H. von Klüber · Harold Knox-Shaw · Zdeněk Kopal · A. Kopff · V. Kourganoff

Max Krook · Gerard P. Kuiper · [illegible] · Gustav Land · [illegible]

Bertil Lindblad · E. M. Lindsay · E. H. Linfoot · A. B. Lovell

W. J. Luyten · R. A. Lyttleton · M. R. Madwar · [illegible] · W. H. McCrea

Dean B. McLaughlin · Donald H. Menzel · Paul W. Merrill · Joseph Meurers · M. Minnaert

W. W. Morgan · [illegible] · J. J. Nassau · Peter Naur · Joseph Needham

O. Neugebauer · H. W. Newton · D. O'Connell · [illegible] · J. H. Oort · P. Th. Oosterhoff

Michael W. Ovenden · Bernard Pagel · F. A. Paneth · Ant. Pannekoek · R. M. Petrie · E. J. Pierce

H. H. Plaskett · [illegible] · Derek Price · [illegible] · J. A. Ratcliffe

R. O. Redman · G. Righini · [illegible] · J. A. Rottenberg · Keith Runcorn

Henry Norris Russell · [illegible] · Martin Ryle · [illegible] · Ed Salpeter

George Sarton · F. Schmeidler · [illegible] · Harlow Shapley · [illegible] · [illegible]

SF Singer · Charlotte E. Moore-Sitterly · [illegible] · F. Graham Smith · [illegible]

[illegible] · [illegible] · [illegible] · Robert Stoneley · R. H. Stoy · Hans Straßl

Bengt Strömgren · Otto Struve · [illegible] · P. Swings · Tcheng Mao-Lin

C. David Thackeray · Richard N. Thomas · Jaakko Tuominen · Albrecht Unsöld

Harold C. Urey · G. Van Biesbroeck · H. C. van de Hulst · Peter van de Kamp · W. H. van den Bos

[illegible] van den Dungen · [illegible] · Julie H. Vinter Hansen · M. Waldmeier

C. Desmond Walshaw · Patrick Wayman · Harold Weaver · [illegible] Weidemann

[illegible] · Peter Wellmann · Fred L. Whipple · J. P. Wild · Emil Wolf

Bradshaw Wood · [illegible] · [illegible] · Herman Zanstra

VISTAS IN ASTRONOMY

Volume 1

My subject disperses the galaxies, but it unites the Earth.

SIR ARTHUR EDDINGTON, at the Meeting of the International Astronomical Union, Cambridge, Mass., U.S.A., September, 1932.

F. J. M. Stratton

THESE TWO VOLUMES ARE DEDICATED TO

F. J. M. STRATTON

C.B.E., D.S.O., T.D., M.A., HON.LL.D., D.PHIL., F.R.S.

EMERITUS PROFESSOR OF ASTROPHYSICS
IN THE UNIVERSITY OF CAMBRIDGE

BY HIS FRIENDS AND ADMIRERS
EVERYWHERE

VISTAS IN ASTRONOMY

IN TWO VOLUMES

EDITED BY

ARTHUR BEER

Volume 1

CO-OPERATION AND ORGANIZATION
HISTORY AND PHILOSOPHY
DYNAMICS
THEORETICAL ASTROPHYSICS
INSTRUMENTS
RADIO ASTRONOMY
SOLAR PHYSICS

PERGAMON PRESS
OXFORD · LONDON · NEW YORK · PARIS

PERGAMON PRESS LTD
Headington Hill Hall, Oxford
4 & 5 *Fitzroy Square, London, W.*1

PERGAMON PRESS INC.
122 *East 55th Street, New York 22 N.Y.*
1404 *New York Avenue, N.W., Washington 5, D.C.*
P.O. Box 47715, *Los Angeles, California*

PERGAMON PRESS S.A.R.L.
24 *Rue des Écoles, Paris Ve*

PERGAMON PRESS G.m.b.H.
Kaiserstrasse, 75, *Frankfurt am Main*

Special Supplement No. 3 to the Journal of Atmospheric and Terrestrial Physics

First printed in 1955
Second impression 1960

Printed in Great Britain by the Pitman Press, Bath, Somerset

CONTENTS OF VOLUME 1

1. CO-OPERATION AND ORGANIZATION

2. HISTORY AND PHILOSOPHY

3. DYNAMICS

4. THEORETICAL ASTROPHYSICS

5. INSTRUMENTS

6. RADIO ASTRONOMY

7. SOLAR PHYSICS

CONTENTS OF VOLUME 2

8. SOLAR-TERRESTRIAL RELATIONS

9. GEOPHYSICS

10. PLANETARY SYSTEM

11. STELLAR ASTRONOMY

12. PHOTOMETRY

13. SPECTROSCOPY

14. SPECTRAL PECULIARITIES AND NOVAE

15. GALAXIES

16. COSMOGONY AND COSMOLOGY

INTRODUCTION

ONE of the results of F. J. M. STRATTON's imaginative understanding of human affairs has been to bring about harmonious co-operation between the very different personalities populating the sphere of astronomical studies. Inspired by his magnificent example, I have tried in this book to bring together and to present in an organized manner the ideas that form the structural units of present-day astronomy.

Everyone familiar with our subject knows that it is a field in which controversy cannot be avoided; indeed, at the frontiers of expanding knowledge, controversy is most desirable. Contradictory views will therefore sometimes be found here side by side with one another.

It can be questioned whether a synthesis such as was achieved in the seventeenth century through the work of GALILEO, KEPLER, NEWTON, and others, is attainable in the twentieth century. But astronomical science can only advance, as it did then, by boldly combining observational and experimental work with the tireless quest for theoretical unification. As long as we realize that theoretical interpretations can only be held valid until further notice, that is until incontrovertible evidence gained by observation compels us to modify them, there can be no objection to the construction of hypotheses. NEWTON's own motto "*hypotheses non fingo*" was, in a sense, disregarded by NEWTON himself: he rejected hypotheses only where they violated his own "*regula philosophandi*", that is to say, his principle of their strict parsimony. In terms of present-day methodology, we reject hypotheses as scientifically meaningless if they are incapable even of indirect test; and we reject them as superfluous or as implausible if they are too complex and artificial to conform with well-established canons of inductive probability. But freedom of scientific theorizing must be preserved wherever the conditions of meaningfulness and of economy appear to be satisfied.

Moreover, we cannot appraise the validity of theories or hypotheses unless a great deal of observational material has been accumulated, scrutinized, checked, and counter-checked. Astronomical and astrophysical hypotheses, involving as they do tremendous extrapolations, must remain controversial for a long time. It is my hope that the rich material here assembled, both of observation and of theory, will promote the synthesis of our knowledge, leading us *per noctem ad lucem*, no matter how long and tortuous the path may be.

* * *

When I first approached our colleagues all over the world to sound them about a "Stratton Volume", any possible doubts as to the feasibility of the project were quickly dispelled by their warm and encouraging response.

In my invitation to contributors, I emphasized that the purpose of the book was to present original articles and reviews, reflecting fully and critically the present situation and recent advances in astronomy and related fields. The authors were also asked to frame their contributions with a view not only to their value for the expert, but also to their use by less specialized readers wishing to gain a general idea of the latest developments. The average length of an article was to be two to three thousand words, of which the first few hundred words were to be a statement of the problem with a brief summary of previous work, while the concluding paragraphs were to be devoted to a survey of unsolved questions and possible "vistas" of future research. In this way, each article was designed to be a self-contained unit, combining the author's own researches with his views on the historical roots of his problem and on its future beyond the limits of to-day.

We thus intended to develop something new, neither a "*Handbuch*"* nor a "*Festschrift*"—but a generous and thorough cross-section through the whole of contemporary astronomy and the allied sciences, always emphasizing the lines along which research is particularly active at the moment, its new techniques and methods, and their interaction with theoretical developments. Arrangements were made to ensure that every article would add a new and significant facet to the whole.

It is hoped that this structure of the book may make its contents more accessible to those whose interests lie in adjacent fields—such as physics, mathematics, and geophysics—than is normally possible with more specialized astronomical texts.

* * *

The original set of collaborators came from the wide range of Professor STRATTON'S immediate colleagues, pupils, and friends. The momentum of the work carried it forward until ultimately 215 authors joined to produce the 192 contributions: 179 of them are astronomers; 36 are physicists, geophysicists, mathematicians, and historians. They cover the span between the ages of twenty-four and ninety-one, and the more significant one between twenty-six nations.

The aim was to include one of the leading representatives in every special field. Overlapping has been avoided, except where several authors were deliberately invited to write on the same topic (a fact which was known to the editor only)—approaching it from different angles, with different points of view, and with divergent results—thus illuminating the situation with the spotlights of opposing new ideas.

Some of those who were originally approached found themselves unable to contribute because of the time-limit necessarily imposed. Happily, they were only few in number, and I should like to record my gratitude to them for their good wishes for the success of the venture. Their warm words were among the many sources of inspiring encouragement extended to me. It is deeply regretted that death has prevented the contributions by BERNARD LYOT, EDWIN HUBBLE, and WALTER GROTRIAN from being included in these volumes, whilst it is sad that the one by THADDÄUS BANACHIEWICZ, who died in November, 1954, must remain his last work.

Practically every country in the world with an active astronomical life has contributed and many languages were involved. For the sake of uniformity it was decided to publish everything in English. The editor had, therefore, to provide some 600 printed pages of translation; in most cases the whole text had to be turned into English; in others the author's provisional English version was revised. The problem facing every translator is the search for a suitable compromise between the demands of style and those of accuracy. "*Les traductions sont comme les femmes: quand elles sont belles, elles ne sont pas fidèles; et quand elles sont fidèles, elles ne sont pas belles.*" On the whole, I have tried to retain the flavour of the original language

* "While the '*Handbuch*' is obviously important to every astronomer, it is clear that such enormous editions cannot appear frequently. Other means must, therefore, be devised to meet the need of scientific summaries. It is to be hoped that a suitable organization will undertake to supplement the '*Handbuch*' by publishing from time to time relatively short summarizing articles containing reviews of new developments in various fields".—OTTO STRUVE, reviewing the first six volumes of the *Handbuch der Astrophysik* in the *Astrophysical Journal*, vol. 80, p. 73, 1934.

rather than to produce a uniform and perhaps colourless version. Thus, while every effort was made to ensure a smooth flow of language, equal care was taken not to impress an impersonal style on the work of colleagues from abroad, each of whom wrote in the lively idiom of his own tongue. The final responsibility for the result is mine.

* * *

I cannot resist the comment that in attempting this cross-section of modern astronomy I have been vividly confronted with a cross-section of contemporary astronomers in all their delightfully individual ways. The suggested length of the papers was some 2500 words: the actual mean length turned out to be 4780 words, and the standard deviation 2920; the minimum was 360 words, the maximum 28,000. The average delay in submitting the manuscripts was 175 days, and the standard deviation 95; the extreme values were − 70 days and + 680 days. Many of the longer delays were, of course, due to circumstances beyond the authors' control, and the overall impression left in my mind is that of the sincere effort made by the contributors to keep to the schedule.

There was also quite a lively exchange of letters. Often these were concerned with entirely necessary points. They added to the labour, but did not detract from the enjoyment of editing. A total of some 6000 letters proved sufficient to ensure smooth relations between contributors, editor, and publisher.

The suggestion that each article should conclude with a "vista" has been interpreted in many ways, and not everywhere is such an account presented in words. One of our most distinguished astrophysicists, for instance, concluded his article with a capital Ω, and another one with the word "false". I was sometimes surprised by the range of topics which made their appearance; on p. 89, for example, there is a discussion of the variability of the human gestation period, and reference 14 on p. 490 concerns a decerebrated albino rabbit. On the other hand, it was decided to reduce discussion of some familiar topics where recent accounts or new textbooks have been published or are in preparation. This is, for instance, the case with certain theoretical problems of stellar atmospheres and stellar interiors, the planetary system, aurorae, and cosmical aerodynamics.

Every effort has been made to keep the articles up to date by additional work and amendments during the growth of the book. However, the general framework of each article had to be fixed before printing could begin, and the mean epoch of final revision is 1954·5.

While the simultaneous publication, in a fixed order, of 192 contributions from twenty-six different countries has necessarily been slower than if they had been dealt with individually like papers in a scientific journal, it is hoped that none of the articles is so ephemeral that its interest will have diminished during the time of printing. The generous inclusion by authors of accounts of their latest researches has involved some sacrifices in the speed of announcement, but has added a refreshing note to the impression which is given of the contemporary astronomer's working day.

No doubt a number of mistakes and misprints have escaped scrutiny; but it is hoped that serious ones have been avoided. In one or two cases manuscripts left on my desk were found to have mysteriously grown unsolicited additions, not always of a strictly scientific nature; it was discovered just in time that a medieval drawing had been provided with the caption "The Editor, half-size". I should be grateful to receive notice of any omissions and errors found by readers.

The system of abbreviations of references is that used in *Science Abstracts* (of the Institution of Electrical Engineers, London; in association with the American Physical Society, *etc.*), and I am much indebted to Dr. B. M. CROWTHER who kindly gave permission to adopt this system and to use it throughout the whole book. The request to the authors that "references should be given generously" has resulted in the accumulation of a thorough bibliography covering a very wide range of astronomy and allied fields, which it is hoped may be not the least valuable part of the work.

* * *

"I owe debts of gratitude to many more people than can conveniently be named, people of all degrees and many nationalities. He who befriends a traveller is not easily forgotten, and I am very grateful indeed to everyone who helped me on a long journey". These words of PETER FLEMING in his *Travels in Tartary* very well suit the case of the editor of *Vistas*.

I should like to mention some of the many friends and colleagues who have always been ready to assist me.

First of all, I wish to say how sincerely grateful I am to Professor R. O. REDMAN, without whose help and understanding forbearance the task of editing this book could not have been accomplished.

I am most obliged also to Sir EDWARD V. APPLETON, Dr. G. BING, Prof. T. G. COWLING, Dr. D. W. DEWHIRST, Prof. H. FEIGL, Dr. R. M. GOODY, Prof. W. HARTNER, Prof. W. H. MCCREA, Prof. M. MINNAERT, Dr. J. NEEDHAM, Dr. D. H. SADLER, Dr. M. J. SMYTH, and Dr. A. D. THACKERAY, for their active interest and support. Nor shall I ever forget the stimulating friendship and devoted help which I received from Dr. PETER FELLGETT and Dr. BERNARD PAGEL.

My wife and my daughter NOVA (whose name was proposed by Professor STRATTON in one of those happy nights when he and I joined in the observations of Nova Herculis) have helped and encouraged me in many ways.

The Mount Wilson and Palomar Observatories have been very helpful: Dr. I. S. BOWEN and Dr. M. L. HUMASON provided some of the latest photographs used in the preparation of the book. I am also much obliged to Dr. WALTER BAADE, who placed his original photograph of the radio source in Cassiopeia at the disposal of *Vistas* (p. 566), as well as his new and previously unpublished record of the peculiar galaxy NGC 2623, which is shown in the frontispiece to Volume 2, and is now known to be a radio source in Cancer. In the same frontispiece is shown a photograph of the interesting double galaxy in Pisces, also believed to be a radio source. For this last photograph I am much indebted to Dr. FRITZ ZWICKY, and for that of NGC 7293, on p. 259, to Dr. R. MINKOWSKI.

Thanks are due to the Master and Fellows of Gonville and Caius College, Cambridge, for permission to reproduce their fine portrait of Professor STRATTON, painted by Sir OSWALD BIRLEY, which forms the frontispiece of this volume; and particularly to the Master, Sir JAMES CHADWICK, for his kindness in making the necessary arrangements.

I gratefully acknowledge the unfailing helpfulness and disinterested care which the publishers devoted to the production and publication of this book, especially that of Dr. P. ROSBAUD, Mr. E. J. BUCKLEY, and Mr. J. G. MASON, who always gave every possible consideration to my suggestions.

I should really have mentioned some 200 more names: I can only say how sad

I feel that the ways of editor and collaborators must now part again, after such a pleasant time together.

* * *

It can rarely, if ever, have happened before that so many leading scholars in a single discipline have co-operated in one book on such a wide international scale. That this was possible is a fitting tribute to the international spirit which has permeated Professor STRATTON's life-work in our science.* I am glad to be able to include here the following passages from a letter recently received from Dr. WALTER S. ADAMS, of Mount Wilson Observatory:

> Of the many important contributions to the fields of solar and stellar spectroscopy which Professor STRATTON has made, I am inclined to rate among the highest his series of investigations of the spectra of novae which began in 1912 with Nova Geminorum. He brought to this great insight, and I think of him as a pioneer in the modern research which has given us so much information regarding the physical state and constitution of these extraordinary stars.
>
> I also want to mention as probably the finest relatively short treatise in its field his *Astronomical Physics*, first published in 1925. The choice of material, the conciseness with which it is handled, and the simple and admirable style of writing still make this a reference book for astronomers and laymen in spite of the quarter century since it appeared.
>
> STRATTON was elected General Secretary of the International Astronomical Union at the Cambridge Meeting in 1925, succeeding Professor ALFRED FOWLER, and continued to serve until 1938. During this time he prepared and edited the reports of the I.A.U. Meetings at Leiden, Cambridge in Massachusetts, and Paris. An enormous amount of labour was involved in these volumes, and STRATTON's pre-eminent qualities of energy, tact, and friendliness were tested to the full in dealing with contributors scattered all over the face of the Earth, many of them with highly individualistic ideas of what should go into a report. In this respect STRATTON had much the same experience as Professor ARTHUR SCHUSTER, who as President of the International Union for Co-operation in Solar Research (the predecessor of the I.A.U.) was moved to say at the close of the Mount Wilson Meeting in 1910: "Gentlemen, the reports of the sections and committees will be printed in the Transactions of the Union. I wish to call your attention to the fact, however, that in the past some of these reports as submitted have borne little or no resemblance to what was said or done at the meeting".
>
> The contribution which STRATTON made to the efficiency and smoothness of operation at the meetings cannot be overestimated. A large group of international scientists, each an individualist in his own right, often presents a serious problem of organization, but STRATTON handled it with remarkable skill and diplomacy. He could make allowances for questions of national prestige which occasionally arose even in a scientific meeting, and his linguistic ability was of immense value on these and similar occasions. I always remember the delightful atmosphere of good fellowship which accompanied the close of the Stockholm Meeting in 1938 after STRATTON had presented to Sweden the greetings and thanks of some twenty different nations, each in its native tongue.
>
> I have many personal memories of STRATTON, associated with all sorts of occasions. His orbit was quite unpredictable as was illustrated by a telegram received on Mount Wilson during one of the years of the Second World War. It was from San Francisco and read: "Arriving Los Angeles 9.30, (signed) STRATTON". Since I had no reason to think that STRATTON was less than 5000 miles distant at the time, one can imagine the thrill this message produced. We met

---

* STRATTON, Frederick John Marrian; D.S.O., 1917; O.B.E., 1929; T.D., 1924; F.R.S., 1947; M.A. (Cantab.); Hon. LL.D. (Glasgow); D.Phil. (Copenhagen); D.L. Cambs., 1924; V.L., 1945; Professor of Astrophysics, Cambridge University, and Director of the Solar Physics Observatory, 1928–47; *b.* Birmingham, 16 Oct. 1881; *s.* of late Stephen Samuel Stratton and Mary Jane, *d.* of John Marrian. Educ.: King Edward's Grammar School, Five Ways, and Mason University College, Birmingham; Gonville and Caius College, Cambridge; 3rd Wrangler, 1904; Isaac Newton Student, 1905; Smith's Prizeman, 1906; formerly Lecturer in Mathematics and Tutor Gonville and Caius College, Cambridge; President of Gonville and Caius College, 1946–48; Halley Lecturer, Oxford, 1927; Secretary International Astronomical Union, 1925–35; General Secretary British Assoc., 1930–35; Pres. Roy. Astronomical Soc., 1933–35; Pres. Cambridge Philosophical Soc., 1930–31; Gen. Sec. Internat. Council of Scientific Unions, 1937–52; Pres. Soc. for Psychical Research, 1953–55; Signal Service, R.E., 1914–19 (D.S.O., Legion of Honour, Bt. Lt.-Col.); Royal Corps of Signals, 1940–45; Deputy Scientific Adviser to the Army Council, 1948–50; Membre Correspondant de l'Académie des Sciences et du Bureau des Longitudes, Paris; Fellow of the Institute of Coimbra.—Publications: *Astronomical Physics*, 1925; papers in the *Monthly Notices* and *Memoirs of the Royal Astronomical Society*.—Address: Gonville and Caius College, Cambridge.—Club: "Savile". (From *Who's Who*, 1955.)

at many places on the European Continent, in his rooms at Caius College in Cambridge, in London, and in California.

Two out of many such occasions remain especially in my memory. The first followed the organization meeting of the I.A.U. at Brussels in the Summer of 1919. STRATTON took a small group of us to the battlefields of Ypres and under his pilotage we went through the dug-outs in which he had lived so long and given all he had to the cause of Britain in this desperate struggle. We climbed Mount Kemmel under a grey sky, and among the trenches and the broken guns and refuse of war STRATTON outlined to us the course of the great battles which marked a turning-point of history.

A second memory is of the I.A.U. Meeting in Paris in 1935. It was Bastille Day in France, and after dinner on the Eiffel Tower, STRATTON, who is a true cosmopolitan and knows Paris as well as he does London, took a few of us to see the dancing in the streets. He knew just where to go, and the Americans in the party received one of the thrills of their lives and a knowledge of the French people they could have learned in no other way. . . .

Another sidelight is contained in the following account, taken from a letter from Dr. A. D. THACKERAY, of the Radcliffe Observatory; it refers to the Cambridge expedition to Japan for the 1936 total solar eclipse (see also pp. 708–10):

On the evening of June 19th, 1936, after the bitter experience of a cloudy eclipse, the following telegram, which we were told had been transmitted in Japanese characters, was placed in our hands:

NOBANIYA EPUSHIRON SEFUEISADOMA GUNICHIYUDO.

STRATTON cheerfully sat down to disentangle the meaning with members of the expedition, and in less than half an hour turned out the message as "Nova near epsilon Cephei third magnitude". (There was no means of indicating beginnings or ends of words in the message.) On this occasion the press reported that we took half an hour to find the nova, without reference to the decoding problem. . . .

* * *

I should like to add a personal footnote to my INTRODUCTION. Circumstances have given me the opportunity of long-continued contact with Professor STRATTON's radiant and inspiring personality; it is an experience which I cherish more deeply than I can readily express and which I consider one of the greatest gifts of my life.

If the work on these volumes, dedicated to him, involved many difficulties and complications, I have been more than rewarded by the spirit in which a group of brilliant and creative minds has co-operated. We should regard the international band of 200 astronomers working together to produce this book in honour of a great scholar - as an example of what can and should be done in a wider sphere.

If it be possible for this book to strengthen the appreciation of human efforts unhampered by national barriers, and to serve as a symbol of what good will can achieve in our troubled world, then, indeed, it will have fulfilled all our hopes.

The Observatories,
University of Cambridge.
March, 1955.

ARTHUR BEER.

Hommage Respectueux, Cordial, et Reconnaissant à Monsieur le Professeur F. J. M. Stratton

E. Delporte

Directeur Honoraire de l'Observatoire Royal de Belgique à Uccle, Associate R.A.S.

Je désire que ma petite planète 1942 XB qui a reçu le numéro 1560 porte le nom de "Strattonia" en l'honneur du présent Jubilaire.

Des voix autorisées retraceront la carrière si touffue, si vivante, si profondément marquée dans la vie astronomique internationale tant par ses travaux personnels que par la part active qu'il a prise dans la naissance et la vie de l'Union Astronomique Internationale.

Pour la Belgique, le Professeur F. J. M. Stratton fut l'un de ses sauveteurs lors de la première guerre mondiale. Il fut le représentant combien autorisé de la Royal Astronomical Society aux cérémonies du Centenaire de l'Observatoire Royal de Belgique en 1935.

Personnellement, je tiens à rappeler la part grande qu'il assuma lors de la publication par l'Union Astronomique Internationale de ma "Délimitation Scientifique des Constellations" et surtout la généreuse collaboration qu'il daigna m'accorder pour la publication de l'*Atlas Céleste* paru sous les mêmes auspices de l'U.A.I., avec les nouvelles limites.

Je m'honore d'avoir pu être considéré par M. le Professeur F. J. M. Stratton comme un ami.

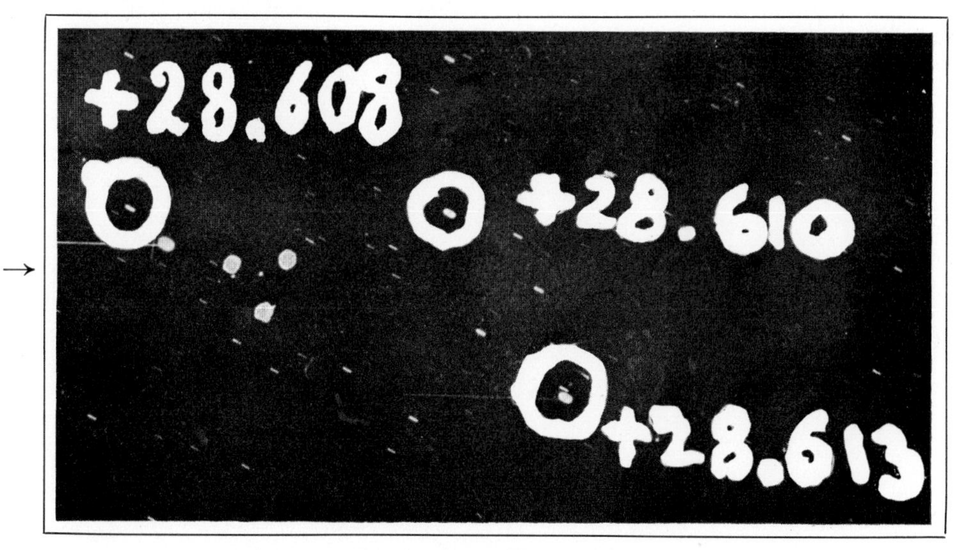

Fig. 1. Planet STRATTONIA

La petite planète 1942 XB = 1560 a les éléments suivants:

Equinoxe 1950,0

m_0	g	1943	M	ω	Ω	i	φ	μ	a
14,4	11,0	III 17	52°,234	92°,208	290°,172	6°,283	12°,236	805″,959	2,6861

d'après l'éphéméride des petites planètes de Leningrad pour 1953 après application des perturbations spéciales.

Les premiers éléments en ont été calculés par M. MUSEN au moyen des positions du 3 décembre 1942, 4 janvier 1943, et 1 février 1943. La numérotation 1560 a été annoncée dans la circulaire 2468 de l'Astronomisches Rechen-Institut. La planète a été réobservée aux oppositions de 1946 et 1950.

J'ai trouvé la petite planète sur la plaque No. 1617 prise le 3 décembre 1942 au soir, au double astrographe Zeiss par la méthode de Metcalf, pose de $1^h\ 50^m$ à $2^h\ 43^m$, par légère brume. La dite méthode a été employée pour la recherche d'une autre petite planète d'éclat faible. Confirmée par la plaque 1620 (astro Agfa) posée le 10 décembre au soir, par la même méthode de METCALF, avec une durée de 56 minutes, la planète fut suivie par les plaques 1621 (10 décembre soir), 1623 de la même nuit (ciel peu clair), 1627 du 27 décembre au soir, 1628 du 4 janvier 1943 au soir, 1630 du 5 janvier au soir, 1633 du 8 janvier au soir, 1638 du 12 janvier, 1639 du 13 janvier et 1640 du 1 février 1943.

L'éclat assez faible de l'astre et surtout le ciel peu propice pendant toute la période a obligé l'observateur à se servir constamment de la méthode METCALF, concentrant la lumière renvoyée par l'astre en un point, alors que les étoiles traçaient des traits sur la plaque. La copie d'une partie de la plaque de découverte reproduite ici, accuse la difficulté de la recherche; les 3 étoiles ayant servi à la mesure de la position sont entourées d'un cercle, tandis que l'image de la planète se trouve entre 3 points.

SECTION 1

CO-OPERATION AND ORGANIZATION

"That tho' a Man were admitted into Heaven to view the wonderful Fabrick of the World, and the Beauty of the Stars, yet what would otherwise be Rapture and Extasie, would be but a melancholy Amazement if he had not a Friend to communicate it to."

Attr. to ARCHYTAS by CHRISTIANUS HUYGENS.
The Celestial Worlds Discover'd, Book 1, pt. 4, 1722,

International Co-operation in Astronomy

M. MINNAERT

Sterrewacht, Utrecht, Netherlands

SUMMARY

Scientific co-operation between the nations is found already in Antiquity and the Middle Ages and has proved a strong stimulus to the development of astronomy. Different forms of modern international co-operation in astronomy may be distinguished: (1) co-ordinated observations at widely separated stations; (2) collective achievement of a great amount of work; (3) creation of international centres; (4) unification of notations and terminology. The increasing need for co-operation in astronomy was the reason for the constitution of international bodies, among which the I.A.U. acquired the greatest importance; the history of the Union shows that scientific co-operation must be kept outside political implications. International meetings, colloquia, travels, and exchanges should be encouraged. The introduction of an auxiliary international language would be highly desirable. International co-operation is a necessary complement to the national development of science.

1. EARLY FORMS OF CO-OPERATION

FROM the earliest periods of civilization, contact between the nations has stimulated the development of science. In Antiquity, it was especially the interaction between the Oriental and the Greek world which had important consequences. THALES and PYTHAGORAS, later PLATO and EUDOXOS, travelled to Egypt and were initiated into the wisdom of the priests. In 280 A.D. the Babylonian BEROSSOS taught astronomy to Greek scientists at Kos. The transmission of astronomical learning from the Greeks to the Romans and the Arabs, later from the Arabs to Western Europe, was made possible by numerous laborious translations of astronomical textbooks, and it saved the continuity of the growth of astronomy.

At the end of the Middle Ages, the national groups were taking shape, and in each country centres of science developed with their particular schools of learning, influenced by exchanges and scientific relations. So, as early as in the thirteenth century, we find JOHN HOLYWOOD, of York, as a professor of astronomy at the Paris University. And when the Renaissance stimulated all sciences and arts to a wonderful flowering, it was the international contact which diffused the new concepts, brought fresh ideas, and new impulses to a rejuvenating world. At most universities, students of many nations assembled, easily following the teaching in Latin. It became a custom that scientists in their youth should visit foreign countries in order to complete their education. PURBACH travelled to France and Italy; REGIOMONTANUS to Vienna, Rome and Hungary; NICOLAUS of CUSA to the Netherlands, Germany, and Italy; COPERNICUS to Bologna, Rome, and Padua; RHETICUS from N. Italy to Copernicus at Frauenburg; the young TYCHO to Germany; SIMON MARIUS from Francony to Tycho and to Italy. Brilliant scientists were called as professors to foreign universities, in the same spirit of internationalism which is our pride nowadays. The printed book conveyed knowledge from one country to another and became the main instrument of international scientific co-operation: no other is so effective, so easy, so permanent.

However, the diffusion of new discoveries by books takes too much time and is not efficient for quick scientific intercourse. In hundreds of years, the real contact between actively working scientists was entertained by direct correspondence. The letters of HUYGENS, for example, comprise nearly one-half of his whole scientific production and have been published, together with the replies, in ten quarto volumes of about 600 pages each; they are addressed to scientists all over Europe. It was only in 1679 that the *Connaissance des Temps* appeared as the first important astronomical year book. From 1800 to 1813 the first astronomical periodical was edited by Baron von Zach, the *Monatliche Korrespondenz*, followed after a short interruption by the *Zeitschrift für Astronomie* (1816–18) and the *Correspondance Astronomique* (1818–26); astronomers of several nations contributed papers and the editors endeavoured to give information about the development of astronomy all over the world. The important astronomical reviews which have originated since then have each kept their national character more or less, though papers of foreign colleagues are in general welcomed. This international contact has been considerably increased by the publication of regular observatory annals which are freely distributed to all foreign observatories, according to a system almost unique among scientists and testifying of the generosity of the great institutes in favour of the minor ones.

2. Modern Co-operative Enterprises

The aspects of international co-operation thus far described do not yet include the form which is considered nowadays as the most typical: an organization, consciously planned for team work. In 1736, when the French Academy of Sciences wished to compare an arc of the meridian at different latitudes, two expeditions were prepared, one to Lapony, the other to Ecuador; nobody considered the possibility of asking foreign countries to make independent observations which could be compared afterwards. Similarly, in 1671, the parallax of Mars was determined from the results of two French astronomers, Richer at Cayenne and Cassini at Paris.

Probably the first really co-operative international enterprise in astronomy was undertaken in 1761 and 1769, when the transit of Venus was observed by numerous expeditions with a view to determining the solar parallax. The preliminary calculations were derived from an intercomparison of observations from a few distant stations; the final discussion, published by Encke in 1822 and 1824, was based on the individual results of all the stations. This then may be called the first type of worldwide astronomical co-operation: when observations are needed, made at distant stations, it is quite natural to combine the efforts of observatories of different nationalities. Such combined efforts for the determination of the solar parallax were repeatedly made later. For the Venus transits of 1874 and 1882, the results were obtained independently by the French, by the German, by the English, and by the American observers, each from a comparison of their own national expeditions. Really international, however, was the co-operative observation of the minor planets Iris, Victoria, and Sappho, organized by Gill (1888–9); and so was the well-planned and fruitful Eros campaign of 1930–1, in which forty observatories co-operated. Of this same "first type" is also the International Latitude Service, created in 1898 and now centralized at Turin; it has become clear that this will have to be extended, since the number of co-operating observatories is not sufficient for the acquiring of complete information. Curiously enough, recent eclipse expeditions for the determination of longitudes by the observation of the beginning totality have nearly always been made by several parties of one and the same nation; it seems that international co-operation for this subject has not been considered as yet. Recent co-operative schemes are concerned with the continuous observation of solar phenomena: the sudden disturbances, especially the flares, are studied by a number of observatories, located at different geographical longitudes, each observatory being responsible for a daily one-hour watch. The progress of radiophysics has necessitated a similar organization for the observation of the radio noise-bursts and outbursts, though in this case a smaller number of stations is sufficient, because these phenomena are automatically recorded, independently of the weather. Three or four transmitting stations are broadcasting daily in code form a survey of the solar activity, the ionospheric perturbations, and the cosmic radiation in the last 24 hours ("ursigrams").

A second type of international co-operation has been organized in those cases where a scientific enterprise involves such a great amount of work that it could not be handled by one observer alone. Such collaboration is especially important for astronomy, which deals with an immense amount of material while depending on a limited number of scientists and instruments. Probably the earliest attempt in this direction was the conference convened by von Zach at Lilienthal, in the autumn of 1800, where twenty-four astronomers of different nations assumed the task of undertaking a systematic search for the hypothetical planet between Mars and Jupiter. To the same category belong: the project of the Catalogue of the Astronomische Gesellschaft, initiated by Argelander in 1867; the celebrated enterprise of the Carte du Ciel (Paris, 1887); and, more recently, the Plan of Selected Areas (1905). In the theoretical field, computational work is sometimes so extensive that even there a co-operation between the computing centres of different countries has proved very effective. For several tables of the astronomical almanacs the computations are carried out either in England, in Germany, in France, in Spain, in the U.S.A., or in the U.S.S.R. and the results are afterwards exchanged. Quite recently, the International Astronomical Union has planned a very interesting co-operative calculation of fundamental data concerning stellar atmospheres: the ionization, the absorption coefficient, and similar data will be computed at increasing depths for a great variety of model atmospheres, differing in their chemical composition; a dozen observatories have taken an interest in this work. The assistance of international computing centres with electronic machines will be of great value for similar problems in the future.

Mutual co-operation in industry leads to specialization and to distribution of the work. The same has proved true in astronomical research. The modern big telescopes produce photographic records at such a tremendous rate that it would be impossible for the astronomers of such an observatory to run their instruments continuously and to work out all these invaluable documents. It has become more and more frequent that colleagues of all nations receive the photographs which are necessary for their work or get the opportunity to take them themselves, while the measurements, the microphotometrical investigation of the plates and the theoretical development are made at their own institutes. Such a co-operation was inaugurated by Kapteyn and Gill, when they agreed that the plates of the southern skies, obtained at the Cape Observatory (1885–90), would be measured at Groningen. The result was the Cape Photographic Durchmusterung. As one of the very numerous recent examples we may perhaps quote the Photometric Atlas of the Solar Spectrum, a result of a co-operation between the Mount Wilson and the Utrecht Observatories, and from which a Catalogue of Fraunhofer lines is being derived by another co-operation between some American and Dutch astronomers specialized in a particular branch of this subject.

A third type of co-operation has a more centralized character. The co-operation of a great number of observations is entrusted to a central bureau, which makes the combined results available to all. In about 1850 Wolf, at Zürich, had already begun his computation of sunspot relative numbers from observations made in many countries. We have now a centre for the observation of solar flares at Meudon; a centre for solar radio noise data at Sydney-Canberra; the Zürich centre for the study of sunspot development and for the publication of a Quarterly Bulletin of Solar activity; the Cincinnati, Heidelberg, and Leningrad centres for the ephemerides of minor planets; the Moscow Catalogue of Variable Stars. The same principle is applied to astronomical bibliography, embodied in the Astronomischer Jahresbericht at Heidelberg, the Bulletin Analytique, and the Science Abstracts, and also to the publication of the volume *Observatoires et Astronomes*, by the Uccle Observatory. An important institution of this kind is also the central Bureau for Astronomical Telegrams, which has been working at Copenhagen since 1922, and communicates by wire and by circulars the latest discoveries about novae, comets, or exceptional planetoids. The Bureau International de l'Heure, set up at Paris in 1913 and reorganized in 1919, intercompares the time determinations of several observatories and broadcasts time signals of high precision.

A far reaching plan for co-operative enterprise is the project of an international observatory, to be erected in the south of Europe, in excellent climatic conditions, and to be financed by those nations which desire to have the benefit of the telescopes there erected. This proposition was originally presented by the Polish delegation at the preparatory Copenhagen conference in 1947, and afterwards more fully developed by Professor Shapley, preference being given to a location in the southern hemisphere. It was put on a list of similar projects, submitted to the UNESCO, but relatively to these it was not considered to merit a high priority. In the meantime, a plan of the same character, though less ambitious, had been realized in Switzerland, where it became a decided success. The high altitude station at Jungfraujoch was organized there, not only for astronomical research, but also for the study of biological and cosmic ray problems; it receives contributions from several countries and is gradually developing its instrumental equipment in a most promising way. Very recently, a number of European observatories which had planned the establishment of small southern stations, have considered the possibility of building in common effort a first-class observatory in the southern hemisphere.

Compared to these ambitious enterprises, a fourth category of international co-operative work might look quite insignificant: the unification of notations, terminology, and units. However, all actively working astronomers know the confusions which arise if there is no general accord on such matters. Quite recently the definitions of time had to be again modified, a new unit of Fraunhofer line strength had to be introduced and a new terminology for radio-astronomy had to be found, these innovations being made necessary by new scientific advances. It was also important to reach agreement on the normals of wavelength to be used in spectroscopic work; or on the frequencies at which the solar radio noise will be observed, so that the measurements become comparable.

3. Organizations for International Co-operation in Astronomy

The increasing international co-operation in astronomical work made necessary the organization of congresses and the creation of a body where such co-operation could be systematically planned. One of the first astronomical meetings with an international character was the congress at Gotha in 1796. Because of the presence of the Frenchman LALANDE, the Austrian astronomers were not allowed by their government to take part at this meeting, and the court at Gotha was warned: "*il pourrait bien s'agir d'autres révolutions que des révolutions célestes*". The Duke and Duchess of Gotha did not care very much and were personally present. In 1863 the German astronomers founded the Astronomische Gesellschaft at Heidelberg, which from the start had a more or less international character, due to the membership of numerous foreigners. In 1904, systematic co-operation was started in the new field of astrophysics, when HALE succeeded in founding an International Union for Co-operation in Solar Research. The first meeting was held in St. Louis, in connection with the International Congress of Science. In a remarkable speech he emphasized that in co-operative scientific work special importance should be attached to the encouragement of individual initiative, no less than to the accomplishment of big projects for routine work.

The Union for Solar Research gave the inspiration for the constitution of a much more far-reaching and systematic organization of astronomical research. Immediately after the First World War, three meetings were held at London, Paris, and Brussels (1918–19), where an International Research Council was created, with several International Unions for the various sciences. Originally, the foundation of the Unions was not laid in a truly international spirit: interallied and neutral countries only were allowed to adhere. This may be frankly recognized: the Union has amply corrected this *vitium originis*, and ". . . 'tis thirty years since!" Already at the first meeting of the International Astronomical Union at Rome, in 1922, Professor CERULLI opened the general assembly by a speech in which he emphasized the necessity for uniting the astronomers of all nations, without any exception. After some years of hesitation, limitations to the membership were removed in 1926 and invitations for co-operation were addressed to Germany, Austria, and Hungary. But the seed of resentment bears evil fruits. It was only in 1947 that Hungary became a member, and it was 1952 when Germany and Austria joined. There could be no clearer demonstration that scientific co-operation must be kept outside all political implications, that it should never be used as a means of uniting one group of nations against another group. The I.A.U. has quickly developed a considerable and very stimulating activity. There are now thirty-three member countries. The work is distributed over forty-two Commissions, which act in a most efficient way and may truly be said to give inspiration to individual initiative, according to HALE's recommendation. The successive meetings of the Union, separated by intervals of three years, are each the result of intense scientific work and at the same time the starting point of new research. Besides these general meetings, small symposia for specialists on selected subjects have proved of great use.

In this whole organization, the General Secretary is the man carrying the heavy burden and the greatest responsibility; we shall never forget how Professor STRATTON devoted more than ten years of his life to this important task and contributed more than anyone else to the vigorous development of the Union.

The international feeling has now become so strong among astronomers that very quickly after the Second World War scientific co-operation was resumed, and no new dissensions were allowed to disturb this work. It is a source of pride and happiness to the members of the Astronomical Union that among them the Russian, Polish, and Czech astronomers as well as the American colleagues give each other a full-hearted co-operation, that they regularly exchange their publications and fraternize at their meetings. A critical event was the preparation of the 1952 meeting, for which a Russian invitation had been received but finally was not accepted because of the international tension. Neutrality was saved, for the decision taken applied in the same way to meetings in the U.S.A. and in the U.S.S.R. But it is a sad thought that a more positive demonstration of the universality of science could not yet be realized.

International co-operation is more than the planning of a common scientific programme. It must never be forgotten that scientific research is made by men, and therefore it is not sufficient to exchange ideas—the scientists themselves should meet and discuss and work together. This is the reason why visits of astronomers to their colleagues abroad are so highly important and have developed to such an extent in recent years. It is a privilege when we are able to welcome great astronomers from distant parts of the world, to have them working, lecturing, discussing among us. But it is perhaps equally wonderful that young astronomers, in the springtime of their life, are able to visit foreign observatories, to enjoy the stimulus of fresh contact and new surroundings, to see excellent astronomers at work in their institutes and in the midst of their collaborators. Temporary assistantships for foreigners have become available at many observatories. If the financial means of an institute are limited, an exchange may be easily organized between a junior staff member and one of his colleagues at another institute, the salary of each of them being available for the maintenance of the visiting astronomer. The recent success of such arrangements is in a considerable degree due to the activity, practical spirit, and never-failing helpfulness of Professor STRATTON, now the president of the I.A.U. Commission for Exchange of Astronomers.

In all international co-operation, the language differences are a major difficulty. At one time, Latin was the only vehicle of science and the common tongue of scientists all over the world. However, in the eighteenth and the beginning of the nineteenth century, when the influence of science on economic life and on society at large became increasingly important, the cryptic language of the learned was felt as an unendurable barrier between them and the nation. Temporarily, French was used in the learned societies all over Europe; then the vernacular superseded the Latin. But how about international relations? At first, French, German, and English were the languages of the great scientific periodicals and standard works. Gradually science developed in many other countries, which began to publish in Italian, Spanish, Japanese. Recently the remarkable rejuvenation of astronomy in Russia has put the problem before us in an acute form. While science is growing, it requires more and more labour to master the established disciplines and less time is left to learn foreign languages. There is thus an increasing need for reconsidering the whole problem, and radical solutions, such as the introduction of Esperanto as an auxiliary scientific language, should be seriously examined. Let the sceptical reader ask himself, whether he is able to propose a better solution.

In all its varied ways international co-operation in astronomy has developed quite naturally out of the requirements of scientific life itself. It has adapted itself to our modern way of living, it has become increasingly important, even to such an extent that it really could not be done without any more. Let us for a moment ponder about the significance of this international contact as a complement to national differentiation. It is sound that science should develop within each country as a part of the national activity and in narrow connection with the local circumstances, the production, the special interests of the people. By the selection of the problems, by the philosophical background on which these are treated, by the special qualities of the nation, it will have a character of its own, even if it is an "exact" science like astronomy. The interaction of the work of all these nations automatically eliminates errors due to preconceived ideas; refinement of mind is complemented by stubborn labour or by practical common sense. But it is the faith of our life that ultimately there will never be contradictions between the findings of all these scientists, varied in their personalities and nationalities, since Nature is unique and since Truth is unique. And, finally, there emerges, in a purified form, what we may call *International Science*, which is no more the work of individuals but the work of the community, of humanity as a whole, the noblest expression of the human mind.

International scientific co-operation demonstrates to all that there is a way of living together on Earth in peace and mutual aid and happiness; a way of living which has not been found by politicians, but which has developed out of the simple desire for truth, and because we relied upon each other, and because we loved each other.

. . . *Ye Heavens, whose pure dark regions have no sign*
Of languor, though so calm, and though so great
Are yet untroubled and unpassionate:
Who though so noble share in the world's toil,
And though so task'd keep free from dust and soil . . .

MATTHEW ARNOLD, 1852.[1]

The International Astronomical Union

P. TH. OOSTERHOFF
Sterrewacht Leiden, Netherlands

INTERNATIONAL co-operation in astronomy as well as international organization of certain specific problems which are too extensive or costly to be undertaken by a single observatory, have a long history. In the previous article Professor MINNAERT has given a general outline of the historical development of international co-operation in astronomical science. The same topic was treated by Professor STRATTON when he delivered his Presidential Address at a meeting of the Royal Astronomical Society.[2]

Nevertheless it was not until 1919 that the many different and separate efforts were combined into a single organization. In common with some other Unions the International Astronomical Union was founded at the Constitutive Assembly of the International Research Council, which was held at Brussels in July 1919. During the thirty-four years which have elapsed since its establishment the Union has developed a great many activities in different fields and it has no doubt become an institution of unique value for astronomical science. Interesting facts about the early history of the I.A.U. can be found in an article by Dr. W. S. ADAMS (*Publ. Astron. Soc. Pacific*, **61,** 5, 1949).

[1] For full text see the fascinating collection "Dichters over Sterren" by M. MINNAERT (van Loghum Slaterus, Arnhem, 219 pp., 1949).—The Editor.
[2] F. J. M. STRATTON; *M.N.* **94,** 361–372, 1934.

This development however has not always been smooth and in the course of years many difficult problems had to be solved. Most of these problems paralleled the international political situation. It is very encouraging that the Union despite these political setbacks, is to-day as strong as, and probably even stronger than it ever was before. This result is not only due to the efforts, the good-will and the wise decisions of the members of the Executive Committee, who had a large share in determining the general policy of the Union, but also to the conviction of all the individual members that a strong international organization like the Union is essential for a sound and fruitful development of science in general and of astronomy in particular. Many astronomers of widely different nationalities have spoken to this effect and have given expression to their belief that the international relations between astronomers are too strong to be broken or even affected for long by political forces. We may quote here from a message by Sir ARTHUR EDDINGTON which was read at the concluding meeting of the General Assembly of Stockholm in 1938:

> "But, if in international politics the sky seems heavy with clouds, such a meeting as this at Stockholm is as when the sun comes forth from behind the clouds. Here we have formed and renewed bands of friendship which will resist the forces of disruption."

This firm belief has never changed and it has succeeded in making the Union the strong organization it is now.

Before dealing with the present-day activities and problems of the Union, we shall pay tribute to those astronomers who have given much of their thought and time for the benefit of the Union, and we shall give some numerical data to demonstrate the continuous growth of the Union.

The Union is governed by the Executive Committee, consisting of a President, five Vice-Presidents, and a General Secretary who is also the Treasurer. The President stays in office for one term, normally of three years, whereas the other officers are elected for two terms. The Executive Committee is responsible for the administration of the affairs of the Union and acts in accordance with the decisions of the General Assembly. As a rule the General Assembly meets once every three years. For obvious reasons it has not always been possible to keep to this rule and altogether eight General Assemblies have been held so far. Without derogating the great merits of the many Vice-Presidents who have served the Union, we give here a list of the Presidents and General Secretaries who have been in office since the foundation of the Union.

Presidents	
Mr. B. BAILLAUD	1919–1922
Prof. W. W. CAMPBELL	1922–1925
Prof. W. DE SITTER	1925–1928
Sir FRANK DYSON	1928–1932
Prof. F. SCHLESINGER	1932–1935
Prof. E. ESCLANGON	1935–1938
Sir ARTHUR EDDINGTON	1938–1944
Sir HAROLD SPENCER JONES	1944–1948
Prof. B. LINDBLAD	1948–1952
Prof. O. STRUVE	1952–

General Secretaries	
Prof. A. FOWLER	1919–1925
Prof. F. J. M. STRATTON	1925–1935
Prof. J. H. OORT	1935–1948
Prof. B. STRÖMGREN	1948–1952
Prof. P. Th. OOSTERHOFF	1952–

Among the names on these lists two stand out clearly because of their very long service. They are Prof. F. J. M. STRATTON, who fulfilled the task of General Secretary for ten years, over three consecutive terms, and Prof. J. H. OORT, who kept the same office for the usual two terms, but on account of the war these terms covered a period of thirteen years. It can safely be said that these two prominent astronomers, who have sacrificed on behalf of the Union so much time which they might have used for their own research, know the Union and its affairs better than anyone else. Their prudence and careful handling of the Union's affairs have contributed greatly to the continual growth and strength of this organization. For this work alone they have earned lasting gratitude from astronomers all over the world. We shall see later that Professor STRATTON did not discontinue in 1935 his efforts to further international collaboration and international scientific organization.

The growth of the Union can be demonstrated by some simple figures.

Year	*Number of standing commissions*	*Number of members*	*Number of adhering countries*
1922	32	207	19
1938	31	553	26
1952	39	814	33

The figures indicate the situation just after the first General Assembly at Rome, after the sixth at Stockholm and after the eighth and latest at Rome. The figures in the last column are especially important, showing that the adjective "international" is continually gaining weight. The figures in the third column prove that astronomy is very much alive at present—though they may cause some worry to the General Secretary. The table, however, gives a very incomplete picture of the growth of the Union. Although the increase in the number of commissions already indicates an extension of its activities, a number of important developments have taken place especially since the end of the war.

Although some of the regular activities of the Union, many of which were started immediately after its foundation, have been enumerated in the preceding article by Professor MINNAERT, they are too important to omit for this reason in an article which should give a description of the I.A.U. Much work has been done by the standing commissions and it is difficult to estimate how many times discussions between members of these commissions have paved the way for international collaboration. In this connection should be mentioned the stimulating influence of the Union on work bearing upon the "Carte du Ciel" and the plan of Selected Areas. The work on nomenclature, the co-ordination of solar research, with publication at regular intervals of solar observations, the work on ephemerides and Minor Planets, the naming and cataloguing of variable stars are only a few of the items which require international collaboration and which therefore have received the full attention, sometimes including financial aid, of the Union. Furthermore the Union is scientifically and financially concerned in a number of permanent services, like the International Latitude Service, the International Time Bureau, and the telegram Bureau at Copenhagen. The first two of these services work under the auspices of the I.A.U. and of the International Union of Geodesy and Geophysics. The Union also took a very active interest in the International Longitude Campaigns of 1926 and 1933, and it is now engaged with other Unions, under the auspices of the International Council of Scientific Unions, in the preparations for an International Geophysical Year and another Longitude Campaign.

Of the developments which became important after the end of the last war we should mention first the organization of an increasing number of international symposia, some of them taking place at the occasion of a General Assembly, others in the interval between such meetings. It is not unusual for a considerable part of the meetings of the standing commissions during a General Assembly to be devoted to rather technical matters and the discussions are naturally restricted within the province of the commission. Consequently the introduction of symposia opened new possibilities for an exchange of opinions on more general scientific topics. Since the war the following symposia have been held:

During General Assembly,
Zürich, 1948

Infra-red Spectrophotometry
The Spectral Sequence and its Anomalies
The Abundances of the Chemical Elements in the Universe

Paris, 1949

Problems of Cosmical Aerodynamics
(Organized by the International Union of Theoretical and Applied Mechanics and the I.A.U.)

During General Assembly,
Rome, 1952

Stellar Evolution
Astrometry of Faint Stars
Astronomical Instrumentation
Spectra of Variable Stars

Groningen, 1953

Co-ordination of Galactic Research

Cambridge (England), 1953

Gas Dynamics of Interstellar Clouds
(Organized by I.A.U. and I.U.T.A.M.)

These symposia have proved so successful that it may be assumed that the Union will continue to organize them.

The other most important development, though of a completely different character, was the establishment of UNESCO, which has taken an active interest in international organization. Since 1947 the Union has received annually considerable sums from this body, which have enabled the Union to increase its activities to a great extent. The grants allotted by UNESCO to the I.A.U. are as follows:

	$
1947	11,740
1948	21,880
1949	14,000
1950	13,105
1951	17,900
1952	14,300
1953	13,750

As the income from the annual dues paid to the Union by the adhering countries amounted to $24,018 in 1951 and to $23,451 in 1952, it is clear that the financial aid from UNESCO plays an essential role in the budget of the Union. The International Latitude Service, le Bureau International de l'Heure, several symposia, many publications and other activities of the Union have profited from this financial aid.

This larger budget led also to the formation of Commission 38 for the exchange of astronomers. Under the able presidency of Professor STRATTON this commission has been very active and many young and promising astronomers have obtained through the efforts of Professor STRATTON and his colleagues the opportunity to work for shorter or longer periods at renowned institutes in foreign countries. This form of exchange is so important and has proved so successful that the Union will continue to support Commission 38 financially, even though UNESCO has discontinued its financial aid towards this special purpose.

It is realized that this survey of the activities of the Union is very incomplete. The choice of the subjects treated is a personal and rather arbitrary one, but a fuller treatment would go beyond the scope of this article. However, some of the problems which the Union has still to solve should be mentioned. The first of these difficulties is probably met in all international organizations. It is the problem of languages. The fact that many tongues are spoken by the members of the Union usually does not seriously hamper conversation and the verbal exchange of ideas between them, as often a language can be found which is understood by both parties. But in large meetings, like General Assemblies, meetings of commissions and symposia, long translations are sometimes required which impede an effective and expeditious course of the proceedings. The problem is most urgent however with respect to scientific publications. Here the Union has always stressed the desirability that abstracts in another main language be added to the original articles. It seems impossible that the Union can solve this problem of languages completely. It will be the task of several generations to come to remove the barriers which at the present time impede the free intercourse between the different nationalities.

The second problem which may be mentioned here is one of an "organizational" character. Since its establishment the Union has divided its scientific task over a number of standing commissions. It is evident that any subdivision of astronomical science into a number of specialized fields must be arbitrary and artificial. Probably for this reason some other big Unions have followed a less drastic course and work through a small number of large sections. During its early history the number of members of the Union was so small that the standing commissions could be considered as working groups of a relatively small number of experts in the field. At present an astronomer can only be a member of the Union if he is nominated as member of a commission. As a consequence of this rule and of the present growth of astronomy, several commissions have become very large, and it is well known that the efficiency of a commission is certainly not proportional to the number of its members. The difficulties indicated here have been studied and discussed more than once by the Executive Committee, but no other form of organization has been suggested so far which would guarantee an improvement over the present system. During the last General Assembly a number of combined commission meetings were arranged which proved to be quite useful and Professor LINDBLAD suggested that in large commissions small working parties should be formed. Although we do not know what the future course of the Union will be, it is reasonable to expect that the Union which has already made itself indispensable to astronomy during the few decades of its existence, will succeed in solving such organizational difficulties.

This article would be incomplete without a few words about the International Council of Scientific Unions (I.C.S.U.). This body, the central organization of the International Scientific Unions, was established in 1931 as a continuation of the International Research Council, which was founded by the Allies in 1919 after the First World War. At present, eleven Unions adhere to this Council. Space does not permit describing in any detail the structure and the activities of this important organization. Although I.C.S.U. and UNESCO are both completely autonomous organizations, an agreement of mutual recognition was drawn up between them in 1947. As a consequence, the granting of financial support by UNESCO to the International Scientific Unions adhering to I.C.S.U. takes place through the intermediary of I.C.S.U. Furthermore, I.C.S.U. has organized and finances a number of special research stations of which the High Altitude Research Station at the Jungfraujoch has proved to be of great importance for astronomical research. Another form of activity of I.C.S.U. is the formation of Joint Commissions between members of two or more Unions in order to co-ordinate efforts in special fields which fall within the domain of more than one Union. At the moment, members of the I.A.U. are active in four of these Joint Commissions, *viz.* on High Altitude Research Stations, the Ionosphere, Solar and Terrestrial Relationships, and Spectroscopy.

The board of I.C.S.U. consists of a Bureau, an Executive Board, and a General Assembly. No doubt it will interest the reader to know that Professor STRATTON, who has served the I.A.U. for so many years, has also been the General Secretary of I.C.S.U. from 1937 until 1952. Few scientists have taken such a very large share in the efforts to improve international organization and collaboration as has Professor STRATTON.

The International Astronomical News Service

JULIE VINTER HANSEN

Universitetets Astronomiske Observatorium, Copenhagen

SUMMARY

This is the story of the development of the International Astronomical News Service from the wishful thinking of TYCHO BRAHE up to the present organized global news service by telegrams and circulars, under the auspices of the International Astronomical Union. The location of the present International Telegram Bureau is: University Observatory, Copenhagen, Denmark.

THE feeling of the importance of co-operation between scientists has been steadily growing through the years. No doubt most scientists have always felt an urge to contact other learned men to discuss problems, discoveries, and inventions. In ancient times such contacts were not easily made; travelling was hazardous and mail-service non-existent, hence progress was slow. The lack of intercourse between scientists sometimes led to bitter quarrels about the priority of theories, discoveries, or inventions, and it was not easy to pass judgment on the various claims.

In astronomy we see how the discoveries and calculations of the ancient Eastern countries spread to Europe where particularly the early Greeks developed theories about the motions of the celestial objects around us, and this accumulated knowledge was kept alive by the Arabs during the decline of science in Europe during the Middle Ages. With the Renaissance astronomy, as all other sciences and arts, awoke to new life and eminent astronomers even voiced ideas of organized co-operation. TYCHO BRAHE, for instance, in his description of his life* and work (written about 1598) does some wishful thinking in this respect. Writing about his own catalogue of 1000 stars, he points out the desirability of observing the Southern stars—both those visible from Egypt and those further South—and he adds: "So if some mighty nobleman would care to fulfil our own and others' wishes in both these respects, they do a very good deed that would be ever gloriously remembered. Up to now no one has even tried to do a thing like this in the right way, let alone carried it out, as far as I know. I would be willing to provide the necessary instruments and tools if somebody could organize the work and get the right people for such a deserving enterprise". A little later in the same paper, talking about the importance of determining correct geographical positions of localities on the Earth, he once more calls upon "kings and princes and other mighty noblemen in widely separated parts of the world" to make suitable and generous preparations in this respect for "then they would really be doing a good deed, and in this way astronomy, which is in need of widely different terrestrial horizons, would develop towards greater perfection". It took over 200 years before any organized work as that proposed by TYCHO BRAHE was undertaken. That TYCHO also keenly felt the importance of closer intercourse between fellow astronomers is seen from a paragraph in the same paper in which he tells that before King FREDERIK II of Denmark offered him the island Hven as site for an observatory he had had plans for settling in Basel, one of the reasons for this choice being, that "Basel is located so to speak at the point where the three biggest countries in Europe, Italy, France, and Germany meet, so that it would be possible by correspondence to form friendships with distinguished and learned men in different places. In this way it would be possible to make my inventions more widely known so that they might become more generally useful". Into these remarks may be read a desire for a centre, from which useful astronomical news could be distributed. Such a centre, however, took its time to materialize.

The first start was made in Denmark in connection with a comet medal that the Danish King FREDERIK VI instituted on 17th December, 1831,† this gold medal to be given to the person who was the first to announce the discovery of a new telescopic comet. All discoverers had to write immediately about the new comet to H. C. SCHUMACHER, professor of astronomy in the University of Copenhagen, but resident in Altona, Holstein, and editor of the *Astronomische Nachrichten.* In case of more than one claimant Schumacher was to decide who was to be given the medal. The first of these medals was presented in January, 1833, to GAMBART, Marseille, for the discovery of a comet on 19th July, 1832 (GAMBART, 1833).

A few years afterwards co-operation was established with FRANCIS BAILY (1835), in such a way that discoverers of comets "if in any part of Europe except Great Britain must send *immediate* notice to Professor SCHUMACHER of Altona; and if in

* TYCHO BRAHE'S description of his instruments and scientific work, *Det. Kgl. Danske Videnskabernes Selskab*, Copenhagen, 1946.

† Announcement published in *Astron. Nachr.*, **10**, No. 221, 1832.

Great Britain, or any other part of the globe except Europe, must send *immediate* notice to FRANCIS BAILY, Esq., of Tavistock Place, London. Professor SCHUMACHER and Mr. F. BAILY are to determine whether a discovery is to be considered as established or not; but should they differ in opinion, Dr. OLBERS, of Bremen, is to decide between them". This arrangement seems to have worked very well.

In addition to publishing news of astronomical discoveries in the *Astronomische Nachrichten*, SCHUMACHER also printed, when needed, circulars to spread the news of such discoveries as quickly as possible. These circulars contained not only discoveries of comets and cometary orbits and ephemerides but also news about asteroids, for those were the days when the discovery of a minor planet was considered quite a remarkable event. Famous names like that of Gauss are sometimes found among the contributors to the circulars, and these circulars were not always as matter of fact as present-day circulars from the Central Bureau for Astronomical Telegrams. For instance, in a circular of 22nd October, 1847 (reprinted in *A.N.*, **26** No. 616), announcing the discovery by HIND of a minor planet, a proposal by Sir JOHN F. W. HERSCHEL to name the next discovered asteroid Flora is mentioned, and SCHUMACHER quotes in full the following lyrical paragraph from HERSCHEL's letter:

> "Pallas, Juno, Ceres, and Vesta, as sober and majestic Duennas will abundantly provide for the respectability of the group between Mars and Jupiter, while Astraea, Iris, Hebe, and Flora will attract all eyes and fill all imagination with sweet and graceful images".

A quicker way to spread astronomical news was provided when telegraph service came into being. In 1869 the Imperial Academy of Sciences in Vienna instituted a gold medal or the corresponding value thereof to be given to the person, who first announced a new telescopic comet. The announcement was to be sent, preferably by telegram to Vienna and from there the news was to be distributed by telegrams to selected observatories. The first two medals went to TEMPEL (1873), of Marseille, for comets discovered in 1869. In 1873 the intercourse between astronomers on opposite sides of the Atlantic Ocean was made more easy when the Associated Trans-Atlantic Cable Companies granted the Smithsonian Institution of Washington a limited number of astronomical cablegrams free of charge between America and Europe, more strictly between the Smithsonian Institution and the Astronomer Royal of England.* In the first place the idea was to transmit discoveries of comets and planets, but later† it was stated that other astronomical phenomena, for instance, disturbances on the Sun, outbursts of some variable stars, unexpected meteor showers, would be proper subjects for telegrams. Although great satisfaction was felt in this quick new way of obtaining news from distant observatories, no real organized news service was yet in existence.

At a congress of the Astronomische Gesellschaft in Berlin, Professor FÖRSTER (1879) pointed out that the existing arrangement for the distribution of astronomical news was not quite satisfactory, particularly the form of the telegrams did not permit any check on the correctness of the imparted numbers except through repetition; also national vanities had showed up. FÖRSTER thought it desirable that a central news bureau be established. Although astronomers took kindly to this idea, nothing much was done about it. In 1881 FÖRSTER once more aired his ideas about such a

* Announcement in *M.N.*, **33**, 369, 1873.
† Announcement in *M.N.*, **34**, 185, 1874.

bureau. He considered the telegraphic transmission of astronomical news free of charge a doubtful blessing, because experience showed that people when not having to pay were apt to be too hasty in transmitting their supposed discoveries without checking them carefully. He also wished for a telegraphic code that would permit immediate check on the cabled figures. At the congress of the Astronomische Gesellschaft in Strasbourg, in 1881, FÖRSTER's plans for a central bureau were debated (FÖRSTER, 1881). Some favourable resolutions were passed, and there the matter rested—presumably because some astronomers were reluctant, considering the proposals too radical and fearing national vanities might be hurt.

In 1882 the appearance of the bright September comet brought on a crisis and made the astronomers realize that the arrangement for the transmission of news was thoroughly inadequate. FÖRSTER (1882) took the initiative and had a central bureau established in the German town of Kiel, under the management of Professor A. KRUEGER, the editor of the *Astronomische Nachrichten*. This bureau had at its start thirty-nine subscribers, thirty-eight in Europe, one in Asia (TASHKENT). For the U.S.A. RITCHIE, Boston, was to act as intermediary between American observatories and the bureau. The first telegram code used was the so-called "Science Observer Code", a code that had been warmly recommended to the Astronomische Gesellschaft at the meeting in Strasbourg by S. C. CHANDLER and JOHN RITCHIE, the editors of *Science Observer*. This code was in use in the U.S.A. and it is based on a dictionary: *A Comprehensive Dictionary of the English Language*, by JOSEPH E. WORCESTER, L.L.D., Boston, 1876. In this code it is possible to communicate five-figure numbers, and thus telegraph positions or orbits and ephemerides together with a check-number, being the sum of the preceding numbers; for instance, if $\omega = 354° \ 9'$ $= 35409$ you had to seek the 9th word on page 354 of the dictionary and came out with the word pyrrhic. This code, however, was not quite satisfactory; some of the pages did not contain the needed 100 words, and none of them had a word for the figure 00, and furthermore several words listed in the dictionary were the same, so that uncertainties arose when decoding. KRUEGER (1882) therefore very soon decided to change code and adopted a five-figure code that in principle was the same as that still in use, although minor changes have taken place. This new code was an amendment of a figure code, proposed by KARLINSKI (1866), of Cracow. At the start the Vienna Observatory had agreed to help the Central Bureau in providing orbits and ephemerides to be telegraphed from the bureau to its subscribers. In 1883 the Harvard College Observatory undertook the distribution in the United States of astronomical information received from the bureau; for many years it favoured another code than the bureau in its telegrams, a syllabic code (Gerrish system).

The new telegram bureau functioned well, not only by telegrams but also, in less urgent cases, through circulars. From a notice in the *Astronomische Nachrichten*, dated 5th December, 1883, about meteor observations, it can be read that KRUEGER had been in for the same worries that befell later leaders of central bureaus, namely, how much to convey by telegrams and to how many of the subscribers. All went smoothly until the fateful year of 1914, when World War I started, and telegraphic and postal communications between warring nations came to a stop. At that time Professor H. KOBOLD was head of the Central Bureau in Kiel and on 3rd November, 1914, he made an arrangement with Professor ELIS STRÖMGREN, in neutral Denmark, to the effect that the Copenhagen Observatory took over the management of the Central Bureau for astronomical telegrams during hostilities. ELIS STRÖMGREN was well suited for this work, as in earlier years he had been an assistant in Kiel and was familiar with the workings of the Central Bureau. It was possible for him all through the war to keep up astronomical intercourse and satisfactory news service between astronomers all over the world. After the war when great bitterness existed between various peoples, it proved impossible to return to the status of having only one Central Bureau. The international relations were so strained that some astronomers even refused to deal with the Copenhagen Bureau, because it was considered too closely connected with Kiel. For a short while Professor B. BAILLAUD, of the Paris Observatory, undertook to act as intermediary between such astronomers and the Copenhagen Observatory. In 1919 the International Research Council was founded in Brussels and from this Council sprang the International Astronomical Union (I.A.U.), at first allowing only astronomers from the *Entente* powers to adhere, but soon astronomers from neutral countries were also invited to join.

The I.A.U. established its own central bureau for astronomical telegrams in Uccle, with Professor G. LECOINTE as its director. This bureau started work on 1st January, 1920, and three central bureaus were thus in existence: Uccle and Kiel, with Copenhagen acting as an intermediary between the two. This peculiar state of affairs lasted until 1st October, 1922, when the I.A.U. Bureau was transferred from Uccle to Copenhagen, with E. STRÖMGREN as its director. The number of Central Bureaus were thus reduced to two, and friendly relations existed between these two bureaus.

The Second World War brought the next great upheaval in the story of the astronomical news service. Denmark tried to stay out of the debacle but was invaded by the Germans in April, 1940. ELIS STRÖMGREN, being a great diplomatist and being well-known in German circles, was despite all odds fairly successful in keeping up international astronomical intercourse; he even obtained permit to send code telegrams, via the Lund Observatory, Sweden, to subscribers in allied countries, and the I.A.U. circulars, although often late in arriving, generally did reach their wide-spread destinations. The I.A.U. Bureau having functioned all during the war could at the end of hostilities immediately resume its activities in a normal way. The old Kiel Bureau did not fare so well; it had been transferred to the Astronomisches Rechen-Institut in Berlin, but in the present political difficulties of that city the location as news centre was most unfavourable, and the bureau was moved to Heidelberg with Professor KOPFF as its director. When telegraph service was again established between Denmark and West Germany the two bureaus could once more co-operate in the old routine.

ELIS STRÖMGREN remained director of the I.A.U. bureau in Copenhagen until his death in 1947, when the writer took over. The bureau is steadily acquiring new subscribers; at present it has sixty subscribers to its telegrams and circulars and eighty-five that receive circulars only. Its services really cover the whole of our globe from Moscow to New Zealand and from Tokyo all around the world to Rio de Janeiro and Santiago, Chile. In U.S.A. the Harvard College Observatory receives the I.A.U. telegrams and distributes the news to the Americas and in the U.S.S.R. the Sternberga Observatory, Moscow, likewise acts as intermediary for telegrams between the bureau and Soviet institutions. The I.A.U. circulars, however, go to numerous subscribers in both the above-mentioned countries. Since 1922 and up to the end of 1951 the I.A.U. Bureau has transmitted 6409 single telegrams and 1339 circulars were printed. The bureau is supported by grants from the I.A.U. and from the

Danish Rask-Örsted foundation which greatly helps to reduce the costs of subscription.*

Life at the bureau is quite variegated. At times when nothing much is happening in the sky, all is quiet; at other times life grows very hectic when a bright comet or nova or several comets are discovered at practically the same time. The clerical work of decoding and rewriting the telegraphic messages is so well organized that in an hour or so we are able to have the telegrams to all our sixty subscribers ready to go to the telegraph office, but this clerical work represents the smallest part of the bureau work. The deliberations before it is decided to transmit a message often claim several hours. It has to be investigated whether the announced object really is new and whether the announcement may be considered trustworthy, and if the decision is favourable the question arises whether telegrams have to go to all subscribers or only to a selected few or whether circulars, perhaps sent by air-mail, will be sufficient. This probing of the news is by far the most important and difficult part of our work, and we certainly have our surprises, when for instance, an announced bright nova proves to be one of the major planets. Besides transmitting the news of discoveries in the sky the bureau also has the responsibility of following up this news, so that the new object may be properly investigated, that is, for comets we have to provide our subscribers with sufficient orbits and ephemerides. In busy times the staff of the Copenhagen Observatory is far too small to take care of this task but fortunately a number of astronomers and observatories come forward to help us. Very often the discoverer himself or his near colleagues take a patriotic pride in providing the necessary computations, and the Leuschner Observatory, particularly Dr. Cunningham, is also keen on such computations. For the short-period comets the Computing Section of the British Astronomical Association does extensive and appreciated work in supplying predicted elements and ephemerides.

I may conclude in adding that the activities of the bureau are well-known outside the astronomical world, which in a way is gratifying, but also has its drawbacks; we can, for instance, thank non-astronomers for good observations of meteors but we have also had our share of announcements of "flying saucers"; unfortunately it has not added to our popularity in a sensation-greedy world that so far we have staunchly refused to spread such announcements through the channels at our disposal.

References

Baily, F. . . . . 1835 *M.N.*, **3**, 132.
1870 *Astron. Nachr.*, **76**, No. 1809.
Förster, W. . . . . 1879 *Vierteljahrsschrift d. Astron. Gesellschaft*, **14**, 345.
1881a *Astron. Nachr.*, **100**, No. 2386.
1881b *Vierteljahrsschrift d. Astron. Gesellschaft*, **16**, 350.
1882 *Astron. Nachr.*, **103**, No. 2472.
Gambart, J. F. A. . . . . 1833 *Astron. Nachr.*, **10**, No. 238.
Karlinski, F. . . . . 1866 *Astron. Nachr.*, **66**, No. 1562.
Krueger, A. . . . . 1882 *Astron. Nachr.*, **104**, No. 2481.
Stratton, F. J. M. . . . . 1934 *M.N.*, **94**, 361.

* For a summary dealing with international astronomical co-operation, see also Stratton (1934).

The Committee for the Distribution of Astronomical Literature and the Astronomical News Letters

BART J. BOK
Harvard College Observatory, Cambridge, U.S.A.
and
V. KOURGANOFF
Laboratoire d'Astronomie de Lille, France

IN 1940, shortly after the invasion of the Low Countries, it became evident that for several years to come there would be great obstacles to the continued exchange on a world-wide basis of astronomical literature. At the Wellesley meeting of the American Astronomical Society (September, 1940) the Council appointed a Committee for the Distribution of Astronomical Literature (C.D.A.L.), with H. R. MORGAN and JAMES STOKLEY as members and BART J. BOK as Chairman. The C.D.A.L. was charged with the responsibility of promoting as far as possible the continued world-wide flow of astronomical literature. WILHELM BRUNNER, in Switzerland, BERTIL LINDBLAD, in Sweden, KATHLEEN WILLIAMS, in Great Britain, and G. NEUJMIN, of the U.S.S.R., offered to assist in the work of this Committee. Much help was also received from J. H. OORT, in Holland, and A. KOPFF, in Germany. About a dozen copies of each astronomical publication from the United States, Great Britain, and Canada were distributed in this fashion and in return the C.D.A.L. received from abroad many publications intended for distribution and circulation in the United States, Great Britain, and Canada.

With the entry of the United States into the war (December, 1941), there developed increasing obstacles to the sending abroad of books and publications and in the summer of 1942 the C.D.A.L. decided upon the publication, in co-operation with the U.S. Department of State and the Office of War Information, of a monthly Astronomical News Letter, reporting on current research activities at home and abroad. These News Letters received a very wide distribution and provided for many astronomers abroad an invaluable continuing link with astronomical activity in parts of the world otherwise almost completely cut off from communication. The News Letters proved also to be very helpful in keeping astronomers in Allied military service, or engaged upon military research, informed about the current happenings in their peace-time field of research. Altogether thirty-six of these war-time monthly News Letters were distributed before the project was abandoned. Some of our colleagues abroad, in Europe, South Africa, Australia, and the U.S.S.R., were extremely helpful in promoting the wide distribution of these News Letters.

During the war years BRUNNER and LINDBLAD did everything possible to provide the C.D.A.L. with copies of current astronomical literature from the occupied countries and Germany. The few available copies of each publication were first abstracted, and the abstracts were then collected in a separate series of C.D.A.L. "Bulletins," which were then distributed in mimeographed form to interested astronomers in the United States, Canada, and occasionally Great Britain.

At the end of the war, the Council of the American Astronomical Society requested the C.D.A.L. to do what it could for the rehabilitation of devastated observatory libraries in war-torn countries. Through the services of an enlarged Committee of American astronomers and a group of representative astronomers abroad, the needs were assessed and a significant contribution toward the rehabilitation problem was made. Since the funds available to the C.D.A.L. were small, we had to depend largely on gifts from publishers and observatory directors, but help was freely given and most of the requests could be satisfied.

When in 1948 all western-language abstracts of astronomical publications of the U.S.S.R. were discontinued, a new need developed and the Astronomical News Letters were resurrected, as the medium for the publication of abstracts in English of astronomical papers published only in Russian or in another language of the Soviet Union. OTTO STRUVE became the Editor of the new series, which began with News Letters No. 37 and which has now reached No. 72; his principal collaborators in the preparation of the abstracts were A. N. VYSSOTSKY and S. GAPOSCHKIN, with others occasionally assisting in the work. As in the past the mimeographing and distribution of the News Letters was handled by MARGARET OLMSTED at Harvard Observatory. Early in 1953, STRUVE had to resign as Editor and this post has now been taken by V. KOURGANOFF, who will report below in detail on current activities, and plans for the future.

When I look back upon my fourteen years of intimate association with the C.D.A.L. and the News Letters, I am very much aware of the fact that the work could never have been done without the whole-hearted co-operation from the astronomical fraternity the world over. It is only because of this co-operation, freely given by all, that in astronomy we have had a record of which we can all be proud of continued exchange of information during and after the war. I have named already the astronomers and organizations who were principally involved in the work. But I should add here that success could never have been achieved without the enthusiastic support of JEAN RENDALL ARONS, ELIZABETH CHAPMAN, and MARGARET OLMSTED, who successively acted as Executive Secretary to the C.D.A.L.

It is interesting to reflect how low the actual financial outlay for the whole programme has been. During the past fourteen years, the costs of the programme have been met, first of all, through three grants from the Foreign Relations Division of the U.S. National Research Council; these total $700. In addition, we have received grants totalling $200 from the American Astronomical Society and quite recently a grant of $300 from the International Astronomical Union. Astronomers in the United States and Canada have paid a small subscription fee for the recent series of News Letters, but there have been no other sources of subscription income. All this goes to prove that the astronomical fraternity can do wonders with limited funds, provided the project is one in which all are ready to assist.

B. J. B.

* * *

The use of Latin in "learned works" at the time of COPERNICUS and NEWTON, when all cultured people could read this language, was very fortunate, and contributed in a most powerful way to the progress of science.

No major difficulties seem to have been encountered, when French took progressively the place of Latin as the "official language" of scientists.

Extension to German and English, in the nineteenth century, was smooth enough, and was facilitated by the increased interest in foreign languages for general purposes, including literature, trade, and travel.

Thus a stable state had been reached at the beginning of our century when the knowledge of three "official" languages was sufficient for every scientist of any nation. Distinguished contributions from Dutch, Swedish, Danish, Norwegian, Italian, Russian, Japanese, Polish, Spanish, and other scientists, could get an international diffusion, being all published in French, German or English.

Real difficulties begun in the problem of "linguistic barriers" when some nations started scientific publications in their own "unofficial" languages. This was, apparently, not only an effect of the development of the "national feeling", but also very often the result of a deep change in the structure of scientific research itself. For the progressive differentiation and the extreme specialization, made it increasingly difficult for those attracted to scientific research to maintain the old standard of general culture and to fit themselves to write good papers in any but their own language.

So long as only *isolated* scientists broke with the tradition of "three languages", international co-operation was in no real danger. Authors who, for some reasons, did not obey the general rule knew very well that their paper would be "buried", and would receive less diffusion, but they took the risk. They often published, however, an abstract in French, German or English.

The sudden decision of the U.S.S.R., in 1948, to suppress the use of any foreign language in scientific publications was a hard blow to those who—having painfully mastered French, German, and English—imagined that they then had access to the whole of the international scientific production.

Astronomers are deeply indebted to Professor OTTO STRUVE for the invaluable service he rendered to astronomy by the creation, in 1948, of the "Russian Section" of the Astronomical News Letters. This obviously met an urgent need.

With the help of A. N. VYSSOTSKY and also that of S. GAPOSCHKIN, he undertook the truly gigantic enterprise of making the vast astronomical literature published in the U.S.S.R. available as quickly as possible to the great majority of astronomers.

Through the efforts of STRUVE and his collaborators, the Astronomical News Letters (from No. 37 to No. 71) give a faithful picture of the astronomical production of the U.S.S.R. between 1948 and 1953. This is very clearly reflected by the Subject Index of the Astronomical News Letters, which is to be published in one of the next issues.

Into the huge amount of summarizing work done in the past seven years STRUVE introduced *a new method* of bibliographical analysis, especially well-suited to his purpose; these are its main features:

(1) The notations used by the author in formulae, numerical tables, figures, *etc.*, are translated and explained. This enables the reader, without any knowledge of Russian, to use the original paper for research purposes.

(2) The reviewer expresses, besides discussing the *physical content* of the paper, his appreciation of ideas expressed by the author. He gives both a summary and a review.

(3) All "delicate" topics, for instance, those involving the general and political outlook of Soviet scientists, receive an integral translation, as faithful as possible, in order to avoid any addition to the misunderstandings which at present make difficult the relations between the "East" and the "West".

STRUVE's method was so perfectly adapted to this aim and to present circumstances that, for the moment, there is no reason to introduce any change. It will be best to try to maintain the high standard of the past Astronomical News Letters, compensating the lack of universality of the present Editor by the division of the work between more specialists.

To-day's organization of the Astronomical News Letters differs from the former one only in three points:

(*a*) KOURGANOFF takes the place of STRUVE as Editor.

(*b*) Extension of the former team, to include several European astronomers.

(*c*) Some of the summaries will be written as previously in English, while others, according to the preference of the authors, will be written in French.

Work on the summaries is centralized at Lille, where most of the manuscripts are typed and revised by the Editor, and then mailed to Harvard College Observatory, U.S.A. There Professor BART J. BOK and Miss OLMSTED take care of the printing and of the distribution.

The modest financial help received by the Astronomical News Letters from the International Astronomical Union, the American Astronomical Society, and the French "Centre National de la Recherche Scientifique", is entirely absorbed by secretarial work, printing, and mailing expenses.

For the present (March, 1954), American and Canadian observatories are paying $3 for five issues of the Astronomical News Letters, while all other institutes and observatories, on account of the difficulty of getting dollars, can receive, on application to the Editor, *one* copy of the Astronomical News Letter for their library. They can, of course, subscribe to more copies at the same rate as American and Canadian observatories.

Though the Astronomical News Letter service has at present insufficient means for ambitious plans to be made, we expect to place the whole enterprise on a better financial basis to allow:

(*a*) Simultaneous appearance of each issue in two languages: one entirely in English, and another entirely in French.

(*b*) Complete translations of important papers and textbooks.

(*c*) Extension to other languages than Russian.

We hope that the description given above clearly indicates the spirit of the Astronomical News Letters: that of international friendship, service, and sincere political neutrality.

V. K.

Some Problems of International Co-operation in Geophysics

J. M. Stagg

Meteorological Office, Air Ministry, London*

Summary

To provide for new and rapidly developing subjects of great importance n geophysical science it is desirable that the constituent associations of the International Union of Geodesy and Geophysics should be modernized. This might also serve to keep the size of assemblies and the general administration of the Union within such bounds as can be effectively handled by an honorary secretary. The arrangements for administering grants-in-aid could also be simplified. An important aspect of international co-operation is to maintain efficiently those permanent services of the Union whose function it is to collect and publish observational data and to provide standards, and individual countries should be encouraged to take responsibility for particular services.

In geophysics as in astronomy international co-operation means more than the exchange of ideas and the encouragement of research: it includes the fostering of arrangements by which all countries interested in the physics of the Earth, its oceans, or its atmosphere, adopt common procedures of observation and make the observations available to all who wish to use them. In particular branches of study and between particular countries co-operation goes still farther; but the stimulation of research by discussion and by exchange of publications, and the organization of the wherewithal to do it are the main objects of co-operation in geophysics on a world-wide scale.

1. Present Organization

Is the present machinery adequate for achieving these objects? Because of its special need for rapid interchange of data in standard form at fixed times each day over a network of stations in every country, meteorology is alone among the geophysical sciences in having a separate organization, devoted primarily to the organizational side of the science. For geophysics as a group the principal medium is the International Union of Geodesy and Geophysics (U.G.G.I.) with a semi-autonomous association for each of its seven main branches. So long as the boundaries between these branches remained fairly clear, the mechanism for ensuring international collaboration worked well enough. But when new techniques and instruments of exploration, such as radar and high altitude rockets, and new lines of thought, like convection currents within the Earth, began to override the boundaries, some of the associations were found to have become too limited in their interests and responsibilities for adequate encouragement of the new fields of investigation. Ways of providing for these new branches had to be found.

2. New and Borderline Subjects

One solution has been sought in joint meetings of the associations most directly concerned. These serve the purpose of ventilating views, but they can be satisfactory in the larger sense only if the responsibility for developing the new subject and for organizing the necessary investigations lies specifically with one or other of the

---

* Lately General Secretary of the International Union of Geodesy and Geophysics (U.G.G.I.).

associations. Joint meetings cannot adequately provide the continuity of stimulus and support needed by new and rapidly expanding branches of the science. Nor does the formation of additional sections within existing associations or even the formation of additional associations offer an adequate solution. Without forceful handling, sections having their own officers and their own programmes of activity are liable to become progressively more self-centred and self-contained, to the detriment of the parent association and its collaboration with other associations. And apart from the difficulty of finding accommodation at the Union's general assemblies for many associations and their committees in such close proximity as to preserve a community of spirit and opportunity for both formal and informal meetings, the present number of associations is about the limit that can satisfactorily be handled by an honorary secretariat, more especially since collaboration with other international bodies (*e.g.* UNESCO) has increased the amount of correspondence that must be dealt with centrally through the Union bureau.

The most likely solution seems to lie in combining the smaller associations so as to make room for more vigorously expanding branches of the science; or, less drastically, the alternative may lie in modernizing them so as to change the emphasis of their hitherto restricted interests and activities under new titles, and in widening their responsibilities accordingly.

3. Permanent Services

Another important problem concerns the maintenance of the permanent services which are so essential in geophysics, and of which U.G.G.I. is wholly or partly responsible for seven. Though most countries can contribute their quota of basic information about some aspect of geophysics, only a few have the facilities for applying the collected data to operational use or research on a world-wide scale. In meteorology the machinery required for this kind of collaboration is provided by World Meteorological Organization, which, being a governmental body and also a specialized agency of United Nations with a full-time secretariat of about forty persons, has considerable resources through which it can function. The other geophysical sciences are less well provided. As membership of U.G.G.I. is by academic agreement, there is no legal obligation by member countries to implement the Union's resolutions or even to contribute to its funds. One indirect result is that, for a group of subjects each so dependent for development on measurements and observations made all over the world, geophysics has inadequate resources in bureaus for collecting, reducing, and publishing these basic data, and the bureaus that have been established lead an uncertain and precarious life.

4. Affiliation with UNESCO

As one possible way of making the permanent services more effective it has been proposed from time to time that membership of the Union should be by government convention. Since a change of this kind would undoubtedly alter the whole character of the Union as well as introduce serious difficulties for some member countries, it was not surprising that the proposal was allowed to lapse when UNESCO offered affiliation to the International Council of Scientific Unions and through I.C.S.U. to each of its unions, including U.G.G.I., particularly since the offer was made on the understanding that private (non-governmental) agreement would continue to be the basis of membership. As affiliation carried with it financial support, some of the services sponsored by the Union (*e.g.* the compiling and publication of the *International Seismological Summary*) have been very materially helped, and others have been allowed to extend their important functions. Further, by relieving the Union and its associations of part of the increased cost of their publications and other expenses, the grants from UNESCO have permitted it to stimulate investigations which need assured support on a long-term basis. Financial assistance has also led to a rejuvenation of some of the older joint commissions and to the establishment of new joint commissions for the study of topics of mutual interest in borderline subjects.

5. Grants-in-Aid

With so much generous assistance on the credit side, other aspects of affiliation with UNESCO are introduced only because they form part of the whole picture. The acceptance of grants naturally puts the recipient unions under obligation to conform with UNESCO's procedures, and since the honorary officer of the union can seldom compete with the paid official of UNESCO, some of the unions have found it necessary to employ staff, and the others, including U.G.G.I., find it increasingly difficult to avoid following their example. This in itself is not evil, but it seems bound to lead to a tendency for secretariats to become anchored in particular countries and also to an extension of the bureaucratic element in the functioning of the unions. Furthermore the grants themselves tend to lead the associations to think in terms of budgets expanded to a degree which the normal resources of the union would have difficulty in maintaining should UNESCO policy change and the grants stop. To operate the scheme of grant allocation effectively would require each union to become more rigorously organized than is appropriate to the traditions of U.G.G.I., and even then an element of unseemly scramble might remain. But perhaps the recent changes in procedure by which UNESCO gives a block grant to I.C.S.U. will improve these less welcome aspects, especially if I.C.S.U. is in turn allowed to give a block allocation to each of its adhering unions for approved but not rigidly defined schemes, the extent of the grant to be determined more by the essentially international character of the use to be made of it and less by criteria of present fashion and popular appeal. If to this were added some assurance of continuity of UNESCO policy so that plans could be made for terms of three or even five years, the unions and U.G.G.I. in particular could assure UNESCO that a more effective use could be made of the grants allotted.

6. Joint Commissions

A similar devolution of responsibility might with advantage be extended to the unions concerned with each of the joint commissions. At present these are set up by I.C.S.U. for the study of specific subjects which overlap the interests of two or more unions. I.C.S.U. provides grants for meetings and for the travelling and expense allowances of members: but because she cannot supervise the activities of all the joint commissions that have come into being—U.G.G.I. alone has representatives on six—I.C.S.U. nominates a mother union for each. Without question some of the commissions function admirably, but others develop a tendency to extend their interests, objects, and terms of responsibility farther than their terms of reference should permit. The weakness probably lies in the division of responsibility between I.C.S.U., who pays the piper, and the unions, who are required to call the tune. For it is not always easy for unions to consort together about the guidance to be given to a body whose activities are defined and paid for by I.C.S.U., and the mother union is often reluctant to seem to interfere, even though she may consider that the funds might be spread among other deserving projects. Perhaps the present arrangements for joint commissions would be improved if I.C.S.U. were to shed part of her financial responsibility on to the unions most intimately concerned with the work of each commission.

7. International Obligations

The best international co-operation is achieved when administration and finance either in the form of grants from bodies like UNESCO or even from central union funds are least obtrusive. As soon as payments are made to any commission or service, the others expect similar treatment and before long contributions have to be made to activities which individual countries had formerly helped to finance as part of their international obligations to science. In this way the machinery of administration becomes more complex and expensive and resources intended for encouraging new projects and investigations are used up. The machinery becomes an end in itself. This process has not yet gone far in geophysics, though the signs are already there. Perhaps the best way to counteract it is to stimulate again the older-fashioned concept of co-operation as self-help in furthering common interests. This spirit in geophysics was at its best in the international polar years of 1882–83 and 1932–33, and is still potent between particular countries: perhaps it will be resuscitated on a wider basis in the forthcoming international geophysical year.

Some Educational Aspects of Astronomy

V. Barocas

Jeremiah Horrocks Observatory, Preston, England

Summary

The interest in astronomy has much increased in recent years among people of all ages. The Preston Municipal Observatory has ventured the experiment of organizing the study of the subject in schools and by encouraging it also in evening classes, in lectures for the adult public, in local astronomical society work, and by active participation of advanced students at the observatory's activities. Some of the encouraging results obtained in this important field are discussed.

Astronomy was in earlier centuries an accepted part of education, but has for a long time been either abandoned or relegated to a position of secondary importance by the general public. In recent years, however, a new interest in astronomy has arisen, and it is important that this interest should be fostered and directed along the right lines. This problem has already been recognized with regard to the interest shown by adults, but there is still a wide scope for work in this field among young people.

The reasons for the increased interest shown by people of all ages are perhaps difficult to assess. Town-dwellers, who lived through the last war with its prolonged black-out and many night duties out of doors, became aware of the starry sky for the first time. Their curiosity was aroused and this led them to try to learn more about the heavenly bodies. During those years a number of simple books on star identification and similar subjects were published and sold very rapidly.

The interest of the younger people has a completely different origin. It arises from science fiction films and "comics" dealing with space travel. The subject has proved very fascinating to young boys and girls and it is even reflected in their games. Unfortunately, the information obtained from these sources is very seldom accurate. As these interests seem to be increasing, it is more than ever desirable that young people should receive some sort of guidance to enable them at least to distinguish true scientific facts from fiction. Nor does this apply to young people alone.

In the last few years a little time has been given to some aspects of astronomical problems in the syllabus of some grammar schools, where it is treated mainly from the point of view of applied mathematics. Little time can be afforded for it and it affects only a few of the pupils. Yet to-day, as astronomy expands in many new fields and new techniques and large instruments are being introduced, one of the main difficulties experienced in observatories all over the world is the lack of trained people to man fully these large instruments and to be able to deal with all the observations obtained. Apart, therefore, from research and routine work, a certain amount of educational work is desirable, if we wish to obtain a steady supply of new recruits for our science. This has already been realized in some schools fortunate enough to have either some kind of equipment, or what is more important, some enthusiastic teacher. The results have always been worth the amount of work and time spent by the teacher. Where small telescopes are available, a definite programme of work is often carried out, while in other schools, where there is no equipment, the astronomical societies, run by the pupils under the supervision of a master, keep interest alive.

After leaving school, young people find an outlet for their interest in the many local astronomical societies, often founded in recent years, which generally have a junior section. The programme of these societies varies. Some are fortunate enough to have a telescope either bought or built by their members, enabling them to follow an observational programme; others limit their activities to lectures, discussions, publication of a news sheet, and hope one day to be able to acquire a small observatory of their own. These societies are doing very good pioneering work, but many of them find that their funds are not sufficient to cover the many activities that they would like to undertake. A few societies are fortunate in having secured the interest of the local education authorities and receive some help from them under the Further Education Scheme.

In most districts evening classes in astronomy are organized by the Workers' Educational Association, which provide a carefully graduated syllabus under the guidance of specialists. Finally we must not forget the Extra Mural Departments of the Universities and their valuable work.

In this way a young person who is interested in astronomy has opportunities of extending his knowledge and furthering his studies. If he is unable to follow astronomy as a career, it may well become a very important hobby in his life. Amateur astronomers have done, and are still doing, very good work in this country. Very often they take a keen interest in certain aspects of astronomy which require patient and continuous observations, and which are generally not carried out by large professional observatories which are naturally occupied in more specialized work. Moreover, in a country with a varied and variable climate like Great Britain, the possibility of having a large number of observers scattered in all parts of the country is a great asset. The work done in this field by the B.A.A., the British Astronomical Association, is too widely known to need any description here.

It is unfortunate that not more young people are aware of these opportunities. Schools can seldom afford as much as they would like to awaken interest. An observatory maintained by the local education authorities and available to the schools would appear to be an answer to this difficulty. There are only a few municipal observatories in the country but the educational work they do is of great value in the educational field.

There is no doubt that a number of instruments are available in the country, which although not large enough for modern research work, could nevertheless be extremely useful for training purposes and for adult educational work. The main difficulty is the expense of acquiring and maintaining such equipment. An organized plan is needed for the erection and maintenance of a number of small observatories in the largest centres of the country. These observatories can do very varied work, as is illustrated by the activities of the municipal observatory, run by a local education committee as an institution for further education, and to which the writer is attached.

This observatory has started an experimental educational programme aimed not only at increasing the interest in the subject of people who have already a general knowledge of astronomy, but also at giving sound scientific facts to people (child or adult) who are simply curious and who must otherwise rely for their information either on newspapers or on space travel films.

The work has been concerned specifically with schools, the general public, the local astronomical society, and with more advanced students. The schools are visited periodically and simple lectures using film strips, slides, photographs, and models are given to young people (age thirteen to sixteen). The teachers also help by allowing their classes time to listen to the course of talks on astronomy for schools given by the B.B.C. After these talks, discussions usually follow and some of the points which may have presented difficulties are elaborated. When the class has a basic knowledge of elementary facts, it is then allowed to visit the observatory in the day-time so as to see the main equipment of an observatory and to receive a practical demonstration on how the instruments are used. Finally, the pupils visit the observatory, in small groups, at night, in order to see for themselves how celestial objects appear through a telescope.

The experiment has shown that the young people are very interested and receptive. Some simple test-papers set at the end of the course have shown a general high standard. For the grammar school type of student, since one is talking to young people who have already obtained the G.C.E., and in some cases to pupils who have already gained admission to a University, the approach to the subject is different. The course generally consists of lectures dealing with particular aspects of astronomy. As the pupils have a knowledge of physics and mathematics, the lectures can be of a more advanced type. Here too the results obtained, judging by the standard of the questions asked, are very encouraging. Those students who show a definite interest in the subject are encouraged to come to the observatory to ask advice on further reading or on practical work. It is found that particularly among the boys there is a desire to do something themselves. Some observe simple phenomena with the naked eye or by means of binoculars, while one particular boy, who possesses a 3-in. refractor, takes regular observations of the planets, making sketches of them and is now learning to take observations of occultations. This boy hopes to gain admission to a University and intends to make astronomy his career.

For boys and girls who are older and who have already left school and are working, a series of lectures is arranged at the Day Continuation College which they attend. The attendance to these lectures is optional, but I have found that they are attended by all the students.

The education of adults must be approached in a different way. The observatory is opened to the public one evening a week and, by arrangement, groups from societies and organizations can visit it on other nights. The number of people who visit the observatory during the year is very large. In these groups we often find people who express a wish to learn something more. To satisfy this demand evening classes are held where the more serious enquirers can learn some fundamentals of astronomy. They often join the local astronomical society and also the B.A.A. Apart from this, lectures for the general public, given by eminent astronomers, are organized every year.

Three years ago an astronomical society was formed in the town. The members meet regularly for lectures and they visit the observatory periodically, where they can obtain information and can consult periodicals and publications of other observatories. A few members have purchased or constructed small telescopes, and an observational programme is being prepared.

Finally, the observatory is also available for more advanced students. Some are people who are preparing for an external degree of London University and who, having chosen astronomy as a subsidiary subject, require practical knowledge of the use of astronomical instruments. Others are preparing for examinations in surveying, and they too require a knowledge of field astronomy and practical observations.

On the whole it is felt that the new experiment has been very successful. It provides the schools and the residents of the town and district with something that was lacking before. I feel that although the work may not be so immediately rewarding as research work, it does perform a necessary service to the science of astronomy. It helps to foster an interest in astronomy, provides a starting point for those who wish to acquire some elementary knowledge which can later be developed, and enables them to appreciate the work and the problems that face the astronomer to-day. It is interesting to note that the general reaction of adult visitors to our observatory is one of envy that young people to-day have this opportunity of learning about astronomy which was lacking in their young days.

The introduction of this subject to young people brings a realization of new and wider horizons. It teaches and encourages them to think for themselves instead of accepting passively the evidence their eyes see about them, and it provides, especially for the more thoughtful child, a source of inspiration that is so often lacking in this materialistic age. It is, indeed, an introduction to the power of the human mind and lays the foundations for a fuller life bringing with it the pleasure and appreciation that comes from reading and sharing the experiences of our great scientists, who through centuries of patient work and their interpretation of the observations have gradually built up the knowledge of the Universe that we have to-day.

Recollections of Seventy Years of Scientific Work

J. Evershed

Ewhurst, Surrey, England

Summary

This article records some of the researches in solar physics undertaken during my life. These have included the following: prominence observations covering sixteen years; spectroheliograph work with a direct vision prism; experiments on the radiation of heated gases; eclipse expeditions in 1898 and 1900; studies of the relation between emission and absorption spectra, and of the continuous spectrum of hydrogen. In *India* (1906–1923) my work included: the discovery of radial motion and the estimate of pressure in sunspots; a study of the exceptional observing conditions in Kashmir; measurements of the red-shift in connection with Einstein's prediction; great magnetic storms during flares; the motion in the tail of Halley's comet and its transit over the Sun; observations of novae. At *Ewhurst* (1923–1954) I was concerned with: high-dispersion work with liquid prisms; the study of the Zeeman effect and the discovery of a particularly sensitive line for its determination; measurements of minute line-shifts due to horizontal motions; and the decrease of wavelength of solar lines. Reference is made to scientific men I have met and to whom I am indebted for interest and encouragement.

---

1. Early Memories

My interest in astronomy began over eighty years ago, when, as a small boy, there happened a partial eclipse of the Sun, and I walked, or ran most of the way, from Gomshall to Shere to see it in the doctor's telescope. An added excitement and one which was destined to occupy my thoughts for the rest of my life was a spot on the Sun's disk.

I can remember also, at the age of six, seeing in an illustrated paper a picture of German shells falling in the streets of Paris, and it was during this siege by the Germans that Janssen, the French astronomer, escaped by balloon from the besieged city to observe a total eclipse of the Sun. Was it not on this occasion, when astonished by the brilliance of the red line of hydrogen he determined to look for it again in full sunshine? This was the beginning of the daily spectroscopic observation of prominences without waiting for an eclipse.

About the year 1875, my eldest brother, who was a student at the School of Mines in London, became acquainted with Raphael Meldola, F.R.S., who later became a friend of the family and brought us news of the scientific world. He was a friend of Charles Darwin, with whom I had the great honour of speaking when Meldola took me to the Royal Society. He also introduced me to the naturalist, Alfred Russell Wallace. My interest in evolution and in natural history was thereby greatly stimulated.

In later years I was impressed by Lockyer's articles in *Nature* about solar prominences, and desiring to see these marvels myself, I made a spectroscope with a tiny prism of 1 cm aperture and a pair of lenses taken from a disused opera-glass. This revealed the solar spectrum with its wonderful array of dark slit images, but was useless for the prominences. Later I constructed a spectroscope with a train of small prisms after the pattern of Lockyer's instrument. With this attached to a 3-in. telescope I got beautiful views of the prominences; and so began a long series of observations and records with a view to determining their distribution in solar latitude and their connection with sunspots.

The discovery by Hale in 1891 of the reversals of the lines H and K in scattered regions over the surface of the Sun and in the prominences, and his subsequent work at the Kenwood Observatory with the beautiful instrument he designed for recording these phenomena, led me to abandon some attempts I had made to photograph prominences and disk markings in the hydrogen line β and to see what I could do with the calcium lines. This led to some interesting correspondence between Kenwood and my home at Kenley.

In the year 1893 I became acquainted with Mr. Cowper Ranyard, F.R.A.S., and he introduced me to Hale, who was on a visit to England. Ranyard died in December, 1894, and left to me his astronomical equipment which included an 18-in. reflecting telescope, and a small spectroheliograph which he lent to Hale for use on Etna in an attempt to photograph the corona in full daylight. This instrument had the disadvantage of giving distorted spectral images of the Sun, owing to the action of the prisms in giving curved spectrum lines. I erected the telescope and its dome at Kenley, but being dissatisfied with the distorted images given by the spectroheliograph, I designed an instrument using a large direct vision prism giving straight lines and undistorted solar images. This type of spectroheliograph was adopted and successfully used at the Tortosa Observatory in Spain (Fig. 1).

Fig. 1. My first spectroheliograph of 1899 (with direct-vision prism; attached to the 18 in. telescope).

In the year 1895 it was generally believed that the radiation of gaseous elements could not be produced by heat alone, but only by electrical or chemical stimulation could their characteristic line spectra be produced, and this, of course, applied to the spectrum of the Sun. The researches of Pringsheim, Paschen, and Smithells in this connection induced me to undertake some interesting experiments on the radiation of heated gases. I was able to show first that the coloured vapours of iodine and bromine heated to the temperature of a red heat glowed with a continuous spectrum, at the same time giving a discontinuous absorption spectrum by transmitted light.

Other coloured vapours, including chlorine, also gave a continuous spectrum. I was also able to show that vaporized sodium could be made to emit its characteristic *D* radiation by heat alone under conditions where there could be no action other than heat. These results were published in *The Philosophical Magazine* (EVERSHED, 1895).

2. ECLIPSE EXPEDITIONS

In 1898 I joined an eclipse expedition to India organized by the British Astronomical Association. I hoped to get photographs of the spectrum of the reversing layer to study the relation of the emission spectrum to the Fraunhofer dark line spectrum. I obtained good spectra extending far into the ultra-violet. A new feature was shown in the hydrogen spectrum of prominences, consisting of a continuous spectrum beginning at the limit of the hydrogen series of lines at $\lambda 3646$ and extending to the end of the plate. This proved to be the counterpart of the continuous absorption spectrum discovered by Sir WILLIAM HUGGINS in stars having very strong hydrogen lines.

The results of measures of wavelength and intensity in the flash spectra at this eclipse were published in *The Philosophical Transactions of the Royal Society* (EVERSHED, 1902) and also those of the eclipse of 28th May, 1900 (EVERSHED, 1903). The general result of these two eclipses showed that the flash spectrum represents the higher, more diffused portion of the gases which by their absorption give the Fraunhofer dark line spectrum, and that the flash spectrum is the same in all solar latitudes. Previously, it was considered that the flash and the Fraunhofer spectra were separate and distinct phenomena.

At the 1900 eclipse in Algeria a station was chosen near the limit of totality instead of, as is usual, on the central line of eclipse. Near the limit the duration of the "flash" is increased from about 2 sec. to 30 sec. or more. Actually, owing to irregularities in the contour of the Moon the eclipse was not quite total, a small point of the Sun's disk remaining visible at mid-eclipse.

Some Arabs at my station were arguing about whether the eclipse was total or not; the *berger* stated that some little piece of "el Simpsh" remained, as much, he said, as a "garro" (cigarette), but this was contradicted by some men who were hoeing maize 500 metres north-east of my camp, these all declared that the whole Sun was obscured for a moment, as it no doubt was.

The results obtained from the spectra of these eclipses led to an interesting correspondence with Sir WILLIAM HUGGINS, then President of the Royal Society. It was, I think, in the year 1897 that I first had the privilege of visiting Sir WILLIAM at Tulse Hill, where he showed me his instruments and explained how he had discovered the gaseous nature of the Orion nebula. In the year 1905 it was largely through his influence that the India Office offered me the post of assistant to Mr. MICHIE-SMITH, director of the Kodaikanal Observatory.

I must here refer to the very helpful advice given me by the late Professor H. H. TURNER, who arranged for me to travel to India via the United States and Japan, and provided me with introductions to the leading American astronomers.

After interesting and instructive visits to Harvard and Yerkes Observatories, my wife and I arrived at Mount Wilson, where we stayed for some time, and I had excellent opportunities for studying their instruments and methods and the work being carried out under the inspiring direction of Professor HALE.

3. India, 1906–23

My early work at Kodaikanal was largely concerned with the Cambridge spectroheliograph which I brought into working order early in 1907; and so began the long series of photographs of the Sun's disk and the prominences.

The observatory possessed an excellent grating of about 70,000 lines, and with this I constructed a high dispersion spectrograph. I used this at first to estimate the pressure in the reversing layer and in sunspots, and to find out whether there was any motion of ascent or descent in sunspots. The first result showed that there could be no appreciable pressure in the reversing layer or in sunspots. An excellent opportunity presented itself for this work on 5th January, 1909, with two large spots on the Sun, and excellent definition after heavy rainstorms. The spectra revealed a curious twist in the lines crossing the spots which I at once thought must indicate a rotation of the gases, as required by HALE's recent discovery of strong magnetic

Fig. 2. An early stage of a spectacular solar eruption, recorded in Kashmir on 26th May, 1916

fields in spots, but it soon turned out from spectra taken with the slit placed at different angles across a spot that the displacement of the lines, if attributed to motion, could only be due to a radial accelerating motion outward from the centre of the umbra. Later photographs of the calcium lines H and K and the hydrogen line α revealed a contrary or inward motion at the higher levels represented by these lines (EVERSHED, 1910a).

In January, 1911, Mr. MICHIE-SMITH retired and I became director of the observatory.

My late wife was greatly interested in the prominences I had recorded at Kenley and in those we had studied together with the fine equipment at Kodaikanal. She was much occupied at this time in writing her important work on "Dante and the Early Astronomers". She nevertheless was able to find time to study in detail the prominences recorded in the years 1890–1914 inclusive. The results were published in a Memoir of the Kodaikanal Observatory.

In 1914 HALE informed me that he was getting very interesting spectroheliograms using the red hydrogen line α, and now that red sensitive plates were available I attached to the Cambridge instrument a second spectroheliograph, using a grating and special arrangements for getting disk photographs in hydrogen light, utilizing the very perfect movement of the Cambridge instrument. Henceforward, spectroheliograms were taken daily in both calcium and hydrogen light. Notable examples were obtained on the 10th August, 1917. Several spectacular eruptions were recorded, a very interesting one on 26th May, 1916, in which the speed of recession of the flying fragments was measured. This was photographed at Kodaikanal and in Kashmir (EVERSHED, 1914); (Fig. 2).

Owing to the excellent conditions prevailing in the Valley of Kashmir a temporary observatory was established there in 1915, where I installed a large spectroheliograph

Fig. 3. Temporary observatory established in 1915 in the Valley of Kashmir

(Fig. 3) and co-operated with Dr. T. ROYDS who had charge, in my absence, of the Kodaikanal Observatory.

The Valley of Kashmir is a level plain containing a river and much wet cultivation of rice. It is 5,000 ft above sea level and is completely surrounded by high mountains. Under these conditions the solar definition is extremely good at all times of the day, and unlike most high level stations it is best near noon and in hot summer weather. After experiencing these remarkable conditions, observations were made of solar definition in various localities, including one on an ocean liner in tropical waters far from land. The best solar definition was found in low level plains near the sea, or on small islands surrounded by extensive sheets of water. At sea, the definition appeared to be perfect so far as could be judged, with a moving solar image.

A large section of our work at Kodaikanal was devoted to the measurement of the

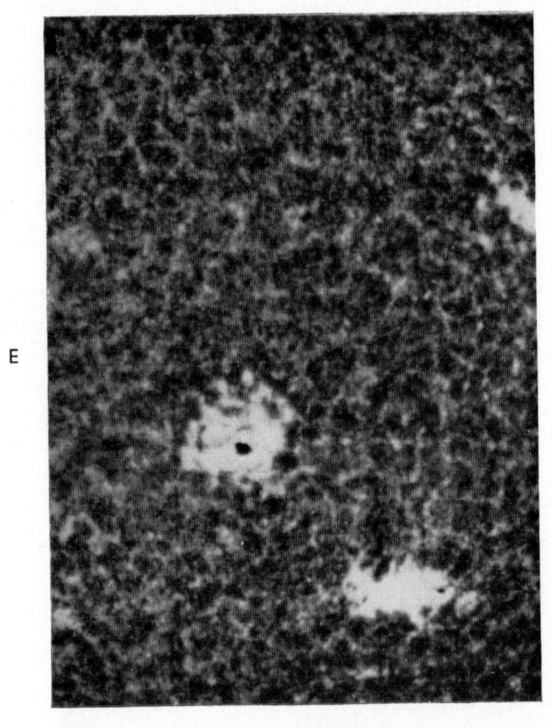

Fig. 4. Enlargement of a calcium spectroheliogram, obtained on 24th September, 1909. This large sunspot was responsible for the great magnetic storm on the 25th, which upset the Indian telegraph system

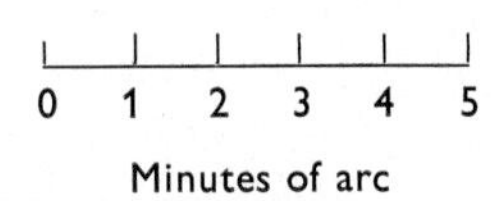

minute shifts of the solar lines towards the red, especially the iron lines, when compared with spectra obtained with the electric arc. In general, in light from the centre of the Sun's disk, the shifts were found to agree more or less with the shift predicted by EINSTEIN, but at the Sun's limb the shift was greater, almost by a factor of two, and this excess remains an unsolved problem.

The integrated light of the whole disk of the Sun also gives spectrum lines shifted to red in excess of the predicted shift, and it occurred to me to put the planet Venus into my optical train in order to see whether or not this excess shift would appear in light reflected from the far side of the Sun. Yet here again when the planet was nearing superior conjunction the line shifts were unaltered.

Our time in India was characterized by some interesting and, indeed, exciting events:

(1) A great spot in September, 1909 (Fig. 4), gave a spectrum in which the Zeeman splitting of the iron lines in the region near H and K indicated a magnetic effect of about 10,000 gauss during what would now be called a "flare" (EVERSHED, 1910b). The accompanying magnetic storm upset the Indian telegraph system, for which we appeared to be held responsible by the Director of Telegraphs!

(2) On the morning of 17th January, 1910, when we were starting the daily series of spectroheliograms, there appeared a great sun-grazing comet shining in the blue sky with a brilliance almost like a detached piece of the Sun itself. In April and May of this same year HALLEY's comet gave us a memorable display, with its tail extending for 100° far up towards the zenith. Many interesting photographs were obtained

and these revealed an accelerating outward movement of the tail. It appeared as though the nucleus had thrown off an entire tail and was forming a new one.

On May 19th the nucleus was computed to be in transit over the Sun. We attempted to photograph this event with the spectroheliograph, using the ultra-violet light of cyanogen which is a specially prominent radiation of the head of comets, but no trace of the comet could be seen in the photograph, nor could it be seen on the Sun in an ordinary telescope. At this time the tail must have swept over the Earth and a sensation was caused in the newspapers by the suggestion of a poisonous gas entering our atmosphere.

(3) On 9th June, 1918, a brilliant new star appeared and the resources of the observatory had to be mobilized at very short notice. In this I had the assistance of the late Mr. R. J. POCOCK of the Nizamiah Observatory, who was visiting Kodaikanal at the time, and it was due to his very resourceful help that we were able to get all the arrangements perfected on 12th June. It is interesting to recall some of the results obtained. The hydrogen emission lines assumed extraordinary forms, Hα appeared as a block of brilliant red light, the β and γ lines were more extended blocks of blue and violet, well defined at their edges. Interpreting these wide bands as due to motion in the line of sight, an expansion of the hydrogen was indicated with velocities up to 1580 km per sec. On the violet sides of these bands two absorption lines appeared, one on the edge of the bright bands; these indicated velocities of the absorbing hydrogen on 12th June of 1586 and 2193 km per sec. respectively. After 13th June the more refrangible line died out, but the other remained, indicating by its decreasing wavelengths on successive days an accelerating outward motion. This motion outward from the star was shared also by the iron vapour (EVERSHED, 1919).

Our somewhat isolated life at Kodaikanal was relieved occasionally by visitors from home or other parts of India. The Director of Indian Observatories, Sir GILBERT WALKER, came on his official visits from Simla once every year with Lady Walker, a most welcome interlude.

4. EWHURST, 1923–55

On retiring from the Indian Meteorological Service in 1923, and establishing a private observatory near Ewhurst, I can recall again the interest and encouragement of H. H. TURNER, F.R.S., and of his friend, Sir CHARLES PARSONS. Sir CHARLES supplied a large block of optical glass which was essential, when made into a 6 in. prism, for much of the work undertaken. This included daily spectroheliograms in hydrogen light, and other spectrograms which confirmed and extended to the prominence region the very remarkable fact that the angular rotation of the Sun increases with the height above the photosphere (EVERSHED, 1935).

My early experiments with a hollow prism filled with ethyl cinnamate revealed great possibilities for high dispersion work if certain formidable difficulties could be overcome. Here I was fortunate to have the co-operation and very skilled work of Mr. F. J. HARGREAVES, who not only perfected the first prism used (1938), but designed and made larger prisms giving very high resolving power. These liquid prisms have to their credit many beautiful spectrum photographs which have revealed some interesting facts; in particular in the magnetic field of sunspots (EVERSHED, 1939a and 1944), the significant fact was disclosed that lines which are Zeeman triplets over the umbrae are doublets in the penumbrae. They also made possible the detection of minute shifts indicating movements parallel to the Sun's surface over

Fig. 5. Spectra showing the D-lines of sodium in the Sun and in a vacuum tube, and illustrating the perfection of the liquid prism (ten transmissions through a single prism). On *top*: photograph taken on 25th November, 1937, at 8^h 23^m GMT. *Centre and bottom:* these two narrow spectra, *b* and *a*, were photographed on the same day (at about 11^h 16^m) at the west and east limbs at the solar equator, respectively. They show also the bright lines D_1 and D_2 given by a vacuum tube containing sodium

wide areas of the photosphere (1933), and gave evidence of decreasing wavelengths during recent years, of iron and nickel lines in spectral regions where atmospheric lines of oxygen and water vapour formed fixed standards of reference (Evershed, 1949); see Fig. 5.

Throughout my time in India and at Ewhurst in later years I have owed much to the friendship of the late Professor H. F. Newall, F.R.S., and I recall with great pleasure the many visits to his beautiful home at Madingley Rise, Cambridge. In 1933 a memorable event was a visit from Newall and Hale to Ewhurst, when Hale brought me his original spectra which were thought to give evidence of the Sun's general magnetic field. These were measured by the positive on negative method with the final result that any minute shifts that could be detected indicated Doppler and not Zeeman effects (Evershed, 1939b).

Finally, it is interesting to recall that I began observing the Sun with a tiny prism of half-inch aperture, and have ended with giant solid and liquid prisms of five to six inches effective aperture.

References

Evershed, J. . . . . 1895 *Phil. Mag.*, (May), p. 460.
1902 *Phil. Trans.*, **197**, 381.
1903 *Phil. Trans.*, **201**, 457.
1910a *M.N.*, **70**, 220.
1910b *Kodaikanal Obs. Bull.*, No. 22.
1914 *Kodaikanal Obs. Bull.*, No. 42.
1919 *M.N.*, **79**, 468.
1933 *M.N.*, **94**, 96.
1935 *M.N.*, **95**, 509.
1938 *M.N.*, **99**, 127.
1939a *M.N.*, **99**, 217.
1939b *M.N.*, **99**, 438.
1944 *Observatory*, **65**, 190.
1949 *M.N.*, **109**, 595.

Astronomical Recollections

MAX BORN

Department of Mathematical Physics, University of Edinburgh*

SUMMARY

Reminiscences of the author's experiences, fifty years ago, as a student of astronomy at two German universities, Breslau and Göttingen, and of his teachers, FRANZ and SCHWARZSCHILD.

I AM not an astronomer, nor have I done any work in physics applicable to astronomy. Yet I cannot resist the wish to be included amongst those who offer their congratulations to Professor STRATTON by an article in this volume. There was a time in my life when I was very near to devoting myself to the celestial science; but I failed. May I offer, as a substitute for a more serious contribution, the story of my wrangling with astronomy and some recollections of remarkable astronomers who were my teachers.

I have to begin with Professor FRANZ, the director of the observatory of my home city, Breslau. My father, who died just before I finished school, had left me the advice to attend lectures on various subjects before choosing a definite study for a profession. In Germany of that period this was possible because of the complete "academic freedom" at the university.

There was in most subjects no strict syllabus, no supervision of attendance, no examinations except the final ones. Every student could select the lectures he liked best; it was his own responsibility to build up a body of knowledge sufficient for the final examinations which were either for a professional certificate or for a doctor's degree, or both. Thus I made up a rather mixed programme for my first year, including physics, chemistry, zoology, general philosophy and logic, mathematics and astronomy. At school I had never been very good nor interested in mathematics, but at the university the only lectures which I really enjoyed were the mathematical and astronomical ones. The greatest disappointment were the philosophical courses; there we heard a lot about the rules of rational thinking, the paradoxes of space, time, substance, cause, the structure of the universe, and infinity. Yet it seemed to me an awful muddle. Now the same concepts appeared also in the mathematical and astronomical lectures, but instead of being veiled in a mist of paradox they were formulated in a clear way according to the case. For that was the important discovery I then made: that all the high-sounding words connected with the concept of infinity mean nothing unless applied in a definite system of ideas to a definite problem.

Astronomy was attractive in another way. There the problems of cosmology are related to the infinity of the physical universe. But little about these great questions was mentioned in the elementary lectures of our Professor FRANZ. What we had to learn was the careful handling of instruments, correct reading of scales, elimination of errors of observation and precise numerical calculations—all the paraphernalia of the measuring scientist. It was a rigorous school of precision, and I enjoyed it. It gave one the feeling of standing on solid ground. Yet actually this feeling was not quite justified by facts. The Breslau Observatory was not on solid ground, but on

* Now at Bad Pyrmont, Germany.

the top of the high and steep roof of the lovely university building, in a kind of roof pavilion, decorated with fantastic baroque ornaments and statues of saints and angels. The main instrument was a meridian circle, which a hundred years ago had been used by the great Bessel; although it was placed on a solid pillar standing on the foundations and rising straight through the whole building, it was not free from vibrations produced by the gales blowing from the Polish steppes. The whole outfit of this observatory was old-fashioned and more romantic than efficient. There were several old telescopes from Wallenstein's time, like those Kepler may have used. We had no electric chronograph but had to learn to observe the stars crossing the threads in the field of vision by counting the beats of a big clock and estimating the tenths of a second. It was a very good school of observation, and it had the additional attraction of an old and romantic craft.

I remember many an icy winter's night spent there in the little roof pavilion. We were only three students in astronomy, and we took the observations alternately. When my turn was finished I enjoyed looking down on the endless expanse of snow-covered, gabled roofs of the ancient city, the silhouettes against the starry sky of the massive towers of the churches around the market place and of the Cathedral further away beyond the river. There on the narrow balcony amongst the stucco saints and old-fashioned telescopes, one felt like an adept of Dr. Faustus and would not have wondered if Mephistopheles had appeared behind the next pillar. However, it was only old Professor Franz who came up the steps to look after his three students—he had not had so many for a long time—and who carried with him the soberness of the exact scientist, checking our results and criticizing our endeavours with mild and friendly irony.

These, our results, I rather think were not very reliable; it was not so much our fault as that of the exalted but exposed position of the observatory. Professor Franz himself, therefore, abstained from doing research, which needed exact measurements, and restricted himself to descriptive work, a thorough study of the moon's surface which he knew better than the geography of our own planet. He made strenuous efforts, however, to obtain a modern observatory but never succeeded. During my student time there were great hopes. The firm Carl Zeiss, Jena, had sent a set of modern instruments to the World's Fair at Chicago. After the end of the show these were purchased by the Prussian State for its university observatories. Breslau obtained an excellent meridian instrument and a big parallactic telescope; yet no proper building was granted, and the meridian circle was installed in a wooden cabin on a narrow island of the Oder River, just opposite the university building. This island was in fact an artificial dam between the river and a lock through which many barges used to pass. The time service for the province of Silesia, which had been practised for scores of years with the help of the old Bessel circle, was transferred to the new Zeiss instruments, but the results remained highly unsatisfactory. Eventually we discovered a correlation between the strange irregularities of the time observations with the changing level of the water in the lock; the island suffered small displacements through the water pressure. Professor Franz's hopes of a more efficient observatory had broken down again.

We youngsters took this disappointment rather as a funny incident. It did not diminish the fascination which astronomy exerted on my mind. This fascination was, however, shattered by the horrors of computation. Franz gave us a lecture on the determination of planetary orbits, connected with a practical course where we had to learn the technique of computing, filling in endless columns of seven decimal logarithms of trigonometric functions according to traditional forms. I knew from school that I was bad at numerical work, but I tried hard to improve. It was in vain, there was always a mistake somewhere in my figures, and my results differed from those of the class mates. I was teased by them, but that made it worse. I do not think that I ever finished an orbit or an ephemeris, and then I gave up—not only this calculating business but the whole idea of becoming an astronomer. If I had known at that time that there was in existence another kind of astronomy which did not consider the prediction of planetary positions as the ultimate aim, but studied the physical structure of the universe with all the powerful instruments and concepts of modern physics, my decision might have been different. But I came in contact with astrophysics only some years later, when it was too late to change my plans.

At that period German students used to move from one university to another, from different motives. Sometimes they were attracted by a celebrated professor or a well-equipped laboratory, in other cases by the amenities and beauties of a city, by its museums, concerts, theatres, or by winter sport, by carnival and gay life in general. Thus I spent two summer semesters in Heidelberg and Zürich, returning during the winter to the home university. The observatory of Heidelberg was on the Königstuhl, a considerable, wooded hill, where the astronomers lived a secluded life remote from the ordinary crowd. I had then definitely changed over to physics, and not even the celebrated name of WOLF, the professor who has discovered more planetoids than anybody else, deflected me from my purpose.

The observatory in Zürich was more accessible, and the name of the professor was WOLFER, which could be interpreted as a comparative to WOLF. But even that did not attract me.

The following summer I went to Göttingen for the rest of my student time. There KARL SCHWARZSCHILD was director of the famous observatory which had been for many years under the great GAUSS. SCHWARZSCHILD was the youngest professor of the university, about thirty years of age; a small man with dark hair and a moustache, sparkling eyes and an unforgettable smile. I joined his astrophysical seminar and was for the first time introduced to the modern aspect of astronomy. We discussed the atmosphere of planets, and I had to give an account of the loss of gas through diffusion against gravity into interstellar space. Thus I was driven to a careful study of the kinetic theory of gases which then, in 1904, was not a regular part of the syllabus in physics. But this is not the only subject which I first learned through SCHWARZSCHILD's teaching. His was a versatile, all-embracing mind, and astronomy proper only one field of many in which he was interested. About this time he published deep investigations on electro-dynamics, in particular, on the variational principle from which LORENTZ's equations for the field of an electron and for its motion could be derived. In the following year (1905) there appeared the first of his great articles on the aberrations of optical instruments; these are, in my opinion, classical investigations, unsurpassed in clarity and rigour, by later work. I have presented this method in my book *Optik* (Springer, 1932), and it is again to be the backbone of a modernized version which will appear soon as an English book on optics (in collaboration with E. WOLF*). SCHWARZSCHILD applied his aberration

* Pergamon Press, London. *To be published in 1956.*

formulae to the actual construction of new types of optical systems; but I am not competent to speak about this part of his activities. Nor can I discuss his astronomical work, experimental or theoretical. Personally he was a most charming man, always cheerful, amusing, slightly sarcastic, but kind and helpful. He once saved me from an awkward situation. I had intended to take geometry as one of my subjects in the oral examinations for the doctor's degree, but was not attracted by the lectures of FELIX KLEIN, the famous mathematician, and attended somewhat irregularly. This fact did not escape KLEIN's observation and he showed me his displeasure. A disaster at the orals, only six months ahead, seemed to be impending. But SCHWARZSCHILD said that half a year was ample time to learn the whole of astronomy. He gave me some books to read and tutored me a little, in exchange for my training him in tennis. When the examination came his first question was: "What do you do when you see a falling star?" Whereupon I answered at once: "I have a wish"—according to an old German superstition that such a wish is always fulfilled. He remained quite serious and continued: "Yes, and what do you do then?" Whereupon I gave the expected answer: "I would look at my watch, remember the time, constellation of appearance, direction of motion, range, *etc.*, go home and work out a crude orbit". Which led to celestial mechanics and to a satisfactory pass. SCHWARZSCHILD differed from the ordinary type of the dignified, bearded German scholar of that time; not only in appearance, but also in his mental structure, which was thoroughly modern, cheerful, active, open to all problems of the day. Still he had his hours of professorial absent-mindedness. There was a "Stammtisch", a certain table in a restaurant where a group of young professors and lecturers used to meet for lunch. SCHWARZSCHILD was one of them until his marriage. A few weeks after the wedding he was again at his accustomed place at the lunch table and plunged in his usual way into a lively discussion about some scientific problem, until one of the men asked him: "Now, SCHWARZSCHILD, how do you like married life?" He blushed, jumped up, said: "Married life—oh, I have quite forgotten—", got his hat and ran away. But I think this kind of behaviour was not typical of him. He always knew what he was doing. His life was short, his achievements amazing, his success great—his end tragic. When the great war of 1914–18 broke out he was employed as a mathematical expert in ballistics and attached to the staff of one of the armies on the Eastern front. There, in Russia, he contracted some rare infectious disease. It was said that he refused to be sent home, until it was too late. On his way home, he visited me in my military office in Berlin; he was still cheerful, but he looked terribly ill. Soon afterwards he died. Now his son, Martin, keeps up the astronomical tradition, thus founding another one of those hereditary lines of astronomers, the HERSCHELS, the STRUVES, and so on.

I have met many other distinguished astronomers and been intimate with some of them; but as most of them are still wandering on this globe, I had better refrain from telling stories about them.

May I conclude by wishing Professor STRATTON a happy future and by adding the request that he too may present us with some recollections of astronomical personalities out of his long experience.

SECTION 2

HISTORY AND PHILOSOPHY

"Glaubt ihr denn, dass die Wissenschaften entstanden und gross geworden wären, wenn ihnen nicht die Zauberer, Alchimisten, Astrologen und Hexen vorangelaufen wären als die, welche erst Durst, Hunger und Wohlgeschmack an verborgenen und verbotenen Mächten schaffen mussten?"

FRIEDRICH NIETZSCHE, *Fröhliche Wissenschaft*, IV, 1882.

The Egyptian "Decans"

O. NEUGEBAUER
Brown University, Providence, R.I., U.S.A.

SUMMARY
It is shown that the 10° sections of the ecliptic, called decans by the Greeks, were originally constellations rising heliacally 10 days apart, and invisible for 70 days. Such stars belong to a zone south of the ecliptic and include Sirius and Orion. The use of the decans for time measurement at night leads to a twelve-division of the period of complete darkness. From this is eventually derived the twenty-four division of day and night.

---

1. THREE different systems of astronomical reference were independently developed in early antiquity: the "zodiac" in Mesopotamia, the "lunar mansions" in India, and the "decans" in Egypt. The first system alone has survived to the present day because it was the only system which at an early date (probably in the fifth century B.C.) was associated with an accurate numerical scheme, the 360-division of the ecliptic. The lunar mansions, *i.e.* the twenty-seven or twenty-eight places occupied by the Moon during one sidereal rotation, were later absorbed into the zodiacal system which the Hindus adopted through Greek astronomy and astrology. With Islamic astronomy the mansions returned to the west but mainly as an astrological concept. A similar fate befell the decans. When Egypt became part of the Hellenistic world the zodiacal signs soon show a division into three decans of 10° each. As "drekkana" they appear again prominently in Indian astrology, and return in oriental disguise to the west, forming an important element in the iconography of the late Middle Ages and the Renaissance.

2. We shall be concerned not with these wanderings of early astronomical concepts but with the much discussed problem of the localization of the decans. I think we now can satisfactorily solve this problem and simultaneously gain an insight into the origin of the twelve-division of night and day in Egypt, from which eventually our twenty-four-division was derived (SETHE, 1920). This progress has been made possible by utilizing information contained in a Demotic papyrus of the Roman period, purchased by the Carlsberg Fund about twenty years ago for the Egyptological Institute of the University of Copenhagen. The late H. O. LANGE recognized the importance of this text, now called "P. Carlsberg 1", which was then published by him and the present writer (LANGE-NEUGEBAUER, 1940). In recent years, my colleague, Professor R. A. PARKER, and I assumed the study of this text in connection with our plans for a comprehensive publication of all available astronomical texts from Egypt. It was from our discussions that it became clear that the Egyptian texts of the Middle and New Kingdom contain all the information required for determining, at least qualitatively, the position of the decans and their use for time measurement.

3. The "decans" make their appearance in drawings and texts on the inner side of coffin lids of the tenth Dynasty (around 2100 B.C.). Here we find thirty-six constellations arranged in thirty-six columns of twelve lines each in a diagonal pattern of which the following scheme represents the right upper corner (the columns proceed

from right to left, as is customary in Egyptian inscriptions):

day 21	day 11	day 1	
S_3	S_2	S_1	hour 1
	S_3	S_2	hour 2
		S_3	hour 3

The constellations "S" are our thirty-six "decans". Among them figure Sirius and Orion, which, except for the Big Dipper, are the only two identifiable asterisms of the Egyptian sky.

The use of these "diagonal calendars" was first explained by Pogo (1932). Each vertical column serves as a star clock during the particular decade the first day of which is quoted at the top of the column. For example, the rising of decan S_3 indicates during the first decade the third hour of the night; in the second decade, the second, *etc.* When a decan rose in the first hour of the night, it was obviously near its acronychal setting ten days later. (Fig. 1 may serve as an interesting illustration of the decans.)

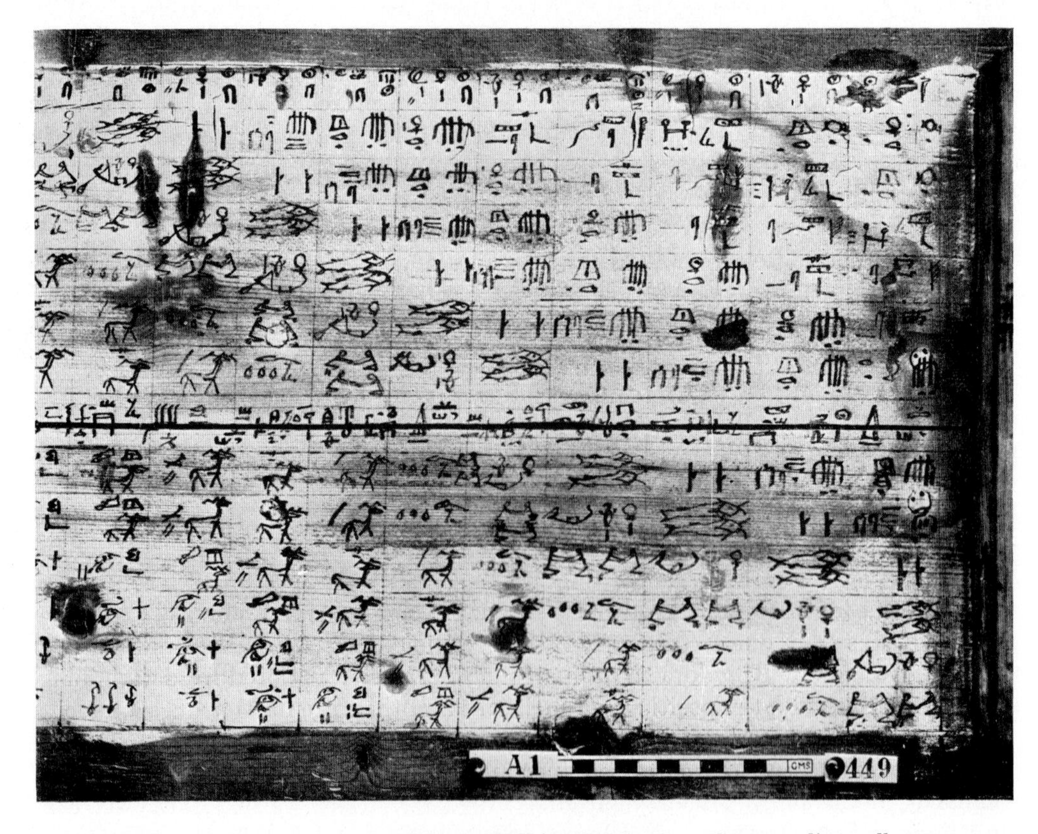

Fig. 1. The Decans on the Coffin of Ḥ K ꜣ-t, showing how these are diagonally arranged. (Cairo Mus. 28127)

We shall not discuss the details of these texts, particularly those modifications of the scheme which became necessary because of the five epagomenal days at the end of the Egyptian year. On the contrary, we shall simplify our discussion by being quite unhistorical and replacing ten-day intervals by 10°-segments of the ecliptic. The error thus committed has no influence on the interpretation of such crude schemes and can be remedied at once if the need should arise.

Using the rising of stars as indications of the beginning of "hours" means that all stars which rise simultaneously—synanatellonta in Greek terminology—are in principle equally serviceable. Thus the diagonal calendars tell us only that the decans were stars located at thirty-six positions of the eastern horizon, such that these horizons intersect the ecliptic at points that are about 10° apart.

4. The next bit of information comes from monuments of Seti I and Ramses IV (about 1300 and 1170 B.C.), where the decans are represented on the body of the sky goddess Nut, and from P. Carlsberg 1, which is an extensive commentary to the often very cryptic inscriptions on the monuments. Again we shall not describe details but utilize only one fact which became clear only through the ancient commentator: all decans are invisible for 70 days between acronychal setting and heliacal rising. In other words, all the decans have (at least ideally) the same duration of invisibility as their leader, Sirius. This shows clearly the origin of the whole concept of the decans. The heliacal rising of Sirius marks, in principle, the beginning of the year, and similarly one chose other stars whose rising indicated the beginning of the consecutive decades of the Egyptian civil calendar. The gradual removal of each such constellation away from heliacal rising was used to mark intervals of the night—we shall call them the "decanal hours". And in order to make the whole scheme as uniform as possible the selection of the decans from among all simultaneously rising stars was made such that they had been seen for the last time 70 days earlier, just as Sirius had spent 70 days in the nether world before rising again at the end of the last hour of the night.

5. Before discussing the resulting decanal hours we shall combine what we know from the diagonal calendars and from P. Carlsberg 1. The diagonal calendars told us that the decans were located on thirty-six horizons which meet the ecliptic at intervals of 10°. But a similar condition is imposed on the thirty-six positions of the horizon with respect to the ecliptic when the stars set. The two groups must be in the relation to each other such that 70 days elapse between corresponding settings and risings. The thirty-six intersections between such pairs give the places of the thirty-six decanal stars.

It is easy to carry out this construction graphically. We make, say, a cylinder projection of the celestial sphere with the equator as circle of contact (see Fig. 2). Let B and A be two positions of the Sun, 70 days apart. Let EH be the position of the horizon when a star S rises heliacally, WH the position of the western horizon when S sets acronychally. For a star of given brightness, it is known how far the Sun in A and B must be distant from the horizon. Thus for given A and B, 70 days apart, we can find S at the intersection of the two proper horizons. Moving A and B 10 days ahead, we get a new pair of horizons and, at their intersection, the next decan. And because Sirius is one of the decans, we are given an initial position AB from which to start. Thus all the other decans can be found, at least in principle.

Of course, this presupposes that we know the brightness of the stars—which is obviously not the case except for Sirius. Nevertheless it is clear that less brightness must remove the two horizons from A and B and thus bring S closer to the ecliptic. And the opposite holds for bright stars. Thus we obtain for the decans a zone, instead of a curve, following by and large the ecliptic toward the south, with Sirius located at its farthest boundary.

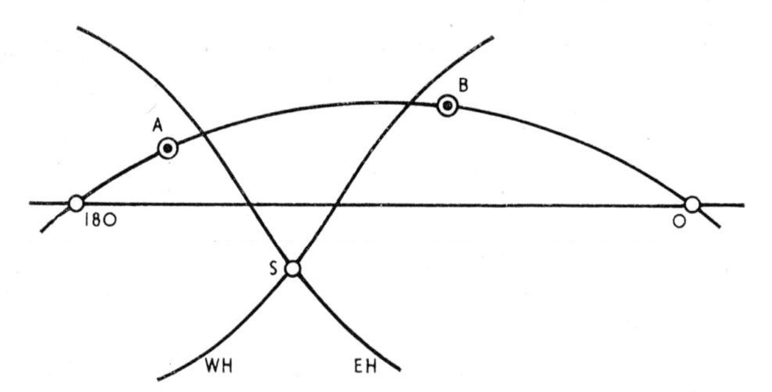

Fig. 2. Graphical determination of the Decans

6. There would be no point in trying to push the identification of the decans any further. What we now know definitely is that both hypotheses which have found support among scholars, namely, that the decans are either ecliptical or equatorial stars (considering the only certain identifications, Sirius and Orion, as "exceptions" to confirm the rule), are equally wrong. It is clear that the 70-day invisibility cannot be taken absolutely literally, not to mention the idealizations made by the Egyptians in order to maintain the relationship between stars and rotating calendar. All these effects soon enough rendered unusable the whole device of the diagonal calendars, and it was probably already obsolete in practice when we meet it as the traditional time instrument for the use of the dead person on his coffin. Indeed, as Parker has recognized, the star tables of the temples of the Ramessides abandon decanal risings in favour of transits of quite different constellations. But the decans maintain to the end of Egyptian history their rôle as representatives of the consecutive decades of the year and as such they were readily absorbed into the Hellenistic zodiacs.

7. We must now come back to the "decanal hours". Obviously these "hours", determined by the rising of stars on horizons with constant longitudinal difference, are neither constant nor even approximately 60 min in length. They vary as the oblique ascensions of the corresponding sections of the ecliptic and are about 45 min long because during each night eighteen decans rise and set. Thus twelve decanal hours seem too short to measure time at night. In so arguing, one forgets, however, that each decan has to serve for ten days as indicator of its hour. If we furthermore require that these "hours" of the night never should be part of twilight, we get a satisfactory covering of the time of total darkness by means of only twelve decans during ten consecutive days, especially for the shorter summer nights.* Again for the sake of consistency a simple scheme which held true for Sirius near the shortest summer nights was extended to all decans and all nights, thus making the number of

* This can be checked, *e.g.* by using the tables of oblique ascensions in the Almagest II, 8.

"hours" twelve for all seasons of the year. And finally the symmetry of night and day, of upper and nether world, suggested a similar division for the day. This parallelism between day and night is still visible in the "seasonal hours" of classical antiquity. It is only within theoretical astronomy of the Hellenistic period that the Babylonian time-reckoning with its strictly sexagesimal division, combined with the Egyptian norm of 2 × 12 hours led to the twenty-four "equinoctial hours" of 60 min each and of constant length.

REFERENCES

LANGE, H. O. and NEUGEBAUER, O. . . . . . 1940 Papyrus Carlsberg No. 1, ein hieratisch-demotischer kosmologischer Text, Danske Videnskabernes Selskab Hist.-filol. Skrifter 1, No. 2.

POGO, A. . . . . 1932 Calendars on coffin lids from Asyut, *Isis* 17, pp. 6–24.

SETHE, K. . . . . 1920 Die Zeitrechnung der alten Aegypter, Nachr. d. Ges. d. Wiss. zu Göttingen, Philol.-hist. Kl., p. 28–55; p. 97–141.

The Astral Religion of Antiquity and the "Thinking Machines" of To-day

GEORGE SARTON

Harvard University, Cambridge, Mass., U.S.A.

1. THE ASTRAL RELIGION

THE theory of homocentric spheres invented by EUDOXOS of Cnidos (IV–1 B.C.)[1] made it possible "to save the phaenomena" (σώζειν τὰ φαινόμενα), that is, to give a mathematical explanation of the motions of the "erring" bodies, or planets. For example, the spherical lemniscate (ἱπποπέδη) which Mercury (or Venus) describes in heaven could be explained as the cinematical resultant of the motions of four homocentric spheres, which are linked together, and each of which rotates around its own axis with a different speed. This theory was a magnificent illustration of the rationalism of the Greeks, their mathematical genius and their boldness; they were not more shy of introducing incredible postulates in order to give a rational account of observations than are the astrophysicists of to-day.[2]

What is stranger still, they were inclined to combine rational and workable theories, such as that of the homocentric spheres, with oriental fantasies. Pythagorean astronomers were largely responsible for such combinations. Two of their fundamental ideas were very probably of Babylonian or Mazdean origin, to wit, the dualism of the universe (sublunar and translunar) and the divinity of the astra,[3] especially the

[1] Symbols like (IV–1 B.C.) or (V–2) mean first half of the fourth century B.C., second half of the fifth century after Christ.

[2] Sometime later the theory of homocentric spheres was replaced by other theories making use of epicycles and eccentrics, but that did not change the situation. It was simply the replacement of one kinematic explanation by another more convenient. The theory of epicycles and, possibly, also that of eccentrics, was originated by APOLLŌNIOS of Perga (III–2 B.C.).

[3] The word *astra* is used because of its generality. The *astra* include stars and planets.

planets. The Pythagoreans inheriting other oriental ideas and, translating them into the Greek idiom, added two new conceptions, first the excellence of circular and spherical motions, second, the return of men's souls to heaven after death. On account of their divinity, celestial motions must be circular; the erratic trajectories of the planets across the stars must be explained in terms of cinematic compositions of circular movements. In the second place, there is a relationship between astra and souls. The immortality of the soul thus became part and parcel of astronomical doctrine.

The endeavour to explain celestial mechanics had led to the formulation of what might be called an astronomical theology. The earliest, if not the clearest, formulation of it was the *Epinomis*, which as its title suggests, is an appendix to the *Laws*. Was it written or drafted by PLATO? Was it composed or edited by PLATO's secretary, PHILIP of Opus (IV–1 B.C.)? We have no means of knowing.

A great historian of ancient religion, the Belgian FRANZ CUMONT, said of the *Epinomis* that it was "the first gospel preached to the Hellenes of the stellar religion of Asia".[4] That description is striking, but a little delusive. The astral religion had oriental roots, but it was definitely a Hellenic creation, Pythagorean and Platonic.

I do not claim to understand the *Epinomis* which is chock full of irrationality and is as good an example of pseudo-scientific writing as the *Timaios*. The gist of it is as follows. The aim of wisdom is to contemplate numbers, especially celestial numbers. The five regular solids are equated with the five elements (ether being the fifth). The most beautiful things are those revealed to our understanding by our own souls, or by the cosmic soul and the celestial regularities. The cult of the astra must be legalized. Astronomy is not only the climax of scientific knowledge; it is a rational theology. The supreme magistrates are not to be philosophers but rather astronomers, that is, theologians.

The popularity of that astronomical theology and its acceptance for centuries by men of great intelligence and wisdom was justified by the following circumstances. In the first place, the old mythology had become untenable. The majority of good men respected the traditional rites and valued the myths as a kind of national and sacred poetry, but they could no longer accept them as true.[5] Hence, there was in their hearts a religious vacuum which had to be filled. Secondly, after the loss of independence and subjection to Macedonian rule, every Greek was profoundly disillusioned. To political servitude were added the tumult of passions and no end of economic difficulties. The sublunar world was as chaotic as it could be; it was the home of decay, disorder, corruption, instability, disease, and death. It was insufferable. There must be another world where order and justice obtain. Thirdly, the heavens offer an excellent image of the cosmos. Everything in heaven is regulated. The motion of the fixed stars is majestic, the complex periodicities of planetary motions are even more awful. This splendor is revealed by mathematical analysis which allows astronomers to build up a theoretical reconstruction of the whole system, the verisimilitude of which is proved by the comparison of observed positions with calculated ones. The concordance is imperfect, but such as it is, it could not be accidental; it is imperfect but indefinitely perfectible.

---

[4] F. CUMONT; *Astrology and Religion among the Greeks and Romans* (New York, Putnam, 1912), p. 51.

[5] LUCIAN of Samosata made fun of the myths, but he was very exceptional, a child of the Euphrates, and came much later (fl. c. 120–200). He has been called the Voltaire of antiquity. The average Greek gentleman (καλοκἀγαθός) attended the ceremonies and mysteries and kept silent, but he could not believe the stories.

Hence, the final conclusions were readily assented to. The stars and planets are the homes of the Gods and the final retreats of human souls.

We may still add that the *Epinomis* was an intrinsic part of the Platonic canon and partook in its authoritativeness and its glory. This was the result of a series of misunderstandings and unconscious prevarications. Non-mathematicians, that is the majority of the people, accepted Plato's conclusions as corollaries of his mathematical work[6]; the mathematicians accepted it as a kind of metaphysics beyond their own ken. Plato's reputation for wisdom was so great that everybody assumed he had established the validity of his statements, however fanciful they might be.

Hence, the astral religion was countenanced by almost every thinker of the Hellenistic and Roman ages. It was incorporated into the fabric of Stoicism and explained by Poseidōnios of Apameia (I–1 B.C.); Poseidōnios' disciple Cicero (I–1 B.C.) re-explained it to Latin readers. Cicero wrote good accounts of it in the *Somnium Scipionis* and in the *De natura deorum*. The astral religion was the religion of every educated man who could not swallow any longer the traditional mythology and had not been converted to another faith. In particular, we may say that it formed the natural intermediary between the doctrines of Paganism and those of Christianity.

The *Somnium Scipionis* preserved by Macrobius (V–1) had a deep influence upon Christian thought.[7] It did not transmit the whole of astral theology to the western world, but it offered a kind of proof of the immortality of the soul which was as highly acceptable to Christian theologians as it would have been to Jewish or Muslim ones. That is another story, however, which does not concern us at present.

For some six centuries (three before and three after Christ) the astral religion was the religion of the elite of the Pagan world. It was accepted by the wisest men, such as Marcus Aurelius. One may wonder why Jewish or Christian doctrines did not reach the Pagan world earlier. The answer is simply that the Pagans, the Christians, and the Jews hardly mixed even in the great cities where they lived together. For example, Galen was a very learned man, the kind of man who knew everything; yet, his knowledge of Judaism and Christianity was so rudimentary that he was hardly able to distinguish between them.[8] This is the more astonishing, because he must have come across many Christian and Jews in Pergamon, Alexandria, and Rome, and he actually used the commentary on *Epidemics* by Rufus of Samaria, who flourished in Rome at the same time as himself. Similar remarks might be made about Marcus Aurelius. Jews and Christians did not mix with Pagans any more than they could help and the revulsion was mutual. They were minority groups under suspicion of disloyalty; the intolerant considered it their duty to denounce and persecute them, while the more tolerant preferred to steer clear of them. The unfortunate circumstances of our own time make it easier for us to understand that situation.

It is not difficult for us to appreciate and even to consider with sympathy the attitude of mind of those enlightened Pagans who found solace in celestial mechanics. It did not occur to them (and it would be unfair to blame them) that the trajectory of

---

[6] For example, the cosmological meaning of the five regular solids, which is reaffirmed in *Epinomis*, was accepted as a piece of scientific knowledge. In spite of the fact that the dovetailing of the five solids with the five elements was fantastic and irrelevant, for the non-scientific minds it clinched the argument.

[7] May I seize this opportunity to draw attention to the English translation of Macrobius' commentary well annotated by William Harris Stahl (292 pp., 2 figs., New York, Columbia University Press, 1952; Isis 43, 267–8).

[8] Richard Walzer: *Galen on Jews and Christians* (102 pp., Oxford University Press, London, 1949).

the Earth observed by a Venerian astronomer would have been as pure and sublime as the trajectory of Venus as seen by themselves. The main trouble with the sublunar world is that we are too close to it for abstractions and illusions. We understand their feelings of sacredness which were comparable to KANT'S, writing in 1788, "Two things fill one's conscience with ever increasing wonder and awe, the stars in heaven and the moral law in oneself". Yet, KANT was deeply moral and rational, while they were demoralized and overwhelming miseries had driven them into irrationality.

There is one thing, however, which puzzles me very much. The astronomical theologians admired the celestial order so much that they identified it with intelligence, and, on the other hand, they tended to identify disorder and freedom with stupidity (*Epinomis* 982–3). The planets reveal divine intelligence by the eternal accuracy of their motion. Now we might admit with Plato that the planetary motions reveal God, but not that the planets themselves are gods. Think of the popular argument of the clock. Its mechanism and regular motions reveal the existence of a clockmaker. Nobody ever said that the clockmaker was in the clock, or that the clock was itself the clockmaker. Yet, according to the new astrologic religion, the planets did not simply reveal God, they were themselves gods, each planet regulating its own motion with divine intelligence and repeating it eternally to evidence its own wisdom. Does that make sense? Yet the argument was accepted by the New Academy and by the Stoics and we find it very clearly stated by CICERO (*De natura deorum II*, 16). The confusion of thought was probably caused by a wrong generalization; the soul or intelligence of an animal is within itself; we may say that the animal has intelligence or that he is an intelligent being; his intelligence, however, is revealed not by the regularity and precision of his motions, but rather by their unexpectedness.

How could these wise people confuse automatism with intelligence and freedom with foolishness? Is it not tragic to witness Greek philosophy ending in such a shameful impasse?

A similar confusion occurring under our own eyes may help us to judge that paradoxical situation with indulgence, but before speaking of that it is worth while to end our account of the astral religion with a brief discussion of the week.

The best proof of the popularity of the astral religion during the decline and fall of Paganism is given by the establishment of the week.[9] The number of days in the week is seven, because of the recognition of seven planets which were generally listed in the following order: Saturn, Jupiter, Mars, Sun, Venus, Mercury, Moon. This is proved by the fact that the names of the days are derived from planetary names in many languages, and because the order of the week days is easily deductible from the normal order just given.[10] What is astounding and could not have been foreseen is

---

[9] For the history of the week, see FRANCIS HENRY COLSON: *The Week* (133 pp., Cambridge University Press, 1926) and SOLOMON GANDZ: "The Origin of the Planetary Week in Hebrew Literature" (*Proc. American Academy of Jewish Research*, **18**, 213–54, 1949).

[10] It is the order of what they believed to be decreasing distances from us; the order was correct except, of course, for the Sun. The week order is deduced from the normal order by taking the first and dropping two, then the second and dropping two, *etc.* One finally gets the series, Saturn, Sun, Moon, Mars, Mercury, Jupiter, Venus. The Jewish week was essentially different from the astrological week, except that both had seven days. The first day of the Jews was Sunday and the last Sabbath. The first day of the astrological week was Saturday, an unlucky day, while the last Jewish day, Sabbath was a day of blessing.

The days of the week were called after the Planets or Gods which dominated their first hour; the following hours were dominated by the following planets in the normal order. Thus, the first Pagan day, Saturday (*Saturni dies*) was so called, because its first hour was dominated by Saturn; the second hour was dominated by Jupiter; the third, tenth, seventeenth, and twenty-fourth hours were dominated by Mars. The first hour of the second day (*Solis dies*) was thus dominated by the Sun, *etc.* If one continues the calculation, one obtains the astrological order of the days: Saturday, Sunday, Monday, Tuesday, Wednesday, Thursday, and Friday (the astrological origin of the days' names is more obvious in many other languages, *e.g.* in Italian).

that the astrological week was diffused unofficially throughout the Roman empire not long before Christ and about the beginning of our era.

The six days of creation followed by the sabbath equalled in number the seven planets. That was a pure coincidence, but it favoured the success of the astrological week. At any rate, the hebdomadary period, if not always the names of the days, spread gradually all over the world.

Next to the silent adoption of the decimal basis of numbers, it is the most remarkable example of unconscious convergence and unanimity in the whole history of mankind.

The establishment of both periods was certainly facilitated by the fact that it was not deliberate. If one had tried to establish it by means of international congresses (assuming that that would have been possible or even conceivable in those early days), the chances of failure would have been considerable, because the two periods unconsciously agreed upon, ten and seven, are good enough but not absolutely so. It is clear that a week of six, eight, or ten days would have been equally acceptable, and the number bases eight or twelve have as much to recommend themselves as the basis ten. On the other hand, it is just as well that the convergences were not carried too far; logicians might have been tempted to promote a decadic week instead of the hebdomadary one in order to agree with the decimal system, and we would have lost three holidays out of every ten.

In India the hebdomadary succession of days was accepted together with much else in Greek astrology. Witness the Sūrya-siddhānta which dates probably from the fourth or fifth century.[11] In early China, however, where Greek astrology did not penetrate, there seems to have been a decimal week, and much later, under the T'ang (621–907) officials were allowed one holiday in every decade of days (*hsün hsiu*).[12] The seven-day week was introduced into China about the eighth century by Manichaeans, by Nestorians or by Hindus[13]; it did not find as wide an acceptance as in the western countries.

Bearing in mind the Chinese anticipation, the decimal week introduced by the French Revolution was not a complete novelty. It did not last very long, only fifteen years (1792–1807); its unpopularity was probably aggravated, if not caused, by its very length. The period seven is more in agreement with physiological needs; a stretch of six working days is ample,[14] we would hate to have it increased 50 per cent for the sake of decimal consistency.

The cultural convergence which caused the establishment of the week and the decimal basis is even more mysterious than the linguistic convergence making possible the birth of a language. In the linguistic cases only a relatively small group localized in a relatively small region was concerned while the diffusion of the week and of the decimal basis implied unconscious collaboration on an international, interracial, interreligious, and polyglottic scale. On the other hand, it must be admitted that any language, even the least developed, is a very complex fabric implying agreement on a very large number of words and rules.

---

[11] SARTON: Introduction (1, 386–8).

[12] *Hsün* (Giles No. 4864) is a period of ten days, but later it came to mean also seven days; perhaps, because the period was thought of rather than its length (as if a Frenchman spoke of a "semaine de dix jours" when trying to describe the Revolutionary decade). *Hsiu* (No. 4651) means to rest, to stop. I am indebted for Chinese information to my friend, L. CARRINGTON GOODRICH, of New York, and for Sanskrit information to my friend, PAUL EMILE DUMONT, of Baltimore.

For Chinese decimalism in cartography, see my Introduction (2, 225).

[13] SARTON: Introduction (1, 504, 513–14).

[14] For people who really work! A missionary told the story of primitive men converted to Christianity who rested with great affectation on Sundays in order to suggest that they did actually work on the other days. RAOUL ALLIER: *La Psychologie de la Conversion chez les Peuples Non-civilisés* (2 vols., Paris, Payot, 1925; Isis 30, 306–10).

2. "Thinking Machines"

The history of science, and even the history of mankind, could be written in terms of tools and artifacts. The stone axes of our Palaeolithic ancestors were tools, and so were the wheels, pulleys, microscopes and telescopes, thermometers, dynamos of later times. The arms and armour of all nations from the club to the atomic bomb are tools of another kind.

One might claim that the history of science is a history of ideas, and that claim is correct; yet every scientific idea (with the possible exception of very abstract or mathematical concepts) led to the creation of a new instrument or of a better one. If books and prints are included among the instruments, there is no exception of any kind.

Every tool is an extension of a human organ or the amplification of a function, material or spiritual. Many of them amplify our mechanical power, some amplify it exceedingly; they may multiply it a million times or more. Others increase the acuity of our vision; telescopes and microscopes increase it so much that they reveal new worlds to us, infinitely large or infinitely small. If photographic plates are properly attached to those instruments, they serve to amplify immeasurably our memory. Of course, every book or print helps to do the same; a good library is the memory of mankind.

No tool is worth anything unless there be somebody able to use it, and when a tool or, let us say, a machine is sufficiently complex, it must be guided or restrained by a workman. Ingenious people, however, found means of contriving the machine to help or guide itself. Humphrey Potter, the lazy boy who got tired of opening and closing valves in an early steam engine made the engine self-acting by causing the beam to open and close them with suitable cords and catches. That was as early as 1713. Not only was the work done, but no human being could have done it as faithfully. Another early example was Watt's governor (1784) controlling automatically the speed of an engine, and this type of governor (pendulum governor) was followed by many others. An admirable kind of servo-mechanism was invented in 1799 by Joseph Marie Jacquard (1752–1834) of Lyon. Jacquard looms made it possible to weave a fabric with the most complicated pattern as easily as a plain one. His invention opened endless possibilities, for his cards and looms could be modified indefinitely.

The mechanical governors represented a type of contrivance which was extended later to many other fields; the purpose of all of them was to establish a kind of homeostatic control, for example, thermostats and voltage stabilizers.

To return to the eighteenth century, many machines had been invented before the end of it; and a variety of servo-mechanisms had been introduced. There were also calculating machines and slide rules of various kinds. In our own days computing mechanisms and linkages have been carried to such a high stage of perfection that innocents are beginning to speak of them as "thinking machines". The extraordinary speed and accuracy of those machines and their fast increasing "memory" suggest comparisons with human brains.

The reference to them as "thinking machines" (a reference which was never made with regard to Jacquard looms) was caused partly by the fact that ignorant people, mistaking computation with mathematics, assumed it to be a highly intellectual function,[15] partly by the lightning speed and precision of those machines. From that

[15] For example, lightning calculators or outstanding chess-players are often believed to be men of great intellectual genius.

particular point of view, electronic computers are not human but superhuman, for they approach a perfection hopelessly out of man's reach. (But are the strength of a motor, the visual power of a telescope or microscope not superhuman in the same way?)

No machine is able to do anything by itself without human initiative. In that respect, an electronic computer is on the same level as any other tool. The confusion exists, however, and it is pernicious. Some enthusiasts are dreaming of machines which would play chess, translate from one language into another, or compile bibliographies. To return to reality, it is clear enough that machines could be invented to accomplish the most servile parts of any undertaking, machines which could do everything, except the essential.

Outside of the mathematical aspect (which is always full of mystery and glamour for the non-mathematician), the feature which impresses the popular mind most deeply is the contrivance of a mechanical memory, comparable to human memory, though with an infinitely greater capacity and precision. Such a memory may be realized by (1) magnetic drums (any of 81,920 digits can be stored or recalled in an average of $\frac{1}{25}$ of a second), (2) cathode ray tubes (any of 10,240 digits can be stored or recalled in 10^{-5} sec., (3) magnetic tapes (any of 2,000,000 digits can be stored on one tape and recalled from it at the rate of 12,500 a second).[16]

Dr. VANNEVAR BUSH was so impressed by various mnemonic contrivances that he coined the ingenious phrase "memex instead of index".[17] Memory is certainly one of the fundamental components of intelligence, but it has gradually lost much of its importance, because mankind has created artificial treasures of information which are so much superior to natural memories that the latter may be disregarded. Men endowed with phenomenal memories such as existed in the past and still exist in countries like India,[18] are no longer looked up to as if they were men of miraculous wisdom, but looked down upon as idiots. Memory will always be indispensable as a general guide, but the emphasis has changed from literal memory to a broader and vaguer kind. A man of science relies on his memory but does not trust it overmuch; he does not try to remember special data, but rather to know the books wherein those data are exactly recorded.

There are in any kind of material or intellectual work an endless number of repetitive processes which could (and will) be done by machines incomparably better than by men. Not only are many operations susceptible of being carried out by a machine, but servo-mechanisms may enable us to accomplish series of operations of indefinite length and complexity, and to control groups of simultaneous operations or of simultaneous series of operations. There is no limit to such a development. As soon as the greatest and most efficient machine has been invented,[19] it is relatively easy to invent one that is more complex, more versatile, and more efficient.

The ancients worshipped the planets because of the awful regularity of their motions. It is well that our contemporaries are satisfied to boast of their "thinking machines" (which are far more regular than any planet could ever be) and have not yet ventured any apotheosis.

---

[16] The figures quoted represent the realizations of to-day (May, 1953), as given by International Business Machines, New York. Present possibilities can be and will be indefinitely increased.

[17] V. BUSH: *Endless Horizons* (p. 31, 1946).

[18] Examples and references given in my *History of Science* (vol. 1, 132, 1952).

[19] "Invented", not "built". A better machine will be invented before the preceding one is actually built. As LE CORBEILLER puts it, "Well before the machine is finally built, the designers have in mind an entirely different principle and twenty improvements of detail which, when incorporated into the next type, will, they tell you, make the present one 'look like peanuts'."

3. Conclusions

Every "thinking machine" needs a human brain and human hands behind it or it cannot think at all. Many problems may be solved with lightning speed and unerring precision, but who will formulate them? Machines can answer questions exceedingly well, but who will ask the questions? If a long series of operations is determined by a sequence of holes punched in a ribbon, there must still be somebody to punch the holes. Some machines are exceedingly versatile, but versatility should not be confused with freedom.[20]

As far as performance goes, the wonder of our time is less the giant computer than the means of reproducing music, synchronizing it with pictures and transmitting the combination to any distance. Curiously enough, the tools which make television possible are never assumed to be "thinking"; that supreme quality is assumed to be restricted to the computing machines, because a great number of people mistake computation for thought. Let us repeat that real thought is free, and that computation is the opposite of free; it is servile.

In both cases—the ancients adoring the planets and the children of our time standing in awe before the "thinking machines"—there is an identical confusion between regularity and precision, on the one hand, and intelligence on the other.[21]

Let us reverse the argument and say: everything that is perfectly regular betrays a lack of intelligence. The essential difference between the average man and the perfect robot is that the former can and will think, however little, while the latter cannot think at all.

The paradox which misleads people lies in this. Human thought is free, spontaneous, unpredictable; yet we expect it to develop along logical patterns. It must be logical and consistent, free of arbitrariness, but not too much so. The same is true of everything beautiful; beauty implies complex symmetries, not perfect ones.

Most of the paradoxes of life, art, and virtue hinge on that one. Think of the main treasure house of any culture, and if one may use the metaphor, its main power house, language. Each language is the combination of innumerable rules and words; the words hang together in capricious sequences; the rules are dovetailed in many ways, but the meanings of many words are erratic and the rules are full of exceptions. The elaboration of thought in a single head is mysterious enough, but how were the languages developed, each of them simultaneously in thousands of heads, without deliberation and almost without consciousness? Now, that miracle, the creation of a language, implying the web and woof of capricious rules and the combination of logical with illogical patterns, has not happened once but thousands of times.

Or look at it from the point of view of a student. If it be our maternal language, we learn its rudiments intuitively, without analysis. If our mother spoke it purely, we learn to speak it with equal purity. The Greeks wrote their masterpieces before any grammar had been composed; indeed, the early grammars were built on the basis of those masterpieces, not vice versa. If we have to learn a foreign language later in life, we proceed more methodically. We try to build up a sufficient vocabulary

---

[20] As does Le Corbeiller. The automatic telephone is very versatile but has no freedom; all the freedom is in the head of the person who dials a given number. That number is a definite order which the telephone has no choice but to obey; that is why it is called automatic.

[21] One might claim that the ancients were aware of the fact that planetary motions are not perfectly regular. The planets of heaven do not move as regularly as those of an orrery. The ancient astronomers were aware of irregularities when computed and observed places did not coincide; their reaction, however, was to conclude, not that the planets were irregular but that their cinematic explanations were imperfect.

and to master the rules of morphology, accidence, and syntax Yet, as long as we have to think of the dictionary and the grammar of a language, we can hardly use it in a fluent way. We are beginning to know a language only when all the words and rules have been driven into our unconsciousness; we know it only when we have reached a high degree of automatism in our use of it. Automatism in this case is a measure of progress, but not for very long, for no one can use a language supremely well unless he does so with great deliberation.

Consider the problems of administration. Every administrator dreams of regularizing his business as much as possible, of automatizing it, but he could never succeed unless the business entrusted to his care were almost dead or he were himself exceptionally stupid. Such insensitive and dictatorial administrators exist, but they are as amiable as Procrustēs. The paradox lies in the fact that administration must be very orderly yet human, which means disorderly; it must be impartial yet sensitive, and this implies love and partiality.

Consider art. Gilbert Murray hit the nail on the head when he wrote: "The unwillingness to make imaginative effort is the prime cause of almost all decay in art. It is the caterer, the man whose business it is to provide enjoyment with the very minimum of effort, who is in matters of art the real assassin".[22] Now, any imaginative effort or creative effort is an escape from automatism, a rebellion against it, a defence of personality and of freedom. Dictators in art as well as in politics arise only because too many men are spiritless and timorous and prefer to eschew the difficulties of choice and the anxieties of decision; such men, incipient robots, betray humanity; they are more like sheep and pigs than soulful beings.

Consider education. It is to some extent a school of automatism. Educators try to inculcate good habits. We all need good habits. The business of life is so complex that we must learn to do many things instinctively and to reserve our thinking power for things of greater worthwhileness. Our physiological functions are largely automatic; if they were not so, we could not exist. We must try to automatize other functions, material and spiritual; this will enable our mind to soar off from a higher level. Habits are activities which have become gradually unconscious; they are necessary but very insufficient. A man whose life is too regular and who is dominated by habits is not admirable at all but contemptible.

Success in life, in art, in virtue always implies a victory of freedom over automatism, of new creation over dead-like repetition.

It is well that we have with us enough engineers who are obsessed with the vision of new gadgets, for gadgets may be very useful, but the real business of life is independent of them.

There will always be a need for "thinking machines", better ones and more of them, but the greatest need is for thinking men, not robots nor even "yes-men" but independent men, honest and free.

Short Bibliography

1. *Astral Religion*

Cumont, Franz (1868–1947): *Les Religions Orientales dans le Paganisme Romain* (4th edition, 355 pp., 16 pl., 13 figs., Paris, Paul Geuthner, 1929; *Isis* 15, 271).

Cumont, Franz (1868–1947): *Lux Perpetua* (562 pp., Paris, Geuthner, 1949; *Isis* 41, 371).

Epinomis: Greek text and English translation by W. R. M. Lamb in the Plato edition of the Loeb series (vol. 8, 426–87, 1927).

[22] *Religio Grammatici* (London, Allen, 1918), p. 34.

ROUGIER, LOUIS: *L'origine Astronomique de la Croyance Pythagoricienne en l'Immortalité Céleste des Ames* (152 pp., Recherches d'archéologie, de philologie et d'histoire, 6; Institut français d'archéologie orientale, Le Caire, 1933; *Isis* 26, 491).
SARTON, GEORGE: *A History of Science: Ancient Science through the Golden Age of Greece* (672 pp., 103 figs., Harvard University Press, Cambridge, Mass., 1952), pp. 447–54.
SARTON, GEORGE: *Introduction to the History of Science* (3 vols. in 5, Baltimore, Williams & Wilkins, 1927–48).

2. "*Thinking Machines*"

BERKELEY, EDMUND C.: *Giant Brains, or Machines that Think* (286 pp., New York, Wiley, 1950).
BUSH, VANNEVAR: *Endless Horizons* (192 pp., Washington, D.C., Public Affairs Press, 1946; *Isis* 37, 250).
DIEBOLD, JOHN: *Automatism.* The advent of the automatic factory (182 pp., New York, Van Nostrand, 1952).
LE CORBEILLER, PHILIPPE: "Large-scale Digital Calculating Machinery" (*American Journal of Physics*, **16**, 345–7, 1948).
LEMOINE, JEAN GABRIEL: "La 'Machine à Penser' de Raymond Lull et l'Astrologie Arabe" (*Bulletin de la Société de philosophie de Bordeaux*, 5e année, No. 25, pp. 55–64, Août 1950).[23]
Proceedings of two symposia on large-scale digital calculating machinery (vols. 16 and 26 of the *Annals of the Computation Laboratory of Harvard University*, Harvard University Press, 1948, 1951).
SVOBODA, ANTONIN: *Computing Mechanisms and Linkages* (No. 27 of Massachusetts Institute of Technology Radiation Laboratory series, 372 pp., New York, McGraw-Hill, 1948).
WILLERS, FRIEDRICH ADOLF: *Mathematische Maschinen und Instrumente* (340 pp., 258 figs., Berlin, Akademie-Verlag, 1951).

[23] This item is mentioned, because the curious use of the phrase "thinking machine" in its title. There is a difference, however, which it is worth while to explain. Every branch of logic or mathematics might be called a "thinking machine", because it helps us to think somewhat automatically. For example, given a system of equations, one proceeds to solve it according to the rules of algebra or analysis and without bothering about their physical meaning. Whether these equations be solved with paper and pen or with more complicated tools is immaterial. The main point is that their solution may lead eventually to the solution of a physical problem and to a discovery. RAMON LULL (XIII-2) was trying to create such a machine, logical rather than algebraical (Introduction, 2, 901–4).

Ptolemy's Precession

A. PANNEKOEK

Formerly Sterrekundig Instituut, Universiteit van Amsterdam, Netherlands*

ACCORDING to Ptolemy's great astronomical work, the *Mathematical Composition*, the movement of the equinoxes relative to the stars was discovered by Hipparchus. Probably it had been suggested by his result that the length of the year, as a cycle of seasons determined by the equinoxes, was shorter than the return to the same phenomena of the stars. This was explained by the assumption that the sphere of the fixed stars had a movement along the ecliptic in the same direction as the movement of the planets.

To confirm this explanation Hipparchus made use of observations of the stars themselves. Ptolemy says: "In his treatise 'On the variation of the solstices and equinoxes', Hipparchus by comparing lunar eclipses observed in his time with others observed formerly by Timocharis, arrives at the result that Spica in his time stood at 6° distance, and in Timocharis' time stood at 8° distance before the autumnal equinox. . . . And also for the other stars which he compared he shows that they have proceeded by the same amount in the direction of the zodiacal signs" (Book VII, Chapter 2). The advantage of using lunar eclipses is that the direct comparison of the star with the moon gives the place relative to the Sun which, through the tables or by direct comparison, can be easily reduced to the equinox. The years between which the change of 2° took place are not indicated in the quotation. Total lunar

* Now at Wageningen, Netherlands.

eclipses near Spica happened in Timocharis' time in 283 B.C. on 7 March, and in Hipparchus' time in 135 B.C. on 21 March. In Book III, Chapter 1, dealing with the length of the year, the latter is mentioned with the addition that a distance of $5\frac{1}{4}°$ between Spica and the equinox resulted therefrom; in addition, another eclipse in 146 B.C. on 21 April is mentioned with a resulting distance of $6\frac{1}{2}°$. If these are the source of the rather cursory statement of the quotation, then a change of 2° in an interval of, say, 283 — 140 = 143 years, means a precession of 50″ per year.

Ptolemy does not give this computation. Immediately after the sentence quoted above he continues by describing his own measurements which, he says, are in accordance with Hipparchus' result. With his "astrolabon" (armilla) he determined the longitude difference between the Sun and a star, first measuring the distance between the setting Sun and the Moon, and then when darkness had fallen, between the Moon and the star. "As an instance" he gives the data of an observation of Regulus in A.D. 139; he finds its longitude to be 122° 30′ which, compared with the longitude of 119° 50′ found by Hipparchus, affords a displacement of 2° 40′ in about 265 years or 1° in 100 years. "This was also assumed with reserve by Hipparchus, who in his treatise 'On the length of the year' says '. . . when therefore the solstices and equinoxes in one year are retrograding at least 1/100 degree against the direction of the signs, they must have receded in 300 years by at least 3°'." Ptolemy continues that "After having observed in the same way also Spica and the brightest stars near to the ecliptic . . ." he found from them the same amount. Thus with the omission of the restriction "at least" he assumes 1° in 100 years (36″ per year) to be Hipparchus' value of the precession.

It has often been assumed that Ptolemy simply copied and took over this value from Hipparchus and doctored his own observations to bring them into accordance with it. It has been made clear, however, by DREYER* that the wrong amount of 2° 40′ can quite well have been the genuine result of observations affected by an accumulation of systematic errors (refraction, solar tables, instrumental errors) which, moreover, were working on different stars in a similar way.

Ptolemy's Chapter 3 of Book VII is devoted to a more detailed test of Hipparchus' assumption that it was a movement of all the stars about the pole of the ecliptic. Good results could be expected because he had observations from three epochs; those of the early Alexandrian astronomers Timocharis and Aristyllus, those of Hipparchus, and finally his own, thus extending over a far longer interval. The distances in latitude from the ecliptic he found to be nearly equal to what they were at the time of Hipparchus. The distance to the equator, however, measured along a circle through the celestial poles, was different for the three times of observation. For the stars nearer the vernal equinox the new positions were farther to the north than the older ones, and for those nearer the autumnal equinox, farther to the south. To show this clearly he gives, for eighteen stars, the distance to the equator (our declination) as measured by the observers at the three epochs.

In the ensuing discussion he says that the precession of 2° 40′ in 265 years between Hipparchus and himself will appear clearly in the differences (of declination) for stars in the vicinity of the equinoxes. He chooses six stars for which he gives the observed difference and for each of them states, without giving details, that the same difference of declination is found for two points of the ecliptic situated about the star's longitude at a mutual distance of 2° 40′. The latter quantity can be easily

* J. L. E. DREYER, *M.N.*, **77**, 536 (1917).

read from the table in Book I, Chapter 15, giving the declination for every degree of longitude of the ecliptic. We find for:

	η Tauri	Capella	Bellatrix	Spica	η Ursae Majoris	Arcturus
Observed difference .	+ 65′	+ 46′	+ 42′	+ 66′	+ 65′	+ 70′
Computed difference.	+ 56′	+ 39′	+ 42′	+ 64′	+ 56′	+ 64′

The concordance must be deemed satisfactory, especially if we consider that the computed values hold for stars in the ecliptic and should be increased for stars of higher declination.

The catalogue of declinations of eighteen stars given by Ptolemy deserves a closer examination, because it can be put to a wider use than that of confirming qualitatively the course of the precession and to test Ptolemy's supposition for some few cases. It can provide us with a more quantitative knowledge of the precession as well as of the character and accuracy of Greek observational astronomy. It represents wellnigh the only collection of first-hand data left in the literature which can give information on the measuring work of the first Alexandrian astronomers as well as of Hipparchus. Whereas we know almost nothing about instruments used in those ancient times, so that it could even remain doubtful whether they had done anything more than observe the phenomena of eclipses and equinoxes, we have here the results of measurements with graduated circles and expressed as "distances in latitude to the equator on a great circle through its poles".

In Table 1 after the names of the stars in the first column the six stars selected by Ptolemy are marked by an asterisk; the change of declination by the precession is given in the second and third column, first between the early Alexandrians and Hipparchus, and secondly between Hipparchus and Ptolemy. They are positive at

Table 1. Comparison of declination differences

Star	Observed declination difference		α (for 100 B.C.)	cos α	Computed declination difference		Errors (num.)	
	H.-Al.	Pt.-H.			H.-Al.	Pt.-H.	H.-Al.	Pt.-H.
1. Altair . .	0′	+ 2′	271° 58′	+ 0·034	+ 2′	+ 3′	− 2′	− 1′
2. η Tauri* . .	+ 40	+ 65	27 13	+ 0·889	+ 48	+ 79	− 8	− 14
3. Aldebaran . .	+ 60	+ 75	39 53	+ 0·767	+ 41	+ 69	+ 19	+ 6
4. Capella* . .	+ 24	+ 46	42 38	+ 0·736	+ 40	+ 66	− 16	− 20
5. Bellatrix* . .	+ 36	+ 42	53 41	+ 0·592	+ 32	+ 53	+ 4	− 11
6. Betelgeuse . .	+ 30	+ 55	60 47	+ 0·488	+ 26	+ 44	+ 4	+ 11
7. Sirius . .	+ 20	+ 15	78 9	+ 0·205	+ 11	+ 18	+ 9	− 3
8. Castor . .	+ 10	+ 14	79 31	+ 0·182	+ 10	+ 16	0	− 2
9. Pollux . .	0	+ 10	83 31	+ 0·113	+ 6	+ 10	− 6	0
10. Regulus . .	− 40	− 50	123 7	− 0·546	− 29	− 49	+ 11	+ 1
11. Spica* . .	− 48	− 66	174 21	− 0·995	− 54	− 89	− 6	− 23
12. η Ursae* . .	− 45	− 65	184 42	− 0·997	− 54	− 89	− 9	− 24
13. ζ Ursae . .	− 45	− 90	177 17	− 0·999	− 54	− 89	− 9	+ 1
14. ε Ursae . .	− 54	− 81	166 32	− 0·973	− 53	− 87	+ 1	− 6
15. Arcturus* . .	− 30	− 70	189 54	− 0·985	− 53	− 88	− 23	− 18
16. α Librae . .	− 36	− 94	194 53	− 0·966	− 52	− 86	− 16	+ 8
17. β Librae . .	− 48	− 84	201 56	− 0·928	− 50	− 83	− 2	+ 1
18. Antares . .	− 40	− 75	216 40	− 0·802	− 43	− 72	− 3	+ 3

the vernal, negative at the autumnal side. To the modern astronomer it offers a set of data from which he can directly derive the constant of precession, since $\Delta\delta$ = number of years $\times$ $n \cos \alpha$, where n = precession $\times$ $\sin \varepsilon$ (obliquity). There is an uncertainty in the number of years, which is hardly relevant; in what follows we took 265 years (between + 137 and − 128) from Ptolemy, and took 287 B.C. as a mean earliest date. In Columns 4 and 5 the right ascension α for 100 B.C., and $\cos \alpha$, are given. Taking the sum total of all $\Delta\delta$ and all $\cos \alpha$, separately for the positive and the negative side, and then combined, we have:

Stars	$\Sigma\Delta\delta$		$\Sigma \cos \alpha$	$160n$ H.-Al.	$265n$ Pt.-H.	n	
	H.-Al.	Pt.-H.				H.-Al.	Pt.-H.
1–9	+ 220′	+ 324′	+ 4·006	55′	81′	0′.34	0′.31
10–18	− 386′	− 675′	− 8·191	47′	82′	0′.29	0′.31
1–18	606′	999′	12·197	49′.7	81′.9	0′.311	0′.309

The results n = 18″.7 for the first, n = 18″.5 for the second interval, correspond to precessional constants of 46″.4 and 46″.0.

This is the value of the precessional constant derived from the observational data presented in Ptolemy's book. Ptolemy himself could not make such a derivation, since exact formulae giving the dependence of $\Delta\delta$ on the position were lacking, as well as the right ascension of the stars. There is no sense in making a solution by the method of least squares to see what errors remain, for we are able to derive the real errors of the observed changes in declination, since we know the true precessional constant. With the true n = 20″.24 = 0′.337, $160n$ = 54′.0, $265n$ = 89′.4, we find the values in column 6 and 7; the next two columns give the errors in the numerical values of the observed differences of declination. The resulting square mean is $\sqrt{109}$ for the first interval of time, $\sqrt{135}$ for the second, so $\sqrt{122}$ = 11′ is found as the mean error of the difference between two ancient observers, and 8′ as the mean error of a declination measured by one of them.

The distribution of the values and the signs shows that we have here a normal set of observational data, apparently unbiased and undoctored, and not marred by systematic errors from faulty tables of the Sun or Moon. We can study this apart from the precession problem, by comparing the observed declinations themselves with the real values deduced from modern data. For the years 289 B.C., 129 B.C., and A.D. 137, assumed to represent the mean dates, the values were interpolated from NEUGEBAUER's tables (1912). The comparison is given in Table 2.* The differences in columns 8–10 are the errors made by the observers. What strikes the eye here is the smallness of the errors and the absence of any systematic character; the signs + and − are rather evenly distributed. One abnormal deviation of 46′ occurs in Timocharis' result for Arcturus, where a mistake might be presumed; otherwise his errors do not exceed 15′, Hipparchus' largest error is 18′, and Ptolemy's largest 28′. The mean error for Hipparchus is 7′ only, for Ptolemy it is 13′, and for the ancient Alexandrians Timocharis and Aristyllus it is 14′ (or by exclusion of Arcturus 9′). As far as a conclusion can be based on this small amount of material—we have no

* P. V. NEUGEBAUER, *Sterntafeln von 4000 vor Chr. bis zur Gegenwart*, Leipzig, 1912. Inconsistencies between the computed values in Tables 1 and 2 are due to second-order effects and large proper motions. It may be added that the above article and these calculations were completed in 1952.

other—the excellence of Hipparchus' work is seen to surpass that of both former and later Greek astronomers.

Returning now to the problem of Ptolemy's value for the precession, the question may be asked: when this set of observations is so consistent with the modern true data, how could Ptolemy find therein a confirmation of his far too small constant of precession? The answer is given by column 7 of Table 1, containing the errors of the

Table 2. Comparison of declinations

Star	Declinations observed			Declinations computed			Errors of observation		
	Tim.-Ar.	Hipp.	Ptol.	− 288	− 128	+ 137	Tim.-Ar.	Hipp.	Ptol.
1. Altair	+ 5° 48′	+ 5° 48′	+ 5° 50′	+ 5° 40′	+ 5° 41′	+ 5° 47′	+ 8′	+ 7′	+ 3′
2. η Tauri	+ 14 30	+ 15 10	+ 16 15	+ 14 31	+ 15 20	+ 16 38	− 1	− 10	− 23
3. Aldebaran	+ 8 45	+ 9 45	+ 11 0	+ 9 0	+ 9 42	+ 10 48	− 15	+ 3	+ 12
4. Capella	+ 40 0	+ 40 24	+ 41 10	+ 39 46	+ 40 26	+ 41 27	+ 14	− 2	− 17
5. Bellatrix	+ 1 12	+ 1 48	+ 2 30	+ 1 15	+ 1 47	+ 2 39	− 3	+ 1	− 9
6. Betelgeuse	+ 3 50	+ 4 20	+ 5 15	+ 3 48	+ 4 15	+ 4 57	+ 2	+ 5	+ 18
7. Sirius	− 16 20	− 16 0	− 15 45	− 16 12	− 16 4	− 15 53	− 8	+ 4	+ 8
8. Pollux	+ 33 0	+ 33 10	+ 33 24	+ 33 4	+ 33 15	+ 33 28	− 4	− 5	− 4
9. Castor	+ 30 0	+ 30 0	+ 30 10	+ 29 58	+ 30 5	+ 30 11	+ 2	− 5	− 1
10. Regulus	+ 21 20	+ 20 40	+ 19 50	+ 21 8	+ 20 40	+ 19 50	+ 12	+ 0	0
11. Spica	+ 1 24	+ 0 36	− 0 30	+ 1 26	+ 0 33	− 0 56	− 2	+ 3	+ 26
12. η Ursae	+ 61 30	+ 60 45	+ 59 40	+ 61 35	+ 60 41	+ 59 12	− 5	+ 4	+ 28
13. ζ Ursae	+ 67 15	+ 66 30	+ 65 0	+ 67 30	+ 66 37	+ 65 6	− 15	− 7	− 6
14. ε Ursae	+ 68 30	+ 67 36	+ 66 15	+ 68 37	+ 67 45	+ 66 18	− 7	− 9	− 3
15. Arcturus	+ 31 30	+ 31 0	+ 29 50	+ 32 16	+ 31 18	+ 29 43	− 46	− 18	+ 7
16. α Librae	− 5 0	− 5 36	− 7 10	− 4 46	− 5 38	− 7 4	− 14	+ 2	− 6
17. β Librae	+ 1 12	+ 0 24	− 1 0	+ 1 5	+ 0 15	− 1 7	+ 7	+ 9	+ 7
18. Antares	− 18 20	− 19 0	− 20 15	− 18 22	− 19 6	− 20 17	+ 2	+ 6	+ 2

changes of declination between Hipparchus and himself. The six stars selected by Ptolemy (indicated by an asterisk in Table 1) are those which present strongly negative errors, thus exhibiting too small a change of declination. If we derive the precession from these six stars alone we find $\Sigma\Delta\delta = 354'$; $\Sigma \cos\alpha = 5{\cdot}194$, $265n = 68'\!.2$; $n = 0'\!.257 = 15''\!.4$; precessional constant $= 38''$, whereas the remaining twelve stars give $\Sigma\Delta\delta = 645'$, $\Sigma \cos\alpha = 7{\cdot}003$, $265n = 92'\!.0$, $n = 0'\!.348 = 21''\!.0$, precessional constant $= 52''$.

There can be no doubt that Ptolemy selected these six stars *because* they were favourable to his assumed value of the precession and could be quoted as confirmations, and that other stars were omitted because they did not confirm his assumption. Yet we cannot speak of an attempt to deceive his readers; he presents to them the full material with the unfavourable cases also. It comes down to saying: "my result is confirmed by a number of data; the other data which do not conform to it do not count".

In the third part of the same chapter Ptolemy gives "a still more clear demonstration of the phenomena" by determining the change in longitude of some stars. From two occultations of the Pleiades by the Moon, one observed by Timocharis in 283 B.C. and the other by Agrippa in Bithynia A.D. 92, he finds the longitudes 29° 30′ and 33° 15′ and hence an increase in longitude of 3° 45′ in 375 years. From three occultations of Spica, two observed by Timocharis in 294 B.C. and 283 B.C., and one by Menelaus at Rome in A.D. 98, he finds the longitudes 172° 20′, 172° 30′, and 176° 15′, giving an increase of 3° 45′ in 379 years and 3° 55′ in 391 years. From two occultations of β Scorpii, one seen by Timocharis in 295 B.C. and one by Menelaus in A.D. 98, he finds the longitudes 212° and 215° 55′, and hence an increase of 3° 55′ in 391 years. The real increase in these cases must have been between 5 and $5\frac{1}{2}$ degrees.

Of this derivation DREYER says: "A worse selection of material for the object in view than the seven conjunctions of the Moon with stars it would be difficult to conceive". Because the star places here are based upon Ptolemy's lunar theory with all its imperfections they may contain grave errors. DREYER finds the errors of the older and the later longitude to be: for η Tauri − 73′ and + 14′, for Spica − 20′ and + 70′, and for β Scorpii − 41′ and + 49′. "By a curious piece of ill luck the longitudes for the time of Timocharis were all too great and those for the end of the first century too small, which produced the faulty precession of 1° in 100 years". Was it ill luck indeed? Certainly it must be deemed highly improbable that three cases, as the only available ones, should by mere chance all be affected by the same large error. Considering his dealing with the declinations we must assume rather that here also Ptolemy had at his disposal more abundant material of other star occultations which he deemed it useless to exhibit in detail since their testimony, not fitting in his theory, was of no account. Selection of the data in this way is, of course, strictly condemned by modern scientific standards. In condemning Ptolemy we should not forget, however, that the principle of selecting data and rejecting deviating results as unreliable was followed up to almost modern times; not until the seventeenth and eighteenth century did it become habitual to derive and use the average of all the observed data. Even in the nineteenth century, scientists felt themselves warranted in excluding strongly deviating values, and they established an exact criterion for exclusion. It is well known that Ptolemy has been harshly criticized and suspected to have falsified or even invented the observations used in his deductions, *e.g.* in a detailed way by DELAMBRE (1817).* When at the same time he is praised as one of the greatest scientists of antiquity, it is clear that his work cannot be judged by the present standards of scientific research. His way of looking at the phenomena and their explanation by theory was different from ours; and this holds for the scientists and philosophers of antiquity at large.

A few cases of ancient eclipses excepted, it is only rarely that the study of ancient astronomy can contribute to our present knowledge of the celestial motions. Rather, conversely, the numerical data derived from modern research are so accurate that by computing the then positions of the celestial bodies, we can examine the accuracy of the statements made by the ancient astronomers. Thus in studying the history of astronomy we are not learning something about the stars, but something about the astronomers. What we are informed about is the psychology of the ancient scientists; we are confronted with what was in their minds, and not only with their opinions on the stars and the world, but also with their logic and their mode of thinking and arguing in establishing facts and theories.

Ptolemy, as the author of the most extensive textbook on Greek astronomy, is the most prominent object of such study. His prominence appears, for example, in the fact that he was the only one among the later authors who took notice of Hipparchus' discovery of precession. Struck by the statement, in cautious form, that there was a regular displacement certainly as large as 1° per century and perhaps larger, he tried to determine it himself in a direct way. His determination by means of Regulus,

* J. B. J. DELAMBRE, *Histoire de l'Astronomie Ancienne*, Tome I, pp. XXV, XXXI, XXXII; Tome II, pp. 250, 599. (1817).

through the chance concurrence of systematic errors, afforded him the same numerical value. Not content with this single result he looked for confirmation by other methods; and in a list of old and new measurements he found six stars whose motions agreed with the value derived. Should he not have been cautious, seeing other stars that disagreed? It is a common experience in all later scientific progress that often in the first enthusiasm of a new discovery the disagreeing facts are simply disregarded. Science was not built up as we do it in our textbooks, where the materials are given and we know exactly what is relevant and what is not. In early times, when nature was so little understood and there was such abundance of unknown, mysterious forces in the world, it did not seem surprising that so many cases did not fit; the happy surprise that some few cases did fit well, and therefore appeared more important, was sufficient to give confidence that theory was right. Considered in this light, Ptolemy's treatment of the precession may be helpful in understanding the character of scientific thinking in antiquity.

A Medieval Footnote to Ptolemaic Precession

DEREK J. PRICE
Christ's College, Cambridge

IT is interesting that when Ptolemaic astronomy returned to Europe via Islam and Spain the false value of precession was corrected to a remarkably accurate one. The Alfonsine astronomers considered the motion of the equinoxes to be compounded from a steady and an oscillatory part. The steady motion took 49,000 years for a complete rotation and corresponded to the difference between the tropical and the calendar years; the modern Gregorian correction omits three intercalary days in four centuries, or $365\frac{1}{4}$ days in 48,700 years. The Alfonsine value is equivalent to an error in date of equinox of 7 hr in 1000 years—a very commendable accuracy for thirteenth-century observations.*

The oscillatory part of precession, the trepidation, had to account for the residual displacement of the fixed stars with respect to the tabulated places of the Sun. It was considered to go through $\pm\ 9^\circ$ with a period of 7000 years and a zero on 17th May A.D. 15. At the date of the Alfonsine Tables, A.D. 1272, the trepidation was $8^\circ\ 7'$ and the calendrical precession in the same interval of 1257 years from the zero was $9^\circ\ 12'$. The total Alfonsine precession was therefore only $11'$ less than the amount $17^\circ\ 30'$ calculated from modern tables, and well within the limits of medieval accuracy in observation. Unfortunately the trepidation was then approaching its maximum, and within a few centuries it became evident that the anticipated decrease in rate of precession had not occurred.

Possibly PTOLEMY was led astray by inaccurate values for calendrical precession. At all events the authority of his incorrect constant of precession probably led the Alfonsine astronomers to believe they had detected a secular variation of that constant and permitted them to introduce the erroneous and misleading notion of trepidation.

* A fuller account of Alfonsine precession theory is given by D. J. PRICE, *The Equatorie of the Planetis*, Cambridge, 1955, pp. 104–7.

The Peking Observatory in A.D. 1280 and the Development of the Equatorial Mounting

JOSEPH NEEDHAM

Caius College, Cambridge, England

SUMMARY

One of the focal points in the intercourse between civilizations was the year + 1267, when the Persian astronomer JAMĀL AL-DĪN was sent by the Ilkhan from the Marāghah Observatory to confer with the astronomers of the observatory at Peking directed by KUO SHOU-CHING. The dynastic history of the Yuan preserves accounts both of the designs or models of instruments which the Persian brought with him, and of the instruments which KUO SHOU-CHING set up about the same time. It is suggested that the "Simplified Instrument," essentially identical with the equatorial mounting of modern telescopes, was KUO SHOU-CHING's modification of the earlier mediaeval Arabic and European instrument known as the torquetum. It was called "simplified" because the ecliptic components had been removed, in accordance with the system of equatorial co-ordinates, classically Chinese, and adopted generally in the West after the time of TYCHO BRAHE.

OF the armillary spheres of the late Sung dynasty little is known, but some of the + 13th-century instruments of that great scientist and engineer KUO SHOU-CHING [1][1] are still extant,[2] and kept at the present time at the Purple Mountain Observatory of Academia Sinica, north-east of Nanking (Figs. 1, 2, and 2*a*). They were still in use at the time of the arrival of the Jesuits about + 1600. Here is what MATTHEW RICCI wrote about them[3]:

> "Not only in Peking, but in this capital also (Nanking) there is a College of Chinese Mathematicians, and this one certainly is more distinguished by the vastness of its buildings than by the skill of its astronomers. They have little talent and less learning, and do nothing beyond the preparation of almanacs on the rules of calculation made by the ancients—and when it chances that events do not agree with their computations they assert that what they had computed was the regular course of things, and that the aberrant conduct of the stars was a prognostic from heaven about something that was going to happen on earth. This something they made out according to their fancy, and so spread a veil over their blunders. These gentlemen did not much trust Fr. MATTEO, fearing, no doubt, lest he should put them to shame; but when at last they were freed from this apprehension they came and amicably visited him in the hope of learning something from him. And when he went to return their visit he saw something that really was new and beyond his expectation.
>
> There is a high hill at one side of the city, but still within the walls.[4] On the top there is an ample terrace, capitally adapted for astronomical observation, and surrounded by magnificent buildings which form the residences of the astronomers. . . . On this terrace are to be seen astronomical instruments of cast metal, well worthy of inspection whether for size or for beauty, and we certainly had never seen or read of anything in Europe like them. For nearly 250 years they have stood thus exposed to the rain, the snow, and all other atmospheric inclemencies, yet they have lost nothing of their original lustre. . . .
>
> First we inspected a great (celestial) globe, graduated with meridians and parallels; we estimated that three men would hardly be able to embrace its girth. . . . A second instrument was a great (armillary) sphere, not less in diameter than that measure of the

[1] Figures in square brackets refer to the table of Chinese characters on p. 68.

[2] The present writer had the great pleasure of examining some of them in 1946. They were, of course, at Peking when the Jesuits first went there, but were transported to Germany at the time of the Boxer Rebellion, whence in due course the German Government returned them to China from Potsdam. Good photographs of them taken during their period at Potsdam have been published by R. MÜLLER.

[3] *Opere*, ed. VENTURI, i, pp. 24, 315. Quoted by TRIGAULT (1615), tr. GALLAGHER, p. 329 ff., and by YULE.

[4] This must have been the Cockcrow Temple Hill, at the top of which the Institute of Meteorology of Academia Sinica now stands, close to the temple in which the Liang emperor, Wu Ti, a Buddhist, piously starved himself to death.

1 郭守敬	11 景符	21 札馬魯丁	32 苦來亦阿兒子	43 窺衡
1a 耶律楚才	12 闚几	22 咱禿哈剌吉	33 地理志	43a 端為圭首
2 玲瓏儀	13 日月食儀	23 渾天儀	34 兀速都兒剌	43b 横耳
3 皇甫仲和	14 星晷	24 咱禿朔八台	35 北極雲架	44 赤道環
4 簡儀	15 定時儀	25 測驗周天星曜之器	36 規環	45 界衡
5 渾天象	16 正方案	26 魯哈麻亦渺凹只	37 龍柱	45a 宿
6 仰儀	17 候極儀	27 冬夏至晷	38 南極雲架	46 定極環
7 四海測驗	18 九表懸	28 魯哈麻亦木思塔餘	39 百刻環	47 陰緯環
8 高表	19 正儀	29 春秋分晷	40 四游雙環	48 立運環
9 立運儀	19a 蘇頌	30 苦來亦撒麻	41 直距	
10 證理儀	20 西域儀象	31 斜丸渾天圖	42 横	

Table of Chinese characters referred to in this article, and indicated in the text by the figures in square brackets outstretched arms which is commonly called a geometric pace. It had a horizon (-circle) and poles; instead of circles it was provided with certain double hoops (armillae), the void space between the pair serving the same purpose as the circles of our spheres.[5] All these were divided into 365 degrees and some odd minutes. There was no globe representing the earth in the centre, but there was a certain tube, bored like a gun-barrel, which could readily be turned about and fixed to any azimuth or altitude so as to observe any particular star, just as we do with our vane-sights—not at all a despicable device. . . .

The third machine was a gnomon, the height of which was twice the diameter of the former instrument, erected on a very large and long slab of marble, on the northern side of the terrace. The stone slab had a channel cut round the margin, to be filled with water in order to determine whether the slab was level or not, and the style was set vertical as in hour dials. We may suppose this gnomon to have been erected that by its aid the shadow at the solstices and equinoxes might be precisely noted, for in that view both the slab and the style were graduated.

Fig. 1. Armillary sphere (equatorial) of Kuo Shou-Ching, about + 1276 (made for the latitude of Phing-Yang in Shansi, but later at Peking and now at Nanking). The photograph is one by SAUNDERS which YULE obtained from WYLIE. Another good one is in THOMSON, vol. iv

The fourth and last instrument, and the largest of all, was one consisting of, as it were, 3 or 4 huge astrolabes in juxtaposition; each of them having a diameter of such a geometrical pace as I have specified. The fiducial line, or Alhidada, as it is called, was not lacking, nor yet the Dioptra. Of these astrolabes, one having a tilted position in the direction of the south, represented the equator; a second, which stood crosswise on the first, in a north and south plane, the Father took for a meridian, but it could be turned round on its axis; a third stood in the meridian plane with its axis perpendicular, and seemed to represent a vertical circle, but this also could be turned round to show any vertical whatever. Moreover, all these were graduated and the degrees marked by prominent studs of iron, so that in the night the graduation could be read by touch without any light. All this compound astrolabe instrument was erected on a level marble platform with channels round it for levelling.

[5] This was to allow of the swinging of the sighting-tube within these circular guides. It is noteworthy that such a device was strange to RICCI.

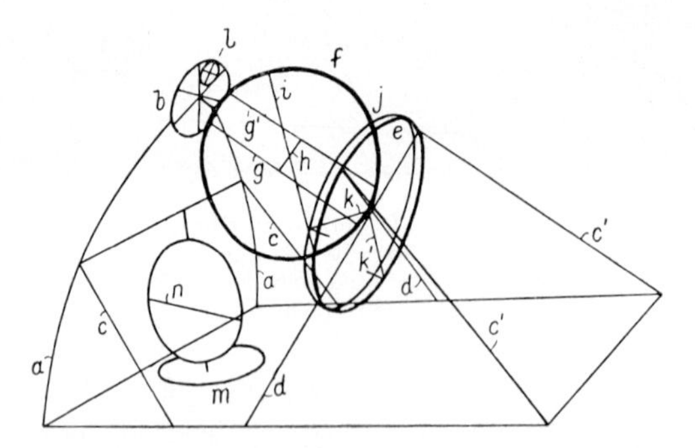

Fig. 2. "Simplified instrument" (really an equatorial torquetum) of Kuo Shou-Ching, about + 1276 (made for the latitude of Phing-Yang in Shansi, but later at Peking and now at Nanking). The photograph is one by SAUNDERS which YULE obtained from WYLIE; it was taken at Nanking

Fig. 2 (*a*). See Fig. 2; a more recent photograph of the "Simplified Instrument" (taken at Nanking) showing the "pole-determining circle" at the top, with its crossbars and central hole, attached to the "normal circle"

The table of identifications given below refers to the sketch on the lower part of the opposite page, Fig. 2.

a, *a*	North pole cloud frame standards (pei chi yün chia [35]).
b	Normal circle (kuei huan [36]), diameter 2 ft 4 in. (fixed).
c, *c*, *c′*, *c′*	Dragon pillars (lung chu [37]).
d, *d*	South pole cloud frame standards (nan chi yün chia [38]).
e	(fixed). Diurnal circle (pai kho huan [39]), graduated in "the twelve hours and the hundred divisions", each division containing 36 sub-divisions; it carries four rollers for the better rotation of the equatorial circle. Diamater 6 ft 4 in.
f	(revolving). Meridian double circle (ssu yü shuang huan [40]), 6 ft in diameter, graduated in degrees and minutes. For declination.
g, *g′*	Stretchers (chih chü [41]).
h	Brace (hêng [42]).
i	Diametral alidade (khuei hêng [43]), with pointed ends (like ceremonial tablets) for accuracy (tuan wei kuei shou [43a]), and sighting-vanes (hêng erh [43b]).
j	(revolving). Equatorial circle (chhih tao huan [44]), 6 ft in diameter, graduated in degrees and minutes. For right ascension.
k, *k′*	Independently movable radial pointers with pointed ends (chieh hêng [45]). It is not clear whether these carried sighting vanes. Their name ("boundary bars") indicates that they were used to mark off the boundaries of the hsiu [45a], *i.e.* the equatorial lunar mansions.
l	Pole-determining circle (ting chi huan [46]). Diameter equivalent to 6°. Attached to the upper part of the normal circle. It cannot be seen in Fig. 2 but appears clearly in Fig. 2 (*a*). This circle had a cross-piece inside it, with a central hole, and seems to have been used for determining the moment of culmination of the pole-star itself. Observation was effected through a small hole in a bronze plate attached to the south pole cloud frame standards. The main polar axis through the centre of the diurnal and normal circles was also provided with holes constituting a sighting-tube.
m	(fixed). Earthly co-ordinate circle (yin wei huan [47]), for azimuths.
n	(revolving). Vertical circle with alidade (li yün huan [48]), for altitudes.

On each of the instruments explanations of everything were given in Chinese characters, and there were also engraved the 24 zodiacal constellations which answer to our 12 signs, 2 to each.[6] There was, however, one error common to all the instruments, *i.e.* that in all the elevation of the pole was taken to be 36°. Now there can be no question about the fact that the city of Nanking lies in lat. $32\frac{1}{4}$°, whence it would seem probable that these instruments were made for another locality and had been erected at Nanking, without reference to its position, by someone ill-versed in mathematical science.[7]

Some years afterwards Father MATTEO saw similar instruments at Peking, or rather the same instruments, so exactly alike were they, insomuch that they had unquestionably been made by the same artists. And indeed it is known that they were cast at the period when the Tartars were dominant in China; so that we may without rashness conjecture that they were the work of some foreigner acquainted with our studies."[8]

Thus RICCI was greatly impressed by the astronomical instruments of the Yuan dynasty, though holding a poor opinion of those Chinese contemporaries whom it was his strategy to supplant, and venturing a particularly erroneous guess about the original constructor of the instruments.

The authentic Chinese texts from which we gain information about the equipment of KUO SHOU-CHING's observatory of + 1276 include, of course, the *Yuan Shih* (History of the Yuan Dynasty).[9] The instruments are there listed as follows:

(1) Ling Lung I [2] . . Ingenious Armillary Sphere (Fig. 1).
RICCI'S "second instrument". The instrument of this kind still preserved at the Purple Mountain Observatory near Nanking is considered to be the copy made by HUANGFU CHUNG-HO [3] in + 1437, and not the original one of KUO himself.

(2) Chien I [4] . . Simplified Instrument (Fig. 2, 2a).
RICCI'S "fourth instrument".[10]

(3) Hun Thien Hsiang [5] . Celestial Globe.
RICCI'S "first instrument".

(4) Yang I [6] . . . Upward-looking Instrument.[11]
A hemispherical sun-dial intermediate in size between earlier Chinese types and the much larger Jai Prakāś instruments of the Indians.

(5) Kao Piao [8] . . Lofty Gnomon.[12]
Undoubtedly the 40 ft gnomon, especially that at Yang-chhêng.[13]

(6) Li Yün I [9] . . Vertical Revolving Circle.
The vertical circle in (2) described by RICCI, and seen in Fig. 2. This, with (13) below, would be equivalent to modern altazimuths and theodolites. (SPENCER JONES (1), p. 83.)

(7) Chêng Li I [10] . . Verification Instrument.
It is not clear what this was, but perhaps it was a component of (1) which permitted exact determinations of the positions of the sun and moon near the equator; such at any rate was its stated purpose. Perhaps a sighting-tube.

---

[6] This was a mistake of RICCI'S.

[7] This point can be cleared up at once. The instruments which RICCI saw belonged to a set which had originally been made for the astronomical college founded by the great statesman and astronomer YEHLÜ CHHÜ-TSHAI [1a] about + 1220 at Phingyang in Shansi (latitude just over 36°). In the scientific decadence of the Ming they had been removed to Nanking (JOHNSON). That the astronomers of KUO SHOU-CHING's time were very conscious of the importance of the latitude is shown by the fact that the *Yuan Shih* (Chapter 48, p. 12b, ff) reproduces a table of the latitudes of some twenty-five important centres, some of which undoubtedly had astronomical instruments; it is entitled "Ssu Hai Tsche Yen" [7]. See GAUBIL, p. 110.

[8] Tr. YULE, i, p. 451; cit. BERNARD-MAITRE, p. 59; WYLIE, p. 14.

[9] Chs. 48 and 164, paraphrased and partly translated by WYLIE, who lists the other sources.

[10] *Yuan Shih*, Chapter 48, p. 2b.

[11] P. 6a. [12] P. 8b.

[13] RICCI'S "third instrument" must have been one of the usual 8-ft gnomons.

(8) Ching Fu [11] . . Shadow Definer.[14] A device for determining more precisely the end of the sun shadow on (5).

(9) Khuei Chi [12] . . Observing Table. Apparently an adaptation of the gnomon and shadow-definer to lunar shadows.

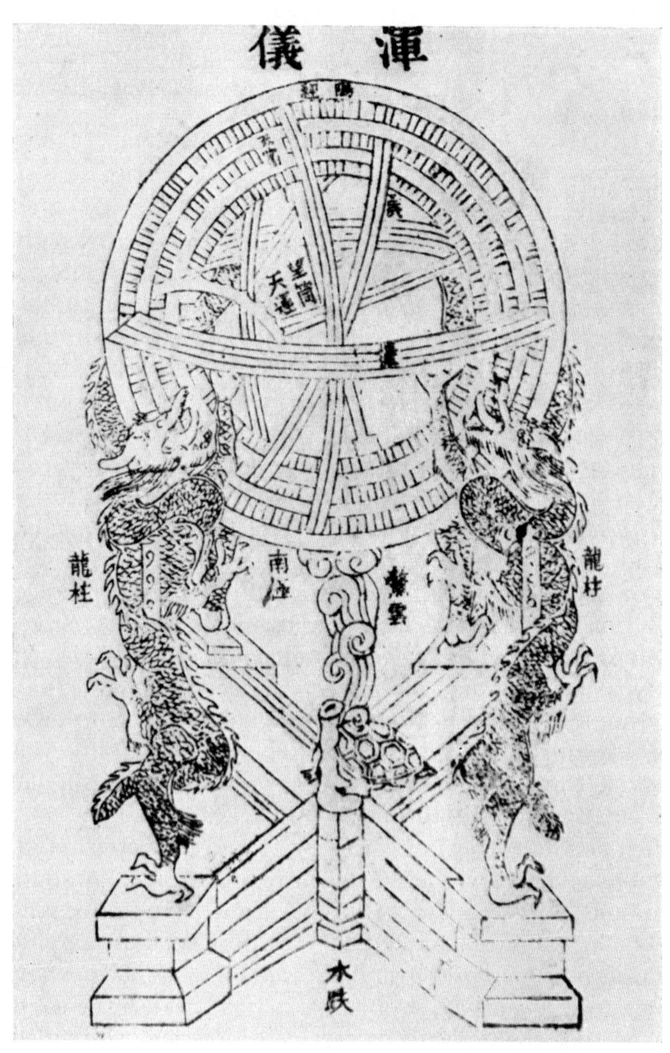

Fig. 3. Armillary Sphere of Su Sung (fl. + 1070/+ 1095), as described in his book *Hsin I Hsiang Fa Yao* (New Description of the Armillary Sphere) of + 1094 (from the *Shou Shan Ko Tshung Shu* edition)

(10) Jih Yüeh Shih I [13] . Instrument for Observation of Solar and Lunar Eclipses. Not clear and not explained.

(11) Hsing Kuei [14] . . "Star-dial". Could this have been a forerunner of the + 16th century nocturnal?

(12) Ting Shih I [15] . . Time-determining Instrument. Probably the "diurnal circle" of (2) above.

(13) Chêng Fang An [16] . Direction-determining Table. This must have been an azimuthal circle, probably the "earthly co-ordinate circle" of (2) above.

---

[14] P. 9b.

(14) Hou Chi I [17] . . Pole-observing Instrument.
Presumably a quadrant for obtaining the polar altitude and hence the latitude. Or it may have been the polar sighting-tube embodied in (2) above.

(15) Chiu Piao Hsüan [18] . Nine Suspended Indicators.
Though details are not given, this probably refers to the "groma" or hanging plumb-lines whereby the trueness of the instruments, especially the gnomons, was checked.

(16) Chêng I [19] . . Rectifying Instrument.
Purpose uncertain.

Apart from the minor devices and those about which we have not sufficient information, the chief interest lies in the great armillary sphere (1), and the "Simplified Instrument" (2). There is little need to describe the former more fully, since it did not differ in any fundamental way from that which SU SUNG [19a] had already used in + 1090 (Fig. 3), though no doubt of finer and exacter workmanship. But the second of these two, which RICCI could only describe as a collection of "astrolabes" set in different axes, is something new. I believe that it was a variant of the mediaeval instrument known as the "torquetum"[15] which consisted of a series of discs and circles not, like those of the armillary sphere, placed concentrically. The Arabic and European versions had a disc mounted in the plane of the equator, and another one revolving at any angle to it in the plane of the ecliptic; this then carried a celestial longitude circle at right angles to itself[16]. The equipment was completed by a half-disc or protractor and plumb-line for reading off altitude. Probably the apparatus was used mainly for computations, as it permitted the direct conversion of ecliptic to equatorial co-ordinates and *vice versa* as well as other comparisons. The invention of this unwieldy instrument has often been ascribed to KUO SHOU-CHING's elder contemporary, NAṢĪR AL-DĪN AL-ṬŪSĪ of Marāghah[17] but more probably goes back to the Spanish Muslim, JĀBIR IBN AFLAḤ (b. about + 1130).[18] Europeans were using it at the same time as the Marāghah astronomers in Persia (THORNDIKE, 1, 2) and there were treatises on the torquetum by Regiomontanus and Apianus about + 1540,[19] but TYCHO BRAHE spoke scornfully of it,[20] and after him no one employed it except the Indians who perpetuated the Arabic tradition. An outdoor example is still extant at Jaipur, the Krāntivṛitti Valaya Yantra[21] (see Fig. 4).

Now the great point of interest about KUO SHOU-CHING's "simplified instrument" is that although (as a "dissected" armillary sphere) it is recognizably related to the torquetum, it is a true *equatorial*. Doubtless it was because the ecliptic components had been removed that the instrument received its name "simplified". This means that though Arabic influence may have been responsible for suggesting its construction, KUO adapted it to the specific character of Chinese astronomy, namely equatorial co-ordinates. And in so doing, he fully anticipated the equatorial mounting so widely used for modern telescopes. This is generally attributed to JAMES SHORT, F.R.S. (+ 1710/+ 1768) in + 1749.[22] Here we reach the situation which several recent historians of science have recognized as being of such great interest,

[15] To be distinguished carefully from the "triquetrum", see on.
[16] See R. WOLF, ii, p. 117; HOUZEAU, p. 952; GUNTHER (1), vol. 2, pp. 35, 36; ANON., p. 18, No. 348 and opp. p. 30; ROHDE, pp. 79 ff; MICHEL (2), p. 68 and Pl. XIII. One of the best expositions is that of MICHEL (3), who showed that the Rectangulus of Wallingford, which GUNTHER (1), vol. 2, p. 32, had not understood, was a skeleton torquetum.
[17] SARTON, ii, p. 1005. [18] SARTON, ii, p. 206.
[19] The oldest European specimen is that preserved at Cues near Treves; Nicholas of Cusa bought it in + 1444 (see HARTMANN).
[20] RAEDER, STRÖMGREN, and STRÖMGREN tr., p. 53.
[21] KAYE, pp. 32, 33, and Fig. 58; SOONAWALA, p. 38. This instrument now lacks several components.
[22] Cf. CHAUVENET, p. 367, who favours FRAUNHOFER; and A. WOLF, pp. 136 ff.

namely (*a*) the faithfulness of the Chinese from the beginning of their history to what afterwards became the co-ordinates universally used, and (*b*) the influences which may have acted upon TYCHO BRAHE to cause him to abandon the characteristic Graeco-Arabic-European system of ecliptic co-ordinates.

Before examining this, however, it will be more logical to explore somewhat further what is known of the Arabic influence upon the astronomers of the Yuan dynasty.[23] The task is easy owing to the special studies which HARTNER; YABUUCHI; and TASAKA[24] have devoted to the subject. They have been able to identify the diagrams of seven astronomical instruments which reached China from Persia in + 1267. In

Fig. 4. The only existing outdoor torquetum, the Krāntivṛitti Valaya Yantra at the Jaipur Observatory (from KAYE)

the *Yuan Shih*[25] a couple of pages are devoted to the "Plans (or Models) of Astronomical Instruments from the Western Countries" (Hsi yü i hsiang [20])—these were sent by HŪLĀGU KHAN or his successor to KHUBILAI KHAN through the hands of one of the Marāghah astronomers CHA-MA-LU-TING [21][26] (JAMĀL AL-DĪN) in + 1267. The identity of this man is somewhat obscure but he may have been the JAMĀL AL-DĪN IBN MUḤAMMAD AL-NAJJĀRĪ,[27] who had declined to take on the full responsibility of building the Marāghah observatory in + 1258. The Chinese names of the instruments about which he was deputed to inform the Chinese astronomers, together with brief explanations of them, are given in the *Yuan Shih* text.

These instruments were the following:

Chinese transcription[28]	*Persian-Arabic original*	*Chinese translation and explanation*
(1) Tsa-thu ha-la-chi [22]	Dhātu al-ḥalaq ("the owner of the rings").	Hun Thien I [23] Armillary sphere.

[23] WAGNER has described two interesting manuscripts preserved in his time in the Pulkowo Observatory in Russia, one in Arabic or Persian, the other in Chinese. They are tables of the motion of sun, moon, and planets, calculated from an epoch starting at + 1204, and written about + 1261. As probable relics of the collaboration of JAMĀL AL-DĪN and KUO SHOU-CHING, they would be precious indeed, and it is to be hoped that they were not destroyed when the Observatory was burnt during World War II. In the following century, there were numerous Arabic-Chinese scientific contacts.

[24] Cf. FUCHS. In the following identifications of Arabic terms and their interpretations we accept those of HARTNER rather than those of TASAKA.

[25] Chapter 48, p. 10b. [26] Cf. SARTON, ii, p. 1021.

[27] Or more probably AL-BUKHĀRĪ (note from Dr. W. HARTNER).

[28] These are not the only extant transcriptions, but they are the most correct. Chhien-Lung editions of the *Yuan Shih* were subjected to revision by a commission of learned linguists who mongolicized all foreign words even when these had been transliterations from other languages such as Arabic (note from Dr. W. HARTNER).

(2) Tsa-thu shuo-pa-thai [24]	Dhātuʿsh shuʿbatai ("the instrument with two legs").	Tshê Yen Chou Thien Hsing Yao chih Chhi [25] ("instrument for observing and measuring the rays of the stars of the celestial vault"). Certainly PTOLEMY's "organon parallacticon", 'όργανον παραλλακτικόν, *i.e.* the long ruler, or "triquetrum", for determining zenith distances of stars at culmination.[29] The conjectures of ZINNER, p. 236, that it was a divided circle, and of FUCHS, p. 4, that it was the Jacob's staff of the surveyors, are not to be retained.
(3) Lu-ha-ma-i miao-wa-chih [26]	Rukhāma-i-muʿwajja	Tung Hsia Chih Kuei [27] "Solsticial Dial", *i.e.* plane sun-dial for unequal hours.
(4) Lu-ha-ma-i mu-ssu-tha-yü [28]	Rukhāma-i-mustawīya	Chhun Chhiu Fên Kuei [29] "Equinoctial Dial", *i.e.* plane sun-dial for equal hours.
(5) Khu-lai-i sa-ma [30]	Kura i-samāʿ	Hsieh Wan Hun Thien Thu [31] ("obliquely set globe with map of the stars"). Celestial Globe.
(6) Khu-lai-i a-erh-tzu [32]	Kura i-arḍ	Ti Li Chih [33] Terrestrial Globe.
(7) Wu-su-tu-erh-la [34]	al-Usṭurlāb	Astrolabe. The text says: "The Chinese name (for this) has not been worked out. The instrument is to be made from bronze, on which the times (hours) of the day and night are engraved". Certainly not a clepsydra, as ZINNER, p. 236, supposed.

The list is an interesting one. The first suggestion was certainly no novelty for the Chinese,[30] but Jamāl al-Dīn's instrument was surely adapted for ecliptic measurements, and as we saw above, KUO SHOU-CHING paid no attention to this. It would also have used a graduation of 360° but KUO SHOU-CHING retained the $365\frac{1}{4}°$ system.[31] Nor was the fifth instrument, the celestial globe, anything new. On the other hand, the terrestrial globe was perhaps a novelty; there is no previous record of one before the time of MARTIN BEHAIM (+ 1492)[32] except the ancient globe of Crates of Mallos[33] in the — 2nd century, which had been entirely forgotten. The Chinese text describes the new instrument as "a globe to be made of wood, upon which seven parts of water are represented in green, three parts of land in white, with rivers, lakes, *etc.* Small squares are marked out so as to make it possible to reckon the sizes of the regions and the distances along roads". There seems no evidence, however, that the Chinese took this up. As for the sun-dial, they were probably puzzled at the conception

[29] The best description is TYCHO BRAHE'S (RAEDER, STRÖMGREN, and STRÖMGREN, p. 44). See also GUNTHER (1), ii, p. 15.

[30] HARTNER goes far astray in his suggestion that armillary spheres were unknown in China until the + 13th century—an under-estimate of perhaps seventeen centuries. We have details of some twenty-five important instruments from the — 2nd century onwards, and good reasons for thinking that the astronomers of the — 4th also used them, at any rate in simple form. This is said not in any criticism of HARTNER, since, as he informs us, his facilities were lamentably inadequate, and the work of MASPERO (1, 2) was not known to him; but in order to correct views which might otherwise claim the authority of *Isis* in 1950.

[31] This seems the only justification, and not at all a strong one, for JOHNSON'S emphasis on what he calls the "tragic conservatism" of the Chinese. BOSMANS, himself a Jesuit, has on the contrary drawn attention to the imposition of the sexagesimal graduation of degrees and minutes on the Chinese by VERBIEST at a later date. The Chinese had always graduated them decimally in tenths and hundredths. BOSMANS freely admits that this change was a retrograde one.

[32] RAVENSTEIN. [33] SARTON, i 185; cf. STEVENSON; SCHLACHTER and GISINGER.

of unequal hours, and it is fairly clear that their traditional type of sun-dial persisted unchanged.

The instrument most strikingly absent from the list is the torquetum, though it would have been expected more than any other, if KUO SHOU-CHING's "Simplified Instrument" arose from the stimulus of contact with Arabic science as on all the circumstantial evidence it did. Moreover, no torquetum is listed in AL-'URḌI's account of the equipment at the Marāghah Observatory (SEEMANN (1); JOURDAIN).

As for the second and seventh of the instruments, if they were not adopted it was surely because they did not fit into the characteristic system of Chinese astronomy, polar and equatorial. The parallactic ruler for determination of zenith distance[34] could hardly interest astronomers in whose work the zenith played no part. And the astrolabe, so universal in Arabic and mediaeval European astronomy, was primarily intended to measure altitude, and to compute ecliptic co-ordinate positions, which the Chinese did not particularly want. HARTNER considers that what the Marāghah astronomers offered was well chosen;[35] they did not send the apparatus for determining sines, azimuths, and versed sines,[36] because it was probably known that the Chinese astronomers were unfamiliar with spherical trigonometry. JAMĀL AL-DĪN had an overwhelming task before him if he intended to explain to the Chinese the whole system of Arabic gnomonics and the mathematics required for the stereographic projection on which the astrolabe markings were based, and if he tried he certainly did not succeed. But what has been unperceived, even by HARTNER, is that the measurements and computations which the rulers and the astrolabe could yield were simply not wanted in the Chinese polar-equatorial system.

The astrolabe is a very complex instrument upon which mediaeval Arabic and European astronomers lavished all their mathematical art. It might be called a "flattened" armillary sphere,[37] combining the armillary rings of HIPPARCHUS and the theodolite of THEON with a projection of the zodiac and the starry hemisphere. The luminous treatise of MICHEL (1), published a few years ago, makes recourse to older explanations of the theory and construction of astrolabes unnecessary, while a massive compendium by GUNTHER (2) gives elaborate details of the principal surviving instruments. The place of origin of the instrument is uncertain,[38] but its first known user was the Byzantine Ammonios (ca. + 500), though the earliest dated astrolabe is Persian, that of "Aḥmad and Maḥmūd the sons of Ibrāhīm the astrolabist of Iṣpahan", + 984. Venerable also is the Byzantine astrolabe of + 1062, described by DALTON. The earliest extant treatise on its use derives from JOANNES PHILIPONUS (+ 525), the Byzantine physicist, a pupil of Ammonios, and in the next century the Syrian bishop SEVERUS SEBOKHT also wrote on it.[39] Not a single example is known from China, either by textual references or actual preservation.

It was DREYER (1) who was perhaps the first to appreciate the historical importance of KUO SHOU-CHING's retention of the equatorial system in his Simplified Instrument. "We have here", he said, "two remarkable instances of how the Chinese people often came into possession of great inventions many centuries before the western

[34] Illustration in GUNTHER (1), ii, p. 15.

[35] It will be remembered that the Marāghah Observatory had on its staff at least one Chinese astronomer. His name was apparently FU MÊNG-CHI, or as some think, FU MU-CHAI. Cf. SARTON, vol. 2, p. 1005.

[36] Cf. JOURDAIN and SEEMANN.

[37] Hence the name "astrolabium planisphaerium", R. WOLF, ii, 45. There was an intermediate form, in which flat rings were made so as to slide over a solid celestial sphere, but this was never widespread. One such, made for ALFONSO X, is figured in SINGER and SINGER, p. 227. See SEEMANN (2).

[38] Cf. NEUGEBAUER. What PTOLEMY called an astrolabe was an armillary sphere, a fact which has caused some terminological confusion in modern times. RICCI's use of the word astrolabe in our opening quotation was, of course, quite unjustified.

[39] Another old treatise, by the Jewish astronomer MANASSEH of Baghdad (d. + 815) (Messahalla) has been translated in GUNTHER (1), vol. v. See also J. FRANK.

nations enjoyed them. We find here in the + 13th century the equatorial armillae of TYCHO BRAHE, and better still, an equatorial instrument like those 'armillae aequatoriae maximae' with which TYCHO observed the comet of + 1585 as also the fixed stars and planets".[40]

JOHNSON considers that "the Yuan instruments exhibit a simplicity which is not primitive, but implies a practised skill in economy of effort, and in this sense compares favourably with the Graeco-Muslim tendency to rely on separate instruments for each single co-ordinate to be measured—neither Alexandria nor Marāghah[41] exhibit

Fig. 5. TYCHO BRAHE'S *Armilla Aequatoria Maxima* (from H. RAEDER, E. STRÖMGREN, and B. STRÖMGREN)

any device so complete and effective and yet so simple as the 'simplified' instrument of KUO SHOU-CHING. Actually our present-day equatorial mounting has made no further essential advance".[42] JOHNSON adds that the gun-barrel sighting-tubes rotating in the double rings were much preferable to the open alidades of the Arabs.[43]

An urgent question thus arises. What was it that led TYCHO BRAHE in the + 16th century to abandon the age-old Graeco-Muslim ecliptic co-ordinates and ecliptic armillary spheres in favour of the equatorial co-ordinates which the Chinese had had all along? GUNTHER[44] expressed the greatest surprise that the Chinese had

[40] See Fig. 5. The difference was that TYCHO retained a half-circle of the equator embracing the hour-circle centrally.

[41] For lists of the Marāghah equipment see JOURDAIN and SEEMANN (1), summarized in SARTON, ii, 1013.

[42] It is interesting to reflect that MARCO POLO was in China just at the time when all this was going on. But he noticed only the astrological aspects of Chinese state-supported astronomy. Some of his texts contain a chapter (Chapter 33; see YULE (1), i, p. 446), "Concerning the Astrologers in the City of Cambaluc". He says that no less than 5000 of them were entertained by the Great Khan with an annual maintenance and clothing, and that "they have a kind of astrolabe on which are inscribed the planetary signs, the hours and critical points of the whole year". All three sects of star-clerks, Cathayan, Saracen, and Christian (presumably Uighur Nestorian), used these instruments for prognostications, on which they were consulted by many people. Moreover, they prepared "certain little pamphlets called Tacuin", *i.e.* Ar. Taqwīm, calendars or ephemerides, which were published by the government in surprising numbers. For example, in + 1328 more than three million copies were printed and issued. Some of these (*e.g.* one for + 1408) later came into the possession of such men as ROBERT BOYLE, ROBERT HOOKE, and SAMUEL PEPYS, arousing their curiosity concerning Chinese astronomy.

[43] Another matter in which the Renaissance astronomers of Europe adopted Chinese practices was the greater use of meridian transit observations. There was TYCHO'S great mural quadrant, followed by the first mounting of a telescope permanently in the meridian by ROEMER in + 1681 (DREYER (1), GRANT, pp. 461 ff.; SPENCER JONES (2)).

[44] (1), ii, pp. 145 ff.

used equatorial armillae, which were considered one of the chief advances of European Renaissance astronomy, and concluded that KUO SHOU-CHING had anticipated TYCHO by three centuries. Now TYCHO tells us himself in his book on instrument-construction[45] that he found zodiacal or ecliptic armillary spheres very unsatisfactory

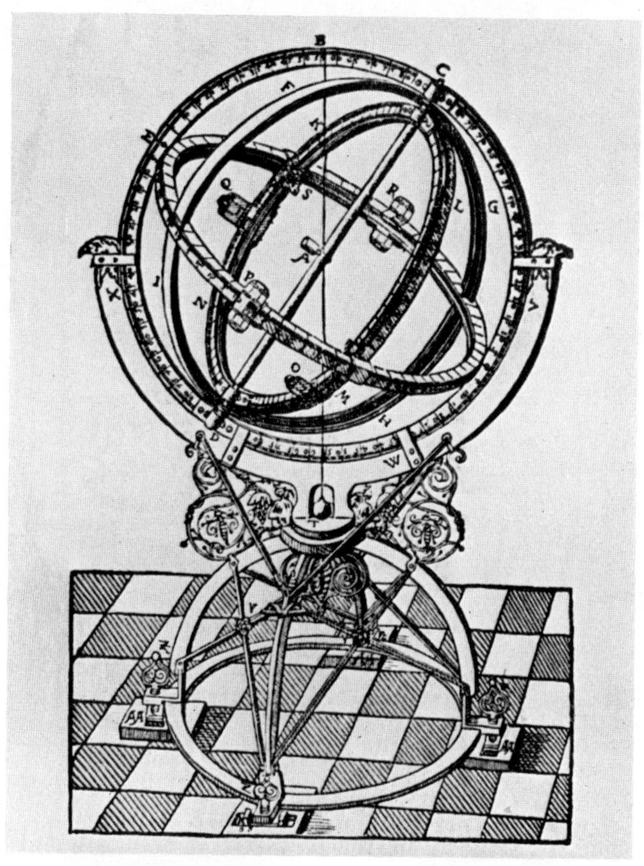

Fig. 6. TYCHO BRAHE's smaller equatorial armillary sphere (from H. RAEDER, E. STRÖMGREN, and B. STRÖMGREN)

because since they are not always in a position of equilibrium (*i.e.* since their centre of gravity shifts according to the position of the equinoctial junctions) they become deformed owing to the weight of the metal, introducing errors of as much as a couple of minutes of arc. For this reason, he preferred to construct equatorial armillary spheres (Fig. 6). But a purely technical reason of this kind seems insufficient for a change in the basic method of expressing celestial co-ordinates. DREYER (2) therefore raised the question of whether there had been some Arabic influence sapping the assured foundations of ecliptic methods. SÉDILLOT[46] (1) suggested that equatorial armillary spheres had been known to the Arabs, and he provided more evidence later (2), quoting, among other sources, BETTINI's seventeenth-century view that AL-HAITHAM had used them in the late + 10th or early + 11th century.[47] In such a place as this it is impossible to pursue the matter further, but its importance would

[45] Tr. RAEDER, STRÖMGREN, and STRÖMGREN, pp. 55 ff. [46] P. 198.

[47] P. cxxxiv. BETTINI (Apiaria, II, progym. iii, p. 41) had written: "Adhibuit TYCHO armillare quoddam instrumentum quod tamen comperi ego positum et adhibitum olim fuisse ante TYCHO ab Alhazeno"

Fig. 8. View of the Peking Observatory, as reconstructed by the Jesuits (from VERBIEST)

Fig. 7. Ecliptic Armillary Sphere at the Peking Observatory, dating from its reconstruction by the Jesuits in the + 17th century (photograph from the WHIPPLE Collection, Cambridge)

seem to justify a special investigation. There exists the possibility that in spite of the general trend of Muslim astronomy, there were isolated instances of the use of equatorials, and that these might have been derived from Arab-Chinese contacts (at any time indeed back to the Han).[48] The idea might then have stimulated a few European astronomers, such as GEMMA FRISIUS, who in + 1534 first described a small portable equatorial armillary sphere,[49] and then in turn Tycho Brahe himself, whose reasons for making the change can surely hardly have been confined to the purely technical ones mentioned above.

How paradoxical it is, in the light of all this, that when the Jesuits proceeded to enlighten the Chinese in scientific matters, they erected (+ 1673/+ 1714) *ecliptic* armillary spheres at the Peking observatory[50] (Figs. 7[51] and 8[52]). And Verbiest, in his book of + 1687 about astronomy in China, had nothing to say about the Yuan instruments, save that they had been the products of a "ruder Muse".

[48] If we may include pre-Muslim times and the Syrian and Persian predecessors of Arabic science.
[49] DREYER (2), p. 316.
[50] LECOMTE, p. 67. They are still in good condition (1952) and have retained their sighting-tubes (square-sectioned outside and tubular inside) with fiducial cross-wires.
[51] Other good photographs were published by B. F. ROBINSON. [52] From VERBIEST, copied by LECOMTE and others.

NOTE

This study, which it has given the author much pleasure to contribute to a collection honouring Professor F. J. M. STRATTON, will form part of the third volume of a work *Science and Civilisation in China* (now in course of publication by the Cambridge University Press), the first volume of which appeared in the autumn of 1954.

He wishes to thank Mr. WANG LING and Dr. DEREK PRICE for valuable assistance.

REFERENCES

ANON. . . . . *Les Instruments de Mathématiques de la Famille Strozzi faits en 1585–1586 par Erasmus Habermehl de Prajue* (Frederik Muller, Amsterdam), 1911.

BERNARD-MAITRE, H. . . . . *Matteo Ricci's Scientific Contribution to China* (Vetch, Peiping), 1935.

BETTINI, M. . . . . *Apiaria* (Bononia), 1645.

BOSMANS, H. . . . . "Une Particularité de l'Astronomie Chinoise au XVIIème Siécle", *Ann. Soc. Scientifique de Bruxelles*, 1903 (1), **27**, 122.

CHAUVENET, W. . . . . . . *Manual of Spherical and Practical Astronomy* (Lippincott, Philadelphia), 1900.

DALTON, O. M. . . . . . . "The Byzantine Astrolabe at Brescia [of A.D. 1062]", *Proc. Brit. Acad.*, 1926.

DREYER, J. L. E. (1) . . . . . "The Instruments in the Old Observatory in Peking", *Proc. Roy. Irish Acad.*, 1883 (2nd series), **3**, 468; *Copernicus*, 1881, **1**, 134.

(2) . . . . . *Tycho Brahe; a Picture of Scientific Life and Work in the XVIth Century* (Black, Edinburgh), 1890.

FRANK, J. . . . . *Die Verwendung des Astrolabs nach al-Chwarizmi* (Mencke, Erlangen), 1922. (Abhdl. z. Gesch. d. Naturwiss. u. d. Med., No. 3.)

FUCHS, W. . . . . *The 'Mongol Atlas' of China by Chu Ssu-Pên and the Kuang Yü Thu* (Fu-Jen University Press, Peking), 1946. (Monumenta Serica Monograph Series, No. 8.)

GAUBIL, A. . . . . "Histoire Abregée de l'Astronomie Chinoise", in *Observations Mathématiques, Astronomiques, Géographiques, Chronologiques, et Physiques, tirées des anciens Livres Chinois, ou faites nouvellement aux Indes, à la Chine, et ailleurs, par les Pères de la Compagnie de Jésus*, ed. E. Souciet (Rollin, Paris), 1732, vol. ii.

GRANT, R. . . . . *History of Physical Astronomy, from the Earliest Ages to the Middle of the XIXth Century* (Baldwin, London), 1852.

GUNTHER, R. T. . . . . . . *Early Science in Oxford*, 14 vols. (Oxford), 1923–45, especially vols. ii and v.

GUNTHER, R. T. . . . . . . *The Astrolabes of the World*, 2 vols. (Oxford University Press, Oxford), 1932.

HARTMANN, J. . . . . "Die astronomischen Instrumente des Kardinals Nikolaus Cusanus", *Abhdl. d. Gesellsch. d. Wiss. z. Göttingen* (*Math.-Phys. Kl.*), 1919 (NF) **10**, No. 6.

HARTNER, W. . . . . "The Astronomical Instruments of Cha-Ma-Lu-Ting, their Identification, and their Relation to the Instruments of the Observatory of Maragha". *Isis*, 1950, **41**, 184.

HOUZEAU, J. C. . . . . *Vade Mecum de l'Astronomie* (Hayez, Brussels), 1882.

JOHNSON, M. C. . . . . "Greek, Muslim, and Chinese Instrument Design in the Surviving Mongol Equatorials of 1279 A.D.", *Isis*, 1940–47, **32**, 27. Reprinted in *Art and Scientific Thought; Historical Studies towards a Modern Revision of their Antagonism* (Faber & Faber, London), 1944.

JOURDAIN, A. . . . . "Mémoire sur l'Observatoire de Meragha et les Instruments Employés pour y Observer", *Magasin Encyclopédique*, 1809 (6ème ser), **84**, 43; and sep., Paris, 1810.

KAYE, G. R. . . . . *The Astronomical Observatories of Jai Singh* (Government Printing Office, Calcutta), 1918 (Archaeol. Survey of India, New Imp. Ser., No. 40).

LECOMTE, LOUIS . . . . "Memoirs and Observations Topographical, Physical, Mathematical, Mechanical, Natural, Civil, and Ecclesiastical, made in a late Journey through the Empire of China, and published in several Letters, particularly upon the Chinese Pottery and Varnishing, the Silk and other Manufactures, the Pearl Fishing, the History of Plants and Animals, *etc.*, translated from the Paris edition, *etc.*", London, 1698.

MASPERO, H. (1) . . . . . . "L'Astronomie Chinoise avant les Han", *Toung Pao*, 1929, **26**, 267.

(2) . . . . . . "Les Instruments Astronomiques des Chinois au Temps des Han", *Mélanges Chinois et Bouddhiques* (Bruxelles), 1939, **6**, 183.

MICHEL, H. (1) . . . . . . *Traité de l'Astrolabe* (Gauthier-Villars, Paris), 1947.

(2) . . . . . . *Introduction à l'Etude d'une Collection d'Instruments Anciens de Mathématiques* (de Sikkel, Antwerp), 1939.

(3) . . . . . . "Le Rectangulus de Wallingford, précédé d'une Note sur le Torquetum", *Ciel et Terre*, 1944, (Nos. 11 and 12), 1.

MÜLLER, R. . . . . "Die Astronomischen Instrumente des Kaisers von China in Potsdam", *Atlantis*, 1931, 120.

NEUGEBAUER, O. . . . . "The Early History of the Astrolabe", *Isis*, 1949, **40**, 240.

RAEDER, H., STRÖMGREN, E. and STRÖMGREN, B. . . . . . . *Tycho Brahe's Description of his Instruments and Scientific Work, as given in his "Astronomiae Instauratae Mechanica* (Wandesburgi, 1598), (Munksgaard, Copenhagen), 1946. (Publication of K. Danske Videnskab. Selskab).

RAVENSTEIN, E. G. . . . . . *Martin Behaim; his Life and his* [*Terrestrial*] *Globe* (London), 1908.

ROBINSON, B. F. . . . . . . "The Astronomical Observatory in Peking", *Art and Archaeology* (Washington), 1930.

ROHDE, A. . . . . *Die Geschichte d. wissenschaftlichen Instrumente vom Beginn der Renaissance bis zum Ausgang des* 18. *Jahrhunderts* (Klinkhardt & Biermann, Leipzig), 1923.

SARTON, GEORGE . . . . . . *Introduction to the History of Science*, 5 vols. (Carnegie Institution of Washington, Washington D.C.), 1927–47 (Publication No. 376).

SCHLACHTER, A. and GISINGER, F. . . *Der Globus, seine Entstehung und Verwendung in der Antike* (Teubner, Leipzig and Berlin), 1927. (Stoicheia, Stud. z. Gesch. d. antik. Weltbildes u.d. griechischen Wiss., No. 8.)

SÉDILLOT, L. P. A. (1) . . . . "Mémoire sur les Instruments Astronomiques des Arabes, pour servir de complement au Traité d'Aboul Hassan", *Mémoires Présentés par Divers Savants à l'Acad. Roy. des Inscr. et Belles-Lettres* (Paris), 1844, **1**; also sep. at Impr. Roy. 1841.

(2) . . . . *Prolégomènes des Tables Astronomiques d'Oloug Beg* (Notes, Variantes, et Introduction) (Didot, Paris), 1839 and 1847.

SEEMANN, H. J. (1) . . . . . "Die Instrumente der Sternwarte zu Maragha nach den Mitteilungen von al-Urdi", *Sitzungsber. d. physik. med. Soc. Erlangen*, 1928, **60**, 15.

(2) . . . . . *Das kugelförmige Astrolab* (Mencke, Erlangen), 1925. (Abhdl. z. Gesch. d. Naturwiss. u.d. Med.).

SINGER, C. and SINGER, D. W. . . "The Jewish Factor in Mediaeval Thought", article in *Legacy of Israel* (Oxford University Press, Oxford), 1927.

SOONAWALA, M. F. . . . . . *Maharaja Sawai Jai Singh II of Jaipur and his Observatories* (Jaipur Astronomical Society, Jaipur), n.d. (1953).

SPENCER JONES, H. (1) . . . . *General Astronomy* (Arnold, London), 1946.

(2) . . . . "The Royal Greenwich Observatory", *Proc. Roy. Soc.* B, 1949, **136**, 349.

STEVENSON, E. L. . . . . *Terrestrial and Celestial Globes; their History and Construction*, 2 vols. (Hispanic Society of America, Yale University Press, New Haven), 1921.

TASAKA, K. . . . . "About an Aspect of Islamic Culture Moving Eastwards" (in Japanese), *Shigaku Zasshi*, 1942, **53**, 404.

THOMSON, J. . . . . *Illustrations of China and its People, etc.*, 4 vols. (Sampson Low, London), 1873–4.

THORNDIKE, L. (1). . . . . . "Franco de Polonia and the Turquet", *Isis*, 1945, **36**, 6.

(2). . . . . . "Thomas Werkworth on the Motion of the Eighth Sphere", *Isis*, 1948, **39**, 212.

TRIGAULT, NICH. . . . . . . *Histoire de l'Expédition Chrétienne au Royaume de la Chine Entrepris par les PP. de la Compagnie de Jésus, comprise en cinq livres, . . . tirée des Commentaires du P. Matthieu Riccius, etc.* (Lyon), 1616. Partial English translation in *Purchas His Pilgrimes* (London), 1625, iii, 380. Full Eng. tr. L. V. GALLAGHER (Random House, New York), 1953.

VENTURI, P. T. (ed.) . . . . . *Opere Storiche di Matteo Ricci, S. J.*, 2 vols. (Macerata), 1910–13.

VERBIEST, F. . . . . *Astronomia Europaea sub Imperatore Tartaro-Synico Cam Hy [Khang-Hsi] appellato ex Umbra in Lucem Revocata . . .* (Bencard, Dillingae), 1687. This is a quarto volume of 126 pp., edited by P. Couplet. A folio volume with approximately the same title had appeared in 1668, consisting of 18 pp. of Latin text, and 250 plates of apparatus on Chinese paper, only one of which, the general view of the Peking Observatory, was re-engraved in small format for the 1687 edition. The large version is very rare and I have not seen it. Cf. Bosmans; and Houzeau, p. 44.

WAGNER, A. . . . . "Über ein altes Manuscript der Pulkowaer Sternwarte" (with additional note by J. L. E. Dreyer), *Copernicus*, 1882, **2**, 123.

WOLF, A. . . . . *History of Science, Technology, and Philosophy in the XVIIIth Century* (Allen & Unwin, London), 1938.

WOLF, R. . . . . *Handbuch d. Astronomie, ihrer Geschichte und Literatur*, 2 vols. (Schulthess, Zürich), 1890.

WYLIE, A. . . . . "The Mongol Astronomical Instruments in Peking" (Travaux de la IIIeme Congress des Orientalistes, 1876); incorporated in *Chinese Researches* (Shanghai), 1897; photolitho reproduced Peiping, 1936.

YABUUCHI, K. . . . . "The Introduction of Islamic Astronomy to China in the Yuan Dynasty" (in Japanese), *Toho Gakuho*, 1950, No. 19, 65.

YULE, H. . . . . *The Book of Ser Marco Polo the Venetian, concerning the Kingdoms and Marvels of the East*, translated and edited, with notes, by H. YULE. Subsequently edited by H. Cordier (Murray, London), 1903 (reprinted 1921), with a third volume *Notes and Addenda to Sir Henry Yule's Edition of Marco Polo*, by H. Cordier (Murray, London), 1920.

ZINNER, E. . . . . *Geschichte d. Sternkunde, von den ersten Anfängen bis zur Gegenwart* (Springer, Berlin), 1931.

The Mercury Horoscope of Marcantonio Michiel of Venice

A Study in the History of Renaissance Astrology and Astronomy

WILLY HARTNER

Institut für Geschichte der Naturwissenschaften,
Frankfurt a.M., Germany

SUMMARY

A statuette of Mercury (1527) wrought by ANTONIO MINELLI and consecrated by a Venetian patrician, MARCANTONIO MICHIEL, carries a horoscope combined with a graphic demonstration of the planet Mercury's motion according to the Ptolemaic system.

The first part of the study deals with the dating and astrological significance of the horoscope proper. It is shown that a uniquely propitious moment: 15th June, 1527, 8 a.m., was chosen to mark some important event in Michiel's life. Judging from the general character of the horoscope, it is likely that it was destined to mark the conception of Michiel's first child. As his son Vettore was born only 7 months later, the possibility of a premature birth must be envisaged. The concluding note (at end of study) evidences that the data of the horoscope were directly taken from JOHANNES STÖFFLER's ephemerides for the years 1499–1531.

The second part of the paper discusses the history and development of PTOLEMY's theory of Mercury. In a digression the algebraic curve described by the centre of Mercury's epicycle is analysed. It is shown that this curve is practically interchangeable with an ellipse. The curve appears for the first time in a treatise by AZARQUIEL (eleventh century) which is preserved in a Spanish translation incorporated in the *Libros del Saber*. European historians, with the exception of A. WEGENER (1905), have failed to recognize the true significance of this curve, which is by no means an anticipation of KEPLER's ellipses though it may have been one of the stimuli that led to his experimenting with oval (elliptiform) curves. G. PEURBACH (fifteenth century) almost totally depends on the Islamic astronomers' (AZARQUIEL, ALHAZEN) interpretation of the Ptolemaic theory; he deals in extenso also with the elliptiform curve. It is finally shown that the geometrical designs filling the interior of the horoscope were directly copied from PEURBACH's *Theoricae Novae Planetarum*.

---

1. INTRODUCTION

SOME three years ago, the Victoria and Albert Museum in London was presented, by Dr. W. L. HILDBURGH, F.S.A., with a marble statuette of the god Mercury,[1] carved by a sixteenth-century Venetian sculptor, ANTONIO MINELLI (Figs. 1 and 2). From a study by JOHN POPE-HENNESSY[2] we learn that little is known about the life and work of MINELLI, whose fame ranks lower than that of his father, Giovanni, of Padua. Antonio's name first occurs in a contract signed in the sacristy of the Santo at Padua, on 21st June, 1500. In 1525 he went to Venice, where a few of his works are preserved. There is no other evidence about him, and even the year of his death remains unknown.

If we content ourselves, for the moment, with this brief statement on the place held by MINELLI's Mercury statuette in the history of cinquecento art, we may turn to one peculiarity which makes it remarkable from an astronomical point of view. On the left-hand side of the figure, the artist has placed a small altar bearing on its inner side an inscription which refers to the completion and dedication of the statuette: it reads as follows:

MERC. SIMVLAC.
M. ANT. MICHAEL. P.V.
ANN. VRB. VENET.
MC. VI.F.C.
CONSECRAVITQ.

---

[1] No. A_{44}—1951.

[2] A statuette by ANTONIO MINELLI, in *The Burlington Magazine*, January, 1952, pp. 24–28.

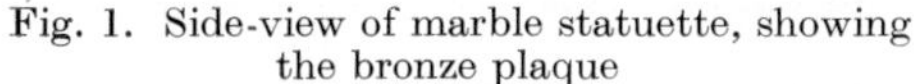

Fig. 1. Side-view of marble statuette, showing the bronze plaque

Fig. 2. Other side of the marble statuette, showing the inscription (not visible on photograph)

ANT. MINELL. SCVLPT.
PATAVINVS.
XVI. KAL. MARTI. COEPTVM.
XVII. KAL. QVINT.
PERFECIT
ANN. SAL.
M.D.XXVII.

On the two narrow sides of the altar appear a crane and a bearded, flat-nosed, horned deity, probably representing Pan (the son of Hermes-Mercury and the nymph Dryope) whose characteristic features are frequently confused with those of Silenus, another of Hermes' descendants. On the outer side (under a relief where the caduceus of Mercury in the centre is flanked on the left by a lyre placed over a cock, and on the right by a knapsack[3] over a long-legged bird, apparently also a crane), a circular bronze plaque with an engraved horoscope is suspended; see Fig. 3. It is this horoscope with which we are principally concerned.[4]

The "horoscope" as a whole is of particular interest because, contrary to the normal type, it stresses the predominance of one single planet, Mercury, over all of his companions. This is proved, on the one hand, by its combination with the Mercury statuette and the various Mercurial emblems; on the other, by the fact that the interior of the plaque gives a complete illustration of the mathematical theory of

[3] Not a scorpion, as interpreted by POPE-HENNESSY. The lyre and the knapsack are the two indispensable attributes of Mercury, the god of poets, scholars, and merchants (including thieves). The cock and the crane also frequently occur in combination with Mercury. It would take too long to go into a discussion of their origin. They are commonly interpreted as symbols of vigilance and learning.

[4] Owing to the kindness of my friend Dr. ARTHUR BEER, and through his good offices, a photograph of the horoscope was sent me by the WARBURG INSTITUTE of the University of London. The result of my computation from the data of the horoscope (date 15th June, 1527) was found to agree with the inscription cited; see POPE-HENNESSY, *l.c.* p. 28.

Mercury, as conceived in early Hellenistic times and perfected by PTOLEMY and Islamic astronomers, representing the extravagant motion of the most capricious of all the seven planets.

I. ASTROLOGICAL PART

2. THE HOROSCOPE PROPER

The plaque has a diameter of 102 mm. In Fig. 4 all its characteristic details are found represented on an enlarged scale. The outer rim, consisting of three circular rings, carries the ordinary division of the zodiac into twelve signs of 30° each, starting from the vernal point, shown on the right half of the horizontal diameter, and running counter-clockwise from Aries to Pisces. The longitudes of six of the seven planets as well as that of the ascending node of the Moon, are indicated by the usual planetary symbols followed by Roman numerals. Only the position of Mercury is not listed. This apparent omission, however, is readily explained by the circumstance that the position of Mercury can be obtained by sighting from the centre of the horoscope over that of the disk of Mercury in the upper half of the horoscope, and reading off the graduation of the rim. It yields 2° Cnc = 92°. Thus we have the following positions:

Sun = Jupiter .	2° Cnc = 92°
Mercury . .	2° Cnc = 92°
Moon . . .	20° Cap = 290°
Venus . . .	26° Cnc = 116°
Mars . . .	23° Sco = 233°
Jupiter (see Sun)	
Saturn . . .	1° Tau = 31°
Ascending Node .	24° Sgr = 264°

As is evident from these figures, our horoscope presents the unusual case of a simultaneous conjunction of the Sun with Mercury and Jupiter in 2° Cnc, Venus standing 24° east in the same sign, and the Moon—about 35^h after her full—close to opposition with Venus. Mars, holding his "domicile" Scorpio, has passed his opposition with Saturn, but the two ill-omened planets still stand in diametrically opposite signs, whereby their disastrous influence, according to most contemporary astrological theories, is increased. On the other hand, the danger of the situation may be considered extenuated by a favourable coincidence: Saturn stands in the auspicious sextile aspect with the Sun, Jupiter, and Mercury; and Venus, in the equally favourable trigonal aspect with Mars; secondly, the Ascending Node, no less evil than Mars and Saturn, holds the sign of Sagittarius which astrologers regard as its "dejectio", *i.e.* the place where its action is minimized.[5] More will have to be said about the astrological significance of the horoscope in connection with an attempt at clearing up its history (see below, Sections 5–7).

[5] For further information, see WILLY HARTNER, "The Pseudoplanetary Nodes of the Moon's Orbit in Hindu and Islamic Iconographies", in *Ars Islamica*, Vol. V, Pt. 2, pp. 114–154, Ann Arbor, 1939; and A. BOUCHÉ-LECLERCQ, *L'Astrologie Grecque*, Paris, 1899, Ch. VII, pp. 180–240.

3. THE DATE OF THE HOROSCOPE

The computation of the date is most conveniently effected with the aid of P. V. NEUGEBAUER's tables.[6] The longitude of the Sun indicates a day about the middle of June (in the period 1200–1600, the date corresponding to a mean solar longitude of 92° recedes from 17th June to 13th June, Jul. style[7]). On the other hand, a rough calculation of the positions of the slow planets, Jupiter and Saturn, and of the Ascending Node (period of revolution 18·6 years) immediately points to the summer of 1527. Thus we have to look for a date in the middle (close to the 15th) of June, 1527, which lies about 35 hours after the true full-moon. According to NT, II, 83–88, the mean full-moon of June, 1527, occurred on 14th June, about 2^h a.m. (G.M.T.), and the true opposition less than 1^h later, the mean anomaly of the Moon being close to 0°. By adding 35^h, we arrive at the date of 15th June, 1527, about 1^h p.m. Hence it is this day, if any at all, that fulfils the conditions defined by the horoscope. Evidently, however, the hour of the day as found by our rough calculation, cannot be expected to be correct, for neither do the Moon-tables in use during the sixteenth century warrant an exactness to the degree, nor can a rough estimate like the one applied here be regarded as conclusive. But there is another way of determining the hour with a higher degree of accuracy, *viz.* by computing it from the longitude of the ascending point of the ecliptic (see below, Section 5). Anticipating that the result is about 8^h a.m.[8], the theoretical *true* geocentric longitudes of the seven planets and the Ascending Node are given below as computed from NT, II, for the date 15th June, 1527, 8^h a.m. The second line indicates the data derived from the horoscope, the third, the difference "Horoscope minus computed"[9].

The congruence is indeed remarkably good. It not only excludes any doubt of the correctness of our date, but also proves that the horoscope was cast by a skilled astronomer, and probably with the aid of the best astronomical ephemerides or tables

	Sun	Moon	Mercury	Venus
Computed . . .	92°40	287°57[10,11]	89°33[12]	116°44[13]
Horoscope . . .	92	290	92	116
Horoscope — Computed .	0°	+ 2°	+ 3°	0°

	Mars	Jupiter	Saturn	Ascending Node
Computed . . .	232°08	92°76[11]	28°83[11]	264°1
Horoscope . . .	233	92	31	264
Horoscope — Computed .	+ 1°	— 1°	+ 2°	0°

[6] *Tafeln zur Astronomischen Chronologie*, Vol. II: Sonne, Planeten und Mond, Leipzig, 1914 (cited "NT, II"); III: Hilfstafeln zur Berechnung von Himmelserscheinungen, 2nd ed., 1925 (cited "NT, III"); and *Astronomische Chronologie*, Vols. I–II, Berlin and Leipzig, 1929.

[7] All dates are given in Julian style.

[8] Evidently, in this context, the exact hour of the day is relevant only in the case of the Moon.

[9] Thus in analogy with the modern "O — C". The data of the horoscope were definitely not observed but computed in advance or taken from ephemerides then in use. See Concluding Notes at the end of the present study, p. 135.

[10] Reduced to the meridian of Venice ($\lambda = + 12°20$).

[11] In the case of the Moon, Jupiter, and Saturn, perturbations in the mean heliocentric longitudes as well as in the eccentricity and the longitude of the perihelion, were taken into account. Their effect is practically negligible. For the Moon, the main perturbations in longitude are comprised in the position indicated.

[12] $1–2^d$ after inferior conjunction.

[13] $60–61^d$ after first appearance as evening star.

obtainable. It seems likely that these were either JOHANNES STÖFFLER's ephemerides for the years 1499–1531 (*Almanach nova plurimis annis venturis inservientia*, Ulm, 1499) or else the Alphonsine Tables[14], the fame of which had not yet been eclipsed by those compiled at the time immediately after the death of Copernicus: ERASMUS REINHOLD'S *Tabulae Prutenicae* ("Prussian Tables") and the tables of TYCHO BRAHE. But even these new tables were so far from perfect that it is dubious whether they really marked a decisive advance on the Alphonsine Tables. One repeatedly encounters complaints from the contemporary astronomers about the unreliability of the Prussian Tables which were frequently found to give less accurate results than TYCHO BRAHE'S and even the Alphonsine Tables; thus KEPLER states that, in the case of the planet Mars, their deviation from the positions observed in 1625 amounts to 4°, or even 5°[15]. Unfortunately, STÖFFLER's ephemerides as well as the Alphonsine Tables, which played a predominant part in the history of astronomy of the fourteenth to the seventeenth century (according to ZINNER[16] they were printed in ten editions, 1488–1649, and besides, hundreds of handwritten copies are preserved to-day), were inaccessible to me so that I was unable to verify their assumed connection with the horoscope[17].

4. ORIGIN AND PURPOSE OF THE HOROSCOPE. THE INSCRIPTION ON THE ALTAR

The date derived from the horoscope by purely astronomical means is fully confirmed by the second part of the inscription incised on the back of the altar (see Section 1), which reads in translation: "The sculptor ANTONIO MINELLI of Padua finished the work, which he had started on 14th February, on 15*th June*, 1527 *A.D.*"[18]. In other words, the date for which the horoscope was cast coincides with the day on which the artist put the last hand to his work. This obviously implies that not only the statuette, but also the plaque carrying the horoscope had been commissioned in advance to be ready on the day indicated on both, which means that the horoscope must have been precalculated in order to celebrate one single day of decisive importance in the future life of the person who commissioned the statue.

The possibility of an ordinary birth horoscope thus being excluded, the question arises as to what may have been the purpose of working out a horoscope valid for a day in the near future—about four months in advance. It is well to remember that precalculated horoscopes of this kind were hardly less frequent than the ordinary *ex post facto* specimens like birth horoscopes. A Persian king, in the Islamic Middle Ages, would hardly have dared to mount a horse without previously having consulted his court astrologer, and still in the seventeenth century (Thirty Years' War, Wallenstein) the dates of battles could depend just as much on astrological forecasts as on purely strategic needs. During the Renaissance it also became customary for private persons of rank and importance to follow the example of statesmen and generals, in order to know what steps to take and which to avoid. Hence we may safely assume that MARCANTONIO's horoscope either had an apotropaic significance, *viz.* serving

[14] Or perhaps the tables of GIOVANNI BIANCHINI (ca. 1458), based on the Alphonsine tables, and reduced to the meridian of Ferrara. See R. WOLF, *Geschichte der Astronomie*, Munich, 1877, p. 79, and E. ZINNER, *Geschichte der Sternkunde*, Berlin, 1931, p. 371.

[15] Cf. E. ZINNER, *l.c.*, p. 462. [16] *l.c.*, pp. 369–70. [17] See "Concluding Notes" at the end of this study (Section 13).

[18] Concerning the first part of the inscription, which gives the commissioning and consecration date "1106 Urbis Venetorum", I was at a loss to find any reference to the era of Venice in the books it might be expected to be. The only passage pertinent to the question, I found eventually in the *Enciclopedía Universal Ilustrada Europeo-Americana*, Tomo LXVII, Bilbao, 1929, p. 946, col. 1: "La historia de Venecia . . . puede dividirse en cuatro períodos: 1°. Desde la *fundación de la ciudad* (italicized by me) hasta la elección del dux ENRICO DANDOLO, lo que pudiera llamarse período de desarrollo (421–1192), . . . ". The foundation date indicated there (421 A.D.) tallies perfectly with the inscription on the altar: 421 + 1106 = 1527.

the purpose of averting threatening danger, or that it was destined to mark a day on which the fates were peculiarly propitious.

It is understandable that MARCANTONIO MICHIEL—one of the central figures of Venetian humanism and, as a patrician of his home town, undoubtedly a man of considerable wealth—should have chosen the planet Mercury as his protector and regulated his life according to the caprices of his astrological patron. Unfortunately, however, although we possess a study on the life and works of MARCANTONIO[19], our knowledge does not suffice to allow of a definite answer to our problem; but the little we happen to know about MARCANTONIO's private life during the years 1527 and 1528 does throw some light on the significance of his horoscope.

From a playful remark in a letter to SADOLETO, written by GIROLAMO NEGRO in 1527, we learn that MARCANTONIO in that same year had married a girl without dowry: *Ceteri communes amici bene habent praeter M. Michaelum nostrum qui uxorem duxit, quamquam in hoc quoque philosophum egit, indotatam enim accepit, ne, ut est Plautinum illud, dote imperium venderet.* The statement is confirmed by the *Genealogie del Barbaro*[20] according to which the marriage of MARCANTONIO to MARIA SORANZO took place in February, 1527. Recalling to our memory that the work of the Mercury statuette was begun on 14th February of the same year and anticipating that there is sufficient astrological evidence to assume that the marriage was celebrated at about the same time (see below, Section 8), it seems more than a mere guess to postulate a connection between the two events. Indeed, there is a fairly high probability of the horoscope's bearing directly upon the married life of the young couple. The first thing to be thought of, then, would be that it is in some way connected with MARCANTONIO's eldest child, Vettore, who was born on 13th January, 1528.[21] But in this case no other interpretation appears to offer itself than that the horoscope was set for a day particularly auspicious to conception. Conception horoscopes are known to have been quite a common phenomenon from the days of late antiquity down to the period concerned. Therefore, it would hardly surprise us to learn that MARCANTONIO, obviously a firm believer in astrology, had allowed himself to be guided by astrological considerations even in the most intimate steps of his private life.[22] However, counting back from the day of Vettore's birth by an average normal period of 280 days, would bring us down to 8th April, 1527, and for the probable date of conception to about 15th April, which is two months earlier than indicated by the horoscope. The possibility of a premature birth might be thought to explain the discrepancy;[23] if new documentary material on Vettore should bear witness to an unusual weakness or untimely death of the child this might support our hypothesis. On the other hand, of course, a second possibility cannot be excluded, namely, that nature prevailed over astrological considerations, and that Vettore was conceived six weeks or two months before the propitious day.

---

[19] E. A. CICOGNA, "Intorno la vita e le opere di M. MICHIEL patrizio veneto della prima metà del secolo XVI," in *Memorie dell' Istituto Veneto*, IX (1861), p. 383 (cited after POPE-HENNESSY, not consulted by the writer of this article).

[20] MS. Venice, *Museo Correr*, Cic. 515, MSS. II, 174, Vol. V.

[21] According to a private information of Dr. GIOVANNI MARIACHER of the *Museo Correr*, cf. POPE-HENNESSY, *l.c.* p. 25, footnote.

[22] I am, of course, aware that the lapse of four months between the marriage and the horoscope date is rather long. However, there was no lack of medical devices to comply with the needs of astrology.

[23] According to modern statistics, the average period of post-conceptional pregnancy is 273^d. The actual period can be considerably shorter, but the probability rate for the birth of a mature child before the 250th day *post conceptionem* decreases rapidly. One out of 4299 children is born before the 246th day, and one out of 3,333,333 before the 234th day. This, of course, concerns only the question of *maturity*, not of *viability*. The lower limit of viability is assumed to be 181^d *post conceptionem*, which is a good deal less than the period between the day of the horoscope and that of Vettore's birth (212^d). Yet, even in our days, the cases of viable children born after less than 215 days' pregnancy *post conceptionem* seem extremely rare, and it is doubtful whether such a possibility could even be considered in the Renaissance. Cf. L. NÜRNBERGER, "Abnorme Schwangerschaftsdauer", in *Biologie und Pathologie des Weibes*, Vol. VII, part 1, Berlin and Vienna, 1927, pp. 365–406.

Fig. 3. Details of the plaque

Finally, a third possibility must be examined, namely, that the date of the horoscope is meant to mark the moment of animation of the embryo. Against TERTULLIAN, *De anima*, Chapter XXVII, who assumes the soul to come into being simultaneously with conception, St. AUGUSTINE, *Quaest. in Exod. quaest. LXXX*, distinguishes between the unanimated and the animated embryo. Relying on this passage, Canon Law[24] assumes the male embryo to be animated on the 40th, and the female on the 80th day, whereas Secular Law indiscriminately assumes the 40th day for both sexes. But, obviously, there is no reason to believe that the question of animation is applicable to our case; on the contrary, the fact alone that a definite *hour* is indicated by the horoscope, is sufficient to refute such an interpretation.

[24] For references, cf. L. NÜRNBERGER, "Fehlgeburt und Frühgeburt", *op. cit.*, pp. 407–646 (see in particular p. 414).

5. THE HOUR OF THE HOROSCOPE. THE ASCENDANT AND THE TWELVE HOUSES (LOCI). DOMICILES AND EXALTATIONS. APHETIC POINTS AND APHETS (HYLEG). THE WHEEL OF FORTUNE

The relative positions of the planets in the zodiac determine only the astrological situation in general. In order to establish a complete forecast of future events, which

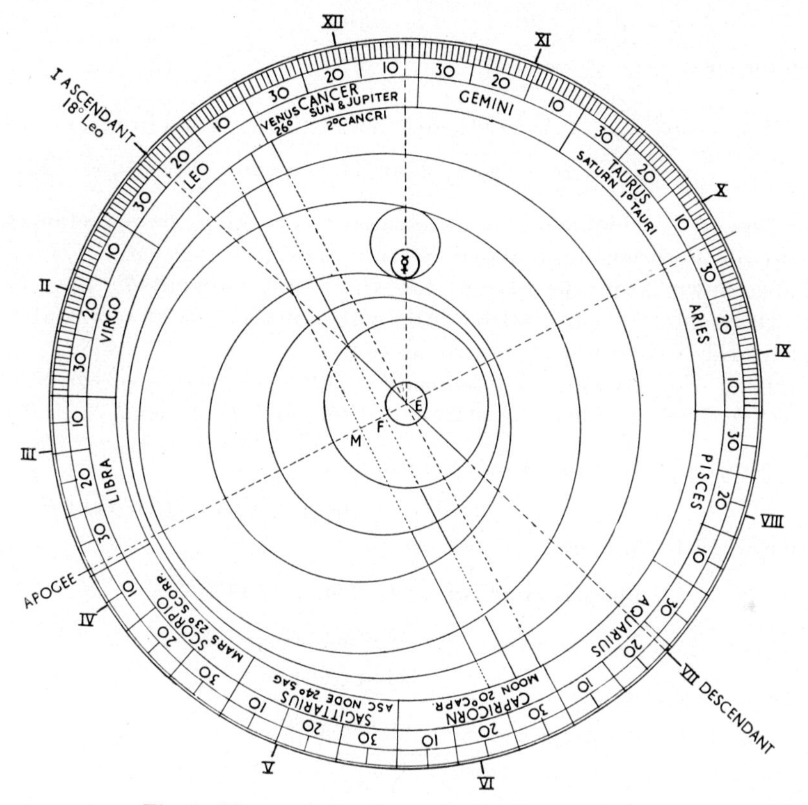

Fig. 4. MARCANTONIO MICHIEL'S horoscope (redrawn)

is the true purpose of all horoscopes, the exact knowledge of the hour, and even the minute, is indispensable. For only if the day and the hour are known is it possible to compute the decisive point of the horoscope, namely, the degree of the ecliptic rising above the horizon at the moment concerned, which serves as a starting point of a most peculiar, and no less illogical and abstruse, division of the zodiac into twelve *houses* (Lat. *loci*) of unequal extent.

In our case, this ascending point, or short, *Ascendant* (Greek ὡροσκόπος, Lat. *pars horoscopans*, whence "horoscope"), is found by producing the sloping straight line (see Figs. 3 and 4) through the centre of the horoscope (marking the horizon of Venice) and reading on the rim, in the left upper quadrant, which yields 18° Leo, or 138°. The point diametrically opposite, 18° Aqu, or 318°, is called *Descendant* (Lat. *occasus*) as it sets at the same moment that the Ascendant rises[25].

[25] Theoretically, the inverse interpretation according to which the ascendant and the descendant would have to be interchanged, is possible. But in horoscopes the ascendant nearly always stands on the left, so that the upper half of the circle represents day, the lower, night. Moreover, such an inverse interpretation would not tally with the position of the Moon, nor would it lead to any reasonable astrological result—so far as the adjective *reasonable* can be used in connection with astrology.

(a) The Hour

The equatorial co-ordinates corresponding to an ecliptical longitude of $\lambda = 138°$ are $\alpha = 140\overset{\circ}{.}5$, or $9\overset{h}{.}36$[26], and $\delta = +\ 15\overset{\circ}{.}5$. To $\delta = +\ 15\overset{\circ}{.}5$ corresponds, for the latitude of Venice ($\Phi = +\ 45\overset{\circ}{.}43$), the semi-diurnal arc $t = 7\overset{h}{.}15$. The sidereal time at mean noon of 14th June, 1527, is $\vartheta = 6\overset{h}{.}09$[27]. Hence

$$\alpha - t - \vartheta = 20\overset{h}{.}12$$

Correction for mean time $= -\ 0{\cdot}05$

Hour of horoscope $T_0 = 20\overset{h}{.}07$ Mean Time Venice, counted from noon, 14th June

$T_1 = 8^h\ 4^m$ a.m. Mean Time Venice, 15th June, 1527[28].

It is this value of T_1 which we had anticipated in checking the planetary positions indicated by our horoscope (see above, Section 3).

As will be shown later, the reasons for which this moment T_1 was chosen are manifold. Only one of them need be dealt with now because it serves at the same time to prove the correctness of our calculation.

In converting T_1 into *unequal* ("*temporal*") *hours*, *i.e.* sixths of the Sun's semi-diurnal arc $S = 7\overset{h}{.}83$, we obtain (equation of time on 15th June, $J = +\ 0\overset{h}{.}02$):

Sunrise, 15th June, 1527, Venice, at $12^h - S + J$

$= 4\overset{h}{.}17 = 4^h\ 10^m$ Mean Time Venice

$\frac{1}{6} S = 1\overset{h}{.}305 = 1^h\ 18\overset{m}{.}3$, hence

unequal hour No.	I starts at	4^h	10^m
	II starts at	5	28
	III starts at	6	47
	IV starts at	8	5
	V starts at	9	23
	VI starts at	10	42
	VII starts at	12	0

This means that the moment T_1 marks the beginning of the 4th unequal hour of the day (*i.e.* after sunrise) of 15th June, 1527 (*Saturday*). According to the theory of *chronocratories*[29] ("planetary week") each of the 168 hours of the week is governed by one of the seven planets which succeed one another in the following sequence: Sun, Venus, Mercury, Moon, Saturn, Jupiter, Mars. Starting with the *Sun* as chronocrator of the first hour of the *day*, on *Sunday*, the 12th hour of the *day* will be ruled by Saturn, the 1st of the *night* by Jupiter, the 12th of the *night* by Mercury, and the 1st of the next day (*Monday*) by the *Moon*. Similarly, the 1st hour of *Tuesday* (dies *Martis*) belongs to *Mars*, of *Wednesday* (dies *Mercurii*) to *Mercury*, *etc.*, so that each day is initiated by the planet after which it is called.

[26] Fraction of an hour, not minutes.
[27] According to *NT* III, T. 1; hours counted in the old-fashioned way, *i.e.* from noon of preceding day.
[28] Accuracy of our calculation $\pm\ 3^m$. As the hour of the horoscope was probably determined by means of an astrolabe (see below, Section 6), we have to allow of an additional reading error of $\pm\ 1° = \pm\ 4^m$. Hence the hour intended, T, is defined by $7^h\ 57^m < T < 8^h\ 11^m$.
[29] Cf. BOUCHÉ-LECLERCQ, *op. cit.*, pp. 476–480.

Thus, on the day in question, a *Saturday*, the chronocrator of the 1st hour of the *day* is *Saturn*, of the 2nd *Jupiter*, of the 3rd *Mars*, and of the 4th, which has our particular interest, the *Sun*.

I venture to say that this is not due to a pure game of chance because, in our horoscope, the Sun is practically equal in rank to the predominant planet, Mercury. It is the Sun that stands in conjunction with Mercury and Jupiter, marking the point second in importance to the Ascendant; moreover, the Ascendant itself is in the sign of Leo, which from remote antiquity had been regarded as the solar dominion *par excellence*, and which scientific astrology, by the beginning of Hellenism, attributed to the Sun as his "domicilium". When the Sun stands in Leo, he exercises his maximum action. But also by itself, Leo has the qualities of the great luminary and plays the rôle of his deputy.

(*b*) *The Twelve Houses*

Concerning the method of dividing the ecliptic into twelve *houses*, the disagreement which rules among mediaeval astrologers is remarkable indeed. It would carry us too far to enter upon a detailed discussion here. Therefore, referring the reader to BOUCHÉ-LECLERCQ's standard work[30] and, particularly, to C. A. NALLINO's excellent historical summary[31], we limit ourselves to a few remarks.

All astrologers agree on the four cardinal points of the horoscope: *Ascendant*, *Descendant* (*Occasus*), "*Medium Coelum*" (abbreviated *MC*, *i.e.* the point of the zodiac being in upper culmination), and "*Imum* [*Medium*] *Coelum*" (abbreviated *IMC*, *i.e.* the point in lower culmination). Counting counter-clockwise (with increasing longitudes), the Ascendant marks the beginning of the 1st house, IMC the 4th, Occasus the 7th, and MC the 10th. The further subdivision of the four quadrants, however, is effected according to methods or hypotheses that differ widely from one another; strangely enough, late mediaeval and Renaissance astrologers, instead of trying to reconcile conflicting assumptions, even competed in inventing new systems, thus adding to the confusion inherited from their predecessors. Yet it appears that one system, which will be described here, was considered superior to all others. It is the one that PTOLEMY seems to have had in mind when dealing with the subject in his *Tetrabiblos*, and which all the later famous Arabic and Jewish astrologers made use of in their computations, such as AL-BATTĀNĪ (Albategnius), AL-QABĪSĪ (Alcabitius), ABRĀHĀM BEN 'EZRĀ (Avenezra), ABŪ'L-WAFĀ', and ULŪGH BEG. Through ALPHONS X of Castile's *Libros del saber de astronomía* and through the writings of early Italian astrologers, such as GIOVANNI CAMPANO's (thirteenth century), it found its way also into European astrology and maintained a predominant position during the subsequent centuries[32].

According to this system, the semidiurnal and the seminocturnal arcs of the Ascendant are each divided into three equal parts. Then the sections of the ecliptic limited by the great circles laid through the poles and the divisions of the equator represent the six eastern houses, running from No. X (which starts at the meridian) through XI and XII (ending at the horizon), and on from I (below the horizon) through II to III (ending at the northern half of the meridian).

The six western houses, Nos. IV–IX, are situated symmetrically with the former, the point of symmetry being the centre of the Universe (*i.e.* the centre of the Earth).

[30] *Op. cit.*, Ch. IX (pp. 256–310).
[31] In AL-BATTĀNĪ, *Opus Astronomicum*, Part I (Milan, 1903), pp. 246–249.
[32] See "Concluding Notes" at end of study (Section 13).

Thus we obtain six pairs of diametrically opposite houses each of equal extent (No. VII = No. I + 180°, *etc.*), but varying in size according to the position of the vernal point relative to the horizon[33]. The absurdity of this method could not be better characterized than by the words of J.-B. DELAMBRE, in his *Histoire de l'astronomie du moyen âge*[34], where he deals with the system of ALCABITIUS:

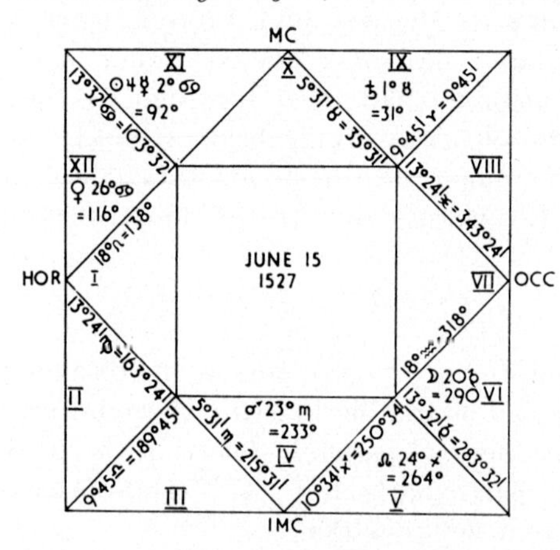

Fig. 5. The twelve astrological houses according to PTOLEMY's system

"Les six maisons dernières sont toujours diamétralement opposées aux six premières; il en résulte cependant une espèce d'absurdité. Le quart de l'équateur, entre le méridien et l'horizon occidental, se trouve divisé suivant les arcs nocturnes, quoiqu'il appartienne au jour; le quart entre l'horizon occidental et le méridien inférieur, est divisé suivant les heures du jour, quoiqu'il appartienne à la nuit. Au reste, le calcul de cette méthode est extrêmement simple, et c'est peut-être ce qui a fait passer sur l'absurdité que nous venons de remarquer".

In applying the theory of the twelve houses as outlined here, to our special case, we arrive at the following result:

$$\text{Longitude of Ascendant } \lambda = 138^{\circ};$$
$$\text{AR of Ascendant } \alpha = 140^{\circ}.5;$$
$$\text{Semidurnal arc of Ascendant } t_d = 7^{h}.15;$$
$$= 107^{\circ}.25;$$
$$\tfrac{1}{3}t_d = 35^{\circ}.75;$$
$$\text{Seminocturnal arc } t_n = 4^{h}.85;$$
$$= 72^{\circ}.75$$
$$\tfrac{1}{3}t_n = 24^{\circ}.25.$$

House No.	α	λ		
I (*Ascendant*)	α = 140°.5	λ = 138°.0 =	Leo	18°.0
II	164·7	163·4 =	Virgo	13·4
III	189·0	189·8 =	Libra	9·8
IV (*IMC*)	213·2	215·5 =	Scorpio	5·5
V	249·0	250·6 =	Sagittarius	10·6
VI	284·7	283·5 =	Capricornus	13·5
VII (*Occasus*)	320·5	318·0 =	Aquarius	18·0

[33] Twice a day (not once, as erroneously stated by NALLINO, *l.c.*, p. 247), the four quadrants of the zodiac will be equal, namely when the vernal or the autumnal point coincides with the horizon. But, evidently, an equality of all of the twelve houses can never occur.

[34] Paris, 1819, p. 502.

House No. VIII . . .	344·7	343·4 = Pisces	13·4
IX . . .	9·0	9·8 = Aries	9·8
X (*MC*) . .	33·2	35·5 = Taurus	5·5
XI . . .	69·0	70·6 = Gemini	10·6
XII . . .	104·7	103·5 = Cancer	13·5

In Fig. 4, the beginnings ("cusps") of the houses corresponding to the values of λ indicated, are marked by roman figures (at the outer rim). Fig. 5 represents the same houses in a more conventional schematic way as a square subdivided into twelve small triangles, leaving a small square open in the middle of the figure. There the values of λ are marked, in signs and degrees, at the edges of the triangles, while the positions of the planets are inscribed inside them.

The astrological significance of these imaginary "houses" is described by the following mediaeval distych:

Vita, lucrum, fratres, genitor, nati, valĕtudo.

Uxor, mors, pietas, regnum, benefactaque, carcer.

In other words, House No. I, commonly called *Horoscope*, decides on the life of the newly-born (or the future life of the foetus) in general; II, on pecuniary circumstances ("lucre"); III to V, on brothers, parents, and children; VI, on health and disease; VII, on marriage; VIII, on death and inheritances; IX, on religion (and also travels); X, on domicile and State (also honours, arts, character, and conduct of life); XI (*Bonus Genius*), on benefit, charity, and friends; XII (*Malus Genius*), on enemies, captivity, and other afflictions.

Not all of the twelve houses are attributed equal weight. The four cardinal ones (I, IV, VII, X) are, of course, considered predominant, though IV (*IMC*) and VII (*Occasus*) are often regarded of lesser interest. Next to these in importance are those standing in propitious aspects to the Ascendant, *viz.* the trigonal (V and IX) and the sextile (III and XI). The four remaining (II, VI, VIII, XII) are regarded as "lazy" and least effective.

Remembering that the hour of MARCANTONIO's horoscope was precalculated, it seems astonishing at first sight that the Ascendant was chosen by the astrologer in such a way that, with the exception of Mars in the relatively unimportant House No. IV, no other planet stands in the cardinal houses. As I am going to demonstrate, however, this apparent gap is filled by the vicarious action of the planetary *domiciles* and *exaltations* that become effective, not only at the four cusps of the zodiac, but at the beginning of each of the twelve houses.

(c) *Domiciles and Exaltations* (*Dejections*)

In astrology, there are two competitive systems of attributing the signs of the zodiac to the seven planets, the "*domicilia*" and the "*exaltationes et dejectiones*". According to the first, each planet governs two signs, one during the day and one at night, with the exception of the Sun and the Moon, each of whom governs one sign only. According to the second, each planet has one sign in which he is "exalted", *i.e.* exerts his maximum action, and a sign diametrically opposite, in which he is "depressed" or "dejected", *i.e.* where his influence is least. As the co-existence of the two systems necessarily led to intolerable contradictions, it was agreed[35] that not

[35] Thus already in early astrological papyri (Mich. pap. No. 149, XVI, 2nd cent. A.D., and others, see *Mich. Papyri*, Vol. III, ed. J. G. WINTER, Ann Arbor, 1936, p. 116, and my review, in *Isis*, Vol. 27 (1937), pp. 337ff). PTOLEMY, at the same time, seems to disregard the degrees.

the whole sign, but only one single degree of it, and its close surrounding, should be counted as exaltation or dejection. The distribution is as follows:

Planets	Domicilia		Exaltations		Dejections	
	Day	Night				
Sun . . .	Leo	—	Aries	19°	Libra	19°
Moon . . .	—	Cancer	Taurus	3	Scorpio	3
Mercury . .	Virgo	Gemini	Virgo	15	Pisces	15
Venus . . .	Libra	Taurus	Pisces	27	Virgo	27
Mars . . .	Scorpio	Aries	Capricornus	28	Cancer	28
Jupiter . .	Sagittarius	Pisces	Cancer	15	Capricornus	15
Saturn . . .	Capricornus	Aquarius	Libra	21	Aries	21
Ascending Node .	—	—	Gemini		Scorpio	

A graphic illustration of the two systems is given in Figs. 6 (*a*) and (*b*).

As mentioned before, the significance of the *domicilia* and *exaltationes* is not completely covered by the statement that the influence of the planets standing *in* or near to them becomes particularly powerful. For also alone, the signs and degrees in question are regarded as efficient, assuming the qualities of their lords. In horoscopes, therefore, they have to be taken into account and regarded as vicarious planets.

Thus, examining the cusps of the twelve houses and comparing them with the above list, we find:

House No.	Sign	Degree	
I .	Leo	18° .	Sun's *day domicile*
II .	Virgo	13 .	Mercury's *day domicile*, and Mercury exaltation − 2°
III .	Libra	10 .	Venus' *day domicile*
IV .	Scorpio	6 .	Mars' *day domicile*, and Moon dejection + 3°
V .	Sagittarius	11 .	Jupiter's *day domicile*, and Ascending Node dejection (no degree ascribed)
VI .	Capricornus	14 .	Saturn's *day domicile*, and Jupiter dejection − 1°
VII .	Aquarius	18 .	Saturn's *night domicile*
VIII .	Pisces	13 .	Jupiter's *night domicile*, and Mercury dejection − 2°
IX .	Aries	10 .	Mars' *night domicile*
X .	Taurus	6 .	Venus' *night domicile*, and Moon exaltation + 3°
XI .	Gemini	11 .	Mercury's *night domicile*, and Ascending Node exaltation (no degree ascribed)
XII .	Cancer	14 .	Moon's *night domicile*, and Jupiter exaltation − 1°

As may be seen, the distribution is such that the greater part of the day domiciles lie under the horizon, in the night half of the zodiac, and of the night domiciles, in the day half, so that only Leo and Aquarius, which are approximately bisected by the horizon, become really effective. By this choice, the astrologer has evidently succeeded in reducing the influence of the domiciles altogether in particular the evil effect of Mars at IMC, thereby giving full weight to the exaltations and dejections. Thus, from the point of view of vicarious planetary action, without respect to the actual positions of the planets themselves, of all the twelve houses only the following will be of astrological interest:

I (Vita): Sun's domicile, reinforced by rising.

II (Lucrum): Mercury's exaltation.

VII (Uxor): Saturn's domicile, extenuated by setting.

X (Regnum): Moon's exaltation.

XI (Benefacta): Ascending Node's exaltation.

XII (Carcer): Jupiter's exaltation.

The astrological inferences to be drawn are the best possible ones: the royal lord of the planets, the Sun, governing the first house indicates good fortune and happiness in general. The cusp of the second house near the exaltation of Mercury, who is the

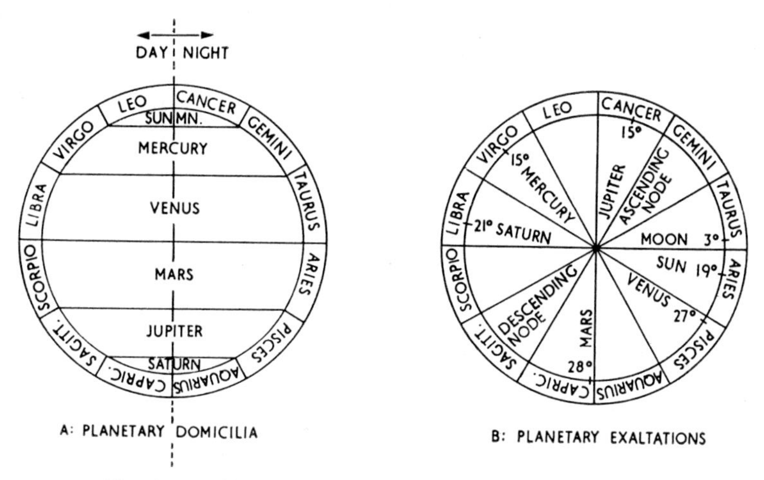

Fig. 6. (*a*) Planetary domicilia; (*b*) planetary exaltations

dominating planet of the horoscope, undoubtedly means wealth. Saturn's evil influence on the future marriage, in the seventh house, is reduced by the setting of the planet's domicile, Aquarius. The most important tenth house is determined by the exaltation of the benevolent Moon. The beneficent eleventh house is threatened by the exaltation of the evil Node (which, however, at the same time is heavily counterbalanced by the planetary double conjunction occurring in it). The danger of the maleficent twelfth house, finally, is averted by the exaltation of the good and friendly Jupiter.

Now, passing over to the actual positions of the planets, whose influence, of course, prevails over that of their "deputies", we find that:

IV (Genitor): Mars' evil influence extenuated (see above).

VI (Valetudo): determined by the propitious Moon.

IX (Pietas): determined by the ill-omened Saturn, whose effect, however, may be considered reduced by his proximity to dejection (distance 10°).

XI (Benefacta): overwhelming cumulation of luck[36]: the royal Sun and the

[36] It is true that planetary conjunctions, especially those with the Sun, are often regarded as unprosperous (but see below, Section 7). "The Sun *burns* or *paralyzes* the planet with which he stands in conjunction" (BOUCHÉ-LECLERCQ, *op. cit.*, p. 245f). At the same time, however, "the planets in conjunction communicate to one another their good or bad qualities". In our case, Mercury being ambiguous (good with the good, and bad with the bad), the Sun being the symbol of creativeness and the source of the sensorial life—*i.e.* so-to-say beyond good and evil—and Jupiter, exclusively propitious, this double conjunction is hardly capable of an inauspicious interpretation. (As a parallel case, I refer to the laying of the first stone of TYCHO BRAHE's observatory "Uraniborg" on the Isle of Hven, on 8th August, 1576, at sunrise, which date had been chosen expressly because it marked a conjunction of Jupiter with the Sun (ca. Leo 25°) near Regulus, both standing in trigonal aspect with Saturn (ca. Sagittarius 24°), the Moon occupying ca. Aquarius 22°, *i.e.* being ca. 3° before her full. Though BRAHE, in *Explicatio Partium majoris et praecipuae domus* (*Opera omnia*, ed. I. C. L. DREYER, Vol. V, Copenhagen, 1921, p. 143) mentions only the positions of the Sun, Jupiter, and the Moon ("Exoriente Sole unâ cum Jove juxta cor Leonis, Lunâ occiduum cardinem in Aquario occupante"), he evidently had cast the whole horoscope so as to find a particularly auspicious day and hour. In any case, it leaves no room

(*See bottom page 98*)

auspicious Jupiter in the House of Benefit, reinforced by Mercury, moving towards the exaltation of Jupiter (which is 13° distant, standing at the cusp of the next house).

XII (Cancer): the house of sorrow and trouble dominated by Venus—the most pleasant consolation that can ever be conceived!

(*d*) *Aphetic Points and Aphets* (*Hyleg*)

The theory of genethlialogy, as established by PTOLEMY and developed by the Arabs, pays special attention to the so-called *aphetic points* (τόποι ἀφετικοί) occupied by a planetary *aphet* (ἀφέτης)[37]. BOUCHÉ-LECLERCQ[38] calls this an "assimilation du Zodiaque à une roulette dans laquelle la vie des individus est lancée avec une force plus ou moins grande d'un certain point de départ (τόπος ἀφετικός) et se trouve arrêtée, ou risque d'être arrêteé, par des barrières ou lieux destinateurs (τόποι ἀναιρετικοί), sans pouvoir en aucun cas dépasser un quart du cercle". From the position of an *aphet* in an *aphetic point*, most essential conclusions are drawn as to the life of the individual, particularly its duration. Of the twelve houses only five can become aphetic points, namely, in the following sequence:

X (*MC*), I (*Ascendant*), XI (*Bonus Genius*), VII (*Occasus*), IX (called *Deus*).

For births or conceptions occurring during the day, the gradation of aphets is:

(1) The Sun, if standing in an aphetic point.

(2) The Moon, if standing in an aphetic point.

In case neither of the two fulfill the condition:

(3) A planet standing in an *aphetic point*. If there are two or more competitors, special rules are to be observed to decide on the question of priority.

(4) In the last resort, the house of the Ascendant is regarded as *aphetic point*. In that case, again, special prescriptions exist for deriving the *aphet* from the Ascendant (he may be identified with the lord of the *domicile*, or of the *exaltation*, *finis*, or the like).

For births occurring at night, the Sun and the Moon, in the above gradation, are interchanged.

In the present case, evidently, the Sun occupying the *locus boni genii* (House No. XI) is to be regarded as *aphet of the horoscope*.

(*e*) *The Wheel of Fortune* (κλῆρος τύχης, *Fortuna*)[39]

It seems astonishing that this chimera of early astrological (allegedly Egyptian) fancy does not occur on the horoscope. The "*Wheel of Fortune*" is an imaginary point to which auspicious planetary qualities are ascribed; it is commonly represented by the figure of a cross in a circle ("wheel"), ⊕. It holds the degree of the ecliptic whose angular distance from the Moon (counting with increasing longitudes) is the same as

(*Continuation of footnote 36 from page 97*)
for doubt that the sixteenth century astrologers regarded conjunctions of the beneficent planets with the Sun as strictly favourable.) In the case of Marcantonio's horoscope, the conjunction takes place in the *locus boni genii* (XI), the auspiciousness of which is increased by the Sun (who is the *aphet* of the horoscope, see below, Section 5 (*d*)). But it should not be forgotten that most of the rules of astrology are vague enough to allow of different, often strictly contradictory, interpretations. Two astrologers dealing with the same case will seldom, if ever, arrive at identical results. Therefore, it is hardly necessary to state that my inferences from the given facts have no claim to absolute certainty. I may have overlooked details to which the Renaissance astrologer ascribed particular importance, and I may have unduly stressed others which, according to the taste of the time, were considered negligible.

[37] Renaissance astrologers use the term *hyleg* or *alhyleg*, derived from the Persian *hailāj*, "the one who lets loose", a literal translation of the Greek ἀφέτης (Persian—HIL—, = Greek ἀφίημι). The stress laid on the *hyleg* in early European astrology is obviously due to Islamic influence.

[38] *Op. cit.*, p. 411; see also AL-BĀTTANĪ, *Opus Astronomicum* (ed. NALLINO), Vol. I, p. 313f.

[39] Cf. BOUCHÉ-LECLERCQ, *op. cit.*, pp. 288–310.

that of the Ascendant from the Sun ($\lambda_{\oplus} - \lambda_{☽} = \lambda_{\text{ASC.}} - \lambda_{\odot}$). The house occupied by it becomes prosperous even if it be that of death or evil spirits. *Et si fuerit in loco mali fortuna verte sentenciam et dic loco mali bonum*, we read in the *Flores Albumasaris*[40].

In the present horoscope, the *Wheel of Fortune* would be in 336° = 6° Pisces (*below* horizon), in the *night domicile* of Jupiter, in the *finis* governed by Venus, and in the *facies* of Saturn. As it is simultaneously in the House of Marriage (VII, *Occasus*), additional information on the future marriage of MARCANTONIO's child can be drawn from it. The prognostication is evidently propitious, because the evil influence of Saturn is heavily counterbalanced by the accumulation of auspicious elements.

6. SECONDARY ELEMENTS OF PROGNOSTICATION: LIMITS (FINES), DECANS, AND TRINES (TRIGONA)

Even the statements made hitherto do not exhaust the possibilities of astrological prognostication[41]. The "secondary elements" may sometimes be just as important as those treated above, and it would, therefore, be a serious omission if we did not take some of them into consideration.

(*a*) *Limits* (Lat. *fines*)

These result from a subdivision of each of the twelve signs of the zodiac into five *unequal* parts, each of which is governed by one of the five planets (taking no account of the Sun and the Moon). There are three main systems competing with one another, the "Egyptian", the "Chaldean", and that worked out by PTOLEMY. Only the first of them, however, seems to have been used in practical astrology of the Arabic Middle Ages and the Renaissance[42]. According to these Egyptian rules the first *finis* of Aries, comprising 6°, is governed by, or stands as a substitute for, Jupiter; the second, comprising 6°, belongs to Venus; the third (8°) to Mercury; the fourth (5°) to Mars; the fifth (5°) to Saturn, *etc.*, as is seen from the following table.

	I	II	III	IV	V
Aries	Jupiter 6°	Venus 6°	Mercury 8°	Mars 5°	Saturn 5°
Taurus	Venus 8	Mercury 6	Jupiter 8	Saturn 5	Mars 3
Gemini	Mercury 6	Jupiter 6	Venus 5	Mars 7	Saturn 6
Cancer	Mars 7	Venus 6	Mercury 6	Jupiter 7	Saturn 4
Leo	Jupiter 6	Venus 5	Saturn 7	Mercury 6	Mars 6
Virgo	Mercury 7	Venus 10	Jupiter 4	Mars 7	Saturn 2
Libra	Saturn 6	Mercury 8	Jupiter 7	Venus 7	Mars 2
Scorpio	Mars 7	Venus 4	Mercury 8	Jupiter 5	Saturn 6
Sagittarius	Jupiter 12	Venus 5	Mercury 4	Saturn 5	Mars 4
Capricornus	Mercury 7	Jupiter 7	Venus 8	Saturn 4	Mars 4
Aquarius	Mercury 7	Venus 6	Jupiter 7	Mars 5	Saturn 5
Pisces	Venus 12	Jupiter 4	Mercury 3	Mars 9	Saturn 2

(*b*) *The Thirty-six "Faces" or "Decans"* (Lat. "*facies*")

This is another subdivision of each of the twelve signs into three *equal* parts ("decans"), which can be traced back to early Egyptian prototypes, being the main

[40] Fol. 7v–9r (*passim*), Augsburg, 1488 (facsimile reprint by Deutscher Verein für Buchwesen und Schrifttum, Leipzig, 1928).

[41] "Les débris de toutes les fantaisies qui n'avaient pas trouvé place dans la répartition des οἶκοι ("houses") et des ὑψώματα ("exaltations") ont servi à fabriquer le système des trigones planétaires. On s'achemine peu à peu vers l'incompréhensible, qui atteint sa pleine floraison dans le système des ὅρια ("limits")", says BOUCHÉ-LECLERCQ, *op. cit.*, p. 198f.

[42] Thus, for example, on astrolabes; cf. W. HARTNER, "The Principle and Use of the Astrolabe", in *A Survey of Persian Art* (ed. A. U. POPE), Vol. III, Oxford, 1939, Ch. 57, p. 2548ff.

contribution of pre-Hellenistic Egypt to mediaeval astrology[43]. Again, each of the decans is governed by one of the planets—this time including the Sun and the Moon—as illustrated by the following table.

	I	II	III		I	II	III
Aries . . .	Mars	Sun	Venus	Libra . .	Moon	Saturn	Jupiter
Taurus . . .	Mercury	Moon	Saturn	Scorpio . .	Mars	Sun	Venus
Gemini . . .	Jupiter	Mars	Sun	Sagittarius .	Mercury	Moon	Saturn
Cancer . . .	Venus	Mercury	Moon	Capricornus .	Jupiter	Mars	Sun
Leo	Saturn	Jupiter	Mars	Aquarius . .	Venus	Mercury	Moon
Virgo . . .	Sun	Venus	Mercury	Pisces . .	Saturn	Jupiter	Mars

The practical application of the two systems to our case is shown in the following table, where the longitudes of the cusps of the twelve houses are listed in the left-hand column, the corresponding *fines* with their lords in the middle, and the *decans* with their lords in the right-hand column.

				Finis	Facies
House No.	I (*Ascendant*) .	Leo	18°	IV, Mercury	II, Jupiter
	II . .	Virgo	13	II, Venus	II, Venus
	III . .	Libra	10	II, Mercury	I, Moon
	IV (*IMC*) . .	Scorpio	6	I, Mars	I, Mars
	V . . .	Sagittarius	11	I, Jupiter	II, Moon
	VI . . .	Capricornus	14	III, Venus	II, Mars
	VII (*Occasus*) .	Aquarius	18	III, Jupiter	II, Mercury
	VIII . . .	Pisces	13	II, Jupiter	II, Jupiter
	IX . . .	Aries	10	II, Venus	II, Sun
	X (*MC*) . .	Taurus	6	I, Venus	I, Mercury
	XI . . .	Gemini	11	II, Jupiter	II, Mars
	XII . . .	Cancer	14	III, Mercury	II, Mercury

Any astrologer who contrived to select a similarly auspicious Ascendant would, indeed, be worthy of praise. Of the two malevolent planets, Saturn is completely banished from the *fines* as well as the *facies*, whereas Mars occurs only once in the former, and three times in the latter. All of the rest are governed either by Mercury or by the two benevolent planets (Jupiter and Venus), the Sun occurring only once, as lord of the second *decan* of Aries, at the cusp of House No. IX (*Pietas*).

In particular, the conjunction of the Sun with Mercury and Jupiter is "repeated" by the Ascendant being simultaneously in the *domicilium* of the *Sun*, in the *finis* of *Mercury*, and in the *facies* of *Jupiter*. Leo 18° marks exactly the beginning of the fourth *finis* (extending from 18°–24°); hence the first 25^{m} approximately (corresponding to 6° in longitude, or a little more in AR) subsequent to the moment for which the horoscope was set, will be dominated by the lord of the horoscope, Mercury. Again, Mercury, together with Jupiter, appears at the Descendant, and, with Venus, at the *MC*. Finally, he concludes the "circulus geniturae" as the double lord of the *finis* and the *facies* of the cusp of the twelfth house, whereby the evil influence of this *locus mali genii*, already extenuated by Jupiter's exaltation (see above, Section 5 (*c*)), is minimized.

[43] Cf. W. GUNDEL, "Dekane und Dekansternbilder" (*Studien der Bibliothek Warburg*, Vol. XIX), Glückstadt and Hamburg, 1936.

Only the danger threatening on the part of Mars cannot be completely disregarded. The fact that he is Lord of the *finis* and the *facies* standing at the cusp of the fourth house (*IMC*, *Genitor*), moreover that this house coincides with Scorpio, his *domicilium*, and that it is actually occupied by the planet himself, necessarily implies that the father of the child to be born will have to face martial hardships. The other threat, which arises from Mars' being connected with the decisive eleventh house, where the great conjunction takes place, is counterbalanced to a great extent by the auspicious conjunction itself, but is probably not completely negligible because of the Ascending Node's exaltation. Nevertheless, as a whole, the astrological situation seems extraordinarily, not to say uniquely, auspicious.

The reader who has followed my analysis thus far is entitled to an explanation of the really amazing fact that the astrologer, in selecting the day and the hour, contrived to make a number of heterogeneous circumstances tally so perfectly that the propitious symptoms of the horoscope by far prevail over the unpropitious ones. One part of the answer is given by referring to the fact that, above all, the auspicious nature of the day offers itself almost automatically, for not only the double conjunction but also all the other actual planetary aspects (see below, under (*c*), and Section 7) appear to be as favourable as they can be. Hence particular astrological skill is indeed required only in selecting the Ascendant in such a way that the propitious sections or points of the zodiac coincide with the cusps of the twelve houses, while the unpropitious ones are banished from the critical points or at least counterbalanced by one or the other accessory element. Obviously, it seems a laborious, not to say "sisyphean", work to find the most appropriate solution by way of calculation, because for each ascendant we obtain a different distribution of houses and, consequently, also different lords of *fines* and *facies*, as well as different effects of the *domiciles* and *exaltations*. It is true that the distribution of the planets among the *fines* (cf. the above table) is apt to facilitate the choice, because the three first *fines* of each sign (*i.e.* approximately the first 20° of each) are prevalently governed by auspicious planets (exceptions being only the following four out of thirty-six: Cancer I, *Mars*; Libra I, *Saturn*; Scorpio I, *Mars*; and Leo III, *Saturn*), whereas the ill-famed planets appear in the two last *fines*. But as there is no similar regularity in the *facies* and because, on the other hand, the deviation of the houses from the average extent of 30° can be considerable, the astrologer's task still seems difficult enough.

The explanation, no doubt, is that a rough approximation was obtained with the aid of an astrolabe. This most ingenious instrument of the Middle Ages enabled the astrologer to solve these or similar problems simply by way of systematic trial. The work of days or weeks could thus be done in less than an hour. A first approximation being found, the astrologer would resort to the manifold tables at his disposal[44] in order to establish the definitive value of the Ascendant. Thus actual calculation could be reduced to a minimum or even disposed of completely.

(*c*) *Trines* (Lat. *trigona* or *triquetra*)

They consist of four groups of three signs of the zodiac lying 120° apart; each trine is attributed to one of the four elements, and governed by one lord of the day, one of the night, and one companion.

[44] Tables of right ascensions for each degree of the ecliptic, of oblique ascensions in the seven climates and for the latitudes of important cities, *etc.*; cf. AL-BATTĀNĪ, *op. cit.*, Vol. II, pp. 61–71, also *Die astronomischen Tafeln des Muḥammed ibn Mūsā al-Khwārizmī* (ed. H. SUTER), Copenhagen, 1914, Tables 79–90 (pp. 194–205).

	Signs	Element	Lord of		Companion
			Day	Night	
Trine	I Aries, Leo, Sagittarius	Fire	Sun	Jupiter	Saturn
	II Taurus, Virgo, Capricornus	Earth	Venus	Moon	Mars
	III Gemini, Libra, Aquarius	Air	Saturn	Mercury	Jupiter
	IV Cancer, Scorpio, Pisces	Water	Venus	Mars	Moon

In the present case, we find:

Trine I: not occupied by planets (the Ascending Node in Sagittarius does not count).

Trine II: (1) Saturn in Taurus. This is of no interest because Saturn is neither lord nor companion. (2) Moon in Capricorn. Of decisive importance, because the auspicious Moon which governs the *night* stands in the *night* part of the *trine* (Capricorn below the horizon). Highly propitious symptom for prognostication.

Trine III: not occupied by planets.

Trine IV: (1) Sun, Jupiter, Venus, and Mercury, in Cancer (above horizon). Very propitious because Venus as lord of the *day* stands in the *day* part of the trine, supported by three other benevolent planets. (2) Mars as lord of the *night* in the *night* part, is powerful but not obnoxious because he is neutralized by the propitious cumulation of planets in Cancer.

7. Actual Planetary Aspects

It is due to practical considerations only and not to intrinsic reasons, that I am treating this question only after having dealt with the influence of the various fictitious divisions and subdivisions of the zodiac (*domiciles*, *exaltations*, *houses*, *fines*, *facies*, *trines*). Evidently, the alleged effect produced by the relative positions (angular distances) of the planets ranks above that of vicarious points and sections, and will, therefore, have to be considered first by the astrologer.

The rules are simple. An angular distance (difference in longitude) of 180° is called *opposition*, as in astronomy; 120° make the *trigonal aspect*; 90° the *quadrature*; and 60° the *sextile aspect*. Oppositions and quadratures are considered unfavourable, especially when ill-omened planets are involved, and the latter more so than the former, while the trigonal and sextile aspects are regarded as propitious phenomena.

The conjunction, in classical astrology, was not counted among the aspects, strictly speaking, but[45] starting with the early Islamic period it played an important, even a decisive rôle. The famous Arabic astrologer Albumasar (Ja'far b. Muḥammad b. 'Umar Abū Ma'shar al-Balkhī), who died at the age of 100, in 886 a.d., was one of the first to compile a bulky book specially devoted to conjunctions, which was copied and commented on innumerable times by later astrologers. In the translation of Johannes Hispalensis (fl. about 1135–1153), this work of Albumasar's was among the first to be published in print[46]. A first edition appeared simultaneously in Augs-

[45] Cf. footnote 36, page 97.
[46] *Albumasar de magnis conjunctionibus et annorum revolutionibus ac eorum profectionibus octo continens tractatus. The Flores Albumasaris* ("*Tractatus Albumasaris florum astrologie*"), mentioned in Section 5 (*e*) which were printed by Ratdolt in 1488, are possibly an extract from the great work.

burg (printing office of ERHARD RATDOLT) and in Venice; a second was published in Venice in 1515.

The fact that the City of Venice, in the course of one generation, witnessed the publication of two editions of this work, obviously bears upon the present problem. Or perhaps it is only chance that twelve years after the publication of the second edition, the patrician MARCANTONIO of Venice ordered a horoscope to be worked out in which a conjunction of two planets with the Sun holds the central place[47]? It would seem a promising task to examine Albumasar's great work from this point of view to find out what prognostication, according to him, can be inferred from Marcantonio's double conjunction. This new problem, however, will not be treated here; it may be made the subject of another study, inshā'llāh.

In the present case, the mutual angular distances between the planets are as follows.

Sun–Moon (198°)	162°	*Moon*–Venus (186°)	174°	*Venus*–Mars	117°
–Mercury	0	–Mars (303°)	57	–Saturn (275°)	85
–Venus	24	–Saturn	101	–Ascending Node	148
–Mars	141	–Ascending Node (334°)	26		
–Jupiter	0			*Mars*–Saturn (202°)	158°
–Saturn (299°)	61	*Mercury* and *Jupiter* (see Sun)		–Ascending Node	31
–Ascending Node	172				
				Saturn–Ascending Node (233°)	127

Apart from the conjunction, only the following are of interest from the point of view of aspects:

Sun (with Jupiter and Mercury)–Saturn: sextile aspect + 1°. Strictly favourable, because binding the dangerous Saturn.

Sun–Ascending Node: 8° *past* opposition. The danger threatening from opposition has gone.

Moon–Venus: 6° *before* opposition. The beneficent effect of the two planets is not yet blocked by the opposition occurring ca. 11^h later.

Moon—Mars: 3° *before* sextile aspect. The Moon will reach the favourable aspect within 6 or 7 hours. The evil influence of Mars will, therefore, be banished until the afternoon of the subsequent day[48].

Moon–Saturn: 11° *before* quadrature. The Moon has just passed the limit of 13°, where the evil effect of her quadrature with Saturn may become perceptible. But the danger is still minimal.

Venus–Mars: 3° *past* trigonal aspect[49]. The prosperous effect of the aspect still lasting.

Venus–Saturn: 5° *before* quadrature. The evil effect of the aspect not yet perceptible.

In other words, this last of the main elements of prognostication (primary and secondary) is just as auspicious as all the preceding ones. It would probably be hard

---

[47] But the doctrine of conjunction played a big part all through the fifteenth century. (See also A. WARBURG'S studies on MARTIN LUTHER in "Heidnisch-antike Weissagung in Wort und Bild zu Luther's Zeiten" in *Sitzungsberichte d. Heidelberger Akad. d. Wiss., Philol.-histor. Kl.*, 26. Abh., 1919, *e.g.* pp. 12 and 80; furthermore G. BING (Editor), A. WARBURG. *Gesammelte Schriften*, Band II, Teubner, Leipzig, 1932. See also F. BOLL and C. BEZOLD, *Sternglaube und Sterndeutung; die Geschichte und das Wesen der Astrologie*, 4th Edition; *e.g.* pp. 34 and 117; Teubner, Leipzig, 1931.)

[48] According to BOUCHÉ-LECLERCQ, *op. cit.*, p. 245, the effect of conjunctions is perceptible in an area extending 3° to both sides of the point of contact as far as the Sun and the five planets (except the Moon) are concerned. It seems reasonable to apply the same rule to the other aspects. In the case of the Moon, the sphere of influence extends to ± 13° from the critical points (aspects), corresponding to the distance travelled in the course of one day.

[49] Venus in direct motion, between superior conjunction and maximum eastern elongation.

to find a second day and hour in MARCANTONIO's life that marks a similarly propitious constellation. It would seem perfectly reasonable, therefore, that our Venetian patrician considered this horoscope worthy of being engraved on the marble statuette representing his astrological patron and protector.

8. THE PROBABLE DATE OF MARCANTONIO'S MARRIAGE

Once our curiosity has been roused, we find it difficult to stop at this point. For does it not seem probable, nay certain, that MARCANTONIO would also allow himself to be guided by astrological considerations in fixing the date of his marriage?

All we know is that the marriage was celebrated in February, 1527[50]. Considering the fact that the horoscope of 15th June was dominated by the conjunction of Mercury with the Sun and Jupiter, we may duly expect another striking phase of Mercury—possibly also a conjunction—to play an analogous part in the marriage horoscope. Of course, a double conjunction would be highly unlikely (indeed, there is none during the whole month). The only astrological phenomenon to be considered, therefore, would be a conjunction of Mercury either with the Sun or with Venus, which is even more auspicious than Jupiter. In fact, both occurred in February, 1527, the former about 7th February, the latter, 11th February. In checking the astrological situation of the two days, it has been found by means of approximate calculations that only the second one is capable of a favourable interpretation.

	7th February, 1527 (noon) (Conjunction Mercury-Sun)	11th February, 1527 (noon) (Conjunction Mercury-Venus)
Sun	328° (Aquarius 28°)*	332° (Pisces 2°)
Moon	45 (Taurus 15)	98 (Cancer 8)
Mercury	328 (Aquarius 28)*	325 (Aquarius 25)
Venus	320 (Aquarius 20)	325 (Aquarius 25)
Mars	236 (Scorpio 26)	238 (Scorpio 28)
Jupiter	72 (Gemini 12)	73 (Gemini 13)
Saturn	13 (Aries 13)	13 (Aries 13)
Ascending Node	271 (Capricornus 1)	271 (Capricornus 1)

* Exact values (mean noon Venice): Sun 327° 52′, Mercury 328° 7′.

The difference between the two consists mainly in the changed position of the Moon. On 7th February, her good influence is blocked by the approaching opposition with Mars, which will occur within 24^h, while on 11th February, she not only occupies her *domicilium* (Cancer), but also stands close to the trigonal aspect with the Sun. It is true that the danger of her quadrature with Saturn is imminent, but it could be averted by choosing the later part of the preceding night for the marriage ceremony, when the Moon entering Cancer is still 13° distant from the critical point. Equally unprosperous, in both cases, is the proximity to quadrature with Mars, on 7th February, of the Sun and Mercury, on 11th February, of Mercury and Venus. But as the figures given above are only rough approximations, and as undoubtedly they do not necessarily agree with those derived from contemporary astronomical tables, it would be unjustified to draw further conclusions.

The most powerful argument for 11th February as the probable wedding day—or rather a day near to 11th February on which, according to the unknown tables used,

[50] See Section 4, page 89.

the conjunction of Mercury and Venus took place—is that no believer in astrology who has chosen Mercury as his patron, will fail to celebrate his marriage on the day on which his planet-god celebrates his conjunction with the goddess of love, Venus[51]. And should there be evil aspects threatening to destroy the harmony of the day, a skilled astrologer will always find ways to explain them away or twist them so as to make them form a friendly picture. For, here as elsewhere, it would appear that the real aim of man is not to learn the truth but to have confirmed what he would like to be the truth. *Mundus vult decipi*[52].

II. ASTRONOMICAL PART

9. The Ptolemaic Theory of Planetary Motion

(a) Venus, Mars, Jupiter, and Saturn

It is not without pride that PTOLEMY (ca. 140 A.D.), in his *Mathematical Syntaxis*[53] (which we are accustomed to call by its corrupted Arabic name, *Almagest*) claims to have been the first to work out a complete mathematical theory of the planetary motions. In the second chapter of the 9th book[54], he pays tribute to his predecessor, HIPPARCHUS (ca. 150 B.C.), "this great lover of Truth", who, occupying himself very thoroughly with the theory of the Sun and Moon, had proved that the orbits of these two luminaries can be represented on the basis of the (Aristotelian) postulate of uniform circular motion in the eccentric deferent as well as in the epicycle. For the planets, however, HIPPARCHUS had only shown that this postulate does not suffice to explain their complicated motions, as determined by observation.

For this reason, PTOLEMY introduces the new hypothesis that the centre of the epicycle moves on the eccentric with varying velocity, in such a way that only when seen from the *punctum aequans*—*i.e.* the point on the line of apsides whose distance from the Earth is the double of the linear eccentricity—the motion in the eccentric appears to be uniform. The characteristic features of the Ptolemaic hypothesis as applied to the planets Venus, Mars, Jupiter, and Saturn, are illustrated by Fig. 7, where

E = Earth, centre of ecliptic, $AXYA_1D$,

F = centre of eccentric deferent, GHK,

E' = *punctum aequans*,

$EF = FE'$ = linear eccentricity[55],

AA_1 = line of apsides,

G = apogee of deferent,

K = perigee of deferent,

[51] See "Concluding Notes" at end of study.

[52] The period under discussion here is only some sixty years later than that of the Chigi-Horoscope of 1466, on the ceiling of the Sala di Galatea of the Farnesina in Rome, the investigation of which was carried out in 1934 by FRITZ SAXL and ARTHUR BEER (Reale Accademia d'Italia, Collezione "La Farnesina", Rome, 1934, XII): Part I, by F. SAXL, "La fede astrologica di Agostino Chigi"; Interpretazione dei dipinti di Baldassare Peruzzi nella Sala di Galatea della Farnesina (57 pp.). Part II, by A. BEER, "Il significato astronomico e la data dei dipinti della volta della Sala di Galatea" (7 pp.). It is interesting to compare the historical and personal aspects as they appear from this interpretation of CHIGI'S horoscope with those set out in the present paper.

[53] Greek text: *Cl. Ptolemaei Syntaxis Mathematica*, ed. J. L. HEIBERG, Part I (Books I–VI), Leipzig, 1898; Part II (Books VII–XIII), 1903. German translation: *Des Cl. Ptolemäus Handbuch der Astronomie, aus dem Griechischen übersetzt von K. Manitius*, Vol. I, Leipzig, 1912; Vol. II, 1913. Quotations (*Almagest*) are made from the German translation.

[54] *Almagest*, Vol. II, p. 96ff.

[55] Expressed in parts (60th) of the radius FG of the deferent (all other linear quantities, such as the radius of the epicycle, *etc.*, are expressed in the same way).

H = centre of epicycle, gTk,
g = mean apogee of epicycle,
k = mean perigee of epicycle,
T = planet,

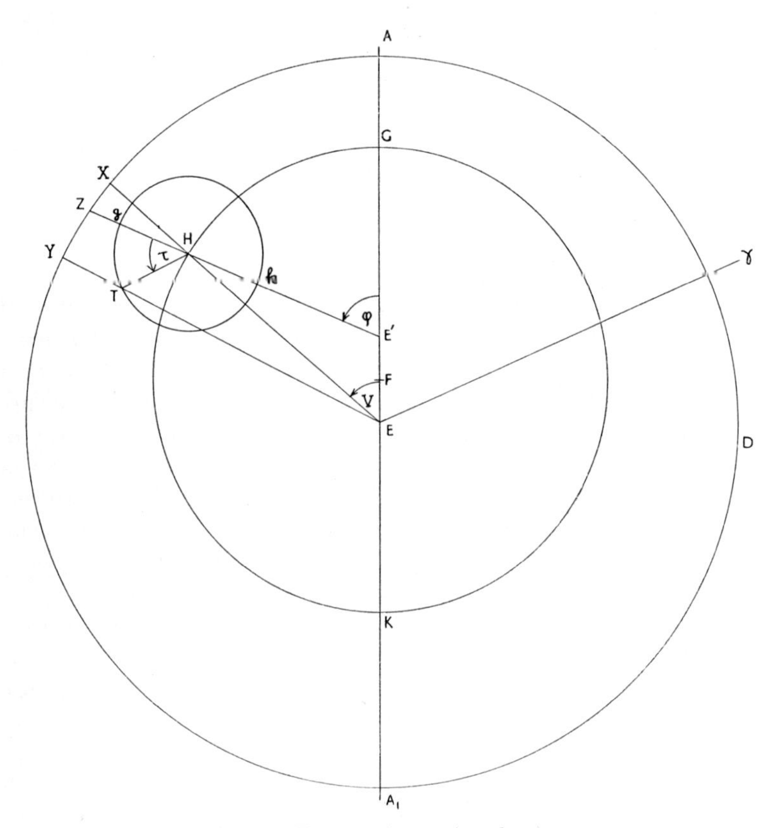

Fig. 7. Venus and superior planets

$\angle \Upsilon EA$ = longitude of apogee G of deferent,
$\angle AE'Z = \Phi$ = mean anomaly of centre H of epicycle,
$\angle AEX = v$ = apparent[56] anomaly of centre H of epicycle,
$\angle XHZ = \angle EHE'$ = "zodiacal inequality" (PTOLEMY: ἡ ζωδιακὴ ἀνωμαλία, in Arabic: *Taʿdīl al-ḫāṣṣa waʾl-markaz*, whence mediaeval Lat. *aequatio anomaliae et centri*, and in modern usage, "*equation of the centre*"),
$\angle gHT = \tau$ = (mean) anomaly of planet T in the epicycle,
$\angle XEY$ = *aequatio anomaliae* (ἀνωμαλίας προσθαφαίρεσις, "prosthaphairesis").

The apogee of the deferent is supposed to have a uniform direct motion round E, its amount being the same as that of precession (1° in 100[a]); in other words, the line of apsides is fixed relative to the fixed stars. The revolution of H on the deferent, as

[56] *Apparent* always to be understood as "appearing to the eye", *i.e.* as actually observed from the Earth.

mentioned before, is uniform in regard to the *punctum aequans*, E', not the centre of the deferent, F. Finally, the planet itself revolves uniformly—and equally counter-clockwise, as all other points—in the periphery of the epicycle[57]. Thus the *tropical* revolution of the planet is represented by the motion of H in the deferent (called μῆκος, [motion in] longitude)[58], whereas the *synodic* is taken care of by that of the planet in the epicycle (ἀνωμαλία, *anomaly*). In Alm. IX.4[59], the longitudes and anomalies are tabulated for periods of 18^a, single years, months, days, and hours, of the Egyptian calendar ($1^a = 12$ months of $30^d + 5$ epagomenal days, without intercalation), the epoch being Thoth 1 of the 1st year of NABONASSAR, *noon* = 26th February, 747 B.C. The reduction of *mean* to *apparent* longitudes is effected with the aid of special tables, one for each planet (Alm. XI.12[60]), arranged in such a way that the *aequatio centri* as well as the *aequatio anomaliae* can be obtained from them by double interpolation.

PTOLEMY's theory of *latitudes* is developed in the six first chapters of Alm. XIII[61]. It will be disregarded here as it does not bear directly on our subject.

(b) *Mercury*

The mathematical hypothesis demonstrated above yields fairly good results in the case of the four planets, Venus, Mars, Jupiter, and Saturn. It is, however, not sufficient to represent the motion of Mercury, which, on account of the considerable eccentricity of its orbit, was of crucial importance for ancient and mediaeval astronomers. They were particularly puzzled by the circumstance that, in the course of one revolution, the centre of the epicycle passes twice through a perigee point, but only once through an apogee (see Section 10).

Being aware of the impossibility of obtaining satisfactory results by operating with a deferent having a constant eccentricity, PTOLEMY devises the following modification of his simple hypothesis. In Fig. 8 the line AA_1, corresponding to the line of apsides in the preceding figure[62], follows the precession of the equinoxes, *i.e.* is considered immovable relative to the fixed stars.

On it, E marks the Earth, and E' the *punctum aequans* (being the centre of the *circulus aequans*, or "equant", a), from which the revolution of the centre of the epicycle appears uniform. The centre of the movable deferent lies on the periphery of a small circle $MM'E'$ the centre of which, F, has the same distance from E' as E' from E.

Starting from a certain moment, t, when the centre H of the epicycle coincides with the apogee G of the deferent in its initial position on the apse-line (dotted circle, d, with centre M), the centre of the deferent, in the course of one tropical year, revolves with uniform *retrograde* motion on the small circle around F, while the radius vector $E'H$ simultaneously carries around, with uniform *direct* motion, the centre H of the epicycle. At the moment t', the centre of the deferent (circle d') being in M', and its apogee in G', the centre of the epicycle will occupy H', so that $\angle GFG' = \angle GE'H' = \phi$. After one half-year, the centre of the deferent coincides

[57] *Almagest*, IX.6 (Vol. II, p. 123).
[58] In the case of Venus, as with Mercury, the period of revolution of H is the tropical year because the longitude of H is always equal to the Sun's mean longitude; see below, Section 9 (*b*).
[59] *Almagest*, Vol. II, pp. 104–118.
[60] *Almagest*, Vol. II, pp. 261–265.
[61] *Almagest*, Vol. II, pp. 325–380.
[62] Contrary to the other planets, the point A_1 is *not* the perigee point, whence the term "line of apsides" is not applicable to this case. Only for the sake of brevity, the line AA_1 will henceforth be called "apse-line."

with the *punctum aequans*, E', and the deferent itself, with the "equant" a, and thus the centre H'' of the epicycle and the movable apogee, G'', will again meet on the apse-line, but this time between E and A_1.

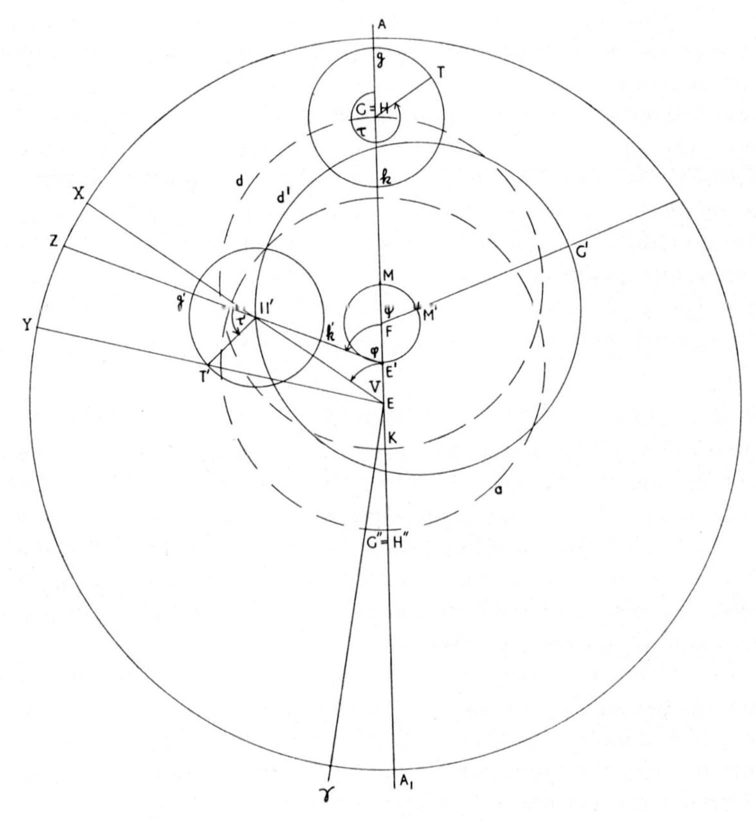

Fig. 8. Mercury

As in the preceding case, the planet itself revolves uniformly in the epicycle. The period of revolution of the centre of the epicycle is the tropical year, as also in the case of Venus, and thus the centre of the epicycle always has the same mean longitude as the Sun[63]. This, however, does not necessarily imply that the Sun itself is regarded as the centre of revolution. On the contrary, PTOLEMY and his followers obviously take no notice of the (so-called) "Egyptian" hypothesis as propagated by HERACLEIDES, according to which Mercury and Venus are believed actually to revolve around the Sun. They disregard or reject it with good reason indeed, because evidently this naively devised hypothesis becomes improbable and useless as soon as the true Sun is replaced by the mathematical point representing the Sun's mean longitude.

The equation of the centre ($\angle E'H'E$) and that of anomaly ($\angle H'ET'$) are computed and tabulated in the same way as in the case of the other planets. From his observations, PTOLEMY derives the following values for the constants by which the

[63] *Almagest*, IX.3 (Vol. II, p. 102, l. 17–19), see footnote, 58, page 107.

orbit of Mercury is determined, taking again no account of its inclination as not relevant to our subject):

Linear eccentricity $\varepsilon = 3^p\ 0'$[64] $= 0{\cdot}05R$ ($R = 1$).

Radius of epicycle $\rho = 22^p\ 30' = 0{\cdot}375R$.

Longitude of apogee $A_0 =$ Libra $1°\ 10' = 181°\ 10'$.

Mean longitude of centre of epicycle $L_0 =$ Pisces $0°\ 45' = 330°\ 45'$ (= mean longitude of Sun).

Anomaly (counted from mean apogee of epicycle) $\tau_0 = 21°\ 55'$.

(Epoch of A_0, L_0, τ_0 = 1st year of Nabonassar, Thoth I, *noon* = 26th February, 747 B.C.[65].)

10. EXCURSUS: THE CURVE DESCRIBED BY THE CENTRE OF THE EPICYCLE. IBN AL-SAMḤ AND AZARQUIEL, THE LIBROS DEL SABER, AND PEURBACH

The Almagest contains no statement concerning the nature of the curve which the centre of the epicycle actually describes according to PTOLEMY's theory of Mercury. To the best of my knowledge, the first European author to speak about it explicitly is GEORG PEURBACH (1423–61). In his *Theoricae novae planetarum*[66], he refers to it with the words: "Sexto ex dictis apparet manifeste centrum epicycli Mercurii propter motus supra dictos non ut in aliis planetis fit: circumferentiam deferentis *circularem sed potius figurae habentis similitudinem plana ovali* periferiam describere".

This statement is perfectly true, for under the given conditions $\left(\text{"linear" eccentricity } \varepsilon = 3^p = \frac{R}{20}\right)$, as will be seen, the curve is practically identical with an ellipse having F as its centre, and $a = R + \varepsilon$ and $b = R - \varepsilon$ as its axes (eccentricity according to modern definition $e = \frac{2\sqrt{R\varepsilon}}{R+\varepsilon} = 0{\cdot}4259180\ldots$). Theoretically, however, it is an algebraic curve represented by the polar equation

$$r = \varepsilon(\cos\phi + \cos 2\phi) + \sqrt{R^2 - \varepsilon^2(\sin\phi + \sin 2\phi)^2}, \qquad \ldots.(1)$$

where the punctum aequans, E', marks the origin, and $E'A$, the axis, of a system of polar co-ordinates, r and ϕ (see Fig. 9).

My assertion that, for the Ptolemaic ratio of its two parameters, $R = 20\varepsilon$, this curve becomes practically identical with an ellipse, is not a matter of course and has to be proved. Since a complete analysis of its mathematical properties is beyond our scope, I limit myself to the following observations.

For small values of the ratio $\frac{R}{\varepsilon}$, the curve is pear-shaped, as shown in Fig. 9[67]. Letting $u = \sin\phi + \sin 2\phi$, and $\frac{du}{d\phi} = \cos\phi + 2\cos 2\phi$, we find, from $\frac{du}{d\phi} = 0$ for $\phi_M = \pm 53{\overset{\circ}{.}}625$, that u reaches its maximum, $u_M = +1{\cdot}76017$, for this value of ϕ[68].

---

[64] $1^p = 60' = \frac{1}{60}$ of radius of deferent, $R = 60^p$, see footnote 55, page 105.

[65] NALLINO, in AL-BATTĀNĪ, *Opus Astronomicum*, Vol. I, p. 241, erroneously assumes the year 137 A.D. to be the epoch to which Ptolemy reduces his observations, though it is expressly stated, *Almagest*, IX.11 (Vol. II, p. 155 *et passim*), that the data are consistently referred to the beginning of the era of Nabonassar. Therefore the differences between PTOLEMY'S values for the apogees of Saturn, Jupiter, and Mars, and, on the other hand, those computed from LEVERRIER'S elements are about 10° smaller than indicated by NALLINO (2°–5°, instead of 12°–15°).

[66] Editio princeps, printed by Regiomontanus, Nuremberg, 1472, fol. 21f.

[67] As the curve evidently has the line AE as its axis of symmetry, only the left half of it is drawn, for various ratios $\frac{R}{\varepsilon}$.

[68] u attains its minimum $u_m = -0{\cdot}3690$ near $\phi = \pm 147{\overset{\circ}{.}}5$. Because $|u_m| < |u_M|$, this value is of no interest to us.

Hence the curve will become undefined in the neighbourhood of ϕ_M, for $R < \varepsilon \, . \, u_M$.

The equation

$$\frac{dr}{d\phi} = -\varepsilon(\sin\phi + 2\sin 2\phi) - \frac{\varepsilon^2\{\sin\phi(\cos\phi + \cos 2\phi) + \sin 3\phi + \sin 4\phi\}}{\sqrt{R^2 - \varepsilon^2(\sin\phi + \sin 2\phi)^2}} = 0 \quad \ldots.(2)$$

has the trivial solutions $\phi = 0°$ and $\phi = 180°$, indicating that r has two maxima[69], *viz.* $r = R + 2\varepsilon$ for $\phi = 0°$ and $r = R$ for $\phi = 180°$.

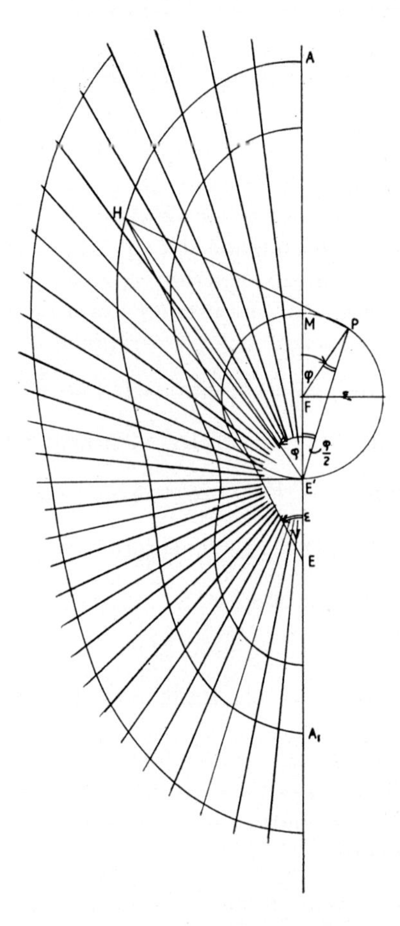

Fig. 9. The Ptolemaic curve for small $\dfrac{R}{\varepsilon}$

$EE' = E'F = \varepsilon$ $\quad \angle AEH = v$

$MA = PH = R$ $\quad E'H = r$

$\angle AFP = \angle AE'H = \phi$ $\quad EH = s$

As we may expect r to reach its minimum near $\phi = \pm 90°$, we first compute that value of the ratio $\dfrac{R}{\varepsilon}$ which satisfies our equation for this value of ϕ. From $\sqrt{R^2 - \varepsilon^2} = 2\varepsilon$ we obtain $\dfrac{R}{\varepsilon} = \sqrt{5}$. For $\dfrac{R}{\varepsilon} < \sqrt{5}$, as can easily be seen, the minima of r will occur at values of $|\phi| < 90°$, whereas for $\dfrac{R}{\varepsilon} > \sqrt{5}$ the minima of r occur at $|\phi| > 90°$.

[69] The position of the maxima and minima is evident. I therefore omit a discussion of the second differential quotient, which is a rather clumsy expression:

$$\frac{d^2r}{d\phi^2} = -\varepsilon(\cos\phi + 4\cos 2\phi) - \varepsilon^2\{\cos 2\phi + 4(\cos 3\phi + \cos 4\phi) - \sin\phi\sin 2\phi\} - \frac{\varepsilon^4\{\sin\phi(\cos\phi + \cos 2\phi) + \sin 3\phi + \sin 4\phi\}^2}{R^2 - \varepsilon^2(\sin\phi + \sin 2\phi)^2}.$$

The upper limit of ϕ is defined by the equation $\sin\phi + 2\sin 2\phi = 0$, which is satisfied by $|\phi| = 104^\circ\!.477$. Hence, for all $\frac{R}{\varepsilon} > \sqrt{5}$, r will reach its minimum at $90^\circ < \phi < 104^\circ\!.477$.

In the following table, some corresponding values of ϕ and $\frac{R}{\varepsilon}$ are given for which r becomes a minimum.

ϕ	$\frac{R}{\varepsilon}$
90°0	2·24
101·0	5·39
102·0	7·10
103·0	11·03
104·0	31·3
104·477	undefined

It shows, in general, that $\frac{R}{\varepsilon}$ grows very rapidly near the critical value of $\phi = 104{\cdot}477$, and, in particular, that for PTOLEMY's value of $\frac{R}{\varepsilon} = 20$, r reaches its minimum between $\phi = 103^\circ$ and 104°. Indeed, this statement has only theoretical interest because PTOLEMY's and our own concern is not the minimum of the radius vector r, but that of the geocentric distance, which we shall henceforth designate by s (radius vector EH, in Figs. 9 and 10). Before entering upon a discussion of the latter quantity, however, I have to come back to my statement concerning the interchangeability of the Ptolemaic curve with an ellipse, in the case of a sufficiently large value of $\frac{R}{\varepsilon}$.

In Fig. 11, as in the preceding case, we have E = Earth, E' = punctum aequans, F = centre of a small circle with radius $\varepsilon = FE' = E'E$, in whose circumference a point P revolves. The angle AFP which corresponds to a certain moment t will be called ψ here. Availing ourselves of a well-known method, we construct the isosceles triangle FPS with apex P. By producing PS over S and making $PK = R$, the point K will evidently lie on an ellipse with the centre F and the axes $R + \varepsilon$ and $R - \varepsilon$. For $t = t_0$, we choose $\psi = 0$, so that K_0, on the apse-line, coincides with A.

Our problem then is to express the radius vector $E'K = \rho$, in analogy with the preceding, as a function of the angle $\phi = AE'K$. In raising the perpendicular TK on AE, we recognize that $\cos\psi = \frac{\rho\cos\phi - \varepsilon}{R + \varepsilon}$, and $\sin\psi = \frac{\rho\sin\phi}{R - \varepsilon}$. On the other hand, we have

$$(R - \varepsilon)^2 = \overline{KT}^2 + \overline{TS}^2 = \rho^2\sin^2\phi + (R - \varepsilon)^2\cos^2\psi.$$

By substitution we find

$$(R - \varepsilon)^2 = \rho^2\sin^2\phi + \left(\frac{R - \varepsilon}{R + \varepsilon}\right)^2(\rho\cos\phi - \varepsilon)^2$$

or
$$O = \rho^2\left\{\sin^2\phi + \left(\frac{R - \varepsilon}{R + \varepsilon}\right)^2\cos^2\phi\right\} - 2\rho\varepsilon\left(\frac{R - \varepsilon}{R + \varepsilon}\right)^2\cos\phi + \varepsilon^2\left(\frac{R - \varepsilon}{R + \varepsilon}\right)^2 - (R - \varepsilon)^2.$$

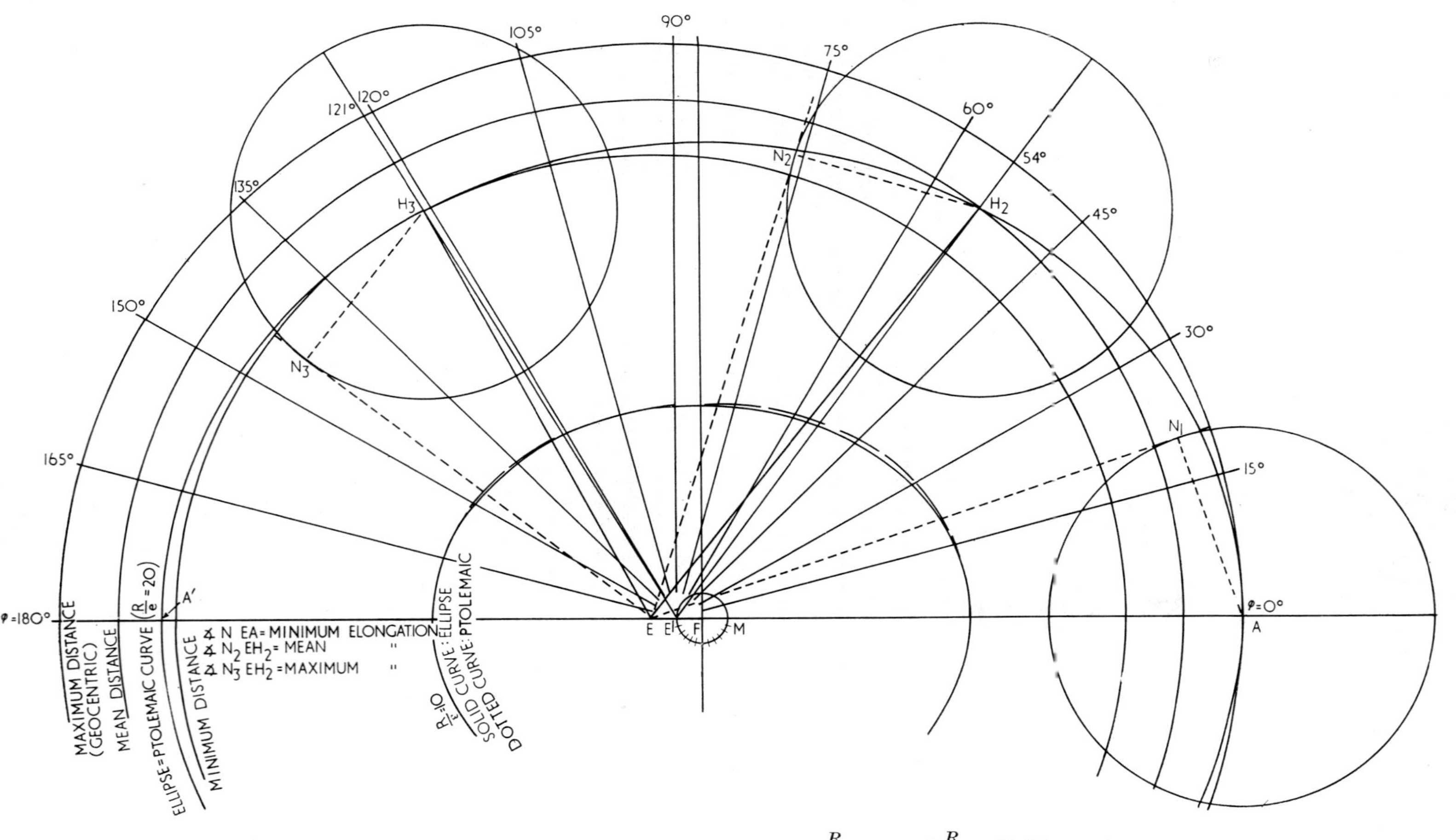

Fig. 10. The Ptolemaic curve compared with ellipse for $\frac{R}{\varepsilon} = 10$ and $\frac{R}{\varepsilon} = 20$ (Mercury)
Correct proportions

The solution of this quadratic equation is given by the rather cumbersome expression

$$\rho = \pm \frac{(R - \varepsilon)\sqrt{2}}{2(R^2 + \varepsilon^2 - 2R\varepsilon \cos 2\phi)}$$

$$\left\{\sqrt{2(R^2 + \varepsilon^2)[(R + \varepsilon)^2 - \varepsilon] + \varepsilon^2(R - \varepsilon)^2 - [4R\varepsilon(R^2 + 2R\varepsilon) - \varepsilon(R - \varepsilon)^2] \cos 2\phi}\right\}$$

$$+ \frac{\varepsilon(R - \varepsilon)^2 \cos \phi}{R^2 + \varepsilon^2 - 2R\varepsilon \cos 2\phi}, \qquad \ldots .(3)$$

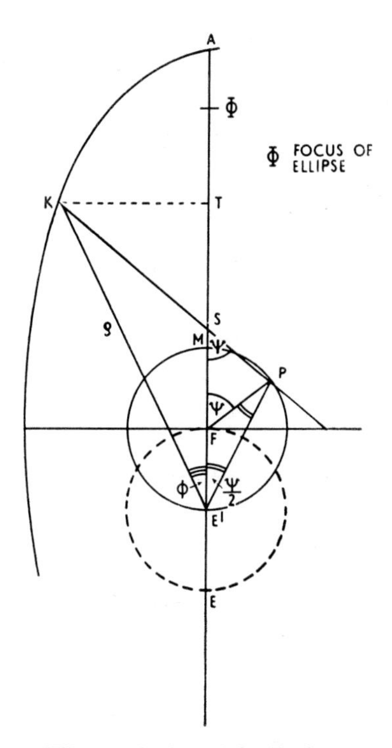

Fig. 11. Ellipse substituted for Ptolemaic curve

where, of course, only the positive sign of the first term is applicable to our case.

Unfortunately, there is no way to show that, for a sufficiently large value of $\frac{R}{\varepsilon}$, this formula for the radius vector ρ of the ellipse passes over into the one for r as given by equation (1). We therefore have to content ourselves with proving it numerically for the Ptolemaic ratio $\frac{R}{\varepsilon} = 20$.

In letting $\varepsilon = 1$, our formula (3) becomes

$$\rho = \frac{19\sqrt{2}}{2} \cdot \frac{\sqrt{353{,}241 - 34{,}839 \cos 2\phi}}{401 - 40 \cos 2\phi} + \frac{361 \cos \phi}{401 - 40 \cos 2\phi}. \qquad \ldots .(4)$$

With a very small error, this formula can be simplified as follows:

$$\rho \sim \frac{19 \cdot 10^2}{2} \cdot \frac{\sqrt{2(35 \cdot 324 - 3 \cdot 484 \cos 2\phi)}}{40(10 - \cos 2\phi)} + \frac{9 \cdot 00 \cos \phi}{10 - \cos 2\phi}$$

$$\sim \frac{23 \cdot 75\sqrt{70 \cdot 65 - 6 \cdot 97 \cos 2\phi} + 9 \cdot 00 \cos \phi}{10 - \cos 2\phi}.$$

In order to simplify the radicant $70 \cdot 65 - 6 \cdot 97 \cos 2\phi$, it seems practical to reduce it to the form $c \,.\, (10 - \cos 2\phi)$, where $7 \cdot 065 > c > 6 \cdot 97$. Remembering then that, for $\phi = 0$ and $\cos \phi = \cos 2\phi = 1$, the numerical value of ρ has to be 22·00, we obtain from

$$22 \cdot 00 = \frac{23 \cdot 75\sqrt{c(10 - 1)} + 9 \cdot 00}{10 - 1}$$

$$c = \left(\frac{63 \cdot 00}{23 \cdot 75}\right)^2 = 7 \cdot 035,$$

which, in fact, lies between the limits indicated.

Hence we finally obtain the simplified formula

$$\rho \sim \frac{63 \,.\, \sqrt{10 - \cos 2\phi} + 9 \cos \phi}{10 - \cos 2\phi}. \qquad \ldots.(5)$$

It can easily be shown that the total error committed by this series of simplifications does not exceed 0·1 per cent. This means that, in computing the values of ρ to the second decimal place, the last digit may in extreme cases be affected with a maximum error of $\pm$ 2, for $R = 20$ and $\varepsilon = 1$[70].

Our Table 1 serves in the first place to compare r and ρ; in the second, to show the variation of r in the neighbourhood of its minimum (between $\phi = 103°$ and $104°$, see above). In the last column, the values of the geocentric distance s, in which we are particularly interested, are listed (three decimals in general, four for the interval $116° < \phi < 122°$, to show the variation of s near its minimum, as well as the situation of the latter). The table is computed for $\phi = 0, \pm 5°, \pm 10°, \ldots, \pm 180°$, except for the critical interval $95° < \phi < 125°$, in which fall the minima of r and s. r and ρ are computed to the second decimal place. Only for $101° < \phi < 106°$, the values of r are given with a greater accuracy (three or four decimals). The geocentric distance is computed from $s = \sqrt{r^2 + \varepsilon^2 + 2r\varepsilon \cos \phi}$.

The table demonstrates with sufficient clarity that the "Ptolemaic curve" (about which no word can be found in PTOLEMY's *Almagest*) is practically interchangeable with the ellipse as defined above, which it has been our aim to prove. The geocentric distance, s, occurs only as a denominator in the formula for the equation of the centre, $\phi - v$[71],

$$\sin(\phi - v) = \frac{\varepsilon \sin \phi}{s}. \qquad \ldots.(6)$$

As this quantity, in PTOLEMY's table of anomalies[72], is given only to the minute of a degree, it makes practically no difference whether we compute s from the radius vector r, as in our table, or from the approximate value ρ, which corresponds to the ellipse.

---

[70] Such an extreme case occurs for $\phi = 90°$, where the difference "approximate minus exact" amounts to + 0·017; see note to our Table 1.

[71] $\phi = \angle AE'H$ (Figs. 9 and 10) = *mean* anomaly; $v = \angle AEH$ = *true* anomaly, or *apparent* anomaly, according to PTOLEMY (cf. footnote 56, page 106).

[72] *Almagest* XI.11(Vol. II, p. 265).

In order to demonstrate the low degree of accuracy attained by PTOLEMY, and later by AL-BATTĀNĪ[73], and thus to exclude the last doubt as to the validity of my

Table 1

ϕ	r	ρ	$r - \rho$	s
0°	22·00	22·00	0·00	23·000
5	21·98	21·98	0·00	22·976
10	21·93	21·93	0·00	22·916
15	21·82	21·80	+ 0·02	22·788
20	21·68	21·65	+ 0·03	22·623
25	21·51	21·48	+ 0·03	22·420
30	21·32	21·26	+ 0·06	22·192
35	21·10	21·03	+ 0·07	21·927
40	20·88	20·82	+ 0·06	21·655
45	20·63	20·57	+ 0·06	21·349
50	20·39	20·33	+ 0·06	21·047
55	20·15	20·10	+ 0·05	20·740
60	19·92	19·88	+ 0·04	20·438
65	19·71	19·68	+ 0·03	20·153
70	19·52	19·48	+ 0·04	19·884
75	19·34	19·36	− 0·02	19·623
80	19·18	19·18	0·00	19·379
85	19·06	19·01	+ 0·05	19·173
90	18·97	19·00*	− 0·03	18·996
95	18·91	18·98	− 0·07	18·850
96	18·90			18·821
97	18·89			18·794
98	18·88			18·767
99	18·88			18·750
100	18·88	18·91	− 0·03	18·732
101	18·8726			18·707
102	18·8703			18·688
103	18·8691			18·669
103·50	18·8688			18·661
103·75	18·8688			18·656
104	18·8689			18·652
105	18·8699	18·91	− 0·04	18·636
106	18·8719			18·621
107	18·88			18·607
108	18·88			18·594
109	18·89			18·582
110	18·89	18·92	− 0·03	18·572
111°	18·90			18·562
112	18·91			18·553
113	18·91			18·545
114	18·92			18·538
115	18·93	18·97	− 0·04	18·535
116	18·9455			18·5290
117	18·9578			18·5253
118	18·9711			18·5227
119	18·98527			18·5211
120	19·00000	19·05	− 0·05	18·5203
121	19·01549			18·5176
122	19·03171			18·5212
123	19·048			18·522
124	19·064			18·524
125	19·084	19·08	0·00	18·528
130	19·18	19·17	+ 0.01	18·553
135	19·29	19·29	0·00	18·596
140	19·41	19·38	+ 0·03	18·655
145	19·52	19·52	0·00	18·710
150	19·63	19·63	0·00	18·771
155	19·74	19·73	+ 0·01	18·839
160	19·83	19·83	0·00	18·893
165	19·90	19·91	− 0·01	18·936
170	19·95	19·95	0·00	18·966
175	19·99	19·99	0·00	18·994
180	20·00	20·00	0·00	19·000

* Exact value $\rho_{90}^{(e)} =$ 18·978
According to approximative formula $\rho_{90}^{(a)} =$ 18·995
$\rho_{90}^{(a)} - \rho_{90}^{(e)} = + 0·017$.

statement, I give in Table 2 a comparison of the theoretical values of the equation of the centre, $\phi - v$, computed from the above formula (6), for $\varepsilon = 1$, with those found in PTOLEMY's and AL-BATTĀNĪ's tables (columns 1–4). For all $\Delta \neq 0$, the values of s corresponding to PTOLEMY's wrong figures are computed from the latter and compared with those of Table 1 (columns 5–7).

As is seen, the minimum change in s caused by an error of 1′ in $(\phi - v)$ amounts to $0{\cdot}08\varepsilon$, whereas we have found that, on the other hand, the maximum error committed by substituting the ellipse for the Ptolemaic curve was $0{\cdot}07\varepsilon$ (attained only twice: for $\phi = 35°$ and $\phi = 95°$, cf. column "$r - \rho$" in the preceding table).

[73] AL-BATTĀNĪ's tables (*Op. Astr.*, II, pp. 132–137) still have the same degree of accuracy as PTOLEMY'S, but are more elaborate; they list the equation of the centre for every degree, while PTOLEMY, for $0° < \phi < 90°$, lists it only from 6 to 6°, and for $90° < \phi < 180°$ from 3 to 3°. The sum of PTOLEMY'S third and fourth columns is equal to AL-BATTĀNĪ'S third column. A difference of ± 1′ occurs only in a few cases, and only one such case appears in our list, *viz.* $\phi = 165°$. In order to avoid interpolations the figures in column 3 of my table are taken from AL-BATTĀNĪ.

We now have to remember (cf. above, Section 9 (*b*)) that PTOLEMY's really ingenious theory of Mercury owed its origin to the circumstance that, in comparing the maximum elongation of Mercury (from the Sun's mean longitude), he had found that this

Table 2

ϕ	$\phi - v$		Δ (Ancient minus Theoretical)	s		δ (Ancient minus Theoretical)
	Theoretical	Ancient*		Theoretical	Ancient	
0°	0°·000 = 0° 0′	0° 0′	0′	—	—	0·00
5	0·217 = 0 13	0 15	+ 2	22·98	19·98	− 3·00
10	0·435 = 0 26	0 28	+ 2	22·92	21·31	− 1·61
15	0·651 = 0 39	0 40	+ 1	22·79	22·23	− 0·56
20	0·866 = 0 52	0 53	+ 1	22·62	22·19	− 0·43
25	1·080 = 1 5	1 5	0	—	—	0·00
30	1·291 = 1 17	1 17	0	—	—	0·00
35	1·499 = 1 30	1 31	+ 1	21·93	21·66	− 0·27
40	1·701 = 1 42	1 43	+ 1	21·66	21·45	− 0·21
45	1·898 = 1 54	1 54	0	—	—	0·00
50	2·086 = 2 5	2 6	+ 1	21·05	20·91	− 0·14
55	2·264 = 2 16	2 16	0	—	—	0·00
60	2·429 = 2 26	2 25	− 1	20·44	20·54	+ 0·10
65	2·572 = 2 35	2 35	0	—	—	0·00
70	2·709 = 2 43	2 43	0	—	—	0·00
75	2·821 = 2 49	2 49	0	—	—	0·00
80	2·913 = 2 55	2 54	− 1	19·38	19·47	+ 0·09
85	2·978 = 2 59	2 58	− 1	19·17	19·25	+ 0·08
90	3·018 = 3 1	3 1	0	—	—	0·00
95	3·029 = 3 2	3 2	0	—	—	0·00
100	3·014 = 3 1	3 1	0	—	—	0·00
105	2·971 = 2 58	2 58	0	—	—	0·00
110	2·900 = 2 54	2 54	0	—	—	0·00
115	2·803 = 2 48	2 49	+ 1	18·54	18·44	− 0·10
120	2·680 = 2 41	2 41	0	—	—	0·00
125	2·534 = 2 32	2 32	0	—	—	0·00
130	2·366 = 2 22	2 22	0	—	—	0·00
135	2·179 = 2 11	2 11	0	—	—	0·00
140	1·975 = 1 59	2 0	+ 1	18·66	18·42	− 0·24
145	1·757 = 1 45	1 47	+ 2	18·71	18·43	− 0·28
150	1·526 = 1 32	1 32	0	—	—	0·00
155	1·285 = 1 17	1 18	+ 1	18·84	18·63	− 0·21
160	1·037 = 1 2	1 3	+ 1	18·89	18·66	− 0·23
165	0·783 = 0 47	0 48*	+ 1	18·94	18·54	− 0·40
170	0·525 = 0 32	0 32	0	—	—	0·00
175	0·263 = 0 16	0 16	0	—	—	0·00
180	0·000 = 0 0	0 0	0	—	—	0·00

* "Ancient" = PTOLEMY and AL-BATTĀNĪ, whose figures are identical for the values of ϕ listed, except for $\phi = 165°$, where PTOLEMY has $\phi - v = 0° 47'$, whence $\Delta = 0$.

planet, in the course of one tropical revolution, passes twice through a perigee position, whereas it has only one apogee.

He refers to the following two pairs of corresponding observations[74]:

I { 2nd February, 132 A.D. (mean long. of Sun = Aquarius 10° = 310° = mean long. of centre of epicycle): max. *east* elong. = 21° 15′

2nd February, 141 A.D. (mean long. of Sun = Aquarius 10° = 310°): max. *west* elong. = 26° 30′ }

47° 45′

[74] *Almagest*, IX.8 (Vol. II, p. 139f).

II

4th June, 134 A.D. (mean long. of Sun = Gemini 10° = 70°):
max. *west* elong. = 21° 15′

4th June, 138 A.D. (mean long. of Sun = Gemini 10° = 70°):
max. *east* elong. = 26° 30′

47° 45′

This means that, in both perigee positions, the angle subtended by the epicycle is 47° 45′, and hence the angle subtended by its radius, $\sigma_P = 23° 52' 30''$, whereas in the apogee position the maximum elongation is found to be $\sigma_A = 19° 3'$.

With the aid of elementary trigonometrical considerations PTOLEMY then contrives to prove that his theory tallies perfectly with his observations if one sets $\varepsilon = 3^p$, and the radius of the epicycle, $q = 22^p\ 30'$ (for $R = 60^p$), or $R = 20$ and $q = 7{\cdot}5$ (for $\varepsilon = 1$). Accepting PTOLEMY's assumption that the longitude of the apogee, about 140 A.D., is ca. Libra 10° = 190°, the anomalies of the centre of the epicycle in the above cases will be 120° and 240° respectively. In both cases we have $s = 18{\cdot}5203$ (see Table 1), whence

$$\sin \sigma_P = \frac{7.5}{18{\cdot}5203} = 0{\cdot}404961; \quad \sigma_P = 23^{\circ}{\cdot}8887 = 23° 53' 19''$$

(PTOLEMY 23° 52′ 30″, error < 50″).

For the apogee we find

$$\sin \sigma_A = \frac{7{\cdot}5}{23{\cdot}0000} = 0{\cdot}326087; \quad \sigma_A = 19^{\circ}{\cdot}0314 = 19° 1' 53''$$

(PTOLEMY 19° 3′, error < 70″).

The agreement of theory and observation thus must be said to be extremely good.

As regards the value of ϕ for which the geocentric distance becomes a minimum, it appears that PTOLEMY naively assumes it to be ± 120° exactly. The same naive assumption is still found thirteen centuries later in PEURBACH's *Theoricae Novae Planetarum*. This however is not true, although it is a very close approximation. I refer to our Table 1, where the lowest value of s listed (18·5176) corresponds to $\phi = 121°$[75].

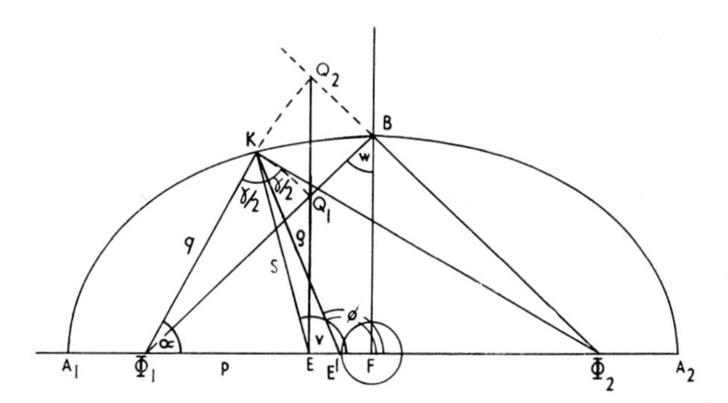

Fig. 12. Determination of minimum geocentric distance (ellipse)

[75] It is, of course, of no practical interest to compute the values of s and ϕ for the theoretical minimum. In any case, this would involve rather elaborate considerations and calculations. That the ellipse substituted for the Ptolemaic curve, which indeed offers itself to direct mathematical treatment, does not furnish a satisfactory answer in this case, may be seen from the following, illustrated by Fig. 12.

As mentioned before, the first European author I know of who expressly stated the similarity of the curve described by the centre of the epicycle with an ellipse was PEURBACH, and even he contents himself with saying that it is a "kind of oval". In the Islamic world, however, as will be seen, the discovery is of a much earlier date; to prove this, I refer to the *Libros del Saber de Astronomía*[76], composed in 1276–77 by order of King ALPHONSE X of Castile, where we find all necessary information in the section bearing the title *Libros de las láminas de los VII planetas*[77]. This section consists of Spanish translations of two western Arabic treatises, one (Book I) by ABŪ 'L-QĀSIM AṢBAGH IBN MUḤAMMAD IBN AL-SAMḤ of Cordova (d. 1035), the other (Book II), which was written ca. 1081, by IBRĀHĪM IBN YAḤYĀ AL-NAQQĀSH ABŪ ISHĀQ IBN AL-ZARQĀLĪ, commonly cited by his latinized name AZARQUIEL (d. 1100). The problem dealt with is the same in both treatises, namely, to construct instruments (*láminas* = "disks") reminiscent in their general aspect of planispheric astrolabes, and consisting of a series of graduated circles or rings, which permit the quick determination of the position of the planets without computation[78].

IBN AL-SAMḤ, who constructs one particular disk for each of the seven planets, devises two different constructions for the motion of Mercury, one very crude based

---

(Continuation of footnote 75 from page 117)

Fig. 12 shows an ellipse A_1BA_2, with centre F, foci Φ_1 and Φ_2, the major axis $2a$, and angle of eccentricity $\Phi_1BF = w$. As in Fig. 11, E (Earth) marks a point between Φ_1 and F, whose distance from F is 2ε. Our problem is to find the minimum distance $EK = s$ and the corresponding angle $A_2EK = v$ (apparent anomaly), from which, finally the angle $A_2E'K = \phi$ (mean anomaly) has to be computed.

s being the normal through E by which $\angle \Phi_1K\Phi_2 = \gamma$ is bisected, we have

$$\frac{\Phi_1K}{\Phi_1E} = \frac{\sin(180^\circ - v)}{\sin\frac{\gamma}{2}} = \frac{\sin v}{\sin\frac{\gamma}{2}} = \frac{\Phi_2K}{\Phi_2E}$$

or, calling $\Phi_1K = q$, $\Phi_2K = 2a - q$, $\Phi_1\Phi_2 = 2f$, $\Phi_1E = p$

$$\frac{p - 2f}{p} = \frac{q - 2a}{q},$$

whence

$$\frac{2f}{p} = \frac{2a}{q} \text{ or } \frac{q}{a} = \frac{p}{f}.$$

In other words, the perpendicular raised in E on A_1A_2 will meet $\Phi_1B = a$ in a point Q_1 such that $\Phi_1Q_1 = q$; it will similarly meet the line Φ_2B produced in Q_2, where $\Phi_2Q_2 = 2a - q$.

By introducing $\angle w = \Phi_1BF = \Phi_1Q_1E$, we can write the last relation

$$q = a - \frac{2\varepsilon}{\sin w}$$

and, correspondingly,

$$2a - q = a + \frac{2\varepsilon}{\sin w}.$$

From

$$q^2 + (2a - q)^2 - 2q\,.\,(2a - q)\,.\,\cos\gamma = (2f)^2$$

we then obtain by substitution

$$\cos\gamma = \frac{a^2\sin^2 w(1 - 2\sin^2 w) + 4\varepsilon}{a^2\sin^2 w - 4\varepsilon}$$

In the present case ($\cos w = \frac{19}{21}$, $w = 25^\circ.2087$, $a = 21$, $\varepsilon = 1$) the computation yields

$$\cos\gamma = 0{\cdot}7233593 \text{ and } \gamma = 43^\circ.6675.$$

Then from

$$\sin v = \frac{\sin\frac{\gamma}{2}}{\sin w}$$

we obtain $(v) = 60^\circ.8340$ and $v = 180^\circ - (v) = 119^\circ.1660$.

Finally, the formula $s = \dfrac{p\,.\,\sin\alpha}{\sin\frac{\gamma}{2}}$ yields (α being $= 97^\circ.3322$): $s = 18{\cdot}5189$.

By transforming the formula for the equation of the centre, $\sin(\phi - v) = \dfrac{\varepsilon\sin\phi}{s}$,

we obtain $\operatorname{ctg}\phi = \operatorname{ctg} v - \dfrac{1}{s\sin v}$,

which furnishes

$$\phi = 121^\circ.797$$

as the anomaly for which the "*elliptic*" geocentric distance, s_E, becomes a minimum.

According to the formula of the Ptolemaic curve (1), however, we obtain for this value of ϕ, $r = 19{\cdot}0283597$ and for the "*Ptolemaic*" geocentric distance $s_P = 18{\cdot}5210$, which is larger than the smallest value listed in our table (approximate minimum, $s_{121} = 18{\cdot}5176$).

Thus the minimum of s_E is a little larger than that of s_P and occurs at a value of $|\phi|$, which is about 1° larger than that corresponding to the minimum of s_P.

[76] *Libros del Saber de Astronomía del Rey D. Alfonso X de Castilla*, ed. M. RICO Y SINOBAS, 5 vols. in folio, Madrid, 1863–67.

[77] Vol. II, pp. 241–284. [78] Cf. G. SARTON, *Introduction to the History of Science*, Vol. II, Baltimore, 1931, p. 837, No. 14.

on a fixed circular eccentric (Ch. VII, p. 253), and one elaborate (Ch. XV, p. 263), saying twice that it is *grieue de fazer*, *i.e.* "hard to make"). This latter, as far as I can see from the obviously mutilated text, is mainly based on PTOLEMY's theory as described above and shows no feature of particular interest.

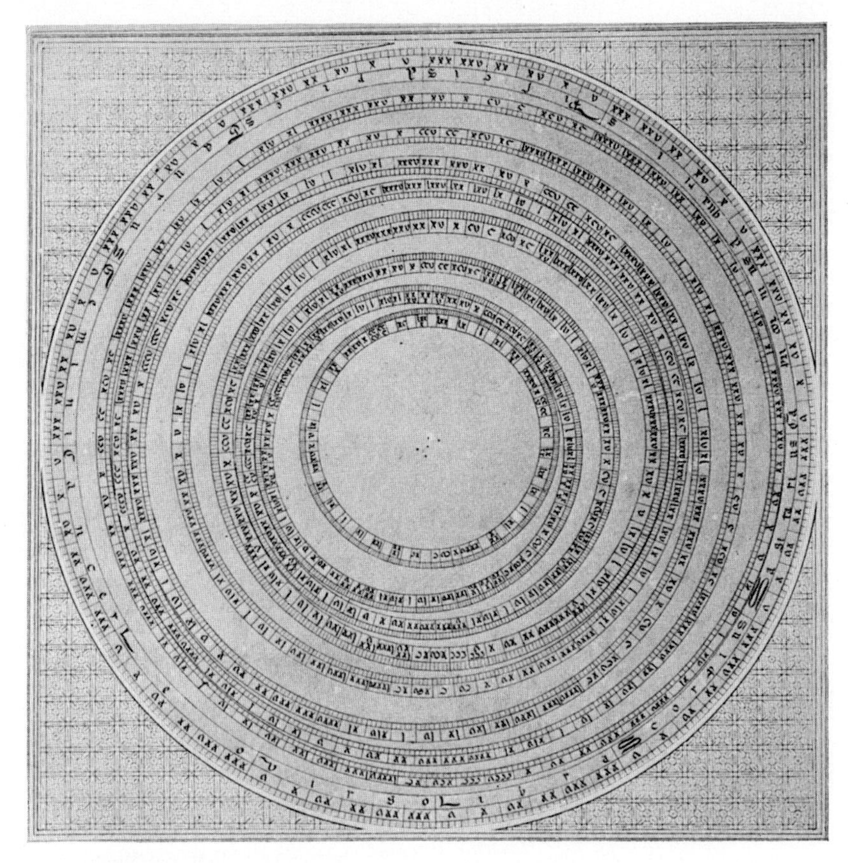

Fig. 13. *Libros del Saber*, vol. III, p. 281: the circles of Venus and of the superior planets

AZARQUIEL's treatise, however, as far as Mercury is concerned, is of first-rate importance. Unlike that of his predecessor, his device serves the purpose of representing the motions of all the seven planets on the two surfaces of one and the same disk, namely Venus and the superior planets on one (see Fig. 13), and the Sun, the Moon, and Mercury on the other (see Fig. 14). Unfortunately, the drawings are as incorrect as they are beautiful, and as the text accompanying them also bristles with errors it is often no easy task to make out the true meaning of the author. Yet from the wording of the chapter referring to the "Circles of Mercury" (Book II, Ch. IX, pp. 278–80), it is perfectly clear in what way Azarquiel wants those "circles" to be drawn.

It is, of course, the elliptiform curve with the axes 90 and 76 mm in the middle of Fig. 14 that evokes our curiosity. It has been paid attention to by earlier historians[79], but those I know of apparently failed to recognize its historical context and true

[79] R. WOLF, *Geschichte der Astronomie*, Munich, 1877, who mentions it on p. 207, adds the remark: ". . . und wenn es auch etwas gewagt erscheint, in dieser Ellipse, deren *Mittelpunkt* das Zeichen der Sonne zeigt, einen Vorläufer der Kepler'schen Ellipsen sehen zu wollen, so documentirt sie dagegen, wie Mädler richtig hervorhebt, 'dass man schon früh die Unmöglichkeit

significance[80]. The text of Chapter IX, however, which will be summarized here, leaves no doubt that it is nothing but the curve resulting from PTOLEMY's theory, which we discussed above in detail:

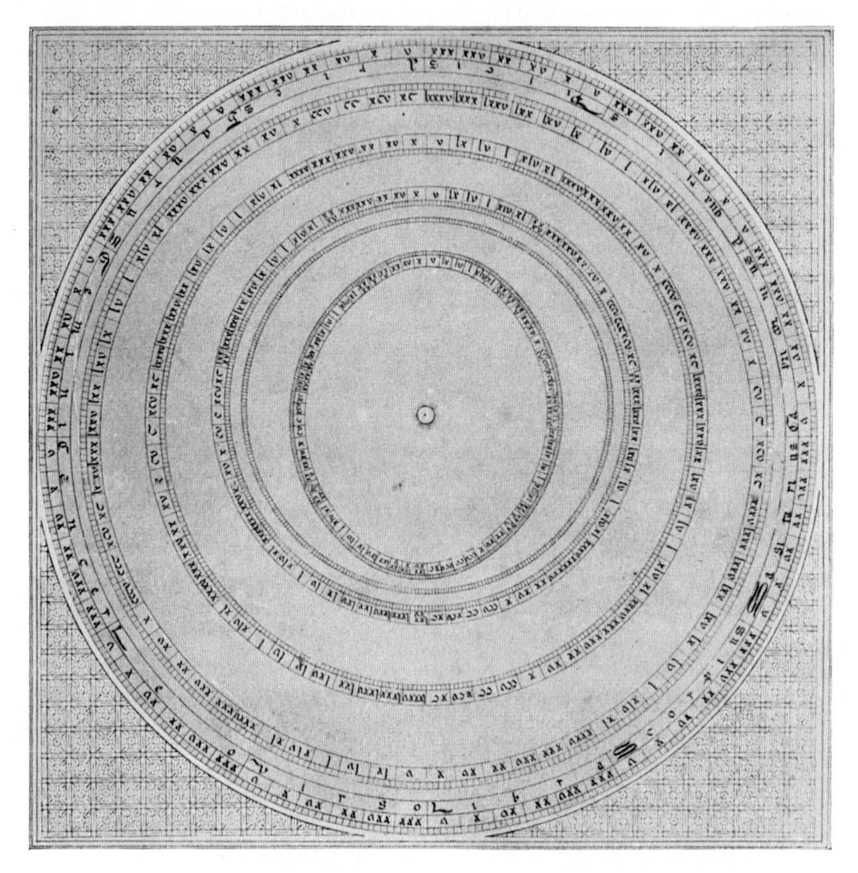

Fig. 14. *Libros del Saber*, vol. III, p. 282: the circles of Sun and Moon, and the elliptiform curve of Mercury

Mark a point (our point F, see Fig. 9) which is $4^p\ 42'$ distant from the centre of the disk (E) and make it the centre of a circle with the radius $2^p\ 21'$ (ε), and call this latter the "deferent of the centre of the deferent of Mercury" (*el çerco leuador*[81] *del centro del leuador de mercurio*). Then draw a "hidden" circle (*i.e.*

Continuation of footnote 79 from page 119)
eingesehen hat, mit dem excentrischen Kreise in allen Fällen auszureichen'." As demonstrated above, Mädler's "schon früh" has to be interpreted as "from the very beginning", because the elliptiform curve exists, at least by implication, already in Ptolemy.

SARTON, *Introduction* II, p. 838: "A figure of the deferent of Mercury in the form of an ellipse with what looks like a Sun in the centre in the second treatise on the plates of the seven planets (no. 14 in my list above, Madrid edition, vol. 3, 282, 1864) is purely accidental; it is not in any sense an anticipation of KEPLER'S discovery of the ellipticity of celestial orbits". JOSÉ M. MILLÁS VALLICROSA, *Estudios sobre historia de la ciencia española*, Barcelona, 1949, p. 137, footnote 48, refers to his treatment of the *Libros de las láminas de los siete planetas*, in Chapter X of his *Estudios sobre Azarquiel*, then in print. As I do not possess this work, I can only refer to G. SARTON'S review, in *Isis*, 42, p. 152f, where a discussion of the ellipse is not mentioned. No negative conclusion, however, can be drawn from this *argumentum e silentio*.

[80] *Note added in proof.* It was only after having concluded the present study that the author found a correct interpretation of the curve in question in ALFRED WEGENER, "Die astronomischen Werke Alfons X", in *Bibliotheca Mathematica*, 3. Folge, 6. Band, Leipzig, 1905, pp. 129–85. There, on pp. 156–61, WEGENER demonstrates its construction according to Ptolemy's theory (his Fig. 2, p. 157), though without entering upon a mathematical analysis and comparison with the ellipse. He then refers to ERASMUS REINHOLD'S commentary on PEURBACH (1542), URSTISIUS (WURSTEISEN, 1568), and RICCIOLI'S *Almagestum Novum* (1551). SARTON (*Introduction II*, p. 841) emphasized the importance of WEGENER'S study, yet did not mention its most important point (cf. the preceding footnote). WEGENER already points out the really amazing errors committed, and the fantastic theories put forward in this connection by earlier historians, particularly N. HERZ, *Geschichte der Bahnbestimmung von Planeten und Kometen*, Leipzig, 1887–94. [81] Old Spanish *levador* = "deferent", cf. modern Spanish *llevar* = "to carry".

which will later be erased) with radius 81ᵖ round F and divide it *clockwise* into seventy-two equal parts, starting from the apse-line. Moreover, draw another "hidden" circle with radius 80ᵖ round a point on the apse-line (E') which is 2ᵖ 21′ distant from E, divide it *counter-clockwise* into seventy-two parts, and call it the "circle of equal motion of Mercury" (*el çerco dell yguador del mouimiento de mercurio*). Then transfer the division of the large circle round F on the small concentric circle and, on the other hand, draw the radii (as "hidden lines") from E' to the seventy-two divisions (*cinculares*, *i.e.* sections comprising 5°) of the circle round E'. Thereafter, mark the point on the apse-line which is 49ᵖ 21′ distant from the point of intersection (M) of the small circle round F with the apse-line, and, placing subsequently the point of your compasses with a constant opening of 49ᵖ 21′, in the subdivisions of the small circle, mark the points of intersection with the corresponding radii through E'. Finally, join every three of the points thus marked, by an arc, and there will result a *figura pinnonada* (a curve similar to a pignon). *Et quando fizieres los çercos de mercurio assí cuemo te mostré aquí en este capítolo. salirtá so logar con ellos bien cierto. mas que dotra manera.* ("And when you have made the circles of Mercury as I have shown here in this chapter, its position will result from them very accurately, more so than in any other way".)

Thus the first explicit description of the curve of Mercury's true deferent, as well as its practical application, is undoubtedly Arabic; it precedes the first European mentioning by nearly 400 years. As IBN AL-SAMḤ, half a century before AZARQUIEL, does not yet mention it, we may moreover safely assume that AZARQUIEL himself is entitled to the honour of having discovered it.

The only difference between AZARQUIEL and PTOLEMY lies in the change of the numerical values of the parameters. Instead of PTOLEMY's $\varepsilon = 3^p$ and $\frac{R}{\varepsilon} = 20$, AZARQUIEL lets $\varepsilon = 2^p\ 21'$ and $\frac{R}{\varepsilon} = 21$. The question whether this was due to new observations cannot be decided upon off hand: (AL-BATTĀNĪ still has the Ptolemaic values). The radius of the epicycle, however, is not changed: AZARQUIEL makes it 18ᵖ 30′ for his major semi-axis of $(49^p\ 21' + 2^p\ 21') = 51^p\ 42'$, which corresponds almost exactly to PTOLEMY's 22ᵖ 30′ for a major semi-axis of $(60 + 3) = 63^p$.

Concerning the plate illustrating AZARQUIEL's text, it may be well to note that it was obviously not carried out in accordance with the author's prescription. The small circle in the middle (which looks like a Sun and, therefore, has deceived many interpreters) is nothing but the small circle with radius ε round F. The rays apparently emanating from it indicate its division into equal parts, but there are not seventy-two rays as prescribed, but only sixty. In the original, ε is 2·5 mm, whence the major semi-axis should be $22 \times 2{\cdot}5 = 55$ mm, and the minor, approximately $20 \times 2{\cdot}5 = 50$ mm, whereas we have stated above that they actually are 45 and 38 mm, respectively (the outer, not the inner, ring counts, hence the different values given by WOLF and MÄDLER: 82 and 67 mm for the *whole* axes). From $R + \varepsilon = 45$ and $R - \varepsilon = 38$ results $R = 41{\cdot}5$, $\varepsilon = 3{\cdot}5$, and $\frac{R}{\varepsilon} = 11{\cdot}85$, which is completely wrong and must yield false results. Thus, evidently, the curve was not constructed correctly with the aid of the small circle, but probably drawn at random, at the whim of the designer. For $\frac{R}{\varepsilon} \approx 12$, moreover, the curve would no longer have (practically) two axes of symmetry, as in the figure, but be noticeably narrower in its lower part; cf. the dotted curve corresponding to $\frac{R}{\varepsilon} = 10$ in Fig. 10.

Finally, the orientation of the apse-line of Mercury is as wrong as it can be, *viz.* 25° Pisces instead of ca. 24° Libra (IBN AL-SAMḤ's text, Book I, Ch. XIII, p. 262, has 23° 40′ Libra for the epoch 416 A.H. = 1025 A.D.).

Considering the general dependency of PEURBACH on Arabic astronomy (cf. below, Section 12), it is highly improbable that his statement concerning the oval shape of Mercury's deferent should have no connection with AZARQUIEL and the *Libros del Saber*. But it will not be easy to find out through what channels he got acquainted with the achievements of his Arabic predecessor[82].

There is, of course, no doubt that COPERNICUS as well as KEPLER were thoroughly conversant with the contents of PEURBACH's treatise, and perfectly aware of the oval shape of Mercury's deferent. But in spite of this, COPERNICUS did not think of introducing ovals or ellipses into his heliocentric system[83]. And, certainly there appears to be no natural or obvious way leading from the oval (or elliptiform) geocentric deferent to an elliptic heliocentric orbit. As for KEPLER, it is not impossible that the first idea of using ellipses may have come to his mind through PEURBACH's treatise. As a matter of fact, his first attempt, after having proved the impossibility of a circular orbit, was to introduce an oval curve which was wider near the apogee, and narrower near the perigee. Yet the independence of his procedure is so striking that even the definite proof of one or the other impulse received from others would not impair the greatness of his discovery[84].

Another question, however, is the later development of instruments similar to those of AZARQUIEL. It would be worth while investigating whether the planispheres of PETER APIANUS[85], which he constructed for the same purpose (KEPLER called his efforts an "industria miserabilis"), show a definite dependence on the ones described in the *Libros del Saber*. In particular, it would be interesting to know if he, too, makes use of an ellipse in connection with Mercury. An answer to both questions can easily be found in his *Astronomicum Caesareum* of 1540. I should expect it to be to the affirmative.

11. MATHEMATICAL THEORY VERSUS PHYSICAL REALITY. THE NATURE OF THE SPHERES ACCORDING TO ALHAZEN

In the early fourth century B.C., EUDOXUS of Cnidos devised his theory of homocentric spheres which, to the best of our knowledge, is the first attempt made in

[82] Here reference is also made to the experiences made in an early Islamic zodiacal interpretation: ARTHUR BEER, "The Astronomical Significance of the Zodiac of Qusayr'Amra", in K. A. C. CRESWELL's *Early Muslim Architecture*, Vol. I, pp. 296–303, Clarendon Press, Oxford, 1932. (See also F. SAXL, *l.c.*, pp. 289–295.)

[83] As for COPERNICUS, the famous passage about the ellipse cancelled by the author's own hand (*De Revolutionibus*, Book III, Ch. 4, original Ms., fol. 75r, German translation by C. L. MENZZER, Thorn, 1879, p. 139) is not, as claimed time and again (for instance, by J. HOPMANN, in his Preface to the anastatic reprint of MENZZER's translation, Leipzig, 1939, p. VII) an anticipation of KEPLER's discovery, but only a mathematical triviality. Those wishing to maintain the myth are recommended to read the whole passage, and not only the five or ten lines in question. It would help them to understand that COPERNICUS does not refer to planetary motion at all. In V.4 (ed. MENZZER, p. 272) a deviation from the perfect circle is mentioned in connection with the planets, but it is not said that the resultant curve is an ellipse. KEPLER discusses this chapter extensively in *De motibus stellae Martis*, I.4, q.v.

[84] E. ZINNER, *Geschichte und Bibliographie der astronomischen Literatur in Deutschland zur Zeit der Renaissance*, Leipzig, 1941, p. 33, states that ERASMUS REINHOLD (1511–43), "the best theoretician of his time, by referring to the oval orbit of the centres of the Moon's and Mercury's epicycles, initiated or adumbrated ('hat angebahnt') future considerations". ZINNER does not mention the passage found in PEURBACH's treatise. It is, of course, the source of REINHOLD's statement, as he wrote himself a commentary on it (published in 1542).

[85] See R. WOLF, *op. cit.*, p. 265.

Greek antiquity at subjecting the complicated motions of the planets to mathematical treatment. From the reports on which we have to rely it seems to result that EUDOXUS himself did not bother about the physical reality of his construction (to each planet he attributes a series of spheres, one revolving inside the other around

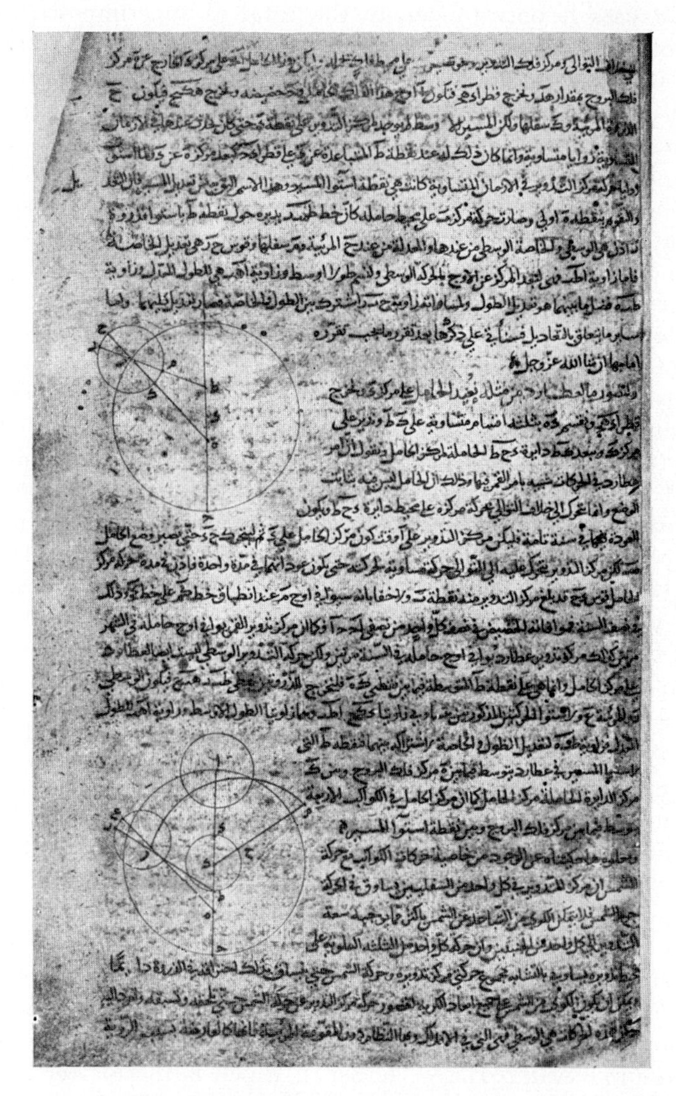

Fig. 15. A page from AL-BĪRŪNĪ's *Mas'ūdic Canon*: Ms. Berlin Qu. 1613, Fol. 190r. Arabic text with two inserted sketches; the upper showing the motion of Venus and the superior planets, and the lower that of Mercury

different axes, the planet itself being fixed to the equator of the innermost sphere, while the outer sphere, in every case, was thought to have the same rotation as that of the fixed stars).

It was ARISTOTLE who tried to raise this theory of homocentric spheres to the rank of a system based on physical reality, by inserting sets of reacting spheres between the single, and originally independent, planetary systems so as to reduce the complex motion of the innermost sphere of one planet to the simple motion of the outermost sphere of the next one.

In the *Almagest* the question of physical reality is not discussed. Although in Greek the word *σφαῖρα* may refer to the solid sphere as well as the circle, PTOLEMY in his *Almagest* uses it only to denote the total of superimposed circles (*κύκλος*, *ἐπίκυκλος*) representing a planet's orbit, and thus the orbit itself[86]. In his *Hypotheseis*, however, he takes a different attitude; there, obviously, the term *sphere* has to be understood in our modern sense of the word, *i.e.* as referring to solid spheres or spherical shells.

During the Islamic Middle Ages we encounter both interpretations, varying according to the greater interest of the authors in mathematical and astronomical, or in physical and philosophical matters. Thus while astronomers such as AL-BATTĀNĪ and AL-BĪRŪNĪ[87] show but little interest in the physical reality of their system, the question plays a predominant part in the writings of physicists, among whom I mention first and foremost the great ALHAZEN (AL-ḤASAN IBN AL-ḤUSAIN IBN AL-HAITHAM of Basra, c. 965–1039); in the late Middle Ages, probably because of the enormous reputation of ALHAZEN's writings, the physical interpretation obtained a footing also in certain treatises of a purely astronomical character. We find it not only in QAZWĪNĪ's famous *Cosmography*[88] but also in an astronomical treatise by AL-JAGHMĪNĪ[89]. The latter seems to have been highly esteemed in Islamic countries. There exist numerous manuscripts of his treatise; it was translated into Persian and commented upon by the Persian astronomer and philosopher 'ALĪ IBN MUḤAMMAD AL-JURJĀNĪ (1340–1413). The dependency of early Renaissance astronomers on ALHAZEN and AL-JAGHMĪNĪ is beyond doubt. Yet I am unable to tell at the moment from which of the two (possibly from both), and through which channels, they drew their information.

The theory set forth by ALHAZEN in his treatise *Fī hai'at al-'ālam* ("On the Shape of the Universe")[90] is in its main outlines as follows:

> The universe, which is spherical in shape, consists of nine spherical shells (*aflāk*, plural of *falak* = *σφαῖρα*)[91] gliding inside one another; each of these shells again is composed of a set of concentric or eccentric shells or complete spheres. *There is no empty space or "void"*. The Earth with its water is surrounded by air, which again is surrounded by fire. The sphere of fire is limited by that of the Moon, after which come the spheres of the six other planets and of the fixed stars. Finally, the outer limit of the universe is the "sphere of spheres" (Caelum Empyreum). While the four elements of which consist the earthy, watery, airy, and fiery spheres are either heavy or light, the fifth element (Arabic *aithīr* = ARISTOTLE's "aether"), which is the matter filling the translunar world, is neither the one nor the other. Unlike the four terrestrial elements it has the essential quality of eternal circular motion.

---

[86] Thus, in III.3 (ed. HEIBERG, I, p. 216, 1–12) he speaks about the *θέσεις καὶ τάξεις τῶν ἐν ταῖς σφαίραις αὐτῶν κύκλων* ("the situations and positions of the circles lying in their spheres").

[87] For AL-BĪRŪNĪ's (973–1048) purely mathematical treatment, see Fig. 15, showing a page from one of the earliest manuscripts known of his *Mas'ūdic Canon* (AL-QĀNŪN AL-MAS'ŪDĪ), written less than a century after the author's death.

[88] ABŪ YAḤYĀ ZAKARĪYĀ IBN MUḤAMMAD IBN MAḤMŪD AL-QAZWĪNĪ, born 1203/04, d. 1283. Arabic editions of his *Cosmography* by F. WÜSTENFELD (2 vols.), Göttingen, 1848–49; German translation of the first part of Vol. I (containing the astronomical part), by H. ETHÉ, *Die Wunder der Schöpfung*, Leipzig, 1868.

[89] MAḤMŪD IBN MUḤAMMAD IBN 'UMAR AL-JAGHMĪNĪ (d. 1344/45), see G. RUDLOFF and A. HOCHHEIM, "Die Astronomie des Ǵagmînî", in *Zeitschrift der Deutschen Morgenländischen Gesellschaft*, Vol. 47 (Leipzig, 1893), pp. 213–275.

[90] See K. KOHL, "Über den Aufbau der Welt nach Ibn al Haiṯam", in *Sitzungsberichte d. Physik. Med. Sozietät in Erlangen*, Vol. 54/55 (1922/23), Erlangen, 1925 pp. 140–179.

[91] The ambiguity of the term *falak*, which is exactly the same as that of *σφαῖρα*, is expressly discussed in the beginning of the 3rd part. In ALHAZEN's practical application, however, it nearly always denotes the globe or the spherical shell, not the circle.

(*a*) In the first place, following ALHAZEN'S example, I shall describe the system of the Sun, for the sole reason that it is simpler than the others (see Fig. 16[92]). The sphere of the Sun is a *corporeal* (material and perfectly transparent) spherical shell concentric with the centre of the universe (the Earth). It surrounds the sphere of Venus and is itself surrounded by that of Mars. In this spherical shell, called *al-falak al-mumaththal* = "assimilated (*i.e.* concentric or parecliptic) sphere", another such

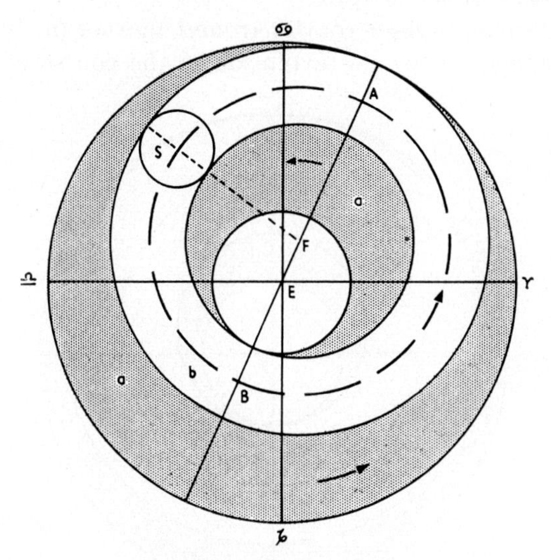

Fig. 16. The spheres of the Sun

E = Earth
A = apogee of Sun's centre
B = perigee of Sun's centre
F = centre of eccentric sphere
a = concentric sphere
b = eccentric sphere
S = Sun

shell is embedded eccentrically, in such a way that its interior surface touches from outside the interior surface of the first (concentric) shell, while its exterior surface touches from inside the exterior of the concentric one. This second shell, called *al-falak al-khārij al-markaz* = "eccentric sphere" (which in this case is identical with *al-falak al-ḥāmil* = "bearing sphere" or "deferent") rotates from west to east round an axis through the poles of the ecliptic, carrying along the Sun, which is a solid globe fitting exactly between the two surfaces of the eccentric shell. The apogee (Arabic *awj*, whence mediaeval Latin *aux*, gen. *augis*, as consistently used by PEURBACH[93]) precedes the point of summer solstice by $24\frac{1}{2}°$ (*i.e.* has a longitude of $65\frac{1}{2}°$)[94].

(*b*) Before entering upon the discussion of the problem of the planet which has our particular interest, it will be useful to review the less complicated system devised by ALHAZEN to describe the motion of Venus and the superior planets.

The sphere of Venus is embedded between those of Mercury and the Sun. It consists of a spherical shell concentric with the Earth and an eccentric shell analogous

[92] After KOHL, *l.c.*, Fig. 2 (p. 154).
[93] The *Libros del Saber*, too, employ the terms *auxe* or *alaux*.
[94] ALHAZEN here gives the same value as PTOLEMY, *Almagest*, III.4 (Vol. I, p. 167), although AL-BATTĀNĪ some time before had discovered the displacement of the apogee. Yet in another place he refers to the new discovery.

to that of the Sun (see Fig. 17). Between the two surfaces of the latter a rotating (solid) sphere (the sphere of the epicycle, *falak al-tadwīr*) is embedded, in whose equator the spherical body of Venus is fixed.

The concentric sphere has a slow rotation (in the direct sense) of 1° in 100 years (PTOLEMY's value of the precession of the equinoxes; *i.e.* the apse-line, in the case of all planets, is considered immovable relative to the fixed stars[95]); the rotation takes place in the plane of the ecliptic.

The eccentric, or bearing, sphere rotates around another (inclined) axis, likewise in the direct sense, making one whole revolution in the course of one tropical year[96].

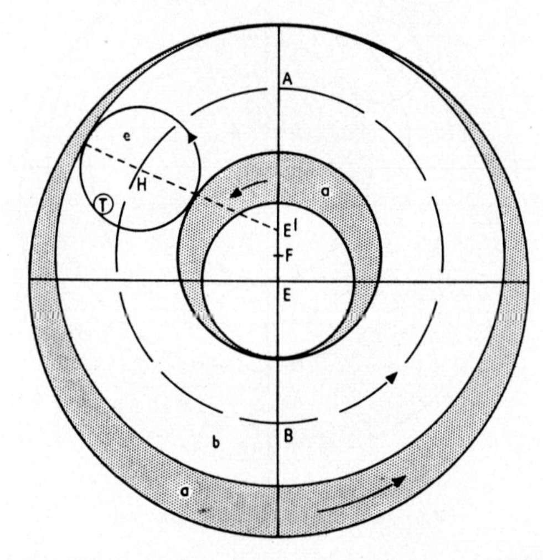

Fig. 17. The spheres of Venus and superior planets

E = Earth	a = concentric sphere
A = apogee of centre of epicycle	b = eccentric sphere
B = perigee of centre of epicycle	e = sphere of epicycle
F = centre of eccentric sphere	H = centre of epicycle
E' = *punctum aequans*	T = planet

The rotation is uniform when seen from the *punctum aequans*, E', which lies on the apse-line beyond the centre of the eccentric, its distance being twice that of the latter from the Earth.

The sphere of the epicycle, again, turns in the same sense around an axis that is inclined to that of the concentric as well as to that of the eccentric. Its period is that of the planet's synodic revolution.

(*c*) After this, there will be no difficulty in understanding the system of Mercury.

Again, the sphere of Mercury (Figs. 18 and 19) is embedded between those of the Moon and Venus. In perfect analogy with the preceding, the concentric sphere moves (directly) 1° in 100 years, in the plane of the ecliptic. A first eccentric, called *al-falak al-mudīr*, "the turning sphere", is embedded in the concentric, while a second eccentric, the deferent proper (*al-falak al-ḥāmil*), is embedded in the first. For the centres of these three spherical shells I refer to our Figs. 9 and 10. In the initial position (centre of the epicycle in the apogee, see Fig. 18) the three centres, E, F

[95] Cf. above, Section 9 (*a*).
[96] True also in the case of Mercury. For the other planets, the time of one complete rotation of the deferent is equal to the planet's tropical revolution.

and M lie on the line which joins the apogee with the Earth, called by us "apse-line"[97]. For other positions of the centre of the epicycle (see Fig. 19), the centre of the deferent (P in Figs. 9 and 10) lies on the circle with radius ε round F. Again E', situated, in the case of Mercury, in the middle between E and F, marks the *punctum aequans* from which the revolution of the centre of the epicycle appears uniform.

The sphere of the epicycle carrying the planet itself in its equator is fitted between the two surfaces of the deferent.

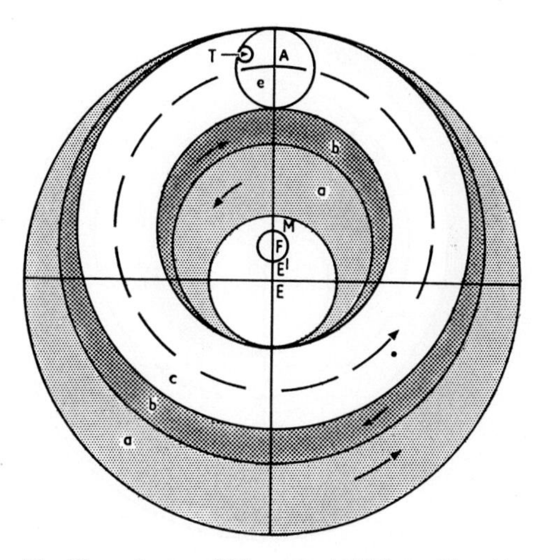

Fig. 18. The spheres of Mercury, initial position ($\phi = 0$)

E = Earth
E' = *punctum aequans*
F = centre of turning sphere (Mudīr, b)
M = centre of deferent (Ḥāmil, c)
A = apogee of centre of epicycle
a = concentric sphere
b = first eccentric sphere or turning sphere (Mudīr)
c = second eccentric sphere or deferent (Ḥāmil)
e = sphere of epicycle
T = planet

The turning sphere rotates retrogradely around an axis which is inclined to that of the ecliptic; the period of revolution is one tropical year. The deferent rotates in the direct sense around an axis parallel to that of the turning sphere; its period is also the tropical year.

The sphere of the epicycle rotates around an axis which is inclined to that of the ecliptic as well as the parallel axes of the turning sphere and the deferent[98]. Its period of revolution is that of the synodic revolution of Mercury.

12. PEURBACH AND MARCANTONIO MICHIEL

As a last link in the chain leading from PTOLEMY through the Arabs to the horoscope of MARCANTONIO MICHIEL, I shall now briefly discuss the theory of Mercury according to PEURBACH'S *Theoricae Novae Planetarum*[99]. This work, which is indeed far superior to SACROBOSCO'S exceedingly poor composition[100], has exercised a considerable impact on later Renaissance astronomy (REGIOMONTANUS, also COPERNICUS, and ERASMUS REINHOLD, see footnote 84, page 122). The statement of its superiority to SACROBOSCO,

[97] On account of the occurrence of two perigees, there is strictly speaking, no line of apsides as in the other cases. Cf. footnote 62, page 107.

[98] Here, too, I refrain from discussing the theory of latitudes. It has obviously escaped the attention of PTOLEMY and his disciples that the inclination of the epicycle (equator of the sphere of the epicycle, according to ALHAZEN) to the plane of the ecliptic must be regarded as constant. Cf. K. KOHL, *l.c.*, p. 167, footnote 40.

[99] Cf. footnote 66, page 109.

[100] SACROBOSCO'S *Sphaera Mundi* does not deal with the theory of planets at all.

however, has only a relative import. Anyone familiar with the history of Islamic astronomy will recognize at once that there is very little in PEURBACH that is not borrowed or directly copied from the Arabic masters, and nothing at all that would entitle us to speak of an independence and freedom from prejudice, such as is considered characteristic of the spirit of the Renaissance.

I shall begin by giving the Latin original of PEURBACH'S relevant theory; the illustrations will be discussed below.

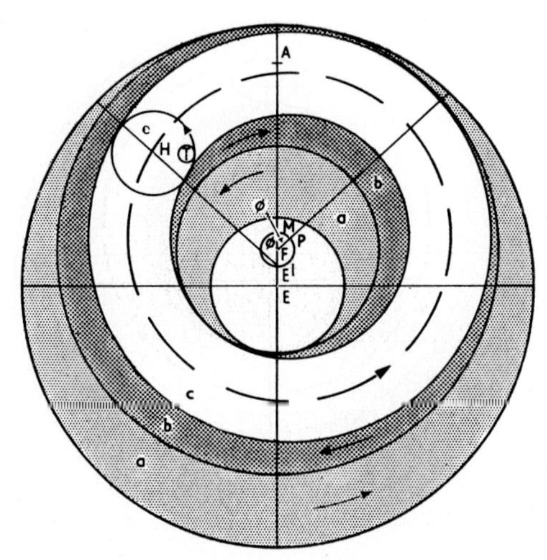

Fig. 19. The spheres of Mercury ($\phi \neq 0$)
P = centre of deferent (Ḥāmil, c)

DE MERCURIO

(GEORGII PURBACHII, *Theoricae Novae Planetarum*, Nuremberg, 1472 or 73, fol. 9r.)

Mercurius habet orbes quinque et epicyclum. Quorum extremi duo sunt eccentrici secundum quidem. Superficies nanque convexa supremi et concava infimi mundo concentricae sunt. Concava autem supremi et convexa infimi eccentricae mundo: sibi ipsis tamen concentricae. Et centrum earum tantum a centro aequantis quantum centrum aequantis a centro mundi distat. Et ipsum est centrum parvi circuli quem centrum deferentis ut videbitur describit. Vocantur autem deferentes augem aeqantis. Et moventur ad motum octavae sphaerae super axe zodiaci. Inter hos extremos sunt alii duo similiter difformis spissitudinis intra se quintum orbem scilicet epicyclum deferentem locantes. Superficies nanque convexa superioris et concava inferioris idem cum parvo circulo centrum habent. Sed concava superioris et convexa inferioris una cum utrisque superficiebus quinti orbis aliud centrum habent mobile: quod centrum deferentis dicitur. Hi duo orbes augem eccentrici deferentes vocantur. Et moventur regulariter super centrum parvi circuli contra successionem signorum tali velocitate ut praecise in tempore quo linea medii motus Solis unam facit revolutionem et orbes isti in partem oppositam similiter unam perficiant.

(fol. 9v.) Et fit motus iste super axe quandoque aequidistante axi zodiaci et per centrum parvi circuli transeunte. Motum autem horum orbium sequitur ut centrum orbis deferentis epicyclum circumferentiam quandam parvi circuli similiter in tanto tempore regulariter describat. Huius vero semidiameter est tanta quanta est distantia qua centrum aequantis a centro mundi distat. Unde haec circumferentia per centrum aequantis ibit. Sed orbis quintus epicyclum deferens intra duos secundos locatus movetur in longitudinem secundum successionem signorum centrum epicycli deferendo regulariter super centro aequantis. Quod quidem in medio est inter centrum mundi et centrum parvi circuli. Hanc tamen habet velocitatem ut centrum epicycli in eo tempore semel revolvatur in quo linea medii motus Solis unam complet revolutionem. Habet se nanque Mercurius in hoc ad Solem ut Venus. Fit enim semper ut medius motus Solis sit etiam medius motus horum duorum.

Ex his igitur et dictis superius manifestum est singulos sex planetas in motibus eorum aliquid cum Sole communicare: motumque illius quasi quoddam commune speculum et mensurae regulam esse motibus illorum. Huius autem orbis epicyclum deferentis motus fit super axe imaginario cuius extremitates sicut aparuit in Venere (fol. 10r) propter motum alium quem habet in latitudinem similiter accedunt ad polos zodiaci et ab eis recedunt. Axis tamen iste secundum se totum mobilis est secundum motum centri deferentis in circulo parvo. Patet itaque sicut in Luna centrum epicycli bis in mense lunari deferentes augem eccentrici pertransit: ita in Mercurio centrum epicycli bis in anno deferentes augem epicyclum deferentis peragrare. Non tamen est in auge deferentis nisi semel. Aux enim deferentis Mercurii non circulariter movetur circulares revolutiones complendo sicut in Luna contingit. Sed propter motum centri deferentis in parvo circulo nunc secundum successionem signorum nunc contra procedit. Habet nanque limites certos quos egredi ab auge aequantis recedendo non valet: sed continue sub arcu zodiaci a duabus lineis circulum parvum contingentibus a centro mundi ad zodiacum ductis comprehenso: ascendendo et descendendo volvitur atque revolvitur. Quotienscumque enim centrum epicycli fuerit in auge deferentis ipsum etiam motuum similitudine erit in auge aequantis et centrum deferentis in auge sui parvi circuli. Quare tunc centrum epicycli in maxima remotione a centro mundi fiet: et centrum deferentis in duplo plus distabit a centro aequantis quam centrum aequantis a centro mundi.

(fol. 10v.) Deinde vero cum centrum deferentis per motum orbium duorum secundorum movebitur ab auge sui circuli versus occidentem: centrum epicycli per motum deferentis movebitur ab auge aequantis tantundem versus orientem. Unde centrum deferentis ad centrum mundi incipit accedere. Et aux deferentis ab auge aequantis versus occidentem recedit continue donec centrum deferentis fuerit in linea contingente circulum occidentali. Id autem fit cum ab auge parvi circuli quatuor signis distiterit. Et tunc similiter centrum epicycli ab auge aequantis versus orientem distabit quatuor signis. Aux autem deferentis erit in maxima sua ab aequantis auge versus occidentem remotione. Atque in hoc situ centrum epicycli fiet in maxima sua quam solet habere ad centrum mundi accessione. Non tamen tunc erit in opposito augis deferentis: nec in linea ad parvum circulum contingenter per centrum mundi producta. Post enim descendente centro deferentis versus centrum aequantis aux deferentis incipit reaccedere versus augem aequantis: centrum autem epicycli proportionaliter descendet in altera medietate versus oppositum augis aequantis. Unde magis removebitur a centro mundi: nec perveniet ad oppositum augis deferentis nisi cum ipsum fuerit in opposito augis aequantis. Id autem fiet cum centrum deferentis perveniet in centrum aequantis. Et tunc aux deferentis erit etiam cum auge aequantis. Et tam deferens quam aequans ex quo aequales in quantitate constituuntur: erunt circulus unus. Et plus distabit a centro mundi centrum epicycli tunc quam distabat cum erat in situ ab auge aequantis per signa quatuor. Hinc autem cum centrum deferentis recedet a centro aequantis in suo circulo ascendendo centrum epicycli recedet ab opposito augis aequantis et deferentis et continue magis centro mundi propinquabit. Sed aux deferentis removebitur ab auge aequantis versus orientem continue donec perveniet centrum deferentis ad lineam contingentem circulum parvum a parte orientis. Qui punctus contactus etiam ab auge parvi circuli versus orientem quatuor signis distat. Tunc enim aux deferentis fiet in maxima remotione ab aequantis auge versus orientem. Et centrum epicycli iterum erit in maxima eius ad terram accessione quam habere solet: non tamen erit in opposito augis deferentis. Ab hoc vero loco ascendente centro deferentis versus augem parvi circuli aux deferentis continue revertetur ad augem aequantis. Et centrum epicycli magis elongabitur a centro mundi versus augem aequantis ascendendo usque dum centrum deferentis ad augem parvi circuli perveniet. Nam tunc aux deferentis erit cum auge aequantis: et centrum epicycli similiter tam in auge deferentis quam aequantis. Unde iterum erit in maxima remotione a centro mundi sicut primo. Rursusque deinde similis ut iam dicta est mutatio redibit.

Ex his primo videtur in anno tantum semel centrum deferentis esse idem cum centro aequantis. Alias autem semper deferentis centrum a centro mundi distantius esse quam aequantis centrum. Quare sequitur contrarium ei quod in superioribus et Venere accidit: ut scilicet quanto centrum epicycli vicinius augi aequantis fuerit tanto velocius: et quanto vicinius eius opposito tanto tardius moveatur. Secundo licet centrum epicycli tantum semel in maxima remotione fuerit in anno a centro mundi: bis tamen in maxima propinquatione quam habere solet ipsum esse contingit. Similiter quamquam bis in anno sit in maxima accessione: tamen tantum semel in anno in opposito augis deferentis reperitur. Tercio necesse est ut oppositum augis deferentis centro epicycli extra augem aequantis aut oppositum eius existente (fol. 11r) inter centrum epicycli et oppositum augis aequantis semper versetur: aliquando quidem versus centrum epicycli aliquando ab eo tam praecedendo quam sequendo sese devolvens. Quarto sicut aux deferentis ad certos limites utrinque ab auge aequantis removetur ita etiam se habet oppositum augis deferentis respectu oppositi augis aequantis. Maior tamen est arcus huiusmodi motus augis deferentis quam arcus

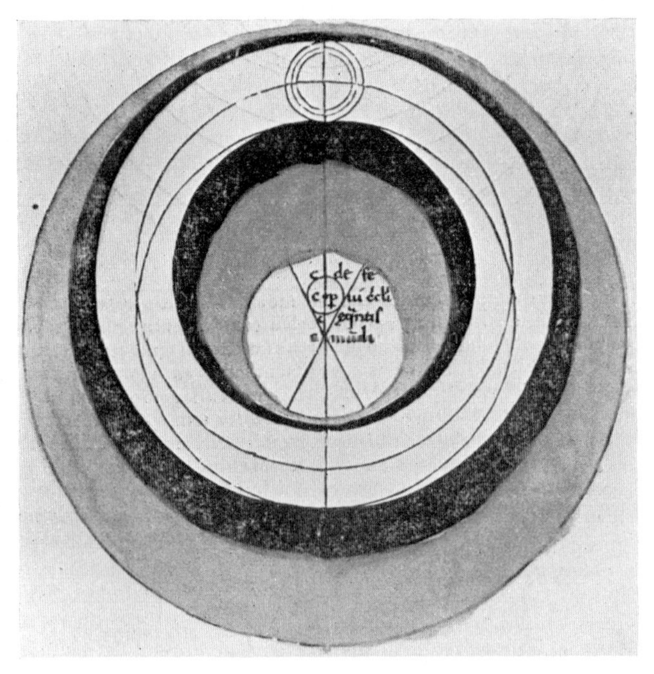

Fig. 20. *Peurbach Theoricae*, Fol. 9r: the circles of Mercury

motus oppositi eius. Unde motus unius motu alterius velocior erit. Quinto etsi centrum epicycli contingat esse in puncto deferentis a centro mundi remotissimo nunquam tamen est in puncto deferentis quem centro mundi vicinissimum esse contingit. Nam dum centrum epicycli fuerit in auge deferentis talis est habitudo deferentis ut oppositum augis eius sit centro mundi ita vicinum quod in quacunque alia deferentis quam habet habitudine nullus punctus eius vicinior aut tam vicinus centro mundi reperiatur. In tali autem puncto quem vicinissimum esse contingit: centrum epicycli non est eo tempore quo propinquissimum eum esse contingit: sed in eius opposito. Sexto ex dictis apparet manifeste centrum epicycli Mercurii propter motus supra dictos non ut in aliis planetis fit: circumferentiam deferentis circularem sed potius figurae habentis similitudinem (fol. 11v) cum plana ovali periferiam describere. Epicyclus vero in longitudinem movetur sicut epicyclus Veneris. Revolutionem tamen unam in quatuor mensibus solaribus fere super centro suo perficit. Termini autem tabularum hic sicut in superioribus declarantur: nisi quod diversitas in minutis proporcionalibus aliqualis existit. Aequationes enim argumentorum Mercurii quae in tabulis scribuntur sunt quae contingunt dum centrum epicycli fuerit in mediocri eius a terra remotione. Haec autem accidit centro epicycli ab auge aequantis per duo signa quatuor gradus et XXX minuta distante. Sed in aliis planetis centro epicycli in longitudinem media deferentis existente fiebat. Item minima centri epicycli Mercurii a centro mundi remotio fit dum centrum epicycli ab auge aequantis eius quatuor signis distiterit. Haec autem in aliis centro epicycli in opposito augis aequantis existente contingebat.

Minuta igitur proporcionalia longiora sunt excessus remotionis centri epicycli maximae super mediocrem eius remotionem in sexaginta partes aequales divisus. Sed minuta proporcionalia propiora dicuntur excessus remotionis centri epicycli mediocris super remotionem eius minimam similiter in LX particulas aequales divisus. Et secundum hoc duplex diversitas diametri diffiniatur. Quia tamen a loco maximae accessionis centri epicycli versus oppositum augis aequantis minuta proporcionalia propiora minuuntur quae prius a loco mediocris remotionis usque ad locum maximae accessionis continue augebantur: ideo dicitur in Mercurio minuta proporcionalia tripliciter se habere: quae tamen in Venere atque tribus superioribus dupliciter: in Luna vero simpliciter ut manifeste patuit: se habere solent.

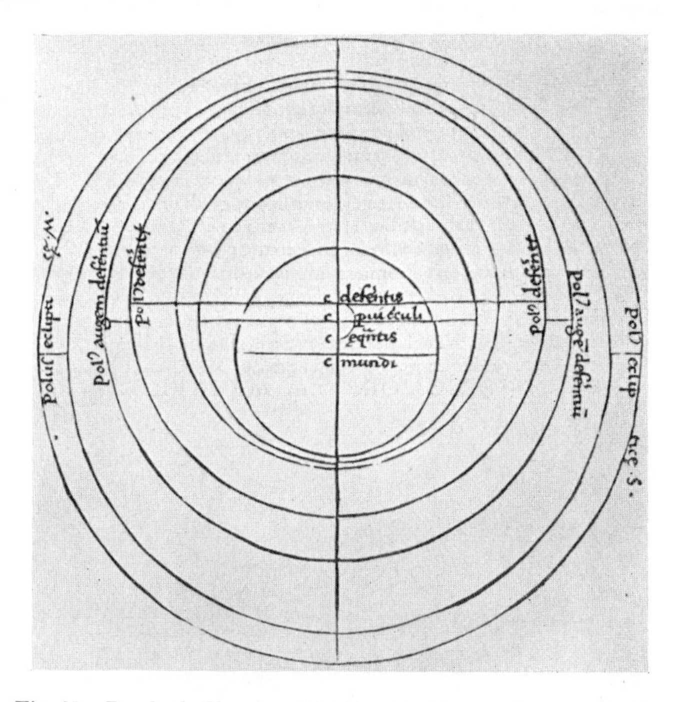

Fig. 21. *Peurbach Theoricae*, Fol. 9v: the theory of axes and poles

The first of PEURBACH'S figures (Fig. 20) is identical with our Fig. 17. The second (Fig. 21), an interesting supplementary representation of ALHAZEN'S theory, is a cross-section laid through the poles of the ecliptic and the apogee of the centre of the epicycle. Contrary to ALHAZEN (see above, under Section 11 (*c*), and footnote 98, page 127), PEURBACH recognizes the necessity of making the turning sphere revolve in the plane of the ecliptic, and of taking care of the latitudes only by attributing an appropriate inclination to the plane of Mercury's epicycle. Thus by introducing the conception of absolute direction in space, which was not yet sufficiently developed in the Middle Ages, the nodes of Mercury's orbit are defined as the points of intersection of the plane of its epicycle with that of the ecliptic. This evidently involves a considerable simplification of the numerical treatment of the problem. The plate, which shows the north pole of the ecliptic to the right, illustrates the moment when the axis of the bearing sphere (deferent), being in its apogee position, lies in the same plane as that of the ecliptic and of the first eccentric. Subsequently, it will describe the surface of a circular half-cylinder with radius ε above the plane of projection, until after one-half synodic revolution, when passing through the *punctum aequans*, it again falls in the plane of projection.

The third plate (Fig. 22) illustrates the motion of the apogee of the *deferent* (*aux deferentis*)[101], and similarly that of its perigee (called by PEURBACH "*oppositum augis deferentis*"), with regard to the fixed apse-line. According to PEURBACH's figure both seem to describe discontinuous lune-shaped curves fitted between two circles, one of which corresponds to the deferent in its apogee position, the other to the deferent when its centre coincides with the *punctum aequans*. Starting from $\phi = 0°$ and $\phi = 180°$, the two points in question will first move retrogradely on the outer halves of the lunes until they reach the tangent drawn from the Earth on the small

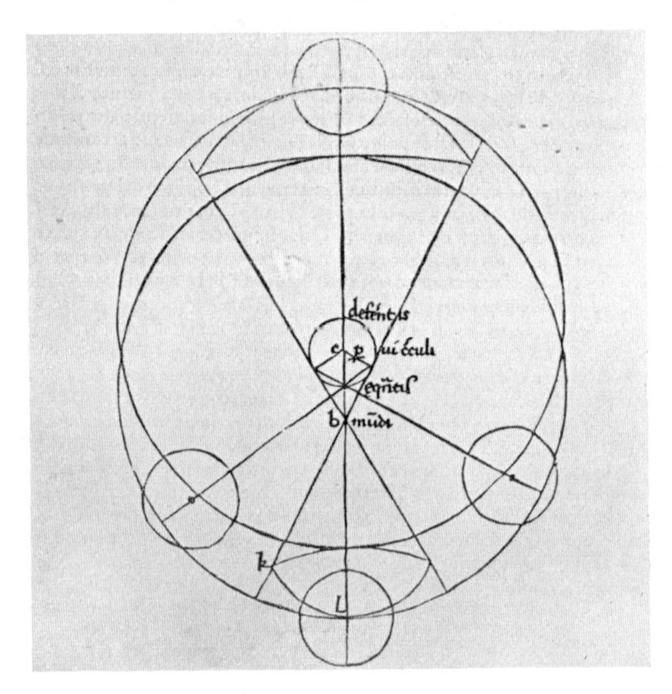

Fig. 22. *Peurbach Theoricae*, Fol. 10r: the motion of the apogee of the deferent

circle with radius ε. Then they will move directly on the inner halves till they coincide with the other tangent; whereafter they again move retrogradely on the outer halves (period 1 tropical year). Fig. 23 shows that in truth the curves are not discontinuous. They have, of course, only one tangent also in the points of maximum elongation from the apse-line, and this tangent $Q_1P_1EQ_1'$ coincides with the one drawn from the Earth on the small circle with radius ε[102]. Although the conception of continuity of curves is alien to fifteenth century mathematics, it is hardly understandable that PEURBACH would content himself with such an incorrect and misleading figure. A discussion of the mathematical properties of these interesting curves is beyond our scope.

The fourth plate (Fig. 24) shows the "oval" curve described by the centre of the epicycle, which we have sufficiently discussed in the preceding chapters. Again PEURBACH's drawing evokes the erroneous impression of discontinuity for the points corresponding to $\phi = 0°$ and $\phi = 180°$. In fact, as can easily be seen from our

[101] PEURBACH seems to have been the first to pay attention to the curve described by the apogee and the perigee of the deferent.

[102] The two tangents drawn from the Earth ("*Centrum Mundi*") on the circle with radius ε are shown also in Fig. 20.

demonstrations, the curve has only one tangent, which is perpendicular to the apse-line, in each of the two points.

I have mentioned that PEURBACH's *Theoricae*, composed in 1460, were edited and printed for the first time by REGIOMONTANUS about 1473. They enjoyed an enormous reputation and popularity. At least four reprints were made during the fifty subsequent years: Frankfurt an der Oder, K. Baumgardt, 1507; Basel, 1509; Vienna, Johann Singriener, 1518 (together with SACROBOSCO's *Sphaera Mundi*); Basel, 1523 (again together with SACROBOSCO)[103]. There is no doubt that it was regarded as one

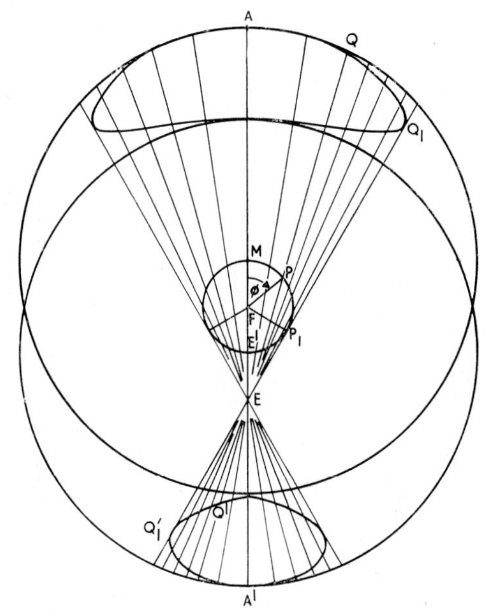

Fig. 23. The curves described by the apogee and perigee of the deferent

of the most important works on planetary astronomy in learned circles of all European countries.

Thus we may safely expect also that the interior of MARCANTONIO's horoscope, which originated five years after a second edition of the *Theoricae* had been issued in Basel, was based on this work of PEURBACH's. As a matter of fact, a single glance will put in evidence the very close relationship between the two, and therefore a few explanatory remarks will suffice to conclude our study.

The designer evidently intended to represent the orbit of Mercury as seen from the north pole of the ecliptic, in accordance with PEURBACH's first plate (Fig. 20). Indeed there is no other way, because only thus can the longitudes be read directly on the outer rim. However, his astronomical skill was obviously somewhat deficient. He therefore contented himself with copying the three spheres in the same situation (valid for the centre of the epicycle in the apogee) and approximately in the same proportions as found in PEURBACH's treatise (and these proportions are wrong, like all similar ones extant in earlier works). Then he placed the epicycle carrying the disk of Mercury in such a way that the straight line joining the Centre of the Universe (Earth) with that of the disk of Mercury indicates the "apparent" longitude of the planet. The

[103] Cf. ZINNER, *Geschichte und Bibliographie der astronomischen Literatur in Deutschland zur Zeit der Renaissance*, Leipzig, 1941: Nos. 890, 918, 1098, and 1216.

situation of the planet in the epicycle is also indicated schematically, because, at the moment in question, Mercury did not occupy the perigee of the epicycle. Moreover, the designer committed the error of placing the small circle with radius ε in the centre of the whole disk, with the Earth as its centre, whereas he ought to have placed it eccentrically. Apart from this obvious lapse the designer could not possibly arrive at a satisfactory solution of his problem except by drawing the figure in correct proportions, as is done in our Fig. 25.

There is one more very characteristic symptom of the designer's insufficient training. The three parallel broken straight lines perpendicular to the apse-line (this latter is not drawn in MARCANTONIO's figure) make no sense here. We do not find

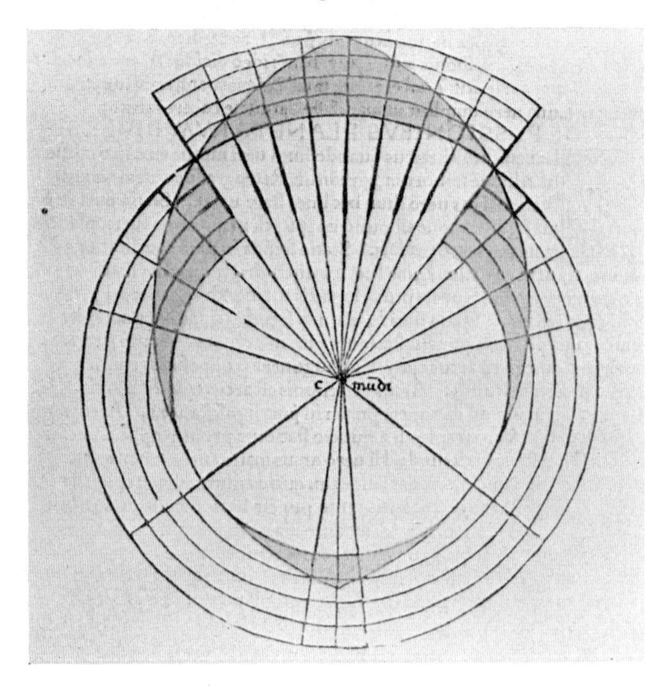

Fig. 24. *Peurbach Theoricae*, Fol. 11r: the oval curve of the centre of the epicycle

them in PEURBACH's first plate, but they play an important part in the second (Fig. 21), being the *axes* of the spheres! In other words, our designer has here confused the projection of the ecliptic with the cross-section in the plane of the circle of longitudes through the apogee.

The last question concerns the longitude of the apogee itself, which in the figure of the horoscope is ca. Libra 29° = 209°. We do not know whether this figure was really computed for the year 1527 or directly taken over without computation from an earlier source. It is almost exactly 20° larger than PTOLEMY's figure (which refers to the epoch ca. 134 A.D.[104]). As the interval between the two is nearly 1400 years, this would indicate a precession constant of 1° in ca. 70 years, which would correspond to the best value known in the Middle Ages: *viz.* the one determined, about 1260 A.D., by NAṢĪR AL-DĪN AL-ṬŪSĪ. A comparison with other values of the longitude of the

[104] See footnote 65, page 109. PTOLEMY's exact figure, for 134 A.D., is Libra 9° 15′ = 189° 15′. The difference of 8°, which NALLINO (AL-BATTĀNĪ, *Op. Astr.*, I, p. 241) regards as unexplained, approximately corresponds to the time elapsed between the era of NABONASSAR and that of PTOLEMY, which is ca. 900 years. NALLINO mistakenly believes the figure 181° 10′, which is PTOLEMY's value computed for the first year of NABONASSAR, to refer to PTOLEMY's own period.

apogee that I know of yields less satisfactory results: 875 A.D., AL-BATTĀNĪ: 201° 28′ (1° in 87ᵃ); 1029, AL-BĪRŪNĪ: 203° 43′ (1° in 94½ᵃ); 1204, AL-JAGHMĪNĪ: 206° 23′ 33″ (1° in 127ᵃ).

If however MARCANTONIO's astronomer really had employed NAṢĪR AL-DĪN's value of the precession, it would be inconceivable that he did not also employ his value of the longitude of Mercury's apogee. I have been unable to find the latter, but it must evidently be ca. 48′ larger than AL-JAGHMĪNĪ's, *viz.* ca. 207° 10′. To the remaining 1° 50′ would correspond 130 years, which would bring us to the year 1390,

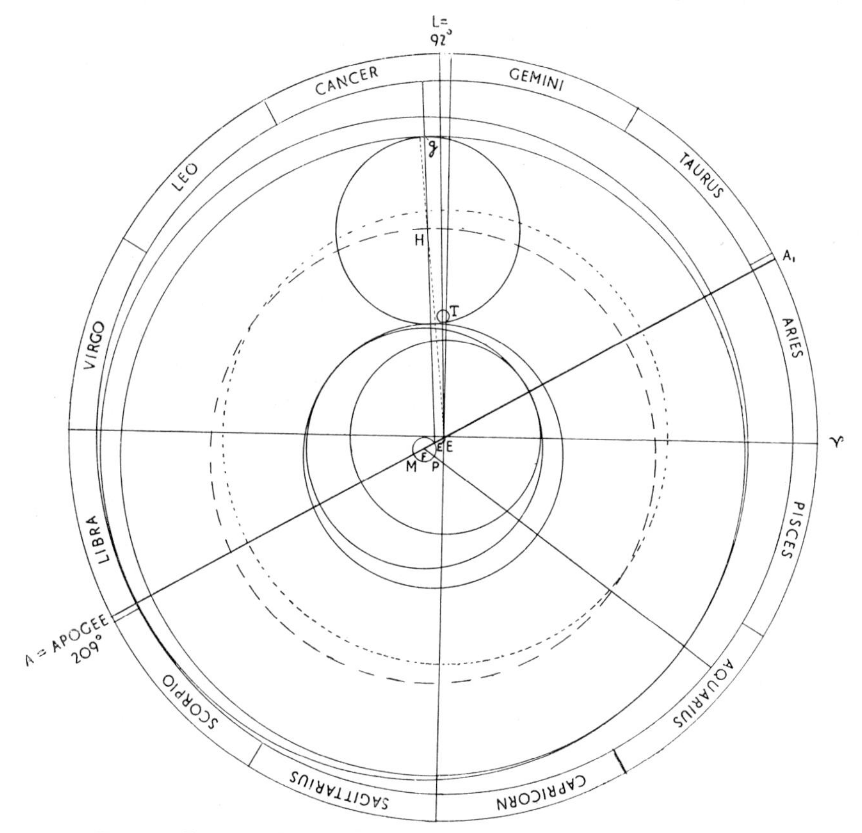

Fig. 25. Interior of MARCANTONIO's horoscope in correct proportions
Longitude of apogee, $A = 209°$
Mean anomaly ($\angle AE'H$), $\phi = 244°$
"App." anomaly ($\angle AEH$), $v = 247°$
(Mean) anomaly of planet in epicycle ($\angle gHT$), $\tau = 187°$
"App." longitude of Mercury ($\angle \Upsilon ET$), $L = 92°$
Dotted circle (centre E) = circle indicating mean geocentric distance

whereas for the year 1527 we would have to expect a longitude of 207° 10′ + 3° 48′ = ca. 211°, or 2° more than indicated by the horoscope.

The exactness of MARCANTONIO's figure is not sufficient, and the discrepancy not large enough, to allow of a reliable conclusion on this point.

13. CONCLUDING NOTES

Thanks to the generous co-operation of the Director of the Bayrische Staatsbibliothek and the good offices of the Central Administration of the Frankfurt Libraries, one of the extremely rare copies of JOHANNES STÖFFLER'S *Almanach nova plurimis annis venturis inservientia*[105] containing planetary ephemerides for the years 1499–1531, was placed at the author's disposal just in time to make the following additions and corrections to the present article.

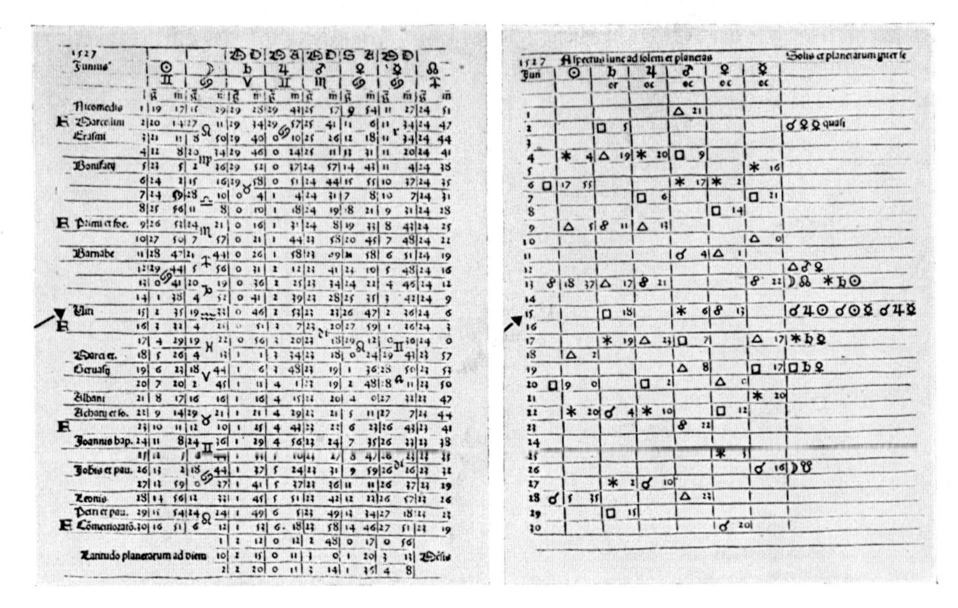

Fig. 26. Two pages from STÖFFLER'S *Almanach Nova*: Ephemerides for June, 1527

(*a*) Under the date 15th June, 1527, the planetary positions listed on the left-hand page (cf. Fig. 26, showing the two pages for June, 1527) are as follows:

Sun	Moon	Saturn	Jupiter	Mars	Venus	Mercury	Ascending Node
Cnc 2° 35′	Cap 19° 33′	Tau 0° 46′	Cnc 2° 53′	Sco 23° 23′	Cnc 26° 47′	Cnc 2° 36′	Sgr 24° 6′

In comparing these with the positions indicated on the horoscope:

Sun	Moon	Saturn	Jupiter	Mars	Venus	Mercury	Ascending Node
Cnc 2°	Cap 20°	Tau 1°	Cnc 2°	Sco 23°	Cnc 26°	Cnc 2°	Sgr 24°

we see that the latter were directly taken from the former, not by interpolation but by rounding off in a very arbitrary and unsystematic way. As the positions given by STÖFFLER refer to mean noon on the 15th June, while the horoscope was cast for 8 a.m. of the same day, the correctly interpolated values would have been as follows:

Sun	Moon	Saturn	Jupiter	Mars	Venus	Mercury	Ascending Node
Cnc 2° 25′	Cap 17° 6′	Tau 0° 45′	Cnc 2° 51′	Sco 23° 24′	Cnc 26° 35′	Cnc 2° 47′	Sgr 24° 6′

[105] Printed by JOHANNES REGER, Ulm, 1499, see Section 3.

and when rounded off correctly (prescriptions for this procedure are given in STÖFFLER's introduction), as evident,

Sun	Moon	Saturn	Jupiter	Mars	Venus	Mercury	Ascending Node
Cnc 2°	Cap 17°	Tau 1°	Cnc 3°	Sco 23°	Cnc 27°	Cnc 3°	Sgr 24°

It seems impossible to decide whether the incorrect figures of the horoscope are due to the astrologer's lack of mathematical insight alone, or whether he wanted the

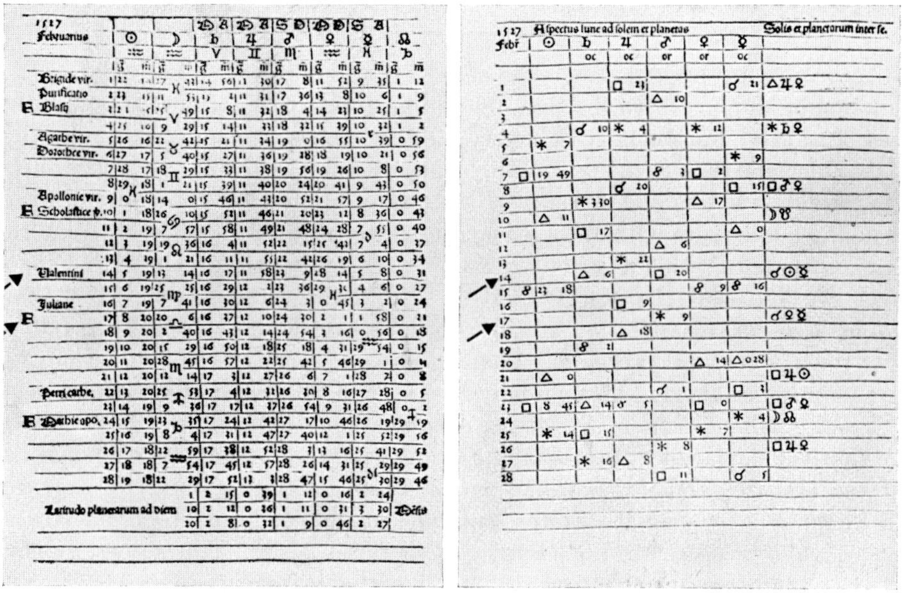

Fig. 27. Two pages from STÖFFLER: Ephemerides for February, 1527

double conjunction to take place in Cnc 2° instead of 3°, which would have been more correct. I can see no astrological reason for it, nor do I see any for neglecting the Moon's motion of ca. 2°.5, or for making Venus stand in Cnc 26° instead of 27°.

(*b*) From the circumstance that STÖFFLER's *Almanach* contains tables for the computation of the twelve astrological *houses* for the latitudes 42°–54°, which are based on REGIOMONTANUS', not PTOLEMY's, system, we may conclude that also the *houses* figuring in MARCANTONIO MICHIEL's horoscope were based on REGIOMONTANUS[106]. The distribution of the *houses*, *fines*, and *facies* then is as shown in table overleaf.

In comparing this table with the ones given in Sections 5 (*b*) and 6 (*a*), (*b*)[107], we find that three out of the twelve cusps (*Houses* No. III, IV, and VI) fall on unpropitious *fines*, against one according to PTOLEMY's system, and four (Nos. IV, VIII, IX, XI) on unpropitious *facies*, as against three according to PTOLEMY. The general character of the horoscope, however, is not altered thereby, especially because the four main *houses* (I, IV, VII, X) are the same in all existing systems.

[106] REGIOMONTANUS subdivides each of the four equal divisions of the *equator* which are confined between the horizon and the meridian into three equal sections of 30°. The *houses*, then, are the unequal sections of the ecliptic cut out by the "circles of position" (with poles in the north and south points of the horizon) laid through the twelve equal sections of the equator.

[107] See in particular the second table on p. 100.

				Finis	Facies
House No.	I . . .	Leo	18°	IV, Mercury	II, Jupiter
	II . . .	Virgo	9	II, Venus	I, Sun
	III . . .	Libra	3	I, Saturn	I, Moon
	IV . . .	Scorpio	6	I, Mars	I, Mars
	V . . .	Sagittarius	16	II, Venus	II, Moon
	VI . . .	Capricornus	22	IV, Saturn	III, Sun
	VII . . .	Aquarius	18	III, Jupiter	II, Mercury
	VIII . . .	Pisces	9	I, Venus	I, Saturn
	IX . . .	Aries	3	I, Jupiter	I, Mars
	X . . .	Taurus	6	I, Venus	I, Mercury
	XI . . .	Gemini	16	III, Venus	II, Mars
	XII . . .	Cancer	22	IV, Jupiter	III, Moon

(*c*) According to STÖFFLER's ephemerides for February, 1527 (Fig. 27), the conjunction of Mercury with the Sun did not occur on 7th February as computed by me (see Section 8), but on 14th February, and the one with Venus, not on 11th, but on 17th February (see STÖFFLER's last column). On 14th February, the Moon is listed as standing in trigonal aspect with Saturn at 6 p.m. (col. 3) but simultaneously approaching quadrature with Mars (occurring 20 hours after noon, on 15th February, at 8 a.m.); therefore, this day could not be found appropriate. On 17th February, however, we find that the Moon stood in the propitious sextile aspect with Mars (at 9 p.m., see col. 5) and that no unpropitious aspect disturbed the marriage of MARCANTONIO's celestial protector. Hence there can hardly be a doubt that MARCANTONIO himself celebrated his own marriage with MARIA SORANZO on the same day.

(*d*) From a comparison of STÖFFLER's ephemerides for 7th and 11th February we see that the deviations of the planets from the true positions, as calculated from NT II and III, are very considerable. In the case of Mercury, the error amounts to no less than 12°; for Venus, Jupiter, and Saturn the positions are given fairly correctly, as could be expected, but Mars again deviates ca. 6°. This shows, on the one hand, that there are problems of dating, like the one treated in Section 8, that cannot be solved with the aid of modern computation but only by resorting to the tables or ephemerides actually used at the time in question; and, on the other hand, that the incorrectness of the tables must have been felt so disturbing that a complete revision of the foundations of Ptolemaic astronomy seemed inevitable. This revision, as will be remembered, was almost completed by the time MARCANTONIO's horoscope was cast. Only sixteen years later, the scientific world witnessed the dawn of the new era of astronomy.

Observatories and Instrument Makers in the Eighteenth Century

D. W. Dewhirst
The Observatories, Cambridge

1. During the last decade a renewed interest has been taken in the study of early scientific instruments, not only for their intrinsic interest as examples of the craftsman's art, but also for the light they throw on the instrumental limitations imposed on the ability of the experimental scientist. The exacting requirements of positional astronomy have always made greater demand on the skill of the instrument maker than almost any other science, and this interplay between the observer's requirements and the ability of the instrument maker to satisfy them is especially significant in the rapid development of this branch of astronomy during the eighteenth century. The topic is here considered with reference to Bernoulli's account of the instrumental equipment of the various observatories in Europe in the latter half of the eighteenth century.

2. The Jean Bernoulli here referred to was Jean Bernoulli III, born at Basle in 1744. He was the eldest son of Jean Bernoulli II (1710–1790, professor of mathematics at Basle) and a grandson of the celebrated Jean Bernoulli I (1667–1748). He inherited the distinguished mental ability of this remarkable family, took his doctor's degree in philosophy at the age of thirteen, and was called to Berlin as astronomer by the Berlin Academy when only nineteen. During October, 1768, to July, 1769, he undertook an extensive tour of the observatories of Europe and wrote an account of what he had seen (Bernoulli, 1771). This account is published in the form of letters to a hypothetical correspondent, which (as he explains in the preface) enables him to give, by means of the dates of the letters, the exact time at which he found things as he describes them, and also to beg his readers' indulgence for the style of writing. This latter is very informal, and the letters contain impressions and personal anecdotes expressed in a frank simplicity which makes them not only entertaining to read, but sometimes a valuable source of otherwise inaccessible background material. The *Lettres Astronomiques* are of greater importance, however, in that they describe the interest and reactions of a young but knowledgeable astronomer on visiting the principal observatories of Europe in his day. Although known to such bibliographers as Houzeau and Lancaster the book does not seem to have received much attention from previous historians of the period, perhaps because copies are now rather scarce.

3. The revival of observational astronomy in Europe in the sixteenth century took place largely in observatories established by the zeal of individuals and financed either from private sources or by noble patronage. Tycho Brahe and Hevelius both expended private fortunes on their observatories, whilst the Landgraf of Hessen-Cassel was not only an able observer but a munificient patron of the science. The foundation of such bodies as the "Invisible College" in England in the 1640's, which

became the Royal Society in 1662, and the Academy of Sciences in Paris, gave great impetus to the experimental sciences, both by providing a medium for the exchange of information and the diffusion of learning throughout Europe, and in part by providing further encouragement and patronage for the individual of ability.

4. During the latter part of the seventeenth and in the eighteenth century many of the universities also established observatories which were used both for instruction and research. These also, however, usually owed their existence to the zeal of individuals rather than to any determination to prosecute astronomy as an integral part of the work of the university, and were frequently little more than an adornment to the *apparatus philosophicus* of the institution. Professor STRATTON (1949) has described the early history of observational astronomy in Cambridge, which well exemplifies these points. During the hundred years from 1703 there were three separate observatories in Cambridge which flourished each but a few years, whilst many private individuals had collections of instruments in the several colleges: no original contributions of great moment came from any of them, although they were occasionally used for elementary instruction. PRICE (1952) has given a detailed description of the instruments of the first of these observatories, namely that at Trinity College, many of which are still extant. The instruments acquired in 1703 included a Gunter's quadrant of 13 in. radius "with a nocturnal on ye backside", a telescopic level, a universal ring dial of 10 in. diameter, and a 16 ft telescope with two object glasses. There were numerous minor items, "a plain table with its furniture" (used in surveying), "a concave burning metal", "Mr. MOLYNEUX his telescopic dial", *etc.* It is evident that an institution with such equipment, although referred to by a contemporary as "a stately astronomical observatory, well stor'd with the best instruments in Europe" was not really equipped to prosecute valuable research even if a constant supply of able observers had been forthcoming. In fact the observatory fell into disuse within a few years of its completion, and although the building existed until 1797 BERNOULLI did not regard it as worth a visit when he was in Cambridge in 1769.

Whilst taken only as an example, this history typifies many of the university observatories described by BERNOULLI later in the century. In England, it is true, the tide of intellectual activity in the universities was perhaps at its lowest ebb in the mid-eighteenth century, while in the mathematical sciences in particular the development of NEWTON's work was stifled by the continued use of the master's fluxional calculus, and by a rigid adherence to methods of investigation which NEWTON had virtually exhausted. This is perhaps a contributing factor to the lack of inspiration in observational astronomy at least in the English universities of the time. There were, of course, a few notable exceptions, as exemplified by TOBIAS MAYER, who became director of the Göttingen Observatory in 1751 and carried out, with a mural quadrant of 6 ft radius constructed by BIRD, a notable series of accurate observations for use in the construction of his solar and lunar tables. We may also note a series of accurate observations made by HORNSBY in Oxford with a similar instrument some years later.[1] But these exceptions resulted usually from the combination of an individual of ability and a good instrument; if nothing else, the changing population and uncertain support of university observatories were in

---

[1] *See* Dr. KNOX-SHAW's contribution to this volume, p. 144.

general not conducive to the sustained work of high quality necessary for good results in positional astronomy.

5. We may consider briefly the sources of the instruments used by observers prior to and during the eighteenth century. The main centres of instrument making in the sixteenth century had been in Germany (especially in Augsburg and Nürnberg), the Low Countries, and in London. The principal products were small portable dials, surveying and navigational instruments, of high artistic merit and skilled craftsmanship. When BRAHE first started to use the divided instruments of large dimensions which characterized later work in positional astronomy, he at first sought the assistance of the Augsburg instrument makers—as exemplified by the great Augsburg quadrant (BRAHE, 1598). But in general the specialised skill of the instrument makers were not immediately adaptable to the astronomer's needs, and most of the great observers of the early seventeenth century relied on their own skill to produce mountings and divided circles. At this time, too, the larger optical components were made by the observers themselves: when JAMES GREGORY took his plans for the reflecting telescope, which he had described in *Optica Promota* (1663), to London, he was unable to find an optician to polish the necessary surfaces. Both NEWTON and HOOKE made the first reflecting telescopes with their own hands (BIRCH, 1756).

6. The growth of the Royal Society in London and similar academies in Europe at about the same time had two further influences on the development of observational astronomy. On the one hand, the increased demand for philosophical apparatus stimulated the clock makers and spectacle makers of the capital cities to take an increased interest in the demand of new clients. HOOKE's diary (1672–80) contains frequent references to his visits to such famous clock makers as THOMAS TOMPION (1639–1713) and opticians like CHRISTOPHER COCK (sometimes COCKS, COX, *etc.*), fl. ca. 1660–1680, who was, indeed, referred to by OLDENBURG, Secretary of the Royal Society, as "our perspective maker Coxe . . .". Many of these craftsmen were men of considerable ability outside the requirements of their profession, and like GEORGE GRAHAM (1673–1751), the clockmaker, were later to become Fellows of the Royal Society in their own right.

On the other hand, influential support could be given to a new type of observatory: the national observatory, supported by State funds for the prosecution of observations generally recognized to be urgently needed. The Royal Observatory of Paris was completed in 1671; in England a number of Fellows of the Royal Society, considering the long standing difficulty of the determination of longitude at sea, suggested the construction of a small observatory where observations could be directed to this problem, and JOHN FLAMSTEED commenced his duties as the "Astronomical Observator" in the new Observatory at Greenwich in 1675. Despite the smallness both of the building and of FLAMSTEED's salary, the observatory was better designed and regulated than that of Paris, which had an unfortunate history in its early years.

At first, FLAMSTEED sought through HOOKE the assistance of the clockmaker TOMPION in the construction of his apparatus, but he eventually employed the talented mathematician and then amateur instrument maker ABRAHAM SHARP to produce a mural arc of 140° and 79 in. radius which was undoubtedly the most accurate divided arc ever produced up to that time. This was the beginning of a distinguished partnership at Greenwich between instrument maker and astronomer, the main course of which may be outlined as follows (Table I).

Table I

Astronomer Royal	Instrument	Maker	Notable Observations
JOHN FLAMSTEED (1676–1719)	50 in. mural quadrant 1683		Obliquity of ecliptic, *etc.* Fundamental right ascensions.
	79 in. mural arc 1689	ABRAHAM SHARP	Catalogue of the "Historia Coelestis".
EDMUND HALLEY (1720–1742)	66 in. focus transit instrument 1721	ROBERT HOOKE?	Zodiacal stars. Places of Moon for determination of longitude.
	8 ft mural quadrant 1725	GEORGE GRAHAM	
JAMES BRADLEY (1742–1762)	1721 transit repaired	JONATHAN SISSON	Latitude and place of equinox.
	1725 quadrant modified	GEORGE GRAHAM	Refraction.
	12 ft zenith sector 1727	GEORGE GRAHAM	First general transit observations of high order.
	8 ft radius mural quadrant 1750	JOHN BIRD	(Aberration) and Nutation.
	8 ft focus transit 1750	JOHN BIRD	
NATHANIEL BLISS (1762–1764)	—	—	Continued BRADLEY's work.
NEVILLE MASKELYNE (1765–1810)	Commenced work on new mural circle	—	Fundamental stars. Places of Sun, Moon, and planets. Nautical Almanac from 1767.
JOHN POND (1811–1835)	Mural circle 1812	EDWARD TROUGHTON	Catalogue of 1112 stars (1833).
	10 ft focus transit 1816	EDWARD TROUGHTON	

The successive instruments by SHARP, GRAHAM, SISSON, BIRD, and TROUGHTON may each be said to mark an epoch in the art of instrument design and circle dividing, as the accuracy of the Greenwich observations through this period bear witness. Missing from the list are JESSE RAMSDEN, F.R.S., fl. c. 1762–1800, whose greatest contribution was the dividing engine, and, of course, such opticians as DOLLONDS (1750 onwards) and JAMES SHORT, fl. c. 1735–1768, who was the first person to make successful reflecting telescopes commercially, and whose labours amassed him a fortune of £20,000.

7. Space has not permitted reference to the contributions of OLE RÖMER (1644–1710) and JOHANN TOBIAS MAYER (1723–1762) to the construction and design of divided instruments, and there were, of course, developments in instrument manufacture on the continent allied to those in England in the eighteenth century. But it is clear that the continental instruments makers did not vie with those of London in the provision of the capital instruments of the world's observatories until the nineteenth century.

That this view is not a result of faulty historical perspective or prejudice is confirmed when we revert to BERNOULLI's *Lettres*. A Swiss astronomer, interested in instruments, he notes in almost all the cities he visits instruments by SISSON, BIRD, DOLLOND, and SHORT, and especially regards the divided instruments as the best available in his time. After visiting Greenwich for three days, he wrote on 26th March, 1769, "Je finis ici cette lettre; elle est bien longue, mais c'est aussi du principal

Observatoire d l'Europe que je vous ai parlé, et je crains bien de n'avoir pas traité le sujet comme il meritoit" (p. 99). But despite his enthusiasm for what he saw at Greenwich, his first concern on arriving in England was to visit not observatories but the instrument workshops: he saw the "very considerable" one of the Dollonds in St. Paul's Churchyard, most of the shops of the numerous philosophical instrument makers, and wrote later on 3rd May, 1769, about the instruments of Bird, Sisson, and Ramsden. The precision of the divided circles of these artists he places in that order, "dans le quels ils sont mis par le plus grand nombre de ceux qui connoissent leurs ouvrages". Bird's workshop made instruments of all types, including the necessary lenses (unlike some of the other makers, who relied on other firms for their lenses), and Bernoulli notes that his divided instruments were the most expensive: a 5 ft mural quadrant cost 250 guineas. Of Ramsden, unfortunately, he does not tell us much, but Ramsden had completed his first dividing engine only three years previously, and was yet to achieve the fame that enabled him to equip the observatories of Palermo, Florence, Padua, Mannheim, and St. Petersburg—to name a few—with their well-known transit instruments and vertical circles.

Apart from the individual genius of these artists and the stimulus of the astronomers' needs, there are perhaps contributing factors to their success: some were related by marriage and shared technical information within the family, whilst others had been trained as apprentices by their predecessors: always there was the inducement of the rewards offered by the Board of Longitude for such improvements in instruments, clocks, or the art of circle dividing as might lead to the ready determination of longitude at sea. But whatever the cause, the observatories of Europe clamoured for their productions, and frequently the makers had difficulty in meeting sufficiently quickly the orders placed with them.

Bernoulli concludes his first letter from London: "Me voici à *Londres* depuis 8 jours; je ne puis vous parler encore d'Astronomes ni d'Observatoires; mais je vous ferai part de la surprise agréable où est jetté un Astronome en parcourant les rues de cette Capitale. Vous avés sûrement oui parler de la richesse et de l'éclat des boutiques de *Londres*, mais je doute que vous vous représentiés combien l'Astronomie contribue à la beauté du spectacle: *Londres* a un grand nombre d'Opticiens; les Magasins de ces artistes sont remplis de Telescopes, de Lunettes, d'Octans, &c. Tous ces instruments, ranges et tenus proprement, flattent l'oeil autant qu'ils imposent par les reflexions auxquelles ils donnent lieu". (8th December, 1768.)

References

Bernoulli, Jean . . . . 1771 *Lettres Astronomiques, où l'on donne une idée de l'état actual de l'Astronomie pratique dans plusieurs villes de l'Europe.* Berlin. Chez l'auteur. (8vo., 175 pp. and 2 plates.)

Birch, Thomas . . . . 1756 *History of the Royal Society*, London, vol. III, p. 122.

Brahe, Tycho . . . . 1598 *Astronomiae Instauratae Mechanica.* (English translation by H. Raeder, E. Strömgren and B. Strömgren, Copenhagen, 1946, p. 89.)

Hooke, Robert . . . . 1680 *Diary of Robert Hooke*, 1672–1680. Edited by H. W. Robinson and W. Adams; London, 1935 (*passim*).

Price, D. J. . . . . 1952 *Annals of Science* (London), vol. 8, p. 1.

Stratton, F. J. M. . . . . 1949 "The History of the Cambridge Observatories", *Annals of the Solar Physics Obs.*, vol. I.

The Radcliffe Observatory

H. Knox-Shaw

Formerly Radcliffe Observer, Pretoria, South Africa[1]

As Professor Stratton has been for several years one of the Radcliffe Trustees, it is perhaps not inappropriate to include in this volume a short history of the Radcliffe Observatory.

When Dr. John Radcliffe, who had been London's leading physician during the reign of Queen Anne, died in 1714 he left a considerable fortune. His will directed that the residue of his estate, after several specific bequests, should be employed at the discretion of the trustees for charitable purposes. Now in English law any institution the sole purpose of which is the pursuit of knowledge is classed as a charity. So in 1771 the Radcliffe Trustees agreed, on the application of the University of Oxford, to build an observatory for the use of the Savilian Professor of Astronomy. A magnificent building on nine acres of land on the then confines of the city of Oxford was erected in the course of the next twenty years, the architects being Henry Keene, who had been responsible for the Radcliffe Infirmary, and James Wyatt. In addition to library, lecture room, and accommodation for the astronomical instruments there was a commodious house for the director of the Observatory, who became known as the Radcliffe Observer. The Observatory was thenceforth entirely controlled and financed by the Radcliffe Trust, and until nearly 160 years later, when the question of moving the observatory to the southern hemisphere was being considered, there were no astronomers among the Trustees. Since then, however, the Trust has had the advantage of the technical advice of Sir Frank Dyson, Sir Arthur Eddington, Sir Harold Spencer Jones, and Professor F. J. M. Stratton.

The Rev. Thomas Hornsby, Savilian Professor of Astronomy since 1763, to whose energetic action the founding of the Observatory was due, had at his disposal an 8-ft transit, two 8-ft mural quadrants, and a 12-ft zenith sector, which were the last and probably the best instruments made by that famous craftsman John Bird; and very good use Hornsby made of them, for he observed the positions of the Sun, Moon, planets, and brighter stars assiduously and almost continuously for nearly thirty years until stopped by ill health in 1803, when he was seventy years of age. He made all these observations, some 80,000 transits and 20,000 zenith distances, himself for he had no assistant and, as a reduction of them made 120 years after Hornsby's death has shown, they were of a high order of accuracy for those days. It is astonishing that one man should have been able to observe a star in both co-ordinates at one transit, as he frequently did, taking the time (by the eye-and-ear method, of course) over the first three wires of the transit, then going to the next room to observe the star's zenith distance and returning to the transit in time for the last three wires. This remarkable man also held the posts of Radcliffe Librarian and Sedleian Professor of Natural Philosophy for most of the time he was observing; and he managed to produce the first volume of Bradley's observations. It is thus

[1] Now at Elgin, C. P., South Africa.

not surprising that he had to leave to other hands the publication of his own observational work.

When HORNSBY died in 1810 he was succeeded by the Rev. ABRAM ROBERTSON, who continued to observe with the same instruments, though with the aid of an assistant, one of them taking the transit and the other the quadrant. The next Observer and Savilian Professor was STEPHEN PETER RIGAUD, who was appointed in 1827. He observed very little himself, as his interests lay in the direction of the literature and history of science. He amassed a large library of early scientific works, which on his death was purchased by the Trustees. Three years before his death in 1839 a 6-ft meridian circle by THOMAS JONES was erected to take the place of the two quadrants. In design it was transitional between the mural circle and the meridian circle proper and seems to have been unique.

These first three Radcliffe Observers held also the Savilian Professorship of Astronomy, and the appointments to the two posts were made by agreement between the Radcliffe Trustees and the Board of Electors to the Savilian Chair. However, in 1840 these two bodies failed to agree on a successor to RIGAUD. G. H. SACHEVERELL JOHNSON was elected to the Professorship, but as his subsequent career showed (he exchanged soon after to the Whyte Professorship of Moral Philosophy and later became Dean of Wells) he had no particular interest in astronomy. The Trustees were naturally anxious to have someone with experience of astronomical instruments and observation as their Observer and appointed MANUEL JOHNSON, who as an officer in the St. Helena Artillery had been in charge of the observatory on that island and had compiled a catalogue of six hundred southern stars. From this time onwards there was no direct link between the University and the Observatory, the use of which was lost to the Savilian Professor. On the other hand, the Observer, freed from the duty of teaching, could give his whole attention to the needs of the Observatory.

MANUEL JOHNSON thoroughly justified his selection, for he proved himself an energetic and resourceful director. In the nineteen years before he died at the age of fifty-four he introduced several new activities into the Observatory's routine. Amongst these was the publication of an annual volume of *Radcliffe Observations* containing astronomical, and a few years later also meteorological, results. The transit was remodelled by WILLIAM SIMMS in 1842 and was in constant use along with the JONES meridian circle. In 1848 a $7\frac{1}{2}$-in. heliometer by REPSOLD, the only example of an instrument of this kind ever erected in England, was installed in a separate dome. It was used by JOHNSON and his assistants for measuring double stars and the positions of minor planets. A second assistant was added to the staff in 1851. He was NORMAN POGSON, the originator of the proposal that the "light-ratio" defining the scale of stellar magnitudes should be the fifth root of one hundred. The devoted services as a computer of WILLIAM LUFF, given for many years gratuitously, commenced in 1844 and were to last until 1889. The longitude of the Observatory was determined by the Rev. RICHARD SHEEPSHANKS in 1842 by the transport of chronometers from Greenwich, and in 1854 apparatus for the continuous photographic recording of atmospheric pressure and temperature was installed by Mr. (afterwards Sir) WILLIAM CROOKES.

From the inception of the Observatory its chief concern had been with fundamental meridian work. Though HORNSBY had at his disposal an equatoreal sector, and

10-ft and 3½-ft Dollond achromatic telescopes which could be wheeled out of the large room in the tower on to the balcony, he only used them on occasions for such events as eclipses, occultations and the phenomena of Jupiter's satellites; he obtained approximate positions of Uranus with the sector a few days after its discovery by WILLIAM HERSCHEL. And although later on a 10-ft reflector by HERSCHEL, the heliometer and a 10-in. equatorial were added to the equipment, the reputation of the Observatory rested on its meridian work until that ceased early in the present century.

The meteorological observations, which started with the reading of pressure and temperature for the calculation of refraction, developed gradually until a full set of observations of all the elements, including the continuous recording of pressure, dry and wet bulb temperatures and rainfall, was made. The instruments were naturally changed from time to time, and so indeed were their exposures, but it was found possible to construct reasonably homogeneous series of values for the elements over long periods (for rainfall and mean temperature from 1815 onwards), and these results were published in 1932 as an appendix to Volume LV of *Radcliffe Observations.*

When MANUEL JOHNSON died in 1859 his place was taken by the Rev. ROBERT MAIN, who had been Chief Assistant at Greenwich for twenty-five years. He himself observed mainly with the heliometer, measuring over 800 double stars and the diameters of the major planets. His assistants, at first two and later three, observed with the TROUGHTON AND SIMMS 5-in. transit circle, which had belonged to RICHARD CARRINGTON and was purchased in 1861. This instrument was in regular use until 1903. Main published the *First* and *Second Radcliffe Catalogues*, containing respectively the results of the meridian observations made between 1840 and 1853 and from 1854 to 1861. The *Third Catalogue*, compiled in Germany from the observations made between 1862 and 1876, was not published until 1910.

MAIN was succeeded in 1879 by EDWARD JAMES STONE, who came to Oxford from the Cape Observatory, where he had served for nine years as Her Majesty's Astronomer. He did not observe himself except with the heliometer, which he employed for determining the solar parallax by measuring the positions of minor planets; but his activities were many-sided. He produced the *Radcliffe Catalogue for* 1890, as a continuation up to the equator of the *Cape Catalogue for* 1880, which had been his main concern while at the Cape; he organized the British expeditions to observe the transit of Venus in 1882, training the observers beforehand in Oxford; and he successfully observed the total solar eclipse on 8th August, 1896, in Novaya Zemlya. In 1886 the Observatory received as a gift from Mr. GURNEY BARCLAY a very fine 10-in. equatorial by COOKE, which later ousted the heliometer from its dome and was used on occasions as long as the Observatory remained in Oxford.

On STONE's death in 1897 ARTHUR ALCOCK RAMBAUT, Royal Astronomer of Ireland, became Radcliffe Observer. His first work was to introduce some improvements into the transit circle, and the observations made with this instrument during the last ten years that it was in regular use were published in 1906 as the *Radcliffe Catalogue for* 1900. In the meantime the new Grubb twin telescope with 24-in. photographic and 18-in. visual objectives had been erected in a dome furnished with a hydraulically-operated rising floor. It was at first employed in testing an idea suggested by J. C. KAPTEYN for determining simultaneously the parallaxes of a large number of faint stars. This method was not a success owing to the parallaxes being too small in relation to the magnitude equation affecting the results. So in

1909 the telescope was turned to photographing the northern and equatorial Selected Areas for the proper motions of stars down to the fourteenth magnitude. There were 115 such Areas in the programme and the plates were taken in duplicate, one being stored for re-exposure after an interval of at least ten years and the other being developed at the time for eventual comparison with another plate taken at the second epoch. All the first epoch plates were secured in the next nine years, but although the efforts of the whole staff were concentrated on this work a further fifteen years were to pass before those at the second epoch were completed.

RAMBAUT died in 1923, and in the following year HAROLD KNOX-SHAW, Director of Helwan Observatory and of the Egyptian Meteorological Service, was appointed in his place. The new Observer's first duty was clearly to complete at as early a date as possible the observing programme on the Selected Areas and to arrange for the heavy computation involved in the derivation of proper motions for 30,000 stars. By the introduction of calculating machines and the appointment of two extra computers the work was completed in a further ten years, and in 1934 the *Radcliffe Catalogue of Proper Motions*, representing the unremitting labour of the Observatory for a quarter of a century, was published.

The two previous Observers had each wished to reduce the meridian observations made by HORNSBY, but had been unable to obtain sufficient money for that purpose. However, an application now made to the Royal Society was most sympathetically received, and generous grants over a period of six years provided the salaries of two computers. The work was done in collaboration with Dr. J. JACKSON, then Chief Assistant at Greenwich, who made himself responsible for the tabular positions of the Sun and stars. The results appeared in 1932 as *Hornsby's Meridian Observations, 1774–1798*.

But as is soon to be recorded, the Observatory was ere long to bid farewell to Oxford and at the same time to one of its most faithful servants. WILLIAM HENRY ROBINSON, after three years at Greenwich, was in 1879 appointed by STONE to be his Second Assistant. He became First Assistant in 1920 and retired in 1935 after serving the Observatory with the utmost loyalty and devotion for fifty-six years.

For some time the Radcliffe Infirmary being in urgent need of room for expansion had looked with envious eyes at the Observatory's nine acres next door, and had indeed sounded the Trustees on whether they would sell part of the land. It was decided, however, that the Observatory could not carry on efficiently if it were surrounded by buildings any more closely than it was. So with the backing of its new President, Sir WILLIAM MORRIS (now LORD NUFFIELD) the infirmary offered to buy the whole site. It was clear that if the Observatory was to move it should be to a place with a better climate and preferably to the southern hemisphere. So as the British Association was meeting in South Africa in 1929, Sir FRANK DYSON and the Observer were commissioned to look for a possible site there. Their enquiries met with a ready response, and one of the sites offered to the Trust, on the hills outside Pretoria, was provisionally chosen; W. H. STEAVENSON went out for six months and made exhaustive tests of the atmospheric conditions there. These proved so good that the generous offer of the Pretoria Municipality to present 57 acres of land to the Trust, and to lay on water and electric power, was accepted. The Trustees thereupon agreed to sell the whole site to Sir WILLIAM MORRIS for the use of the Infirmary and the University Medical School, and were granted a lease of the buildings and the immediately surrounding land to enable the proper-motion programme to be completed.

There were, however, legal difficulties ahead of the project. The Court of Chancery had to sanction not only the sale of the land but also the expenditure of any of the capital of the Trust outside its own jurisdiction, and for such an action there was no precedent. The scheme providing for the removal to South Africa eventually came before the Court in 1934. The Attorney-General was there to see that the interests of the general public were not outraged, and the University of Oxford was given leave to intervene and plead that the Trust's funds could be put to better use nearer home. The Trustees' scheme, which included an offer of close co-operation with the Oxford University Observatory, was backed by affidavits from a regular galaxy of astronomers: Sir FRANK DYSON, Sir ARTHUR EDDINGTON, Dr. J. S. PLASKETT, and Professors WILLEM DE SITTER, FRANK SCHLESINGER, and HARLOW SHAPLEY, the last three testifying to the great value to the home University of a southern station in South Africa. The University of Oxford filed affidavits from Professor F. A. LINDEMANN (now LORD CHERWELL) and Professor ALBERT EINSTEIN. After a sitting of three days Mr. Justice BENNETT sanctioned the scheme in principle, and it was approved in detail a year later. The scheme included provision by the Trustees of funds for a Travelling Fellow in Astronomy, who would spend half his time in Oxford and half in Pretoria, the selection of the Fellow to lie with the University.

It was then possible to order from Grubb, Parsons & Co. the 74-in. reflector the plans for which had been under consideration for several years. The Oxford site was vacated in June, 1935, and the building of the new Observatory was commenced early in 1937. The mechanical parts of the telescope were erected in the following year, but the large mirror was delayed by two failures in casting the disk and by the war, and was not installed until 1948. The Cassegrain spectrograph ordered in 1937 was not completed until 1951. Though so long delayed the telescope could still claim to be the most powerful in the southern hemisphere.

On the removal to Pretoria new assistants were appointed, who for the first time in the history of the Observatory were University graduates; R. O. REDMAN, and E. G. WILLIAMS whose early death was a grievous loss. During the ten years that the telescope was without its mirror, observations could be made only through its finders and two wide angle cameras lent by the Cape Observatory, which were attached to the tube. WILLIAMS used the former for determining the magnitudes and colours of early-type stars by the Fabry method, and REDMAN the latter for photometric work on the Harvard *E* regions. He also photographed the change in the solar spectrum on transition from chromosphere to photosphere at the total eclipse in 1940.

With the completion of the telescope the new Observatory was at last able to embark on the programme of work which had been planned in outline so many years ago. In the forefront of this programme was the measurement of the radial velocities of the O and early B type stars in that third of the galaxy too far south to be reached from the northern hemisphere. Photography of the far southern nebulae also had its place in the original plan. Before this work was well under way KNOX-SHAW had retired, the first Observer to do so, and had handed over to ANDREW DAVID THACKERAY. He was able to bring into action almost at once a scheme that had been brewing for some little time, whereby in return for an annual grant from the British Admiralty a third of the observing time with the 74-in. reflector is allotted to staff from the Cape

Observatory. Thus full use can be made of the good climate of Pretoria in a way that would not have been possible by the staff of the Radcliffe Observatory alone. There is a tremendous field of opportunity open to the new Observatory, and there is good reason for thinking that it will be both energetically and successfully explored.

A System of Quantitative Astronomical Observations

C. W. ALLEN
University of London Observatory

1. THE PROBLEM

ASTRONOMY is an experimental science, but, since the objects of astronomical inquiry cannot be handled or influenced in any way, the experiments of astronomy become a matter of observation only. All the practical evidence relating to the heavenly objects comes to us in the form of electromagnetic radiations, visible or invisible, which may be detected by various methods. An observing astronomer is therefore occupied entirely with devices for detecting and measuring these radiations.

Since it is possible to describe an observer's activities so simply one feels that there should be corresponding simplicity about his measurements and results. We may introduce some formal simplicity by regarding practical astronomy as a series of observations on a number of celestial objects which bear highly significant points of similarity and difference.

The observer's first attention is given to a primary classification of celestial bodies by means of their qualitative characteristics. Individual objects are then found to have attributes that can be measured quantitatively and the results may be given as a set of numbers—one for each attribute. Other attributes are more complex and in the first instance can best be represented by a pictorial illustration. However, it may be found that much of this illustration can be resolved into a collection of entities each to be measured by a further set of numbers. Without doubt further analysis could reduce every observable detail to numerical expression but at some stage this would become inefficient both to the observer and the user. Thus astronomical observations become reduced to (*a*) a quantitative part expressed by sets of numbers, and (*b*) a qualitative part expressed in illustrations or verbal descriptions. In many branches of astronomy both parts are needed, the quantitative for comparative analysis, and the pictorial to decide what types of objects exist and to provide the ideas.

We are concerned at present with quantitative measurements which are urgently required for all celestial objects in order that an accurate intercomparison might be possible. The objects occur in great variety of forms and do not all lend themselves very obviously to uniform measurement. There has been no general agreement as to how such measurements should be made and each observer has tended to use his own system, leading to results that are not directly comparable with others. We must, therefore, face the serious question of whether it is in fact possible to make uniform quantitative measurements of all astronomical objects, and, if the measurements cannot be made uniformly, what is the best approach to uniformity that we can offer.

We may make some progress with this question by considering the requirements of an ideal set of uniform astronomical measurements. With the list of requirements before us it should be possible to determine the degree of compromise that must be adopted.

2. Requirements

(*a*) The measurements should be within the power of existing observing techniques at several observatories and for several objects, but should be capable of improvement to any accuracy required by future techniques.

(*b*) Any measurement should be available for the whole range of astronomical objects without requiring their qualitative classification. It should be suitable for describing, measuring, and labelling.

(*c*) The measurements should be independent of one another. They should be chosen in such a way that each has a useful meaning in its own right and without reference to other observations. If a measurement is slightly dependent on another factor that factor should be chosen centrally in the range that might exist in practice.

(*d*) The set of measurements should be complete in the sense that all available general quantitative information should be obtainable from them.

(*e*) Each measurement should express a physical quantity and should be given in units appropriate to that quantity. They should be readily convertible to analogous measurements or quantities.

(*f*) Defining rules should be in terms of fundamental physical units and should contain no arbitrary factors. If arbitrary numbers are required they should be carefully chosen from 2, *e*, 10, 100, 1000, *etc.*

(*g*) Measurements should possess stability in the sense that if the associated quantity is stable the measurement is a constant. They should be on an absolute scale but suitable also for differential measurements.

(*h*) Measurements should be suitable both for those engaged in specialized researches and those engaged in accumulating standard observations.

If all observations could be made on a scheme that obeyed these requirements a very considerable economy might be effected both in observing and expressing results. Plentiful observations could be expected each one adding directly to the pool available and ready for use in either a theory or statistical analysis. New values could be readily compared with old and a mean or "best" value rapidly assessed.

However, the possibility of developing such a scheme is dependent firstly on whether one can set down a list of measurements that can represent the main quantitative observations of astronomy. This is not as difficult as it might appear and the list below of twenty-five measurements, together with their correlations, will be found to be fairly complete. In the list each independent measurement is labelled with a number, and the correlation parameters are collectively labelled C.

3. Measurements

[1] Number of astronomical objects of particular categories in whole, or a defined portion, or the sky.

[2, 3] Position in the sky.

[4, 5] Total apparent magnitude, and colour.
[6, 7] Polarization.
[8, 9] Size and ellipticity.
[10, 11] Spectrum classification and luminosity class.
[12] Time.
[13] Period (of periodic phenomena).
[14] Duration of phenomena.

C Variations of [1–11] with time may be expressed in many correlation parameters other than [13, 14].

[15] Multiplicity (= number of entities making up whole object).
[16, 17] Position within object (or departure from mean position) usually relative to [2, 3] (includes orientation).
[18] Surface brightness.
[19] Number density of entities.

C Variations of [5, 6, 7, 10, 11, 13, 14, 18, and 19] with [16, 17].

[20, 21] Wavelength of a spectrum line, and its shift from normal position.
[22] Intensity or equivalent breadth of a line.
[23] Line breadth.
[24] Line multiplicity.
[25] Spectral intensity of a continuum.

C Variations of [20–25] with other measurements.

We must consider how well this list of measurements suits the requirements. The measurements are, of course, not entirely complete, but they cover the main factors relating to objects in the sky and their spectra, and include radio astronomy. Some measurements cannot be applied in all cases—for example, some objects have no clearly defined period for the measurement [13]. However, there are not many inconsistencies in the scheme of applying [1–19] to all celestial objects, and [20–25] to all spectra. It is mainly technical limitations that prevent us from making all measurements on all objects. The items in the list are not entirely independent, for example, [5] and [10] or [4, 5] and [25] could give the same information, and [13] could be regarded as a correlation of [12] with other factors, but these are rather unimportant inconsistencies caused by an attempt to make the list simulate the main observations as closely as possible. We might call these the primary measurements of astronomy and restrict any further discussion to the domain covered by them.

The important question to be decided is whether each measurement can be adequately expressed by one primary value, and how this can be done by uniform observations without loss of vital informations. We take the point of view that, whatever other observations might be made for the needs of a special inquiry, the listed primary observations should be made in a standardized form on every object or spectrum line that is of interest.

The problems to be faced in standardizing each of these measurements are quite varied and must be considered individually. This is done below, and in each case an opinion is given as to the most appropriate definition for the primary measurement in view of the many facts and requirements.

[1] It is not possible to detect and thence measure the number of objects of any particular category in the whole universe nor even in the galaxy. The most useful measurement is number of objects down to a selected limiting magnitude. However one cannot select one standard limiting magnitude to suit all categories, and for the observations to be useful it is desirable to make the two dimensional measurement of total number against limiting magnitude. For the most general representation the counts should refer to the whole sky, while for regional studies it is more appropriate to make counts per square degree.

While most celestial statistics will be measured in this form there is scope for a single primary measurement to express the populations of different categories as seen from the earth. Any such measurements must necessarily be obtained from the brighter objects of the particular category since the faint ones often cannot be classified. Probably the most direct and stable measurement would be *the magnitude of the tenth brightest object in the particular category*. Such an observation could be referred to the population of all objects to the same magnitude (a stable relation dominated by stars) to give the proportion of that object relative to all others. If the mean absolute magnitude of the object were also known it could be interpreted to estimate population in space.

[2, 3] The measurement of absolute position by right ascension and declination offend the stability requirement (g) unless a standard epoch is selected. It would appear that 1900·0 shows promise of being adopted as a permanent standard epoch for absolute reference of celestial positions.

[4] The bolometric magnitude would be the ideal standardized measurement of apparent magnitude if this were capable of being measured directly. However, the practical choice falls between the photographic and photovisual system, of which the photovisual appears to have the slight advantage as a primary standard because the effective wavelength is more stable and more central among the values that might be selected. In the case of extended or composite objects the magnitude must refer to the total light in excess of the radiation from the background and field stars.

[5] With the photovisual system chosen as the magnitude standard the colour index $P - V$ (where P is photographic and V the photovisual magnitude) becomes the natural choice for expressing colours. The weakness of the system is that the base-line of colour difference changes with the colour itself, however there is still a one-to-one relation between colour index and any other general measurement of colour, so that there is every opportunity of standardizing these relations.

[6, 7] The degree of polarization is unambiguously measured from the ratio I_x to I_y, where I is intensity and the subscripts x and y represent directions at right angles for linear polarization and rotations in different sense for circular polarization. The ratio might be expressed for preference as $(I_x - I_y)/(I_x + I_y)$. The preferred conventions for expressing the direction of polarization are (a) for linear polarization, the direction in which the electric vector is maximum, (b) for circular polarization, the optical convention whereby for R polarization the rotation of a vector in a fixed plane when looking towards oncoming light (so that it enters the eye) should be clockwise. At present radio astronomers adopt the convention opposite to (b).

[8, 9] For the primary measurement of size there appears to be every reason to use apparent radius (in seconds of arc) for stars, planets, and satellites, but to seek another definition for other objects with no sharply defined edge. There are great

difficulties in deciding on a measurement for extended astronomical objects such as nebulae and clusters. Not only is there no defined edge but the fall of intensity often trails off very slowly and irregularly with the result that size measurements may differ by a factor of more than ten. The definition of the edge should be related in some way to the brightness of the rest of the object. The central intensity cannot be used as standard since it may be an unresolved stellar nucleus, and we are left with the most appropriate definition, *the diameter of that circle (or ellipse) which will contain half the total light*. The selection of half for the adopted fraction should give diameters comparable with the usual conception of size. It will be noticed that although radius is used for stars and planets, diameter is recommended for other objects which have no assurance of radial symmetry.

When objects have measurable ellipticity (*i.e.* a long and a short axis) it is usual in astronomy to quote the major diameter a as a first measurement, and then ellipticity $(a - b)/a$ from which the minor diameter b may be obtained. The mean diameter for most purposes would be $\sqrt{ab}$. The procedure for measuring any object (not in the star or planet category) would be to decide first on the ellipticity and orientation of the major axis. Ellipses of this shape and orientation are then fitted to the object and the size varied until it contains half the total radiation. The ellipticity may need adjustment to simulate the intensity contours when the final size is known. The primary size of the object would be defined by the major diameter of this ellipse and its ellipticity. As in [4] only that radiation that is in excess of the background radiation should be included.

[10, 11] The two-dimensional system of spectrum and luminosity classification is maintained by standard stars. However a certain amount of classification error should be allowed for these standards so that they do not prevent the establishment of smooth relations between spectral class and other factors such as colour. To establish such relations one must know whether the domain of classification extends from 0·0 to 9·9 for each class (O to M) or whether there are some sections definitely missing. Existing systems (mainly the Yerkes classification) are best illustrated graphically when the domain K8·0 to M0·0 is completely omitted. It would be preferable, however, if the K classification could be adjusted to cover the whole range as for the other classes.

[12] The U.T. will remain the primary measurement of time, although the actual observations might be made against a sidereal time clock, or the result may be converted to ephemeris time, heliocentric time, *etc.* for special purposes.

[13] The representation of periods in seconds, days or years presents no formal difficulty.

[14] The measurement of the duration of a phenomenon that commences or ends suddenly is quite straightforward, but difficulties arise when one attempts to devise a scheme that will give full duration for such events and still give an appropriate value for a phenomenon that commences or ends gradually or sporadically. A rule that will suit all cases can hardly be expected but some satisfaction may be obtained from measurement of the interval between the two instants (a) when 0·1 of the total flux of the phenomenon has passed, and (b) when 0·9 of the total flux has passed. However, these ratios are too arbitrary for a primary definition and it would appear that for this measurement the definition itself will have to be varied to suit the circumstances. The measurement of the interval for which the phenomenon exceeds an arbitrary threshold level will be used in many cases.

[15] The measurement of the multiplicity of an object is formally equivalent to [1], the counting of objects. Again there is no way to select a standard limiting magnitude (or limiting resolution) and usually the multiplicity is expressed by giving the number of entities as a function of some limiting factor that controls their detectability (such as the magnitude). It is only when the multiplicity converges to a finite value as the other factor is extended indefinitely that a stable primary value of multiplicity can be given.

[16, 17] These measurements can represent either (*a*) the displacement of the object from some mean or standard position, or (*b*) the measurement of components of the object from some fiducial position. The primary co-ordinates would be the radial distance (usually in seconds of arc from the central fiducial position represented by [2, 3]), and the orientation (anticlockwise from N) at epoch 1900·0.

[18] The practice of expressing surface brightness as an apparent magnitude of a defined area has the advantage that great differences of brightness can be expressed without change of units. This being so it is desirable that the unit of area be selected once for all, and the square second of arc would appear to be the suitable choice for primary measurement. However, when components of radiation have to be added the logarithmic magnitude scale is inconvenient and one might adopt as a second choice the number of $m_v = 0$ or $m_v = 10$ stars per square degree.

[19] The measurement of number density of entities or components of an object is similar to the measurement of its multiplicity [15].

[20] The custom of expressing wavelengths shorter than λ 2000 in vacuum units and those longer than λ 2000 normal dry air at 15°C is likely to be continued. In the radio spectrum frequencies are more usually adopted for accurate work and the ambiguity avoided.

[21] No ambiguity is involved in expressing wavelength shifts in Ångstroms or kaysers (cm^{-1}).

[22] The intensity of an emission or absorption line that is associated with a stable continuous spectrum may be expressed as an equivalent breadth of that continuum. This is convenient both for measurement and analysis. However, if no continuum is present, measurements must be given directly in physical units, either absolute or relative. For the sake of uniformity it would be preferable to express primary absorption line intensities as equivalent breadths, and emission intensities in physical units such as erg $cm^{-2}s^{-1}$.

[23] Line broadening is usually symmetrical and regular and hence its measurement does not involve the difficulties of [14], which is analogous. The accepted primary measurement is the total wavelength distance between the points where the intensity is half the maximum.

[24] The measurement of line multiplicity in astronomical sources presents no difficulty provided the spectrographic resolution is finer than the line width. Components that would have been resolved if the line width were much smaller are not to be regarded as adding to the multiplicity.

[25] Absolute measurements of spectral intensity f_λ may readily be expressed in physical units such as erg $cm^{-2}s^{-1}Å^{-1}$. However, actual measurements are usually relative and it is recommended that these should be expressed in the same physical units relative to the value at 5400 Å. Such measurements could be readily related to visual magnitudes.

Reviewing this discussion we find that standard practice provides suitable primary measurements in most cases. A new measurement to represent the population of objects seen from the earth is suggested in [1], and a suggestion for standardizing the size of nebulae, *etc.*, is revived in [8, 9]. On the other hand, we have found no suitable primary measurement of the duration of irregular events.

If astronomers in the course of their many and varied observations could add as many as possible of the above primary measurements to their programme the results would always be welcome. In particular, if the work has other intentions the primary measurements should be added as a by-product. There are, of course, many other primary factors (such as distance) which are also needed from observers and which fall outside the scheme of observations considered here. But the same arguments apply—if a clear-cut, unambiguous, and well-accepted method of expressing the factor is available there will be much better prospects of obtaining usable results.

Fact and Inference in Theory and in Observation

H. BONDI

Trinity College, Cambridge*

SUMMARY

A critical examination is made of the relative reliabilities of theoretical and observational work in astronomy. It is found that, contrary to widely held opinions, observational papers are no less liable than theoretical papers to reach erroneous conclusions.

THE aim of the present paper is an examination of the relative reliabilities of so-called "theory" and so-called "observation" in astronomy. They may simply be regarded as different tools used by research workers. Such examinations of relative reliabilities are common enough in all fields of science. Measurements made by different instruments are generally found not to agree amongst themselves, and the subject of the combination of observations attempts to deal rationally with such difficulties. Statistical weights are used to describe the relative reliabilities of similar instruments. In the event of a contradiction between two instruments it is helpful to have some knowledge of the degree of certainty to be assigned to them. In ordinary laboratory practice this assessment is generally based chiefly on past experience with the instruments concerned, combined with comparisons of their constructions and estimates of how "direct" the measurements are.

This purely empirical approach has been found far more satisfactory in laboratory work than reliance on emotional prejudices, however real they may seem without critical examination. Empiricism is vital to all scientific progress, and there seems to be good reason to apply this empirical approach, which has been so successful in the laboratory, to a rather wider field.

It is fashionable nowadays to divide astronomical research into two categories, "theoretical" and "observational". These terms are ill-defined and lack sure philosophical significance, but it is nevertheless true that most astronomers would have

* Now at King's College, University of London.

little doubt or disagreement on how to classify most published papers. All references in this essay to "theory" and to "observation" are intended only in this colloquial sense and should not be understood as more than a rough and ready classification.

There is undoubtedly a widespread opinion that astronomical theory is, by and large, airy speculation, that it changes from day to day in its most fundamental tenets, that it is based on ill-considered hypotheses and that its reliability should accordingly be assessed to be low. By contrast the results of observational work are, according to this opinion, solid incontrovertible facts, permanent and precise achievements, that will never change and whose reliability is accordingly high.

In my view these opinions are unfounded and false, and their prevalence does great harm to the progress of astronomy. If they were not injurious, there would be little point in attempting to dispel these views, but, in the present state of astronomy, conflicts between observation and theoretical results continually arise. In the past, credence in such a conflict has generally been given to the observational result, and theoretical advances of considerably significance have been held up by undue reliance on the observations. The chief purpose of this paper is to show that in such a conflict it is far wiser to keep an open mind or even to lean to the side of the theory. Such an attitude, which can be firmly based on a critical examination of past experience, is likely to be far more helpful to the progress of astronomy than reliance on prejudices based on the emotional significance of such outworn phrases as "observational fact" and "theoretical speculation".

Before an examination of the relative reliabilities of these so-called theoretical and observational results can be attempted, it is desirable to show that theoretical work falls into two quite distinct classes. One is essentially an attempt to classify observational material; the *ad hoc* assumption is made that the type of classification is based on intrinsic qualities without making these particularly clear or (and this is the chief characteristic of this type of structure) linking up these qualities with known phenomena. An example of this type of structure is the old RUSSELL "theory" of the luminosity-type diagram. As will be remembered the principal suggestion of this theory was that the different appearance of giant and dwarf stars was due to the fact that they were made of different materials, "giant stuff" and "dwarf stuff" as the terms were. Any further explanation beyond this, any link with the known properties of material was left for the future. As is well known the suggestion was later disproved, but it is a typical example of such suggestions.

An example of more recent date concerns the ingenious observation by McKELLAR (1949) of the abundances of carbon isotopes on certain very red giant stars. He suggested that the stars in which the isotope ratio agreed with that produced in the Bethe cycle were well mixed stars, and those where the ratio was different, were not mixed. The sole purpose of any such suggestion can only be merely to fit the immediate observational classification. Indeed it is little more than a re-wording. To anyone who has the slightest knowledge of stellar structure, McKELLAR's suggestion can only sound utterly unconvincing, since it raises many more questions than it solves.

An entirely different type of logical structure to which frequently the same name of "theory" is given, is in fact a development of terrestrial physics. An attempt is made to apply the laws of physics as obtained in the laboratory to the conditions of the astronomical problem in question. Naturally it is sometimes necessary in the present state of knowledge to make a number of additional assumptions and the value of the attempt will depend greatly on the clarity and explicit nature of these assumptions. In my view misunderstandings have frequently arisen through confusion between the two types of approach both of which have been referred to as "theory". Different names are required, and I suggest "observational inference" for the first type of structure, and "physical theory" for the second type. An example of this second type of structure is the theory of stellar constitution.

In attempting to assess the relative reliabilities of theoretical and observational work in the past it is necessary to establish a criterion of failure. This is not quite as simple as one would expect it to be.

It is an essential feature of science that everything in it may stand in need of revision as a result of new and unexpected evidence—that it contains no "ultimate truths". Such truths are certainly outside the scope of science, though they may appertain to philosophy and to religion. Whenever we examine any approach to science we will not be surprised to find that changes occasionally occur, and this is true both of theory and of observation. Some such changes arise indeed merely from a widening of the field of enquiry. Classical mechanics was found wanting when it was applied to atomic dimensions and had to be replaced by quantum mechanics. The validity of classical mechanics in the macroscopic field is, however, not affected by this change. The changes that are to be examined here, the errors that should be taken account of, are of a far grosser and less subtle kind. They are simply statements that were later shown to be incorrect in the very field to which they were originally intended to apply within the accuracy originally claimed. Such errors have almost always been a hindrance rather than a help to the progress of science and we shall understand by the word error this its usual conventional meaning (as, for example, in the expression "probable error") without in any way entering upon the meta-scientific problem of a definition of the term.

Three questions seem then to arise:

(i) Is observational knowledge by its very nature more or less certain than theoretical knowledge?

(ii) As an empirical test, have theories in the past been shown to be in error more or less frequently than observational results?

(iii) Is an error in observation likely to persist for longer or shorter than an error in a theory?

Where the first of these is concerned the extreme complexity of present-day observational methods hardly need be stressed. To derive any significant astronomical result from the blackening of a photographic plate or the simple reading of a meter a tremendous amount of intervening work has to be done. Corrections may have to be applied, calculations and reductions may have to be carried out, and above all interpretations requiring a great deal of theoretical background may have to be made. Consider, for example, such an apparently straightforward matter as the determination of the masses of an eclipsing binary. Gravitational theory, tidal theory, theoretical reflection factors, and other theoretical notions have all to be brought in and used, frequently to the limits of their power. Or consider work on spectroscopic abundances where not only the most accurate photometry, but also particularly complex aspects of quantum theory are involved. Nobody can fail to admire such work, but by no stretch of imagination can it be termed purely factual or purely observational. And yet this is what some people seek to do!

In a recent review of Mr. HOYLE's broadcast talks by Dr. WILLIAMSON (1951), a review possibly written more in anger than in earnest, the reviewer states that he has attempted to establish the percentage of astronomical "facts" as compared with "theory" in Mr. HOYLE's talks. But what is an astronomical fact? At most it is a smudge on a photographic plate! Does he expect Mr. HOYLE to give a broadcast talk on smudges?

The purely factual part of the vast majority of observational papers is small. It is also important to realise that these basic facts are frequently obtained at the very limit of the power of the instruments used, and hence are of considerable uncertainty. To refer to observational results as "facts" is an insult to the labours of the observer, a mistaken attempt to discredit theorists, a disservice to astronomy in general and exhibits a complete lack of critical sense. Indeed I would go so far as to say that this sort of irresponsible misuse of terminology is the curse of modern astronomy.

Present-day observational astronomy may fairly be called the science of extracting the maximum information from the fundamentally meagre data that can be obtained about outer space, an endeavour to stretch both observation and interpretation to the very limit.

A similar examination of physical theories in astronomy reveals that their primary basis is very sound indeed, since it rests on established terrestrial physics. But in order to apply this knowledge to astronomy inferences of considerable range have to be made, sometimes with the aid of additional assumptions. In the theory of stellar structure, for example, results obtained in thermodynamic systems of temperatures of at most 4000° or in non-thermodynamic systems (particle accelerators) employing extremely tenuous matter have to be applied to dense matter at millions of degrees. It is only because of the comprehensive nature of laboratory physics that such extrapolations are possible. Other examples of a similar nature could be given, but it would probably be fair to sum up the situation by saying that in *observational work long chains of inferences are based on frequently somewhat uncertain data, whereas in physical theories of astronomy, though long chains of inferences are also used, they are generally based on much more reliable experimental data.* There is, therefore, no reason to expect any marked difference in the degrees of reliability of so-called theory and of observation, unless indeed it is that theoretical results are of greater reliability.

We can now turn our attention to the second question, that of empirical test. The question is, have theories been disproved more or less frequently than observational results? Clearly it is difficult to give an exhaustive list on either side. On the theoretical side I might mention JEANS' proof (1925, 1927) that stars would be unstable if the subatomic generation of energy depended very sensitively on temperature, a proof that confused the development of the theory of the constitution of the stars until COWLING (1934, 1935) showed it to be fallacious. Again JEANS' "long time scale" for the age of the galaxy (1929) was widely accepted until it was disproved by many arguments (see BOK, 1946). Finally, one might mention EDDINGTON's estimate of the hydrogen content of the stars as about 35 per cent (1930). This last error must be shared between theoretical and observational astronomy, since spectroscopists were of the same opinion. The recognition that the hydrogen content was far higher came more recently (DUNHAM, 1939; HOYLE, 1947).

Although in my own work I am more in contact with theoretical than with observational research, yet I find that I have more often met observational errors. One might recall VAN MAANEN's observations of the proper motions in extragalactic nebulae (1916, 1921, 1922, 1923, 1925, 1927), observations that have been proved to be incorrect not by factors of 2 or 3, but factors of 100 or 1000 (HUBBLE, 1935). Then one might mention ADAMS' "discovery" of the EINSTEIN shift of the spectral lines of Sirius B (1925). This shift is now not only considered to be practically impossible to measure, but is known through modern quantum theory to be very seriously obscured by almost incalculable pressure shifts (LINDHOLM, 1941; ADAM, 1948). But the shift "observed" in 1925 was supposed to agree as well as could be expected with relativity theory without account being taken of these pressure shifts.

Yet another example in this field is formulated by the Trumpler stars (TRUMPLER, 1935; An unjustified interpretation of an uncertain and difficult observation was widely accepted as an established fact, as a proof that extremely massive stars existed with known luminosities and radii that were apparently in contradiction with the theory of stellar structure. Confusion was caused by reliance on this result, but now it is regarded as having been based on erroneous interpretations (STRUVE, 1950).

In an earlier period, astronomers in this country, after failing to discover Neptune, endowed this planet with a ring and satellites (LASSELL, 1847). More recently, HUBBLE and HUMASON (1931) inferred from their data that the constant of the red-shift of the nebulae was $4{\cdot}967 \pm 0{\cdot}012$ and soon afterwards, from almost the same data, that it was $4{\cdot}707 \pm 0{\cdot}016$ (1931, 1934; HUBBLE, 1936). Similarly the last determination of the solar parallax (SPENCER JONES, 1941) is well outside the three-fold stated probable errors of earlier determinations which vary amongst themselves far more than their individual stated errors.

In a field with which I have been in close contact, Struve's recent work on Capella (STRUVE, 1951) has come as something of a shock. We were told in the most meticulous survey (KUIPER, 1938) that the masses of the components of Capella were the second best known ones, that, except for two cases, all other mass determinations were at least twice as uncertain as Capella, and, except for four other cases, more than three times as uncertain. And now it appears that even in the case of Capella there has been all along an error of more than 25 per cent. EDDINGTON in particular used Capella as his standard when he developed the theory of stellar structure. More recently, but for STRUVE's timely discovery of the mistake, the subject of the structure of Red Giant stars would have been put on the wrong track by reliance on this supposedly so precise and permanent fact.

These examples, though by no means exhaustive, will illustrate sufficiently the thesis of this essay. It seems that, by the empirical test, errors in theories are if anything less frequent than in observational work. A detailed numerical test is difficult, but what I have said is enough to refute the view that theories are airy pieces of guesswork disproved every few days whereas observational results are "hard", "incontrovertible" facts.

We come then to the third question: are errors likely to persist for longer in theory or in observation? The answer to this question is clear. Observational equipment is so scarce, is devoted to so many tasks, and is so difficult to set up, that repetitions and checking are not as common as one would hope, particularly in the case of many theoretically important measurements. There is also a widespread opinion, an opinion that I hope will not persist, that the certainty of observational work is so great that no repetitions are required. This uncritical attitude is greatly to be deplored. If more effort were devoted to repeating observations, the gain in certainty would be of great value.

The attitude and climate of opinion in which theoretical astronomy is conducted are entirely different. Almost every paper is received with sceptical interest and many papers immediately stimulate work connected with their proof or disproof. While in observational work it is unfortunately considered somewhat impolite for one observer to criticize the observations and immediate inferences of another observer, similar criticism between theorists is luckily considered perfectly natural. There is therefore a considerable likelihood of an error being rectified speedily. As a recent example of this I might mention a paper by RICHARDSON and SCHWARZSCHILD (1948) & SCHWARZSCHILD (1948) on Red Giants which was refuted by GOLD (1949), his paper being submitted less than two months after the publication of the first paper. An error in a paper on stellar structure by CHANDRASEKHAR and SHOENBERG (1942) was shown up speedily by HOYLE and LYTTLETON (1946). Of course, sometimes the rectification of an error takes much longer. JEANS' statement (1925) about the instability of stars with temperature sensitive energy production was finally disbelieved only after COWLING's work in 1934, twelve years after the claim had appeared. But even this interval is not long compared withthe cases quoted in observational astronomy where the intervals are usually more like twenty years. The intricate nature of observational equipment makes intervals of several years almost unavoidable, but the great lengths of interval occurring are probably due to the unfounded prejudice of regarding many observational results as facts not requiring confirmation.

The persistence of observational errors being generally much greater than of theoretical errors implies that if the average fraction of errors produced in the two branches is roughly equal then the fraction of incorrect current observational work is considerably greater than the corresponding fraction for theoretical work. Careful statistical analysis would be required to confirm this statement and put it into quantitative form, but the arguments given here seem to support this conclusion strongly.

So far attention has been confined to physical theories rather than observational inferences, since there is a considerable structural difference between them from the point of view of scientific methodology. The fate of observational inferences is not encouraging to anyone thinking of relying on this method, many of them having turned out to be quite incorrect. Nevertheless they still continue to be made, mainly, though not entirely, in the context of observational papers. This seems to be the result of a deep human prejudice, that if only one continues to look at an object for long enough its nature will become apparent. In science this is, of course, nonsensical. One could stare at a piece of wood for years if not generations without discovering its atomic nature, or being able to infer its properties in any way from appearances. I must refer again to Dr. WILLIAMSON's article (1951) which is so usefully revealing in presenting current prejudices without any attempt at veiling or rationalizing them. He clearly considers it to be a valid argument that people who have never done actual observational research are not entitled to discuss astronomy. A more preposterous statement is hard to imagine. It is on the same plane as the statement that only plumbers and milkmen have the right to pronounce on questions of hydrodynamics.

As an example of an observational inference one can quote from STRUVE's book (1950, p. 116): "It looks as though the K-type and M-type dwarfs represent something in the nature of the final stage. . . . This part of the H-R diagram resembles a sink into which many stars drop . . .".

This is not the place for discussing such an idea in detail but I cannot help suspecting that the feeling described is at least partly due to a prolonged study of HERTZSPRUNG-RUSSELL diagrams drawn in the conventional way with the red dwarfs *near the bottom* of the picture.

The final point I wish to make concerns stated probable errors. All too often subsequent work has shown that they bear little relation to the actual errors made but are at most an indication of the internal consistency of the methods used. It would be a tremendous help if more observational papers were to contain (as some do now) a reasonable assessment of the errors that may have arisen. On the theoretical side it would be similarly of great advantage if more papers could contain clear explicit statements of the assumptions made, of every appeal to observation, and of every subsidiary hypothesis. Then the observational disproof of a theory would convey immediately valuable information. It would become clear that one of the bases of the theory was wrong, and such discoveries have frequently been very valuable, as for example, when the MICHELSON-MORLEY experiment showed that the velocity addition formula underlying the original theory was wrong. If these habits of clearly stating uncertainties and assumptions became general, and if prejudices regarding observational results as facts and theories as bubbles were overcome, astronomy would greatly benefit. Both so-called theory and so-called observation are liable to error, and critical appreciation and impartial scepticism are the best foundations for progress.

REFERENCES

ADAM, M. G. . . . . 1948 *M.N.*, **108**, 446.
ADAMS, W. S. . . . . 1925 *Proc. Nat. Acad. Sci.*, U.S.A., **11**, 382.
BOK, B. J. . . . . 1946 *M.N.*, **106**, 61.
CHANDRASEKHAR, S. and SHOENBERG, M. . . . 1942 *Ap. J.*, **96**, 161.
COWLING, T. G. . . . . 1934 *M.N.*, **94**, 768.
1935 *M.N.*, **96**, 42.
DUNHAM, T. F., Jr. . . . . 1939 *Proc. Amer. Phil. Soc.*, **81**, 277.
EDDINGTON, A. S. . . . . 1930 *The Internal Constitution of the Stars*, p. 159 (Cambridge).
GOLD, T. . . . . 1949 *M.N.*, **109**, 115.
HOYLE, F. and LYTTLETON, R. A. . . . . . 1946 *M.N.*, **106**, 525.
HUBBLE, E. and HUMASON, M. L. . . . . . 1931 *Ap. J.*, **74**, 43.
1934 *Proc. Nat. Acad. Sci.*, U.S.A., **20**, 264.
HUBBLE, E. . . . . 1935 *Ap. J.*, **81**, 334.
1936 *Ap. J.*, **84**, 270.
JEANS, J. H. . . . . 1925 *M.N.*, **85**, 914.
1927 *M.N.*, **87**, 400, 720.
1929 *Astronomy and Cosmogony*, p. 381 (Cambridge).
KUIPER, G. P. . . . . 1938 *Ap. J.*, **88**, 472.
LASSELL, W. . . . . 1847 *M.N.*, **7**, 157, 167, 297, 307.
LINDHOLM, E. . . . . 1941 *Ark. Mat., Astron. Fys.*, **28**B, No. 3.

MAANEN, A. VAN	1916	*Ap. J.*, **44**, 210.
	1921	*Ap. J.*, **54**, 237, 347.
	1922	*Ap. J.*, **56**, 200, 208.
	1923	*Ap. J.*, **57**, 49, 264.
	1925	*Ap. J.*, **61**, 130.
	1927	*Ap. J.*, **64**, 89.
MCKELLAR, A.	1949	*Publ. Astron. Soc. Pacific*, **61**, 199.
RICHARDSON, R. S. and SCHWARZSCHILD, M.	1948	*Ap. J.*, **108**, 373.
SCHWARZSCHILD, M.	1948	*Ap. J.*, **107**, 1.
SPENCER JONES, H.	1941	*M.N.*, **101**, 356.
STRUVE, O.	1950	*Stellar Evolution*, pp. 18–20 (Princeton).
	1951	*Proc. Nat. Acad. Sci.*, U.S.A., **37**, 327.
TRUMPLER, R. J.	1935	*Publ. Astron. Soc. Pacific*, **47**, 254.
WILLIAMSON, R.	1951	*J. Roy. Astron. Soc. Canada*, **45**, 185.

Philosophical Aspects of Cosmology

HERBERT DINGLE

Department for History and Philosophy of Science, University College, London

1. THE purpose of this paper is to indicate, but not to solve, certain philosophical problems peculiar to the study of cosmology. Cosmology and cosmogony—no attempt is made here to distinguish them, and they are in fact inseparable—are at the present time primarily scientific subjects; that is to say, we are no longer forced to approach them on grounds of pure reason alone but have in our possession a growing body of observed facts which must form the data on which our reason begins to operate. Nevertheless, they have the peculiarity that in them we treat the whole field of investigation as having characteristics of its own, independent of the characteristics of any of its parts, and it is those universal characteristics that we seek to discover. But, up to the present at least, the part of the universe that we can observe is at most only a very small portion of what we have reason to believe exists. We have therefore to introduce considerations over and above the ordinary scientific process of inductive generalization, and those considerations are philosophical in character and so give to cosmology a philosophical aspect which the other departments of science do not show in the same degree.

An example will make the point clearer. When NEWTON demonstrated that within the solar system the movements of bodies everywhere conformed to the law that every piece of matter attracted every other piece of matter with a force varying directly as the masses of the bodies and inversely as the square of the distance between them, this law was generalized to apply to matter everywhere, and so became known as a *universal* law. But it was not a law of the universe; it was a law that, supposing it to be true, was exemplified wholly and completely in every part of the universe, but it had nothing to do with the universe as a whole. The universe might be large or small, finite or infinite, eternal or temporary, homogeneous or heterogeneous—in fact, it might, as a whole, have any conceivable characteristics at all, and the Newtonian law of gravitation would be the same in all cases. Similarly, the laws of thermodynamics are universal laws but not laws of the universe. Again assuming them to be true, they characterize any closed system whatever, small or large, irrespective of whatever else the universe might contain; we therefore learn from them nothing at all about the character of the universe *as a whole*. On the other hand, when you say that space has a positive or a negative or a zero curvature, you are stating a law of the universe, but you are saying nothing at all about any particular part of the universe. The curvature in a given region may be anything; it is only the universe *as a whole* that is supposed to have the curvature you postulate. Accordingly, the assertion that the universe has this curvature must rest on something other than direct generalization from observation or experiment in limited regions; it must involve reasoning over and above the ordinary type of scientific reasoning. It is the problems connected with that kind of reasoning that are most properly called philosophical problems of cosmology.

2. By its very definition the universe is necessarily unique. The first question that arises, therefore, is whether its uniqueness is, so to speak, essential or accidental. That is to say, can we properly imagine different kinds of universes that might have been formed, and so present ourselves with the problem of showing why it was in fact the actual one that came into existence; or must we take it as a primary axiom that any conceivable alternative to the actual universe must necessarily be impossible? This question must be answered before we can decide whether or not familiar scientific methods of attack are legitimate when applied to the universe as a whole. For example, in the statistical mechanics of GIBBS, when we wish to study the behaviour of a sample of gas, we first of all suppose the sample to be composed of a very large number of molecules, each behaving in an unspecified—or at most an incompletely specified—manner, and then consider a very large number of such samples. Owing to the incomplete specification we cannot deduce with certainty how this *ensemble* of samples will behave, but we can deduce its probable behaviour. We then draw conclusions, from this probable behaviour of the *ensemble*, about the actual behaviour of the sample in which we are interested.

Now whatever philosophical doubts may be aroused as to the validity of deductions made in this way, there can be no doubt concerning the possibility of making the investigation. Undoubtedly a very large number of samples of gas of the kind postulated do exist, and undoubtedly, if they are composed of molecules, we must leave unspecified the instantaneous positions and momenta of those molecules and therefore have no right to assume that such positions and momenta in the various samples bear any relation to one another. But have we any right to apply the same kind of treatment to the universe? Can we consider an *ensemble* of universes, in each of which the stars and nebulae can form any configuration whatever, and then draw conclusions about our actual universe from the probability of behaviour of this imaginary *ensemble*? It is at least a plausible proposition that we have no right to talk of the probability of an event before we are satisfied that the event is possible. We have, so far as I know, no assurance that any universe other than one *is* possible.

3. Consider now another example—not an imaginary one, but one that has revealed itself in the recent history of cosmology without, apparently, arousing the philosophical questions that should have been asked. The notion of "stability of equilibrium" is a very familiar one in mechanics. A system is said to be in *equilibrium* when, in the absence of disturbances from the outside, its state remains unchanged for an indefinite time; the equilibrium is said to be *stable* when, if the system is momentarily disturbed by an indefinitely small external impulse, it automatically

tends to return to its former state; if, on the other hand, its reaction to such an impulse is to remove itself further from its former state, it is said to have been in *unstable* equilibrium. Without the external impulse we cannot tell whether the equilibrium is stable or unstable; it is simply equilibrium. Now suppose we are considering the universe, and suppose we have reason to think that it is in equilibrium; how can we tell whether its equilibrium is stable or unstable? Only by disturbing it from the outside. But by hypothesis there is no outside; the universe, whether finite or infinite, comprises all that is. Hence the distinction between stability and instability ceases to have any meaning when applied to the universe. Now it is well known that, according to the cosmology of the late Sir ARTHUR EDDINGTON, the universe was once in equilibrium; it was the static EINSTEIN universe familiar in the early days of the relativity theory. But EDDINGTON showed that this universe was unstable—that is to say, if it was disturbed it would not return to its original state but would start and continue to expand or contract—and in this way he accounted for the present observed recession of the extra-galactic nebulae. But what, then, disturbed it? If something outside, the system considered was not the universe; if something inside, the system considered was not in equilibrium. Actually EDDINGTON chose the latter alternative, but he did not face the problem of explaining how the original universe could have been both in equilibrium and not in equilibrium. Yet the problem remains, and is clearly not one to be settled by observation; it is a problem of the philosophy of science.

4. A more general aspect of problems of this kind is presented by the following question: can we, in a description of the universe, maintain the separation that has been implicit in science at least from the seventeenth century onwards, between universal laws of nature and the particular material system that actually exists and behaves in accordance with those laws? It is probable that, at any rate up to very recent times, few would have questioned the legitimacy of this separation. For instance, suppose it had been asked: "If there had been another major planet between Mars and Jupiter, would the law of gravitation have been different?" The almost unanimous answer, I think, would have been: "No, the law of gravitation is true independently of the actual bodies which happen to be in the universe. The introduction of another planet would affect in some degree the motion of every body in the universe, but the new motions would still obey the same law of gravitation". The reverse question: "If the law of gravitation had been different, would the number of bodies in the solar system have been different?" would, I think, have been regarded as scarcely permissible; the law of gravitation would have been accorded a stronger element of necessity, so to speak, than the actual population of the universe. But if the question had been pressed, the answer would probably have been that one could not say with confidence, but that probably there would have been a difference. The process of formation of the planets, taking place according to a different law, would certainly have been different in detail, and that might very well have affected the number of planets formed. The general picture was that of eternal and unchangeable laws, independent of any material creation that might or might not exist to display them; and superposed on that, a particular arrangement of bodies, whose number and disposition must result from the particular process that brought them into being.

Such a picture has proved invaluable for the study of isolated systems—that is, systems bounded in space and time. We choose an arbitrary moment at which we observe their state, and then calculate what the universal laws will do with bodies in that state; and in innumerable instances our calculations are justified by the event. It is therefore tempting to apply the same way of thinking to the problem of the universe. But is this justifiable? Apart from the impossibility, which relativity has revealed, of uniquely associating a particular spatial boundary with a particular instant of time—an impossibility whose effects become more and more serious the larger the system we study—we find that we must choose our "arbitrary" instant of time at the beginning of things, since the further out in space we go, the earlier is the time at which we must necessarily observe the state of the bodies there. In other words, we have no longer, in fact, an *arbitrary* instant to start from; the moment of creation itself, if that is the proper name for it, is the only possible one. Similarly, our spatial "boundary" must be removed to an indefinitely great distance, for we cannot screen off a portion of the universe from the rest and generalize from that. We are thus forced to start at the origin of the whole of things, and how can we be sure that the laws which we deduce from the contemporary behaviour of small regions are necessarily the laws that operated everywhere then?

5. The question is closely bound up with the view we take of the source of our knowledge. The traditional scientific belief has been that both the laws of nature and the structure of the material system that is the visible body of nature are knowable only through observation, through experience. If that is so, then it becomes very difficult to maintain that the laws are essentially independent of the structure. The observed behaviour of the universe now is the single source of both elements of our knowledge; if that were different—as we all believe it would be if the configuration of bodies were different—the proposition that we would infer the same laws becomes very dubious. And, indeed, the modern relativistic approach no longer assumes the old independence of laws and structure. Relativistic cosmology gives us in the same formula the structure of a system and the motions occurring in it, and we cannot have one without the other. The EINSTEIN universe contains so much matter, and it is necessarily static; the DE SITTER universe is expanding, but necessarily has no matter in it; the LEMAÎTRE universe is expanding, and necessarily has a homogeneous distribution of matter having certain characteristics and no others. This is quite consistent with a purely empirical approach in which we make no presuppositions but describe as a whole the kind of universe we observe. A separation between laws and structure would indeed be plausible, if not inevitable, if one had a rational and the other an empirical basis—if, that is to say, the laws of nature, but not the appearances of the constellations, could be derived by reason without recourse to experience. This, in fact, was maintained by EDDINGTON, and on other grounds by MILNE, though EDDINGTON at least accepted the relativity principle that bound the two inseparably together. It is somewhat paradoxical that when such a rational origin for the laws of nature would have justified the separation that was universally assumed, no scientist admitted it, and now that the separation is discredited, cosmologists appear who hold a view that would demand its acceptance. Here is perhaps a philosophical problem of psychology, but cosmology affords scope enough for our present troubles.

6. Space forbids a consideration of the problems concerning time and the origin of

the universe. It must suffice to mention, and to question, one assumption that has been tacitly made in some recent theories, namely, that, provided a hypothesis is consistent with experience and contains no self-contradictions, it is valid, and therefore that anything postulated to have happened before human experience began is exempt from the requirement that it must conform to our possibilities of imagination. It would perhaps not be easy to give a brief direct refutation of this, but let us look at its implications. Nearly 100 years ago PHILIP GOSSE, in order to reconcile the facts of geology with the Hebrew scriptures, advanced the theory that, in his son's words, "there had been no gradual modification of the surface of the earth, or slow development of organic forms, but that when the catastrophic act of creation took place, the world presented, instantly, the structural appearance of a planet on which life has long existed".* The beginning of the universe on this theory occurred some 6000 years ago. There is no question that the theory is free from self-contradiction and is consistent with all the facts of experience we have to explain; it certainly does not multiply hypotheses beyond necessity since it invokes only one; and it is evidently beyond refutation by future experience. If, then, we are to ask of our concepts nothing more than that they shall correlate our present experience economically, we must accept it in preference to any other. Nevertheless, it is doubtful if a single person does so. It would be a good discipline for those who reject it to express clearly their reasons for such a judgment, but the matter is raised here merely to show that we cannot grant the inaccessible past freedom to sow its wild oats as it pleases provided that it takes care to observe the proprieties when experience begins; it must to some extent conform to the pattern on which we organize the present behaviour of the universe.

Modern Cosmology and the Theologians

M. DAVIDSON
The College of St. Mark, Audley End, Essex, England

1. STATEMENT OF THE PROBLEM

SOME recent pronouncements on cosmology have left a number of theologians in a state of bewilderment and perplexity. Many of them know well that a restatement of traditional doctrines which rest on a discredited cosmology would lead to a serious disturbance in the faith of a large number of people, and for this reason they often maintain a discreet silence. Some are disposed to adopt a suspicious attitude towards theories which apparently undermine faith in the Christian dogmas, but they welcome pronouncements from men of science which seem to support such dogmas. It is suggested that, while discredited cosmologies should not be used to bolster up the Christian Faith, nevertheless theologians should not hastily accept the latest pronouncements of cosmologists as the final word on the subject and should exercise great caution before accepting them as necessarily supporting, or proving destructive

* *Father and Son*, EDMUND GOSSE, Chapter V.

to, religion. An instance is given of the same law—the Second Law of Thermodynamics—leading to diverse theological conclusions, and the inference is that theologians may find it necessary to adopt a more independent attitude towards cosmology. The same remark applies to other sciences but these are not considered in this discussion.

A short summary is given of the views of some cosmologists on theological questions, but a *caveat* on this matter is considered advisable and there follows a quotation from a book by the Astronomer Royal as a warning that astronomers are not necessarily better equipped than others to make pronouncements on the purpose behind the universe. Towards the end a reminder is given to men of science that their explanations are not necessarily exhaustive and that there may be many unmanifested properties in nature. Further, just as the physicist can be said to make use of "myths"*—using the word with a more comprehensive meaning than usual—which serve a useful purpose, so there may be some justification for myths in theology, for example, the myth of the Fall which attempts to explain the origin of evil in the world.

Dr. Inge (1932) has pointed out that there are three positions which the Church may adopt in dealing with astronomical development. The first is to condemn astronomical science as impious—an attitude which no one except an extreme Fundamentalist is likely to adopt. The second is to admit that traditional doctrines do not belong to the natural world with which science deals, but this does not necessarily deny that they may possess a higher truth outside the reach of science and hence may be regarded as symbolic of eternal truths. The third is to recognize the necessity for recasting all theological doctrines which rest on the geocentric theory of the universe. Perhaps it was unfortunate that Dr. Inge did not develop a scheme for recasting these theological doctrines, though he expressed the opinion that anything was better than trying to conceal an open sore which destroys our joy and peace in believing. A complete recasting could, however, and possibly would, lead to devastating effects on the faith of many earnest Christians.

Many to-day accept theological doctrines which apparently rest on a discredited cosmology and it is equally true that many others have lost interest in theological cosmologies, chiefly on the grounds of their naïve anthropocentrism and—partly as a consequence of this—of their anthropomorphism. It is remarkable that the anthropocentric attitude still prevails amongst some eminent men of science whom we should truthfully describe as deeply religious, and they experience less disturbance to their faith than do many theologians whose lives have been devoted to a study of some of the deeper problems of life. Nevertheless, theologians are confronted by serious difficulties and we shall now turn our attention in particular to those that arise in consequence of the impact on theology of developments in cosmology.

2. Necessity for Independence of Views amongst Theologians

In 1948 a paper with the title "Science and Christian Modernism" was read at the Modern Churchmen's Conference at Oxford and a brief reference will now be made to a few of the points discussed in this paper (Davidson, 1948).

It was emphasized that the hasty acceptance of scientific theories and using them ot support cherished theological views could be detrimental to theology itself. On the other hand, the rejection of scientific theories merely on the grounds that they

* It is not stretching the meaning of the word too far if it is accepted that mythology is an attempt to account for facts in the natural order, and that it is more like primitive philosophy than primitive science.

seemed to be opposed to certain tenets of the Church (the word "Church" is used in its most comprehensive sense) could be even more disastrous. Within comparatively recent times certain pronouncements were made on problems of indeterminancy, freedom of the will, *etc.*, in the light of the developments in atomic physics, which were eagerly welcomed by some theologians—not always to the advantage of theology —but as this paper is restricted to problems of cosmology these will now be considered.

In the paper previously mentioned an example was taken from the law of the increase of entropy on which some Christian apologists have been disposed to build an imposing edifice, and it was pointed out that some day this law might be shown to be valid only under local and restricted conditions and inapplicable in dealing with the universe. If this should happen—and there is the possibility that it may, as shown by R. C. TOLMAN (1934)—the structure erected on this foundation would be like the house in the parable that was built on the sand. The conclusion from this example was, that while the Church should not shrink from presenting the Christian Faith in terms of contemporary thought, at least as far as it is competent to do so, nevertheless it was essential that it should preserve a certain amount of independence of opinion. It is very unwise for theologians to hang on to the mantles of men of science merely because some of their pronouncements seem to support certain cherished theological tenets. In addition, diametrically opposite views in the realm of theology have been deduced from the acceptance of scientific theories, and this will now be shown to have occurred on the basis of the Second Law of Thermodynamics.

In a recent work by C. F. VON WEIZSÄCKER (1951), we are informed that the theory of evolution and the Second Law of Thermodynamics were put forward about the middle of last century as purely scientific theories, and the former quickly became the battle cry of every modern mind. On the other hand, the Second Law of Thermodynamics remained a technical detail of physics, and subterfuge was used to evade its application to the world as a whole. The reason given for this subterfuge is interesting —"For the prospect of the heat death of the world, however far off in the future, would have shaken the faith that life has meaning".

It is remarkable that, while this gloomy prospect was disturbing to the faith of some, it has had exactly the opposite effect with others as the following instance in recent times will show.

About ten years ago Sir EDMUND T. WHITTAKER (1952), dealing with the law of the increase of entropy, made the following statements: "The knowledge that the world has been created in time, and will ultimately die, is of primary importance for metaphysics and theology; for it implies that God is not Nature, and Nature is not God; and thus we reject every form of pantheism, the philosophy which identifies the Creator with creation, and pictures him as coming into being in the self-unfolding or evolution of the material universe. For if God were bound up in the world, it would be necessary for God to be born and to perish".

Later on WHITTAKER speaks about the evolutionary process in which the individual counts for nothing amongst the lower forms of life, whereas in humanity the race loses its value and the individual acquires the supreme value. Nevertheless, though mankind and all his works must vanish away (and everyone will agree that this must happen if we accept the validity of the Second Law of Thermodynamics), yet something remains: "The goal of the entire process of evolution, the justification of creation, is the existence of human personality: of all that is in the universe, this alone is final and has abiding significance: and we believe that this has been granted, in the eternal purpose of God, in order that the individual man, born into the new creation of the Church, shall know, serve, and love Him for ever".

Here we have two almost antithetical views emanating from the same law; the first offers little inducement to mankind to survive; the second, anthropocentric though it may seem, does supply a motive for persevering against the most formidable obstacles, and if accepted, would go a long way towards dispelling the gloom and pessimism which overshadow the world to-day. Without offering any opinion on the validity of either argument, one may merely add that it would serve no useful purpose to disturb the faith of those who believe that there is purpose in the universe and an ultimate goal for mankind. The two examples mentioned above (and others could be given) should be a warning to theologians. They should beware of accepting the latest pronouncements of cosmologists as final and of accepting them for destructive or constructive purposes, and equally they should avoid all manifestations of undue alarm at such pronouncements. That this last warning is not out of place may be shown by the following instance.

At the Modern Churchmen's Conference at Cambridge in 1950 a number of the members were very disturbed by some recent pronouncements of a cosmologist who had expressed the opinion that Christianity offers merely an eternity of frustration. Their main anxiety was due to their dread that the faith of many might be undermined by such a statement. The writer of this article, who did not share their view in the matter, endeavoured to reassure them (not, it is to be feared, with much success). It is very difficult to understand why theologians should be perturbed by a personal view of this kind which, after all, cannot be regarded as authoritative. Incidentally, it may be remarked that the Founder of Christianity warned His followers that frustration—temporary at any rate—would be their lot, and frustration is not necessarily an evil thing. That the cosmology of the book (HOYLE, 1950) is very interesting, no one will deny, nor will anyone deny that previous cosmologies have also been most interesting; but the final words on the subject have not yet been spoken. Regarding the pronouncements of cosmologists on some of the deeper issues of life, the following quotation from the end of a work by the Astronomer Royal, Sir HAROLD SPENCER JONES (1952), is relevant: "The task of the astronomer is to learn what he can about the universe as he finds it. To endeavour to understand the purpose behind it and to explain why the universe is built as it is, rather than on some different pattern which might have accorded better with our expectations, is a more difficult task; for this the astronomer is no better qualified than anyone else".

These words should be a standing rebuke to all cosmologists who venture to make pronouncements in realms in which they have no special qualifications to do so.

3. RECENT COSMOLOGICAL PRONOUNCEMENTS

We shall now consider the most recent views of a well-known cosmologist, though it must be admitted that many of his theories have not met with general approval. The late Prof. E. A. MILNE, who wrote extensively on the subject of cosmology, was a deeply religious man, and in one of his works which appeared nearly twenty years ago, MILNE (1935) claimed that, in the view of the universe that he advocated, he could say he had found God. "For the universe seems to be a perfect expression of those extra-temporal, extra-spatial attributes we should like to associate with the nature of God". Later he admits that the physicists and cosmologists need God only once, to ensure creation, but in a recent—and posthumous—work MILNE is more definite, and a short summary of his views expressed in this book (MILNE, 1952) now follows.

The main thesis of the book is that God formed the visible universe, that it bears the marks of a created article, and that He has also created the laws of nature. These are not special *ad hoc* creations but, following the view of ERNST MACH, the laws of nature are a consequence of the structure of the universe. MILNE found evidence of the creation of the universe by a rational Creator from the rationality of the universe itself. We are able to know something about the rules that nature obeys from the study of our own rational thought-processes which inform us of the limitations placed on nature if it is to be rational. He rejects the hypothesis of BONDI and GOLD (1948) (see also BONDI, 1952) on continuous creation of matter which, he says, is irrational "because it requires the specification of a rate of creation of matter per unit volume of space". He thinks that no reason can be given for the choice by God of any particular value of this quantity, but asserts that his own hypothesis that the universe was created as a point-singularity is not open to such objection.

Many will probably think that this argument is by no means convincing and they will not feel that much assistance is afforded by MILNE's definition of "rational", which is as follows: "We say that the universe is rational when laws of nature, predicted *a priori*, are found observationally and experimentally to be obeyed in nature". This definition is largely based on MILNE's view that science should aim at becoming less empirical and more deductive, but even if it is admitted that physics should become more deductive, we cannot but ask how far the deductive method is applicable to biology. Obviously MILNE foresaw a difficulty here because he admits that he has not excluded the possibility of divine interference in the details of biological evolution. Readers will probably agree that a serious difficulty arises once such divine interference is admitted, because the deductive method is then inapplicable unless it is capable of predicting the minute details of the mind of the Creator. Limits of space prevent an exhaustive examination of this book but one other point will be referred to.

In the last chapter MILNE regards the infinity of the galaxies as an infinite number of scenes of experiment in biological evolution, but he is perplexed by the difficulty felt by many Christians in connection with the infinite number of planets that have evolved from the infinite mass brought into existence at the epoch $t = 0$. He even discusses the question of the incarnation—whether it was a unique event on the Earth or whether it has been re-enacted on each of a countless number of planets. He gives reasons—not very convincing from the theological point of view—for favouring the former view and suggests that some day we may be able to signal the news to other planets. He thinks it is not outside the bounds of possibility that radio-signals apparently proceeding from the Milky Way are genuine signals from intelligent beings from other planets, and that "in principle, in the unending future vistas of time, communication may be set up with these distant beings".

There are, of course, other cosmologies, but as they do not deal so fully with the theological implications as MILNE'S, they do not lie within the sphere of the present discussion. As a personal opinion only, which must not be regarded as expressing the views of theologians in general, it might on the whole be better if cosmologists confined themselves to their own particular subject and left the philosopher and the theologian to draw their own conclusions.

4. Suggestions on a Constructive Policy

Can anything constructive be built on the results of modern research in physics and cosmology (it is impossible to separate the two in such an enquiry)? First of all, physicists and cosmologists should realize—as indeed most of them do—that their scientific explanations are by no means exhaustive, and while they can observe such things as masses, lengths, times, *etc.*, there may be an almost unlimited number of other properties which are not only unmanifested but which, for the greater number at least, will ever remain so. Then they may be prepared to give some sympathetic consideration to certain religious tenets—even to those that are based on mere mythology which finds a place in the Christian religion. Physicists themselves are not averse to the use of "myths"—if the word may be used to signify the basis of some of their theories. It is only necessary to mention such myths as the ether of space which, with all the extraordinary properties with which it was endowed, once played an important rôle. Are the "solar system"-atoms, the various fundamental particles, the differential equations of wave mechanics, and so on, anything more than myths used to describe a number of puzzling phenomena? They are not permanent myths but serve their purpose until some less crude myths supplant them. In the same way, is the theologian not justified in using the myth of the Fall of Man to explain the fact of evil in the world—a fact which few men of science will deny? Admittedly it is a crude explanation, but there is a deep moral significance in it all which even biological evolution has not yet destroyed. Other similar instances of the use of mythology or of legendary lore could be cited but sufficient has been said to suggest that men of science and theologians, each in their own sphere, can come more and more to see each other's point of view. This is all the more necessary amid the perplexities of the present time when the eternal questions, Whence? Whither? Wherefore? still remain unanswered.

5. Some Quotations

In conclusion, the following quotations from men of science and philosophers, are relevant to certain points discussed in this chapter.

"Science is a certain kind of knowledge, notable for its high degree of reliability within its self-imposed limitations; so it must necessarily contribute to philosophy in the wide sense in which we have used the word. Religion is, in part, also a kind of knowledge, relating but little to things, and more to men—both being considered in their relation to God". (Taylor, 1951.)

"Philosophy and science do not answer our deepest questions, nor do they solve effectively our most pressing problem, that of human conduct". (Taylor, 1951.)

"Examples of legitimate philosophies are numerous. One thinks, for example, of various systems of theology for which men have been ready to die, yet which, with larger experience, have been discarded. Their holders have held them indispensible for rationalizing their own limited religious experience, and it would have been the height of folly for them to have exchanged such theologies for a not yet achieved inclusion of religious experience within the scientific scheme". (Dingle, 1952.)

"I cannot, however, help feeling it is likely to be more important for religion in the future to have a theology that is founded on the reality of religious experience, than to have one that builds its doctrines upon supposed events in the past: supposed events which some of the best scholars of history are unable to regard as established beyond doubt by the rules of evidence accepted in other fields of historical research". (HARDY, 1951.)*

REFERENCES

BONDI, H. . . . . 1952 *Cosmology* (Cambridge University Press).
BONDI, H. and GOLD, T. . . . . 1948 *M.N.*, **108**, 252.
DAVIDSON, M. . . . . 1948 *The Modern Churchman*, XXXVIII (September 3rd) (Blackwell, Oxford).
DINGLE, H. . . . . 1952 *The Scientific Adventure* (Pitman, London).
HARDY, A. C. . . . . 1951 *Science and the Quest for God* (Lindsey Press, London).
HOYLE, F. . . . . 1950 *The Nature of the Universe* (Blackwell, Oxford).
INGE, W. R. . . . . 1932 *The Church and the World* (Longmans, London).
MILNE, E. A. . . . . 1935 *Relativity, Gravitation, and World Structure* (Oxford University Press).
1952 *Modern Cosmology and the Christian Idea of God* (Oxford University Press).
SPENCER JONES, H. . . . . 1952 *Life on Other Worlds* (English Universities Press, London).
TAYLOR, S. . . . . 1951 *Man and Matter, Essays Scientific and Christian* (Chapman and Hall, London).
TOLMAN, R. C. . . . . 1934 *Relativity, Thermodynamics, and Cosmology* (Oxford University Press).
WEIZSÄCKER, C. F. VON . . . . 1951 *The History of Nature* (Routledge and Kegan Paul, London).
WHITTAKER, E. T. . . . . 1942 *The Beginning and End of the World* (Oxford University Press).

* A few weeks after the manuscript of this paper had been sent to the Editor, an article by RUDOLF BULTMANN, the substance of his broadcast, "What is Demythologizing?", appeared in *The Listener* of 7th February, 1953, which has some points similar to this article. This similarity, however, is entirely accidental, and the writer of this paper has not altered his original draft in consequence of Prof. BULTMANN's interesting broadcast which he has since read, but which he did not hear.

SECTION 3

DYNAMICS

"Nous devons donc envisager l'état présent de l'Univers comme l'effet de son état antérieur, et comme la cause de celui qui va suivre. Une intelligence qui, pour un instant donné, connaîtrait toutes les forces dont la nature est animée, et la situation respective des êtres qui la composent, si d'ailleurs elle était assez vaste pour soumettre ces données à l'analyse, embrasserait dans la même formule les mouvements des plus grands corps de l'Univers et ceux du plus léger atome; rien ne serait incertain pour elle, et l'avenir comme le passé, seraient présents à ses yeux. L'esprit humain offre, dans la perfection qu'il a su donner à l'astronomie, une faible esquisse de cette intelligence".

PIERRE-SIMON LAPLACE, *Essai Philosophique sur les Probabilités*, Paris, 1814.

Vistas in Celestial Mechanics

D. H. Sadler

H.M. Nautical Almanac Office, Royal Greenwich Observatory,
Herstmonceux Castle, Hailsham, Sussex, England

Summary

A brief review of the present status of celestial mechanics, in relation both to the need for increased accuracy and to the new methods made possible by the use of electronic digital computing machines.

---

1. Problems

The aim of that branch of astronomy known as celestial mechanics is, in its most general form, the determination of the motions of celestial bodies under their mutual gravitational attraction and under such other forces as may exist. The solar system has provided so far a sufficient number of problems, as well as the observations necessary for their numerical solution and subsequent verification. Most of these practical problems have been solved to a high degree of accuracy, enabling the positions of the Sun, Moon, planets, satellites, and comets to be predicted for many years in advance. But these solutions, although generally of the same or superior order of accuracy to the observations with which they are to be compared, are not entirely satisfactory.

There is no theoretical or literal solution to the n-body problem, or at least none admitting of practical development. The classical solution consists of the expression of the co-ordinates of the celestial bodies as trigonometrical series, with numerical coefficients, and with time as the independent variable; even so it has been applied rigorously only in the case of the Moon-Earth-Sun system and, in most other cases, the time is allowed to appear as a factor outside the trigonometrical expressions, so limiting the period of validity. Such solutions represent the culmination of mathematical and analytical skill developed over more than a century's effort by some of the world's foremost mathematicians; but the amount of labour involved, both theoretical and computational, is enormous and an extension of the accuracy, by these classical methods, is quite prohibitive in computational effort.

Thus the state of the subject before the second World War was an uneasy balance. On the one hand the immediate requirements of the practical astronomer had been met and there was little inducement for new work in a subject in which few new lines of attack were possible; on the other hand the unsatisfactory nature of the solutions, and their unsuitability for further extension, was recognized by the few workers in the field.

The great technological progress during the war has led to two, or possibly three, developments that have transformed the whole subject of celestial mechanics, in respect of both demands of accuracy and methodology. Firstly, the development of quartz crystal (and other) clocks and the possibility of improved observational techniques due to electronic observing and recording gave warning that the present standards of accuracy in the fundamental ephemerides of the Sun, Moon, and planets might soon be inadequate. Simultaneously, the development of electronic automatic digital computing machines not only provided the tool by which the required extension in accuracy could be made possible, but also gave rise to a new method of approach to the classical problems. The third development is in a very different field, namely that of "space-travel"; little serious work has been done on the many problems of celestial mechanics that will arise in the "navigation" of a "space-ship", capable of supplying a limited motive force to itself, in the gravitational field of the solar system. It is, however, probable that the practical problems will have to be solved in an *ad hoc* fashion, and it is certain that celestial mechanics will be able to supply all necessary solutions to the general problems concerned whenever the technological difficulties of rocket-travel have been overcome.

2. Ephemeris Time

The first and most important practical problem to be solved by the aid of celestial mechanics is the accurate determination of time. Clocks can keep time over short periods more accurately than astronomical observations can be made, but the latter still provide the only means over long periods. The rotation of the Earth can no longer be regarded as uniform, being subject to unpredictable erratic changes in rate in addition to the gradual retardation due to tidal friction and to small seasonal fluctuations. Yet day-by-day determinations of time must be made by observation of the directions of stars, relative to the rotating Earth; time so determined is Universal Time (U.T.), and is not uniform. A new time, Ephemeris Time (E.T.), defined by means of the revolution of the Earth round the Sun, has accordingly been introduced* and this is uniform, to the best of our present knowledge.

Ephemeris Time, being uniform, can be chosen so that its origin and rate can be identified with the independent variable ("time") of the theories, tables and ephemerides of the Sun, Moon, and planets. Actually it is in practice defined by reference to Newcomb's tables of the Sun, and the ephemeris of the Moon is modified† to bring both ephemerides into accord with this common argument. Ephemeris Time can be determined observationally by comparison of the positions of the Sun, Moon, planets (and satellites) in their orbits with those predicted by gravitational theory, without reference to the rotation of the Earth. Naturally, its value at any particular instant cannot be found by such observations to the high accuracy associated with Universal Time, but the average value of the difference between the two times can be obtained accurately over a period of, say, one year, and, with the better observing techniques now envisaged, perhaps in as little as a month. Owing to its rapid motion and in spite of the difficulty of accurate observation, the Moon offers the best means for the practical determination of E.T. by observation, and a precise ephemeris of the Moon is thus required.

3. Ephemeris of the Moon

A very accurate knowledge of the Moon's position and motion is also required for the determination of geodetic distances on the Earth's surface by means of observations of lunar occultations and solar eclipses.

This requirement is being met by the direct calculation of the lunar ephemeris from the numerical trigonometrical series developed by Brown (1919), thus avoiding

* Recommendation No. 6 of the Paris Conference on the Fundamental Constants of Astronomy, *Bulletin Astronomique*, tome XV, fascicule 4, p. 291 (1950). Adopted by the I.A.U. as Resolution No. 6 of Commissions 4 and 4*a*.

† Recommendation No. 5 of the Paris Conference (also Resolution No. 2 of Commissions 4 and 4*a*). See Sadler (1951).

the approximations necessarily involved in the use of the *Tables*. The synthesis of some 1500 separate terms of the series would have been quite impracticable without the aid of electronic computing machines, and is a considerable task even with such machines. The ephemeris, to $0^{s}.001$ in R.A. and $0''.01$ in Dec., is being computed from 1952 onwards, but will not appear in the almanacs in its proper place until 1960.

An artificial satellite of high albedo and of stellar appearance, revolving round the Earth in a period of a few hours, would serve as the hand of a clock of far greater long-term precision than anything available to-day. Such a body, even if revolving round the Earth in a stable orbit without self-generated forces, would provide problems in celestial mechanics of considerable difficulty if its position were to be calculated to an accuracy sufficient for the determination of time. It is possible that such artificial satellites will eventually be launched, but it is certain that the requirements of direct observation will be a secondary consideration; both observation and the computing problem will thereby be made more difficult.

One of the fundamental problems of celestial mechanics is to determine the motions, for all time, of a number of bodies of known mass moving under their mutual gravitational attraction, given at any one instant their positions and velocities in space. The very simplicity of its statement emphasizes its generality and its difficulty; it has attracted the attention of many eminent mathematicians, but, in spite of great mathematical achievements and ingenuity, no solution has been found which does not involve almost prohibitive numerical computation. Two entirely different methods have been used—those of "general perturbations" and of "special perturbations".

In the first of these methods the co-ordinates of the bodies are expressed as trigonometrical series with time as the independent variable. The series converge slowly even in the more favourable cases, while the possibility of a close approach of two of the bodies will usually make the method impracticable. But the availability of computing machines, which can be made, for instance, to multiply trigonometrical series of some hundreds of terms, is changing the significance of a given level of impracticability. In fact consideration is being given by BROUWER, CLEMENCE, and ECKERT, in the United States, to a comprehensive numerical solution of the motions of all the planets in the solar system. The great advantage of this classical method is that the positions of the bodies, consistent with the adopted masses, can be obtained, relatively simply, for any time merely by substituting that time in the trigonometrical series.

4. SPECIAL PERTURBATIONS

The alternative method consists of the direct numerical integration, by step-by-step methods, of the differential equations defining the motions of the bodies concerned. It has the great merit of simplicity of principle and involves no mathematical analysis; but it suffers from the disadvantage inseparable from numerical work of this nature—namely the accumulation of building-up errors. This means that no solution can be valid, to the required precision, for more than a limited time; the number of significant figures required to be kept in the integrations increases too rapidly with the number of steps. Until the advent of the automatic computing machines, the method had been limited to various special problems generally in which the positions of the perturbing bodies were assumed known, so that only three second-order equations had to be solved. The position is very different now:

this method has been used by BROUWER, CLEMENCE, and ECKERT (1951) on the I.B.M. Selective Sequence Electronic Calculator to compute the positions of the five outer planets (Jupiter to Pluto) for the period 1653–2060, with such success that the results have been adopted by the International Astronomical Union as the basis for the fundamental ephemerides from 1960 onwards. NAUR (1951) has demonstrated

Fig. 1. A general view of the Selective Sequence Electronic Calculator (SSEC) of the International Business Machines Corporation, by whose permission this photograph is reproduced. This machine was used for the calculation, by numerical integration, of the fundamental ephemerides of the five outer planets, and for the direct calculation of the lunar ephemeris from BROWN's theory as modified according to the resolution adopted at the Rome meeting, 1952, of the I.A.U.

the power of EDSAC (an automatic computing machine in Cambridge) in computing the very precise orbits of minor planets required as fundamental standards of reference for right ascensions. Many other integrations (Jupiter's outer satellites, Icarus and other special minor planets), now being laboriously done by hand, will doubtless soon be done on these automatic machines (Fig. 1 and Fig. 2).

The method of special perturbations thus now offers a practical solution to many of the problems of celestial mechanics; but it can never replace entirely the classical method, since it fails explicitly to exhibit the dependence of the perturbations on the masses and periods of the perturbing bodies. The principal terms of the expansions of the classical method are still essential to the full understanding of the physical (and mathematical) basis of the theory. Moreover, the short time-interval necessary for the inner planets makes numerical integration unsuitable for fundamental orbits, for which an essential condition is consistency and continuity over the whole period of observation. A new theory of Mars has accordingly been undertaken by

Fig. 2. A general view of the EDSAC (Electronic Delay Storage Automatic Calculator) at the University Mathematical Laboratory, Cambridge, by whose permission this photograph is reproduced. A programme has been worked out and used on this machine for computing the orbit of a minor planet with high precision under the attractions of the Sun and perturbing planets

CLEMENCE (1949), and the first-order theory has already been published. In this theory, which shows up some significant errors in NEWCOMB's theory, all periodic terms are included whose coefficients are as large as 0″0001; it must surely be the most carefully planned and accurately executed planetary theory ever undertaken, and could hardly have been possible with the calculating equipment available in NEWCOMB's day. Good progress is being made with the second-order theory, for which the published work furnishes such a sound basis; it is already clear that the perturbations of the second order will introduce further substantial corrections to the present theory.

No reference has so far been made to the by-no-means inconsiderable theoretical work that is in progress in celestial mechanics. Although in practice numerical answers are nearly always required, there are many questions to which the final ephemeris does not give the answer directly—for instance, secular perturbations of the elements of the orbit. It is a measure of the rising interest in the subject, fostered largely by the "school" in the United States and now actively encouraged by the ability to undertake computations that were previously impracticable, that no less than three major textbooks on celestial mechanics are being written, after a gap of nearly thirty-five years!

After being in a state of suspended progress for many years, celestial mechanics thus is faced simultaneously with new problems and with the mechanical means of solving them. There is no doubt that the theory will keep pace with the demand and that, even in the difficult field of man-made bodies, it will continue to provide, as it must, solutions to an accuracy greater than that of observation.

References

Brown, E. W. . . . . 1919 *Tables of the Motion of the Moon*, New Haven.

Clemence, G. M. . . . . 1949 *Astronomical Papers Prepared for the Use of the American Ephemeris and Nautical Almanac*, Vol. XI, Part II.

Eckert, W. J.; Brouwer, Dirk; and Clemence, G. M. . . . . 1951 *Ibid.*, Vol. XII. See also the review in *Observatory*, 1952, **72**, 35.

Naur, P. . . . . 1951 *M.N.*, **111**, 609.

Sadler, D. H. . . . . 1951 *M.N.*, **111**, 624.

An Introduction to the Eclipse Moon

R. d'E. Atkinson

Royal Observatory, Greenwich, London, S.E.10

Summary

Observations of the Moon's place by different methods differ in such important systematic ways that it is advisable in practice to think of the Earth as accompanied by a number of different, but incomplete, Moons, whose positions, though nearly identical, are in many ways logically independent. The seven most important of these Moons are briefly described, and the various methods which have been used to study the place (and diameter) of the "Eclipse Moon" are summarized, including a method devised by the author and successfully tried out in 1948 and 1952. Either a total or an annular eclipse can now give the (Eclipse) Moon's place more accurately than a year's occultations can give that of the Occultation Moon; the important problem of fusing these two (and others) of the aggregate of Moons is briefly touched on.

The *Nautical Almanac* will shortly start giving the Moon's place to a hundredth of a second of arc, and it already gives the semidiameter and parallax to this accuracy. The precision of individual observations is, of course, considerably less; but it is not at all the case that the observed values differ from the computed ones by amounts which are simply random. The Moon's outline is full of features which vary systematically, and it departs fairly often by more than two full seconds from the best circular arc that can be fitted to it at any given moment; there is also evidence that the best semicircles for the east and west limbs have slightly different radii (and perhaps also different centres), and that these differences may change as the libration changes [1]. The libration does in fact change rapidly in a two-dimensional pattern, with two incommensurable periods, and over a range large enough, in each co-ordinate, to sink even the largest mountain entirely below the limb; thus the apparent place inferred for the Moon from any given observation will depend greatly on the librational situation at the moment, and on the point or points actually observed, but also on the extent to which the particular arc of a circle, to which those points could in principle be reduced by applying limb-corrections, differs from the complete circle that may be considered characteristic of the mean whole Moon. In practice, limb-corrections hardly ever are applied, as yet, to any observations of position (except in eclipses); moreover, different methods of observation are necessarily subject to different systematic errors; and since in addition they give systematically different weights to different parts of the limb at different times, including zero weight for large lengths of it during large (and systematically different) fractions of each month, it really is the case in principle that the earth is at present accompanied not by one Moon, but by a sort of out-of-focus aggregate of several incomplete Moons, which are logically distinct in the sense that a correction to the place of one of them is not bound to apply also to another. Any large correction, of course, such as that due to the E.T.-U.T. difference, can approximately be applied to all, and indeed no predicted place of the "Eclipse Moon" is ever based on previous eclipse observations at all, but only on observations of some other member of the aggregate; systematic disagreements with the places observed in eclipses are thus only to be expected, but reliable data are still somewhat scanty for assessing their actual values. The purpose of the present note is first to summarize the properties of these different Moons, and then to review briefly the observations which have dealt directly with the Eclipse Moon; it will appear that this is in many ways peculiarly isolated from all the others.

(1) *The Moon of Brown's Tables*

If the Empirical Term is removed, this Moon's place is based on pure gravitational theory; the disposable constants of the equations have been adjusted to suit long series of past observations as well as possible, but even so this Moon differs systematically from all the observational ones in two important respects: the place is given in terms of a strictly uniform time (E.T.) instead of a time involving the earth's rotation (U.T.); and the mean (orbital) latitude is necessarily zero, whereas the mean latitudes of the observational Moons are generally slightly negative. The latter fact may imply that the centre of apparent figure lies slightly south of the centre of gravity, presumably owing to the mountainous nature of the south polar regions; from the former consideration it follows [2] that the comparison of "Brown's Moon" with any observational one gives an indication of the difference between the two kinds of time. Brown's Moon has sometimes been supplemented by direct integration ("special perturbations") if its behaviour over a short interval was required with extreme accuracy, *e.g.* in an eclipse [3].

(2) *The Moon of Photographic Limb-Surveys*

(See Hayn [4], Watts [9].) This Moon's place is not observed at all, and its diameter is doubtful; but an accurate knowledge of its other characteristics is essential if any real improvement in the place and diameter of any other Moon is to be made. Only the bright limb exists, of course; and in fact even at "full" Moon there is usually "defective illumination" over a large fraction of one semicircle. Defective illumination is serious, not only because the (mean) terminator then lies a little inside the (mean) limb, but because the terminator *is* the limb as seen from the Sun, and the libration (and thus the limb-contour) differs, as between a solar and a terrestrial observer, by quite large amounts even for small extents of the defect. ($0''\cdot 1$ means $0\cdot 84^\circ$ of libration-difference.) This Moon exists only during hours of darkness, and most observations are thus made within a week of full moon; but even so it is clear that it will seldom be possible to use more than 200° or 220° of limb on any one photograph, without running into these libration-differences, and the complete 360° contour characteristic of any one pair of libration-co-ordinates (l, b) can only be built up out of pieces. There are, however, obstacles to this; the absolute radius depends on the seeing and transparency and cannot be used to control the fit, and this lack can be directly made up only if one obtains substantial overlaps, between successive regions of the limb, on photographs taken at the same l and b. In practice, the same pair of l, b values does not usually recur again at all closely for a long time [5], so that one really has to build up a surface rather than a set of circles. Moreover, the complete circular outline for any one pair of l, b values would in any case remain incompletely observed at one point; photographs near full moon give a good overlap with both east and west limbs, but (at least as far as most outlines are concerned) they do so near one pole only. The "Cassini rules" have the effect that whichever pole is tipped towards us at full moon, the limb beyond it is always the defectively illuminated one, and the combination of the successive overlapping arcs (whether into circles or into a single sphere) cannot therefore be checked by proper closure-tests. Watts is at present engaged on a very thorough programme of constructing the best mean sphere characteristic of all observed arcs, but this work is necessarily very difficult.

(3) *The Heliometer Moon*

This somewhat resembles the former, except that observations are visual, and that points on the limb are also related to craters on the surface (usually Mösting A.) It is this work which has recently given information on the difference between east and west radii, and their dependence on libration; the mean inclination and node of the Moon's equator, and the constants of the "physical" libration, are also inferred from it [6].

(4) *The Moon of the Meridian-circle Observer*

This Moon does not exist within about three days either way from new moon, but it can exist in full daylight; its east "limb" (really only one point, at any one transit) does not exist during the first half of the month nor its west one during the second; and although both these points, or both the north and south pair, are occasionally listed in the *Almanac* as "observable" at the same transit, one in each pair is always (necessarily) shown as "defectively illuminated". Further, since the position-angle of the axis cannot exceed about $\pm$ 25°, there are four regions, each about 40° long, which do not exist at all; meridian observations cannot well be corrected for limb-features and must be taken as they stand, and thus the "Meridian Moon" may differ systematically from all Moons located by means of their complete contours, whether or not corrections are applied in their case. (The meridian observer can apply corrections only if he can remember, and specify accurately from memory, just how he laid the wires up against the irregularities with which he found himself suddenly confronted at transit.) The diameter of the Meridian Moon is seldom directly observed and is in any case falsified by irradiation; the much more extensive data which have been obtained for the Sun [7] suggest that it may also be subject to large personality-effects, and that these may themselves depend strongly on the zenith distance. In contrast to the "Meridian Sun", an error in the diameter usually causes an error in the place. On all counts, therefore, there is little hope of discovering the precise short-period behaviour of this Moon; but its general behaviour is well known.

(5) *The Occultation Moon*

This is the most precisely known Moon nowadays, but it too is very unevenly observed. It scarcely exists within about three days either way from new moon, nor during daylight or bright twilight; in addition, dark-limb observations are easier, and can be made with many more stars, than bright-limb ones, and disappearances are generally more reliable than reappearances, and also much more often observed. For all practical purposes indeed, the west limb does not exist when bright, and the main weight is in the dark-limb disappearances (age 5 to 13 days, say); the corrections derived for the Moon's place are often meaned by lunations, and these means may sometimes therefore be discussed as if the orbital and libration elements for all the observations had been what they were at the age of 9 days or thereabouts. The values so meaned [8] show oscillations in latitude which it would certainly be impossible to explain by oscillations in U.T.; WATTS has shown [9] that the application of individual limb-corrections would reduce them considerably, but they may also be due in part to errors in the star-places, which necessarily affect them. The Occultation Moon's diameter exists, and is entirely unfalsified by irradiation; it cannot be based on the simple comparison of dark-limb observations near first and last quarters, since it would then hardly be separable from the parallactic inequality, but it has been determined by disappearances and reappearances, and also by pure dark-limb disappearances, preferably in regions rich in stars [10]. It follows that there is a tendency for the results to be tied to special values of b; for example, occultations in the Pleiades involve $b \approx -5\frac{1}{2}°$, and occultations (faint stars) in lunar eclipses involve $b \approx 0$ [11].

(6) *The Astrometric Moon*

This hardly exists at all, as yet, but it is likely to become very important. The difficulty in getting a place by direct photography, as is regularly done for minor planets, has been that the Moon's motion is appreciable even in the short time needed to photograph suitable comparison-stars; the new technique (MARKOWITZ) of countering this motion by means of a slowly-tilting glass plate, covering only the Moon's image, seems most promising. What is obtained is a place for the *centre* of the entire arc which is photographed, since although the radius is falsified by irradiation the centre is unaffected; this is a very great advantage. With the exception of the Eclipse Moon, this is the only one for which the apparent places of a really large number of limb-points are observed simultaneously, before the librations have had time to change, and each photograph can in principle supply its own limb-corrections, including a direct and valuable check on their position-angles. Only the bright limb exists, of course, but observations just before and after full moon should help to show how far the centres of figure of the east and west limbs coincide. As with the

Occultation Moon, systematic errors in the star-places go straight into the Moon's; but their random errors are less serious.

(7) *The (Solar) Eclipse Moon*

This exists only at an age (or elongation) entirely outside the range of all other observational Moons, namely zero. Its libration in latitude is always nearly zero also, and its contours thus form almost a one-parameter array, being nearly determined by the libration in longitude alone. It is the only Moon whose *complete* contour can be observed before the librations have changed. This may obviously be done at an annular eclipse, but it can be nearly as well done at a total one too; if one has cameras a little outside the track, on both sides, and takes photographs for about 30–40 min near mid-eclipse, the rapid change in position-angle allows the entire limb to be covered, with good overlaps. This method was tried at the Sudan eclipse of 25th February, 1952, but the seeing in the desert was poor. The place of the Eclipse Moon is tied to that of the Sun; its diameter is separable from the Sun's in principle, but usually not in practice. The approximate place is known back to great antiquity, and gives some information on the secular change in the Earth's rotation, as well as useful checks on early chronology. The following are the principal modern observational methods; regrettably, work has been confined almost entirely to total eclipses, and the apogee half of this Moon's orbit is nearly unrepresented.

(*a*) From timed position-angles of the line of cusps, in the partial phase, one can derive corrections to both co-ordinates of the Moon's place, provided one observes long enough to obtain a good change in P.A. The diameter clearly cannot be obtained in this way (or at least not without knowledge of the limb-detail), because for two circular outlines the line of cusps is perpendicular to the line of centres whatever the diameters. If, however, one also measures the separation of the cusps, the diameters of both Sun and Moon are in principle obtainable; in practice, one of the two still has usually to be assumed, and in addition there is obviously more risk of an error due to irradiation than in the case of position angles. Most early work of this kind was done at established observatories, where the eclipse was only partial anyway; some of the best visual results were obtained with heliometers, but filar micrometers, and even measurements on the two cusps separately, were also successfully employed. Some important cases, with the probable errors in right ascension and declination, were: Wichmann, 1851, $\pm$ 0″.3 and $\pm$ 0″.2 [12]; Kobold, 1890, 1891, and 1893, about $\pm$ 0″.18 in each co-ordinate [13]; also 1900, $\pm$ 0″.26 and $\pm$ 0″.28 [14]; Merlin, 1905 (using a projected image), $\pm$ 0″.1 and $\pm$ 0″.2 [15]; also 1908, $\pm$ 0″.4 and $\pm$ 0″.5 [16]; Naumann, 1912 (applying limb-corrections to the cusps), $\pm$ 0″.07 and $\pm$ 0″.06 [17]; also 1914, $\pm$ 0″.12 and $\pm$ 0″.08; and 1927, $\pm$ 0″.26 and $\pm$ 0″.17 [18]. Photographic work was started in 1912, particularly by Hayn, $\pm$ 0″.08 and $\pm$ 0″.11 [19]; also 1914 [20]; in 1927 [21] he adopted the procedure of determining the centres of the lunar and solar limb-arcs from measurements on twenty points of each (not using the actual cusps at all), and he also measured the distances and position angles of each cusp from the Sun's centre, so determined; his probable errors were about $\pm$ 0″.07.

The above results were among those which claimed the smallest probable errors [22], but systematic errors (*e.g.* in timing) may be quite as large; timing was in fact rather doubtful in some cases. And the whole theory of the method breaks down if the Moon is elliptical (unless limb-corrections based on a circle are applied) or if its two halves are corrected to circular arcs whose centres and radii are not identical, since one cusp is pretty sure to be in the eastern half and one in the western.

(*b*) Cinematography was first used at the "limiting" eclipse of 1912, when it was devoted mainly to deciding whether the eclipse was total or not, and where the centre-line ran. Several observers set up one or more cine cameras near the centre-line [23], and decided where the line had been, and when mid-eclipse had occurred, by mere inspection, picking those photographs which showed the most uniform intensities of three or four Baily's beads that were well distributed round the Moon. This method is very seldom applicable; it is not really objective; and it obtains the place of a Moon which is different from all those discussed above, namely the Moon of beads due to the deepest valleys only. (The "centre of gravity" of a bead is an uncertain distance above the valley floor, even if the latter is known.)

(*c*) A much better cinematographic method was devised by Banachiewicz, and was tried out, under his direction, by Kordylewsky at the 1927 eclipse [24]. This was to use the beads quantitatively, by comparing their times of disappearance, and of appearance, with those calculated from the appropriate Hayn profile. The times when a thin thread of light was first (or last) broken by a peak were also used. The probable errors were $\pm$ 0″.04 and $\pm$ 0″.06; the method was also employed in 1936, 1945, 1947, and 1948 [25] but results do not seem to be available.

(*d*) Banachiewicz's method still used only the bottoms of valleys and the tops of mountains, though it used a considerable number of them and allowed for all their different depths and heights; an essentially continuous use of the limb-contour can be made in the "spectroscopic" method proposed by Lindblad, and used by expeditions directed both by him and by Lundmark in 1945. In this, slitless spectrograms were taken as rapidly as possible of the thin crescent near second and third contacts, and to measure them Kristenson [26], following Lindblad, ran a photometer across the spectrogram at a region where there were no appreciable chromospheric lines. The change of wavelength (and so of intensity) involved in travelling a fairly small fraction of one solar radius in the direction of the dispersion is small, and can in any case be determined; accordingly, running the photometer in a straight line across the spectrum (which is convenient and safe) amounts to the same as running it round in an arc of a circle parallel to the image of a spectral line. The method does not eliminate errors due to bad seeing (as has sometimes been claimed for it); but the fact that the continuum is spread out by dispersion does allow a sort of rectification of the crescent, as well as neutralizing "photographic spreading" of the image; the photometer readings (calibrated, of course, to reduce them to intensities) thus give essentially the total intensity as a function of position along the crescent. If it were not for solar limb-darkening, the intensity would be simply proportional to the thickness of the crescent (reckoned parallel to the dispersion) and should show variations directly dictated by the lunar limb-detail; in fact, by comparing the intensity at the same limb-point at different times, the limb-darkening itself was also determined, in good agreement with Wildt's theory [27], so that the results did give a series of true pictures of the actual distance of the lunar limb from the solar one, point by point. The probable error of the Moon's place is given as $\pm$ 0″.023,

$\pm$ 0.″09. This method, as well as BANACHIEWICZ's, should apparently help to determine the extent to which the east and west limbs of the Moon can be fitted to a single circle.

(*e*) Another method, which has not this particular advantage, but which seems to leave very little room for ambiguity in its interpretation, and which also uses a "continuous" run of lunar limb-points, was devised by the writer [28], and tried out at Mombasa in 1948, and, on a more extensive scale, in 1952. This consists in making a timed cinema record, at a station a little outside the track of totality, of the rapid swing in position-angle shown by the crescent near mid-eclipse. The swing may be made so rapid that a wide range of angles is obtained before the image has moved out of the field of a stationary camera, so that the diurnal motion can give the zero of position angles very precisely (Fig. 1). Owing to the sharpness of the cusps, the

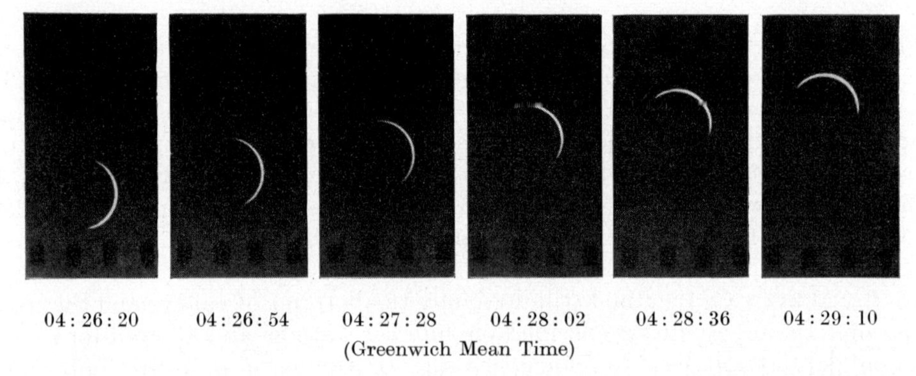

Fig. 1. Six of the 3100 pictures obtained at the Mombasa eclipse of 1st November, 1948. The times shown are only approximate, and actual times are good to a hundredth of a second or better, for all 3100 exposures. The camera was stationary, the Sun was rising almost vertically, and the Moon was rising less fast and moving south. The minimum separation of centres of Sun and Moon was just over 1′, which accounts for the rapid rotation

fluctuations in the run of position-angles, due to lunar irregularities, are marked; but the Mombasa fluctuations follow in considerable detail the fluctuations inferred from two appropriate limb-traces which were kindly supplied by WATTS. The closeness of the fit in detail is sensitive to the value assumed for the Moon's diameter, and may even provide a check on its value. The systematic revision which WATTS is still undertaking has resulted already in a marked improvement in the systematic fit, as against that obtained with his first analysis, and it seems probable that the method can locate "WATTS' Moon" more accurately, in both co-ordinates, than he himself can yet be sure of the true heights of that Moon's mean spherical surface, relative to its observed contour. The possible geodetic accuracy is thus greater (as may also be true with LINDBLAD's method), than the accuracy with which the Moon's place can be derived from the results, essentially because one cannot yet sufficiently specify the Moon itself. A detailed analysis of this work is now nearing completion.

In discussing the "probable error" of the place of the Moon obtained by "continuous" observations along its contour, one is brought up against a difficulty which is also inherent (though less evidently so) in the other methods. The Gaussian

Theory of Errors assumes that the errors of successive measurements of a quantity are independent; but if one is trying to determine the best mean circle for a given lunar profile, and one makes a very large number of radius-measurements, so that a hundred of them (say) all fall within the compass of a single mountain, the departures from the mean radius are clearly not independent at all; a hundred consecutive residuals are all positive. For determining the minutiae of a mountain's shape, a hundred measurements within a short length of limb are, of course, better than ten; but as far as the best mean radius for the whole profile goes, they are very little better, and they certainly have not got ten times the weight of ten. If we determine the co-ordinates of the centre from (say) twenty points equally spaced along the whole arc (HAYN, *loc. cit.*), the twenty residuals may reasonably be considered independent (though their distribution will still not be Gaussian); but since we know they will run up to $\pm 2''$ or more, the "probable error" of our result must be expected to be about $\pm 0''.15$, in good agreement with the values actually claimed. One cannot fairly claim to get a much smaller value merely by making many more measurements, because (to put it roughly) a change in libration could so easily abolish one mountain, or push another up, over a range which would then be long enough to have a *systematic* effect on one's measurements. In the author's Mombasa work, 3100 timed position-angles were obtained; but it would quite clearly be wrong to take the mean scatter of the measured position-angles from a smooth run, and to divide it by $\sqrt{3100}$ and infer a probable error of the Moon's place from the figure so obtained; if one did, one would quote $\pm 0''.004$ or thereabouts, but a slight change in libration would give a result different by many times this figure.

One can, however, legitimately claim a smaller probable error than the $0''.15$ quoted above if one is comparing one's observations with those to be expected for a Moon of known profile (KORDYLEWSKY, KRISTENSON, ATKINSON). For (assuming that the profile is in fact suitable) the differences between the computed and observed values can now with much more freedom be called random: measurements at successive points along the limb can quite evidently be made much closer together than before without the residuals becoming strongly correlated with their neighbours. Moreover, the mean residuals will now themselves be appreciably smaller also, which would in itself reduce the probable error. It seems difficult to formulate any rigorous criterion for the maximum number of observations which it would be fair to regard as independent, and certainly any rigorous criterion could only depend on an appropriate theory of (dependent) errors [29]. But bearing in mind how often the Gaussian theory is formally applied to cases in which its fundamental assumptions certainly do not hold, one might perhaps formally apply it here also, subject to a rule that the number of "independent" observations can be taken as no more than twice the number of times that the residuals change sign; this rule evidently holds whenever the residuals are random (provided their median is zero), since for each residual there is then a 50 per cent chance that the next one will be of opposite sign. The "probable error" of the place of "WATTS' Moon" (say), as located by "ATKINSON's method" (say) would then follow from the mean residual divided by the square root of twice the number of its changes of sign, in much the same way as probable errors are generally derived from mean residuals and numbers of residuals themselves; neither the coefficient 0·8453 (if the mean residual is taken) nor 0·6745 (if the root mean square) would be strictly applicable; but that has frequently not prevented their use in the past.

(8) *Conclusion*

This paper has stressed the considerable chances which exist at present for systematic differences between the places of the different Moons. But the application of a really uniform system of limb-corrections, such as may be hoped for from WATTS' work, should greatly help to fuse some of these Moons into one; indeed, it seems probable that a most significant observational advance would be made if the Limb-contour Moon and the Astrometric Moon were at once fused observationally (as is clearly possible in principle), and were jointly observed as widely as possible from now on. The Occultation Moon, if reduced to WATTS' sphere, should clearly fall into agreement with the Astrometric so reduced, and a rigorous revision of the star-places would then allow a reliable comparison with the Eclipse Moon; meanwhile, this last one must remain definitely in need of continual further observation in its own right. It falls in a quarter of the month that is otherwise quite unobserved; if it can nevertheless be relied on to be consistent with the others, it provides a check on the Sun's place (and diameter), free from the risk of falsification by refraction. It can provide geodetic information [30]; however, this is true also of the Occultation Moon [31], and probably of the Astrometric one in addition. Indeed, the Astrometric Moon may well supplant both the others before long in this field; it is not too much to say that if MARKOWITZ's method works out as it appears likely to, then as far as geodetic work is concerned we shall in future be as well off as if we could order our eclipses by the dozen, whenever and wherever we wished.

NOTES AND REFERENCES

[1] *I.A.U. Draft Reports*, 1952, pp. 94–98; I. V. BELKOVICH, *Engelhardt Obs. Publ.*, **24**, 242, 1949; A. A. YAKOVKIN, *Kiev Obs. Publ.*, **3**, 17, 1950 and **4**, 71, 1950. However, see also K. KOZIEL, *Acta Astron.* (a), **4**, 61, etc., 1948–9.
[2] H. SPENCER JONES, *M.N.*, **99**, 541, 1939.
[3] K. F. SUNDMAN, *l'Activité de la Comm. Géod. Balt.*, **1944–7**, 63, 1948.
[4] F. HAYN, *Abh. d.k. sächs. Ges. d. Wiss. Leipzig*, **30** and **33**, 1907 and 1914. (Selenographische Koordinaten, III and IV.) C. B. WATTS and A. N. ADAMS, *A.J.*, **55**, 81, 1950.
[5] According to T. BANACHIEWICZ, the Saros is the smallest period for an approximate repetition of both co-ordinates together. *Acta Astron.* (c), **4**, 160, 1951.
[6] F. J. M. STRATTON, *Mem. Roy. Astron. Soc.*, **59**, 257, 1909, and references there. Also [1] above.
[7] R. T. CULLEN, *M.N.*, **86**, 344, 1926.
[8] F. M. MCBAIN, *Greenw. Obs.*, 1939 Appendix (re-discusses 1943–7); *Astron. J.*, **53**, 163, 1948 and **55**, 7, 47, 247, 1949–51; D. BROUWER, *Astron. J.*, **51**, 144, 1945, etc.
[9] C. B. WATTS, *Astron. J.*, **48**, 170, 1940.
[10] For example, H. SPENCER JONES, *M.N.*, **85**, 11, 1925. For a general discussion, with other important references, see H. ILLIGNER, *Astron. Nachr.*, **253**, 293, 1934.
[11] For the Pleiades, F. KÜSTNER, *Nova Acta*, **41**, 1, 1879; J. PETERS, *Astron. Nachr.*, **138**, 113, 1895; H. ILLIGNER, *loc. cit.*; for lunar eclipses, L. STRUVE, *Astron. Nachr.*, **135**, 169, 1894.
[12] M. WICHMANN, *Astron. Nachr.*, **33**, 309, 1852.
[13] H. KOBOLD, *Strasbourg Ann.*, II, Annexe A, 1–39, 1899.
[14] *Ibid.*, III, Annexe A, 1–36, 1901.
[15] J. MERLIN, *C.R. Acad. Sci.* (*Paris*), **144**, 20, 1907.
[16] *Ibid.*, **148**, 146 (and 263), 1909.
[17] H. W. NAUMANN, *Astron. Nachr.*, **193**, 123, 1913.
[18] *Ibid.*, **201**, 217, 1915, and **233**, 189, 1928.
[19] F. HAYN, *Astron. Nachr.*, **193**, 117, 1913.
[20] *Ibid.*, **201**, 185, 1915.
[21] *Ibid.*, **233**, 183, 1928.
[22] Some others were: C. W. WIRTZ, *Astron. Nachr.*, **171**, 102, 1906; E. BECKER, *Astron. Nachr.*, **180**, 387, 19; A. WILKENS, *Astron. Nachr.*, **193**, 139, 1913; G. N. NEUJMIN, *Mitt. Nik. Obs. Pulkowo*, **5**, 77, 1913; M. SIMONIN, *Ann. Obs. Paris*, 1914; H. BATTERMAN, *Astron. Nachr.*, **202**, 23, 1915; C. LE MORVAN, *Bull. Astron.* (*Paris*), **32**, 337, 1915; Moscow Obs., *Russian Astron. J.*, **2** (2), 70–76, 1925; (various) *Bull. Astron. Inst. Netherlands*, **3**, 266, 1927; S. ISHII, *Tokyo Astron. Bull.*, **5–6**, 1927; (various) *Bull. Astron. Inst. Netherlands*, **4**, 71, 1927; E. PRZYBYLLOK, *Astron. Nachr.*, **231**, 293, 1928; E. PEREPELKIN, *Astron. Nachr.*, **232**, 127, 1928; S. ISHII, *Japan J. Astron. Geophys.*,

11, 37, 1933; H. G. LENGAUER, *Proc. of Eclipse Expeditions* 1936, pp. 129–144, published by Russian Academy of Science (in Russian); J. E. WILLIS, *Astron. J.*, **47**, 109, 1939; S. ISHII, *Tokyo Astron. Bull.*, **426–7**, 851, 1940; G. G. LENGAUER, *Reports of Moscow-Leningrad Eclipse Expeditions*, vol. II, 137, 1941.
[23] F. VLÈS and J. CARVALLO, *C.R. Acad. Sci.* (*Paris*), **154**, 1142, 1912, and **155**, 545, 1912; DA COSTA LOBO, *C.R. Acad. Sci.* (*Paris*), **154**, 1396, 1912; A. DE LA BAUME-PLUVINEL, *ibid.*, 1139; E. CARVALLO, *ibid.*, 1072; R. SCHORR, *Astron. Nachr.*, **191**, 429, 1912.
[24] K. KORDYLEWSKY, *Acta Astron.* (b), **1**, 133, 1932.
[25] T. BANACHIEWICZ, *Acta Astron.* (c), **3**, 16, 1936; I. BONSDORFF, *l'Activité de la Comm. Géod. Balt.*, 1943, 12 (1944); B. LINDBLAD, *ibid.*, 17; H. KRISTENSON, *Stockholm Obs. Ann.*, **17**, pp. 15–16, 1951; I. BONSDORFF *et al.*, *Proc. Finn. Acad.* 1948, 99 (1950); W. KINNEY, *Nat. Geogr. Mag.*, **95**, 325, 1949; T. N. PANAY, *Trans. Amer. Geophys. Union*, **31**, 809, 1950.
[26] H. KRISTENSON, *Stockholm Obs. Ann.*, **17**, 3, 1951.
[27] R. WILDT, *Ap. J.*, **105**, 36, 1947.
[28] R. D'E. ATKINSON, *M.N.*, **113**, 18, 1953, and later papers.
[29] H. JEFFREYS, *Theory of Probability*, Cambridge University Press, pp. 229–245, 1939.
[30] I. BONSDORFF, *l'Activité de la Comm. Géod. Balt.* 1942–3, 12 (1944).
[31] J. A. O'KEEFE, *Astron. J.*, **55**, 177, 1950.

The Moon's Principal Librations in Rectangular Co-ordinates

SIR HAROLD JEFFREYS
St. John's College, Cambridge

SUMMARY
The Moon's librations of the first and second orders are discussed by a method that involves direct use of rectangular co-ordinates and LAGRANGE's equations of motion. The secular stability in presence of dissipation of the type that might arise from internal tidal friction is also discussed.

ALL the textbooks on celestial mechanics that I know deal with the Moon's librations by expressing its angular position in terms of EULER's angles, using EULER's dynamical equations, and then making various more or less cumbrous transformations to reduce the problem to one of small oscillations. As the direct use of rectangular co-ordinates seems to be gaining popularity in planetary theory, and has already gained it in lunar theory, I think that it may be of interest to show how the chief results for the librations can be found by using rectangular co-ordinates and Lagrange's equations from the start.

The main difficulty comes from the fact that at least one second-order libration is on the verge of being observable, and consequently it is necessary to have the kinetic energy T and the work function W to the third order of small quantities. Since the principal rotation is not small, it needs to be known to the third order; the other angular velocities are needed to the second order.

The quantities observed most directly are the displacements with respect to the limb of a point near the centre of the disk. A third co-ordinate is provided by the change of position angle of a known mark near the limb with respect to the centre. If we take the principal axes at the centre of the Moon, OA pointing nearly to the Earth, and OC the principal axis nearly normal to the orbit, we can refer them to a set OX, OY, OZ, where OZ is fixed in direction and OX, OY rotate about it with angular velocity n, OX pointing nearly to the Earth. The usual practice is to choose Euler's angles so as to specify the position of C first and then introduce a further turn about C; thus the displacements of A in latitude and longitude, which are directly measurable quantities, are each the sum of two parts. It would be more natural to specify the position of A first and then B. We need the matrix of direction cosines of OA, OB, OC with respect to OX, OY, OZ, expressed in terms of the small quantities of the problem, and some at least of the elements are needed to the third order. For the usual Euler angles θ, ϕ, ψ this matrix is exactly, (JEFFREYS, 1950):

$$\begin{array}{l} \\ B = A' \\ \\ C = B' \\ \\ A = C' \end{array} \begin{array}{c} \begin{array}{ccc} \quad X \quad & \quad Y \quad & \quad Z \quad \end{array} \\ \begin{pmatrix} \cos\phi\cos\theta\cos\psi - \sin\phi\sin\psi & \sin\phi\cos\theta\cos\psi + \cos\phi\sin\psi & -\sin\theta\cos\psi \\ -\cos\phi\cos\theta\sin\psi - \sin\phi\cos\psi & -\sin\phi\cos\theta\sin\psi + \cos\phi\cos\psi & \sin\theta\sin\psi \\ \cos\phi\sin\theta & \sin\phi\sin\theta & \cos\theta \end{pmatrix} \end{array} \quad \ldots (1)$$

A', B', C' are the axes as given in the book. Take $\sin\phi = \chi$, $\theta = \frac{1}{2}\pi + \sin^{-1}\xi$, $\psi = \frac{1}{2}\pi + \sin^{-1}\eta$, expand to the third order in χ, ξ, η, and rearrange. We have

$$\begin{array}{l} \\ A \\ B \\ C \end{array} \begin{array}{c} \begin{array}{ccc} \quad X \quad & \quad Y \quad & \quad Z \quad \end{array} \\ \begin{pmatrix} 1 - \frac{1}{2}\chi^2 - \frac{1}{2}\xi^2 & \chi - \frac{1}{2}\chi\xi^2 & -\xi \\ -\chi + \xi\eta + \frac{1}{2}\eta^2\chi & 1 - \frac{1}{2}\chi^2 - \frac{1}{2}\eta^2 + \xi\eta\chi & \eta - \frac{1}{2}\xi^2\eta \\ \xi + \chi\eta - \frac{1}{2}\xi(\chi^2 + \eta^2) & -\eta + \chi\xi + \frac{1}{2}\eta\chi^2 & 1 - \frac{1}{2}\xi^2 - \frac{1}{2}\eta^2 \end{pmatrix} \end{array} \quad \ldots (2)$$

It is simple to start with the elements of zero and first order and apply corrections in succession to make the matrix orthogonal to the third order; but there are some ambiguities, which are most conveniently resolved by choosing the higher terms to that the velocity components do not contain χ explicitly (this being the chief good point of the Euler co-ordinate ϕ). This object is achieved automatically by direct transformation of the exact matrix.

The space velocities of C with respect to X, Y, Z are to the second order

$$\dot{\xi} + \chi\dot{\eta} + \dot{\chi}\eta - n(-\eta + \chi\xi), \; -\dot{\eta} + \dot{\chi}\xi + \chi\dot{\xi} + n(\xi + \chi\eta), \; -\xi\dot{\xi} - \eta\dot{\eta},$$

and resolving along OA, OB we have two components of angular velocity

$$\omega_2 = \dot{\xi} + (n + \dot{\chi})\eta, \quad -\omega_1 = -\dot{\eta} + (n + \dot{\chi})\xi. \quad \ldots (3)$$

The velocity components of A are

$$-\chi\dot{\chi} - \xi\dot{\xi} - n(\chi - \tfrac{1}{2}\chi\xi^2), \; \dot{\chi} - \tfrac{1}{2}\xi^2\dot{\chi} - \chi\xi\dot{\xi} + n(1 - \tfrac{1}{2}\chi^2 - \tfrac{1}{2}\xi^2), \; -\dot{\xi},$$

and resolving along OB gives, to the third order,

$$\omega_3 = n + \dot{\chi} + \tfrac{1}{2}\chi^2\dot{\chi} - \tfrac{1}{2}(n + \dot{\chi})(\xi^2 + \eta^2) - \eta\dot{\xi}. \quad \ldots (4)$$

Then $$2T = A\dot{\eta}^2 + B\dot{\xi}^2 + C\dot{\chi}^2 - 2nA\xi\dot{\eta} - 2n(C-B)\dot{\xi}\eta - (n^2 + 2n\dot{\chi})\{(C-A)\xi^2 + (C-B)\eta^2\} - 2A\xi\dot{\eta}\dot{\chi} - 2(C-B)\eta\dot{\xi}\dot{\chi}. \quad \ldots.(5)$$

Constants and derivatives with regard to t have been dropped.

The direction cosines of the Earth with respect to OX, OY, OZ are taken to be

$$-\{1 - \tfrac{1}{2}\sin^2 i \sin^2(v+pt)\}\cos(v-nt), \quad -\sin(v-nt), \quad -\sin i \sin(v+pt),$$

where v is the Moon's longitude, $-p$ the motion of the node, and i the inclination. $v - nt$ is small of the order of the eccentricity e. It is easily verified that the sum of squares differs from 1 by a quantity of order i^2e^2. Then we are treating i and e also as small quantities of the first order. Then with respect to OA, OB, OC the direction cosines are, (l' to the third order and m', n' to the second),

$$\left.\begin{aligned} l' &= -\{1 - \tfrac{1}{2}\sin^2 i \sin^2(v+pt)\}\cos(v-nt)(1 - \tfrac{1}{2}\chi^2 - \tfrac{1}{2}\xi^2) \\ &\quad - \chi\sin(v-nt) + \xi\sin i\sin(v+pt) \\ m' &= \chi - \xi\eta - \sin(v-nt) - \eta\sin i\sin(v+pt) \\ n' &= -\xi - \chi\eta + \eta\sin(v-nt) - \sin i\sin(v+pt) \end{aligned}\right\} \quad \ldots.(6)$$

l' can be eliminated at once from the work function. If R is the actual distance from the Earth and a the mean distance, the relevant part of the work function is

$$W = \frac{n^2}{2(1+\mu)}\left(\frac{a}{R}\right)^3\{A + B + C - 3(Al'^2 + Bm'^2 + Cn'^2)\}$$
$$= \frac{n^2}{2(1+\mu)}\left(\frac{a}{R}\right)^3\{B + C - 2A - 3(B-A)m'^2 - 3(C-A)n'^2\}. \quad \ldots.(7)$$

Dropping terms independent of ξ, η, ζ we have

$$W = -\frac{3n^2}{2(1+\mu)}\left(\frac{a}{R}\right)^3[(B-A)\{\chi - \sin(v-nt) - \eta(\xi + \sin i\sin(v+pt))\}^2 + (C-A)\{\xi + \sin i\sin(v+pt) + \eta(\chi - \sin(v-nt))\}^2]$$
$$= -\frac{3n^2}{2(1+\mu)}\left(\frac{a}{R}\right)^3[(B-A)\{\chi - \sin(v-nt)\}^2 + (C-A)\{\xi + \sin i\sin(v+pt)\}^2 + 2(C-B)\{\xi + \sin i\sin(v+pt)\}\{\chi - \sin(v-nt)\}\eta] \quad \ldots.(8)$$

Forming LAGRANGE's equations to the first order, and putting

$$C - A = B\beta, \; C - B = A\alpha, \; B - A = C\gamma \quad \ldots.(9)$$

with $$\alpha + \gamma - \beta - \alpha\beta\gamma = 0, \quad \ldots.(10)$$

we have the equations

$$\ddot{\xi} - n(1-\beta)\dot{\eta} + \left(4 - \frac{3\mu}{1+\mu}\right)n^2\beta\xi = -\frac{3n^2\beta}{1+\mu}\sin i\sin(v+pt) \quad \ldots.(11)$$

$$\ddot{\eta} + n(1-\alpha)\dot{\xi} + n^2\alpha\eta = 0 \quad \ldots.(12)$$

$$\ddot{\chi} + \frac{3n^2\gamma}{1+\mu}\chi = \frac{3n^2\gamma}{1+\mu}\sin(v-nt) \quad \ldots.(13)$$

To this order $v + nt = (n + p)t$, $v - nt = 2e \sin (n - g)t$, where g is the motion of perigee. Since $1 \gg p/n \gg \alpha, \beta, \gamma$, the first approximation is, as usual,

$$\xi = P \sin (n + p)t, \; \eta = - Q \cos (n + p)t, \; \chi = \frac{6n^2\gamma e}{(1 + \mu)(n - g)^2} \sin (n - g)t, \qquad \ldots.(14)$$

where

$$P = \frac{3\beta \sin i}{(1 + \mu)(2p/n - 3\beta)}, \quad Q = \frac{3\beta \sin i}{(1 + \mu)(1 + p/n)(2p/n - 3\beta)}. \qquad \ldots.(15)$$

On account of the small divisor $2p/n - 3\beta$, P and Q are of the same order of magnitude as i. χ has no small divisor (we are not at present considering the annual term in $v - nt$). The speeds of the free motions are approximately $n(1 + \frac{3}{2}\beta)$, $2n\sqrt{(\alpha\beta)}$, and $n\sqrt{(3\gamma)}$.

In proceeding to a second approximation we use

$$v - nt = 2e \sin (n - g)t, \; (a/R)^3 = 1 + 3e \cos (n - g)t, \qquad \ldots.(16)$$

$$\sin (v + pt) = \sin (n + p)t - e \sin (p + g)t + e \sin (2n + p - g)t \qquad \ldots.(17)$$

Second-order terms may rise in importance if their speeds are small or near n. In T, the term $- A\xi\dot{\eta}\dot{\chi}$ does not contain the small factors α, β, γ. Substituting the first approximations to ξ and η we find that it contributes to the left of (13) a term $- n(n + p)P^2 \sin 2(n + p)t$. The contribution of the terms on the right is

$$\frac{- 3n^2}{1 + \mu} A\alpha Q(P + i) \sin 2(n + p)t, \qquad \ldots.(18)$$

which is much smaller on account of the factor α. Then the second-order part of χ is

$$\tfrac{1}{4}P^2 \sin 2(n + p)t. \qquad \ldots.(19)$$

In the third-order terms in W, χ is much less than $v - nt$, and may be dropped in forming the equations for ξ and η. Terms in e in $\sin (v + pt)$ must be retained.

The extra terms in W are

$$\begin{aligned} - \frac{3n^2e}{2(1 + \mu)} & [3B\beta\{\xi + \sin i \sin (n + p)t\}^2 \cos (n - g)t \\ & + 2B\beta\{\xi + \sin i \sin (n + p)t\}\{\sin (2n + p - g)t - \sin (p + g)t\} \sin i \\ & - 4A\alpha\{\xi + \sin i \sin (n + p)t\}\eta \sin (n - g)t]. \end{aligned} \qquad \ldots.(20)$$

The new terms in $\dfrac{\partial \omega}{\partial \xi}$ are

$$- \frac{3n^2e}{1 + \mu} (\tfrac{3}{2}B\beta P + \tfrac{1}{2}B\beta i - A\alpha Q) \sin (p + g)t \qquad \ldots.(21)$$

and in $\dfrac{\partial W}{\partial \eta}$

$$\frac{3n^2e}{1 + \mu} A\alpha(P + \sin i) \cos (p + g)t. \qquad \ldots.(22)$$

Terms of speed $2n + p - g$ have been dropped because, being fortnightly, they are not subject to magnification.

The term $-A\xi\dot{\eta}\dot{\chi}$ in T still needs consideration. It adds to the left of the ξ equation

$$A\dot{\eta}\dot{\chi} \doteqdot \frac{3n^2}{1+\mu} AQ\gamma e \sin(p+g)t \qquad \ldots.(23)$$

and converts the $A\alpha$ of (22) into $A\beta$, which we can replace by $B\beta$.*

The last change is comparable with the effects of terms in $C-B$ and $B-A$ on the right, which we have not considered. The effect on the η equation contains the derivative of the slowly varying factor, and is negligible. If we also, as usual, neglect the difference between P and Q the additional terms on the right of (11) and (12) reduce to

$$-\frac{3}{2}\frac{n^2 e\beta}{1+\mu}(P+\sin i)\sin(p+g)t, \quad \frac{3n^2 e\alpha}{1+\mu}(P+\sin i)\cos(p+g)t. \qquad \ldots.(24)$$

The largest terms on the left are those containing $\dot{\xi}$ and $\dot{\eta}$, and integration gives the Poisson terms

$$\xi = \frac{3ne\alpha}{1+\mu}\frac{P+\sin i}{p+g}\sin(p+g)t, \qquad \ldots.(25)$$

$$\eta = -\frac{3}{2}\frac{ne\beta}{1+\mu}\frac{P+\sin i}{p+g}\cos(p+g)t. \qquad \ldots.(26)$$

ξ and η are usually defined as displacements of C toward X and Y, and therefore correspond to the present $\xi+\chi\eta$, $-\eta+\chi\xi$; but the second-order terms do not contain the small divisor $p+g$ and are therefore negligible.

The amplitudes of the Poisson terms in ξ and η are of the order of 1'.4; TISSERAND gives 947″ and 972″, apparently through a misplaced decimal point, and the mistake is not indicated in the errata. PLUMMER'S *Dynamical Astronomy* gives the correct value.

The annual term in γ, of speed n' (which will not be confused with the direction cosine), leads to second-order terms in ξ and η. The most important come from the $A\xi\dot{\eta}\dot{\chi}$ term in T, but the differentiation of χ introduces the small factor n', and the terms in ξ and η have speeds $n+p\pm n'$ and are less magnified than terms in $(n+p)t$. Hence these terms are negligible.

The conditions for ordinary stability are $\gamma > 0$, $\alpha\beta > 0$, so that the Moon would be stable either for $C > B > A$ or for $B > A > C$. If, however, there is small damping proportional to the velocities relative to the mean velocity of rotation, such as might occur through tidal friction, the equations of free motion become

$$\ddot{\xi} + nk\dot{\xi} - n(1-\beta)\dot{\eta} + \left(4 - \frac{3\mu}{1+\gamma}\right)n^2\beta\xi = 0,$$

$$\ddot{\eta} + nk\dot{\eta} + n(1-\alpha)\dot{\xi} + n^2\alpha\eta = 0,$$

$$\ddot{\chi} + nk\dot{\chi} + \frac{3n^2\gamma}{1+\mu} = 0.$$

* Euler's equations, of course, give $B\beta$ directly at this point, essentially because they refer to rotations about exactly rectangular axes.

The time factors in ξ and η are, approximately, $\exp \lambda t$, with

$$\frac{\lambda}{n} = -k \pm i\sqrt{(1+3\beta)},\ -\tfrac{1}{2}k(\alpha+4\beta) \pm \tfrac{1}{2}i\{16\alpha\beta - k^2(\alpha+4\beta)^2\}^{\frac{1}{2}}.$$

That in χ gives

$$\frac{\lambda}{n} = -\tfrac{1}{2}k \pm i\sqrt{(3\gamma)}.$$

$\alpha\beta > 0$ for ordinary stability; but if α, β were both negative there would therefore be secular instability in ξ and η. Hence the Moon could be secularly stable only in the actual condition $C > B > A$.

The direct use of the matrix (2) reduces much of the work to routine; actually only the direction cosines of A are needed to the third order. Once the angular velocities and l', m', n' are known in terms of ξ, η, χ, EULER's equations are probably slightly simpler to use than LAGRANGE's. LAGRANGE's have the advantage that (8) is still available if perturbations of the orbit are to be considered. The problem does not look hopeful for the application of "vector methods".

REFERENCE

JEFFREYS, H. and B. S. . . . . 1950 *Methods of Mathematical Physics*, Cambridge, p. 123.

A General Expression for a Lagrangian Bracket

W. M. SMART
University Observatory, Glasgow

SUMMARY
In the paper, based on an earlier investigation by A. Y. G. CAMPBELL, the general expression—in terms of any functions of the orbital elements—for any one of the Lagrangian Brackets which are constituents of LAGRANGE's planetary equations, is derived.

(1) ONE of Professor STRATTON's earliest researches (1909a, 1909b) was in the strenuous discipline of celestial mechanics on the subject of *The Constants of the Moon's Physical Libration*; in joining with so many of his friends and admirers in saluting him on his retirement from the Chair of Astrophysics, I think it not inappropriate to offer a small contribution in the field of his original interests.

(2) A Lagrangian Bracket, $[u, v]$, is defined by the expression

$$[u, v] = \sum_{x,y,z} \frac{\partial(x, \dot{x})}{\partial(u, v)}, \quad \ldots.(1)$$

where x, y, z are the rectangular co-ordinates of a planet and u, v are usually taken to be two of the six elements of the orbit—a, e, i; Ω, $\tilde{\omega}$, and ε in the well-established notation; each co-ordinate is a function of the time and the six elements. The fifteen brackets (1) are constituents of the six Lagrangian equations of planetary motion and their evaluation, when u and v are elements, is found in the principal works on celestial mechanics.

The present note is based on a method first suggested by CAMPBELL (1897), over half a century ago, and is substantially in the form in which I have given it in lectures for some time past. We shall assume that, in general, u and v are any independent functions of the six elements.

(**3**) Let

$$F_u = \dot{x}\frac{\partial x}{\partial u} + \dot{y}\frac{\partial y}{\partial u} + \dot{z}\frac{\partial z}{\partial u}, \qquad \ldots .(2)$$

with a similar expression for F_v; then, from (1),

$$[u, v] = \frac{\partial F_u}{\partial v} - \frac{\partial F_v}{\partial u}. \qquad \ldots .(3)$$

Now

$$\frac{dF_u}{dt} = \sum_{x,y,z}\left\{\ddot{x}\frac{\partial x}{\partial u} + \frac{1}{2}\frac{\partial}{\partial u}(\dot{x}^2)\right\};$$

also

$$\dot{x}^2 + \dot{y}^2 + \dot{z}^2 = \mu\left(\frac{2}{r} - \frac{1}{a}\right) = 2V + 2V_0,$$

where

$$V = \frac{\mu}{r}, \; V_0 = -\frac{\mu}{2a} \text{ and } \mu = n^2a^3. \qquad \ldots .(4)$$

The equations of motion in the elliptic orbit are

$$\ddot{x} + \frac{\mu x}{r^3} = 0, \textit{ etc.}, \text{ or } \ddot{x} = \frac{\partial V}{\partial x}, \textit{ etc.}$$

Hence

$$\frac{dF_u}{dt} = \sum_{x,y,z}\left\{\frac{\partial V}{\partial x}\cdot\frac{\partial x}{\partial u}\right\} + \frac{\partial}{\partial u}(V + V_0) = \frac{\partial}{\partial u}(2V + V_0)$$

$$= 2\frac{\partial}{\partial u}(V + V_0) - \frac{\partial V_0}{\partial u}.$$

Integrate between τ and t, where τ is the time of perihelion passage; then,

$$F_u - F_u(\tau) = 2\int_\tau^t \frac{\partial}{\partial u}(V + V_0)dt - (t - \tau)\frac{\partial V_0}{\partial u}.$$

Then, if

$$I = \int_\tau^t (V + V_0)dt,$$

we have, since $\tau = -\frac{1}{n}(\varepsilon - \tilde{\omega})$,

$$\frac{\partial I}{\partial u} = \int_{\tau}^{t} \frac{\partial}{\partial u}(V + V_0)dt - \{V(\tau) + V_0\}\frac{\partial \tau}{\partial u}.$$

Hence

$$F_u - F_u(\tau) = 2\frac{\partial I}{\partial u} + 2\{V(\tau) + V_0\}\frac{\partial \tau}{\partial u} - (t - \tau)\frac{\partial V_0}{\partial u}. \qquad \ldots (5)$$

Now, in terms of the eccentric anomaly E, r is given by

$$r = a(1 - e \cos E);$$

hence, since $E = 0$ when $t = \tau$, $V(\tau) = \mu/\{a(1 - e)\}$, and (5) becomes

$$F_u - F_u(\tau) = 2\frac{\partial I}{\partial u} + \frac{\mu(1 + e)}{a(1 - e)} \cdot \frac{\partial \tau}{\partial u} - (t - \tau)\frac{\partial V_0}{\partial u},$$

which we write as

$$2\frac{\partial I}{\partial u} - t\frac{\partial V_0}{\partial u} - F_u = C_u, \qquad \ldots (6)$$

where

$$C_u = - F_u(\tau) - \frac{\mu(1 + e)}{a(1 - e)} \cdot \frac{\partial \tau}{du} - \tau\frac{\partial V_0}{\partial u}. \qquad \ldots (7)$$

Here C_u is a constant, being a function of the elements. Similarly,

$$2\frac{\partial I}{\partial v} - t\frac{\partial V_0}{\partial v} - F_v = C_v, \qquad \ldots (8)$$

where C_v is defined in an analogous way to (7).

Differentiate (6) with respect to v and (8) with respect to u, and subtract; then, from (3),

$$[u, v] = \frac{\partial C_v}{\partial u} - \frac{\partial C_u}{\partial v}. \qquad \ldots (9)$$

Since the right-hand side of this equation is constant, the characteristic property of a bracket is deduced, namely, that $[u, v]$ does not involve the time explicitly.

The evaluation of $[u, v]$ requires the evaluation of $F_u(\tau)$ in (7).

(**4**) The three angular elements of the planetary orbit with which it is convenient to begin are:

$$\Omega = \Upsilon N,\ \omega = NA, \text{ and } i \text{ (the inclination)},$$

where N is the ascending node and A is the direction of perihelion on the sphere, the Sun being at the centre O.

Let (ξ, η) be the co-ordinates of the planet with respect to axes OA, OB in the orbital plane ($NB = 90° + \omega$).

If (l_1, m_1, n_1) and (l_2, m_2, n_2) are the direction-cosines of OA and OB with respect to the original axes, then

$$x = l_1\xi + l_2\eta,\ y = m_1\xi + m_2\eta,\ z = n_1\xi + n_2\eta \qquad \ldots (10)$$

We have:

$$l_1 = \cos \Omega \cos \omega - \sin \Omega \sin \omega \cos i,$$
$$m_1 = \sin \Omega \cos \omega + \cos \Omega \sin \omega \cos i,$$
$$n_1 = \sin \omega \sin i,$$

and

$$l_2 = - \cos \Omega \sin \omega - \sin \Omega \cos \omega \cos i,$$
$$m_2 = - \sin \Omega \sin \omega + \cos \Omega \cos \omega \cos i,$$
$$n_2 = \cos \omega \sin i.$$

Also, if (l_3, m_3, n_3) are the direction-cosines of OC, the normal to the orbital plane, then

$$l_3 = \sin \Omega \sin i, \; m_3 = - \cos \Omega \sin i, \; n_3 = \cos i.$$

From these

$$\frac{\partial l_1}{\partial \Omega} = - m_1, \quad \frac{\partial m_1}{\partial \Omega} = l_1, \quad \frac{\partial n_1}{\partial \Omega} = 0;$$

$$\frac{\partial l_1}{\partial \omega} = l_2, \quad \frac{\partial m_1}{\partial \omega} = m_2, \quad \frac{\partial n_1}{\partial \omega} = n_2;$$

$$\frac{\partial l_1}{\partial i} = l_3 \sin \omega, \quad \frac{\partial m_1}{\partial i} = m_3 \sin \omega, \quad \frac{\partial n_1}{\partial i} = n_3 \sin \omega.$$

Hence

$$\frac{\partial l_1}{\partial u} = \frac{\partial l_1}{\partial \Omega} \cdot \frac{\partial \Omega}{\partial u} + \frac{\partial l_1}{\partial \omega} \cdot \frac{\partial \omega}{\partial u} + \frac{\partial l_1}{\partial i} \cdot \frac{\partial i}{\partial u}$$

$$= - m_1 \frac{\partial \Omega}{\partial u} + l_2 \frac{\partial \omega}{\partial u} + l_3 \sin \omega \frac{\partial i}{\partial u}.$$

Similarly,

$$\frac{\partial m_1}{\partial u} = l_1 \frac{\partial \Omega}{\partial u} + m_2 \frac{\partial \omega}{\partial u} + m_3 \sin \omega \frac{\partial i}{\partial u},$$

$$\frac{\partial n_1}{\partial u} = n_2 \frac{\partial \omega}{\partial u} + n_3 \sin \omega \frac{\partial i}{\partial u}.$$

From these and the relations between direction-cosines,

$$\sum_{l,m,n} l_2 \frac{\partial l_1}{\partial u} = n_3 \frac{\partial \Omega}{\partial u} + \frac{\partial \omega}{\partial u}. \qquad \ldots.(11)$$

Also, since $\Sigma l_1^2 = \Sigma l_2^2 = 1$, and $\Sigma l_1 l_2 = 0$, then

$$\Sigma l_1 \frac{\partial l_1}{\partial u} = \Sigma l_2 \frac{\partial l_2}{\partial u} = 0 \qquad \ldots.(12)$$

and

$$\Sigma l_1 \frac{\partial l_2}{\partial u} = - \Sigma l_2 \frac{\partial l_1}{\partial u}. \qquad \ldots.(13)$$

(5) From (2) and (10) we have

$$F_u = \sum_{l,m,n} (l_1\dot{\xi} + l_2\dot{\eta})\left(l_1 \frac{\partial \xi}{\partial u} + \xi \frac{\partial l_1}{\partial u} + l_2 \frac{\partial \eta}{\partial u} + \eta \frac{\partial l_2}{\partial u}\right)$$

$$= \dot{\xi} \frac{\partial \xi}{\partial u} + \dot{\eta} \frac{\partial \eta}{\partial u} + (\xi\dot{\eta} - \dot{\xi}\eta)\, \Sigma l_2 \frac{\partial l_1}{\partial u},$$

by means of (12) and (13).

But,

$$\xi\dot{\eta} - \dot{\xi}\eta = h = \sqrt{\{\mu a(1 - e^2)\}} = na^2\sqrt{(1 - e^2)},$$

where h is twice the rate of description of area in the elliptic orbit; hence by means of (11),

$$F_u(\tau) = \left[\dot{\xi} \frac{\partial \xi}{\partial u} + \dot{\eta} \frac{\partial \eta}{\partial u}\right]_\tau + h \frac{\partial \omega}{\partial u} + h \cos i \frac{\partial \Omega}{\partial u}. \qquad \ldots (14)$$

Now,

$$\xi = a(\cos E - e),\ \eta = a\sqrt{(1 - e^2)} \sin E;$$

also, Kepler's equation is

$$E - e \sin E = n(t - \tau),$$

from which

$$\dot{E} = \frac{na}{r} \text{ and } \dot{E}(\tau) = \frac{n}{1 - e}.$$

Then,

$$\dot{\xi} = -a \sin E \,.\, \dot{E}, \text{ so that } \dot{\xi}(\tau) = 0.$$

Also,

$$\dot{\eta} = a\sqrt{(1 - e^2)} \cos E \,.\, \dot{E},$$

from which

$$\dot{\eta}(\tau) = \frac{na\sqrt{(1 - e^2)}}{1 - e}.$$

Again,

$$\left(\frac{\partial \eta}{\partial u}\right) = a\sqrt{(1 - e^2)} \left(\frac{\partial E}{\partial u}\right)_\tau.$$

From Kepler's equation

$$\frac{r}{a} \frac{\partial E}{\partial u} - \sin E \,.\, \frac{\partial e}{\partial u} = (t - \tau) \frac{\partial n}{\partial u} - n \frac{\partial \tau}{\partial u};$$

hence

$$\left(\frac{\partial E}{\partial u}\right)_\tau = -\frac{n}{1 - e} \cdot \frac{\partial \tau}{\partial u}.$$

Inserting these results in (14) we obtain, with μ given by (4),

$$F_u(\tau) = -\frac{\mu(1 + e)}{a(1 - e)} \cdot \frac{\partial \tau}{\partial p} + h \frac{\partial \omega}{\partial u} + h \cos i \frac{\partial \Omega}{\partial u}.$$

Hence (7) becomes

$$C_u = -\tau \frac{\partial V_0}{\partial u} - h \frac{\partial \omega}{\partial u} - h \cos i \frac{\partial \Omega}{\partial u}, \qquad \ldots (15)$$

or, in terms of $\tilde{\omega}$, ε, and replacing V_0 by $-\dfrac{\mu}{2a}$,

$$C_u = \frac{\mu^{\frac{1}{2}}}{2a^{\frac{1}{2}}}(\varepsilon - \tilde{\omega})\frac{\partial a}{\partial u} - h\frac{\partial \tilde{\omega}}{\partial u} + h(1 - \cos i)\frac{\partial \Omega}{\partial u}. \qquad \ldots.(16)$$

CAMPBELL's procedure for evaluating the Lagrangian Brackets when u and v are any two of the elements a, e, i, Ω, $\tilde{\omega}$, and ε may be interpolated here. From (16),

$$C_a = \frac{\mu^{\frac{1}{2}}}{2a^{\frac{1}{2}}}(\varepsilon - \tilde{\omega}), \quad C_e = C_i = 0,$$

$$C_\Omega = h(1 - \cos i), \; C_{\tilde{\omega}} = -h, \; C_\varepsilon = 0.$$

Then, for example,

$$[a, e] = \frac{\partial C_e}{\partial a} - \frac{\partial C_a}{\partial e} = 0.$$

In this way the non-zero brackets are found to be, on writing $e = \sin\phi$,

$$[a, \varepsilon] = -\tfrac{1}{2}na, \; [a, \Omega] = \tfrac{1}{2}na\cos\phi(1 - \cos i), \; [a, \tilde{\omega}] = \tfrac{1}{2}na(1 - \cos\phi),$$
$$[e, \Omega] = -na^2\tan\phi(1 - \cos i), \; [e, \tilde{\omega}] = na^2\tan\phi, \; [i, \Omega] = na^2\cos\phi\sin i.$$

(**6**) The general expression for a bracket can be derived at once from (9) and (15), u and v being any functions of the elements; it is, with $V_0 = -\dfrac{\mu}{2a}$,

$$[u, v] = \frac{\partial\left(-\tau, -\dfrac{\mu}{2a}\right)}{\partial(u, v)} + \frac{\partial(\omega, h)}{\partial(u, v)} + \frac{\partial(\Omega, h\cos i)}{\partial(u, v)}, \qquad \ldots.(17)$$

or, in terms of $\tilde{\omega}$ and ε,

$$[u, v] = \frac{\partial\left(\dfrac{\varepsilon - \tilde{\omega}}{n}, -\dfrac{\mu}{2a}\right)}{\partial(u, v)} + \frac{\partial(\tilde{\omega} - \Omega, h)}{\partial(u, v)} + \frac{\partial(\Omega, h\cos i)}{\partial(u, v)}. \qquad \ldots.(18)$$

In particular, if u and v are any two elements the evaluation of the corresponding bracket can be quickly obtained.

(**7**) The formula (17) enables us to derive the canonic equations expressed in the usual forms.

Introduce α_r and β_r ($r = 1, 2, 3$) given by

$$\alpha_1 = -\frac{\mu}{2a}, \; \alpha_2 = h \equiv \sqrt{\{\mu a(1 - e^2)\}}, \; \alpha_3 = h\cos i,$$

$$\beta_1 = -\tau, \quad \beta_2 = \omega, \qquad \beta_3 = \Omega.$$

Then (17) becomes

$$[u, v] = -\sum_{r=1}^{3}\frac{\partial(\alpha_r, \beta_r)}{\partial(u, v)}.$$

Hence, since the α's and β's are independent,

$$[\alpha_r, \beta_r] = -1 \qquad \ldots.(19)$$

and

$$[\alpha_r, \alpha_s] = [\beta_r, \beta_s] = [\alpha_r, \beta_t] = 0, t \neq r. \qquad \ldots.(20)$$

If R is the disturbing function, Lagrange's planetary equations in their general form are:

$$\sum_{s=1}^{3} [\alpha_r, \alpha_s]\dot{\alpha}_s + \sum_{s=1}^{3} [\alpha_r, \beta_s]\dot{\beta}_s = \frac{\partial R}{\partial \alpha_r}, \qquad \ldots.(21)$$

$$\sum_{s=1}^{3} [\beta_r, \alpha_s]\dot{\alpha}_s + \sum_{s=1}^{3} [\beta_r, \beta_s]\dot{\beta}_s = \frac{\partial R}{\partial \beta_r}. \qquad \ldots.(22)$$

Then (22) and (21) become, by means of (19) and (20),

$$\dot{\alpha}_r = \frac{\partial R}{\partial \beta_r}, \ \dot{\beta}_r = -\frac{\partial R}{\partial \alpha_r}.$$

These are the canonic equations in which α_r and β_r are a pair of conjugate variables.

References

CAMPBELL, A. Y. G. . . . . 1897 *M.N.*, **57**, 118.
STRATTON, F. J. M. . . . . 1909a *Mem. Roy. Astron. Soc.*, **59**, Pt. IV.
1909b *M.N.*, **69**, 568.

The Rôle of Cracovians in Astronomy

T. BANACHIEWICZ

Astronomiczne Obserwatorium, Kraków, Poland

SUMMARY

The importance and usefulness of cracovians—matrices obeying special rules of multiplication—is made evident in the treatment of astronomical problems. To show this, simple applications are dealt with in the fields of spherical astronomy, transformation of co-ordinates, and the theory of least squares. The discussion emphasizes the theoretical interest of cracovians, and brings out the simplicity, ease, and reliability of cracovian operations.

1. Introduction

THE rise of cracovians in computational astronomy is to a certain degree connected with the decline of logarithms. Of course, logarithms continue to play a conspicuous rôle both in the definition of stellar magnitudes and in theoretical developments. As regards computations, however, a revolutionary development has taken place, which is mainly due to the use of calculating machines (desk calculators). This happened because the desk calculator serves better than logarithms in the formation of products of numbers, especially in expressions such as

$$p = aA - bB + cC, \qquad \ldots (1)$$

for which the desk calculator does not demand the writing out of the separate aA, $-bB$, cC. This has had the consequence that linear expressions like (1), formerly almost tabooed in astronomical writings (and often hidden, as, for example, in Gauss' *Theoria Motus*, in the thick of trigonometrical formulae) have now risen to an honourable place. We come here to another factor in this development. It has turned out, in fact, that further progress may be achieved by a suitable mathematical interpretation of the linear expressions, together with a corresponding theory and symbolism. For instance, the formulae of common algebra

$$\begin{array}{lll} l = a\xi - b\eta & x = cl - dm & X = ex - fy, \\ m = a'\xi - b'\eta & y = c'l - d'm & Y = e'x - f'y, \end{array} \qquad \ldots (2$$

which give us the relations between X, Y and ξ, η, are not very convenient and demand a wearisome attention in their evaluation. They may, however, be essentially simplified by the use of collective numbers, such as the matrices invented by HAMILTON and CAYLEY about a hundred years ago, or the cracovians introduced by the writer since 1916. These two types of collective numbers differ theoretically only in the definition of the product, and it is natural that mathematicians in their desire to deal with a minimum of mathematical entities, would like *a priori* to drop the new numbers, or to treat both types in a common theory. The second alternative, however, would lead inevitably to errors in the calculations, as it is essential for the computer to have single fixed rules. As regards the choice between matrices and cracovians, we may quote here some lines from the recent authoritative Polish treatise by HAUSBRANDT (1953): "The use (in the book) of cracovians . . . instead of matrices . . . does not appear to demand detailed explanations. Everybody engaged in practical mathematics knows that the (cracovian) multiplication by parallel lines (columns by columns or rows by rows) is in practical computation more convenient than the (matrix) multiplication by perpendicular lines. The apparent defect of cracovian multiplication, *i.e.* the absence of the associative law, $(ab)c = a(bc)$, is (really) no defect . . . because for the cracovians it is $\boldsymbol{abc} = \boldsymbol{a}(\boldsymbol{c} \,.\, \boldsymbol{\tau b})$, $\boldsymbol{\tau b}$ being the transpose of $\boldsymbol{b}$ . . . . The solution of different problems by cracovians has preceded, sometimes by more than ten years, their solution by matrices. This appears to indicate the superiority of cracovians as a tool of research".

Also a number of astronomers from different countries have given preference to computations with cracovians, which in Poland are used exclusively. We quote here VILLEMARQUÉ (1936) in China, ECKERT and BROUWER (1937) in U.S.A., AREND (1941–1950) in Belgium, HERGET (1948) in U.S.A., and SAMOILOWA-JACHONTOWA (1949) in U.S.S.R. Hence only the applications of cracovians will be considered here.

In 1923 the author of this paper introduced the cracovians in print, and developed in the following years their theory (BANACHIEWICZ, 1923a, 1949). Multiplication of cracovians is carried out according to the following rule: the element p_{ij} in the ith column and the jth row of the product $\boldsymbol{p} = \boldsymbol{a} \,.\, \boldsymbol{b}$ of two cracovians $\boldsymbol{a}$ and $\boldsymbol{b}$ is obtained by multiplying the ith column (denoted by $\boldsymbol{a}_i$) of the first factor $\boldsymbol{a}$ by the jth column (denoted by $\boldsymbol{b}_j$) of the second factor $\boldsymbol{b}$. Thus

$$p_{ij} = \boldsymbol{a}_i \,.\, \boldsymbol{b}_j. \qquad \ldots (3)$$

On the basis of this definition the relation (2) between X, Y and ξ, η, becomes:

$$\begin{Bmatrix} X \\ Y \end{Bmatrix} = \begin{Bmatrix} \xi \\ \eta \end{Bmatrix} \begin{Bmatrix} a & a' \\ -b & -b' \end{Bmatrix} \begin{Bmatrix} c & c' \\ -d & -d' \end{Bmatrix} \begin{Bmatrix} e & e' \\ -f & -f' \end{Bmatrix}, \qquad \ldots .(4)$$

denoting the cracovians by the symbols { } and supposing that the cracovians in the right member of (4) are to be multiplied in turn: the first by the second, their product by the third, and so on. For the numerical values $a = +4$, $a' = -1$, $b = +5$, $b' = -6$, $c = -5$, $c' = +7$, $d = -1$, $d' = +2$, $e = +1$, $e' = +3$, $= +4, f' = +2, \xi = +3, \eta = +2$, it follows from (2) or from (4) that $X = +15$, $Y = +5$. In employing (4) the problem is arranged according to the second precept of DESCARTES (1636): the given values of a, a', . . . are inserted into the scheme (4) and then the operations are carried out with the desk calculator (BANACHIEWICZ, 1929a). Although the arithmetical operations are the same whether we use (2) or (4), it is easy to convince oneself by trial that the use of (4) saves much mental work. In such an arrangement of the work lies to a large extent the importance of the cracovians as a tool for reducing the mental strain of the computer.

2. SPHERICAL ASTRONOMY

The cracovians lead quickly, rigorously, and probably more directly than any other way, to the fundamental formulae of spherical trigonometry for a general triangle, as well as to those of spherical polygononometry (BANACHIEWICZ, 1923b).

The underlying concept is the method of "wandering axes". Take a system of perpendicular co-ordinate axes OX, OY, OZ with O in the centre of a sphere. When this system rotates about the axis OZ through the angle α, or about the axis OX through the angle β, the co-ordinates (x, y, z) of any fixed point M on the sphere become:

$$\begin{Bmatrix} x \\ y \\ z \end{Bmatrix} \boldsymbol{r}(\alpha) \quad \text{or} \quad \begin{Bmatrix} x \\ y \\ z \end{Bmatrix} \boldsymbol{p}(\beta),$$

where

$$\boldsymbol{r}(\alpha) = \begin{Bmatrix} \cos\alpha & -\sin\alpha & 0 \\ \sin\alpha & \cos\alpha & 0 \\ 0 & 0 & 1 \end{Bmatrix}, \quad \boldsymbol{p}(\beta) = \begin{Bmatrix} 1 & 0 & 0 \\ 0 & \cos\beta & -\sin\beta \\ 0 & \sin\beta & \cos\beta \end{Bmatrix}. \qquad \ldots .(5)$$

Consider a closed spherical polygon with vertices $\boldsymbol{1}$, $\boldsymbol{2}$, . . . $\boldsymbol{n}$, and let the axis OX be directed initially towards the vertex $\boldsymbol{1}$ and the axis OY to a point lying on the great circle through the vertices $\boldsymbol{1}$ and $\boldsymbol{2}$. Let s_i denote the length of the side $(\boldsymbol{i}, \boldsymbol{i}+\boldsymbol{1})$ and a_i the angle between the sides $(\boldsymbol{i}, \boldsymbol{i}+\boldsymbol{1})$ and $(\boldsymbol{i}+\boldsymbol{1}, \boldsymbol{i}+\boldsymbol{2})$. Let the co-ordinate axes rotate in turn about the axes OZ and OX, first through the angle s_1 about OZ, then through a_1 about the new position of OX, and afterwards through the angle s_2 about the new position of OZ, and so on. After all these rotations round the whole polygon the axes $OXYZ$ will return to their initial positions, and the initial and the final co-ordinates of all points M will coincide.

If we take as M in turn the points (1, 0, 0), (0, 1, 0), and (0, 0, 1), we obtain from (5) the fundamental formula of spherical polygonometry:

$$\boldsymbol{\tau} \,.\, \boldsymbol{r}(s_1) \,.\, \boldsymbol{p}(a_1) \,.\, \boldsymbol{r}(s_2) \,.\,.\,.\, \boldsymbol{r}(s_n)\boldsymbol{p}(a_n) = \boldsymbol{\tau}, \qquad \ldots .(6)$$

where $\boldsymbol{\tau}$ denotes the unit cracovian

$$\boldsymbol{\tau} = \begin{Bmatrix} 1 & 0 & 0 \\ 0 & 1 & 0 \\ 0 & 0 & 1 \end{Bmatrix}. \qquad \ldots .(7)$$

In the particular case of a spherical triangle the formula (6) leads directly to the familiar eight formulae of GAUSS and to the formula of CAGNOLI.

Moreover, the formula (6) is valid *mutatis mutandis* also if the cracovians $\boldsymbol{r}$ and $\boldsymbol{p}$ are replaced by certain other cracovians $\boldsymbol{R}$ and $\boldsymbol{P}$ of the fourth order (*i.e.* cracovians having 4 columns and 4 rows, and depending on the arguments $\frac{1}{2}s_i$ and $\frac{1}{2}a_i$ (BANACHIEWICZ, 1927). This result furnishes two methods of solving polygonometric problems. It extends to spherical polygons the four well-known formulae of DELAMBRE of the form: $\sin \frac{1}{2}(A - B) \, . \, \sin \frac{1}{2}c = \sin \frac{1}{2}(a - b) \, . \, \cos \frac{1}{2}C$.

KOEBCKE (1937), with reference to the above equation (6), wrote as follows: "The solution of general spherical triangles was once performed by their decomposition into right-angled triangles; only considerably later was the solution of general triangles carried out directly. Nevertheless, spherical polygons have previously been solved by resolving them into triangles. In this respect the cracovians . . . bring about an essential change, and in this domain lies their main theoretical significance".

Formula (6) is frequently applied to calculations concerning the Moon, particularly to the determination of the selenographic co-ordinates P and D for occultations or solar eclipses (BANACHIEWICZ, 1930). It formed the basis of the differential formulae of spherical polygonometry deduced by KOZIEL (1949).

Detailed expressions for spherical polygons of 12 or less elements have been given by the writer (BANACHIEWICZ, 1948).

3. THEORETICAL ASTRONOMY

The rotational cracovians $\boldsymbol{p}$ and $\boldsymbol{r}$ figuring in (6) are especially useful in practice in connection with the transformation of co-ordinates. The vectorial constants of an orbit, *e.g.* are given by the formula (BANACHIEWICZ, 1929b):

$$\begin{Bmatrix} A_x & B_x \\ A_y & B_y \\ A_z & B_z \end{Bmatrix} = a \begin{Bmatrix} \cos \omega & -\sin \omega \\ \sin \omega & \cos \omega \end{Bmatrix} \begin{Bmatrix} 1 & 0 & 0 \\ 0 & \cos i & \sin i \end{Bmatrix} \begin{Bmatrix} \cos \Omega & \sin \Omega & 0 \\ -\sin \Omega & \cos \Omega & 0 \\ 0 & 0 & 1 \end{Bmatrix} \begin{Bmatrix} 1 & 0 & 0 \\ 0 & \cos \varepsilon & \sin \varepsilon \\ 0 & -\sin \varepsilon & \cos \varepsilon \end{Bmatrix} \qquad \ldots .(8)$$

This formula contains 47 symbols (besides the zeros) against 137, *i.e.* almost three times as many symbols in the equivalent common forumlae (see, *e.g. Planetary Co-ordinates*, 1939). The latter are not only less transparent, but they also do not contain in themselves the handy computing scheme for formula (8); in contrast to the latter they also cannot be written down directly. The transcription of the formula (8) using the cracovians $\boldsymbol{R}$ and $\boldsymbol{P}$ was given by BANACHIEWICZ (1928).

More important are the formulae for computing the differential coefficients of geocentric co-ordinates of a planet or comet with respect to the elements of the orbit. The notable simplification of the problem brought about by the cracovians is here very attractive; see the papers by KEPIŃSKI (1927), ZAGAR (1928), ORKISZ (1931), SZELIGOWSKI and KOEBCKE (1934), PRZYBYLSKI (1939), ECKERT and BROUWER

(1937), p. 132), MOŠKOVA (1949), and others. For the case of the elements ☊, i, and ω, the theory was developed by the present writer (1929c) and for OPPOLZER'S elements by American astronomers (see ECKERT and BROUWER, 1937, p. 132).

For the application of the cracovians to the theory of special perturbations see KOEBCKE (1937, p. 13). Different applications to the theory of orbits are to be found in the papers of ŠTEINS (1950) and the writer (in connection with the problem of the accuracy of an orbit; BANACHIEWICZ, 1950) and in many other places.

4. Least Squares

$1954 = 1 \,.\, 10^3 + 9 \,.\, 10^2 + 5 \,.\, 10^1 + 4$. This universally adopted positional system of writing the numbers considerably facilitates the four arithmetical operations on them. In all operations on numbers we do not need to trouble about the different powers of 10; we forget them. A similar mental simplification may be achieved in the solution of linear equations. Since astronomers are interested mainly in the solution of the normal equations of the theory of Least Squares, such equations only will be considered here.

Take a system of two normal equations for two unknowns

$$\begin{aligned} 4x + 2y &= 10. \\ 2x + 10y &= 14. \end{aligned} \qquad \ldots.(9)$$

By introducing the cracovians

$$\boldsymbol{A} = \begin{Bmatrix} 4 & 2 \\ 2 & 10 \end{Bmatrix} \quad \text{and} \quad \boldsymbol{L} = \begin{Bmatrix} 10 \\ 14 \end{Bmatrix}$$

we present the solution of (9) in the form

$$\begin{Bmatrix} x \\ y \end{Bmatrix} = \boldsymbol{L} : \boldsymbol{A}. \qquad \ldots.(10)$$

The problem is thus reduced to the division of two cracovians. In such a solution we have to deal only with the elements of $\boldsymbol{L}$ and $\boldsymbol{A}$, but not with the equations and the unknowns, thus attaining the simplification indicated.

The division in (10) is made by a resolution of $\boldsymbol{A}$ into a product of two triangular cracovians consisting of the upper right-hand half of the square array (BANACHIEWICZ, 1938), *i.e.* with zeros below the leading diagonal. This operation, made solely on the basis of the single equation $\boldsymbol{G}_i \,.\, \boldsymbol{H}_j = A_{ij}$, is very easily performed with a desk calculator. One supposes usually $\boldsymbol{G} = \boldsymbol{H} = \sqrt{\boldsymbol{A}}$ ("method of the square root"). One can also take $\boldsymbol{G}$ as having units on its leading diagonal ("method of proportional factors"); the method of DOOLITTLE then leads, less conveniently, to the same numerical operations.

The numbers ρ_1 and ρ_2, satisfying $\begin{Bmatrix} \rho_1 \\ \rho_2 \end{Bmatrix} \sqrt{\boldsymbol{A}} = \boldsymbol{L}$, are also found. In our example, where $\sqrt{\boldsymbol{A}} = \begin{Bmatrix} 2 & 1 \\ 0 & 3 \end{Bmatrix}$, $\rho_1 = 5$, $\rho_2 = 3$, it follows that

$$2x + y = 5,\ 3y = 3, \quad \text{and hence} \quad x = 2,\ y = 1. \qquad \ldots.(11)$$

This is the determinate solution, which was never a matter of any great difficulty. It was presented by CHOLESKY (see BENOIT, 1924) in ordinary algebraic language, and therefore without the advantages of the operations with collective numbers.

Denoting by $\boldsymbol{q}$ the inverse of $\boldsymbol{r} = \sqrt{\boldsymbol{A}}$, so that $\boldsymbol{q} = \boldsymbol{r}^{-1}$, one obtains the indeterminate solution of (10), *i.e.* the cracovian $\boldsymbol{Q}$ in the equation $\begin{Bmatrix} x \\ y \end{Bmatrix} = \boldsymbol{L} \,.\, \boldsymbol{Q}$, namely

$$\boldsymbol{Q} = \boldsymbol{q}^2 \qquad \ldots .(12)$$

in our example

$$\boldsymbol{Q} = \begin{Bmatrix} \frac{1}{2} & 0 \\ -\frac{1}{6} & \frac{1}{3} \end{Bmatrix}^2 = \tfrac{1}{18} \begin{Bmatrix} 5 & -1 \\ -1 & +2 \end{Bmatrix}.$$

GAUSS was much occupied with a problem which was equivalent to the determination of $\boldsymbol{Q}$, but his results have proved in practice to be somewhat difficult to apply. In spite of some misstatements an equation equivalent to (12) was never obtained by CHOLESKY. We note that a still shorter method of calculating $\boldsymbol{Q}$ is based on the equation $\boldsymbol{Q} = \boldsymbol{q} : \boldsymbol{r}$, demanding a knowledge of the diagonal elements of $\boldsymbol{q}$ only (BANACHIEWICZ, 1939).

Besides facilitating computation, the cracovians also simplify the theory of least squares. The expression (10), for instance, leads at once to the formulae

$$\begin{Bmatrix} x \\ y \end{Bmatrix} = \boldsymbol{L} \,.\, \tau\boldsymbol{q} \,.\, \boldsymbol{q} = (\boldsymbol{L} : \tau\boldsymbol{r}) \,.\, \boldsymbol{q} = (\boldsymbol{L} : \tau\boldsymbol{r}) : \boldsymbol{r},$$

each of which would demand for its derivation a lengthy procedure using ordinary algebra. The numerical determination of the unknowns may be carried out in a variety of ways.

"In Poland the Gauss algorism has been almost wholly replaced by the cracovian calculus", writes LEŚNIOK (1953).

5. "THE CRACOVIAN IDEA"

The simplification of the solutions of linear problems and the reduction of the mental work involved in computations, as achieved by the cracovian calculus, has encouraged some workers to introduce other tabular numbers or symbols (relating to tables) for the solution of non-linear computational problems. KOCHMANSKI (1952) established his "nuclear algebra", dealing with operations on power series in two or more variables. KOZIEL (1954), using the concept of the "nuclei", solves with respect to x and y the equations

$$\xi = \Sigma A_{ij} x^j y^i, \quad \eta = \Sigma B_{ij} x^j y^i,$$

where A_{ij} and B_{ij} denote the known coefficients. HAUSBRANDT (1953, p. 67) invents special auxiliary symbols for geodetic computations.

Everybody who familiarizes himself with cracovians, and proceeds to practise them, will become their friend, often further developing their basic ideas.

REFERENCES

AREND, S. . . . . 1941a "Voies nouvelles dans le calcul scientifique", *Ciel et Terre*, **57**, No. 12, pp. 497–515.

1941b "Etablissement par voie raccourcie des formules de Thiele-Innes, relatives aux orbites d'étoiles doubles, en recourant aux principes de l'affinité", *Annales de l'Observatoire de Toulouse*, **16,** 109–113.

1947 *Bull. Obs. Uccle*, **4,** pp. 64–99.

1949 "Cracoviens rationnels. Leur utilisation dans les transformations fondamentales de coordonnées astronomiques, dans la théorie de la lunette méridienne et dans celle du théodolite", *Bulletin Astronomique, Uccle*, **4,** 64–70.

1950 "Orientation des orbites d'étoiles doubles visuelles par rapport à la galaxie", *Communications de l'Observatoire Royal de Belgique*, No. 20, pp. 17–29.

BANACHIEWICZ, T. . . . . 1923a "On a Certain Mathematical Notion", *Bull. Acad. Sci.* (*Poland*), A.

1923b "La trigonométrie sphérique et les voies nouvelles de l'astronomie mathématique", *Bull. Acad. Sci.* (*Poland*), A.

1927 "Les relations fondamentales de la trigonométrie sphérique", *C.R. Acad. Sci.* (*Paris*), **185,** 21, XI; *Circ. Obs. Crac.*, **25.**

1928 "Determination of the Position of an Orbit", *M.N.*, **89,** 215–17.

1929a *Acta Astronomica* (*Kraków*), Ser. c, **1,** 64.

1929b *Acta Astronomica* (*Kraków*), Ser. c, **1,** 66.

1929c "Méthodes arithmométriques de la correction des orbites", *Acta Astronomica* (*Kraków*), Ser. c, **1,** 71–86.

1930 "Coordonnées sélénographiques relatives aux occultations", *Acta Astronomica* (*Kraków*), Ser. c, **1,** 127–37.

1938 *Bull. Acad. Sci.* (*Poland*), A, pp. 393–404.

1939 "Computation of Inverse Arrays", *Acta Astronomica* (*Kraków*), Ser. c, **4,** 30.

1948 "Formulas fundamentale pro hexagono spherico et polygonos spherico", *Rocz. Astr. Obs. Krak.* (SAC), **19.**

1949 "Les cracoviens et quelques-unes de leurs applications en géodésie, *Cracow Obs. Reprint*, **25.**

1950 *Acta Astronomica* (*Kraków*), Ser. a, **5,** 37–50.

BENOIT, E. . . . . 1924 *Bull. géod. internat.*, No. 2.

DESCARTES, R. . . . . 1636 *Discours de la méthode; Deuxième partie.* Paris.

ECKERT, W. J. and BROUWER, D. . . . . 1937 *Astron. J.*, **46,** No. 13.

HAUSBRANDT, S. . . . . 1953 *Rachunki geodezyjne* (*Geodetic Computations*), Warszawa, pp. 6, 53.

HERGET, P. . . . . 1948 *Computation of Orbits.* Ann Arbor.

KEPIŃSKI, F. . . . . 1927 *Bull. Acad. Sci.* (*Poland*), A, p. 740.

KOCHMAŃSKI, T. . . . . 1952 *Acta Astronomica* (*Kraków*), Ser. a, **5,** 51–114.

KOEBCKE, F. . . . . 1937 *O zastosowaniu krakowianów w astronomii* (*On Astronomical Applications of the Cracovians*). Poznan, p. 1.

KOZIEL, K. . . . . 1949 *Bull. Acad. Sci.* (*Poland*), A, pp. 1–16.

1954 *Acta Astronomica* (*Kraków*), Ser. a, **5** (in press).

LEŚNIOK, H. . . . . 1953 *Rocznik Geodezyjny* (*Geodetic Almanac*), p. 260.

MOŠKOWA, W. S. . . . . 1949 *Bull. Inst. Astron. Acad. Sci., U.S.S.R.*, **4,** No. 6.

ORKISZ, L. . . . . 1931 "Definitive Bahn des Kometen 1925 I", *Mém. Acad. Sci.* (*Poland*).

Planetary Co-ordinates . . . . 1939 (For the years 1940–1960, referred to the equinox of 1950·0), 150 pp., H.M. Nautical Almanac Office, London.

PRZYBYLSKI, A. . . . . 1939 "Definitive Bahn des Kometen 1919 V", *Acta Astronomica* (*Kraków*), Ser. a, **4,** 34–60.

ŞAMOILOWA-JACHONTOWA, N. S. . . . . . 1949 *Bull. Inst. Astron. Acad. Sci. U.S.S.R.*, **4,** No. 6.

ŠTEINS, K. . . . . 1950 *Acta Astronomica* (*Kraków*), Ser. *a*, **5,** 19–36.

SZELIGOWSKI, S. and KOEBCKE, F. . . . 1934 *Acta Astronomica* (*Kraków*), Ser. *a*, **3,** 57.

VILLEMARQUÉ, E. . . . . 1936 "Calcul numérique par la méthode des bandes mobiles", *Bull. Univ. l'Aurore*, p. 1, Shanghai.

ZAGAR, F. . . . . 1928 *Mem. R. Istituto Veneto d. Sci.*, **29,** No. 8, p. 40.

Regularization of the Three-Body Problem

G. LEMAÎTRE
Université de Louvain, Belgium

1. INTRODUCTION

THE motion of three point-masses under their mutual attraction, according to NEWTON'S law, is one of the oldest and most celebrated problems of celestial mechanics, and is known as the "Three-Body Problem". When two of the bodies approach one another, the attractive force becomes very large and for an actual encounter it is infinite. In that case the equations of motion lose their meaning: encounters are singularities of the problem. In the case of binary collision, that is to say when only two bodies collide, it has been possible to define a modified problem which has no singularity for binary collisions and is completely equivalent to the former problem where this was regular. Such a substitution of a regular problem for the singular one is called a regularization of the problem.

An older example of such a regularization is afforded by the problem in which one body only moves under the Newtonian law due to the attraction of the two other bodies, which are constrained to remain fixed. This problem has been solved by EULER, and JACOBI refers to it as the masterpiece of this great mathematician. It is better known in the form in which LAGRANGE has presented the solution. This is essentially a change of co-ordinates, which introduces the so-called elliptical co-ordinates. It is a conformal transformation of the network of rectangular co-ordinates of which each centre is a singular point. Its effect is that when a trajectory of the original problem makes a sharp bend passing near to a centre in a parabolic orbit, its transformed image follows nearly a straight line.

This would not be sufficient to regularize the problem because this straight line would be described with infinite velocity. To complete the regularization it is necessary to transform the time by introducing a kind of spurious regraduation of it, so that from the point of view of this regraduation the motion would be slowed down when approaching the singularity, the new speed remaining finite. The new variable introduced in this way is called a regularizing time.

This process of LAGRANGE has been extended by THIELE to some particular cases of the Three Body Problem and it has been extensively used in practical computation in the work of the Copenhagen school under E. STRÖMGREN. A similar process has been introduced by LEVI-CIVITA and applied to the motion of three bodies in a plane. It is the purpose of this paper to extend such a regularization process to the general case.

SUNDMAN has shown that from a pure-mathematical point of view, the regularization of time was the essential part of the process, the rectification of the trajectory being a mere luxury. But nobody has been able to apply SUNDMAN'S theory to an actual numerical problem. It is possible that it is the introduction of the superfluous conformal transformation which makes the whole difference between a mere mathematical theorem and a powerful numerical tool.

2. Geometrical Aspects of the Regularization

The nature of the regularization of the binary collisions in the Three-Body Problem can be conveniently exhibited in the simpler case of the Two-Body Problem of Keplerian motion. The Hamiltonian for that case,

$$H = \tfrac{1}{2}p_x^2 + \tfrac{1}{2}p_y^2 - \frac{\mu}{r}$$

is transformed by a conformal transformation, written

$$z = \zeta^2$$

in complex-variable notation, or

$$x = \xi^2 - \eta^2,$$

$$y = 2\xi\eta$$

in real Cartesian co-ordinates.

Then

$$dx^2 + dy^2 = 4(\xi^2 + \eta^2)(d\xi^2 + d\eta^2)$$

and therefore the Hamiltonian becomes

$$H = \tfrac{1}{2}\frac{p_\xi^2 + p_\eta^2}{4(\xi^2 + \eta^2)} - \frac{\mu}{\xi^2 + \eta^2} = h.$$

In order to regularize we introduce a new Hamiltonian

$$H' = (H - h)4(\xi^2 + \eta^2) = \tfrac{1}{2}(p_\xi^2 + p_\eta^2) - 4\mu - 4h(\xi^2 + \eta^2) = 0,$$

for which the canonical equations will be valid if they are written with a new time τ, in place of t, defined by

$$\tau = 4\int \frac{dt}{r}.$$

We have, in fact,

$$\frac{d\xi}{d\tau} = \frac{\partial H'}{\partial p_\xi} = p_\xi, \quad \frac{dp_\xi}{dt} = \frac{d^2\xi}{d\tau^2} = -\frac{\partial H'}{\partial \xi} = 8h\xi.$$

This is the device introduced by Levi-Civita.*

It connects the given problem with a new one by a transformation of co-ordinates involving the introduction of a new time τ. This transformation is regular everywhere, except at the singularity of the first problem. Therefore regularization means replacement of the given problem by another which is regular everywhere and is equivalent to the first one except at the point of singularity.

There are two aspects in this regularization process: first the introduction of a regularizing time τ and secondly a geometrical aspect of the transformation which realizes a simplification of the shape of the trajectories in the neighbourhood of the singularity. Mathematically speaking, the new graduation of time by τ is the only essential point. Sundman's realization of this enabled him to regularize the general Three-Body Problem, at least when three-body-collisions are excluded and he has shown that this always occurs when the total angular momentum does not

* See, for example, Happel (1941), p. 516, and Whittaker (1937), p. 424, where references to the original papers of Levi-Civita and Sundman are given.

vanish. Nevertheless, his process has found no practical use in numerical computations. This may be related to the fact that, although SUNDMANN accounted for the required regraduation of the time he completely neglected the secondary aspect of, LEVI-CIVITA's regularization, *i.e.* the geometrical simplification of the trajectories in the neighbourhood of the singularity.

A confirmation of his point of view may be afforded by the practical interest of the process of regularization, introduced by LAGRANGE in the problem of two fixed centres (cf., *e.g.*, WHITTAKER (1937), p. 91), which has been used with so much success by THIELE and the Copenhagen school in the problem of two rotating centres.

This process makes use of the conformal transformation

$$z = \cosh \zeta$$

or

$$x = \cosh \xi \cos \eta,$$

$$y = \sinh \xi \sin \eta.$$

Using WHITTAKER'S notation with $c = 1$, it leads to the introduction of a new Hamiltonian,

$$\begin{aligned} H' &\equiv (H - h)(\cosh^2 \xi - \cos^2 \eta) \\ &\equiv p_\xi^{\,2} - (\mu + \mu') \cosh \xi - 2h \cosh^2 \xi + p_\eta^{\,2} + (\mu - \mu') \cos \eta + 2h \cos^2 \eta = 0, \end{aligned}$$

to be used with the regularizing time

$$\tau = \int \frac{dt}{\cosh^2 \xi - \cos^2 \eta}.$$

Both collisions with each of the two centres at $\xi = 0$ and $\eta = \pm \pi/2$ are regularized.

The object of the present paper is to make a regularization of the binary collisions which would be efficient for the general case of the Three-Body Problem and would include both aspects of the device of LEVI-CIVITA. This has been applied only to the problem when the three bodies remain in a fixed plane.

3. A TRANSFORMATION FIXING THE SINGULARITIES

The method we propose to apply consists in introducing co-ordinates in such a way that the singularities for binary collisions would be fixed, *i.e.* would be represented by fixed values of some of the co-ordinates. In fact, they will be mapped on three points equally spaced on the equator of a sphere and the regularization will be achieved by mapping these three points conformally on the three vertices of a spherical triangle with three right angles.

The transformation does not imply infinite series, it can be expressed in finite terms and the formulae needed may be considered as being relatively simple.

The connection between the sphere we have just introduced and the three distances of the three bodies may be described in the following way. Let γ_1, γ_2, γ_3 be the three angular distances between a general representative point on the sphere and the three points on the equator which represent the binary collisions. Then the three distances r_k between the bodies are given by

$$r_k = r\sqrt{2} \sin \frac{\gamma_k}{2}.$$

More precisely, introducing polar co-ordinates L and B as longitude and latitude on the sphere and writing

$$L_k = L - \frac{2\pi}{3} k,$$

we have

$$r_k^2 = r^2(1 - \cos B \cos L_k).$$

It is obvious that when the distances r_k are given these equations permit the computation of a kind of mean distance r and of the position of the representative point with polar co-ordinates L, B. The representation is twofold, symmetrical representative points being equivalent. Equilateral configurations lead to indeterminacies which are precisely those which occur at the pole: $B = \pi/2$ and L indeterminate.

It is possible to associate with the triangle formed by the three bodies a set of rectangular axes such that the six Cartesian co-ordinates of these bodies referred to these axes would be

$$x_k = r \sqrt{\frac{2}{3}} \cos \frac{B}{2} \cos \frac{L_k}{2},$$

$$y_k = r \sqrt{\frac{2}{3}} \sin \frac{B}{2} \sin \frac{L_k}{2}.$$

This can be checked by assuming these values of the co-ordinates and computing the mutual distances.

The values of the masses of the bodies can be defined in a manner similar to the foregoing definition of the distances by three quantities m, μ, ν such that the three masses m_k are

$$m_k = m \left[1 + 2\mu \cos \left(\nu - \frac{2\pi}{3} k\right)\right].$$

Then the co-ordinates of the centre of mass referred to the axes just defined are

$$x = \mu r \sqrt{\frac{2}{3}} \cos \frac{B}{2} \cos \left(\nu + \frac{L}{2}\right),$$

$$y = \mu r \sqrt{\frac{2}{3}} \sin \frac{B}{2} \sin \left(\nu + \frac{L}{2}\right).$$

To check this computation it is convenient to obtain and to apply the relation

$$\Sigma m_k \cos \frac{L_k}{2} = 3m\mu \cos \left(\nu + \frac{L}{2}\right)$$

and the corresponding one with sines in place of cosines. We shall need later corresponding relations such as

$$\Sigma m_k \cos L_k = 3m\mu \cos (L - \nu).$$

In order to obtain the kinetic potential (Lagrangian) and the corresponding Hamiltonian of the problem, we must consider rectangular axes parallel to the foregoing ones but drawn through the centre of mass. The position of these axes as regards a set of three fixed rectangular axes will be defined in the ordinary way by the three Eulerian angles (cf., *e.g.*, WHITTAKER, p. 9) and the angular velocity $\omega_x\omega_y\omega_z$ will be expressed in the usual way.*

The resulting kinetic energy is expressed in terms of the coefficients of the momental ellipsoid† by

$$T_e = \tfrac{1}{2}(A\omega_x^2 + B\omega_y^2 + C\omega_z^2 - 2D\omega_x\omega_y).$$

We need also the relative kinetic energy T_r and the relative angular momentum K_r in order to obtain the total kinetic energy

$$T = T_r + \omega_z K_r + T_c.$$

Each of these computations requires the calculation, for α, β, γ independent of L, summations, such as

$$\sum_k m_k(\alpha + \beta \cos L_k + \gamma \sin L_k) - 3m\mu^2[\alpha + \beta \cos (L + 2\nu) + \gamma \sin (L + 2\nu)].$$

They are found to be

$$3m(1 - \mu^2)(\alpha + \beta\Gamma + \gamma\Delta),$$

where we write

$$\Gamma = \frac{1}{1 - \mu^2}[\mu \cos (L - \nu) - \mu^2 \cos (L + 2\nu)]$$

and

$$\Delta = \frac{1}{1 - \mu^2}[\mu \sin (L - \nu) - \mu^2 \sin (L + 2\nu)],$$

or, defining μ' and ν',

$$\Gamma = \mu' \cos (L - \nu'),$$

$$\Delta = \mu' \sin (L - \nu').$$

We write also

$$m_0 = m(1 - \mu^2), \quad m' = m_0(1 - \mu'^2).$$

We find in this way

$$T_e = \tfrac{1}{2}m_0 r^2\left[(1 - \Gamma)\omega_x^2 \sin^2\frac{B}{2} + (1 + \Gamma)\omega_y^2 \cos^2\frac{B}{2} + (1 + \Gamma \cos B)\omega_z^2 - 2\Delta \sin B\omega_x\omega_y\right]$$

also

$$T_r = \frac{m_0}{2}[(1 + \Gamma \cos B)\dot{r}^2 + \tfrac{1}{4}r^2(1 - \Gamma \cos B)(\dot{L}^2 + \dot{B}^2)$$
$$- \Gamma r\dot{r}\dot{B} \sin B - \Delta r\dot{r}\dot{L} \cos B + \tfrac{1}{2}\Delta r^2\dot{B}\dot{L} \sin B]$$

and finally

$$K_r = \tfrac{1}{2}m_0 r^2(\dot{L} \sin B + \dot{B}\Delta).$$

The kinetic potential (Lagrangian) is then

$$L = T + V$$

---

* WHITTAKER, p. 16, with due allowance for an obvious misprint in the expression of $\omega_2 = \omega_y$. The formula is *correct in the first edition*.

† WHITTAKER, p. 124; we write D in place of his F.

for $m_{k+3} = m_k$, with

$$U = \frac{G}{r} \sum_k \frac{m_{k+1} m_{k+2}}{\sqrt{1 - \cos B \cos L_k}}.$$

4. Elimination of the Nodes

We must proceed to the "elimination of the node" in the classical way (Whittaker, p. 344).

We notice that when the Z axis is taken along the fixed total angular momentum $K = p_\varphi$ we have

$$p_\psi = p_\varphi \cos \theta, \quad p_\theta = 0$$

and, therefore, according to the canonical equations,

$$\frac{\partial H}{\partial \theta} = 0.$$

It follows that if θ is eliminated and replaced by its value in p_φ and p_ϕ, the modified Hamiltonian will still satisfy the canonical equations for the remaining variables.

The ω's can be eliminated by

$$A\omega_x - D\omega_y = - p_\varphi \sin \theta \cos \psi,$$

$$B\omega_y - D\omega_x = p_\varphi \sin \theta \sin \psi,$$

$$C\omega_z + K_r = p_\psi,$$

and this last equation has to be introduced in the expression for the remaining canonical momenta. For instance we have

$$p_B = \frac{\partial L}{\partial \dot{B}} = \frac{\partial T_r}{\partial \dot{B}} + \omega_z \frac{\partial K_r}{\partial \dot{B}} = \frac{\partial}{\partial \dot{B}} \left(T_r - \frac{1}{2C} K_r^2 \right) + \frac{p_\psi}{C} \frac{\partial K_r}{\partial \dot{B}}.$$

It is therefore convenient to write

$$A_B = \frac{p_\psi}{C} \frac{\partial K_r}{\partial \dot{B}} = \tfrac{1}{2} \frac{\Delta . \, p_\psi}{1 + \Gamma \cos B}$$

and similarly

$$A_L = \frac{p_\psi}{C} \frac{\partial K_r}{\partial \dot{L}} = \tfrac{1}{2} \frac{p_\psi \sin B}{1 + \Gamma \cos B},$$

and p_B and p_L occur only in the expressions

$$\tilde{\omega}_B = p_B - A_B, \quad \tilde{\omega}_L = p_L - A_L$$

for which we have

$$\tilde{\omega}_B = \frac{\partial}{\partial \dot{B}} \left(T_r - \frac{1}{2C} K_r^2 \right),$$

$$\tilde{\omega}_L = \frac{\partial}{\partial \dot{L}} \left(T_r - \frac{1}{2C} K_r^2 \right),$$

and similarly

$$p_r = \frac{\partial}{\partial \dot{r}} \left(T_r - \frac{1}{2C} K_r^2 \right).$$

The last steps of the computations may be greatly simplified if we use, in place of the variable r, a new variable r_0, defined by

$$r_0 = r\sqrt{1 + \Gamma \cos B}.$$

Then it is found that

$$T_r - \frac{1}{2C} K_r^2 = \tfrac{1}{2} m_0 \dot{r}_0^2 + \tfrac{1}{8} \frac{m' r_0^2}{(1 + \Gamma \cos B)^2} (\dot{B}^2 + \dot{L}^2 \cos^2 B)$$

reduces to a sum of squares and the transition to the Hamiltonian is straightforward.

In this way we obtain finally

$$H \equiv T_r - \frac{1}{2C} K_r^2 + \tfrac{1}{2} \frac{p^2_\psi}{m_0 r_0^2} + I - U$$

with

$$T_r - \frac{1}{2C} K_r^2 = \frac{1}{2m_0} p_{r_0}^2 + \frac{2(1 + \Gamma \cos B)^2}{m' r_0^2} \left(\tilde{\omega}_B^2 + \frac{1}{\cos^2 B} \tilde{\omega}_L^2 \right)$$

and

$$I = \tfrac{1}{2} \frac{p_\varphi^2 - p_\psi^2}{AB - D^2} [B \cos^2 \psi + A \sin^2 \psi - 2D \sin \psi \cos \psi]$$

$$= \frac{p_\varphi^2 - p_\psi^2}{2m' r_0^2} (1 + \Gamma \cos B) \left[(1 + \Gamma) \frac{\cos^2 \psi}{\sin^2 \frac{B}{2}} + (1 - \Gamma) \frac{\sin^2 \psi}{\cos^2 \frac{B}{2}} - 2\Delta \frac{\sin 2\psi}{\sin B} \right].$$

In U, r must be replaced by its expression in r_0.

5. CONFORMAL TRANSFORMATION

Now that we have found the Hamiltonian expressed in terms of co-ordinates for which the binary collisions are represented by three fixed points on a sphere with polar co-ordinates $B = 0$ and $L = \frac{2\pi}{3} k$, we must introduce a conformal transformation which maps these three points on the three vertices Q_1, Q_2, Q_3 of a rectangular triangle.

We first introduce a complex variable by a stereographic projection

$$\zeta = \tan \left(\frac{\pi}{4} - \frac{B}{2} \right) e^{Li}$$

and we have

$$dB^2 + \cos^2 B dL^2 = 4 \frac{d\zeta d\bar{\zeta}}{(1 + \zeta\bar{\zeta})^2}.$$

If z is any function of ζ and if $B_0 L_0$ are polar co-ordinates related to z as B and L are related to ζ, then ζ' being the derivative and $\bar{\zeta}\bar{z}$ the conjugates, we shall have

$$dB^2 + \cos^2 B dL^2 = \zeta' \bar{\zeta}' \frac{(1 + z\bar{z})^2}{(1 + \zeta\bar{\zeta})^2} (dB_0^2 + \cos^2 B_0 dL_0^2),$$

and accordingly we shall have to put in the Hamiltonian

$$\tilde{\omega}_B^2 + \frac{\tilde{\omega}_L^2}{\cos^2 B} = \frac{(1 + \zeta\bar{\zeta})^2}{(1 + z\bar{z})^2} \frac{1}{\zeta' \bar{\zeta}'} \left(\tilde{\omega}_{B_0}^2 + \frac{\tilde{\omega}_{L_0}^2}{\cos^2 B_0} \right)$$

where, of course,

$$\tilde{\omega}_{B_0} = p_{B_0} - A_{B_0}, \quad \tilde{\omega}_{L_0} = p_{L_0} - A_{L_0}.$$

The transformation of the A's must be such that

$$A_L dL + A_B dB = A_{L_0} dL_0 + A_{B_0} dB_0.$$

The zeros and the poles of the function z correspond to the points which map the points $B = \pi/2$ (zero) and the antipodal point $B = -\pi/2$ (pole).

They are: the point, say P_0, with $B_0 = \pi/2$ as zero and as poles the points P_1, P_2, P_3, as symmetrical images of this point relative to the side of the triangle Q_1, Q_2, Q_3, the vertices of which map the singularities of the problem and therefore also of the transformation. The points P_0', P_1', P_2', P_3' are also poles or zeros.

The function

$$\zeta = z \frac{\sqrt{8} - z^3}{1 + z^3\sqrt{8}}$$

has the required zeros and poles and it will not be difficult to show that the multiplicative constant which is still left undetermined has been chosen correctly.

The derivative of ζ is

$$\zeta' = \frac{\sqrt{8} - 20z^3 - z^6\sqrt{8}}{(1 + z^3\sqrt{8})^2}$$

and it can be written

$$\zeta' = \frac{-\sqrt{8}}{(1 + z^3\sqrt{8})^2} \prod_{k=1}^{3} \left(z - e_k\right)\left(z + \frac{1}{e_k}\right),$$

where we have written

$$e_k = -\frac{\sqrt{3} - 1}{\sqrt{2}} e^{\frac{2\pi}{3}ki}$$

for the values of z corresponding to the singular points Q_1, Q_2, Q_3.

We have also

$$e^{\frac{2\pi}{3}ki} - \zeta = \frac{1}{1 + z^3\sqrt{8}} \left(z - e_k\right)^2 . \left(z + \frac{1}{e_k}\right)^2$$

and this shows that the points Q_k are really the images of the singularities.

When M_1 and M_2 are two arbitrary points on the sphere we have for the corresponding values of z

$$\sin^2 \frac{M_1 M_2}{2} = \frac{(z_1 - z_2)(\bar{z}_1 - \bar{z}_2)}{(1 + z_1\bar{z}_1)(1 + z_2\bar{z}_2)}.$$

From this formula we may deduce

$$\zeta'\bar{\zeta}' = \prod_{k=1}^{3} \frac{\sin^2 Q_k M}{\sin^2 MP_k/2}$$

and similarly for the gravitational potential

$$U = 2\sqrt{3}\frac{G}{r}\sqrt{\frac{1 + \zeta\bar{\zeta}}{1 + z\bar{z}}} \prod_{k=1}^{3} \sin \frac{MP_k}{2} . \sum_{k=1}^{3} \frac{m_{k+1}m_{k+2}}{\sin^2 MQ_k}.$$

This shows that singularities would disappear if the Hamiltonian H were replaced by a new Hamiltonian H',

$$H' \equiv (H - h) \sin^2 MQ_1 \sin^2 MQ_2 \sin^2 MQ_3 = 0,$$

to be used with a new time

$$\tau = \int \frac{dt}{\sin^2 MQ_1 \sin^2 MQ_2 \sin^2 MQ_3}$$

and this gives a regularization of the Three-Body Problem which, besides introducing a regularizing time τ as in SUNDMANN's process, introduces also the conformal transformation and the resulting simplifications of the co-ordinates characteristic of LEVI-CIVITA's device of regularization.

In this paper we have treated the question in a geometrical way. In a preceding paper* the question was treated analytically and the computation carried out up to the end, including the transformation of the A's and the expression of the non-planar terms.

The classical methods of celestial mechanics are very efficient when the trajectories of two of the bodies relative to the third one are well separated, as they are in the lunar problem or for the planetary problem with a ratio of the axes not too near to unity. These methods could not be applied, for instance, to work out the mutual perturbation of planets with encroaching orbits, so that one of them penetrates in the inside of the other. The difficulty of such problems is clearly connected with the occurrence of the singularity for binary encounter and it may be hoped that it will be removed if the regularization process can be sufficiently mastered and used in a practical and effective way.

REFERENCES

HAPPEL, H. . . . . 1941 *Das Dreikörperproblem* (Koehler, Leipzig).

LEMAÎTRE, G. . . . . 1952 "Coordonnées Symétriques dans le Problème des Trois Corps", *Bulletin de l'Académie Royale de Belgique* (classe des sciences), **38**, 582–92 and 1218–34.

WHITTAKER, E. T. . . . . 1937 *Analytical Dynamics* (Cambridge University Press).

* LEMAÎTRE (1952). For comparison with the present paper the reader is advised to look first at the formulae (10) and (79).

Recent Developments in Stellar Dynamics

G. L. Camm

Department of Mathematics, University of Manchester

Summary

The dynamical theory of stellar systems, as formulated forty years ago, was apt to be insufficiently precise in its treatment of the physical problem. For example, it rarely made use of Poisson's equation to relate the star density to the gravitational field in which the stars move. At the same time the theory was often too rigid in its mathematical assumptions, as in the adoption of the generalized Maxwellian distribution of velocities. In recent years these two points have received much attention from several authors. There has also been a growing emphasis on non-steady systems, though this problem has not so far yielded any very precise results. The stochastic effects of the gravitational forces between neighbouring stars have also been considered.

---

1. Introduction

From the beginning of the present century until 1940 the theoretical approach to stellar dynamics advanced steadily in a single direction. The statistical approach of the kinetic theory of gases was modified to suit the different circumstances, but from kinetic theory the Boltzmann equation of continuity and the Maxwellian distribution of velocities were carried over and tended to dominate stellar dynamics. The position of the theory in 1940 is readily seen in the books of that period by Smart (1938) and by Chandrasekhar (1942). Since that time there has been a very noticeable tendency to widen the approach, sometimes by introducing a new principle, and sometimes by discarding one which had long been accepted. This, then, is a suitable opportunity at which to draw attention to the changes that have recently occurred. It is first necessary to summarize the fundamental problem, and the older methods of attacking it.

Statistically, the state of a star-system at a time t is described by its velocity-distribution function. Suppose (x_1, x_2, x_3) are cartesian space co-ordinates, and (u_1, u_2, u_3) are the corresponding components of velocity. Then the velocity-distribution function, $f(t, X_1, X_2, X_3, U_1, U_2, U_3)$, is defined in such a way that $f(t, X_1, X_2, X_3, U_1, U_2, U_3)dX_1dX_2dX_3dU_1dU_2dU_3$ is the number of stars for which, at time t, $X_i < x_i < X_i + dX_i$, $U_i < u_i < U_i + dU_i$, $(i = 1, 2, 3)$. Thus the function f can be regarded as the number-density of stars in the six-dimensional phase space (X_i, U_i).

It should be pointed out that a velocity-distribution function is usually associated with a selected group of stars—most frequently with stars of a certain spectral type. Different groups may have different distribution functions, but all such functions satisfy the same mathematical equation of continuity. It is generally agreed that collisions between stars are so infrequent that they may be neglected. On this assumption considerations of continuity require the function f to satisfy the Boltzmann differential equation,

$$\frac{\partial f}{\partial t} + \sum_{i=1,2,3} \left\{ U_i \frac{\partial f}{\partial X_i} + \frac{dU_i}{dt} \frac{\partial f}{\partial U_i} \right\} = 0.$$

It is usual to suppose that the acceleration, of which the components dU_i/dt occur in this equation, is due to the gravitational field of some smoothed-out distribution of matter.

The actual density of matter in a star system is extremely high in the immediate neighbourhood of certain points (corresponding to the stars), and extremely low in interstellar space. The gravitational field of such a distribution will obviously have large fluctuations in the vicinity of a star, but since close encounters between two stars are so rare, it is more convenient to imagine the matter more uniformly distributed, so that the density and the gravitational field are smoothed.

If $\Omega(t, X_i)$ is the gravitational potential of this distribution, the fundamental equation becomes

$$\frac{\partial f}{\partial t} + \sum_{i=1,2,3} \left\{ U_i \frac{\partial f}{\partial X_i} + \frac{\partial \Omega}{\partial X_i} \frac{\partial f}{\partial U_i} \right\} = 0.$$

It was pointed out by JEANS* that this equation simply requires that f is an arbitrary function of the six integrals of the equations of motion,

$$dX_i/dt = U_i,\ dU_i/dt = \partial\Omega/\partial X_i \quad (i = 1, 2, 3).$$

If Ω is independent of t, then $(2\Omega - U_1^2 - U_2^2 - U_3^2)$ is one of these integrals. If Ω has axial symmetry about the X_3-axis, there is also an integral of angular momentum, namely $(X_1U_2 - X_2U_1)$. If Ω has spherical symmetry, there are three such integrals of angular momentum. Other integrals can only be determined when the explicit form of Ω is given.

In the case of a steady system possessing complete spherical symmetry the function f depends on three variables only, namely the radius, and the radial and transverse components of velocity. Consequently, there are only two integrals of motion, and these are the energy and the square of the angular momentum. The most general form of f for this problem is therefore an arbitrary function of these two integrals.

In other respects, JEANS's result is rather less helpful. It has frequently been supposed that the gravitational field of the galactic system itself has axial symmetry, and that the velocity-distribution function depends solely on the energy integral and the integral of angular momentum about the axis. On these assumptions, the distribution of velocities in planes through the galactic axis should be independent of direction. This is not the case in practice. The mean-square of the velocity components in the direction of the galactic centre is significantly greater than that in the direction of the galactic pole.

2. THE ELLIPSOIDAL DISTRIBUTION OF VELOCITIES

An alternative treatment, originally due to EDDINGTON, has been considered in great detail by CHANDRASEKHAR (1942). By analogy with the Maxwellian velocity distribution in the kinetic theory of gases, it is reasonable to suppose that the function f has the form,

$$f = \exp\{-(aU_1^2 + bU_2^2 + cU_3^2 + 2fU_2U_3 + 2gU_3U_1 + 2hU_1U_2 + pU_1 + qU_2 + rU_3 + s)\},$$

where the terms $a, b, c, f, g, h, p, q, r, s$ are functions of X_1, X_2, X_3 and t, to be determined. If this form of f is substituted in the Boltzmann equation, we obtain an

* Where specific references are not given, the work is reproduced in the books of SMART (1938) and CHANDRASEKHAR (1942).

identity in U_1, U_2 and U_3. For systems in which t is not explicity involved, the functional form of the coefficients can readily be obtained.

In the case of a non-steady system this method has been used by CHANDRASEKHAR (1942) to obtain a rather special solution, but no general solution has been found.

3. THE USE OF POISSON'S EQUATION

The outstanding omission in the development of the theory up to 1940, as outlined above, was the use of POISSON's equation, connecting the gravitational potential with the density distribution which produces it. This was not an oversight, but was in recognition of the observed fact that the velocity distribution is not the same for all types of star. Yet it is possible to formulate the problem in such a way that POISSON's equation can be applied. If the function f is defined so that

$$fdX_1dX_2dX_3dU_1dU_2dU_3$$

is the *total mass* of the stars which at time t are in the element of phase space $dX_1dX_2dX_3dU_1dU_2dU_3$, then f still satisfies the same Boltzmann equation, and in addition the integral of f over all velocities is the mass-density which gives rise to the gravitational potential Ω.

The addition of POISSON's equation has provided only disappointing results. It has been shown (CAMM, 1941) that the steady ellipsoidal distribution which possesses the two most obvious properties of the galactic system—namely, a flattened density distribution and a rotation about an axis—cannot satisfy POISSON's equation.

It also appears that the special non-steady solution given by CHANDRASEKHAR is unsatisfactory. SCHÜRER (1943) has shown that this exceptional case is directly obtained from a steady solution by a change of variables, and KURTH (1949a, 1949b) has shown that when POISSON's equation is introduced, this system is either uniform in density, or constant in time.

The case against the ellipsoidal distribution has also been presented by FRICKE (1951a, 1951b). He has shown that by abandoning this restrictive form, it is possible to explain another feature of the observations which the theorists had rather ignored. This is the well-known asymmetry of the high-velocity stars. It was first pointed out by OORT that the vector velocities of those neighbouring stars whose speeds relative to the Sun exceed 63 km per sec. are not distributed symmetrically about the line from the Sun towards the galactic centre. Such a state of affairs cannot be explained by a quadratic distribution, but FRICKE has found that a simple polynomial in the two integrals of energy and angular momentum about the galactic axis will serve remarkably well as a distribution function for the observed asymmetry. Of course, this function still fails to explain the relatively small observed velocity dispersion in the direction towards the galactic pole, but it shows that the ellipsoidal distribution can usefully be replaced by something slightly more complicated to explain one more feature of the observed stellar movements.

4. FINITE SYSTEMS

It is rather remarkable that three quite independent investigations, pursued more or less simultaneously in different countries (CAMM, 1941, 1950; FRICKE, 1951a; KURTH, 1949a), all put increased emphasis on the Poisson equation. It is still more strange that all should also consider the necessity of finiteness in a mathematical model of a stellar system (CAMM, 1950, 1952; FRICKE, 1951a, 1951b; KURTH, 1949b).

The original generalization of the Maxwellian distribution proposed by SCHWARZSCHILD cannot describe a system which is finite in extent. With this model, however large a velocity we choose, there is a non-zero number of stars moving with this velocity. Thus there is no limit to the kinetic energy of a star, and so the distribution in space is not bounded. If, on the other hand, we wish to have a model which is finite in extent, then we must ensure that the potential difference between any two points of the system is not infinite, and consequently the kinetic energy must have a definite upper bound. To obtain a finite system it is therefore necessary—but not sufficient—that the form of velocity-distribution function should impose a limit on the range of velocity components.

Examples of this kind have been given by CAMM (1950, 1952). The aim in both these investigations was to find non-steady solutions of the Boltzmann equation which were also compatible with POISSON's equation. The systems considered were deliberately simplified so as to reduce the number of independent variables. In one case, the system was stratified in plane parallel layers, so that only the space- and velocity-components normal to the layers were introduced. There is the further advantage that the density is simply proportional to the gradient of the gravitational intensity. The other simplified model had complete spherical symmetry, so that the radius was the only space co-ordinate involved, and there were two components of velocity, along and perpendicular to the radius. Again the density and the gravitational intensity were simply related.

In each case the mathematical problem could be reduced to a single non-linear partial differential equation, but no non-steady solution of either equation has been found. Steady solutions of finite radius (or finite thickness in the stratified problem) were discovered, and good agreement, especially in view of the over-simplification, was found between observed and theoretical density distributions.

5. STOCHASTIC FORCES IN STELLAR DYNAMICS

An entirely new concept was introduced into the theory of stellar dynamics by the papers of CHANDRASEKHAR and VON NEUMANN (1942, 1943) on stochastic forces. Although stellar encounters are very infrequent in regions of the galaxy such as that in which the Sun is situated, they are by no means negligible in their effect. As we have explained earlier, it was previously assumed to be sufficient to represent the gravitational forces of the whole star system by the gradient of the potential function due to a smoothed-out distribution of matter. In this work, CHANDRASEKHAR and VON NEUMANN have determined the probability distribution of the gravitational force at a point in a field of stars, and also the distribution of the rate-of-change-of-force. The application of this work to find the probability distribution of force and rate-of-change-of-force on a star moving in the field is bound to be much more difficult, since the test-star itself will contribute to the gravitational field, and will alter the paths of the field stars. Now the circumstances in which the stochastic forces are most important are those rare occasions when a field star passes near the test-star. These are precisely the occasions where the present analysis breaks down. In general, the test-star will have little effect on the field stars, but for a close encounter the effect, both on the path and on the speed, may have remarkable results in changing the gravitational field. Obviously the present theory by CHANDRASEKHAR and VON NEUMANN overlooks this kind of situation, for the probability distributions do not contain any reference to the mass of what we have called the test-star.

In two further papers CHANDRASEKHAR (1943a, 1943b) puts aside the notion of stochastic forces due to a field of stars, and considers instead the effect on a test-star of a series of binary encounters, developing in this way a theory of dynamical friction. Naturally, the masses of the stars enter this kind of treatment. The drawback of this method is that the regions where the forces due to random encounters are most important are the regions of high density—so that the frequency of encounters is raised. But if the density is high, it is extremely difficult to simplify the problem to a series of binary encounters. Near the centre of a globular cluster, for example, one star may be experiencing fairly large attractions from several neighbouring ones simultaneously.

This work has still to be followed up. The situation with which it deals must occur, and will affect the whole velocity distribution, but at the present time the theory raises more problems than it solves.

6. NON-STEADY SYSTEMS

The various developments in the theory of stellar dynamics which have occurred in the past ten years all emphasize the need to consider non-steady systems. A study of the observational material points in the same direction. For example, it has already been mentioned that different types of stars have very different velocity distributions. This cannot be attributed to the constitution of the stars, nor yet to their evolution; but it seems likely to be related to the circumstances of their formation, and more especially to their age. This is certainly evidence that the velocity distribution is not independent of the time.

There are other reasons for stressing the importance of non-steady systems. We have seen that the effect of binary encounters is not completely negligible; such encounters will certainly produce changes in the velocity distribution, even though the process may be extremely slow. Similarly the process of accretion of interstellar matter will also cause a slow change. Both these effects have been overlooked in formulating the fundamental equation. In a non-steady solution, the motion and smoothed gravitational field may produce changes in the velocity distribution which are far more rapid than those caused by encounters or accretion, and these may then reasonably be neglected. But in so-called steady solutions, these slow changes are the only ones which occur, and they cannot therefore be disregarded.

The search for non-steady solutions is really just beginning. The example given by CHANDRASEKHAR proved, as mentioned earlier, to be a trivial one, with no real bearing on the general problem. CAMM's treatment also revealed a trivial solution, but it may be possible to find more general solutions by the methods he has used.

VON DER PAHLEN (1947) and KURTH (1951b) have considered the evolution of a spherical system, by expressing the distribution function as a power series in the time. KURTH (1952) has examined in great detail the mathematical principles involved in the general problem, and has shown that in theory a solution can be obtained by successive approximations.

The possibility of periodic solutions has yet to be examined. Since the total energy of the system is conserved, an oscillating system seems to be quite feasible. It might well be considered with the aid of the virial theorem. Since this theorem is based on the equations of motion of each star and the inverse square law of gravitation, it seems likely to be directly deducible from the Boltzmann and Poisson equations, but this does not seem to have been considered. Clearly the virial theorem gives much less information than the combination of those two equations, but it so reduces the complexities that there may be some advantage gained. It has been used in special problems with some success by FREUNDLICH (1945, 1947) and by KURTH (1950, 1951a).

The whole subject, it will be seen, has undergone a remarkable change in recent years. The change has not removed the inherent difficulties (and, indeed, we have only recently become aware of some of those difficulties), but many new lines of attack have been introduced, and progress has certainly been made. Not least important is the revival of interest in a branch of mathematical astronomy which for twenty years had appeared moribund.

REFERENCES

CAMM, G. L.	1941	*M.N.*, **101**, 195.
	1950	*M.N.*, **110**, 305.
	1952	*M.N.*, **112**, 155.
CHANDRASEKHAR, S.	1942	*The Principles of Stellar Dynamics.* (Chicago.)
	1943a	*Ap. J.*, **97**, 255.
	1943b	*Ap. J.*, **97**, 263.
CHANDRASEKHAR, S. and VON NEUMANN, J.	1942	*Ap. J.*, **95**, 489.
	1943	*Ap. J.*, **97**, 1.
FREUNDLICH, E. F.	1945	*M.N.*, **105**, 237.
	1947	*M.N.*, **107**, 268.
FRICKE, W.	1951a	*Naturwissenschaften*, **19**, 438.
	1951b	*Astron. Nachr.*, **280**, 193.
KURTH, R.	1949a	*Z. Astrophys.*, **26**, 100.
	1949b	*Ibid.*, **26**, 168.
	1950	*Ibid.*, **28**, 60.
	1951a	*Ibid.*, **29**, 26.
	1951b	*Ibid.*, **29**, 33.
	1952	*Ibid.*, **30**, 213.
PAHLEN, E. VON DER	1947	*Z. Astrophys.*, **24**, 68.
SCHÜRER, M.	1943	*Astron. Nachr.*, **273**, 230.
SMART, W. M.	1938	*Stellar Dynamics.* (Cambridge.)

"Intrinsic" Studies of Stellar Movements in the Milky Way

PAUL BOURGEOIS

Observatoire Royal de Belgique, Uccle-Bruxelles

---

1. INTRODUCTION

STELLAR statistics provides evidence of systematic stellar movements in the Milky Way. Stellar dynamics, based on this knowledge, furnishes a simple hypothetical basis on which the laws governing the structure and evolution of the galactic system may be deduced from observations (COUTREZ, 1949, 1951).

The work of LINDBLAD and the demonstration by OORT of the differential rotation of the Milky Way, shed new light on a series of investigations which had been going on since the end of the last century; in particular, the well-known phenomena of

STRÖMBERG's asymmetrical pattern of stellar motions and those of KAPTEYN's preferential motions have thus been explained.

In the absence of sufficient observational data, attempts to discover the statistical laws governing systematic motions within the galaxy were generally based on the components of the individual stellar movements, *i.e.* on proper motions and radial velocities. Space velocities were used only in exceptional cases. In these investigations use was made of simplifying hypotheses, such as the adoption *a priori* of a Gaussian distribution of velocities. Although such hypotheses are capable of representing the observations in a first approximation, they tend to obscure local effects, such as the deviations from the galactic differential rotation, which one would logically expect in a system as complex as the Milky Way. We know further that the distribution of stars in space is not as uniform as one might suppose. The existence of stellar agglomerations and of clustering in the dispersive track again show (AMBARZUMIAN, 1949; BLAUUW, 1952) that there are deviations from the general pattern of differential rotation. In this connection we may also mention the K-effect discovered at the beginning of this century.

2. METHOD

Following a study of stellar dynamics by COUTREZ (1944), which led to the interpretation of the classical K-effect as the radial components of a spatial K-effect in spiral nebulae, we have been engaged in the last few years in a more extensive study of stellar motions in space. A general synopsis of work which we have published on this subject has been given by BOURGEOIS (1951). We should like to complete it here and to discuss the character and possibilities of the method used, without dwelling on the details of the results obtained so far.

The principle of this method consists of an *intrinsic* statistical analysis of groups of stars of substantially the same mass, the spatial movements of which are well known and homogeneous; the term *intrinsic* implies that in the course of the analysis no use is made of any numerical data not derived from the catalogue itself. Although this mode of procedure involves a certain loss of information in consequence of the *a priori* rejection of previously obtained data such as the coefficient of differential rotation, it assures great reliability in the analysis of the phenomena studied, just because it does not depend on anything outside the system of observational results taken as a basis.

In the course of the investigation the statistics of individual space velocities, of barycentric space velocities, and of barycentric space velocity residuals were obtained in succession. These different stages permit: (1) the study of the systematic motions peculiar to each group; (2) the study of the barycentric velocity field in the neighbourhood of the Sun; (3) the study of the peculiar relationship between each group and the general field of stellar velocities in the neighbourhood of the Sun.

3. OBSERVATIONAL DATA

The success of such an investigation depends essentially on the homogeneity in radial velocity, proper motion, and parallax of the catalogues used in the determination of space velocities. Thus no heterogeneous element can be kept in the system; but very refined criteria must be applied to avoid any systematic effect that might result from the elimination of some of the stars. This method may reduce the number of usable stars, but it increases the accuracy of the results and their representative value. One may judge this by calculating well-known systematic effects depending on a single quantity, such as the classical K-effect (radial velocities) and the differential rotation of the galaxy (proper motions); one should then obtain the same result from a study of the space motions. Another important indication of the representative value of the method can be obtained by comparing the results obtained in passing from one system of homogeneous data to another; for example, for proper motions, from the system of the General Catalogue to that of the FK3 (BOURGEOIS and COUTREZ, 1948a). The homogeneity in mass of the stars in a group is checked by the homogeneity in intrinsic luminosity, assuming the mass-luminosity relation. Finally, we always attempted to make an estimate of the errors in the calculated values.

4. INDIVIDUAL SPACE VELOCITIES

The statistical study of individual space velocities is made by analysing the distribution of the representative points of the stars in velocity-space. This distribution contains the effect of all the systematic movements of the group. The distribution is

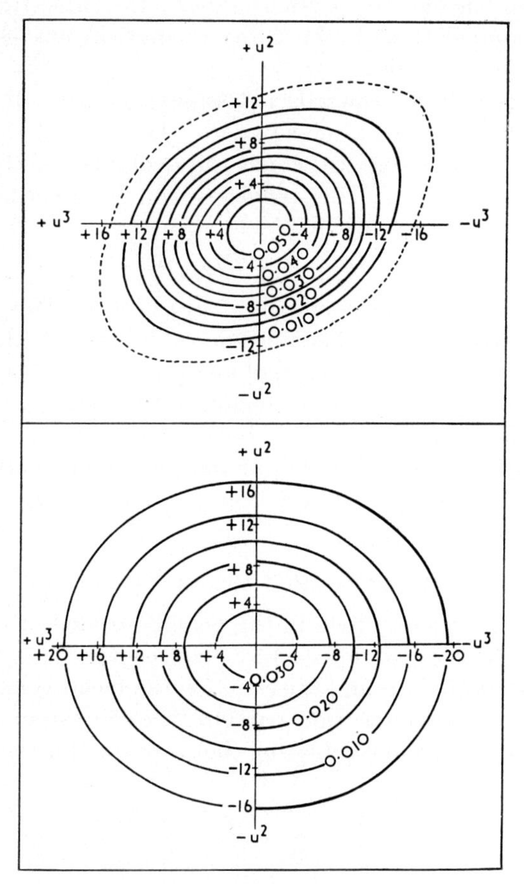

Fig. 1. A-type stars

Intersection of the surfaces of equal point-density in velocity-space with a plane perpendicular to the direction of the galactic centre. The axes of the diagrams are the mean axis and the minor axis of the representative ellipsoid, graduated in km per sec. The figures on the lines of equal density are numbers of stars per (km per sec.)^3. The *upper* diagram refers to a representation by a distribution of CHARLIER's type A, the *lower* one to an ellipsoidal representation. The difference between these two descriptions is very marked.

represented by a frequency function, from which a series of surfaces of equal point density is deduced. A study of these surfaces then permits an easy analysis of the distribution. In the work discussed here we have adopted a frequency function of CHARLIER's type A, and we have calculated the coefficients in such a way as to keep the moments of the distribution up to the fourth order. A simple Gaussian frequency function (SCHWARZSCHILD's ellipsoidal distribution) is insufficient, as may be seen from an examination of Fig. 1, which shows the deformation in the surfaces of equal density that occurs when one passes from a simple Gaussian function to a function of this type, characterized by coefficients of asymmetry (skewness) and excess.

We insist in particular on the necessity for numerical checking in the course of the reductions, since the mechanical procedures which are indispensable in such work are open to the danger of computational errors.

To judge the representative value of the frequency distribution obtained, there is no question of actually calculating the differences between theoretical and observed frequencies. This difficulty is circumvented by successive projection of the observed "statistical cloud" on the three co-ordinate axes; the three frequency curves thus obtained are then compared with the three theoretical curves deduced from the representative frequency function.

Since series of type A tend towards a Gaussian density distribution when the coefficients of order higher than 2 tend to zero, the value of these coefficients here gives us a measure of the extent to which the evolution of the system studied is affected by random influences. This tendency can be described (BOURGEOIS and COUTREZ, 1948b) by the expression:

$$D^2 = 1 + \Sigma(\alpha_{ijk})^2,$$

where α_{ijk} are the coefficients of asymmetry and excess. This expression tends to unity when the effective velocity distribution approaches the Gaussian form; but, it should be mentioned that the evolution of a group of stars is here considered from a statistical point of view in a very small region of the galaxy and may thus present peculiar characteristics in relation to the whole. Hence D is not necessarily an index of the dynamic evolution of the system. Further, a comparison between the D values obtained for stellar groups with different homogeneous characteristics should be regarded with caution.

5. "BARYCENTRIC" SPACE VELOCITIES

The study of the stellar velocity field in the neighbourhood of the Sun is facilitated if individual stellar velocities are replaced by "barycentric" mean velocities of groups of stars regularly distributed around the Sun, *i.e.* the mean velocities with respect to the centre of gravity of the groups, thus reducing the effect of individual fluctuations.

The classical relations used in calculating the constants for the differential rotation of the galaxy are:

$$\begin{aligned} u &= -xV + y(U - B), \\ v &= +x(U + B) + yV, \\ w &= 0, \end{aligned}$$

where u, v, w are the galactic components of barycentric velocities; x, y, z are the galactic co-ordinates of the centres of gravity of the regions; and where U and V are given by $U = A \cos 2l_c$ and $V = A \sin 2l_c$, (l_c = longitude of the galactic centre). These relations imply *a priori* a certain form for the coefficients, depending on the hypotheses used (*i.e.* mean movements of groups of stars parallel to the galactic plane, and rotating about the centre with an angular velocity which depends uniquely on the distance from the centre). It is interesting to express these equations in a general form of the type

$$u_i = \Sigma a_{ij} x_j \quad (i, j = 1, 2, 3),$$

so that on the average the velocities u_i of the stars in the neighbourhood of the Sun depend in the simplest possible manner on their galactic co-ordinates x_i (BOURGEOIS and COUTREZ, 1950b). As no restrictive hypothesis is made here concerning the

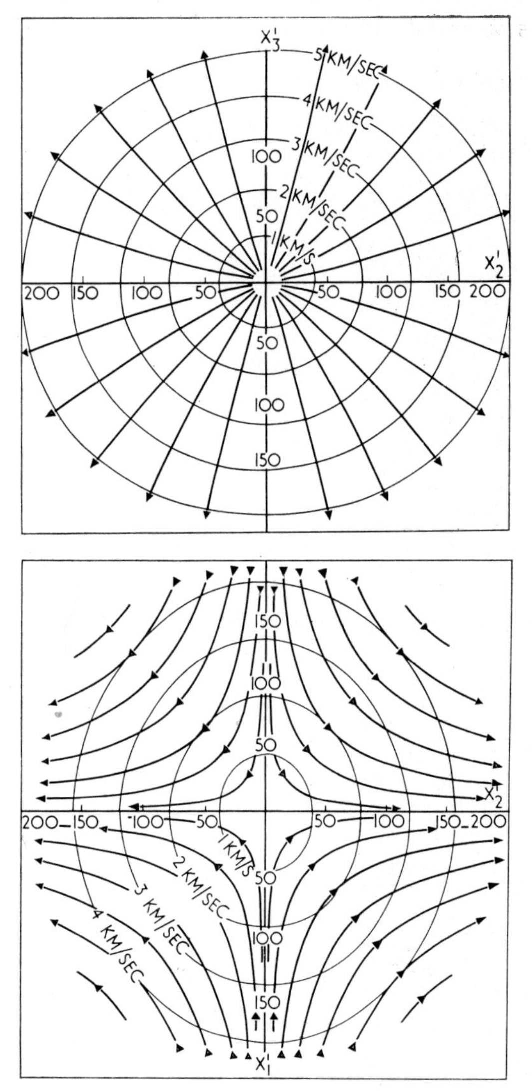

Fig. 2. A-type stars

Representation of the field of barycentric velocities of deformation. The heavy curves are line of flow, the sense of motion being shown by the arrows. The lighter curves are the intersections of surfaces on which the velocity has a given value, respectively by the galactic plane (*below*) and by the plane normal to the galactic plane and to the direction of the centre (*above*). The x_1'-axis is directed towards the galactic centre. The axes are graduated in parsecs, and the curves of constant velocity in km per sec.

coefficients a_{ij}, the latter appear as the components of a tensor which represents the linear relation (valid in the first approximation) between the velocities and the co-ordinates. These components represent the group's local rotation and local deformation, the latter giving rise to a linear dilatation and to a shear. By calculating this tensor a_{ij} one may obtain a precise idea of stellar movements in the vicinity of the Sun in the linear approximation. In particular, one may judge the extent to which the stars in a group which has been studied follow the classical differential rotation (BOURGEOIS and COUTREZ, 1950a, 1950b). Fig. 2 illustrates, for A-type stars, the characteristics of the velocities of deformation in the local group, taken with respect to the centre of gravity of the system.

6. Residual Space Velocities in the Barycentric System

Another method of studying the special characteristics of the stellar velocity field in the neighbourhood of the Sun consists in separating the barycentric space velocities

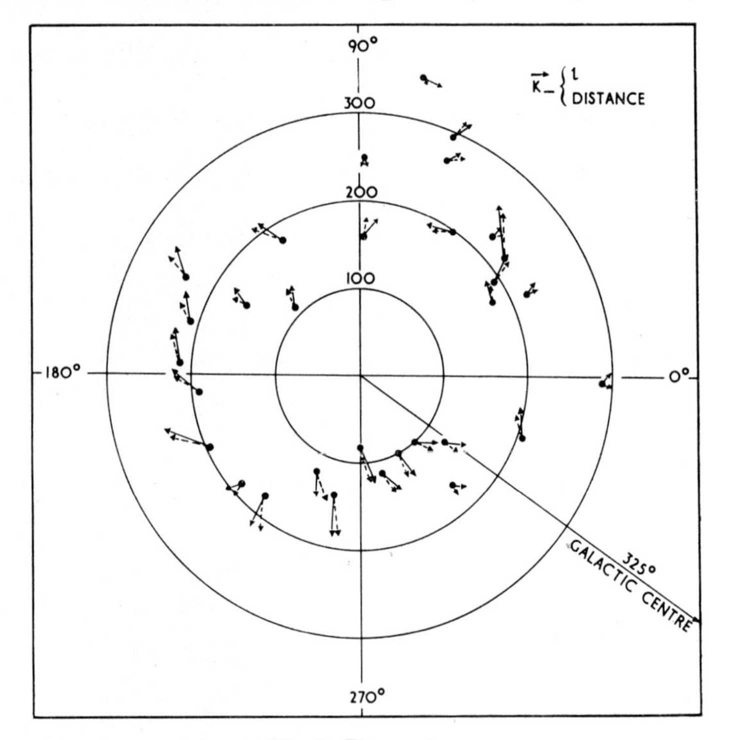

Fig. 3. B-type stars

Relation between residual barycentric velocities (spatial K-effect), distances, and longitudes, represented in the galactic plane. The vectors have a marked tendency to align themselves parallel to the direction from the Sun to the galactic centre, thus indicating the tendency of the galaxy, taken as a whole, to show an expansion effect. The diameters of the points give the accuracy. The full-line vectors were calculated on the basis of the FK3 system, those shown by broken lines used the system of the General Catalogue. It will be noticed that the results are essentially unaffected by transition from one system to the other.

from the effects of the apex and of the classical differential rotation. In this way one obtains the barycentric velocity residuals as a function of longitude and distance. Fig. 3, obtained from a study of stars of spectral type B, illustrates the distribution in the galactic plane.

Similarly, one can separate the barycentric space velocities from the general effects of a linear dependence on the co-ordinates as discussed in Section 5. The barycentric residual space velocities found in these two cases are closely correlated (BOURGEOIS and COUTREZ, 1950a), which shows that the linear approximation used above gives a very satisfactory description of the region of space studied, without the use of any special hypothesis.

7. CONCLUSIONS

The results obtained in the course of the studies which we have pursued in this field, in collaboration with A. COUTREZ, are concerned with stars of spectral types B and A (BOURGEOIS, 1951).

For these spectral types the concentration of the representative points near the centre of gravity in velocity-space has turned out to be particularly strong, and it may be attributed to the presence of streaming, probably in a spiral arm of the Milky Way. The elongation of the "distribution clouds" in a direction close to that of the galactic centre and their flattening normal to the galactic plane are a strong indication of the presence of differential galactic rotation, and follow from the position of the Sun in our stellar system.

The method described in Section 5 reveals a local rotation of the two groups studied about axes inclined to the galactic plane; this seems to be due to turbulence in the neighbourhood of the Sun.

The study of barycentric residual velocities for stars of spectral type B shows, for the groups studied, a tendency to radial expansion parallel to the galactic plane and to concentration towards this plane.

These few examples are given here merely to indicate the interest of this method, which is very promising for the detailed study of stellar movements in the vicinity of the Sun.

Recent work in radio astronomy promises to provide knowledge of the Milky Way in depth. The radial velocities obtained from observations of the radiation from interstellar hydrogen at 21 cm, related to the distances through the adoption of a scheme for the motion of the entire system, already provide a proof of the spiral character of the galaxy in agreement with MORGAN's model.

Progress in stellar dynamics depends essentially on an ever-increasing knowledge of the structure of our galaxy. The development of complementary investigations of this kind is the only means of securing this. Such studies should therefore be actively pursued.

REFERENCES

AMBARZUMIAN, V. A. . . . . 1949 "Associations stellaires", *Astron. J. U.S.S.R.*, **26**, 3.

1950 "Les associations stellaires et l'origine des étoiles", *Acad. Sci. U.S.S.R.*, **14**, No. 1, 15.

BLAAUW, A. . . . . 1952 "The Age and Evolution of the ζ Persei Group of O- and B-type Stars", *Bull. Astron. Inst. Netherlands*, **9**, 433.

BOURGEOIS, P. . . . . 1951 "Les mouvements spatiaux des étoiles dans l'étude de la Galaxie", *Ciel et Terre*, **1–2**; *Commun. de l'Observatoire Royal de Belgique*, No. 24 (with additional references).

BOURGEOIS, P. and COUTREZ, R. . . . 1948a "Comparison des composantes tangentielles des vitesses spatiales résiduelles dans les systèmes du General Catalogue et du FK3 pour les étoiles de type spectral B", *Ann. Astrophys.*, **11,** fasc. 3; also *Commun. de l'Observatoire Royal de Belgique*, No. 7.

1948b "Etude statistique intrinsèque des étoiles de type spectral A à mouvement spatial connu", *Ann. de l'Observatoire Royal de Belgique*, 3e Série, tome III, fasc. 4, p. 209.

1950a "Mouvements systématiques à l'approximation linéaire et vitesses spatiales résiduelles pour les étoiles de type spectral B", *Ann. Astrophys.*, tome 13, No. 2; also *Commun. de l'Observatoire Royal de Belgique*, No. 17.

1950b "Mouvements systématiques à l'approximation linéaire et vitesses spatiales résiduelles pour les étoiles de type spectral A," IIIe Congrès National des Sciences, *Compte-rendus et Commun. de l'Observatoire Royal de Belgique*, No. 18.

COUTREZ, R. . . . . 1944 "La dynamique des nébuleuses spirales", *Ann. de l'Observatoire Royal de Belgique*, 3e série, Tome III, pp. 76–80.

1949 "Contribution à l'étude de la Dynamique des systèmes stellaires", *Ann. de l'Observatoire Royal de Belgique*, 3e Série, Tome IV, fasc. 3.

1951 "Méthodes actuelles de la dynamique stellaire appliquées aux problèmes de la Voie Lactée et des nébuleuses extragalactiques", *Scientia*, January; also *Commun. de l'Observatoire Royal de Belgique*, No. 22.

The K-Effect in Stellar Motions

HAROLD F. WEAVER

University of California, Leuschner Observatory, Berkeley, U.S.A.

SUMMARY

The large K term associated with bright early B stars signifies expansion, but does not indicate any fundamental characteristic of the local region of the galaxy. It is a chance property of the sample population traceable to nearby community motion groups within the sample. Removal of chance and real community motion groups reduces the K term to approximately $+ 2{\cdot}5$ km per sec. for the O to B2 stars. This term may be attributed principally, if not entirely, to gravitational red shift.

A method of investigating on the basis of galactic structure the significance of the negative K term found for faint B stars is suggested. Tests indicate:

(1) No variation of the K term within the local spiral arm of the galaxy.

(2) No relative motion of the local and inner spiral arms, hence the same K term holds for both.

(3) While stars in the outer arm also tend to confirm the normal K term, the stars in the Perseus-Cassiopeia region possess large negative peculiar radial motions. Inclusion of the numerous B stars in this region will bias any solution towards a small or negative K term and a large OORT A-constant.

1. DISCOVERY OF THE K-EFFECT

FIFTY years ago FROST and ADAMS (1904), from a discussion of twenty measured radial velocities corrected for solar motion, discovered the two principal characteristics of the motions of the B stars.

(1) The average peculiar motion of the B stars is small, approximately 7 km per sec.

(2) The average residual with regard to sign is positive and unexpectedly large, $+ 4{\cdot}6$ km per sec., according to FROST and ADAMS' pioneering investigation.

In 1910 these findings were confirmed by KAPTEYN and FROST (1910) and simultaneously by CAMPBELL (1911, 1913), who had available radial velocities of 138 B stars reasonably well distributed over the sky. CAMPBELL found the average value of the difference observed radial velocity minus computed solar motion component to be + 4·93 km per sec. To represent this clearly present but unexplained systematic motion of the B stars, CAMPBELL introduced into the usual solar motion equation an empirical constant K,

$$v_r = K - u_{\odot} \cos \alpha \cos \delta - v_{\odot} \sin \alpha \cos \delta - w_{\odot} \sin \delta. \qquad \ldots.(1)$$

Here v_r represents the expected radial velocity of a star situated at position α, δ; $u_{\odot}$, $v_{\odot}$, $w_{\odot}$ represent the components of solar motion in a rectangular co-ordinate system oriented in the customary manner in the α, δ frame. Thus introduced, a non-zero K term indicates only that the average velocities of B stars in opposite parts of the sky are not equal in numerical value.

2. OBSERVATIONAL CHARACTERISTICS OF THE K-TERM

The K term is large for O and early B stars; for later spectral types it is small or negligible. Numerical values of the K term found by CAMPBELL and MOORE (1928), and more recently by SMART and GREEN (1936), are shown in Table 1.

Table 1. Numerical values of the K term

Spectral type	Campbell and Moore		Smart and Green	
	K term (km per sec.)	Number stars	*K* term (km per sec.)	Number stars
B	+ 4·9	284	+ 4·7	645
A	+ 1·7	500	0·0	742
F	+ 0·3	199	− 0·6	523
G	− 0·2	244	− 1·0	433
K	+ 0·3	687	− 0·2	1118
M	+ 0·7	234	0·0	222

The numerical value of the K term is a function of the apparent magnitude range of the B stars investigated. This correlation, first pointed out by J. S. PLASKETT (1930), is illustrated by the data of Table 2.

Table 2. Variation of K with apparent magnitude
(Data from J. S. PLASKETT, 1930)

Type	Magnitude range	Average magnitude	Number of stars	*K* (km per sec.)
O–B2	$< 5·5$	3·98	78	+ 4·3
O–B2	$> 5·5$	6·60	139	+ 0·1

3. INTERPRETATIONS OF THE K-EFFECT

Various explanations of the K term have been suggested.

(1) *Erroneous Wavelengths or Unknown Blends of Lines Used for Radial Velocity Determinations.* CAMPBELL (1911, 1913) pointed out that an average increase of only 0·07 Å in the wavelengths of all lines would eliminate the K term. Errors of this size do not exist in modern wavelength scales, yet the K term persists; the explanation fails. The triplet series of He is composed of very closely spaced lines which are

blended on small dispersion spectra. Modern data do not support the early suggestion that the individual line intensities are abnormal in stellar spectra and hence that the wavelengths of the blends used for radial velocity determination are incorrect.

(2) *Pressure Shift of Spectral Lines.* A few decades ago the principal source of line broadening in B stars was thought to be pressure, which was also assumed to shift the lines longward (CAMPBELL, 1911, 1913). To-day it is clear that the pressure in B-star atmospheres cannot produce the broad lines or the K effect.

(3) *Atmospheric Currents.* In 1914 CAMPBELL, basing his remarks on EVERSHED'S measurements of convection in the solar atmosphere, suggested that much greater descending currents in the atmospheres of B stars might account for the K term. It is difficult entirely to eliminate this suggestion on the basis of scanty modern data.

(4) *Expansion of the Galaxy.* PILOWSKI (1931), OGRODNIKOFF (1932), MILNE (1935), and earlier investigators have suggested that the K term represents a real expansion of the entire galaxy or of the local cluster in which the Sun is presumably located. Expansion is generally taken to imply that K should increase with distance from the Sun. Failure to find an increase of K with an increase of apparent magnitude of the stars observed is evidence against the interpretation. It does not rule out expansion of the local cluster.

(5) *Relativity Red Shift.* FREUNDLICH (1915) attributed the K effect to the gravitational red shift of spectral lines in the massive O and B stars. Calculations indicate (PLASKETT, 1930; PLASKETT and PEARCE, 1936) that for O–B2 stars the relativity K term is between 2 and 3 km per sec. While the relativity red shift is undoubtedly present in the B stars, it cannot account for the entire K term shown in Table 1.

(6) *Stream Motions and Moving Clusters.* PLASKETT and PEARCE (1929) suggested that the K term can be traced to groups of stars in community motion. In particular, they asserted that removal of the Scorpio-Centaurus Stream from the data decreases the K term to the amount to be expected from relativity shift.

SMART (1936) has criticized the particular explanation and certain specific questionable mathematical procedures used by PLASKETT and PEARCE. In spite of these specific criticisms, the original suggestion of moving clusters loses none of its applicability. Suitably generalized, it offers the most promising means of explaining the effect.

4. GENERALIZATION OF THE MOVING CLUSTER CONCEPT

(1) *Real or Chance Community Motion Groups.* Solar motion solutions made with equation (1) contain the implicit assumption that the stars surveyed constitute a homogeneous sample population in random motion. A departure from random motion in any subvolume of the sample implies that the stars in the subvolume possess community motion. No physical association of the involved stars is necessarily implied by such community motion, which may be solely a chance phenomenon.

If an appreciable fraction of the sample stars possesses community motion, the precise values of the derived K term and solar motion will not necessarily have fundamental significance. They will represent only properties of the sample population studied, not fundamental properties of the region of the galaxy surveyed. If the stellar population as a whole is inhomogeneous and various subpopulations possess community motion, then, as the observed sample is varied and community motion groups are gained or lost, the numerical value of the K term and the solar motion may be expected to change.

Percentagewise, the change in the K term may be large. The numerical value of the K term is determined by:

(*a*) physical properties of the stars (M/R, for example, determines the gravitational red shift), and

(*b*) the state of motion of the sample population.

In this later connection, the K term indicates whether the sample population is, on the average, expanding, contracting, or stationary. B-star samples are generally small; a few stars in community motion will give the sample the property of expanding or contracting and will greatly influence the value of K.

(2) *Fictitious Community Motion Groups.* Galactic rotation imposes a field of differential motions in the sample volume. The stars in any subvolume of the sample space will, in general, have group motion with respect to the stars in any other subvolume. To allow for the effects of galactic rotation, the equation for solar motion is usually written

$$v_r = K - u_\odot \cos \alpha \cos \delta - v_\odot \sin \alpha \cos \delta - w_\odot \sin \delta + \bar{r}A \sin 2(l - l_0) \cos^2 b, \quad \dots (2)$$

where the symbols have their customary definitions. Equation (2) is valid only if $\bar{r}$ is not correlated with l or b; its use implies that the stars are distributed uniformly in space, or for the B stars, at least distributed uniformly in the galactic plane. Further, it may be noted that the galactic rotation term in (2) is valid only for nearby stars, though it is often applied to rather distant ones. A more precise equation is desirable. The expression (see TRUMPLER and WEAVER, 1953)

$$v_r = K - u_\odot \cos \alpha \cos \delta - v_\odot \sin \alpha \cos \delta - w_\odot \sin \delta + R_0[\omega(R) - \omega_0] . \sin (l - l_0) \cos b \quad \dots (3)$$

is preferable. For circular galactic orbits the rotation term of (3) is exact.

Equation (3) applies to an object at distance R from the galactic centre; $\omega(R)$ represents the angular velocity of a body moving in a circular orbit of radius R; R_0 is the value of R for the Sun; ω_0 represents $\omega(R_0)$. As in equation (2) we assume that the object under discussion is so close to the galactic plane that $\omega(R, z)$ may be taken as $\omega(R, 0)$ or simply $\omega(R)$. For B stars the assumption is adequate.

Very early B stars (B2 and earlier) sharply define the spiral arms of the galaxy (MORGAN, WHITFORD, CODE, 1953; WEAVER, 1953a). Stars of later spectra type appear to delineate the arms much less clearly. The principal arm structures defined by stars B2 and earlier in the solar neighbourhood are indicated in Fig. 1. The distribution of the early-type stars within the arms is very spotty; major distribution trends are indicated by shading in Fig. 1. Use of equation (2) requiring no correlation of $\bar{r}$ and l is clearly inadmissible for the B2 and earlier stars, particularly when great distances are involved. Use of equation (2) will introduce fictitious star drifts in those directions in which the average distance differs from the $\bar{r}$ value used in the equation. These fictitious drifts will distort the results for solar motion and may introduce a spurious K term, the numerical value of which will depend upon the space distribution of the stars. If equation (3) is employed in the discussion, and if each star is treated individually for galactic rotation, space distribution irregularities and approximations in the galactic rotation term of (2) will not affect the results. However,

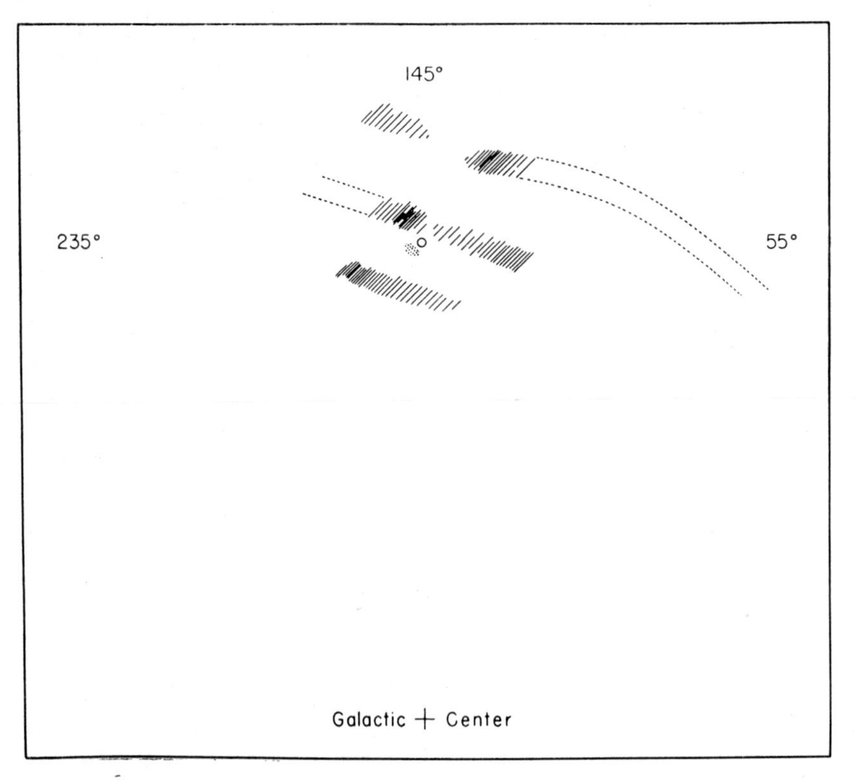

Fig. 1. Schematic representation of principal spiral structures in the solar neighbourhood. The Sun is indicated by the circle; the Scorpio-Centaurus aggregate is shown as a dotted area.

star groups of real or chance community motion will still influence the values derived. Such localized effects must be detected and treated individually.

5. Examples of K-Term and Solar Motion Solutions

We consider principally stars B2 and earlier and magnitude 5·0 and brighter. Radial velocities are available for eighty stars satisfying these criteria; nearly all lie within a few degrees of the galactic plane. For purposes of exhibiting velocity results graphically, all stars will be treated as though they were in the plane. No loss of detail or accuracy will occur as a result, only a small additional fictitious scatter will be introduced in the graphical representation. All calculations are carried out precisely; no such approximation is made. The galactic concentration of the B stars will permit only a very weak determination of B_0, the latitude of the apex. However, it will play no part in the discussion.

(1) Fig. 2 shows the eighty stars projected on the galactic plane. Fig. 3 exhibits the observed radial velocity, v_r, of the eighty stars plotted as a function of galactic longitude, l. Solar motion appears in this diagram as a single sine wave, galactic rotation as a double sine wave of at most a few kilometres per second amplitude, since the stars are nearby. A K term representing a *general* expansion or gravitational red shift appears as a constant and raises the entire group of points. Community motion groups appear as distortions of these three features and need follow no regular pattern.

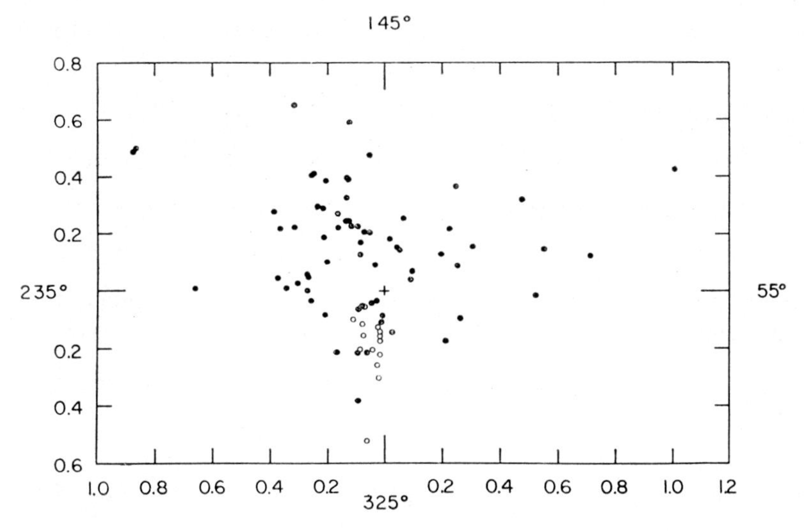

Fig. 2. Galactic plane distribution of O–B2 stars for which radial velocities are available. Unfilled circles represent members of Scorpio-Centaurus aggregate

(i) Solar motion, galactic rotation, and K term do not permit a satisfactory representation of the observed points in Fig. 3, which define a distorted, asymmetric curve. The points lie predominately above the zero axis; they attain a maximum at approximately $l = 220°$, a minimum at $l = 15°$. The curve requires 205° to rise, 155° to fall.

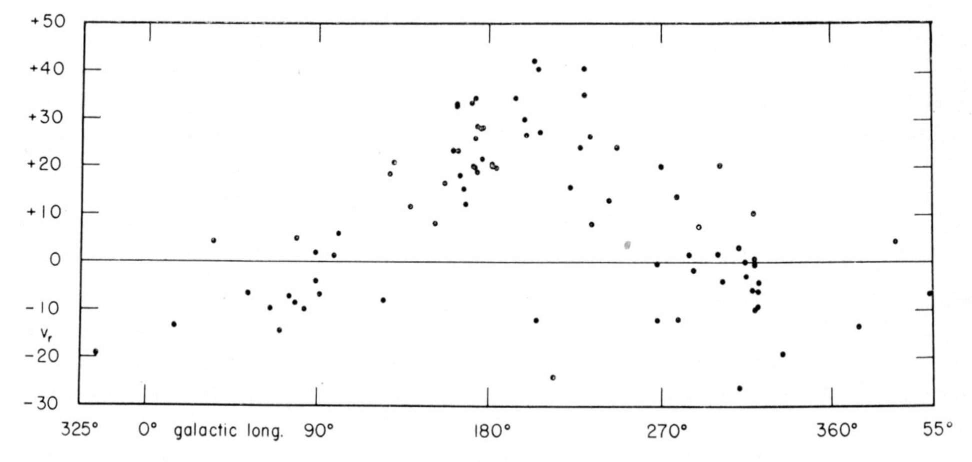

Fig. 3. Observed v_r, l - relation for stars shown in Fig. 2

(ii) The points to the left and to the right of $l = 195°$ show a difference in dispersion. The reader may demonstrate this for himself by covering first one half of the diagram and then the other: the points to the left of 195° represent stars lying within the local arm pictured in Fig. 1; the points to the right represent stars outside the local arm in the region occupied by the Scorpio-Centaurus group. The distortion of the v_r, l relation appears to be caused primarily by the points to the right of 195°.

Discussed on the basis of equation (1), the points of Fig. 3 lead to the parameter values

$$K = +\ 4{\cdot}75 \text{ km per sec.},\ S_{\odot} = 19{\cdot}46 \text{ km per sec.},\ L_{\odot} = 16{\overset{\circ}{.}}2,\ B_{\odot} = +\ 9{\overset{\circ}{.}}3.$$

(2) The points shown in Fig. 3 may be individually corrected for galactic rotation:

$$\begin{aligned} v_r' &= v_r - R_0[\omega(R) - \omega_0] \sin (l - l_0) \cos b \\ &= K - u_{\odot} \cos \alpha \cos \delta - v_{\odot} \sin \alpha \cos \delta - w_{\odot} \sin \delta. \qquad \ldots .(4) \end{aligned}$$

For this calculation R_0 has been taken as 8·75 kiloparsecs; the function $\omega(R) - \omega_0$ has been determined by WEAVER (1954) on the basis of radial velocities and the

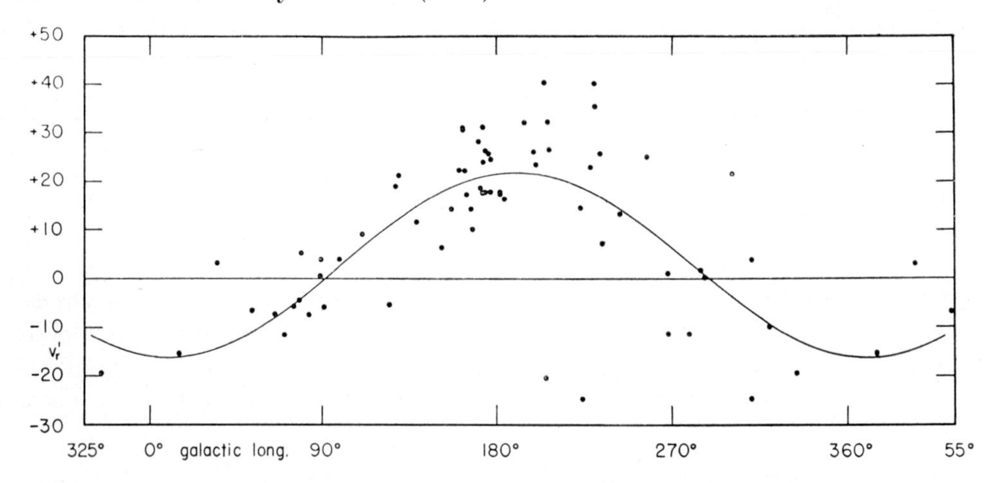

Fig. 4. Observed v_r', l relation for stars shown in Fig. 2; Scorpio-Centaurus members omitted. The curve represents solution number 5

revised distance scale of the cepheids. The modifications caused by galactic rotation are slight since the stars are all nearby. The peak at $l = 220°$ is slightly lowered. The points in the range $l = 235°$ to $325°$ are raised; this effect increases the asymmetry. The radial velocities v_r' discussed by equation (4), lead to the results

$$K = +\ 4{\cdot}88 \text{ km per sec.},\ S_{\odot} - 19{\cdot}37 \text{ km per sec.},\ L_{\odot} = 16{\overset{\circ}{.}}8,\ B_{\odot} = +\ 9{\overset{\circ}{.}}3.$$

Correcting for galactic rotation has slightly increased the K term, decreased the solar motion, and shifted the apex to a greater longitude. These effects can be traced principally to the raising of the points in the l-range 235° to 325° and to the fact that there are essentially no early B stars in the quadrant 325° to 55° to balance the change.

(3) The disturbance in the solar motion curve arises mainly in longitude range 235° to 325°. Thirteen Scorpio-Centaurus stars identified in this l-range by BLAAUW (1946) are plotted in Fig. 2 as open circles. We remove these objects from the data and find for the remaining stars

$$K = +\ 3{\cdot}35 \text{ km per sec.},\ S_{\odot} = 18{\cdot}73 \text{ km per sec.},\ L_{\odot} = 9{\overset{\circ}{.}}5,\ B_{\odot} = +\ 14{\overset{\circ}{.}}6.$$

The Scorpio-Centaurus stars thus account for 1·5 km per sec. of the K term.

A plot of the v_r', l relation of the sixty-seven stars to which the solution applies and the representation afforded by the solution are shown in Fig. 4.

(4) The most pronounced difference between the plotted points and the curve in Fig. 4 occurs in the longitude range 190° to 235°, where the points lie predominantly above the computed curve. If the slightly large velocity found for these stars is of significance in galactic motions, we should expect to obtain the same result from other B stars in the same direction and region of space. If the same high velocity is not found from these other stars, we shall be inclined to judge that the bright B stars form a community motion group.

In Fig. 5 (*a*) the solid dots show the galactic plane distribution of the bright B stars for which the v_r', l relation is plotted in Fig. 4. Fainter, generally more distant early

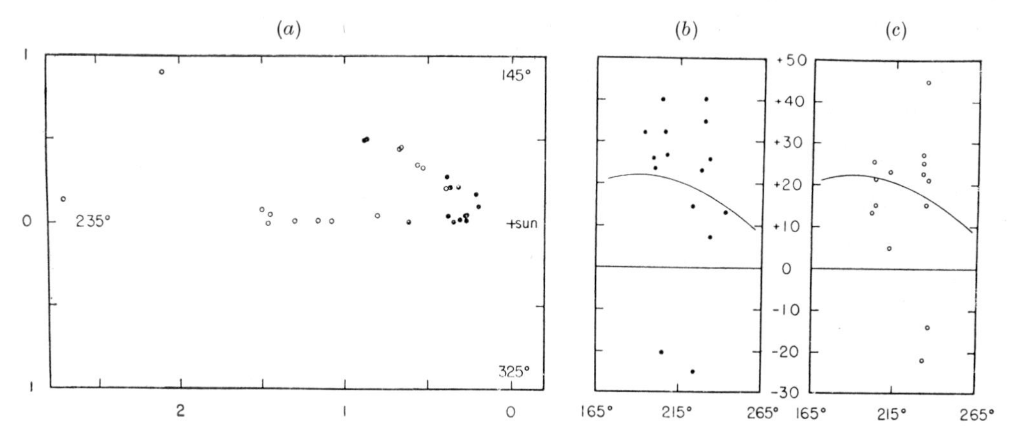

Fig. 5 (*a*). Galactic plane distribution of O–B2 stars in longitude interval 195°–235°. Filled circles, stars brighter than magnitude 5·0

(*b*) The v_r', l relation for stars represented as filled circles in Fig. 5 (*a*). Curve is same as that drawn in Fig. 4

(*c*) The v_r', l relation for stars represented as unfilled circles in Fig. 5 (*a*). Curve is same as that drawn in Fig. 4

B stars for which velocities are available are shown as open circles. Fig. 5 (*b*) exhibits the v_r', l relation for the stars indicated by solid dots in Fig. 5 (*a*). The curve drawn is a section of that shown in Fig. 4. Fig. 5 (*c*) shows the v_r', l relation for the stars represented by open circles in Fig. 5 (*a*) and repeats the curve of Fig. 5 (*b*).

(i) The B stars pictured in Fig. 5 (*c*) do not exhibit high velocities as do the stars in Fig. 5 (*b*). We take this to indicate that the stars pictured in Fig. 5 (*b*) form a chance community motion group.

(ii) The stars outside the local arm in the 235° direction show the large velocity dispersion characteristic of stars in the inner arm of the galaxy (WEAVER, 1953b) or of some of the stars in the region of the Scorpio-Centaurus group.

(5) In the sample population treated in solution (3) we replace the stars for which we found chance community motion (Fig. 5 (*b*)) by the stars shown in Fig. 5 (*c*). We find for this altered sample that does not contain the chance community motion group

$$K = +\ 2{\cdot}89 \text{ km per sec., } S_{\odot} = 18{\cdot}94 \text{ km per sec., } L_{\odot} = 9^{\circ}\!.9,\ B_{\odot} = +\ 22^{\circ}\!.0.$$

The K term has been further diminished; the solar speed is not significantly changed.

(6) The disturbance in the v_r', l relation in Fig. 4 appears to arise primarily from those stars outside the local arm. If we make use only of those stars in the local arm (the stars in solution (5) minus those in the l-range 235° to 0°) we find for the forty stars available.

$$K = 2{\cdot}07 \text{ km per sec., } S_{\odot} = 19{\cdot}71 \text{ km per sec., } L_{\odot} = 8^{\circ}\!.0,\ B_{\odot} = +\ 23^{\circ}\!.3.$$

The K term is still further diminished. We note that the solution is not a strong one, however, since only 235° of longitude are covered. Nevertheless, the trend is clear: the local-arm stars show a small K-term.

6. The Faint B Stars

We investigate the decrease in the K term with increasing magnitude by considering two related questions.

(*a*) Do distant early B stars located in the local arm show, after they have been individually freed from the effects of galactic rotation, the same v_r', l relation as the nearby local-arm stars?

(*b*) Do early B stars located in the inner and outer arms indicate, after they have been individually freed from the effects of galactic rotation, the same v_r', l relation as the nearby local-arm stars?

(1) To investigate the first problem we make use of a number of early B stars and galactic clusters of type 1B2 or earlier, distributed in the galactic plane as shown in Fig. 6 (*a*). Radial velocities for the clusters have been kindly supplied by Dr. R. J. Trumpler. The v_r', l relation derived for the distant B stars and clusters is illustrated in Fig. 6 (*b*), where the results of solution (5) are shown as a curve. Though the scatter in the v_r', l relation found for these more distant objects is somewhat larger than that found for the nearby B stars (owing to distance uncertainties and hence galactic rotation uncertainties) visual inspection of the diagram indicates no significant differences in the results obtained from the nearby and distant local-arm stars. The stars within the local arm thus appear consistent in their motions.

(2) Fig. 6 (*a*) also shows projected on the galactic plane several galactic clusters (1B2 or earlier in type) and individual B stars located in the inner and outer arms. The v_r', l relation derived from these is shown in Fig. 6 (*b*).

Inner Arm. Visual inspection discloses no discrepancy between the v_r', l relation of the inner arm and that found for the local arm. Examination of the inner arm is made generally difficult by the large velocity dispersion found for stars in that arm. Significant values can be derived only if observations are available for a considerable number of stars.

Outer Arm. The v_r', l relation of that portion of the outer arm studied agrees generally with the v_r', l relation of the local-arm stars. The Double Cluster in Perseus (and all field stars in the region of the Double Cluster) are notable exceptions to this rule. That the abnormal velocity of the rich stellar association in the Per-Cas region represents community motion rather than a direct effect of galactic rotation can be shown by comparing the value of $\omega(R) - \omega_0$ (see equation (4)) derived from local-arm stars with that derived from Per-Cas stars of the same R. We find $[\omega(R) - \omega_0]_{\text{Per-Cas}} \approx 1{\cdot}5[\omega(R) - \omega_0]_{\text{local arm}}$; the abnormal velocity of the stars in the Per-Cas region is of the order of -20 km per sec.

Community motion in the Per-Cas region is not confined to B stars; it is also shown by the cepheids (Weaver, 1954).

Inclusion of the extensive Per-Cas community motion group in the determination of the K term, solar motion, or galactic rotation constant will strongly bias the results. Velocities of many Per-Cas stars are included among the data discussed by Plaskett and Pearce; these Per-Cas stars have decreased the K term and increased the Oort constant A. However, discussion of such details is beyond the scope of the present paper; they will be considered in a more extensive discussion elsewhere.

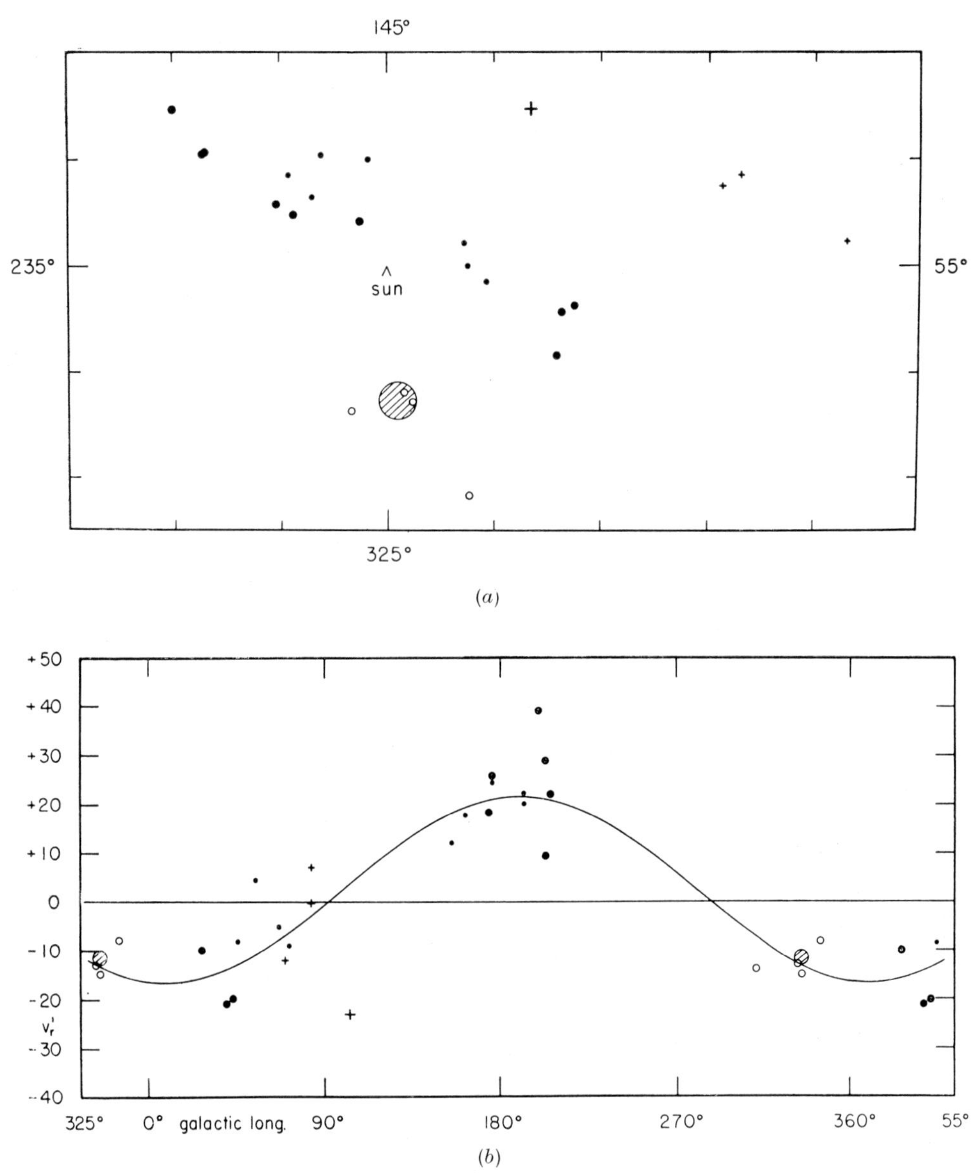

Fig. 6 (a). Galactic plane distribution of O–B2 stars and clusters. Filled circles: members of local arm; unfilled circles: members of inner arm; crosses: members of outer arm. Large symbols represent clusters; small symbols represent individual stars. The shaded area represents the location of thirty-seven stars treated as one point in the discussion. The large cross represents the double cluster in Perseus

(b) The v_r', l relation for objects pictured in Fig. 6 (a). Symbols as in Fig. 6 (a).

7. Conclusions

The large K term associated with the bright early B stars is real and signifies expansion, but does not indicate any fundamental characteristic of the local region of the galaxy. It is a chance property of the sample population traceable to nearby community motion groups within the sample. The largest of the community motion groups is the Scorpio-Centaurus cluster. Its influence on the K term was first pointed out by PLASKETT and PEARCE. When community motion groups are removed from the data, a K term of approximately $+ 2{\cdot}5$ km per sec. remains for the O to B2 stars. This may be attributed principally if not entirely to gravitational red shift.

A method of investigating on the basis of galactic structure the significance of the small K term found for faint B stars is suggested. Brief, incomplete tests that can be made with available data indicate:

(1) Motions within the local arm are consistent; there is therefore no variation of K term within the arm.

(2) No relative motion of the local and inner arms is found, hence the same K term holds for both arms.

(3) While stars in the outer arm also tend to confirm the normal K term, the stars in the Per-Cas region possess large negative peculiar motions. Inclusion of the numerous B stars in this region will bias any solution towards a small K term and a large OORT constant.

Though it is not discussed in detail, there is mentioned the danger of introducing fictitious community motion groups among distant stars by making inadequate corrections for galactic rotation. Inspection of former investigations shows that at times this has been a powerful source of bias.

REFERENCES

BLAAUW, A. . . . . 1946 *Publ. Kapteyn Astron. Lab., Groningen*, No. 52.

CAMPBELL, W. W. . . . . 1911 *Lick Obs. Bull.*, **6,** 101, 125.
1913 *Stellar Motions* (Yale University Press, New Haven).
1914 *Lick Obs. Bull.*, **8,** 82.

CAMPBELL, W. W. and MOORE, J. H. . . . . 1928 *Publ. Lick. Obs.*, **16,** xxxviii.

FINLAY-FREUNDLICH, E. . . . . 1915 *Astron. Nachr.*, **202,** 17.

FROST, E. B. and ADAMS, W. S. . . . . . . 1904 *Dec. Publ. Univ. of Chicago*, **8,** 105.

KAPTEYN, J. C. and FROST, E. B. . . . . . 1910 *Ap. J.*, **32,** 83.

MILNE, E. A. . . . . 1935 *M.N.*, **95,** 560.

MORGAN, W. W., WHITFORD, A. E. and CODE, A. D. 1953 *Ap. J.*, **118,** 318.

OGRODNIKOFF, K. . . . . 1932 *Z. Astrophys.*, **4,** 90.

PILOWSKI, K. . . . . 1931 *Z. Astrophys.*, **3,** 53, 279, 291.

PLASKETT, J. S. . . . . 1930 *M.N.*, **90,** 616, 621.

PLASKETT, J. S. and PEARCE, J. A. . . . . . 1929 *Amer. Astron. Soc.*, **6,** 277.
1936 *Publ. Dominion Astrophys. Obs. Victoria*, **5,** 227 (gives an excellent historical account of the problem of the K-term).

SMART, W. M. . . . . 1936 *M.N.*, **96,** 568.

SMART, W. M. and GREEN, H. E. . . . . . . 1936 *M.N.*, **96,** 471.

TRUMPLER, R. J. and WEAVER, H. F. . . . . . 1953 *Statistical Astronomy* (University of California Press, Berkeley), Chapter 6.21.

WEAVER, H. F. . . . . 1953a *Astron. J.*, **58,** 177.
1953b *Publ. Astron. Soc. Pacific*, **65,** 132.
1954 *Astron. J.*, **59,** 375.

On the Empirical Foundation of the General Theory of Relativity

E. FINLAY-FREUNDLICH
University Observatory, St. Andrews, Scotland

1. INTRODUCTION

I FIRST had the pleasure of meeting Professor STRATTON at an eclipse expedition in Sumatra twenty-eight years ago. On the occasion of writing this summary of my eclipse work, I hope he will share my pleasant recollection of the time we spent there together.

During the past years the task of putting the general theory of relativity to the test has receded into the background; and thus among many scientists the feeling seems to prevail that the theory has been successfully tested, or that the astronomical observations have not yet reached the necessary accuracy to decide definitely whether the new effects predicted by the general theory of relativity exist or not. I need only refer to P. G. BERGMANN, 1946, p. 221.*

Both attitudes are wrong. Since various new attempts are in progress to settle this problem, which is of fundamental importance, it is justified to summarize the present situation.

One of the three possibilities of testing the general theory of relativity refers to the orbital motion of a small mass in the gravitational field of the Sun; the improved law of motion, as compared with Kepler's law, predicts an advance of the perihelion of Mercury moving in the gravitational field of the Sun alone to the amount of 42″.9 per century. This effect is, as such, not of a fundamentally new character but a refinement of Newton's law of gravitation by the theory of relativity. The latest exhaustive discussion of all observational material, covering the time from 1765 to 1937, by G. M. CLEMENCE, Washington, results in a complete agreement between theory and observation. As to the planet Venus, the eccentricity of its orbit is too small to allow a sufficiently accurate determination of the motion of the perihelion of its orbit, and for the still more distant planets the predicted effect becomes too small.

The two other possibilities of testing the theory, the red-shift of the solar lines and the deflection of light passing near the Sun, concern fundamentally new phenomena, namely an influence of a gravitational field upon the propagation of electromagnetic energy. These two predictions of the theory of relativity have not yet been confirmed; not, however, because the predicted effects are too small to be determined with the necessary accuracy. In both cases, as will be shown, a determination of the predicted effect is possible with a standard deviation of not more than about 5 per cent of the predicted value. Both effects either do not agree with the theoretical prediction, or else they are strongly falsified by secondary effects of unknown origin and character.

* See also the Notes in the last paragraph of this article (p. 246).

2. General Red-shift of the Solar Lines

Let us first consider the case of the *general red-shift of the solar lines.*

Compared with the wavelength of the corresponding lines, produced in a terrestrial light source, all solar lines taken from any point of the surface of the Sun, allowing, naturally, for the influence of the Sun's rotation, should appear displaced towards the red end of the spectrum by an amount $\Delta\lambda$ equal to $+ 2{\cdot}12 \times 10^{-6}$ times the wavelength. This shift corresponds to a loss of the energy $\Delta(h\nu)$ of the light quanta when rising from the Sun to the Earth's surface, equal to their gain in potential energy in the gravitational field of the Sun. Fig. 1 reproduces the results of the latest observations by M. G. Adam (1948) of Oxford, compared with observations made many years

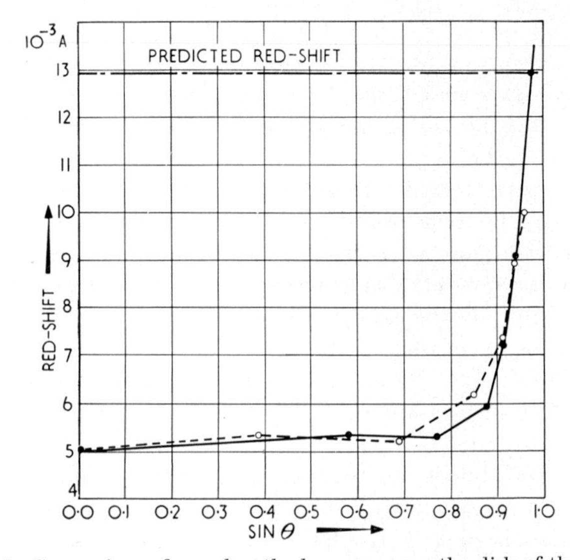

Fig. 1. Comparison of wavelength changes across the disk of the Sun
• = Oxford (Miss Adam, 1948), λ6100, 14 lines, mean intensity 6
∘ = Potsdam (Freundlich *et al.*), λ4420, 9 lines, mean intensity 3, mean of 12 radii.

earlier in Potsdam. The latter observations were only relative measurements of wavelength changes along twelve radii on the Sun's disk; for the centre of the Sun the observations of Oxford and of Potsdam are made to coincide. All lines should reveal a shift equal to the constant amount $12{\cdot}9 \times 10^{-3}$Å represented in Fig. 1 by the parallel line. The probable error of each measurement represented by a black dot in the graph is about 10 per cent. The complete coincidence of the two independent sets of observations shows clearly that the red-shift is not just "barely observable"; nor is it justified to add "but appears to be in agreement" with the theoretical predicted value. Fig. 1 shows quite clearly that, if the general red-shift were existent according to the theoretical prediction, the observations would reveal it with absolute certainty; what they reveal, however, is (1) that indeed a certain general red-shift is indicated (for the values do not drop at the Sun's centre below $+ 5 \times 10^{-3}$Å), but they definitely do not indicate a constant red-shift of the amount $2{\cdot}12 \times 10^{-6} \times \lambda$; (2) near to the Sun's limb the red-shift rises suddenly steeply to a value which seems to lie near to the predicted effect.

The possibility exists therefore that the theoretically predicted effect is existent, but that it is eclipsed by other effects. Disturbing effects that may be made responsible are Doppler effects due to radial currents in the solar atmosphere and pressure effects in the solar atmosphere.

Many attempts have been made to bring the observational results into agreement with the theory by taking account of both possibilities; but neither an acceptable model for radial currents in the solar atmosphere could be constructed, nor is there a generally accepted theory of pressure effects which make it possible to extract a constant red-shift of the demanded amount from the observational data.

So far, the marked increase of the red-shift near to the Sun's limb is the only effect firmly established. At the limb Doppler effects due to radial currents will not affect the wavelengths of the solar lines; similarly, the influence of pressure effects should become insignificant near the limb when the lines arise from the highest layers of the solar atmosphere.

Therefore, observations restricted to the Sun's limb should yield a reliable proof for the existence of the predicted red-shift, if the observed red-shift converged at the limb towards the theoretically predicted value. This, however, according to the present results, is definitely not the case.

EVERSHED (1936) has obtained extensive material using lines from the reversing layer near to the limb of the Sun, as well as from the chromosphere and from eruptions rising well above the Sun's limb. The results of these observations are very consistent, but they yield a red-shift at the Sun's limb considerably larger than the predicted value. The following table gives all necessary data concerning these observations by EVERSHED (1936).

Table 1

	$\Delta\lambda_{obs}$ ⊙ limb minus vacuum arc, Mt. Wilson	$\Delta\lambda_{obs}$ ⊙ limb minus vacuum arc, EVERSHED	Excess over relativistic effect	Number of lines and spectra used
Reversing Layer: Fe lines: 3932 Å (max. of 14 lines)	+ 0·0141 Å	+ 0·0146 Å	+ 0·0063 (62 per cent)	14 lines, 50 spectra
Reversing Layer: Fe lines: 4433 Å (max. of 11 lines)	+ 0·0148 Å	+ 0·0139 Å	+ 0·0045 (50 per cent)	11 lines, 25 spectra
Reversing Layer: Fe lines: 6203 Å (max. of 5 lines)	+ 0·0248 Å	+ 0·0252 Å	+ 0·0120 (90 per cent)	5 lines, 25 spectra
Chromosphere: H and K lines		+ 0·0151 Å	+ 0·0068 (80 per cent)	2 lines, 22 E. and W. spectra
Prominences: H and K lines		+ 0·0150 Å	+ 0·0067 (79 per cent)	2 lines, 180 spectra

The obvious disagreement with the theoretical prediction led to the *ad hoc* hypothesis of an additional "limb effect". Since no explanation for such a limb effect can be given, it is more consistent to summarize the results as follows:

At the Sun's limb, where neither radial currents nor pressure effects in the atmosphere of the Sun will seriously affect the wavelengths of solar lines, a red-shift is observed which, however, does not have the value predicted by the theory of relativity, but a value considerably larger than the one predicted by it.

Frequently the results obtained from observations of the companion of Sirius are cited as a reliable confirmation of the theory of relativity. These observations definitely do indicate the existence of a red-shift; the value may even roughly agree with the predicted amount; this is by the way also the case with the solar observations; for the average red shift which would result if one derived the mean value by integrating over the whole solar disc, would not fall far short of the predicted value. The possibility of a detailed investigation along the Sun's disc reveals, however, that the problem is not so simple. The experimental proof of the theory of relativity asks for more than a crude confirmation of the theoretical prediction. The results which can be derived from observation of Sirius B are far too uncertain to yield a sound foundation for an empirical test of the theory of relativity.

3. The Light-Deflection

The situation with regard to the deflection of a light beam passing near to the Sun is very different from that concerning the general red shift discussed in the preceding paragraph. While in the latter case the existence of a general constant red shift of the solar lines relative to terrestrial lines has not yet been safely established, the existence of a light deflection of the expected order, decreasing apparently proportional to $1/\Delta$ (Δ denoting the smallest distance at which the beam passes the Sun) is beyond doubt. It remains undecided, however, whether the observed and predicted values of the deflection agree with each other. With the increasing accuracy of the determinations of the light deflection, a systematic difference between observations and theory has been disclosed with increasing certainty, the observed values exceeding the theoretical prediction by nearly 25 per cent. The whole effect exceeds considerably the observational errors; it is therefore quite misleading to say that it "is just outside the limits of experimental error" (Bergmann, 1946, page 221). The standard deviation of a determination of the light deflection need not exceed 5 per cent of the predicted value. The difficulties in measuring accurately the light deflection are partly of a practical, partly of a theoretical nature.

The practical difficulty in measuring the light deflection with sufficient accuracy lies in the fact that the observations have to be performed under the exceptional conditions during the short and rare moments of a total solar eclipse. The duration of the last total eclipse in Sweden in 1954, for instance, was only about 150 sec. The theoretical difficulty lies in the fact that the method of deriving accurate stellar positions from astrographic plates cannot be applied to this problem, unless special precautions are taken. This fact was not recognized at one time and has reduced the weight of the earlier determinations of the light deflection. Table 2 gives the five determinations for which the mean error does not exceed $\pm$ 0".3. The aim should be to determine the light deflection with an accuracy of $\pm$ 0".1 or less; according to the theory the value should be equal to 1".75 for light passing the Sun's limb. Observations made in Potsdam with the powerful apparatus specially constructed for this problem, and made under normal night conditions, show that it should be possible to bring the standard deviation even below the value of $\pm$ 0".1; v. Brunn and v. Klüber (1937).

Apart from the result of the Lick expedition, given in the second line of Table 2, all values exceed considerably the predicted value of 1".75; also in the case of the Lick expedition the value actually observed, before having been specially corrected, was equal to 2".05.

The only method which promises to yield the value of the light deflection with the necessary accuracy is a differential comparison of the star field surrounding the eclipsed Sun during a total eclipse with the same field under normal night conditions. On the eclipse plate the stars should appear shifted away from the Sun's centre when compared with their position on the night plate. Chosing for a numerical comparison the proportions of the Horizontal Twin Camera of 850 cm focal length, specially devised for this problem, a star the light from which grazes the Sun's limb should appear shifted by 0·07 mm. In practice, stars will hardly ever be visible through the solar corona as near as that to the Sun; the actually observable shifts will not surpass 0·03–0·04 mm. If, however, all stars in a field of $3° \times 3°$ numbering more than, say, 20, show a systematic general shift of this amount, this shift is determinable with great certainty; and if the reduction is correctly performed, *i.e.* with complete knowledge of the theoretical implications, the value for the deflection at the limb can be derived with a standard deviation not exceeding 5 per cent of the theoretically predicted value.

By using, as an intermediate reference plate, one exposed through the glass to the star field surrounding the Sun during the eclipse, the positions of all stars visible on an eclipse plate and on a night plate can be compared with each other, so to speak, emulsion in contact with emulsion; and relative co-ordinates Δx_i, Δy_i can be measured. The two values Δx_i and Δy_i for every star i ($i = 1, 2, 3, \ldots n$, if n is the total number of stars on the plate) arise from various sources and can be expressed by the equations:

$$\Delta x_i = Zx + O \,.\, y_i + S \,.\, x_i + p x_i^2 + q x_i y_i + L \frac{x_i}{r_i^2} + Rx_i + Ax_i$$

$$\Delta y_i = Zy - O \,.\, x_i + S \,.\, y_i + p x_i y_i + q y_i^2 + L \frac{y_i}{r_i^2} + Ry_i + Ay_i.$$

Here the x_i, y_i are the rectangular co-ordinates of the star i relative to the centre of the Sun which is supposed to coincide with the centre of the plate; r_i is the distance of the star from the Sun's centre; Zx and Zy denote the small relative shifts of the centres of the two plates, one eclipse and one night exposure, when brought into contact and measured in the comparator; O is a small difference in their orientation relative to the reference system of co-ordinates; S denotes the unavoidable difference of scale on both plates. Due to the fact that, since the exposures of the night sky have to be made either a few months before or after the eclipse, these exposures are made under different conditions as to temperature, *etc.*, and the focal length of the telescope will not have been accurately the same at both exposures. The coefficients p and q account for possible small differences in the inclination of the photographic plates in the plate-holder relative to the optical axis of the telescope during their exposures. L represents the expected influence of the light deflection on the positions of the stars on the eclipse plate; and finally A and R the influences of aberration and refraction upon the relative positions of the stars; both corrections naturally change with the atmospheric situation and the conditions at the time of the exposures.

The last two effects can be accurately determined, if the time of the exposure and if the temperature and barometric conditions during the exposure are noted. These effects can consequently be calculated and subtracted, before formulating the above equations. Similarly, in order to simplify the following discussion, we shall drop the terms with p and q, for in rigidly constructed telescopes of steel the position of the plates relative to the optical axis can be kept accurately under control.

The remaining $2n$ equations have to be solved for the five unknowns Zx, Zy, O, S, L. If $2n \gg 5$ a least squares solution should yield accurately their most probable values. This, however, is only true if the determinant of this set of equations does not vanish; when this is the case, two of the unknowns must be linearly coupled and one of them can be freely chosen. It is a special misfortune for the problem under consideration that this singular case really endangers the accurate determination of the two unknowns L and S. For a difference in scale of two plates that are compared, will, assuming always the Sun's centre to coincide with the plate's centre, increase in a linear manner in all directions away from the centre. The light deflection on the, other hand, is supposed to decrease proportionally to the distance from the Sun; that means along a hyperbola. But in practice the hyperbola is indistinguishable from its asymptote. In the area from 4–10 solar radii from the Sun—this is the region in which the majority of stars lie that contribute measurably to the light deflection—the hyperbola differs from the linear decline by no more than $\pm 0''.04$; the mean error of one co-ordinate measurement is about eight to ten times this amount. Thus L and S are quasi-linearly coupled. That means L can only be determined accurately if S is independently determined by a special set of observations and not introduced into the above set of equations.

The weight of the determination of L is given by the expression

$$w_1 \simeq n \left(\frac{1}{h} - \frac{1}{a}\right)$$

if L and S are both derived from the least squares solution; the weight is, however, $w_2 \simeq n \cdot \frac{1}{h}$, if S is determined independently. These two expressions refer to the simple case that the stars are symmetrically distributed around the Sun, the centre of which is supposed to coincide with the centre of the plate; h denotes the harmonic mean, a the arithmetic mean of the r_i^2. The second value w_2 reveals the unfavourable conditions under which L has always to be derived. This is due to the fact that the value of the light deflection decreases with r_i; and since the majority of the stars lie in the outer region of the plate and consequently do not contribute to the determination of L with sufficiently large, accurately measurable, shifts the light deflection cannot be derived from a least squares solution as an unknown of large intrinsic weight. The few stars lying near to the Sun will always be the backbone of the determination. If, however, S is left in addition to L, to be derived from the least squares solution, the weight of the determination of L is further seriously reduced by the smallness of the factor $\left(\frac{1}{h} - \frac{1}{a}\right)$.

Thus L can only be accurately determined if S is derived by special independent observations, and only if the determination of S itself is extremely accurate. For any error made in the determination of S, say ΔS, will produce, due to the tight coupling of L and S, an error ΔL in the determination of L, equal to $\Delta L = - h\Delta S$; here the possible effect of an asymmetrical distribution of the stars around the Sun is neglected. Since h is of the order of $\bar{r}_i^2$, and $\bar{r}_i$ rarely is smaller than 4–5 solar radii, an error in the determination of S, producing an error of the assumed value for the Sun's radius equal to $0''.01$, or, expressed in other terms, an error of the scale corresponding to no more than 1/100,000 of the focal length of the telescope, will increase or reduce the value obtained for L by $0''.2$.

So far only in the case of the Potsdam expedition of 1929 was the scale correction independently derived with sufficient accuracy to reduce the mean error of L to

$\pm$ 0″1. That this accuracy was really attained is proved by the fact that observations with a twin telescope, during the eclipse, of a second star field so distant from the Sun that no influence of a light deflection could be expected—this field being reduced in the same way as the eclipse field—gave, as they should, no shifts of the star positions. The mean value of the measured changes of star positions, in this second field, taken irrespective of sign, did not exceed 0″13.

The two preceding determinations of L (see Table 2) are not based on independent observations to control S; of the two later determinations the last one failed in its attempt to derive S independently, while of the Russian expedition no detailed results are available.

Table 2

Expedition	L	Standard deviation	Remarks
Greenwich, 1919 .	1″98	$\pm$ 0″16	A revised calculation gives 2″16
Lick, 1922 . .	1″72	$\pm$ 0″15	Original, uncorrected, result: 2″05
Potsdam, 1929 .	2″24	$\pm$ 0″10	Scale correction derived by special observations
Moscow, 1936 .	2″71	$\pm$ 0″26	—
Yerkes, 1947 .	2″01	$\pm$ 0″27	Scale correction *not* derived independently

However, all these determinations point towards a larger value of the light deflection; they obviously do not scatter around the predicted value of $L = 1''75$, but around a value of the order of 2″2.*

To summarize: while the prediction of the general theory of relativity, concerning the more accurate law of the orbital motion of a small body around the Sun, appears to be in full agreement with the observational results, the predictions of effects of a fundamentally new character, concerning the motion of photons in the gravitational field of the Sun, have not yet been safely confirmed. The existence of a general, systematic red-shift of all solar lines, relative to the corresponding terrestrial lines, amounting to $2{\cdot}12 \times 10^{-6}$ times λ has not yet been established. The observations indicate a small red-shift of the solar lines, but the value changes across the disk of the Sun, apparently affected by superimposed effects of a different character. At the Sun's limb, however, where the disturbing influences of Doppler effects and pressure effects should become insignificant, the observed red-shift does not converge towards the value predicted by the theory of relativity. As to the light deflection, the existence of it is absolutely safely established. Its quantitative value, however, is not yet ascertained. Observations point towards a value nearly 25 per cent larger than the theoretically predicted one. The increasing accuracy of the determinations does not tend to reduce this disagreement, but on the contrary, seems to bring it more clearly to light.

Both effects are well beyond the limits of observational errors, so that research should be once more concentrated on a final solution of these problems. A repetition of the experiment to measure the light deflection with the powerful special telescope developed in Potsdam is in preparation. The *Deutsche Akademie der Wissenschaften* had generously given the necessary funds to improve this instrument, so that in case of favourable conditions of observation it should be possible to redetermine the value

* In this connection reference should be made to a recent result obtained by G. VAN BIESBROECK, in *Astron. J.*, **58**, 87–88, 1953, who derives a value of 0″70 $\pm$ 1″10 from his observations at the Khartoum eclipse of 25th February, 1952. See also the author's previous discussion with VAN BIESBROECK in the *Astron. J.*, **55**, 49, 1950; **55**, 245, 1951; **55**, 247, 1951.

of L with a standard error not exceeding $\pm 0''.1$. The observation of the total eclipse of 30th June, 1954, in Öland, Sweden, was prevented by clouds; we are now hoping for 1955 in Ceylon. Also new observations concerning the general red shift are planned.

Notes added in proof

A new hypothesis concerning the origin of the red-shift of spectral lines in stellar atmospheres has recently been put forward by the author in the *Nachrichten der Akademie der Wissenschaften in Göttingen*, Mathem.-Phys. Klasse, 1953, No. 7, pp. 96–102 (also in *Contrib. Obs. St. Andrews*, No. 4); see also the subsequent article *l.c.*, pp. 102–8 by MAX BORN; furthermore, a note by the author in *Proc. Phys. Soc. (London)*, A, **67**, pp. 193–4, 1954; and the more detailed paper in *Phil. Mag. Ser.*, 7, **45**, pp. 303–19, 1954 (also in *Contrib. Obs. St. Andrews*, No. 5).

REFERENCES

ADAM, M. G. . . . . 1948 *M.N.*, **108**, 446.
BERGMANN, P. G. . . . . 1946 *Introduction to the Theory of Relativity*. Prentice Hall, New York.
EVERSHED, J. . . . . 1936 *M.N.*, **96**, 152.
BRUNN, V. and KLÜBER, V. . . . . . . 1937 *Z. f. Ap.*, **14**, 242.

SECTION 4

THEORETICAL ASTROPHYSICS

LECTURER: ". . . Fundamentally, a star is a pretty simple structure . . ."
VOICE FROM THE AUDIENCE: "*You* would look pretty simple, too, at a distance of ten parsecs."

Colloquium in Cambridge University, 1954.

Fraunhofer Lines and the Structure of Stellar Atmospheres

A. Unsöld and V. Weidemann

Institut für Theoretische Physik und Sternwarte der Universität,
Kiel, Germany

Summary

A review is given on the development of our ideas concerning stellar atmospheres and their spectra. After some brief remarks concerning the early comparisons, between stellar and laboratory spectra it is emphasized that quantitative analysis of stellar spectra must involve essentially three points: (1) spectrophotometric measurements; (2) the theory of radiative and possibly convective transfer; and (3) the atomic theory of absorption coefficients. After following them for the period 1905–39, the more recent developments are characterized by treating radiative transfer in spectral lines mostly according to the scheme of "true absorption" and by giving closer attention to the temperature-stratification, using essentially non-grey models. Finally, the importance of more and better theoretical, and particularly laboratory work on transition probabilities, line broadening, *etc.*, is stressed. The problem of damping constants in solar type stars is given special attention.

1. Direct Comparison of Stellar and Laboratory Spectra

In the early days of stellar spectroscopy astrophysicists mostly attempted to interpret the *spectra of celestial bodies* by comparing them directly with those of *laboratory sources*. That procedure was very successful as to the comparison of wavelengths and identification of elements. Also, it happened to lead to a correct interpretation of sunspot spectra in terms of lowered temperature while, for instance, the comparison of the spectra of novae with those of underwater-sparks led to quite erroneous ideas. Quantitative conclusions could by no means be derived from a direct comparison of stellar and laboratory spectra since one essential characteristic of all cosmical light sources can never be reproduced on earth: namely their *enormous dimensions*.

2. Quantitative Interpretation of Stellar Spectra in Connection with Spectrophotometric Measurements, Theory of Radiative Transfer and the Atomistic Theory of Absorption Coefficients

Quantitative interpretation of stellar spectra became possible only by connecting *spectrophotometric measurements* on solar and stellar spectra with the increasing knowledge of *atomic physics* through the intermediary of *theory*. The latter in this connection has a twofold function: it must consider (*a*) the transfer of energy through stellar atmospheres by radiation and possibly convection, and (*b*) the atomic theory of absorption and re-emission coefficients for continuous and line radiation under the conditions prevailing in stellar atmospheres. Let us recall briefly the essential steps on these two lines of research up to about 1939.

(*a*) *Radiative transfer and radiative equilibrium**

After some preliminary work by Schuster and others the theory of radiative transfer and radiative equilibrium was established in two fundamental papers (1905 and 1914)

* We propose to speak of radiative *transfer* quite generally, when energy is transported as radiation, of radiative *equilibrium* however when *all* the energy flux is carried by radiation only.

by K. SCHWARZSCHILD. In the second paper he fully realized the importance of spectrophotometric measurements on line profiles and briefly indicated also that of the theory of line absorption. Most of SCHWARZSCHILD's further work was connected with BOHR's theory of spectral lines, and he would certainly have returned to the subject of Fraunhofer lines had not an untimely death prevented him from doing so. Ten years later the Bohr theory was firmly established and in 1924 RUSSELL and STEWART reviewed various mechanisms that might contribute towards the broadening of stellar absorption lines. In 1927 UNSÖLD tried to connect specially made spectrophotometric measurements of solar line profiles with the theory of radiative transfer *and* the quantum theory of line absorption, in order to determine the electron pressure P_e and—at least partly—the quantitative chemical composition of the solar atmosphere. But let us first follow the further development of the theory of radiative transfer.

In *The Internal Constitution of the Stars* (1926) EDDINGTON had considered various "*models*", *i.e.* various assumptions concerning the ratio of line and continuous absorption as a function of depth in the atmosphere. But it was only in 1928 that MILNE pointed out clearly the rôle of continuous absorption as limiting the depth of the layer which contributes towards the formation of Fraunhofer lines.* Practically speaking, that meant an important step toward the explanation of the absolute magnitude effects which are used for determining spectroscopical parallaxes. A general theory of radiative transfer in the case of "weak absorption", *i.e.* weak lines or the wings of strong lines, was established in 1932 by UNSÖLD under the title of "theory of weight functions"; the contribution of atoms located at a given depth towards the depression in a line being characterized by certain functions of depth which can be calculated once for all. MINNAERT (1936) rightly remarked that the variation with depth should be taken into account accurately not only for the ratio of line to continuous absorption but also for the Kirchhoff-Planck-function determining the re-emission of radiation. Calculations in that direction were made by MINNAERT himself and later by UNSÖLD, KURZ, and HUNGER (1949–50). Perhaps we should note that EDDINGTON's old idea, that the intensity $I_\nu(0, \vartheta)$ leaving the solar surface at inclination ϑ to the normal should roughly equal the *Kirchhoff-Planck*-function B_ν for an optical depth $\tau_\nu = \cos \vartheta$, was put on a firmer mathematical basis by BARBIER in 1943 and later applied also to line-problems by UNSÖLD (1948).

The one-sided emphasis laid in the literature upon radiative transfer should not distract us from giving more attention also to problems of convection. After the fairly obvious thermodynamical foundations had been given by UNSÖLD in 1930, BIERMANN first established the principles of convective energy transport in stellar atmospheres. Progress in the application of these ideas to the actual interpretation of spectra is difficult and slow, chiefly owing to aerodynamical difficulties.

Before going into the modern developments concerning the theory of radiative and convective transfer as well as structure of stellar atmospheres we should first report the earlier history of line absorption and related problems.

(*b*) *Atomic theory of line absorption and emission*

In his 1927 paper UNSÖLD had still used a semi-classical theory of *radiative damping*, taking into account quantum-theoretical values for the oscillator strength f, but not

* At that time many astrophysicists considered as the important point in MILNE's papers his "generalized theory of ionization" taking into account the increase of pressure with depth. That, however, was an illusion, since actually the corresponding increase of temperature (which had been ignored) is just as important.

yet for the damping constants γ of the lines. Fortunately for the chiefly considered resonance lines, the error was not significant. A correct quantum-mechanical theory of radiative damping was given in 1930 by WEISSKOPF and WIGNER. In 1929 STRUVE clearly demonstrated that the hydrogen and helium lines of early type spectra were broadened by the Stark effect due to electric fields in the ionized matter. MINNAERT (1931) recognized the importance of the *thermal Doppler effect* and STRUVE and ELVEY (1934) that of *turbulence*. They all used very effectively for the analysis of solar and stellar spectra the "curve of growth"-method whose history goes back to an important paper by RUSSELL, ADAMS, and MOORE (1928). MINNAERT's finding of unexpectedly large damping constants and his observation (with GENARD) on the strange appearance of the diffuse Mg-series in the Sun finally led UNSÖLD in 1936 to apply also the quantum theory of *collision damping* to astrophysics.

Here we should interrupt our report for a moment to recall that in 1938 our quantitative ideas concerning the pressure and the hydrogen-abundance in later type atmospheres underwent a considerable change through WILDT's remarkable discovery of the continuous absorption by the H^- ion, followed by the well-known papers of STRÖMGREN and CHANDRASEKHAR.

About 1939 most workers in the field of stellar atmospheres had the impression that we had essentially the necessary fundamental knowledge for going into a detailed and "final" analysis of solar and stellar spectra. But it was just work of that type that led to an almost complete reversal of our ideas during the second world war.

3. MODERN DEVELOPMENT OF THE THEORY OF RADIATIVE TRANSFER; LOCAL THERMODYNAMICAL EQUILIBRIUM; STRUCTURE OF NON-GREY STELLAR ATMOSPHERES

Hitherto the radiative transfer of line radiation had been mostly treated following K. SCHWARZSCHILD's scheme of "scattering" or "monochromatic radiative equilibrium", using MILNE's terminology. In 1942, HOUTGAST in connection with the interpretation of his observations on "Variations in the profiles of strong Fraunhofer lines along a radius of the solar disc", carefully studied the mechanism of incoherent scattering. HOUTGAST's and SPITZER's work indicated that actually the wings of these lines should closely follow SCHWARZSCHILD's scheme of "true absorption" or "local thermodynamical equilibrium", in MILNE's language. At the same time UNSÖLD noticed that—judging from quantum-theoretical arguments—practically all the lines of early type stars should be formed by "true absorption" and confirmed that conclusion by measurements of the central intensities of all the sufficiently strong lines.

To-day we are convinced that—perhaps with the exception of some features in the resonance lines of late type stars—the re-emission in practically all Fraunhofer lines can be calculated fairly accurately simply by applying KIRCHHOFF's law. That relieves the astrophysicist from a heavy mathematical burden. The transfer problem for true absorption then admits the well known exact solution which is so simple that even in the most complicated cases it can be handled numerically rather easily.

However, *naturam expellas furca, tamen usque recurret . . . !* Up to 1942 it was almost generally agreed that the increase of temperature with depth in stellar atmospheres could be calculated with an accuracy sufficient for spectroscopic purposes using the simple theory of the "grey" atmosphere in connection with the Rosseland average over the absorption coefficient. However, in early type stars the variation of the continuous absorption coefficient with wavelength amounts to several powers of ten near the Lyman-limit of hydrogen and in later type stars the densely crowded absorption lines involve huge variations of the absorption coefficient over considerable parts of the spectrum. Such deviations from the previously assumed greyness lead to a steep decrease of the temperature in the highest layers of the atmospheres. For instance, the boundary temperature of the Sun, which had been given in 1938 generally as 4830°K is now definitely known to be as low as 3800°K.

The mathematical theory of radiative equilibrium in non-grey atmospheres is very cumbersome indeed. It is easy to write down their general equations, but the numerical evaluation is beset with difficulties. A first idea as to their behaviour could be derived from a most useful paper, which CHANDRASEKHAR had published as early as 1935 on the influence of lines on the temperature stratification of the Sun. In 1947 STRÖMGREN and UNSÖLD proposed the so-called "Λ-iteration" procedure in which—starting from a reasonable zero approximation of the source function—one calculates first the radiation intensity, then again (taking into account the conservation of energy) the source function and so on. This method, however, gives reasonably fast convergency only for the topmost layers. Therefore in 1951 UNSÖLD proposed the "flux-iteration" procedure. Here, again starting from some zero approximation, one calculates exactly the corresponding radiative flux as a function of depth and tries then to correct its deviations from constancy by means of a relation between source function and flux which had been first published by R. v. d. R. WOOLLEY in 1941. The second method works well in the deeper layers but is not so good in the high ones. So probably the thing to do will be to apply first the flux-iteration and then the Λ-iteration.

For the deeper layers the models of later type atmospheres must moreover be improved by taking into account the convection. Unfortunately, the practically most important layers exhibit energy transfer by convection *and* radiation. It is, however, just the transition from the radiative to the convective zone of a stellar atmosphere that is theoretically most difficult to handle. Possibly it will be necessary also to allow for the fact that in a convection zone one has at one and the same optical depth quite a considerable range of temperatures. In the accessible layers of the solar atmosphere that range of temperatures connected with the boiling of the granulation may be about ± 500°K.

Calculation of model atmospheres on this new standard, of course, takes much time. In Kiel we have tried to give the thermodynamical and atomistical fundamentals in a series of papers "Der Aufbau der Sternatmosphären". The solar atmosphere has—with particular emphasis on convection—been treated by E. VITENSE in two papers published in *Z. Astrophysik*. Work on earlier spectral types is being done in connection with the analysis of large dispersion plates kindly taken for us by O. STRUVE. At other institutes similar problems have been tackled in somewhat different ways; we should mention especially the work done by STRÖMGREN in Copenhagen, DE JAGER (Diss. Utrecht, 1952), J. C. PECKER (Diss. Paris, 1951), *etc.*

4. NUMERICAL VALUES OF CONTINUOUS AND LINE ABSORPTION COEFFICIENTS AND DAMPING CONSTANTS, ETC., CALCULATED FROM QUANTUM MECHANICS AND MEASURED IN LABORATORY LIGHT-SOURCES

All the work which we have briefly indicated so far can, however, only lead to actual extension and higher precision of our knowledge about the stars if it is supplemented by a better knowledge of the involved atomistical constants: absorption coefficients, oscillator strengths, damping constants, *etc.*

Important progress in this direction has been made using quantum mechanical methods. Let us recall only the calculation of transition probabilities by HARTREE, BIERMANN, GREEN, BATES, and their collaborators, and the exceedingly laborious calculation of the H^- absorption coefficient by CHANDRASEKHAR.

However, many of the astrophysically important atomic or ionic transitions are very unsuitable for quantum theoretical calculation. So it will be necessary that astrophysics take up again much closer connection with laboratory spectroscopy. The aim of such work will not be—as in the early days—to duplicate astrophysical spectra in the laboratory, but to measure the atomistic constants entering into the theory of solar and stellar spectra. Nevertheless it is necessary to develop light-sources approaching more or less stellar conditions. In any case they must have a definite temperature and often—what is more difficult to realize—also a known electron pressure.

Oscillator strengths or transition probabilities for many spectra of astrophysical interest have been measured in the electric furnace by R. B. and A. S. KING (1935) in absorption and by CARTER (1949) in emission. Absolute values have also been determined by KOPFERMANN and WESSEL (1949–51) using an ingenious atomic beam method avoiding the use of a vapour pressure curve. However, in the most important case of iron the Göttingen and the Pasadena measurements differ by a factor 3 whose origin could not yet be traced in spite of considerable efforts. More laboratory work is evidently needed.

In the Institute for Experimental Physics of Kiel University W. LOCHTE-HOLTGREVEN and H. MAECKER with their collaborators have developed a high current arc which, being stabilized by a whirl, burns very steadily and in which temperature *and* electron pressure can be measured quite accurately. By drastic cooling the arc can be contracted so that with currents up to 1200 A the temperature of its column rises to $\sim 50{,}000^\circ$.

Under conditions, which previous calculations by VITENSE had indicated as favourable, LOCHTE-HOLTGREVEN and his students (1951) found WILDT'S H^- continuum in the laboratory. G. JÜRGENS (1952) tested HOLTSMARK'S theory for the broadening of the Balmer lines as well as INGLIS and TELLER'S formula for the termination of the series by micro-fields under conditions of $\sim 12{,}700^\circ$K and an electron pressure of $\sim 0{\cdot}15$ atm, corresponding about to a white dwarf atmosphere! It is hoped that similar light sources will allow measurements of transition probabilities for higher stages of ionization. Also the broadening of helium and other lines as well as the inducing of forbidden components by the electric fields of ions and possibly electrons offers a most promising field of research.

One of the most disappointing problems in the theory of stellar spectra used to be the calculation of *damping constants* caused by collisions between the radiating atoms and neutral hydrogen atoms in later type atmospheres.

Recently, however, progress became possible by suitably combining theoretical and experimental methods. Quantum mechanics indicate that the van der Waals energy of interaction ΔE between the radiating atom in a state k and the perturbing neutral hydrogen atom over a distance r can be calculated from the formula

$$\Delta E_k = -\frac{hC_k}{r^6},$$

where the interaction constant C_k is given approximately by

$$C_k = \frac{e^2}{h} \cdot \alpha \overline{R_k^2}.$$

Here e, h have the usual meaning; α is the polarizability of the perturbing atom, and $\overline{R_k^2}$ the matrix element of the (radius)^2 for the considered quantum state. Following BATES and DAMGAARD (1949) the $\overline{R_k^2}$ matrix elements can be calculated with fair accuracy even for complex spectra using the well-known hydrogen-like matrices in connection with the effective principal quantum number n^*. We have

$$\overline{R_k^2} = a_0^2 \cdot \frac{n^{*2}}{2Z^2} \{5n^{*2} + 1 - 3l(l + 1)\},$$

where a_0 means the Bohr radius, $Z = 1, 2, \ldots$ for arc-, spark, . . . spectra, and $l = 0, 1, 2$, for s, p, d . . . states.

The simplifying assumptions made in deriving the formula for ΔE_k are essentially: (*a*) the energy diagram of the perturbing atom should have a large gap between the ground state and the excited levels near the ionization limit; (*b*) the energy differences between the level of the radiating atom and the nearer combining levels should be small compared with the ionization energy of the perturbing atom; and (*c*) the average over all possible relative orientations of the two colliding particles is formed classically. In general, these assumptions are fulfilled reasonably well. WEISSKOPF's quantum theory of collision damping further says that the broadening of a line should be determined by the difference of the interaction constants for its upper (*a*) and lower (*b*) level, *i.e.* $C = C_a - C_b$. It would be rather hopeless to investigate in the laboratory collision broadening of, say, Fe-lines by atomic hydrogen and to measure the corresponding collisional cross sections either for direct astrophysical application or for testing the described theory. However, if we substitute some rare gas, preferably He, for H, we have only to change the well-known α's in the formula for the interaction energies ΔE_k while everything else remains the same. So it appears possible to test the theory by measurements made with rare gases. One must only make sure that the applied pressure is low enough so that one is still dealing with two particle collisions, *i.e.* the ratio of the average distance r_0 between two perturbing particles and the Lorentz collision radius ρ_0 should be $\gg 1$. The table opposite gives a compilation of all the published measurements on line broadening by rare gases; the figures for which the last-mentioned condition is well satisfied are printed in bold type. In general, the difference between measurements and theory keep within quite reasonable limits. More experimental work on line broadening, especially in complex spectra and by helium under not too high pressures (for avoiding multiple collisions) would be highly desirable.

In concluding this brief survey we should apologize for having left aside several aspects of our problem. We have not dealt with variable and peculiar stars and we have not spoken about magnetic fields and their importance for the dynamics of cosmical plasms. These fields, which in course of time will no doubt lead to important new vistas, will, however, be taken care of by more competent authors in this book.

Table 1. Collision broadening by rare gases

1	2	3	4	5	6	7	8	9	10	11
Rare gas $\alpha \cdot 10^{25}$ [cm³]	Broadened line	Term value [eV]		Interaction constants $\times 10^{32}$				r_0/ρ_0	n	Literature
		(*a*) Upper	(*b*) Lower	Observed C_{exp}	Calculated					
					C_a	C_b	$C_a - C_b$			
He 2·06	Na 3^2S–3^2P	3·04	5·14	**0·8**	0·8	0·4	**0·4**	**20**	10	SCHÜTZ, W.: *Z. Phys.*, **45,** 30, 1927.
	K $4^2S_{\frac{1}{2}}$–$5^2P_{\frac{3}{2}}$ $4^2S_{\frac{1}{2}}$–$5^2P_{\frac{1}{2}}$	1·29	4·34	29 13	4·9	0·5	4·4	2	2 2	FÜCHTBAUER, Chr. and H. J. REIMERS: *Z. Phys.*, **97,** 1, 1935
	Rb $5^2S_{\frac{1}{2}}$–$6^2P_{\frac{3}{2}}$ $5^2S_{\frac{1}{2}}$–$6^2P_{\frac{1}{2}}$	1·22 1·23	4·17	25 70	5·9 5·5	0·6 0·6	5·3 4·9	2	1 1	NY TSI-ZÉ and CH'EN SHANG-YI: *Phys. Rev.*, **52,** 1158, 1937.
	Cs 6^2S–7^2P	1·17	3·89	120	6·2	0·7	5·5	2	3	FÜCHTBAUER, Chr. and F. GÖSSLER: *Z. Phys.*, **87,** 1, 1934
	Ag 5^2S–5^2P	3·85	7·58	0·9	0·43	0·18	0·26	2	5	CLAYTON, E. D. and CH'EN SHANG-YI: *Phys. Rev.*, **85,** 68, 1952.
	Hg 6^1S_0—6^3P_1	5·56	10·44	**0·03** **0·08**	0·18	0·10	**0·08**	**10** **8**	3 5	ZEMANSKY, M. W.: *Phys. Rev.*, **36,** 219, 1930 KUNZE, P.: *Ann. Phys.*, (5), **8,** 500, 1931
Ne 3·98	Na 3^2S–3^2P	3·04	5·14	**0·75**	1·50	0·74	**0·76**	**17**	20	SCHÜTZ, W.: *Z. Phys.*, **45,** 30, 1927
	K $4^2S_{\frac{1}{2}}$–$5^2P_{\frac{3}{2}}$ $4^2S_{\frac{1}{2}}$–$5^2P_{\frac{1}{2}}$	1·29	4·34	8·0 5·7	9·6	1·0	8·6	2	— —	FÜCHTBAUER, Chr. and H. J. REIMERS: *Z. Phys.*, **97,** 1, 1935
	Rb $5^2S_{\frac{1}{2}}$–$6^2P_{\frac{3}{2}}$ $5^2S_{\frac{1}{2}}$–$6^2P_{\frac{1}{2}}$	1·22 1·23	4·17	32 8·5	11	1	10	2	1 1	NY TSI-ZÉ and CH'EN SHANG-YI: *Phys. Rev.*, **52,** 1158, 1937
	Hg 6^1S_0–6^3P_1	5·56	10·44	**0·37**	0·34	0·19	**0·15**	**8**	4	KUNZE, P.: *Ann. Phys.* (*Leipzig*) (5), **8,** 500, 1931
A 16·3	Na 3^2S–3^2P	3·04	5·14	6·7 5·9 5·4 **4·5** **2·0** **5·0**	6·2	3·0	**3·2**	2 2 **12** **5** **5**	2 2 1 15 — —	KLEMAN, B. and E. LINDHOLM: *Ark. Mat. Astron. Fys.*, **32**B, No. 10, 1945 MARGENAU, H. and W. W. WATSON: *Phys. Rev.*, **44,** 92, 1933 SCHÜTZ, W.: *Z. Phys.*, **45,** 30, 1927 MINKOWSKI, R.: *Z. Phys.*, **93,** 731, 1935
	K $4^2S_{\frac{1}{2}}$–$4^2P_{\frac{3}{2}}$ $4^2S_{\frac{1}{2}}$–$4^2P_{\frac{1}{2}}$	2·73	4·34	10·0	8·0	4·1	3·9	2	2	HULL, G. F.: *Phys. Rev.*, **50,** 1148, 1936
	Ag $5^2S_{\frac{1}{2}}$–$5^2P_{\frac{3}{2}}$ $5^2S_{\frac{1}{2}}$–$5^2P_{\frac{1}{2}}$	3·80 3·91	7·58	10 4	3·6 3·4	1·4	2·2 2·0	2	1 1	CLAYTON, E. D. and CH'EN SHANG-YI: *Phys. Rev.*, **85,** 68, 1952
	Hg 6^1S_0–6^3P_1	5·56	10·44	**1·1** **1·1** **0·9**	1·40	0·77	**0·63**	**5** **5** **4**	6 5 —	ZEMANSKY, M. W.: *Phys. Rev.*, **36,** 219, 1930 KUNZE, P.: *Ann. Phys.* (*Leipzig*) (5), **8,** 500, 1931 KUHN, H.: *Proc. Roy. Soc.*, **158,** 212, 1937

Explanation of column numbering:
1. Rare gas and its polarizability α.

2–4. Broadened line; upper and lower (ground) term in electron-volts.

5. Interaction constant C_{exp} derived from the *measured* damping constant γ of the line according to LINDHOLM'S formula

$$\gamma = 17{\cdot}0 \, . \, C_{exp}^{2/5} \, v^{3/5} N,$$

where v means the average relative velocity of the colliding particles and N the number of rare gas atoms per cubic centimetre.

6–8. Interaction constants for the upper and lower term as well as their difference entering into WEISSKOPF'S theory.

9. Ratio of the average distance between the perturbing atoms $r_0 = 0{\cdot}62 N^{-1/3}$ and the LORENTZ radius $\rho_0 = \sqrt{12C/v}$.

10. n = number of measurements for which $r_0/\rho_0 > 2$.

11. Literature.

Where $r_0/\rho_0 > 2$ and column 10 ensures some accuracy of the measurements (bold type figures) the agreement between C_{exp} and $C = C_a - C_b$ is about as good as might have been expected.

References

It is not intended to give here extensive references to literature. Instead we mention the following modern treatises:

Astrophysics, edited by J. A. Hynek (McGraw-Hill Book Co., 1951), especially Chapter 5, by B. Strömgren.

Basic Methods in Transfer Problems, by V. Kourganoff (with collaboration of I. W. Busbridge) (Oxford University Press, 1952).

Radiative Transfer, by S. Chandrasekhar (Oxford University Press, 1950).

A revised edition of *Physik der Sternatmosphären*, by A. Unsöld, with extensive literature references is in course of preparation (to be published in 1955).

On the Formation of Condensations in a Gaseous Nebula

H. Zanstra

Sterrekundig Instituut, Universiteit van Amsterdam, Netherlands

Summary

In order to get an insight into the process of formation of condensations in an ionized gas of high kinetic temperature, two main principles are applied:

(1) that the pressure in the less dense medium 1 should be about the same as in the denser medium 2, which means that, in the denser medium, the kinetic or electron temperature is correspondingly lowered, and

(2) that this lowering of electron temperature is brought about by electron impacts exciting low level lines. (If the optical depth of the condensations is large, so that the ionizing radiation is nearly completely absorbed, this effect may be enhanced.)

The approximate balancing of pressure is verified in the case of prominences considered as condensations in the solar corona (Section 1).

In order to account for condensations as observed by Baade in the planetary nebula NGC 7293 in Aquarius (Section 2) a very simplified model of a nebula is worked out, consisting of one substance only, occurring as ions A^+ (for instance N^{++}) and "atoms" A (for instance N^+), the latter having a low metastable level, while the ionization of A into A^+ is fairly complete. Introducing the liberation temperature $\bar{T}$ corresponding to the average energy of the photo-electrons when freed from A, it is shown that separation into two phases 1 and 2 will occur *above* a certain critical liberation temperature $\bar{T}_{cr}$ of about 50,000°, the curves of constant $\bar{T}$ having a shape resembling those of the van der Waals theory. For $\bar{T} \simeq 110,000°$, or stellar temperature 170,000°, the volumes per unit mass in phases 1 and 2 have the ratio $v_1/v_2 = 15$, while the electron temperatures are $T_{\varepsilon 1} = 130,000°$ and $T_{\varepsilon 2} = 9000°$, the latter being about the value derived in regions of planetary nebulae where forbidden lines are excited. Similar results can be expected for a nebula consisting of hydrogen with the substances A and A^+ intermixed.

It is suggested that the cometary shape of many of the condensations observed by Baade in NGC 7293 might be due to aggregates of matter already present in interstellar space, serving as nuclei for condensation, the tails being formed by dragging along of matter by the outward moving gases of the nebula.

Introduction: The Mechanisms of Ionization and Recombination and of Electron Collision for a Gaseous Nebula

The spectrum of many nebulae shows a number of bright lines, which have practically all been identified as being produced by atoms and their ions, so that they can be described as luminous clouds of gas. Thus a planetary nebula consists of a rather regular mass of gas (disk, ring, *etc.*) at the centre of which a hot star is situated. Of this, the well-known ring nebula in Lyra is a typical example; also the helical nebula in Aquarius, discussed in Section 3 of the present paper, belongs to this class. According to present theories, the far ultraviolet radiation of the hot central star is absorbed by the gases in the surrounding envelope and ionizes the hydrogen atoms. These absorbing atoms are practically all in the ground state, since the nebula is much larger than the star, so that the radiation incident on it is very diluted. By this process the ultraviolet stellar radiation beyond ν_0, the ionizing frequency of hydrogen, may be absorbed, and, for every absorbed quantum, one ionization occurs. Subsequently, the photo-electron, thus freed, recombines on the various levels of the hydrogen atom and, in returning to the ground state, emits the various line spectra of hydrogen, among others the lines of the Balmer series, while captures on the second level produce the continuous spectrum at the head of this series. It is clear that, from the observed intensity of the hydrogen spectrum, one can estimate the number of ultraviolet quanta beyond ν_0 emitted by the star, and thus the stellar temperature, for which values of 20,000° and much higher are found (the star is approximated by a black body and the absorption of ionizing quanta assumed to be complete, though this may not be quite correct). This *mechanism* 1 of luminosity, *ionization* and *recombination*, applies to the spectra H, He I, He II, and others.

Generally, the strongest lines in the nebular spectrum are however low level forbidden lines, like the two green "nebulium lines" of the O III spectrum, or the two red lines of the N II spectrum. These cannot be sufficiently produced by mechanism 1, but are believed to be excited by electron impact of the ion (O^{++} or N^{+} for the two examples) with the photo-electrons freed by mechanism 1. In other words, before recombining, the photo-electron may lose the major portion of their energy which passes into forbidden lines. This is stated as the *principle of cooling* in Section 2. This *mechanism* 2, excitation of low level lines by *impact with photo-electrons*, is therefore a by-product of mechanism 1. It is also clear that the observed energy emitted in total by all forbidden lines is indirectly determined by the star's ultraviolet energy and gives another determination of stellar temperature. The fact that the stellar temperatures obtained by mechanisms 1 and 2 agree approximately furnished a check of the theory.

Historically, the discussion of observations on diffuse nebulae (like the great nebula in Orion) and planetary nebulae by HUBBLE (1922) has been fundamental. His empirical rules induced the writer (ZANSTRA, 1926, 1927) to verify mechanism 1 for hydrogen. This mechanism was also conceived independently by MENZEL (1926). After his very important identification of "nebulium lines" as forbidden lines of ions, the mechanism 2 was suggested by BOWEN (1927, 1928) and checked quantitatively in the above manner by the writer (ZANSTRA, 1928, 1931a, 1931b) from his observations of Victoria spectra of planetary nebulae and by BERMAN (1930) from similar observations of Lick spectrograms taken by him.

1. EQUALITY OF PRESSURE AT DIFFERENT ELECTRON TEMPERATURES: PROMINENCES IN THE CORONA

In 1946 the writer put forward and verified the idea that the pressure in a prominence of the quiescent type should be approximately the same as that of the surrounding solar corona (see ZANSTRA, 1947). Moving picture films of such prominences as were made by MACMATH, LYOT, and the Harvard observers at Climax clearly show a predominantly downward motion of luminous material, which somehow must originate by condensation in the surrounding coronal material. This process of condensation is shown in a convincing manner in the coronal streamers (PETTIT) which start from a point in the corona above the solar surface, and, equally convincing, in two-sided streamers, where the luminous material moves from a point above the solar surface in two opposite directions along an uninterrupted arc. This being the case, the pressure in the upper more extended portion of the prominence, where the spectrum for determining concentration and temperature is generally taken, should be balanced by the surrounding coronal material.

If, in general, the less dense medium is indicated by the suffix 1, and the condensation by 2, we have the requirement

$$p_1 = p_2, \quad n_1 k T_{\varepsilon 1} = n_2 k T_{\varepsilon 2}, \quad \frac{T_{\varepsilon 1}}{T_{\varepsilon 2}} = \frac{n_2}{n_1}, \qquad \ldots .(1)$$

where T_ε represents the kinetic temperature or electron temperature and n the number of particles per cm^3, which may be taken equal to $2n_\varepsilon$, the concentration of electrons or protons. At an altitude of 50,000 km, the approximate height of a prominence, the observed values for the corona are about $n_{\varepsilon 1} = 2 \times 10^8$, $T_{\varepsilon 1} = 7 \times 10^5$ and, for the prominence (REDMAN and ZANSTRA, 1952) $n_{\varepsilon 2} = 2 \times 10^{10}$, $T_{\varepsilon 2} = 5 \times 10^3$, so that $n_2/n_1 = 100$, $T_{\varepsilon 1}/T_{\varepsilon 2} = 140$, which shows that the requirement of equal pressure is approximately satisfied.

In the next sections attention will be confined to condensations in planetary nebulae, where the same principle (1) will be applied.

2. Condensations in the Planetary Nebula NGC 7293 (Baade). Cooling by Excitation of Low Level Lines

Dr. BAADE has observed a great number of very remarkable bright condensations in the planetary nebula NGC 7293, the helical nebula in Aquarius. They were mentioned by Dr. MENZEL in a colloquium at the Zurich meeting of the International Astronomical Union in 1947 (MENZEL, 1950).

The temperature T_* of the central star, the distance D and radius r_n of this nebula are approximately (ZANSTRA, 1932)

$$T_* > 85{,}000°, \quad D = 350 \text{ parsecs}, \quad r_n = 105{,}000 \text{ astr. units} \qquad \ldots .(2)$$

The stellar temperature is definitely a lower limit and may be a few times 100,000°.

The photographs were taken with the 100-in. Mount Wilson reflector in the red region of $H\alpha$ and [N II] $\lambda\lambda$6548, 6584 with exposures of $\frac{3}{4}$ and 4 hours. We are much indebted to Dr. BAADE for having placed prints at our disposal and for his comments. Fig. 1 shows an enlarged reproduction of such a red photograph, which was taken in 1952 at the 200 in. Mount Palomar reflector by Dr. R. MINKOWSKI. Like Dr. BAADE's photographs it shows the ring, in reality three turns of a helix to be largely made up of condensations. In particular towards the inside of the ring there is a radial structure with respect to the central star, and at the inside blurred cometary shapes consisting of a head and a tail directed away from the central star are recognizable. In the nearly dark space inside the ring there are a fair number of small condensations with a typical cometary shape: a round head with a fairly long tail directed radially away from the central star. They do not occur in the neighbourhood of the central star. These "comets" in the dark space form the most startling feature of the photograph. They contribute however only a very minute fraction of the total brightness of the nebula on the photograph, which is practically given by the outer ring. Table 1 shows the emission lines in the photographic region observed by MINKOWSKI (1942) for the brightest part of the ring north-east of the central star with intensity, excitation potential, ionization potential and the ionization potential of the next higher ion. In Dr. BAADE's opinion it seems likely that in the red region

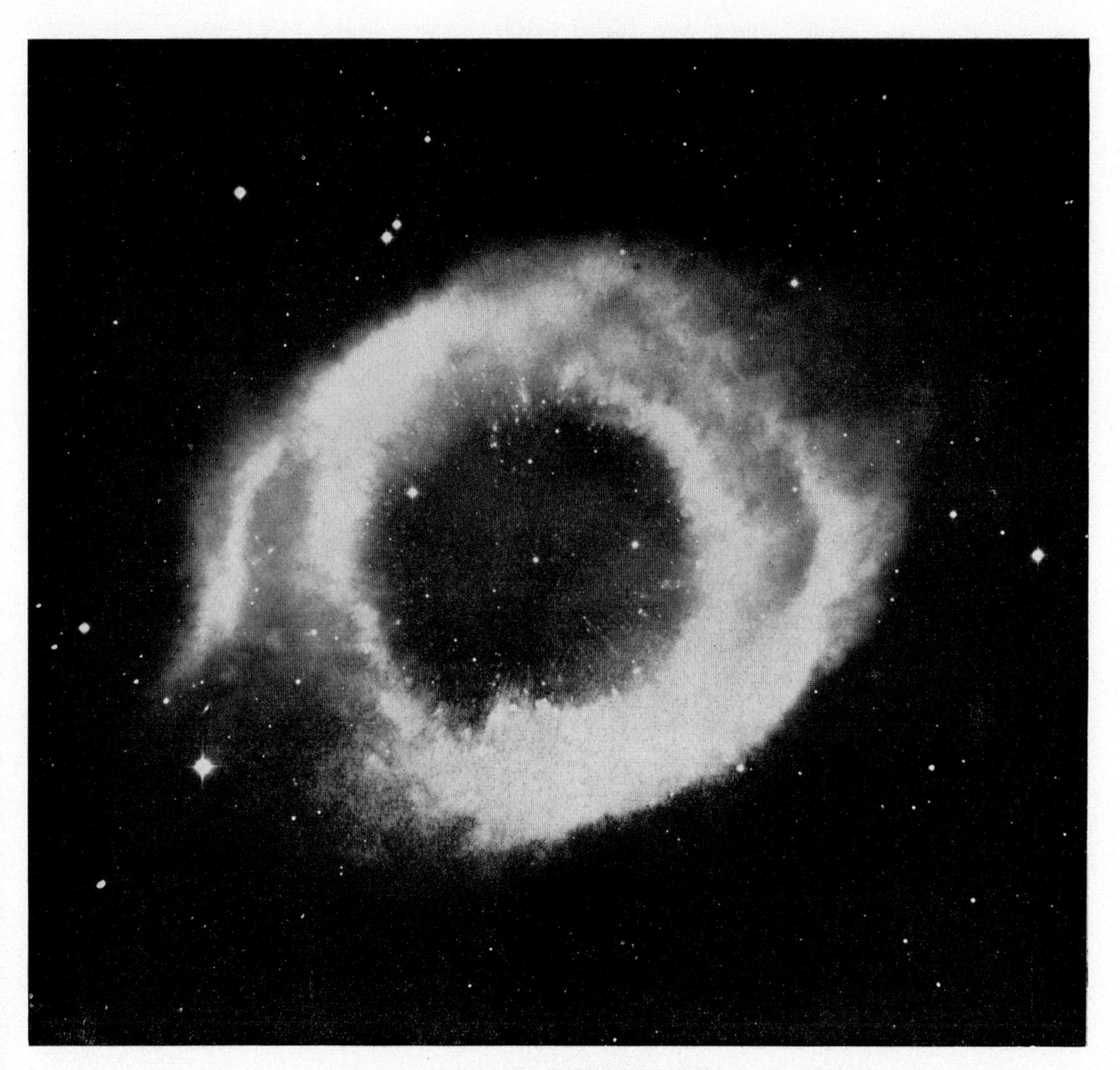

Fig. 1. The planetary nebula NGC 7293
(Red photograph by R. Minkowski, Mount Wilson and Palomar Observatories.)

Table 1. NGC 7293

Line	H_β	H_γ	H_δ	[O III]	[O III]	[O II]	[Ne III] + H_ε	[Ne III]	Red H_α	Red [N II]	Red [N II]
λ.	4861	4340	4101	5007	4959	3727	3790	3869	6563	6548	6548
Int.	10	4·1	2·1	60	21	31	3·1	8·4	—	—	—
Ex. P.	12·07	12·76	13·05	2·51	2·51	3·33	3·20	3·20	10·20	1·90	1·90
I.P.	13·60	—	—	54·93	—	35·15	64	—	13·60	29·60	—
I.P.$^+$	—	—	—	77·39	—	54·93	97·2	—	—	47·43	—

in which his photographs were taken most of the light would originate from the two [N II] lines also given in the table. Since the intensity of Hα is generally about five times Hβ, or fifty units, this statement indicates that the light of these [N II] lines together might well be in excess of the total light of the forbidden lines in the photographic region.

Altogether this indicates that most of the light in the ring or the condensations (medium 2) originates from the forbidden lines of various kinds. These lines are all low level lines excited by the mechanism of collision by free electrons, and this means that the electron temperature in the condensations (medium 2) will be considerably lowered with respect to medium 1:

$$T_{\varepsilon 2} < T_{\varepsilon 1} \qquad \ldots (3)$$

and then equality of pressure (1) leads to $n_2 > n_1$, which would explain qualitatively the formation of condensations *viz.* regions of high density.

In the next section this principle of cooling by excitation of low level forbidden lines will be formulated quantitatively for a simple model and indeed will be shown to lead to the formation of two phases.

3. Simplified Model of a Nebula with Condensations

The ideas developed in the previous section will now be applied to a very simplified model of a planetary nebula. The treatment is quantitative, but the model is idealized and the assumptions approximate to such an extent that a direct quantitative comparison with observations is hardly justified, the main purpose being to show the possibility of the formation of two phases in the nebular material, provided the temperature of the central star is high enough.

The nebula is considered to consist of one substance only, say nitrogen, occurring only in two stages of ionization: the lower ions or "atoms" A and the next higher ions A^+. The higher ions $A^+ = N^{++}$ have no low levels, but $A = N^+$ has a low metastable level B excitable by electron impact, giving [N II] lines. (In the case of oxygen, $A^+ = O^{+++}$, $A = O^{++}$, emitting [O III] lines.)

The radiation incident on the element of the nebula considered has an intensity $I_\nu d\nu$ ergs per cm² per sec. within the frequency interval ν to $\nu + d\nu$ and is but incompletely absorbed by the nebular material under consideration, if the element is not too extended.

Let n_A be the number of "atoms" per cubic centimetre, and α_ν the atomic absorption coefficient for the frequency ν. The number of ionizations per cubic centimetre per second is then

$$N = n_A \int_{\nu_0}^{\nu_0'} \alpha_\nu \frac{I_\nu}{h\nu} d\nu, \qquad \ldots (4)$$

where ν_0 is the ionizing frequency of A and ν_0' is either infinity or the ionizing frequency of the next state A^+.*

According to HEBB and MENZEL (1940) the number of exciting collisions per cubic centimetre per second which raise the atoms A from the ground level to the metastable level B are

$$\mathscr{F}_{AB} = \frac{\Omega_{AB}}{g_A} \mathscr{F}_0 n_A n_\varepsilon \frac{1}{T_\varepsilon^{\frac{1}{2}}} e^{-\frac{\chi_{AB}}{kT_\varepsilon}}, \qquad \ldots.(5)$$

where $\mathscr{F}_0 = 8{\cdot}54 \times 10^{-6}$ a universal constant, Ω_{AB} a factor dependent on the kind of atom and excited level of excitation energy χ_{AB}, g_A the weight factor for the ground level, n_ε the number of electrons per cubic centimetre and T_ε the electron temperature.

The average energy of the electrons which are present in 1 cm³ is then $\frac{3}{2}kT_\varepsilon$, which might also have served as a definition of the electron temperature. Let, in a similar way, $\frac{3}{2}k\bar{T}$ represent the average energy of the photo-electrons when freed from the ground state of A, and $\frac{3}{2}k\bar{T}_\varepsilon$ the average energy of the free electrons when recombining on all levels of A^+, then $\bar{T}$ will be called the *liberation temperature* and $\bar{T}_\varepsilon$ the *capture temperature*. (We may expect $\bar{T}$ to be of the same order as the temperature T_* of the central star, and $\bar{T}_\varepsilon$ of the same order as T_ε, see formulae (15) and (16).) The average energy lost per electron between ionization and recombination is then $\frac{3}{2}k(\bar{T} - \bar{T}_\varepsilon)$.

The energy lost per cubic centimetre per second, disregarding loss by electron switches, all goes into the excitations of the forbidden line, so that

$$N\tfrac{3}{2}k(\bar{T} - \bar{T}_\varepsilon) = \mathscr{F}_{AB}\chi_{AB}, \qquad \ldots.(6)$$

provided de-exciting collisions are negligible.

After substitution of (4) and (5) in (6), the electron concentration is expressed by

$$n_\varepsilon = \frac{\frac{3}{2}k \int_{\nu_0}^{\nu_0'} \alpha_\nu \frac{I_\nu}{h\nu} d\nu}{\frac{\Omega_{AB}}{g_A} \mathscr{F}_0 \chi_{AB}} \,.\, T_\varepsilon^{\frac{1}{2}}(\bar{T} - \bar{T}_\varepsilon) e^{+\frac{\chi_{AB}}{kT_\varepsilon}}. \qquad \ldots.(7)$$

Let the "atom" A be an ion of charge $i - 1$, so that A^+ has charge i. Then $n_\varepsilon = in_{A+} + (1 - i)n_A$. It will be assumed that, for all concentrations considered, the ionization is nearly complete, so that approximately

$$n_\varepsilon = in_{A+} \gg n_A, \quad n_\varepsilon + n_{A+} = \left(1 + \frac{1}{i}\right) n_\varepsilon. \qquad \ldots.(8)$$

Let m_A be the mass of the neutral atom. The concentration ρ in grams per cubic centimetre, or the volume occupied by 1 g is then, using (7) and (8), given by

$$\frac{1}{v} = \rho = m_A n_{A+} = \frac{m_A}{i} n_\varepsilon = \frac{m_A}{i} K \,.\, \left(\frac{k}{\chi_{AB}}\right)^{\frac{3}{2}} T_\varepsilon^{\frac{1}{2}}(\bar{T} - \bar{T}_\varepsilon) e^{+\frac{\chi_{AB}}{kT_\varepsilon}}, \qquad \ldots.(9)$$

where

$$K = \frac{\frac{3}{2}\chi_{AB}^{\frac{1}{2}} \int_{\nu_0}^{\nu_0'} \alpha_\nu \frac{I_\nu}{h\nu} d\nu}{k^{\frac{1}{2}} \frac{\Omega_{AB}}{g_A} \mathscr{F}_0}. \qquad \ldots.(10)$$

---

* For a planetary nebula, the incident radiation is short wavelength ultraviolet, supplied by the central star, but in general it may be ultraviolet radiation of any origin. For incident corpuscular radiation the results of the following derivation are equally valid, provided one replaces the integral in equation (4) by the number of ionizations it produces per cubic centimetre per second on one atom A. Interstellar gas colliding with an expanding nebula may be considered as such corpuscular radiation, incident with the relative velocity (OORT, 1946).

The pressure in dynes per square centimetre $p = (n_\varepsilon + n_{A+})kT_\varepsilon$ similarly becomes

$$p = \left(1 + \frac{1}{i}\right) n_\varepsilon k T_\varepsilon = \left(1 + \frac{1}{i}\right) \chi_{AB} K \left(\frac{k}{\chi_{AB}}\right)^{\frac{5}{2}} T_\varepsilon^{\frac{3}{2}}(\bar{T} - \bar{T}_\varepsilon) e^{+\frac{\chi_{AB}}{kT_\varepsilon}}. \qquad \ldots .(11)$$

Introducing the unit of temperature u_T degrees Kelvin, the unit of mass u_m grams and the unit of pressure u_p dynes per cm² defined by

$$u_T = \frac{\chi_{AB}}{k}, \quad u_m = \frac{1}{i} m_A K, \quad u_p = \left(1 + \frac{1}{i}\right) \chi_{AB} K, \qquad \ldots .(12)$$

with K given by equation (10), the equations (9) and (11) assume the simple form

$$\frac{1}{v} = \rho = T_\varepsilon^{\frac{1}{2}}(\bar{T} - \bar{T}_\varepsilon) e^{+\frac{1}{T_\varepsilon}}, \qquad \ldots .(13)$$

$$p = \frac{T_\varepsilon}{v} = T_\varepsilon^{\frac{3}{2}}(\bar{T} - \bar{T}_\varepsilon) e^{+\frac{1}{T_\varepsilon}}, \qquad \ldots .(14)$$

where, according to the definitions, ρ represents the number of mass units u_m per cm³ and v the volume in cm³ occupied by the mass u_m.

For a given incident diluted stellar radiation, the average energy of the photo-electrons when freed is fixed, so that also the liberation temperature $\bar{T} =$ *constant*. Then, since $\bar{T}_\varepsilon$ must be a function of T_ε (see below), equations (13) and (14), by means of this variable parameter T_ε, express p as a function of v, which is represented in the p–v diagram as a *curve of constant liberation temperature*, but of course variable electron temperature. From considerations based on KRAMERS' absorption law EDDINGTON (1926) has shown that, for stellar black body temperature T_*,

$$T_* = \tfrac{3}{2}\bar{T}, \text{ if } k\bar{T} \ll \text{ionization energy}, \qquad \ldots .(15)$$

and that in general T_* is a function of $\bar{T}$ such that $T_* \geqslant \frac{3}{2}\bar{T}$.

Considering similarly KRAMERS' law for capture of electrons, one may show that

$$T_\varepsilon = \tfrac{3}{2}\bar{T}_\varepsilon, \text{ if } k\bar{T}_\varepsilon \ll \text{ionization energy}, \qquad \ldots .(16)$$

and that in general T_ε is a function of $\bar{T}_\varepsilon$ such that $T_\varepsilon \geqslant \frac{3}{2}\bar{T}_\varepsilon$. Now, in virtue of the cooling effect by the excitation of the forbidden line, $\bar{T}_\varepsilon$ and consequently T_ε decreases as the density increases or the volume decreases, and eliminating this variable parameter from equations (13) and (14) the equation of state $p = p(v,\bar{T})$ is obtained. Using the approximation (16) $T_\varepsilon = \frac{3}{2}\bar{T}_\varepsilon$, the equation of state becomes

$$p = (\bar{T} - \tfrac{2}{3}pv) p^{\frac{3}{2}} v^{\frac{3}{2}} e^{+\frac{1}{pv}}. \qquad \ldots .(17)$$

For actual computation p and v are best calculated from equations (13) and (14) for a set of T_ε values at a given $\bar{T} = \frac{2}{3}T_*$ (16).

In Figs. 2 to 4 the curves $\bar{T} =$ constant thus obtained for various values of $\bar{T}$ are plotted. The value of the electron temperature T_ε is also indicated for each point.

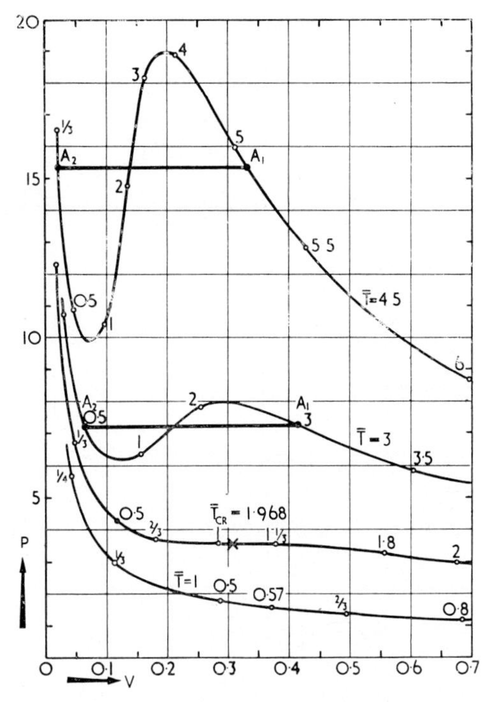

Fig. 2. p-v diagrams for liberation temperatures $\bar{T} = 1$, 1·986, 3, and 4·5. The volume per unit mass is v. The units of temperature, mass, and pressure (12) are used, which vary from curve to curve. The electron temperature T_ε varies along each curve and is indicated at various points by the corresponding figure. *Above* the critical liberation temperature 1·968 separation in two phases 1 and 2, represented by the points A_1 and A_2 occurs

At higher $\bar{T}$ (or stellar temperature $T_* > \frac{3}{2}\bar{T}$) each curve shows a maximum and a minimum, which coincide at a certain critical liberation temperature

$$\bar{T}_{cr} = 1{\cdot}968,\ p_{cr} = 3{\cdot}53,\ v_{cr} = 0{\cdot}310,\ \text{at } T_\varepsilon = 1{\cdot}086, \qquad \ldots.(18)$$

as may be obtained analytically from equations (13) and (14) with (16). For lower values of $\bar{T}$ the maximum and minimum have disappeared. For a liberation temperature above $\bar{T}_{cr}$ the part between the two maxima is unstable, since dv/dp is positive: with increasing volume the pressure increases also and the expansion continues, similarly a decrease of volume results in a collapse by decreasing pressure. Stable equilibrium can only occur when the pressure is the same throughout the system, which means that only two phases, a dilute phase 1 and a denser phase 2, co-exist (see Section 1). The volumes per unit mass v_1 and v_2 of these two phases are found as the extreme points of intersection A_1 and A_2 of a horizontal line with the curve of constant $\bar{T}$ such that the areas cut off above and below this line are equal.* The

* One cannot apply thermodynamical considerations. But, assuming continuity of the various states according to the continuous curve of constant $\bar{T}$ when passing through the boundary region between the two phases, passage of the mass dm from phase 2 to phase 1 through the boundary layer results in a mechanical energy freed $dm\int_2^1 p\,dv$. If in a hypothetical experiment this is brought about by having the gas in a cylinder transparent to radiation and moving the piston outward, the work done on the piston is $dmp(v_1 - v_2)$, where p is the equilibrium pressure. Since there is equilibrium, these two quantities should be equal, which results in the above equality of areas. If in this hypothetical experiment p is larger than its equilibrium value, which means that the horizontal line is drawn higher, one sees from the figure that the mechanical energy freed upon expansion is smaller than the work on the piston required to keep up this larger pressure, so that the pressure must go down upon expansion and thus come closer to the equilibrium value. Similarly one can show that also for too low initial p or compression one goes towards the equilibrium. Thus the equilibrium is always stable. However, the above assumption regarding the continuity of states requires further justification, since the variable pressure in the boundary layer could only be kept up by some field of force.

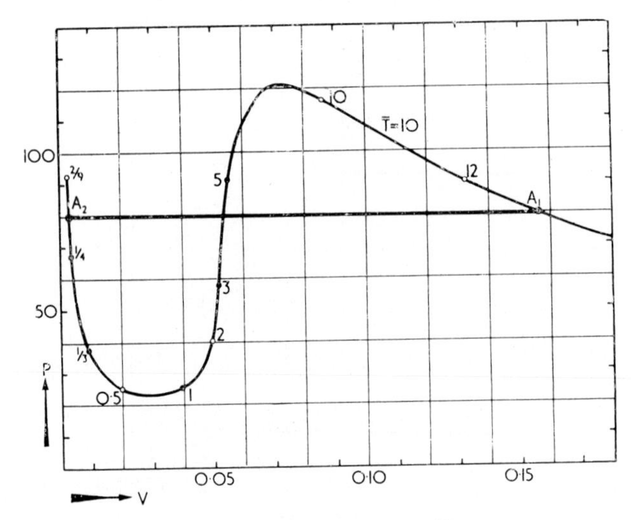

Fig. 3. The same as Fig. 2, but for $\overline{T} = 10$ units

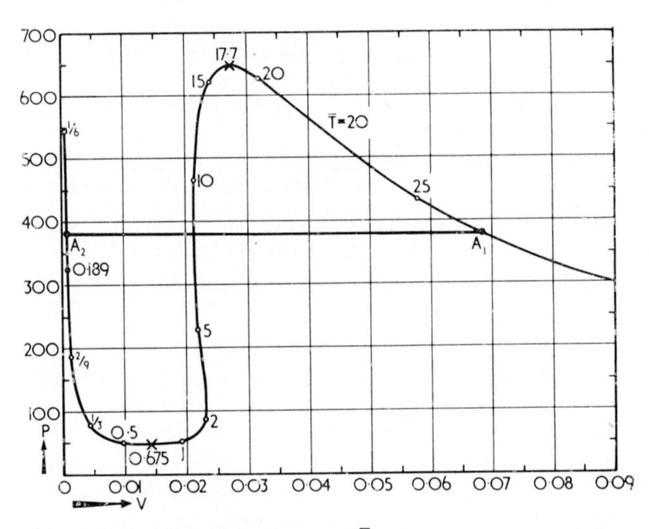

Fig. 4. The same as Fig. 2, but for $\overline{T} = 20$ units. Maximum and minimum are marked by crosses

whole situation is analogous to the isotherms of the van der Waals Theory, except that the separation into phases occurs *above* a critical temperature of liberation and that there is no thermodynamical equilibrium; in fact throughout there is dissipation of energy and, though the pressures in the two phases are equal, they are at different kinetic temperatures.

Let v be the volume of one unit mass of mixture consisting of x units of phase 1 and $1 - x$ units of phase 2. Then $v = xv_1 + (1 - x)v_2$, so that one gets the well-known expression

$$\frac{1-x}{x} = \frac{v_1 - v}{v - v_2} = \frac{A_1A}{A_2A},$$

where A represents the point on A_1A_2 corresponding to v.

Table 2 contains, for a given $\bar{T}$, the electron temperatures and ratio of the volumes per unit mass for the two phases.

Table 2. Electron temperatures and volume ratios for the two phases

The unit of temperature χ_{AB}/k is 21,800° for [N II], λ6572 and 28,800° for [O III], λ4991 (see Table 1).

$\bar{T}$	3	4·5	10	20
$T_{\varepsilon 1}$	3·0	5·1	12·5	26
$T_{\varepsilon 2}$	0·46	0·35	0·23	0·18
v_1/v_2	6·5	15	50	140

The temperature unit is about 25,000°, being the average for the green [O III] and the red [N II] lines. Above the critical liberation temperature (18) of about 50,000° separation in two phases may occur. While $\bar{T}$ rises from 75,000° to 500,000°, the electron temperature $T_{\varepsilon 2}$ of the condensations goes *down* from 11,500° to 4500°. This is not far from the electron temperatures around 10,000° derived by MENZEL, ALLER, and HEBB (1941) from the relative intensities of [O III] lines in planetary nebulae. These authors in the case of the nebula in Draco NGC 6543 find a value even as low as 6000°. For the highest values $\bar{T}$ of 10 or 20, that is $k\bar{T}$ (in ordinary units) 10 or 20 times the excitation energy χ_{AB}, the application of equation (5) may be somewhat doubtful. It is better founded on $\bar{T} = 4{\cdot}5 = 110{,}000°$, which corresponds to a stellar temperature of about 170,000°, a reasonable value for the central star of NGC 7293 (Section 2), though it might be higher. The theoretical electron temperatures according to the table are then about 130,000° for the dilute medium and 0·35 units or about 9000° for the condensations, which would be fifteen times as dense as the medium. Though not much stress should be laid on the actual values, the theory as developed in the foregoing seems to account for the *formation of dense and very luminous condensations in NGC* 7293, provided that the concentration of matter is such that v in Fig. 2 is between A_1 and A_2.

For an actual nebula, the assumption of only one substance A, A^+ is probably too idealized, since it also contains much *hydrogen*. Assuming hydrogen H, H^+, mixed with the substance A, A^+, equation (4) is replaced by

$$N = n_A \int_{\nu_0}^{\nu_0'} + n_H \int_{\nu_0(H)}^{\nu_0'(H)}, \qquad \ldots (4a)$$

the second integral referring to hydrogen. In equation (7), $\int^{\nu_0}_{\nu_0}$ is then to be replaced by $\int_{\nu_0}^{\nu_0'} + \frac{n_H}{n_A}\int_{\nu_0(H)}^{\nu_0'(H)}$. From considerations based on the Kramers theory, or, less generally, on the ionization equation for diluted black-body radiation, one may then show that, for not too high T_ε, the ratio of the two terms is independent of T_ε, and this again leads to equations (13) and (14), however with definitions of the units u_m and u_p which differ from equation (12), provided that the ionization, both for hydrogen and the substance, is nearly complete. *The curves of constant $\bar{T}$ given in Figs. 2 to 4 then remain valid.*

4. Influence of Absorption of Ionizing Radiation. Remarks on the Tail Formation for the Cometary Condensations in NGC 7293. Possible Future Applications

According to the mechanism proposed in Section 3, separation in two phases can only occur if the smoothed out density ρ is such that v, or $1/\rho$, is between the points A_1 and A_2 of Figs. 2 to 4, say the point halfway. But even then a condensation, $\rho_2 = 1/v_2$, can only be formed on a nucleus, the dimensions of which are large compared with the mean free path and also large enough for thermal conductivity to be inappreciable. One must therefore assume that, when the nebula first originates, there are such accidental regions of increased density of sufficient dimensions. Once such a nucleus is present, the condensation will grow in size until it gets so large that all ionizing radiation forming photo-electrons is practically exhausted by absorption. If the photo-electrons originate from hydrogen (with which some material A, A^+ is intermixed) the condensation consists of nearly completely ionized hydrogen (H^+ region) passing rather abruptly into an inner core where the hydrogen is practically neutral (neutral H region), and of very low electron temperature.* Assuming equality of pressure of the condensation (phase 2) and the core, the latter (region 3) would have a high density or small volume per unit mass, since $v_1 : v_2 : v_3 \simeq T_{\varepsilon 1} : T_{\varepsilon 2} : T_{\varepsilon 3}$. In any case the formation of this neutral H core would mean that the condensation might progress a good deal further.

In this way more or less globular condensations might be formed, so that this would account for the *heads* of the cometary structures observed by Baade in NGC 7293 (see Section 2). Near the inside of the ring his photograph shows these heads to have a diameter of about 1000 astronomical units, farther inward they are smaller. But the condensation mechanism in itself is not capable of explaining the *tails*, which have a length about 3000 astronomical units, nor can it account for the radial structure of the outer ring. Now Biermann (1951) has recently explained long straight tails of comets by the action of solar corpuscular radiation (known by magnetic storms) which imparts momentum to the gases in the heads. It seems that a similar explanation might apply to the tails of the cometary structures in NGC 7293, assuming an *outward movement of the nebular gases*, which consists of an expansion with a velocity increasing with the distance from the central star. Such an expansion for planetary nebulae was put forward by the writer (Zanstra, 1931c, 1932) on the basis of broadening and doubling of emission lines observed by Campbell and Moore and theoretical considerations, and it is now generally recognized to occur for many, if not all, planetary nebulae. If one assumes aggregates of matter to be present in interstellar space before the planetary nebula has developed to its present state, such aggregates might serve as nuclei for condensation as considered above, when the nebular gases reach them, and at the same time matter dragged along from them by these gases might form a streak on which the tail might further condense. Head and tail would further develop by condensation from the moving medium, carrying with it its momentum, so that finally the whole cometary structure would share the outward motion of the medium. The condensations should thus move radially outward with

* These H^+ and H regions are well-known from Strömgren's treatment (1939) of interstellar hydrogen. A similar development for planetary nebulae had been given earlier by the writer (1932, pp. 142, 143). The ionizing radiation incident on the condensation originates primarily from the central star. But it should not be forgotten that in our case the nebula (medium 1) surrounds the condensation and that recombinations in this nebula on the ground level of the hydrogen atoms produce an emission of L_c, the continuous spectrum beyond the head of the Lyman series, which amounts to about 40 per cent of the radiation absorbed by them. This stray L_c radiation, incident in all directions on the condensation, is added to the direct stellar radiation.

a considerable velocity, perhaps of the order 50 km per sec., so that the suggestion might be submitted to an observational test. For this explanation the original aggregates would probably have to be small clouds of gas, or else consist of solid particles like ice crystals which easily can be converted into a gas, either by the stellar radiation or by the impact of the oncoming gases. For actually working out the problem, use could be made of the treatment by Miss KLUYVER (1951) of the collision of an expanding shell of gas with an interstellar cloud.

At present the two main mechanisms for condensation of matter into a denser cloud, leading eventually perhaps to the formation of stars and planets, are *gravity* of the cloud and *radiation pressure on the solid particles* in it by stellar radiation from outside stars, the latter according to WHIPPLE or SPITZER. The present treatment adds to this another mechanism: *formation of two phases of different density and electron temperature* in an originally homogeneous medium exposed to radiation, the gas pressure being operative in producing the equilibrium.

Besides the application to planetary nebulae given above, one might think of possible applications to clouds formed in interstellar matter, where very low levels can be excited by electron collisions, as considered by SPITZER and SAVEDOFF (1950). SPITZER (1951) has shown that the gas pressure of such clouds may be approximately the same as that of the surrounding interstellar medium of much higher kinetic temperature and much lower density. For this reason an interpretation in terms of two phases would be rather tempting, but the writer, who made some preliminary attempts, is rather doubtful whether this would be successful.

Future applications to prominences might also be considered (see Section 1) but it seems probable that this problem is more involved. It is even conceivable that in this case the system is not exposed from outside to strongly ionizing radiation I_ν, but that material of high kinetic temperature is continually supplied to the corona and removed as the denser and colder phase in the prominences, so that there would be a stationary state of moving matter rather than an equilibrium.

The writer is greatly indebted to Dr. W. BAADE for having placed information regarding NGC 7293 at his disposal and for criticizing his views at an earlier stage, and who, in particular, stressed the idea that the condensations occurred in the nebular material itself rather than being there ready made before the nebula was formed.

REFERENCES

BERMAN, L. . . . . 1930 *Lick Obs. Bull.*, No. 430.

BIERMANN, L. . . . . 1951 *Z. Astrophys.*, **29**, 274.

BOWEN, I. S. . . . . 1927 *Nature (London)*, **120**, 473.
1928 *Ap. J.*, **67**, 1.

EDDINGTON, A. S. . . . . 1926 *The International Constitution of the Stars*, p. 377 (Cambridge).

HEBB, M. B. and MENZEL, D. H. . . . . . 1940 *Ap. J.*, **92**, 408.

HUBBLE, E. . . . . 1922 *Ap. J.*, **56**, 162, 400.

KLUYVER, HELEN A. . . . . 1951 Chapter 12 of *Problems of Cosmical Aerodynamics* (Proceedings of the symposium on the motion of gaseous masses of cosmical dimensions, held at Paris on 16th to 19th August, 1949) (Central Air Documents Office, Dayton 2, Ohio).

MENZEL, D. H. . . . . 1926 *Publ. Astron. Soc. Pacific*, **38**, 259.
1950 *Trans. Internat. Astron. Union*, **7**, 468.

MENZEL, D. H., ALLER, L. H. and HEBB, M. H. . 1941 *Ap. J.*, **93**, 230.

MINKOWSKI, R.	1942	*Ap. J.*, **95,** 243.
OORT, J. H.	1946	*M.N.*, **106,** 159.
REDMAN, R. O. and ZANSTRA, H.	1952	Circular No. 6 of the Astron. Inst. of the University of Amsterdam, Table 3 (from *Proc. K. Ned. Akad. Wet.*, Ser. B, **55,** 598).
SPITZER, L.	1951	Chapter 3 of *Problems of Cosmical Aerodynamics* (see Reference KLUYVER).
SPITZER, L. and SAVEDOFF, M. P.	1950	*Ap. J.*, **111,** 593.
STRÖMGREN, B.	1939	*Ap. J.*, **86,** 526.
ZANSTRA, H.	1926	*Phys. Rev.*, **27,** 644.
	1927	*Ap. J.*, **65,** 50.
	1928	*Nature (London)*, **121,** 790.
	1931a	*Publ. Dominion Astrophys. Obs., Victoria* **4,** 209.
	1931b	*Z. Astrophys.*, **2,** 1.
	1931c	*Z. Astrophys.*, **2,** 329.
	1932	*M.N.*, **93,** 131, Table I.
	1947	*Observatory*, **67,** 10.

The Calculation of Atomic Transition Probabilities

R. H. GARSTANG

University of London Observatory, London

SUMMARY

Recent calculations on the strengths of spectral lines are reviewed. Transitions permitted in Russell-Saunders coupling are discussed with particular reference to absolute strengths. Mention is made of studies of the effect of departures from Russell-Saunders coupling. Calculations on forbidden lines are outlined. Some of the difficulties of the subject are discussed. Appendices contain detailed references to recent work on transition probabilities and atomic wave functions.

1. INTRODUCTION

DURING the last twenty years immense progress has been made in the study and interpretation of astronomical spectra. The identification of forbidden lines in the spectra of the gaseous nebulae, solar corona, and upper atmosphere opened up a new phase of spectroscopic analysis. The development of high dispersion spectrographs and the increasing number of large telescopes available have led to the detailed analysis of the spectra of many individual celestial bodies. The development of the theory of stellar atmospheres, and especially the discovery and exploitation of the curve of growth technique have enabled quantitative information to be extracted from the mass of observational data which can now be obtained. In order to relate the observed intensities of spectral lines to the number of atoms responsible for their production a knowledge is required of the atomic parameters involved. In particular, reasonably accurate values of the atomic transition probabilities are needed for very many lines. Such data can be obtained in some cases by laboratory experiment, while in others recourse must be had to theoretical calculations based on quantum mechanics. In this paper we shall review the calculations which have so far been carried out.

2. PERMITTED LINES: RUSSELL-SAUNDERS COUPLING

The general theory of the calculation of the strengths of permitted lines has been fully described by CONDON and SHORTLEY (1952). Under the assumption of Russell-Saunders coupling it has been shown that the "strength" S of an electric dipole line is given by

$$S = \mathscr{S}(M)\mathscr{S}(L)\sigma^2, \qquad \ldots .(1)$$

where $\mathscr{S}(M)$ is a numerical factor depending on the particular multiplet of a transition array and $\mathscr{S}(L)$ is a numerical factor depending on the particular line of a multiplet. $\mathscr{S}(M)$ can be evaluated from two sets of tables computed by GOLDBERG (1935, 1936) and $\mathscr{S}(L)$ from the tables of WHITE and ELIASON (1933) or of RUSSELL (1936). The f-values and spontaneous transition probabilities A (sec.$^{-1}$) associated with a line of wavelength λ (Ångstroms) and strength S are given by the formulae

$$f = \frac{304S}{\omega_1 \lambda},$$

$$A = \frac{2{\cdot}02 \times 10^{18} S}{\omega_2 \lambda^3},$$

where ω_1 and ω_2 are the statistical weights of the lower and upper levels of the line, respectively.

If we adopt the central field approximation (in which we regard the total wave function of the atom as a suitable combination of one electron central field wave functions) then it can be shown that

$$\sigma^2 = \frac{1}{4l^2 - 1}\left(\int_0^\infty rP_iP_f dr\right)^2, \qquad \ldots .(2)$$

where l is the greater of the two azimuthal quantum numbers involved and $\frac{1}{r}P_i, \frac{1}{r}P_f$ are the radial parts of the wave functions of the jumping electron. The determination of line strengths in Russell-Saunders coupling is thus reduced to the calculation of σ^2.

Many attempts have been made to determine σ^2. Among early work may be mentioned that of TRUMPY and others based on the WKB method (CONDON and SHORTLEY, 1935, Chapter XIV). The most important calculations have been based on the self-consistent field procedure devised by HARTREE and described by him in a comprehensive review (1947). An immense amount of labour is required to carry through calculations of this type and even when it is done the results are not always as satisfactory as might be desired. The principal weakness of the self-consistent field method is its neglect of correlation between electrons. A number of attempts to improve the approximation have been made by BIERMANN and LÜBECK (1948, 1949) and by BIERMANN and TREFFTZ (1949a), who modified the central field by the introduction of a polarization potential proportional, for large r, to r^{-4}. The final energies obtained agree more closely with observation than those from self-consistent fields.

Opportunity is taken to refer to some recent unpublished investigations by DOUGLAS (1953). Using "EDSAC", the electronic calculating machine at the Cambridge University Mathematical Laboratory, he has developed methods of calculating atomic wave functions and has applied these to Si IV, an ion with a single electron outside a closed core. The novel feature of his work is the use of a polarization potential given not analytically (as used by BIERMANN) but as a tabulated function. The use of EDSAC enabled over seventy trial functions to be investigated and a function was eventually obtained which reproduced the observed energies of six levels to within the accuracy set by the assumed field. This is very satisfactory as far as it goes. It is not yet clear, however, whether the potential used is uniquely determined and much further work on other simple systems will be needed before the method can be extended to more complex cases. It will be interesting to examine transition probabilities calculated from his wave functions when these become available.

GREEN (1949) and GREEN and WEBER (1950) have made extensive calculations of f-values for Ca II deriving wave functions by mechanical integration of the wave equation on relay calculating machines. It is hoped that this work (and other similar investigations) can be extended to other complex systems and more use made of modern calculating machines as these become available.

The most important of all recent work on the calculation of atomic transition probabilities is that of BATES and DAMGAARD (1949). They have shown that in calculating the transition integrals (σ) it is permissible to neglect the departure of the potential of an ion from Coulomb form. This enables the transition integrals to be systematically tabulated in terms of (essentially) the energy parameters of the initial and final terms. For the simpler atomic systems (especially those with one electron outside a closed shell) the method gives remarkably accurate results, and it seems probable that the results are as good as or better than those obtained by the use of more elaborate methods referred to earlier. Lack of suitable comparison data makes it difficult to assess the accuracy of the method for more complex systems for which the Coulomb approximation would be at best of doubtful validity. Such information as is available, however, suggests that the method may be quite useful even for the heavier atoms.

3. PERMITTED LINES: DEPARTURES FROM RUSSELL-SAUNDERS COUPLING

All the calculations on permitted lines so far described have been based on the assumption of Russell-Saunders coupling. Most of the atoms whose spectra are of astrophysical interest show substantial departures from Russell-Saunders coupling. It is desirable to investigate these departures which are usually discussed in two parts, intermediate coupling and configuration interaction.

(a) *Intermediate Coupling*

In intermediate coupling we include in the Hamiltonian terms arising from electron spin. It is normally sufficient to include only the spin-orbit interactions. The general theory of line strengths in intermediate coupling has been given by CONDON and SHORTLEY (1935). The first major attempt at a detailed calculation was made by SHORTLEY (1935) for the $2p^53s$–$2p^53p$ array in Ne I. The results were not completely satisfactory because the final strengths were sensitive to the adopted values of various parameters and these were not accurately known. Nevertheless the final results were in much better agreement with the available observational data than were values computed on the basis of Russell-Saunders coupling.

Similar calculations for a complex spectrum of great astrophysical importance have been completed by GOTTSCHALK (1948). He has calculated line strengths for the $3d^7(^4P)4s$–$3d^7(^4P)4p$ and $3d^7(^4F)4s$–$3d^7(^4F)4p$ arrays in Fe I using intermediate coupling theory and neglecting configuration interaction and the interactions between terms based on different parents. Although some discrepancies remain, the agreement obtained between the calculated strengths and the available laboratory data is remarkably good.

Much further work by these methods could and should be undertaken. GARSTANG has investigated the line strengths of the arrays $2p^23s$–$2p^23p$ in O II, $2p^43s$–$2p^43p$ in Ne II, $3p^24s$–$3p^24p$ in S II, $3p^44s$–$3p^44p$ in A II, $2p^23p$–$2p^23d$ in O II, and $2p^43p$–$2p^43d$ in Ne II. In the first three of these arrays the departures from Russell-Saunders coupling are quite small. In the other three arrays there are a number of large departures from Russell-Saunders coupling and many intercombination lines appear. Satisfactory strengths have been computed for many of these lines.

It is important to note that all intermediate coupling calculations give relative line strengths normalized to the same total as those Russell-Saunders values from which they were derived. This is equivalent to saying that, in formula (1), a new value of $\mathscr{S}(M)\mathscr{S}(L)$ is found for each individual line, while $\Sigma\mathscr{S}(M)\mathscr{S}(L)$ for the whole array and the value of σ^2 remain unaltered.

(b) *Configuration Interaction*

The effect of configuration interaction on line strengths has been investigated in a series of very important papers by BIERMANN and TREFFTZ (1949b) and TREFFTZ (1950) for Mg I and by TREFFTZ (1951) for Ca I. These calculations, like those for intermediate coupling, have to be performed for one atom at a time and generalization is difficult. For the resonance lines of Mg I and Ca I for which earlier results calculated by self-consistent field methods (with exchange) are available, the general effect is to lower the f-values and transition probabilities by factors of the order of 30 per cent. Much additional work is required on calculations of this type. TREES, GREEN, BOYS, and others have made extensive studies of the effect of configuration interaction on the energy levels of various atoms. Detailed extensions of their work to transition probability computations will be awaited with much interest.

4. FORBIDDEN LINES

It should be stated that in this connection the term "forbidden" is used as synonymous with "multipole", the lines with which we are concerned arise from magnetic dipole or electric quadrupole radiation. We do not include intercombination lines which arise from departures from Russell-Saunders coupling but which are nevertheless due to electric dipole radiation.

The identification of many of the most important lines in the spectra of the gaseous nebulae provided a strong stimulus to study the theory of forbidden radiation. Many attempts were made to calculate the transition probabilities of these lines, none of the interesting lines being susceptible to laboratory production. The most important tabulation of numerical data was by PASTERNACK (1940), whose work superseded previous calculations. He studied the configurations p^2, p^3, p^4, d^2, and d^3 which produce most of the astronomically interesting lines. These lines are strictly forbidden for electric dipole radiation by the Laporte Rule (the transitions are allowed only between states of opposite parity). Some transitions are allowed for magnetic dipole and electric quadrupole radiation even in Russell-Saunders coupling and others appear in intermediate coupling. The detailed theory of these transitions was worked out by SHORTLEY (1940) and numerical developments given by SHORTLEY, ALLER, BAKER, and MENZEL (1941) enable transition probabilities to be calculated for the p^n configurations. The work of PASTERNACK remains the only source of data on the

d^2 and d^3 configurations. OSTERBROCK (1951) has computed transition probabilities for a number of additional lines of astronomical interest.

The calculations so far discussed were based on the theory of intermediate coupling taking as the perturbation the spin interaction of an electron with its own orbit. In general the results so obtained gave intensity ratios of various lines in good agreement with astronomical observations. In one important case, however, the theory failed to agree with observation. In O II the ratio of the intensities of the pair of nebular lines at $\lambda3727$ due to the transitions ^4S–^2D was widely different from the calculated intensity ratio. This discrepancy was explained by ALLER, UFFORD, and VAN VLECK (1949). In the p^3 configuration the lines ^4S–^2D are forbidden for both magnetic dipole and electric quadrupole radiation to the first order in the spin-orbit interaction, but have non-vanishing intensities when the second order spin-orbit effects are included. ALLER, UFFORD, and VAN VLECK showed that for this particular transition the effect of the spin-other-orbit and spin-spin interactions between pairs of electrons is large and comparable with the second order effect of spin-orbit interaction. Only the latter effect had been included in the earlier calculations of PASTERNACK and others. Other ions in the p^3 configuration which have been studied with special reference to spin-other-orbit and spin-spin effects include N I by UFFORD and GILMOUR (1950) and S II, O II, and N I by GARSTANG (1952a). GARSTANG (1951b), and also NAQVI (1951), have examined the effects in the p^2 and p^4 configurations of a number of ions. It has been shown that while the energy levels are appreciably affected there is little change in the transition probabilities. Similar calculations for other ions have been made by OBI (1951) and GARSTANG (1952b).

These investigations seem to provide a satisfactory basis for the calculations of the strengths of forbidden lines. So far as the writer is aware no attempt has been made to examine what effect, if any, would be caused by the extension of the calculations to include the effects of configuration interaction. There do not appear to be any discrepancies between theory and observation which might be ascribed to configuration interaction, but the absolute values of the line strengths might be affected. The quadrupole lines are comparatively sensitive to the adopted quadrupole moments, and these in turn to the wave functions used. They would be more susceptible to perturbing interactions than magnetic dipole lines. It should be emphasized that all the transition probability calculations for forbidden lines of the magnetic dipole type give absolute values, subject to the uncertainties of certain parameters which are normally obtained by an empirical study of the energy levels. The electric quadrupole lines are subject to the same small uncertainties in addition to the very approximate quadrupole moment. Uncertainty of the quadrupole moment appears to be the most doubtful factor in the calculations of the strengths of forbidden lines. The most important task remaining is the computation of the strengths of some lines in more complicated configurations. Some work has been begun on the d^4 and d^6 configurations in Fe V and Fe III by GARSTANG but no results are yet available.

5. GENERAL DISCUSSION

Reference must be made to two points of difficulty in the calculation of transition probabilities. The formula (2) for σ^2 involves the dipole length integral

$$\int_0^\infty rP_iP_f dr.$$

It can be shown that this can be transformed into the dipole velocity form

$$\frac{2}{E}\int_0^\infty P_{l-1}\left(\frac{l}{r}P_l + \frac{dP_l}{dr}\right)dr,$$

where E is the energy (in Rydbergs) of the transition and l is the larger azimuthal quantum number involved, and a third form (dipole acceleration) can also be obtained. These formulae would be exactly equivalent if P_i, P_f were exact solutions of the wave equation. When P_i, P_f are approximate the dipole velocity and dipole length integrals will have different numerical values. The importance of this difference was pointed out for H^- by CHANDRASEKHAR (1945) and many subsequent investigators have calculated σ^2 from the two formulae. It is not at all clear which value of σ^2 is the most reliable, or what mean value might be chosen. It is not even certain that, in general, the true value of σ^2 lies between the two calculated values. The correctness of this last statement has been shown by BATES (1951) for the case of the $1s\sigma$–$2p\sigma$ transition in H_2^+. Exact wave functions for this molecule have been obtained and used to compute exact f-values (as a function of internuclear distance R). Approximate f-values have been obtained by applying the dipole length and dipole velocity formulae to wave functions obtained as a Linear Combination of Atomic Orbitals approximation. It was found that the dipole length formula gave greater and the dipole velocity smaller values than the true values. In many cases, unfortunately, the last conclusion appears to be uncertain or perhaps definitely false (BATES and DAMGAARD, 1949; Table 7).

A second difficulty arises in several calculations. If, for example, we compute a self-consistent field for an atom we obtain as the appropriate eigenvalue of the Schrödinger equation a parameter which, for a system consisting of one electron outside a closed core, gives the best theoretical approximation to the energy of the corresponding spectroscopic level. This differs from the observed value. Should the theoretical or experimental energies be used in calculations of f-values? If in a particular case, as for example the intermediate coupling calculations mentioned earlier, the labour involved prohibits obtaining theoretical values for the energies of the centres of gravity of the various terms, will the use of the observed energies give satisfactory transition probabilities, bearing in mind the well-known disagreement between the observed multiplet intervals and any calculations which neglect configuration interaction?

This problem has been investigated in an interesting paper by GREEN, WEBER, and KRAWITZ (1951). They calculated oscillator strengths for all important transitions from the $3d$ state of Ca II using wave functions with and without exchange and using both calculated and observed energies. No experimental data was available for comparison, and recourse was had to the f-sum rule and partial f-sum rule. It was shown that the use of the calculated rather than the observed energies gave results in much closer agreement with the sum rules. This conclusion held whether wave functions with or without exchange were used, and whether dipole length, velocity, or acceleration formulae were used. The results obtained using observed energies were very poor for some transitions.

It should be remembered that the f-sum rule is only an approximation based on the assumption that the series electron moves in a fixed field. The general agreement of the sum rule with the sum of the f-values calculated both with and without exchange and using calculated energies, together with the marked disagreement between the results with and without exchange when observed energies are used, suggests that, at least for a one electron spectrum, the sum rules are valid with fair accuracy. It is important to note that while failure to satisfy the sum rules implies unreliability of the f-values, the satisfaction of the sum rule does not necessarily imply that the f-values are accurate. This is well shown by the results of GREEN, WEBER, and KRAWITZ, whose individual f-values using calculated energies with and without exchange differ greatly although the f-sums are nearly equal. See SEATON (1951) for further discussion of f-sum rules.

These last remarks illustrate what, in the writer's opinion, is the greatest difficulty in the whole subject of transition probabilities, whether theoretical or experimental. We refer to the almost complete absence of any transition probabilities of absolutely certain accuracy for any complex atoms. We have discussed some of the theoretical difficulties and uncertainties. Many of the experimental determinations, while of immense assistance to astrophysicists are nevertheless subject to serious uncertainties. The "absolute" determinations of ESTABROOK (1952) and HULDT and LAGERQUIST (1952) for λ4290 of Cr I, for example, differ by a factor of 100. This may be a particularly bad case, but we cannot regard as satisfactory a situation in which such discrepancies exist. An experimental programme of especial interest is one being developed at Kiel. New high-temperature stabilized arcs have been constructed. Transition probabilities for carbon by MAECKER (1953) are among the first results of this work. Much more work must be done to improve the accuracy of both theoretical and experimental determinations, to supply accurate experimental comparison data for checking theoretical work for atoms of moderate complexity (such as Ca I) and, above all, to supply numerical data which is greatly desired by all astronomical spectroscopists.

ACKNOWLEDGMENT

The writer would like to thank Mr. A. S. DOUGLAS for communicating his results in advance of publication.

APPENDIX I

ATOMIC TRANSITION PROBABILITY DATA

For a summary of the information available in 1947 see UNSÖLD (1938, pp. 128 and 191) and MINNAERT (1950). For recent work on forbidden lines see: ALLER, UFFORD, and VAN VLECK (1949); GOLD (1949); UFFORD and GILMOUR (1950); OBI (1951); NAQVI (1951); OSTERBROCK (1951); GARSTANG (1951b, 1952a, 1952b). BATES and SEATON (1949) list work on continuous absorption up to 1949. BATES and DAMGAARD (1949) give absolute line strengths for a number of atoms. We list below, with references, all other work on permitted transitions, theoretical and experimental, known to the writer which is not mentioned by these authors.

Continuous absorption

Ion	References	Ion	References	Ion	References	Ion	References
O I	49	Ne I	49, 39	Mg II	8	K II	48
F I	49	Na I	50, 16	Si II	8	Ca II	31, 32, 50

Line absorption

Ion	References	Ion	References	Ion	References	Ion	References
H	64	Mg I	9, 10, 56	S III	1	Mn I	36
Li I	55	Mg II	1, 7,	Cl II	1	Fe I	12, 29, 38
Be I	9	Al I	7	A II	1, 25	Fe XIV	26
C I	39a	Al III	1	K I	7, 54, 61	Ni I	18, 37
C II	1, 7, 39a	Si II	7	Ca I	57	Rb I	54
N II	1	Si III	1	Ca II	31, 32, 50, 58	Ba I	62
O II	1, 25	Si IV	1, 17, 33	V I	46	Hg I	11, 41
Ne II	20, 25	P III	1	Cr I	18, 19, 35, 36	Tl I	54
Na I	7, 50, 54	S II	1, 25				

APPENDIX II

CALCULATION OF WAVE FUNCTIONS

A bibliography of wave functions has been given by HARTREE (1947). We append a short supplementary list of wave functions published since 1947.

Ion	References	Ion	References	Ion	References	Ion	References
C II	7	Mg II	7	Si II	7, 8	Ca II	57
N II	50a	Al I	7	K I	7, 41, 61	Ca XIII	15
Ne III	21	Al III	36a	K III	48	Fe XIV	26
Na I	7	Al IV	36a	Ca I	57	Hg I	41
Mg I	9, 10, 56						

Unpublished wave functions are available from A. S. DOUGLAS (of the Cambridge University Mathematical Laboratory) for S II and Si IV, from D. R. HARTREE (Cavendish Laboratory, Cambridge) for Ne I computed by B. H. WORSLEY and from the present writer for Ne II.

REFERENCES

[1] ALLER, L. H. . . . . 1949 *Ap. J.*, **109**, 260.
[2] ALLER, L. H., UFFORD, C. W. and VAN VLECK, J. H. . . . . 1949 *Ap. J.*, **109**, 42.
[3] BATES, D. R. . . . . 1946 *Proc. Roy. Soc.* A., **188**, 350.
[4] BATES, D. R. and DAMGAARD, A. . . . . . 1949 *Phil. Trans.*, **242**, 101.
[5] BATES, D. R. and SEATON, M. J. . . . . . 1949 *M.N.*, **109**, 698.
[6] BATES, D. R. . . . . 1951 *J. chem. Phys.*, **19**, 1122.
[7] BIERMANN, L. and LÜBECK, K. . . . . . 1948 *Z. Astrophys.*, **25**, 325.
[8] BIERMANN, L. and LÜBECK, K. . . . . . 1949 *Z. Astrophys.*, **26**, 43.
[9] BIERMANN, L. and TREFFTZ, E. . . . . . 1949a *Z. Astrophys.*, **26**, 213.
[10] BIERMANN, L. and TREFFTZ, E. . . . . . 1949b *Z. Astrophys.*, **26**, 240.
[11] BROSSEL, J. and BITTER, F. . . . . . . 1952 *Phys. Rev.*, **86**, 308.
[12] CARTER, W. W. . . . . 1949 *Phys. Rev.*, **76**, 962.
[13] CHANDRASEKHAR, S. . . . . 1945 *Ap. J.*, **102**, 223.
[14] CONDON, E. U. and SHORTLEY, G. H. . . . 1935 *Theory of Atomic Spectra* (Cambridge, reprinted 1952).
[15] DAVIS, W. H. . . . . 1949 *Ap. J.*, **110**, 17.
[16] DITCHBURN, R. W. and JUTSUM, P. J. . . . 1950 *Nature* (*London*), **165**, 723.
[17] DOUGLAS, A. S. . . . . 1953 (Not yet published.)
[18] ESTABROOK, F. B. . . . . 1951 *Ap. J.*, **113**, 684.
[19] ESTABROOK, F. B. . . . . 1952 *Ap. J.*, **115**, 571.
[20] GARSTANG, R. H. . . . . 1950 *M.N.*, **110**, 612.
[21] GARSTANG, R. H. . . . . 1951a *Proc. Cambridge Phil. Soc.*, **47**, 243.

[22] Garstang, R. H. . . . . 1951b *M.N.*, **111**, 115.
[23] Garstang, R. H. . . . . 1952a *Ap. J.*, **115**, 506.
[24] Garstang, R. H. . . . . 1952b *Ap. J.*, **115**, 569.
[25] Garstang, R. H. . . . . 1953 Unpublished.
[26] Gold, M. Tuberg . . . . 1949 *M.N.*, **109**, 471.
[27] Goldberg, L. . . . . 1935 *Ap. J.*, **82**, 1.
[28] Goldberg, L. . . . . 1936 *Ap. J.*, **84**, 11.
[29] Gottschalk, W. M. . . . . 1948 *Ap. J.*, **108**, 326.
[30] Green, L. C. . . . . 1949 *Ap. J.*, **109**, 289.
[31] Green, L. C. and Weber, N. E. . . . . . 1950 *Ap. J.*, **111**, 582, 587.
[32] Green, L. C., Weber, N. E. and Krawitz, E. 1951 *Ap. J.*, **113**, 690.
[33] Hartree, D. R., Hartree, W. and Manning, M. F. . . . . 1941 *Phys. Rev.*, **60**, 857.
[34] Hartree, D. R. . . . . 1947 *Reports on Progress in Physics*, XI, 113.
[35] Hill, A. J. and King, R. B. . . . . . . 1951 *J. Opt. Soc. Amer.*, **41**, 315.
[36] Huldt, L. and Lagerquist, A. . . . . . 1952 *Ark.Fys.*, **5**, 91.
[36a] Katterbach, K. . . . . 1953 *Z. Astrophys.*, **32**, 165.
[37] King, R. B. . . . . 1948 *Ap. J.*, **108**, 87.
[38] Kopfermann, H. and Wessel, G.. . . . . 1951 *Z. Phys.*, **130**, 100.
[39] Lee, Po and Weissler, G. L. . . . . . . 1952 *J. Opt. Soc. Amer.*, **42**, 214.
[39a] Maecker, H. . . . . 1953 *Z. Phys.*, **135**, 13.
[40] Minnaert, M. . . . . 1950 *Trans. I.A.U.*, **7**, 388.
[41] Mishra, B. . . . . 1952 *Proc. Cambridge Phil. Soc.*, **48**, 511.
[42] Naqvi, A. M. . . . . 1951 *Astron. J.*, **56**, 45.
[43] Obi, S. . . . . 1951 *Pub. Astron. Soc. Japan*, **2**, 150.
[44] Osterbrock, D. E. . . . . 1951 *Ap. J.*, **114**, 469.
[45] Pasternack, S. . . . . 1940 *Ap. J.*, **92**, 129.
[46] Righini, G. . . . . 1949 *Mem. Soc. Ast. Italy*, **20**, 303.
[47] Russell, H. N. . . . . 1936 *Ap. J.*, **83**, 29.
[48] Seaton, M. J. . . . . 1950 *M.N.*, **110**, 247.
[49] Seaton, M. J. . . . . 1951a *Proc. Roy. Soc.* A., **208**, 408.
[50] Seaton, M. J. . . . . 1951b *Proc. Roy. Soc.* A., **208**, 418.
[50a] Seaton, M. J. . . . . 1953 *Proc. Roy. Soc.* A., **218**, 400.
[51] Shortley, G. H. . . . . 1935 *Phys. Rev.*, **47**, 295.
[52] Shortley, G. H. . . . . 1940 *Phys. Rev.*, **57**, 225.
[53] Shortley, G. H., Aller, L. H., Baker, J. G. and Menzel, D. H.. . . . . 1941 *Ap. J.*, **93**, 178.
[54] Stephenson, G. . . . . 1951a *Proc. Phys. Soc.* A. (*London*), **64**, 458.
[55] Stephenson, G. . . . . 1951b *Nature* (*London*), **167**, 156.
[56] Trefftz, E. . . . . 1950 *Z. Astrophys.*, **28**, 67.
[57] Trefftz, E. . . . . 1951 *Z. Astrophys.*, **29**, 287.
[58] Trefftz, E. and Biermann, L. . . . . . 1952 *Z. Astrophys.*, **30**, 275.
[59] Ufford, C. W. and Gilmour, R. M. . . . . 1950 *Ap. J.*, **111**, 580.
[60] Unsöld, A. . . . . 1938 *Physik der Sternatmosphären.* Berlin.
[61] Villars, D. S. . . . . 1952 *J. Opt. Soc. Amer.*, **42**, 552.
[62] Wessel, G. . . . . 1951 *Z. Phys.*, **126**, 440 and **130**, 106.
[63] White, H. E. and Eliason, A. V. . . . . 1933 *Phys. Rev.*, **44**, 753.
[64] Wild, J. P. . . . . 1952 *Ap. J.*, **115**, 206.

Atomic Collision Processes in Astrophysics

H. S. W. Massey

Department of Physics, University College, London

1. Introduction

The development of modern astrophysics owes a great deal to the great extension of knowledge of atomic phenomena. Although the great majority of the applications of atomic physics in this direction have been concerned with radiative processes of one kind or another there are a number of circumstances in which a knowledge of the rates of collision processes involving electrons, ions and/or neutral atoms and molecules is required.

One very important example of the need for accurate knowledge of collision cross sections associated with electron impact is provided by the theory of the solar corona. Apart from its own intrinsic interest a knowledge of the quality and quantity of the radiation emitted from the outer layers of the solar atmosphere is necessary for the interpretation of solar-terrestrial relations. The identification by Edlén (1942) of the coronal lines as due to the $^2P_{\frac{1}{2}} - {}^2P_{\frac{3}{2}}$ transition in the ground configuration of Fe XIV showed that the kinetic temperature of the corona must be of the order 10^6 °K. Further evidence for this remarkable result has since been provided from observation of the radio-noise spectrum emitted from the quiet sun (Martyn, 1946). An atmosphere at this high kinetic temperature must emit radiation of very short wavelength, probably in the soft X-ray region. Hoyle and Bates (1948) have shown that a suitable emission of such radiation could well produce terrestrial atmospheric ionization similar to that observed as the F layer. Radiation from deeper layers of the solar atmosphere, such as the chromosphere, may also be important in producing ionization in the earth's atmosphere in excess of that expected from the usual black body model of the Sun as radiator, and at a different altitude. To investigate the possibilities theoretically it is necessary to have available reliable information on cross sections, not only for radiative recombination of electrons to highly ionized atoms, but also for ionization and excitation of such atoms by electron impact. This may be seen by reference to one of the pioneer investigations of the solar corona, by Woolley and Allen (1948).

They consider first the nature of the equilibrium in the corona where the kinetic and radiation temperatures differ by several orders of magnitude. The ionization equilibrium must arise from a balance between ionization by electron impact and radiative recombination. If there are xn atoms per cubic centimetre in the $(s + 1)$th and $n(1 - x)$ in the sth stage of ionization the equilibrium equation is

$$nxn_e\alpha = n(1 - x)n_e\gamma,$$

where α is the coefficient of radiative recombination of electrons to the $(s + 1)$-fold ionized atom and γ that for further ionization (by electron impact), n_e being the number of electrons per cubic centimetre. This gives

$$\frac{x}{1 - x} = \frac{\gamma}{\alpha}.$$

Hence if γ and α are known as functions of electron energy it is possible to determine under what conditions x approaches unity for a particular ion. In the solar corona this must be so for Fe XIV and this means that, from γ and α, the kinetic temperature of the corona can be determined. With this information available it is possible in principle, from a knowledge of the appropriate excitation cross sections and transition probabilities, to calculate the intensity of the various characteristic radiations emitted from coronal atoms in terms of their concentrations and that of the electrons.

A programme of this kind was carried out by WOOLLEY and ALLEN (1948) using the best available information about the collision cross sections. They obtained many results of importance; the coronal kinetic temperature comes out to be close to 10^6 °C, evidence that the main contribution to the coronal line emission in the ultraviolet comes from the principal lines of Mg X was obtained, and the intensity of

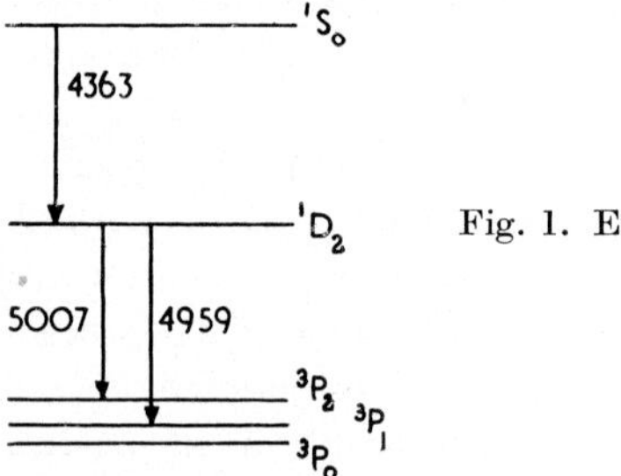

Fig. 1. Energy level diagram (O III)

the continuous emission was estimated to be less than the number of Lyman quanta emitted by the chromosphere. It is therefore important to verify that the assumptions made about the atomic constants are valid, including particularly the ionization and excitation cross sections.

The development of a theory of the chromospheric emission offers much greater difficulties. A start has been made by GIOVANELLI (1949) and by WOOLLEY and ALLEN (1950). The former author investigated the hydrogen spectrum of the sun by attempting an analysis of the population of the different atomic hydrogen levels in the chromosphere. It is impossible to carry out this analysis without knowledge of the cross sections for production of various transitions between the low-lying levels of the hydrogen atom by electron impact. Whereas in the coronal problem the transitions are optically allowed, in the chromospheric case optically disallowed transitions such as $1s$–$2s$ are important. In addition, at the kinetic temperature of the chromosphere, the exciting electrons have energies only slightly in excess of the threshold for the transition concerned. The distinctive feature of such collisions is that electron exchange effects are often of dominating importance whereas in optically allowed transitions this does not apply.

An accurate knowledge of the population distribution among the low-lying hydrogen states is also important in connection with the possibility of observing the $2^2s_{\frac{1}{2}} - 2^2p_{\frac{1}{2}}$ transition as a line emission in solar radio-noise at 9882 Mc per sec (WILD, 1952). The effectiveness of collisions in determining the distribution of atoms between the $2s_{\frac{1}{2}}$, $2p_{\frac{1}{2}}$ and $2p_{\frac{3}{2}}$ states is the decisive factor in this case.

The effectiveness of electron impact in producing transitions from the ground state to the metastable states of atoms is also a major factor in the determination of the intensity of emission of many principal lines in nebular spectra. The classical example is provided by the "auroral" and nebular lines of O III. Fig. 1 illustrates the levels arising from the ground configuration of this ion. The green nebular lines λ 5007 Å and λ 4959 Å arise from the $^1D_2 - {}^3P_2$ and $^1D_2 - {}^3P_1$ transitions and the auroral line λ 4363 Å from $^1S_0 - {}^1D_2$. Under nebular conditions effective electron collisions are sufficiently frequent to ensure that the population of the three levels of the ground ^3P term follows a Boltzmann distribution about the electron temperature (HEBB and MENZEL, 1940). The two metastable levels $^1S_0$ and $^1D_2$ are populated by electron excitation from the ^3P states. Deactivation occurs by radiation and by superelastic collisions. It follows that the relative intensity of the auroral and nebular lines is determined by the relevant transition probabilities and collision cross sections and by the electron temperature (MENZEL and ALLER, 1941). The absolute intensity of either line depends further on the concentration of O III ^3P ions and of electrons in the nebula. The electron concentration may be obtained from the intensity of the Balmer continuum emitted by the nebula.

Hence, if the radiative transition probabilities and collision cross sections are known from atomic theory it follows that measurement of the absolute intensities of the nebular and auroral O III lines from a nebula enables one to determine the O III abundance in that nebula. If this procedure could be applied to all states of ionization the atomic constitution of the nebula could be obtained. Even though this is not possible for all states of ionization it can be carried out for many atomic ions and the results extrapolated in a reasonable way. The pioneer work in this field is that of MENZEL and his co-workers (1941, 1945). We shall see, however, that some of the collision cross sections they employed in the analysis of the O III forbidden line intensities were too large.

The green and red forbidden lines of O I which arise from similar transitions to those of O III (see Fig. 1 (*b*)) are a prominent feature of airglow and auroral spectra. In the auroral case electron impact undoubtedly plays an important part in determining the relative intensities at different heights in the earth's atmosphere but in the airglow other processes of excitation are responsible for populating the S and D states.

Although, in general, collisions between systems which are both of atomic dimensions are not of such importance under astrophysical conditions as are electron impacts, there are phenomena in which such collisions play a vital role. The light emitted and the ionization produced by meteors in the earth's atmosphere are due to inelastic impacts between atoms evaporated from the meteor and atmospheric atoms and molecules. Again, protons are more effective than electrons in the chromosphere in producing transitions between the $2s$ and $2p$ levels of hydrogen (PURCELL, 1952).

The recent observations of GARTLEIN (1951) and by MEINEL (1950) of Doppler-shifted hydrogen lines in auroral spectra have confirmed the expectation that energetic protons are among the primary particles whose entry into the earth's atmosphere gives rise to auroral displays. This makes it important to study the various processes, including particularly charge exchange, which can occur as the incoming protons penetrate into the atmosphere. Charge transfer on impact between ions and neutral molecules may also profoundly modify the ionic constitution of the ionosphere and it may be important in a similar way in the radiating regions of comets. Other types of collisions between atomic systems are also very important in the ionosphere.

It is clear therefore that the development of atomic collision theory is important in many applications to astrophysical (including geophysical) phenomena. We shall now examine briefly how far this development has proceeded paying special attention to electron collisions for which the greatest progress has been made.

2. The Theory of Inelastic Collisions of Electrons with Atoms

There is a profound difference between the state of the theory of energy transfer on collision between material particles and that of radiative transition probabilities. In the latter case the only difficulty is that of determining with sufficient accuracy the wave functions which describe the initial and final states; the formula expressing the probability in terms of these wave functions is quite reliable. On the other hand, when the electron energy is close to the threshold for the inelastic collisions concerned, the usual formulae for the collision cross section are quite unreliable and cannot be used with confidence even for collisions with atomic hydrogen for which the wave functions are known exactly.

The usual, perturbation, formula which is employed in the first instance to calculate collision cross sections is that of Born. This is derived on the assumption that the motion of the impinging electron relative to the atomic system can be described by plane waves, both before and after the collision. A modification of this approximation due to Oppenheimer allows for the possibility that electron exchange may take place during the collision but still employs the plane wave approximation. We shall refer to these two as the Born and OB approximations respectively.

A detailed examination of the validity of these Born and OB approximations by comparison with experimental data has been made by Bates, Fundaminsky, Leech and Massey (1950) (B.F.L.M.). They also employed, as a further check, a general conservation theorem which states that the maximum possible contribution to the cross section for any inelastic collision from encounters in which the relative angular momentum is $\{l(l+1)\}^{1/2}\hbar$ is $\lambda^2(2l+1)/4\pi^2$, where λ is the wavelength of the relative motion. It is seen from this analysis that a distinction must be made between inelastic collisions involving optically allowed and optically disallowed transitions. In the former case the Born approximation does not appear to give results which are grossly in error at any electron energy. At low energies it overestimates the cross section to an increasing extent but usually by not more than a factor of 2 (B.F.L.M., 1950). The observed behaviour may be represented empirically by expressions of the form adopted by Woolley and Allen (1948) in their study of the solar corona and it is unlikely that any very serious error was introduced in their analysis from their assumptions in this regard. Exchange seems to be relatively unimportant even at very low energies. It is fortunate that the simplest approximations work fairly well for these cases as it is very difficult to improve them without very great labour.

The situation is very different for optically disallowed transitions, particularly those in which there is no change in azimuthal quantum number of the active atomic electron. Inelastic or superelastic electron collisions of this kind are often among the most important in applications as they include the excitation or deactivation of metastable states. For collisions of this kind, electron exchange effects are usually of dominating importance and it is found that the OB approximation, the simplest that can be used, is usually wholly unreliable (B.F.L.M., 1950). Whereas in optically allowed transitions contributions come from so many relative angular momenta that the conservation theorem cannot be employed as a check, for the encounters we are now considering the major contribution, near the threshold where the cross section is large, comes from impacts with a definite and known relative angular momentum. The conservation theorem may therefore be employed as a convenient and reliable check. It was found in this way that many of the earlier calculations carried out by the OB method were invalid, giving cross sections in excess of the limit $\lambda^2(2l+1)/4\pi^2$. This included those used by MENZEL and ALLER (1941) in their determination of the abundance of O III in nebulae. Unsatisfactory results were also obtained when the test was applied to the excitation of the 2^3S level of helium. It has recently been found that the OB method also gives invalid results for the excitation of the metastable level of hydrogen (ERSKINE and MASSEY, 1952).

Although the simple approximations are so very unsatisfactory for these cases there is much more scope for introducing improved methods. This is largely because most of the contributions come from one relative angular momentum. The most important modification which can be made is that of replacing the plane wave representation of the colliding electron by a wave distorted by the mean atomic field in which the electron moves. In determining this distortion exchange effects may or may not be included. If they are so included we refer to the approximation as the D.W.B.O. (distorted wave Born-Oppenheimer), if not, as the D.W.B. The calculation of the appropriate distorted waves has been greatly facilitated by the introduction of variational methods (ERSKINE and MASSEY, 1952; HULTHÉN, 1944).

Both of these approximations still assume that the coupling between the initial and final atomic states is not strong. To investigate this MASSEY and MOHR have discussed a schematic model for which exact solutions may be obtained even if the coupling is strong, and compared with the predictions of the corresponding D.W.B.O. and B.O. approximations. It is found that, whereas the B.O. approximation is always unreliable, no matter how weak the coupling, the D.W.B.O. gives good results even when the coupling is so strong that the cross section rises to as much as one-half the limiting value. Their calculations show, however, that the distorted wave approximation is only reliable when the distortion is accurately allowed for. In short it will usually be necessary to employ the D.W.B.O. approximation in which exchange effects in producing distortion are included, and not content oneself merely with the D.W.B.

Detailed calculations using the D.W.B.O. approximation have been carried out by ERSKINE and MASSEY (1952) for the excitation of the $2S$ level of hydrogen. Their results profoundly modify the predictions of the OB approximation. This may be seen by reference to Table I.

Table I. Comparison of cross sections for 1S–2S excitation of hydrogen calculated by the OB *and* D.W.B.O. *methods*

Electron energy (e-volts)	Cross section in units πa_0^2		Limit set by Conservation Theorem
	OB	D.W.B.O.	
10·2	0	0	—
13·5	1·590	0·178	1·000
19·4	0·503	0·094	0·694
30·4	0·104	0·035	0·444
54·0	0·020	0·011	0·250

Note.—The cross sections given are those for excitation by electrons with zero angular momentum only. The contribution from higher angular momenta is very small at energies below 20 e-volts, but is important at higher energies. It can be calculated adequately by the OB approximation.

Detailed analysis of the data shows that in this case the D.W.B.O. results, while not violating the conservation theorem and representing a great improvement in the OB results, may still be somewhat in error near the maximum. In other words, the coupling between the $1S$ and $2S$ levels may be too strong. Some evidence supporting this is forthcoming from some preliminary calculations by MASSEY and MOISEIWITSCH (I), in which a variational procedure is employed but no assumption made about the strength of the coupling. It is hoped to carry out a still more accurate calculation in which the coupled integro-differential equations concerned are solved numerically.

The D.W.B.O. approximation is also being applied in detail, by MASSEY and MOISEIWITSCH (II), to the excitation of the 2^3S and 2^1S levels of helium. Preliminary results for the 2^3S case are encouraging, whereas the OB method gives a maximum cross section near the threshold of $1{\cdot}5\pi a_0{}^2$, the D.W.B.O. reduces this to $0{\cdot}04\pi a_0{}^2$ which is quite close to that derived from the rather imperfect experimental data.

Even for much more complicated atoms such as oxygen it has been possible to obtain greatly improved results. SEATON has been able to calculate cross sections for the excitation of metastable states of p^2 and p^4 configurations by a method which is virtually an improved version of the D.W.B.O., applicable to transitions within a configuration, which allows for close coupling. In all cases the conservation theorem is no longer violated and the cross sections can be used with some confidence for the study of nebular abundances and auroral and airglow phenomena.

It seems probable therefore that reasonably reliable data on electron collision cross sections will soon be available for many astrophysical applications. An experimental research programme is also being initiated at University College, London, to provide additional reliable observational checks on the theory.

3. Collisions Involving Atomic Systems (Massey and Burhop, 1952)

The situation is much less satisfactory when both colliding systems are of atomic dimensions. If the velocity of relative motion is high compared with that of the internal motion of the electrons in the atomic systems which are concerned in the transition there is no difficulty in principle as BORN's approximation may be safely applied. Even then the computation of the integrals which appear in charge exchange cases is a tedious and lengthy task. When the opposite conditions apply, the velocity of relative motion being small compared with that of the atomic electrons concerned or, in other words, the time of collision is long compared with the effective period of the electronic motions, the problem is much more difficult. Under these conditions the probability of the particular inelastic collision concerned is small, the process being nearly adiabatic, but it is impossible to say at present from theoretical arguments how small it is for a collision with any particular relative velocity. Unfortunately in many cases of practical importance, as for meteor ionization, the collisions are of this nearly adiabatic type. Progress in predicting cross sections for processes of this kind must necessarily be slow. An extensive research programme devoted to this end is now in progress both at University College, London, and Queen's University, Belfast. The theoretical work, which is being carried out at both establishments, includes a detailed study of the excitation of hydrogen atoms by slow protons in which the perturbation producing the energy transfer is the kinetic energy of relative motion. The experimental work, being carried out in London, includes a systematic study of charge exchange between positive ions and neutral atoms and molecules and of electron detachment from negative ions on impact with rare gas atoms. This will be extended to study ionization by impact of ions and of neutral atoms. It is hoped that systematic research on these lines will eventually make possible the prediction of the rates of nearly adiabatic processes of sufficient accuracy for application.

REFERENCES

ALLER, L. H. and MENZEL, D. H. . . .	1945	*Ap. J.*, **102**, 239.
BATES, D. R., FUNDAMINSKY, A., LEECH, J. W. and MASSEY, H. S. W.	1950	*Phil. Trans.* A., **243**, 93.
EDLÉN, B.	1942	*Ark. Mat. Astr. Fys.*, **28**, No. 1B.
ERSKINE, G. and MASSEY, H. S. W.. . .	1952	*Proc. Roy. Soc.*, **212**, 521.
GARTLEIN, C. W.	1951	*Phys. Rev.*, **81**, 463.
GIOVANELLI, R.	1949	*M.N.*, **109**, 298.
HEBB, M. H. and MENZEL, D. H. . . .	1940	*Ap. J.*, **92**, 408.
HOYLE, F. and BATES, D. R.	1948	*Terr. Mag. and Atmos. Elect.*, **53**, 51.
HULTHÉN, L.	1944	*K. fysiogn. Söllsk. Lund. Förk*, **14**, No. 21.
MARTYN, D. F.	1946	*Nature* (London), **158**, 632.
MASSEY, H. S. W. and BURHOP, E. H. S. .	1952	*Electronic and Ionic Phenomena* (Oxford University Press), Chaps. VII and VIII.
MASSEY, H. S. W. and MOHR, C. B. O. . .	1952	*Proc. Phys. Soc. (London)* A., **64**, 545.
MASSEY, H. S. W. and MOISEIWITSCH, B. L.	1953	I. *Proc. Phys. Soc. (London)* A., **56**, 406. II. Work in progress.
MEINEL, A. B.	1950	*Phys. Rev.*, **80**, 1096.
MENZEL, D. H. and ALLER, L. H. . . .	1941	*Ap. J.*, **94**, 30.
PURCELL, E. M.	1952	*Ap. J.*, **116**, 467.
SEATON, M. J.	1953	*Phil. Trans.* A., **245**, 469.
WILD, J. P..	1952	*Ap. J.*, **115**, 206.
WOOLLEY, R. v.d. R. and ALLEN, C. W. .	1948	*M.N*,, **108** 292.
	1950	*M.N.*, **110**, 358.

Energy Production in Stars

E. E. SALPETER

The Australian National University, Canberra, Australia*

SUMMARY

Recent work and trends in the field of stellar energy production and related topics are reviewed. Thermonuclear reactions are discussed in general, and a list is given of those reactions which might be important in the interior of different types of stars, together with the temperature at which these reactions set in. It is shown that the proton-proton chain almost certainly is the source of energy of the Sun and of cooler main sequence stars, and the carbon-nitrogen cycle most probably the energy source for the hotter main sequence stars (A, B, O). Some current ideas on the evolution of stars, particularly of subgiants in globular clusters, are discussed.

---

1. INTRODUCTION

LET us consider first of all our best-known star, the Sun. We know accurately the amount of energy the Sun loses in the form of electromagnetic radiation per second. The rate of energy loss per unit mass of the Sun is not very large, about 2 ergs per gm per sec., modest even compared with the rate of heat production during animal metabolism. On the other hand, geological evidence tells us that the Sun has been radiating at approximately its present rate for an extremely long time. This rules

* At present at Australian National University, on leave of absence from Cornell University, Ithaca, New York.

out chemical sources of energy, which could have supported the Sun's radiation for not much longer than the age of the Egyptian Empire. The energy released in the gradual gravitational contraction of the Sun is much more impressive and could have served as the Sun's fuel for tens of millions of years. Such a life-span seemed adequate not too many decades ago. But nowadays geologists and astronomers agree that the Earth and Sun are several thousand million years old and the only source of energy potent enough for such a long time is the energy released in nuclear transformations.

During the 1930's ATKINSON and HOUTERMANS, GAMOW and TELLER, BETHE and CRITCHFIELD, and v. WEIZSACKER laid the foundation for the study of nuclear reactions in stars. They pointed out how the thermal energy of nuclei in the interior of stars took the place of man-made accelerators to initiate nuclear reactions and discussed which species of nuclei might be involved. In 1939 BETHE wrote a classic paper, analysing in great detail all the possible reactions involving fairly light nuclei. He proved that in stars not too different from the Sun only two reaction chains are of importance, the carbon-nitrogen cycle and the proton-proton chain. On the nuclear side the biggest advance since 1939 has been a vast increase in the experimental information on nuclear resonance levels and on cross-sections of nuclear reactions. Thus the numerical values for the rates of many relevant reactions are different and much more accurate than in BETHE's paper. But BETHE's main conclusions about the reactions transforming hydrogen into helium and supplying the star's energy for most of its lifetime still hold. At least this is the case for main sequence stars, which consist mainly of hydrogen and are in a steady state. On the astrophysical side, too, our quantitative knowledge of the chemical composition of main sequence stars and notably of the opacity in their interior has improved greatly, but without altering BETHE's picture qualitatively. One new trend, however, has been an increased interest in stellar evolution and hence in stars which do not lie on the main sequence. Of particular interest are very young stars which are still under gravitational contraction and very old stars which have exhausted a considerable fraction of their hydrogen supply.

In Section 2 we give a brief review of the principles and results of calculations on the rates of various thermonuclear reactions of importance in different types of stars. In Section 3 we take stock of the present position of the study of the energy sources and interior structure of main sequence stars. In Section 4 we discuss briefly, in a highly tentative and oversimplified way, some current ideas on the evolution of stars. More detailed reviews of earlier work on some of the topics treated here will be found in: BETHE, 1938; BETHE and MARSHAK, 1939; GAMOW and CRITCHFIELD, 1949; JOHNSON, 1950; CHANDRASEKHAR, 1951; SALPETER, 1953; ALLER, 1954. The bibliography mainly contains recent papers which give references to earlier work.

2. THERMONUCLEAR REACTIONS

We know that all atomic nuclei consist of neutrons and protons (nucleons), bound together by the very strong nuclear forces. We give below a table of binding energy (B) per nucleon in mega-electron-volts for a few typical nuclei.

	H^1	He^3	He^4	B^{11}	O^{16}	Fe	U
B	0	2·8	7·1	6·9	8·0	8·5	7·5

One feature of such a table is that the binding energy B tends to increase with atomic weight until atomic weight 40–80 (Fe-region) is reached. Thus energy will be released, in general, in a reaction building up a heavier nucleus out of two lighter ones (up to the Fe-region). Another important feature is the fact that the He^4-nucleus (alpha-particle) has a large binding energy compared to neighbouring nuclei. Considering this fact and the fact that hydrogen is the most abundant element in the Universe, it is not surprising that main sequence stars live solely on the conversion of hydrogen into helium.

All the reactions of stellar interest involve collisions between positively charged nuclei, close enough for the short range nuclear forces to come into play. At these

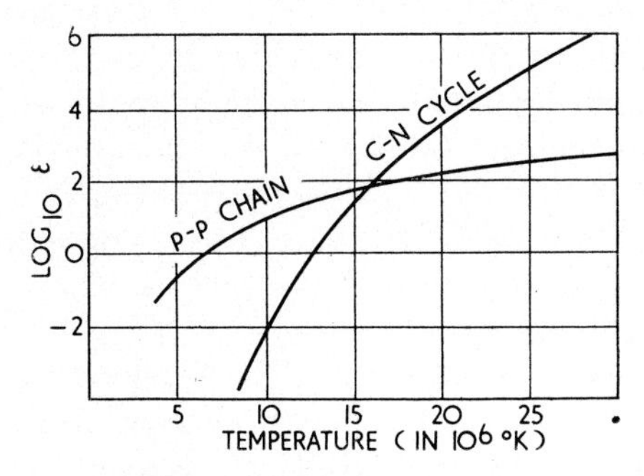

Fig. 1. The rate of energy production ε (in ergs per gramme per second) as a function of temperature for the proton-proton chain and for the carbon-nitrogen cycles. The curves are plotted for $(\rho x_H{}^2) = 100$ and $(\rho x_H x_{CN}) = 1$, where ρ is the density in grammes per cubic centimetre and x_H, x_{CN} are the abundances (by mass) of hydrogen and carbon plus nitrogen, respectively

very small separations the two nuclei experience a strong electrostatic repulsion, corresponding to a Coulomb barrier of about a million electron volts (MeV). At temperatures of the order of 10^7 °K the mean thermal energy is only about a thousand electron-volts. Most of the nuclear reactions actually occurring, involve nuclei whose kinetic energy lies in the "high velocity tail" of the Maxwell-Boltzmann distribution, but is still small compared with the Coulomb barrier energy. The main feature of thermonuclear reactions then is that the reaction rate involves two exponential factors, one a Maxwell distribution factor depending on temperature, the other a Coulomb barrier penetration factor depending on the nuclear charge. Consequently the reaction rate increases strongly with temperature and decreases strongly with increasing atomic charge (BETHE, 1938; BETHE and MARSHAK, 1939). Further, the temperature dependence is stronger for higher atomic charge.

As we mentioned before there are two chains of reactions, each having the same net result, namely,*

$$4H^1 \rightarrow He^4 + 2\nu + 26 \text{ MeV}. \qquad \ldots.(1)$$

The carbon-nitrogen cycle (BETHE, 1938) involves successive reactions between protons and the isotopes of carbon and nitrogen, which merely act as "catalysts" without being used up. The cross-sections for all the reactions involved in this cycle have now been measured in the laboratory, but all for kinetic energies larger than occur in stellar interiors. In Fig. 1 the present value for the rate of energy production on this cycle is plotted as a function of temperature (FOWLER, 1954). These values

* ν stands for neutrino.

are probably correct to within a factor of two, but there is still a small possibility that one particular nuclear resonance level has been missed in the laboratory experiments, in which case the rate would be about twenty times larger than given in Fig. 1.

The proton-proton chain involves the collision between two protons, accompanied by a beta-decay resulting in a deuteron (BETHE and CRITCHFIELD, 1938). Such a beta-decay is very improbable, but, on the other hand, the penetration of the Coulomb barrier is much easier for singly charged protons than for the reactions of the C-N cycle. This *p-p* reaction cannot be observed in the laboratory but its rate of energy production can be calculated from the known properties of the two-nucleon system and from beta-decay theory (SALPETER, 1952a; FRIEMAN and MOTZ, 1953). This rate, accurate to within about 20 per cent is also plotted in Fig. 1. Note that at low temperatures the *p-p* chain predominates, at high temperatures the C-N cycle.

At temperatures appreciably below 10^7 °K the conversion of hydrogen into helium proceeds at a very slow rate. But some other light nuclei, in the presence of hydrogen, react appreciably even at these lower temperatures. A deuteron, in particular, will absorb a proton and release energy in the form of a gamma ray already at about 10^6 °K. The various isotopes of Li, Be, and B absorb protons at slightly higher temperatures, an alpha particle being emitted in each process. The net result is the destruction of Li, Be, and B and the building up of He^3 and He^4. The He^3 produced is in turn converted into He^4. Many of the relevant reactions have already been investigated to some extent in the laboratory and more thorough investigations are under way. It is hoped that in the near future all these reaction rates will be known, at least to within a factor of two.

He^4 cannot react with a proton, since Li^5 does not exist. Furthermore two He^4-nuclei cannot react with each other, since Be^8 is unstable. The only process by means of which He^4 nuclei can be destroyed, is one requiring very high temperature and density (ÖPIK, 1951a; SALPETER, 1952b), namely

$$3He^4 \rightarrow C^{12} + \gamma + 7{\cdot}3 \text{ MeV}. \qquad \ldots.(2)$$

The detailed mechanism of this reaction is rather complicated, proceeding via intermediate steps involving the unstable Be^8 and a resonance level in C^{12} which was only discovered quite recently. Once C^{12} has been produced, it can absorb another He^4-nucleus giving O^{16} and a gamma ray. O^{16} in turn can be converted into Ne^{20}, Ne^{20} into Mg^{24}, *etc.* At present the rates of these reactions can only be calculated very crudely (any of the results could be in error by a factor of more than 100). But present indications are that C^{12}, O^{16}, and Ne^{20} are produced in roughly comparable proportions, whereas the production of Mg^{24}, *etc.*, is much too slow.

At much higher temperatures still even C, O, and Ne nuclei can react with each other, even in the absence of hydrogen and helium (SALPETER, 1952b; HOYLE, 1946). A large number of different reactions can take place, including for instance the collision between two carbon nuclei. Most important are reaction chains in which some very energetic photons (from the "high energy" tail of the Planck distribution of black-body radiation) break up some nuclei which have a relatively low binding energy. In most of these processes alpha-particles or protons are produced which are again absorbed by other nuclei. The end result of such reaction chains is that less stable nuclei are destroyed and more stable ones built up. There is thus tendency to build up the nuclei which have the largest binding energy per particle, namely those in the iron region.

In Table I we summarize some of the groups of reactions we have discussed. The temperatures stated are roughly the ones at which these reactions proceed at an appreciable rate. It should be noted that, although a large number of reactions are involved in the groups following the formation of helium, the total energy release in all these reactions is small compared with that of the H-He conversion.

Table I. Required temperature and energy release of typical reactions

Initial nuclei	D^2	He^3, Li, Be, B	H^1	He^4	C^{12}, O^{16}, Ne^{20}
Final nuclei	He^3	He^4	He^4	C^{12}, O^{16}, Ne^{20}	Fe-region
Energy release per nucleon (in MeV)	3	2	7	1	0·5
Temperature (in 10^6 °K)	1	5	8	100	(1 to 2) $\times 10^3$

3. Main Sequence Stars

Let us assume we know the mass M of a star, the law of energy production in its interior (as a function of temperature, density, and chemical composition), and that the star is in complete equilibrium (temperature, radius, *etc.*, steady). We can then calculate all the remaining physical properties of the star, including its radius R, absolute luminosity L, and central temperature T_c, provided that we assume a specific "stellar model" and chemical composition. For a main sequence star the model involves the assumptions that most of the energy is produced in a small region near the centre of the star and that the chemical composition is roughly uniform throughout the star. The chemical composition of stellar atmospheres is known at least roughly, the concentration (by mass) Y of helium being 5–30 per cent, the concentration Z of all heavier elements about $\frac{1}{3}$–3 per cent and the bulk of the star consisting of hydrogen.

For the Sun such calculations have been carried out by several authors (Epstein and Motz, 1953; Naur and Strömgren, 1954), using the known solar mass $M_\odot$ and the law of energy production for the proton-proton chain (see Fig. 1). The calculated values for R and L agree surprisingly well with the accurately known observational values. These results certainly confirm one's faith in the theory of stellar interiors and in the *p-p* chain as the energy source. In fact one can now invert the above argument, assume the observed values of M, R, and L and calculate the composition parameters Y and Z. The latest calculation for the Sun gives (Naur and Strömgren, 1954):

$$T_c = 13{\cdot}5 \times 10^6\ °\mathrm{K},\ Y = 0{\cdot}25,\ Z = 0{\cdot}01. \qquad \ldots(3)$$

The calculated values for Y and Z are, unfortunately, very sensitive to any small errors made in the computation. Thus the values in, for example, (3) are not yet very accurate, but it is hoped that reliable values will be available in the not-too-distant future—when accurate formulae for the opacity coefficients become available for numerical solutions on electronic computing machines.

For other main sequence stars the picture is not quite as complete as for the Sun, largely due to the lower precision of the observational data. Some calculations have been done for red dwarf stars whose physical properties are fairly well known.

Earlier calculations (ALLER *et al.*, 1952), using the accepted rate of energy production for the *p-p* chain, gave a slight but systematic discrepancy with observation, the calculated luminosity being higher than the observed one. Recent theoretical work on the effect of convective envelopes of red dwarfs (OSTERBROCK, 1953), however, predicts lower central temperatures (and hence lower calculated luminosity) than a simple calculation. According to this work theory and observation agree also for red dwarfs, to the accuracy expected.

As will be seen from Fig. 1, the *p-p* chain dominates the energy production at low temperatures, the C-N cycle at high temperatures. For the Sun (NAUR and STRÖMGREN, 1954), the *p-p* chain exceeds the carbon cycle's energy production by a factor of about 100, and for the cooler red dwarfs by an even larger factor. But for the hotter and more luminous main sequence stars (all O and B and most A stars) the carbon cycle should predominate. Only a few quantitative calculations have been done so far in this region of the main sequence. The physical properties of Sirius A can be fitted by assuming the rate of energy production given by the carbon cycle (SWIHART, 1953). Qualitatively, at least, observations on even more luminous main sequence stars are also compatible with the carbon cycle.

4. STELLAR EVOLUTION AND DISCUSSION

There is at present no generally-accepted theory of the evolution of stars. However, the following crude picture, based mainly on oversimplified theoretical considerations, may at least serve as the basis of comparison with observation.

In regions containing gas and dust under "suitable" conditions, condensations of various masses form. Such a condensation (initially still with fairly low density and temperature) begins to radiate and to contract gravitationally, its temperature, temperature gradient, and luminosity rising. The time taken from such a proto-star to shrink to its dimensions on the main sequence depends on its gravitational energy content and its luminosity throughout the contraction. Presumably a lower limit to this time is obtained by assuming a luminosity throughout the contraction equal to the main sequence value. This time is about $2{\cdot}5 \times 10^7$ years for a proto-star of solar mass $M_\odot$, about 10^6 years for $3M_\odot$, *etc.* If the proto-star contained an appreciable amount of D, Li, Be, or B it will burn these elements at different stages of its contraction, increasing the total contraction time slightly.

Once the star reaches its main sequence position, the energy production from the H $\rightarrow$ He conversion keeps pace with the radiative energy loss and the contraction stops. The time taken for the conversion of only 1 per cent of the stellar mass from hydrogen into helium is as long as 10^9 years for a star of mass $M_\odot$, a few times 10^7 years for $3M_\odot$.

Let us assume the star is poorly mixed, which seems likely at least for a considerable fraction of stars (ÖPIK, 1951b). Hydrogen is then converted into helium almost exclusively in the hottest central regions of the star. Thus a core of different composition (and mean molecular weight) from the rest of the star is formed. When this core contains about 10 per cent of the mass of the star, its configuration ceases to be stable (SCHÖNBERG and CHANDRASEKHAR, 1942). Its core will contract and heat up, while its envelope expands and cools (SANDAGE and SCHWARZSCHILD, 1952). At least in the early stages of this development the star will become much redder and slightly more luminous (red giant or subgiant). Details of this development are not yet known, but it seems likely that in the centre of at least some such stars high enough temperatures and densities are reached for the conversion of He into C, O, Ne (and possibly even for the further reactions outlined in Section 2). It should be remembered that even after leaving the main sequence the bulk of the star still consists of hydrogen and the bulk of the energy production still comes from the H-He conversion in a thin shell just outside the core. But it is rather likely that the onset of further nuclear energy production in the centre of the core can alter the structure of a star. Such processes may be involved in some types of variable stars or supergiants.

It also seems likely that at some late stage in the development of a star, it will become unstable enough to lose mass from its outer layers, either gradually or catastrophically. This loss of mass will presumably stop when the mass of the remaining core is less than the Chandrasekhar limit and this core will eventually become a white dwarf. A white dwarf contains essentially no hydrogen in its interior (MESTEL, 1952), has too low a temperature for nuclear reactions involving helium or heavier nuclei, and gravitational contraction is prevented by the high pressure of the degenerate electron gas. Having no possible sources of energy left, the white dwarf will cool down very slowly and, like old soldiers and some famous generals, merely fade away.

The above outline of stellar evolution at best represents a sort of average picture, since the evolution of different stars may well depend strongly on various factors. In particular, the structure of a star (as well as the timescale of the evolution) depends strongly on its mass and its development will be affected by its original rotational velocity, its magnetic field and by the presence or absence of accretion from nearby gas clouds (HOYLE and LYTTLETON, 1949). But the structure of main sequence stars is well understood, the proton-proton chain certain as energy source for the Sun and cooler main sequence stars, the carbon cycle very probable for hotter ones. Much theoretical work has been done on the structure of red giants by SCHWARZSCHILD and his co-workers, HOYLE, LYTTLETON, H. and C. M. BONDI, and others (for references see a recent paper by ROY (1952)). There probably are different types of red giant stars and the theory of SCHWARZSCHILD and SANDAGE outlined above may only apply to subgiants in globular clusters (and other systems belonging to the type II population). The absence of very luminous main sequence stars and the luminosity and colour of subgiants present in globular clusters (ARP, BAUM, SANDAGE; 1952) agree well with this theory.

REFERENCES

ALLER, L., *et al.* . . . . 1952 *Ap. J.*, **115**, 328.

ALLER, L. H. . . . . 1954 *Astrophysics*, Vol. 2 (Ronald Press, New York).

ARP, W., BAUM, W. A. and SANDAGE, A. R. . . 1952 *Astron. J.*, **57**, 4.

BETHE, H. A. . . . . 1938 *Phys. Rev.*, **54**, 248.

BETHE, H. A. and CRITCHFIELD, C. L. . . . . 1938 *Phys. Rev.*, **54**, 248.

BETHE, H. A. and MARSHAK, R. E. . . . . . 1939 *Rep. Progr. Phys.*, **6**, 1.

CHANDRASEKHAR, S. . . . . 1951 Chap. 14 in *Astrophysics, A Topical Symposium* (A. HYNEK, Ed., McGraw-Hill, New York).

EPSTEIN, I. and MOTZ, L.. . . . . 1953 *Ap. J.*, **117**, 311.

FOWLER, W. A. . . . . 1954 *Rev. mod. Phys.* (in press).

FRIEMAN, E. and MOTZ, L. . . . . 1953 *Phys. Rev.*, **89**, 648.

GAMOW, G. and CRITCHFIELD, C. L. . . . . 1949 *Theory of Atomic Nucleus*, p. 264 (The Clarendon Press, Oxford).

HOYLE, F. . . . . . 1946 *M.N.*, **106**, 343.
HOYLE, F. and LYTTLETON, R. A. . . . . . 1949 *M.N.*, **109**, 631.
JOHNSON, M. . . . . . 1950 *Astronomy of Stellar Energy* (Faber, London).
MESTEL, L. . . . . . 1952 *M.N.*, **112**, 583 and 598.
NAUR, P. and STRÖMGREN, B. . . . . . 1954 *Ap. J.*, **119**, 365.
ÖPIK, E. J. . . . . . 1951a *Proc. Roy. Irish Acad.*, Å**54**, 49.
1951b *M.N.*, **111**, 278.
OSTERBROCK, D. E. . . . . . 1953 *Ap. J.*, **118**, 529.
ROY, A. E. . . . . . 1952 *M.N.*, **112**, 484.
SALPETER, E. E. . . . . . 1952a *Phys. Rev.*, **88**, 547.
1952b *Ap. J.*, **115**, 326.
1953 *Ann. Rev. Nucl. Sci.*, **2**, 41.
SANDAGE, A. R. and SCHWARZSCHILD, M. . . . . 1952 *Ap. J.*, **116**, 463.
SCHÖNBERG, M. and CHANDRASEKHAR, S. . . . . 1942 *Ap. J.*, **96**, 161.
SWIHART, T. L. . . . . . 1953 *Ap. J.*, **118**, 577.

The Effective Pressure Exerted by a Gas in Turbulent Motion

G. K. BATCHELOR

Trinity College, Cambridge

SUMMARY

This paper examines the way in which the effective mean pressure exerted by turbulent motion in a gas varies when the gas is subjected to an adiabatic homogeneous distortion. The results depend on the directional properties of the distortion and of the turbulence, and on the speed with which the distortion is carried out. In certain simple cases the effective normal pressure due to the turbulence is proportional to the nth power of the mean density, where n has a value which sometimes is as small as unity and sometimes as large as three. The dependence of total pressure (thermal plus turbulent) on mean density may therefore change when the turbulent pressure becomes greater than the thermal pressure, *i.e.* when the turbulent velocity fluctuations become supersonic.

1. INTRODUCTION

A SIGNIFICANT feature of current research in theoretical astrophysics is the emphasis laid on the existence of turbulent motion in large-scale gaseous bodies. This very proper emphasis is based on the notion that non-turbulent relative motion of a fluid is almost always unstable and that the energy of the random turbulent motion usually grows until it is an appreciable fraction of the total kinetic energy. The general effect of the turbulence is to act as an exchange mechanism between different parts of the gas and to produce fluctuations in the values of macroscopic quantities—that is, to produce all those effects also produced by the random thermal motion of the discrete particles of the gas. The magnitude of the effect of turbulent motion of the gas will sometimes be larger than that of thermal motion of the particles, depending on the nature of the effect concerned. The intention in this note is to consider briefly the effective pressure exerted by the gas as a result of its turbulent motion, and to show that in certain circumstances turbulent fluid behaves under adiabatic compression as a polytropic gas. The corresponding index in the pressure-density relation seems in some cases to be larger than both the index for the thermal motion of the particles and the effective index for radiation, so that in such cases variations of the state of the gas when the density is large will be dominated by the turbulent pressure.

We begin with a demonstration, in the manner of REYNOLDS' classical work on turbulence, that the effective turbulent pressure is directly related to the kinetic energy of the turbulent motion in unit volume of the gas. Let v_i be the i-component of the macroscopic, or continuum, velocity at any point in the gas, and ρ and p the gas density and pressure. Then the force equation for a small volume element of the gas acted on by an external force with components X_i per unit mass is

$$\frac{\partial \rho v_i}{\partial t} + \frac{\partial \rho v_i v_j}{\partial x_j} = \rho X_i - \frac{\partial p}{\partial x_i}, \qquad \ldots.(1)$$

(repeated suffixes being summed) on the assumption that viscous stresses are negligible. If the motion is turbulent, we can write

$$v_i = U_i + u_i,$$

where U_i is the average value of v_i and u_i is the turbulent fluctuation in velocity. (The averages involved in turbulent motion are best thought of as probability averages, *i.e.* as averages over the values which the quantity would take if the whole dynamical development of the system occurred a large number of times; occasionally averages over a large volume of fluid or over a large time will give the same results and will be more useful.) Then on taking the average of all terms in (1) we find

$$\frac{\partial \bar{\rho} U_i}{\partial t} + \frac{\partial \bar{\rho} U_i U_j}{\partial x_j} = \bar{\rho} X_i - \frac{\partial}{\partial x_j}(\bar{p}\delta_{ij} + \overline{\rho u_i u_j}) + \varepsilon, \qquad \ldots.(2)$$

where the term

$$\varepsilon = -\frac{\partial \overline{\rho u_i}}{\partial t} - \frac{\partial}{\partial x_j}(U_i \overline{\rho u_j} + U_j \overline{\rho u_i}),$$

will usually be small compared with the other terms since the mechanical connection between fluctuations in density and velocity is weak. If ε is neglected, equation (2) shows that the effect of the turbulence on the distribution of mean velocity is to add a term $\overline{\rho u_i u_j}$ to the stress tensor. It is in this sense that the turbulence adds to the effective "pressure" exerted by the gas; we note, however, that the thermal motion of the discrete particles of the gas is without appreciable directional preference so that the diagonal components of the thermal stress tensor are dominant, whereas the turbulent motion may not be isotropic and the non-diagonal components of $\overline{\rho u_i u_j}$ may be important.

This result is probably familiar to everyone. Let us now go on to ask two questions which are of interest in astrophysics. First, when is the turbulent pressure more important than the thermal pressure?; and second, how does the turbulent pressure change when the gas is expanded or compressed? The answer to the first question emerges from the relation

$$\frac{\overline{\rho u_i u_i}}{\bar{p}} \approx \frac{\bar{\rho}\, \overline{u_i u_i}}{\bar{p}} = \gamma M^2, \qquad \ldots.(3)$$

(γ = ratio of specific heats of the gas), which shows that the two pressures have comparable magnitudes when the Mach number of the turbulence M (= ratio of root-mean-square velocity fluctuation to the velocity of sound corresponding to the mean density $\bar{\rho}$ and mean gas pressure $\bar{p}$) is of order unity. Thus, as expected, the turbulent pressure is larger than the gas pressure when the turbulence is supersonic, *i.e.* when the turbulent velocity fluctuations are larger than the thermal velocities of the gas particles.

The answer to the more important second question is not as evident. Two complicating factors occur in the problem. First, it is to be expected that the turbulence will be changed by compression of the gas in a way which depends on how the compression is carried out, so that an imposed strain of general type (although possibly still spatially homogeneous) must be considered in place of an isotropic expansion or compression. Second, the relaxation time of turbulent motion, *i.e.* the time required for turbulence to adjust itself to a new mean velocity distribution or to new boundary conditions, is not negligibly small as it is in the case of thermal motion of the gas particles, so that some account must be taken of the speed with which the distortion or strain is carried out. With regard to this latter difficulty, it does not seem to be possible to consider the general case, but the two extremes of very rapid and very slow homogeneous distortions have certain simple features. When the distortion takes place so quickly that the simultaneous relative displacement of elements of the gas produced by the initial velocity distribution (or produced by the external force acting on the new arrangement of the fluid) is small compared with the relative displacement produced by the distortion, all elements of the fluid are subjected to the same distortion since the gas behaves approximately as though it were initially at rest. The change in the velocity distribution (both mean and fluctuating) is then due to the consequent rearrangement of the vorticity, or angular velocity, distribution. On the other hand, when the distortion is very slow, there will be sufficient time for complete adjustment of the turbulence to the new conditions presented at each stage of the distortion. We proceed to consider these two extreme cases separately.

2. Cases of Rapid Distortion

The simplest kind of distortion is a homogeneous isotropic strain, *i.e.* a uniform expansion or compression, in which all line elements of the fluid are multiplied by a factor e, say. The direction of the angular velocity vector at each point of the fluid is unchanged, and the requirement that the angular momentum of each portion of the gas shall remain unchanged implies that the velocity of each element of the gas is divided by the factor e. Hence if suffix 0 denotes the initial state of the gas,

$$\begin{aligned} \overline{\rho u_i u_j} &\approx \bar{\rho}\, \overline{u_i u_j} \\ &= e^{-5}(\bar{\rho})_0(\overline{u_i u_j})_0 \\ &\approx \left(\frac{\bar{\rho}}{\bar{\rho}_0}\right)^{5/3} (\overline{\rho u_i u_j})_0, \end{aligned} \qquad \ldots .(4)$$

showing that all components of the turbulent stress tensor behave, for adiabatic changes, like the thermal stress in a monatomic gas. This is not surprising, since there are no "internal" degrees of freedom of the fluid continuum assumed in discussions of turbulent motion, and the energy of compression is distributed among the translational degrees of freedom in proportion to the energy they already possess.

It is possible to work out the change in the turbulent stress tensor consequent upon a rapid distortion which is spatially homogeneous but which is not the same in all directions (Ribner and Tucker, 1952; Batchelor and Proudman, 1953). However, the details of the general case are rather complicated, and will not be reproduced here since the turbulent pressure does not vary with density in the simple manner of a polytropic gas. The reason for the complexity is that in general the direction of the angular velocity vector at any point in the fluid is changed by the distortion, so that the changes in the three components of the velocity are not independent and cannot be considered separately. Two further special cases, additional to the one leading to (4), in which the direction of the angular velocity vector at each point is unchanged, are (*a*) one principal extension ratio large compared with the other two, and (*b*) two principal extension ratios equal and large compared with the third extension ratio. We have to consider advanced stages of these distortions, at which the angular velocity vector has already been turned considerably and has approximately attained its asymptotic direction. With further increase in the relative magnitude of the one principal extension in case (*a*) or of the two ratios in case (*b*), conservation of the angular momentum of any portion of the gas requires the velocity at any point to change according to the relation

$$u_i \approx (u_i)_0/e_i \text{ (no summation)}, \qquad \ldots .(5)$$

where the initial instant specified by suffix 0 must now be an instant at an advanced stage of the distortion and e_i is a (principal) extension ratio in the i-direction. Hence

$$\begin{aligned} \overline{\rho u_i u_j} &\approx (\overline{\rho u_i u_j})_0/e_1 e_2 e_3 e_i e_j \\ &\approx \frac{(e_1 e_2 e_3)^{2/3}}{e_i e_j}\left(\frac{\bar{\rho}}{\bar{\rho}_0}\right)^{5/3}(\overline{\rho u_i u_j})_0, \end{aligned} \qquad \ldots .(6)$$

so that the relation between turbulent pressure and mean gas density can be a power law with an index either greater or less than 5/3, according to the choice of direction of the normal stress or pressure force. In the interesting case of a disk-shaped mass of fluid for which e_1 and e_2 are equal and large compared with e_3, the normal stress in the direction of the axis of the disk is $\overline{\rho u_3{}^2}$, and

$$\overline{\rho u_3{}^2} \approx \left(\frac{e_1}{e_3}\right)^{4/3}\left(\frac{\bar{\rho}}{\bar{\rho}_0}\right)^{5/3}(\overline{\rho u_i u_j})_0.$$

Thus if the disk is flattened still further by decrease of e_3 with e_1 and e_2 remaining constant, we shall have $\bar{\rho} \propto e_3{}^{-1}$, and the turbulent pressure offered to the boundary at right angles to the direction of the compression is proportional to the cube of the mean gas density. This large value of the index is made possible by the fact that the velocity components u_1 and u_2 in the plane of the disk receive a share of the energy provided by compression of the gas only inasmuch as the mean density increases.

3. CASES OF SLOW DISTORTION

At the other extreme, we can consider a distortion which is slow enough to permit complete adjustment of the turbulence to the new mean velocity distribution or boundary conditions presented at each stage of the distortion.* This requires specification of the source of the energy of the turbulence (there must be a continual replenishment of the turbulence energy for the case of a slow distortion to have any significance, since unreplenished turbulence would simply dissipate itself and disappear), so that it is difficult to find general conclusions. Two possible continuing sources of energy of the turbulence suggest themselves. The first is convection from sources of heat within the fluid, but there seems to be too little known about turbulence accompanying convection for consideration of this case to be useful. Another possible source is the mean velocity distribution. In the simple and interesting case

* The argument here will be rougher than in the cases of rapid distortion.

in which the mean velocity distribution takes the form of a differential rotation about an axis of symmetry, and the distortion has symmetry about the same axis, the mean velocity of each circular filament of fluid will change in the manner required for preservation of angular momentum (assuming that no appreciable change in the spatial distribution of mean angular momentum is brought about by means of turbulent exchange). We can also make use of the well-known empirical fact that the equilibrium magnitude of turbulent velocity fluctuations is proportional to the maximum mean velocity difference in the fluid, for boundaries of similar shape, and with the neglect of any effects due to change of the Mach number of the motion. Hence if the axis of symmetry is a principal axis of the strain, with extension ratio e_3, we shall have for the change in the mean velocity

$$U_i = (U_i)_0/e_1 \quad (e_1 = e_2), \qquad \dots.(7)$$

and then since all three components of the turbulent velocity come to equilibrium with the new mean velocity distribution,

$$\overline{u_1^2}/(\overline{u_1^2})_0 = \overline{u_2^2}/(\overline{u_2^2})_0 = \overline{u_3^2}/(\overline{u_3^2})_0 = e_1^{-2},$$

and

$$\begin{aligned} \overline{\rho u_i u_j} &\approx \bar{\rho}\, \overline{u_i u_j} \\ &= \frac{\bar{\rho}_0}{e_1 e_2 e_3} \frac{(\overline{u_i u_j})_0}{e_1^2} \\ &\approx \left(\frac{e_3}{e_1}\right)^{2/3} \left(\frac{\bar{\rho}}{\bar{\rho}_0}\right)^{5/3} (\overline{\rho u_i u_j})_0. \end{aligned} \qquad \dots.(8)$$

The behaviour of all components of the stress tensor is the same in this case, and the turbulent pressure varies with $\bar{\rho}$ and with the ratio e_3/e_1. If the fluid is being contracted in the axial direction, with no change in the lateral plane (*i.e.* $e_1 = e_2 = 1$), the turbulent pressure is proportional to $(\bar{\rho}/\bar{\rho}_0)^n$ and $n = 1$ (in this case no change occurs in the mean velocity distribution), whereas if the fluid is expanding in the lateral plane without change in the axial direction, the index is 2. If the distortion is isotropic we have the value 5/3 for the index; the same value was found for an isotropic distortion which takes place rapidly, so that it seems very likely that this will be the approximate value of the index for isotropic distortions of all speeds in the case of a rotationally symmetric mean velocity distribution.

4. Application of the Results

With the aid of results of the above kind it may be possible to analyse the history of masses of gas which are being subjected to distortion, for example, by explosive expansion or by gravitational contraction. Many factors other than turbulent pressure are involved, and a complete discussion requires a wider knowledge of astrophysics than the author possesses. One or two features of the problem are perhaps worthy of comment and may lead to further developments. It has been shown that in some cases the turbulent pressure varies with mean density, under adiabatic distortion, in the same way as for a polytropic gas, *i.e.* turbulent pressure is proportional to $(\bar{\rho}/\bar{\rho}_0)^n$. Such cases were special only in the sense that the kind of distortion considered was of a fairly simple kind, so that it seems reasonable to regard the qualitative features of such cases as having general significance. It was also found that the value of the index n could be more than 5/3. 5/3 is the maximum possible value of the index arising from the thermal motion of the particles of the gas, and the effective value of the index for radiation is 4/3 (EDDINGTON, 1927). Thus there exist cases in which the rate of increase of turbulent pressure with increase of mean density, as brought about by adiabatic compression of a specified kind, is greater than that of gas thermal pressure or of radiation pressure. In such cases, and provided the turbulent pressure is not negligible (*i.e.* provided the Mach number of the turbulence is not small compared with unity), questions of equilibrium between gravitational force and pressure gradient will need to be decided with due account of the existence of the turbulent pressure.

The case of a disk-shaped mass of gas which has a rotationally symmetrical distribution of mean velocity, and which is being flattened still further by gravitational self-attraction (the gravitational force in the plane of the disk being resisted by the centrifugal force of the rotation) is one of particular interest in that it may represent one part of the process of formation of galaxies, and possibly of other aggregates of interstellar matter. According to the analysis of this note, the value of the index (for the turbulent pressure) when further flattening of the disk occurs is 3 if the whole flattening process takes place more-or-less "instantaneously" (using the mean velocity of the fluid as a standard), and is 1 if the flattening occurs so slowly that the turbulence is in balance with the mean velocity distribution at each stage of the flattening. The order of magnitude of the duration of an axial contraction which is neither slow nor fast in the above senses will be given by the time required for an element of fluid in a typical position to make one circuit round the axis. Possibly some of the axial contractions that occur as a result of self-gravitation are closer to being "rapid" than to being "slow". If so, the index n will approximate to 3, and will surely be greater than 5/3. If the values of the index n for the turbulent pressure and the index γ for the gas thermal pressure are widely different, there will be a strong tendency for the equilibrium state of the disk of gas to be such that the two pressures are of the same order of magnitude, for this is the state at which further contraction would generate a very large turbulent pressure which in turn would resist the contraction. Thus there may exist a tendency for flattened disks of gas to settle down to a state in which the root-mean-square of the turbulent velocity fluctuations is of the same order as the thermal velocities in the gas. The occurrence of a Mach number of the turbulence in the neighbourhood of unity may have some general significance, if these ideas are correct.

REFERENCES

BATCHELOR, G. K. and PROUDMAN, IAN . . . 1953 "The Effect of Rapid Distortion of a Fluid in Turbulent Motion", *Quart. J. Mech. appl. Math.*, **6**.

EDDINGTON, A. S. . . . . 1927 *The Internal Constitution of the Stars.* University Press, Cambridge, p. 29.

RIBNER, H. S. and TUCKER, M. . . . . . . 1952 "Spectrum of Turbulence in a Contracting Stream". National Advisory Committee for Aeronautics, U.S.A., Technical Note No. 2606.

"Turbulence", Kinetic Temperature, and Electron Temperature in Stellar Atmospheres*

P. L. Bhatnagar, M. Krook, D. H. Menzel, and R. N. Thomas

Harvard College Observatory, Cambridge, Massachusetts U.S.A.

Summary

The phenomenological use of the term "astronomical turbulence" is reviewed and earlier conclusions that the physical nature of the phenomenon is more likely anisotropic mass-motion, or jet-prominences, than the customary aerodynamic turbulence are restated. The primary problem under such conditions is the relative importance of mechanical energy-transport and momentum transport in perturbing the structure of the atmosphere. The problem of the difference between kinetic temperatures of the atoms and electrons is treated, and it is concluded that the difference is negligible in those parts of the stellar atmosphere which are in a statistically-steady state.

1. Introduction

Theoretical investigations of the outer layers of a star have been based, for the most part, on the "standard" model of a stellar atmosphere. The following assumptions specify this model:

(A1) Radiative equilibrium.
(A2) Hydrostatic equilibrium of the matter.
(A3) Local thermodynamic equilibrium for the matter.
(A4) Neglect of electric and magnetic fields.

The physical structure of a standard model atmosphere is then determined completely (at least in principle) when its effective temperature, surface gravity, and chemical composition are specified. In particular, these data determine unambiguously the density and temperature distribution for the model. A prediction of the spectrum follows from additional assumptions on the mode of line formation. The term "turbulence" has been used in a rather vague way to characterize certain discrepancies between actual observations and predictions from the model.

At least six types of observational results provide the background for the discussion of "turbulence" in stellar atmospheres. These are:

(T1) The extent of the atmosphere.
(T2) The density distribution.
(T3) A random, line-of-sight velocity deduced from the so-called Doppler segment of the curve of growth.
(T4) A random, line-of-sight velocity deduced from measures of line profiles.
(T5) The directly observed, or the inferred, presence of some form of prominence activity.
(T6) The occurrence of anomalously high excitation; this feature deserves more attention than it has received to date.

* The research reported in this document has been made possible through support and sponsorship extended by the Geophysics Research Division of the Air Force Cambridge Research Centre, under Contract No. AF19(604)-146. It is published for technical information only, and does not necessarily represent recommendations or conclusions of the sponsoring agency.

As regards (T1–4), the significant features in the present context are:

(i) The existence of atmospheres considerably more distended than in the standard model (T1, 2).

(ii) The existence of random velocities other than, and in addition to, thermal ones predicted by the standard model.

What astronomers have usually called "turbulence" in a stellar atmosphere consists of a purely phenomenological interpretation of the residual random-velocity fields obtained in (T3, 4) when the standard-model thermal motions are subtracted out. No detailed model of an atmosphere actually containing such velocity fields has so far been presented. In a rather more qualitative way, "turbulence" is also invoked to explain the anomalous distension of an atmosphere (T1, 2).

The physical self-consistency of such atmospheric velocity fields has not as yet been investigated. Thus far investigations have not gone beyond an attempt to apply the statistical theory of isotropic turbulence (in the aerodynamic sense), to macroscopic motions in a stellar atmosphere. However, the conviction appears to be growing that a system of anisotropic mass-motions, or jet-prominences (T5), represents astronomical "turbulence" more closely than does the ordinary concept of aerodynamic turbulence.

In discussions of astronomical "turbulence" it is common practice to neglect the effects of the coupling between the macroscopic turbulent velocity field and the microscopic thermal motion. The energy of the turbulence dissipated in this way has implicitly been assumed to be negligible. The few exceptions to this remark have considered the possibility that the coupling between turbulent motion and thermal motion may be significant only in the outermost layers—chromosphere or corona—and hence could not easily be observed. Most generally, it has been assumed that the existence of a distended atmosphere does not imply the existence of appreciable temperature anomalies in the regions responsible for the formation of absorption lines.

Most studies of "turbulence" in stellar atmospheres depend on an assumption of isotropy. We believe that this assumption is unjustified and oversimplified. One should consider carefully and in detail the source of the astronomical "turbulence" in order to relate it to the conventional turbulence observed in physical laboratories. Physically, the onset of turbulence is associated either with thermal instability in a quiescent medium, or with dynamical instability in fluid flow. The earliest suggestion (ROSSELAND, 1929) of the stellar occurrence of hydrodynamic turbulence was in connection with stellar rotation, the high Reynolds number of the flow in the atmosphere leading to dynamical instability. On the other hand, turbulence associated with the star's convective zone results from thermal instability. From the standpoint of mechanisms for generating turbulence, we note then that the axes of preferred motion are at right angles in these two cases. Consequently, we believe that the general problem should eventually be investigated from the standpoint of origin.

In preceding discussions (THOMAS, 1947, 1948, 1949) one of us has suggested that astronomical "turbulence" consists of strongly anisotropic mass-motions, preferentially in the radial direction, associated with high kinetic temperatures induced by this mass-motion. Any such system of mass-motion will produce atmospheric distention. If energy transfer predominates over momentum transfer, the distension will effectively result from a high kinetic temperature. On the other hand, for sufficiently large momentum transfer, the distension results primarily from dynamic support. We believe that the Wolf-Rayet atmosphere represents an extreme case of the dynamic support. The solar chromosphere-corona forms an example of a case where the relative importance of momentum and energy transfer is not yet clear. Over a period of years, another of us has urged that a prominence model of a stellar atmosphere deserves more attention (MENZEL, 1930, 1931, 1939, 1946). If small jet-prominences comprise essentially the entire atmosphere, momentum transfer will be more significant. However, if the jets eject into a gaseous medium, to which they supply energy, then energy transfer may be the more important. One may well question whether such a system of prominences could exist without creating a gaseous substratum. Then, given such a substratum, even if the main process is momentum transfer, one must investigate the extent to which the coupling of a jet-prominence with the medium may produce a high kinetic temperature.

The translational kinetic energy of a jet-prominence is converted into thermal energy of the atmosphere by elastic collisions between jet particles and atmospheric particles. A jet atom loses its excess kinetic energy after a few collisions with atmospheric atoms. On the other hand, because of the small ratio of electron mass to atom mass, a jet atom would require many more collisions with atmospheric electrons to lose its excess kinetic energy. Thus the jet loses its translational energy by raising the kinetic temperature T_k of the atmospheric atoms (which, hereafter, we call Process *a*). The elastic collisions of atmospheric electrons with atmospheric atoms and ions then tend to raise the electron temperature T_e (Process *b*). This last process competes with such processes as radiative capture, collisional excitation with ultimate radiation, which tend to lower the electron temperature (Process *c*). The question then arises whether a steady-state atmosphere with such mass-motions could exist in which the atom kinetic temperature T_k is significantly greater than the electron kinetic temperature T_e.

From the observational aspects, the above picture would be clarified if the atmosphere kinetic temperature, T_k, could be measured. Working within the "standard" model, one customarily assumes that T_k is equal to the electron temperature, T_e, and identifies the latter with the excitation temperature, T_{ex}. The uncertainty in this last identification has been discussed (THOMAS, 1949). In the solar chromosphere, however, the largest current source of uncertainty seems to be the discrepancy between T_k and T_e. T_k is largely fixed by optical observations. T_e comes from radio observations. It appears that $T_e < T_k$. We must, however, always remember that the radio and the visual observations refer to very different layers of the solar atmosphere, and the comparison of T_k and T_e is by extrapolation.

Criticism has been directed at the eclipse observations of line profiles (MIYAMOTO, 1953), largely because self-absorption may cause the observed profiles to appear wider than the true Doppler profile at temperature T_k. We believe it significant that these same line profiles place an upper limit on the tangential component of any turbulent velocities. This limit lies considerably below the value required for the interpretation of the chromospheric density gradient as arising from an isotropic turbulence. Thus we return to the question whether a purely radial set of motions can exist without raising T_k. We have attempted, as yet unsuccessfully, to investigate how a departure from Maxwellian velocity distribution for the electrons may affect the value of T_e inferred from the radio observations. The question then arises whether there could actually exist a difference between T_e and T_k.

The literature on discharge-tube phenomena contains reference to situations with $T_k < T_e$. In investigating the structure of very strong shock-waves (non-steady state), we have encountered a situation with $T_k > T_e$. The question then arises whether $T_k > T_e$ might occur in a steady-state in a stellar atmosphere.

2. INVESTIGATION OF THE POSSIBILITY OF THE STEADY-STATE CONFIGURATION $T_k > T_e$

With reference to the processes of energy transfer mentioned above, a balance of process (a), energy transfer from atoms in a jet-prominence to those in the atmosphere, against process (b), energy transfer from atmospheric atoms to atmospheric electrons, should give the atom kinetic temperature in terms of the energy dissipation from the mechanical energy source. A balance of (b) against (c), energy dissipation by the electrons, should give the above discussed comparison between electron and atom kinetic temperatures. The "kind" of energy dissipation, (a), to be expected from a system of jets, or prominences, has previously been discussed in a rough manner for the solar spicules (THOMAS, 1948). Here, we consider the T_e, T_k distinction. We consider a hydrogen atmosphere in the following.

Process b: elastic collisions in the medium transfers energy from the atoms to the electrons at the rate:

$$\boldsymbol{E}_b = \sigma_b \frac{m_e}{m_a} N_e N_a v_e k(T_k - T_e), \qquad \ldots.(1)$$

where the subscripts a and e refer to atom and electron, respectively; where N is the particle density; m, the mass; v_e the mean electron velocity; T_e the kinetic temperature of the electrons and T_k of the atoms; and k, BOLTZMANN's constant.

We have derived the cross sections σ_b, for energy transfer in elastic collisions, for the respective cases of neutral atoms and of ions interacting with electrons. We shall omit the details of the derivation in this paper.

Atom-electron:

$$\sigma_b = 8\sqrt{\frac{2}{3\pi}}\left[1 + \frac{m_e}{m_a}\frac{T_k}{T_e}\right]^{1/2} \sigma \sim 4.10^{-16}, \qquad \ldots.(2)$$

where σ is the ordinary kinetic cross-section, of the order of 10^{-16} cm^2.

Ion-electron:

$$\sigma_b = 2\sqrt{\frac{2\pi}{3}}\left(1 + \frac{m_e}{m_a}\frac{T_k}{T_e}\right)^{-3/2} (kT_e)^{-2}\varepsilon^4 \ln\left[1 + (4kT_e\varepsilon^{-2}N_e^{-1/3})^2\right]$$

$$\simeq 1{\cdot}2 \,.\, 10^{-4} T_e^{-2}\left[1 + 0{\cdot}13 \ln T_e N_e^{-1/3}\right], \qquad \ldots.(3)$$

where ε is the electronic charge.

Process c: the electrons lose energy in three ways. (1-c): by interaction with the ions they may show an excess of emission over the absorption of radiation in free-free (ff) transitions. Defining:

$$\lambda = \frac{T_n}{T_e}; \quad P_1 = 3{\cdot}2 \,.\, 10^{-6} k N_i N_e T_e^{-1/2}. \qquad \ldots.(4)$$

We obtain the rates of loss and gain by these two processes (MENZEL, 1937):

$$\text{emission} \quad = \boldsymbol{E}_{ff} = P_1 \frac{kT_e}{2hR} \qquad \ldots.(5)$$

$$\text{absorption} = \boldsymbol{A}_{ff} = \boldsymbol{E}_{ff} \, . \, \overline{W} \, . \, \lambda \left[1 + \left(\tfrac{1}{2} - \frac{1}{1+\lambda}\right) + \left(\tfrac{1}{3} - \frac{1}{2+\lambda}\right) \ldots \right]. \qquad \ldots.(6)$$

where R is the RYDBERG constant, h is PLANCK'S constant, and the subscript i refers to ions. In the derivation of (5) we have taken the incident radiation field to consist of black-body emission at a temperature T_r, with a mean dilution factor $\overline{W}$.

(2-c): the electron component of the medium loses energy by free-bound and gains energy by bound-free transitions (MENZEL, 1937):

$$\text{emission} \quad = \boldsymbol{E}_{fb} = P_1 n^{-3} - \boldsymbol{F}_{fb} E_n, \qquad \ldots.(7)$$

$$\text{absorption} = \boldsymbol{A}_{bf} = \overline{\overline{W}} P_1 n^{-3} X_n e^{X_n} Y_n^{-1} \sum_{a=1}^{\infty} e^{-aY_n}[b_n a^{-1} - (a+\lambda)^{-1}] - \boldsymbol{F}_{bf} E_n \qquad \ldots.(8)$$

where the $\boldsymbol{F}$'s represent numbers of transitions per cubic centimetre per second:

$$\boldsymbol{F}_{fb} E_n = P_1 n^{-3} X_n e^{X_n} E_i(-X_n), \qquad \ldots.(9)$$

$$\boldsymbol{F}_{bf} E_n = P_1 n^{-3} X_n e^{X_n} \sum_{a=1}^{\infty} \{b_n E_i(-aY_n) - E_i(-Y_n[a+\lambda])\}. \qquad \ldots.(10)$$

These formulas refer to hydrogenic atoms, with n as the principal quantum number:

$$X_n = \frac{E_n}{kT_e}, \quad Y_n = \frac{E_n}{kT_r}, \quad E_n = \frac{E_1}{n^2}, \quad E_1 = hR.$$

The quantity b_n specifies the degree of departure from thermodynamic equilibrium.

$$b_n = N_n \left[h^3 (2\pi m_e k T_e)^{-3/2} \frac{\omega_n}{2} e^{X_n} N_i N_e\right]^{-1}$$

where $\qquad \omega_n = 2n^2$

the statistical weight of level n. We have introduced the exponential integral,

$$E_i(-u) = \int_u^{\infty} x^{-1} e^{-x} dx.$$

(3-c): inelastic collisions represent the third process of energy loss by the electrons. These collisions may ionize the atom or merely excite it. The net energy loss in ionization processes is (MATSUSHIMA, 1952):

$$\boldsymbol{C}_{ni} = C_{ni} E_n = 3{\cdot}6 \, . \, 10^{-21} P_1 N_e T_e^{1/2} b_n (1 - b_n^{-1}) X_n (1 + X_n) \omega_n Q_{ni} \quad \ldots.(11)$$

where the subscripts ni refer to the ionization from level n. Q_{ni} is the mean cross-section for ionization, expressed in atomic units. For collisional excitation from level n to n'' and the reverse, we have the net loss (MATSUSHIMA, 1952):

$$\boldsymbol{C}_{nn''} = C_{nn''}(E_n - E_{n''}) = C_{nn''} E_{nn''}$$

$$= 3{\cdot}6 \, . \, 10^{-21} P_1 N_e T_e^{1/2} b_n \left(1 - \frac{b_{n''}}{b_n}\right) e^{X_{n''}} (1 + X_{nn''}) X_{nn''} Q_{nn''} \omega_n. \qquad \ldots.(12)$$

Thus, equating energy gains by process b to energy loss by processes c, we have

$$\boldsymbol{E}_b = (\boldsymbol{E}_{ff} - \boldsymbol{A}_{ff}) + \sum_n [(\boldsymbol{E}_{fb} - \boldsymbol{A}_{bf}) + \boldsymbol{C}_{ni} + \sum_{n''>n} \boldsymbol{C}_{nn''}]. \qquad \ldots (13)$$

We now introduce the conditions that the atmosphere be in a statistically steady state, an assumption requiring that N_n, N_i, and N_e be constant. Given T_e, T_n, $\overline{\overline{W}}$, and N_e, we may employ the constancy of N_n to determine the b_n. Constancy of the numbers of ions and electrons supplies the auxiliary condition:

$$\sum_n (C_{ni} + \boldsymbol{F}_{bf} - \boldsymbol{F}_{fb}) = 0 \qquad \ldots (14)$$

balance of the total number of transitions up and down.

If, now, we set:

$$\boldsymbol{E}_{fb}{}^* = \boldsymbol{E}_{fb} + \boldsymbol{F}_{fb} E_n; \quad \boldsymbol{A}_{bf}{}^* = \boldsymbol{A}_{bf} + \boldsymbol{F}_{bf} E_n$$

and use (14), (13) becomes:

$$\boldsymbol{E}_b = (\boldsymbol{E}_{ff} - \boldsymbol{A}_{ff}) + \sum_n [(\boldsymbol{E}_{fb}{}^* - \boldsymbol{A}_{bf}{}^*) + E_{1n}(\boldsymbol{F}_{fb} - \boldsymbol{F}_{bf} - C_{ni}) + \sum_{n''>n} \boldsymbol{C}_{nn''}] \quad \ldots (15)$$

Introducing the explicit forms, we obtain:

$$\left.\begin{aligned}
&\sigma_b \frac{m_e}{m_a} N_e N_a v_e k(T_k - T_e) \\
&= P_1 \frac{kT_e}{2hR} \left[1 - \overline{\overline{W}}\lambda \left\{1 + \left(\tfrac{1}{2} - \frac{1}{1+\lambda}\right) + \left(\tfrac{1}{3} - \frac{1}{2+\lambda}\right) \ldots \right\}\right] \\
&+ P_1 \sum_n [n^{-3}\{[1 - \overline{\overline{W}} X_n Y_n{}^{-1} e^{X_n} \sum_{a=1}^{\infty} e^{-aY_n}(b_n a^{-1} - [a+\lambda]^{-1})] \\
&+ X_{1n} e^{X_n} E_i(-X_n) \\
&\quad [1 - E_i{}^{-1}(-X_n) \sum_{a=1}^{\infty} \{b_n E_i(-aY_n) - E_i(-Y_n[a+\lambda])\}] \\
&+ X_{1n} \,.\, 3{\cdot}6 \,.\, 10^{-21} N_e T_e{}^{1/2} \omega_n b_n [-(1 - b_n{}^{-1})(1 + X_n) Q_{ni} \\
&+ \sum_{n''>n} (1 - b_n \,.\, b_n{}^{-1})(1 + X_{nn''}) X_{nn''} X_{1n}{}^{-1} e^{X_{n''}} Q_{nn''}]].
\end{aligned}\right\} \quad \ldots (16)$$

Now, employ (4), introduce

$$v_e = (3kT_e/m_e)^{1/2}$$

and insert numerical values:

$$\begin{aligned}
T_e = T_k - 8{\cdot}7 \,.\, 10^{-9} T_e{}^{-1} \sigma_b{}^{-1} N_i N_a{}^{-1} \{3{\cdot}2 \,.\, 10^{-6} T_e [1 - \ldots] \\
+ \sum_n [n^{-3}(1 - \ldots) + 7{\cdot}2 \,.\, 10^{-21} N_e T_e{}^{1/2} X_{1n} n^2 b_n (< 10)]\} \quad \ldots (17)
\end{aligned}$$

Consider now the term $\sigma_b{}^{-1} N_i N_a{}^{-1}$. This term represents elastic collisions of electrons with both ions and neutral atoms. Thus, by (3) and (4):

$$\begin{aligned}
\sigma_b{}^{-1} N_i N_a{}^{-1} &= \left[\sigma_b(\text{ion}) + \sigma_b(\text{atom}) \frac{N_a}{N_i}\right]^{-1} \\
&= [1{\cdot}2 \,.\, 10^{-4}(1 + 0{\cdot}13 \ln T_e N_e{}^{-1/3}) + 4 \,.\, 10^{-32} N_e T_e{}^{1/2} \textstyle\sum n^2 b_n e^{X_n}]^{-1} T_e{}^2
\end{aligned} \quad \ldots (18)$$

where we have used the dissociation equation in the form modified for departure from thermodynamic equilibrium:

$$N_a = N_i N_e \left[\frac{h^2}{2\pi m_e k T_e}\right]^{3/2} \sum_n n^2 b_n e^{X_n}. \quad \ldots .(19)$$

For the solar chromosphere, we can set the limits:

$$5 \cdot 10^3 < T_e < 10^5; \quad 10^{10} < N_e < 10^{12}$$

and all computations (MATSUSHIMA, 1952) give $b_1 < 10^6$; $b_{30} \sim 1$. Observations show the effective hydrogen continuum to begin at about H_{35}. The excluded-volume principle allows the summation in the partition function to be terminated at this point. In such cases the ion-electron term in (18) always predominates. The range of values specified above fits the inequality:

$$\sigma_b^{-1} N_i N_a^{-1} < 10^4 T_e^2.$$

Further, the bracketed quantity in (17), in general, does not exceed 10. Thus:

$$T_k - T_e < 10^{-3} T_e.$$

This analysis thus leads to the significant fact that the difference between T_e and T_k is indeed negligible. This result appears to remove all possibility of reconciling the radio and optical observations in terms of differences between kinetic and electron temperature, at least for the case of a homogeneous atmosphere in a statistically-steady state. One further possibility of reconciliation lies in the investigation mentioned earlier of the influence upon the free-free emission of a non-Maxwellian velocity distribution for the electrons. Another, lies in an atmosphere with marked non-uniform distribution of temperatures, such as the models proposed by GIOVANELLI (1949) and by HAGEN (1953) of a hot chromosphere containing cool columns of gas.

3. COMMENT ON THE RESULT $T_e \sim T_k$

In the literature on electric discharge in gases one finds frequent reference to situations in which $T_e > T_k$. Physically, our above conclusion, that $T_k \sim T_e$ in a steady state, is a consequence of the fact that in electron-atom collisions the elastic cross-section is much larger than any of the inelastic cross-sections. Consequently, to obtain a steady state with $T_e > T_k$, an efficient mechanism for the atom to dissipate energy must be introduced. In the discharge tube, wall-effects may provide such an energy sink. However, this mechanism must be explicitly introduced, and the rate at which it dissipates energy must be calculated. Thus, for example, ALFVÉN (1950) balances the energy gained by an electron drifting in an electric field against its loss in elastic collision with ions and atoms. The possibility of a steady-state is simply assumed, and the conditions then holding are determined by *assuming* the above energy balance to be exact. No mention, however, is made of the mechanism of energy loss by the atoms. Unless this mechanism is introduced, the assumed situation clearly violates the conservation of energy.

REFERENCES

ALFVÉN, H.	1950	*Cosmical Electrodynamics*, p. 45 (Oxford University Press).
GIOVANELLI, R. G.	1949	*M.N.*, **109**, 298.
HAGEN, J. P.	1953	Private communication.

MATSUSHIMA, S.	1952	*Ap. J.*, **115**, 544.
MENZEL, D. H.	1930	*Amer. Astron. Soc. Abstracts*, **6**, 370.
	1931	*Popul. Astron.*, **39**, 16.
	1937	*Ap. J.*, **85**, 330.
	1939	*Popul. Astron.*, **47**, 6, 66, 124.
	1946	*Physica*, **12**, 768.
MIYAMOTO, S.	1953	Private communication.
ROSSELAND, S.	1929	*M.N.*, **89**, 49.
THOMAS, R. N.	1947	*Ap. J.*, **52**, 158.
	1948	*Ap. J.*, **108**, 130.
	1949	*Ap. J.*, **109**, 500.

The Intensity of Emission Lines in Stellar Spectra

P. WELLMANN

Hamburger Sternwarte, Hamburg-Bergedorf, Germany

SUMMARY

The conditions in an extended stellar atmosphere may be adequately explained by the theory of diluted radiation if it is postulated that the optical depth in the lines is small enough.

In order to check this basic postulate the dependence of emission intensities on the numbers of atoms is developed by analogy with the curve of growth for absorption lines. This relation, called curve of growth for emission lines, is used in analysing observed emission spectra and reveals that the assumption of small optical depths is permissible in many cases.

Moreover, it is seen that even a strong emission spectrum may be affected by the shell absorption. If these two phenomena, *i.e.* spectra, are treated as a unit, the curves of growth for absorption and emission lines appear as limiting cases of a more general relation between line intensity, transition probability, and characteristic data of the atmosphere. This relation describes the difference between various shell spectra and the changes in the spectrum of an expanding atmosphere, and it can be used to study the structure of a shell.

A generalized equation of ionization has to be introduced for the investigation of emission spectra.

1. THE ORIGIN OF EMISSION LINES

IT is now almost unanimously recognized that emission lines visible in the light of a star as a whole are caused by the action of diluted radiation in extended atmospheres. The observations of the solar chromosphere, of Be-stars, WR-stars, and novae, prove that the monochromatic emission is generated in the outer low-density parts of the stellar envelopes, and it is obvious that on the average the more extended atmospheres have the stronger emission.

Formerly the presence of emissions comparable in intensity with the continuum was attributed entirely to the relatively large mass of gas in the shells. This explanation implicitly presupposed normal excitation and local thermodynamical equilibrium. In this case, the flux of continuous radiation in a line frequency may suffer absorption and scattering after emerging from the deeper parts of the star. At any rate it is diminished by absorption. Under the most favourable circumstances, the radius of the star being infinitely small, the scattering does not influence the total energy output. But normally a fraction of the back flux of scattered light will be re-absorbed in the interior of the star and this part is also lost in the line frequency. If we assume radial symmetry of star and envelope, absorption lines always result irrespective of the extent and mass of the atmosphere. Therefore, if an emission line exists, some energy has been transferred from one frequency to another by monochromatic processes. This rather general argument proves that the principle of detailed balancing is no longer valid, and that it must be replaced by the more general principle that a steady state exists. We write down the steady-state conditions for a gas consisting of atoms with n energy levels and m allowed transitions between these levels, the atoms being in a radiation field of density $u(\nu)$. Let the population of a given level be N atoms per cm^3, then we have

$$\frac{dN}{dt} = 0 \text{ and } \frac{du(\nu)}{dt} = 0. \qquad \ldots.(1)$$

The members of the first set of equations have the form

$$\sum_i N_i B_{ij} u_{ij} + \sum_k N_k (A_{kj} + B_{kj} u_{kj}) = N_j [\sum_i (A_{ji} + B_{ji} u_{ji}) + \sum_k B_{jk} u_{jk}] \quad \ldots.(2)$$

where A, B = transition probabilities; i, k = levels lower or higher than level j. Only $n - 1$ of these equations are independent of one another. They contain $n - 1$ ratios N_i/N_k because they are homogeneous in N. But they contain also as unknowns the functions $u(\nu)$. The equations of the second set are equivalent to the equations of transport of radiation. Explicitly written they run

$$\cos\theta \frac{dI_{ik}}{ds} = \frac{h\nu}{4\pi} \left[-N_i B_{ik} \frac{4\pi}{c} I_{ik} + N_k (A_{ki} + B_{ki} u_{ki}) \right], \qquad \ldots.(3)$$

with

$$u_{ik} = \frac{1}{c} \int I_{ik} d\omega.$$

The absolute values of the N's now appear as unknowns, and we have to add the condition

$$\sum_i N_i = N_t$$

that the total number of atoms is final.

There are $n + m$ equations and $n + m$ unknown functions. On principle it is possible to solve the problem. These considerations are valid in the same form if continuous states are included. The continuum will be divided into elements of infinitely small width dE which will be treated as an infinite number of discrete states. This consideration also holds for a mixture of different atoms and molecules. Furthermore, the result is the same if collisional excitation is introduced in (2). The solution is completely independent of any concept of temperature. The distribution with frequency of the radiant energy and the energy distribution among the free electrons are not assumed, but are deduced from the theory.

The practical solution of this system of integro-differential equations is extremely complicated, but we can reduce the problem to simpler forms by adequate approximations.

It has been proved that the population of the continuous states, that is the frequency of electrons with certain velocities, is very well described by the Maxwell distribution, even when the departures from thermodynamic equilibrium in the populations of the discrete states are rather strong. By introducing the Maxwell law an electron temperature is defined and determined.

The equation of radiation transport can be transformed into

$$u(\nu) = \frac{1}{c}\int_{\omega} I^* e^{-\tau} d\omega + \frac{1}{c}\int_{\omega}\int_{s} \varepsilon e^{-\tau} ds d\omega, \quad \tau = \int_{s} \kappa N ds, \qquad \ldots .(4)$$

where ε is the energy emitted by the unit of volume in the shell. I^*, the intensity of radiation at the photosphere, is given as a boundary condition. It is customary to write

$$I^*(\nu, \omega) = \frac{2h\nu^3}{c^2} \cdot \frac{1}{e^{\frac{k\nu}{kT_s}} - 1} = I_P(\nu),$$

i.e. to neglect the darkening to the limb of the star and to introduce a radiation temperature T_s. The next simplification implies a significant loss of generality. Instead of using the transport equations (3), we substitute in (4) an appropriate assumed functional form.

First it is assumed that the second term is negligible in consequence of the low density in the shell. Hence

$$u(\nu) = \frac{4\pi I_P}{c}\int_{\omega} e^{-\tau}\frac{d\omega}{4\pi} = u_P(\nu) \,.\, W = u_P(\nu) \,.\, e^{-\bar{\tau}} \,.\, \frac{\omega^*}{4\pi}, \qquad \ldots .(5)$$

ω^* is the solid angle subtended by the star;

$$W_g = \frac{\omega^*}{4\pi} \approx \left(\frac{r^*}{2r}\right)^2 \qquad \ldots .(6)$$

is the geometrical dilution of the radiation from the star, $e^{-\bar{\tau}}$ is the dilution by absorption in the medium between the star and the point in question. In most computations of the populations of discrete levels, however, W is assumed to be constant, *i.e.* only the geometrical part is considered. This is permissible if the optical depth of the atmosphere is small in all wavelengths. In some cases W was assumed to be a certain constant at wavelength longer than the Lyman limit, and a constant considerably smaller below this limit, in order to allow for a strong absorption in the Lyman continuum.

If we introduce arbitrary functions we are no longer sure that the solution is correct with regard to the energy balance. Therefore it is necessary to introduce the condition of constant energy content. This yields a relation between T_e and T_s (BAKER, MENZEL, ALLER, 1938; WOOLLEY, 1947)*.

2. THE POPULATION OF DIFFERENT ATOMIC LEVELS

The problem is now reduced to a rather simple form: a gas of temperature T_e is excited by a radiation of temperature T_s, the radiation being diluted by the factor W. How are the populations changed relative to the equilibrium at T_e?

The first treatment of this problem was published by ROSSELAND (1926) in his classical papers on the theory of cycles. He began with three discrete states and showed that the transition 1—3 from the ground state to the highest level was not compensated by 3—1, but partly by 3—2—1, and vice versa. The probability of the cycle 1—2—3—1 is smaller than that of the reverse cycle by a factor equal to W.

* Strictly, the results of the next section should be known in establishing the relation between T_e and T_s. But it is actually possible to solve the problem in advance with sufficient accuracy by a process of successive approximations.

This means that the emission in the two lines of longer wavelength was strengthened considerably at the expense of the line of shortest wavelength. In most cases W is very small: therefore one of the cycles nearly ceases to operate and the lines of longer wavelength appear as pure emissions. Since this work by ROSSELAND it has been possible to understand the existence of emissions. Further application of his method showed that the first results give a qualitative guide for many other cases. The steps of small energy difference always gain emission at the expense of transitions of larger or of the largest energy difference. The emission caused by recombination is a limiting case. It is qualitatively understood on the same basis why in a series the emission becomes weaker with increasing quantum number.

Another important rule follows from these early computations of dilution effects by ROSSELAND (1926), AMBARTSUMIAN (1933), STRUVE (1935a), and other authors. When the ground state combines with higher levels, the populations of these levels are reduced, approximately by a factor W, relative to that of the ground state. But, if one of the higher states is metastable, its population is nearly independent of the dilution and bears the normal relation to that of the ground state. This holds approximately in more complicated cases with a larger number of levels if one or more metastable levels are included. These metastable levels reduce the effects of the cycles; their high populations cause a relative strengthening of all the lines due to transitions starting from these levels. This is the key to the explanation of a great many observations.

In special cases this general, qualitative reasoning has been replaced by numerical solutions of the steady-state conditions assuming a larger number of energy levels with individual values of frequencies and transition probabilities.

This has been done by CILLIÉ (1936) and by MENZEL and his collaborators for hydrogen, all atomic constants of which are known functions of the quantum numbers (MENZEL, 1937; MENZEL, BAKER, 1937; BAKER, MENZEL, 1938; MENZEL, ALLER, BAKER, 1938). The solution of the complete infinite system of linear equations (2) is given in series development. The results are relative populations N_i/N or certain coefficients b_i, defined by MENZEL in the following way.

In thermodynamic equilibrium the number of atoms in the state i may be written

$$N_{i0} = \frac{\tilde{\omega}_i}{2\tilde{\omega}^+} N_0^+ N_e \frac{h^3}{(2\pi m k T_e)^{3/2}} e^{\frac{\chi_i}{kT_e}},$$

since the Saha equation is valid. In the general problem the departure from this expression is measured by the factor b, where

$$N_i = b_i N_{i0}, \quad N^+ = b_k N_0^+.$$

Since the system is homogeneous in N and in b, the b-values may and must be normalized. This is done by putting $b_k = 1$ for the ion such that

$$N_i = b_i \frac{\tilde{\omega}_i}{2\tilde{\omega}^+} N^+ N_e \frac{h^3}{(2\pi m k T_e)^{3/2}} e^{\frac{\chi_i}{kT_e}}. \qquad \ldots (7)$$

The natural definition would perhaps have been to introduce $b = 1$ for the ground state, but MENZEL's definition has many advantages for practical computation.

MENZEL and his collaborators calculated the b's for the first fifteen levels of hydrogens as functions of T_s and T_e for the limiting case of very low W or very strong dilution. In this case b_i/b_1 is approximately proportional to W, hence all the b_i's except b_1 are independent of W, and b_1 is proportional to $1/W$.

The values of b for He II, valid for T_s and T_e, are the same as the hydrogen values for $T_s/4$ and $T_e/4$.

The b-values of He I have been calculated, but it is not possible here to use serial laws in order to establish a manageable infinite system. Hence STRUVE (1935b), later STRUVE and WURM (1938), used only six levels. GOLDBERG (1941) used eight levels and improved the method by introducing integrals as an approximation to the sum of the higher states, but he restricted the solution to $W \approx 0$. The present author used GOLDBERG's method and computed the b-values of thirteen levels for different W and of twenty-one levels for $W \approx 0$ (WELLMANN, 1952a). The introduction of the higher levels up to the quantum number 4 enables us to deal with some of the important He I-emission lines in the visual and photographic region of the spectrum. Furthermore, the extension of the work to larger W made the results applicable to atmospheres of intermediate heights and to the lower parts of high atmospheres. The He-spectrum is particularly interesting. Because of the two term systems and the two metastable levels, it shows many lines having very different behaviour. This makes He more important than hydrogen, the lines of which are too uniform in their reaction to dilution and temperature.

The results may be described by some general rules:

(1) The relative population of two levels is nearly independent of dilution: the population of the ground state and the metastable 2^1S-state.

(2) The higher singlet levels are depopulated proportionally to W.

(3) The population of the metastable level 2^3S gains a factor of from 10 to 20 relative to the ground state, but is independent of W if the dilution exceeds a certain limit.

(4) The population of the higher triplet states is proportional to W. Most of the triplets behave like the singlet states, except the n^3P-levels in which the atoms are ten to twenty times more frequent.

3. THE CURVE OF GROWTH FOR EMISSION LINES

The b-values furnish the basis for the construction of the curve of growth. When we derive a curve of growth for absorption lines, we plot a certain function of the line intensity against a theoretical intensity, which depends on the oscillator strength of the line and the population of the lower level given by the Boltzmann distribution. We expect the emission intensity, the oscillator strength f, and the b of the upper state, to be related. But, as emission does not exist without more or less reabsorption, the b-value of the lower level will also affect these functions.

In the following we have to remember that we are only trying to obtain a first idea of the relation. We have a complicated theory of the formation of absorption lines, but in dealing with absorptions in integrated starlight, use was generally made of an interpolation formula for the profiles (UNSÖLD, 1938; MENZEL, 1939) and only in more recent papers the theory of atmospheric models has been considered (WRUBEL, 1949, 1950, 1954). To the same degree of first approximation, we regard the emission of a homogeneous gas filling a cylindrical volume of height H. The intensity directed outward at the front surface is

$$I = \int_{\nu} \int_{s=0}^{H} \varepsilon e^{-\tau} ds = \frac{2h\nu^3}{c^2} \left(1 + \frac{W}{e^{\frac{h\nu}{kT_s}} - 1}\right) \frac{\tilde{\omega}_i}{\tilde{\omega}_k} \int_{\nu} \int_{\tau} \frac{N_k}{N_i} e^{-\tau} d\tau d\nu.$$

Assuming $N_{i,k}$ to be constant, we get

$$I = \frac{2h\nu^3}{c^2} \cdot \frac{\tilde{\omega}_i}{\tilde{\omega}_k} \frac{N_k}{N_i} \cdot A_\nu = \frac{2h\nu^3}{c^2} \frac{b_i}{b_k} \cdot A_\nu,$$

with
$$A_\nu = \int_{\nu=0}^{\infty} (1 - e^{-\kappa N_i H}) d\nu,$$

and after re-arranging

$$\frac{I}{\nu^4} \cdot \frac{\tilde{\omega}_k}{\tilde{\omega}_i} \frac{N_i}{N_k} = \text{const.} \frac{A\nu}{\nu}. \qquad \ldots .(8)$$

The theory of absorption gives a relation between

$$\frac{A\nu}{\nu} \text{ and } \frac{N_i f H}{\nu},$$

the well-known curve of growth for absorption in an absorption tube.

Because of (8) this may be regarded as a relation between

$$\frac{I}{\nu^4} \frac{\tilde{\omega}_k}{\tilde{\omega}_i} \cdot \frac{N_i}{N_k} \text{ and } \frac{N_i f H}{\nu}, \qquad \ldots .(9)$$

and this is the required curve of growth for the emission intensity. By this procedure we avoid the repetition of laborious integrations already discussed in the theory of absorption lines. The resulting curve is easily interpreted. The linear part expresses the fact that the emission is proportional to the transition probability and the number of atoms in the upper state, as long as the optical depth is small. When it becomes larger the reabsorption increases sensibly and causes a bending of the curve. The reabsorption limits the intensity in the middle of the emission profile to $W \cdot I_p$. When the number of atoms grows further, then only the width of the line can increase and the intensity follows a square-root law.

As an example of a Be star with strong emissions of H and He I, γ Cass has been investigated (WELLMANN, 1952c). From the curve of growth, constructed according to (9), the values $T_s \sim T_e \sim 10{,}000°$K and $W \sim 10^{-2}$ were derived. The figures are not defined very precisely, but as a first attempt to deduce these values from emission lines they may be sufficient. The Balmer lines fall on the linear part of the curve, as do most of the He lines, but some He lines in the upper part show the beginning of the bending. Hence we conclude that the lines do not suffer from appreciable reabsorption: some He lines with very strong population of the lower states are the only ones to be slightly affected. The optical depth is, therefore, small for most of the emissions, though the whole atmosphere participates in producing them. This results from the strong Doppler broadening of the lines.

The emission He I λ3888 is exceptional in that it is too weak compared with the prediction of the curve of growth. This is the strongest transition between the metastable level 2^3S and the higher triplet states, and the extremely high population of this level which results from the dilution will cause a very strong absorption in the line λ3888. In the spectra of shells with moderate height showing weak emissions, this line is often the only line which remains when the normal absorption spectrum is blotted out by dilution. However, the reabsorption is already allowed for in the curve of growth and the observed reduction must be the effect of another superposed absorption spectrum. That is easily explained by the absorption of the light of the star in the atmosphere.

4. THE COMPLETE SHELL SPECTRUM OF A BE-STAR

We have now to change our theoretical assumptions by including the shell absorption spectrum. This is at first sight a serious matter because we are apparently forced to seek the solution of a problem of radiation transfer. But fortunately the re-absorption is negligible, and we thereby escape all the difficulties. The equation of radiation transport reduces to a simple expression, and immediately leads to a formula for the flow of radiation in the emission lines at the outer boundary of an atmosphere with radial symmetry. We get, assuming a homogeneous envelope (WELLMANN, 1952b),

$$E = \alpha' h\nu(b_k D_{ki}\eta - b_i D_{ik}) N^+ N_e V,$$

with the function $\eta(W)$ according to the following table:

W . .	1·0	0·5	0·10	0·01	0
η . .	0·5	0·5	0·81	0·97	1

The product $b_i \,.\, D_{ik}$ is the frequency of the transition $i \rightarrow k$. E is the energy expressed in equivalent width, positive for emission, negative for absorption. V is the volume, α' a certain constant.

After some transformation we have

$$\frac{E}{\nu^4} \cdot \frac{\tilde{\omega}_k}{\tilde{\omega}_i} \cdot \frac{N_i}{N_k} = \text{const.} \frac{N_i f H}{\nu} \cdot \psi, \quad \psi = 1 - \frac{b_i D_{ik}}{b_k D_{ki}\eta}.$$

If no shell absorption is present, as in cases of extreme dilutions, $b_i \,.\, D_{ik}$ vanishes, $\psi = +1$ and the result is the linear part of our curve of growth for pure emission. The other limit is given by thermodynamical equilibrium, $W = +1$, $\eta = 0{\cdot}5$, $b_i = b_k = 1$, $D_{ik} = D_{ki}$, hence $\psi = -1$, and the result is the linear part of the normal curve of growth of absorption lines. Imagine a low atmosphere. It will show the normal absorption spectrum. If this atmosphere expands for some reason, the dilution factor diminishes and ψ grows. Consequently, the absorption lines are diminished by superposed emission. If the expansion is large enough, the absorption will be masked completely, later the emission will dominate and, finally, when $\psi = +1$ is reached, the emission will show the undisturbed maximum value.

The growth of ψ with dilution, however, differs from line to line. The lines He I λ6678, λ5875, and λ3888 are examples of a normal singlet line, a normal triplet line and a line of which the lower state is metastable. With $T_s = T_e = 25{,}000°$, we find

λ	W			
	1	0·1	0·01	0
6678	− 1	+ 0·65	+ 0·96	+ 1
5875	− 1	+ 0·09	+ 0·59	+ 1
3888	− 1	− 0·15	+ 0·09	+ 0·13

We see that in an atmosphere with moderate dilution (W from 0·5 to 0·1) the singlets have the strongest tendency to emission, the triplets have weaker emission or stronger absorption, while $\lambda 3888$ is still an absorption line. Even with extreme dilution, $\lambda 3888$ remains fainter than the curve of growth for pure emission predicts.

This is, at least on principle, what we observe in the spectra of different shell stars or in the spectra of stars with outer atmospheres of variable extent.

Finally, this formula not only describes the mean quality of an atmosphere, but helps in attacking the problem of stratification. If we want to allow for the variability of density and dilution with distance from the star, we must write:

$$E = 4\pi\alpha' h\nu \int_{r=r^*}^{\infty} (b_k D_{ki} \eta' - b_i D_{ik}) N^+ N_e r^2 dr.$$

The function in brackets is given by the theory and differs from line to line. If

$$N^+ N_e = f(r, a_1, a_2, \ldots)$$

is a given function containing n unknown parameters a, it is possible to deduce these parameters from the measured intensities of n lines. Because not all of the lines are suitable for actual use, the number of parameters is limited, but we can secure preliminary information on the structure of such an emitting atmosphere. The first tentative calculations show that in the case of γ Cassiopeiae an atmosphere with an exponential outward decrease of density fits the observations. The density gradient has been found, and the density itself has been derived from the absolute intensities of the emissions integrated over the whole atmosphere. The result is

$$\log N_e = 11{\cdot}8 - 0{\cdot}74 \frac{r}{r^*}$$

at the time of greatest extension of the gases. The total density is proportional to the electron density, because hydrogen and helium are almost completely ionized. The density decreases rather rapidly with the height, but nevertheless the effective atmosphere reaches to a radius of about $10r^*$. The emission chiefly originates in the outer parts, the absorption in an inner shell. Fig. 1 represents this model.

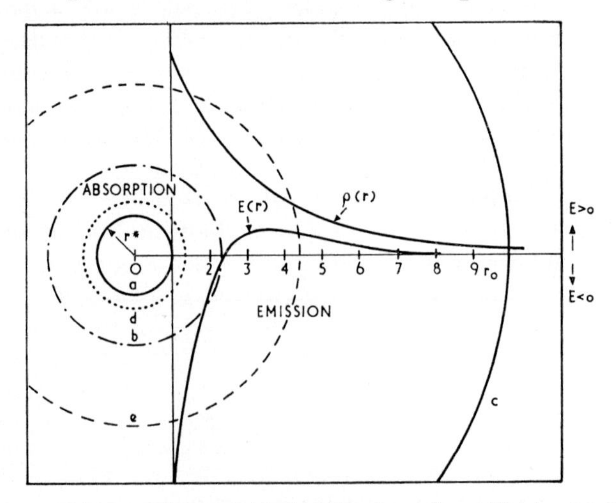

Fig. 1. Be star model-atmosphere. $\rho(r)$ is density of envelope, $E(r)$ is contribution to line intensity ($E > 0$ emission, $E < 0$ absorption). (*a*) Photosphere, (*b*) limit of absorbing region, (*c*) limit of effective emitting region, (*d*) effective radius of absorbing layer, (*e*) effective radius of emitting shell

5. IONIZATION BY DILUTED RADIATION

In the same manner as in the construction of the curve of growth for absorption, the individual curves for different ions or elements will be shifted parallel to the log (NfH/ν)-axis and combined into one final function. According to (9) and (7) these shifts depend on the abundance and degree of ionization of the elements. For instance, it is possible to find N(He II)/N(H II), N(He III)/N(He II), and other ratios which demand an interpretation in terms of T_s, T_e, and W.

The frequency of transitions between successive states of ionization, i and $k = i + 1$, may be expressed with the aid of the atomic absorption coefficient κ as follows.

The number of ionizations from level n of ion i, produced by radiation of frequency ν, is

$$4\pi N_{i,n} I_\nu(T_s) \,.\, \kappa_n \frac{d\nu}{h\nu} \text{ cm}^{-3}\text{-sec.}^{-1};$$

the corresponding number of recombinations is:

$$\frac{8\pi}{c^2} \cdot N_{k,1} \cdot N_e \frac{\bar{\omega}_{i,n}}{\bar{\omega}_{k,1}\bar{\omega}_e} \frac{h^3}{(2\pi mkT_e)^{3/2}} \kappa_n e^{\frac{\chi_n - h\nu}{kT_e}} \left(1 + \frac{c^2 I_\nu(T_s)}{2h\nu^3}\right) \nu^2 d\nu.$$

We now introduce the same assumptions concerning the departure from thermodynamic equilibrium as in the preceding sections, and the usual expression for the absorption by hydrogen-like atoms. After integration over all frequencies and summing up the contributions from all levels of the ion with the lower ionization, we have for the number of ionizations:

$$Z_{ik} = W \frac{8\pi}{c^2} \kappa_0 \,.\, N_i \frac{2}{B_i} e^{-\frac{\chi_i}{kT_e}} \,.\, \sum_n \frac{(Z+s)^4}{\bar{n}^3} \,.\, b_n e^{\frac{\chi_n}{kT_e}} \,.\, \phi_{ns} \text{ cm}^3\text{-sec.}^{-1}.$$

The analogous expression for the recombination is much more complex, therefore we restrict it to $W = 1$ and $W \ll 1$:

$$Z_{ki} = \frac{8\pi}{c^2} \kappa_0 \,.\, N_k N_e \,.\, \frac{2b_{k,1}}{\bar{\omega}_e B_k} \cdot \frac{h^3}{(2\pi mkT_e)^{3/2}} \cdot \sum_n \frac{(Z+s)^4}{\bar{n}^3} e^{\frac{\chi_n}{kT_e}} \begin{cases} \cdot\, \phi_{ne} & W = 1 \\ \cdot\, K\left(\dfrac{\chi_n}{kT_e}\right) & W \ll 1 \end{cases}$$

where we write

$$\kappa_0 = \frac{2^6\pi^4}{3\sqrt{3}} \frac{me^{10}}{ch^6}, \quad \phi_{ns} = \int_{\nu_n}^{\infty} \frac{1}{e^{\frac{h\nu}{kT_s}} - 1} \frac{d\nu}{\nu} = \sum_{\mu=1}^{\infty} K\left(\frac{\mu h\nu_n}{kT_s}\right), \quad K(x) = -Ei(-x),$$

B_i, B_k are the generalized partition functions

$$B = \sum b_n \bar{\omega}_n e^{\frac{\chi_n - \chi_i}{kT_e}},$$

$N_{i,k}$ is the number of all atoms in state i or k, Z is the atomic number, s the screening constant, and $\bar{n}$ the effective quantum number.

Now the generalized equation of ionization is

$$Z_{ik} = Z_{ki}$$

or, with $W \ll 1$,

$$\frac{N_{i+1}N_e}{N_i} = W \frac{B_{i+1}\bar{\omega}_e}{b_{i+1,1}B_i} \cdot \frac{(2\pi mkT_e)^{3/2}}{h^3} e^{-\frac{\chi_i}{kT_e}} \,.\, \frac{\sum_n \frac{(Z+s)^4}{\bar{n}^3} b_n e^{\frac{\chi_n}{kT_e}} \phi_{ns}}{\sum_n \frac{(Z+s)^4}{\bar{n}^3} e^{\frac{\chi_n}{kT_e}} \cdot K\left(\frac{\chi_n}{kT_e}\right)} \quad \ldots\ldots(10)$$

It is easily seen that equations of the same kind given by previous authors are equivalent to (10) or special cases of this formula (see STRÖMGREN, 1948). A numerical comparison with the earlier approximations (WELLMANN, 1952a), for instance in the case of He, shows that an error amounting to a factor of about 10 can result from disregarding the singular population of the metastable levels and the recombination in the higher levels, especially if T is large. Therefore the use of (10) is unavoidable in some problems of emission line intensity.

Most elements appear in more than two different states of ionization. We may ask whether it is permissible to calculate the abundances of higher ions by successive applications of (10). In principle, the degree of ionization is affected by transitions between all states of ionization, that is by the normal process as well as by the simultaneous escape or capture of several electrons. The probability of these multiple processes is very small, however. From a detailed discussion of the simplest case, the equilibrium between He I, He II, and He III, where the original Rosseland theory of three states applies exactly, it is found that the removal of the two electrons at a time has no appreciable influence.

In the case of high ionization, for instance of Fe, the situation is not so clear, but it may be estimated that the contribution of multiple ionization only affects the number of ions when this number is too small to give observable lines. Therefore we can use (10) throughout the analysis of an emission spectrum.

As an example, we may choose the spectrum of a nova. The intensities of H, He I, He II, [O I], [O III], [Fe VII], and [Fe X] had been measured in the spectrum of Nova Serpentis 1948 in its early nebular stage, 14th–18th May, 1948 (WELLMANN, 1954). From the He-lines the abundance ratio He III/He II is deduced, using the calculated b-values. The ratios O III/O I and Fe X/Fe VII are found from more qualitative estimates of the dilution effects, which at least give the right order of magnitude.

The observed intensities of the H, He-, and O-lines all refer to the whole volume of the main shell because they belong to the principal spectrum. The dilution factor is given by the geometrical relations according to (6). The radius of the main shell is the velocity of expansion multiplied by the time elapsed since the outburst, the radius of the star was about $1{\cdot}0\, r_{\odot}$. Hence in the average $W \sim 10^{-9}$. This equation of ionization contains W, N_e, T_s, and T_e, but the latter is cancelled almost completely by the dependence of b on the electron temperature. The value $N_e = 10^7$ is found from a model of the nova shell, which reproduces the observed spectrum-time relation. The ratio of He III/He II yields $T_s = 40{,}000° \pm 3000°$; the ratio H/He III yields $T_s = 46{,}000° \pm 3000°$, assuming normal He-abundance. The same procedure with the [O III] and [O I] lines had the result: $T_s = 23{,}000° \pm 2000°$. T_e is found from a relation between the intensity ratio [O III]($N_1 + N_2$)/[O III]λ4363, the ratio [O I]$\lambda\lambda$6300, 6363/[O I]λ5577, the electron temperature, and the density; (MENZEL, ALLER, HEBB, 1941). It seems to be comparatively high, namely, $T_e = 18{,}000° \pm 5000°$. The ratio Fe X/Fe VII, however, points to a much higher radiation temperature, about 120,000°, if these ions were located in the main shell. But this discrepancy is partly removed if it is realized that the radius of the region containing ions of such high energy will be much smaller than the region of lower ionization (WURM, SINGER, 1953). In addition, this is supported by a discussion of the observed intensity ratio [Fe VII]/[O III]. Hence W must be much larger than in the case of emissions originating in the main shell. The strength and the intensity ratio of Fe VII- and

Fe X-lines can be explained by assuming $T_s = 80{,}000° \pm 4000°$ and, (in an atmosphere immediately surrounding the star), a density 10^{-3} times that of the main shell. It should be noted that the resulting radiation temperature increases with the highest ionization potential involved, namely: O II, 35 eV, 23,000°; He II, 54 eV, 40,000°; Fe IX, 235 eV, 80,000°.

REFERENCES

AMBARTSUMIAN, V.	1933	*Poulkovo Obs. Circ.* No. 6.
BAKER, J. G. and MENZEL, D. H.	1938	*Ap. J.*, **88**, 52.
BAKER, J. G., MENZEL, D. H. and ALLER, L. H.	1938	*Ap. J.*, **88**, 422.
CILLIÉ, G. G.	1936	*M.N.*, **96**, 771.
GOLDBERG, L.	1941	*Ap. J.*, **93**, 244.
MENZEL, D. H.	1937	*Ap. J.*, **85**, 330.
	1939	*Popul. Astron.*, **47**, 6 and 66.
MENZEL, D. H., ALLER, L. H. and BAKER, J. G.	1938	*Ap. J.*, **88**, 313.
MENZEL, D. H., ALLER, L. H. and HEBB, M. H.	1941	*Ap. J.*, **93**, 230.
MENZEL, D. H. and BAKER, J. G.	1937	*Ap. J.*, **86**, 70.
ROSSELAND, S.	1926	*Ap. J.*, **63**, 218.
STRÖMGREN, B.	1948	*Ap. J.*, **108**, 252.
STRUVE, O.	1935a	*Ap. J.*, **81**, 66.
	1935b	*Ap. J.*, **82**, 252.
STRUVE, O. and WURM, K.	1938	*Ap. J.*, **88**, 93.
UNSÖLD, A.	1938	"Physik der Sternatmosphären", Berlin.
WELLMANN, P.	1952a	*Z. Astrophys.*, **30**, 72.
	1952b	*Z. Astrophys.*, **30**, 88.
	1952c	*Z. Astrophys.*, **30**, 96.
	1954	*Phys. Verhandlungen*, **5**, 126.
WOOLLEY, R. v. d. R.	1947	*M.N.*, **107**, 308.
WRUBEL, M. H.	1949	*Ap. J.*, **109**, 66.
	1950	*Ap. J.*, **111**, 157.
	1954	*Ap. J.*, **119**, 51.
WURM, K. and SINGER, O.	1952	*Z. Astrophys.*, **30**, 153.

Dynamo Theories of Cosmic Magnetic Fields

T. G. COWLING

Department of Applied Mathematics, University, Leeds

SUMMARY

The development of the dynamo theory of cosmic magnetic fields during the last twenty years is critically surveyed. After a general introduction to the theory, separate sections are devoted to the steady-state theory and the theory of turbulent magnetic fields. The problems raised but not answered by earlier work are generally surveyed, and the important questions needing a decisive answer are enumerated.

---

1. INTRODUCTION

IN 1930, Professor STRATTON was one of the examiners of my D.Phil. thesis at Oxford. In this thesis I discussed the dynamo theory, in the belief that it could provide the explanation of sunspot magnetic fields and possibly of the sun's general magnetic field. It therefore seems appropriate for me here to trace the development of the dynamo theory since then.

The dynamo theory considered here is that which ascribes the maintenance of a cosmical magnetic field to currents induced in moving conducting fluid by the already existing field. The maintenance mechanism thus resembles that in a self-exciting dynamo. The recent development of such theories was initiated by LARMOR (1919), in an attempt to explain sunspot fields and the sun's general field. He assumed the field, and the material flow, to be symmetric with respect to an axis, the lines of force and stream lines lying in planes through the axis. If material flows in towards the axis across the lines of force at great depths induced currents flow in circles round the axis here. Such currents are in the direction required to maintain the field, and LARMOR suggested that they might in fact explain the origin of the field. If in the upper layers the material flows outward away from the axis across the lines of force, currents flow in the reverse sense here; since, however, the material here is less conducting, the currents at greater depths can be expected to be dominant. A rough calculation suggests that the material motions necessary for maintenance of the field are very small—less than 10^{-2} cm per sec for sunspot fields, and less than 10^{-6} cm per sec for the sun's general field.

However, I was able in 1934 to show that a dynamo mechanism cannot work in the axially symmetric case. In this particular case the field possesses one or more "neutral rings". These are circles on which the field vanishes, with the axis of symmetry as axis; near the rings the lines of force are closed loops threaded on to the rings. Since there is no magnetic field on a neutral ring, no current can be induced in the moving material at the ring; moreover, since near the ring the material cannot move wholly inward (towards the ring) or wholly outward, the induced current cannot be in the same sense round the axis of symmetry at all points near the ring. But a field whose lines of force are loops threaded on the ring requires for its maintenance a current which does not vanish on the ring, and which is directed in the same sense round the axis of symmetry at all points near the ring. It follows that in the axially symmetric case dynamo maintenance of the field is impossible.

2. General Theory

This result was regarded for some time as barring the possibility of dynamo maintenance of cosmic magnetic fields. The dynamo theory was indeed invoked again by FRENKEL (1945) and by GUREVICH and LEBEDINSKY (1945), but their work appeared to have been done in ignorance of the result in question. However, repeated failures to provide an alternative mechanism led to a deeper examination of dynamo theories. My 1934 argument applied only to axially symmetric fields, or fields possessing a general similarity to these. The possibility that more general fields might be maintained by a dynamo mechanism now began to receive attention.

The new attack on the problem was led by ELSASSER (1946; 1947). He considered fields due to motions in a uniform incompressible sphere. Displacement currents can be neglected in the discussion; thus if $\boldsymbol{B}$ is the magnetic induction, $\boldsymbol{j}$ the current-density, and μ the magnetic permeability

$$\operatorname{curl} \boldsymbol{B} = 4\pi\mu\boldsymbol{j}. \qquad \ldots.(1)$$

As a consequence of this div $\boldsymbol{j} = 0$, so that any piling up of charge exerts no appreciable influence on the current flow. If $\boldsymbol{E}$ is the electric intensity and σ the conductivity (both measured in E.M.U.), and $\boldsymbol{v}$ the material velocity,

$$\boldsymbol{j} = \sigma(\boldsymbol{E} + \boldsymbol{v} \times \boldsymbol{B}) \qquad \ldots.(2)$$

and

$$-\frac{\partial \boldsymbol{B}}{\partial t} = \operatorname{curl} \boldsymbol{E}. \qquad \ldots.(3)$$

Eliminating $\boldsymbol{j}$ and $\boldsymbol{E}$ between these equations, and using the fact that div $\boldsymbol{B} = 0$, we find

$$\nabla^2 \boldsymbol{B} = 4\pi\mu\sigma \left(\frac{\partial \boldsymbol{B}}{\partial t} - \text{curl}\,(\boldsymbol{v} \times \boldsymbol{B})\right). \qquad \ldots.(4)$$

This equation indicates how the field varies in the presence of an assigned motion $\boldsymbol{v}$. The motion is regarded as arbitrary, save that it satisfies the continuity equation

$$\text{div}\ \boldsymbol{v} = 0. \qquad \ldots.(5)$$

ELSASSER obtained a formal solution of equation (4) as follows. When $\boldsymbol{v} = 0$, the equation indicates how a magnetic field decays as the energy of the currents flowing is converted into Joule heat. There are certain normal modes of decay in which the field decays exponentially; putting $\boldsymbol{B} = \boldsymbol{B}_r e^{-\lambda_r t}$, the field in these normal modes is given, inside the sphere, by

$$\nabla^2 \boldsymbol{B}_r = -\,4\pi\mu\sigma\lambda_r \boldsymbol{B}_r. \qquad \ldots.(6)$$

The different possible fields $\boldsymbol{B}_r$ have been considered by various authors, beginning with LAMB (1883). They form a complete set; that is, any field $\boldsymbol{B}$ generated by currents inside the sphere can be expressed in terms of a series

$$\boldsymbol{B} = \sum_r b_r \boldsymbol{B}_r. \qquad \ldots.(7)$$

ELSASSER assumed the field $\boldsymbol{B}$ satisfying (4), in the general case $\boldsymbol{v} \neq 0$, to be expanded in a series of form (7), in which the coefficient b_r are functions of the time. Equation (4) is then used to indicate how the coefficients vary with the time.

The fields $\boldsymbol{B}_r$ satisfy the orthogonality relation

$$\int \frac{1}{\mu} \boldsymbol{B}_r \,.\, \boldsymbol{B}_s d\tau = 0 \quad (r \neq s) \qquad \ldots.(8)$$

the integration being one over the whole of space. For if $\boldsymbol{E}_r e^{-\lambda_r t}$ and $\boldsymbol{j}_r e^{-\lambda_r t}$ are the electric intensity and current-density corresponding to $\boldsymbol{B}_r$,

$$\begin{aligned}\frac{4\pi}{\sigma}\int \boldsymbol{j}_r \,.\, \boldsymbol{j}_s d\tau &= 4\pi \int \boldsymbol{E}_r \,.\, \boldsymbol{j}_s d\tau \\ &= \int \boldsymbol{E}_r \,.\, \text{curl}\left(\frac{1}{\mu}\boldsymbol{B}_s\right) d\tau \\ &= \int \text{div}\left(\frac{1}{\mu}\boldsymbol{B}_s \times \boldsymbol{E}_r\right) d\tau + \int \text{curl}\,\boldsymbol{E}_r \,.\, \frac{1}{\mu}\boldsymbol{B}_s d\tau.\end{aligned}$$

The first of these two integrals vanishes by Green's Theorem, since the tangential components of $\boldsymbol{B}_s/\mu$ and $\boldsymbol{E}_r$ are continuous at the surface of the sphere. Hence, using an equation similar to (3),

$$\frac{4\pi}{\sigma}\int \boldsymbol{j}_r \,.\, \boldsymbol{j}_s d\tau = \lambda_r \int \frac{1}{\mu} \boldsymbol{B}_r \,.\, \boldsymbol{B}_s d\tau.$$

By symmetry, these also equal

$$\lambda_s \int \frac{1}{\mu} \boldsymbol{B}_r \,.\, \boldsymbol{B}_s d\tau.$$

Since $r \neq s$, equation (8) follows. The fields $\boldsymbol{B}_r$ are also supposed to be normalized to satisfy

$$\int \frac{1}{\mu} \boldsymbol{B}_r^2 d\tau = 1. \qquad \ldots.(9)$$

In terms of $\boldsymbol{j}_r$, equations (8) and (9) are equivalent to

$$\int \boldsymbol{j}_r \,.\, \boldsymbol{j}_s d\tau = 0 \qquad (r \neq s)$$
$$= \frac{\sigma\lambda_r}{4\pi} \quad (r = s). \qquad \ldots.(10)$$

Now by arguments similar to those used in deriving the orthogonality relations, if $\boldsymbol{j}$, $\boldsymbol{E}$, and $\boldsymbol{B}$ are the general vectors satisfying equations (2), (3) and (4),

$$\frac{4\pi}{\sigma} \int \boldsymbol{j} \,.\, \boldsymbol{j}_r d\tau = 4\pi \int (\boldsymbol{E} + \boldsymbol{v} \times \boldsymbol{B}) \,.\, \boldsymbol{j}_r d\tau$$
$$= - \int \frac{1}{\mu} \frac{\partial \boldsymbol{B}}{\partial t} \,.\, \boldsymbol{B}_r d\tau + 4\pi \int (\boldsymbol{v} \times \boldsymbol{B}) \,.\, \boldsymbol{j}_r d\tau.$$

In this substitute the expansion (7) and a similar expansion for $\boldsymbol{j}$. Then simplifying by equations (8), (9), and (10), we get

$$\lambda_r b_r = - \frac{db_r}{dt} + 4\pi \sum_s b_s \int (\boldsymbol{v} \times \boldsymbol{B}_s) \,.\, \boldsymbol{j}_r d\tau$$

or

$$\frac{db_r}{dt} = - \lambda_r b_r + \sum_s a_{rs} b_s \qquad \ldots.(11)$$

where

$$a_{rs} = 4\pi \int (\boldsymbol{v} \times \boldsymbol{B}_s) \,.\, \boldsymbol{j}_r d\tau. \qquad \ldots.(12)$$

Equation (11) indicates how the field given by (7) varies due to the joint action of electrical resistance and the motion.

3. Stationary Fields

The discussion now bifurcates. On the one hand, one may ask if steady motions of special kinds exist, able to maintain stationary fields; on the other, the question is whether less regular motions may maintain fields which, though not precisely stationary, none the less possess certain statistically steady properties. We discuss first the possibility of the maintenance of stationary fields; this has been considered especially by Bullard (1949).

For a stationary field, equation (11) becomes

$$\lambda_r b_r = \sum_s a_{rs} b_s. \qquad \ldots.(13)$$

Putting $r = 1, 2, 3, \ldots$ we get an infinite number of equations to determine the infinite set of quantities $b_1, b_2, b_3, \ldots$ These equations in general possess only the trivial solution $b_1 = b_2 = \ldots = 0$. However, a simple argument suggests that it should not be difficult to find motions for which the equations possess a non-zero solution, so that such motions can maintain the field corresponding to these values of $b_1, b_2, \ldots.$

The argument is as follows. Suppose that the motion $\boldsymbol{v}$ is increased in the ratio

indefinitely. This means that the field grows, but not necessarily that it grows regularly. It may simply grow more and more irregular, lines of force in one direction being continually brought into close juxtaposition with lines of force in other, and perhaps opposed, directions. For the steady-state dynamo theory to be worth while, the field must be "stable" against such increases in the motion; that is, such increases must produce increases in the field which preserve certain regular characteristics until the field has grown so large that it is able to react on and limit the motion by its mechanical effects. Thus it is not sufficient only to know that a given motion maintains a steady field; when increased, it must not at once destroy the regular features of this field.

Such conditions are, of course, satisfied for the dynamos in common use. BONDI and GOLD (1950) have suggested that the reason why these dynamos work is that the currents flow in a multiply-connected volume (a closed circuit). They have considered dynamo effects in the case of infinite conductivity, which is equivalent in many respects to assuming finite conductivity and infinitely great motions. They found that, in this case, the external field cannot increase indefinitely unless the currents flow in a multiply-connected space. Their results depended on the fact that in perfectly conducting material no new lines of force can leak across the surface of the material, and to that extent are unrepresentative of the results for finite conductivity. But they are important as indicating that in a simply-connected region one does not get an indefinite regular amplification of the field simply by increasing the dynamo motions indefinitely. Rapid motion inside a conducting sphere does not lead to a correspondingly rapid increase in the external field; it leads to rapid changes in the internal field, but since the only rapid changes in the external field with which these can be correlated are irregular ones, any increased internal field must itself be irregular.

4. TURBULENT FIELDS

The dynamo maintenance of non-steady fields has usually been considered in connection with turbulent motion. Since moving material tends to drag the lines of force with it, turbulent motion tends to twist and stretch the lines of force until these are distorted into an irregular tangle. However, just as turbulent motion possesses certain statistically steady features, the resulting turbulent magnetic field can also be expected to possess statistically steady properties. It is, moreover, suggested that from time to time the external field may possess a fairly appreciable dipole component.

However, the suggestion most frequently made (FERMI, 1949) is that a steady state tends to be set up, in which there is equipartition of energy between the turbulent motion and the turbulent magnetic field, for all "wavelengths" of the turbulence "spectrum" of each. Equipartition is supposed to be set up most rapidly in the short wavelengths, which contribute only a small fraction of the energy, and to extend later to progressively longer and longer wavelengths. Equipartition cannot be expected save between quantities which behave similarly and interchange energy on equal terms. ELSASSER (1950) has put forward the following argument to suggest that material motions and magnetic forces do, in fact, behave very similarly.

Using the equations div $\boldsymbol{B} = 0$, div $\boldsymbol{v} = 0$, equation (4) can be put in the form

$$\frac{\partial \boldsymbol{B}}{\partial t} + (\boldsymbol{v} \cdot \nabla)\boldsymbol{B} = (\boldsymbol{B} \cdot \nabla)\boldsymbol{v} + \frac{1}{4\pi\mu\sigma}\nabla^2\boldsymbol{B}. \qquad \ldots.(18)$$

The equation of motion of the material is

$$\rho\left(\frac{\partial \boldsymbol{v}}{\partial t}+(\boldsymbol{v}\,.\,\nabla)\boldsymbol{v}\right)=-\nabla p+\boldsymbol{F}+\kappa\nabla^2\boldsymbol{v}+\boldsymbol{j}\times\boldsymbol{B} \qquad \ldots.(19)$$

where ρ is the density, p the pressure, $\boldsymbol{F}$ the body force, and κ the viscosity. Put

$$\left.\begin{aligned} &\boldsymbol{v}+\boldsymbol{B}/\sqrt{(4\pi\rho\mu)}=\boldsymbol{P}, &&\boldsymbol{v}-\boldsymbol{B}/\sqrt{(4\pi\rho\mu)}=\boldsymbol{Q}\\ &\kappa_1=\tfrac{1}{2}\kappa+\rho/(8\pi\mu\sigma), &&\kappa_2=\tfrac{1}{2}\kappa-\rho/(8\pi\mu\sigma).\end{aligned}\right\} \qquad \ldots.(20)$$

Then (18) and (19) can be combined to give

$$\rho\left(\frac{\partial \boldsymbol{P}}{\partial t}+(\boldsymbol{Q}\,.\,\nabla)\boldsymbol{P}\right)=-\nabla\left(p+\frac{B^2}{8\pi\mu}\right)+\boldsymbol{F}+\kappa_1\nabla^2\boldsymbol{P}+\kappa_2\nabla^2\boldsymbol{Q} \qquad \ldots.(21)$$

$$\rho\left(\frac{\partial \boldsymbol{Q}}{\partial t}+(\boldsymbol{P}\,.\,\nabla)\boldsymbol{Q}\right)=-\nabla\left(p+\frac{B^2}{8\pi\mu}\right)+\boldsymbol{F}+\kappa_1\nabla^2\boldsymbol{Q}+\kappa_2\nabla^2\boldsymbol{P}. \qquad \ldots.(22)$$

Remembering that $B^2/8\pi\mu$ is the magnetic "pressure", these equations are seen to be closely analogous with the form taken by the equation of motion (19) in the absence of a magnetic field, the variables $\boldsymbol{P}$ and $\boldsymbol{Q}$ corresponding to the velocity $\boldsymbol{v}$. Remembering the definitions of $\boldsymbol{P}$ and $\boldsymbol{Q}$, it follows that $\boldsymbol{B}/\sqrt{(4\pi\rho\mu)}$ should fill a role closely analogous to that of $\boldsymbol{v}$. This suggests that the magnetic energy $B^2/8\pi\mu$ per unit volume fills a place essentially like that of the material energy $\frac{1}{2}\rho v^2$.

BATCHELOR (1950) and CHANDRASEKHAR (1950), combating this point of view, point out that equation (18), satisfied by $\boldsymbol{B}$, is identical in form with the equation satisfied by the vorticity in the motion of a viscous liquid, $1/4\pi\mu\sigma$ corresponding to the kinematical viscosity. They therefore suggest that the correct analogy for $\boldsymbol{B}$ is with the vorticity. In turbulent motion, as is well-known, the vorticity becomes progressively more important, compared with the velocity, as one passes to smaller and smaller wavelengths in the turbulence spectrum. Thus if $\boldsymbol{B}$ behaves like the vorticity equipartition is unlikely for all wavelengths; it may prevail for very small wavelengths, but for greater the magnetic energy is likely to be the smaller.

The defect of this argument is that the vorticity is functionally connected with the velocity; the vector $\boldsymbol{B}$ is not. Moreover, it rests only on equation (18). Equation (19), indicating how mechanical forces of magnetic origin affect the motion, is not used, and it is through this equation that equipartition might be expected to be set up. The main importance of the argument is in indicating how dangerous it is to build simply on analogy. Equipartition of energy is not proved by any analogy; it is proved only by actually establishing that magnetic energy and material kinetic energy interact with each other on equal terms.

One obvious difference exists between magnetic energy and material kinetic energy. Turbulent motion can be generated and maintained by large-scale shearing motion; a large-scale magnetic field cannot, by itself, generate a turbulent magnetic field. Thus some limitations of equipartition must exist; how serious these limitations can be demands closer investigation.

The equation of motion in the absence of a magnetic field is

$$\frac{\partial \boldsymbol{v}}{\partial t} = -(\boldsymbol{v}\,.\,\nabla)\boldsymbol{v} + \frac{1}{\rho}(\boldsymbol{F} - \nabla p + \kappa \nabla^2 \boldsymbol{v}).$$

In discussing the application of this equation to turbulence, the following argument is often used. If the turbulent motion is divided into an energy spectrum, the term $-(\boldsymbol{v}\,.\,\nabla)\boldsymbol{v}$ in the equation involves the products of different spectrum elements. These products contribute to the spectrum of $\partial \boldsymbol{v}/\partial t$ interaction terms which may have longer wavelengths than the elements which compose the products, but more often have shorter wavelengths. Thus there is a continuous tendency for large eddies to break up into smaller and smaller eddies. Energy is continually being fed into large eddies from the kinetic energy of mass motions; it passes from these to progressively smaller and smaller eddies, and is destroyed by viscosity when it reaches the smallest eddies.

A similar argument can be applied to equation (18). The terms $(\boldsymbol{v}\,.\,\nabla)\boldsymbol{B}$, $(\boldsymbol{B}\,.\,\nabla)\boldsymbol{v}$ contribute to the spectrum of $\partial \boldsymbol{B}/\partial t$ interaction terms which more often than not have shorter wavelengths than the elements of $\boldsymbol{B}$ and $\boldsymbol{v}$ which interact. In ordinary language this means that turbulent motion tends to twist existing lines of force into an increasingly tangled confusion, rather than to generate large-scale fields from small-scale ones. Thus, just as in turbulent motion small eddies are supplied by the interaction of large eddies, so in turbulent magnetic fields small-scale elements of the field are supplied by the interaction of larger elements with the eddies. If the energy of large-scale elements of the field is small compared with the kinetic energy of large eddies, the energy of small-scale elements of the field must equally be smaller than the kinetic energy of small eddies.

If this argument is sound, it implies that there is no reason to expect equipartition of energy between turbulent magnetic fields and turbulent motion, unless the conditions for equipartition have already been met by imposing a large-scale field whose energy equals that of the large-scale motion. There is, of course, no automatic reason why equipartition of energy should obtain in any particular problem. In the kinetic theory of gases it obtains between molecules of different gases in a mixture because these can interchange energy on equal terms, without loss, at collision. In turbulence, on the other hand, there is no "equilibrium" state; energy is continuously being passed from long wavelengths to shorter. Elements of the magnetic field do not interchange energy simply with turbulence elements of the same size, and so there is no obvious reason to expect equipartition between the two.

Only in one case is equipartition known to occur; this is the case of Alfvén's magnetohydrodynamic waves, where there is equipartition between material kinetic energy and the energy of the disturbance field $\boldsymbol{b}$ due to the wave. In this case equipartition is to be expected on general grounds; dissipation of energy is ignored, and the second-order terms like $(\boldsymbol{v}\,.\,\nabla)\boldsymbol{v}$ in (19), or $(\boldsymbol{v}\,.\,\nabla)\boldsymbol{b}$ in (18), cancel out. Thus the whole process of degradation of energy into shorter wavelengths is stopped; the only interaction terms in the equations are those due to interaction of $\boldsymbol{v}$ or $\boldsymbol{b}$ with the undisturbed field $\boldsymbol{B}_0$. Since $\boldsymbol{B}_0$ is assumed to be uniform, these last terms introduce into $\partial \boldsymbol{v}/\partial t$ and $\partial \boldsymbol{b}/\partial t$ only those wavelengths already present. Hence in this case kinetic energy is freely interchanged with magnetic energy of similar wavelengths and equipartition follows. However, the factors securing equipartition in this instance are not met in more general circumstances.

The advocates of equipartition can easily criticize the above arguments, since, like the arguments in support of equipartition, they rest on analogy rather than detailed analysis. I personally find them more convincing than the arguments in favour of equipartition. But clearly any conclusion must remain doubtful until confirmed by detailed analysis.

5. Conclusions

To sum up, the steady-state theory is under somewhat of a cloud at present. A proof is needed of the possibility of dynamo maintenance of steady fields by currents in a simply-connected region. When this has been given, a further proof must be given that the field is not twisted out of all recognition if the dynamo motions are slightly increased. Until such proofs are given, the validity of steady-state dynamo theories must be regarded as open to question, and the proofs appear to require some altogether new idea.

The non-steady dynamo theory is in a much less developed state. Not even has it been definitely proved that a sufficiently violent turbulent motion can spontaneously generate a turbulent magnetic field from nothing, and any more ambitious assertion is open to criticism. Nevertheless, the possibilities open if non-steady dynamo mechanisms work are so interesting in many directions that any positive results in this field will be very worth while.

References

Batchelor, G. K. . . . . 1950 *Proc. Roy. Soc.*, **201**, 405.
Bondi, H. and Gold, T. . . . . . 1950 *M.N.*, **110**, 607.
Bullard, E. C. . . . . 1949 *Proc. Roy. Soc.*, **197**, 433.
Chandrasekhar, S. . . . . 1950 *Proc. Roy. Soc.*, **204**, 435.
Cowling, T. G. . . . . 1934 *M.N.*, **94**, 39.
Elsasser, W. M. . . . . 1946 *Phys. Rev.*, **69**, 106; **70**, 202.
1947 *Phys. Rev.*, **72**, 821.
1950 *Phys. Rev.*, **79**, 183.
Fermi, E. . . . . 1949 *Phys. Rev.*, **75**, 1169.
Frenkel, J. . . . . 1945 *C.R. Acad. Sci. U.S.S.R.*, **49**, 98.
Gurevich, L. E. and Lebedinsky, A. I. . 1945 *C.R. Acad. Sci. U.S.S.R.*, **49**, 92.
Lamb, H. . . . . 1883 *Phil. Trans.*, **174**, 519.
Larmor, Sir J. . . . . 1919 *Brit. Assoc. Reports*, p. 159.
1934 *M.N.*, **94**, 469.

On the Interpretation of Stellar Magnetic Fields

S. K. Runcorn

Gonville and Caius College and Department of Geodesy and Geophysics, Cambridge University

Summary

The present knowledge of stellar magnetic fields is reviewed. Their large variation with time in many stars presents a difficulty, the explanation of which in terms of mechanical oscillations is not satisfactory. In this paper a new attempt is made to examine under what conditions the measurements can be reconciled with the assumption of a steady field. This "oblique rotator" theory can only be true if the fields contain strong multipole components and a suggestion is made as to why this is probable. In addition certain elements must be concentrated in certain areas of the stellar surface to account for the intensity and Doppler displacement variation observed for certain spectral lines.

---

1. Introduction

The explanation of the magnetic fields associated with the Earth, the Sun, and certain stars seems likely to be found in the flow of electric current in the interior of these bodies. At various times it has been held that these fields were a new property of massive bodies in rotation but experimental tests of certain simple consequences of such "fundamental theories" have latterly proved this view to be unfounded. Runcorn, Benson, Moore, and Griffiths (1951) have measured in deep mines the radial gradient of the geomagnetic field and they find that the magnetic field just inside the Earth is irrotational, which it would not be if matter acted as a virtual current, *i.e.* a source of magnetic field. Also Blackett (1952) has proved by a laboratory experiment the non-existence of magnetic fields near dense bodies of the magnitude demanded by this hypothesis. The original objection felt by Schuster (1912) to explanations of the magnetic fields of cosmical bodies in terms of current flow was that the maintenance of an axial magnetic field demands electromotive forces directed round the lines of latitude, the existence of which in a spherically symmetrical body does not admit of any ready explanation. Now it had long been known that inside a stationary electrically conducting sphere magnetic fields may exist of "toroidal" type which do not appear outside the conductor. Elsasser (1947) raised the possibility that such fields exist within the Earth's core, thus, in effect, allowing the simple axial symmetry of the magnetic field, which led to the difficulty mentioned above, to be removed. Bullard (1948) and Runcorn (1954) have shown how such fields might be excited by simple physical processes in the Earth's interior.

Further progress in the understanding of cosmical magnetic fields would seem to depend on the correct appraisal of the evidence now available of the apparent variability of the fields of the Earth, the Sun, and the early-type stars. The gradual recognition of the short time scale, geophysically speaking, of the geomagnetic secular variation has enabled inferences concerning the fluid motions in the Earth's core to be made, which even now can be seen to be quite fundamental to the explanation of the main field itself. There seems little doubt, for instance, that the irregular fluctuations in the length of day, the proof of the existence of which forms a fascinating chapter of astronomical literature, have their origin in the same motions of the Earth's core, which are a key factor in the understanding of the generation of the geomagnetic field. In time inferences drawn from the astronomical data will add considerably to our knowledge of the physical properties of the interior of the Earth. Further, there is now strong evidence, notably that of HOSPERS (1953, 1954) on the remanent magnetization of the lavaflows in Iceland, that the geomagnetic field has actually reversed several times since early in the Tertiary epoch. If this interpretation of the reversed polarization of igneous rocks is finally maintained it may well prove of decisive importance in choosing between the various theories of the cause of the electromotive forces in the Earth's core. HALE's original measures of the Sun's field as a dipolar one with a polar magnitude of 50 gauss seem, on internal evidence of a convincing type (VON KLÜBER, 1952), to be not mistaken. Yet various observers find a value not greater than about 5 gauss over the last decade. Theoretical progress is somewhat hampered by doubt as to the reality of these considerable variations of the terrestrial and solar magnetic fields.

But perhaps the behaviour of the magnetic fields of the early-type stars detected in pioneer work by H. W. BABCOCK, presents the most considerable and, from the point of view of theory, most urgent problem of interpretation. BABCOCK (1953) finds among stars later than type B8 fourteen in which the magnetic fields actually reverse, ten in which the fields do not reverse but show large fluctuations in magnitude and eleven in which the fields are steady. Many of the magnetic variables are apparently spectrum variables, in which certain lines undergo fluctuations of intensity of the same period as the magnetic field, the lines usually belonging to two groups showing oscillations in antiphase. The stars with steady fields likewise show no evidence of variability in spectrum. Two of the stars in which the measured fields reverse have been investigated in detail; HD 125248 by BABCOCK (1951) and α^2 Canum Venaticorum by BABCOCK and BURD (1952). Fluctuations in the velocities in the line of sight for the various elements are found with periods equal to that of the magnetic field, though the velocity curves are not sinusoidal, second harmonics being notably present in the velocities of the Cr II and Ti II lines of α^2 Canum Venaticorum. A further notable feature is that in the star HD 125248 the intensities and also the velocities in the line of sight of the rare earth elements (*e.g.* Eu II) and the chromium lines (Cr II) oscillate in antiphase. The velocity differences observed in the lines of these stars are of the order of a few kilometres per second, representing movements of the elements of the order of the stellar diameter during the period. It is, of course, most essential that a satisfactory physical model of these magnetic variables should be found and in particular to decide whether these reversals are to be taken as wholesale alterations of the distribution of the magnetic lines of force.

On the latter view it seemed impossible, in view of the high electrical conductivity of stellar material, that the magnetic fields at a point in the star could actually change with respect to an element of the atmosphere in such short times and this consideration prompted the suggestion of RUNCORN (1948) that the form of the magnetic field was being periodically altered by mechanical oscillations of the star. Using the well-known device, appropriate for materials of high conductivity, of considering the lines of magnetic force attached to particles of matter, so that motion of the fluid may stretch lines of force but not create new ones, it can be shown that motions in the star may easily redistribute the lines of force over the surface. That this might lead under certain circumstances to an apparently reversing field is due to the fact that, in the interpretation of a measured Zeeman Shift, BABCOCK assumes an axial dipolar magnetic field and coincidence between the axis of rotation and the line of sight. Now the magnetic field actually measured is the mean component in the line of sight of the field integrated over the visible hemisphere, allowing for limb darkening, and on the above assumptions is the difference between the (larger) contribution from the polar regions and that from the equatorial regions. It was suggested that the effect of a radial pulsation of the star would be to sweep the lines of force alternatively towards and away from the poles, when the star is expanding and contracting respectively, thus making the polar and equatorial contributions alternatively dominant. In the papers of SCHWARZSCHILD (1952), FERRARO and MEMORY (1952), GJELLSTAD (1952), and COWLING (1952) are to be found detailed mathematical consideration of the "magnetic oscillator" theory. These discussions are not entirely favourable to it for SCHWARZSCHILD'S suggestion that the oscillation is a free one under its own magnetic field is shown by FERRARO and MEMORY to imply impossibly large magnetic fields in the interior of the star and though COWLING shows that gravity oscillations are more likely to occur and could lead to sufficiently long periods, if the motions had very strong horizontal components, he is able to demonstrate that the resulting redistribution of the lines of magnetic force is not of the type required to lead to apparent reversals in the field.

2. OBLIQUE ROTATOR THEORY OF MAGNETIC VARIABLES

It is consequently of interest to reconsider the alternative explanation; termed the oblique rotator theory, discussed by BABCOCK (1951), on the whole unfavourably at that time. This assumes that the field of the star is steady but that due to the rotation of the star the field presented to the observer changes with time. It is easily seen that for this to happen it is necessary for the axis of rotation of the star, the line of sight, and the magnetic axis to have different directions in space. Suppose first that the magnetic axis and the line of sight are inclined at angles α and β respectively to the axis of rotation. If the surface field of the star is dipolar in character then SCHWARZSCHILD (1950) has shown that the effective field strength H_e, as measured by the observer, including the effect of limb darkening given by the coefficient u, is

$$H_e = \frac{15 + u}{15 - 5u} \cdot \frac{H_p}{4} \cos i$$

where H_p is the field strength at the magnetic pole and i the inclination of the magnetic axis to the line of sight.

Now for the oblique rotator by the use of a simple formula of spherical trigonometry we obtain

$$\cos i = \cos \alpha \cos \beta + \sin \alpha \sin \beta \cos 2\pi t/T,$$

where T is the period of rotation and t is the time measured from the moment when the three axes are coplanar.

Then

$$H_e = 0{\cdot}303 H_p (\cos \alpha \cos \beta + \sin \alpha \sin \beta \cos 2\pi t/T),$$

taking $u = 0{\cdot}45$, the value given by BABCOCK (1951), as typical of A0 stars.

This demonstrates an important apparent objection to the oblique rotator theory, rather similar to one raised by BABCOCK (1951), *i.e.* there is in the curves of H_e against t for the magnetic variables a distinct second harmonic. For instance, for the star H.D. 125248 BABCOCK (1951) finds

$$H_p = + 2000 + 6600 \cos 2\pi t/T - 1600 \cos 4\pi t/T \text{ (gauss)}$$

and for the star α^2 Canum Venaticorum from the mean curve given by BABCOCK and BURD (1952) the expression for the apparent polar field is

$$H_p = - 1100 - 3800 \cos 2\pi t/T + 1800 \cos 4\pi t/T \text{ (gauss)}$$

where t is measured in each case from the Eu II maximum.

Thus assuming the axis of rotation is fixed in space, the oblique rotator theory can only be maintained by assuming that the surface magnetic fields contain higher harmonics. However, it seems for mechanical reasons to be difficult to justify the existence of a purely dipolar field. For the production of an axial dipole field outside a conducting sphere it is necessary to have a distribution of electric current flow over concentric spherical surfaces directed around lines of latitude, the intensity of flow per unit distance measured along lines of longitude being proportional to the sine of

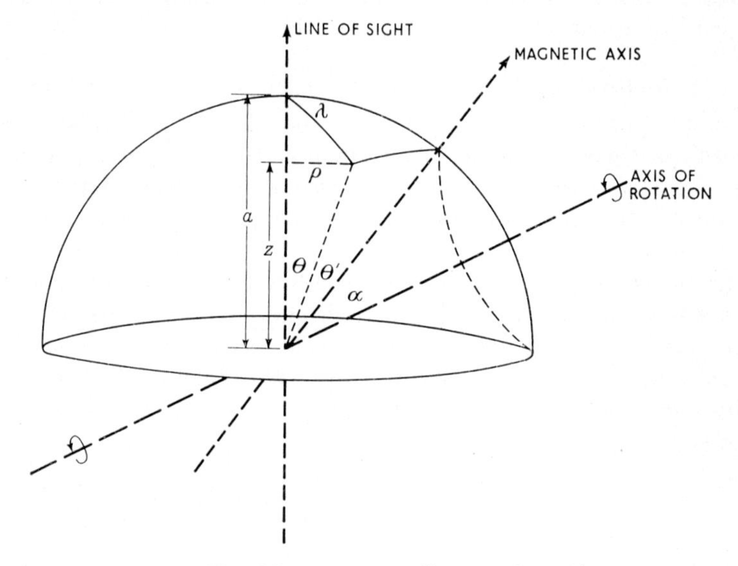

Fig. 1. The oblique rotator: diagram of notations

the angle of colatitude (θ). Particularly if it is assumed that the current flow producing the observed field is concentrated in the outer layers of the star, considerable mechanical forces would be necessary to preserve this distribution against the resulting electromagnetic forces. Suppose that the magnetic field is produced by a surface distribution of current ($I_0 \sin \theta$). The radial component of the resultant magnetic field will be continuous through the current sheet and will be given by $H_r = (8\pi/3)I_0 \cos \theta$. Thus the electromagnetic force on the atmosphere per cm^2 at colatitude θ will be of magnitude $(3/8\pi)H_r^2 \tan \theta$ and will be directed towards the equator. In the case of the stars this force will clearly be of the order of 10^8 dynes. If the currents are distributed only over the whole depth of the atmosphere, say 100 km, the electromagnetic forces per cubic centimetre of gas will be 10 dynes. Now the gravitational forces in a stellar atmosphere are of the order of 10^{-6} dynes per c.c., so that there is a strong tendency for the current to be concentrated in the equatorial regions, *i.e.* the state of minimum magnetic energy. This enhanced equatorial current would, of course, give rise to a strong axial octupole component in the surface magnetic field and on the oblique rotator theory result in a magnetic field variation markedly different from a purely sinusoidal one.

The potential ϕ of an axial octupole in spherical co-ordinates is given by

$$\phi = \frac{H_p \, . \, a}{4} \left(\frac{a}{r}\right)^4 P_3(\cos \theta'),$$

where H_p is the polar field. This may be converted to spherical co-ordinates about the line of sight by the addition theorem of spherical harmonics. The terms for $m \neq 0$ do not contribute to the integral and are therefore omitted. Thus

$$\phi = \frac{H_p \, . \, \mathrm{a}}{4} \left(\frac{a}{r}\right)^4 P_3(\cos \theta) P_3(\cos i) + \, . \, . \, . \, .$$

Converting to cylindrical co-ordinates about the line of sight, the component of the field in the line of sight may easily be found by differentiation of ϕ with respect to z. This is found to be

$$H_z(\rho) = \frac{(-\,8a^4 - 35\rho^4 + 40a^2\rho^2)}{8a^4} \, . \, H_p \, . \, P_3(\cos i).$$

The field measured by the observer is then

$$H_e = \int_{\rho=0}^{\rho=a} 2\pi\rho \, . \, H_z(\rho) \, . \, (1 - u + u\sqrt{a^2 - \rho^2}/a) d\rho$$

$$= \frac{1 + 7u}{3 - u} \cdot \frac{H_p}{8} \, . \, P_3(\cos i).$$

The apparent field will fluctuate with second and third harmonic components of the period of rotation.

An important study by DEUTSCH (1952a) of the spectrum variable H.D. 124224, which has a period of 0·52 days, rather strongly suggests that spectrum variability involves a distribution of the sources of certain absorption lines with longitude with at the same time a rotation about an axis inclined to the line of sight. In a general discussion of twelve spectrum variables of type A, DEUTSCH (1952b) is able to show that the differences in profile of a certain line (Mg II 4481), common to all these stars and of constant intensity and therefore presumably uniformly distributed over the surface, are predictable from estimates of the peripheral velocities of the individual stars as deduced from their period of variability. This evidence, together with the direct verification discussed elsewhere in this book* that the sources of certain spectral lines can be concentrated in certain areas of the stellar surface, provides strong indication that the oblique rotator theory of magnetic variables must be reconsidered, as DEUTSCH himself suggests (1952b).

---

* See Section 14 (Spectral Variations and Novae), Volume 2.

3. Spectrum Variability

To make a first attempt to discuss quantitatively the differences observed at different phases in the intensities and Doppler shifts of certain of the elements, as, for example, in H.D. 125248, we note that the velocity in the line of sight at a point on the stellar surface arises from the component of the angular velocity $(2\pi/T)$ perpendicular to the line of sight. Thus, taking the line of sight as the axis of spherical co-ordinates, the velocity in the line of sight at colatitude θ and azimuth λ, measured from the plane containing the line of sight and axis of rotation, equals $(2\pi/T) \sin\beta \sin\theta \sin\lambda$, taking the radius of the star as unity. We will further assume that the sources of certain lines are not distributed evenly over the stellar surface. Deutsch (1952b) has suggested that the magnetic field affects the excitation by inhibiting convection and so disturbing the temperature gradient. It thus seems generally reasonable to take a distribution of sources given by a zonal harmonic with respect to the magnetic axis. In the model of the magnetic field distribution, which we considered earlier, the magnetic field is mainly vertical at the poles and horizontal at the equator, thus it seems reasonable to represent the source distribution as a sum of even zonal harmonics, say, $a_0 + a_2 P_2(\cos\gamma)$, where γ is the angle of colatitude with respect to the magnetic axis and a_0 and a_2 are constants with a_0 greater than $a_2/2$. We convert this to a sum of tesseral harmonics along the line of sight as the axis of co-ordinates. Then by the additional formula of spherical harmonics

$$P_2(\cos\gamma) = P_2(\cos i)P_2(\cos\theta) + 2\sum_{m=1}^{2}\frac{(2-m)!}{(2+m)!}P_2^m(\cos i)P_2^m(\cos\theta)\cos m(\lambda' - \lambda),$$

where λ' is the azimuth of the magnetic axis at any time t, given by the expression

$$\sin\lambda' = (\sin\alpha \,.\, \sin 2\pi t/T)/\sin i.$$

By integration over the visible hemisphere the mean intensity (I) of lines which are distributed in the above way over the stellar surface and the mean velocity in the line of sight (v) as given by the Doppler shift may be derived. The surface brightness is taken as proportional to the expression $(1 - u + u\cos\theta)$,

$$\begin{aligned}I &= \int_{\lambda=0}^{\lambda=2\pi}\int_{\theta=0}^{\theta=\pi/2}(1 - u + u\cos\theta)(a_0 + a_2P_2(\gamma))\cos\theta\sin\theta d\theta d\lambda\\ &= (1 - u/3)a_0 + 1/4(1 + u/15)a_0(3\cos 2i - 1)\\ &= (3 - u)a_0/3 + (15 + u)a_2/60 + (15 + u)a_2(2\cos^2\alpha\cos^2\beta + \sin^2\alpha\sin^2\beta)/40\\ &\quad + \frac{15 + u}{40}\,.\, a_2 \,.\, \sin 2\alpha\sin 2\beta[\cos 2\pi t/T + 1/4\tan\alpha\tan\beta\cos 4\pi t/T].\end{aligned}$$

As a_2 cannot exceed $2a_0$ the greatest fluctuation in the light from this harmonic cannot exceed 2·4 : 1, if $u = 0{\cdot}5$. This is perhaps rather less than the fluctuation observed and indicates that in a fuller discussion it might be necessary to include higher harmonics.

$$\begin{aligned}v &= (1/I)\int_{\lambda=0}^{\lambda=2\pi}\int_{\theta=0}^{\theta=\pi/2}(2\pi a_2/3T)\sin\beta P_2^1(\cos i)P_2^1(\cos\theta)\sin\lambda'(1 - u + u\cos\theta)\\ &\qquad \cos\theta\sin^2\theta\sin^2\lambda\, d\theta d\lambda\\ &= (1/I)\,.\,(1/5 - 3u/40)(\pi^2a_2/T)\sin 2\alpha\sin 2\beta[\sin 2\pi t/T + 1/2\tan\alpha\tan\beta\sin 4\pi t/T].\end{aligned}$$

A detailed comparison of these results with the observational data will not be attempted here but an inspection of these expressions for I and v show up two points of interest. Firstly, the line intensity at certain phases varies rather rapidly due to the presence of the second harmonic: this is in accord with DEUTSCH's observations. Secondly, v may vary rapidly when I is small: the rapid acceleration of the Eu II lines while strong just before the phase of maximum magnetic intensity is a notable feature of BABCOCK's results on H.D. 125248. It is clear also that the oblique rotator theory provides an explanation for the light curve and for the cross-over effect found in H.D. 125248, from which BABCOCK has inferred that different areas of the star have different polarities and radial velocities. The theory also accounts in a general way for the systematic difference in Zeeman shift measured in the constant and varying lines.

BABCOCK (1951) objected further to the oblique rotator theory on the grounds that the antiphase fluctuations in intensity of the rare earth and chromium group of lines require each group to be concentrated at one magnetic pole and he points out the real difficulty of thinking of a process of separation which would take account of the polarity rather than the intensity of the field or its inclination to the stellar surface. Further, in H.D. 125248 and α CVn the rare earth lines and Cr lines vary in antiphase, but the relationship of magnetic polarity to the respective elements is opposite in the two stars. The above calculation of the line intensity shows that a source distribution can be used which is not open to this objection.

The F0 type star, γ Equilei, found to have an unvarying magnetic field by BABCOCK (1948), has extremely fine metallic lines and so presumably rotates about an axis inclined little to the line of sight. There is, of course, no evidence as to whether its magnetic axis is inclined appreciably to the line of sight.

4. CONCLUSION

Thus it seems possible now to explain the apparent variability of the magnetic fields of stars in terms of the various aspects presented by the magnetic and rotational axes to the observer. This view might to some extent have been inferred from the fact that of the stars in which magnetic fields have been detected, some have measured fields which are constant, some which are variable and some which actually reverse.

Little progress has, of course, been made towards the understanding of the origin of the stellar magnetic fields, but it is not out of the question in view of the long free decay time that these fields are residual fields and have their origin in some event in stellar evolution. The reason for the non-axial character of the fields we have inferred above is also obscure. But there seems no reason for the coincidence of the magnetic and rotational axes as there is in the case of the Earth, see RUNCORN (1954).

The understanding of cosmical magnetic fields is a considerable challenge to the extra-laboratory sciences. It is clear that in both the geophysical and astrophysical examples we are still at the stage of providing a reasonable physical model to take account of all the facts available. Satisfactory theories as to the cause of the electric current flow postulated to explain the fields must probably await the completion of this task.

REFERENCES

BABCOCK, H. W. . . . . 1948 *Ap. J.*, **108**, 191.
1951 *Ap. J.*, **114**, 1.
1953 *M.N.*, **113**, 357.
BABCOCK, H. W. and BURD, S. . . . . 1952 *Ap. J.*, **116**, 8.
BLACKETT, P. M. S. . . . . 1952 *Phil. Trans.*, A **245**, 309.

COWLING, T. G. . . . . 1952 *M.N.*, **112**, 527.
DEUTSCH, A. J. . . . . 1952a *Ap. J.*, **116**, 536.
1952b *Proc. Joint Comm. on Spectroscopy*, I.A.U. Trans., vol. VIII.
ELSASSER, W. M. . . . . 1947 *Phys. Rev.*, **72**, 821.
FERRARO, V. C. A. and MEMORY, D. J. . . . 1952 *M.N.*, **112**, 361.
GJELLESTAD, G. . . . . 1952 *Ann. Astrophys.*, **12**, 148.
HOSPERS, J. . . . . 1953 *Proc. K. Ned. Akad. Wetensch.*, B, **56**, 467.
1954 *Proc. K. Ned. Akad. Wetensch.*, B, **57**, 112.
KLÜBER, H. VON . . . . 1952 *M.N.*, **112**, 540.
RUNCORN, S. K. . . . . 1948 *Trans. Oslo Meeting, Assoc. of Terr. Mag. and Elect.*, I.A.T.M.E. Bulletin No. 13, p. 421.
1954 *Trans. Amer. Geophys. Un.*, **35**, 49.
RUNCORN, S. K., BENSON, A. C., MOORE, A. F. and GRIFFITHS, D. H. . . . . 1951 *Phil. Trans.*, A **244**, 113.
SCHUSTER, A. . . . . 1912 *Proc. Phys. Soc. (London)*, **24**, 121.
SCHWARZSCHILD, M. . . . . 1950 *Ap. J.*, **112**, 222.
1952 *Ann. Astrophys.*, **12**, 148.

Theories of Variable Stellar Magnetic Fields

V. C. A. FERRARO
Queen Mary College, University of London

(1) As early as 1892 SCHUSTER conjectured that massive celestial bodies in rotation might, like the Earth, be the seat of magnetic fields (see also SCHUSTER, 1912). It is said that one of the reasons which led HALE to consider the project of the 100-in. telescope at Mount Wilson was the hope that it might reveal the existence of stellar magnetic fields. HALE (1908) was successful in establishing the existence of sunspot magnetic fields, but the discovery of stellar magnetic fields was reserved for his compatriot, H. W. BABCOCK (1947), who discovered that the star 78 *Virginis* possessed a general magnetic field of the order of 1500 gauss at the pole.

Since then BABCOCK has discovered that several other stars possess magnetic fields exceeding 1000 gauss; furthermore, he found that in some the magnetic field was variable, whilst in others, notably HD 125248, the field was not only variable but, what is remarkable, its polarity reversed once every $9\frac{1}{4}$ days.

All magnetic stars appear to be spectral variables of the types A and F in rapid rotation. The variability of the star HD 125248 has been the subject of a careful study by BABCOCK (1950) who found that the polar magnetic field varies from about 7000 gauss to − 6000 gauss. Radial velocity measurements show an harmonic variation which seems coupled with changes in the magnetic field, the velocity amplitude varying from 3–10 km per sec. The variations are characterized by a rapid deceleration.

(2) There are at least three possible ways of accounting for the variability and reversal of stellar magnetic fields. The obvious explanation that the star possesses a

true variable magnetic field, in the sense that it is caused by alternating currents in the deep interior, has not been entertained seriously by astrophysicists on the grounds that the high electrical conductivity of the stellar material would enable the mantle of the star to act as an electromagnetic screen and so prevent all but an imperceptible portion of the field from penetrating to the surface.

The second possibility is to suppose that the magnetic axis of the star is inclined to the axis of rotation of the star, which is fixed in space, but which in turn is inclined to the line of sight. Babcock (*loc. cit.*) has rejected this model (the oblique rotator) on various grounds and chiefly because he maintains that it could not account for the rapid deceleration in the observed radial velocity curves mentioned earlier in section (1).

The third possibility, and so far the one to which attention has been chiefly directed, supposes that the variability and reversal of the magnetic field is produced by oscillations of the star in its own (otherwise stationary) magnetic field. The fact that the period of the variation of the magnetic field very nearly coincides with the period of the variability in the spectrum appears to lend support to this suggestion. It was first considered by Schwarzschild (1949), who made a number of simplifying assumptions to render the problem tractable mathematically. He assumed a uniform liquid star, with a uniform magnetic field (in the absence of the oscillations) and supposed that the mechanical forces due to gravity were negligible compared with the electromagnetic forces. To get a nine-day period, as found by Babcock, Schwarzschild deduced that a uniform field of 10^6 gauss was necessary. He interpreted this to correspond to the magnetic field near the centre of the star, the field falling off to a value comparable with 10^4 gauss near the surface. Since this is of the same order as the observed field, Schwarzschild concluded that this model would lead to a possible explanation.

The agreement was shown by Ferraro and Memory (1952) to be fortuitous; by making the same simplifying assumptions as Schwarzschild, but taking account of the *increase* of the magnetic field with depth, Ferraro and Memory showed that to get a nine-day period, a field far in excess of 10^6 gauss at the centre was necessary. They concluded that the reversal of the field could not be explained by considering pure magnetic oscillations of the star alone and suggested that this might be due to the excitation by a mechanical oscillation of the star, without, however, formulating any definite theory. Another objection to Schwarzschild's theory, to which Cowling (1952) had drawn attention, is that his theory indicates a field which varies periodically but without reversing sign.

(3) To explain the reversal of the field, Cowling (1952) has suggested that the oscillations must be horizontal and nearly non-divergent, as in the case of tidal oscillations. The reason is that the material of a star has a very high electrical conductivity and hence the magnetic lines of force are "frozen" in the material and are carried along with it. Because rotational broadening of spectral lines is generally absent in magnetic stars, this is interpreted as implying that such stars are oriented with their axis towards the observer. Hence during the phase of the oscillation in which the material is moving (horizontally) away from the pole, the magnetic lines of force are directed prominently away from the observer and during the opposite phase, half a period later, they are directed towards him. This creates the appearance of a reversal of polarity to an observer viewing the field from a point on the magnetic

axis as illustrated in Fig. 1, which shows the appearance of the magnetic lines of force at the extremes of displacement. Such oscillations have nevertheless been shown by COWLING to be not wholly favourable to the theory. The argument given above, though plausible, is vitiated by the fact that it is impossible to fit smoothly the variable magnetic resulting from the oscillation to the potential field which must exist outside.

(4) We now consider the general properties of oscillations of a gaseous star under the combined action of the electromagnetic and gravitational forces. The argument

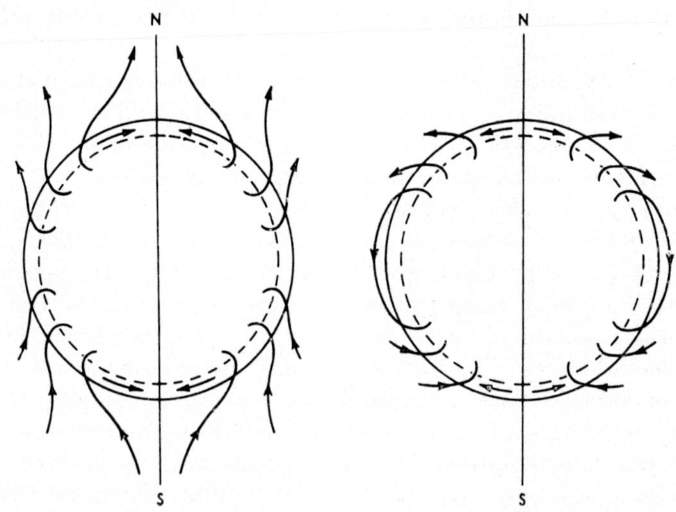

Fig. 1. Illustrating the bending of the magnetic lines of force of the permanent stellar magnetic field by (nearly) horizontal oscillations of the star. At one extreme of the oscillation they are bent towards the pole (left), at the other they are bent towards the equator (right)

follows along the lines of COWLING's analysis (*loc. cit.*). Consider the case of a non-rotating star possessing a permanent magnetic field which has the character of the field of a dipole, the lines of force being in planes through the axis. As we have already mentioned, the electrical conductivity of the star may be taken as infinite. We denote by ρ and p respectively the density and pressure at any point of the star, by g the acceleration of gravity, by j the current density, and by H the permanent magnetic field. MAXWELL's equation giving the relation between j and H is

$$\text{curl } \boldsymbol{H} = 4\pi \boldsymbol{j}, \qquad \ldots.(1)$$

the equations of mechanical equilibrium is

$$\text{grad } p = \rho \boldsymbol{g} + \boldsymbol{j} \wedge \boldsymbol{H}, \qquad \ldots.(2)$$

and the gravitational equation

$$\text{div } \boldsymbol{g} = -4\pi G\rho, \qquad \ldots.(3)$$

where G is the constant of gravitation.

When the star performs small oscillations, the changes in any variable ϕ from its value in the position of equilibrium will be denoted by $\delta\phi$. The velocity of a gas element v is given in terms of its displacement $\boldsymbol{\xi}$ by

$$\boldsymbol{v} = \frac{\partial \boldsymbol{\xi}}{\partial t}. \qquad \ldots.(4)$$

As in all problems of small oscillations, squares and products of small quantities, such as $\delta\rho$, δp, . . . , $\boldsymbol{v}$, $\boldsymbol{\xi}$, . . . are neglected.

Since the conductivity is infinite, the electric field* $\boldsymbol{E}$ is given by

$$\boldsymbol{E} + \boldsymbol{v} \wedge \boldsymbol{H} = 0 \qquad \ldots.(5)$$

expressing the fact that the total electric field on the moving material must vanish.

Substituting this equation in MAXWELL'S equation,

$$\text{curl } \boldsymbol{E} = -\frac{\partial \delta \boldsymbol{H}}{\partial t},$$

we have

$$\text{curl } (\boldsymbol{v} \wedge \boldsymbol{H}) = \frac{\partial \delta \boldsymbol{H}}{\partial t} \qquad \ldots.(6)$$

and integrating with respect to the time we have

$$\delta \boldsymbol{H} = \text{curl } (\boldsymbol{\xi} \wedge \boldsymbol{H}). \qquad \ldots.(7)$$

This equation expresses the fact that the changes in the magnetic field are produced by displacement of the lines of force "frozen" in the moving material.

Assuming oscillations of period $2\pi/\sigma$, the equation of motion is

$$\rho\sigma^2\boldsymbol{\xi} = \text{grad } \delta p - \rho\delta\boldsymbol{g} - \delta\rho\boldsymbol{g} - \boldsymbol{j} \wedge \delta\boldsymbol{H} - \delta\boldsymbol{j} \wedge \boldsymbol{H}. \qquad \ldots.(8)$$

Also (1) and (3) give

$$\text{curl } \delta\boldsymbol{H} = 4\pi\delta\boldsymbol{j}, \ \text{div } \delta\boldsymbol{g} = -4\pi G\delta\rho. \qquad \ldots.(9)$$

The equation of continuity,

$$\frac{\partial \delta\rho}{\partial t} + \text{div } \rho\boldsymbol{v} = 0,$$

becomes after partial integration with respect to the time,

$$\delta\rho + \text{div } \rho\boldsymbol{\xi} = 0. \qquad \ldots.(10)$$

Finally, the equation of state, $p \propto \rho^\gamma$, gives

$$\Delta p/p = \gamma\Delta\rho/\rho, \qquad \ldots.(11)$$

where Δ signifies a change in the variables following the motion of the mass-element; (10) and (11) lead to

$$\Delta\rho = -\rho \text{ div } \boldsymbol{\xi}. \qquad \ldots.(12)$$

Thus, $\Delta\rho$ vanishes if div $\xi = 0$, *i.e.* if the oscillations are non-divergent, as in the case of torsional oscillations.

(5) We can now show that unless the magnetic field of the star at the centre greatly exceeds the value of 10^5 gauss, inferred from surface observations, electromagnetic forces are negligible compared with the mechanical forces. It will suffice to consider orders of magnitude only. Suppose that the oscillations corresponds to waves of

* Electromagnetic units are used throughout for the electric and magnetic variables.

length L; clearly L cannot exceed the radius of the star R. It then follows that, in general, for any function G, $|\text{grad } G| \sim G/L$.

From (7), (10), (12) we then have

$$\delta \boldsymbol{H} \sim \boldsymbol{\xi} \wedge HL^{-1},\ \delta\rho \sim \rho L^{-1} \sim \Delta\rho,\ \delta p \sim pL^{-1} \sim \Delta p.$$

Also, (3) and (9) give

$$g \sim 4\pi G R\rho,\quad \delta g \sim 4\pi G L \delta\rho,$$

and (1) and (9) give

$$4\pi Rj \sim H,\ 4\pi L\delta j \sim \delta H.$$

Using these results we see that in (8), $\rho\delta g$ is at most of the same order as $g\delta\rho$. Also

$$|\text{grad } \delta p| \sim (\delta p)L^{-1} \sim p\xi L^{-2}, \qquad \ldots.(13)$$

Hence, $\rho\delta g$, $g\delta\rho$ are at most of the same order as grad δp, so that the first of the last two terms in (8) is at most comparable with the first. Thus in the right side of the equation of motion (8), the dominant terms are grad δp and $\delta \boldsymbol{j} \wedge \boldsymbol{H}$. The latter is of the order

$$H\delta H/4\pi L \sim \xi H^2/4\pi L^2.$$

Using (13), it follows that the ratio of the mechanical forces to the electromagnetic forces is as p is to $H^2/4\pi$, *i.e.* comparable with the ratio of the gas pressure to the magnetic pressure, as might have been expected. Taking the mass and radius of HD 125248 to be comparable with that of Sirius A, we find $p \sim 10^{14}$ dynes per cm^2, so that the magnetic pressure becomes comparable with the gas pressure when $H \sim 10^7$ gauss or greater. Since, as we have seen, the central value of the magnetic field is unlikely to exceed 10^5 gauss, the electromagnetic forces are negligible. This, however, supposes that the motion is neither practically horizontal ($\boldsymbol{\xi}$. grad p small) or non-divergent (div $\boldsymbol{\xi}$ small). Otherwise the mechanical forces are much reduced.

Neglecting the electromagnetic forces and considering orders of magnitude, equation (8) gives the period of mechanical oscillation to be $2\pi/\sigma$, where

$$\sigma^2 \sim gR/L^2,$$

or at least of order $g/R = GM/R^3$, where M is the total mass of the star. This is in good agreement with known results. For the Sun, $\sigma^2 \sim 4 \times 10^{-7}$ sec.$^{-2}$, giving an oscillation of period about 2 hr. For BABCOCK'S star, the corresponding period is probably longer on account of its greater radius (about twice the solar radius), but still much too rapid to give a period of a few days.

To get agreement with observations, an oscillation must be found in which the effect of the mechanical forces is diluted. As noted earlier, this demands that the oscillations should be nearly horizontal and non-divergent, so that the terms of order div $\boldsymbol{\xi}$, $\boldsymbol{\xi}$. grad ρ, and $\boldsymbol{\xi}$. grad p are small, and so $\delta\rho$ and δp are small.

Such oscillations were considered by COWLING (1941) and called by him *gravity oscillations*. An appropriately long period is easily obtained in the case of such oscillations, but there at once arises a difficulty in that there are *many* free periods not very different from the observed period and the question arises as to why a particular period should be excited in preference to a neighbouring one.

Nevertheless, COWLING has carried out the analysis in such cases and finds that the oscillations are confined to the surface layers.

(6) Whilst gravity oscillations would be of the right period we have already mentioned at the end of section (3) that the oscillations are unlikely to produce the desired reversal of the magnetic field. The difficulty concerns the fitting of the solutions for the magnetic field inside and outside the star at the surface of the star. For continuity of the normal resolute of the magnetic flux requires the continuity of the radial resolute of the magnetic field $\delta \boldsymbol{H}$ at the surface. Once this has been assigned at the surface, the potential field outside the star is completely determined, and the continuity of the horizontal resolutes of $\delta \boldsymbol{H}$ on opposite sides of the surface cannot be effected without introducing a surface discontinuity. This would imply the existence of a surface current sheet and for this there is no justification. Thus gravity oscillations cannot be expected to produce a reversal of the field.

(7) The observed variation of the radial velocity of BABCOCK'S star strongly suggest an oscillation, and the failure of the oscillation theory to explain the reversal of the field suggests that further observations are necessary.

As COWLING suggests, the oscillation theory cannot be lightly set aside; yet in its present form one can expect fluctuations of the mean field of 13 per cent of the undisturbed value at most. This is clearly insufficient to account for a reversal of the field. Moreover, the theory gives no direct explanation of the differences between lines of iron and chromium, which, as BABCOCK suggests, implies that the lines of chronium and of the rare earth are emitted in different regions of the star's atmosphere. Clearly, further investigation of this model are required and the possibility of the oblique rotator as a model of a magnetic variable should also be reconsidered.

REFERENCES

BABCOCK, H. W. . . . . 1947 *Ap. J.*, **27**, 315.
1940 *Nature* (London), **166**, 249.
COWLING, T. G. . . . . 1941 *M.N.*, **101**, 367.
1952 *M.N.*, **166**, 249.
FERRARO, V. C. A. and MEMORY, D. J. . . . . 1952 *M.N.*, **112**, 361.
HALE, G. E. . . . . 1908 *Ap. J.*, **105**, 105.
SCHUSTER, A. . . . . 1892 *Report B.A.A.S.*
1912 *Proc. Roy. Soc.* A, **24**, 121.
SCHWARZSCHILD, M. . . . . 1949 *Ann. Astrophys.*, **12**, 148.

Theories of Interstellar Polarization

LEVERETT DAVIS, JR.

Physics Department, California Institute of Technology,
Pasadena, California

SUMMARY

The experimental data to be explained by any theory of interstellar polarization are summarized. The expression for the ratio of polarization to absorption produced by a cloud of partially aligned small dust grains is put in a form applicable to all current theories of alignment. A semi-quantitative treatment is given of GOLD's proposal that alignment is due to relative motion of dust through gas and this mechanism is shown to be inadequate to produce the observed polarization. Magnetic alignment theories are preferred. Equations are given for use when starlight passes through several clouds with various orientations. Any orienting mechanism can be tested by a statistical study of the observed distribution of polarization over neighbouring stars. Thus it appears tentatively that it is more plausible to assume that a galactic magnetic field lies along a spiral arm than that the field is randomly directed in the plane of the galaxy.

1. INTRODUCTION

AN investigation of the possible causes of the interstellar polarization discovered by HILTNER (1949, 1951) and by HALL (1949, 1950) holds out the promise of giving important information on the interstellar gas, dust, and magnetic fields whose properties are of such importance in any study of galactic structure. For example, the magnetic field required by some theories affects the motion of the gas as much as does gravitation. Quantitative treatments of the various theories are needed in order to provide quantitative information and to explore accurately their merits and difficulties. Formulae that can be applied to any of the polarization mechanisms that have been proposed are given in Sections 2 and 3. In Section 4 a new semi-quantitative treatment of GOLD's theory is summarized and the results that have been obtained from the other theories are described. Topics upon which more work needs to be done will be emphasized and, particularly in Section 5, ways of getting information bearing on galactic structure will be pointed out.

The main observational results to be explained by any theory may be listed as follows (further treatment of the observational side of the problem being left to HILTNER's article in this book).* First, the plane in which the electric vector of a star's light is a maximum, here called the plane of vibration, shows a preference for the galactic plane. In some regions the degree of alignment is quite high; for homogeneous groups near galactic longitude $l = 100°$ and latitude $b = 0°$ the root-mean-square deviation in orientation was found by STRANAHAN (1954) from HILTNER'S (1951) data to be of the order of

$$\alpha = 0{\cdot}1 \text{ radian.} \quad \ldots.(1)$$

(A smaller, but still impressive, degree of uniformity extends from $l = 70°$ to $l = 170°$ in the galactic plane. Second, the polarization, p, measured in magnitudes as defined by HILTNER, is partially correlated with the colour excess, E_1, and hence with the total absorption, A_{pg}, which is about $9(E_1 + 0{\cdot}05)$ if the zero point correction of $0{\cdot}05$ is used for E_1. The average value of $9p/A_{pg} = p/(E_1 + 0{\cdot}05)$ is about $0{\cdot}3$ with

* See W. A. HILTNER on "Interstellar Polarization" in Section 11 (Stellar Astronomy), Volume 2.

standard deviation 0·1 for good regions near $l = 100°$, $b = 0°$. The maximum value that need be accounted for by any theory is

$$9p/A_{pg} = p/(E_1 + 0{\cdot}05) \leqslant 0{\cdot}5 \text{ mag. per mag.} \qquad \ldots.(2)$$

Third, the polarization in the red is about 80 per cent of that for the same star in the blue.

Since colour excess and interstellar absorption are due to dust, equation (2) implies that polarization is also due to dust and that the total absorption of plane polarized light varies a maximum of about 6 per cent as the plane of vibration is rotated. The dust can produce polarization if the grains are not spherical and if they are partially aligned by some interstellar field with their long axes predominantly normal to the galactic plane. Since the distance to appreciable quantities of dust is 300–500 parsecs in the direction $l = 100°$, the uniformity of orientation listed above implies a corresponding uniformity of the field over distances of 500–1000 parsecs. The only fields that can align dust grains and can maintain rough uniformity, perhaps only statistically, over such large distances seem to be magnetic fields or velocity fields of dust and gas. Variations in the properties of the dust, the effectiveness of the aligning mechanism, and its orientation with respect to the line of sight can account for the fact that p is not a unique function of E_1.

2. THE POLARIZATION PRODUCED BY A CLOUD OF PARTIALLY ALIGNED DUST GRAINS

Consider now an idealized model in which all dust grains are supposed to be identical spheroids which are small compared to the wavelength of the light considered. The classical theory of GANS (1912) then gives the extinction cross-section of such a grain in terms of its radii, index of refraction, orientation, and the wavelength of the light. Assume an unspecified orienting mechanism which, for each dust cloud, has an axis of symmetry called the cloud axis. This means that it does not disturb the uniform distribution of the azimuthal angles ϕ_A, where θ_A, ϕ_A are the spherical polar co-ordinates of the axis of symmetry of a grain referred to the cloud axis as polar axis, but that it does produce a distribution over θ_A that is not proportional to $\sin \theta_A$. If the extinction produced by the cloud in various beams of plane polarized light is then computed,* one finds that when a beam of originally unpolarized light passes through the cloud in a direction that makes the angle $(\pi/2) - \nu$ with the cloud axis, it will emerge partially plane polarized and p/A_{pg}, the ratio of the polarization to the total absorption, both in magnitudes, will be

$$\frac{p}{A_{pg}} = \frac{p}{9(E_1 + 0{\cdot}05)} = \frac{9}{2} RF \cos^2 \nu, \qquad \ldots.(3)$$

Here F is the distribution parameter

$$F = \tfrac{1}{3} - [\cos^2 \theta_A]\text{av}, \qquad \ldots.(4)$$

obtained by averaging $\cos^2 \theta_A$ over all grains in the cloud, and R is

$$R = \frac{\sigma_A - \sigma_T}{\sigma_A + 2\sigma_T}, \qquad \ldots.(5)$$

* For details and a more extensive treatment of the material in this section see DAVIS and GREENSTEIN (1951).

where σ_A and σ_T are the extinction cross-sections obtained from the GANS theory for grains whose axes of symmetry are parallel and perpendicular, respectively, to the electric vector in a beam of plane polarized light. Some values of R are given below in Table 1. If $RF > 0$, the plane of vibration contains the cloud axis; if $RF < 0$, the negative value of p in (3) indicates a rotation through 90° of the plane of vibration. If the grains are prolate ellipsoids $R > 0$ but for oblate ellipsoids $R < 0$. If the grain axes are all normal to the cloud axis F has its maximum value of $\frac{1}{3}$, while if they are all parallel to the cloud axis, F has its minimum value $-\frac{2}{3}$.

The assumption that the grains are identical spheroids is not important as long as the grains are small, it merely simplifies the averaging procedures and the rigid body dynamics. But the assumption that the grains are small compared to the wavelength of the light involved is an essential defect of equation (3) that cannot be met rigorously merely by using a value of R somewhat smaller than that given by the GANS theory. For large grains, no workable theory is now available that gives σ_A and σ_T, and these quantities do not determine the extinction cross-section of a grain having a general orientation. The colour dependence of the total absorption and of the polarization shows that the larger grains dominate these effects.

For want of a more accurate theory, equations (3) and (4) can be used to give a first approximation to a quantitative discussion of any theory of interstellar polarization. One determines the value of F produced by the orientation mechanism under consideration and uses R as given by the GANS theory. This procedure should give considerable insight but may predict too high a numerical value of p/E_1 since for very large grains $p = 0$. It should not be expected to explain the difference in the colour dependence of p and of A_{pg} since this is determined by the larger grains.

3. THE EFFECT OF MANY CLOUDS

The lack of complete uniformity in alignment of the planes of polarization implied by equation (1) suggests consideration of a model in which each dust cloud is regarded as uniform with a well defined cloud axis, there being some variation in alignment from cloud to cloud. It is usual to assume that the light by which we observe a distant reddened star showing considerable polarization has passed through several such thin clouds, say of the order of five or ten. If one prefers a galactic structure containing fewer but larger dense clouds having a non-uniform structure, one need only say that the small clouds are the elements of volume into which the large cloud can be divided. To determine the polarization produced when light passes through several clouds, partially plane polarized light is described (CHANDRASEKHAR, 1950) by the STOKES' parameters and the effect on the parameters of an arbitrarily oriented cloud is computed. By repeated application of the formula, the effect of many clouds is obtained in the following form. Let p_i be the polarization that would be produced if natural light passed through the i'th cloud normal to its axis. Let ξ_i, η_i, ζ_i be the direction cosines of the cloud axis in a co-ordinate system in which OY is in the direction of the light ray. Let the absorption of the i'th cloud be A_i magnitudes; assume that A_i is independent of ξ_i, η_i, ζ_i. The total absorption produced by a number of clouds is then

$$A = \Sigma A_i \qquad \ldots.(6)$$

and the polarization as observed is

$$p = (Q^2 + U^2)^{1/2}, \qquad \ldots.(7)$$

where $$Q = \Sigma p_i(\xi_i^2 - \zeta_i^2), \quad U = \Sigma p_i 2\xi_i\zeta_i. \qquad \ldots(8)$$

The angle in the OXZ plane, measured from OX toward OZ, made by the plane of polarization is

$$\chi = \tfrac{1}{2}\sin^{-1}(U/p) = \tfrac{1}{2}\cos^{-1}(Q/p). \qquad \ldots(9)$$

These formulae are valid even when some p_i, as given by equation (4), are negative. They are approximate since quadratic terms in p_i have been dropped. If n clouds, all the same except for orientation, obscure each star, then $A = nA_1$ and the mean value of p is np_1 times the mean of the combinations of direction cosines. Thus the mean p/A is independent of n but depends on the distribution of the direction cosines. The distribution about the mean depends both on n and on the distribution of direction cosines.

To illustrate the above formulae, note that if two clouds have axes along OX and one has its axis along OZ, then the three produce $p = p_1$, $A = 3A_1$, $\chi = 0°$. If two clouds have their axes along the lines $(3^{1/2}/2, 0, \pm \frac{1}{2})$, which make angles of $\pm 30°$ with OX, then they produce $p = p_1$, $A = 2A_1$, $\chi = 0°$.

4. MECHANISMS FOR THE ALIGNMENT OF GRAINS

A galactic magnetic field could align prolate spheroidal dust grains either if they were permanently magnetized, as treated by SPITZER and TUKEY (1949, 1951), or if small non-conservative torques act due to paramagnetic relaxation, as treated by DAVIS and GREENSTEIN (1951). Formulae are given for F in the two cases (SPITZER and TUKEY use $M = -3F/2$) and the polarization is treated by means of equations that are equivalent to (3). The following discussion is not adequate as a summary or review of these papers; it merely forms a basis for comparison with GOLD's theory.

SPITZER and TUKEY assume each grain to be a single, highly magnetized ferromagnetic domain whose magnetization is parallel to the long axis of the grains. The MAXWELL-BOLTZMANN distribution law then allows a direct, accurate calculation of F. The following are the points of major difficulty: (1) to concentrate the small amounts of Fe and Ni present in ordinary ice grains into ferromagnetic grains requires a somewhat specialized mechanism involving collisions of grains; (2) the ordinary, unoriented ice grains dilute the polarizing effect of the ferromagnetic grains; (3) the magnetic field required seems to be quite large, of the order of 2×10^{-3} to 10^{-4} gauss; (4) since $F < 0$, the magnetic field must be predominantly normal to the plane of the galaxy; (5) a substantial fraction of the clouds must have kinetic temperatures as low as 30°K.

DAVIS and GREENSTEIN consider a mechanism whose details are less familiar. It depends on the fact that, as the grains rotate in a weak magnetic field, the magnetization lags behind the field by an amount that can be determined from the paramagnetic absorption (GORTER, 1947). This produces a dissipative torque that gradually tends to leave the grains spinning with their long axes normal to the field in both H_I and the much warmer H_{II} regions. Since $F > 0$, the required magnetic field should lie in the plane of the galaxy. The major difficulties with this theory include: (1) although both theory and extrapolation of experimental data on several substances indicate that all grains of any reasonable composition should have enough paramagnetic relaxation to be partially aligned, the subject is new enough and the grain composition uncertain enough that a surprise is possible; (2) since the torques are not conservative, the MAXWELL-BOLTZMANN theory does not apply, and only a very crude estimate is as yet available for the dependence of F on field strength and the properties of the surrounding gas; (3) the magnetic field required seems to lie somewhere in the range of 10^{-4} to 10^{-5} gauss, although a somewhat smaller field might possibly suffice.

A fair summary of the situation seems to be that if no better mechanism can be found and if the existence of a galactic magnetic field does not seem too unreasonable, either of these mechanisms can be regarded as giving a plausible explanation of the origin of interstellar polarization and a means of investigating the dust grains and the galactic magnetic field. BIERMANN and SCHLÜTER (1951) have shown that the electromotive forces of the galaxy would initiate a magnetic field which, because of the turbulent motion of the gas clouds, should have grown to a field that is in the galactic plane, is more or less homogeneous over distances of 300 parsecs, and has a strength of 10^{-6} to 10^{-5} gauss. CHANDRASEKHAR and FERMI (1953) have pointed out that the equilibrium of the gas in a spiral arm of the galaxy seems to require an axial magnetic field of the order of 10^{-5} gauss. On this basis, the paramagnetic absorption theory seems to be preferred.

GOLD (1952, 1953) states without proof that such magnetic fields are questionable and suggests that an alignment mechanism based on a streaming of the interstellar dust clouds through the gas clouds would have fewer difficulties than do the magnetic theories. If the velocity of the dust relative to the gas is supersonic, the angular momentum imparted to a grain by collisions with gas atoms tends to lie normal to the relative velocity. And an elongated grain tends to be normal to its angular momentum. Thus, as GOLD points out, there is a tendency for the long axes of the grains to be parallel to the relative velocity, which defines the cloud axis. Quantitative estimates can be made on the basis of simple models. If the grains are infinitely slender, each grain spins with its axis normal to **H**, its angular momentum. And if the relative velocity is high enough, each **H** becomes normal to the cloud axis. Then $F = -1/6$ and equation (3) shows that (2) is satisfied for reasonably small values of R. A more realistic model gives

$$F = F_\theta(\gamma)F_\beta(r), \qquad \ldots.(10)$$

where $F_\theta(\gamma)$ allows for the fact that the grains are not infinitely slender and $F_\beta(r)$ allows for the fact that, due to the non-zero kinetic temperatures of the gas and dust clouds, each **H** is not normal to the cloud axis.

To examine the first of these factors, assume that the grains are axially symmetric with finite thickness but that each **H** is normal to the cloud axis. The distribution of the angle between **H** and the grain axis depends on the grain shape and is difficult to determine exactly. To estimate F, assume that this angle has the equipartition distribution characteristic of thermal equilibrium. The same distribution would be obtained if the grains were subject to completely random bombardment by the gas while mechanically constrained so that the direction of **H** is fixed. Appendix 2 and equations (36) to (44) of DAVIS and GREENSTEIN (1951) then enable one to evaluate the distribution function, getting

$$F = F_\theta(\gamma) = -\frac{1}{6} + \frac{1}{2(\gamma - 1)}\left[\frac{\gamma^{1/2}\sinh^{-1}(\gamma - 1)^{1/2}}{(\gamma - 1)^{1/2}} - 1\right], \qquad \ldots.(11)$$

where γ^{-1} is the moment of inertia about the axis of symmetry divided by that about a transverse axis through the centre of mass. If the grains are assumed to be small homogeneous spheroids, to which equations (3) to (5) apply, straightforward calculation gives F_θ, R, and p/A_{pg} as functions of $x = a_A/a_T$, the axial radius divided by the transverse radius. The results are listed in Table 1 for various values of x running from 0 for a disk through 1 for a sphere to ∞ for a line. R and p/A_{pg} depend on the index of refraction of the grain; $n = 2^{1/2}$ is appropriate for dielectric grains and $n = 2^{1/2}(1 - i)$ gives an estimate of R for grains having metallic absorption.

Table 1. Values of R for small grains from the Gans theory for any alignment mechanism, and values of F_θ and p/A_{pg} for Gold's mechanism when $F_\beta = 1$

x	$F_\theta(\gamma)$	$n = 2^{1/2}$		$n = 2^{1/2}(1 - i)$	
		$R(x)$	p/A_{pg}	$R(x)$	p/A_{pg}
0·0	0·048	− 0·333	− 0·072	− 0·454	− 0·098
0·2	0·045	− 0·241	− 0·049	− 0·419	− 0·085
0·6	0·026	− 0·098	− 0·012	− 0·239	− 0·028
1·0	0·000	0·000	− 0·000	0·000	− 0·000
1·2	− 0·013	0·036	− 0·002	0·104	− 0·006
1·5	− 0·031	0·080	− 0·011	0·226	− 0·032
2·0	− 0·056	0·130	− 0·033	0·353	− 0·089
3·0	− 0·090	0·189	− 0·077	0·459	− 0·186
5·0	− 0·124	0·239	− 0·133	0·507	− 0·282
10·0	− 0·150	0·275	− 0·185	0·519	− 0·350
∞	− 0·167	0·294	− 0·221	0·520	− 0·390

In order that each $\mathbf{H}$ be normal to the cloud axis, it is necessary that all velocities of gas atoms relative to dust grains be parallel to the cloud axis and that the angular momentum added by collisions be much larger than the initial angular momentum of the grains. Departures from this condition reduce F, and the values of p/A_{pg} in Table 1, by the factor F_β which may be estimated as follows. Assume that the three components of $\mathbf{H}$ are normally distributed, the standard deviation of the component along the cloud axis being s_P and that for any normal direction being $s_N = s_P/r$. Assume that the angle between the grain axis and $\mathbf{H}$ has the equipartition distribution. Then F is given by equation (10) with

$$F_\beta = 1 - 3\frac{r^2}{1 - r^2}\left[\frac{\tanh^{-1}(1 - r^2)^{1/2}}{(1 - r^2)^{1/2}} - 1\right] \qquad \ldots.(12)$$

For $r = s_P/s_N = 1$, $2^{1/2}/2$, $2^{1/2}/4$, and 0, F_β is, respectively, 0, 0·26, 0·65, and 1. Inspection of Table 1 shows that for reasonable values of x the maximum allowable reduction in p/A_{pg}, if condition (2) is to be satisfied, corresponds to $F_\beta > \frac{1}{4}$ and hence to $r < 0{\cdot}7$. To discover the circumstances under which this condition is satisfied, consider a dust cloud that strikes a gas cloud. Assume that the dust cloud is very diffuse and does not contain enough evaporated gas to make a hot shock wave where the two clouds interpenetrate. Before collision, $\mathbf{H}$ in the dust cloud will be randomly distributed and F_β will be zero. After the dust grains have penetrated a certain distance the anistropy in angular momentum will be built up and F_β will become reasonably large. But when the relative velocity is reduced to v_H, the mean thermal velocity of the atoms in the gas cloud, enough of the collisions will have been oblique to reduce F_β to about $\frac{1}{4}$, as determined from calculations for spherical grains. For these supersonic velocities, the drag is proportional to the square of the velocity and the distance L travelled by the dust grain while its velocity is reduced from v_0 to v_H is

$$L = (m_g/\pi a^2 m_H n_H)\ln(v_0/v_H) \approx 0{\cdot}1 \text{ parsec}, \qquad \ldots.(13)$$

where m_g and m_H are the masses of the grain and a hydrogen atom, respectively, a is the grain radius, and n_H is the number of gas atoms per unit volume. The numerical value corresponds to $a = 10^{-5}$ cm, grain density unity, $n_H = 10$, $v_H = 1{\cdot}2$ km per sec. (corresponding to 60°K), and $v_0 = 12$ km per sec. Hence appreciable polarization is produced only in a very thin layer at the edge of a gas cloud and the average effect throughout the cloud is much smaller. Although somewhat crude models have been used in getting these results, it does not seem likely that a more accurate theory would give a substantially larger polarization and it therefore appears that GOLD's mechanism cannot produce the observed polarization of starlight.

ZIRIN (1952) has suggested that with differential motion of dust and gas the alignment would be with the long axis normal to the direction of relative motion. This seems based on a hydrodynamic argument that does not apply here since the grain diameter is less than the mean free path of gas atoms. His point that oblate grains should be considered is well taken; actually they have been included in all the analyses referred to above and, except perhaps in the case of ferromagnetic grains, give effects of the same sign but of smaller magnitude than do prolate grains.

5. APPLICATIONS TO GALACTIC STRUCTURE

If any one of these theoretical explanations of interstellar polarization is accepted, interpretation of the observations provides information bearing on the structure of the galaxy. For example, if GOLD's mechanism were accepted and the polarization were assumed to occur in a single cloud, then by equation (1) the cloud axes near galactic longitude 100° would have to be nearly normal to the galactic plane, with a dispersion of only 0·1 radian. This implies that the cloud velocity normal to the galactic plane should be about ten times the 8 km per sec. dispersion in cloud velocity observed in the plane. But such high normal velocities are inconsistent with the observed concentration of dust and gas within a few hundred parsecs of the galactic plane. If it is assumed that analysis of the dynamics of gas clouds in the gravitational field of the galaxy would limit the r.m.s. velocity normal to the plane to twice the r.m.s. velocity in the plane, then the resulting dispersion in orientation of the cloud axes could still be consistent with equation (1) if the light went through enough thin clouds to average out most of the fluctuation. However, calculations based on equations (6) to (9) indicate that the number of clouds would have to be of the order of eighty and that the mean p/A_{pg} would be reduced by a factor of 0·3. These implications of GOLD's theory are further evidence of its unacceptability.

Next, consider a type of analysis which may eventually give some information on the uniformity of the clouds and the orientation of their axes. For example, if the mechanism of DAVIS and GREENSTEIN produces the interstellar polarization, the magnetic lines of force lie predominantly in the galactic plane. But they could either make random whirls in this plane or they could constitute a reasonably uniform field, perhaps along a spiral arm since the region of maximum uniformity is that in which we look normal to the arm. Accordingly, apply the treatment of Section 3 to a model in which all p_i are equal, the Z-axis is normal to the galactic plane, all $\zeta_i = 0$, and $\cos^{-1} \xi_i$ is uniformly distributed, corresponding to random whirls. Then the plane of vibration is the galactic plane, as observed, and the computed p/A_{pg} shows a statistical variation that depends on n, the number of clouds through which the light is assumed to pass. Superficial comparison with the observed distribution of

$p/(E_1 + 0 \cdot 05)$ near galactic longitude 100°, as taken from HILTNER'S (1951) data, indicates that an n of the order of 8 is required if the theoretical distribution is to resemble the observed; $n = 1$ is certainly excluded and $n = 4$ seems to be. If the p_i have a statistical distribution, then n must be larger. If further work indicates that the value of n required is unreasonably large, then the roughly uniform magnetic field along the spiral arm will seem more appropriate since it can easily give the observed distribution of $p/(E_1 + 0 \cdot 05)$. In this way the recent assumption of CHANDRASEKHAR and FERMI (1953) that the galactic magnetic field lies along the spiral arm could receive additional support. This argument, if it holds, also will tend to discredit any theory, such as ZIRIN'S, in which $p_i > 0$ and in which the cloud axes would be randomly distributed in the galactic plane. For mechanisms that give $p_i < 0$, the corresponding implication is that the cloud axes are roughly normal to the galactic plane rather than randomly oriented normal to the axis of a spiral arm.

Consider next the magneto-hydrodynamic formula

$$B = (4\pi\rho/3)^{1/2} v/\alpha_1 \text{ gauss} \qquad \ldots.(14)$$

used by DAVIS (1951) and by CHANDRASEKHAR and FERMI (1953) to determine the strength of the magnetic field from ρ, the density of the gas, v, its root-mean-square velocity, and α_1, the root-mean-square angular deviation of the lines of force from a uniform direction. This assumes that the field is roughly uniform. If the cloud axes are determined by the magnetic field and if the observed polarization is produced in going through n clouds that are separated by distances larger than the wavelength of the variations in B, then the root-mean-square variation in the orientation of the planes of polarization will be

$$\alpha_p = \alpha_1 n^{-1/2}, \qquad \ldots.(15)$$

provided $\alpha_1 \ll 1$. Since α_p is given by equation (1), B can be estimated. If one takes $n = 6$, $v = 8 \times 10^5$ cm per sec., $\rho = 2 \times 10^{-24}$ gm per cm^3, and $\alpha_p = 0 \cdot 1$ radian; one gets $B = 10^{-5}$ gauss. Evidently it is important that appropriate values of n and α_p be used in this argument.

In these illustrative calculations only order of magnitude accuracy has been attempted. If more extensive calculations seem worth while, they should be based on extensive measurements of p, E_1, and the orientation of the plane of vibration for stars in a number of regions. These observations should be compared with the distributions predicted with the aid of equations (6) to (8) for models in which the magnetic field has some reasonable assumed structure that determines both the cloud axes and p_i as given by the paramagnetic absorption theory.

Although it would be unfair to ascribe to Dr. J. L. GREENSTEIN any of the astronomical mistakes made by a physicist who ventures into an unfamiliar field, he does deserve much credit, which it is a great pleasure to acknowledge, for his constant helpfulness in supplying information, suggestions, and constructive criticism.

REFERENCES

BIERMANN, L. and SCHLÜTER, A.	1951	*Phys. Rev.*, **82,** 863.
CHANDRASEKHAR, S.	1950	*Radiative Transfer*, Oxford, p. 24.
CHANDRASEKHAR, S. and FERMI, E.	1953	*Ap. J.*, **118,** 113.
DAVIS, L., Jr.	1951	*Phys. Rev.*, **81,** 890.
DAVIS, L., Jr. and GREENSTEIN, J. L.	1951	*Ap. J.*, **114,** 206.

GANS, R. . . . . 1912 *Ann. Phys. (Leipzig)*, **37**, 881.
GOLD, T. . . . . 1952 *Nature (London)*, **169**, 322.
1953 *M.N.*, **112**, 215.
GORTER, C. J. . . . . 1947 *Paramagnetic Relaxation* (Elsevier Publ. Co., New York).
HALL, J. S. . . . . 1949 *Science*, **109**, 166.
HALL, J. S. and MIKESELL, A. H. . . . . . 1950 *Publ. U.S. Naval Obs., Washington*, **17** (Pt. I).
HILTNER, W. A. . . . . 1949 *Science*, **109**, 165.
1951 *Ap. J.*, **114**, 241.
SPITZER, L., Jr. and TUKEY, J. W. . . . . . 1949 *Science*, **109**, 461.
1951 *Ap. J.*, **114**, 187.
STRANAHAN, G. . . . . 1954 *Ap. J.*, **119**, 465.
ZIRIN, H. . . . . 1952 *Bull. Harvard Coll. Obs.*, **921**, 26.

The Gravitational Instability of an Infinite Homogeneous Medium when a Coriolis Acceleration is Acting

S. CHANDRASEKHAR

Yerkes Observatory, University of Chicago

SUMMARY

It is shown that JEANS's criterion for the gravitational instability of an infinite homogeneous medium remains unaffected even if the system is partaking in rotation and Coriolis forces are acting.

1. INTRODUCTION

ONE of the fundamental concepts in modern cosmogony is that of gravitational instability due to JEANS (1902). Its importance lies in the fact that it enables one to estimate the scale of the condensations which may take place in an extended gaseous medium; and the possibility of estimating this will give precision to the ideas one may have regarding the formation of stars and of galaxies from an "original primeval nebula". To give only two examples of recent applications of gravitational instability and which at the same time illustrate the fruitfulness of JEANS's concept, we may refer to SPITZER's efforts (1951) to interpret and incorporate in a picture of the formation of stars, the highly irregular distribution of the interstellar matter and GAMOW's speculations (1950) on the origin of the galaxies.

The principle underlying JEANS's estimate of the scale of the condensations which may take place in an extended gaseous medium is the following: starting from an initial state of homogeneity and rest, we consider the velocity of propagation of a wave of density fluctuation through the medium. If the gravitational consequences of the density fluctuations are ignored, the problem is the classical one of the propagation of a sound wave; and, as is well known, the velocity of sound in a gaseous medium is independent of the wavelength and is given by

$$c = \sqrt{(\gamma p/\rho)}. \qquad \ldots .(1)$$

where p denotes the gas pressure, ρ the density and γ is the ratio of the specific heats. On the other hand, if the change δV, in the gravitational potential consequent to the density fluctuation, $\delta\rho$, is taken into account in the equations of motion in accordance with Poisson's equation,

$$\nabla^2 \delta V = - 4\pi G \delta\rho \qquad \ldots.(2)$$

then the velocity of propagation of a wave is no longer independent of its wavelength, λ. It is given by

$$v_J = \sqrt{(c^2 - G\rho\lambda^2/\pi)}. \qquad \ldots.(3)$$

The velocity of wave-propagation, therefore, becomes maginary when

$$\lambda > \sqrt{\frac{\pi c^2}{G\rho}} = \sqrt{\frac{\pi\gamma p}{G\rho^2}} = \lambda_J \text{ (say)}. \qquad \ldots (4)$$

But an imaginary velocity of wave-propagation means, only, that the amplitude of the wave can increase exponentially with time. Accordingly, if an arbitrary initial perturbation in density is represented by a Fourier integral, then the amplitudes of the components in the Fourier representation which have wavelengths greater than λ_J will increase exponentially with time; λ_J is therefore a measure of the linear dimensions of a "condensation" which may take place in the medium. This is JEANS's result.

Now doubts have been expressed (cf. SPITZER) regarding the applicability of JEANS's criterion (4) to a system partaking in rotation. And it has been stated that if

$$\Omega^2 > \pi G\rho \qquad \ldots.(5)$$

where Ω denotes the angular velocity of rotation, then gravitational instability in the sense of JEANS cannot take place. If this were true, the usefulness of the concept of gravitational instability would, indeed, be limited. Certainly, its applicability to the interstellar medium could be questioned. For, with the known angular velocity of galactic rotation ($\sim 10^{-15}$ per sec) the limit on the density imposed by (5) is

$$\rho < 4 \times 10^{-24} \text{ gm per cm}^3. \qquad \ldots.(6)$$

Since the average density of interstellar space is appreciably less than this we might conclude that gravitational instability offers or suggests no clue. However, an examination of the underlying dynamical problem shows that no such inequality as (5) limits the applicability of JEANS's criterion. It is the object of this paper to show this.

2. THE SOLUTION OF THE PROBLEM

Consider an extended gaseous medium which is partaking in a rotation with angular velocity $\mathbf{\Omega}$. The equations of motion and continuity governing the motions consequent to infinitesimal fluctuations in density ($\delta\rho$) and pressure (δp) are:

$$\rho \frac{\partial \mathbf{v}}{\partial t} = - \text{grad } \delta p + \rho \text{ grad } \delta V + 2\rho \mathbf{v} \times \mathbf{\Omega} \qquad \ldots.(7)$$

and

$$\frac{\partial}{\partial t} \delta\rho = - \rho \text{ div } \mathbf{v}. \qquad \ldots.(8)$$

If the changes in pressure and density are assumed to take place adiabatically,

$$\delta p = c^2 \delta \rho \qquad \ldots.(9)$$

where c denotes the velocity of sound (equation (1)). In addition to the foregoing equations we have also Poisson's equation (2).

We shall consider a solution of equations (2), (7), (8), and (9), which correspond to the propagation of a wave in the z-direction (say); for such solutions, $\partial/\partial z$ is the only non-vanishing component of the gradient and the equations become

$$\left.\begin{aligned} \frac{\partial v_x}{\partial t} + 2\Omega_y v_z - 2\Omega_z v_y &= 0 \\ \frac{\partial v_y}{\partial t} + 2\Omega_z v_x &= 0 \\ \frac{\partial v_z}{\partial t} + \frac{c^2}{\rho}\frac{\partial}{\partial z}\delta\rho - \frac{\partial}{\partial z}\delta V - 2\Omega_y v_x &= 0 \\ \frac{\partial}{\partial t}\delta\rho + \rho\frac{\partial v_z}{\partial z} &= 0 \\ \text{and} \qquad \frac{\partial^2}{\partial z^2}\delta V + 4\pi G\delta\rho &= 0. \end{aligned}\right\} \qquad \ldots.(10)$$

In writing these equations we have supposed that the orientation of the co-ordinate axes has been so chosen that $\mathbf{\Omega}$ lies in the (y, z) plane and that,

$$\mathbf{\Omega} = (0, \Omega_y, \Omega_z). \qquad \ldots.(11)$$

For a solution which represents the propagation of a wave in the z-direction,

$$\frac{\partial}{\partial t} = i\omega \text{ and } \frac{\partial}{\partial z} = -ik, \qquad \ldots.(12)$$

where ω denotes the frequency and k the wave-number. Substituting (12) in equations (10), we obtain a system of linear homogeneous equations which can be written in matrix notation in the form:

$$\begin{vmatrix} 2i\Omega_z & \omega & -2i\Omega_y & 0 & 0 \\ \omega & -2i\Omega_z & 0 & 0 & 0 \\ 0 & +2i\Omega_y & \omega & -c^2k/\rho & k \\ 0 & 0 & -\rho k & \omega & 0 \\ 0 & 0 & 0 & 4\pi G & -k^2 \end{vmatrix} \begin{vmatrix} v_y \\ v_x \\ v_z \\ \delta\rho \\ \delta V \end{vmatrix} = 0. \qquad \ldots.(13)$$

The condition that the foregoing system of equations has a non-trivial solution is that the determinant of the matrix on the left-hand side should vanish. On expanding the determinant, we find that it can be reduced to the form

$$\omega^4 - (4\Omega^2 + \Omega_J^2)\omega^2 + 4\Omega_z^2\Omega_J^2 = 0 \qquad \ldots.(14)$$

where (cf. equation (3))

$$\Omega_J = \sqrt{(c^2k^2 - 4\pi G\rho)}. \qquad \ldots.(15)$$

From equation (14) it follows that there are, in general, two modes in which a wave can be propagated through the medium. If ω_1 and ω_2 are the frequencies of these two modes, then (cf. equation (14))

$$\omega_1^2 + \omega_2^2 = 4\Omega^2 + \Omega_J^2 \qquad \ldots.(16)$$

and

$$\omega_1\omega_2 = 2\Omega_z\Omega_J. \qquad \ldots.(17)$$

Hence *if Ω_J is imaginary, then either ω_1 or ω_2 must be imaginary.* In either case, there is a mode of wave-propagation which is unstable. Now the condition for Ω_J to be imaginary is

$$k^2 < 4\pi G\rho/c^2; \qquad \ldots.(18)$$

but this is precisely JEANS's criterion. The condition for gravitational instability is, therefore, unaffected by the presence of Coriolis forces. There is only one exception to this rule, namely, when the wave is propagated in a direction at right angles to the direction of $\mathbf{\Omega}$. In that case $\Omega_z = 0$, and equation (14) gives

$$\omega^2 = 4\Omega^2 + \Omega_J^2 = 4\Omega^2 + c^2k^2 - 4\pi G\rho. \qquad \ldots.(19)$$

Accordingly, in this plane there is only one mode of wave-propagation; and if the condition (5) should in addition be fulfilled, $\omega^2 > 0$ and gravitational instability cannot arise. This may explain how the inequality (5) came to be current. However, as we have already stated JEANS's criterion for gravitational instability applies for wave-propagation in every other direction which is not at right angles to $\mathbf{\Omega}$.

REFERENCES

GAMOW, G. and CRITCHFIELD, C. L. . . 1950 *Theory of Atomic Nucleus and Nuclear Energy-Sources*, Oxford, p. 336.

JEANS, J. H. . . . . 1902 *Phil. Trans. A.*, **199,** 1; see p. 49.

1929 *Astronomy and Cosmogony*, Cambridge, see pp. 345–7.

SPITZER, L. . . . . 1951 *J. Washington Acad. Sci.*, **41,** 309.

SECTION 5

INSTRUMENTS

"Der Spiegel dient, Dir selbst die Flecken zu entdecken:
Am Spiegel wische nicht, an Dir wisch' ab die Flecken!"
FRIEDRICH RÜCKERT (1788–1866).

Modern Developments in Telescope Optics

E. H. LINFOOT

The Observatories, Cambridge

SUMMARY

The paper falls into two main parts.

In Part 1 a brief account is given of the development of Schmidt and Schmidt-Cassegrain systems from 1930 to the present day.

Part 2 describes an optical development of contemporary interest and its application to the problem of aberration balancing in fast Schmidt cameras.

CONTENTS

1. GENERAL SURVEY

1.1. *Introduction*

TWENTY-THREE years ago, BERNHARD SCHMIDT devised and constructed the new type of optical system which bears his name. Perhaps it is not too early to look back and review briefly the progress in astronomical instrument design which has resulted from his invention.

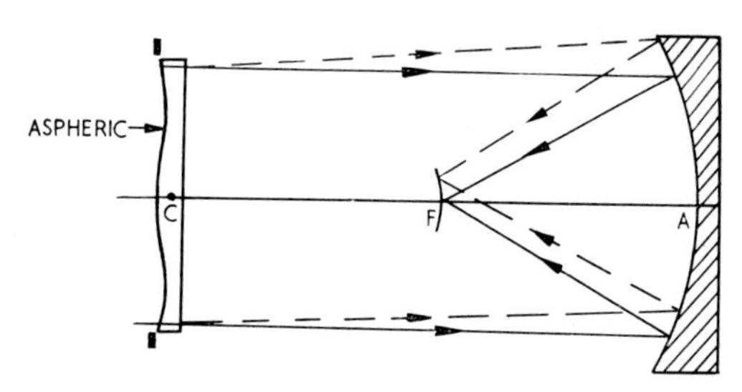

Fig. 1. The Schmidt camera (the asphericity on the front surface of the corrector plate is exaggerated about 100 times)

SCHMIDT'S new idea was to correct the spherical aberration of a concave mirror by means of a surface deformation or "figuring" carried out, not on the mirror itself, but on a thin plane-parallel glass plate situated at its centre of curvature (see Fig. 1).

It can be seen, without any detailed calculation, that a corrector plate in this position can remove the aberrations in the off-axis images nearly as completely as in the on-axis image. In particular, the system consisting of sphere and corrector plate at its centre of curvature will be free from the off-axis coma which spoils the optical performance of a paraboloid reflector. The system may therefore be expected to give good photographic images over a relatively wide field. In SCHMIDT'S first model, an $f/1{\cdot}75$ camera of 44 cm aperture working over a 16° field, this expectation was brilliantly fulfilled.

How did SCHMIDT come upon his novel and ingenious idea? From what has been put on record by R. SCHORR (1936) concerning SCHMIDT'S methods of work it is possible to see his invention as a natural development of ideas known to be already in his mind.

SCHORR mentions that when making Cassegrain cameras SCHMIDT generally used a spherical mirror as the primary and removed spherical aberration of the system by figuring the secondary. Such a procedure was entirely in keeping with his practice of working all surfaces by hand, not with machines, and of using light, rather thin glass tools for grinding and polishing. So acute an observer could hardly fail to notice that the off-axis coma of a Cassegrain system is made worse when the "figuring" ordinarily carried by the primary is transferred in this way to the secondary. Thus it appears that SCHMIDT was familiar with the notion of transferring a "figuring" from one surface to another, was almost certainly aware that such a transfer alters off-axis coma, and, possibly for reasons connected with his own mirror-making technique, preferred systems in which the large primary mirror is spherical. The stage was therefore well set for the advent of the new idea of using a glass plate to carry the "figuring" which removes the spherical aberration of a spherical primary mirror and of controlling the coma by adjusting the position of the plate. SCHMIDT'S own explanation of his idea (SCHMIDT, 1932) is in substantially these terms.

It is also interesting to note that, even after his idea had fully taken shape, SCHMIDT did not embark on the construction of the corrector plate until he had thought out a method which reduced the problem to that of grinding and polishing a spherical surface, of hardly perceptible curvature,* on one side of the elastically deformed glass plate. In this way he was able to avoid the necessity of figuring a relatively large aspheric surface with sub-diameter tools.

1.2. *Variants of the Schmidt Camera*

The field surface in the Schmidt camera is not flat but spherical and concentric with the mirror. In large Schmidt cameras, the inconvenience of the curved field is not serious; thin glass photographic plates can be held pressed against a spherical backing surface during the exposure. On removal from the holder, the plates spring flat again. A more serious drawback arises from the fact that the overall length of the system is a little more than twice its focal length; this much increases the cost of mounting and dome.

F. B. WRIGHT (1936) proposed a modified system, in which the strength of the corrector plate is doubled and its distance from the mirror halved, while the mirror is given a deformation of figure equal and opposite to that which would parabolize it. The field surface of this system is flat, and the geometrical images are, to a first approximation, uniformly lit disks. The system is not anastigmatic and its theoretical

* P. C. HODGES (1948). HODGES says that the surface of the deformed plate was ground flat, but this seems to be incorrect.

performance is much inferior to that of the classical Schmidt camera. A better method of flattening the field is to compensate the Petzval curvature by adding a plano-convex lens immediately in front of the focal plane. The resulting system, usually known as the field-flattened Schmidt camera, can be designed so as to give an optical performance not much inferior to that of the plain Schmidt.* Thus it is the short tube, not the flat field, which is the strong point of WRIGHT's design.

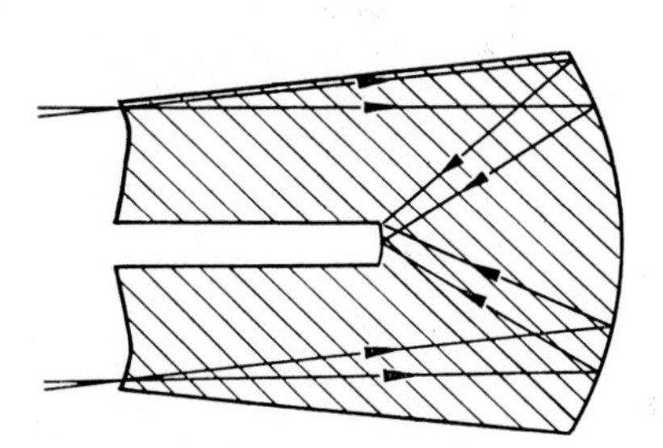

Fig. 2. Solid Schmidt camera

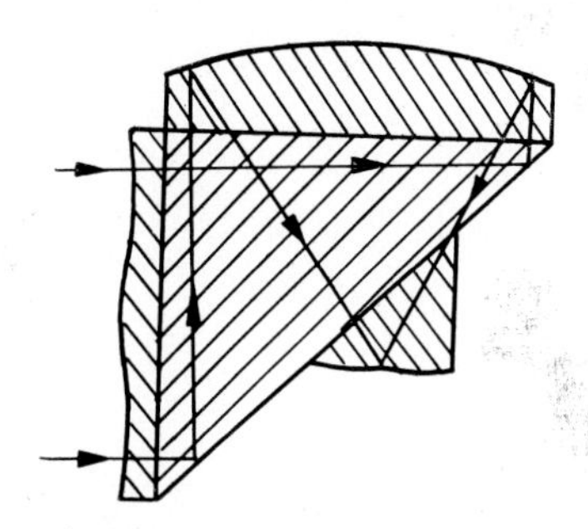

Fig. 3. Folded solid Schmidt camera (D. HENDRIX)

VÄISÄLÄ (1935) pointed out the existence of a whole range of systems with aspherized mirror and corrector plate, which include the Wright camera as a special case. The other cases are not flat-fielded, and performance falls off seriously as the tube length decreases. But when a Schmidt camera of preferred focal length just fails to go into an already existing dome, VÄISÄLÄ's systems may provide the best compromise design.

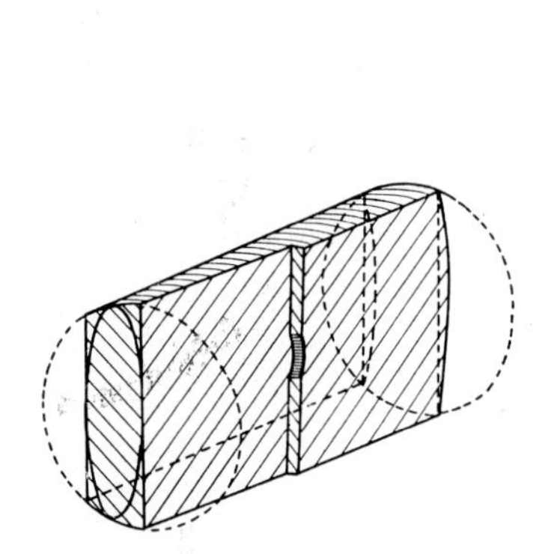

Fig. 4. Solid Schmidt camera (J. G. BAKER)

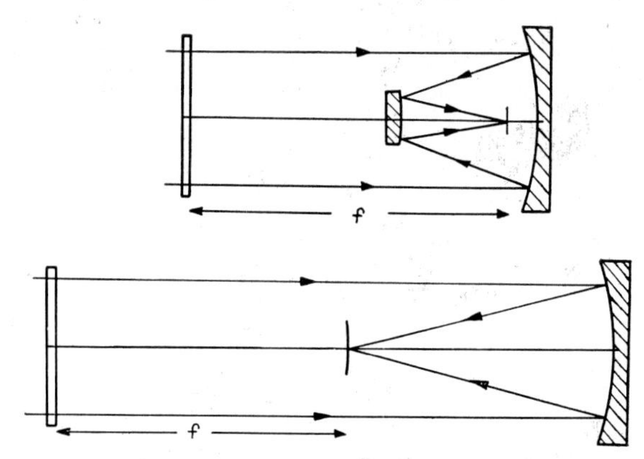

Fig. 5. (a) Schmidt-Cassegrain camera. (b) Schmidt camera of the same aperture and focal length

At aperture ratios shorter than $f/1$, the aberrations of the Schmidt camera can no longer be ignored, and they begin to be troublesome even in small systems at the still higher ratios used in Schmidt spectrograph cameras. For these the use of a solid Schmidt system (see Fig. 2) is sometimes advisable. Constructed in glass of refractive index n, the solid Schmidt is n^2 times faster than the ordinary Schmidt of similar

* A practical drawback to the system in astronomical photography is the formation of halos round the photographic images of bright stars. They arise from reflections at the two surfaces of the field flattener, and can be practically eliminated by "blooming" these surfaces.

Fig. 6 (*on facing page*). The 15-in. Schmidt-Cassegrain telescope of St. Andrew's University Observatory. A 30-in. Baker-Schmidt anastigmat is in course of construction

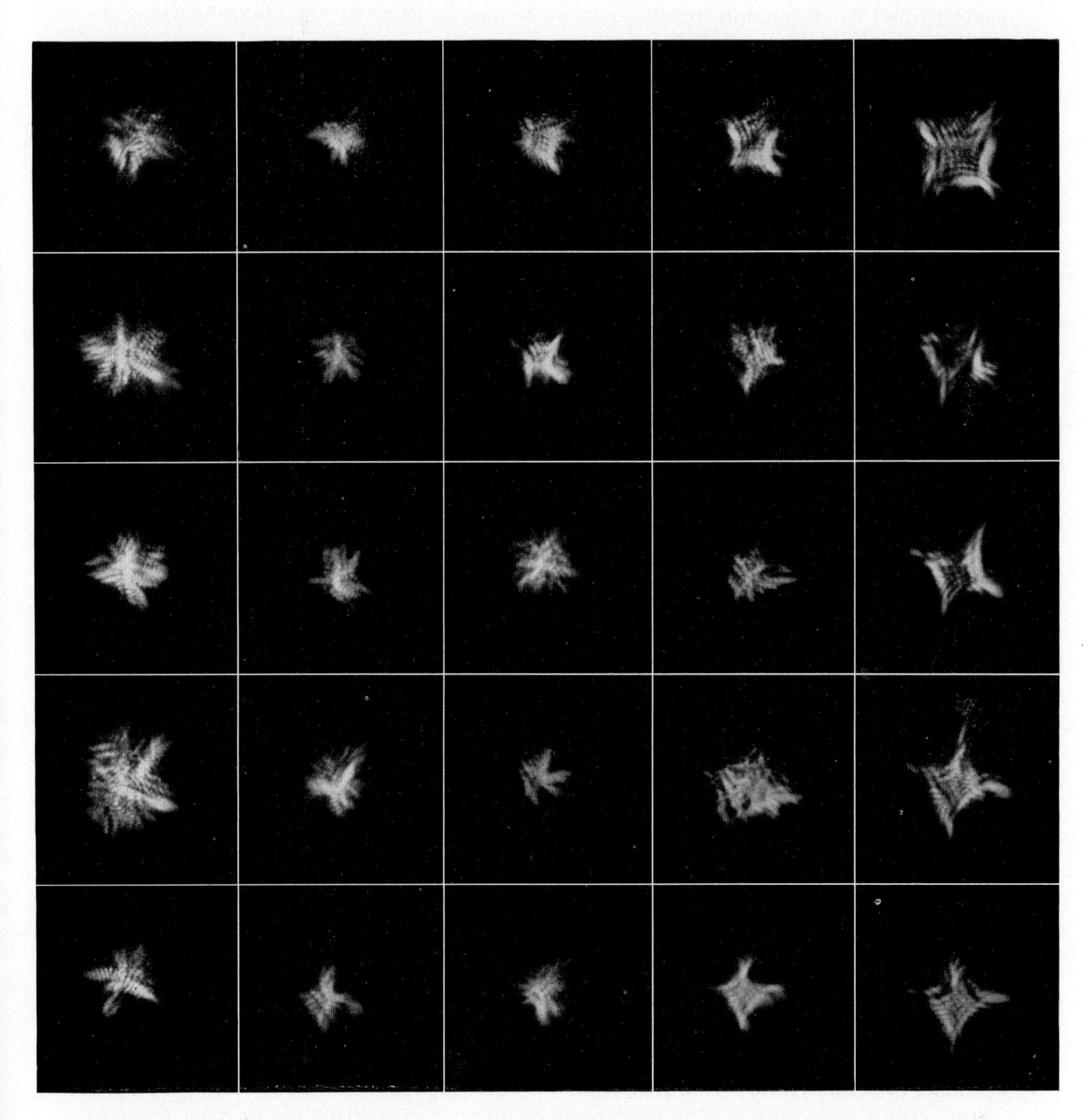

Fig. 7. *Microphotographs (enlarged 200×) of images of an artificial star at selected points in the field of the St. Andrew's Pilot Telescope (15-in. f/3 Schmidt-Cassegrain aplanat).* The twenty-five selected points form a square lattice covering the field; each image was separately adjusted to best focus, so that the images do not provide a test for the absence of residual field-curvature. A pin-hole, imaged down by a micro-objective, was placed at the focus of a 19-in. paraboloid collimator mirror. The pilot telescope on its mounting could be moved about in the parallel beam issuing from the collimator, to produce a star image in any desired part of the field. This image was photographed through a 6 mm micro-objective. Fuller details are given in a paper in preparation by Dr. J. CÍSAŘ and the writer.

dimensions, and effective speeds of $f/0{\cdot}35$ are practicable in small sizes. The glass is cut away behind the field surface to allow access to the photographic film, which is oiled on to the spherical field surface. Because the initial refraction into the glass is accompanied by dispersion, resulting in a chromatic difference of magnification, the astronomical usefulness of the system is restricted to spectroscopy.

An elegant and ingenious variant of the design, due to HENDRIX (1939), is shown in Fig. 3. Here the aspheric first surface and the mirror are cemented to the perpendicular faces of a right-angled prism; a smaller prism cemented to the centre of the diagonal face carries the accessible field surface. Another variant, due to J. G. BAKER (1940a), is shown in Fig. 4. It is designed to work with a prism train from which the collimated beam in each wavelength emerges with an elliptical cross-section, the major axis being parallel to the direction of spectral resolution.

1.3. *Schmidt-Cassegrain Cameras*

1.31. *Baker-Schmidt Systems; Schmidt-Cassegrain Aplanats*—The most effective way of flattening the field of the Schmidt camera is by the addition of a convex secondary mirror to the system, so as to compensate the Petzval curvature. Of course, this involves changes in the strength and position of the corrector plate. Not only is the new system flat-fielded, but the ratio of its overall length to the focal length is also improved. (See Fig. 5.) J. G. BAKER (1940b) showed that flat-fielded anastigmats can be obtained in this way provided that at least one of the mirrors is aspherized in addition to the corrector plate. The same discovery was made independently by C. R. BURCH (1942). The ADH telescope at Bloemfontein, a Baker-Schmidt anastigmat of 30 in. aperture and 120 in. focal length, has a spherical primary mirror and a "figured" secondary.

H. SLEVOGT (1942) and E. H. LINFOOT (1944) pointed out independently that the use of two spherical mirrors and a corrector plate gives a system in which the off-axis astigmatism is small enough to be harmless in many practical cases. One of these cases is that of the 15-in. $f/3$ St. Andrew's pilot telescope, shown in Fig. 6; microphotographs obtained by J. CÍSAŘ and the writer of the diffraction images formed by this system are shown in Fig. 7.

1.32. *Monocentric Schmidt-Cassegrain Systems*—An even more favourable ratio of tube-length to focal length is obtained in the monocentric Schmidt-Cassegrain shown in Fig. 8. Here primary and secondary mirrors are concentric spheres, and the corrector plate is situated at their common centre of curvature. There is no difficulty in arranging that the concave field surface takes a conveniently accessible position just behind the pierced primary mirror. The optical aberrations of an $f/3{\cdot}5$ system, approximately six times larger than those of an ordinary Schmidt camera of the same aperture and focal length, are still small enough to give good performance over a $4\frac{1}{2}°$ field (WAYMAN, 1950). Used with a simple plano-concave field flattener, the system still gives a sufficiently good performance; it is worthwhile in most cases to modify the plate profile so as to offset the chromatism of the field flattener.

1.33. *Two-plate Schmidt-Cassegrain Systems*—At aperture ratios of $f/3$ and longer, chromatism becomes the dominating optical aberration in Schmidt and Schmidt-Cassegrain systems working over the fields of diameter not exceeding about 5° to which they are more or less restricted in astronomical use by considerations of vignetting. It can be shown (LINFOOT, 1945) that chromatism can be greatly reduced, and anastigmatism can be obtained with both mirrors spherical, provided that the single corrector plate is replaced by two separated aspheric plates, of different glasses and of opposite asphericities, a short distance apart; see Fig. 9. Less general systems of this kind had already been considered by A. WARMISHAM (1941, 1943).

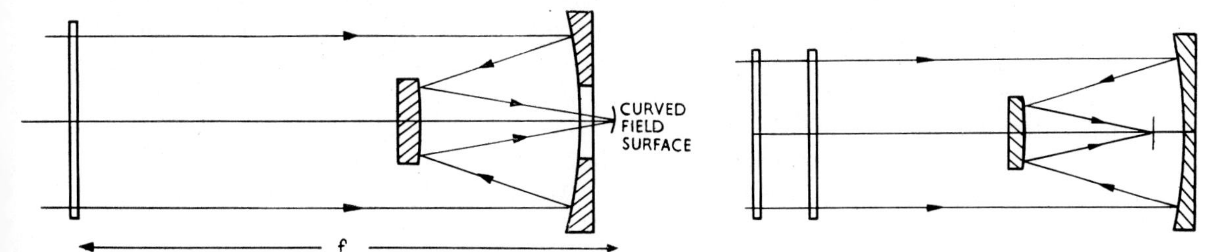

Fig. 8. Monocentric Schmidt-Cassegrain system

Fig. 9. Two-plate Schmidt-Cassegrain system

A small-scale system, of aperture 7·75 in. and focal length 14·5 in., was constructed by the writer some years ago in Bristol; the mounting was later improved in Cambridge and the plates "bloomed" to reduce reflection-ghosts. Both the aspheric plates are of ordinary plate glass; at this short focal length fuller colour-compensation can be dispensed with. The nominal aperture-ratio is $f/1{\cdot}9$.

Trials made by Dr. G. MERTON at the Oxford University Observatory showed that this system gives photographic images which appear completely uniform over a flat photographic plate 2 in. square; the smallest fully-exposed images being only

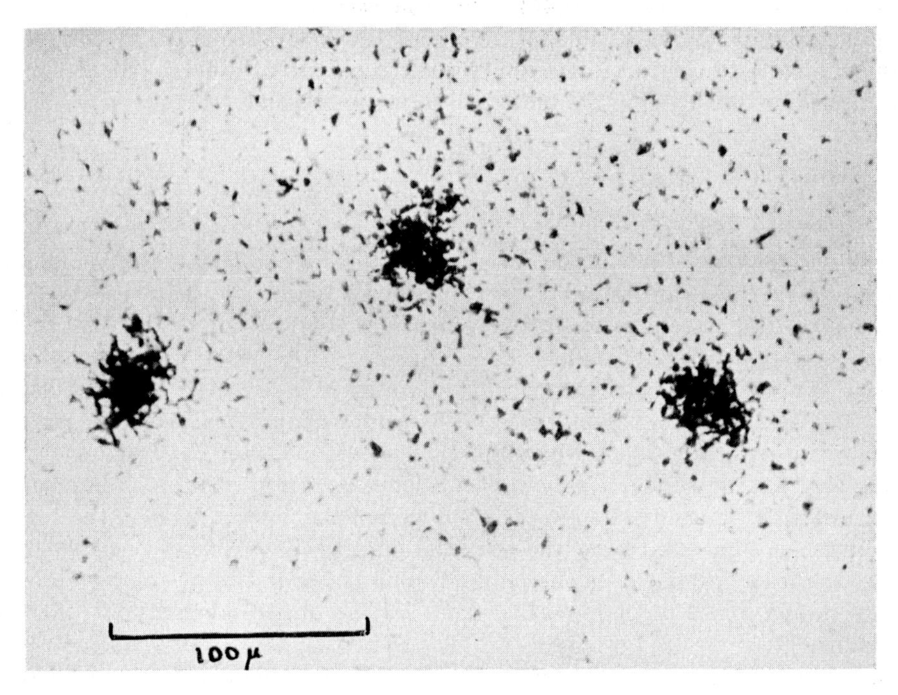

Fig. 10. Showing an enlarged portion of a focus-test plate taken by Dr. MERTON with this camera. Three images of the same star are recorded at focal settings differing by 0·001 in.

15μ in diameter. Fig. 10 shows an enlarged portion of a focus-test plate taken by Dr. MERTON with this camera. Three images of the same star are recorded at focal settings differing by 0·001 in.

1.34. *Meniscus Schmidt Systems*—Extremely good performance over a wide field at aperture ratios near $f/1$ can be obtained if the spherical aberration of the mirror is first compensated by a concentric meniscus lens, as shown in Fig. 11, and the residual zonal aberration then corrected by figuring a plane air-glass surface passing through the centre of curvature. That is to say, the mirror-meniscus system is provided with a corrector plate analogous to a Schmidt plate, but of only about one-sixteenth the strength. (HAWKINS and LINFOOT, 1945.) An essentially similar design, due to

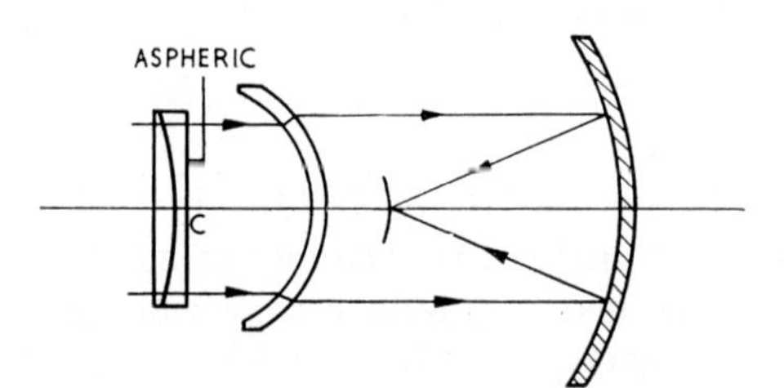

Fig. 11. Meniscus Schmidt camera (HAWKINS and LINFOOT)

J. G. BAKER, uses two concentric menisci, one on each side of the corrector plate (see *Sky and Telescope*, **10**, 219, 1951). In both cases, some colour-error is introduced by the menisci, which act as weakly divergent lenses, and this can be corrected by making the plate in the form of a doublet, with glasses of different dispersions but nearly the same mean refractive index, as shown in Fig. 11.

2. ABERRATION-BALANCING IN SCHMIDT CAMERAS AND RELATED SYSTEMS

2.1. *Preliminary Remarks*

In a well-designed anastigmat, the Seidel aberrations are not made to vanish identically but small traces are retained, in controlled amount, to "balance" the higher aberrations of the system as well as possible over the working field. The Schmidt system is also an anastigmat, but because of its very high performance in comparison with the older types of astrographic objective, error-balancing has so far received very little attention from the makers and users of astronomical Schmidt cameras.

At focal ratios near $f/3{\cdot}5$, for example, the dominant error in a Schmidt camera working over a 3° or 4° field is sphero-chromatic aberration and the main problem is to balance this by suitably choosing the position of the neutral zone on the corrector plate. When this has been done, the practical limit to the performance of the camera is set by the resolving power of the photographic plate, and by residual astigmatism of figure in the corrector plate, rather than by the optical aberrations inherent in the design.

The situation is different in the fast Schmidt cameras used in modern spectrographs. At an aperture ratio of $f/1$ or shorter, it is the monochromatic aberrations which dominate the images in the outer parts of the field, and at sufficiently high apertures it is the optical aberrations rather than plate-grain which set a limit to the photographic resolution of the system.

In such cameras, it is worthwhile to improve on the classical Schmidt design, in which the corrector plate is figured to give axial stigmatism in light of a selected wave-

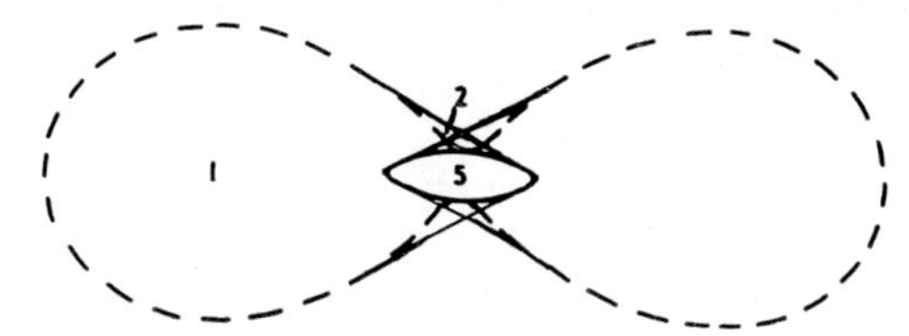

Fig. 12. Monochromatic images of axially stigmatic Schmidt camera. The numbers state the degree of many-sheetedness of the image. The dotted lines correspond to the edge of the aperture. (After LINFOOT, 1949)

length near the middle of the working spectral range. Some idea of the balancing problem can be obtained from Figs. 12, 13, and 14, which show the monochromatic aberrations in the classical design. In Fig. 12, the degree of folding of the many-

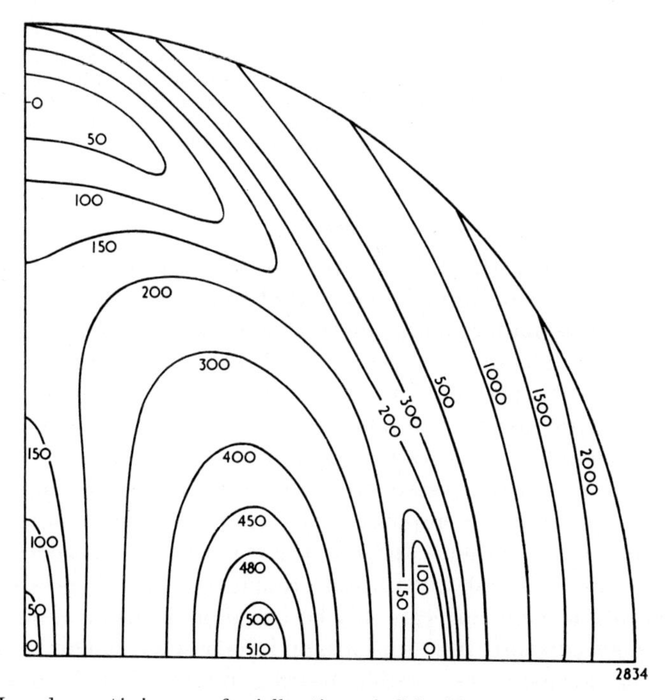

Fig. 13. Monochromatic images of axially stigmatic Schmidt camera. The figure represents the aperture stop, but since the complete diagram is bisymmetrical only one quadrant is shown. Numbers indicate, for any entering pencil of parallel light, the (normalized) deviations with which rays through the different points of the aperture meet the image-surface; for example, a ray through any point of the curve marked 200 meets the receiving surface at a (normalized) distance 200 from the centre of the image. (After HAWKINS and LINFOOT, 1945)

sheeted geometrical images is indicated by numerals, while the dotted line corresponds to the edge of the aperture. Fig. 13 represents a quadrant of the aperture stop; it shows which parts of the aperture send rays to the central part of the image and which parts to the outer wings. Fig. 14 gives a "spot diagram" of the monochromatic image, from which a rough idea can be formed of the illumination density in different parts of the image in those cases where the effects of diffraction are small compared with those of the geometrical aberrations. Of course, in the actual balancing problem we have to consider chromatism as well as the monochromatic aberrations. One way of doing this is described in Section 2.3.

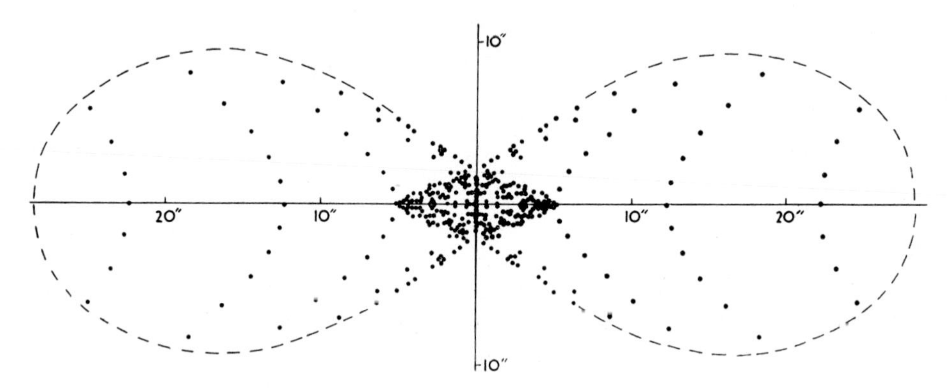

Fig. 14. Spot diagram showing the monochromatic image formed 6° off axis in an axially stigmatic $f/1{\cdot}2$ Schmidt camera. The individual dots represent the intersections with the receiving surface of rays through the corner points of a lattice of small squares filling the aperture stop. The dotted line, as in Fig. 12, corresponds to the edge of the aperture. (After HAWKINS and LINFOOT, 1945)

2.2. *Image-assessment in Astrographic Cameras*

The notion of optimizing the performance of an optical system by aberration-balancing only acquires a meaning when related to a particular definition of image-quality. No really satisfactory general method of assessment of optical image-quality has yet been devised. The problem is simplified if the discussion is restricted to the imaging of bright points spread at random over a dark background, and this covers some astronomical applications. But so far no one has succeeded in analysing even the simplified problem in such a way as to allow the prediction of the photographic resolving power of an optical system from its own geometrical-optical aberrations on the one hand and from the empirically determined characteristics of the photographic emulsion on the other. Indeed, the concept of photographic resolving power is probably not sufficiently definite at present to make this possible. Until it has been done, aberration balancing can only be based on a theory of image-assessment which involves some arbitrary assumptions, and the best we can do is to try to choose these assumptions as sensibly as possible, bearing in mind that the most appropriate choice of assumptions may differ from one application to another.

Elementary considerations are enough to show that neither of the two methods of aberration-balancing most commonly used by optical designers can be expected to lead to optimum resolution, in any acceptable sense of the term, when applied to astrographic cameras.

The first of these methods, which aims at minimizing the overall image spreads of the system, can only lead to an optimized (or very nearly optimized) design in cases where the geometrical image-spreads are large compared with both the Airy diffraction disk and the photographic image-spread of the smallest fully developed images.

In all other cases some resolving power is wasted through a mismatch between the structure of the theoretical image and the combined effects of plate grain and of atmospheric turbulence.

The second method, used mainly for the design of systems in which field and aperture are pushed to the limit, consists in keeping the wave-front deviations well below one wavelength over most of the clear aperture, without regard to the large deviations which then develop over the remainder of the aperture in the outer parts of the field. Cameras designed in this way are well suited to pictorial photography, since their aberrations affect the picture mainly through a loss of contrast in fine detail near the edges of the field and this often improves the aesthetic effect of the picture. They are usually not very suitable for scientific work, because of the large variation in the characteristics of the image between the centre and edge of the field.

The above two methods of aberration balancing correspond to two very different methods of image assessment. In the first, the correspondence is clearly defined; the quality of the physical images is assessed by means of the diameter of the corresponding geometrical images. In the second, the design procedure ensures that the images shall each consist of a bright nucleus, comparable in size to the Airy disk of the system and accompanied by faint halos, flares, or streamers which extend for some distance but are too faint to have much effect either on resolution or on the general appearance of the image-field. No formulation of a definition of polychromatic image quality corresponding to this type of correction seems to have been attempted in the literature, though the "interferometer quality" of COLEMAN, CLARK, and COLEMAN (1947) provides one for monochromatic images. These authors studied the interferometer patterns in green mercury light of telescopes focused visually on large numerals (presumably of high contrast) placed optically at a great distance. They state that, at the visually best focal adjustment, the interferometer pattern is found to have the largest possible diameter covered by a single light or dark interference area; the size of this area depends on the aberrations present. The diameter of the largest circle that can be inscribed in the area, expressed as a percentage of the diameter of the axial entrance pupil, is called by them the interferometer quality (I.Q.) of the system and is stated by them to be "in agreement with quality evaluations based on resolving power measurements". They add: "For practical purposes, the interferometer quality might be thought of as the percentage of the entrance pupil of the system under test that is free from aberrations". (Since the system is tested by double transmission, "free from aberrations" here means "with wave-aberrations less than one-quarter of a wavelength".)

COLEMAN *et al.* also point out that the best visual and the best interferometer focus for primary astigmatism both lie midway between the sagittal and tangential foci, and that the diameter of the I.Q. circle is always greater, but the aberration-free area often less, at this mean focus than at the sagittal or tangential focus. Thus the shape of the aberration-free area is relevant as well as its size, a result which was to be expected on theoretical grounds.

Since COLEMAN, CLARK, and COLEMAN'S concept of interferometer quality depends only on the properties of the system in light of one given wavelength, it requires some extension before it can form the basis of even a provisional definition of polychromatic image quality.

A natural way of extending it is as follows: at a given focal setting P, the interferometer, supplies a separate I.Q. value $q_P(\lambda)$ for each wavelength λ. The weighted λ-mean of these values is the quantity

$$\bar{q}_{P,\rho} = \frac{1}{L}\int q_P(\lambda)\rho(\lambda)d\lambda, \qquad \ldots .(1)$$

where $\rho(\lambda)$ measures the effective spectral density distribution of the light and where the normalizing constant $L = \int\rho(\lambda)d\lambda$.

Let $Q(\rho)$ denote the maximum of $\bar{q}_{P,\rho}$ as the point P varies.* Then it seems possible that $Q(\rho)$ might serve as a workable definition of image quality for optical systems of this type. It agrees tolerably well with optical design practice, takes account of chromatism as well as of the monochromatic aberrations, and includes "interferometer quality" as the special case where the light is monochromatic. To find out how well it agrees with the results of the customary resolution tests would require an experimental investigation going beyond the one so ably carried out by COLEMAN, CLARK, and COLEMAN, but on essentially similar lines. The prospects of approximate agreement seem good, since these authors already report close agreement between interferometer quality in mercury green light and the so-called K.D.C. efficiency values† in the case of an Ic Tessar photographic objective.

The present situation in this field well illustrates the point that there can be no systematic theory of aberration balancing except in relation to a precise definition of image-quality. Designers of high grade optical systems have at present no general theory to guide them and have to base their assessments of quality on the results of resolution tests which are not always mutually consistent. These private troubles of the optician are, however, of minor importance here in comparison with the fact that no design which produces images consisting of a small bright nucleus and an extensive faint outer region can reasonably be described as optimized for astrographic work, since images of this kind are intensely disliked by the more experienced observational astronomers.

On the mathematical side, it appears that a general analytical theory of image optimizing based on either of the above two definitions of image quality must be regarded as impracticable because of the complexity of the formulae involved.

Partly for this last reason, and partly because it seems intrinsically better suited to the application in view than either of those already described, a third type of image assessment will now be considered.

2.21. *R.m.s. Image Assessment (Monochromats)*‡—The geometrical aberrations of a centred optical system can be completely specified by two functions

$$\xi = \xi(u, v;\ \Theta), \quad \eta = \eta(u, v;\ \Theta), \qquad \ldots .(2)$$

which express the aberration displacements ξ, η in a particular receiving surface as functions of normalized rectangular co-ordinated u, v in the exit pupil and of a field-parameter Θ which runs from 0 at the centre of the field to 1 at its edge.

The co-ordinate axes Ou, Ov are oriented so that the u-axis lies in the plane containing the object point and the optic axis (principal plane) and scale-normalized so that the exit pupil aperture is represented by the region $u^2 + v^2 \leqq 1$; (see Fig. 15).

* P is the centre of a comparison sphere relative to which the wave-front deviations are made visible by the interferometer.
† See H. S. COLEMAN and SAMUEL W. HARDING, *J. Opt. Soc. Amer.*, **37**, 267 (1947).
‡ The contents of this and the following sections will be found set out in more detail in Chapter I of the author's forthcoming book *Recent Advances in Optics* (Clarendon Press, Oxford), 1955.

We denote the focal length of the system by f and its numerical aperture, after multiplication by an appropriate normalizing constant, by μ. For present purposes we may suppose the object surface at infinity. The receiving surface is not necessarily flat, but may be any "smooth" surface of revolution about the optic axis lying in the so-called image layer of the system.* In the case of a general centred system, this means that the receiving surface lies everywhere within a distance $O(f\mu^2)$ of the Petzval surface, which plays the part of a natural reference surface. In the case of an anastigmat, it means that the receiving surface lies everywhere within a distance $O(f\mu^4)$ of the Petzval surface.

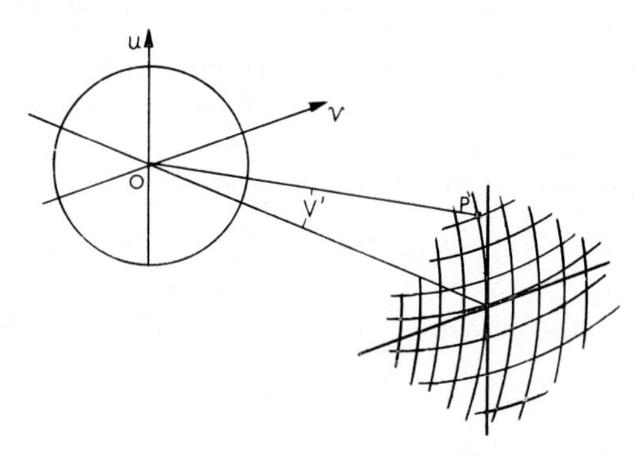

Fig. 15. Co-ordinates in the exit pupil and receiving surface

The working field will be supposed not to exceed $O(\mu)$ in angular radius. This assumption is justified in astrographic systems where the photographic plate is roughly comparable in size with the lens aperture; it holds *a fortiori* in Schmidt and Schmidt-Cassegrain telescopes.

Rays issuing from a particular object point P and passing through the system meet the receiving surface in a set of points called the *image patch* of P. To each of these rays belongs a pair of *co-ordinate numbers* (u, v), namely the normalized co-ordinates of the point in which, after reflection or refraction by the different surfaces of the system, it intersects the exit pupil. The ray OP' with the co-ordinate numbers $(0, 0)$ is called the *principal ray*; its intersection P' with the receiving surface is called the *principal point* of the image patch. The angle V' which OP' makes with the optic axis is called the *angular off-axis distance of the image patch*. We suppose that the working field is defined by the unequality $V' \leqq V_0'$. Then a practically convenient definition of the field-parameter Θ is given by the equation

$$\Theta = V'/V_0'. \qquad \ldots.(3)$$

Let the ray with co-ordinate numbers u, v intersect the receiving surface in P''. We define the aberration displacements of the ray as the components of the vector $P'P''$ in the directions parallel to Ou, Ov respectively and denote them by $\xi(u, v;\ \Theta)$ and $\eta(u, v;\ \Theta)$ respectively. Because the normal at P' to the "smooth" receiving surface makes an angle at most $O(\mu)$ with the direction of the optic axis, ξ and η differ only by factors $1 + O(\mu^2)$ from the aberration displacements as ordinarily defined in the tangent plane at P' to the curved field surface. Even in an $f/2$ system, this represents an error of only a few per cent.

* Receiving surfaces which do not satisfy these conditions are of less practical importance.

In the case of a general centred system with normal figurings,* imagining monochromatic light, the deviations ξ, η can be most usefully represented in the form of an approximation consisting of a leading term and an error term. If we introduce scale-normalized polar co-ordinates, r, ϕ, in the exit pupil in accordance with the equations

$$u = r \cos \phi, \quad v = r \sin \phi, \qquad \ldots.(4)$$

then ξ, η can be written in the form

$$\xi = \xi_0^* + O(f\mu^5), \quad \eta = \eta_0^* + O(f\mu^5), \qquad \ldots.(5)$$

where $$\xi_0^* + i\eta_0^* = f\mu^3[a_1 r^3 e^{i\phi} + ia_2 r^2\Theta(2 - e^{2i\phi}) + a_3 r\Theta^2(2e^{i\phi} - e^{-i\phi}), \qquad \ldots.(6)$$

and the real constants a_1, a_2, a_3 are all $O(1)$. In equation (6) a_1 is the coefficient of spherical aberration, a_2 that of coma, a_3 that of astigmatism.

In the case of an anastigmat with normal figurings, imaging monochromatic light, the equations corresponding to (2), (3) are respectively

$$\xi = \xi_0^* + O(f\mu^7), \quad \eta = \eta_0^* + O(f\mu^7), \qquad \ldots.(7)$$

where $$\xi_0^* + i\eta_0^* = f\mu^5.\frac{f\mu}{H}\left(\frac{\partial}{\partial u} + i\frac{\partial}{\partial v}\right)\Phi_0^*(u, v; \Theta), \qquad \ldots.(8)$$

H is the radius of the entry (not exit) pupil, and

$$\begin{aligned}\Phi_0^*(u, v; \Theta) = {} & (\tfrac{1}{2}a'' r^2 - m''\Theta v) \\ & + (\tfrac{1}{4}b' r^2 - f'\Theta v r^2 + c'\Theta^2 v^2 + \tfrac{1}{2}d'\Theta^2 r^2 - e'\Theta^3 v) \\ & - (S_2\Theta^4 r^2 + S_3\Theta^5 v + S_4\Theta^2 r^4 + S_5\Theta^3 v r^2 + S_6\Theta^4 v^2 \\ & + S_7 r^6 + S_8\Theta v r^4 + S_9\Theta^2 v^2 r^2 + S_{10}\Theta^3 v^3). \qquad \ldots.(9)\end{aligned}$$

In equation (9), the real constants a'', m'', b', f', c', d', e', S_2, S_3, . . . S_{10} are all $O(1)$ and the contents of the three brackets on the right correspond to the first-, third-, and fifth-order aberrations of the system. (Only small traces of the first- and third-order aberrations can be present, of course, in an anastigmat.)

When approximate expressions, of the form (5) and (7) respectively, have been obtained for ξ, η on a spherical or flat receiving surface S lying in the image layer, it easy to deduce the corresponding approximate expressions for other receiving surfaces in the image layer. Such receiving surfaces can be specified by a displacement function $f\mu^2\varepsilon(\Theta)$ in the first case, or $f\mu^4\varepsilon(\Theta)$ in the second, which for each value of Θ measures, to a sufficient approximation, the focal shortening in the new field surface relative to the surface S, which then takes on the function of a field reference surface. By choice of the smooth function $\varepsilon(\Theta)$ subject to the condition $\varepsilon(\Theta) = O(1)$, the receiving surface can be varied at will within the image layer. In a general centred system, the new values for the aberration displacements ξ, η are given by the equations

$$(\xi, \eta) = (\xi^*, \eta^*) + O(f\mu^5), \qquad \ldots.(10)$$

$$(\xi^*, \eta^*) = (\xi_0^*, \eta_0^*) + f\mu^3\varepsilon(\Theta)(u, v); \qquad \ldots.(11)$$

in an anastigmat by the equations

$$(\xi, \eta) = (\xi^*, \eta^*) + O(f\mu^7), \qquad \ldots.(12)$$

$$(\xi^*, \eta^*) = (\xi_0^*, \eta_0^*) + f\mu^5\varepsilon(\Theta)(u, v). \qquad \ldots.(13)$$

---

* See LINFOOT (1955) for the precise meaning of this term.

The last four equations allow the determination, with sufficient accuracy, of the best field surface corresponding to a given definition of image quality. A procedure which is both analytically convenient and physically acceptable in many practical applications is to define the effective radius of a single monochromatic image patch as the square root of the expression

$$\rho^2 = \frac{1}{\pi} \iint_{u^2 + v^2 < 1} (\xi^2 + \eta^2)\, du\, dv \qquad \ldots.(14)$$

and the effective monochromatic image-radius over the whole working field $\Theta \leqq 1$ on a given receiving surface in the image layer as the square root of

$$E = \frac{2}{\pi} \int_0^1 \Theta d\Theta \iint_{u^2 + v^2 < 1} (\xi^2 + \eta^2)\, du\, dv. \qquad \ldots.(15)$$

This amounts to defining the effective radius of each monochromatic image as the radius of gyration of its ray-density distribution about its principal point, and the effective monochromatic image radius over the working field as the root mean square average of these effective radii over the field area $\Theta \leqq 1$.

2.22. *Best Field Surface* (*Monochromats*)—When ξ, η are replaced by ξ^*, η^* in (14) and (15), the resulting changes in value of ρ and of $E^{\frac{1}{2}}$ are $O(f\mu^5)$ in a general centred system, $O(f\mu^7)$ in an anastigmat, and so are negligible in the present approximation. We define the best field surface as that for which $\varepsilon(\Theta)$ minimizes the quantity

$$E^* = \frac{2}{\pi} \int_0^1 \Theta d\Theta \iint_{u^2+v^2<1} (\xi^{*2} + \eta^{*2}) du dv. \qquad \ldots.(16)$$

In a general centred system, where ξ^*, η^* are given by (11) and (6), the best field surface is that for which

$$\varepsilon(\Theta) = -\frac{f\mu}{H}[\tfrac{2}{3}a_1 + 2a_3\Theta^2] \qquad \ldots.(17)$$

and is therefore spherical to the approximation accuracy used, *i.e.* to within $O(f\mu^4)$. The value of E^* in the best field surface is

$$E_1^* = \tfrac{1}{36} f^2\mu^6(a_1^2 + 30a_2^2 + 6a_3^2). \qquad \ldots.(18)$$

If we re-define the effective radius of each image as the radius of gyration of its ray-density distribution about the centre of gravity of the image patch, instead of about the principal point,* the best field surface turns out to same as before, while the mean square of the effective radii over the working field is now

$$E_2^* = \tfrac{1}{36} f^2\mu^6(a_1^2 + 12a_2^2 + 6a_3^2). \qquad \ldots.(19)$$

From (18), (19) it appears that on both assessments coma (with coefficient a_2) is the aberration which has the worst effect on image quality, while spherical aberration (with coefficient a_2) is relatively harmless.

* More accurately: if we shift the (ξ, η)-origins so that the mean values of ξ and η over the circle $u^2 + v^2 < 1$ are always zero.

In an anastigmat, where ξ^*, η^* are given by (13), (8), and (9) and we suppose the (ξ, η)-origins to correspond to the centroids of the image patches, the best field surface is (to within $O(f\mu^6)$) that for which

$$\varepsilon(\Theta) = -\frac{2f\mu}{H}[(\tfrac{1}{2}a'' + \tfrac{1}{3}b' - \tfrac{3}{2}S_7) + \Theta^2(\tfrac{1}{2}d' + \tfrac{1}{2}c' - \tfrac{4}{3}S_4 - \tfrac{2}{3}S_9) + \Theta^4(-S_2 - \tfrac{1}{2}S_6)] \quad \ldots.(20)$$

and the best spherical field surface that for which

$$\varepsilon(\Theta) = -\frac{2f\mu}{H}[(\tfrac{1}{2}a'' + \tfrac{1}{3}b' - \tfrac{3}{2}S_7) + \Theta^2(\tfrac{1}{2}d' + \tfrac{1}{2}c' - \tfrac{4}{3}S_4 - \tfrac{2}{3}S_9 - S_2 - \tfrac{1}{2}S_6)], \quad \ldots.(21)$$

where the coefficients a'', b', c', d', S_2, S_4, S_6, S_7, S_9 are evaluated on the field reference surface S.

2.23. *Aberration Balancing (Monochromats)*—From (8), (9), (13), (16), (21) it follows that when an anastigmat images monochromatic light on to the corresponding best spherical field surface, the mean square effective image radius E^* is a homogeneous quadratic in the variables a'', m'', b', f', c', d', e', S_2, S_3 . . . S_{10}.

Because E^* has this relatively simple form, it is possible to obtain a general answer to the question: what amounts of residual Gauss and Seidel aberrations will give to a monochromat with known fifth-order aberration coefficients the smallest r.m.s. image radius over its best spherical field surface?

Analysis shows that for the "optimized" system

$$\left\{\begin{aligned} a'' &= -(\tfrac{1}{6}S_2 + \tfrac{2}{3}S_4 + \tfrac{1}{12}S_6 + \tfrac{9}{10}S_7 + \tfrac{1}{3}S_9), \\ m'' &= -(\tfrac{2}{3}S_5 + \tfrac{1}{2}S_8 + \tfrac{1}{2}S_{10}). \end{aligned}\right.$$

$$\left\{\begin{aligned} b' &= \tfrac{1}{2}S_4 + \tfrac{4}{5}S_7 + \tfrac{1}{4}S_9, \\ f' &= \tfrac{2}{3}S_5 + \tfrac{3}{2}S_8 + \tfrac{1}{2}S_{10}, \\ c' &= S_9 + \tfrac{3}{4}S_6, \\ d' &= S_2 + \tfrac{4}{3}S_4 + \tfrac{1}{8}S_6 + \tfrac{1}{6}S_9, \\ e' &= S_5 + \tfrac{3}{4}S_{10}, \end{aligned}\right. \quad \ldots.(22)$$

and (because the aberration deviations are measured from the centroids of the image patches)

$$S_3 = 0. \quad \ldots.(23)$$

The practical importance of this result lies in the fact that it indicates which small changes in the design-parameters of an anastigmat will suffice to optimize its performance. For these small changes alter the eight fifth-order coefficients S_2, S_4, . . . S_{10} only by $O(\mu^2)$, and S_2, S_4 . . . S_{10} may therefore be supposed, in the present approximation, to remain fixed during the optimization process. (See LINFOOT, 1955, Chapter I, Section 1.7.)

2.24. *Polychromatic Images*—In a general centred system, the effect of a variation $O(\mu^2)$ in refractive index is to introduce variations $O(f\mu^2)$ in the focal position of the image and in the focal length (plate scale), and variations $O(\mu^2)$ in the coefficients of the third-order monochromatic aberrations on a given receiving surface.

We restrict the discussion to systems for which the variation in the refractive indices of any refracting components is $O(\mu^2)$ over the effective spectral range. This includes all the systems mentioned in the first part of this survey, as well as many others which find applications in astronomy.

Then the chromatism may be expressed, to the Seidel order of approximation, by adding to $\xi + i\eta$ the expression

$$f\mu^3[A_1 re^{i\phi} + iA_2\Theta] + O(f\mu^5), \qquad \ldots.(24)$$

in which the coefficients A_1, A_2 are smooth functions of λ which are $O(1)$ over the effective wavelength range and vanish at the wavelength λ_0 in which the monochromatic aberrations were calculated, while the effects of the chromatic variation in the Seidel coefficients are covered by the term $O(f\mu^5)$.

The aberration displacements in light of wavelength λ are then

$$(\xi_\lambda, \eta_\lambda) = (\xi_\lambda{}^*, \eta_\lambda{}^*) \pm O(f\mu^5), \qquad \ldots.(25)$$

where $$\xi_\lambda{}^* + i\eta_\lambda{}^* = \xi_0{}^* + i\eta_0{}^* + f\mu^3\left[\left(\frac{H}{f\mu}\varepsilon(\Theta) + A_1\right)re^{i\phi} + iA_2\Theta\right] \qquad \ldots.(26)$$

and $\xi_0{}^*$, $\eta_0{}^*$ are given by (6).

We now define the effective polychromatic image radius over the working field as $(\overline{E_\lambda{}^*})^{\frac{1}{2}}$, where $\overline{E_\lambda{}^*}$ is the weighted λ-mean of

$$E_\lambda{}^* = \frac{1}{\pi}\int_0^1 2\Theta d\Theta \iint_{u^2+v^2<1} (\xi_\lambda{}^{*2} + \eta_\lambda{}^{*2})dudv \qquad \ldots.(27)$$

and the weighting corresponds to the assigned spectral density distribution of the light.

Calculation shows that, when ξ_λ, η_λ are measured from the principal point of the image patch as origin, $E_\lambda{}^*$ is minimized by taking

$$\varepsilon(\Theta) = -\frac{f\mu}{H}[(\tfrac{2}{3}a_1 + \overline{A_1}) + 2a_3\Theta^2] \qquad \ldots.(28)$$

and that on this best field surface

$$\overline{E_\lambda{}^*} = f^2\mu^6[\tfrac{1}{36}(a_1{}^2 + 12a_2{}^2 + 6a_3{}^2) + \tfrac{1}{2}d_1{}^2 + \tfrac{1}{2}d_2{}^2 + \tfrac{1}{2}(a_2 + \overline{A_2})^2]; \qquad \ldots.(29)$$

here bars denote weighted λ-means, and

$$d_1{}^2 = \overline{A_1{}^2} - (\overline{A_1})^2, \quad d_2{}^2 = \overline{A_2{}^2} - (\overline{A_2})^2 \qquad \ldots.(30)$$

are the mean square deviations of the coefficients A_1, A_2 from their mean values $\overline{A_1}$, $\overline{A_2}$.

If ξ_λ, η_λ are measured from the centroid of the polychromatic image patch as origin, the best field surface is still given by (28) and the minimized value of $\overline{E_\lambda{}^*}$ is given by the equation

$$\overline{E_\lambda{}^*} = f^2\mu^6[\tfrac{1}{36}(a_1{}^2 + 12a_2{}^2 + 6a_3{}^2) + \tfrac{1}{2}d_1{}^2 + \tfrac{1}{2}d_2{}^2] \qquad \ldots.(31)$$

instead of by (29). In this case we can say that

$$\tfrac{1}{36}f^2\mu^6(a_1{}^2 + 12a_2{}^2 + 6a_3{}^2) \qquad \ldots.(32)$$

is the monochromatic contribution and

$$\tfrac{1}{2}f^2\mu^6(d_1{}^2 + d_2{}^2) \qquad \ldots.(33)$$

the chromatism contribution to the mean square image radius.

Anastigmats

In a colour-corrected anastigmat* containing refracting surfaces and imaging on the spherical reference surface S the aberration coefficients $\frac{1}{2}a''$, $-m''$, $\frac{1}{4}b'$, $-f'$, c', $\frac{1}{2}d'$, $-e'$ may vary by $O(1)$ as the wavelength varies through the effective spectral range, while $-S_2$, $-S_3$, . . . $-S_9$ vary only by $O(\mu^2)$ and can therefore be treated as constant in the fifth-order approximation. Denoting these coefficients respectively by a_1, a_2, . . . a_{16} in accordance with Table 1, we can say that a_1, a_2, . . . a_7 may vary by $O(1)$ but a_8, a_9, . . . a_{16} remain effectively constant. The variation of a_1, a_2, . . . a_7 expresses the chromatic aberrations of the system.

Table 1. Coefficient notation

Gauss		Seidel					Schwarzschild								
a_1	a_2	a_3	a_4	a_5	a_6	a_7	a_8	a_9	a_{10}	a_{11}	a_{12}	a_{13}	a_{14}	a_{15}	a_{16}
$\frac{1}{2}a''$	$-m''$	$\frac{1}{4}b'$	$-f'$	c'	$\frac{1}{2}d'$	$-e'$	$-S_7$	$-S_8$	$-S_4$	$-S_9$	$-S_{10}$	$-S_5$	$-S_2$	$-S_6$	$-S_3$

The mean square image radius over the whole field is now

$$\overline{E_\lambda^*} = \int_0^1 \overline{\rho_\lambda^{*2} \, . \, 2\Theta d\Theta} = \int_0^1 (\overline{\rho_\lambda^{*2}}) 2\Theta d\Theta, \qquad \dots (34)$$

where bars denote weighted λ-means as before,

$$\rho_\lambda^{*2} = \frac{1}{\pi} \iint_{u^2+v^2<1} (\xi_\lambda^{*2} + \eta_\lambda^{*2}) du dv \qquad \dots (35)$$

and ξ_λ^*, η_λ^* are the fifth-order aberration displacements in λ-light, referred to the principal point of the λ_0-image patch as origin.

Calculation gives

$$\overline{E_\lambda^*} = f^2 \mu^{10} \sum_{r, s=1}^{10} t_{rs} (\overline{a_r a_s}), \qquad \dots (36)$$

where the t_{rs} are numerical coefficients, independent of λ and of the particular system under consideration, whose values are shown in Table 2.†

Table 2. Non-zero values of t_{rs} in equation (36)

$r \backslash s$	2	4	7	9	12	13	16
2	$\frac{1}{2}$	$\frac{1}{2}$	$\frac{1}{3}$	$\frac{1}{2}$	$\frac{1}{4}$	$\frac{1}{3}$	$\frac{1}{4}$
4	$\frac{1}{2}$	$\frac{5}{6}$	$\frac{1}{3}$	1	$\frac{5}{12}$	$\frac{5}{9}$	$\frac{1}{4}$
7	$\frac{1}{3}$	$\frac{1}{3}$	$\frac{1}{4}$	$\frac{1}{3}$	$\frac{3}{16}$	$\frac{1}{4}$	$\frac{1}{5}$
9	$\frac{1}{2}$	1	$\frac{1}{3}$	$\frac{13}{10}$	$\frac{1}{2}$	$\frac{2}{3}$	$\frac{1}{4}$
12	$\frac{1}{4}$	$\frac{5}{12}$	$\frac{3}{16}$	$\frac{1}{2}$	$\frac{9}{32}$	$\frac{5}{16}$	$\frac{3}{20}$
13	$\frac{1}{3}$	$\frac{5}{9}$	$\frac{1}{4}$	$\frac{2}{3}$	$\frac{5}{16}$	$\frac{5}{12}$	$\frac{1}{5}$
16	$\frac{1}{4}$	$\frac{1}{4}$	$\frac{1}{5}$	$\frac{1}{4}$	$\frac{3}{20}$	$\frac{1}{5}$	$\frac{1}{6}$

The remaining $t_{rs} = 0$.

$r \backslash s$	1	3	5	6	8	10	11	14	15
1	2	$\frac{8}{3}$	$\frac{1}{2}$	1	3	$\frac{4}{3}$	$\frac{2}{3}$	$\frac{2}{3}$	$\frac{1}{3}$
3	$\frac{8}{3}$	4	$\frac{2}{3}$	$\frac{4}{3}$	$\frac{24}{5}$	2	1	$\frac{8}{9}$	$\frac{4}{9}$
5	$\frac{1}{2}$	$\frac{2}{3}$	$\frac{1}{3}$	$\frac{1}{3}$	$\frac{3}{4}$	$\frac{4}{9}$	$\frac{7}{18}$	$\frac{1}{4}$	$\frac{1}{4}$
6	1	$\frac{4}{3}$	$\frac{1}{3}$	$\frac{2}{3}$	$\frac{3}{2}$	$\frac{8}{9}$	$\frac{4}{9}$	$\frac{1}{2}$	$\frac{1}{4}$
8	3	$\frac{24}{5}$	$\frac{3}{4}$	$\frac{3}{2}$	6	$\frac{12}{5}$	$\frac{6}{5}$	1	$\frac{1}{2}$
10	$\frac{4}{3}$	2	$\frac{4}{9}$	$\frac{8}{9}$	$\frac{12}{5}$	$\frac{4}{3}$	$\frac{2}{3}$	$\frac{2}{3}$	$\frac{1}{3}$
11	$\frac{2}{3}$	1	$\frac{7}{18}$	$\frac{4}{9}$	$\frac{6}{5}$	$\frac{2}{3}$	$\frac{13}{24}$	$\frac{1}{3}$	$\frac{7}{24}$
14	$\frac{2}{3}$	$\frac{8}{9}$	$\frac{1}{4}$	$\frac{1}{2}$	1	$\frac{2}{3}$	$\frac{1}{3}$	$\frac{2}{5}$	$\frac{1}{5}$
15	$\frac{1}{3}$	$\frac{4}{9}$	$\frac{1}{4}$	$\frac{1}{4}$	$\frac{1}{2}$	$\frac{1}{3}$	$\frac{7}{24}$	$\frac{1}{5}$	$\frac{1}{5}$

* That is to say, a centred system with polychromatic image-spreads as low as $O(f\mu^5)$ over the whole of its working field. The Schmidt camera, the field-flattened Schmidt camera, and the Schmidt-Cassegrain cameras are anastigmats in this sense.

† This table, in a slightly different notation, appears as Table IV in WAYMAN (1952).

Because $a_8, a_9, \ldots a_{16}$ are effectively independent of λ, (36) can be written in the alternative form

$$\overline{E_\lambda^*} = f^2\mu^{10} \sum_{r,s=1}^{16} t_{rs}\bar{a}_r\bar{a}_s + f^2\mu^{10} \sum_{p,q=1}^{7} t_{pq}d_{pq}, \quad \ldots.(37)$$

where the *chromatism factors* d_{pq} are defined by the equations

$$d_{pq} = \overline{(a_p - \bar{a}_p)(a_q - \bar{a}_q)}, \quad (p, q = 1, 2, \ldots 7). \quad \ldots.(38)$$

(Evidently $d_p = +\sqrt{d_{pp}}$ is the "scatter" or r.m.s. deviation of a_p from its weighted mean $\bar{a}_p$.)

The quantities d_{pq} and the Schwarzschild aberration coefficients both remain effectively unchanged (*i.e.* they change by at most $O(\mu^2)$) under the small variations of the design parameters of the system which are ordinarily called for in the process of balancing the fifth-degree monochromatic aberrations by means of controlled traces of lower-degree monochromatic aberrations in the manner already outlined above. Thus the d_{pq} provide a *chromatic contribution*

$$E_C^* = f^2\mu^{10} \sum_{p,q=1}^{7} t_{pq}d_{pq} \quad \ldots.(39)$$

to $\overline{E_\lambda^*}$ which is effectively invariant under the process of aberration balancing, while the remaining part of $\overline{E_\lambda^*}$, namely the terms $f^2\mu^{10} \sum_{r,s=1}^{16} t_{rs}\bar{a}_r\bar{a}_s$, can be minimized analytically just as though the system were a monochromat, but with $\bar{a}_r$ in place of $\bar{a}_r$ throughout. Its minimized value on the best spherical field surface is found to be

$$E_M^* = f^2\mu^{10}\left[\frac{3}{50}S_7^2 + \frac{1}{20}S_8^2 + \left(\frac{1}{27}S_4^2 + \frac{1}{27}S^4S^9 + \frac{11}{216}S_9^2\right)\right.$$
$$\left. + \left(\frac{11}{192}S_{10}^2 + \frac{1}{36}S_{10}S_5 + \frac{1}{54}S_5^2\right) + \frac{1}{90}S_2^2 + \frac{1}{90}S_2S_6 + \frac{13}{1440}S_6^2\right]. \quad \ldots.(40)$$

It seems appropriate to call E_M^* the *monochromatic contribution* to the mean square image radius of the system optimized in this way.

2.3. *Application to the Design of Fast Schmidt Cameras*

The general theory outlined in Sections 2.21–2.24 can be used to determine what amount of axial undercorrection, what choice of neutral zone radius on the corrector plate, and what choice of spherical receiving surface will minimize the r.m.s. effective radii of the polychromatic images in a fast Schmidt camera working over a prescribed angular field.

It is convenient here to use the refractive index n' of the plate to specify the wavelength λ, instead of conversely, because the fifth-order aberrations of the system are linear in n'. In the classical Schmidt design, there is a value of n' (corresponding to a wavelength near the middle of the effective spectral range) for which the system is axially stigmatic. We call this value of n' the *stigmatic index* and denote it by n. The spherical surface, concentric with the mirror surface, which passes through the axial image point in n-light will be called the *customary field surface.*

The introduction of axial undercorrection can be shown (see LINFOOT and WOLF, 1949) to be equivalent in fifth-order approximation to an increase in the stigmatic index n, accompanied in general by a corresponding change in the customary field surface. In applying the theory of Section 2.2, we take the customary field surface as reference surface.

N denotes an arbitrary fixed value of n' and it is assumed that n', n, and n_0 all differ by $O(\mu^2)$, whether or not n corresponds to a wavelength in the effective spectral range. This assumption is satisfied in the fast Schmidt cameras for which aberration balancing is of practical importance.

We define

$$k_{nn'\mu}{}^{(0)} = \frac{n' - n}{4\mu^2(n_0 - 1)} = O(1), \quad \ldots.(41)$$

$$\bar{k} = \overline{k_{nn'\mu}{}^{(0)}} = \frac{\bar{n}' - n}{4\mu^2(n_0 - 1)} = k_{n\bar{n}\mu}{}^{(0)}. \quad \ldots.(42)$$

Then it is easy to show that

$$\overline{(k_{nn'\mu}{}^{(0)})^2} = \kappa_\rho{}^2 + (k_{n\bar{n}'\mu}{}^{(0)})^2, \quad \ldots.(43)$$

where $$\kappa_\mu{}^2 = \overline{(n' - \bar{n}')^2}/[16\mu^2(n_0 - 1)^2]. \quad (44)$$

It can also be shown (see LINFOOT, 1955, Ch. I, Section 1.75) that for the Schmidt camera

$$\left.\begin{aligned} a_1 &= -2r_0{}^2k_{nn'\mu}{}^{(0)},\ a_3 = k^{nn'\mu(0)},\ a_5 = -2r_0{}^2c^2, \\ a_6 &= -\frac{r_0{}^2c^2}{n_0},\ a_{10} = \frac{c^2}{2n_0},\ a_{11} = 2c^2 \end{aligned}\right\} \quad \ldots.(45)$$

and the remaining a_r are zero. Here r_0 stands for the radius of the neutral zone of the corrector plate expressed as a fraction of the full aperture radius; and c, defined as the ratio of the angular field radius of the system to its numerical aperture, is the central obstruction ratio by the field surface.

From (45) it follows that

$$\bar{a}_1 = -2r_0{}^2\bar{k},\quad \bar{a}_3 = \bar{k},\quad \bar{a}_5 = -2r_0{}^2c^2,\quad \bar{a}_6 = r_0{}^2c^2/n_0 \quad \ldots.(46)$$

and that the values of the chromatism factors d_{11}, d_{13}, d_{33} in (39) are here given by the equations

$$d_{11} = 4r_0{}^4\kappa_\rho{}^2,\quad d_{13} = 2r_0{}^2\kappa_\rho{}^2,\quad d_{33} = \kappa_\rho{}^2, \quad \ldots.(47)$$

all the other d_{pq} being zero.

Calculation then shows that the best spherical surface is obtained by taking

$$\varepsilon(\Theta) = \frac{2f\mu}{H}\left\{2\left(r_0{}^2 - \frac{2}{3}\right)\left[\bar{k} + c^2\Theta^2\left(\frac{1}{2} + \frac{1}{2n_0}\right)\right] - \frac{2}{3}c^2\Theta^2\right\}$$

and that the value of $\overline{E_\lambda{}^*}$ on this surface is

$$f^2\mu^{10}\left\{\frac{4}{9}\left[\bar{k} + c^2\left(\frac{1}{2} + \frac{1}{4n_0}\right)\right]^2 + \frac{1}{6}c^4[4(r_0{}^2 - 1)^2 + 1]\right.$$
$$\left. + \frac{1}{27}c^4\left(1 + \frac{1}{2n_0}\right)^2 + 4\kappa_\rho{}^2\left[2\left(r_0{}^2 - \frac{2}{3}\right)^2 + \frac{1}{9}\right]\right\}. \quad \ldots.(48)$$

From (48) it follows that for given f, μ, c (that is, for given focal length, focal ratio, and angular field) the r.m.s. image radius over the best spherical field surface is least when

$$\bar{k} = -c^2\left(\frac{1}{2} + \frac{1}{4n_0}\right), \qquad \ldots.(49)$$

i.e. when the plate gives axial stigmatism for a "fictitious" refractive index

$$n = \bar{n}' + \frac{1}{4n_0}\phi_0^2(n_0 - 1)(2n_0 + 1) \qquad \ldots.(50)$$

and when

$$r_0^2 = \frac{2}{3} + \frac{1}{3}\frac{1}{1 + 12\kappa_\rho^2/c^4}, \qquad \ldots.(51)$$

where

$$\kappa_\rho^2/c^4 = \overline{(n' - \bar{n})^2}/\phi_0^4(n_0 - 1)^2. \qquad \ldots.(52)$$

When r_0^2 and n are given the optimum values (50) and (51), the mean square image radius (48) over the best spherical field surface becomes

$$f^2\mu^{10}c^4\left\{\frac{1}{27}\left(1 + \frac{1}{2n_0}\right)^2 + \frac{1}{6} + \frac{4}{3}\frac{\kappa_\rho^2}{c^4}\frac{c^4 + 4\kappa_\rho^2}{c^4 + 12\kappa_\rho^2}\right\}. \qquad \ldots.(53)$$

It has been shown (LINFOOT, 1951) that no improvement on the value (53) for $\overline{E_\lambda^*}$ can be obtained by dropping the restriction that a "stigmatic index" n should exist and allowing quite general smooth variations in the plate profile and in the receiving surface. Under these wider conditions, the design given above still remains the best in the above mean square sense. It is interesting that the two constants n' and κ_ρ give all the information about the spectral distribution of the light which is relevant to mean square optimization of the design.

REFERENCES

BAKER, J. G. . . . . 1940a *Proc. Amer. Phil. Soc.*, **82**, 323.
1940b *Proc. Amer. Phil. Soc.*, **82**, No. 3.
BURCH, C. R. . . . . 1942 *M.N.*, **102**, 159.
COLEMAN, H. S., CLARK, D. G. and COLEMAN, M. F. 1947 *J. Opt. Soc. Amer.*, **37**, 671.
HAWKINS, D. G. and LINFOOT, E. H. . . . . 1945 *M.N.*, **105**, 334.
HENDRIX, D. O. and CHRISTIE, W. H. . . . . 1939 *Scientific American* (August).
HODGES, P. C. . . . . 1948 *Amer. J. Roentgenology*, **59**, 122.
LINFOOT, E. H. . . . . 1944 *M.N.*, **104**, 48.
1945 *Proc. Phys. Soc. (London)*, **57**, 199.
1949 *M.N.*, **109**, 279.
1951 *M.N.*, **111**, 75.
1955 *Recent Advances in Optics* (Clarendon Press, Oxford). In press.
LINFOOT, E. H. and WOLF, E. . . . . 1949 *J. Opt. Soc. Amer.*, **39**, 752.
SCHMIDT, B. . . . . 1932 *Mitt. der Hamburger Sternwarte in Bergedorf*, **7**, 15.
SCHORR, R. . . . . 1936 *Astron. Nachr.*, **258**, 45.
SLEVOGT, H. . . . . 1942 *Z. Instrumentenkunde* (October).
VÄISÄLÄ, Y. . . . . 1935 *Astron. Nachr.*, **254**, 361.
WARMISHAM, A. . . . . 1941 British Patent 541650.
1943 British Patent 551082.
WAYMAN, P. A. . . . . 1950 *Proc. Phys. Soc. (London)*, **63**, 553.
1952 Thesis, Cambridge.
WRIGHT, F. B. . . . . 1935 *Publ. Astron. Soc. Pacific*, **47**, 300.

Thermal Distortions of Telescope Mirrors and their Correction

ANDRÉ COUDER
Observatoire de Paris

SUMMARY

It is pointed out that the gain in mechanical stability of telescope mirrors, brought about by the change from bronze to glass as the material used in their construction, has been accompanied by an undesirable increase in the sensitivity of the mirrors to the temperature changes encountered in astronomical observing. A method is described of compensating for the changes so caused by controlled heating applied to the back of the mirror.

1. TEMPERATURE EFFECTS IN TELESCOPE MIRRORS

IT is nearly a century since silvered glass replaced bronze in the construction of mirrors for reflecting telescopes. Among the consequences of this change we shall consider here only the following.

In the first place, elastic flexure of the mirror has become much easier to correct, principally because the low density of glass allows relatively thick disks to be used without their weight becoming prohibitive, and because for a given diameter and thickness the flexure of the disk under its own weight is proportional to the ratio of the density to the Young's modulus of the material of which it is made. This ratio is $3{\cdot}4 \times 10^{-12}$ (c.g.s.) for glass as compared with 11×10^{-12} (c.g.s.) for bronze of composition $SnCu_4$, so that if two disks of glass and bronze are compared, having the same diameter and weight and carried by identical supports, it is found that their flexure will be in the ratio 1 : 10·8. Moreover, the correction of flexure of mirrors has been the subject of systematic research, both theoretical and experimental (COUDER, 1931), and although the flexure increases as the fourth power of the diameter, we are able to solve the problem sufficiently well having regard to the other difficulties which limit the size of telescopes that can be built.

Our resources for overcoming thermal distortions are not so satisfactory. No matter what precautions are taken, the temperature of the mirror varies during the observation, and as the temperature is not constant it follows that it is also not uniform in the volume of glass, and the unequal expansion of different parts of the disk causes a distortion of the optical surface. Let δ be the density, c the specific heat, m the heat conductivity, and α the coefficient of expansion of the mirror. Suppose that different disks, of the same size but of different materials, are subject to a sudden change of temperature at their surface at some instant $t = 0$. The temperature differences which will exist in their interiors at a certain later time t, relative to the new equilibrium state, will be in the same sequential order as the values of the quantity $\delta c/m$; and the residual distortions at time t will be in the same sequence as the values of $K = \alpha\delta c/m$. Some numerical values are given in the following table.

Table 1

Material	K
Ordinary glass .	$K = 152{\cdot}0 \times 10^{-5}$ c.g.s.
"Pyrex" .	$49{\cdot}0 \times 10^{-5}$
Specular metal .	$6{\cdot}3 \times 10^{-5}$
Fused quartz .	$4{\cdot}6 \times 10^{-5}$

The thermal distortions of the old bronze mirrors may well have been imperceptible, since they are hardly greater than those of fused quartz. The distortions are eight to twenty-four times greater in our large glass mirrors and we cannot unfortunately fail to notice them.

Hitherto these distortions have not been opposed except by passive methods (glasses of small coefficients of expansion, partial heat insulation of the disk, and forced ventilation of the surfaces) which, although useful, are insufficient. I have experimented with intervening actively by applying a small source of artificial heat suitably localized.

The method was first applied to the 81-cm telescope of the Observatoire de Haute Provence. In the Cassegrain arrangement its focal length is 12 m; the mirror has a mean thickness of 7·6 cm; the thermal capacity is equivalent to 19 kg of water, and its coefficient of expansion is 8×10^{-6}. During an observing session the temperature of the air falls often by as much as 1°C per hour. If the temperature of the whole instrument fell uniformly, the only effect observed would be a slow displacement of the focal plane by about 0·04 mm per degree, resulting from the difference of expansion of the glass mirror and the steel tube. This would be a practically negligible inconvenience. In order to fall 1° per hour, however, the mirror must loose heat steadily (by conduction, convection, and radiation) at a rate of about 22 W; for this to happen its surface temperature remains steadily 2°4 below that of the middle (COUDER, 1931, p. 294). The deformations resulting from this heat loss will now be discussed.

The loss by radiation and conduction may be considered ideally as being symmetrical about the optic axis. The same is not true (at least not on average) for the convection loss, because the instrument points upwards. If the movement of air in front of the optical surface is studied, using a Foucault knife edge, a turbulent region is seen, having local velocities of the order of a decimetre per second. When the line of sight is inclined to the vertical, an overall upward movement having about the same average speed is observed. Nevertheless, it is remarkable that the deformation of the surface remains to a very high approximation a figure of revolution. At a distance x from the optic axis the error in path length introduced by the thermal effect appears to be of the general form

$$\Delta = ax^2 + bx^4.$$

The term in x^2 corresponds to a change in focal length; its mean value is nearly zero, but it is subject to random fluctuations depending on the local air conditions. These changes can be taken up by altering the focal setting. The term in x^4 represents an appreciable error in the correction of spherical aberration; this is the well-known edge effect. On the 80-cm telescope overcorrection (turned down edge) is always observed varying between zero and three-times the Rayleigh limit. This limit, a quarter of a wave, corresponds in a $f/15$ instrument to a longitudinal aberration of 2 mm for the marginal rays. It is this part of the effect, the only one which is directly objectionable, that I have attempted to correct.

2. THE COMPENSATION OF THERMAL DISTORTION

The method used consists in warming the back of the mirror M (Fig. 1) over a zone lying between radii R_1 and R_2. A resistance of constantan wire (0·3 mm in diameter) was used, wound on to a ring of plastic (1 mm thick). The resistance forms a compact network in good thermal contact with the glass. It is covered on the outside by a heat-insulating envelope D consisting of two thicknesses of felt separated by a layer of paper. A backing plate E, 5 mm thick, pressed by light springs F, holds all in place. Thanks to the insulating layer, the metal cell which contains the mirror absorbs only a very small part of the heat given off by the resistance. This is very important; the base of the cell is reinforced by raised ribs capable of diffusing the heat, and there would therefore be a risk of them producing local cooling so that the isotherms on the back of the mirror might show the pattern of the structure of the cell. The whole assembly (C, D, E) must be sufficiently flexible so that it does not impose any appreciable constraint on the mirror. For this reason

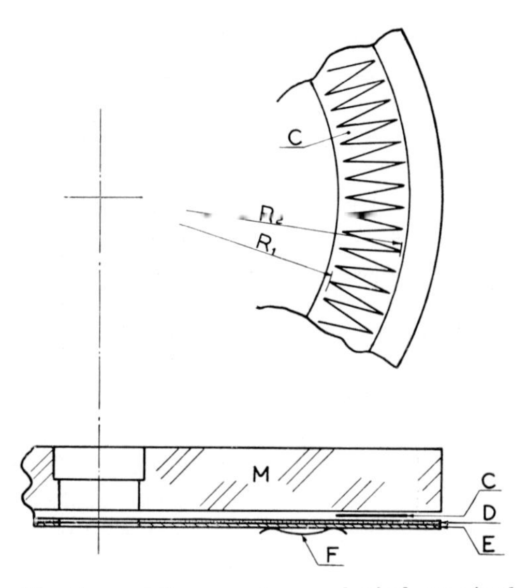

Fig. 1. Diagram of heating-pad for correcting a spherical error in the 80-cm telescope of the Observatoire de Haute Provence

the rigidity of the backing-plate is destroyed almost completely by making saw-cuts across it.

After some trials, the heating ring was made to lie between $R_1 = 28{\cdot}9$ cm and $R_2 = 35{\cdot}4$ cm. It would be better for R_2 to be a little greater, in other words for the glass to be a little warmer at the edge. The presence, however, of the three fixed supports and the six counterpoise levers which form the outside of the back support of the mirror made necessary to limit R_2 to the value given.

The actual form of the mirror has been determined from the extra-focal images of a bright star under high magnification, by comparing the illumination of the diffraction rings as one goes through focus, and also by the knife-edge method. The photographs shown in Fig. 2 are of Foucault shadows. The relief of the wave-surface will be interpreted in the right sense if we image the figure as showing a solid body illuminated from the left. Fig. 2 (*a*) shows the ordinary thermal effect, which at the time the photograph was taken amounted to about half a wavelength. Some asymmetry characteristic of the off-axis aberrations of coma and astigmatism can also be seen; the mirror has been recently replaced and had not yet been perfectly centred.

When the heating circuit is connected to a low-voltage battery, a stable condition is soon reached which is characterized by a very slight diminution of focal length and by relative under-correction of the spherical error; the longitudinal aberration of the marginal rays amounts to — 0·81 mm for each watt dissipated in the winding. The power was 3 W when the photograph of Fig. 2 (*b*) was taken; this gives a satisfactory correction. Not far from the edge of the shadow of the secondary mirror, a zone can be seen on this photograph; this is a defect of the mirror itself, and represents a retardation of about 1/20 of a wavelength. Furthermore, a narrow turned

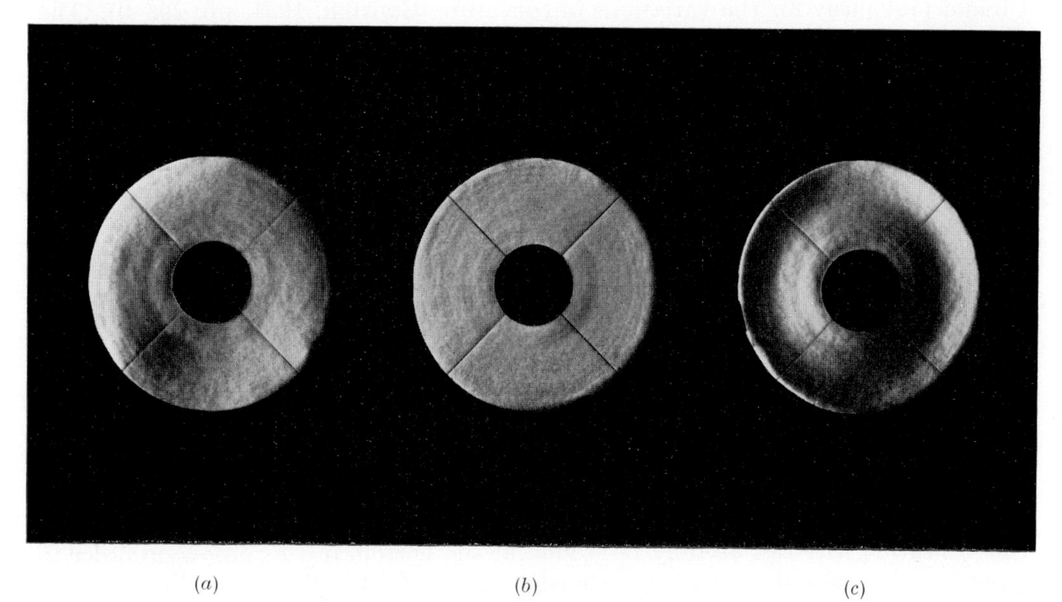

(*a*) (*b*) (*c*)

Fig. 2. The effect of the arrangement shown in Fig. 1
(*a*) Normal condition of the mirror showing over-correction due to falling temperature.
(*b*) Good overall correction with 3 W of heat applied.
(*c*) Over-correction with 25 W of heat.

edge remains; this is an effect of the corrector which does not extend far enough. All the same, the images seen with a short-focus eye-piece were practically perfect and the patterns inside and outside focus were almost exactly alike. It cannot fail to be noticed that the vanes which support the secondary mirror appear to lie at the bottom of narrow valleys in the wave-surface. This is also a thermal effect; in another publication, I have discussed the cause of this effect and ways of reducing it (Couder, 1949).

I wanted to try the effect of an exaggerated heating regime: the heating amounted to 25 W when the photograph of Fig. 2 (*c*) was taken. It shows a very smooth error (except for the extreme edge which remains turned down), comparable in size to that of a spherical mirror. It was this observation that suggested to me the use of intentional thermal deformations for figuring parabolic mirrors or more general aspheric surfaces (Couder, 1952).

We now return to the correction of those thermal distortions which occur during a night's observing. For any uniform rate of cooling there is a corresponding stable deformation of the surface which we can consider as constituting the initial state.

At a certain time (which we shall take to be the origin of our time scale) a particular heating regime is set up, and to this there corresponds a particular final steady state. We note, firstly, that the change in the form of the mirror begins without any appreciable delay after the heat is turned on, and that this is the time at which the change is most rapid. Secondly, the passage from the initial to the final state follows an exponential law. About half of the change occurs, for the mirror considered, during the first 8 min. Thirdly, the time-constant for the variation in spherical aberration found in this way is similar to (but little smaller) than that which I found previously for the variations in curvature (COUDER, 1931, pp. 295–9). The difference is, in fact, of the order of the errors of measurements, which are fairly large in the present case because the initial and final states referred to above are not rigorously fixed under the actual conditions of observations on the sky. Nevertheless, the natural variations are much slower than the effects of artificial heating and, indeed, the errors to be corrected are not very great. For these two reasons the adjustment is easy; after some experience it has always been possible to obtain excellent correction within a quarter of an hour.

It should also be said that the variation indicated by a thermometer, although capricious when the weather conditions are unsettled, shows a quite definite behaviour during nights of good weather without wind, and these are precisely the nights when the quality of seeing makes perfect optical correction of the instrument desirable. For these "normal" nights it is possible to tabulate a schedule which it suffices to follow, without wasting time for empirical adjustment.

Finally, the 120-cm mirror of the Observatoire de Haute Provence (thickness 18 cm, water equivalent 92·5 kg) has been fitted with a heating arrangement similar to that already described. As this mirror has a channel half-way up its cylindrical rim, a heating wire can usefully be wound at the bottom of this groove in order to improve the correction at the extreme edge of the mirror. The thermal errors ordinarily experienced with the 120-cm telescope are relatively small. This is doubtless due to the structure of the dome having been arranged so that the daily warming of the mirror is very small, much less than for the 80-cm telescope. Good correction requires a power of only up to 10 W. The time-constant, which increases as the square of the thickness of the mirror, is here about 40 min.

The additional artificial heat, required to bring about the correction of the edge effect, increases to some extent the temperature excess (which is always present at night) of the mirror relative to the air. It might therefore be thought that the convection currents in the air within the tube of the telescope would become optically more harmful. In fact, such an increase has not been noticed because it is very small. As pointed out, the heat emitted by the mirror normally amounts to 22 W, and this has only been increased by 3 W. The overall effect is found to be an improvement in the quality of the images, which is important when the seeing is good, and still very appreciable when the images are of average steadiness.

The installation of a thermal corrector is planned for the 193-cm telescope of the Observatoire de Haute Provence, which is currently under construction.

It is well known how serious is the convex figure taken up by siderostat mirrors used for observations of the Sun; the image rapidly moves away from the objective and usually has appreciable astigmatism. As the arrangement described above appeared simple and easy to adjust, I had thought that it would also be possible to apply the method *mutatis mutandis* to solar siderostats (COUDER, 1950). It would then be necessary for the entire back of the mirror to be uniformly heated by an amount dependent on the elevation of the Sun, the state of the sky, that of the metallized surface, and also of the angle i of incidence. The power to be brought into play remains quite reasonable. In our climate on a fine summer's day, for $i = 30°$ and with a freshly silvered surface, it would be necessary to provide about 0·4 W per square decimetre; this figure would perhaps be three times greater for an aluminized mirror. I estimate that an approximate adjustment, which would doubtless be sufficient, could easily be maintained after some experiments had been made. On the other hand, an entirely automatic arrangement can be conceived.

References

Couder, A. . . . . 1931 "Recherches sur les déformations des grands miroirs employés aux observations astronomiques", *Bull. astron.* (*Paris*), VII, p. 201 (fasc. VI); p. 283 (fasc. VII); p. 353 (fasc. VIII).

1949 "Sur un effet thermique observé dans les télescopes á réflexion", *L'Astronomie*, **63,** 253.

1950 "Correction des déformations thermiques des miroirs de télescope", *C.R. Acad. Sci.* (*Paris*), **231,** 1290.

1952 "Application d'une déformation thermique intentionnelle á l'exécution d'une surface parabolique", *C.R. Acad. Sci.* (*Paris*), **235,** 491.

On the Interferometric Measurement of Small Angular Distances

André Danjon

Observatoire de Paris

Summary

The applicability of Young's fringes to the measurement of small angular distances is discussed, and an arrangement is described which gives much brighter images. An account is given of the use of a Mach interferometer for the study of scintillation.

---

1. Interference, Seeing, and Size of Source

In 1868, H. Fizeau, to whom astrophysics in indebted for so many interesting ideas, wrote as follows:

"Il existe pour la plupart des phénomènes d'interférence, tels que les franges d'Young, celles des miroirs de Fresnel et celles qui donnent lieu à la scintillation des étoiles d'après Arago, une relation remarquable et nécessaire entre la dimension des franges et celle de la source lumineuse, en sorte que des franges d'une ténuité extrême ne peuvent prendre naissance que lorsque la source de lumière n'a plus que des dimensions angulaires presque insensibles; d'où, soit dit en passant, il est peut-être permis d'espérer qu'en s'appuyant sur ce principe et en formant, par exemple, au moyen de deux larges fentes très écartées, des franges d'interférence au foyer des grands instruments destinés à observer les étoiles, il deviendra possible d'obtenir quelques données nouvelles sur les diamètres angulaires de ces astres."

We say nowadays that the visibility of certain types of interference fringes depends on the diameter of the source. The visibility is defined as the ratio $(I_{max} - I_{min})/I_{max}$ for the system of fringes; it equals unity when the apparent diameter of the source is very small, and falls to zero when this diameter attains a critical value which can be predicted from the theory, knowing the dimensions of the interferometer and the wavelength of the light.

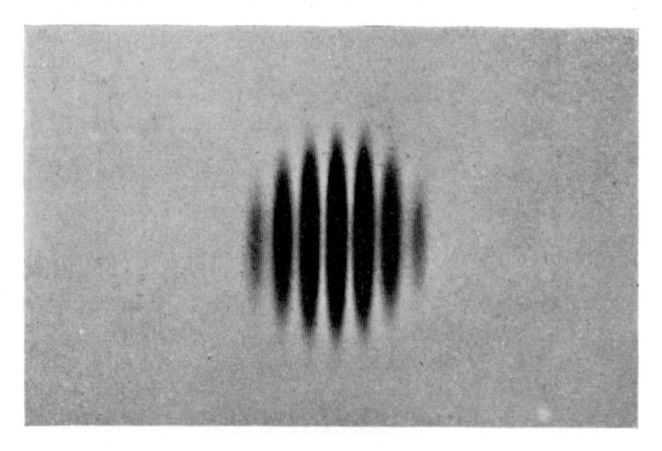

Fig. 1. Young's fringes (negative reproduction)

On FIZEAU'S advice, STÉPHAN (1873, 1874) observed the fringes produced by FOUCAULT'S telescope at the Marseille Observatory. Because of the small aperture of the mirror, 80 cm, the visibility of the fringes remained excellent for all the stars observed, and STÉPHAN concluded that their apparent diameters were appreciably less than 0″.16.

FIZEAU'S method was successfully applied to the satellites of Jupiter by MICHELSON (1891) and HAMY (1899), but it was only in 1920 that MICHELSON and PEASE (1921) obtained the first positive results for stars, using a 20-ft interferometer mounted on the Hooker telescope. During the same period ANDERSON (1920) applied the interferometric method to measurements of the angular separation of the components of Capella. The observation of angular diameters of stars is to-day still a difficult and exceptional procedure, in contrast to the interferometric measurement of double stars which has become part of current astronomical practice. Amongst the latest publications reporting such measures I will cite only the description of a very convenient eye-piece interferometer by VAN DEN BOS (1950, 1951).

All the investigators quoted above observed Young's fringes as produced by an optical system having a double aperture. If the separation of these apertures is progressively increased, the visibility of the fringes decreases. For a circular disk, or for a double star with equally bright components, the fringes completely disappear for a certain critical separation. For a double star with unequal components the visibility passes only through a minimum. The observations have to be made under good atmospheric conditions, but experience shows that a small unsteadiness of atmospheric origin is innocuous.

To reduce the observations, it is necessary to know the effective wavelength of the light used, which is commonly ill-defined; this is the weakness of all such interferometric methods. For bright stars a selective colour-filter can be used, but such cases are rare. It must not be forgotten that, because of its secondary spectrum, the objective of a refractor itself acts as a filter with an effective wavelength always in the vicinity of 5750 Å.

In his statement quoted above, Fizeau mentioned three types of interference phenomena for which the visibility depends on the diameter or the structure of the source:

(1) Young's fringes localized in the diffraction image (the application of which have just been described).

(2) Fresnel's fringes or more generally non-localized fringes, observed in a region common to two beams coming from the same source.

(3) Those fringes which, according to Arago, are the cause of stellar scintillation.

Before discussing the possible application of non-localized fringes, a few remarks will be made on Arago's theory of scintillation; this digression will not take us as far from the subject as might have been thought.

Arago refers scintillation, as we still do to-day (Montigny, Lord Rayleigh), to the inhomogenuity of the atmosphere. The refractive index is different along the various rays reaching the eye, whether or not a telescope is used. In other words, the wave-fronts of the star-light, which were plane outside the Earth's atmosphere, become distorted before reaching ground level, some parts being advanced with respect to others. Consequently, according to Arago (1852), interference is possible between rays arriving at different parts of the pupil or the objective: ". . . il pourra arriver", he wrote, "que les rayons de gauche détruisent en totalité les rayons de droite"; the star might therefore be extinguished. The absurdity of this reasoning is nowadays quite evident; it contradicts the principle of conservation of light, which is the basis of photometry. If a path difference of half a wavelength is established between the two halves of the aperture of an objective, the illumination will vanish at the point of the focal plane which corresponds to the geometrical image of the star, but it will not vanish *everywhere in this plane*. The observed image will differ from the Airy pattern; elementary arguments show that this image will be crossed by a dark fringe parallel to the dividing line between the two halves of the objective, but the total amount of light received by the eye will remain unchanged.

Arago's theory cannot therefore explain scintillation, but it leads to the study of an interference pattern, which incidentally was not considered either by Arago or by Fizeau. It was photographed by me in 1921 (Fig. 2), using a telescope with a Jamin-compensator in front of the objective in order to introduce a half-wave path-difference between the two halves of the aperture. Fig. 3 shows the arrangement. The visibility of the single fringe depends on the apparent diameter of the source in such a way that the apparatus constitutes an interference micrometer. I used it first in 1933 on the satellites of Jupiter (Danjon, 1936, 1944), and later, from 1935 to 1939, for the measurements of visual binaries (Danjon, 1936, 1952).

2. Half-wave Interference Micrometer

In its final form the half-wave micrometer is placed at the eye-end of a large refractor (Fig. 4). A divergent lens (a convergent lens can be used if it is permissible

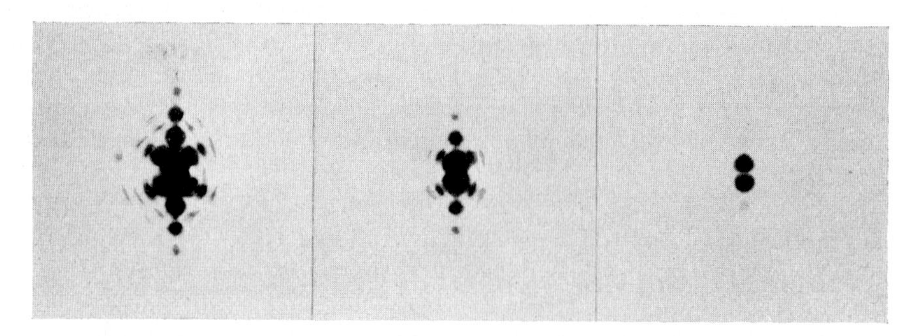

Fig. 2. Diffraction pattern with a dark fringe; circular aperture (negative reproduction)

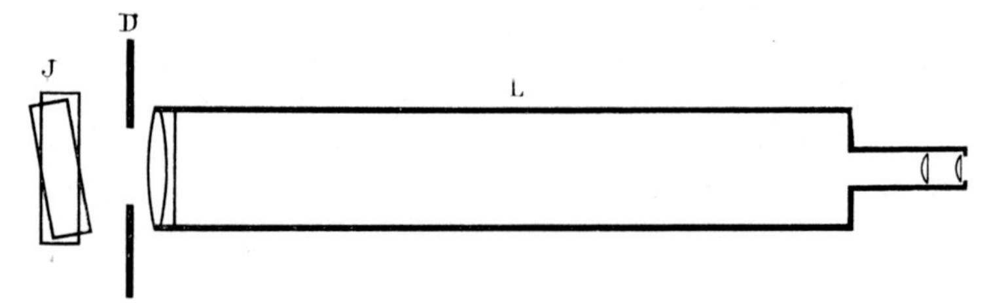

Fig. 3. Lay-out of a half-wave interference micrometer

Fig. 4. Eye-piece interference micrometer at the Observatory of Strasbourg

to lengthen the instrument) forms with the objective a system of zero power. Following this are the Jamin compensator, then a square diaphragm of variable aperture, and finally a small observing telescope. The square diaphragm has the advantage over a circular iris diaphragm that it gives a diffraction image consisting of two patches which are almost exactly circular (Fig. 5). To measure a small disk

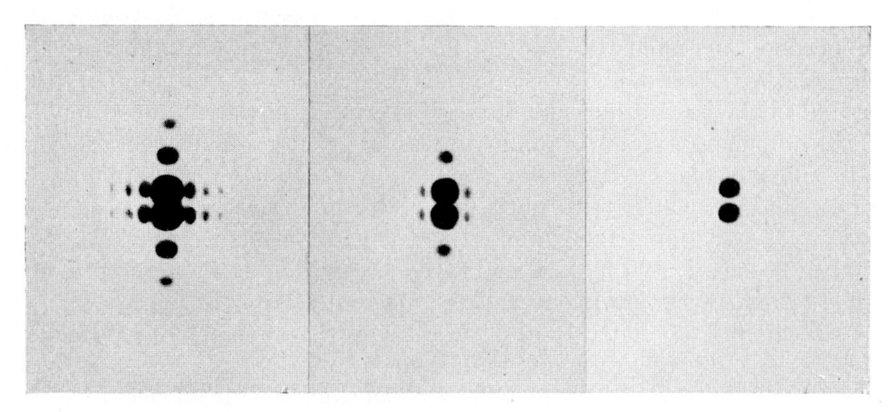

Fig. 5. Diffraction pattern with a dark fringe; square aperture (negative reproduction)

or a double star with almost equal components, the diaphragm is progressively opened until the interference fringe disappears. When the aperture is small, two patches of light are seen (it is necessary to use a high magnification of the order of 1000); the visibility of the fringe diminishes as the diaphragm is opened. For a certain critical value of the size of the square, a pattern having three maxima of light replaces rather suddenly the pattern with two maxima. The transition takes

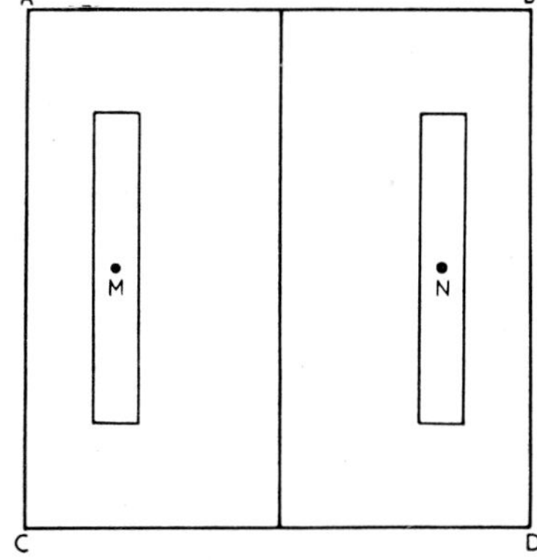

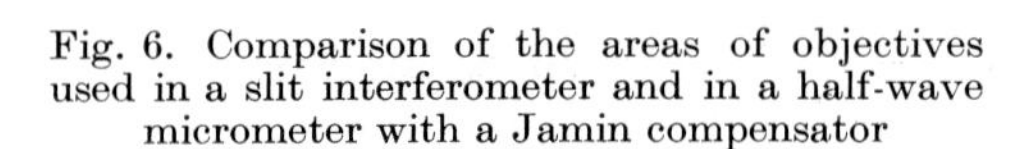

Fig. 6. Comparison of the areas of objectives used in a slit interferometer and in a half-wave micrometer with a Jamin compensator

place over a range of only a few per cent in aperture. When the components are clearly separated, the half-wave micrometer can be used as a double-image device, because it gives two distinct images of each component. If the instrument is turned about the optical axis, and at the same time the diaphragm is adjusted, the four luminous patches can be made to occupy the corners of a square, or indeed they can be made equidistant along a single line.

The half-wave micrometer gives much more light than a slit interferometer, and this enables much fainter stars to be observed, as well as the use of selective filters. Fig. 6 permits a comparison of the surface areas of (1) the slits necessary for the observation of a given binary, and (2) of the square diaphragm needed to observe the same pair. Moreover, the optical performance of the refractor is better used in the second case; in particular, the effect of the second spectrum is reduced.

It will be seen from this that Young's fringes, the only ones used by the first observers, are by no means the best. The interference pattern that I have put forward is not necessarily the best of all, and I would be pleased if some young physicist would undertake the systematic study of this problem. It may be remarked that the methods described depend, above all, on the observation of a minimum of visibility; this is a null in the case of measurements of apparent diameters, or of double stars with equal components, and it is a more or less deep minimum in other cases. When the distance to be measured is very small, it is necessary to obtain interference between two beams which were initially widely separated, and this leads to the construction of very large interferometers. If a quantitative measure of fringe-visibility can be substituted for the qualitative estimation of the minimum, it would be possible to reduce the initial separation of the interfering beams.

In the laboratory it is relatively easy to measure the visibility of a fringe system given by a fixed source in a calm and homogeneous atmosphere. For stars, however, great difficulties arise, caused principally by atmospheric turbulence and irregular guiding. The chief result of the numerous attempts of this kind which I have made, both on artificial sources and on stars, has been to teach me about scintillation. Those who take an interest in this atmospheric phenomenon should be interested in repeating my observations; if ARAGO had known about them he would certainly have given up his theory.

I have come to the conclusion that Young's fringes, and fringes localized in the diffraction image in general, are very badly suited to measures of visibility. It is necessary to return to the second of FIZEAU'S ideas, and to use non-localized white-light fringes, such as are given not only by Fresnel's mirrors, but also by those of Jamin, and by the interferometers of MICHELSON, MACH, and others. These devices enable relatively broad fringes to be obtained over a sufficiently extended field. Their visibility depends only on the initial separation of the beams used, independent of the nature and size of the system producing the fringes, and of the spacing of the fringes. At the time of my first attempts I had at my disposal only visual photometers for measuring fringe-visibility. Later on I was able to photograph them, but I am convinced that the best solution consists in scanning them with a photomultiplier cell.

3. MACH INTERFEROMETER FOR THE MEASUREMENT OF FRINGE VISIBILITY

I have chosen the Mach interferometer as being the most convenient. Fig. 7 shows the form of this apparatus that I have used since 1924 (DANJON, 1926, 1934). The refractor which carries it is arranged as a system of zero-power by means of the divergent lens D. The plates and mirrors of the interferometer are supported, two at a time, by two base plates P and P'. One of these, P, is rigidly connected with the tube of the telescope, and the other, P', can rotate about xy. The interferometer is first set up in the usual way by making all the surfaces plane-parallel, and the two optical paths equal. A uniform tint is then observed over the whole field. The support P' is then turned about xy, and the structure of the emergent beam is observed by means of a condenser lens C which images the star on to the pupil of the eye. In the plane of the condenser one sees a section of the two beams which have been separated by the rotation of P', and in the area common to the two beams a system of white-light fringes is seen, which it is very instructive to examine.

It will be found that the fringes are disturbed by the inhomogenuity of the atmosphere and by the resultant differences in refractive index along the path of any two

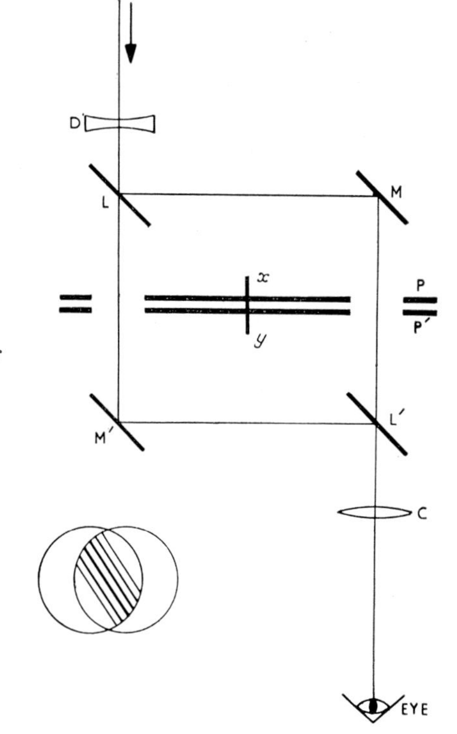

Fig. 7. Mach's interferometer for the study of scintillation

interfering rays. If the distance d between these rays is not greater than several centimetres, the amplitude of the disturbance is proportional to d; but for values larger than a certain limit, which is usually between 10 and 20 cm, the motion reaches a maximum and does not increase further. The variation of the disturbance as a function of d is represented by a saturation curve as shown in Fig. 8. The

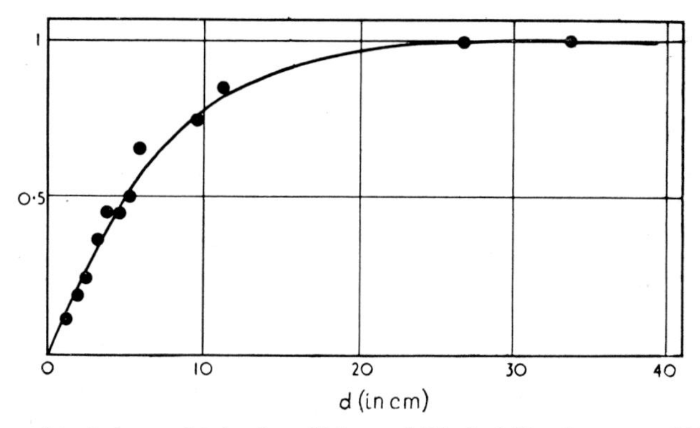

Fig. 8. Curve of turbulence obtained on Sirius and Rigel at Strasbourg on 13th February, 1926. The ordinates show the semi-amplitude of the movement of the fringes expressed in wavelength. The abscissae is the distance d (in centimetres) between the interfering beams

conclusion from this is that the wave-front may be regarded as plane over any small element, and that this plane undergoes fluctuations in direction, the semi-amplitude of which is the *optical turbulence* (DANJON and COUDER, 1935), but that the fluctuations for elements several decimetres apart are uncorrelated. It may be remarked in passing that this fact favours the use of interference methods, since there is no reason to fear that the atmospheric disturbance will increase indefinitely as the distance d between the beams is made larger.

The absolute value of the turbulence can be derived from measures of the movement of the fringes for small values of d (less than 10 cm). Near to the zenith, and in very good weather, values of less than $0''\cdot 05$ have been found very occasionally. Usually, however, the turbulence in good weather is several tenths of a second of arc. When the atmosphere is very disturbed, it can rise to as much as several seconds. In 1924 I used the Mach interferometer described above to calibrate the absolute values given by a method of estimating seeing, which was intended for a study of the steadiness of images in the South-East of France. It was at the end of this exploration that the site was chosen for the construction of the Observatoire de Haute Provence.

The same instrument would also allow the visibility of the fringes to be measured if an appropriate photometer were added to it. I am convinced that it will be possible to perform with a reflector of an aperture of 1 m most of the observations which have required the use of a large Michelson interferometer. I have never given up the idea of making this experiment myself, but until now other commitments have prevented me from doing so.

If the lens C is replaced by a small telescope and the instrument is pointed at an extended illuminated surface, localized white-light fringes at infinity are seen, and these are perfectly stable. The fringe-separation with this arrangement depends only on the distance between the interfering rays, that is to say on the angle through which P' has been turned with respect to P. When one sets on a star, the motion of the fringes which would have been observed at the lens C is by this arrangement replaced by an equivalent motion of the star relative to the fringe-system at infinity. By changing the angle between the emergent beams, the size of the fringes formed at the condenser lens can be conveniently adjusted, and in particular, they can be made as large as one wishes. Atmospheric disturbances are the only hindrance to obtaining a uniform tint; variations in colour are always seen and these, incidentally, are a very beautiful sight.

POKROVSKY proposed in 1912 an optical system whereby polarized light fringes at infinity could be obtained for an extended source, and for a point source a uniform tint can be observed in a plane behind the instrument. Moreover, with a polarizing photometer it should be possible to measure the visibility of fringes from a star having an appreciable apparent diameter (POKROVSKY, 1912, 1915).

In 1926 EDDINGTON drew the attention of astronomers to the difficulties of using POKROVSKY's apparatus, and rightly insisted that it was impossible to keep the star on the central fringe of an interference pattern in infinity with the accuracy required for these measurements. EDDINGTON's (1926) objections apply to the Mach interferometer if one attempts to obtain a uniform tint, but there is no necessity or advantage in doing this. The spacing of the fringes at the lens C can be adjusted at will in such a way as to reduce the displacements caused by bad guiding and by atmospheric turbulence. It is then quite easy to guide the light-sensitive element of a photometer in the field of the fringes. Moreover, the construction and adjustment of Pokrovsky's interferometer seems at first sight to be extremely difficult, whereas Mach's interferometer is a straightforward instrument.

References

Anderson, J. A.	1920	*Ap. J.*, **51**, 263.
Arago, D. F. J.	1852	*Ann. Bureau des Longitudes*, 363.
Danjon, A.	1926	*C.R. Acad. Sci. (Paris)*, **183**, 1032.
	1934	*Réunions de l'Institut d'Optique*, Paris.
	1936	*Ann. Observatoire Strasbourg*, **3**, 181.
	1944	*Ann. Astrophys.*, **7**, 135.
	1952	*J. des Observateurs*, **35**, 85.
Danjon, A. and Couder, A.	1935	*Lunettes et Télescopes*, Paris.
Eddington, A. S.	1926	*M.N.*, **87**, 34.
Fizeau, H.	1868	*C.R. Acad. Sci. (Paris)*, **66**, 934.
Hamy, M.	1899	*Bull. astron. (Paris)*, **16**, 257.
Michelson, A. A.	1891	*Nature (London)*, **45**, 160.
Michelson, A. A. and Pease, F. G.	1921	*Ap. J.*, **53**, 249.
Pokrovsky, S.	1912	*Ap. J.*, **36**, 156.
	1915	*Ap. J.*, **41**, 147.
Stéphan, H.	1873	*C.R. Acad. Sci. (Paris)*, **76**, 1873.
	1874	*C.R. Acad. Sci. (Paris)*, **78**, 1008.
van den Bos, W. H.	1950	*Monthly Notes Astron. Soc. S. Africa*, **9**, 88.
	1951	*Observatory*, **71**, 42.

Partially Coherent Optical Fields

E. Wolf

Department of Mathematical Physics, University of Edinburgh*

Summary

In the first part of this paper the concept of partial coherence is explained and previous researches on partially coherent optical fields are reviewed. A generalized Huygens' principle, applicable to such fields is then formulated, which expresses the intensity $I(P)$ at a typical point P in the field, in the form of an integral taken twice independently over an arbitrary surface. In order to determine $I(P)$ it is necessary to know: (1) the intensity at every point of the surface and (2) a certain correlation factor (called the coherence factor) for all pairs of points on the surface. Next it is shown that apart from a simple geometrical phase factor depending on the positions of the two points, the coherence factor is essentially the Fourier transform of the intensity function of the source.

The results accord particularly well with claims made by Michelson many years ago (and not hitherto related to systematic optical theory) that it is possible in principle to determine the intensity across stellar sources from the measurements of interference fringes in a suitable interferometer.

1. Introduction

In 1890, Michelson (1890, 1891) proposed an interferometric method for determining the separation of double stars and for measuring the apparent size of planetoids, satellites, and stellar diameters, and in the following year confirmed the correctness of his considerations by determining the size of the satellites of Jupiter. The method was actually not used on stellar objects until many years later, probably because it was assumed that it required ideal seeing conditions. However, in 1919, Michelson

* Now at the Department of Astronomy, University of Manchester.

(1920) found that even when the seeing was bad, clear, and relatively steady fringes could be obtained.* Stimulated by this observation, ANDERSON (1920) applied the method to determining the separation of the components of Capella. This attempt was successful, the value found, with less than 1 per cent error was 0".0525. Later measurements by MICHELSON and PEASE (1921) and others confirmed the correctness of the method.

MICHELSON also pointed out that his method could be used for determining the distribution of the intensity of radiation across stellar sources and in his paper with PEASE formulae are given, relating the visibility of the fringes in his stellar interferometer to the exponent of darkening at the limb (MICHELSON and PEASE, 1921).

However, because of difficulties of technical nature, MICHELSON'S method found little favour in astronomical investigations, except in radio astronomy, where in recent years closely related methods for the measurements of the diameters of radio stars have been developed.

Recent researches in optics concerned with partially coherent wavefields have been greatly influenced by MICHELSON'S investigations on the application of interference methods to astronomy, and in turn these researches have thrown new light on the theoretical background of MICHELSON'S methods. It is the purpose of the present article to give a brief account of some of these developments and to direct attention once more to the great potentialities of the Michelson Stellar Interferometer.

2. THE CONCEPT OF PARTIAL COHERENCE

The determination of the intensity at a point P due to the combined effect of light reaching this point via different paths, $p_1, p_2, \ldots$, involves in principle the following three steps:

(1) The determination of the complex amplitudes $V_m(P, t)$ $(m = 1, 2, \ldots)$ of each contribution and the evaluation of their sum:

$$V(P, t) = \sum_m V_m(P, t).$$

(2) The calculation of the instantaneous intensity $I(P, t)$:

$$I(P, t) = V(P, t)V^*(P, t)$$

(asterisk denoting the complex conjugate).

(3) Averaging of $I(P, t)$ over a time interval which is large compared to the period of the light vibrations:

$$I(P) = \overline{I(P, t)}.$$

In the usual treatment of interference and diffraction of light it is implicitly assumed that the light is strictly *coherent*; the resulting time averaged intensity I is then related to the (averaged) intensities $I_m(P) = \overline{V_m(P, t)V_m^*(P, t)}$ associated with the individual contributions by the relation

$$I = \sum_m I_m + 2 \sum_{m \leqslant n} \sqrt{I_m}\sqrt{I_n} \cos \delta_{mn}, \qquad \ldots (2.1)$$

where δ_{mn} is the phase difference between the mth and nth disturbance.

---

* Referring to his tests at Yerkes Observatory and at Mount Wilson, MICHELSON remarks that "the interference fringes in both observations remained remarkably clear and steady, notwithstanding the excessive 'boiling' of the highly magnified images, corresponding to 'seeing' 2 on a scale 10. Subsequent observations showed that the interference bands remain visible even when the seeing is so poor that the usual type of observation is impracticable" (MICHELSON (1920), p. 260).
The experience of other observers is, however, not in a complete agreement with these remarks.

In contrast to (2.1), one has, for light reaching P from *different* sources, the law of combination of *incoherent* disturbances,

$$I = \sum_{m} I_m, \qquad \ldots .(2.2)$$

the remaining terms having disappeared on taking the time average.

(2.1) and (2.2) characterize two extreme cases. Radiation from an actual source of light is neither strictly coherent nor strictly incoherent. For although light vibrations from *different* elements of an extended source may be considered to be completely independent (incoherent) there will in general exist a finite degree of correlation between disturbances reaching any two arbitrary points P_1 and P_2 in the field which are not too far apart from each other.* This is due to the fact that each point receives contributions from the same elements $d\Sigma_1$, $d\Sigma_2$, . . . of the source. Consequently light which reaches P via the points P_1 and P_2 will no longer obey (2.1) or (2.2) and one must then speak of a field which is *partially coherent*. The usual laws of interference and diffraction of light may in a sense be regarded as limiting forms of more general laws governing the behaviour of such fields.

3. Previous Investigations

The first investigations on this subject appear to be due to Verdet (1869), who in about 1860 considered the problem of determining the maximum distance between two points on a screen illuminated by the Sun, at which the vibrations still showed some coherence. He found that the actual length is less than $\frac{1}{20}$ mm. The real foundations of the subject were laid down in investigations of Michelson (1890, 1891, 1920, 1921), but as Zernike (1948) pointed out, Michelson could not, at his time, adequately express his results because of the lack of requisite terms and requisite concepts. To Michelson we owe the important concept of *visibility of fringes*, defined by the expression

$$\text{visibility} = \frac{I_{\max} - I_{\min}}{I_{\max} + I_{\min}}, \qquad \ldots .(3.1)$$

where $I_{\max}$ denotes the intensity at the centre of a bright fringe and $I_{\min}$ the intensity at the centre of the neighbouring dark fringe. It will be seen later than in many cases this expression may, to a good approximation, be identified with the correlation factor which has been introduced in recent investigations as a measure of partial coherence.

Von Laue (1907) studied the thermodynamics of partially coherent beams and was led to the introduction of a certain factor which he showed to be closely related to entropy. Further important advances were made by Berek (1926), who introduced a measure of coherence a so-called "degree of consonance" and applied it in investigations concerning the image formation in the microscope. Investigations along similar lines were also carried out by Lakeman and Groosmuller (1928).

A new stage in the development of the subject was begun in 1934 when Van Cittert (1934) determined the correlation which exists between the disturbances at two points on a screen, illuminated either directly or via a lens by an extended source. A few years later Zernike (1938) introduced as measure of correlation of

* If the difference in the optical paths from the source to the two points exceeds the coherence length of the light the disturbances will be completely incoherent. (For a discussion of the coherence length, see, for example, Born (1933).)

light vibrations a so-called "degree of coherence," defined as the MICHELSON's visibility of the interference fringes, that may be obtained from them under the best circumstances. With the help of this concept, ZERNIKE established a number of interesting new theorems on partial coherence. Shortly afterwards VAN CITTERT (1939) extended his earlier analysis and showed that there is a close formal analogy between his correlation factor and the degree of coherence of ZERNIKE.

The subject was carried forward in an interesting paper by HOPKINS (1951), who defined the correlation factor γ_{12} for vibrations at points P_1 and P_2 in a wave field as

$$\gamma_{12} = \frac{1}{\sqrt{I_1 I_2}} \int_\Sigma U_1 U_2^* d\Sigma. \qquad \ldots.(3.2)$$

$U_1 e^{-i\omega t}$ and $U_2 e^{-i\omega t}$ denote the disturbances at points P_1 and P_2 due to the element $d\Sigma$ of the source, and I_1 and I_2 are the intensities at these points:

$$I_i = \int_\Sigma U_i U_i^* d\Sigma \quad (i = 1, 2). \qquad \ldots.(3.3)$$

The significance of the values of γ_{12} (which is essentially an approximate expression for the "degree of coherence" of ZERNIKE) can best be seen by considering a simple interference experiment. Let a screen $\mathscr{A}$, with two small openings at P_1 and P_2 be placed across the field produced by an extended source and observe the pattern at a second screen $\mathscr{A}'$ (Fig. 1). A simple calculation shows that the intensity

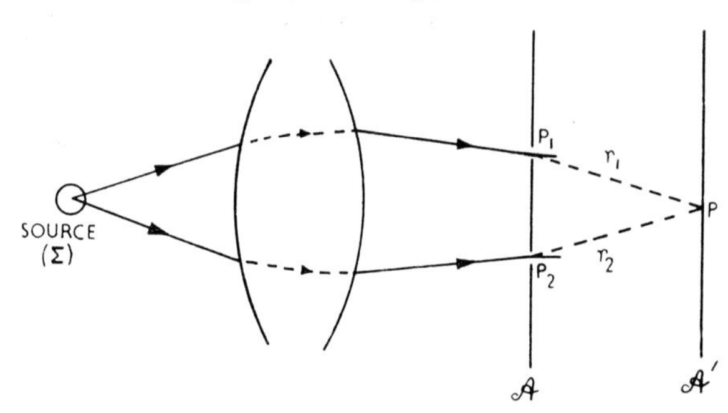

Fig. 1. The significance of the coherence factor

I at a point P on the screen is given by

$$I = I_1 + I_2 + 2\sqrt{I_1}\sqrt{I_2}|\gamma_{12}| \cos [\arg \gamma_{12} + \delta_{12}], \qquad \ldots.(3.4)$$

where $I_1(I_2)$ is the intensity which would be obtained at P if only the aperture at $P_1(P_2)$ alone was open, $\delta_{12} = r_1 - r_2$, r_1 and r_2 being the distances* from P_1 and P_2 to P. For the case of two disturbances, (3.4) is seen to be a generalization of (2.1) and (2.2), the generalization to any number of disturbances being strictly analogous.

From (3.2) and (3.3) it can be deduced that

$$0 \leqslant |\gamma_{12}| \leqslant 1. \qquad \ldots.(3.5)$$

* If the medium between $\mathscr{A}$ and P contains refracting or reflecting surfaces, $r_1 - r_2$ has to be replaced by the corresponding difference $[P_1P] - [P_2P]$ in the optical path lengths.

The two extreme cases $|\gamma_{12}| = 1$ and $\gamma_{12} = 0$ correspond according to (3.4) to perfect coherence and perfect incoherence respectively. Moreover, since by (3.4)

$$I_{\max} = I_1 + I_2 + 2\sqrt{I_1}\sqrt{I_2}|\gamma_{12}|,$$
$$I_{\min} = I_1 + I_2 - 2\sqrt{I_1}\sqrt{I_2}|\gamma_{12}|,$$

it follows that the visibility $\mathscr{V}$ of the fringes is

$$\mathscr{V} = \frac{2}{\sqrt{\dfrac{I_1}{I_2}} + \sqrt{\dfrac{I_2}{I_1}}}|\gamma_{12}|. \quad \ldots .(3.6)$$

In the next section it will be shown that it is possible to formulate a generalized HUYGENS' principle for partially coherent fields. It will also be shown that the coherence factor γ is to a good degree of approximation, and apart from a geometrical phase factor, the Fourier transform of the specific intensity function of the source.* These two results are the basis of a theory of interference and diffraction of light from finite sources, recently proposed by the author (WOLF, 1954a, b).

4. HUYGENS' PRINCIPLE FOR PARTIALLY COHERENT FIELDS

We consider a wave field created by a finite source of natural, nearly monochromatic light of frequency ω. We denote by $\boldsymbol{\xi}_m$ the position vector of the mth element of

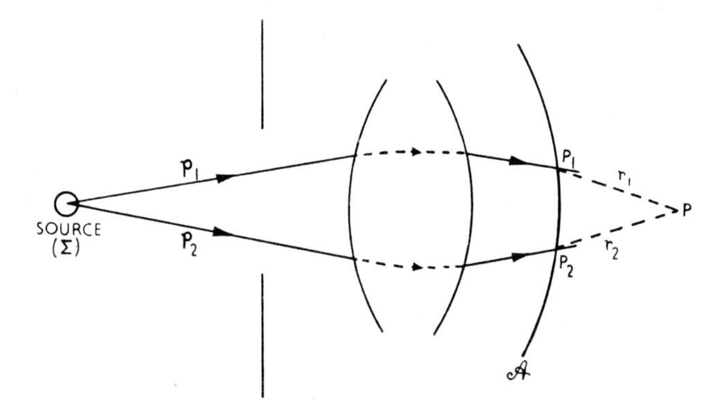

Fig. 2. The generalized *Huygens* Principle

the source and by $\boldsymbol{x}$ the position vector of a point P in the field. The disturbance $V_m(\boldsymbol{x}, t)$ at P due to the element of the source at $\boldsymbol{\xi}_m$ will be written as†

$$V_m(\boldsymbol{x}, t) = U_m(\boldsymbol{x})e^{-i\omega t}. \quad \ldots .(4.1)$$

We take a surface $\mathscr{A}$ cutting across the beam of light and consider the intensity at a point P on that side of $\mathscr{A}$, towards which the light is advancing (Fig. 2). For simplicity we assume that the medium between $\mathscr{A}$ and P is homogeneous.

* This result has also been recently obtained by DUFFIEUX (1953), though under less general conditions. DUFFIEUX also shows that the γ factor is equivalent to one of the "transmission factors" introduced in earlier papers concerned with the application of Fourier transforms to optical images (DUFFIEUX and LANSRAUX, 1945).

Two other papers concerned with partial coherence may also be mentioned: HOPKINS (1953) extended his analysis and applied it to the study of images of different structures and illumination and BAKER (1953) published an account of an experimental investigation on partial coherence and compared the results with predictions of the theory.

† U will actually also change in time, but to the present degree of approximations the resulting effect may be neglected. For a discussion of this point see WOLF (1954a).

Since the radiation from any one element of the source may be regarded as coherent, one can determine $U_m(\boldsymbol{x})$ by an application of Huygens' principle in the usual form,

$$U_m(\boldsymbol{x}) = \int_{\mathscr{A}} U_m(\boldsymbol{x}_1) \frac{e^{ikr_1}}{r_1} \Lambda_1 \, d\boldsymbol{x}_1 \qquad \ldots.(4.2)$$

Here $\boldsymbol{x}_1$ is the position vector of a point P_1 on $\mathscr{A}$, r_1 is the distance from P_1 to P, Λ is the usual inclination factor of Huygens' principle and $k = 2\pi/\lambda$, λ being the wavelength. Hence the intensity $I_m(\boldsymbol{x}) = U_m(\boldsymbol{x})U_m^*(\boldsymbol{x})$ at P, due to the radiation from the element at $\boldsymbol{\xi}_m$ is

$$I_m(\boldsymbol{x}) = \int_{\mathscr{A}} \int_{\mathscr{A}} U_m(\boldsymbol{x}_1)U_m^*(\boldsymbol{x}_2) \frac{e^{ik(r_1-r_2)}}{r_1 r_2} \Lambda_1 \Lambda_2^* d\boldsymbol{x}_1 d\boldsymbol{x}_2, \qquad \ldots.(4.3)$$

the double integration implying that the point $P_1(\boldsymbol{x}_1)$ and $P_2(\boldsymbol{x}_2)$ explore the surface $\mathscr{A}$ independently.

Now the disturbances from different elements of the source may be treated as mutually incoherent. Hence the total intensity at P is

$$I(\boldsymbol{x}) = \sum_m I_m(\boldsymbol{x}). \qquad \ldots.(4.4)$$

Substituting from (4.3) into (4.4) we finally obtain

$$I(\boldsymbol{x}) = \int_{\mathscr{A}} \int_{\mathscr{A}} \sqrt{I_1}\sqrt{I_2}\gamma_{12} \frac{e^{ik(r_1-r_2)}}{r_1 r_2} \Lambda_1 \Lambda_2^* d\boldsymbol{x}_1 d\boldsymbol{x}_2, \qquad \ldots.(4.5)$$

where we set*

$$\gamma_{12} = \frac{\sum_m U_m(\boldsymbol{x}_1)U_m^*(\boldsymbol{x}_2)}{\sqrt{I_1 I_2}}. \qquad \ldots.(4.6)$$

Equation (4.5) may be regarded as a *generalized Huygens' principle for partially coherent fields.* It expresses the intensity at a point P in the field in the form of an integral, taken twice independently over a surface $\mathscr{A}$ and shows that apart from the knowledge of the intensity distribution over the surface, it is in general also necessary to know the values of the coherence factor (4.6) for all pairs of points on the surface.

5. The Coherence Factor

It will be shown that the coherence factor γ may be expressed in terms of the distribution of the intensity across the source.

Let $\tau(\boldsymbol{x}', \boldsymbol{x}'')$ be the transmission function of the medium, defined as the disturbance at $\boldsymbol{x}''$ due to a coherent point source of unit "strength"† at $\boldsymbol{x}'$. To a good degree of approximation we may set

$$\tau(\boldsymbol{x}', \boldsymbol{x}'') = a(\boldsymbol{x}', \boldsymbol{x}'')e^{ikS(\boldsymbol{x}', \boldsymbol{x}'')}, \qquad \ldots.(5.1)$$

where a is an amplitude factor and S the Hamilton point characteristic function of the medium, *i.e.* the optical length of the natural ray between $\boldsymbol{x}'$ and $\boldsymbol{x}''$.

---

* γ_{12} is essentially the coherence factor in the form used by Hopkins but summation is written here in place of integration. The correctness of this procedure follows from dimensional considerations.

† We define a source of unit strength as a source which gives rise to vibrations whose amplitude at a unit distance from it is unity.

If c_m denotes the (time averaged) strength of the mth element of the source, the coherence factor (4.6) becomes

$$\gamma_{12} = \frac{1}{\sqrt{I_1 I_2}} \sum_m c_m{}^2 \tau(\xi_m, \boldsymbol{x}_1)\tau^*(\xi_m, \boldsymbol{x}_2). \quad \ldots . (5.2)$$

It is now permissible to replace the summation (5.2) by an integral involving the intensity function of the source. Denoting by $j(\xi)$ the intensity per unit area of the element at ξ, one obtains, with the help of (5.1):

$$\gamma_{12} = \frac{1}{\sqrt{I_1 I_2}} \int_\Sigma j(\xi) a(\xi, \boldsymbol{x}_1) a(\xi, \boldsymbol{x}_2) e^{ik[S(\xi, x_1) - S(\xi, x_2)]} d\xi, \quad \ldots . (5.3)$$

where
$$I_1 = \int_\Sigma j(\xi) a^2(\xi, \boldsymbol{x}_1) d\xi, \quad \ldots . (5.4)$$

with a similar expression for I_2.

Now in most applications, the dimensions of the source will be small compared to the distance from the source to the surface of integration. It will then be permissible to neglect the variation of the amplitude factor $a(\xi, \boldsymbol{x}_n)$ $(n = 1, 2)$ with ξ and to expand $S(\xi, \boldsymbol{x}_n)$ at a suitable point $(\xi = 0)$ of the source and retain terms up to and including first powers in ξ only*:

$$S(\xi, \boldsymbol{x}_1) - S(\xi, \boldsymbol{x}_2) \sim S(0, \boldsymbol{x}_1) - S(0, \boldsymbol{x}_2) + \xi \frac{\partial}{\partial \xi} [S(\xi, \boldsymbol{x}_1) - S(\xi, \boldsymbol{x}_2)]. \quad \ldots . (5.5)$$

Now by the fundamental property of the S function,

$$\frac{\partial}{\partial \xi} [S(\xi, \boldsymbol{x}_n)]_{\xi=0} = -\boldsymbol{p}_n, \quad (n = 1, 2), \quad \ldots . (5.6)$$

$\boldsymbol{p}_n$ denoting the ray vector† at O of the ray OP.

Substituting from (5.4), (5.5) and (5.6) into (5.3) and replacing $a(\xi, \boldsymbol{x}_n)$ by $a(0, \boldsymbol{x}_n)$, *etc.*, (5.3) becomes

$$\gamma(\boldsymbol{x}_1, \boldsymbol{x}_2) = \sigma(\boldsymbol{p}_1 - \boldsymbol{p}_2) e^{ik[S(0, x_1) - S(0, x_2)]}, \quad \ldots . (5.7)$$

where
$$\sigma(\boldsymbol{p}_1 - \boldsymbol{p}_2) = \frac{\int_\Sigma j(\xi) e^{-ik[(\boldsymbol{p}_1 - \boldsymbol{p}_2) \cdot \xi]} d\xi}{\int_\Sigma j(\xi) d\xi}. \quad \ldots . (5.8)$$

It follows from (5.7) and (5.8) that *apart from the geometrical phase factor* $\exp ik[S(0, \boldsymbol{x}_1) - S(0, \boldsymbol{x}_2)]$, *the coherence factor* γ *is essentially the normalized Fourier transform of the specific intensity function of the source.*

6. DISCUSSION

The results which we have just established accord well with claims made many years ago by MICHELSON, and again, more recently by HOPKINS (1951), that it is possible, in principle, to determine the (relative) distribution of the intensity of

* In the singular case when the source is at infinity, as is effectively the case when radiation from stellar objects is considered, a different expansion must be carried out [cf. WOLF (1954a)], but the final conclusion remains unchanged.

† A ray vector $\boldsymbol{p}$ at a point P is defined as the product of the unit vector along the ray at P, multiplied by the value of the refractive index at that point.

radiation across a source from measurements of interference fringes in a suitable interferometer. For according to (3.4) and (5.7), the intensity in the fringes may be expressed in the form

$$I = I_1 + I_2 + 2\sqrt{I_1 I_2}|\sigma(\boldsymbol{p}_1 - \boldsymbol{p}_2)| \cos [\arg \sigma(\boldsymbol{p}_1 - \boldsymbol{p}_2) + S(0, \boldsymbol{x}_1) - S(0, \boldsymbol{x}_2)], \quad \ldots (6.1)$$

and the visibility $\mathscr{V}$ defined by (3.6) becomes

$$\mathscr{V} = \frac{2}{\sqrt{\frac{I_1}{I_2}} + \sqrt{\frac{I_2}{I_1}}} |\sigma(\boldsymbol{p}_1 - \boldsymbol{p}_2)|. \quad \ldots (6.2)$$

The points P_1 and P_2 may be taken on the same wave-front ($S(0, \boldsymbol{x}_1) = S(0, \boldsymbol{x}_2)$) and it then follows from (6.1) that the factor σ may be determined from the measurements of I_1, I_2, and I. (This is equivalent to the determination of the visibility and position of the fringes.) By (5.8) the normalized intensity distribution in the source may then be obtained by the application of the Fourier inversion formula.

In many applications the intensity across the source will to a good approximation by symmetrical ($j(-\xi) = j(\xi)$) so that σ will be real. Also in most cases $I_1 \sim I_2$ and one then has

$$\mathscr{V} = \pm \sigma(\boldsymbol{p}_1 - \boldsymbol{p}_2), \quad \ldots (6.3)$$

where the sign may be determined from continuity considerations. If one associates with the visibility a positive or negative sign according as to whether σ is positive or negative, $\mathscr{V}$ will become identical with σ and it is seen that the intensity of radiation

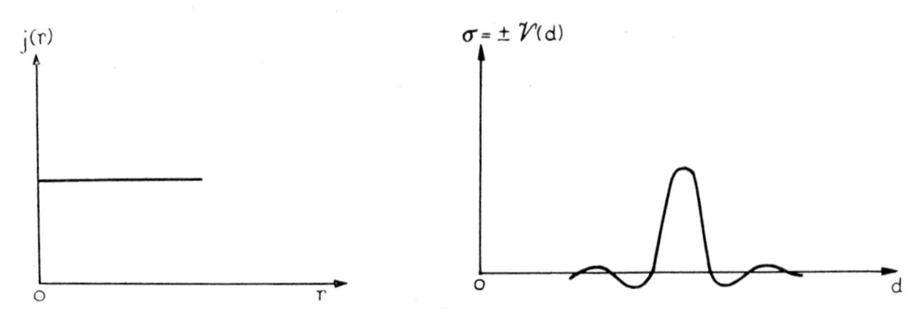

Fig. 3. The specific intensity $j(r)$ across a *uniformly radiating circular disk* as function of the distance r from the centre of the disk and the corresponding visibility curve plotted as function of the path difference d between the interfering beams

across the source may in this case be determined from the measurements of the visibility alone. To illustrate this result the intensity distribution in the source and the corresponding visibility curves for two cases of practical interest are sketched in Figs. 3 and 4. The first represents a uniformly radiating circular disk. In this case only the principal maximum in the interference pattern will as a rule be observable, the secondary maxima being very much weaker than the principal maximum. (The visibility as function of the path difference of the interference beams is in this case identical with the Airy disk curve.) On the other hand, in the case of a double star whose components are of equal brightness and whose angular separation is a moderate multiple of their angular diameters, the visibility will be an almost periodic function of the interferometer setting (Fig. 4).

So far no attention has been paid to the disturbing effect of the atmosphere and an extension of the analysis to take into account the atmospheric turbulences is clearly highly desirable. But it seems safe to conclude that though the disturbing effects of the atmosphere may present serious difficulties of technical nature and may make precise measurements difficult they will not destroy the required information, except for finer details.* In this connection it is of interest to recall that Fürth, Sitte, and Appell (1939) proposed a modification of the Michelson Stellar Interferometer,

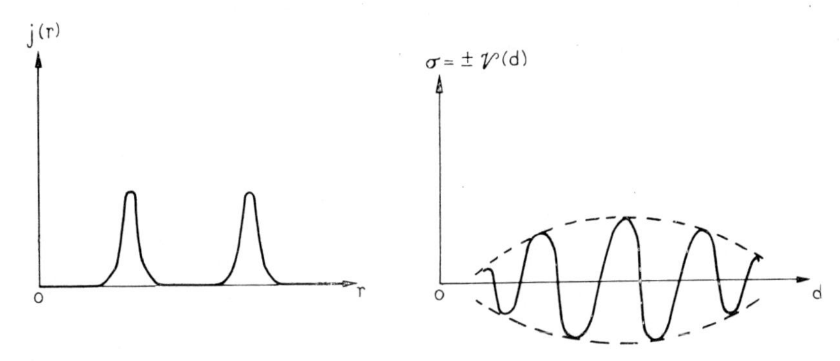

Fig. 4. Specific intensity function $j(r)$ for *double star* and the corresponding visibility curve

the purpose of which is to produce a set of secondary fringes which will remain approximately at rest even when the primary fringes in the focal plane of the telescope are in motion on account of atmospheric turbulences. The principle was tested by laboratory experiments and it was claimed that stellar diameters could be determined from measurements in the secondary fringes alone.† This conclusion has been criticized by Lacroute (1939) and Carroll (1939).

Although this point cannot be completely settled by the present theory until it is extended to take into account the effect of the atmosphere, our analysis suggests that the view advanced by Fürth, Sitte, and Appell is correct. For our analysis shows that the σ factor which characterizes the source is a sort of invariant which is carried with the beam, and which determines substantially the character of every interference pattern obtainable from that particular source. One would therefore expect that the secondary fringes, just as the primary, must contain the required information.

Another possibility for overcoming the disturbing atmospheric effects is suggested by the great developments in recent years in the field of electronics. Although it undoubtedly would be a difficult task to construct an electronic device which would compensate for the motion of the fringe system, it seems not unlikely that with the rapidly increasing knowledge of the principles of servo-mechanism this could indeed

* This is strongly supported by many observations of Michelson, for which he offered the following explanation: "The confusion of the image in poor seeing is due to the integrated effect of elements of the incident light-waves, elements which are not in constant phase relation in consequence of inequalities in the atmosphere due to temperature differences; the optical result being a 'boiling' of the image, closely resembling the appearances of objects viewed over a heated surface.

"In the case of the two elements at opposite ends of a diameter of the objective, the same differences in phase produce a motion of the (straight) interference fringes (and not a confusion) and if, as is usually the case, this motion is not too rapid for the eye to follow, the visibility of the fringes is quite as good as in the case of perfect atmospheric conditions" (Michelson (1927), p. 115).

† The modified method and further experiments are also described in an article by R. Fürth and E. Finlay-Freundlich (1955) in this volume, p. 395.

be done. The construction of such instruments would offer interesting opportunities for future astronomical research.

References

Anderson, J. A. 1920 *Ap. J.*, **51,** 263.
Baker, L. R. 1953 *Proc. Phys. Soc. (London)*, **66**B, 975.
Berek, M. 1926 *Z. Physik.*, **36,** 675, 824; **37,** 287; **40,** 420.
Born, M. 1933 *Optik*, Springer, Berlin.
Carroll, J. A. 1939 *M.N.*, **99,** 735.
Cittert, P. H. van 1934 *Physica*, **1,** 201.
1939 *Physica*, **6,** 1129.
Duffieux, P. M. and Lansraux, G. 1945 *Rev. Opt.*, **24,** 65, 151, 215.
Duffieux, P. M. 1953 *Rev. Opt.*, **32,** 129.
Fürth, R., Sitte, K. and Appell, H. P. 1939 *M.N.*, **99,** 141.
Fürth, R. and Finlay-Freundlich, E. 1955 *Vistas in Astronomy*, Ed. A. Beer, Pergamon Press, London; p. 395.
Hopkins, H. H. 1951 *Proc. Roy. Soc.*, A**208,** 263.
1953 *Proc. Roy. Soc.*, A, **217,** 408.
Lacroute, P. 1939 *M.N.*, **99,** 733.
Lakeman, C. and Groosmuller, J. Th. 1928 *Physica (Gravenhage)*, **8,** 193, 199.
Laue, M. von 1907 *Ann. Phys. (Leipzig)*, **23,** 1.
Michelson, A. A. 1890 *Phil. Mag.*, **30,** 1.
1891 *Phil. Mag.*, **31,** 256.
1920 *Ap. J.*, **51,** 257.
1927 *Studies in Optics*, Chicago University Press.
Michelson, A. A. and Pease, F. G. 1921 *Ap. J.*, **53,** 249.
Lacroute, P. 1939 *M.N.*, **99,** 733.
Verdét, E. 1869 *Léçons d'Optique Physique* (Paris), Tome I 1869; in "Oeuvres . . ." Tome V, 106.
Wolf, E. 1954a *Proc. Roy. Soc.*, A, **225,** 96.
1954b *Il Nuovo Cimento*, **12,** 884.
Zernike, F. 1938 *Physica*, **5,** 785.
1948 *Proc. Phys. Soc.* (London), **61,** 158.

On a Possible Improvement of Michelson's Method for the Determination of Stellar Diameters in Poor Visibility

R. Fürth

Birkbeck College, University of London

and E. Finlay-Freundlich

University Observatory, St. Andrews

Summary

New experimental material is presented in support of a previously proposed improvement of Michelson's interferometric method for the determination of stellar diameters which consists in the production of a "secondary" (extrafocal) fringe pattern. It is claimed that this method is less affected by atmospheric disturbances than the original one. This might make it possible, on the basis of a theorem by E. Wolf (see his article in this volume), to determine the light-intensity distribution over the surface of a star from interferometric measurements, even under conditions of poor visibility.

In a paper published in 1939 one of us (R. Fürth), together with K. Sitte and H. P. Appel*, proposed a modification of Michelson's well-known interferometric method (1918) for measuring stellar diameters which they expected to be less subject to atmospheric disturbances than the original method. The modification consisted in an optical device for producing a system of "secondary fringes" which, according to laboratory experiments, seemed to be less affected by artificially produced air turbulence than the "primary" Michelson fringes appearing in the focal plane of the observing telescope.

The experimental arrangement (Fig. 1) consisted of an "artificial star" (a narrow slit S or a circular aperture in a diaphragm, illuminated by a mercury lamp M

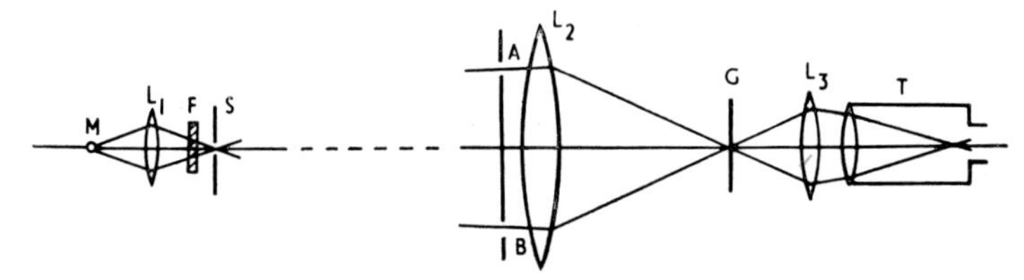

Fig. 1. Schematic diagram of optical arrangement

through a lens L_1 and a filter F), and an astronomical objective lens L_2 (8 in. diameter, 340 cm focal length), with two interferometer slits A, B in front, the width and separation of which could be varied. The primary fringes were formed in the image plane G of S. The secondary interference phenomenon could be observed visually or photographically through a small telescope T with a lens L_3 of 18 cm focal length, whose position could be changed. The air disturbances were produced by a gas heater placed in front of the interferometer slits below the path of the rays.

The results of the laboratory experiments showed clearly that it was possible to determine the size of the artificial star in the usual way (from the disappearance of the fringe pattern at certain critical distances of the interferometer slits) by making

* In the following quoted as FSA.

use of the secondary interference phenomenon, and that the latter could be observed visually and photographically under conditions of turbulence which made the measurements by means of the primary pattern impossible.

Nevertheless, serious objections were raised against the theoretical soundness of the new method in two communications by LACROUTE (1939) and CARROLL (1939) which may be briefly summarized: (1) the fringe pattern, described as "secondary", is in fact only a blurred optical image of the primary pattern, observed in a plane behind the image plane of G in the telescope T, and must therefore necessarily be less distinct than the primary fringes. Furthermore, the two systems must react to air disturbances in exactly the same way. (2) If, by any optical device, a system of "stationary" fringes could be produced which would be insensitive to an irregular movement of the star image it could not be used for determining the size of the light source by MICHELSON's method.

Now in some recent papers of which a résumé is given in his preceding contribution to this volume, WOLF (1953, 1954, 1955) has proved that the measurement of the "visibility" of the interference pattern in *any* arbitrary surface behind the interferometer can in principle provide full information not only on the size but also on the intensity distribution over the surface of the light source as long as there is no air disturbance. (Visibility is defined by the expression

$$V = \frac{I_{\text{max}} - I_{\text{min}}}{I_{\text{max}} + I_{\text{min}}}, \quad \ldots .(1)$$

in which I_{max} and I_{min} are the maximum and the minimum value of the intensity within a period of the fringe pattern.) This fully justifies the use of the secondary (or extrafocal) interference pattern in the modified Michelson method and also invalidates the contention that the pattern must be less distinct than the primary one.

It had already been pointed out in FSA that the secondary pattern had a maximum distinctness for some specific position of the lens L_3 and that for this optimum position the visibility was at least as good as that of the primary system. In order to obtain additional evidence the present authors made some further experiments in 1940 in the Physics Department of St. Andrews University with an arrangement essentially identical with that shown in Fig. 1. In Fig. 2 a series of photographs are given which were taken in calm air for different positions of the lens L_3. The first picture is the "primary" pattern, *i.e.* the pattern produced in the image plane of the artificial star S, which was an illuminated slit of 0·1 mm width. The subsequent photographs show the fringe system in a set of planes behind the image plane, at distances of 1 cm apart, the last at a distance of 20 cm from the image plane. They show that at first the pattern becomes less distinct, but that the distinctness again increases and reaches an optimum at about 10 cm behind the focal plane when the contrast becomes as good as that of the primary system. At the same time the structure of the pattern has changed markedly, so that the term "secondary" pattern seems to be justified.

Since the interference phenomenon is produced by the superposition of the two beams of light passing through the interferometer slits one can naturally expect to observe the pattern clearly only within that region behind the interferometer where

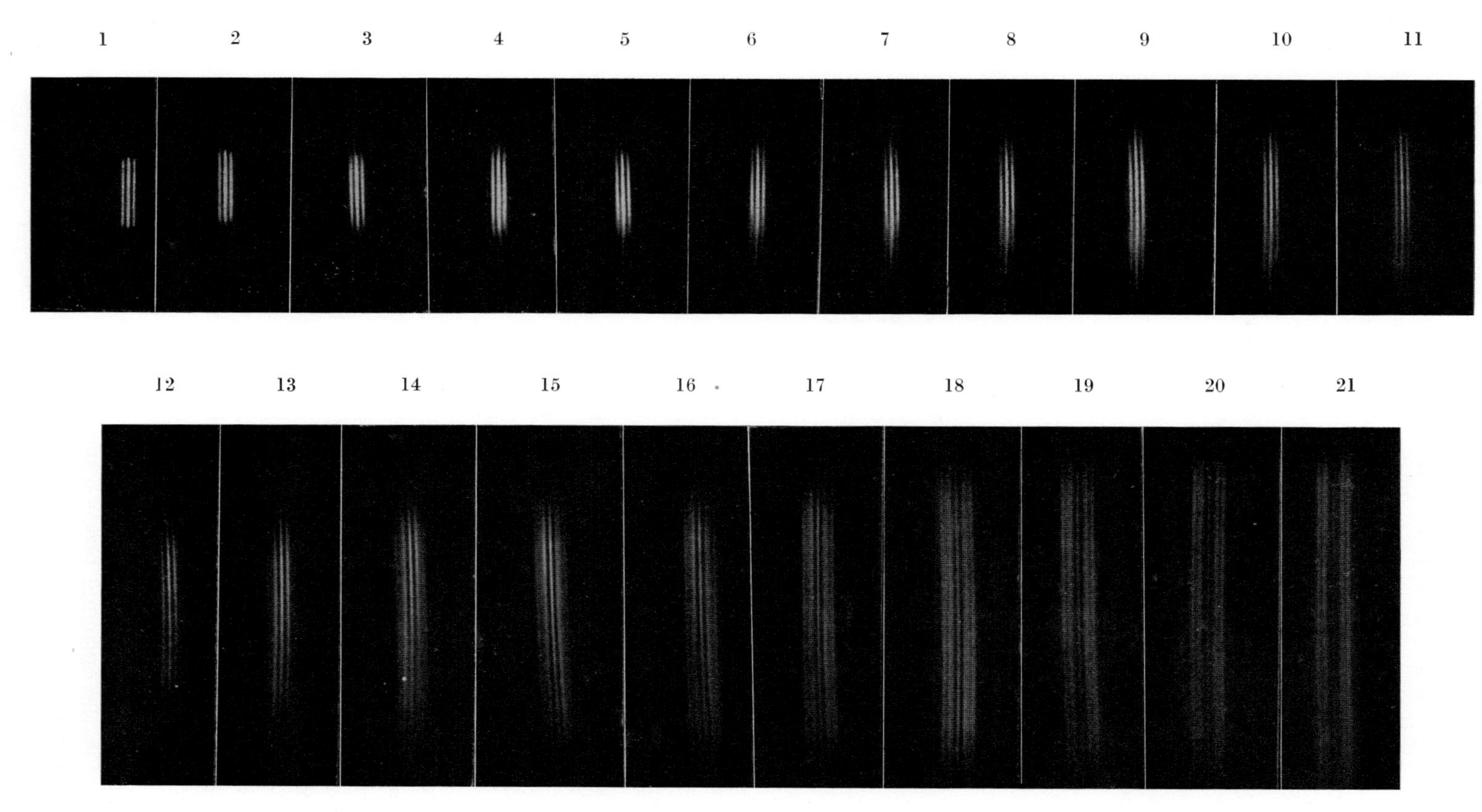

Fig. 2. Development of interference pattern from primary system (1) in image plane of object to secondary systems, produced by successive displacements of lens L_3 by 1 cm. Object slit width 0·1 mm, interferometer slits 15 mm apart

the bulks of the two beams overlap. The diagram of Fig. 3 shows the interferometer plane I and the image plane G of the object at distance L from one another. Let d be the width of the interferometer slits and $2D$ their separation. The effective diameter $2a$ of the image (due to the diffraction of the light of wavelength λ by the slits) is approximately equal to

$$2a \simeq L\lambda/d. \tag{2}$$

The simple geometrical construction of Fig. 3 shows that overlapping of the beams will effectively cease at a distance l behind G which (since l is small compared with L) is given by

$$l \simeq aL/D. \tag{3}$$

From (1) and (2) follows

$$l \simeq \lambda L^2/2dD. \tag{4}$$

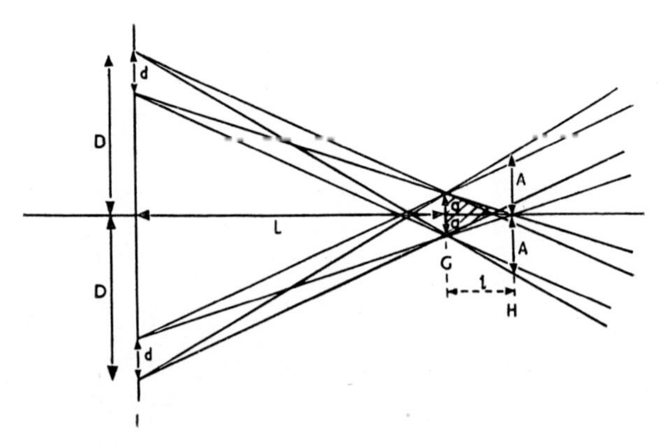

Fig. 3. Region of visibility of secondary pattern

The total width $2A$ of the interference pattern observed in the plane H is easily seen to be equal to

$$2A = 2a(1 + lD/aL); \tag{5}$$

from (3) and (5)

$$A/a \simeq 2 \tag{6}$$

is obtained which shows that the width of the secondary pattern in the plane H is equal to about twice the width of the primary pattern in the plane G.

For the case illustrated by Fig. 2 a value of $l \simeq 11$ cm is found from (4). This is, indeed, the order of magnitude of the shift of lens L_3 from whereon the interference pattern begins to split into two parts separated by a dark space. But even at twice that distance (last picture of the series) a residue of interference pattern can still be distinguished in the dark interspace. It is also seen that the width of the pattern does not increase to more than twice that of the primary pattern, whereas its length increases to about five times its original value.

A further series of photographs, reproduced in Fig. 4, and taken with different separations of the interferometer slits, is meant to demonstrate that the secondary fringe pattern in the optimum position of L_3 is in all cases at least as distinct as the primary pattern, and that both patterns are altered in the same way. It had already been shown in FSA that the secondary pattern disappeared, exactly as did

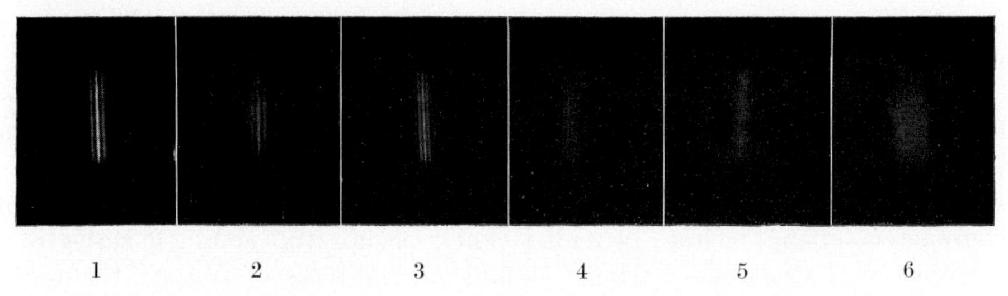

Fig. 4. Change of interference pattern produced by increasing the distance of the interferometer slits from 17 mm (1, 2) to 24 mm (3, 4), and 35 mm (5, 6). 1, 3, 5 primary patterns, 2, 4, 6 secondary patterns

the primary, when the Michelson condition was satisfied. This is further demonstrated by the series of photographs of secondary fringes, shown in Fig. 5, obtained with a fixed setting of the interferometer and varied width of the object slit S; they show how the distinctness of the pattern diminishes as the object approaches the "critical" size.

So far we have dealt only with observations in calm air. According to MICHELSON (1918) atmospheric disturbances produce an irregular movement and distortion of the star image without noticeably impairing the distinctness of the fringe pattern. It

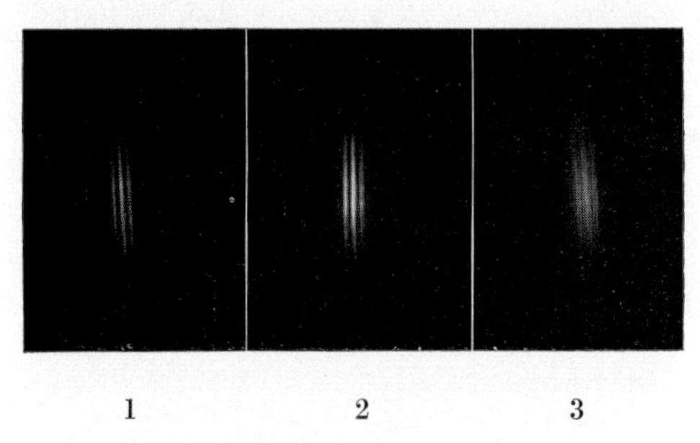

Fig. 5. Decrease of contrast of secondary interference pattern produced by increasing the width of the object slit from 0·1 mm (1) to 0·25 mm (2), and 0·4 mm (3)

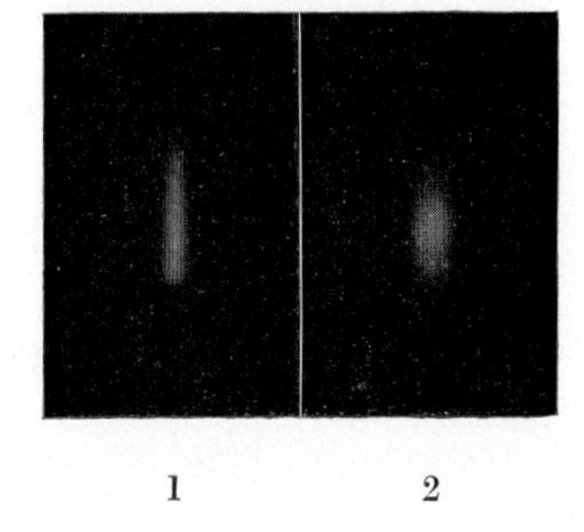

Fig. 6. Primary pattern (1) and secondary pattern (2) for the same object and the same interferometer setting in disturbed air (exposure times 1 min (1) and 2 min (2))

was, therefore, the primary object of the FSA investigation to produce a "stationary" fringe system which would thus be unaffected by air turbulence. There can, however, be no doubt that the secondary system is in fact *not* stationary, as was rightly pointed out by LACROUTE and CARROLL, and was also ascertained in the repetition of the experiments carried out at St. Andrews. Nevertheless, the time exposures reproduced in FSA and additional material obtained at St. Andrews, of which Fig. 6 gives an example, show that the secondary pattern can still be distinguished when the primary pattern is completely obliterated by the disturbances. The reason for this seems to be that, as can be seen from the photographs of Fig. 2 and Fig. 4, the primary pattern consists of broad light bands divided by much narrower black stripes, whereas the secondary pattern in the optimum position is characterized by light and black fringes of about equal width. Thus, if the lateral movement of the pattern is not greater than a fraction of the fringe distance, the secondary pattern will still be preserved, though with diminished sharpness, while the primary one may have become completely blotted out.

A time exposure of the fringe pattern can be used for evaluating the time average $\bar{V}$ of the visibility V, (equ. (1)), for the particular setting of the interferometer. If V_0 is the visibility for the same setting in calm air, it is easily seen that $\bar{V} = \varepsilon V_0$ where the factor $\varepsilon < 1$ depends on the degree of turbulence, but is independent of the interferometer setting. Thus, provided that ε is not too small, a series of time photographs of the secondary fringes should, on the basis of WOLF's theorem, still provide all information necessary for determining the light intensity distribution over the surface of the source under conditions of turbulence which make the observation by the original Michelson method impossible.

It might be worth while to try out this procedure on some celestial objects with a suitably adapted stellar interferometer; if it should turn out to be successful it would open new and unexpected possibilities for the use of this instrument.

REFERENCES

CARROLL, J. A. . . . . 1939 *M.N.*, **99,** 735.
FÜRTH, R., SITTE, K. and APPEL, H. P. . . . 1939 *M.N.*, **99,** 141.
LACROUTE, P. . . . . 1939 *M.N.*, **99,** 733.
MICHELSON, A. A. . . . . 1918 *Ap. J.*, **47,** 283.
WOLF, E. . . . . 1953 *Nature (London)*, **172,** 535.
1954 *Proc. Roy. Soc.*, A, **225,** 96.
1955 *Vistas in Astronomy*, Ed. A. BEER, Pergamon Press, London; p. 385.

Astronomical Spectrographs: Past, Present, and Future*

I. S. BOWEN
Mount Wilson and Palomar Observatories
Carnegie Institution of Washington
California Institute of Technology
Pasadena, California, U.S.A.

AT the beginning of the present century photography was rapidly displacing visual observation for the study of stellar spectra. As a result of this and of the completion of several new observatories at about this time a substantial number of new spectrographs were built and described in the early years of this century (CAMPBELL, 1898; VOGEL, 1900; FROST, 1902; SLIPHER, 1904; ADAMS, 1912; PLASKETT, 1919, 1924). The general pattern of these spectrographs, many of which are still in use, was a lens collimator of 1·5–2·5 in. aperture, a dispersing system of from one to three flint prisms and one or more cameras of focal length from 3 ft down to the shortest that the optical designs of the day would permit for the aperture used. These provided dispersions in the range from 10 to 100 Å per mm.

Somewhat later the demand for higher dispersion resulted in the construction, first at the Mount Wilson Observatory (ADAMS, 1911) and later at the McDonald

* See also the article "Methods in Stellar Spectroscopy" by TH. DUNHAM, Jr., in Section 13 of Volume 2.

Observatory, of large coudé instruments using either prisms or gratings. These had apertures of 3–6 in., focal lengths of up tc 15 ft and yielded dispersions of as much as 2 Å per mm.

With the construction of telescopes of larger and larger aperture, however, it became increasingly evident that the efficiency of these spectrographs, particularly for the higher dispersions, was exceedingly low. Thus the telescope-spectrograph system has a focal length equivalent to that of a single lens having an aperture equal to that of the main telescope mirror and a focal ratio equal to that of the spectrograph camera. If the slit is removed and if the seeing image of a star has a diameter of β radians, the linear diameter of a monochromatic star image on the plate is $A\frac{F}{D}\beta$. A is the aperture of the telescope and F and D are the focal length of the camera lens and the aperture of the beam coming from the collimator, respectively. This value for the image diameter is rigorous for a spectrograph with a symmetrical dispersing system in which a small deviation of the collimator beam produces an equal deviation of the camera beam. For non-symmetrical systems the above formula for the dimension of the image in the direction parallel to the dispersion should be multiplied by a factor, r, equal to the ratio of the deviations of the camera and collimator beams. In general r does not differ greatly from unity.

If this image size is greater than the limit of resolution of the plate, Δ, it is then necessary to introduce a slit to narrow the image to about this limit if optimum resolution is to be attained. The fraction of light that can pass this slit is, therefore,

$$E = \frac{\Delta D}{AFr\beta}. \qquad \ldots .(1)$$

Formula (1) is exact for a square star image. The much more complicated formula for the actual distribution of light in a star image differs from this by an amount much smaller than the uncertainties introduced by large fluctuations in β.

For most fast plates Δ is usually assumed to be about 0·02 mm while under average seeing conditions at the most favourably located observatories the seeing image has a diameter of 1–1·5 sec of arc, or $\beta = 6 \times 10^{-6}$ radians. Under these conditions

$$E = \frac{3300D}{AFr} \qquad \ldots .(2)$$

in which all dimensions are in millimetres. The fraction passed is therefore equal to 3300 mm or 132 in. divided by the effective focal length of the telescope-spectrograph system. This means that to attain 100 per cent efficiency cameras should operate at a focal ratio of two-thirds or less for a 200-in. telescope, of four-thirds for a 100-in., and of 2·2 for a 60-in. Formula (2) also shows that only 2–20 per cent of the light passes the slit of the older high-dispersion spectrographs operating at focal ratios of 10–30 when attached to these large telescopes. In view, however, of the enormous cost of collecting light by means of modern telescopes it is obviously of the greatest importance that the light once collected be used efficiently.

Fortunately optical advances in recent years have made possible the use of apertures and focal ratios of a different order than those hitherto available. These advances are:

(1) The development of the "blazed" grating by WOOD (1944) and by the BABCOCKS (1944; 1951). Gratings of this type throw 60–70 per cent of the incident light into one order. This represents a considerably higher efficiency than even a small-aperture prism train giving the same angular dispersion. Furthermore the grating maintains the same efficiency at all apertures while the efficiency of a prism train falls off rapidly as larger apertures are attempted. Because of this advance all new spectrographs at the Mount Wilson and Palomar Observatories are being built with gratings and all of the older prism instruments are being replaced with gratings as rapidly as shop facilities will permit.

(2) The development of cameras giving critical definition over moderate fields at very low focal ratios. Thus a standard air-type Schmidt camera gives critical definition over a field having a diameter of roughly one-fifth the focal length at focal ratios down to $\frac{F}{D} = \sqrt{\frac{D}{8}}$ in which the aperture D is expressed in inches. In cases where the focal length is not so great that absorption in the glass path becomes prohibitive an even smaller ratio of $\frac{F}{D} = \frac{1}{n}\sqrt{\frac{D}{8}}$ may be obtained with a solid block or thick-mirror-type Schmidt (HENDRIX, 1939; MINKOWSKI, 1944), n being the index of the glass used. If a somewhat smaller field of one-tenth to one-fifteenth of the focal length is permissible still lower focal ratios may be attained with the Schmidt-aplanatic sphere camera (BOWEN, 1952). This means that at the present time one can design spectrographs around cameras that approach the theoretical limit of the focal ratio of one-half for an air system or $1/2n$ for an immersion system in which $n = 1{\cdot}52$, the index of the gelatine of the photographic plate; this index sets the ultimate limit on the focal ratio of any photographic camera.

Obviously a reassessment of spectrograph designs should therefore be made in the light of these new optical possibilities. In equation (1) the quantities A, β and Δ are fixed by the telescope used, by seeing conditions and by the resolution required. Furthermore, F is related to the linear dispersion, K, expressed in Ångstroms per millimetre, and the angular dispersion, α, expressed as Ångstroms per radian, through the relationship

$$F = \frac{\alpha}{K}. \qquad \ldots.(3)$$

From grating theory we may obtain the following relationship:

$$\sin \Theta + \sin \Phi = \frac{n\lambda}{\delta} \qquad \ldots.(4)$$

in which Θ and Φ are the angles of incidence and diffraction, δ is the grating space, and n the order of the spectrum. For a monochromatic image we may differentiate holding λ constant and obtain

$$\cos \Theta \, d\Theta + \cos \Phi \, d\Phi = 0$$

or

$$r = -\frac{d\Phi}{d\Theta} = \frac{\cos \Theta}{\cos \Phi}. \qquad \ldots.(5)$$

To obtain the angular dispersion α we may differentiate holding Θ constant, which gives

$$\cos \Phi \, d\Phi = \frac{n d\lambda}{\delta}$$

or

$$\alpha = \frac{d\lambda}{d\Phi} = \frac{\delta \cos \Phi}{n}. \quad \ldots.(6)$$

Furthermore if L is the length of the ruled surface measured perpendicular to the lines, then D, the largest aperture of collimator beam that can be handled by the grating without loss, is $L \cos \Theta$. Substituting these values of F, r, and α, in (1) and simplifying, one obtains for the efficiency of a grating instrument

$$E = \frac{\Delta K L n}{A \beta \delta}. \quad \ldots.(7)$$

If we substitute for $\frac{n}{\delta}$ from (4) in order to show more clearly the effect of the limitation on the value of n, we obtain

$$E = \frac{\Delta K L(\sin \Theta + \sin \Phi)}{A \beta \lambda}. \quad \ldots.(8)$$

It is evident that any increase in efficiency depends on the possibility of making $L(\sin \Theta + \sin \Phi)$ as large as possible.

In the design of the Coudé spectrograph for the 200-in. Hale telescope L was pushed to 14 in. by resort to the use of a composite made up of four matched gratings each having a ruled area $5\frac{1}{2} \times 7$ in. Since the optical resolving power of an individual grating was greater than that of the photographic plate for all focal lengths used, it was not necessary to bring the four gratings into phase with each other. The much easier condition of so adjusting the gratings that the spectra formed by them were in coincidence within the limit of plate resolution could be satisfied with the aid of proper mechanical devices. Unfortunately it has thus far not been possible to obtain the 60–70 per cent concentration of light in an order higher than the third for a 400 line per mm grating and consequently gratings of this type were used. For the centre of the camera field at λ about 4200 Å this gives a value of $(\sin \Theta + \sin \Phi) = \frac{1}{2}$.

For this spectrograph cameras of focal length 144 in., 72 in., 36 in., 18 in., and 8·4 in., operating at focal ratios of 12, 6, 3, 1·5, and 0·7, are provided. These yield dispersions in the third order violet of 2·3, 4·5, 9, 18, and 38 Å per mm (BOWEN, 1952), and efficiencies of 6 per cent, 13 per cent, 25 per cent, 50 per cent, and 100 per cent, respectively, under the conditions of seeing and slit width mentioned above.

For dispersions in the 110–440 Å per mm range a small grating spectrograph has been constructed for use at the prime focus of the Hale telescope. This has a 3-in. aperture collimator beam and is provided with cameras of the thick-mirror Schmidt type operating at focal ratios of $\frac{F}{D} = 0{\cdot}47$ and $0{\cdot}95$.

Looking to the future for possible ways of still further increasing efficiency, two general possible directions of attack need consideration. The first of these is some scheme such as the image slicer (Bowen, 1938), which changes the shape of the star image to a long narrow one. Such a device by itself increases the fraction of light passing the slit by widening the spectrum rather than by increasing its surface brightness. The brightness may than be increased by narrowing this spectrum with a cylindrical lens placed in front of the photographic plate. A simple cylindrical lens can, without introducing objectionable aberrations, reduce the effective focal ratio in the direction perpendicular to the dispersion to about $F/D = 2$, while the original value of the focal ratio of the camera lens is retained in the direction parallel to the dispersion. With the older cameras operating at focal ratios of $F/D = 10$–30 this made possible a substantial gain in speed. With the much smaller focal ratios now possible for spectrograph cameras the gain in speed is much less, although such an image slicer may often prove useful to widen spectra that otherwise would be unduly narrowed by the small ratios of camera to collimator focal lengths now coming into use.

The second and more direct procedure for increasing efficiency through taking advantage of the factors in equation (7) would appear to have its greatest possibility of success by improving grating ruling techniques in such a way as to increase ($\sin\,\Theta + \sin\,\Phi$) and at the same time retain an efficient blaze. Thus a gain of a factor of over 3, compared to the present 200-in. Coudé spectrograph, is theoretically possible and is probably easier of attainment than by ruling gratings three times as large as the present maximum size. Since the field of view of the present very fast cameras is limited to between one-fifteenth and one-fifth of the focal length, it is desirable that the angular length of the blaze in each order should not exceed this. By combining the grating with a small prismatic cross dispersion all wavelengths may then be recorded on one plate. This would indicate a fairly coarse grating of the order of 100 line per mm. Gratings of this general type have been suggested by Shane and Wood (1947) and by Harrison (1949).

The one serious difficulty encountered in the use of a reflection grating, at large values of Θ and Φ, is the interference of the camera with the collimator beam. Either Θ must differ largely from Φ, in which case a highly elliptical camera beam results, or the camera lens or corrector plate of a Schmidt system must be placed at a large distance from the grating with resultant vignetting difficulties unless an aperture much larger than the beam is used. Either choice results in a very inefficient use of lens aperture and seriously limits the effective focal ratios that may be used.

An ideal solution would be a grating made up of a series of equally-spaced reflecting surfaces arranged like the slats of an open Venetian blind and embedded in a transparent medium, the planes of the individual surfaces being nearly perpendicular to the plane of the grating. By placing the camera and collimator on opposite sides of this grating, Θ and Φ could be large and equal without causing any interference.

Such a grating would have the additional advantage that the position of the blaze in the spectrum could be shifted by a change of Θ and Φ. One procedure for making such a "Venetian-blind" grating would be to rule on a thin layer of plastic bonded to glass with a triangular groove one side of which was held nearly perpendicular to the plastic surface. An aluminium coat would next be evaporated from such a direction that the perpendicular side only of the groove would receive a coat. The grooves would then be filled up with a second application of plastic and possibly covered with a glass optical flat.

REFERENCES

ADAMS, WALTER S.	1911	*Ap. J.*, **33**, 64.
	1912	*Ap. J.*, **35**, 163.
BABCOCK, HAROLD D.	1944	*J. Opt. Soc. Amer.*, **34**, 1.
BABCOCK, HAROLD D. and HORACE W. .	1951	*J. Opt. Soc. Amer.*, **41**, 776.
BOWEN, I. S.	1938	*Ap. J.*, **88**, 113.
	1952	*Ap. J.*, **116**, 1.
CAMPBELL, W. W.	1898	*Ap. J.*, **8**, 123.
FROST, EDWIN B.	1902	*Ap. J.*, **15**, 1.
HARRISON, G. R.	1949	*J. Opt. Soc. Amer.*, **39**, 522.
HENDRIX, D. O.	1939	*Publ. Astr. Soc. Pacific*, **51**, 158.
MINKOWSKI, R.	1944	*J. Opt. Soc. Amer.*, **34**, 89.
PLASKETT, J. S.	1919	*Ap. J.*, **49**, 209.
	1924	*Ap. J.*, **59**, 65.
SLIPHER, V. M.	1904	*Ap. J.*, **20**, 1.
VOGEL, H. C.	1900	*Ap. J.*, **11**, 393.
WOOD, R. W.	1944	*J. Opt. Soc. Amer.*, **34**, 509.
	1947	*J. Opt. Soc. Amer.*, **37**, 733.

Spectroscopy with the Echelle

GEORGE R. HARRISON

Massachusetts Institute of Technology, Cambridge, Massachusetts, U.S.A.

SUMMARY

The echelle is a type of dispersing element intermediate in properties between the blazed plane diffraction grating and the reflection etalon of MICHELSON and WILLIAMS. Echelle spectrographs have now been produced in six forms which for many spectroscopic problems give advantages in speed, compactness, and stigmatic properties, and compete effectively with much larger instruments in resolving power and dispersion. An echelle is crossed with a small high-speed stigmatic prism or grating spectrograph for the production of cyclic spectra, giving the equivalent of a large concave grating instrument. A special advantage of the echelle spectrograph lies in its free spectral range, which is sufficient to reveal hyperfine structures and Zeeman patterns effectively without overlapping.

Echelles are now produced by the Bausch and Lomb Optical Company by three methods, involving respectively ruling in deep aluminium layers on glass as with ordinary gratings, cutting laps in metal with a fairly precise tool room dividing engine and using these to grind into an optical flat the desired set of flat-faced grooves with appropriate procedures to average out periodic errors, and replicating in plastic of either ruled or lapped echelles. It has been found that wavelength precision of about one part in five million, intermediate between that given by the usual large concave gratings and that given by the reflection echelon or etalon, is obtainable with echelles.

1. THE ECHELLE

AT the London Conference on Optical Instruments, held 19th to 26th July, 1950, HARRISON and BAUSCH (1950) reported on the production of a new type of dispersing element, intermediate in properties between the blazed plane diffraction grating (WOOD, 1910) or echellette, and the reflection echelon (MICHELSON, 1898, 1899). This so-called echelle has since been found useful in many fields of spectroscopy, especially where compactness combining high resolution with broad spectral coverage is important. Echelle spectrographs have now been produced in at least six forms, several of which give advantages over concave grating instruments in speed, compactness, and stigmatic properties, and compete effectively with much larger instruments in resolving power and dispersion. The application of the echelle to astronomical spectroscopy holds much promise (HARRISON, 1949).

An echelle consists of a set of flat reflecting optical steps or grooves, worked into a single piece of glass or metal so as to form a multiple-beamed interferometer set in permanent adjustment. Instead of the 2 to 4 steps per inch of the reflection echelon, made perforce by stacking optically-worked flats, which produces a small number of beams that interfere in trains separated by many thousands of orders, it produces a thousand or more beams which interfere in much lower orders. A plane diffraction grating, in turn, combines many thousands of diffracted beams in still lower orders of interference. The echelle can be thought of as a coarse but precisely formed and well-blazed plane grating designed for use at large and equal angles of incidence and reflection, and relying to a minimum on diffraction and spreading for chromatic separation. It is well known, though usually overlooked, that the theoretical resolving powers of a grating and an echelon of equal width are the same, being independent of the number and spacing of grooves, and depending at a given wavelength only on the optical depth. In practice echelons have given resolving powers in excess of one million. Present methods of manufacture give echelles and gratings making possible resolution approaching 500,000.

The essential difference that determines whether one or the other of these types of dispersing units will prove superior in a given case lies in the free spectral range made available. The order of interference m is controlled by the optical depth $2t$ between successive reflected beams, so that $m = 2t/\lambda$. The free spectral range, which is the range of spectrum lying between successive orders of any given spectrum line, is given by the formulae $F_\sigma = 1/2t$ and $F_\lambda = \lambda^2/2t$, in wave numbers and wavelengths respectively, and obviously diminishes as order increases. The free spectral range of an echelon may cover less than the width of a single spectrum line, and asymmetries result which make wavelength measurements extremely difficult, while repeated exposures must be taken to photograph all spectrum lines under proper conditions. In an echelle the free spectral range can be chosen sufficient to reveal hyperfine structures and Zeeman patterns effectively without overlapping. At the same time it offers advantages for wavelength and intensity measurements, and especially for optical design, since it gives dispersion without the wide separation in angle of beams of various wavelengths produced by concave and plane gratings. This ability to produce high dispersion in a collimated beam that diverges only slightly, gives the echelle usefulness for the design of compact spectrographs of high aperture without the drawbacks of the echelon.

2. Design and Production of Echelles

Echelles are designed to be crossed with devices of lower dispersion, such as prisms or gratings used at small angles. An echelle can be crossed externally with any stigmatic spectrograph of suitable aperture, or internally in a design such as those described below. The constants of an echelle may be selected to fit any desired circumstances, and as their manufacture comes under better control it should be possible to obtain echelles suited to any spectrographic need. Some now available can be used to produce very compact and fast spectrographs having resolving powers up to 500,000, which cover extremely broad spectral ranges. The cyclic array of spectra produced makes it possible to cover the spectrum from 2000–7000 Å at high dispersion on a single plate. Dispersions of from 1–10 mm per Å (plate factors from 1–0·1 Å per mm) are not uncommon in echelle spectra.

An example of an echellegram is shown in Fig. 1. The horizontal dispersion is here produced by the echelle, while the vertical dispersion is given by an auxiliary prism or grating which produces from 1/10 to 1/100 as much dispersion as the echelle. Echelles are designed to throw as much of their light as possible, when illuminated in the Littrow manner, directly back toward the illuminating slit.

Fig. 1. Portion of an echellegram produced by crossing a ruled echelle with an ordinary grating. Here the vertical overlapping at short wavelengths is somewhat greater than would be the case when a prism was used for auxiliary dispersion

The ruled width W of the echelle (Fig. 2) contains N parallel grooves, each of optical depth t and reflecting width s. The ratio t/s, called r, controls the intrinsic dispersion, and the larger its value the greater the compactness of a spectrograph producing a given dispersion with the echelle. Also $r = B/A$, the ratio of the optical depth B of the echelle, which controls its resolving power, to its aperture A. Then $W = Na$, where a is the usual grating constant. The angle of illumination α is now usually about 63°, for which $r = 2$. The theoretical resolving power of an echelle for wavelength λ is $\lambda/d\lambda = 2B/\lambda$, while the dispersion in a focal plane at distance P

from the projection lens used with the echelle is $dl/d\lambda = 2rP/\lambda$. The length of an echelle cycle on the plate $l = P\lambda/s$.

When the size and shape of the desired echellegram have been chosen, one can specify the number of echelle grooves N and the groove depth t needed to photograph a spectrum range $d\sigma$ completely on one plate with spectrum lines of height h. If the longest cycle has length l_m on the plate, and the separation between successive cycles is h, one dimension of the plate will be $l_m = fN\lambda_m$, and the other $H = 2ht\Delta\sigma$, where f is the aperture ratio P/A.

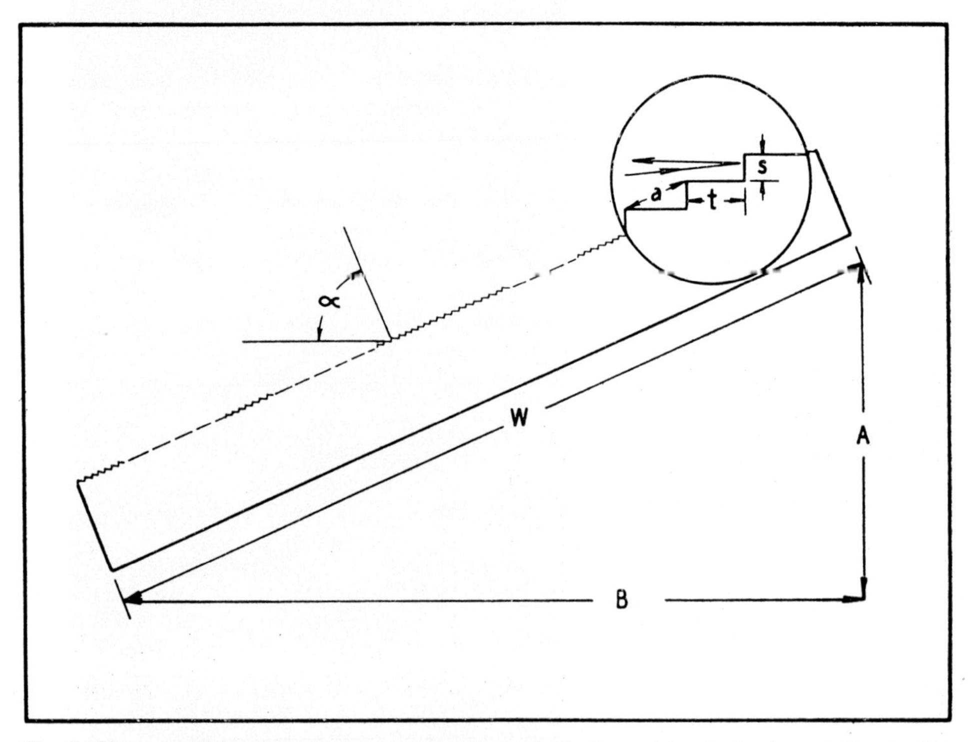

Fig. 2. Diagram of the essential components of an echelle, in position for horizontal illumination

On fixing P, the focal length of the spectrograph, and A its aperture, we determine the size of the spectrograph needed. In theory P could be reduced to a foot or less, so that very small lenses and other optical parts could be used, but this would introduce wide angles and greater aberrations. Echelons have been used at $r/7$, but only at very small free spectral ranges.

Echelles have been produced by the Bausch and Lomb Optical Co. (1954) by three methods. Those having up to 200 grooves per inch are made by cutting laps in metal with a fairly precise tool-room dividing engine, and using these to grind into an optical flat the desired set of flat-faced grooves of triangular cross-section. During this grinding process the lap is moved frequently so that different grooves in lap and echelle come into contact, to average out the periodic errors introduced by the ruling engine when the laps were cut. After the grooves have been ground to the desired depth, they are polished with rouge. During the polishing process, it is necessary to take special pains to keep the grooves flat while preserving their optical alignment to within 1/10 fringe or better. The polished optical steps are then aluminized, giving, in a standard 6-in. echelle, some 1200 equally-spaced parallel mirrors about 3 in. long and 1/8 mm wide.

The theoretical resolving power of such an echelle at wavelength 5000 Å is about 500,000, and at 2000 Å one million. Experimental resolution exceeding 450,000 has been realized in the visible. Rowland ghosts are not, in general, found in echelles produced by this method, though some show satellites very close to or within strong lines, which arise from small local sections of the echelle surface that are displaced slightly from the proper optical plane. These are caused by breakage of the sharp edges of the narrow land between grooves, with embedding of abrasive materials in the broken portions. Techniques have now been developed which greatly reduce this effect.

A second type of echelle, thus far having not fewer than 2000 grooves per inch (because of the thickness of deposited aluminium required for deeper grooves) has been produced by direct ruling on one of the Bausch and Lomb engines. Their Rowland ghosts are likely to be prominent, especially at shorter wavelengths, because of the high angle at which the echelle is used. Bausch and Lomb have also developed a process for making replica echelles of both ruled and ground types. Replicas made from master echelles have shown excellent performance.

3. Echelle Spectrographs

The high intrinsic dispersion of the echelle, coupled with its compactness and luminous effectiveness, make possible design with it of spectrographs having unusual compactness, efficiency, and permanence of adjustment. Resolution at a given speed can be improved with it, or speed at a given resolving power, or some of each.

A type of echelle spectrograph which gives the compactness of the quartz-prism spectrograph, speed only slightly less, and dispersion and resolving power greater than a 21 ft concave grating, is exemplified by the Littrow-Echelle spectrograph manufactured by Bausch and Lomb (1954). This can be used as an ordinary Littrow spectrograph, when it gives the dispersion usually furnished by its quartz prism. By pressing a button the operator can remove a reflecting mirror, which takes the place of the usual coated rear surface of the Littrow prism, from the optical path, as shown in Fig. 3, and the light strikes the echelle. Thus low and high dispersion spectra can be taken on the same plate, as shown in Fig. 4. At 2000 Å, when the prism only is used, the plate factor is 2 Å per mm, while with the echelle the dispersion is increased six-fold. At 5000 Å, the prism gives only 20 Å per mm, while the prism-echelle combination gives 0·68; at 8000 Å, dispersion and the corresponding resolving power are increased fifty-five-fold by the echelle.

Bausch and Lomb also manufacture an echelle attachment, in which two concave spherical mirrors are used to collimate and focus the light on an echelle which has its own slit and mounting; the predispersed but unseparated light from this unit can be thrown on the slit of any desired stigmatic spectrograph, producing the desired cross-dispersion.

Griffin, Loring, Werner, and McNally (1952) have described a spectrograph in which an echelle is crossed with a glass Littrow-prism spectrograph in an external mounting. Harrison, Archer, and Camus (1952) have described a fixed-focus broad-range echelle spectrograph of high speed and resolving power, in which the crossed dispersion is furnished by a concave grating in a Wadsworth type of mounting.

It uses a 6-in. echelle having 200 grooves per inch, and shows about five times the speed of the previously-used 10 m grating with 30,000 grooves per inch in a Paschen-Runge mounting, and double its resolving power. Instead of requiring 750 in. of

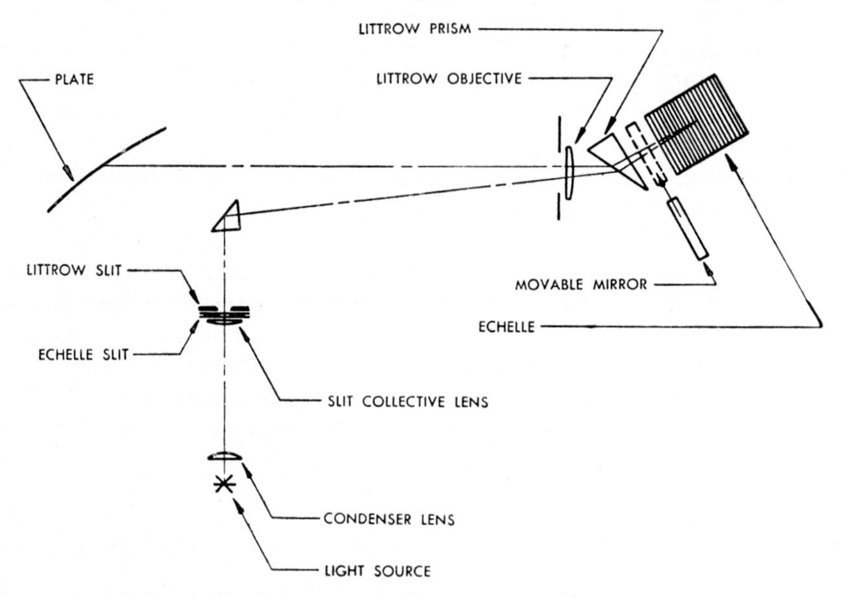

Fig. 3. Optical system of a Littrow-echelle spectrograph

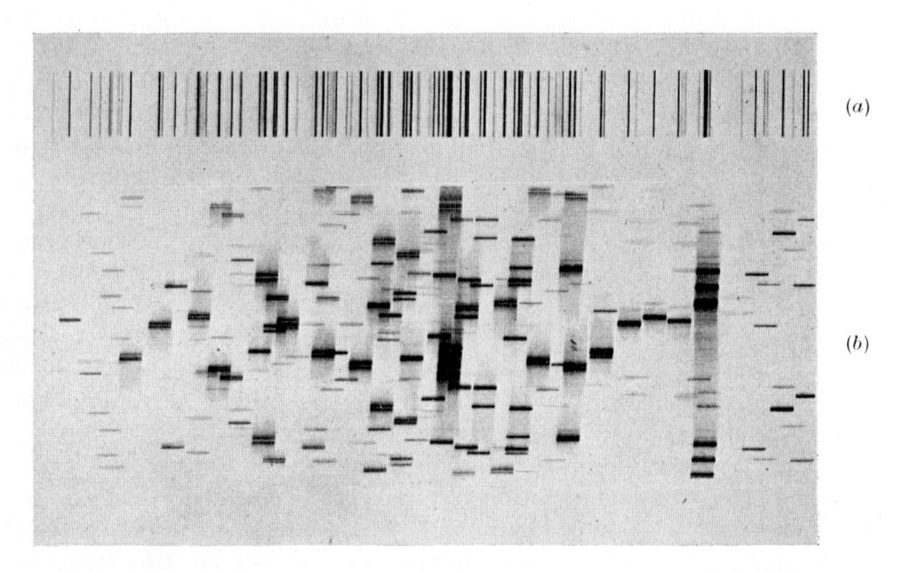

Fig. 4. (a) Spectrum produced with a Littrow-echelle spectrograph without the echelle dispersion. (b) Same spectrum when the light strikes the echelle instead of the backing mirror

plate length to record the spectrum from 2000–7000 Å, it does this on 30 in. of plate at higher dispersion. This echelle spectrograph is stigmatic, and occupies only 35 ft² instead of the 700 ft² required by the previously-used grating mounting. A section of an echellegram taken with this instrument is shown in Fig. 5.

Pierce, McMath, and Mohler (1951) have used an echelle for studying solar spectra, and report it as useful, especially if scattered light and satellites can be reduced beyond what they find in an early echelle. Bausch and Lomb have built a large auroral spectrograph for the Geophysical Directorate of the U.S. Air Force, in which the characteristics of the echelle have been used to combine very high speed with moderate dispersion. This device, with lenses 13 in. in diameter, operates at $f1{\cdot}5$, using a mosaic of four echelles. The compactness, speed, and high intrinsic

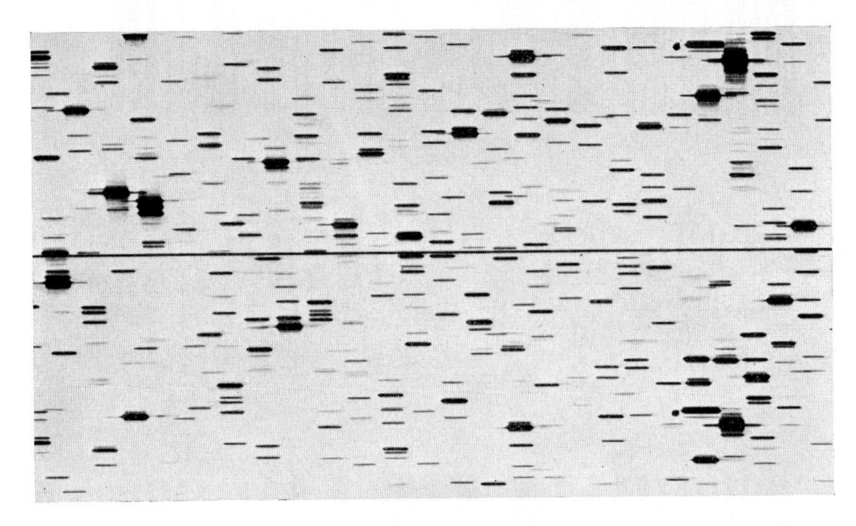

Fig. 5. A portion of a typical echelle spectrum showing fiducial line, with the echelle dispersion vertical

resolving power and dispersion of the echelle should make it extremely useful in many astronomical applications.

4. Measurement of Wavelengths in Echellegrams

Harrison, Davis, and Robertson (1953) have shown that echelle spectrograms can be used to obtain wavelength precision of about 1 part in 5 million, intermediate between the 1 part in 10^6 given by the usual large concave grating, and the 1 part in 5×10^7 given by the echelon or etalon. Fig. 6 shows portions of spectra taken with three such instruments.

A great advantage of the high dispersion fixed-focus spectrographs which the echelle makes possible is that the parameters needed for reducing spectrograms can be maintained constant over long periods. A fiducial line having constant $m\lambda$ can be recorded on each echellegram by reflecting light from a mirror swung into position in front of the echelle, and the exact $m\lambda$ value of this line can be determined from two or three standard lines, such as those of mercury, photographed on the plate. Wavelength reduction is conveniently made in terms of $m\lambda$, which remains nearly constant for a given distance from the fiducial line $m\lambda_0$ in all cycles. A small correction ε can be added for each line, values of ε for the entire fixed-focal surface having been determined in advance with a complex spectrum having many known lines. ε is found to vary by no more than 40 order angstroms over the focal surface of the instrument described (Harrison, Davies, and Robertson, 1953). A simplification in reduction results from the fact that identification of the order of interference m can be made much more quickly than with an etalon or echelon. On division by three-figure integral values of m, wavelengths can be obtained to 0·001 Å or better throughout the visible and ultraviolet spectra.

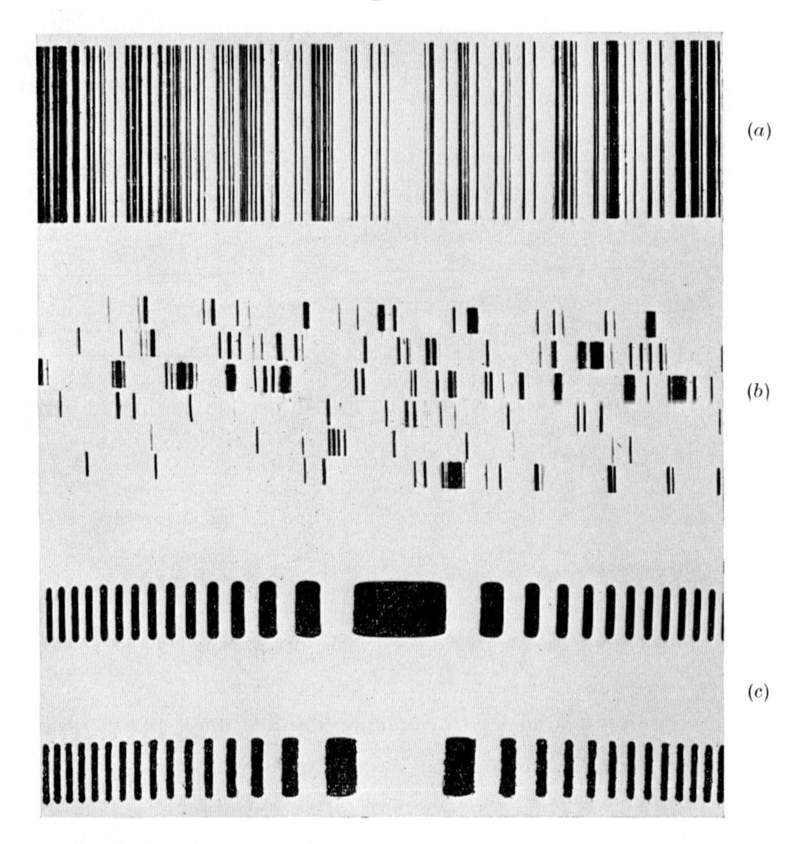

Fig. 6. Sections of spectra taken with: (*a*) a concave grating; (*b*) an echelle spectrograph; (*c*) a Fabry-Perot etalon, showing successively increasing dispersion and diminishing free spectral range

Simplified methods of line identification and less-precise wavelength measurement can be used with the echelle crossed with prism dispersion, as described by FINKELSTEIN (1953).

5. FUTURE POSSIBILITIES

The manufacture of echelles is only in its infancy, and great improvements in resolving power, speed, size, and general effectiveness can be expected. In particular the Bausch and Lomb method of removing periodic errors by groove averaging appears susceptible of development in several different directions, and may well result in removal of the present limitation to available sizes of grooved dispersing instruments. It seems probable that echelles of 20 in. or more in diameter can be produced for astronomical purposes, capable of focusing from 40–80 per cent of the light of a given wavelength in a single order, by using a process of replicating in which a single master echelle is used to extend itself by step-justification.

One limitation of the echelle is that while it may be thought of as blazed for light of all wavelengths, and does not emphasize one range of the spectrum at the expense of another broad range, it does vary the intensity from a maximum at the centre of each cycle to a value which may fall to half as much at the extremes. This effect is minimized, however, by the rounding of the reflecting faces which usually occurs in present-day echelles, which, of course, reduces efficiency, and is not extreme enough to produce the asymmetries of the echelon. Improved techniques of coating the step faces with aluminium should increase the speed of echelles in future, and reduce scattered light. Solution of the problem of coating glass surfaces with thick, homogeneous, strongly reflecting aluminium layers which can be ruled, should also open up the region of echelle spacings between 2000 and 500 grooves per inch. Improved methods of cross-dispersion can be expected also to simplify the astronomical use of echelles.

References

Bausch and Lomb Optical Company, Rochester . . . . 1954 *Catalogue Echelle Spectrographs*, New York.

Finkelstein, N. A. . . . . 1953 *J. Opt. Soc. Amer.*, **42**, 90.

Griffin, Loring, Werner and McNally . 1952 Southeastern Section Meeting of the American Physical Society, North Carolina State College, 10th April.

Harrison, George R. . . . . 1949 *J. Opt. Soc. Amer.*, **39**, 522.

Harrison, George R. and Bausch, Carl L. 1950 *Proc. London Conference on Optical Instruments* (Chapman & Hall, Ltd., London).

Harrison, George R., Archer, J. E. and Camus, J. . . . . 1952 *J. Opt. Soc. Amer.*, **42**, 706.

Harrison, George R., Davis, S. P. and Robertson, H. J. . . . . 1953 *J. Opt. Soc. Amer.*, **43**, 853.

Michelson, A. A. . . . . 1898 *Ap. J.*, **8**, 37.
1899 *Proc. Amer. Acad. Arts Sci.*, **35**, 111.

Pierce, A. K., McMath, R. R. and Mohler, O. . . . . 1951 *Astron. J.*, **56**, 137.

Wood, R. W. . . . . 1910 *Phil. Mag.*, **20**, 770.

Comparison of the Efficiency of Prism and Grating Spectrographs

CH. FEHRENBACH

Observatoires de Marseille et de Haute Provence, France

SUMMARY

The relative luminous efficiencies of spectrographs using prisms and gratings as dispersing elements are discussed; special reference is made to the spectrographic equipment of the Observatoire de Haute Provence. It is shown that grating instruments are almost always the more efficient ones, especially when used with a large telescope or under poor seeing conditions.

THE efficiency of a slit spectrograph used in conjunction with a telescope is measured by the proportion of the incident light which reaches the photographic plate. It is, of course, assumed that the spectra are of good quality and of sufficient height.

The loss of light comes principally from the use of a slit which usually accepts only a very small portion of the spurious disk of the star.

Efficient spectrographs are those which allow a very large slit to be used. In order to prevent this wide slit from giving too large images on the plate, the camera lens must have a very short focal length; more precisely, its numerical aperture must be very large. It is also necessary to have a system of very high dispersion, either a train of prisms, or, better still, a modern grating.

It is the possibility of using very wide slits with gratings which makes modern instruments of this type so effective.

We shall discuss briefly the efficiency of stellar spectrographs and compare systems using prisms with those using gratings.

1. THE LUMINOUS EFFICIENCY OF SPECTROGRAPHS

We shall measure the luminous efficiency of a spectrograph by the illumination produced on the plate by a given star.

The spectrum must have a certain minimum height h of the order of 0·5 mm to be useful. The discussion will be limited to stars with continuous spectra.

We shall denote by E the monochromatic brightness of the star, and by D the diameter of the telescope mirror. The flux received by the objective will be:

$$\Phi d\lambda = \frac{\pi}{4} \,.\, D^2 \,.\, E \,.\, d\lambda.$$

A fraction T will be transmitted to the plate and will occupy on it a portion of the spectrum of height h and length $\frac{dx}{d\lambda}\, d\lambda$, where x is the distance along the plate.

The illumination of the plate will be:

$$e = \frac{\pi}{4} \,.\, E \,.\, T \,.\, \frac{D^2}{h \,.\, \dfrac{dx}{d\lambda}} \,.$$

For a given star and dispersion we write:

$$e = \text{constant} . T . D^2.$$

This quantity depends only on the diameter of the telescope and on the transmission factor T of the spectrograph.

The practical problem to be solved consists therefore in choosing, for a given telescope, a spectrograph having a large transmission factor T.

2. DISCUSSION OF THE TRANSMISSION FACTOR T OF A SPECTROGRAPH

Let us call L the transmission factor of the slit, s that of the system of lenses or mirrors, and S that of the dispersing system (prism or grating). We have

$$T = L . s . S.$$

The importance of the various factors will now be examined.

A. TRANSMISSION L OF THE SLIT

The concept of effective slit width for stars

In order to use all the light of the spectrograph, it is necessary to be able to make the slit sufficiently large, so that the star image falls practically entirely within the slit. The image of the slit which is produced on the plate should have a size of the order of the photographic resolution. It is found that all the available light is used if the f-number of the camera is smaller than O_2 given by:

$$O_2 = \frac{r}{t} \cdot \frac{1}{D},$$

where r is the resolution limit of the plate in centimetres, t the angular diameter of the image in radians, and D as before the diameter of the telescope.

Substituting for s the practical value $25 . 10^{-4}$ cm yields:

$$O_2 = \frac{1}{t''} \cdot \frac{5}{D_m}.$$

Large telescopes seldom give images smaller than 1″.5; they therefore require spectrographs with a f-number less than the following limiting values:

D (meters)	0·40	0·80	1·20	2	5
O_2 . .	7·5	3·7	2·5	1·5	0·6

This table shows clearly the difficulty of equipping large telescopes with spectrographs.

Conversely, for a telescope of given diameter (say, 2 m) full efficiency is obtained only when the turbulence is less than the following limits:

O_2 . .	10	5	2·5	1·25
t'' . .	0″.25	0″.5	1″	2″

This table shows that spectrographs having narrow aperture cameras, that is those giving the largest linear dispersion, are seldom used to full advantage.

B. Transmission Factor *L* of the Slit

Two effects are considered: (1) the slit intercepts only a strip in the seeing image; (2) it diffracts part of the transmitted light, so that it does not enter the objective of the collimator.

(1) *We suppose that the slit lets through an amount of light proportional to the area of the curvi-linear rectangle ABCD in Fig.* 1, where the circle corresponds to the star image.

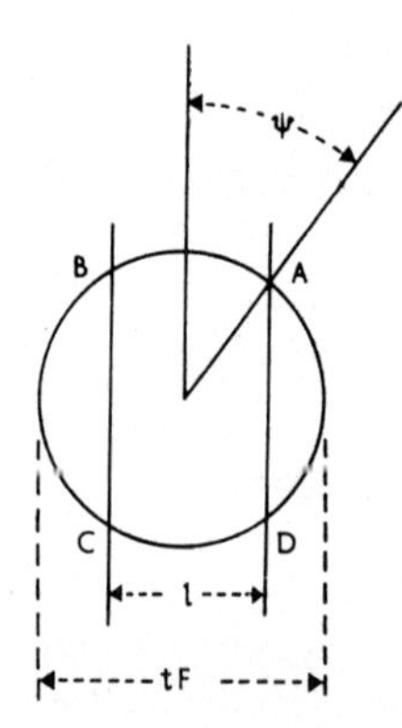

Fig. 1. Geometry of the slit in relation to the star image

Let F be the focal length of the telescope, l the width of the slit, and t the diameter of the image (including turbulence and other defects). We introduce the variables ξ and ψ such that:

$$\xi = \sin \psi = \frac{l}{t\,.\,F}.$$

The proportion of light entering the slit is then

$$L_\xi = \frac{\text{area of rectangle}}{\text{area of disc}} = \frac{2\psi + \sin 2\psi}{\pi}.$$

The assumed distribution of light in the image is clearly somewhat simplified. Weaver* has made a similar calculation, assuming that the distribution is that of a diffraction pattern. Furthermore, we have made the same calculation using a Gaussian distribution.

The results are very similar, provided that the diffuse cases are compared with a sharply-bounded disk which is rather smaller than was taken by Weaver. Table 1 and Fig. 2 refer to the three cases.

The columns labelled "Diffraction image" in Table 1 are taken according to Weaver but with $\xi_e = 1{\cdot}87$. For the diffraction pattern we have thus taken the effective diameter to be that of the first dark ring multiplied by 0·535; and for the Gaussian image it is given by $I = I_0/e$. The difference between the values for the diffraction image and the Gaussian distribution is principally due to the contributions from the illumination outside the central disk of the diffraction pattern. This difference is practicably negligible; it is only the effective diameter which is important; see Fig. 2.

---

* See H. F. Weaver, *Publ. Astron. Soc. Pacific*, vol. 60, p. 79, 1948.

Table 1

Uniformly illuminated circle		Diffraction image		Gaussian distribution	
$\xi = a/R$	L	ξ_e	L	ξ_e	L
0·0	0·000	0·000	0·000	0·0	0·000
0·1	0·127	0·187	0·238	0·1	0·127
0·2	0·253	0·374	0·458	0·2	0·251
0·3	0·376	0·561	0·644	0·3	0·371
0·4	0·495	0·748	0·788	0·4	0·483
0·5	0·609	0·935	0·889	0·5	0·587
0·6	0·715	1·122	0·951	0·6	0·681
0·7	0·812	1·309	0·983	0·7	0·765
0·8	0·896	1·496	0·996	0·8	0·837
0·9	0·963	1·683	1·000	0·9	0·899
1·0	1·000	1·870	1·000	1·0	0·951

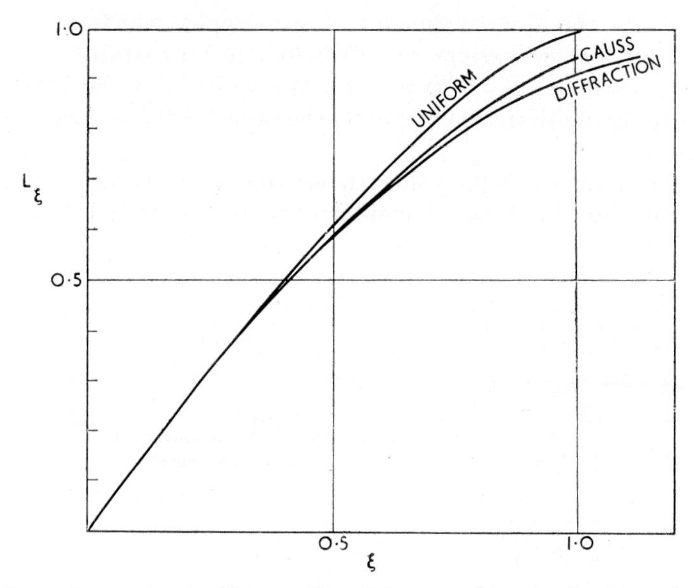

Fig. 2. Proportion L_ξ of the incident light, passed by the slit, for various distributions of light in the star-image

(2) *Diffraction at the slit*

The full treatment of this problem is difficult. However, we can see what happens in the following way:

Consider a beam travelling parallel to the axis of the collimator and penetrating the slit. In the absence of diffraction, the slit would admit a cylindrical beam, the generator of which would have the shape of the slit itself. We shall assume that on account of the diffraction this beam will have an intensity I, in a direction making an angle θ with the original direction, given by the diffraction formula

$$I = \left(\frac{\sin \pi a\theta}{r a\theta}\right)^2,$$

where $a = l/\lambda$.

In the direction $\theta_0 = 1/a = \lambda/l$ the intensity is zero. In the plane of the objective of focal length F_1, this corresponds to a dark ring of radius

$$\rho_0 = \frac{\lambda}{l} \cdot F_1.$$

It is evident that diffraction will cause more light to fall outside the objective the smaller radius R_1 of the objective is relative to ρ_0; we shall therefore take ρ_0 as the unit for expressing R_1. We define m by:

$$m = \frac{R_1}{\rho_0} = \frac{R_1}{F_1} \cdot \frac{l}{\lambda} = \frac{1}{2} \cdot \frac{l}{\sigma} \cdot \frac{1}{\lambda},$$

where σ is the numerical aperture of the collimator, which is equal to that of the telescope.

The ratio l/σ has the same value for spectrographs which are similar except that they are built for the various foci (Newtonian, Cassegrain) of the same telescope. As one goes from $\sigma = 6$ to $\sigma = 15$, the width l of the slit changes by a factor 2·5. The effect of diffraction is exactly the same for these various arrangements of the telescope.

In order to calculate the light collected by the objective of the collimator, we must integrate the light in all the elementary beams of the kind shown in Fig. 3.

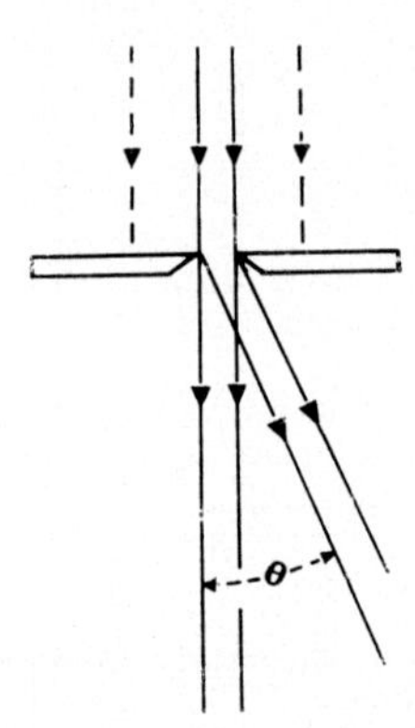

Fig. 3. Geometry of diffraction at the slit of a spectrograph

We shall not give here the details of the calculations, which were made partly with the help of tables of functions, and partly with graphical methods of integration.

Table 2 gives the values for the transparency factor L_m.

Table 2

m	L_m
0·000	0·000
0·405	0·519
0·810	0·706
1·216	0·782
2·431	0·872
4·052	0·913

This theory is really too elementary. It has been possible to compare the results with earlier estimates by MOORE.* The agreement, without being quantitative, is by no means bad.

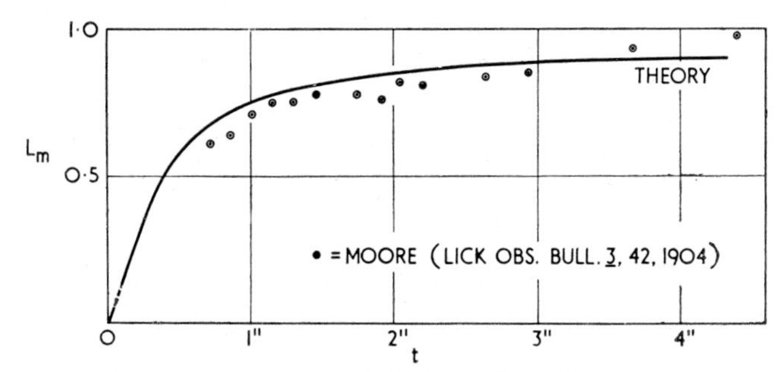

Fig. 4. Comparison of theoretical values of the factor L_m, which represents that proportion of light entering the slit which reaches the collimator despite diffraction at the slit, with the experimental values of MOORE (1904)

It may be noted that the variable m can be written in the form

$$m = \frac{1}{2} \cdot \frac{1}{\Omega} \cdot \frac{l}{\lambda},$$

where Ω is the numerical aperture of the camera.

The amount of light that is actually transmitted will be

$$L = L_\xi \, . \, L_m.$$

An examination of Tables 1 and 2 shows that the amount of the transmitted light increased very rapidly with the width of the slit. Therefore the wider the slit which a spectrograph will allow to be used the more efficient it will be.

Linear dispersion should therefore be effected by having a large angular dispersive power and not by using a long focal length F_2 for the camera. We shall show that a grating has the much larger dispersion and, since the transmission factors of gratings and prisms are nowadays very similar, that thus the use of a grating will give a much greater spectrographic efficiency.

We now compare the dispersive power of a prism with that of a grating. The angular dispersion of the grating is:

$$\frac{\delta D}{\delta \lambda} = K \, . \, n,$$

where K is the order of the spectrum and n the number of lines per centimetre; λ is expressed in centimetres. The angular dispersion of k 60°-prisms is

$$\frac{\delta D}{\delta \lambda} = k \, . \, \frac{2}{n} \, . \, \tan I \, \frac{\delta n}{\delta \lambda},$$

where n is the refractive index of the glass and I the angle of incidence for minimum deviation.

We write:

$$\frac{\delta D}{\delta \lambda} = k \, . \, \nu.$$

* See J. H. MOORE, *Lick Obs. Bull.*, vol. 3, p. 42, 1904.

A prism is therefore equivalent to a grating of ν lines, where

$$\nu = \frac{2}{n} \tan I \, . \frac{\delta n}{\delta \lambda}.$$

The values of ν for a 60°-prism are as follows:

dense flint; H_α . $\nu \sim 1000$ lines per cm
H_γ . $\nu \sim 4000$ lines per cm
quartz; 3500 Å . $\nu \sim 3000$ lines per cm.

The dispersive power is therefore always much smaller than that of a grating of 6000 lines per cm.

3. Comparison between Grating and Prism

We shall now compare the luminous efficiency of a prism spectrograph with that of a grating instrument. In what follows, the subscript 1 refers to the grating and the subscript 2 to the prism spectrograph.

It is reasonable to assume that s_1 and s_2 have the same values, and in this way the efficiency ratio becomes:

$$\frac{e_1}{e_2} = \frac{L_{\xi_1}}{L_{\xi_2}} \cdot \frac{L_{m_1}}{L_{m_2}} \cdot \frac{S_1}{S_2}.$$

The values that will be assumed for S_1 and S_2 are derived from a preliminary study made in connection with spectrographic equipment for the Observatoire de Haute Provence.

Whatever values are taken, the ratio S_1/S_2 is approximately unity when one compares a good modern grating with a prism; it is about 1·5 for the comparison of such a grating with a train of two prisms.

Table 3

Grating

No.	Region	Dispersion (Å per mm)	Order	Kn	S_1	Ω_1	m_1	L_{m1}
1	Hα	60	1	6,000	0·65	4·17	6·0	1·00
2	Hγ	30	2	12,000	0·65	4·17	6·0	1·00
3	Hγ	7·5	3	18,000	0·65	7·42	3·4	0·90
4	3500	30	2	12,000	0·65	5·55	4·5	0·92

Prisms								Comparison		
No.	Material	Number of prisms	$k\nu$	S_2	Ω_2	m_2	L_{m2}	P	$\frac{Kn}{k\nu}$	$P\frac{Kn}{k\nu}$
1	Flint . .	2	2000	0·71	8·33	3·0	0·89	0·97	3·00	2·90
2	Flint . .	2	8000	0·44	4·17	6·0	1·00	1·48	1·50	2·22
3	Flint . .	2	8000	0·44	16·67	1·5	0·82	1·62	2·25	3·64
4	Quartz . .	1	3000	0·83	11·11	2·2	0·86	0·84	4·00	3·36

The efficiencies of grating- and prism-spectrographs are compared in Table 3 for four different cases. The basis of the comparison is as follows:

(1) The grating is taken to be ruled with 6000 lines per cm, and the height of the ruling to be 10 cm. The blaze angle is supposed to be adjusted to the order of diffraction which is used, and the reflectivity is taken to be 0·65.

(2) The prisms are also assumed to be 10 cm high, and their dispersions correspond to the values shown.

The slit is always opened until its image on the plate has a width $s = 25\mu$.

The ninth column of Table 2 gives the value of the product

$$P = \frac{L_{m_1}}{L_{m_2}} \cdot \frac{S_1}{S_2},$$

which has only to be multiplied by

$$R = \frac{L_{\xi_1}}{L_{\xi_2}}$$

in order to obtain the efficiency for given seeing conditions. The variation of this ratio R is fairly easy to understand.

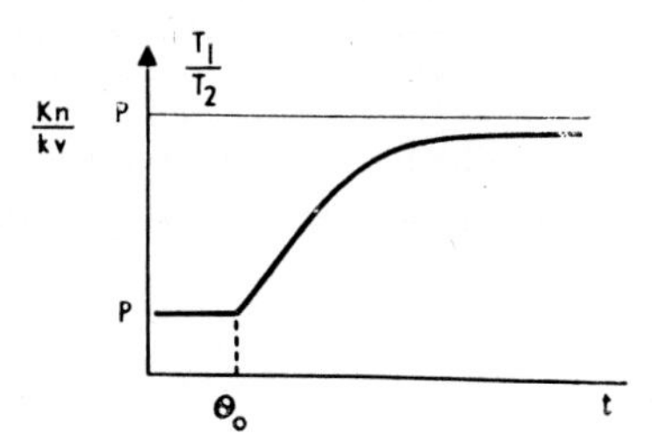

Fig. 5. Ratio of the transmission factor T_1 of a grating to the corresponding value T_2 for a prism, as function of the diameter t of the seeing image

For large values of t, corresponding to small values of ξ, we can develop L_ξ in a power series in ξ:

$$L_{\xi_1} \sim \frac{4}{\pi}\xi_1, \quad L_2 \sim \frac{4}{\pi}\xi_2,$$

so that for the values

$$R = \frac{\xi_1}{\xi_2} = \frac{Kn}{kv}$$

the slit widths l are proportional to the focal length of the camera objective, and therefore inversely proportional to the dispersive power. It is found that for $\xi \rightarrow 0$

$$L = P \,.\, \frac{Kn}{kv}.$$

For small values of t, $\xi > 1$, $L_\xi \simeq 1$, so that for perfect images

$$\frac{L_1}{L_2} = 1, \quad L = P.$$

The variation between these two extremes is shown in Fig. 5.

Fig. 6 shows that under normal conditions of use of a large telescope (turbulence of order 1".5) a grating spectrograph is about three times more efficient than a prism

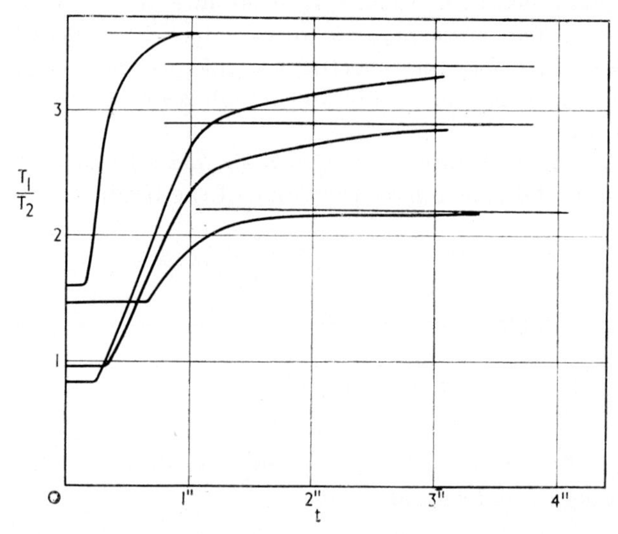

Fig. 6. Comparison of prism- and grating-spectrographs. The co-ordinates are the same as in Fig. 5 and the individual curves refer to the four cases shown in Table 3

instrument. The ratio is so large that small changes in the values of the transmissions of the prisms do not affect the conclusions.

Optical Systems Employed in Spectroheliographs

P. A. Wayman

Royal Greenwich Observatory, Herstmonceux Castle, Hailsham, Sussex

Summary

A survey is made of the properties of optical systems which have been used in spectroheliographs since the inception of that instrument. Particular attention is drawn to those systems in which waste of light is minimized. A method of using a field-lens with a large horizontal spectrograph in an underground chamber is suggested as the solution to the problem of obtaining uniform illumination along the slit and as the scanning motion proceeds.

The spectroheliograph is a combination of two optical instruments, a telescope and a spectrograph. Between the two there is arranged a smooth continuous scanning motion, and an image of the Sun's disk, or a portion of it, in "monochromatic" light, can be built up on a photographic emulsion. The earliest suggestion for an instrument of this type was made by Janssen in 1869 and in the next two decades a number of unsuccessful attempts at building one, especially by Lohse at Potsdam, were made. The first successful construction was evidently that of Hale in 1891, working at Harvard. His success was due, according to himself, to his use of the broad H and K lines of ionized calcium, hardly visible to the eye, but very suitable for the photographic emulsion.

Independently in the years 1891–93, DESLANDRES (1893), in France, and EVERSHED, in England, both constructed successful spectroheliographs. By 1906 small spectroheliographs had been in use continuously for the whole of an eleven-year sunspot cycle, the most complete record being that provided by DESLANDRES' first permanent instrument. In 1906 the Meudon Observatory was planning notable extensions of i facilities for solar physics, while in the United States, the large Snow Telescope had been moved, under HALE's guidance, from Yerkes Observatory to Mount Wilson, and the construction of the first solar tower was being contemplated. At Yerkes HALE left behind the Rumford spectroheliograph of 1899 on the 40 in. telescope and it continued in use there for many years.

1. DESLANDRES' CLASSIFICATION

DESLANDRES (1907) summarized the many possible arrangements for the parts of a spectroheliograph. He enumerated five chief methods for obtaining the scanning motion, as follows.

(1) The diurnal motion of the Sun can be utilized. This was done by EVERSHED in 1892 using a direct-vision spectroscope with two slits and a moving photographic plate, the whole being mounted on an equatorial refractor.

(2) A spectrograph is fixed to an equatorial refractor which is driven at a rate differing from the solar diurnal rate, or is moved in declination, whilst the photographic plate is moved across the second slit at the rate appropriate for obtaining a round image. This arrangement was used on the Rumford spectroheliograph at Yerkes which had its movement arranged in the direction of declination.

(3) A fixed spectrograph is used with a coelostat or siderostat and the objective is moved in its own plane while the photographic plate is moved separately across the second slit of the spectrograph. Up to 1905 this method had not been used but DESLANDRES incorporated it in the new instruments at Meudon with considerable ensuing success and it was also used in the 60-ft tower telescope at Mount Wilson. Many of the finest spectroheliograms in existence have been taken with these two instruments.

(4) The objective and the photographic plate are fixed and so is the main part of the spectrograph but the slits are moved laterally across the optical beam. This arrangement was used by HALE in the first spectroheliograph to be mounted on the Kenwood refractor. Substantially the same system was used in the Mount Wilson 150-ft Solar Tower when arranged as a spectroheliograph.

(5) The whole spectrograph moves while the objective remains stationary. DESLANDRES subdivided this class into the following: (*a*) those cases in which the photographic plate moves independently across the second slit, as in an instrument of his own of 1894, (*b*) those using a rotating direct-vision spectrograph (LOHSE's attempts about 1885 to photograph prominences), (*c*) those consisting of a spectrograph of normal type turning about an axis perpendicular to its own plane and passing through the intersection of the axes of collimator and camera (adopted by EVERSHED (1898), and the basis of a suggestion by STEFANIK (1907)), (*d*) those cases in which the spectrograph is arranged so that the axes of collimator and camera are parallel and the lenses are of equal focal length so that the photographic plate can remain fixed and the second slit can move across it. DESLANDRES' instrument of

1893–1906 was of this kind and so was one of HALE'S as built by TOEPFER of Potsdam in 1893. Subsequently, probably on HALE's recommendation, this type, which he called the "automatic" spectroheliograph, became the most popular. Examples were installed in the Snow Telescope at Mount Wilson, at the Solar Physics Observatory, first in London and later in Cambridge, and at the Kodaikanal Observatory in India.

2. MOUNT WILSON SPECTROHELIOGRAPHS

In his re-erection of the Snow Telescope at Mount Wilson in the years 1904–6, HALE endeavoured to use the results of all his previous experience at Kenwood and Williams Bay. In their account of the Rumford spectroheliograph HALE and ELLERMAN (1903) stated that the best form of spectroheliograph was that in which the photographic plate and objective are fixed and the spectrograph moves bodily

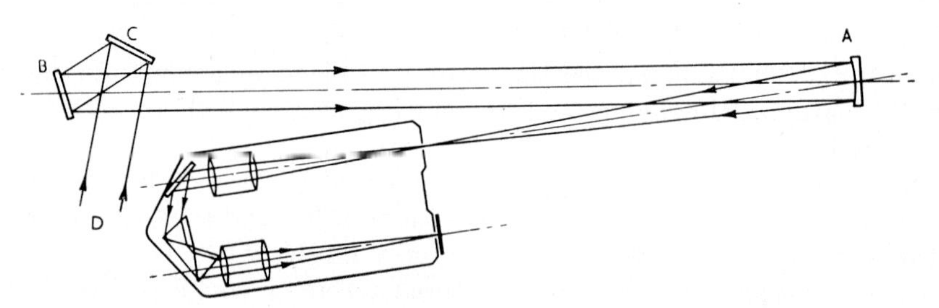

Fig. 1. The five-foot spectroheliograph of the Snow Telescope

(type 5 (*d*) above) but that this arrangement was not possible with a large spectrograph on the 40-in. Yerkes refractor because of the vibration which would be caused by the heavy moving parts. Accordingly, the Rumford spectroheliograph was of type 2 in DESLANDRES' classification. The lenses of the spectrograph were only of $6\frac{1}{4}$ in. aperture, whereas 10-in. lenses were really needed if there was to be no loss of light at the limb of the 7-in. image given by the 40-in. telescope. Furthermore, the spectrum lines were highly curved.

After some preliminary experiments on Mount Wilson with a temporary instrument the principal spectroheliograph of the Snow Telescope was one of type 5 (*d*) to be used with a 24-in. mirror of 60-ft focal length, having collimator and camera lenses of 5-ft focal length and 8-in. aperture (HALE, 1906). These lenses were of "portrait type", as used in the Rumford spectroheliograph. Fig. 1 (not-to-scale) shows the optical system of this spectroheliograph, when used with two prisms. The pencil of light from any one point of the Sun's disk, when intercepting the collimator lens, would be approximately 2 in. in diameter. The size of the collimator lenses permitted pencils from any point of the 6·7-in. solar image to be accepted without very much vignetting; this had not been so in the case of the Rumford spectroheliograph. However, quite a lot of light was lost at the prisms which, although they were 8 in. in height, had faces only 4 in. broad.

In its more usual form with two prisms but also when used with four prisms, the spectrograph used a single reflection in its optical train. In this instrument and in many others like it, this was incorporated so that, by using two slits of equal curvature, a distortion-free image of the Sun could be built up in spite of the strong curvature imparted to the spectrum lines by a prism-train. This was possible in any spectrograph of this type which had an odd number of reflections in its optical train.

(Figs. 2 (*a*) and 2 (*b*).) With an even number of reflections (or none), however, the situation would be as represented by Figs. 2 (*c*) and 2 (*d*). In order to build up a distortion-free image by using two slits of equal curvature as in Fig. 2 (*d*) the photographic plate would have to move over the second slit in the direction opposite to that resulting from having a stationary photographic plate and a moving spectrograph. Thus the simplicity of having only one moving part would have to be sacrificed in order to obtain a distortion-free image. However, as NEWALL originally pointed out, the presence of an even number of reflections is more favourable to the

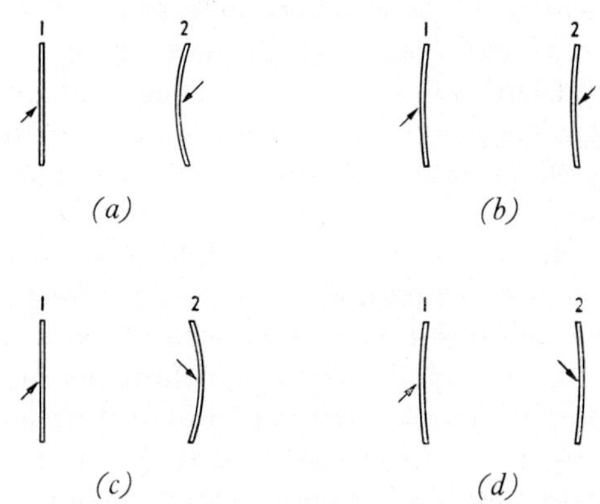

Fig. 2. Curved slits: (*a*) and (*b*) with an odd number of reflections in the spectrograph, (*c*) and (*d*) with an even number of reflections

definition of the image, especially in the case of a wide slit. For, in the cases (*a*) and (*b*) of Fig. 2, as indicated by the arrows which point to corresponding sides of the two slits in the case an odd number of reflections, the image is built up *in reverse*. Hence each point of the solar image, in perfect monochromatic light, would be drawn out, on the photographic plate, into a short line parallel to the dispersion and of length equal to the sum of the two slit-widths. In the case of heterochromatic light a spreading equal to the width of the second slit cannot be avoided but if wide slits are used the extra loss of definition arising from the use of an odd number of reflections may be objectionable. Therefore, HALE made it possible to use two mirrors in the optical train when two prisms were being used sacrificing the distortion-free quality which could be obtained under a single reflection. It is evident that in a spectroheliograph of this sort many different slits were needed, as the curvature is dependent on the wavelength as well as on the optical arrangement. Furthermore, it can be seen that it is not possible *in any system* to obtain a distortion-free image with curved slits which satisfies NEWALL's condition for obtaining the best definition.

It was originally intended, on the Snow Telescope, to incorporate, in addition, a large spectroheliograph using a 17-in. solar image and a spectrograph with a focal length of 30 ft. Preliminary experiments with a temporary spectroheliograph had convinced Hale of the value of larger dispersive power than he had been using before. However, considerable delay in obtaining a suitable train of prisms for this instrument was experienced and in the meantime it was found that an image sufficiently stable for long exposures could not be obtained with an all-mirror telescope because of the rapid change of figure with solar heating. It was accordingly decided to construct a tower 60 ft high, to use a coelostat with glass mirrors 12 in. thick and to use an achromatic lens to obtain the primary image (HALE, 1907). In an underground chamber space was provided for a vertical 30-ft grating spectrograph of Littrow form as well as for the 30-ft spectroheliograph, also vertical, using a train of prisms. The former instrument was used for many of HALE's classic researches on the solar rotation and solar magnetic fields but it seems that the spectroheliograph was only used with a liquid prism (HALE and NICHOLSON, 1938) and was replaced in 1915 by a spectroheliograph of 13-ft focal length. However, the Littrow spectrograph was provided with a spectroheliograph "attachment" and the instrument was used in this form by HALE for obtaining spectroheliograms with iron and hydrogen lines. In this connection HALE (1908) stated that "a large spectroheliograph of the Littrow form will give excellent results", even though previously he had, owing to the increased scattered light, been unsuccessful with an arrangement suggested by NEWALL (1896) incorporating a small Littrow spectrograph suspended on wires and made to swing across the solar image many times—thus "ironing out" variations due to changes in the transparency of the sky.

The spectroheliograph of the 60-ft tower telescope employed a movement of the objective laterally across the parallel beam and a corresponding movement of the photographic plate, the connection being by means of a rotating shaft passing up the tower and the power being supplied by a synchronous motor. The 150-ft tower telescope, built at Mount Wilson in 1910, was provided with a 75-ft vertical spectrograph which could be used as a spectroheliograph, but in this case two moving slits were used and it was possible to cover only a small portion of the 17-in. solar image. The objective and the combined collimator and camera lens were (HALE and NICHOLSON, 1938) of the same focal-ratio—$f/150$. This high value of the focal-ratio precluded the use of very narrow slits, since the exposure-times would then be inordinately long.

In the Solar Laboratory at Pasadena (HALE, 1929) the combined spectrohelioscope and spectroheliograph was an all-mirror design. The spectroheliograph employed an optical system similar to the familiar arrangement used in the Hale spectrohelioscope (in both cases the spectrograph-focal-lengths being 13 ft) and the primary image, obtained from a fixed Cassegrain telescope was moved across the first slit by means of an inverting prism (similar in principle to the Dove prism), moving at half the speed at which the photographic plate moved across the second slit. This arrangement minimized the weight of the moving parts and resulted in a spectroheliograph which could conveniently be used in conjunction with the spectrohelioscope.

3. MEUDON SOLAR OBSERVATORY

Simultaneous with HALE's work at Mount Wilson was the expansion of the Meudon Observatory under the supervision of DESLANDRES. Like HALE, DESLANDRES (1910) had foreseen that a large spectrograph ("with a dispersion at least equal to that of the classical apparatus of ROWLAND") would be of great power in a spectroheliograph. He started to construct such an instrument in 1906 but it was not until 1909 that it reached its final form. As finished, it was a very flexible and versatile instrument and it incorporated many special features originated by DESLANDRES. Principal among these were the following:

(1) The instrument was, in reality, a combination of four spectroheliographs with common objective, first slit and collimator-lens.

(2) Either a grating or a train of prisms could be used at a variety of dispersions.

(3) Mirrors were used in the system of highest dispersion in order to reduce the overall length of the instrument.

(4) The grating or prism-train was separated from the collimator by a distance slightly greater than the focal length of the latter. This condition was more favourable to the inclusion of light from off-axis image-points. (See DESLANDRES (1910), p. 32.)

(5) The scanning motion was obtained, in each case, by moving the objective and the photographic plate across the optical beam by separate synchronous motors and variable reduction-gears.

(6) The largest system incorporated, beyond the second slit, a second spectrograph and a third slit in order that, firstly, the relative intensity of the diffuse light might be reduced and, secondly, a series of camera-focal-lengths could be used (resulting in a permissible reduction of exposure-time with the smaller images) even though the selecting (second) slit remained the same.

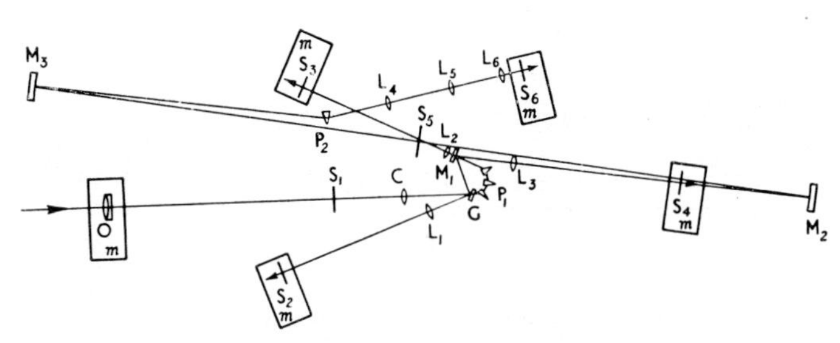

Fig. 3. The large Meudon spectroheliograph of 1909

O = objective; S_1 = first slit; C = collimator; S_2, S_3, S_4, S_5 = second slits; S_6 = third slit; G = grating; P_1 = prism-train; P_2 = supplementary prism; M_1 = plane mirror; M_2, M_3 = concave mirrors (7 m focal lengths); L_1, L_2, L_3, L_4, L_5, L_6 = camera lenses; m = motors

Grating spectroheliograph: $OS_1CGL_1S_2$

Prism spectroheliograph: $OS_1CP_1L_2S_3$

Prism or grating spectroheliograph: OS_1CG (or P_1) $M_1L_3S_4$

Three-slit spectroheliograph with prisms or grating and a variety of camera lenses: OS_1CG (or P_1) $M_1M_2S_5M_3P_2L_4$ (or L_5 or L_6) S_6

Fig. 3 shows the optical arrangement with the different possibilities explained in the caption. This versatile instrument has been in use for many years (D'AZAMBUJA, 1948).

Most other spectroheliographs are similar to those already discussed, although the solar tower at Arcetri employs a vertical spectrograph which can be made to rock about its bottom point, in order to scan the solar disk. If, as in the 150-ft tower of Mount Wilson and the large spectroheliographs of the McMath-Hulbert Observatory*, the slits move across the axes of collimator and camera, there are certain non-linearities in the reflection at the grating which have the result that the proper separation (at the top of the spectrograph) of the two slits is not constant. This defect is much more marked in the case of a train of prisms.

4. A LARGE HORIZONTAL SPECTROHELIOGRAPH

It is not usual for spectroheliographs to provide quantitative measurements of light intensity, although some years ago some experiments in this direction were made at Cambridge (DOBBIE, MOSS, and THACKERAY, 1937) and more recently the instrument at Arcetri has been used photometrically (BALLARIO, 1950). One reason for

* See DIMITROFF and BAKER (1945) for a general description of the extensive solar installation at the McMath-Hulbert Observatory. See also MCMATH (1939).

this is that the total exposure-time for a spectroheliograph is large compared with the exposure-time for each element, and the conditions may change appreciably between the exposure of different portions. Nevertheless, under good atmospheric and instrumental conditions extremely uniform intensities can be obtained. A second factor, however, is that in most spectroheliographs there is a considerable amount of vignetting, both pertaining to points of the slit removed from the central point and to off-axis portions of the solar image, successively exposed as the scanning motion proceeds. HALE's 5-ft spectroheliograph went some way towards obviating this vignetting by using camera lenses whose focal-ratio was some four times smaller than that of the primary beam. This could not be used to full advantage, however,

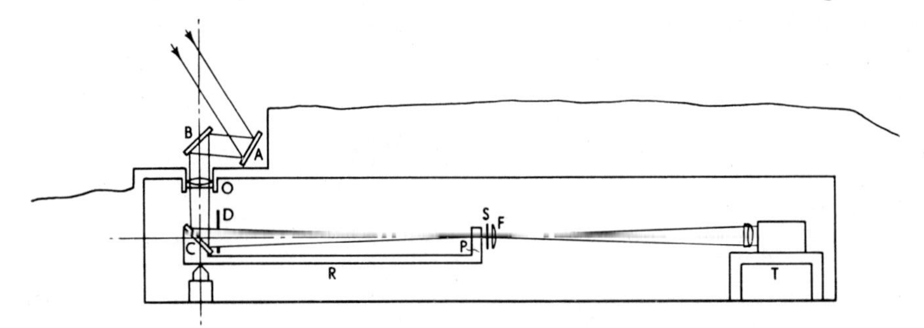

Fig. 4. An underground horizontal spectroheliograph with a field-lens

because of the impossibility of obtaining dispersive elements allowing the use of a circular 8-in. beam. There is clearly a "wastage" of optical elements in any arrangement of this kind and in fact most spectroheliographs have been built with equal focal-ratios in the objective and collimator-lens. DESLANDRES adopted the solution of separating the collimator from the dispersive elements so that the former produced an image of the objective, or entry-pupil of the system, onto the dispersive elements. DESLANDRES used this arrangement to advantage, especially since it made possible the three-slit spectroheliograph, but in the case of a collimator of large focal length, the total size of such a system would be excessive. Furthermore, in the case of a Littrow arrangement, as used by EVERSHED (1911) and ROYDS at Kodaikanal and by EVERSHED (1928) at Ewhurst, it would be impossible to use the same lens as both collimator and camera lens if it were separated so far from the dispersive elements. A suggested solution to this problem is the employment of a field-lens to image the entry-pupil onto the central plane of the dispersive elements where the clear aperture is restricted. A minimum wastage of light is thereby ensured.

Fig. 4 shows a suggested arrangement for an underground horizontal spectroheliograph. The entry-pupil is defined by a diaphragm D which is imaged onto the back-surface of a reflecting train of prisms (or onto the surface of a grating) on the spectrograph-table T by a field-lens F. The arrangement is of the Littrow or autocollimating form and the second slit is parallel to the first and is just beside it. The scanning motion is obtained by rotating the mirror C about a vertical axis of rotation. This motion could be imparted by the rigid frame R, at the further end of which is carried the photographic plate P. One advantage of this arrangement is that of HALE's "automatic" spectroheliograph, in that there is only one moving part, without the disadvantage, in that instrument, of the presence of a heavy moving spectrograph. There would be a very nearly equal exposure given to each part of the solar image, provided the motion were uniform, whilst the clear aperture of the objective need be only very slightly larger than the aperture actually employed for each part of the image. Provided the frame R is of sufficient length the transverse motion at the second slit would not involve any appreciable change of focus and, subject to the same condition, the presence of a slightly rotating image will give rise to only very small distortions in the built-up image. Normally distortions would not exceed 1/150 of the solar radius.

The usefulness of a field-lens in the above arrangement can be clearly seen. In a tower-telescope spectroheliograph of the type at Arcetri a stationary field-lens mounted immediately above the moving first slit would be equally beneficial in preventing the wastage of light.

REFERENCES

BALLARIO, M. C.	1950	*Oss. e Mem. Arcetri*, **66,** 15.
D'AZAMBUJA, M. and Mme. L.	1948	*Ann. Obs. Meudon*, **VI** (7), 1.
DESLANDRES, H.	1893	*C.R. Acad. Sci.* (*Paris*), **116,** 238.
	1907	*Ann. Obs. Meudon*, **III** (2), 94.
	1910	*Ibid.*, **iv** (1), 16.
DOBBIE, J. C., MOSS, W. and THACKERAY, A. D.	1937	*M.N.*, **98,** 606.
DIMITROFF, G. Z. and BAKER, J. G.	1945	*Telescopes and Accessories*, Blakiston, p. 201.
EVERSHED, J.	1911	*M.N.*, **71,** 719.
	1928	*M.N.*, **89,** 175.
HALE, G. E.	1906	*Ap. J.*, **23,** 54. Mt. Wilson Contrib. No. 7.
	1907	*Ibid.*, **25,** 68. Mt. Wilson Contrib. No. 14.
	1908	*Ibid.*, **27,** 1. Mt. Wilson Contrib. No. 23.
	1929	*Ibid.*, **70,** 265. Mt. Wilson Contrib. No. 288.
HALE, G. E. and ELLERMAN, F.	1903	*Publ. Yerkes Obs.*, **III** (1).
HALE, G. E. and NICHOLSON, S. B.	1938	"Magnetic Observations of Sunspots, 1917–24". Carnegie Inst. of Washington Publ. No. 498, Part 1 (= Papers of the Mt. Wilson Obs., vol. **V,** Pt. 1, 1938).
McMATH, R. R.	1939	*Publ. Univ. Obs. Michigan*, **7,** 1.
NEWALL, H. F.	1896	*Proc. Cambridge Phil. Soc.*, **9,** 179.
STEFANIK, P.	1907	*Ann. Obs. Meudon*, **III** (2), 38.

The Solar Installation of Dunsink Observatory

H. A. Brück and Mary T. Brück
Dunsink Observatory, Co. Dublin, Ireland

The Dunsink Observatory was founded in 1783 as a Department of Trinity College, Dublin. After 1921 it served instructional purposes only, and it had been closed altogether for over ten years when in 1947 it was acquired by the Irish Government and re-opened under the auspices of the Dublin Institute for Advanced Studies.

Considerations of climatic and other conditions suggested solar research to be a suitable field of work at the reconstituted observatory. It was decided therefore at the time of re-opening that a combined solar telescope and spectroscope should form part of its new equipment.

In order to allow observation to be extended to the ultra-violet region of the spectrum the solar installation was designed so as to consist of a mirror telescope, fed by a coelostat, and a concave grating spectrograph. It was decided to mount both the telescope and the spectrograph inside the former Meridian Pavilion, a rectangular room, 12 m long, 7 m wide, and, at the apex of its roof, 8 m high. One of the two instruments which had formerly been installed in this room, a famous 8 ft Ramsden circle, now only of historical interest, was dismantled. The two strong piers of lime stone, standing 4 m above the level of the floor, which had supported that circle, and which rest on a solid foundation of rock, were used to carry part of the new solar telescope. This was arranged as a vertical instrument which receives its light from a coelostat, mounted on a platform above the roof of the Meridian Room. This platform, 3 m wide and 6 m long, is at a height of 8·5 m above the ground.

1. The Coelostat

The primary and secondary mirrors of the coelostat are 40·5 cm in diameter. The present mirrors are of glass, but will be replaced by mirrors of fused silica.

The mounting of the primary mirror can be moved east and west up to 1 m off the line of the meridian and over a range of 60 cm north and south. The primary is driven through a worm drive by a synchronous motor. The secondary mirror can be raised or lowered over a range of 1 m and can be moved in azimuth and altitude. The slow motions are operated by electric motors which may be controlled from any position near the slit-end of the spectrograph.

The platform, built on a structure of strong and suitably braced steel girders, has proved to be hardly susceptible to vibration. Except at times of strong wind the definition of the solar image has been found to be determined entirely by the state of seeing. Fig. 1 shows a picture of the platform with the mirrors of the coelostat.

2. The Solar Telescope

The secondary mirror of the coelostat reflects sunlight vertically downwards into the Meridian Room and into the solar telescope, whose main mirror in its cell rests on a solid rock foundation at floor level between the piers of the former Ramsden circle.

The telescope has been designed as an off-axis Cassegrain in which the secondary convex mirror, instead of being vertically above the primary, is slightly shifted to one side of the incoming beam from the coelostat.

A concave mirror whose full aperture is 38 cm and whose focal length is 8·8 m, and which was available when the instrument was first set up, has served as primary up to the present. In combination with a convex mirror of 20 cm aperture and 3·7 m focal length it has an effective focal length of 26·4 m and an aperture ratio of $f/69$, equal to that originally intended for the spectrograph. To fill the 6-in. grating of the

Fig. 1. Platform with coelostat

actual spectrograph eventually installed (aperture ratio $f/47$) the telescope has had to be used as an off-axis Newtonian instrument with a flat replacing the convex mirror.

This arrangement has only been provisional, however, and new concave and convex mirrors are to be installed shortly, whose combined focal ratio matches exactly that of the spectrograph. The original telescope had suffered also from the disadvantage that the primary glass mirror was not of the highest optical quality, and that its shape was that of an ordinary paraboloid (symmetrical about its centre) which, when used obliquely, gives rise to well-known optical aberrations which could only partly be corrected.

The new telescope whose concave and convex mirrors have been made by Messrs. Cox, Hargreaves and Thomson, Ltd., consists of a primary with an aperture of 35 cm and focal length of 6·8 m, to be mounted in the same position at floor level as its predecessor. This primary is to be tilted about $1\frac{1}{2}°$ to the north against the vertical so as to deflect the beam onto the secondary convex of 19 cm aperture which is to be mounted—as in the provisional instrument—face downwards, at a height of about 5 m above floor level. The centre of the secondary will be displaced to the north of the centre of the primary by a distance of 27 cm. Focusing will be done as before by sliding the secondary along a pair of vertical steel bars which are fixed to the two piers mentioned earlier. This part of the telescope mounting can be seen in the upper part of Fig. 2.

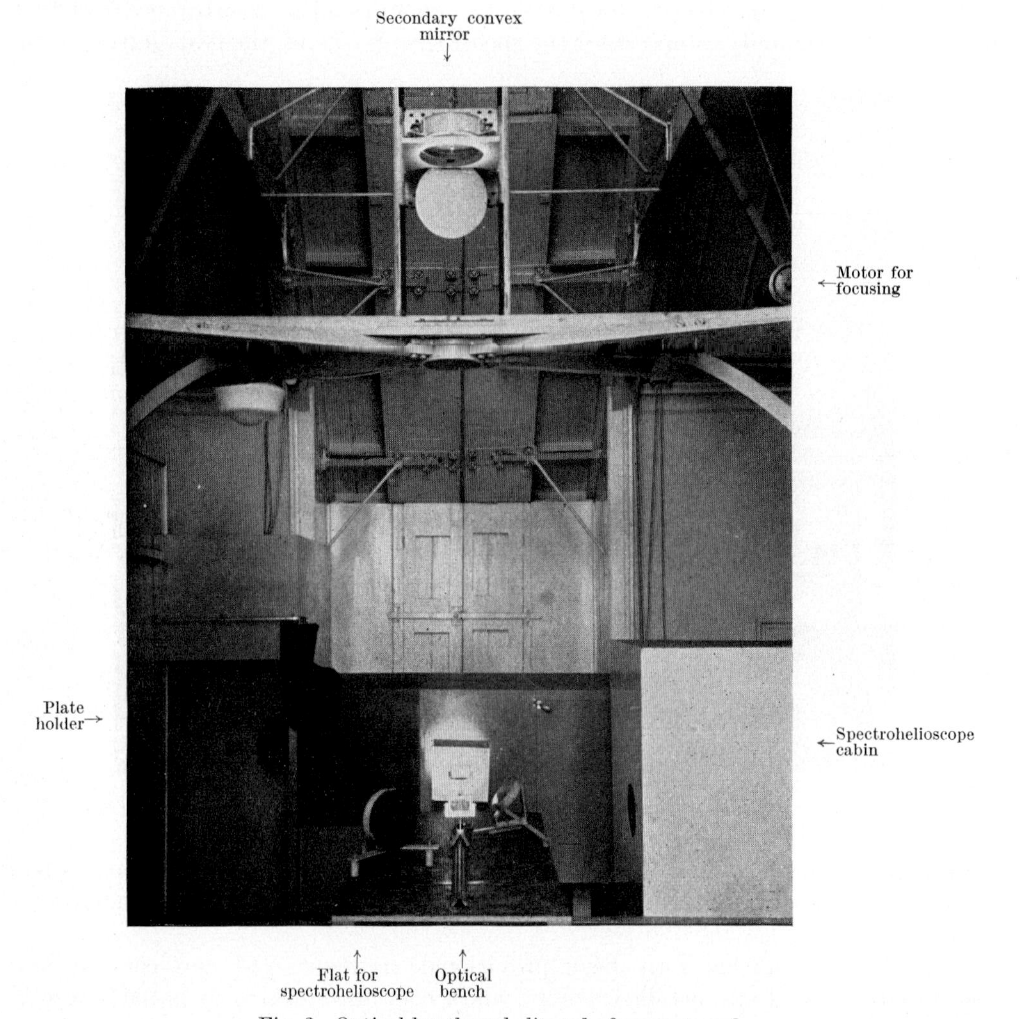

Fig. 2. Optical bench and slit end of spectrograph

The beam coming vertically downwards from the secondary will be deflected by means of a Coudé flat mirror into a horizontal direction and onto the slit of the spectrograph. In front of the slit there is mounted an optical bench on a table which is supported by a rigid girder structure fixed to the piers of the former Ramsden circle. This optical bench, which is nearly 2 m long, allows of ready alignment of optical and other accessories.

All three mirrors in the new telescope are of quartz, and both concave and convex have been made exactly spherical. An ideal instrument of this type should use a primary in the form of a prolate ellipsoid if off-axis aberrations are to be avoided. It is well known, however, that for a mirror of as large a focal ratio as that of the new primary ($f/19$) the oblique aberrations may be considered as pure astigmatism. In designing our telescope, Mr. HARGREAVES has pointed out that, in the event of both component mirrors being truly spherical, such astigmatism may be corrected by having also a slight tilt on the secondary so as to introduce an astigmatism equal in amount, but opposite in sign to that caused by the primary. This method of correction has been in use already with our provisional telescope where it has proved as satisfactory as could be expected, having regard for the imperfections, mentioned earlier, of the primary mirror. Mr. HARGREAVES reports that tests of the new system of mirrors made in his laboratory fully confirm his prediction of the high optical performance obtainable from this type of off-axis arrangement.

3. THE SOLAR SPECTROGRAPH

In the Dunsink spectroscope a 6-in. 7 m concave grating is used in an Eagle mounting. This means that slit and plate-holder are at fixed positions relative to each other. Different regions of the spectrum are thrown on to the plate by rotating the grating, and brought into correct focus by moving the grating along the optical axis of the spectroscope. The plate has to be set along the Rowland circle which is done by rotating the plate-holder around its central axis. The Eagle mounting has the advantage, apart from its compactness, of allowing the use of the higher orders of the grating and of being less affected by astigmatism than, say, the Rowland mounting.

The spectrograph is mounted horizontally on a strong steel girder which rests on two concrete piers in the northern half of the Meridian Room. It is housed within a doubled-walled wooden case, a metre high, more than a metre wide, and 8·5 m long. The case is carefully insulated against variations of outside temperature, and the temperature inside is kept constant to within less than 0·1°C by a thermostat which uses a platinum resistance thermometer and which has been supplied by the Cambridge Instrument Company. In order to reduce scattered light to a minimum, five rectangular diaphragms have been inserted into the case at regular intervals, their dimensions being only just sufficient to pass the main beam to and from the grating.

The slit, produced by Messrs. Adam Hilger, Ltd., which is mounted at the side of the spectrograph, is 2 cm long. It opens symmetrically, the pitch of the screw being 0·5 mm. The slit is provided with a mounting which can be rotated through small angles. The slit unit is mounted on an optical bench, 90 cm long, part of which extends into the spectrograph case, and which also carries the mounting for a small quartz right-angle prism which throws the beam of light from the slit on to the grating. This prism, made also by Messrs. Hilger, is situated on the central axis of the spectrograph case, and its mounting is provided with screws for fine adjustment.

The optical bench carrying the slit is optically aligned with the 2-m bench which, as has been mentioned before, is mounted outside and independently of the spectrograph. Such an arrangement is particularly useful in the case of a normally astigmatic spectrograph such as this one, when stigmatic images are required. It is well known that such images can be obtained of objects which are placed at appropriate distances outside the slit. The positions of these "Sirks points" for various regions of the spectrum and for different orders can be readily marked along the optical bench, and the diaphragm for observations of solar detail or a step-filter can thus be placed without difficulty.

The grating used in this spectrograph was obtained from the Physics Department of the John Hopkins University, Baltimore. It is a 6-in. grating, ruled with 15,000 lines per inch on an aluminized "pyrex" blank, with a radius of curvature of 7 m. The grating is mounted within a cell which can be rotated in its own plane. The grating holder is also provided with an adjustment for tilt so as to make the spectrum fall correctly on to the plate in front of which there is a diaphragm, 15 mm high, and with its centre 2 cm below the level of the centres of slit and grating.

The grating holder is mounted on a carriage which for the purpose of wavelength selection can be turned and for that of focussing can be moved along the axis of the spectrograph. The rotation is done through a worm drive; the focusing is achieved through the use of a precision screw, with a pitch of 2·5 mm, made of a steel bar, 2 m long, which drives the grating carriage along a machined bed. Both movements are controlled electrically from the plate-holder end of the spectrograph, where also are mounted two telescopes by means of which the position of the grating can be read off a millimetre scale for the focussing and a graduated circle for the rotation. The focusing can be set to an accuracy of a quarter of a millimetre; rotation positions can be reproduced within 1′ of arc which, when converted to wavelengths, corresponds to about 10 Å in a second-order spectrum.

The turn-table on which the grating holder is mounted, the machined slide along which it travels and the screw movements have been manufactured by Messrs. Booth Bros., Ltd., Dublin.

The plate holder is attached to a base which can be tilted against the axis of the spectrograph by rotating it around its central vertical axis. In order to exclude false light from the spectrograph the base is in the form of a semicylindrical box which is light-tight with the exception of two slots, 15 mm high and about 80 cm long, through which the spectrum coming from the grating falls on the plate-holder. Part of the plate-holder and base can be seen in Fig. 2. Great care has been taken to make this part of the spectrograph particularly rigid. The tilt of the plate-holder and base can be read off a graduated circle.

The plate-holder can take plates as large as 4 in. × 30 in. (10 cm × 75 cm), which in the second-order cover a range of 875 Å. Standard 4 in. × 10 in. plates are used normally. In producing the plate-holder which was made of mahogany, great attention was paid to shaping its interior in such a way that the photographic plates are accurately curved to the Rowland circle. The plate-holder can be readily raised or lowered so that a number of spectra can be secured on the same plate.

The spectrograph and accessories with the exception of the optics and the parts mentioned earlier have been constructed at the observatory. The same applies to the solar telescope with the exception of the mounting of the primary mirror of the coelostat.

4. The Performance of the Spectrograph

The grating shows relatively bright second- and third-order spectra on one side of the central image. The tests of the performance of the spectrograph which are to be described here, all refer to the blue and violet region of the second-order spectrum where the dispersion is equal to 1·15 Å per mm.

(*a*) *Ghosts*—The ghosts of the Cd line λ4799·9 were photographed with exposure times of up to 3 hr, using a Cd spectrum lamp of the Philips Electrical Co. On long exposure photographs five ghosts can be detected on the violet and four on the red side of the main line, but only the two pairs next to this line are of any appreciable intensity. Even these are so weak that, to compare them with the main line, it was necessary to reduce the intensity of this line by a suitable screen, made up of two superimposed filters, which was placed in front of the plate. The transmission of the screen at λ4800, 0·12 per cent, was determined carefully from photo-electric measurements and also from photographic comparisons with the transmission of a calibrated Hilger step wedge. The plates were standardized, using identical exposure times as those on the ghosts, in a Hilger spectrograph with a step filter in front of the slit. Their densities on the plate were recorded with the Moll Recording Microphotometer of the observatory.

The photometric reduction gave the following values for the percentage intensities of the various ghosts which are displaced by multiples of 3·06 Å from the main line—corresponding to 2·5 Å at λ4000—in terms of the intensity of the main line:

Intensities of Rowland Ghosts
(main line = 100)

	1	2	3	4	5
Violet .	0·14	0·08	$<$ 0·01	$<$ 0·01	$\ll$0·01
Red .	0·13	0·07	$<$ 0·01	$\ll$0·01	—

The total intensity of these ghosts, 0·45 per cent of that of the main line, corresponding to an intensity of 1 per mille in the first order, is satisfactorily small.

(*b*) *Scattered Light*—It has been mentioned before that considerable precautions have been taken to reduce by means of diaphragms and other light traps scattered light within the spectrograph to a minimum. The diaphragms had to exclude from the plate also faint ghost images, 1 in. above and below the focal plane, which are formed by a certain amount of chatter inherent in the grating.

The effect of scattered light on the spectrograms will depend on such conditions as the wavelength region under observation, the order of the grating, the colour sensitivity of the plate, and the transmission of possible colour filters. Estimates of the magnitude of scattered light are provided by two tests which were carried out with sun light in the blue region of the spectrum near λ4300 and in the second order, using blue sensitive (Kodak B4 Half Tone) plates and no colour filter.

The spectrograph slit was masked down to a height of 2 mm. Under the given conditions the spectrum is broadened by astigmatism to a width of about 7 mm. A rough indication of the intensity of the scattered light as observed above and below the actual spectrum is given by the fact that in order to show it up at all on the photographic plate, exposure times had to be given about hundred times greater than those required to produce a well-exposed spectrum of the Sun.

In order to compare the intensity of the scattered light with that of the actual spectrum, a portion of the spectrum was reduced in brightness by placing in front of the plate the same calibrated screen, transmitting 0·12 per cent, which has been used for the measurement of ghosts. To avoid unduly long exposures, the slit of the spectrograph was widened to 0·30 mm, twelve times its normal width, and exposures of 20 min were given. The plates were standardized, using the same exposures, in a Hilger standardizing spectrograph with a step filter.

As can be expected, the photographic density on the plate is greatest near the edges of the spectrum. It reaches a sensibly constant value at distances of 3·5 mm from those edges, and at these distances the photometric reduction of the plates shows the scattered light to amount to 0·1 per cent of the intensity of the spectrum.

The full length of the spectrograph slit of 20 mm was used in a second experiment. In this case a secondary slit, 4 mm wide and 10 mm high, was placed on the optical bench in front of the spectrograph and one metre away from the slit, this being the point of which stigmatic images can be obtained for the region near λ4300 in the second order. Horizontally across the centre of the secondary slit was fixed an opaque bar, 5 mm high, with the result that the spectrogram was cut across the centre by a 5-mm gap. The photographic density within this gap, produced by scattered light, was compared as before with that of the actual spectrum by using suitable screens in front of the plates. With a slit width of 0·05 mm, exposures times of 2 hr were required on the Sun to reveal the scattered light. The photometric reduction of the plates which were standardized in the same way as before, yielded values for the intensity of the scattered light lying between 0·2 and 0·3 per cent of that of the spectrum. It should be mentioned here that to avoid stray light the spectroscope and secondary slits were connected in this experiment by a light-tight telescopic tube such as has been used whenever stigmatic spectra have been taken.

(c) *Resolving Power*—In a search for possible defective areas of the grating such as might impair its resolving power, numerous exposures were taken of the solar spectrum and of spectra of artificial light sources in which different areas of the grating were masked off systematically. With the exception of the very edges of the grating no imperfections could be detected, and as a result of those experiments the grating is now used with a rectangular diaphragm which masks only about one-tenth of an inch of the rulings all round. The effective area of the grating is therefore equal to 5·3 × 2 in. (13·5 × 5 cm).

The theoretical resolving power of the grating in the second order should be 160,000. With an aperture ratio of the spectrograph of $f/47$ a monochromatic line at λ4300 has a theoretical half-value width of 0·022 mm, provided the slit width is set at the van Cittert value for this region (0·020 mm).

In order to determine the instrumental profile experimentally photographs were taken with a Philips spectral lamp of the Kr line λ4319·6 whose half-value width is less than 0·01 Å. The grating was used in the second order, and the slit set at its normal width. On Kodak B4 plates exposure times of up to 30 hr were necessary to reveal the wings of the line, and small mechanical shifts, which tended to broaden the line, could not be entirely avoided. On Ilford Special Rapid plates, however, exposure times of 30 min were sufficient to show the profile to a distance of 0·025 mm from the centre of the line. On the three best plates which were standardized with the Hilger spectrograph and step wedge, and whose densities were recorded by the microphotometer, the profile can be traced accurately between the line centre and points at which the intensity has fallen to about 30 per cent of its central value.

The wings of the line are asymmetrical to the extent that at an intensity of 30 per cent the violet wing is about 5 per cent wider than the red. The intensity distribution in these wings requires further investigation, and at the present time material for the determination of complete instrumental profiles in the first, second, and third orders of the grating, using the same Kr line, is being collected by Dr. G. I. THOMPSON; the results will be published in due course.

The reduction of the second order plates, mentioned above, leads to a half-value width of 0·029 mm. This value is confirmed by a provisional reduction of first and third order plates which give widths of 0·027 mm and 0·035 mm, respectively, showing an increase in width with order, found in most solar grating spectrographs. The difference between actual and theoretical line widths, amounting to about one-third of the latter, is of the same order of magnitude as in the case of well-known solar spectrographs such as those at Mount Wilson, Potsdam, or Göttingen; the second order of the Potsdam grating which practically reaches the theoretical resolution, is a famous exception.

The high resolving power of at least 110,000, or 70 per cent of the theoretical one, which we find in the second order of the Dunsink spectrograph, becomes evident in the quality of the solar spectra. Close pairs of Fraunhofer lines have been examined for resolution within the region extending from $\lambda 4300$ to $\lambda 4500$. It is found that, for example, pairs such as $\lambda 4355{\cdot}9$ and $\lambda 4356{\cdot}0$, $\lambda 4401{\cdot}45$ and $\lambda 4401{\cdot}55$, or $\lambda 4489{\cdot}10$ and $\lambda 4489{\cdot}19$ are clearly resolved. The same pairs are only partly resolved in the Utrecht Atlas, which is based on Mount Wilson spectrograms. To give an idea of the speed of the instrument, it may be mentioned that solar spectra in this region, using the second order, require exposure times of 15 sec. on slow Kodak B4 plates.

5. PROGRAMMES OF WORK

Among programmes which are being undertaken with the instrument, priority is being given to the determination of equivalent widths and the measurement of profiles of Fraunhofer lines in the ultra-violet. The photometry is being done photographically, but will also be carried out soon photo-electrically, using an improved version of the direct recording photo-electric photometer which had first been used at Cambridge.

Other proposed investigations include studies of molecular spectra in sunspots and work on line intensities in prominence spectra.

For the discovery of prominences and other temporary phenomena on the Sun a spectrohelioscope has been erected next to the main solar spectrograph and parallel to its axis. The spectrohelioscope is used in conjunction with the solar telescope operating at a focal length of 8·8 m. An additional flat mirror of 20 cm aperture, placed on the optical bench about 1 m in front of the spectrograph slit, throws the image of the Sun on to the slit of the spectrohelioscope. This flat in its mounting, though not in position, is visible in Fig. 2, which to the right shows the "observing cabin" of the spectrohelioscope.

In the spectrohelioscope there are used a pair of oscillating slits, made at the observatory, and a 10 cm plane grating in autocollimation with a 10 cm lens of 5·8 m focal length. The grating is on loan to the observatory through the kindness of the Director of the Cambridge Observatories.

It is a great pleasure to acknowledge our thanks to Professor T. E. NEVIN of University College, Dublin, for his assistance in designing and setting up the spectrograph, and to Dr. G. I. THOMPSON, Scholar of the Dublin Institute for Advanced Studies, for his help in the taking of spectrograms.

Quartz Crystal Clocks

HUMPHRY M. SMITH and G. B. WELLGATE
Royal Greenwich Observatory, Abinger Common,
Dorking, Surrey, England

SUMMARY

Following a description of the essential components of a quartz crystal clock, a brief survey is given of the salient developments from the first oscillator of CADY in 1922 to the precision standards of time and frequency now employed in America and in the United Kingdom.

A further section deals with current progress, particularly at the Royal Greenwich Observatory. This is followed by assessments of the performance of new operational and experimental standards. The possible trend of future development is reviewed.

A final section deals with molecular and atomic clocks, and their possible use as an alternative to the astronomical standard of time.

PAGE

---

1. INTRODUCTION

RECENT advances in the accuracy of time keeping have been due largely to the outstanding developments in precision clocks. At Greenwich, the introduction of Shortt free pendulum clocks in 1925 marked a significant stage in the evolution of the Time Service: some ten years ago the Shortt clocks were superseded by quartz crystal clocks, which proved superior both as long-term standards and for the day-to-day control of time signals (GREAVES and SYMMS, 1943). Further development has since taken place, and the modern quartz clock attains a standard of precision considerably in advance of that which led to the elimination of pendulum clocks in major time-keeping observatories.

All modern precision clocks employ: (*a*) a resonant element whose natural period of oscillation is, under appropriate conditions, as invariable as possible and which has the lowest possible logarithmic decrement; (*b*) the means of providing the conditions necessary for the resonator to maintain the highest stability; (*c*) the means of keeping the resonator in oscillation by replacing the energy dissipated so as to make the resonator behave as nearly as possible as if it had no losses; (*d*) the means of registering the number of oscillations that have taken place during a specified interval and of producing accurately spaced electrical timing pulses at convenient intervals.

In the quartz clock the resonant element—usually called the crystal—is a bar, plate, or ring of quartz. This is cut to certain chosen dimensions and in a selected orientation to the various axes of the natural quartz crystal, so that it will vibrate at the desired frequency in a longitudinal, shear, flexure, or circumferential mode. In addition the resonator assembly includes the holder which supports the crystal at points or in lines where the motion is small, and the electrodes by which electrical energy is supplied to the crystal in order to maintain the vibration by means of the piezo-electric property of crystalline quartz.

The complete resonator assembly is usually mounted in a container of glass or copper which is evacuated in order to reduce the logarithmic decrement and to render the performance of the resonator independent of atmospheric pressure variations. This container is placed in a chamber called the oven, held within close limits near a chosen temperature.

The crystal is kept in oscillation by an electronic device, the maintaining amplifier, connected to the electrodes of the resonator. The stability of the frequency of oscillation, and thus the uniformity of the rate of the clock, is a function of the phase-stability of the maintaining amplifier; that is to say, the amount by which the voltage applied to the resonator leads or lags the current through the resonator must be as constant as possible. It follows from the work of LLEWELLYN (1931) that the frequency stability is also a function of the linearity of response of the amplifier: in other words, at the resonant frequency the current should ideally be a linear function of the voltage.

In general a standard quartz clock is required to provide a series of electrical timing pulses at intervals of seconds or tenths of seconds; for some applications additional signals are required to mark the minutes, or to indicate a particular time of day. A high crystal frequency facilitates quick and accurate comparisons of rate between standards, but requires extensive ancillary equipment to provide the timing pulses at low repetition rates. Low frequency crystals tend to be bulky and so far have proved inferior in performance. As a convenient compromise, most quartz clocks employ a crystal designed to operate at a nominal frequency of 100 kc/s. Frequency division may be entirely electronic, or electronic in the first stages and then electro-mechanical.

2. HISTORICAL SURVEY

2.1. *Resonators*

The first quartz crystal oscillator was produced by CADY (1922): important contributions towards the better understanding of the behaviour of quartz resonators in electrical circuits were made by DYE (1926). The first quartz clock was described by HORTON and MARRISON (1928). This clock was controlled by a quartz bar oscillating at 50 kc/s. It had a temperature coefficient of 0·35 sec. per day per degree at the temperature of operation. The following years saw the development of special quartz resonators having zero temperature coefficients at certain temperatures, in the U.S.A. by MARRISON (1929), in Germany by SCHEIBE and ADELSBERGER (1934), and in England by ESSEN (1935, 1938).

In the late thirties, MASON (1940) developed the GT-cut crystal. This type of resonator has a very low temperature coefficient over a wide range of temperature and has been successfully made and used by the British Post Office (Booth and Laver, 1946). The electrodes are thin layers of metal deposited on the faces of the crystal and logarithmic decrements as low as $0{\cdot}5 \times 10^{-5}$ have been achieved. Originally these crystal plates were clamped under spring tension between the flat ends of metal pins: in many later models wires soldered to small disks of silver deposited on the surface have been used to support the crystal, and to provide the electrical connection to the electrodes (Greenidge, 1944; Booth and Laver, 1946).

2.2. *Ovens*

As a result of the success achieved in the development of crystals with low temperature coefficients, little attention was paid to temperature control within very close limits. In America a method was developed for controlling the temperature by a self-oscillating bridge circuit (Pierce, McKenzie, Woodward, 1948); in England the Post Office used a mains-driven temperature-sensitive bridge circuit which was a simplified version of a very accurate control devised by Turner (1937); others used contact thermometers. Temperature stability of the order of 0·01°C was usually achieved.

2.3. *Maintaining Amplifiers*

In the earlier period of quartz crystal clock development, the Pierce-Miller circuit was universally used as the basis of the maintaining amplifier (Cady, 1946). Non-linearity is essential to the operation of this circuit and the phase stability is noticeably affected by variations in supply voltages, component values, and valve parameters. In 1938 Meacham published a description of the bridge-stabilized oscillator, which set a new standard of phase stability and linearity. At present it is incorporated, either in the original or in a modified form, in nearly all quartz crystal clocks in this country and in the United States.

2.4. *Dividers*

The first quartz crystal clocks used a type of electronic divider which would now be called a reactance-controlled oscillator, in which a harmonic of the oscillator was synchronized with the input frequency (Horton and Marrison, 1928). The dividers used in later clocks were usually based either on the multivibrator of Abraham and Bloch (1919), which was later shown by Hull and Clapp (1929) to be capable of synchronization at an exact sub-multiple of an applied signal frequency, or on the regenerative frequency divider of Miller (1939). The signal derived from the dividers is usually at a frequency of 1 kc/s and is used to drive a phonic motor. By means of suitable gearing, the phonic motor drives a shaft at one revolution per second and closes a contact every second; further gearing may be used to drive clock dials and subsidiary contacts (Smith, 1949).

3. Current Developments

3.1. *Resonators*

It is characteristic of the ring crystal (Essen, 1938) that the amount by which the clock rate drifts during initial ageing, and the time it takes for this drift to stabilize, are small (Booth, 1951). Recent developments by the British Post Office have accordingly brought this resonator to the fore again, but with certain modifications (see Fig. 1). Instead of being supported on chisel-ended phosphor-bronze screws, the crystal is now suspended by threads held taut by metal springs (ESSEN, 1951). The spaced solid metal electrodes have been replaced by a thin metal film, deposited on the inner and outer faces of the annulus (VIGOUREUX and BOOTH, 1950). The Time Service of the Royal Observatory is based almost exclusively on clocks incorporating these improved ring crystal resonators (SMITH, 1953). In the United States efforts

(*a*)

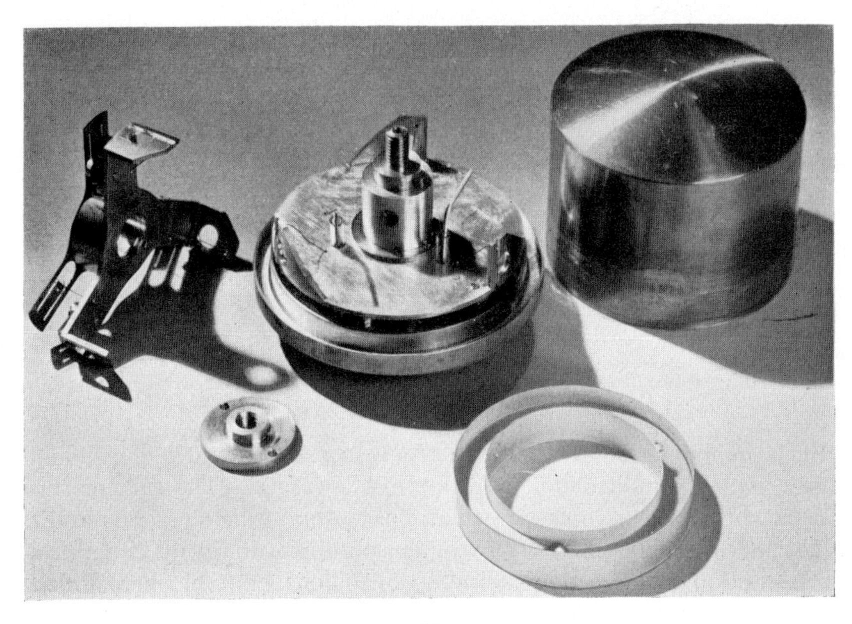

(*b*)

Fig. 1. 100 kc/s quartz ring crystal
(*By courtesy of H.M. Postmaster General*)
(*a*) Ring crystal mounted in crystal holder W.6.
(*b*) Ring crystal and holder W.6 dismantled.

seem to have been concentrated on further improvement of the GT-cut crystal, mainly by taking special precautions during manufacture (GRIFFIN, 1952). Logarithmic decrements as low as 1×10^{-6} have been achieved with both types of crystal.

3.2. *Ovens*

The temperature coefficients of GT-cut crystals may be as high as 0·02 sec. per day per degree. Control to within $\pm$ 0°0005 is thus required to ensure that temperature variations do not give rise to changes in rate in excess of 10 μsec. per day. For a ring crystal resonator the variation of frequency with temperature follows a parabolic law, and the temperature is normally held near the point of zero temperature coefficient: a similar standard of temperature control thus reduces the theoretical rate changes to the order of one or two-tenths of a microsecond per day.

A new oven designed and built at the Royal Observatory, is based on the original design of TURNER (1937), modified to meet the particular requirements of a precision quartz clock. The short-period stability is of the order of $\pm$ 0°0005, but equipment is not yet available for the precise checking of long period changes: provision has been made for readjustment to compensate for a temperature drift if necessary.

3.3. *Maintaining Amplifiers*

The variation in resonant frequency of a typical wire-supported GT-cut crystal as the current through the resonator is varied is shown in Fig. 2. It is evident that,

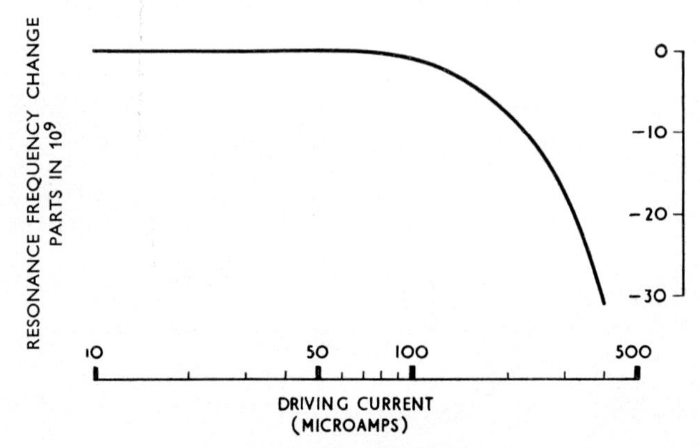

Fig. 2. Variation in resonant frequency of a typical crystal with varying resonator current. (After J. M. SHAULL, *Proc. Inst. Radio Engrs.*, **38**, 6, 1950; by courtesy of the Institute of Radio Engineers, New York)

for currents in excess of 100 μA, the frequency is markedly dependent on the current, *i.e.* on the amplitude of vibration of the crystal. The current through the resonator in the bridge-stabilized oscillator is normally between 500 and 2000 μA: this cannot be reduced without sacrifice of phase stability in the maintaining amplifier. The resonant frequency of a ring crystal is not so dependent on the magnitude of the current, but owing to the low equivalent resistance at resonance, the same driving circuit will produce a higher current through the crystal.

In the conventional bridge-stabilized oscillator, the secular change in the phase shift due to the amplifier gives rise to rate changes of the order of 1×10^{-4} sec. per day: short period changes amount to about one-fifth of this.

Experiments carried out in recent years at the Royal Observatory (WELLGATE, 1951) have resulted in the design of a maintaining amplifier capable of exciting resonators with currents through the crystals of less than 40 μA in the case of a GT-cut plate, and less than 200 μA in the case of a ring. Rate variations resulting from long-term phase instabilities of the amplifier are estimated at 3×10^{-6} sec. per day for a GT plate and 1×10^{-6} sec. per day for a ring: day-to-day variations are about one-fifth of these values.

One experimental clock, G2, employing a GT-cut crystal with a temperature coefficient of 0·011 sec. per day per degree and a logarithmic decrement of 1×10^{-5}, has been in operation for about a year. It incorporates the new design of maintaining amplifier and the improved form of precise temperature control. Its performance is discussed later. Experience gained in the design and adjustment of the experimental clocks has led to modifications to the operational clocks, with consequent improvements in performance.

3.4. *Dividers*

Many dividers of adequate timing accuracy have been developed in recent years. Improved versions of the multi-vibrator have been described (HAMMERTON, 1951). A simple and reliable regenerative divider was designed and installed at the Royal Observatory in 1951: similar instruments have since been developed elsewhere (MORSER, 1953). Dividers of the decade counter type have also been employed at the Royal Observatory, and electronically-derived seconds pulses have superseded the seconds signals from phonic motors for all precision inter-comparisons. Some pulse dividers have been designed to be adjustable in steps of 0·1, 0·01, or 0·001 sec. More recently, methods have been described of division by electron-beam tubes (JONKER, VAN OVERBECK, DE BEURS, 1952; KUCHINSKY, 1953) and cold cathode tubes (ACTON, 1952; HOUGH and RIDLER, 1952), some of which form a complete decade unit in one envelope.

4. PERFORMANCE

The long-term performance of quartz clocks must be related to the fundamental standard of time defined in terms of the rotation of the Earth on its axis (SMITH, 1951): but the rate of rotation of the Earth is not uniform, and the current determination of the irregularities is only possible by using the clocks themselves to define a uniform time system. The estimation of clock performance is thus a complex problem, particularly in view of the standard of accuracy achieved by the modern clock (SMITH, 1952).

The relative rate for a period of 6 months of two clocks, E5 and C6, which both employ silk-suspended Essen ring crystals with modified maintaining amplifiers, is shown in Fig. 3 (*a*). It is convenient to assess performance in terms of the mean absolute monthly second difference of daily rate in milliseconds per day (SMITH, 1953). A value of 0·1 msec. per day per month per month may be taken as typical of the best modern operational clocks. One of the new experimental clocks, G2, which employs a GT-cut crystal, has been regularly compared with both these clocks for a number of months, and it will be seen (Fig. 3 (*b*)) that the performance compares favourably with the ring-crystal standards. G2 has in fact taken its place with the ring crystal clocks in forming the "mean clock" on which the Time Service is based.

In short-term performance the experimental clocks have shown a high standard of uniformity. The day-to-day scatter can be expressed in terms of a daily criterion value which, for the best ring crystal clocks, was found some two years ago to be between 10 and 15 μsec. per day per day per day (SMITH, 1953). The latest values

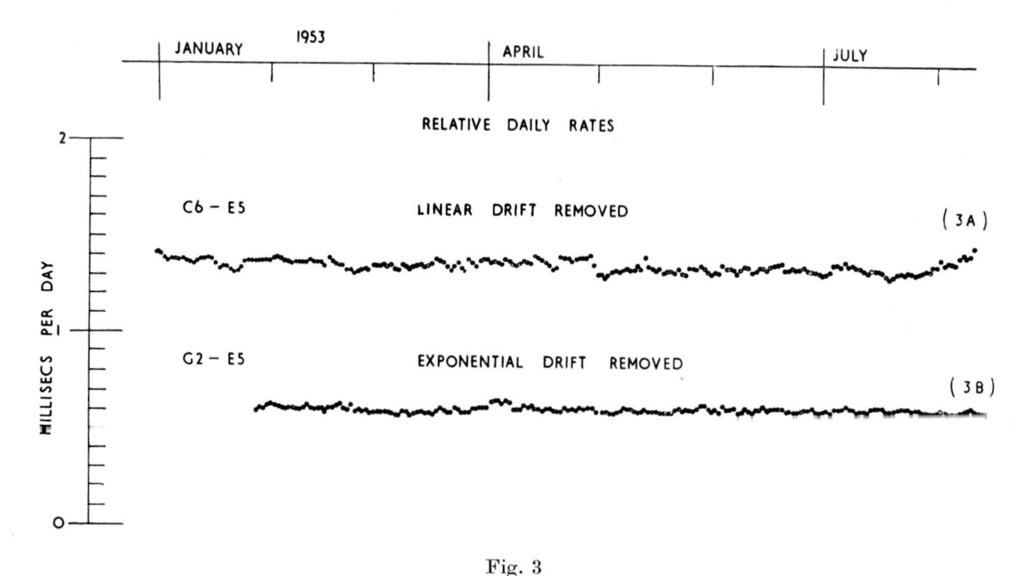

Fig. 3
(A) Relative rate of the crystal clocks E5 and C6 during six months.
(B) Comparison with G2, one of the new experimental clocks.

for the modified clocks E5 and C6 are 12 and 8 respectively over a period of some nine months, the corresponding figure for G2 being 6, which may be compared with a typical figure of 70 for a GT-cut crystal in a modified conventional oscillator.

5. Future Developments

The higher accuracy attained by using GT-cut crystals in new electronic circuits makes it reasonable to suppose that some measure of improvement may be achieved by employing ring crystals in re-designed circuits, but a limit will be reached at which further electronic elaboration produces no commensurate improvement in performance. The quartz crystal itself is thought to be of high inherent stability. It is possible that the conventional methods of finishing, mounting, and plating may prove a serious limitation. Great attention is now paid to surface finish, and some crystals have been designed for mounting at points where both movement and stress are small. Operation at lower temperature may prove advantageous; the effects of variations in the electrodes and supports may be substantially reduced. Investigations at the National Bureau of Standards have in fact shown that near $-150°$C logarithmic decrements of the order of $1{\cdot}6 \times 10^{-7}$, and zero temperature coefficients of rate, may be attained. There are theoretical advantages in using even lower temperatures, for example, the temperature of liquid helium, and experiments on these lines might prove fruitful.

An essential requirement in a standard clock is that it should run for a number of years without interruption. Some of the clocks at the Royal Observatory have been fitted with "long-life" valves, while other circuits have been designed to permit valve replacements to be made without disturbance to the run of the clocks. A description has been published of a clock incorporating a transistor in place of thermionic valves (SULZER, 1953), and magnetic amplifiers may find useful applications in the oven control circuits.

An alternative approach is to dispense with continuously running electronic equipment, and to compare the resonator with other standards by means of daily frequency measurements. It is interesting to note that this procedure does not reduce the "ageing" effect, and the rate seems to drift in a similar manner whether the resonator is continuously excited or not (SHAULL, 1950). To avoid the use of electronic equipment for precise temperature control, the British Post Office propose to install quartz resonators in bore-holes some 50–60 ft below ground level. Temperature tests are already in progress, and show that at this depth diurnal variations do not exceed 0·001°C, and annual variations are of the order of 0·1°C. It is intended to keep continuous records of temperature, and to apply appropriate corrections. In the U.S.A., similar experiments are in progress at the National Bureau of Standards (PENDLETON, 1953).

6. ATOMIC CLOCKS

It has long been regarded as desirable to link the frequencies of oscillating electrical circuits with the frequencies of selected lines in the electromagnetic spectrum of atoms and molecules: in this way electrical frequencies, and hence time intervals, would be directly related to fundamental natural standards. In recent years radio investigations have extended into the centimetric range of wavelengths, *i.e.* to frequencies up to 30,000 Mc/s, and the absorption spectra of gases in this region have been widely investigated (GORDY, 1948). The stabilization of an oscillator by means of the 3·3 ammonia absorption line (23,870·127 Mc/s) was first achieved by W. V. SMITH, DE QUEVEDO, CARTER, and BENNETT (1947).

A different form of control was employed by HERSCHBERGER and NORTON (1948): this was used in a modified form by LYONS in the first molecular clock (see National Bureau of Standards, 1949).

The principle of operation is briefly as follows: the ammonia gas is contained at very low pressure in a section of a waveguide. The frequency of maximum absorption is regularly compared with the frequency of a selected harmonic of the 100 kc/s quartz crystal controlled oscillator. A signal whose sign and amplitude is a function of the frequency difference is used to alter the oscillator frequency so as to reduce this difference to a minimum. The first clock ran only for short periods, and was stable only to some 0·01 sec. per day: the logarithmic decrement was about 6×10^{-5}. Improved models have since been made and work is still in progress (LYONS, 1952).

An alternative and perhaps more promising approach is the employment of molecular-beam magnetic resonances. Transitions between closely-spaced energy levels in some molecules and atoms correspond to electromagnetic radiation in the higher range of radio frequencies. Methods had been developed for exciting these transitions by magnetic resonance in a radio-frequency magnetic field (RABI, MILLMAN, KUSCH, ZACHARIAS, 1939; RAMSEY, 1950), and it was suggested by RABI, in 1945, that such resonances might be utilized as standards of time and frequency. In the experimental equipment so far described, a beam of caesium atoms is magnetically focussed and directed into a chamber in which a radio-frequency magnetic field may be superimposed on a weak homogeneous field. On leaving the chamber, the beam is focussed on a detector. If the radio-frequency field is at the frequency corresponding to the atomic transition, the magnetic moment of some of the atoms is changed, and they fail to reach the detector. A control accordingly operates to adjust the frequency of the radio-frequency field so that the number of atoms reaching the detector is a minimum. The radio-frequency is thus locked to the resonance frequency corresponding to the selected hyperfine structure line of the atomic spectrum. In fact this resonance frequency differs very slightly from the zero-field frequency, but the difference is not critically dependent on the constancy of the homogeneous field. It has been suggested that a clock with a logarithmic decrement of 1×10^{-7} and a rate stability of 0·00001 sec. per day may prove to be practicable. Clocks of this type are under construction in the U.S. at the National Bureau of Standards (Herschberger and Norton, 1948) and in England at the National Physical Laboratory.

In the form now envisaged, an atomic clock is a complex device which may be difficult to maintain in constant operation: but continuity is not essential, as it may be used at regular intervals to check the rate of a quartz clock. In fact, the atomic clock is not a potential successor to the quartz clock, but should be regarded as a new fundamental standard. as such, it is an alternative to the astronomical standard of time. It appears certain that, for many years at least, quartz clocks of the highest quality will be required to make available the full accuracy of the atomic resonator for the precise measurement of radio frequencies and time intervals: they will also still be required for the control of time signals which make Greenwich Mean Time available for survey, navigation, and astronomy.

This report has been prepared by the authors in the course of their official work at the Royal Greenwich Observatory, and is published by permission of the Astronomer Royal.

References

Abraham, H. and Bloch, A. E. . . . . 1919 *C.R. Acad. Sci. (Paris)*, **168**, 1105.

Acton, J. R. . . . . 1952 *Electronic Engng.*, **24**, 48

Booth, C. F. . . . . 1951 *Proc. Instn. Elect. Engrs.*, **98**, III, 1

Booth, C. F. and Laver, F. J. M. . . . 1946 *Proc. Instn. Elect. Engrs.*, **93**, III, 223.

Cady, W. G. . . . . 1922 *Proc. Inst. Radio Engrs.*, **10**, 83.

1946 *Piezo-electricity.* (McGraw-Hill Book Co., Inc.) New York, p. 194.

Dye, D. W. . . . . 1926 *Proc. Phys. Soc. (London)*, **38**, 399.

Essen, L. . . . . 1935 *Nature (London)*, **135**, 1076.

1938 *Proc. Phys. Soc. (London)*, **50**, 413.

1951 *Proc. Instn. Elect. Engrs.*, **98**, II, 154.

Gordy, W. . . . . 1948 *Rev. mod. Phys.*, **20**, 668.

Greaves, W. M. H. and Symms, L. S. T. . 1943 *M.N.*, **103**, 196.

Greenidge, R. M. C. . . . . 1944 *Bell Syst. tech. J.*, **23**, 234.

Griffin, J. P. . . . . 1952 *Bell Lab. Record*, **30**, 433.

Hammerton, J. C. . . . . 1951 *Mullard Res. Rep.*, No. 97. Mullard Radio Valve Co., Ltd., London.

Herschberger, W. D. and Norton, L. E. 1948 *R.C.A. Rev.*, **9**, 38.

Horton, J. W. and Marrison, W. A. . . 1928 *Proc. Inst. Radio Engrs.*, **16**, 137.

Hough, G. H. and Ridler, D. S. . . . 1952 *Electronic Engng.*, **24**, 152, 230, and 272.

Hull, L. M. and Clapp, J. K. . . . . 1929 *Proc. Inst. Radio Engrs.*, **17**, 252.

Jonker, J. L. H., van Overbeek, A. J. W. M. and de Beurs, P. M. . . . . . . 1952 *Philips Res. Rep.*, **7**, 81.

Kuchinsky, S. . . . . 1953 *Inst. Radio Engrs. (New York)*, Convention Record Part 6, p. 43.

Llewellyn, F. B. . . . . 1931 *Proc. Inst. Radio Engrs.*, **19**, 2063.

Lyons, H. . . . . 1952 *Nat. Bureau of Standards, Rep.* 1848.

MARRISON, W. A.	1929	*Proc. Inst. Radio Engrs.*, **17**, 1103.
MASON, W. P.	1940	*Proc. Inst. Radio Engrs.*, **28**, 220.
MEACHAM, L. A.	1938	*Bell Syst. tech. J.*, **17**, 574.
MILLER, R. L.	1939	*Proc. Inst. Radio Engrs.*, **27**, 446.
MORSER, A. H.	1953	*Electronic Engng.*, **25**, 290.
National Bureau of Standards	1949	*Nat. Bureau of Standards, Tech. News Bull.*, **33**, 17.
PENDLETON, T. A.	1953	*Proc. Inst. Radio Engrs.*, **41**, 1612.
PIERCE, J. A., MCKENZIE, A. A. and WOODWARD, R. H.	1948	*Loran, Long Range Navigation.* (McGraw-Hill Book Co., Inc.), p. 237.
RABI, I. I., MILLMAN, S., KUSCH, P. and ZACHARIAS, J. R.	1939	*Phys. Rev.*, **55**, 526.
RABI, I. I.	1945	Richtmyer Lecture before Amer. Phys. Soc.
RAMSEY, N. F.	1950	*Phys. Rev.*, **78**, 695.
SCHEIBE, A. and ADELSBERGER, U.	1934	*Hochfrequenztechnik und Elektroakustik*, **43**, 37.
SHAULL, J. M.	1950	*Proc. Inst. Radio Engrs.*, **38**, 6.
SMITH, H. M.	1949	*British Science News*, **2**, 169.
	1951	*Proc. Instn. Elect. Engrs.*, **98**, II, 143.
	1952	*Proc. Instn. Elect. Engrs.*, **99**, IV, 273.
	1953	*M.N.*, **113**, 67.
SMITH, W. V., DE QUEVEDO, J. L. G., CARTER, R. L. and BENNETT, W. S.	1947	*J. appl. Phys.*, **18**, 1112.
SULZER, P. J.	1953	*Nat. Bureau of Standards Tech. News Bull.*, **37**, 17.
TURNER, L. B.	1937	*J.* (now *Proc.*) *Instn. Elect. Engrs.*, **81**, 399.
VIGOUREUX, P. and BOOTH, C. F.	1950	*Quartz Vibrators and their Applications.* (H.M. Stationery Office, London), p. 139.
WELLGATE, G. B.	1951	*Proc. Instn. Elect. Engrs.*, **98**, II, 167 (discussion).

A Rapid Method of Obtaining Positions from Wide-angle Plates

G. VAN BIESBROECK

Yerkes Observatory, Williams Bay, Wisconsin, U.S.A.

SUMMARY

Positions with an accuracy of two seconds of arc are obtained from settings with a precision theodolite pointing at the plate from a distance equal to the focal length of the telescope with which the plates are obtained.

1. INTRODUCTION

POSITIONS of astronomical bodies on the celestial sphere have always been of great importance to the astronomer. From our present point of view the planetary positions measured by TYCHO in the sixteenth century are very crude. Yet it is from these that KEPLER deduced the empirical laws of planetary motions which formed the basis of NEWTON's gravitational theory. Since then the technique of measuring positions in the sky has been enormously improved: this applies to both fundamental and relative positions. Meridian observations are now accurate enough to detect irregularities in the Earth's rotation and to yield proper-motions of early type stars that confirm their distribution in the spiral arms of our galaxy indicated by other methods. Relative positions deduced from photographic plates obtained with long-focus telescopes are of such accuracy that reliable stellar parallaxes can be secured for many hundreds of stars, thereby calibrating the scale of the visible universe.

However, this high standard of precision involves laborious and time-consuming procedures. Not all problems require such refinement. In asteroid work, for instance, we see many positions published which are only rough values, say to the nearest tenth of a minute of time and the nearest minute of arc. They are generally obtained by plotting the object on BD-charts and bringing the result up to the normal equinox of 1950 by applying roughly the precession. While such positions are generally sufficient for identification purposes computers of orbits mostly require a greater precision and they then ask the observer to deduce from the photographic plate more accurate values. This can become a heavy burden when many objects are involved. A good example is the case of the many hundreds of asteroids which have been recorded in a systematic survey of the ecliptic belt for statistical purposes initiated by KUIPER with a 10-in. Ross type lens at the MacDonald Observatory in Texas. Precise measurement of all these objects on the wide-angle plates is beyond the forces available for such work. In searching for some intermediate solution that would give, on the one hand, an accuracy of some 2″ and, on the other, avoid laborious reductions, it occurred to the writer that a way out could be found by the use of a modern theodolite. The famous Swiss Wild Theodolites, which are on the market at a relatively low cost, are read to 1″ in both azimuth and altitude and guaranteed to give an accuracy of 2″ in spite of their small dimension.

2. INSTRUMENTAL ARRANGEMENT

Let us set up such an instrument at a distance from the plate to be measured equal to the focal length of the telescope with which it was taken. If the plate were held up in such a position that the altitude of its centre read by the theodolite corresponds to the declination of that point we could, by pointing at the stars, retrace the path of the light that went from the telescope lens to the plate in its focus. If, furthermore, the zero-point of the azimuth-circle is set so that the right ascension of the plate centre corresponds to the azimuth-reading it would seem possible, with proper orientation of the plate, to obtain the desired position by direct reading of the circles. This method was in principle used by KAPTEYN when he had REPSOLD build him an equatorial mounting for measuring the Cape Durchmusterung plates (*Ann. Cape Obs.*, Vol. 3, p. 1, 1895). However, he set his polar axis in a horizontal plane for convenience while here the polar-axis corresponds to the vertical axis of the theodolite.

The stars' positions are affected by refraction and aberration which vary appreciably over the extent of the plate taken with a wide-angle camera. In our case the diagonal of the 8 × 10 cm plate corresponds to an angle of 10°. In order to maintain the required precision it is therefore necessary to calibrate the plate by known stars in the vicinity of the asteroid to be measured. If this is done with the star positions of the Yale zones or before long with the new AG positions the result will be referred to the normal equinox of 1950 without the necessity of a reduction for precession. After some preliminary tests it was found easier not to move the plate but to set it always up in a vertical frame while tilting the vertical axis of the theodolite by an angle equal to the declination of the plate centre. The arrangement now in operation at the Yerkes Observatory is illustrated by Fig. 1. The theodolite is mounted on a heavy plate suspended by two tronions. The height of the instrument over that plate is such that the common axis of the tronions passes through the centre of rotation of the theodolite. The plate is permanently set up at a distance equal to the focal length of the telescope and lined up once for all so that the normal to its centre passes through the centre of rotation of the theodolite. The knob K turns a rod that changes the position-angle of the plate in its plane.

Fig. 1. The measuring arrangement in operation at Yerkes Observatory.

3. MEASURING PROCEDURE

The procedure of measurement is now as follows: a known catalogue star being marked on the plate near its centre the altitude of the theodolite is set at the 1950·0 declination of that object. The support plate of the theodolite is then tipped until the star appears under the crosswire and clamped in that position by the knob K. Next the azimuth circle is turned until its reading corresponds to the known right ascension of the star under the crosswire. The plate is then oriented by two stars far apart in declination and not differing too much in right ascension, or vice versa. By trial and error the plate is oriented in its plane by turning the handle H until the observed difference in co-ordinates is equal to the catalogue value. We are now ready to read off the equatorial co-ordinates of any number of asteroids which appear sometimes as many as ten on the same plate. At the same time we read off the co-ordinates of some known stars in their vicinity. The small residuals "observed minus catalogue" can now be interpolated at sight for the asteroids. The only computation required is the transformation of the degrees in time of right ascension, since the theodolite is graduated on the sexagesimal scale. A numerical random example of four asteroids measured on one plate centred at $-15°$ is shown overleaf.

The variation in the value of observed minus catalogue across the plate is mainly due to the combined differential refraction and aberration but it includes also residual errors in scale and orientation. Setting up the instrument and orienting the plates takes about 10 min. These adjustments are approximately made. To make them more accurately and reduce the observed minus catalogue to smaller numbers yet would take much more time and defeat the purpose of the method, since the interpolation from the known stars would have to be made in any case. By the use of two catalogue stars, including the asteroid between them, the observed minus catalogue can be safely interpolated at sight for the asteroid. The above figures are

	Observed		O − C for the stars		Interpolated for the asteroid	
	Right ascension	Declination				
Star	205° 4′ 1″	− 14° 47′ 12″	+ 57″	− 13″		
Asteroid 1	205 18 58	− 14 45 12			+ 55″	− 11″
Star	205 20 53	− 14 39 55	+ 55	− 8		
Star	206 6 55	− 12 51 44	+ 55	+ 1		
Asteroid 2	206 5 39	− 12 54 28			+ 54	+ 1
Star	205 54 1	− 13 46 47	+ 54	− 3		
Star	207 13 8	− 18 7 11	+ 46	− 34		
Asteroid 3	207 27 12	− 17 59 20			+ 41	− 33
Star	207 57 17	− 17 43 35	+ 39	− 31		
Star	210 15 31	− 13 2 57	+ 17	− 2		
Asteroid 4	210 14 37	− 13 13 51			+ 16	− 3
Star	210 29 52	− 13 55 34	+ 16	− 4		

The concluded asteroid positions come out:

Asteroid	Right ascension	Declination
1	205° 18′ 3″	− 14° 45′ 1″
2	206 4 45	− 12 54 29
3	207 26 31	− 17 58 47
4	210 14 21	− 13 13 48

all that had to be written down to obtain the 1950 positions of the four asteroids on the plate.

The same instrument can be used on plates taken with any other telescope. The only modification required is to place the plate at a distance from the theodolite equal to the new focal length.

Nomograms in Astronomy

WITH TWO EXAMPLES

H. STRASSL

Sternwarte der Universität, Bonn, Germany

SUMMARY

Nomograms are valuable tools for many astronomical calculations. Two examples are given to illustrate quite different nomographic representations of the spherical cosine rule, covering (1) the atmospheric absorption at any required wavelength for a given geographical latitude, and (2) the spherical distance of a star from the apex. In the last-mentioned nomogram, which in some respects is not quite perfect, the characteristic features to be emphasized are projective transformations of a regular cosine scale. A full bibliography of astronomical nomograms is added.

---

1. INTRODUCTION

NOMOGRAMS for astronomical computations occasionally made their appearance as early as the beginning of this century. The collection of numerical tables by AMBRONN and DOMKE [3]* published in 1909 contains as an important part of its contents a considerable number of very good nomograms. But whilst nomograms have been much used for technical purposes, they have not up till now been fully applied in astronomy. It is not surprising that astronomers trained in the school of precise classical astronomy have regarded methods of graphical computation with a certain reserve. It would, of course, be a mistake to try to make calculations involving very many digits by purely graphical methods. But there are many occasions in astronomy where one need apply a mathematical formula only approximately. Nomograms are an excellent aid on these occasions, just as the slide rule is used for simple multiplication. Moreover, nomograms can often give valuable aid in the course of accurate calculations, *e.g.* it is possible to speed up enormously the process of solving complicated equations (either algebraic or transcendental) when a nomogram is available which gives an approximate solution as well as the value of the derivative required in the Newtonian method of arriving at more accurate values. As an illustration, the reader is referred to the nomographic treatment of KEPLER'S equation in [53] and [114], [115].

The use of a prepared nomogram is usually very simple, though there is still much prejudice against it, and it is not necessary for the user to understand the theory of the nomogram. On the other hand, the designing and production of a nomogram which will be really useful in numerical calculation is often a very laborious piece of work.

The essence of nomographic charts consists of the representation of functions of two variables, or (as it is more clearly expressed in the language of nomography) of the representation of mathematical relations which connect three variables. Two types of nomograms are the most important for this purpose: alignment diagrams and intersection diagrams (cf. [2], [68]). An alignment diagram consists of three curves (straight lines included) on each of which a scale of one variable is superimposed on the length of the curve and the values of the variable are given. Each

* Numbers in square brackets refer to the reference list at the end of the paper.

triad of connected values is represented by three points which lie on a straight line, and it is easiest when using the diagram to get the result by using a transparent ruler on which a straight "index line" has been marked. An intersection diagram consists of three families of curves, in each family of which one variable varies from curve to curve. Each triple of related values corresponds to a point at which three curves, one from each family, intersect. As the diagram cannot be too closely covered with curves in order that its clarity be not too much impaired, one cannot attain the same degree of accuracy with an intersection diagram as with an alignment diagram. When using an intersection diagram one has always to interpolate by eye a point in two co-ordinates simultaneously, which is much less easy than the corresponding process with an alignment diagram. For these reasons it is advisable, if possible, to represent a given relation by an alignment diagram. But as it is only possible to represent relations of a certain kind by alignment diagrams, one is often forced to give serious attention to a suitable design of intersection diagrams. One often finds intersection diagrams in scientific publications where the author would certainly have given alignment diagrams, had he been familiar with the principle of the latter.

The relationship $\phi(\alpha, \beta, \gamma) = 0$ can be represented by an alignment diagram if and only if it can be written in the form

$$\phi(\alpha, \beta, \gamma) = \begin{vmatrix} f_1(\alpha) & f_2(\alpha) & f_3(\alpha) \\ g_1(\beta) & g_2(\beta) & g_3(\beta) \\ h_1(\gamma) & h_2(\gamma) & h_3(\gamma) \end{vmatrix} = 0. \qquad \ldots.(1)$$

The alignment diagrams for this are obtained from the parametric form of the three curves for α, β, γ in Cartesian co-ordinates:

$$\left.\begin{aligned} x_\alpha &= \frac{Lf_1(\alpha) + Mf_2(\alpha) + Nf_3(\alpha)}{L''f_1(\alpha) + M''f_2(\alpha) + N''f_3(\alpha)} & y_\alpha &= \frac{L'f_1(\alpha) + M'f_2(\alpha) + N'f_3(\alpha)}{L''f_1(\alpha) + M''f_2(\alpha) + N''f_3(\alpha)} \\ x_\beta &= \frac{Lg_1(\beta) + Mg_2(\beta) + Ng_3(\beta)}{L''g_1(\beta) + M''g_2(\beta) + N''g_3(\beta)} & y_\beta &= \frac{L'g_1(\beta) + M'g_2(\beta) + N'g_3(\beta)}{L''g_1(\beta) + M''g_2(\beta) + N''g_3(\beta)} \\ x_\gamma &= \frac{Lh_1(\gamma) + Mh_2(\gamma) + Nh_3(\gamma)}{L''h_1(\gamma) + M''h_2(\gamma) + N''h_3(\gamma)} & y_\gamma &= \frac{L'h_1(\gamma) + M'h_2(\gamma) + N'h_3(\gamma)}{L''h_1(\gamma) + M''h_2(\gamma) + N''h_3()\gamma} \end{aligned}\right\} \qquad \ldots.(2)$$

Here L, M, N, L', M', N', L'', M'', N'' are arbitrary constants, and in each separate case one must choose their numerical values so as to make the chart as useful as possible for its special purpose.

In the following paragraphs, nomograms are given and explained for two different astronomical computation problems. They consist of several alignment diagrams. The spherical cosine rule appears in both nomograms, but is differently represented in each chart, according to the purpose for which it is designed.

2. Atmospheric Absorption as a Function of the Wavelength λ for a Star of Declination δ at Hour Angle t. The Geographical Latitude ϕ is Assumed to be Constant

The first requirement for fixing the amount of atmospheric absorption is to ascertain the zenith distance z of the star. This is obtained from

$$\cos z = \sin \phi \sin \delta + \cos \phi \cos \delta \cos t. \qquad \ldots.(3)$$

It therefore depends on the three variables ϕ, δ, t. In this essay we shall only design the nomogram for a definite numerical value of the geographical latitude. The reader will be able to compute the appropriate nomogram for a different value of the geographical latitude from the formulae given below. In selecting the value ϕ the author has considered it appropriate, in this presentation volume in honour of Professor STRATTON, to select that of the scene of his labours, Cambridge; Fig. 1.

The correction for atmospheric absorption is taken as the value e which one must subtract from the magnitude of a star, observed at zenith distance z, in order to reduce it to the zenith; e depends both on the zenith distance and on the transmission coefficient p which is the ratio of the intensity let through by the atmosphere at the zenith to the extra-terrestrial intensity:

$$m_{z=0} = m_z - e(p, z). \qquad \ldots.(4)$$

It is often convenient (cf. [135]) to introduce, instead of p, the atmospheric absorption coefficient k which represents directly the atmospheric absorption at the zenith, expressed in stellar magnitudes:

$$k = -2{\cdot}5 \log p. \qquad \ldots.(5)$$

If we limit ourselves, as is usual in photometric observations, to $z < 60°$, then e is given with sufficient accuracy by

$$e = k(\sec z - 1). \qquad \ldots.(6)$$

Therefore, the reduction from $m_{z=60°}$ to $m_{z=0°}$ is the same as that from $m_{z=0°}$ to extra-terrestrial magnitude.

The atmospheric absorption k is composed of three parts, all dependent on the wavelength:

$$k = k_R(\lambda) + k_{O_3}(\lambda) + k_D(\lambda). \qquad \ldots.(7)$$

k_R (Rayleigh scattering) is in direct proportion to the barometric pressure and in reverse proportion to λ^4. k_{O_3} is the atmospheric absorption caused by ozone, k_D that caused by mist and dust. The most detailed study of this subject is that by WEMPE [135]. He found from his own observations and a discussion of the results of other observers that the "dust" absorption $k_D(\lambda)$ which varies from place to place and from night to night can within the interval $400 < \lambda < 650\mu$, be approximated by

$$k_D(\lambda) = \text{constant}/\lambda^{\alpha}, \qquad \ldots.(8)$$

where the value of α is characteristic for the place of observation, and is in many cases approximately 1. As no special studies are available for the relationship between wavelength and atmospheric absorption at Cambridge, we shall take $\alpha = 1$. Therefore, if the wavelength of the visual magnitudes $\lambda_v = 550$ is introduced, we can write (8) in the form

$$k_D(\lambda) = \frac{550}{\lambda} \, . \, k_D(550). \qquad \ldots.(9)$$

If we take into account for k_R the unimportant correction for Cambridge of the normal to the local barometric pressure, *i.e.* from 29·92 in. = 760 mm to 29·01 in. = 757 mm, we can write*:

$$k = \frac{757}{760} k_{R,\,760}\,(\lambda) + k_{O_3}\,(\lambda) + \frac{550}{\lambda} k_D\,(550). \qquad \ldots.(10)$$

* It is unnecessary here to go into the question of principle whether a barometric correction should be applied to k_{O_3} as well, as that correction would make practically no difference (k_{O_3} is only $0^m{\cdot}04$ at maximum).

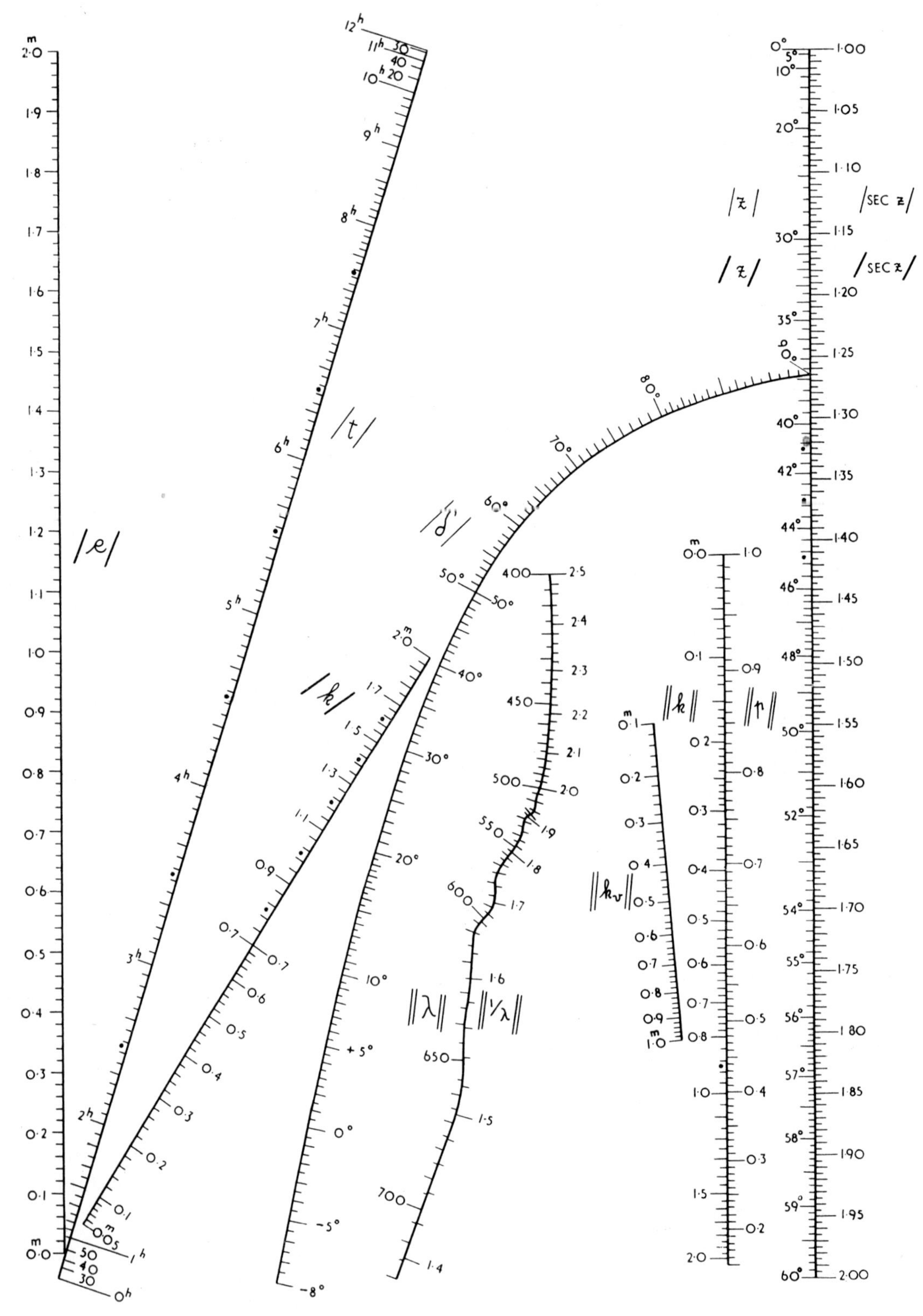

Fig. 1. Nomogram for atmospheric absorption at Cambridge

Applying this formula to the visual wavelength $\lambda_v = 550$ we get

$$k_v = k(550) = 0{\cdot}127 + k_D(550). \qquad \ldots.(11)$$

$k_D(550)$ can be understood as a parameter which characterizes the transparency of the atmosphere on the night of observation. In practice it is probably more convenient to take as a parameter of transparency the coefficient of total visual absorption k_v itself. Accordingly, we can write (10) with the help of (11) as

$$k = \frac{757}{760} k_{R,\,760}\,(\lambda) + k_{O_3}\,(\lambda) + \frac{550}{\lambda}\,(k_v - 0{\cdot}127). \qquad \ldots.(12)$$

With the help of the tables given by Wempe (loc. cit.) it is easy to arrive at the numerical values of the function

$$\mathscr{F}(\lambda) = \frac{757}{760} k_{R,\,760}\,(\lambda) + k_{O_3}\,(\lambda). \qquad \ldots.(13)$$

Thus we get for k the final form:

$$k = \mathscr{F}(\lambda) + \frac{550}{\lambda}\,(k_v - 0{\cdot}127) \quad \text{for } 400 \leq \lambda \leq 700. \qquad \ldots.(14)$$

Here k appears as a function of the two variables λ and k_v. This makes it possible to make at least some allowance for the varying transparency on different nights. From the conclusions of Wempe we know that (with considerable deviations) the average of k_v is about $0^m{.}22$ for Berlin-Babelsberg, about $0^m{.}40$ for Göttingen, about $0^m{.}25$ for Uppsala and about $0^m{.}33$ for Washington.

The whole nomogram for finding the atmospheric absorption (Fig. 1) consists of alignment diagrams for (3), (14), and (6). It is arrived at in the following manner.

On consideration we see we need not so much the value of z as of sec z. In (3) we therefore substitute 1/sec z for cos z and state (3) in the fundamental form (1) by writing (3) as

$$\begin{vmatrix} \cos t & 1 & 0 \\ 1 & 0 & \sec z \\ \sin\phi \sin\delta & -\cos\phi\cos\delta & 1 \end{vmatrix} = 0. \qquad \ldots.(15)$$

The alignment diagram for (3) according to (2) therefore becomes:

$$\left.\begin{aligned} x_t &= \frac{L\cos t + M}{L''\cos t + M''} & y_t &= \frac{L'\cos t + M'}{L''\cos t + M''} \\ x_z &= \frac{L + N\sec z}{L'' + N''\sec z} & y_z &= \frac{L' + N'\sec z}{L'' + N''\sec z} \\ x_\delta &= \frac{L\sin\phi\sin\delta - M\cos\phi\cos\delta + N}{L''\sin\phi\sin\delta - M''\cos\phi\cos\delta + N''} \\ & & y_\delta &= \frac{L'\sin\phi\sin\delta - M'\cos\phi\cos\delta + N'}{L''\sin\phi\sin\delta - M''\cos\phi\cos\delta + N''} \end{aligned}\right\} \qquad \ldots.(16)$$

It consists, therefore, of straight lines for t and z and of an arc of a conic section as scale for δ.

It is so constructed that it takes the fullest advantage of the available rectangular space. The scale for z or sec z stands vertically at the right-hand side and is graduated as a linear scale in sec z. z itself can be read off with considerable accuracy for large zenith distances, but only roughly for small zenith distances.

The unit for the drawing, here and later, has been taken as 1 mm. The arbitrary constants have the following values:

$$\left.\begin{array}{lll} L = +122, & M = -122, & N = 0 \\ L' = +386, & M' = -390, & N' = -192 \\ L'' = +\ \ 1, & M'' = -\ \ 3, & N'' = 0 \end{array}\right\} \quad \ldots .(17)$$

Moreover, $\phi = +52\overset{\circ}{.}21$ (Cambridge, England).

The scales for t, δ, sec z, and z are marked in the reproduction by the thin single rules framing the corresponding letters.

The alignment diagram for (14) is based on formula (14), written according to (1).

$$\begin{vmatrix} (k_v - 0{\cdot}127) & -1 & 0 \\ k & 0 & 1 \\ \mathscr{F}(\lambda) & 550/\lambda & 1 \end{vmatrix} = 0. \quad \ldots .(18)$$

So the alignment chart is obtained from

$$\left.\begin{array}{ll} x_{k_v} = \dfrac{Lk_v - (0{\cdot}127L + M)}{L''k_v - (0{\cdot}127L'' + M'')} & y_{k_v} = \dfrac{L'k_v - (0{\cdot}127L' + M')}{L''k_v - (0{\cdot}127L'' + M'')} \\ x_k = \dfrac{Lk + N}{L''k + N''} & y_k = \dfrac{L'k + N'}{L''k + N''} \\ x_\lambda = \dfrac{L\mathscr{F}(\lambda) + 550M/\lambda + N}{L''\mathscr{F}(\lambda) + 550M''/\lambda + N''} & y_\lambda = \dfrac{L'\mathscr{F}(\lambda) + 550M'/\lambda + N'}{L''\mathscr{F}(\lambda) + 550M''/\lambda + N''} \end{array}\right\} \quad \ldots .(19)$$

It therefore consists of straight lines for k_v and k and a curve for λ. The nomogram has been so contrived that it occupies the free space under the δ scale of the first nomogram. The constants have the values:

$$\left.\begin{array}{lll} L = +108, & M = -148{\cdot}28, & N = +\ \ 95{\cdot}90 \\ L' = -\ \ 43{\cdot}84, & M' = -132{\cdot}43, & N' = +102{\cdot}12 \\ L'' = +\ \ 1, & M'' = -\ \ 1{\cdot}534, & N'' = +\ \ 0{\cdot}888 \end{array}\right\} \quad \ldots .(20)$$

The names of the three scales of this nomogram are framed in the figure with thin double rules.

The k scale is built up as a double-sided scale in the variables k and p according to (5), so that one can also use p instead of k as argument; we have in it a means for converting k (or k_v) into the corresponding p (or p_v) and vice versa. Moreover, the λ scale (λ expressed in millimicrons) bears the values $1/\lambda$ as well (λ expressed in microns), since in spectrophotometric work $1/\lambda$ is often a more suitable argument than λ itself. The waves in the ozone absorption curve (cf. [135]) as a function of wavelength are responsible for the peculiar appearance of the λ curve in the nomogram.

The alignment diagram for (6) is obtained from the determinant

$$\begin{vmatrix} (\sec z - 1) & 1 & 0 \\ 0 & k & 1 \\ e & 1 & -1 \end{vmatrix} = 0. \qquad \ldots .(21)$$

Therefore:

$$\left.\begin{aligned} x_z &= \frac{L \sec z + (M - L)}{L'' \sec z + (M'' - L'')} & y_z &= \frac{L' \sec z + (M' - L')}{L'' \sec z + (M'' - L'')} \\ x_k &= \frac{Mk + N}{M''k + N''} & y_k &= \frac{M'k + N'}{M''k + N''} \\ x_e &= \frac{Le - N}{L''e - N''} & y_e &= \frac{L'e - N'}{L''e - N''} \end{aligned}\right\} \qquad \ldots .(22)$$

The diagram consists of parts of three straight lines. It is so arranged that its z scale is identical with the z scale of the nomogram of formula (3) and the e scale appears at the left-hand side of the drawing space, graduated linearly in e. The constants have the values:

$$\left.\begin{aligned} L &= 0, & M &= +\ 122, & N &= +\ 96/47 \\ L' &= -\ 192, & M' &= +\ 194, & N' &= +\ 576/47 \\ L'' &= 0, & M'' &= +\ \ 1, & N'' &= +\ 96/47 \end{aligned}\right\} \qquad \ldots .(23)$$

The k scale runs from the lower left up towards the right. The letters z (or sec z), k, e of this nomogram are framed by bold single rules.

The use of the nomogram may be explained by an example. Supposing the star α Leo ($\delta = +\ 12\overset{\circ}{.}21$) has been observed photometrically at the hour angle $t = 2^{\rm h}\ 53^{\rm m}$ at Cambridge on a night the quality of which may be characterized by the visual zenith absorption $k_v = 0\overset{\rm m}{.}30$. One wishes to get the correction for atmospheric absorption at Hα ($\lambda = 656{\cdot}3$) and Hδ ($\lambda = 410{\cdot}2$). Solution: in the $\|k_v\|\lambda\|k\|$ nomogram one puts the index line first through the points $k_v = 0{\cdot}30$, $\lambda = 656{\cdot}3$ and so finds $k = 0\overset{\rm m}{.}200$ ($p = 0{\cdot}831$). In the same way one then finds from $k_v = 0\overset{\rm m}{.}30$ and $\lambda = 410{\cdot}2$ the value $k = 0\overset{\rm m}{.}562$ ($p = 0{\cdot}595$). In the $/t/\delta/z/$ nomogram one puts the index line through the points $t = 2^{\rm h}\ 53^{\rm m}$ and $\delta = +\ 12\overset{\circ}{.}21$; in the vertical scale on the right one finds sec $z = 1{\cdot}657$ ($z = 52\overset{\circ}{.}9$). This point is held fixed during the subsequent use of the **/z/k/e/** nomogram. If one turns the index line on this point till it passes through the point $k = 0\overset{\rm m}{.}200$, then one finds for H$\alpha$ $e = 0\overset{\rm m}{.}13$; if one turns it on the same point till it passes through $k = 0\overset{\rm m}{.}562$, then one finds for H$\beta$ $e = 0\overset{\rm m}{.}37$.

As, according to WEMPE, the interpolatory representation of the dust absorption is no longer valid for $\lambda < 400$, it was only possible to design the $\|k_v\|\lambda\|k\|$ nomogram for $\lambda > 400$. It is clear that if one demands a high standard of reliability from the photometric results, then one also requires a more exact basis for the value of k in the region $\lambda > 400$ than that on which formula (14) and its nomographic representation rest. The nomograms for the formulae (3) and (6) are, of course, not affected by this limitation.

In the above nomogram the fact that in equation (10) the mean local barometric pressure $b_0 = 757$ mm should actually be replaced by the individual pressure b at the time of observation is not taken into account, since this difference is practically negligible. If, however, in particular cases it is desirable to take it into consideration, we can put:

$$k' = \frac{b_0}{b} k \text{ and } k_v' = \frac{b_0}{b} k_v. \qquad \ldots.(10a)$$

Equation (14) can then be written:

$$k' = \mathscr{F}(\lambda) + \frac{550}{\lambda}(k_v' - 0{\cdot}127). \qquad \ldots.(14a)$$

The nomogram for (14*a*) is identical with that for (14). In practice one proceeds from k (through multiplication by b_0/b) to k_v', and obtains from the nomogram (with the arguments k_v' and λ) the value of k', which by multiplying by b/b_0 gives k; both these multiplications can be carried out with a slide-rule. In order not to confuse the drawing, no separate nomogram has been introduced for this procedure.

The complete nomogram of Fig. 1 contains as local constants the values of the geographical latitude and the mean barometric pressure of Cambridge. If one wishes to design a corresponding nomogram for another locality, one must substitute in formula (16) the corresponding value for ϕ and calculate the function $\mathscr{F}(\lambda)$ in formula (14) according to the corresponding barometric pressure. Perhaps it would be necessary to take a value other than 1 for α.

If the width of the nomogram is to be increased or decreased in comparison to the nomogram shown here, then one must increase or decrease the three constants L, M, N in formulae (17), (20), (22) accordingly. If the height has to be changed the procedure is the same for the constants L', M', N'. The constants L'', M'', N'' may remain unaltered.

3. Spherical Distance of a Star from the Apex

The spherical distance of a star from the apex plays an important part in the treatment of proper motions and radial velocities. This problem has also been tackled by nomographic means (cf. bibliography). If α_0 and δ_0 are the co-ordinates of the apex and α and δ those of a star, then the spherical distance D of the star from the apex, according to the spherical cosine rule, is obtained from

$$\cos D = \sin \delta_0 \sin \delta + \cos \delta_0 \cos \delta \cos (\alpha - \alpha_0). \qquad \ldots.(24)$$

Therefore, if one considers α_0 and δ_0 as fixed numerical values one can easily construct an alignment chart which contains a regular scale of the values $\cos D$. Such a chart is very useful for finding the solar component of radial velocities, since that component is given by $V_\odot \cos D$ ($V_\odot \approx 20$ km per sec.). An alignment chart by A. F. Dufton [27] gives directly the values $20 \cos D$. But in a scale which is graduated linearly in the argument $\cos D$, D values in the neighbourhood of 0° and 180° appear strongly congested. Therefore, the accuracy with which D values in these regions can be taken from the nomogram is often insufficient when one wishes to know D or $\sin D$ (*e.g.* in the reduction of proper motions).

In the following, a nomogram (Fig. 2) of formula (24) is given from which D values in the neighbourhood of 0° and 180° as well can be read off with a relatively high accuracy. The apex co-ordinates are assumed to be $\alpha_0 = 18^{\text{h}}\ 00^{\text{m}}$, $\delta_0 = +30\overset{\circ}{.}0$.

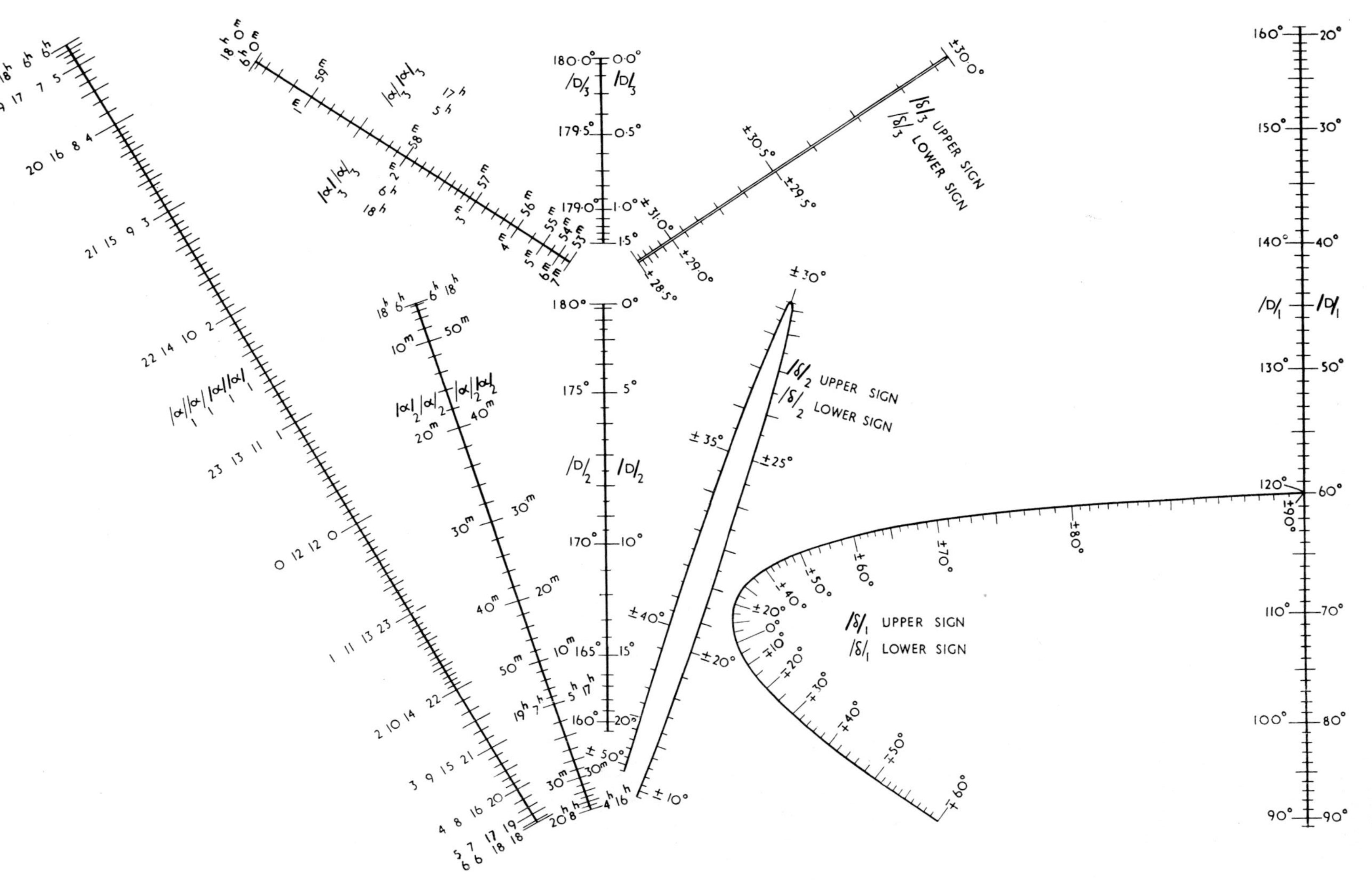

Fig. 2. Nomogram for the spherical distance of a star from the apex (taken at $\alpha = 18^h\,0^m$, $\delta = +\,30\overset{\circ}{.}0$)

If we write (24) in the form (1), we get:

$$\begin{vmatrix} \sin\alpha & 1 & 0 \\ \cos D & 0 & 1 \\ \sin 30^\circ \sin\delta & \cos 30^\circ \cos\delta & 1 \end{vmatrix} = 0. \qquad \ldots.(25)$$

from which follows:

$$\left.\begin{aligned} x_\alpha &= \frac{L \sin\alpha + M}{L'' \sin\alpha + M''} \qquad y_\alpha = \frac{L' \sin\alpha + M'}{L'' \sin\alpha + M''} \\ x_D &= \frac{L \cos D + N}{L'' \cos D + N''} \qquad y_D = \frac{L' \cos D + N'}{L'' \cos D + N''} \\ x_\delta &= \frac{L \sin 30^\circ \sin\delta + M \cos 30^\circ \cos\delta + N}{L'' \sin 30^\circ \sin\delta + M'' \cos 30^\circ \cos\delta + N''} \\ y_\delta &= \frac{L' \sin 30^\circ \sin\delta + M' \cos 30^\circ \cos\delta + N'}{L'' \sin 30^\circ \sin\delta + M'' \cos 30^\circ \cos\delta + N''} \end{aligned}\right\} \qquad \ldots.(26)$$

An alignment diagram of formula (24), therefore, consists of straight-line scales for α and D and a curved scale (arc of a conic section) for δ. A suitable choice must now be made of the arbitrary constants in the D scale. Here a projective transformation of a linear cos D scale should be introduced which neutralizes the congestion of the D values as much as possible. As this congestion occurs at both ends of the D range (0° and 180°), whilst transformations possible within the framework of formula (26) work only in one direction, we design the nomogram only for $0^\circ \leqq D \leqq 90^\circ$. The interval $90^\circ \leqq D \leqq 180^\circ$ can be represented easily by additional lettering of the nomogram. It turned out from numerical experiments that a satisfactory representation of the whole range $0^\circ \leqq D \leqq 90^\circ$ by a single continuous scale is impossible. Therefore, a triple division has been made in the accompanying nomogram. The three partial nomograms apply to the D intervals:

(1) $91^\circ \geqq D \geqq 19^\circ$ and $89^\circ \leqq D \leqq 161^\circ$.

(2) $21^\circ \geqq D \geqq 0^\circ$ and $159^\circ \leqq D \leqq 180^\circ$.

(3) $1^\circ\!.5 \geqq D \geqq 0^\circ$ and $178^\circ\!.5 \leqq D \leqq 180^\circ$.

The arbitrary constants have the values:

	L	M	N	L'	M'	N'	L''	M''	N''
1	− 196	+ 196·0	+490	+ 212·54	+ 212·54	0	− 1	+ 4·3209	+ 2·5
2	− 85	− 85·733	+ 86·70	+ 6·0781	+ 4·94481	− 4·3781	− 1	− 1·013333	+ 1·02
3	− 85	− 85·0015	+ 85·0064	− 88·0846	− 88·09085	+ 88·094	− 1	− 1·000050	+ 1·000075

....(27)

Among them, the numerical values of N''/L'' are of particular interest.

In the completed nomogram the letters α, δ, D are framed in bold single rules for $0^\circ \leqq D \leqq 90^\circ$, in thin single rules for $90^\circ \leqq D \leqq 180^\circ$, and the suffixes 1, 2, 3 are attached to them according to the above division. It is best for the user to look for the given pair of values α, δ first in system $\boldsymbol{/\alpha/_1}$, $\boldsymbol{/\delta/_1}$. If he gets an intersection of the index line with the $\boldsymbol{/D/_1}$ scale, then he has found the value of D. But if one does not find the given δ in $\boldsymbol{/\delta/_1}$ or if otherwise the point of intersection of the index line with the D_1 line occurs *below* the D_1 scale, then one should go over to system $/\alpha/_1$, $/\delta/_1$ (and, consequently, $/D/_1$). If on the other hand, the intersection point occurs *above* the D_1 scale, one goes over to $\boldsymbol{/\alpha/_2}$, $\boldsymbol{/\delta/_2}$ or to $/\alpha/_2$, $/\delta/_2$ respectively. If this gives $D < 1\overset{\circ}{.}5$ or $D < 178\overset{\circ}{.}5$, then it is recommended that one use system $\boldsymbol{/\alpha/_3}$, $\boldsymbol{/\delta/_3}$, $\boldsymbol{/D/_3}$ or $/\alpha/_3$, $/\delta/_3$, $/D/_3$ respectively. The whole process is, as one becomes more accustomed to it, much simpler than one might suppose from reading a description of it.

The reader will notice that within a large interval of the scale D_2, and, even more of D_3, one degree of D is represented by a comparatively greater length than in the scale D_1. This arrangement has been made because in the almost linear D_1 scale one can in favourable circumstances read off the value of D to $\pm\, 0\overset{\circ}{.}1$ or $\pm\, 0\overset{\circ}{.}2$ by linear interpolation by eye, whilst the slightly non-linear scales D_2 and, especially, D_3 allow much less interpolation.

The weakest point of the whole nomogram is probably this, that in system 1 the result is often represented by a strongly "extrapolatory" position of the D point, as this point occurs on the *extension* of the portion of the straight line between the α and δ points. It is not difficult, but it would take too long to explain here an additional nomogram by which one can eliminate this imperfection.

In order to illustrate the use of the nomogram, let us find the distance of Sirius ($\alpha = 6^{h}\,43^{m}$, $\delta = -\,16\overset{\circ}{.}65$) from the apex. Putting the index line in system $\boldsymbol{/\alpha/_1}$, $\boldsymbol{/\delta/_1}$ through the points $\alpha = 6^{h}\,43^{m}$, $\delta = -\,16\overset{\circ}{.}65$ we see that the point of intersection on the D_1 scale occurs below the figure; therefore, we try to use system $/\alpha/_1$, $/\delta/_1$; as the intersection point there occurs above the figure, we have to use system $/\alpha/_2$, $/\delta/_2$; there we find $D = 163\overset{\circ}{.}4$.

If the right ascension of the apex is not, as has been assumed here, $18^{h}\,00^{m}$, but a different value α^*, then one can use the present nomogram by entering the α scale with the value $\alpha + (18^{h}\,00^{m} - \alpha^*)$. But if the declination of the apex differs from $30\overset{\circ}{.}0$, then one cannot use the nomogram. The author intends to publish elsewhere a nomogram for the more general case of a variable apex position.

The accompanying bibliography may be useful to astronomers who are interested in nomograms. In addition to nomograms which are ready for use, it also lists papers from which some information may be taken for the designing of nomograms. It should be added that the bibliography cannot claim to be complete. The author is also well aware that some astronomical nomograms in the non-German literature may have escaped his notice.

In conclusion, the author would very much like to emphasize once more that nomographical methods of calculation can be most useful in different fields of astronomy and should, therefore, be applied to a greater extent than has been customary. In designing the nomograms, however, one would often have to extend the simple representations dealt with most frequently in literature by means of projective transformations.

Bibliography of Astronomical Nomograms

General and Historical Remarks

Luckey [59], [60], [61].

Mathematical Tables

Logarithms and antilogarithms: Lacroix-Ragot [51]. (40 pp.: 5 figures; 6 pp.: 4 figures.)
Logarithms and natural trigonometric functions: Rohrberg [99]. (30 pp., 4–5 figures.)
Elementary and higher functions: Emde [28], [29].

Statistical Computations

Various problems: Koller [49] (15 nomograms).
Auxiliaries for dissecting frequency curves: B. Strömgren [122].

Spherical Trigonometry

Maurer [65]; Becker [9]; Láska [52], Sections 8–10; d'Ocagne [38], [39], Sections 113–18; Friauf [33]; Wood [139].

Calendar

Perpetual calendar: Saldaña [100].

Rising and Setting, especially of Sun and Moon

Hour angle at rising and setting: Perret [6]; Ambronn-Domke [3].
Azimuth at rising and setting: Ambronn-Domke [3].
The Sun's hour angle at rising and setting: Collignon [22]; Luckey [59].
The Sun's azimuth at rising and setting: Wood [138]; Terheyden [124].
Charts of sunrise and sunset in Central Europe: Schütte [104].
Civil twilight anywhere on the Earth: Schütte [102].
Arctic day and arctic night for arbitrary values of geographical latitude and obliquity of ecliptic: Strassl [116].
Auxiliary diagrams for computing moonrise and moonset within a restricted area, especially for Germany (starting point $\phi = 50°$, $\lambda = 0°$): Schütte [103].

Azimuth and Zenith Distance

Transformations between equatorial and horizontal co-ordinates: Favé and Rollet de l'Isle [30]; Florian [32]; d'Ocagne [73]; Littlehales [55]; Maurer [65]; Radler de Aquino [92]; Alessio [1]; K. Schwarzschild [109]; Harms [39]; L. Becker [8]; Schwerdt [110]; E. Kohlschütter [48]; Immler [45]; Maurer [66]; *Deutsche Versuchsanstalt f. Luftfahrt* [26]; Schütte [108]; Slaucitajs [111].
Azimuth: Weir [134]; Littlehales [56]; Perret [83]; Maurer [64]; Reuter [95]; Logan [57]; Bell [10]; Wedemeyer [131], [132]; Henry [40]; Prüfer [90]; Schütte [107]; Wedemeyer [133]; Deutsche Seewarte [25]; Strassl [119].
Orientation with the Sun: Luckey [58], [59]; Schütte [105]; Werner [136].
Zenith distance: d'Ocagne [73]; Luckey [59]; Hamanke [37]; B. Strömgren [123]; Boyer [12]; Burmester [17]; Strassl [119].

Determination of Position

Particular problems, especially in navigation: Pesci [87]; Vital [128], [129]; Perret [84], [85]; Ambronn-Domke [2]; Ogura [76]; Burkewicz [16]; Hydrographic Department Tokyo [44]; Maurer [65]; Feldhusen [31]; Herrick [42]; Aymat [4]; Roelofs [98]; Rodrigues [97]; Terheyden [124].
Horizontal co-ordinates of Polaris: Delville [23]; Brehmer [13]; Kempinski [46]; Roelofs [98].
Longitude and latitude from altitudes of stars (position line methods): Perret [86]; Deutsche Seewarte [24]; Schütte [106]; Steppes [113]; García [35] and [36]; Herrick [41]; Roelofs [98].
See also above: "Azimuth and Zenith Distance".

Pendulum Observations

Air density corrections: Chramow [18].

Spheroids

Equation $r^2 \sin^2 \psi + z^2 \tan^2 \psi = 1$, occurring in computing potential and attractive force of spheroids: PEREK [82].

Refraction

General: WOOD [139] (accuracy needed for navigational purposes); ROELOFS [98].
Differential refraction: MÖLLER [69].

Precession

α and δ:
- Annual values: PIOTROWSKI [88].
- Annual values and values for t years: STRASSL [114], [118], p. 163, [117] (0° to − 24°).
- Orbital elements: MÖLLER [70].

Occultations of stars by the Moon

RIGGE [96]; D'OCAGNE [72]; PERRET [85].

Ephemerides

KEPLER'S equation:

- $0 < e < 0{\cdot}20$: WOOD [139].
- $0 < e < 0{\cdot}25$: LÁSKA [53] (also corrections to approximate solution).
- $0° < \sin^{-1} e < 20°$: BAŽENOW [6].
- $0 < e < 1$: RADAU [91]; CHRÉTIEN [19]; RAND [94]; STRASSL [115] and [114] (also corrections to approximate solution).

Position in an elliptic orbit: FRY [34].

Analogue of KEPLER'S equation for hyperbolic orbits: BAŽENOW [4].

Orbits

Gauss's equation: PIZZETTI [89]; WITT [137]; BAŽENOW [6].
Heliocentric and geocentric distances in elliptic orbits: THERNÖE [125]; BUCERIUS [15].
Equation of eighth degree for heliocentric distances: BAŽENOW [7].
An equation of fourth degree for an auxiliary quantity in computing i and ☊: BAŽENOW [7].
Geocentric distance in parabolic orbits: B. STRÖMGREN [121].
Thiele's transformation: LINDOW [54].

Heliographic Co-ordinates

WALDMEIER [130]; TÜZEMEN [126].

Comets

The data $q = a(1 - e)$ and $P = a^{3/2}$: RAND [93].
Apparent brightness: KRESÁK [50].

Galactic Co-ordinates (Pole at $12^h\ 40^m$, + 28°)

Longitude and latitude:
- PEARCE-HILL [79], OHLSSON [77] — Intersection diagrams.
- BALDWIN [5], PEREK [81], STRASSL [118], pp. 219–22 — Alignment diagrams.

Rectangular co-ordinates: PEREK [44] (x-axis towards galactic centre, assumed at $l = 327°$).

Parallaxes

Conditions for parallax determinations: HINKS [81].

Proper Motions

Computation of proper motions from measurements on photographic plate pairs: LUNDBY [62].

Space Velocity

Computation of space velocity from μ_α, μ_δ, radial velocity, and parallax: PEREK [81].

Apex

Angular distance from star to apex:

SMART [112]	(apex at $18^h\ 0^m$, $+ 34°$)	Intersection diagrams.
PEARCE-HILL [79]	(apex at $18^h\ 4^m$, $+ 28°$)	
BALDWIN [5]	(apex at 28°)	Alignment diagrams.
STRASSL [120]	(apex at $18^h\ 0^m$, $+ 30°$)	

Position angle of apex:

PEARCE-HILL [79] (apex at $18^h\ 4^m$, $+ 28°$) Intersection diagram.
BALDWIN [5] (apex at 28°). Alignment diagram.

Parallactic component of radial velocity:

PEARCE-HILL [80] (apex at $18^h\ 4^m$, $+ 28°$, $V_\odot = 20$ km per sec.). Intersection diagram.
DUFTON [27] (apex at $18^h\ 4^m$, $+ 28°$, $V_\odot = 20$ km per sec.). Alignment diagram.

Stellar Magnitudes and Photometry

Stellar magnitudes and intensity ratios: STRASSL [114], [118], p. 129.
Atmospheric absorption:
Correction of photographic magnitudes (ϕ = constant); BOYER [12].
Corrections at arbitrary wavelengths in the spectrum (ϕ = constant): STRASSL [120].
Apparent magnitude, absolute magnitude, and distance, with uniform interstellar absorption: STRASSL [114], [118], p. 177.

Spectroscopy

Adjusting a diffraction-grating spectroscope: NEWALL [71].

Radiation and Ionization

Black-body radiation (PLANCK's formula):
General review: BOTTLINGER [11].
$10° < T < 50{,}000°K$, $0{\cdot}1\mu < \lambda < 100\mu$; $800° < T < 7{,}000°K$, $0{\cdot}4\mu < \lambda < 1\mu$: HANSEN [38].
Computation of colour temperatures: KIENLE [47].
Thermal ionization:
SAHA's formula: UNSÖLD [127].
SAHA's formula, specified for the Sun's atmosphere: CLAAS [20]; [21].

Double Stars

Mass function $Q = m_2^3/(m_1 + m_2)^2$: PARENGO [78].
Computation of elements of eclipsing binaries: SCHARBE [101]; BROWN [14]; MERRILL [67].

Dark Nebulae

Some problems, connected with total absorption, distance, and radial extension of dark nebulae: LYTTKENS [63].

REFERENCES

[1] ALESSIO, A.; diagram published by Istituto Idrografico, Genoa; description in *Marineblad*, **23,** 895, 1909, De Helder.
[2] ALLCOCK, H. J. and JONES, J. R.; *The Nomogram.* Revised by J. G. L. MICHEL. 4th ed. London, 1950. (pp. x + 238.)
[3] AMBRONN, L. and DOMKE, J.; *Astronomisch-geodätische Hilfstafeln.* Berlin, 1909. (pp. vi + 142, containing 15 nomograms.)
[4] AYMAT, J. M.; *Revista Aeronáutica*, 1946 February, 39–47 (Spanish).
[5] BALDWIN, J. M.; *M.N.*, **89,** 453, 1929. (Description of the chart which was published separately and copies deposited with the R.A.S.)
[6] BAŽENOW, G.; *Astron. Nachr.*, **234,** 425, 1929.
[7] BAŽENOW, G.; *Astron. Nachr.*, **234,** 427, 1929.
[8] BECKER, L.; *Ann. d. Hydrographie u. maritimen Meteorol.*, **58,** 401, 1930; **59,** 36, 1931.

[9] BECKER, L.; "Curves for solving a spherical Triangle". University of Glasgow, 1934. Description in: *M.N.*, **91**, 226, 1930.
[10] BELL, H.; *Proc. Roy. Soc. Edinburgh*, **36**, 192, 1915–16.
[11] BOTTLINGER, K. F.; *Handbuch der Astrophysik*, II_1, 353. Berlin, 1929.
[12] BOYER, R. F.; *Popul. Astron.*, **43**, 14, 1935.
[13] BREHMER; *Ann. d. Hydrographie u. maritimen Meteorol.*, **40**, 192, 1912.
[14] BROWN, O. E.; *Astron. J.*, **47**, 93, 1938.
[15] BUCERIUS, H.; *Astron. Nachr.*, **280**, 73, 1951/52 (Figs. 1 and 2).
[16] BURKEWICZ, G.; *Hydrographic Reports Leningrad* 1934, No. 1 (Russian).
[17] BURMESTER, H.; *Der Seewart*, 1939, 157.
[18] CHRAMOW, D.; *Bull. Astron. Inst. Leningrad*, No. 37, p. 288, 1935.
[19] CHRÉTIEN, H.; *Assoç. Franç. pour l'Avancement des Sciences*, **36**, 83, 1907, Reims.
[20] CLAAS, W. J.; *Bull. Astron. Inst. Netherlands*, **9**, 264, 1942.
[21] CLAAS, W. J.; *Recherches Astronomiques de l'Observatoire d'Utrecht*, XII, Part I, 1951.
[22] COLLIGNON, E.; *Nouvelles Annales de Mathémat.*, 2e série, **18**, 179, 1879; 3e série, **1**, 490, 1882.
[23] DELVILLE, E.; *Proc. Roy. Soc. Canada*, Sect. III, 1906, p. 3.
[24] Deutsche Seewarte; Ortsbestimmung durch astronomische Beobachtung zweier gegebener Fixsterne mit Hilfe von Höhendiagrammen. Oberkommando der Kriegsmarine 1940. (Six volumes with diagrams for latitude zones of 10°.)
[25] Deutsche Seewarte; Azimutdiagramme für alle Breiten, Deklinationen und Stundenwinkel. 2nd ed., Hamburg, 1944.
[26] Deutsche Versuchsanstalt für Luftfahrt: Astronomischer Rechenatlas. Berlin, 1943.
[27] DUFTON, A. F.; *M.N.*, **92**, 688, 1932.
[28] EMDE, F.; *Tables of Elementary Functions*, 2nd ed., Leipzig, 1948, pp. 181.
[29] EMDE, F. (JAHNKE-EMDE); *Tables of Higher Functions*, 4th ed., Leipzig, 1948, pp. 300.
([28]–[29]: Numerical tables, supplemented by numerous graphic representations and intersection diagrams.)
[30] FAVÉ and ROLLET DE L'ISLE; *Ann. Hydrographiques*, 1892.
[31] FELDHUSEN, W.; *Veröff. d. Deutschen Seewarte*, 1944, 115.
[32] FLORIAN, H.; *Mittheilungen aus dem Gebiet des Seewesens*, **28**, 341, 1900, Wien.
[33] FRIAUF, J. B.; *J. Franklin Inst.*, **232**, 151, 1941.
[34] FRY, TH. C.; *A.J.*, **29**, 141, 1916.
[35] GARCÍA, J.; *Introducción a una Geometria de los Valores*. Imprenta del Ministerio de Marina. 1940, pp. 50.
[36] GARCÍA, J.; *Revista general de Marina*, 1940 November, 1940 December, 1941 May.
[37] HAMANKE, E.; *Ann. d. Hydrographie u. maritimen Meteorol.*, **55**, 293, 1927.
[38] HANSEN, G.; *Optik*, **1**, 227, 1946 (Figures 5 and 6).
[39] HARMS, M.; *Ann. d. Hydrographie u. maritimen Meteorol.*, **58**, 98, 1930.
[40] HENRY; *Table graphique d'azimuts à l'usage de la navigation maritime et aérienne*. Vannes, 1930.
[41] HERRICK, S.; *Astron. Papers*, University of California, No. 2, 1943.
[42] HERRICK, S.; *Geogr. Rev.*, **34**, 436 = *Astron. Papers*, University of California, No. 3, 1944.
[43] HINKS, A. R.; *M.N.*, **98**, 440, 1898.
[44] Hydrographic Department Tokyo; *Nomograms for use in navigation*. 1938.
[45] IMMLER, W.; *Der Seewart*, **10**, 117, 1941.
[46] KEMPINSKI, F.; *Inst. d'Astron. Prat. de l'École Polytechn. de Varsovie*, Publ. No. 15, p. 11, 1936.
[47] KIENLE, H.; *Z. Astrophys.*, **20**, 239 = Mitt. Obs., Potsdam, No. 8, 1941.
[48] KOHLSCHÜTTER, E.; *Messkarte zur Auflösung sphärischer Dreiecke*, 3rd ed., Berlin, 1936.
[49] KOLLER, S.; *Graphische Tafeln zur Beurteilung statistischer Zahlen*, 2nd ed., Dresden and Leipzig, 1943; 3rd ed., Darmstadt, 1953.
[50] KRESÁK, L.; *Bull. Astron. Inst. Czechoslovakia*, **4**, 72, 1953.
[51] LACROIX, A. and RAGOT, Ch. L.; *A Graphic Table of Logarithms and Antilogarithms*. New York, 1925.
[52] LÁSKA, W.; *Lehrbuch der Astronomie* I, 2, Aufl., Bremerhaven and Leipzig, 1906.
[53] LÁSKA, W.; *Astron. N.*, **199**, 296, 1914.
[54] LINDOW, M.; *Astron. N.*, **214**, 263, 1921.
[55] LITTLEHALES, G. W.; *Bull. Philos. Soc. Washington*, **14**, 233 = *Proceed. of the United States Naval Institute, Annapolis*, **29**, 953, 1904. (Review by CASPAR in *Ann. d. Hydrographie u. maritimen Meteorol.*, **32**, 242, 1904.)
[56] LITTLEHALES, G. W.; *Pilot Chart of the North Atlantic Ocean*. 1904.
[57] LOGAN, G. W.; *Proc. of the United States Naval Institute, Annapolis*, **34**, 633, 1908.
[58] LUCKEY, P.; *Z. f. math. u. naturwiss. Unterricht*, 1921, pp. 168–75.
[59] LUCKEY, P.; Sirius, **55** (Neue Folge, **50**), 21 and 52, 1922.
[60] LUCKEY, P.; *Unterrichtsblätter f. Math. u. Naturwissenschaften*, **20**, 54, 1923.
[61] LUCKEY, P.; *Astron. Nachr.*, **230**, 1, 1927.
[62] LUNDBY, A.; *Uppsala Astron. Obs. Ann.*, **1**, No. 10 (nomogram on pp. 11–13), 1946.
[63] LYTTKENS, E.; *Uppsala Astron. Obs. Ann.*, **2**, No. 9 (3 nomograms on pp. 35, 40, 46), 1951.
[64] MAURER, H.; *Ann. d. Hydrographie u. maritimen Meteorol.*, **33**, 125 and 323, 1905.
[65] MAURER, H.; *Ann. d. Hydrographie u. maritimen Meteorol.*, **33**, 355, 1905.
[66] MAURER, H.; *Der Seewart*, **10**, 149, 1941.
[67] MERRILL, J. E.; *Astron. J.*, **52**, 26, 1947. (Description of the nomograms only.)
[68] MEYER ZUR CAPELLEN, W.; *Leitfaden der Nomographie*. Berlin-Göttingen-Heidelberg, 1953 (pp. 178).

[69] MÖLLER, J. P.; *Publikationer og mindre Meddelelser fra Köbenhavns Observatorium*, No. 57, 1, 1926.
[70] MÖLLER, J. P.; *Astron. N.*, **249**, 63 = *Publikationer og mindre Meddelelser fra Köbenhavns Observatorium*, No. 88, V, 1933.
[71] NEWALL, H. F.; *M.N.*, **52**, 510, 1892.
[72] D'OCAGNE, M.; *C.R. Acad. Sci. (Paris)*, **130**, 554, 1900.
[73] D'OCAGNE, M.; *C.R. Acad. Sci. (Paris)*, **135**, 728, 1902.
[74] D'OCAGNE, M.; "Edinburgh Lecture of 1914 July 29", published in *Enseignement Mathématique*, **19**, 20, 1917.
[75] D'OCAGNE, M.; *Traité de Nomographie*, 2e éd., Paris, 1921. (pp. xxii + 484.)
[76] OGURA, S.; *Suiro Yôhô* (Hydrographic Bulletin), **5**, 293, 1926.
[77] OHLSSON, J.; *Lund Ann.*, No. 3, 1932. (Diagram as Appendix to numerical tables.)
[78] PARENAGO, P. P.; *Russ. Astron. J.*, **27**, 41, 1950.
[79] PEARCE, J. A. and HILL, S. N.; *Publ. Dominion Astrophys. Obs. Victoria* IV, No. 4, 1927.
[80] PEARCE, J. A. and HILL, S. N.; *Publ. Dominion Astrophys. Obs. Victoria* VI, No. 4, 1931.
[81] PEREK, L.; *Contrib. Astron. Inst. Masaryk Univ., Brno*, **1**, No. 4, 1947. (Contains 9 nomograms in half size; nomograms in original size obtainable from the publishing institute.)
[82] PEREK, L.; *Bull. Astron. Inst. Czechoslovakia*, **2**, No. 5, 1950.
[83] PERRET, E.; *Ann. Hydrographiques*, 1904.
[84] PERRET, E.; *Assoc. Franc. pour l'Avancement des Sciences*, Cherbourg, 1905, pp. 80–102.
[85] PERRET, E.; *Ann. Hydrographiques*, **26**, 170, 1905.
[86] PERRET, E.; *Ann. Hydrographiques*, **47**, 42, 1906.
[87] PESCI, G.; *Rivista Marittima*, 1896, 1897, 1898, 1899.
[88] PIOTROWSKI, S.; *Acta Astron.*, c **33**, 1937.
[89] PIZZETTI, P.; *Tabelle grafiche per la risoluzione approssimata di una equazione di Gauss che si incontra nel calcolo delle orbite*. Pisa, 1910.
[90] PRÜFER, G.; *Ann. d. Hydrographie u. maritimen Meteorol.*, **69**, 331, 1941.
[91] RADAU, R.; *Bull. astron. (Paris)*, **1**, 381, 1884; **2**, 401, 1885; **17**, 37, 1900.
[92] RADLER DE AQUINO; *Proc. of the United States Naval Institute, Annapolis*, **34**, 633, 1908. (Explanation of the nomograms by LAFAY and PESCI.)
[93] RAND, W. C.; *J. Brit. Astron. Assoc.*, **52**, 104, 1942.
[94] RAND, W. C.; *J. Brit. Astron. Assoc.*, **52**, 149, 1942.
[95] REUTER, W.; *Ann. d. Hydrographie u. maritimen Meteorol.*, **34**, 72, 1906.
[96] RIGGE, W. F.; *Astron. Nachr.*, **140**, 81, 1896.
[97] RODRIGUES, O. A. DE AXEREDO; *Graficos usados na navegação*. Marinha de Brasil, Hydrografia & Navegação, 1948.
[98] ROELOFS, R.; *Astronomy applied to Land Surveying*. Amsterdam, 1950. (pp. xv + 259; containing 17 nomograms by N. D. HAASBROEK.)
[99] ROHRBERG, A.; *Graphische Funktionentafeln*. Berlin, 1949.
[100] SALDAÑA, A.; *Urania (Madrid)*, **38**, 20, 1953.
[101] SCHARBE, S.; *Poulkovo Obs. Bull.*, **10**, 31, 1924.
[102] SCHÜTTE, K.; *Meteorol. Z.*, 1936, 1.
[103] SCHÜTTE, K.; *Ann. d. Hydrographie u. maritimen. Meteorol.*, **65**, 422, 1937.
[104] SCHÜTTE, K.; *Wann geht die Sonne auf und unter?* 2. Aufl. Bonn, 1940. (pp. 10 + 37.)
[105] SCHÜTTE, K.; *Ann. d. Hydrographie u. maritimen Meteorol.*, **68**, 215, 1940.
[106] SCHÜTTE, K.; *Archiv der deutschen Seewarte und des Marine-Observatoriums*, **61**, No. 7, 1941.
[107] SCHÜTTE, K.; *Ann. d. Hydrographie u. maritimen Meteorol.*, **70**, 361, 1942.
[108] SCHÜTTE, K.; *Ann. d. Hydrographie u. maritimen Meteorol.*, **71**, 193, 1943.
[109] SCHWARZSCHILD, K.; *Z. f. Instrumentenkunde*, **30**, 75, 1910.
[110] SCHWERDT, H.; *Die Anwendung der Nomographie in der Mathematik*. Berlin, 1931. (pp. vii + 116 + 104 charts.)
[111] SLAUCITAJS, S.; *Contrib. from the Baltic University Hamburg-Pinneberg*, No. 13, 1946.
[112] SMART, W. M.; *M.N.*, **83**, 465, 1923. (Description of charts only; charts on 8 sheets published separately and deposited with the R.A.S.)
[113] STEPPES, O.; *Ann. d. Hydrographie u. maritimen Meteorol.*, **69**, 335, 1941.
[114] STRASSL, H.; *Veröff. d. Univ.-Sternwarte, Bonn*, Nr. 36, 1949. (7 nomograms.)
[115] STRASSL, H.; *Astron. Nachr.*, **279**, 25, 1950.
[116] STRASSL, H.; *Polarforschung (Kiel)*, **2**, 40, 1951.
[117] STRASSL, H.; Additional sheet to the second edition of the Southern BD-Charts. Bonn, 1951.
[118] STRASSL, H.; 7 nomograms in *Landolt-Börnstein: Zahlenwerte und Funktionen*, Band III, Astronomie u. Geophysik. Berlin, 1952.
[119] STRASSL, H.; *Math.-phys. Semesterberichte*, **2**, 271, 1952. (Gottingen.)
[120] STRASSL, H.; "Nomograms in Astronomy" in *Vistas in Astronomy*, Ed. A. BEER, Pergamon Press, London, 1955.
[121] STRÖMGREN, B.; *Publikationer og mindre Meddelelser fra Köbenhavns Observatorium*, No. 66, 1929.
[122] STRÖMGREN, B.; *Publikationer og mindre Meddelelser fra Köbenhavns Observatorium*, No. 93, II, 1934.
[123] STRÖMGREN, B.; *Nordisk Astron. Tidskrift*, **16**, 45, 1935.
[124] TERHEYDEN, K.; *Veröff. d. Univ.-Sternwarte, Bonn*, Nr. 38, 1951. (20 nomograms.)
[125] THERNÖE, K. A.; *Astron. Nachr.*, **260**, 145 = *Publikationer og mindre Meddelelser fra Köbenhavns Observatorium*, No. 110, I, 1936.
[126] TÜZEMEN, E.; *Publ. Istanbul Univ. Obs.*, No. 40, 1941. (Nomogram reprinted in: W. GLEISSBERG; *Die Häufigkeit der Sonnenflecken*, Berlin, 1952, p. 56.)

[127] UNSÖLD, A.; *Z. techn. Physik*, **16**, 460, 1935. (Nomograms reprinted in *Physik der Sternatmosphären*, Berlin, 1938, pp. 66–9.)
[128] VITAL, A.; *Mittheilungen aus dem Gebiet des Seewesens*, **28**, 201, 1900.
[129] VITAL, A.; *Mittheilungen aus dem Gebiet des Seewesens*, **28**, 267, 1900.
[130] WALDMEIER, M.; *Tabellen zur heliographischen Ortsbestimmung*. Basel, 1950. (p. 62, etc.; containing 15 intersection diagrams.)
[131] WEDEMEYER, A.; *Ann. d. Hydrographie u. maritimen Meteorol.*, **46**, 209, 1918.
[132] WEDEMEYER, A.; *Ann. d. Hydrographie u. maritimen Meteorol.*, **47**, 233, 1919.
[133] WEDEMEYER, A.; *Ann. d. Hydrographie u. maritimen Meteorol.*, **70**, 375, 1942.
[134] WEIR, P.; *Proc. Roy. Soc. Edinburgh*, **16**, 354, 1888–1889. (Diagram 76 × 38 cm, published by Potter, London, since 1890.)
[135] WEMPE, J.; *Astron. N.*, **275**, 1, 1947.
[136] WERNER, K.; *Ann. d. Hydrographie u. maritimen Meteorol.*, **71**, 247, 1943.
[137] WITT, G.; *Astron. N.*, **199**, 257, 1914.
[138] WOOD, H. W.; *J. Brit. Astron. Assoc.*, **53**, 61, 1943.
[139] WOOD, H. W.; *J. and Proc. Roy. Astron. Soc. of New South Wales*, **79**, 153. (= Sydney Obs. Paper No. 1, 1946.)

The Use of Electronic Calculating Machines in Astronomy

PETER NAUR

Köbenhavns Universitets Astronomiske Observatorium,
Copenhagen, Denmark

SUMMARY

A short description of the structure of the modern digital electronic computers, and the organization of the work with them, is given. The various types of astronomical computational problems are discussed. The two conclusions are: (1) the electronic computers are likely to be of the greatest importance in astronomical problems which involve a large amount of observational material and a complex theory; (2) if the electronic computers are to be fully utilized for solving astronomical problems the astronomers themselves must learn to use them.

1. THE CHARACTERISTICS OF THE ELECTRONIC COMPUTERS

THE first large electronic calculating machine, the Eniac, was completed in 1946. Since this time the large-scale computing devices have established themselves as extremely useful tools in a wide field of research. The object of the present article is to discuss the computational problems of astronomy with a view of their solution by means of electronic computing machines.

Any appreciation of the vast capabilities of the modern electronic computers must start with some knowledge of their structure and way of operation. Before discussing the astronomical problems we shall therefore give a brief description of the main features of these devices. This task is facilitated by the fact that important similarities exist between most of the machines which are now in operation. Indeed, the following description is believed to apply equally well to any of the machines: Edsac (Cambridge, England), the Ferranti Computer (Manchester, England, and elsewhere), Whirlwind I (Cambridge, Massachusetts), Harvard Mark IV (Cambridge, Massachusetts), The Machine of the Institute of Advanced Study (Princeton, New Jersey), Univac (Philadelphia, Pennsylvania, and elsewhere), Seac (Washington, D.C.), Swac (Los Angeles, California), IBM type 701 (New York City and elsewhere), Cadac 102-A (Hawthorne, California), Commonwealth Computer (Australia), and probably several others.

The structure of such a machine is shown in the diagram. There are four main sections, the storage unit, including arithmetical registers, the output system, the input system, and the control system. The storage unit, sometimes called the "memory" of the machine, is capable of holding a large amount of information, usually or the order of many thousand decimal digits. In addition there are a few storage locations serving special functions during the execution of arithmetical operations, called registers. Thus, in most machines, there is one register holding the accumulated results of additions and multiplications and one is used to hold the multiplier during multiplications, but there may be more, giving additional facilities. The input and output systems are the connections between the computer and the rest of the world. They are mechanisms for converting information as it is held inside

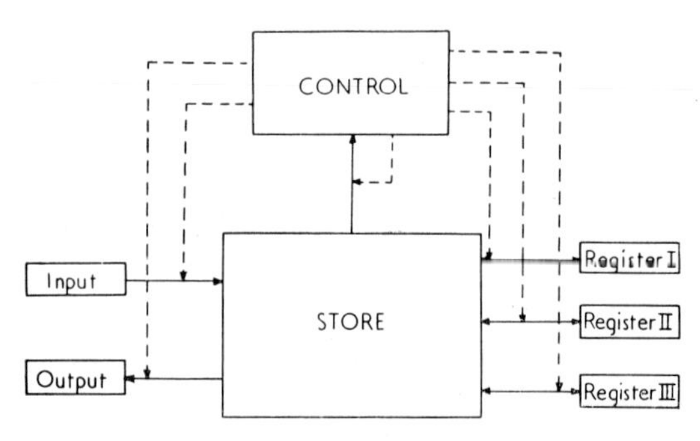

Fig. 1. The logical structure of an electronic computer. The full lines denote paths for information of the kind which can be held in the storage unit. The broken lines denote pulses from the control unit. For the sake of simplicity the connections made during test operations have not been drawn

the computer to a form which is independent of the computer, or vice versa. They differ from one machine to the other, and one machine may be equipped with several systems. The most common input and output media are punched cards, punched tape, magnetic tape, and magnetic wire.

The operation of the storage units and the input and output systems is under control of an automatic control system. This system is connected with the storage unit in a very important manner, which deserves some explanation. For, in fact, besides holding the numerical information operated upon, the storage unit also holds the information which specifies the sequence of operations which the machine has to follow in the particular problem. In other words, a particular piece of information which is present in the storage unit can be interpreted by the machine either as a number or as an operation. Thus, when the computing machine is engaged on a problem, part of the storage space is reserved for the numbers, part of it holds the instructions of operation. The single instructions are of a quite elementary nature. One may cause a single addition to be performed, another a single multiplication, *etc.*

The instructions are usually executed sequentially one after the other as they are held in consecutive locations in the store. Thus the action of the control system is a two-phase process: (1) select from the store the next instruction to be executed, and (2) execute the instruction just selected. This action proceeds at a rate of 50 to 20,000 instructions per second, depending on the particular machine.

In addition to the purely arithmetical instructions there are control, or test, instructions, which permit the action of the computer to depend on the specific values of the results obtained. These test instructions may, for instance, question the sign of a particular number and proceed from one point in the sequence of operations if it is positive, and from another if it is negative. In that way a particular set of instructions may be executed many times during the calculation.

As an illustration we will consider the following problem: suppose that location 10 of the store of the machine contains a positive, but otherwise arbitrary angle, that location 11 contains the quantity 2π, and that it is desired to place the angle, modulo 2π, in location 12. The instructions for doing this job are to be placed from location 50 and onwards. The solution of this problem is supplied by the following five elementary operations*:

50 Place the contents of location 10 in the accumulative register.
51 Subtract the contents of location 11 from that of the accumulator.
52 Test the sign of the number now present in the accumulator: if it is positive, execute next the instruction in location 51. If it is negative, continue with the normal sequence, *i.e.* with the instruction in location 53.
53 Add the contents of location 11 to the accumulator.
54 Transfer the contents of the accumulator to location 12.

Note. The figures preceding the instructions refer to the storage locations in which the orders are placed.
The key-point is the control instruction in location 52. It causes the subtraction in location 51 to be repeated once too often. Therefore one addition is required before the result can be placed in location 12.

Before we discuss the specific applications to astronomical problems we shall mention some details related to the use of the machines for physical problems. The first regards the treatment of common functions such as will often occur in astronomical problems. As mentioned already the automatic control of the machine will usually only produce very simple mathematical functions, sums, products, and perhaps quotients. It might seem that more complicated functions, trigonometric functions, exponentials, *etc.*, would most conveniently be supplied from tables, as in work with ordinary desk calculators. This, however, is by no means the case. In fact, owing to the high speed of the electronic calculators even quite complicated functions will often be generated most easily from their power expansions or by means of iterative processes. These processes will have to be stated in terms of the elementary operations of the machine, but by proper organization it is possible to plan such methods once and for all, ready to be included in larger problems and thus build up a library of prefabricated procedures, subroutines, for the machine. The instructions leading to the generation of a particular function will, of course, still have to be held in the storage of the machine in problems where they are needed, but they can be used almost blindly as they are carefully tested before they are placed in the library. As examples we mention some data relating to the library of subroutines for the Edsac at the Mathematical Laboratory, Cambridge. This machine has at the present time a useful storage capacity of 928 elementary instructions, each equivalent to a five-decimal number. Thus, if in a particular problem 200 five-place

* The machine is supposed to work with what is called an one-address code. The procedure will be quite similar when a three-address code is used.

numbers will have to be available at the same time, there will be left 728 locations for holding the instructions of the problem. The particular structure of the machine also makes operations on ten-place numbers very convenient. A subroutine for generating the cosine of a ten-decimal argument occupies fifty-nine locations, *i.e.* 6 per cent of the total storage capacity, and takes 0·07 sec. each time it is used. The subroutine for generating logarithms occupies thirty-one locations and requires about 1 sec.

The technique of using a library of prefabricated methods is not confined to the evaluation of functions such as those just mentioned, but can equally well be used to relieve the user of the task of specifying the details of the operations involved in the stepwise integration of differential equations, in the treatment of complex numbers or vectors, *etc.* The routine for advancing the variables by one step during the integration of any number of simultaneous differential equations occupies sixty-six locations.

An excellent account of the elements of programming for automatic computers has been given by HARTREE (1952). A detailed description of a library of subroutines and its use is given by WILKES, WHEELER, and GILL (1951).

2. ASTRONOMICAL PROBLEMS

We are now prepared to discuss the astronomical problems. For this purpose the numerical work involved in astronomical research may conveniently be divided into three classes:

(1) Reduction of original observations.
(2) Comparison of partly or completely reduced observations with theory.
(3) Theoretical work of a purely mathematical nature.

(1) *Reduction of Original Observations.* The work in this class is usually comparatively simple, and consists in conversion of pointer readings to digital form, formation of averages, and the application of known systematic corrections. The work of this class is definitely that which is the least suited for treatment by the powerful electronic equipment, firstly because quite often the work necessary to secure the original data, the work at the telescope, will be an appreciable fraction of the total, secondly, because a considerable amount of judgment will often go into the work, and, thirdly, because the preparation of the data to be presented to the automatic calculator from the original data (*e.g.* conversion of pointer readings to digital form) might be a load comparable to the total reduction done by hand. These objections are so serious that there does not seem to be any doubt that most of the work of this kind also in a foreseeable future will be performed with the present-day methods. However, when very large amounts of work become involved this is not so any more. In particular in problems where the last-mentioned objection is the most important one, the problem is entirely solved if it becomes worth while to build special equipment which will bridge the gap between the original data and the automatic computing machine. As an example of such equipment should be mentioned the machine for measuring astrographic plates which is under construction at the I.B.M. Watson Laboratory, New York City. This machine is designed to measure the plates used in the construction of the Yale catalogues and the measurements will automatically be recorded on punched cards. Since the punched cards can be "read" automatically

by calculating machines the complete reduction of the measurements can be carried out automatically. Another example of such a technique which might be useful in astronomy has been given by WELFORD (1952). The apparatus described in this reference is designed for use during psychological experiments and will record the times of certain events as holes in teleprinter tape, of the kind which is used as the input medium for the Edsac. Such a piece of equipment would, of course, be equally useful for recording the times of the meridian passage of stars, and thus the reduction of the observations might be carried out without any manual copying of figures whatever.

(2) *Comparison of Partly or Completely Reduced Observations with Theory.* The value of automatic devices for recording data in the form which can be accepted by an automatic calculator depends very much on what final results it is desired to produce, *i.e.* in most cases, what results one wants to publish. If each observation is going to appear in print it is obviously not economical to construct automatic devices for doing the reduction if the reduction done by hand would require an amount of work comparable to, say, proof-reading. The situation becomes quite different if the observations are to be compared with a theory which is complex enough to make the use of an electronic calculator desirable. In this case the observations will have to be prepared in a form suited to be used by the calculating machine in any case, and then there may not be any reason why the data should not be treated by the machine from the very beginning. This does not necessarily imply that the whole bulk of the work, that is the reduction of the observations and their comparison with theory, has to be done in one single process by the machine. Most modern machines provide the possibility of storing large amounts of intermediate results so that they are later rapidly available to the machine itself, the medium of this storage being, for instance, magnetic tape. Thus a very complicated procedure for the reduction, requiring the whole capacity of the machine, might be used just to prepare the raw data in the form best suited to be compared with theory, and during one or more subsequent runs the theory might be developed and compared with the observations thus prepared. It is only at this stage that the vast capabilities of the machines come into full play, for when deciding what questions to ask, what results to ask the machine to produce, it may often be possible to skip all intermediate steps and simply ask for the final, crucial, result. Two examples of conceivable plans in this direction (though to the writer's knowledge not yet realized) might make the matter clearer.

Assume that a series of photometric measurements has been obtained for a Cepheid variable, and it is desired to perform a harmonic analysis of the light curve. In this case the final questions in the problem might perhaps be the following: how many terms are necessary in the Fourier expansion to satisfy the observations within their probable errors? What are the coefficients in the expansion? What is the root mean square of the residuals? The calculation might be arranged in two parts: the first, reduction of the original measurements for extinction and night differences and evaluation of the probable error of an observation, either on the basis of measurements of the comparison star, or through formation of normal points on the light curve. The results of the reduction might be stored on magnetic tape for later use by the machine. In the second part the machine would have to calculate, one by one, the coefficients of the Fourier expansion and after each term evaluate the residuals from the observations and decide whether they are small enough. When this is the case the process of trials would have to come to a stop and the coefficients and other relevant information be printed.

The classical problem of determining a planetary orbit from three observations, though quite possible, would probably not be the most convenient solution to the problem of orbit determination by means of an electronic computer. For, in fact, it would certainly be more useful, and probably not too complicated to take all available observations into account as soon as a first approximation is available. The results to be presented by the machine would primarily be an ephemeris and sufficient data to enable the machine to continue the process of improving the orbit as soon as new observations are made.

(3) *Theoretical Work of a Purely Mathematical Nature.* Properly speaking, no astronomical work is of purely mathematical nature, since astronomy is an empirical science. However, it is often found convenient to separate the purely mathematical part of a problem from the comparison with observations. Thus we are faced with differential equations such as they occur in celestial mechanics and in the theories of stellar atmospheres and interiors. Taken by themselves these limited mathematical problems are definitely those which can most profitably be solved by the electronic calculators, because a large amount of numerical work follows from a few initial data. It is, therefore, not surprising that most, if not all, astronomical work which has so far been done with the aid of these tools falls in this class. Only one example shall be mentioned, namely the monumental work on the motion of the five outer planets which was done on the SSEC (now demolished) by W. J. ECKERT, D. BROUWER, and G. M. CLEMENCE (1951).

There is little doubt that the problems of theoretical astronomy will be increasingly solved by large automatic calculators. In this situation it seems that the value of the more complete astronomical problem, which includes the observations, should be stressed.

In conclusion, a very important point in the application of electronic computers to astronomical problems shall be touched upon, namely the question of who is going to do the detailed planning of the problems. The alternative is that the problems are handed over to special programmers, who may know little or nothing of the actual problems, or that the astronomers must learn the techniques of using the equipment. In actual practice there may not be any choice, where one particular machine is concerned. If, however, there is a choice, the latter arrangement is, in the present writer's opinion, very much to be preferred. The promise of the electronic computers to astronomy lies in the field of large problems which can nevertheless be solved in their entirety by the machines. It is unlikely that the planning of such an undertaking for the computer could be profitably carried out by someone for whom it is only one problem among many others of varying complexity, and who is not thoroughly familiar with the astronomical aspect of it. The astronomers, on the other hand, should not encounter any difficulties in becoming familiar with the modern electronic calculators, owing to their logical simplicity.

Summarizing, the outstanding features of the modern electronic computers are the large capacity for data, the great flexibility, and the speed of operation. They are eminently suited for solving astronomical problems as soon as these involve at least a moderately large amount of calculation, and it seems probable that the main obstacle in the utilization of them in astronomy will be the reluctance of the astronomers at entering what may seem a new field of research.

REFERENCES

ECKERT, W. J., BROUWER, D. and CLEMENCE, G. M. 1951 *Astronomical Papers Prepared for the Use of the American Ephemeris* and *Nautical Almanac*, Vol. XII.
HARTREE, D. R. . . . . 1952 *Numerical Analysis*, Chapter XII (Oxford).
WELFORD, N. T. . . . . 1952 *J. sci. Instrum.*, **29**, 1.
WILKES, M. V., WHEELER, D. J. and GILL, S. . 1951 *The Preparation of Programs for an Electronic Digital Computer* (Cambridge, Massachusetts).
Articles about many of the modern electronic computers will be found in the later volumes of *Mathematical Tables and Other Aids to Calculation.*

Air Disturbance in Reflectors

W. H. STEAVENSON
Cambridge, England

THE injury to telescopic definition caused by atmospheric turbulence has been the bane of visual observers since the time of HERSCHEL. On the other hand, for a long time after the introduction of photographic methods, the enlargement of star-images from this cause remained small in relation to the inevitable spreading of light in the sensitive emulsion, which with moderate focal lengths amounted often to several seconds of arc. But with the greatly increased image-scale of the large reflectors used to-day, serious loss of effective resolving power may result from enlargements exceeding a single second, so that the elimination of atmospheric effects, where this is possible, has become a matter of great importance.

The effects of turbulence at high altitudes in the atmosphere have long been recognized, and A. E. DOUGLASS pointed out some sixty years ago* how the heights of the disturbing airstreams could be measured. There seems no doubt that the discontinuities of optical density which give rise to the irregularities observed are connected with local and global distributions of atmospheric pressure, as indicated by the formation of cyclones and anticyclones; and in this connection the late W. H. PICKERING showed that steadier conditions are to be expected at places in and near the tropics. In fact latitude is at least as important as altitude in the selection of observatory sites.

But even when the best site has been chosen, so that high turbulence has been reduced to a minimum, there still remain those thermal irregularities which may affect the air in the telescope itself. I have been made painfully aware of these disturbances during the past eight years, while using a 30-in. glass mirror for visual observations of various kinds, and it occurs to me that they must be equally destructive of good definition in photographic work with instruments of comparable and greater size.

* See his article in *Popul. Astron.*, June, 1897.

The nature of the various disturbances is best demonstrated by putting the image of a bright star considerably out of focus. On the large disk thus produced the high disturbances show themselves as a series of parallel straight lines, like a sheaf of telegraph wires, the appearance being an effect of persistence of vision, acting on dark and bright patches moving with great velocity across the line of sight. But in addition to this stream effect there are generally great numbers of discrete dark spots, all in wriggling motion with a tendency to move very slowly towards that part of the disk corresponding to the upper edge of the mirror. Some experiments with a bicycle pump confirmed the suspicion that these disturbances were in actual contact with the surface, and attempts were made to remove them by the action of a large electric fan. The result was disappointing, the effect being merely to break up the disturbances into smaller units and cause them to rush about in greater confusion. They are evidently due to the mirror continually losing heat to the surrounding air, and the appearance was found to change according to the difference of temperature between the latter and the glass disk, which weighs about 200 lb.

After a warm day, when the Sun had been shining on the dome for many hours, the disturbing elements were numerous, perhaps amounting to a hundred or more, and the general effect was that of a seething mass of maggots. But when the difference of temperature between mirror and air was of the order of only 10° or 12°, the numbers would be reduced to one or two dozen, and each would be more obviously elongated and would display a sinuous motion. Finally, when the temperature difference was only 2° or 3°, there would be nothing to see but three or four snake-like disturbances, with lengths comparable with the diameter of the mirror. In the first two cases the blurring and enlarging effect on the focused image of a star would be very marked, and this on nights when the definition with the 25-in. Newall refractor was excellent.

It would seem that the only way of eliminating disturbances of this kind would be so to insulate the walls and dome of the observatory that the daily range of temperature within it was reduced to a very few degrees. Failing this, I feel convinced that no large photographic reflector can ever attain the full resolving power of which it should be capable. Certainly it would be hopeless to expect that a large glass mirror, mounted in an observatory of the normal type, could give the best visual definition under average conditions. In fact it is only in the winter, after a cloudy day, that I have come at all near to getting perfect images with the 30-in. The old observers, such as HERSCHEL, ROSSE, and LASSELL, were probably more fortunate in this respect, as their metal mirrors, used in the open air, must have parted rapidly with their heat at nightfall, so that surface currents would be greatly reduced.

Trouble from air disturbances in the body of the tube is, of course, largely eliminated when the latter, as in the case of my own telescope, is wholly of skeleton construction, right down to the surface of the mirror. On the other hand, when, as with some of the larger reflectors now in use, only the upper part of the tube is of open work, there is a strong tendency for the air to stratify in the solid lowermost section, near the mirror. In particular, in cases where the telescope is mounted on one side of a massive polar axis, there will often be a large temperature gradient across the tube, and this will affect the contained air in such a way as to produce a sort of irregular astigmatism, as shown by the star images. Such an effect has indeed been observed, but has been attributed, wrongly I think, to distortion of the mirror's figure. The continuous stirring of the air by fans would seem to offer the best chance of restoring the circularity of the star images.

Photo-electric Devices in Astronomy

P. B. FELLGETT

The Observatories, Cambridge

SUMMARY

The primary detectors of radiation which are used in astronomy are discussed with special reference to sensitivity. It is emphasized that the performance of good modern detectors is within an order of magnitude of the fundamental limitations to sensitivity set by fluctuations in the radiation incident on the detector. Under any given set of conditions, the sensitivity may be equated to that of an otherwise ideal detector in which only a fraction ε_e of the incident quanta is effective. Then ε_e, called the equivalent quantum efficiency, is a measure of the performance of the detector relative to what is fundamentally possible. This measure can be applied with great generality to both single- and image-detectors; values of interest range from 10^{-4} (certain photographic emulsions) to 0·5 (certain photocells). In terms of this concept, the superiority of photocells for single intensity measurements is emphasized, and is contrasted with the advantages, accruing from their multiple nature, of photographic emulsions in the measurement of extended images. Since photo-electric image devices, such as television camera tubes and image converters, promise to combine the advantages both of the photo-electric and of the photographic method, their significance for the future of astronomy seems to be very great. Examples are given of the performance of an image-orthicon attached to a telescope.

THE term "photo-electric device" is used here to mean those detectors of radiation in which the incident intensity gives rise directly to an electrical output signal; for example, photo-emissive cells, and the corresponding image devices such as are used in television. These detectors are thus distinguished from the human eye and the photographic plate, the detectors traditionally used in astronomy. The difference is perhaps practical rather than fundamental, as it may be argued that the primary process is always photo-electric.

In recent years, photo-electric methods of photometry have made possible improved accuracy, definition of spectral range, and freedom from scale and other systematic errors (STEBBINS, 1950). There is interest in applying photocells more generally, for example, to the measurement of spectra and radial velocities (FELLGETT, 1953a; CODE, 1952) and especially in the possibility of using photo-electric image detectors in astronomy (see, *e.g.*, CODE, 1952). The aim of this discussion is to examine the fundamental reasons why gains may result from the more extensive use of photo-electric devices. It is hoped to show that such extensions promise to open up a wide and interesting vista in the development of astronomy.

1. SENSITIVITY

Since the amount of energy collected by a telescope is small, high sensitivity may be regarded as the primary requirement for an astronomical detector. Nevertheless, the usefulness of a detector is affected by many other factors, and the success of photo-electric photometers has been associated with their being quantitative in a sense in which the eye and photographic plate are only qualitative. Thus a well-developed photo-electric photometer gives a direct measure of relative intensity (KRON, 1952), whereas accurate measures with the eye or photographic plate can be made only by matching the unknown intensity to a known laboratory source; or at most making close interpolations between known intensities. Moreover, the plate suffers from proximity effects, imperfect storage (reciprocity failure), and non-linearity dependent both on time and intensity of exposure and on the spectral region. In addition, the sensitivity varies from one part of the emulsion to another; the corresponding uncertainty in the measurement amounts to a few per cent for the size of image used in astronomy. Yet despite these disadvantages of photography, the photocell could hardly have been used had its sensitivity been poor, especially as it must make measures one at a time, whereas a plate can record many images at once (this will be discussed in more detail later on); and in fact photo-electric photometry was seriously limited in scope until the development of cells capable of measuring in a few seconds stars that need a photographic exposure of several minutes.

The term "sensitivity" is commonly used to denote the output signal caused by a given incident intensity. Recently, there has been a trend to refer to this property by the more specific term "responsivity" (JONES, 1949a, 1949b). Responsivity is mainly of engineering interest; it influences the properties (especially the gain) required in the subsequent amplifier. The responsivity is increased if the amplifier is regarded as part of the detector, and in the Golay cell (GOLAY, 1946, 1949), for example, it is very difficult to consider the amplifier separately.

A more fundamental interpretation is to take sensitivity to mean a measure of the least incident signal that can be detected. If the output of the detector contains random fluctuations (commonly referred to as "noise") of root-mean-square value V volts, and if the responsivity is R volts per watt, then the noise level is equivalent to a fluctuation in the incident power of $E = V/R$ watts. Thus E, which is known as the equivalent noise power, represents the apparent fluctuation that an incident signal must override if it is to be detected. It is invariant to noiseless amplification. The equivalent noise power is smaller the better the detector; its reciprocal is a measure of sensitivity in the required sense. It has been proposed to call this quantity $1/E$ the detectivity (JONES, 1952). A similar definition of equivalent noise energy can be made; and a more general procedure is to express the equivalent noise level in terms of density per unit frequency-time cell (GABOR, 1946; WOODWARD, 1951). These quantities can also be expressed in terms of numbers of photons. It may be helpful to regard responsivity as partly analogous to magnification or dispersion in an optical instrument, and detectivity as analogous to resolution.

2. Fundamental Limitations to Sensitivity

In the last decade a markedly better understanding has been reached of the fundamental limitations to detectivity (MILATZ and VAN DE VELDEN, 1943; GOLAY, 1947; LEWIS, 1947; JONES, 1947; FELLGETT, 1949; MOSS, 1950). A detector exchanges temperature radiation with its surroundings, and must therefore have a noise level at least as great as corresponds to the randomness in the number of exchanged quanta that lie effectively within the spectral sensitivity of the detector. Quantitative details are given in FELLGETT (1949). At room temperature, both the number of quanta and the fluctuation are small at wavelengths shorter than 1μ, and increase very rapidly with wavelength between 1μ and 5μ (Fig. 1). This is the fundamental reason why sensitivity increases as the long wave limit is reduced, for example, in the series *thermocouples and bolometers, lead tulluride, lead sulphide, Cs-Ag-O photocells, Sb-Cs photocells.* It is not that the mechanism of the infra-red detectors is less perfect, or that thermal detectors have special thermodynamic limitations. Actually the noise factor, the ratio of the observed to the fundamental noise level (FELLGETT, 1949b), is much the same for all the members of the series.

A signal incident on the detector gives rise to additional noise corresponding to the randomness in the rate of arrival of the quanta constituting the signal. These signal fluctuations nearly always exceed those of the thermal background in the visible and ultra-violet, but the thermal background usually predominates at the infra-red and radio frequencies. Between 1μ and 3μ, however, the size of the signal may be the deciding factor. In practical terms this may be shown to depend on the signal-to-noise ratio required and the exposure time (FELLGETT, 1951a). Thus signal may

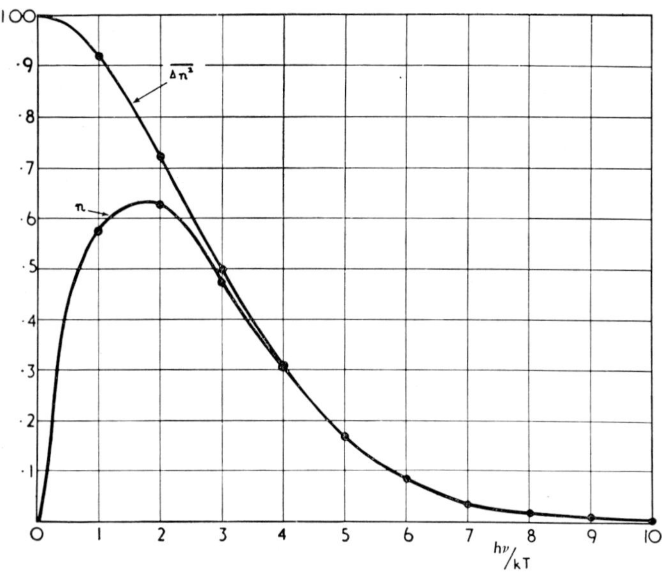

Fig. 1. Number n of quanta, and the fluctuation $\overline{\Delta n^2}$ in this number, as a function of the energy $h\nu$ per quantum divided by the thermal energy per mode $\mathbf{k}T$, for black-body radiation. For unit frequency range $d\nu$, the unit ordinate corresponds to the absolute value $2\pi\mathbf{k}^2T^2d\nu/h^2c^2 = 3{\cdot}05T^2$ sec.$^{-1}$ cm^{-2}, where T is the absolute temperature of the emitter in degrees Kelvin

predominate as far as 3μ for very short times and high accuracy; but under the usual conditions in astronomy of exposures of several seconds and "per cent" accuracy, the transition takes place rather sharply at 1μ because of the steepness of the black body curve at this wavelength.

These causes of fundamental noise are analogous to the "shot" effect (SCHOTTKY, 1918) in an electron tube. If on average N particles arrive in a disordered manner, the mean square fluctuation in the number is kN. For "classical" particle statistics $k = 1$; it is less than unity for a degenerate electron gas in which the energy levels are according to FERMI-DIRAC statistics. In the present case, that of photons, the BOSE-EINSTEIN enumeration of energy levels gives $k \simeq 1$ on the short-wave side of the black-body maximum, and $k > 1$ at the long wavelengths. In terms of energy integrated over all frequencies, the relative error in putting $k = 1$ is only about 0·04 (LEWIS, 1947); and it has sometimes been supposed that the difference from classical behaviour is unimportant. Fig. 1 shows that this is not so; it shows plotted to the same scale (so that the curves coincide for $k = 1$) the distribution as a function of wave number ν of the number N of quanta per unit wave-number range, and the fluctuation $\overline{\Delta N^2} = kN$ (FELLGETT, 1951b). In particular, in the radio region $N \rightarrow 0$ as $\nu \rightarrow 0$, whereas $\overline{\Delta N^2}$ tends to a constant value and this corresponds directly to the classical wave description (RAYLEIGH-JEANS) of the radiation; nevertheless it is also possible to derive the noise in a radio antenna from quantum considerations (FELLGETT, 1949a).

3. FIGURES OF MERIT AND THE EQUIVALENT QUANTUM EFFICIENCY

It is not evident how the sensitivities of detectors of widely different types can be compared. CLARKE JONES (1949a, 1949b) has established a system of classifying detectors, and has shown that for each class the properties can be combined into a "figure of merit" which is particularly appropriate to the comparison of detectors belonging to this class. The figure of merit called in his notation M_2 is of special interest. It takes into account both speed and sensitivity (specifically, the leading factors are steady state detectivity and noise bandwidth), and it can be shown to be a measure of pulse detectivity, the ability to detect a small amount of energy arriving suddenly at the detector. In many cases, M_2 is invariant to changes in design parameters. These figures of merit have been applied mostly to detectors giving single intensity measures, such as thermocouples and photocells. They can be applied to image detectors, but their significance is then largely confined to the performance under non-imaging conditions.

ROSE (1948a, 1948b) has used a very useful and general method of comparing detectors. Under any given conditions, let N quanta be received on average from each resolved element of the target being examined during the time of observation. For the single detectors, the element is the whole target; for the image detectors the element is the effective area of target over which the intensity is averaged by the detector. Then the mean-square fluctuation in the number of incident quanta is kN and their signal-to-noise power ratio is N/k. Let it be supposed that an incident quantum of given energy is either ineffective or else gives rise in the detector to an event of always the same size (this restriction is removed later on), such as the emission in a photocell of an electron which is collected by the anode, or the heating of a thermocouple by the absorption of a quantum. Let a fraction ε of the incident quanta produce such events. Then the average effective signal is εN and certain elementary considerations indicate that the mean square fluctuation is $\varepsilon N + \varepsilon^{(1+a)}N(k - 1)$, giving a signal-to-noise power ratio

$$r = \frac{\varepsilon N}{1 + \varepsilon^a(k - 1)}, \qquad \ldots.(1)$$

but fail to decide between two possible values of a, $a = 0$, or $a = 1$. The theory is more difficult than may appear at first sight, and so far as I am aware the ambiguity has not been resolved experimentally. This uncertainty will not affect the principles to be discussed, and it is unimportant practically in laboratory work (especially in the visible) for which $k - 1 \simeq 0$. For early-type stars, however, k may be of the order of 10 in the visible, and therefore the value of a must be known in order to predict the noise level. Moreover, since k is a function of wavelength and temperature only, this knowledge would in principle enable effective temperatures to be found from noise measurements (FELLGETT, 1951a). The potential interest of this method lies in the possibility of measuring monochromatic temperatures (*e.g.* within a single spectral line) without either knowing the angular size of the source or making any assumptions as to blackness.

Following essentially the method of ROSE, we shall define the equivalent quantum efficiency ε_e of a detector as the value of ε which when substituted in equation 1 gives the same value of the signal-to-noise ratio r as is in fact observed when the N quanta are incident on the detector. Thus the imperfection of the detector is expressed in terms of imperfect quantum efficiency, and ε_e may be regarded as the quantum efficiency that an otherwise ideal detector would have if its performance were the same as that of the one considered. It follows that ε_e cannot exceed the quantum efficiency ε_0 of the primary photo-process. It is particularly to be noted that ε_e, ε_0 are (like the equivalent noise power) unaltered by noiseless amplification,

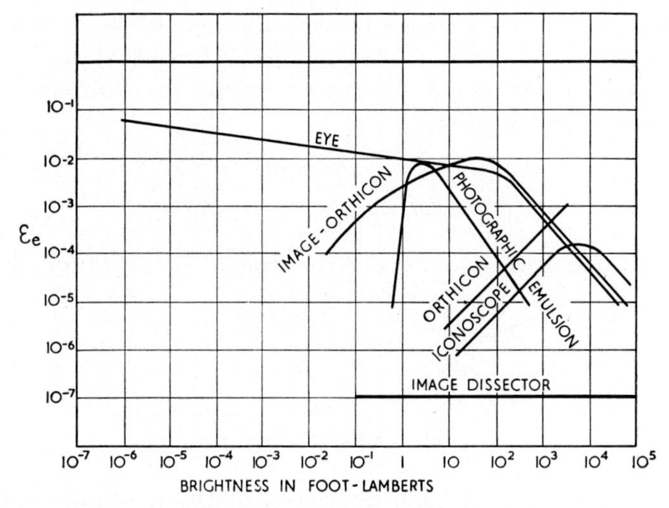

Fig. 2. The variation of equivalent quantum efficiency against brightness of scene viewed, for the human eye, image dissector, iconoscope, orthicon, and image-orthicon. (After ROSE, 1948a; 1948b.)

and by definition they cannot exceed unity. These properties contrast with the behaviour of quantum efficiencies defined in some such way as the number of electrons collected for each incident photon; such definitions may be useful for some purposes, but are largely irrelevant to noise considerations. Sometimes, as in a photomultiplier, the output caused by a quantum effectively absorbed is subject to statistical fluctuations. An effective quantum efficiency ε' can be defined for these detectors by finding the value of ε which represents the performance that would result if all imperfections were removed, in some agreed manner, except the limited primary quantum efficiency and the randomness of amplification. In as much as this randomness gives unequal weight to the incident quanta, the quantum efficiency so defined must be less than ε_0.

So far, the equivalent quantum efficiency has been taken to apply only under one set of conditions. The overall performance is represented by the variation of ε_e with the conditions of use. Thus ROSE has plotted ε_e against the brightness of the object viewed for a number of image detectors, including television camera tubes, photographic emulsions, and the human eye. His diagram showing this variation is the basis of the present Fig. 2. To illustrate the way in which such variations can arise, we shall consider the more simple case of a gas photocell. It will be assumed that $k = 1$.

Let N be the number of quanta incident per second, ε_0 the quantum efficiency of the cathode, M the gas amplification and e the electronic charge. Then $\varepsilon_0 N$ pulses reach the anode per second, each on average containing charge eM. Then if M were constant, the mean-square fluctuation per second would be $e^2M^2\varepsilon_0 N$. Let the randomness of M increase the noise by a factor m^2, giving $e^2M^2m^2\varepsilon_0 N$. An additional source of noise is emission from the cathode which is not dependent on the incident radiation, for example, thermionic emission. If this is at a rate of n electrons per second, the corresponding noise is $e^2M^2m^2n$. Also if the cell is connected to a load resistance R, this gives rise to a thermal agitation noise current having mean square value in 1 sec. of $2\boldsymbol{k}T/R$, where T is the absolute temperature and $\boldsymbol{k}$ Boltzmann's constant. If R is sufficiently large, the noise in the amplifier to which the cell is connected will be negligible at the slow speeds considered here. Then dividing the squared signal current $\varepsilon_0{}^2e^2N^2M^2$ by the total noise current we obtain (for further details see JOHNSON, 1948),

$$r = \frac{\varepsilon_0{}^2e^2N^2M^2}{e^2m^2M^2(\varepsilon_0 N + n) + (2\boldsymbol{k}T/R)} \qquad \ldots.(2)$$

and by equation (1) and the definition of the equivalent quantum efficiency ε_e,

$$r = \varepsilon_e N,$$

when $k = 1$. That is

$$\varepsilon_e = \frac{\varepsilon_0/m^2}{1 + (1/\varepsilon_0 N)(n + 2\boldsymbol{k}T/Re^2M^2m^2).} \qquad \ldots.(3)$$

Thus $\varepsilon_e \to \varepsilon_0/m^2$ as $N \to \infty$. At low intensities the background noise becomes the predominant imperfection, and $\varepsilon_e \to 0$ as $N \to 0$. This condition can be delayed by increasing R and M to such values that $2\boldsymbol{k}T/Re^2M^2m^2$ becomes small compared with n. This is the basis of KRON's work (1952) on the sensitivity of photo-electric photometers. The disadvantage of increasing M is that in practice m and n increase also and a compromise must be made. Increasing R slows the response, and means of compensating for this have been another aspect of KRON's work. The extent to which such compensation is possible is indicated below. Equation (3) applies also to a photomultiplier, but here m tends (within limits) to fall with increasing M (WOODWARD, 1947); but n still increases. One advantage of pulse counting methods (YATES, 1948) over measuring the direct-current output is that the gating circuits ensure that all the pulses which are effective contribute equally, consequently m becomes unity at the expense of a slight loss in the number of pulses observed; another advantage is that thermal agitation noise, and especially the "flicker" component of the leakage current, are discriminated against.

For infra-red detectors, the incident fluctuation includes that portion of the "background" or "dark current" noise which is caused by the temperature radiation received by the detector from its surroundings. The value of ε_e for very large signals $(\varepsilon_e)_\infty$ is a measure of the effective quantum efficiency ε', and the small signal value $(\varepsilon_e)_0$ indicates how much additional imperfection arises from "background" generated in the detector by non-fundamental mechanisms. In the infra-red, the thermal radiation often exceeds the largest signals commonly used so greatly that only $(\varepsilon_e)_0$ can be conveniently measured.

The variation of equivalent quantum efficiency with speed of response may be found by a straightforward analysis. Noise in the amplifier is a leading parameter in the variation, since at a sufficiently high frequency the shunt capacity c across the photocell and its load effectively short-circuits the input of the amplifier. If the equivalent noise resistance of the amplifier is r, then at frequency $\omega/2\pi$ the expression for ε_e is found to be

$$\varepsilon_e(N, \omega) = \frac{\varepsilon_0/m^2}{1 + \dfrac{n_1}{\varepsilon_0 N}\{1 + \omega^2 \tau_1^2\}} \qquad \ldots .(4)$$

where

$$n_1 = n + \frac{2\boldsymbol{k}T}{e^2 m^2 M^2} \cdot \frac{1}{R}\left(1 + \frac{r}{R}\right),$$

$$\tau_1^2 = \frac{R^2 c^2}{\dfrac{R^2}{r}\left\{(n_1 + \varepsilon_0 N)\dfrac{e^2 m^2 M^2}{2\boldsymbol{k}T}\right\}}$$

Here n_1 may be regarded as a fictitious dark current that would generate, at low frequencies and in the absence of a signal, the same noise as is in fact generated by all the imperfections; and τ_1 is the time constant representing the frequency depen-

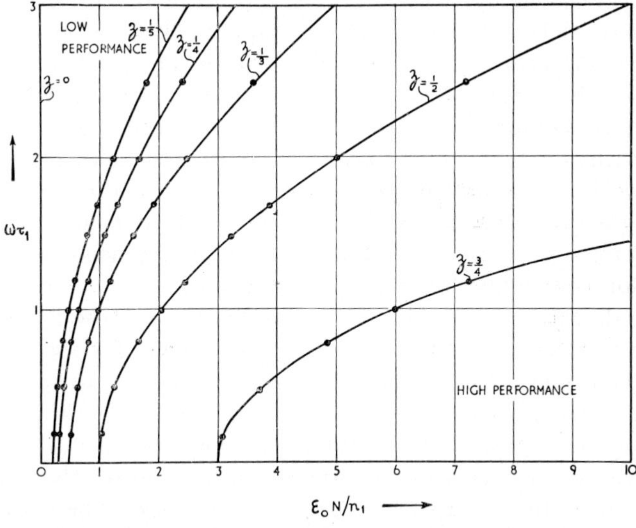

Fig. 3. Contours of performance of a gas-photocell against the normalized intensity and frequency $\omega/2\pi$ of the incident signal (which is assumed sinusoidal). The performance parameter $z = \varepsilon_e/(\varepsilon_0/m^2)$ is the ratio of the equivalent quantum efficiency to its maximum value for any given cell. The ordinate is the angular frequency ω multiplied by the time-constant τ_1 which describes the spectrum of the signal-to-noise ratio. The abscissa is the ratio of $\varepsilon_0 N$, the number of photo-electrons caused by the incident signal in a given time, to the number n_1 of electrons in a fictitious dark-current which would generate (in the same time) the internal noise level which in fact prevails in the absence of a signal.

dence of the signal-to-noise ratio. Lines of equal $\varepsilon_e/(\varepsilon_0/m^2) = \varepsilon_e/(\varepsilon_e)_\infty$ are shown in Fig. 3 plotted against $\varepsilon_0 N/n_1$ and $\omega\tau_1$.

For image detectors, ε_e can be plotted also against fineness of image detail. As the fineness increases, ε_e is at first constant, and then falls rapidly as the "resolution" limit is reached, finer details being greatly attenuated and therefore observed with poor signal-to-noise ratio.

It appears from such examples that a knowledge of the variation of equivalent quantum efficiency gives a detailed account of the sensitivity and range of usefulness of a detector, and provides a method of great generality for comparing detectors. Some applications will now be made of this concept.

4. Some Applications of the Equivalent Quantum Efficiency

We have seen that ε_e is a lower bound for the quantum efficiency ε_0 of the primary process and that $(\varepsilon_e)_\infty$ is a stronger bound; that is $\varepsilon_0 > (\varepsilon_e)_\infty > \varepsilon_e(N) > (\varepsilon_e)_0$. Useful information as to the mechanism of the detector can be obtained in this way. Thus, it is known (Moss, 1950; Fellgett, 1949b) that for the best lead sulphide and telluride infra-red cells $(\varepsilon_e)_0 \simeq 4$ to 5, and therefore at least $\frac{1}{4}$ to $\frac{1}{5}$ of the incident photons must be effective. (Note again that this result is independent of possible multiplying processes in the mechanism of photoconduction.) These facts are adverse to theories of the photoconductivity which postulate that the primary process occurs only at crystal boundaries. Similarly, for the human eye, Rose's results (Fig. 2) show that ε_e remains almost constant up to ordinary levels of indoor illumination. It follows that bleaching of visual pigments cannot be an important part of light adaptation in this range, and this result has been confirmed by direct measurement of the pigment (Pirenne and Denton, 1952; Hagins and Rushton, 1953). What is of special interest is that Rose's simple measure of the noise relations (he actually matched the visual appearance of a test object with the appearance under the same illumination given by a television system for which ε_e was known) could yield the same conclusion as the much more difficult physiological techniques. For astronomy, the most interesting curves in Fig. 2 are the eye, photographic emulsion, and image orthicon. As we have already noted, the range of intensities over which the eye remains efficient is very remarkable. The very sharp drop from the peak of the curve for the photographic emulsion, especially on the low-intensity side, corresponds to the necessity for controlling photographic exposure within narrower limits than suffice for other detectors, and to the practical dictum "expose for the shadows, and the highlights will take care of themselves". The curves for the artificial detectors are drawn for the camera optics which makes the devices equivalent in certain ways (such as depth of focus) which are significant in television; the relative aperture quoted is $f/5$ for the emulsion and $f/10$ for the image orthicon. In astronomy, the relative aperture can be increased and the effect is to move the curve, rigidly and without change in ordinates, to lower brightness values. The shift is limited to a factor of about 100 by optical difficulties, and since the absolute aperture is usually fixed in astronomy, the shortness of focal length that can be used may be limited also by the angular resolution required. The appropriate design procedure would be to draw contours of ε_e in the intensity-resolution plane for the telescope aperture to be used. A further shift to lower light-intensity is possible by increasing the exposure time from the value of 0·2 sec. to which the curves refer. This process is limited for the emulsion by reciprocity failure, and in television tubes by the fact that the whole design is based on television speeds.

It is not yet customary to think of photographic sensitivity in terms of equivalent noise energy, and consequently it is difficult to obtain reliable data for assessing ε_e. From sensitometric and granularity data, I have derived peak values of ε_e for astro-emulsions ranging from 0·06 to 0·01, in substantial agreement with ROSE's diagram. But certain direct grain-counting experiments on 103aO emulsion by BAUM, who has kindly allowed me to quote this result prior to its publication by himself, indicate a value not exceeding 10^{-3}; and values of ε_e exceeding 0·02 are inconsistent with WEBB's estimate that the quantum efficiency is 0·2 and that ten effective quanta are needed to make a grain developable (WEBB, 1948) and *a fortiori* with his earlier estimate of $\approx$ 1000 effective quanta per grain (WEBB, 1941), or SILBERSTEIN and TREVELLI's (1945) quantum efficiencies of 10^{-3}, which, however, do not necessarily refer to the most sensitive materials. ROSE, WEIMAR, and LAW (1946) find that the practical performance of motion picture emulsions is about 200 times less than that of the eye, corresponding to $\varepsilon_e \approx 10^{-4}$; part of this defect is ascribed by them to the sharpness of the ε_e curve, as a consequence of which the least value of ε_e which occurs within the range in brightness of a scene of high contrast is much less than the peak value. It appears, however, that a real discrepancy remains: and certainly elementary experience with an image-orthicon camera suggests that it can reproduce scenes that one would not attempt to photograph in less than several seconds.†

In fairness to the photographic process it must be said that photo-electric devices are commonly used cooled, and it appears that a substantial increase in photographic sensitivity could in principle be realized by controlled cooling to about $-50°C$ (MEES, 1942; BILTZ, 1952). Experiments recently made by Mr. A. N. ARGUE in this observatory have demonstrated, for certain emulsions that might be useful‡ in astronomy, the well-known removal of low-intensity reciprocity failure on cooling, and in one case an increase in sensitivity of ten times was obtained for an exposure time of 30 min. This, however, only restores the sensitivity that prevails, without cooling, for the optimum time of exposure. The older theories of photographic action had indicated that an additional gain might result if the cooled astronomical exposure were required to produce only a sublatent image which requires two effective quanta per grain according to WEBB (1950). This would be made developable by a subsequent long exposure at room temperature to an intensity low enough for the virgin grains to have very strong low-intensity failure. The published data indicate that ε_e might be made as high as 0·1 in this way. Since high-intensity failure is essentially a loss in efficiency of the processes of passing from the sub-latent to the latent condition, any failure of this kind induced by excessive cooling would be innocuous, and indeed post-exposure is successful in removing the effect of such failure in flash photographs taken at ordinary temperatures. Recent work has indicated (MITCHELL, 1951; BILTZ, 1952), however, that the loss of photographic effect on strong cooling is not entirely high-intensity reciprocity failure, but a distinct effect which resembles low-intensity failure in the respect that the initiation of the build-up of the sensitivity speck is more difficult than its continuation.* It follows that post-fogging of cooled exposures is unlikely to be successful, and this is in agreement with our experiments. We have confirmed, nevertheless, that pre-flashing, in which both the time and intensity are separately controlled to give the greatest amount of sub-latent fog, is successful in reducing the low-intensity failure in a subsequent astronomical exposure without cooling; in distinction to the older methods of pre-fogging, the density can usefully exceed that which would correspond, accord-

* But for experiments leading to the contrary conclusion see FARNELL (1952).
† Recent results (1954) modify the values quoted here, but the discrepancy largely persists.
‡ See *Observatory*, vol. 74, p. 213, 1954.

ing to the ordinary characteristic curve, to the sum of the photographic effects of the two exposures. Low-intensity reciprocity failure, although apparently a great nuisance in astronomy, is in fact necessary for stability of the emulsion. Whereas for almost all other detectors the "background" (whether fundamental or accidental) is effective only during the exposure, in photography it is effective during the whole time from manufacture to development. In view of this disadvantage, the performance of modern emulsions is very remarkable. An estimate of the number of ionizations per grain, caused by room temperature quanta of sufficiently high energy and by direct thermal excitation, indicates that fogging would become very severe in a few months were it not that the reciprocity failure protects the emulsions; this certainly applies to red-sensitive materials and is probably true in all cases. It is thus no accident or defect that astro-emulsions must be stored under refrigeration, that "hyper-sensitizing" treatments which remove this failure need to be followed by development within a short time, or that infra-red materials benefit particularly from such treatments.

The quantum efficiency ε_0 of modern commercial photocells and photomultipliers may be as high as 0·2, and values of 0·5 have been reported exceptionally in experimental cells. We have seen that $\varepsilon_e \approx \varepsilon_0$ over a substantial range of conditions for such cells. Yet photocells have not superseded photography, which remains a powerful astronomical tool despite the much smaller value of ε_e even according to the most favourable estimate. The reason is that sensitivity, and signal-to-noise ratio, are not ends in themselves, but a means of collecting information. A 35-mm frame is able to record more than 10^6 independent picture elements, and therefore the total rate at which it collects information may be high despite a restricted performance for each element. The relations governing maximum channel-capacity in information theory (SHANNON, 1948a, 1948b) are not very appropriate to the quantitative evaluation of the effect, since the coding of the information presented to the telescope cannot be adjusted to that postulated in the theory. Instead, we note that for a given total observation time the signal-to-noise power for a photo-electric scanning system varies as ε_e and inversely as the number M of elements to be measured, since the available time must be divided between these. For the emulsion, the number of elements makes no difference, provided $M < 10^6$. If for the cell ε_e is taken as 10^{-1}, and 10^{-3} for an emulsion, the cell will therefore have the better performance if $M < 10^2$.

Thus a photomultiplier might be ten times faster than photography for observing a single spectral line or small group of lines, in a survey of interstellar lines for example, but it would be slower in scanning the whole of a high-resolution spectrum. By orthogonal-modulation techniques, M can be increased to perhaps 10^3 for the cell, but this may necessitate an increase in photo-current or in receiver area which cancels the expected gain. This difficulty does not occur in infra-red spectrometry (FELLGETT, 1953b), and the application of such techniques to the infra-red spectra of stars is being undertaken.

Sometimes a large number of points are customarily measured using an image detector, when the number M of parameters in the final result is small. In principle, therefore, a single cell can be substituted in such cases for the image detector. Means of doing this for the measurement of radial velocity ($M = 1$) and for the classification of spectra into predetermined classes ($M = 1$ for each spectral sequence) have been proposed (FELLGETT, 1953a, 1955).

5. SOME POSSIBILITIES OF IMAGE DETECTORS IN ASTRONOMY

Techniques for using single detectors, when image devices appear at first sight to be necessary, remain valuable notwithstanding the availability of highly developed image detectors, because the greater simplicity of the detector leads in practice to higher performance. For such purposes as examining star fields, however, or for the observation of the details of complex spectra, image detectors are indispensable. The simplest of these is the image-converter tube, in which an optical image at the photo-cathode gives rise to the emission of electrons, which are accelerated and focused on to a fluorescent screen. We can assume $\varepsilon_e \approx \varepsilon_0$; with the grainless evaporated screen, the linear resolution can compare favourably with photography. The screen may be viewed visually, by photography, or by a television tube. If sufficient intensification of brightness can be obtained, the overall ε_e will be that of the image tube. Under these conditions, important gains over the eye are possible if $\varepsilon_0 > 10^{-1}$, and over other image detectors if $\varepsilon_0 > 10^{-2}$; intensification by itself is useless (MORTON, RUEDY, and KRIEGER, 1948). Cathodes are so sensitive to contamination that it will probably be necessary to seal off the screen with an electron-transparent layer. Although several quanta are readily obtained for each one incident, it seems that the number that can be collected by practical optical systems can be made greater than unity only by secondary emission amplification. Cathodes and secondary-emitting surfaces are so similar that they are mutually innocuous, and it is probable that such tubes could be developed for much less than the cost of a major telescope. Alternatively, the electron image can be formed directly on a photographic emulsion. LALLEMAND (1952) has used this method successfully in a continuously pumped system, but the technique must be difficult; possibly an electron-transparent barrier, for example an aluminium pellicule, might be used to separate a demountable vacuum chamber, containing the emulsion, from the sensitive cathode.

Quantitative measures of intensity, in the sense of the introduction to this article, require a television system in the full sense. The complexity of the camera tubes, with the consequent contamination of the cathode and difficulty of obtaining optimum processing of all the components, may always limit the performance to about a tenth of that obtainable in simple photocells, that is at the present time $\varepsilon_e \approx 10^{-2}$ in agreement with Fig. 2, but we have seen that it is likely that image tubes can be developed to go in front of the camera and give an overall performance corresponding to $\varepsilon_e \approx 10^{-1}$. Television systems are costly; and, in as much as the camera tubes are already highly developed, it cannot be expected that they can be much simplified so long as they are required to work at television speeds, as in observation of planetary detail obscured by seeing. Indeed elaboration may be necessary to make the tubes fully quantitative. The associated circuits, however, which are at present designed for the full requirements of broadcasting, can be much simplified when the viewing tube may be directly connected to the camera. Our discussion has shown, moreover, that in appropriate circumstances, the effective light grasp of the telescope may be increased many times, and it would therefore be economical to make for television apparatus an outlay appropriate to a major astronomical instrument.

The primary action of a camera tube is to form a charge distribution corresponding to the optical image. This occurs either by electrons from a photo-cathode being focused on to an insulating plate, or by the charge on a photoconductor decaying in accordance with the intensity of the image; it seems that photoconductors often have high ε_0, but photo-emission has the advantage that amplification can be obtained by secondary emission. The charge distribution is then scanned with an electron beam. For slow speeds, however, it may be possible to simplify the tube considerably by examining the charge distribution electrostatically. This is done in an elementary way, using the electrostatic forces between the surface and charged particles of powder, in xerography (SCHAFFERT and OUGHTON, 1948). A "needle" probe connected to a suitable electrometer might be successful. Another possibility is to focus photo-electrons on to a screen of so-called "infra-red sensitive" phosphor. These phosphors, used for infra-red viewing during the war, can be excited by ultra-violet radiation, store large amounts of this energy for many hours, and emit it as visible fluorescence under the influence of near infra-red radiation (wavelength $\approx 1{\cdot}2\mu$). If such materials could be excited by electron bombardment, the image could be "read off" the screen by scanning it with an intense spot of infra-red radiation. The emitted light would be measured with a photomultiplier cell, and since the screen need not be imaged onto this cell the light might be collected over a large solid angle.

6. CONCLUSIONS

Our discussion began by establishing the usefulness of the equivalent quantum efficiency ε_e as a criterion of performance; since $\varepsilon_e = 1$ represents the maximum performance theoretically possible, the values that we have estimated show that the fundamental limitations are significantly approached by the best detectors. In conclusion, the reasons for expecting a markedly higher performance from photo-electric devices than from the traditional detector will be summarized and somewhat amplified. The primary reason is that ε_e can be higher for the photo-electric detector by a factor of 10 to 100, corresponding to a reduction by the same factor of the observing time necessary to obtain given results. This advantage has been demonstrated by the success of photocells in photometry, but their scope is normally limited by the restriction to one intensity measurement at a time. If photo-electric image devices can be used, combining in effect the high efficiency of cells with the ability of the photographic plate to examine many image points simultaneously, then the astronomer's power to collect information will be markedly increased, and astronomy may enter a new phase. The incidental advantages of these photo-electric devices are also very significant. Both spectral sensitivity and the time of integration may be made to cover ranges not accessible to the eye. It is possible to gain surface brightness at the expense of magnification as cannot be done by purely optical means for the eye, and only to a limited extent for photographs. The more quantitative nature of photo-electric results has been commented on already, and it has been implied that the variation in sensitivity over the photo-cathode can be measured and allowed for.

Through the courtesy of Pye Limited, Cambridge, experiments were recently carried out at the Cambridge Observatories using an image-orthicon attached to a telescope. The photographic analogue of this tube is a motion-picture camera; in comparison with what could have been obtained with such a camera the preliminary results appear to be in good accord with the present discussion. For example, the image of Jupiter and its satellites was similar (apart from greater brightness) to the direct view in an eye-piece. Bright calcium flocculi were also observed in K light on the spectro-helioscope using the same apparatus.

(*a*)

(*b*)

Fig. 4. The Moon, first quarter. Image-orthicon picture taken with minus-blue filter at the prime focus of the Cambridge solar tunnel; lens aperture 12 in., focal length 60 ft. Monitor screen photographed with 0·2 sec. exposure

(*a*) *Above*. Mountainous region, including Maurolycus, towards the South Pole.
(*b*) *Below*. Vicinity of the North Pole, including Aristotle and the Alpine Valley.

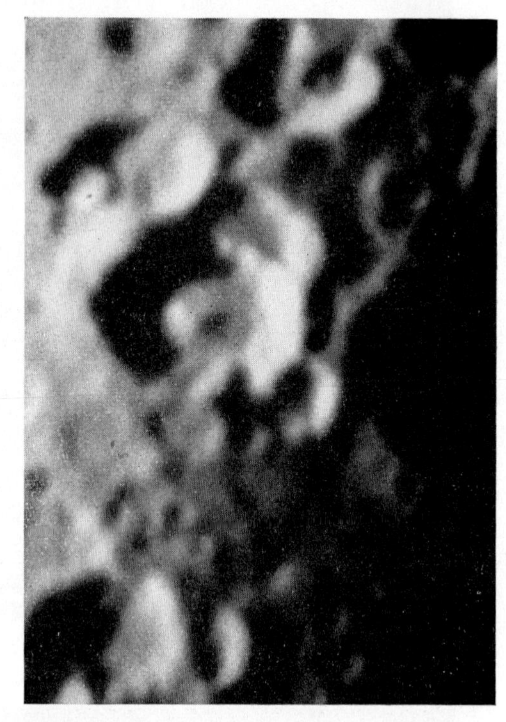

Fig. 5. Comparison of image-orthicon picture with direct photograph

Left. Enlargement of part of Fig. 4 (*a*); exposure 0·2 sec.
Right. Print, to approximately the same scale and contrast of direct photograph on Kodak P.800 plate with minus-blue filter; exposure 4 sec.
The negative was the sharpest obtained in twelve exposures made immediately after the image-orthicon trials. Taking into account the detail shown (for example, the terracing of the outer wall of Maurolycus and the structure of the central peak), the grain, and the resolution, the comparison of these pictures is at least not unfavourable to the television image, despite the fact that the direct photograph needed twenty times as long an exposure. The apparent change in tilt of the surface of the Moon between the two pictures is not, of course, due to libration; it indicates the omission of an adjustment to the monitor cathode-ray tube. The retreat of the shadow between the two exposures can be discerned

The Figs. 4–6 show some of the images obtained of the Moon (Figs. 4 and 5) and of planets (Saturn, Fig. 6). It would, of course, be inappropriate to compare these first attempts either with the earliest endeavours at direct photography in astronomy, or with the best modern photographs; they represent an intermediate level of technique. A solar tunnel, because of the long light path near to the ground, is very sensitive to inequalities of air temperature, and the precautions taken in the design of the building cannot altogether overcome this. The seeing is therefore poorer than can be obtained with a telescope mounted in the usual way. This is not, however, a disadvantage for the purpose of the tests, which was to make a direct comparison of the image-orthicon and photographic plate. Fig. 5 shows a comparison of images obtained successively by the two methods under these conditions of seeing. It appears that the image orthicon has suffered less than the photographic plate from the unsteadiness of the image, because of its ability to work faster than the photographic plate; in the test illustrated it was actually exposed for 1/20 of the time required for the direct photograph.

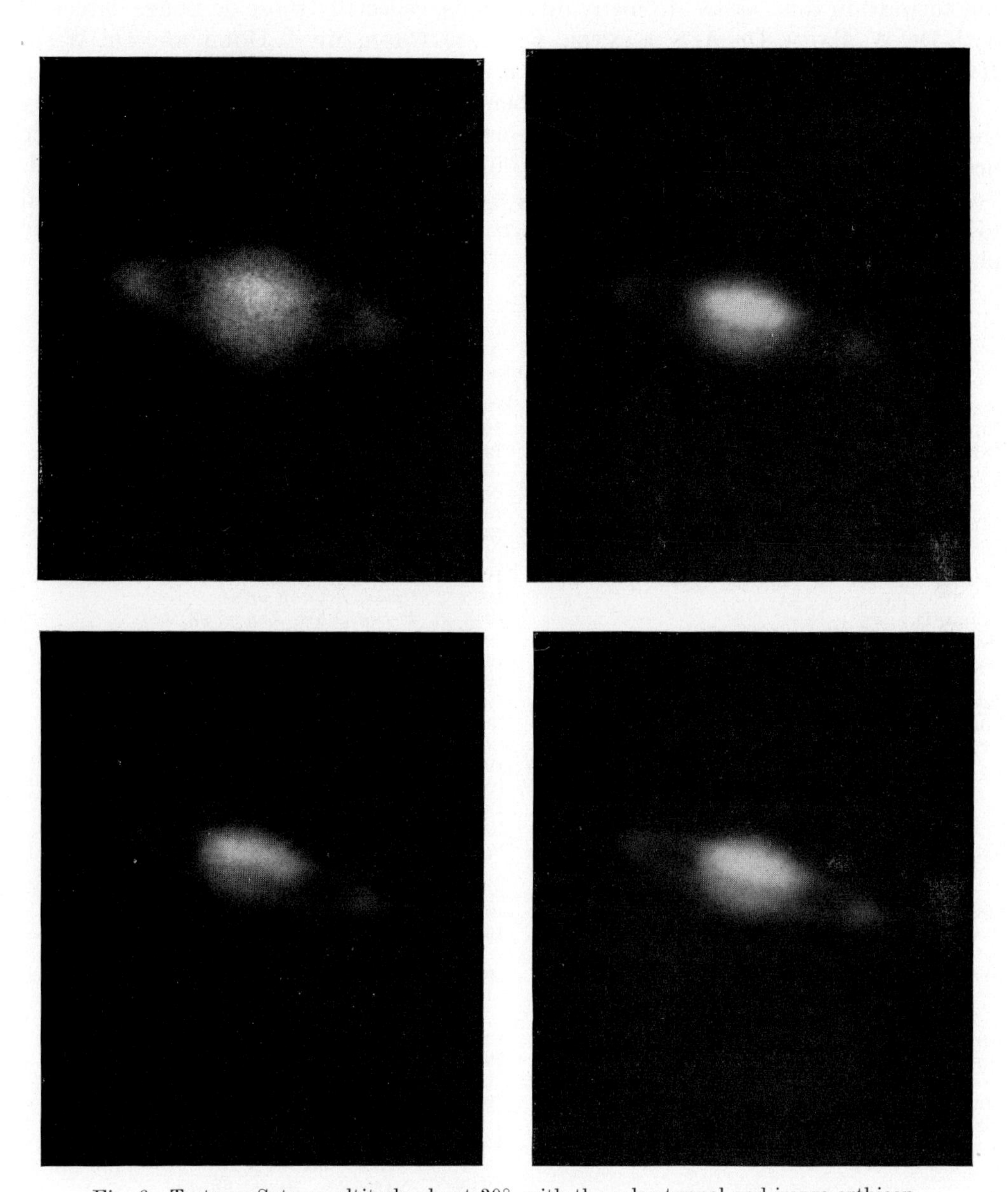

Fig. 6. Tests on Saturn, altitude about 30°, with the solar tunnel and image-orthicon

(*a*) *Above*. Exposure 0·5 sec.; each image is from a super-position of two negatives.
(*b*) *Below*. Exposure 0·2 sec.; each image is from four negatives, and the print was exposed for the region in which the rings cross the ball of the planet.

It is a pleasure to acknowledge my indebtedness for the development of the views put forward in this "vista" to many discussions, especially those on image detectors with Dr. W. Baum, Dr. A. S. Baxter, Dr. A. D. Code, Mr. T. Gold, and Dr. W. A. Hiltner; and to M. A. Lallemand for a most interesting visit to his laboratory.

I am also most grateful to all the members of this observatory and of Pye, who by their helpful participation in the image-orthicon experiments made workable the necessarily improvised arrangements used in the first tests, and especially to Dr. H. von Klüber, who made it possible for the solar tunnel to be used for these tests, and to Dr. D. W. Dewhirst and Dr. Z. Suemoto, who made a special contribution to the photographic comparisons.

References

Biltz, M. . . . 1952 *J. Opt. Soc. Amer.*, **42**, 898.
Code, A. D. . . . 1952 *Observatory*, **72**, 201.
Farnell, G. C. . . . 1952 *Phil. Mag.*, Ser. 7, **43**, 289.
Fellgett, P. B. . . . 1949a *J. Opt. Soc. Amer.*, **39**, 970.
1949b *Proc. Phys. Soc.* (*London*), B **62**, 351.
1951a *Thesis*, Cambridge University.
1951b *M.N.*, **111**, 537.
1953a *J. Opt. Soc. Amer.*, **43**, 271.
1953b *Multichannel Spectrometry*. In prepar.
1955 *Optica Acta*, in press.
Gabor, D. . . . 1946 *J. Inst. Elect. Engrs.*, **93**, 429.
Golay, M. J. E. . . . 1946 *Rev. sci. Instrum.*, **17**, 511.
1947 *Rev. sci. Instrum.*, **18**, 347.
1949 *Rev. sci. Instrum.*, **20**, 816.
Hagins, W. A. and Rushton, W. A. H. . . . 1953 *J. Physiol.*, **120**, p. 61 P.
Johnson, H. L. . . . 1948 *Ap. J.*, **107**, 34.
Jones, R. Clark . . . 1947 *J. Opt. Soc. Amer.*, **37**, 879.
1949a *J. Opt. Soc. Amer.*, **39**, 327.
1949b *J. Opt. Soc. Amer.*, **39**, 344.
1952 *Nature* (*London*), **170**, 937.
Kron, G. E. . . . 1952 *Ap. J.*, **115**, 1.
Lallemand, A. . . . 1952 *Trans. Internat. Astron. Union*, Vol. VIII (Rome Congress, 1952), p. 746. University Press, Cambridge, 1954.
Lewis, W. B. . . . 1947 *Proc. Phys. Soc.* (*London*), **59**, 34.
Mees, C. E. K. . . . 1942 *Theory of the Photographic Process* (Macmillan Co., New York), p. 242.
Mitchell, J. W. (Ed.) . . . 1951 *Fundamental Mechanisms in Photographic Sensitivity*. Butterworth, London. (See especially pp. 61 ff., and pp. 242 ff.)
Milatz, J. M. W. and van de Velden, H. A. . 1943 *Physica*, **10**, 369.
Morton, G. A., Ruedy, J. E. and Krieger, G. L. 1948 *R.C. A. Rev.*, **9**, 419.
Moss, T. S. . . . 1950 *J. Opt. Soc. Amer.*, **40**, 602.
Pirenne, M. H. and Denton, E. J. . . . 1952 *Nature* (*London*), **170**, 1039.
Rose, A., Weimer, P. K. and Law, H. B. . . 1946 *Proc. Inst. Radio Engrs*, **34**, 424.
Rose, A. . . . 1948a *Advances in Electronics* (Editor: Marton; Academic Press, New York), pp. 131 ff.
1948b *J. Opt. Soc. Amer.*, **38**, 196.
Schaffert, R. M. and Oughton, C. D. . . . 1948 *J. Opt. Soc. Amer.*, **38**, 991.
Schottky, W. . . . 1918 *Ann. Phys.* (*Leipzig*), **57**, 541.
Shannon, C. E. . . . 1948a *Bell Syst. tech. J.*, **27**, 379.
1948b *Bell Syst. tech. J.*, **27**, 623.
Silberstein, L. and Trevelli, A. . . . 1945 *J. Opt. Soc. Amer.*, **35**, 93.
Stebbins, J. . . . 1950 *M.N.*, **110**, 416. George Darwin Lecture
Webb, J. H. . . . 1941 *J. Opt. Soc. Amer.*, **31**, 355.
1948 *J. Opt. Soc. Amer.*, **38**, 312.
1950 *J. Opt. Soc. Amer.*, **40**, 3.
Woodward, P. M. . . . 1947 *Proc. Cambridge Phil. Soc.*, **44**, 404.
1951 *Telecommunication Research Establishment Malvern, Memorandum* 425.
Yates, G. G. . . . 1948 *M.N.*, **108**, 476.

The Development of Photo-electric Photometry

C. M. Huffer

Washburn Observatory, University of Wisconsin,
Madison, Wisconsin, U.S.A.

Summary

Photo-electric photometry is to-day an important branch of astronomy and is used wherever accurate measures of light intensity are needed.

The first electric photometry was with a selenium cell, which was first employed by Professor Minchin in England in 1891 for measuring the brightness of stars. It was little used until 1906 when Brown and Stebbins in America revived selenium photometry. In 1911 the photo-electric cell replaced the selenium cell and has been in use ever since.

Developments in the sensitivity and accuracy of photo-electric photometers include the development of electrometers. The string electrometer was used at first for the measurement of the photo-electric current, but was later replaced by the Lindemann electrometer, because the latter could be operated from any position on a moving telescope.

When amplification was successfully accomplished by Whitford in 1932, circuits using vacuum tubes were constructed and the amplified current was measured with a galvanometer. During the war years, a photo-electric multiplier cell (1P21) was invented and perfected which made it possible to eliminate the first stages of amplification.

About the same time, recording meters became available which produced a permanent record of the amplified current, which could be read and studied at leisure. Such installations have made possible the measurement of stars as faint as can be photographed. Distant galaxies also come under observation with the largest reflecting telescopes, with important data resulting from the measurement of the brightness and colour of these objects.

It is hoped that the next step will be the building of a large telescope in the southern hemisphere, where these techniques can be applied to the southern areas of the sky.

Photo-electric photometry is a relatively new branch of astronomy. The invention and perfection of the photo-electric cell has permitted accurate determinations of stellar magnitudes by electronic devices which have the advantage of linear response; it has led to colour measures which have been used in determining the size of the galaxy; and it has provided a long-desired correction to the scale of photographic measures making possible the correlation of the magnitudes of bright and faint stars.

The photo-electric effect was discovered by W. Hallwachs in 1888 and by J. J. Thomson and P. Lenard in 1899, although it was many years before this important discovery was put to astronomical use. They found that when light was allowed to fall on certain metals—the alkali metals, rubidium, sodium, and potassium were best—there was produced a very small but measurable stream of electricity. This small current was proportional to the light and when properly measured could be used for an accurate determination of the amount of the light—in astronomy, starlight.

From the time of the great Greek astronomers Hipparchus and Ptolemy to the modern days of Argelander in Germany and Pickering in the United States, attempts had been made to determine the relative brightness of stars accurately, using the eye as the sensitive instrument. F. W. A. Argelander worked at this problem during the middle part of the nineteenth century and published his great *Durchmusterung* between 1846 and 1869. This work contained the positions of nearly a third of a million stars and their magnitudes, which had been estimated by his "step method" through a small telescope.

Another great work was by Edward C. Pickering, of Harvard University, who invented the meridian photometers which were used in the Harvard Photometry and made more than 1,500,000 observations himself. These photometers permitted the comparison of two stars of different declinations at the same time when on the meridian and used Polaris as a standard. These visual observations and photographic observations made during the same years were published along with the spectral types, which had been classified by Miss ANNIE J. CANNON. It should be pointed out that ARGELANDER's visual observations could be made with an accuracy of only about 0·1 magnitude and PICKERING's with about 0·05 magnitude and were therefore not as accurate as desired. While PICKERING's observations resulted in an internationally accepted scale, part of the scale is still in doubt and is even now being improved by accurate photo-electric methods.

The first application of photography to astronomy was made by JOHN W. DRAPER, of New York, who photographed the Moon on 23rd March, 1840. The first stellar photograph was of the bright star Vega at Harvard in 1850. The daguerre process was used but was insensitive and therefore impractical. An attempt to repeat this process a century later at Harvard failed. However, the rapid development of the dry plate soon made photography very satisfactory, since the exposure time could be long enough to record stars too faint to be seen and the accuracy was considerably improved over the visual observations. With the use of a yellow filter it is also possible to photograph stars in approximately the same spectral region as the maximum response of the human eye. By comparing such "photovisual" observations in the yellow with "photographic" observations in the blue region of the spectrum, the colour-indices of stars could be obtained photographically.

The first electrical measures were made with selenium cells. The fact that selenium changes its electrical resistance when radiated with light was discovered in 1873 by WILLOUGHBY SMITH. In 1891 Professor G. M. MINCHIN in England used selenium for measuring the brightness of stars, but the technique seems to have been forgotten for several years. In 1906 at the University of Illinois, when F. C. BROWN was demonstrating the properties of a selenium cell, the idea of applying it to the measurement of starlight occurred to Dr. JOEL STEBBINS, of the Illinois Department of Astronomy. He applied the selenium cell to the measurement of the light of the Moon at different phases in 1907 and by 1910 the methods of cell production had improved sufficiently to permit the study of certain variable stars, notably the eclipsing star Algol. The secondary eclipse in the light-curve of this star was discovered with the selenium photometer and has since been adequately confirmed with the more sensitive photo-electric photometers.

The selenium cell was unsatisfactory for at least two reasons: it was too insensitive and it had to be refrigerated to be used at all. So selenium photometry was given up with great suddenness when the photo-electric cell was invented and applied to astronomy in 1912.

Attempts to make practical photo-electric cells were begun in Germany by J. ELSTER and H. GEITEL as early as 1893. By 1911 they were producing successful cells and had used them for the first time in astronomy for observing an eclipse of the Sun in 1912. A stellar photo-electric photometer was installed on the 30-cm refractor of the Berlin-Babelsberg Observatory and observations of the variable star β Cephei were reported by P. GUTHNICK that same year. The cell used was of sodium hydride with about 25 volts potential and its spectral sensitivity was similar to that of the photographic plate.

In GUTHNICK's installation the photocurrent was measured by means of a string electrometer, the entire installation having been built by the German firm of Günter & Tegetmeyer to ELSTER and GEITEL's specifications. The electrometer was used by a rate-of-charge method and the observations gave residuals of approximately 1/100 magnitude from a mean curve.

The success of photo-electric experiments by ELSTER and GEITEL reached the ears of Dr. STEBBINS when Professor JAKOB KUNZ came to the University of Illinois in 1911. KUNZ had just come from Germany and had some knowledge of the methods of making cells and of their application. So it was natural to make some experiments, both in the methods of cell preparation and in the construction of a photometer for use with the 12-in. telescope of the University Observatory. STEBBINS spent the academic year of 1912–13 in Germany and KUNZ and W. F. SCHULZ carried on during his absence. They made measures of the bright stars Capella and Arcturus during the winter. A preliminary note entitled "The Use of the Photo-electric Cell in Stellar Astronomy" was published by SCHULZ in the September, 1913, issue of the *Astrophysical Journal*. It is interesting to note that this paper was published three months earlier than the paper of GUTHNICK, but that GUTHNICK published a light-curve while SCHULZ gave only data concerning the rate of deflection of the electrometer and did not attempt to reduce the observations to magnitudes.

The most successful photo-electric cells were made of potassium deposited on a layer of silver in a quartz tube. Potassium was the easiest alkali metal to handle and gave a response in the part of the spectrum which was also a close approximation to the colour response of the photographic plate. The silver was used as a conductor between the alkali metal and the leads into and out of the cell. Quartz was an excellent insulator and prevented the loss of the precious electrons and the leakage of stray electric currents from the high potentials it was necessary to use in making the cells work satisfactorily.

The potassium was first deposited on the silver conducting layer. Then the cell was connected to a vacuum pump, the air was pumped out and an atmosphere of hydrogen introduced. A high voltage across the terminals of the cell produced potassium hydride, the sensitive element, which usually assumed a pale lavender colour. The residual hydrogen was then pumped out and a thin atmosphere of helium was admitted and the cell was sealed and ready for use. The cells made by KUNZ after this fashion were used with a potential of about 300 volts, the purpose of the high potential being to multiply the feeble current from the cathode by ionization in the low-pressure helium atmosphere. If the quartz tube were provided with a guard ring to conduct the surface leakage current on the outside of the tube to the ground, there was very little "dark-current"—a flow of electrons when the cell was not exposed to light.

The first electrometers used to measure the photocurrent were string electrometers. The electrons were conducted to a string made of silvered quartz or, probably better, to a thin platinum wire, which was suspended between two brass plates, one charged positively and the other negatively. As the electrons piled up on the string, it bowed toward the positive electrode, the rate was measured with a stop-watch, and was strictly proportional to the rate of flow of the electrons. This photocurrent was proportional to the brightness of the light being measured, provided the range of intensities was small.

For observations of the stars, some observers used a standard light source for comparison, but the best standards seemed to be the stars themselves. For example, in making observations of the light variations of the eclipsing star Algol, two nearby stars were used as comparison stars and the variations of Algol were compared to the mean brightness of the two standard stars. Thus it was possible to check the two against each other for possible variations and, since they were near the variable star,

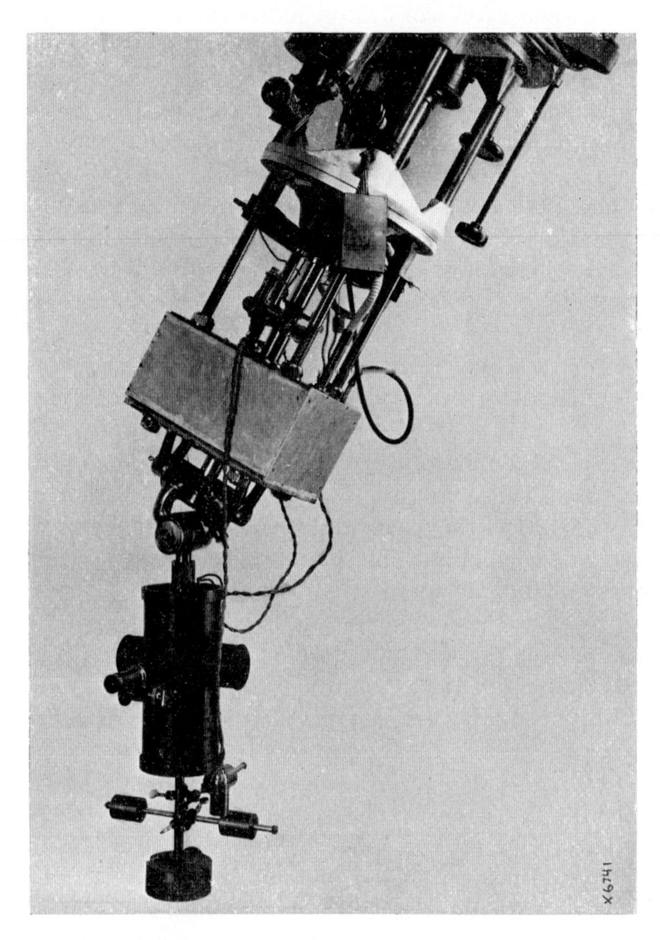

Fig. 1. The first photo-electric photometer of the Washburn Observatory, constructed in 1922 and put into operation early in 1923. The photocell was in the aluminium box and was dried with sulphuric acid. The sensitive surface of the cell was just behind the focus of the 15·6-in. refracting telescope. The string electrometer was hung in gimbals in order that it might hang vertically at all times

all three were subject to approximately the same variations in sky transparency. The two standards had been carefully selected stars of nearly the same colour as Algol. The rate-of-charge measures with the electrometer were converted to differences of magnitude and plotted in the usual manner as magnitudes against time.

Later developments came as improvements in the photometer, since there was little progress in the effort to increase the sensitivity of the photocells. Kunz occasionally made more sensitive cells, but they did not remain sensitive more than a few weeks. The Lindemann electrometer replaced the string electrometer about 1927.

This little instrument had the advantage of being able to operate at any angle, a great help on a moving telescope. The old electrometer had to hang accurately vertical in all positions of the telescope (see Fig. 1). Another electrometer, the cumbersome but highly sensitive Hoffman electrometer, was in use at the Coudé focus of the Mount Wilson telescopes by 1932.

For many years it had been hoped that the photocurrent could be amplified by the use of vacuum tubes and experiments to increase the sensitivity of existing photometers by this method had been made in several places. Such experiments at the University of Wisconsin culminated in 1932 with the successful construction of an

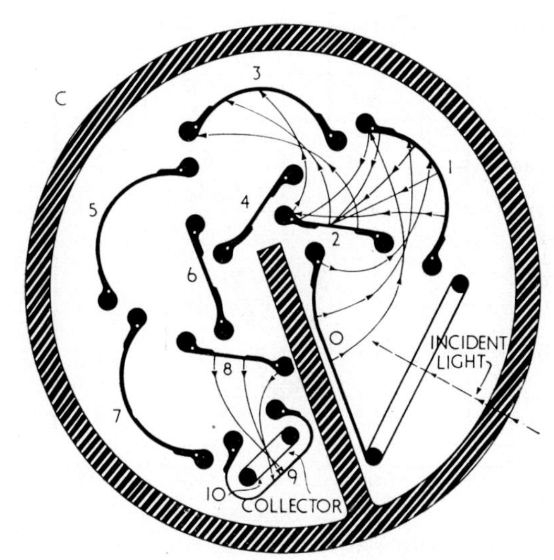

Fig. 2. The 1P21 photomultiplier tube. Left-hand photo is the tube itself. Right-hand diagram shows the arrangement of the elements of the tube

amplifier by Dr. ALBERT E. WHITFORD. Before that time, vacuum tubes were unsatisfactory for the purpose because they all apparently were seriously affected by grid emission larger than the extremely small amounts of current being measured. With the perfection of the FP-54 tube by the General Electric Company in the United States, amplification of direct currents of very small magnitude became possible. These improved tubes operated at low plate potentials and could be used with storage batteries, which were also very steady. The installation of the photometer with amplifier at the University of Wisconsin enabled STEBBINS and his associates to measure stars of the tenth magnitude or fainter. Previously they had been limited to seventh magnitude stars with the 15-in. telescope.

Just before the war the use of the caesium-antimony cathode provided the long-awaited improvement in photocell sensitivity. Also for several years attempts to make phototubes which produce the photocurrent and to amplify it by the principle of secondary emission had been almost successful. The invention of the 1P21 photomultiplier tube combined these two improvements and gave for the first time an amplifier which did not put in noise of its own (see Fig. 2). The 1P21 was used at first with a galvanometer in the circuit, but now there are very satisfactory recording instruments, notably the Brown recording meter. The 1P21 multiplier with an additional amplifier and a Brown recorder permits observations by one observer, instead of two as formerly, and records the data for future examination and computation by the investigator.

The first problem undertaken with the photo-electric photometer was the study of the light-changes of variable stars, particularly of eclipsing binaries. The theory of computation of eclipsing stars had been carefully studied by Professor H. N.

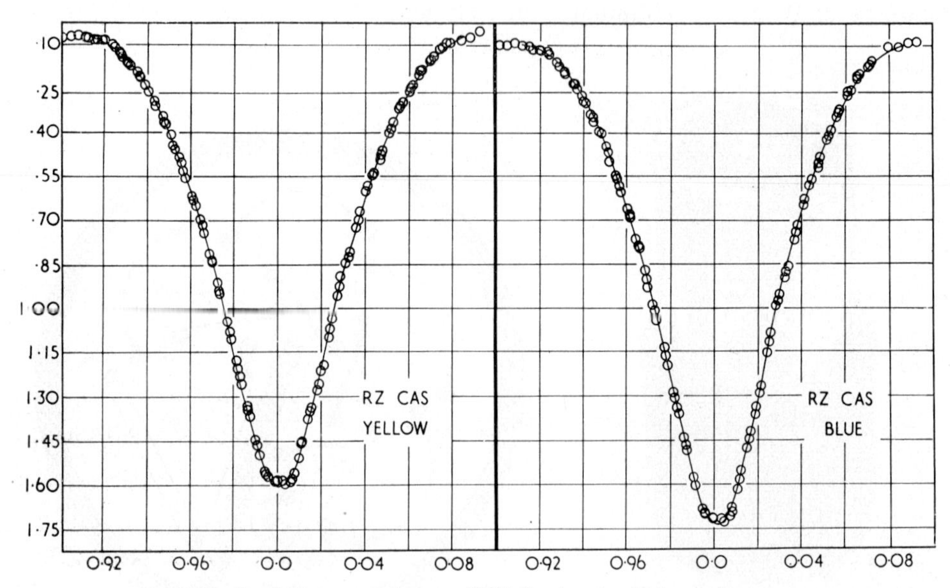

Fig. 3. Primary minimum of RZ Cassiopeiae in two colours

The observations were made at the Washburn Observatory on 15th December, 1952. The left-hand curve was made through a yellow filter, effective wavelength 5409 Å; the right-hand curve through a blue filter at 4282 Å.

The deeper curve in the blue light is a consequence of the lower temperature of the larger star which is eclipsing a star of higher temperature. The eclipses are partial. Probable error of a single observation is about 0·01 magnitude.

The results of the solution are: Ratio of radii . . 0·84 ± 0·01
Inclination of orbit . 81°.9 ± 0°.1

	Smaller star	*Larger star*
Mass	2·00 Sun	0·53 Sun
Radius	1·57 Sun	1·85 Sun
Density	0·52 Sun	0·08 Sun
Light (percentage of total) .	92%	8%

Russell and Dr. Harlow Shapley. The "Russell method" was published in 1912. These two men had computed orbits for many eclipsing stars, but they needed better observations. These were soon provided by the photo-electric measures and over the years many sets of elements were computed by the Russell method. In recent years this method has been further improved and modified by other investigators, especially after the valuable tables of p-functions were computed by Zessewitch in Russia. Leaders in these modern methods of computation have been Drs. J. E. Merrill, J. Irwin, Z. Kopal, and S. Piotrowski, who have been closely associated with each other and with Professor Russell. Their methods include the important addition of partial darkening at the limb of stars, and also improvements for the rectification of the light-curves due to the ellipticity of figure and reflection of light between eclipses. These computations demand observations of high precision such as the light-curve of YZ Cassiopeiae obtained by G. KRON at Lick Observatory and of RZ Cassiopeiae and VV Orionis by C. M. HUFFER at the Washburn Observatory.

Of course, other classes of variable stars have been studied photo-electrically, including the light-curves of Cepheid variables and of some irregular variable stars, but these studies have not had the importance and therefore not the intense interest demanded by the theory of the other class.

At the suggestion of Dr. F. H. SEARES of the Mount Wilson Observatory, STEBBINS and WHITFORD directed their attention to an intensive investigation of the North Polar Sequence. This is a series of stars near the north celestial pole which has been used for many years as standards for the determination of stellar magnitudes and colours by the photographic investigators, of whom Dr. SEARES was one. By 1938 the magnitudes and colours of thirty stars had been measured photo-electrically with the 60-in. reflector at Mount Wilson. This work was extended to fifteenth magnitude in 1950. The magnitudes agreed fairly well with the international photographic scale, except from magnitudes 5·0 to 9·0, where the two scales diverge by nearly 0·2 magnitude. It was also found that the colour scale, which was intended to be Colour-Index = 0·0 for spectral class A0, was somewhat redder for the Polar Sequence. This was in agreement with Seares' belief that there is space absorption and reddening in the region of the pole.

The determination of the colour-index of stars, using two colour filters, was begun by K. F. BOTTLINGER in Germany, although GUTHNICK had published measures of colour-index with one filter in 1920. BOTTLINGER'S catalogue was published in the publications of the Berlin-Babelsberg Observatory in 1920. In 1940 STEBBINS and his associates published colours of 1332 stars of class B and also had measured the colours of all the globular clusters in reach of the telescopes at Mount Wilson. Since those years, many observers have investigated magnitudes and colours of sequences of stars in open custers, such as the Pleiades and Hyades and in several globular clusters. These colour-magnitude arrays have been substituted for the Hertzsprung-Russell diagram and are assuming important roles in the investigations of stellar populations.

Three- and six-colour photometry is also important in astrophysical studies and demand the use of photocells which have a colour sensitivity range further into the red region of the spectrum than was attained by the first potassium cells. Such cells have been developed and used, but require refrigeration. The future will undoubtedly see the perfection of multiplier tubes similar to the 1P21, but with a greater colour range. The use of such multipliers will greatly extend the six-colour photometry and will fill a gap between present measures and those made with the lead-sulphide photo-conductive cell, which is sensitive in the infra-red to 2 or 3 microns. The lead-sulphide cell is being used by L. GOLDBERG and R. R. MCMATH and their associates at the University of Michigan to extend the solar spectrum into the infra-red regions formerly inaccessible.

With multipliers employing the sensitive caesium-antimony surface (such as the 1P21), photometry is now limited only by the background noise from the night sky, a situation which is analogous to the background sky fog that sets the limit in photography. At present, the photo-electric installation on the 200-in. telescope is extending measures to what must be near the limit of instrumental sensitivity. It is to be hoped that the southern hemisphere will soon be as intensively studied as the northern. This will require a larger telescope—100-in. or more—and international co-operation on a large scale, such as had already been started.

Photo-electric photometry is an important branch of modern astronomy and is now receiving the well-earned attention of many investigators.

References

[1] *Objektive Photometrische Methoden*, by BENGT STRÖMGREN, 1937.

[2] "The Development of Astronomical Photometry", by HAROLD F. WEAVER; *Popular Astronomy*, **54,** 1946. 218 footnotes with references.

[3] "Symposium on the Photo-electric Cell in Astrophysical Research", by several authors; *Publications of the Astronomical Society of the Pacific*, **52,** 1940.

[4] "Photo-electric Photometry of Stars", by JOEL STEBBINS; *Publications of the Washburn Observatory*, **XV,** 1928, preceded and followed by many papers in the *Astrophysical Journal* and the *Publications of the Mount Wilson Observatory*, by STEBBINS and associates.

[5] "On the Determination of the Orbital Elements of Eclipsing Binary Stars," by HENRY NORRIS RUSSEL; *Astrophysical Journal*, **35, 36,** 1912. Followed by many other papers.

[6] "Tables for the Computation of the Orbits of Eclipsing Binary Systems", by W. ZESSEWITSCH; *Bulletin de l'Institute Astronomique* (*U.S.S.R.*), **45, 50,** 1940.

[7] "Astronomical Photoelectric Photometry" (Symposium); edited by F. B. WOOD; Publ. Amer. Ass. for the Advancement of Science (Contributors: LINNELL, HALL, BLITZSTEIN, LALLEMAND, LENOUVEL. REDMAN, YATES, WALRAVEN, WHITFORD), Washington, 1953.

On the Application of the Photomultiplier to Astronomical Photometry

R. H. HARDIE

Lowell Observatory, Flagstaff, Arizona, U.S.A.

SUMMARY

A general discussion is made of some of the characteristics of photomultipliers. The relation of these to astronomical measurements is emphasized, especially in regard to obtaining optimum efficiency and stability in photometry in several colours.

WITH the development of the photomultiplier the techniques used in photo-electric photometry have become simplified and improved, with the result that this field, while once of rather specialized and restricted application, has been greatly broadened. However, the multiplier is not without peculiarities, and some attention may profitably be directed to its characteristics in relation to their role in astronomical instruments. For the sake of brevity the discussion will be limited to some of the properties which assume importance in magnitude and colour measurements.

1. Colour Response and Sensitivity of the Cathode

Various substances have been used in the preparation of photocathodes possessing different spectral sensitivities and efficiencies (see ZWORYKIN and RAMBERG, 1949, Chapter 3). In recent years multipliers with antimony-caesium cathodes have been widely used, since this type of cathode has unusual sensitivity through most of the visual spectrum except for the red. Many fruitful investigations, too numerous to cite, of astronomical phenomena in the spectral region from the ultra-violet to the yellow have resulted from the use of such multipliers. Up to the present, investigation in the region from 6000 Å to about 10,000 Å has been dependent on the use of either simple vacuum cells or gas cells with silver-caesium-oxide cathodes (see, for example, HALL, 1941; STEBBINS and WHITFORD, 1943, 1945, *etc.*; KRON and SMITH, 1951). That some of the technical difficulties in making multipliers with silver-caesium-oxide cathodes can be overcome has been demonstrated by the recent construction of some experimental models (HARDIE, 1952; WHITFORD, 1953) and it is hoped that it will be feasible to produce them commercially, as a much accelerated programme of research in the red and infra-red would surely result. Since no photoemissive substances are known to be sensitive beyond about 15,000 Å, there seems to be little promise of exploiting the multiplier techniques for this region, and one must, at least for the present, resort to other devices.

Since the quantum efficiency, or luminous yield, of the cathode determines the response from a given source of illumination, and has a direct bearing on the precision of the measurements, it is desirable to employ a cathode having as high an efficiency as possible in the spectral range to be investigated (LALLEMAND, 1952; KRON, 1948), and to ensure complete collection of the primary electrons by the first dynode. The antimony-caesium types generally available average over 10 per cent efficiency (*i.e.* 10 emitted electrons per hundred incident quanta) in the region of their greatest sensitivity and occasionally may be found to possess about 25 per cent efficiency (LALLEMAND, 1952; ZWORYKIN and RAMBERG, 1949, Chapter 20). Since there is much variation among commercial multipliers, it is most desirable to select one having high cathode efficiency. This may be done by intercomparing several, connected as simple cells (using the first few dynodes as an anode to collect the emitted electrons), and using a common source of illumination with a monochromator or several interference filters. It may be argued, and with some justification, that it is not worthwhile to select multipliers on this basis, since a factor of n in cathode efficiency of two cells results only in a factor of $\sqrt{n}$ in the ratio of signal to statistical fluctuation. One should not fail to realize, however, that by the same token, one effectively increases the aperture in the same ratio if one uses the more efficient cell rather than the poorer one.

2. INFLUENCE OF CATHODE FORM AND POSITION ON RESPONSE

The low efficiency of the silver-caesium-oxide surface (usually less than 1 per cent) leaves much to be desired in the red and infra-red. Among the various factors contributing to this inefficiency is the reflection of some of the light before it can produce electron emission. If the reflected light could be re-directed on to the cathode, an improved response might reasonably be expected. Some experiments were carried out by the writer to test such an effect, using a silver-caesium oxide cathode shaped somewhat like black-body cavity, in which the reflected light was caused to strike the cathode repeatedly. An improvement in the total yield by a factor of three was noted for the response to a lamp at 2870°K; this was principally due to increased response in the infra-red tail of the spectral curve with almost no change for wavelengths shorter than about 8000 Å. In order to make use of this desired effect, however, care would have to be taken in the design of the cathode shape to minimize the effects of position and angle of incidence of the incident light on the stability of sensitivity and colour system.

With regard to minimizing the effects of position and angle of incidence in photometric work, it might be expected that the "end-on" type of cathode would be best. In this type of multiplier, the cathode is deposited on the flat end of the glass envelope and is semi-transparent. The efficiencies of these cathodes are apparently as high as similar solid types and they present fewer difficulties in admitting the light, especially in fast optical systems, than do types like the 1P21 and 931-A. Furthermore, there is no occultation of the light by a wire grid as in these latter types. Some tests carried out on one such "end-on" multiplier indicated that there was, indeed, very little change in the colour system accompanying displacement of the light spot over small distances on the cathode. Similar tests made with a particular 1P21 revealed a rather strong dependence of the colour system on position and moreover the function relating these was not found to pass through a maximum or minimum as in the case of the "end-on" cathode; this was due, no doubt, to large variations in the collection across the cathode. On the other hand, there is also a likelihood of imperfect collection of the emitted electrons in the "end-on" type of multiplier; in the cell that was tested there was found a marked relation between colour system and accelerating voltage. In normal use, however, a fixed high voltage in the first stage would probably prevent any undesired losses or changes in the colour system. Another disturbing characteristic of some multipliers is the defocusing effect of magnetic fields that has been noted by various observers, which probably necessitates the use of effective magnetic shielding.

In general, the cathode area should be as small as possible since the unused portions contribute to the thermal emission (dark current) but not to the photo-electric emission. Refrigeration with dry ice is often resorted to, since it reduces the dark current to a level in which it may often be ignored. This technique is especially important for cathodes having extended red or infra-red sensitivity. It has been found that a cooled cathode has a colour system differing from that of the same cathode in an uncooled condition (JOHNSON, 1953) since the cooling reduces the far red sensitivity somewhat. It may also be found that the sensitivity and colour system of a cathode change over the course of time (STEBBINS and WHITFORD, 1945).

3. SIGNAL-TO-NOISE RATIO

In all these considerations attention has been focused on the cathode; this is because the signal-to-noise ratio for the multiplier is essentially that of the cathode, the multiplying process itself adding only slightly to the noise (LALLEMAND, 1949; ZWORYKIN and RAMBERG, 1949, Chapter 13). This is in fact one of the very attractive features of the multiplier; the task of measuring a minute cathode current is avoided and fortunately not at the expense of much precision. It will be clear, then, that a multiplier, even one possessing high multiplication, cannot make up in precision for an inefficient cathode. It should be said that theoretically a simple cell loaded with an extremely high resistance used in conjunction with an amplifier should perform as well as the corresponding multiplier; however, the problem becomes difficult for technical reasons. The problems of maintaining accurate calibration of very large resistors and high stability of the apparatus, and of avoiding leakage currents and unduly long time constants can be most discouraging. It is not to be inferred that a simple cell cannot be used to advantage in the absence of a multiplier; one has only to refer to the literature to see that these difficulties can be resolved (see, for example, STEBBINS and WHITFORD, 1943, *etc.*; KRON, 1952). What is indicated is the much greater ease and convenience to be found in the use of a multiplier for comparable precision, when one can be found for the spectral region to be studied. However, if the investigation is concerned with light of high intensity, as in the study of bright stars with a large telescope or in some laboratory and solar instruments, the multiplier offers no advantage over the simple cell and amplifier.

A useful expression which describes the noise production in a multiplier is given by LALLEMAND (1949):

$$A = \frac{\text{signal-to-noise ratio of the cathode}}{\text{signal-to-noise ratio of the anode}} = \sqrt{\frac{\delta^{n+1} - 1}{\delta^n(\delta - 1)}},$$

for a multiplier of n stages having a multiplication of δ in each stage. (Consideration is given only to the statistical nature of the multiplication in this expression, all other sources of noise being ignored.) The dependence of A on both δ and n is illustrated in

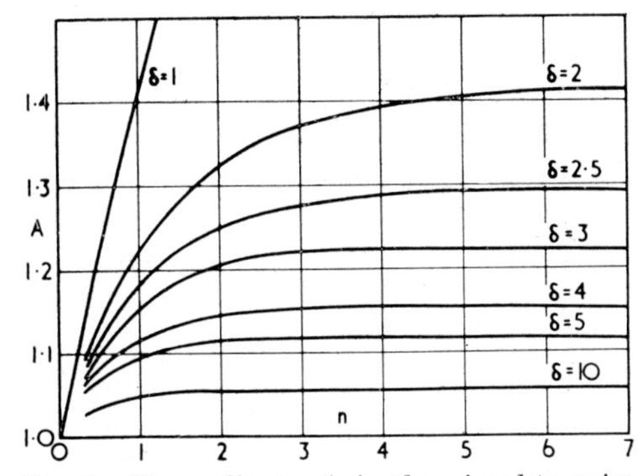

Fig. 1. The ordinate A is the signal-to-noise ratio at the cathode of a photomultiplier, divided by the signal-to-noise ratio at the anode. This is plotted against n, the number of multiplying stages for selected values of the multiplication δ per stage (assumed to be the same for each)

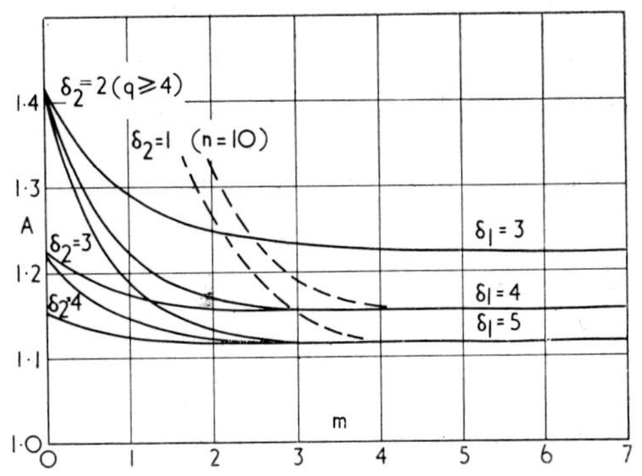

Fig. 2. The relative noise A, as defined for Fig. 1, as a function of the number m of initial stages, each having multiplication δ_1, when these are followed by q ($q \geq 4$) stages having multiplication δ_2 per stage. The curves for unity gain ($\delta_2 = 1$ in) the following stages, which are shown dotted, refer to a total of ten stages

Fig. 1, from which it will be inferred that the initial stages are responsible for almost all of the additional noise. Indeed, it will be noted that it is advisable for these stages to possess high multiplication in order to keep A as near unity as possible. A corresponding expression governing m initial stages of multiplication δ_1, followed by q stages of multiplication δ_2 is given by

$$A = \sqrt{\frac{\delta_1^{m+1} - 1}{\delta_1^m(\delta_1 - 1)} + \frac{\delta_2^q - 1}{(\delta_2 - 1)\delta_2^q\delta_1^m}}.$$

Examination of Fig. 2 will reveal how various values of initial multiplication will improve the anode signal-to-noise ratio.

4. THE FATIGUE EFFECT AND ITS CONSEQUENCES

The multiplier has the property of linear amplification over a very wide range of illumination (ENGSTROM, 1947), that is, the relation between the intensity of the incident radiation and the anode current is very linear. This is especially useful for faint sources, and it has been possible to measure intensities of luminous fluxes much smaller than 10^{-13} lumens (LALLEMAND, 1952; ENGSTROM, 1947). However, the linear relation ceases to apply for large values of the anode current due to an effect known as fatigue. This characteristic probably originates in the last few dynodes, where the current density is sufficiently high in the relatively poor conducting secondary emission layers to reduce the multiplication (LALLEMAND, 1953), and where space charge effects disturb the accelerating potential fields (ZWORYKIN and RAMBERG, 1949, p. 152). This may be avoided by intentionally reducing the multiplication so that high currents are not produced, in which case the linear relation again governs until cathode fatigue sets in at high levels of illumination. It would be of great use to the astronomer if surfaces were developed that could withstand very large current densities and at the same time produce good secondary emission. Many multipliers employ the same kind of surface for the dynodes as is used for the cathode; some others employ a magnesium-silver alloy which has the advantages of not being photo-emissive and having very low thermal emission at ordinary temperatures.

The principal consequence of the fatigue effect, insofar as it enters into the problem of photometry, is the limitation it imposes on the range of magnitudes over which the apparatus will be useful. In general, currents greater than a certain value may not be measured without operating the multiplier in its nonlinear range; hence this places an upper limit on the brightness of the stars to be examined. The lower limit is determined by the sensitivity of the apparatus used to measure this current.

5. MEASUREMENT OF THE OUTPUT CURRENT

The anode current may be measured in several ways: a galvanometer may be used directly, an integrating or pulse-counting technique may be employed, or an amplifier (frequently regarded as an impedance transformer) may be used to operate a meter or pen-recording device. Let us consider the latter technique, since it is the most frequently used and has great flexibility, and moreover the discussion will in general apply to the others.

The sensitivity of the photometer may be varied by changing the load resistor through which the anode current passes in order to develop a voltage to be presented to the amplifier; this is conveniently done by a switching arrangement, the calibration of which may provide a magnitude scale. In addition other means may be used, such as the placing of shunts across the indicating meter. This latter method is feasible, however, for limited ranges only, since it is better not to employ the amplifier over a wide range of operating conditions requiring great overall linearity. Rather the range of such a sensitivity control within the amplifier is confined to perhaps 2·5 magnitudes, and a wider range is achieved by selecting the various load resistors. In essence the linearity of the multiplier's functions over a wide range of conditions is relied on rather than that of the amplifier.

Let us consider a direct-current amplifier having a voltage sensitivity of 0·1 V, a value representative of good present-day amplifiers. If precision wire-wound resistors were to be used for the various anode loads (for example, when the greatest stability of calibration is desired) the minimum anode current which could produce a full scale deflection would be 10^{-8} A, assuming that the largest procurable resistor is 10 megohms. If deflections as small as 10 per cent of full scale could be used, the lower limit for the anode current, then, is seen to be 10^{-9} A when using wire-wound resistors. Further, if the fatigue effect of the multiplier limited the anode current to values less than 10^{-5} A (this, of course, will vary among multipliers), then the maximum range of current is seen to be 10^4, corresponding to 10 magnitudes.

For much photometric work this may be an adequate range (for example, in the study of variable stars, colours, polarization, *etc.*). However, for work such as the establishment of magnitude systems over a wide range, it may be desirable to extend the range to, say, 15 magnitudes without resorting to screens, neutral filters, *etc.*, but relying solely on precisely calibrated wire-wound resistors.

A common solution is to employ composition resistors rather than the wire-wound type, since these may be procured in very high values. In the example of the direct-current amplifier discussed the range would be 15 magnitudes if the largest resistor were 1000 megohms instead of 10. This method has the apparent disadvantage of lacking the high stability desired in the calibration; this may be overcome through frequent calibrations of the resistance ratios, preferably at the telescope.

Another solution which has been very successfully used (for example, by WHITFORD and JOHNSON) is to employ an alternating-current amplifier, since it is relatively easy to extend its voltage sensitivity to 0·001 V, an expedient which provides a further factor of 100 and thus a range of 15 magnitudes altogether. The technique requires interruption either of the light beam with a rotating sector, or of the anode current with a vibrating reed. In either case rectification of the amplified signal is required, and great care is necessary to ensure that the interruption and rectification operations are well synchronized. The fact that a rotating sector wastes half the light is a disadvantage, and this necessitates that the readings be made for intervals twice as long as otherwise necessary for a given precision. However, this technique has the attractive feature of eliminating the dark current (except for the components of the thermal noise lying at the modulating frequency), a convenience when refrigeration is not used. The current interrupting technique does not possess this advantage, but on the other hand does not waste half the light.

Another solution is to provide for various dynode supply voltages; for example, two values of multiplication appropriately selected would permit a range of almost 20 magnitudes under the conditions assumed in the example of the direct-current amplifier discussed, still retaining the wire-wound resistors. The requirements are to use a stable high voltage supply capable of providing several voltage ranges and to select these so that in each range of operation the magnitude range overlaps with that in an adjacent one for purposes of precise calibration. It will be advisable, however, to retain as high a voltage in the first few stages as is consistent with good signal-to-noise ratio in accordance with the earlier discussion and to vary the remaining stages only. It would be imprudent to attempt to use extremely high multiplication in any stages, however, since high voltage gradients are likely to cause ionic currents in the residual gas, thus introducing further noise and possibly damaging the surfaces of the electrodes.

6. SOME APPLICATIONS OF THE PHOTOMULTIPLIER

The importance of any of the foregoing considerations will naturally vary with the particular application of the multiplier and in some cases there will be other characteristics, not discussed here, which will require attention. Some of the notable examples through which substantial progress has been achieved in astronomical problems by the use of the multiplier, other than in direct photometry, may be cited: the recording of transit observations; the automatic guiding of the telescope

(Babcock, 1948); the direct scanning of spectra (Code, 1952); the mapping of the solar corona (Lyot, 1950); the mapping of solar magnetic fields (Babcock, 1953). It may not be unreasonable to expect additional progress in other fields where visual or photographic work may be supplemented by photo-electric methods, for example, in the fields of astrometry and double stars, interferometry, radial velocities, and even spectral classification (as has been reported by Strömgren, 1955). All these and other possibilities await further development, especially as new types of multipliers are introduced.

References

Babcock, H. W. . . . . 1948 *Ap. J.*, **107**, 73.
1953 *Ap. J.*, **118**, 387.
Code, A. D. . . . . 1952 *Observatory*, **72**, 201.
Engstrom, R. W. . . . . 1947 *J. Opt. Soc. Amer.*, **37**, 420.
Hall, J. S. . . . . 1941 *Ap. J.*, **94**, 71.
Hardie, R. H. . . . . 1952 *Publ. Obs. Haute Provence*, **2**, No. 30.
Johnson, H. L. . . . . 1953 Private communication.
Kron, G. E. . . . . 1948 *Harvard Coll. Obs. Monograph No.* 7, "Centennial Symposia", Chapter I, Section II.
1952 *Ap. J.*, **115**, 1.
Kron, G. E. and Smith, J. L. . . . . 1951 *Ap. J.*, **113**, 324.
Lallemand, A. . . . . 1949 *J. Phys. Radium*, **10**, 235.
1952 *Bull. astron.* (*Paris*), **16**, 197.
1953 Private communication.
Lyot, B. . . . . 1950 *Comptes rendus*, **231**, 461.
Stebbins, J. and Whitford, A. E. . . . . . . 1943 *Ap. J.*, **98**, 20.
1945 *Ap. J.*, **101**, 47, etc.
1945 *Ap. J.*, **102**, 320.
Strömgren, B. . . . . 1955 *Vistas in Astronomy*, Ed. A. Beer. (Pergamon Press, Ltd., London), Section 13, Volume 2.
Whitford, A. E. . . . . 1953 Private communication concerning a commercial experimental model.
Zworykin, V. K. and Ramberg, E. G.. . . . . 1949 *Photoelectricity and its Application*, Wiley and Sons, New York.

On the Ideal Reception of Photons and the Electron Telescope

A. Lallemand

Observatoire de Paris

Summary

The importance of the quantum efficiency of the receiver used in observational astronomy is discussed. A description is given of the development of an electron-optical telescope having high quantum efficiency. The original form of this device required a fixed-focus telescope, but a portable version which can be used on any large telescope has now been developed and has given promising results.

---

1. The Rôle of the Photon Structure of Light in Astronomical Observations

Astronomy is an observational science. We cannot perform experiments on the objects which we study; we have to make the best of the information which these objects send to us. There was a time when it could be thought that such information was so rich and extensive that no limit could be envisaged to the possible knowledge of the astronomical universe. The study of double stars, the determination of temperature, radial velocities, apparent diameters, and much else gave rise to this view. In fact, however, the granular nature of light imposes a fundamental limit to our knowledge of the universe. If we suppose that we possess an ideal light-sensitive receiver, and that we can do no more than collect a number of photons from a distant celestial object, then all that it is possible to conclude is that an object exists in a certain direction, but nothing can be said of its form, distance, temperature, or other properties. The amount and quality of our information depends on the number of photons which are detected.

Since the accuracy of our knowledge is determined by the number of photons received, it is important to collect as many photons as possible by constructing large telescopes and by developing a receiver which detects photons without loosing any and without responding to fictitious ones. Moreover, the ideal photon-receiver must be capable of giving an image of the object, since the image contains a great deal of information which a receiver such as a single photo-cell does not transmit. The possible diameter of an astronomical objective is indeed limited, but we can increase its effectiveness, which is to say the number of photons detected, by increasing the exposure time.

It follows that an ideal receiver would have to detect all the photons collected by the objective over as long an interval as may be desirable.

Photographic plates give an image, their exposure times can be made arbitrarily long, and they are fairly free from parasitic background effects. These remarkably sensitive receivers have rendered immense service to astronomy, but they are still far from being ideal receivers. Photographic emulsions deviate from the reciprocity law which an ideal receiver must follow. If the rate at which photons arrive on an element of the plate is decreased, the photographic effect per photon is also decreased. In order to obtain a usable photographic image of an object of low brightness it is therefore necessary to increase the number of photons received by greatly increasing the exposure time. It turns out that one can obtain hardly any photographic effect on very faint objects, however long the exposure time: there is in fact a kind of threshold effect in the sensitivity. This fact shows that our knowledge of the universe is limited because we cannot detect very distant or intrinsically very faint stars which lie below the photographic threshold.

It is, of course, true that the rate of arrival of photons on an element of emulsion can be increased by using objectives of large relative aperture; this is the method adopted in the Schmidt telescope. There are several disadvantages of this method. Not only is the angular resolution of the objective poorly exploited but, in addition, the absolute focal length is short, because it is difficult to construct objectives of very large diameter, so that the small scale of the resultant negatives crowds the details together and fine-structures become confused and indistinguishable from one another. Another phenomenon which limits the use of these otherwise excellent objectives is the luminosity of the night sky. Even on the blackest nights and at the best chosen observing stations, the high atmosphere emits enough photons to fog a photographic plate exposed for more than a fraction of an hour. The use of these objectives gives rise to the important contradiction that in order to study very faint stars and distant nebulae it is necessary to expose for a long time if sufficient photons are to be received, yet the exposure time must be limited in order to avoid general fogging of the plate. From this point of view, the effect of relative aperture of the objective can be summarized by saying that the limiting magnitude for an objective of 25 cm diameter increases from 14 to 19 as the relative aperture goes from $f/1$ to $f/15$; for a 250 cm objective the corresponding magnitudes are 18 and 22. Moreover, if we possessed a more efficient receiver at low light levels than the photographic plate we could afford to analyse the light more finely so as to achieve spectral dispersion, filtering, and selection of photons, and thus obtain the ensuing richness of information which has been exploited for example by DUFAY, BIGAY, and BERTHIER (1952, 1955) in their photography of the centre of the Milky Way in the near infra-red, and by G. COURTÈS (1951) in his observations with Hα-light.

We have attempted to develop a receiver of this kind which would closely approach ideal performance.

2. THE ELECTRON TELESCOPE

In about 1934 our attention (LALLEMAND, 1936a, 1936b, 1937) was arrested by the possibilities of photo-electricity, which is the effect whereby certain surfaces containing alkali metals have the property of emitting electrons when photons impinge on them. This extraordinary phenomenon is inexplicable except in terms of the quantum properties of light. EINSTEIN has given a very simple law for the phenomenon, according to which the energy of a photon is transformed into kinetic energy of the electron after having first overcome a potential barrier corresponding to the escape of the electron from the surface; this energy is known as the work-function. It follows that even at very low light levels one can observe the sudden emission of an electron and it is not necessary to accumulate the energy of the light in order to achieve this phenomenon; the quantum structure of the light itself enables a photon to produce a discrete perceptible effect. This behaviour provides us with a light-sensitive detector which follows the reciprocity law since the cumulative effect of a number of photons is no longer a condition for the production of a measurable effect. But how can these electrons be used? At this time M. LOUIS DE BROGLIE had recently created wave-mechanics and ascribed wave properties to electrons. These new ideas provoked rapid developments in the field and led to the science of electron optics. Electron lenses consisting of electric and magnetic fields enable images to be formed using electrons. Electrostatic lenses can accelerate electrons to high energies, and one can therefore replace photons by high velocity electrons without losing the imaging properties. To do this in an ideal way it is necessary to detect every electron arriving at the target. This is readily possible with electron-sensitive emulsions of the kind used in nuclear physics for the detection of high-velocity atomic particles.

An ideal photon detector would be usable only if there was some means of switching its sensitivity on or off at will; otherwise its manufacture, manipulation, and preservation would present formidable problems. Fortunately, in the arrangement indicated there is no sensitivity to light unless the accelerating voltage is applied.

(a) (b)

Fig. 1. Saturn through the Electron Telescope

(a) Electronic photograph (exposure time $\frac{1}{5}$ sec.)

(b) Conventional photograph on Superfulgur plate (exposure time 10 sec.)

The practical development of an arrangement of this sort involves a number of difficulties, the most important of which is to preserve the sensitivity of the photo-cathode during the necessary manipulations. Photo-cathodes are stable only in high vacuum, and a trace of gas such as oxygen or water vapour quickly destroys the photo-electric response. Moreover, electron-sensitive emulsions occlude a large amount of water and it is necessary to prevent this water from reaching the photo-cathode. The simplest means of doing this is to refrigerate the photographic plate with liquid air, but unfortunately the sensitivity to electrons is rather less at these low temperatures and the plates become very fragile.

We have taken electronic photographs with such an arrangement, constructed in collaboration with M. MAURICE DUCHESNE and have verified that our hopes were justified (LALLEMAND and DUCHESNE, 1951). We have obtained an image in 4 min of an object which, under similar conditions, needed an exposure time of 6 hr by conventional photography using high-speed "Superfulgur" plates. Work has been done at the "Petit Coudé" of the Paris Observatory which has an objective of 26 cm diamter (LALLEMAND and DUCHESNE, 1952). In this way we were able to focus a star image on to the photocathode, despite the complete immobilization at this period of the apparatus by its vacuum-pump connections, the cooling water, and liquid air flasks. This combination enabled us to master the proposed method, and to obtain the first electronic photographs of celestial objects. We obtained in this way images of Saturn in $\frac{1}{5}$ sec., although conventional photography on Superfulgur needed at least 10 sec. Obviously these images are not nearly as good as the best photographs obtained with great telescopes, since the electronic arrangement can only reproduce such details as are effectively present in the optical image given by the telescope. Nevertheless, the electronic images have shown us that the electron-sensitive emulsion is able to render images with exceptionally good quality. Not only does the resolving power seem to be greater than that of ordinary plates, but the electron-sensitive plate can be almost entirely free from silver grains other than those produced by the electrons, and this absence of fogging greatly improves the reproduction of faint images (Fig. 1).

We have now been able to substitute for this primitive arrangement the following apparatus (LALLEMAND and DUCHESNE, 1954). A tube, which can be evacuated by means of a small oil diffusion pump, has a large aperture closed by a glass plate attached by means of a rubber seal. A thin glass capsule containing a photocathode can be inserted into this opening, together with a demountable assembly which includes the electron-sensitive plates in a suitable holder, and the electron optics. The necessary electrical connections are made automatically by plugs and sockets. When a sufficient vacuum has been attained, several suitably placed barium getters are fired, and a side tube containing activated charcoal is refrigerated by liquid air. The parts are then manipulated magnetically so as to break the capsule, and to place the photocathode, which was inside it, into positive relative to the electron optics. The tap connecting the diffusion pump to the tube can now be closed and the pump disconnected. The tube can then easily be placed at the focus of any kind of refractor or reflector, and the vacuum remains sufficiently good for the photocathode to give reasonable performance for about 1 hr; it is not necessary to refrigerate the electron-sensitive plates. We have taken a number of photographs with this arrangement which confirm the previous results.

It may be concluded that the photo-electric effect is well able to fulfill the essential requirements of an ideal photon detector. By its use the limits set by the threshold of sensitivity in a photographic emulsion are overcome, and with long exposures it should be possible to explore the universe at distances which have hitherto seemed inaccessible.

REFERENCES

COURTÈS, G. . . . . 1951 "Étude de la voie lactée en lumière monochromatique Hα de 80° à 180° de longitude galactique," *C.R. Acad. Sci. (Paris)*, **232**, 1283.

DUFAY, J., BIGAY, J. H. and BERTHIER, P. . 1952 "Photographie directe du centre de la voie lactée dans le proche infra-rouge," *C.R. Acad. Sci. (Paris)*, **235**, 120.

1955 *Vistas in Astronomy*, Ed. A. BEER, Pergamon Press, London; Section 15 (Volume 2).

LALLEMAND, A. . . . . 1936a "Application de l'optique électronique à la photographie," *C.R. Acad. Sci. (Paris)*, **203**, 243.

1936b "Sur l'application à la photographie d'une méthode permettant d'amplifier l'énergie des photons," *C.R. Acad. Sci. (Paris)*, **203**, 990.

1937 "La photographie photoélectronique," *L'Astronomie*, p. 300.

LALLEMAND, A. and DUCHESNE, M. . . . 1951 "Sur un recepteur idéal de photons et sa réalisation; resultats preliminaires," *C.R. Acad. Sci. (Paris)*, **233**, 305.

1952 "Application à l'astronomie d'un recepteur idéal de photons," *C.R. Acad. Sci. (Paris)*, **235**, 503.

1954 "Sur le développement d'un recepteur idéal de photons," *C.R. Acad. Sci. (Paris)*, **238**, 325.

Indirect Methods of Star Counting

H. E. Butler

Dunsink Observatory, Co. Dublin, Eire*

Summary

The general problem of star counting consists in the determination from a photographic plate of the distribution of the star images with image size. Since the number of images on one plate can exceed a million, methods which involve counting the stars individually are obviously very limited in scope. The paper describes a method by which, once a certain number of standard stars on the plate have been analysed, the distribution of the images with size can be determined directly from an analysis of all images at once.

Under ideal conditions, the diameters of the different stellar images at the focal plane of an astronomical telescope are all constant. Each is a diffraction pattern whose dimensions are decided, not by the distance or brightness of the individual star, but by the aperture of the instrument. When we photograph such a star field, however, we find on the resulting plate, that the stellar images do vary in diameter and, further, that over any one plate, the apparent diameters can be directly related to the brightnesses of the corresponding stars. This occurrence, fortunate in the particular case, is the result of scattering of the light in the photographic emulsion itself and, the more the light that is available for scattering, *i.e.* the brighter the image, so the larger is the photographed image.

The general problem of star counting consists in the determination, in a certain area of the sky, of the number of stars of different magnitudes. As star counting is normally carried out from a photographic plate, our problem in its simplest form consists of quickly and conveniently grouping all the stellar images on a certain plate, or plates, into different diameter ranges.

It must be made clear that the problem involves many millions of images. The number of stars in the sky brighter than even the twelfth magnitude is already greater than one million, while there are about 500 times that number brighter than the twentieth magnitude.

Conventional methods of star counting inevitably include the determination of stellar co-ordinates as well as magnitudes. As a result even an experienced person can only deal with between, say, 200 and 500 stars per day, and it is evident that complete star counts by this method down to faint magnitudes, and over large areas of the sky, are quite out of the question. Sampling methods are adequate for many purposes, but it is obviously better if complete counts can be made. Star counting has always been a tedious and time-consuming occupation and the development of bigger and better Schmidt telescopes is only making the position more difficult. The ADH 36 in./32 in. Baker-Schmidt telescope at Bloemfontein, for instance, will photograph 18 square degrees on one plate (*i.e.* 1/2300 of the whole sky), with perfect definition, and on a blue-sensitive plate will reach a magnitude of about 19·0 with an hour's exposure. If we have occasion to use a panchromatic emulsion on a region of high star density, it is easily possible to obtain a plate showing a million stars. When

* Now at the Royal Observatory, Edinburgh.

we consider such plates in the light of a reasonable speed of reduction, it is evident that we have reached a stage where detailed statistical research on the faint stars is restricted, not by our inability to photograph them, but by their very numbers on the plates that are now easily obtainable.

There appear to be two possible methods for overcoming this *impasse*. Firstly it should be possible to make an automatic, electronic machine which will explore a photographic plate and produce from it either a complete catalogue, or at least a distribution of the stars with respect to magnitude. Such a machine has apparently not yet been constructed.

There is, however, a second and completely different approach to the problem which appears so far to have been overlooked. Several variants of the method are possible, but in principle each attempts to extract the distribution of stars from a plate with as little reference as possible to individual stars. In theory, provided the images are of uniform definition over the plate and provided there is no vignetting in the telescope, it is only necessary to consider in detail 25 or 50 standard stars on the plate. With the information gained from these, the required distribution over the whole plate, or over chosen sub-areas of it, can be mechanically produced. The actual numbers of stars on the plate become unimportant.

The development of two variants of the method with a view to a critical examination of their limiting accuracy and practicability is being carried out at this observatory and a description of the methods with some discussion of their theoretical limitations is given below.

We will consider first the method which has already been briefly described elsewhere in a paper by the writer.* Take an original, negative, photographic plate of a star field. It consists of a relatively transparent background covered with circular, non-transparent images of different sizes. From this plate make two, enlarged, positive *plates*. These will consist of a relatively black background covered with circular, transparent images of different diameters. Our problem is, of course, to size and count these images. Consider an idealized pair of such plates having completely black backgrounds covered with sharp, circular, transparent images. Allow a uniform beam of parallel light to fall normally on one of them. Evidently the only light transmitted by the plate will be the beams of circular cross-section passed by the various images. If the second plate is now placed behind the first, in a completely general position light will only succeed in passing both of them when an image in the second plate chances to be in one of the beams passed by the first plate. Then some, or all, of the particular beam will be able to continue on its way. In the unique position, however, when the two plates are oriented so that the images are exactly behind each other in pairs, the circular beams passed by the first plate will each pass unaltered through the corresponding images in the second plate, and the presence of this second plate will not in any way affect the amount of light passed. We will call the above-mentioned unique position the "position of correspondence" and we will refer to the first and second plates as such.

Consider now the effect on the total light passed by the two plates when the second plate is displaced in its own plane from this unique position of correspondence.

---

* H. E. Butler, *Observatory*, **73**, 80 (1953).

Evidently the black edges of the images of the second plate all begin to cut into the circular beams passed by the images of the first plate, and the total amount of light passed by both plates begins to decrease. As the displacement of the second plate increases further, the encroachment of the images continues until the smallest beams become completely cut off. Further displacement cuts off larger images and when the plate has eventually been displaced a distance greater than the diameter of the largest stellar image on the plate, all the beams are cut off and the only light passed by both plates occurs when, purely by chance, the image of a star on one plate comes opposite the image of *another* star on the other plate.

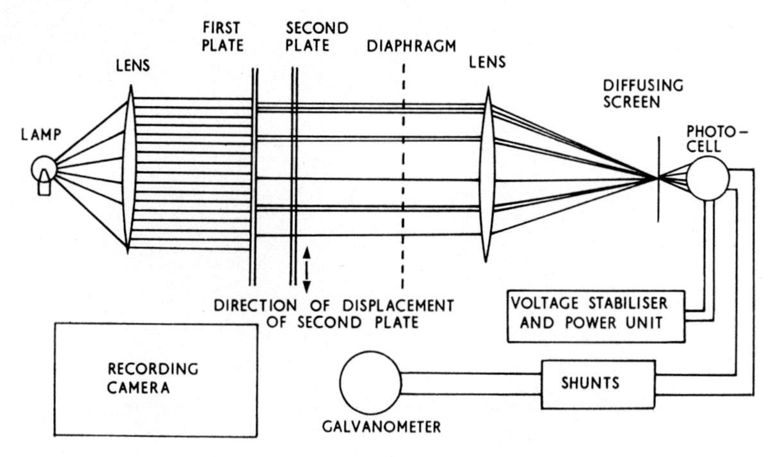

Fig. 1. Schematic diagram of recording instrument

Fig. 1 illustrates the method used to obtain quantitative results during this operation. A battery-run lamp is used with a lens to obtain a uniform, parallel beam of white light. The diameter of the beam in the experimental model is 5 in. The first and second plates are placed in this beam as shown, being about 1 cm apart. A second lens converges the light passed by the two plates on to a photo-cell, the output of which is fed to a galvanometer. The second plate can be moved sideways by a high precision screw and slide, the movement being driven by the motor of the recording camera which records the galvanometer deflection. The arrangement is thus similar to a recording micro-photometer. The rate of movement of the plate is approximately 1 mm per min and the recording paper in the camera moves 50 mm in that time. In the setting up of the instrument so that the images are moved exactly through the position of correspondence, two small relative movements of the two plates are essential in addition to the main sideways motion. The first plate can be rotated in its own plane by a screw and the second plate can be moved in a vertical direction. Once an approximate position of correspondence is obtained, it only remains to make small changes to the three movements while observing the galvanometer deflection. When any translation or rotation of the plate decreases this deflection, we can assume that we have then arrived at the position of correspondence.

A specimen of the type of curve obtained when the second plate is moved through the position of correspondence is shown as the broken line in Fig. 2, and will be referred to as the "all-stars" curve.

The all-stars curve is, of course, made up of the contributions from all the stellar images in the beam. If we now insert after the second plate a diaphragm which can be adjusted so that it only allows light from a single chosen image to pass to the photo-cell, we can then record the similar curve of this star as the second plate is again moved through the position of correspondence. Eight such practical curves corresponding to stars with magnitudes respectively of 10, 11, 12, . . . , 17 are shown in Fig. 2. We will call curves such as these latter "calibration curves" for the different magnitudes on the plate in question, and if we record sufficient of them,

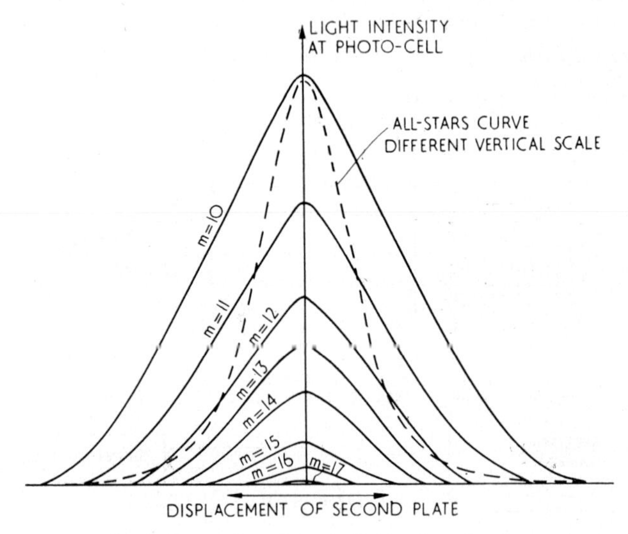

Fig. 2. Specimen curves for "all stars" and for individual stars of magnitudes 10, 11, . . . 17

we can obviously interpolate curves for as many other intermediate magnitudes as we like.

For any one plate, let the all-stars curve be $y = I(x)$, and let a general calibration curve for magnitude m (interpolated or not) be $y = i(m, x)$. Now since the all-stars curve is made up of the contributions from all the stars on the plate, if $N(m) \,.\, dm$ is the number of stars with magnitudes between m and $m + dm$, then

$$I(x) = \int i(m, x) \,.\, N(m) \,.\, dm$$

integrated over all magnitudes on the plate. *The solution of the problem of counting and sizing the images on the plate is thus reduced to solving this equation for* $N(m)$, where $I(x)$ is observed and the $i(m, x)$ are, in general, interpolations between a recorded set of calibration curves.

Experiments have been carried out on three plates taken with the ADH telescope. These were all of the same region of the sky (Harvard E4 region) and all were on blue sensitive emulsion (Kodak 103a–O). The three plates were given exposures of 5, 15, and 60 min respectively, and the distribution of the image diameters with respect to magnitude are plotted in Fig. 3. It will be seen that the limiting magnitudes of the three plates in order are approximately 16·5, 17·5, and 19·0. The steady slope of the curves for the brighter stars shows how well the plates are suited to the determination of magnitude from image diameter. The flattening out of the curves for the fainter stars—when magnitude change has little corresponding diameter change—is most marked for the 5-min exposure and least so for the 60-min exposure.

As a result, quite apart from the fact that this latter plate reaches to stars $2\frac{1}{2}$ magnitudes fainter, it is by far the best plate for measurement. The points on the three curves at which the images, instead of appearing completely black and opaque, show signs of becoming grey and transparent, are indicated.

In practice, the recording instrument itself is not unduly difficult to set up, the most troublesome part being the difficulty of getting uniformity over the parallel beam. Although it was not hard to get a lamp with a reasonably small filament, the indifferent quality of the glass envelope has so far only permitted the intensity over the beam to be constant to $\pm$ 2 per cent. After passing through the plates, the beam converges to an image of the filament of the lamp, and in order to spread

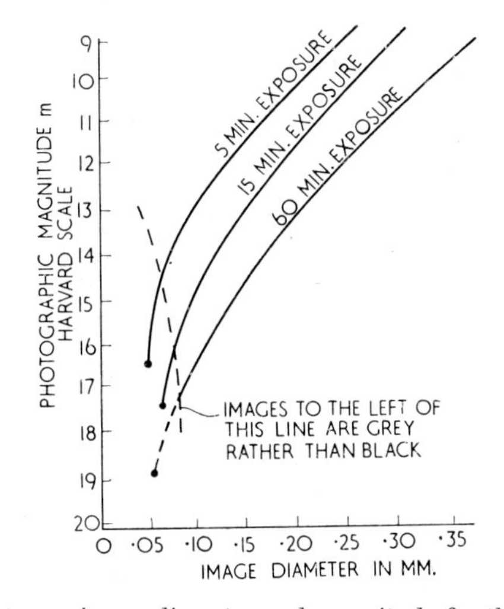

Fig. 3. Relation between image diameter and magnitude for three different plates

this uniformly over the cathode of the photo-cell (a 1P21 photo-multiplier tube), the image is formed on a diffusing screen placed a few inches in front of the cathode.

The light intensity received at the cell can vary between large limits. That received from a faint calibration star can be anywhere between 10^{-4} and possibly 10^{-7} of that received when the all-stars curve is being recorded. As a result a large range of sensitivity control has been needed in order to keep the galvanometer deflection within bounds.

The recording of the calibration curves does not appear to offer any difficulties and the interpolations necessary to give $i(m, x)$ for integral and half-integral values of m can be made by a variety of methods.

So far the integral equation has been solved by least squares applied to its equivalent form in simultaneous equations. It is evident from the curves of Fig. 2 (which are actually for the ADH plate of 60-min exposure mentioned earlier), that if the required solution involved approximately equal numbers of stars of each magnitude, then it would be hopeless to attempt a solution. For instance, the small light contribution from just one star of 17·0 magnitude would be completely insignificant compared with the light from, say, one star of each of the other whole magnitudes. On the other hand, as is generally known, the number of stars in a magnitude interval m to $m + 1$ is between two and three times that in the interval $m - 1$ to m, and so on. As a result, the weakness of the light contributions from the fainter stars is counteracted by the increasing numbers of these stars. In Fig. 4, the curves of Fig. 2 are repeated (only half being given since there is symmetry), with the ordinates of the curves corresponding to magnitudes 10, 11, 12, . . . , 17, being respectively multiplied by the factors, 1, 2, 2^2, 2^3, . . . , 2^7.

It will be seen that on the average, the contributions from the stars of the different magnitudes are roughly of the same order, at least until very near the magnitude limit of the plate. Thus, very approximately, we can say that the method should be able to detect the total contribution from the stars of each magnitude with about the same accuracy. In other words, the numbers found for these magnitude intervals

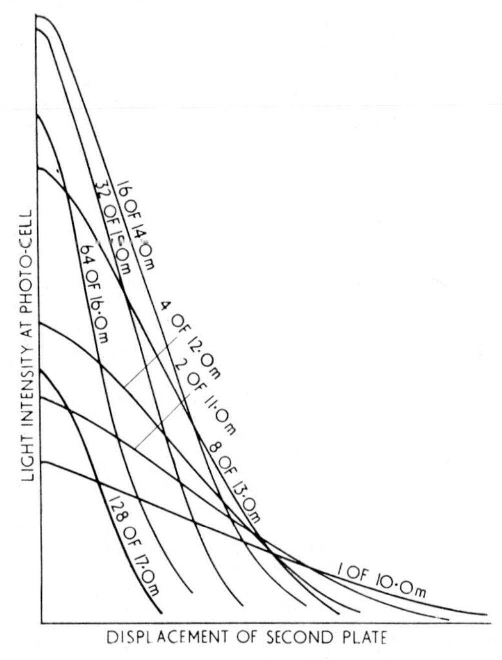

Fig. 4. Specimen curves for stars of different magnitude groups, assuming a realistic distribution of numbers of stars with magnitude

would all have about the same percentage error with respect to the numbers involved. Such a distribution of errors would obviously be acceptable.

It is of interest to examine the curves of Figs. 2 and 4 in the light of the measured diameters of the original star plate—the 60-min exposure of Fig. 3.

For the brighter stars, the fall-off in light with decreasing brightness is very nearly proportional to the square of the image diameter, *i.e.* to the area of the image. As soon as the magnitude gets below about 15·0, however, the fall-off becomes more rapid. There are three obvious reasons for this. Firstly, there is some diffraction at the images of the first plate. This is proportionally greater for the smallest images and as a result there is less light available to pass through the images of the second plate. A second reason is that in the enlarging process the black edges of the images have encroached a little into the clear centres, an effect that also becomes more noticeable the smaller the image may be. Thirdly, the images are becoming grey rather than clear as a result of the grey images that were mentioned as showing up on the original plate.

It should be emphasized that any such changes that may take place to the images in the photographic processes involved in producing the first and second plates, do not in any way affect the consistency of the method, provided that such effects are uniform over the plate. Whatever changes may take place in the diameter or contrast of any image will also equally affect the standard set of stars, and since these remain recognizable, from this point of view we need not concern ourselves further with such photographic effects. On the other hand, it is possible to turn to good use any increases in contrast between the images and the background, particularly in the rather different method shortly to be described. The images on the original star plate are inevitably not sharp at the edges and although the method itself takes care

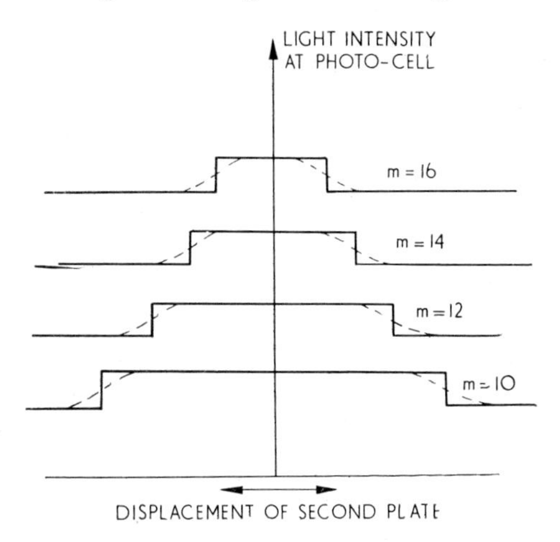

Fig. 5. Examples of flat-topped calibration curves

of this fact, the lack of contrast unnecessarily extends the "wings" of the curves of Figs. 2 or 4. Increasing the contrast minimizes this extension.

A variant to the method, which was mentioned earlier, largely does away with the integral equation, but at the cost of having to do a certain amount of manual work on the second plate.

As before, we use an enlarged positive plate for the first plate, but we change the nature of the second plate so that although the positions of the stellar images are the same as those of the first plate, *the diameters of all the images are now made to be the same.*

Consider the effect of this difference on the curves that the method produces. We will first assume idealized plates in which the first plate has completely clear and sharp images (corresponding in diameter, actually, to those of the 60-min ADH plate mentioned earlier) and the second plate also has clear and sharp images all of whose diameters are equal and negligibly small compared with those of the first plate. Fig. 5 shows a set of calibration curves prepared under these conditions. All are now flat topped, are of the same height and have vertical sides, their widths merely corresponding to the diameters of the stellar images on the first plate. If we consider an all-stars curve made up of individual images having characteristics such as these, as in Fig. 6, then it is obvious that *the curve is itself the distribution of stars on the plate.* The correct horizontal magnitude scale has to be inserted from the widths of the calibration curves and then the scale of ordinates can be made to read numbers of stars. From the curve it is then possible to read off, corresponding to any magnitude, the number of stars brighter than that magnitude.

Needless to say in practice, it is not possible to obtain such sharp curves. In the first place the images in the second plate cannot be of negligible diameter, although they could obviously be made very small indeed. Secondly, the images in the first plate will not normally be completely sharp at the edges. In the only case so far considered, the plates were considerably enlarged and, in order to increase contrast, were copied twice. The resultant calibration curves are found to be rounded off as

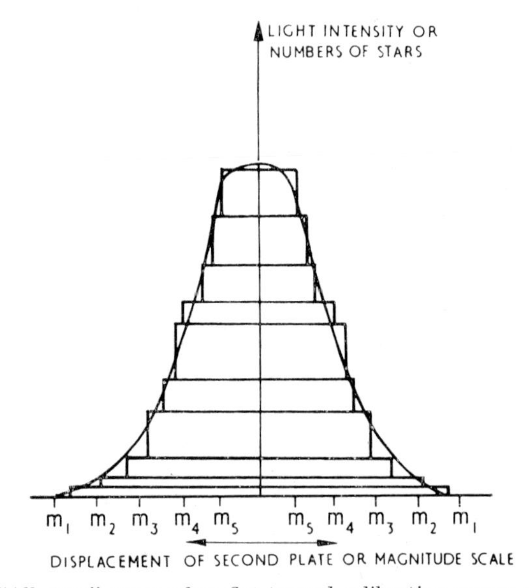

Fig. 6. "All stars" curve when flat-topped calibration curves are in us

shown by the dotted lines in Fig. 5. Evidently once we have allowed this to happen, we have, in theory, to solve the integral equation that we were trying to avoid. On the other hand, if we can keep this rounding off within reasonable limits, we can accept it simply as a limit to the resolution to the method and, if we wish, we may even be able to make some allowance for it by crude methods. From all these points of view, this variant of the fundamental method appears to have every advantage over the original method. On the other hand, it implies the changing of the characteristics of the second plate.

Several practical, but unaesthetic, ways of doing this spring to mind. One way is to fix a sheet of clear glass in front of the first plate and then to view the latter in a measuring machine through the glass. Then, whenever we have centred an image on the crosswires, we imprint a round, controllable ink dot on the upper glass plate. Certain stencil pens were found to make incredibly accurate dots in this way. It would be simple to make a device that would imprint a dot whenever a button was pressed. In that case, the preparation of the second plate would be very quick as it now only remains to make a photographic copy of the inked plate. The time taken to centre the images of a plate individually on to crosswires is very short and the presence of a dot then indicates that the image has already been dealt with, so that no arrangements to avoid duplication or misses are needed. If any dot is imprinted in error where there is no image, there is no need to erase it since the fact that there exists no corresponding image on the first plate means that no light at all will fall on the spurious image at any time when the plates pass through the position of correspondence.

Another possible way of producing the second plate is to make a contact copy of the first plate on to bromide paper, or a similar material, and then, using once again a plate measuring machine, to pierce a hole through the centre of each image. As long as this pierced copy is opaque except for the holes, it can be used as the second plate without any further preparation. Needless to say it would be necessary to devise a similar semi-automatic method for boring the holes. Experience alone will decide the most suitable method. At the same time, we must exploit all the dodges of practical photography in order to produce a first plate of suitably high contrast.

Evidently the latter method will only be of use if it will produce star counts with acceptable accuracy with much less work than the conventional method. Assuming that the fairly straightforward practical points just mentioned can be satisfactorily cleared up, the success or otherwise, of the method will depend on the speed and ease with which the second plate can be produced. It is considered possible in a suitable machine to centre star images at a rate of at least one per 5 sec. A really high rate of the order of 5000 per day by quite unskilled operators is therefore not impossible. The fact that wherever there is any doubt about the image, the operator can assume that there is one, without in any way upsetting the results, makes for the highest possible speed.

The real purpose of this paper is to point out possible, but hitherto unexploited, methods of overcoming the difficulties that exist at present in star counting. Many other methods along similar lines spring to mind and ought to be examined at some stage. Although it is naturally hoped that the methods here presented will be proved practicable, it is not considered that time has been wasted if they do no more than indicate that such unconventional methods are theoretically possible and provoke research along similar lines. Since it is essential to deal statistically with large numbers of stars, it is only reasonable that methods should be developed which are more suitable to present needs than the laborious star by star method now in use.

SECTION 6

RADIO ASTRONOMY

"In the present small treatise I set forth some matters of great interest for all observers of natural phenomena to look at and consider. They are of great interest, I think, first, from their intrinsic excellence; secondly, from their absolute novelty; and lastly, also on account of the instrument by the aid of which they have been presented to my apprehension."

GALILEO GALILEI, Introduction to *The Sidereal Messenger* (translation by E. S. CARLOS, London, 1880); 1610.

Solar Radio Eclipse Observations

J. S. HEY
Ministry of Supply (R.R.E.), Malvern, Worcestershire, England

SUMMARY
The observation of the emission from the Sun at radio wavelengths can provide knowledge of the localized regions of bright radio emission and of the distribution of radio brightness across the quiet Sun. From the latter it is possible to derive the distribution of electron density and temperature in the chromosphere and corona by measurement at centimetre and metre wavelengths respectively. Observationss of solar radio emission at eclipses have been made since 1945 and the results are discussed.

---

1. INTRODUCTION

AMONGST his many activities, Professor STRATTON was for thirty-two years Secretary of the Joint Permanent Eclipse Committee, and it seems appropriate that an article describing one of the most recent developments of eclipse observations should be included in this volume.

The first radio observations of the Sun during an eclipse were made in 1945 by DICKE and BERINGER [1] at a wavelength of 1·25 cm. Subsequent eclipses have been observed at various radio wavelengths. This paper gives a brief general survey of the results achieved and their contribution to our knowledge of the solar atmosphere.

2. PURPOSE OF RADIO ECLIPSE OBSERVATIONS

The most serious practical limitation in radio astronomy is the poor angular resolution fundamentally imposed by the long wavelengths of radio waves. With the low resolution of most radio telescopes so far available for eclipse observations it has only been possible to measure the total intensity radiated from the Sun. During an eclipse the solar radio emission is restricted to the regions unobscured by the Moon and this enables the observer to discriminate between the contributions from different parts of the solar disk. The distribution of radio brightness can then be inferred from the changes in total received intensity during the progress of the eclipse. Whereas in visual observations the eclipse eliminates the glare of scattered light from the terrestrial atmosphere, at radio wavelengths there is no comparable problem of atmospheric scattering; in the radio eclipse the obscuration by the Moon is used to provide an indirect method of increasing the resolution so as to explore the radio Sun.

Two main functions of this exploration of the radio brightness of the Sun may be discerned. One is to distinguish localized regions of enhanced radio brightness, determine their intensity, and find whether they are associated with visually-observed disturbances such as sunspots and prominences, and thus to learn something of the physical processes of radio emission in such regions. Whether these processes are thermal or non-thermal can only be inferred from the characteristics of the emission such as the spectrum, the magnitude and variation of intensity, polarization, and the association with optical data. The other main purpose of exploring the radio brightness of the Sun is to determine the general undisturbed conditions in the solar atmosphere. Here the mechanism of radio emission appears to be classical thermal radiation at radio wavelengths from an ionized atmosphere at high temperature. In order to appreciate the significance of radio eclipse measurements in relation to conditions in the atmosphere of the quiet Sun it is necessary to consider briefly this process of thermal emission.

3. Radio Emission from the Quiet Sun

The solar atmosphere consists mainly of ionized hydrogen, and until the development of radio-astronomy the available knowledge of density and temperature had been derived from spectroscopic lines and optical scattering. From this data, the electron density and its rate of decrease with increasing height seemed reasonably established, but although the temperature in the corona was known to be of the order of 10^6 °K and that in the chromosphere of the order of 10^4 °K there was much uncertainty about the change of temperature with height.

It can be shown that the radio absorption coefficient k is given by

$$k = \frac{AN^2\lambda^2}{\mu T^{3/2}}, \qquad \ldots .(1)$$

where N is the electron density, λ the wavelength, T the electron temperature, and μ the refractive index. A is a slowly varying function of N and T and may be taken

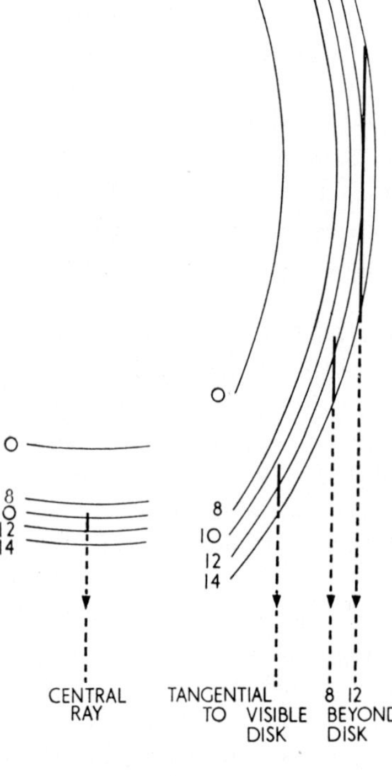

Fig. 1. Illustration of region (heavy line) contributing major part of emission at 3-cm wavelength for different rays. Numbers indicate height in units of 10^8 cm = 1000 km

as constant over extensive ranges of height in the solar atmosphere. The fall of N and rise of T with increasing height in the solar atmosphere causes the absorption coefficient k to diminish rapidly.

The radiation intensity depends on the temperature and on the integration of kdl along the elements of the ray path. From the knowledge of N and T derived from optical data, MARTYN [2] and others [3], [4], [5] have determined the expected distribution of radio brightness. It is found that metre waves undergo considerable refraction and absorption in the corona and it is from this region that emission occurs. At centimetre wavelengths most of the radiation originates from the chromosphere and refraction has negligible effect on ray paths. The radio Sun thus extends beyond the visible to a greater size as the wavelength is increased, and the brightness temperature rises to a maximum at metre wavelengths.

Another interesting feature is the occurrence of limb brightening at short wavelengths. Suppose $\int k dl$ is taken along a ray path emerging from a given height. If $\int k dl$ is large, radiation below this height cannot emerge because it is absorbed. If $\int k dl$ is small the emission above this height is small. Thus it is the height region

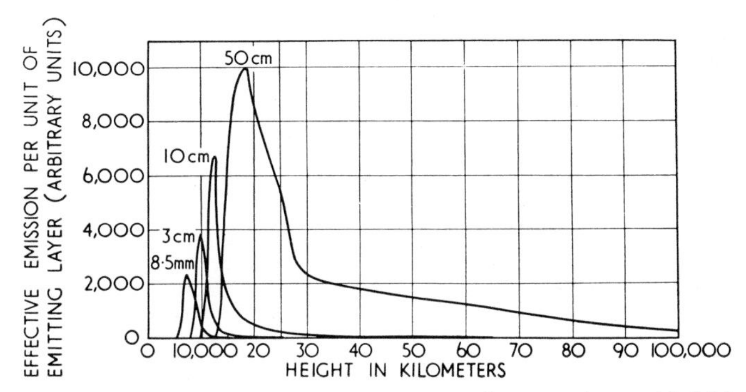

Fig. 2. The contribution of the various layers of the Sun's atmosphere, at heights indicated above the photosphere, to the emission from the centre of the Sun for the four wavelengths 8·5 mm, 3·14 cm, 10·6 cm, and 50 cm. From HAGEN [6]

with intermediate values of $\int k dl$ which contributes most. As the path length increases for a ray nearer the limb $\int k dl$ attains a given value higher in the solar atmosphere, where the temperature rises until the corona is reached. Fig. 1 which is based on values of k and T taken from HAGEN [6] illustrates the height zone contributing the major part of the emission at a wavelength of 3 cm between values of $\int k dl$ arbitrarily chosen at 2·3 and 0·1. As the distance from the centre of the disk is increased the radio brightness reaches a maximum and then falls as $\int k dl$ becomes comparatively small for a ray passing right through the solar corona.

A change of radio wavelength, λ, alters the effective height zone mainly responsible for the radio emission, because the absorption coefficient is proportional to λ^2. Thus the longer the wavelength the higher the region from which effective radio emission occurs, as illustrated by Fig. 2. The effective height zone and consequently the radio brightness increases with wavelength up to metre wavelengths when all the radiation is from the corona.

Although there is general agreement in form between the various calculations of the distribution of radio brightness across the solar disk there are differences in detail according to the assumed distribution of electron density and temperature derived purely from optical data or with the aid of radio data such as the total emission from the quiet Sun at various wavelengths, and Fig. 3 illustrates some of these distributions. The observation of radio brightness at various wavelengths which may be achieved by eclipse measurements offers an important means of helping to verify the correct distribution of electron density and temperature. By combining radio and optical data it should be possible to decide whether there are complicating factors, such as scattering by irregularities, which might influence the observed radio brightness. The presence of a magnetic field modifies k but there is no evidence of an appreciable general field.

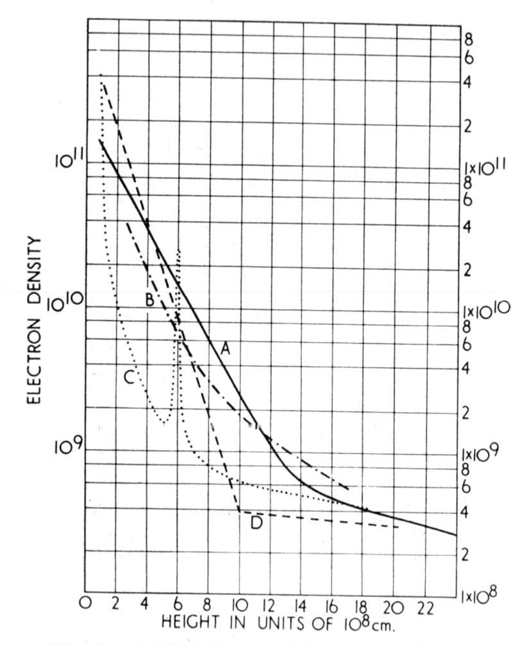

Fig. 3 (a). Distribution of electron density (number per cubic centimetre) with height (in units of 10^8 cm = 1000 km). From (A) HAGEN [6]; (B) PIDDINGTON [26]; (C) WOOLLEY and ALLEN [25]; (D) SMERD [5]

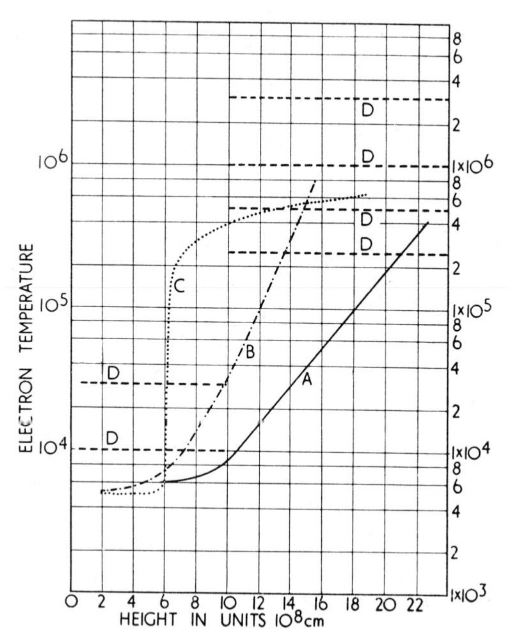

Fig. 3 (b). Distribution of electron temperature (°K) with height (in units of 10^8 cm = 1000 km). From (A) HAGEN [6]; (B) PIDDINGTON [26]; (C) WOOLLEY and ALLEN [25]; (D) SMERD [5]

4. FACTORS AFFECTING ECLIPSE OBSERVATIONS

Ideally, observations should be made during different phases of the solar cycle to ascertain the main changes in solar atmospheric conditions. Unfortunately, at metre wavelengths solar activity can cause localized emission of great intensity and variability. At shorter wavelengths, radiation from such regions is less intense and more stable; their presence is revealed by discontinuities in the total received intensity as they become obscured and subsequently exposed again, and the radiation from the quiet Sun may be deduced by subtracting the contributions of the disturbed regions. Solar flares often have associated radio outbursts over a wide range of wavelengths. Most of them are not of many minutes duration so that the chances that they will upset intensity measurements is not great unless there is exceptional flare activity.

Hence at metre wavelengths in the years near sunspot maximum there is a strong risk of variable disturbances spoiling eclipse observations. At centimetre wavelengths accuracy is impaired if there are many disturbed regions; thus the years near sunspot minimum are the most likely to yield reliable eclipse observations over a wide range of wavelengths.

Radio eclipse measurements, in their independence of weather conditions, have a great advantage over visual eclipse observations. Weather can, however, affect the radio results at very short wavelengths. Atmospheric attenuation due to oxygen and water vapour is very appreciable at millimetre wavelengths. If atmospheric conditions are constant during an eclipse a correction may be applied depending on the altitude of the Sun.

Attenuation due to fog and cloud, or light rain reduces sensitivity, but if constant and homogeneous it does not preclude eclipse measurements. Storms and heavy rain are more serious at millimetre wavelengths, and even at wavelengths of 10 cm can occasionally be troublesome.

It is preferable that the Sun should be at high altitude for eclipse observations, not only because attenuation effects are reduced but there is also less liability of solar radiation reflected by the ground or thermal emission from the ground entering the aerial system. If these effects occur corrections have to be applied by observing the Sun over the same path on days preceding and following the eclipse, and also by measuring the radiation received over this path when the Sun is in a different part of the sky away from the aerial beam. A correction can similarly be made for the change in cosmic noise received over the eclipse path which can be important especially at metre wavelengths. At centimetre and millimetre wavelengths a correction must be applied for the radiation from the Moon which at eclipse totality may represent an appreciable proportion of the received radiation. Another correction often necessary is that for the directional pattern of the aerial, since at narrow beamwidths the reception sensitivity may vary appreciably over the Sun.

5. List of Eclipse Observations

The radio eclipse observations will now be reviewed. The eclipses are listed in Table 1 with remarks indicating briefly the main conclusions concerning the distribution of solar brightness.

6. Discussion of Eclipse Results

The entries in the Remarks column of the above table are an indication only of the conclusions and are not comprehensive. Further results will be mentioned in the following discussion, but reference should be made to the published papers for a complete appreciation of the various eclipse observations.

The study of the localized regions of bright radio emission has yielded interesting information. At the eclipse of 23rd November, 1946, Covington [8], at 10·7 cm, detected two bright areas, one corresponding to an extensive band of prominences of brightness temperature 2×10^7 °K, and the other an area of 2·2 per cent of the disk containing a large sunspot group of brightness temperature $1{\cdot}5 \times 10^6$ °K. Christiansen, Yabsley, and Mills [14] observed the eclipse of 1st November, 1948, at 50 cm, from three well-separated sites. This provides the most precise method of locating bright regions. Assuming there is no sudden fluctuation of emission, the positions of the bright regions must lie along the edge of the Moon's limb whenever abrupt falls of intensity occurred as the regions were obscured or abrupt rises as they were uncovered. The apparent paths of the Moon across the solar disk were appreciably displaced from one another at the different sites, and the localized bright regions could be fixed by the intersections of the position Moon's limb when corresponding discontinuities occurred. In this way Christiansen, Yabsley, and Mills located three bright regions in the positions of sunspot groups, three in positions where sunspots had been observed on the previous solar rotation, and one in the position of a prominence.

Table 1

<table>
<tr><th></th><th>Date of eclipse</th><th>Observer</th><th>λ</th><th>Place of observation</th><th>Type of eclipse</th></tr>
<tr><td>1.</td><td>9th July, 1945</td><td>DICKE and BERINGER [1]</td><td>1·25 cm</td><td>Cambridge, Mass., U.S.A.</td><td>Partial</td></tr>
<tr><td>2.</td><td>23rd November, 1946</td><td>REBER [7]</td><td>1·85 m</td><td>Illinois, U.S.A.</td><td>Partial</td></tr>
<tr><td>3.</td><td>23rd November, 1946</td><td>COVINGTON [8]</td><td>10·7 cm</td><td>Ottawa, Canada</td><td>Partial</td></tr>
<tr><td>4.</td><td>23rd November, 1946</td><td>SANDER [9]</td><td>3·2 cm</td><td>Malvern, England</td><td>Partial</td></tr>
<tr><td>5.</td><td>20th May, 1947</td><td>HAGEN [6]</td><td>3·2 cm</td><td>At sea W20°, N1°</td><td>Total</td></tr>
<tr><td>6.</td><td>20th May, 1947</td><td>KHAYKIN and CHIKHACHEV [10]</td><td>1·5 m</td><td>At sea S12° 45′, W38° 30′</td><td>Total</td></tr>
<tr><td>7.</td><td>28th April, 1949</td><td>LAFFINEUR, MICHARD, SERVAJEAN, STEINBERG, ZISLER [11]</td><td>1·95 m
55 cm
25 cm</td><td>Paris, France</td><td>Partial</td></tr>
<tr><td>8.</td><td>1st November, 1948</td><td>MINETT and LABRUM [12]</td><td>3·18 cm</td><td>Sydney, Australia</td><td>Partial</td></tr>
<tr><td>9.</td><td>1st November, 1948</td><td>PIDDINGTON and HINDMAN [13]</td><td>10 cm</td><td>Sydney, Australia</td><td>Partial</td></tr>
<tr><td>10.</td><td>1st November, 1948</td><td>CHRISTIANSEN, YABSLEY, MILLS [14]</td><td>50 cm</td><td>3 sites in Australia:
(1) Sydney
(2) Rockbank
(3) Strahan</td><td>Partial</td></tr>
<tr><td>11.</td><td>12th September, 1950</td><td>HAGEN, HADDOCK, REBER [15]</td><td>3 cm
10 cm</td><td>Aleutian Islands</td><td>Total</td></tr>
<tr><td>12.</td><td>1st September, 1951</td><td>BOSSON, BLUM, DENISSE, LEROUX, STEINBERG [16], [17]</td><td>3·2 cm
1·78 m</td><td>Markala, W. Africa</td><td>Annular</td></tr>
<tr><td rowspan="2">13.</td><td rowspan="2">25th February, 1952</td><td rowspan="2">BLUM, DENISSE, STEINBERG [17]</td><td rowspan="2">1·78 m</td><td>Dakar, W. Africa</td><td>Partial</td></tr>
<tr><td>Paris, France</td><td>Partial</td></tr>
<tr><td>14.</td><td>25th February, 1952</td><td>COUTREZ et al.</td><td>1·78 m</td><td>Three sites in Africa—
1. Dakar
2. Lwiro
3. Ngili</td><td>Partial</td></tr>
<tr><td rowspan="2">15.</td><td rowspan="2">25th February, 1952</td><td rowspan="2">LAFFINEUR et al. [18]</td><td>55 cm</td><td>Khartoum, Sudan</td><td>Total</td></tr>
<tr><td>117 cm</td><td>Meudon, France</td><td>Partial</td></tr>
<tr><td>16.</td><td>25th February, 1952</td><td>HAGEN et al. [19]</td><td>8 mm
9·4 cm</td><td>Khartoum, Sudan</td><td>Total</td></tr>
<tr><td>17.</td><td>14th February, 1953</td><td>AOKI [20]</td><td>10 cm</td><td>Mitaka, Japan</td><td>Partial</td></tr>
</table>

Table 1

Remarks
Intensity curve not greatly different from that for exposed area of optical disk.
Intensity approximately proportional to disk area exposed.
Intensity approximately proportional to disk area exposed. Sudden decrease of 9 per cent 3 min before first contact as extensive prominences obscured. 25 per cent reduction of intensity as large sunspot obscured.
Intensity curve suggested sources of radiation mainly located about circumference. Rain caused appreciable absorption during part of eclipse.
About 30 per cent of total radiation due to sunspots. After correction for sunspot contribution, intensity curve indicated limb brightening.
Intensity fell to about 40 per cent that of uneclipsed Sun. Intensity curve approximately proportional to exposed area of prominences and flocculi.
At 1·95 m, rapid fluctuations due to variable sources of emission. At 25 cm and 55 cm, intensity curve between that for area of exposed disk and that for area of bright Hα faculae exposed.
Sunspot radiation of low intensity. Intensity curve for quiet Sun consistent with (*a*) uniform distribution over disk diameter 1·1 times that of optical disk; (*b*) uniform distribution over optical disk with bright ring at circumference.
Intensity curve corresponds to 68 per cent contribution proportional to disk area, and 32 per cent from circumference. Radio eclipse commenced 30 sec. before optical contact corresponding to about 10,000 km height. A localized bright region detected where a sunspot group had been observed 27 days previously.
Comparison of intensity changes at three sites fixed positions of localized bright regions corresponding to visible sunspot groups, to positions occupied by sunspots 27 days previously, and to a prominence. After correcting for these, the intensity curves were consistent with either: (1) a uniform distribution of diameter 1·3 times the optical disk; or (2) a uniform distribution of diameter 1·2 times the optical disk with a bright ring at the disk circumference.
At 3 cm and 10 cm, rainstorm caused fluctuations and attenuation. Sunspot emission of order of 50 per cent quiet Sun radiation at 10 cm, and 6 per cent at 3 cm. At 10 cm, results indicated temperature in lower chromosphere less than 30,000°K. Limb brightening indicated at 3 cm.
At 3·2 cm, sunspots contributed about 3 per cent of total emission. At maximum eclipse, residual intensity 16 per cent of total quiet Sun intensity. Limb brightening confirmed by comparison with results of eclipse No. 11 At 1·78 m, spot emission negligible. Residual intensity at maximum eclipse about 48 per cent of total quiet Sun intensity. Intensity curve indicates equatorial diameter at 1·78 m greater than polar diameter.
The results confirm that radio equatorial diameter is greater than the polar diameter.
Asymmetry in the occultation curves indicated brightness distribution not spherically symmetrical. Many bright regions located, the most important at a maximum in emission of coronal line λ5303 and near base of a coronal streamer.
At 55 cm, residual emission of 19·5 per cent at totality. By first optical contact the intensity was reduced 3 per cent. At 117 cm, the residual radiation at totality was 30·5 per cent. Detailed analysis of results not yet published.
At 8 mm, limb brightening as expected, but also evidence that centre is brighter than predicted. A localized region of high emission found near the eastern limb. At 10 cm, two localized active areas present; limb brightening as predicted. Detailed analysis of results not yet published.
Radio diameter about 5 per cent greater than optical diameter. Limb about three times brighter than centre. A region of enhanced radio emission apparently covered about 0·12 of disk area.

Three other sunspot groups and another prominence gave no increased emission. The bright areas subtended on the average 0·004 of the solar disk, at a brightness temperature of 5×10^6 °K. PIDDINGTON and HINDMAN [13], at 10 cm, also detected one of the regions and estimated the area at 0·002 of the solar disk and brightness temperature 10^6 °K; small discontinuities in the eclipse curve might have been due to two other areas at brightness temperatures of $0{\cdot}2 \times 10^6$ °K and $0{\cdot}5 \times 10^6$ °K respectively. HAGEN, HADDOCK, and REBER [15] detected radio emission from a sunspot group during the eclipse of 12th September, 1950, and assuming the emitting area equal to the visible spots the brightness temperature was 4×10^7 °K at 10 cm and 2×10^6 °K at 3 cm. At the eclipse of 1st September, 1951, BLUM, DENISSE, and STEINBERG [16] found that at 3·2 cm wavelength one spot group gave no excess radio emission, but another had an area of emission comparable with that of the visible spots, the brightness temperature being $0{\cdot}6 \times 10^6$ °K. Eclipse measurements such as these, combined with daily intensity measurements have made it possible to derive the spectrum, and other characteristics of these bright regions, from which PIDDINGTON and MINNETT [21] have concluded that the radiation is consistent with thermal origin and that the electron temperature is of the order of 10^7 °K. It may be noted that AOKI [20], at 10 cm, stated that his observations of the eclipse of 14th February, 1953, indicated a bright region of brightness temperature $8{\cdot}9 \times 10^4$ °K over an area of 0·12 of the disk, which included the few small sunspots present on the disk; the area was thus at lower temperature but far more extensive than those reported previously.

Polarization measurements have been included in some of the observations [13], [14]. Sunspot radiation is often circularly polarized as may be expected in the presence of the high magnetic field of spots. At eclipses, changes in the proportions of the circularly polarized components of the total intensity occur during the covering and uncovering of "bright" sunspot regions. This circularly-polarized radiation from sunspots unfortunately tends to mask the attempts to determine whether there is any degree of circular polarization from uneclipsed portions of the quiet Sun and hence to decide whether there is any general solar magnetic field. Depending on the limits of accuracy, eclipse measurements have been able to set an upper limit to such a field, and CHRISTIANSEN, YABSLEY, and MILLS [14] concluded that if there is any field it is less than 8 gauss at the poles.

An important deduction at 1·78 m was reached by BLUM, DENISSE, and STEINBERG [16], [17], when they showed that their eclipse measurements on 1st September, 1951, and 25th February, 1952, were incompatible with a spherical model of the corona. On 1st September, 1951, it was found that at optical contact about 9 per cent of the radiation was occulted. This was incompatible with a residual radiation of 48 per cent at the maximum of the eclipse if spherical symmetry were assumed. The fall of intensity by first optical contact observed on 25th February, 1952, from Paris and Dakar respectively confirmed this and it was concluded that the distribution of radio brightness appeared as an ellipse of nearly uniform brilliance with the major axis in the equatorial plane. The way this fits the reduction of intensity by first optical contact is illustrated in Fig. 4. BLUM [22] has also discussed the intensity reductions by first optical contact in the results at 50 cm obtained on 1st November, 1948, by CHRISTIANSEN, YABSLEY, and MILLS [14] in relation to the residual intensity at totality measured on 25th February, 1952, by LAFFINEUR *et al.* [18], and contradictions which are apparent if spherical symmetry is assumed disappear if the radio

Sun has a greater equatorial extension similar to that deduced at 1·78 m. Optical data have previously shown that the coronal density may be expected to vary with heliographic latitude, but the radio eclipse observations offer a useful means of studying the coronal distribution and its departure from spherical symmetry.

COUTREZ, KOECKELENBERGH, and POURBAIX [27] in co-operation with other observers obtained occultation curves at 169 Mc/s at three separated sites in Africa during the eclipse of 25th February, 1952, and concluded that although the brightness distribution appeared approximately elliptical the axis did not coincide with the rotational axis of the Sun; the distribution appeared more heterogeneous than had been hitherto suggested and this could be associated with the distribution of localized bright areas. Twenty suspected regions, of which eight were definitely confirmed,

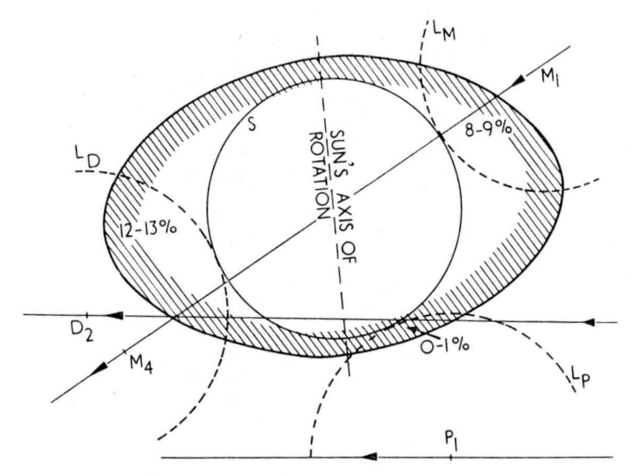

Fig. 4. Model of corona deduced by BLUM, DENISSE, STEINBERG [16, 17] at 1·78 m from eclipse observations 1951, 1952. Figures indicate intensity reductions observed at Paris (L_P), Dakar (L_D), and Markala (L_M)

were determined from the analysis of the slope variations of the eclipse curves at the three sites. The most important, at an altitude of $0{\cdot}4R_{\odot}$ from the limb, coincided with a maximum in the intensity of the coronal line λ5303 and was situated at the base of a large coronal streamer. Two other bright radio regions appeared to be related to maxima of the λ5303 emission. These observations have emphasized the importance, at metre wavelengths as well as centimetre wavelengths, of determining the bright regions which occur even near sunspot minimum. It is possible that the information concerning such regions may help in elucidating the nature of the solar *M* regions.

An alternative method of determining brightness distribution is by means of interferometers with different spacings as used by STANIER [23]. Nevertheless, eclipses provided the first radio results to lead to the elliptical model of the corona. The advantage of the interferometer methods is that the experiments do not have to await the occurrence of an eclipse which may entail an expedition to a distant land. Nevertheless, interferometer methods are also indirect and subject to limitations imposed by the presence of localized bright regions, and there is much to be gained by utilizing both methods.

ALON, ARSAC, and STEINBERG [24] have combined eclipse and interferometer results to derive a detailed distribution of radio brightness of the Sun at 3·2 cm, which is shown in Fig. 5. The choice between the two curves A and B which could both fit the observations, illustrates the difficulty of deriving a unique distribution from measurements of total intensity, and the necessity for a high order of accuracy to help in differentiating between different distributions.

HAGEN [19], in his observations at 8 mm of the total eclipse of 25th February, 1952, has used an aerial with a sufficiently narrow beamwidth to limit observation to a part of the solar disk. A parabolic reflector 16 ft by 4 ft gave a beam 0°.12 by 1° and was so oriented during the eclipse that the long dimension of the beam passed through the centre of the Sun and was aligned along the track of the Moon's centre. This increase of resolution is a most desirable achievement and the publication of the results is awaited with interest.

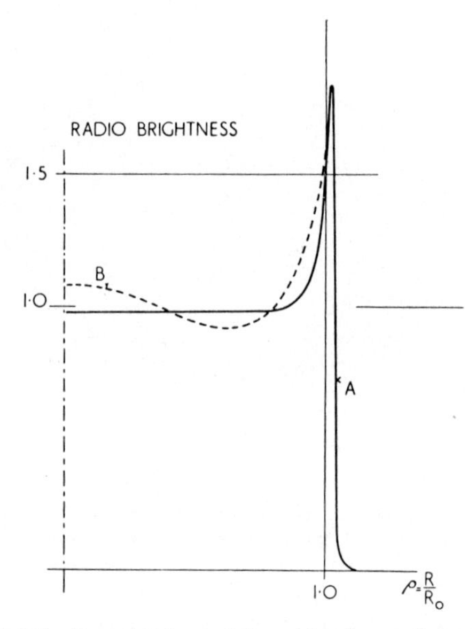

Fig. 5. Brightness distribution at 3·2 cm deduced by ALON, ARSAC, and STEINBERG [24]

The construction of large aerials with sufficient resolution to explore the solar disk will have the important advantage that progressive changes in solar radio brightness may be followed, as compared with the comparatively infrequent opportunities provided by eclipses. Already several high resolution aerials are in use; HAGEN at 8 mm has achieved a beamwidth of 4′ of arc by means of a parabolic reflector, COVINGTON at 10 cm wavelength has obtained a fan beam of width 7·5′ of arc in the narrow dimension with a slotted waveguide array, and CHRISTIANSEN and WARBURTON [28] at 21 cm wavelength have used an array of thirty-two parabolic reflectors to produce fan beams of width 3′ of arc spaced at intervals of 1·7°. Although large aerial systems will tend to reduce the necessity of eclipse observations, an eclipse will be able to improve the resolution by means of the obscuration produced within the part of the solar surface surveyed in the aerial beam. The limit of resolution is then determined by diffraction by the Moon, and also the method of determining the detailed distribution from the total intensity received within the beam. The limit to angular resolution imposed by diffraction is of the order of $(\lambda/d)^{1/2}$ radians, where d is the distance of the Moon from the Earth; at 10 cm wavelength this limit is equal to about 3″ of arc.

7. CONCLUSION

In conclusion, eclipse results have already made a valuable contribution to our knowledge of the solar atmosphere, and may be expected in the future to reveal much detailed information. Observations from one eclipse to another can determine any significant changes during the solar cycle; alternatively, if the solar chromosphere is not much altered, comparison of eclipses of different magnitudes can be of great value in leading to the accurate determination of the limb brightening curve at centimetre wavelengths. Observations at any eclipse should cover as wide a range of wavelengths and with as high accuracy as possible. The eclipse should be observed from different sites so as to discriminate regions of localized emission and to determine the eccentricity of the shape of the corona, and the intensity should be recorded also from an uneclipsed site for comparison purposes. Such elaboration is desirable if the brightness distribution and hence the solar atmospheric conditions are to be uniquely derived.

Acknowledgement is made to Chief Scientist, Ministry of Supply, for permission to publish this communication.

REFERENCES

[1] DICKE, R. H. and BERINGER, R.; *Ap. J.*, **103,** 375, 1946.
[2] MARTYN, D. F.; *Proc. Roy. Soc.*, **193,** 44, 1948.
[3] UNSÖLD, A.; *Naturwissenschaften*, **7,** 194, 1947.
[4] WALDMEIER, M. and MÜLLER, H.; *Mitt. Eidgen. Sternwarte*, No. 154, 1948.
[5] SMERD, S. F.; *Austral. J. sci. Res.*, A, **3,** 34, 1950.
[6] (a) HAGEN, J. P.; Naval Research Laboratory, Washington, Report 3504, 1949.
(b) HAGEN, J. P.; *Ap. J.*, **113,** 547, 1951.
[7] REBER, G.; *Nature* (*London*), **158,** 945, 1946.
[8] COVINGTON, A. E.; *Nature* (*London*), **159,** 405, 1947.
[9] SANDER, K. F.; *Nature* (*London*), **159,** 506, 1947.
[10] KHAYKIN, S. E. and CHIKHACHEV, B. M.; *Isv. Akad. Nauk, U.S.S.R.*, **12,** 38, 1948.
[11] (a) LAFFINEUR, M., MICHARD, R., STEINBERG, J. L., ZISLER, S.; *C.R. Acad. Sci.* (*Paris*), **228,** 1636, 1949.
(b) LAFFINEUR, M., MICHARD, R., SERVAJEAN, R., STEINBERG, J. L.; *Ann. Astrophys.*, **13,** 337, 1950.
[12] MINNETT, H. C. and LABRUN, N. R.; *Austral. sci. Res.*, A, **3,** 60, 1950.
[13] PIDDINGTON, J. H. and HINDMAN, J. V.; *Austral. J. sci. Res.*, A, **2,** 524, 1949.
[14] CHRISTIANSEN, W. N., YABSLEY, D. E., MILLS, B. Y.; *Austral. J. sci. Res.*, A, **2,** 506, 1949.
[15] HAGEN, J. P., HADDOCK, F. T., REBER, G.; *Sky and Telescope*, **10,** No. 5, 1951.
16] (a) BOSSON, F., BLUM, E. J., DENISSE, J. F., LEROUX, E., STEINBERG, J. L.; *C.R. Acad. Sci.* (*Paris*), **223,** 917, 1951.
(b) BLUM, E. J., DENISSE, J. F., STEINBERG, J. L.; *Ann. Astrophys.*, **15,** 184, 1952.
[17] (a) DENISSE, J. F., BLUM, E. J., STEINBERG, J. L.; *Nature* (*London*), **170,** 191, 1952.
(b) BLUM, E. J., DENISSE, J. F., STEINBERG, J. L.; *C.R. Acad. Sci.* (*Paris*), **234,** 1597, 1952.
[18] LAFFINEUR, M., MICHARD, R., PECKER, J. C., D'AZAMBUJA, M., DOLLFUS, A., ATANASIJEVIC, I.; *C.R. Acad. Sci.* (*Paris*), **234,** 1528, 1952.
[19] HAGEN, J. P.; Private communication, 1954.
[20] AOKI, K.; *Rep. Ionospher. Res. Japan*, **7,** 109, 1953.
[21] PIDDINGTON, J. H. and MINNETT, H. C.; *Austral. J. sci. Res.*, A, **1,** 131, 1951.
[22] BLUM, E. J.; *C.R. Acad. Sci.* (*Paris*), **237,** 135, 1953.
[23] STANIER, H. M.; *Nature* (*London*), **165,** 354, 1950.
[24] ALON, I., ARSAC, J., STEINBERG, J. L.; *C.R. Acad. Sci.* (*Paris*), **237,** 300, 1953.
[25] WOOLLEY, R. v. d. R. and ALLEN, C. W.; *M.N.*, **110,** 358, 1950.
[26] PIDDINGTON, J. H.; *Proc. Roy. Soc.*, **203,** 417, 1950.
[27] R. COUTREZ, A. KOECKELENBERGH, and E. POURBAIX; *Commun. de l'Observatoire Royal de Belgique*, No. 60, 1953.
[28] W. N. CHRISTIANSEN and J. A. WARBURTON, *Austral. J. sci. Res.* A, **6,** 190, 1953.

The Application of Interferometric Methods in Radio Astronomy

M. Ryle

Cavendish Laboratory, Cambridge

Summary

Interferometric methods have been applied to a number of different problems in radio astronomy and the increased resolution has allowed important advances in both solar and galactic problems. Observations using spaced-aerial interferometers of variable aperture have enabled the distribution of radio "brightness" across the solar disk to be determined at wavelengths between 60 cm and 7·9 m. Similar methods have been used to determine the detailed structure of the galactic background radiation in directions near the galactic plane; such observations are important in connection with the effect of the ionized regions of interstellar matter.

The use of interferometers, particularly of the "phase-switching" type, has allowed considerable advances in the study of "radio stars". In addition to providing accurate positions, such observations have made it possible to determine the angular diameter of some of the more intense sources.

1. Methods

The extension of astronomical observations to the radio band of wavelengths between 1 cm and 20 m has made possible the further investigation of the solar envelope and the interstellar medium. In addition, the discovery of radio stars and the observation of other features of cosmic radio emission have raised important new theoretical questions. Some of the most important of these investigations have involved observations on the longer wavelengths, between 1 m and 10 m. It is clear that the use of wavelengths about 10^7 times greater than optical wavelengths must introduce very serious restrictions, and in fact nearly all observations in radio astronomy are limited by the relatively poor resolving power which can be achieved in practice.

A small resolving power may introduce a number of separate difficulties. Firstly, it may not be possible to distinguish between the radiation from a source of small angular size and that from the incidence over the remainder of the reception pattern of the general background radiation; an aerial used to observe the radiation from the undisturbed sun at a wavelength of 5 m would have to have an area greater than about 3000 m^2 if the power intercepted from the Sun were to exceed that from the background radiation. Secondly, it limits the accuracy with which the position of a source may be found; the identification of most radio stars with visible objects has been limited in this way by the uncertainty in the radio positions. Thirdly, it may make it impossible to determine the angular diameter or the distribution of radio "brightness" across an extended source; thus even to derive the approximate distribution of "brightness" across the solar disk at a wavelength of 5 m it would be necessary to use an aperture of about 300 m.

Because of these difficulties it is natural that in the development of observational methods in radio astronomy considerable attention should have been paid to the application of interferometric methods. This article is intended to give an outline of the methods which have been used, and to show how they have been applied to a number of different problems.

Two basic types of interferometer have been used: the first, developed in Australia (McCready, Pawsey, and Payne-Scott, 1947), employs an aerial mounted on a high cliff overlooking the sea, and makes use of the interference between the direct and reflected rays, as in a Lloyd's mirror interferometer. The second, developed in Cambridge (RYLE and VONBERG, 1946), uses two separated aerials connected to the same receiver, and is analogous to MICHELSON's stellar interferometer. In both systems the rotation of the Earth causes the source of radiation to pass through the interference fringes; the times at which maxima of the received power occur may then be used to derive the position of the source, whilst the ratio of maximum/minimum received power (the "visibility" of the fringes) enables information to be derived about the angular extent of the source.

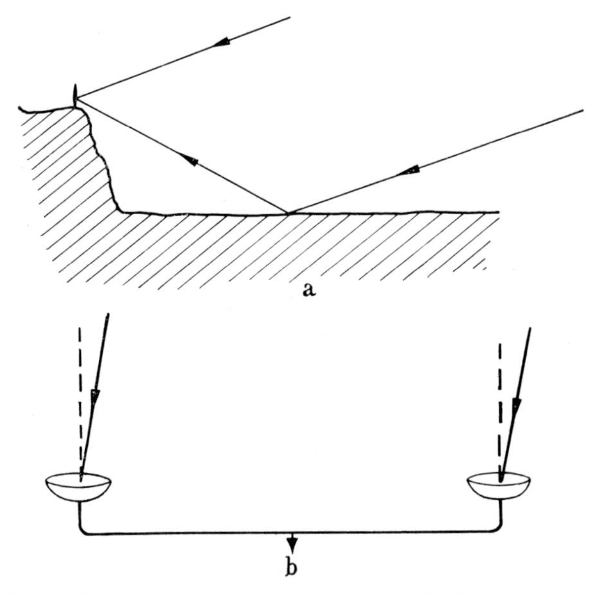

Fig. 1. Basic interferometer systems: (*a*) cliff, (*b*) spaced-aerial

The cliff interferometer has the advantage that only one aerial is needed, without the accurately balanced cables required in the spaced aerial method, but it is usually restricted to observations at large zenith angles, where refractive effects in the atmosphere may be important. The spaced aerial interferometer is frequently constructed with the aerials in an east-west direction, and observations are then made when the source is near the meridian and the zenith angle is least.

A later modification of the spaced aerial interferometer (RYLE, 1952) has a number of important advantages over both earlier systems, and has been applied to many different problems. In this system the receiver is not used to measure the variation of total power intercepted by the two aerials, but is used instead to measure the degree of correlation between the two e.m.f.'s. This operation is achieved by measuring the difference in available power when the aerials are connected alternately in phase and in anti-phase; the two interference patterns which are produced are shown in Fig. 2. An extended source of radiation, such as the galactic background, having no angular structure comparable with the separation of the interference maxima, would produce equal power in the two positions of the change-over switch, and no variation of the power in the receiver would occur. A source of small angular diameter would, on the other hand, produce different powers in the two positions, and the output voltage from the receiver will therefore contain a periodic component at the switching frequency; the amplitude and sign of this component will depend on the intensity and position of the source relative to the interference pattern. If this component is used to deflect a recording milliammeter a sinusoidal trace will be produced as the

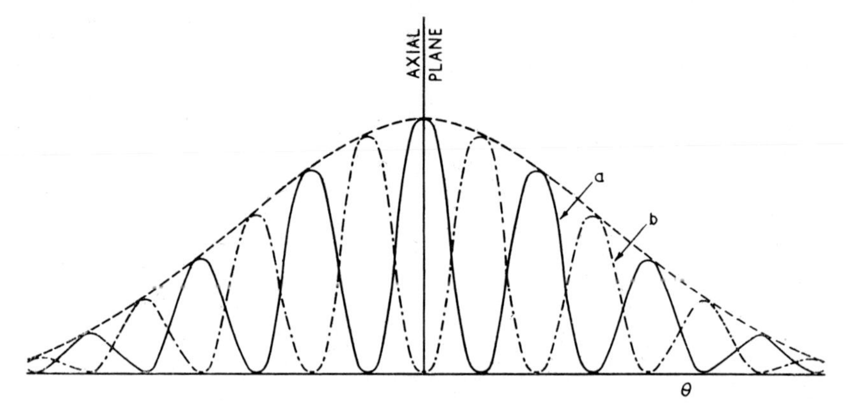

Fig. 2. Reception pattern of spaced aerials connected (*a*) in phase, (*b*) in anti-phase

whole pattern is moved across the source by the Earth's rotation; an example of such a record is shown in Fig. 3.

It can be shown that the amplitude of the recorded trace is proportional to one term of the Fourier transform of the integrated "strip" distribution of brightness across the source; the angular periodicity of this term is determined by the separation of the interference maxima, and hence by the spacing between the aerials.

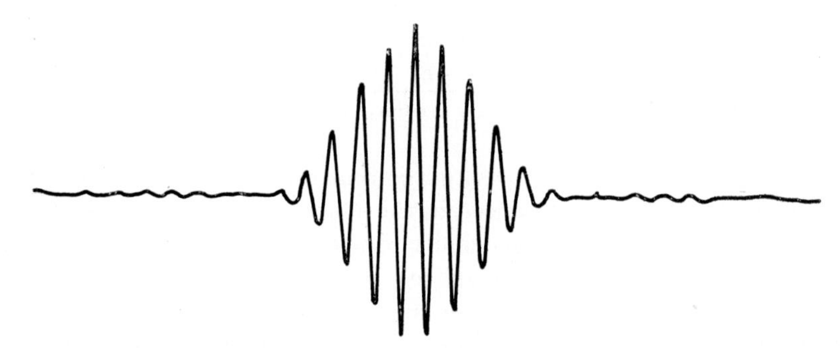

Fig. 3. Record obtained with phase-switching interferometer, showing the intense radio star in Cassiopeia

Since no deflection is produced by the background radiation it is possible to use a very much greater recording sensitivity than with systems based on the measurement of the total aerial power; it has been found possible to detect radio stars which produce an aerial power about 1/1000 of that produced by the background radiation. A further advantage of the system lies in the ability to use simple pre-amplifiers at the aerials, and thus overcome the effect of attenuation in the long transmission lines; the most serious limitation in obtaining large resolving powers is therefore removed.

By deriving the position of a source from the times at which the trace crosses the axis (a condition which corresponds to the signals received at the two aerials being in phase quadrature), improved accuracy is obtained; firstly the observations are made at a time when the rate of change of the recorded output is a maximum rather than a minimum; and secondly, the time of crossing the axis is unaffected by inequality in the gains of the aerials or pre-amplifiers. The "cross-over" time is also unaffected by variations in the intensity of the radiation and the system therefore has particular advantages for accurate measurements of the position of a source of variable intensity.

The application of interferometers, and particularly the phase-switching interferometer, to various problems in radio astronomy will now be discussed.

2. THE RADIATION FROM SUNSPOTS

The study of the intense sources of metre-wave radiation associated with sunspots was the first application of interferometric methods in radio astronomy. Both cliff and spaced-aerial interferometers were used to demonstrate that the radiation originated in a small region, having angular dimensions comparable with the visible spot-group.

A large number of further observations was subsequently made, including a series by LITTLE and PAYNE-SCOTT (1951), who used a modification of the spaced aerial method to observe very rapid changes of the position of the source, and a series by MACHIN with the phase-switching method in which the variation in apparent position of the source at three wavelengths could be determined with an accuracy of about 2 sec. of arc (RYLE, 1952; MACHIN, 1955).

3. RADIATION FROM THE UNDISTURBED SUN

Early observations (PAWSEY, 1946; RYLE and VONBERG, 1947) of the emission of metre-wave radiation from the Sun showed that in the absence of sunspots the intensity fell to a value corresponding to the emission from a source of the diameter of the Sun and having a temperature of about a million degrees. Since different wavelengths should originate at different heights in the solar corona it was suggested that observations of the intensity over a wide range of wavelengths should provide information on the distribution of temperature in the corona. Detailed theoretical work (MARTYN, 1948; SMERD, 1950) has, however, shown that owing to the large refractive effects which occur at metre-wavelengths the total emission from the Sun at different wavelengths does not provide a sensitive measure of coronal temperature. It was found, however, that the distribution of "brightness" across the solar disk depended markedly on the temperature assumed; measurements of the *distribution* at different wavelengths might therefore allow the variation of coronal temperature with height to be derived.

The first attempt to measure the distribution of brightness at metre-wavelengths was based on the observation of the variation of intensity during a total eclipse (HAIKIN and CHICKHACHEV, 1948) and several further observations of this type have since been made (CHRISTIANSEN *et al.*, 1949; BLUM *et al.*, 1952). Unlike the similar observations on centimetric wavelengths, however, no entirely satisfactory results have yet been obtained, partly because of the presence of active areas on the disk at the times of the eclipses, and partly because at wavelengths greater than 1 m the effective diameter of the Sun considerably exceeds that of the Moon; the variation of received power does not then give a sensitive measure of the distribution.

An alternative interferometric method of measuring the distribution has been devised (STANIER, 1950), which does not depend on an eclipse, and which can therefore be carried out on days specially selected for low sunspot activity. This method makes use of the relationship, which has already been mentioned, between the amplitude of the recorded trace and the Fourier transform of the distribution of "brightness" across the source; by observing the Sun with a large number of different separations between the aerials it is possible to derive the complete Fourier amplitude curve and hence to deduce the radial distribution of "brightness".

The experiment was first carried out on a wavelength of 60 cm (STANIER, 1950), and was subsequently repeated on 3·7 m (MACHIN, 1951). More recently an extensive

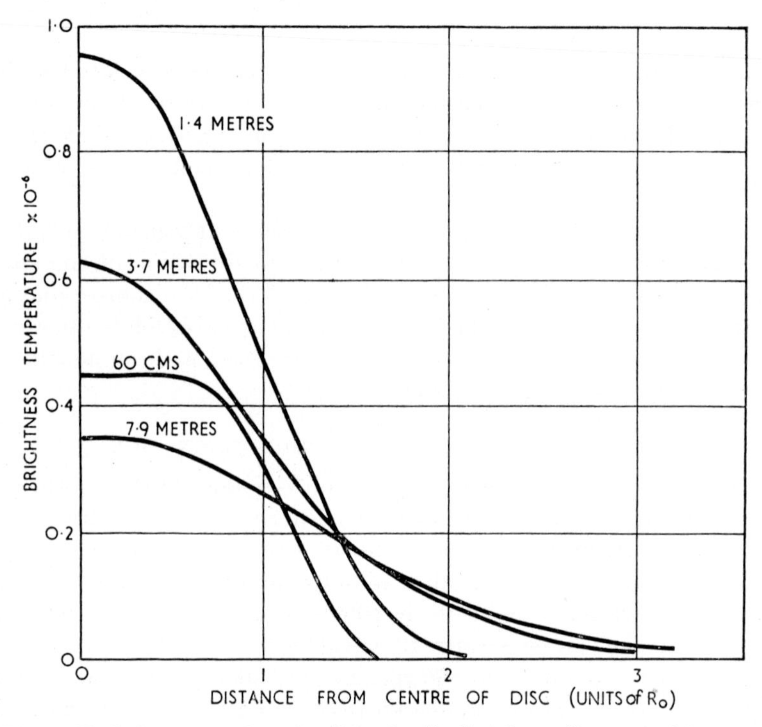

Fig. 4. Variation across the solar disk of radio "brightness" expressed in terms of an equivalent temperature

series of observations on wavelengths of 1·4 m, 3·7 m, and 7·9 m has been made (O'BRIEN, 1953); in some of the latter experiments observations were made with interferometers whose axes made a number of different angles with the solar axis. In this way it was no longer necessary to assume that the Sun had spherical symmetry and the experiments showed that in fact the extent in the equatorial plane was greater than in polar directions. Curves showing the radial variation of "brightness temperature" for the equatorial plane derived from all the Cambridge observations are shown in Fig. 4.

4. RADIO STARS

At the present time the study of radio stars may conveniently be divided into two groups of observation:

(*a*) The detection of a large number of radio stars, so that statistical work on their distribution in direction and apparent magnitude may be made; for this type of observation the accuracy of position measurement need not be very great, but the overall sensitivity of the system must be as great as possible.

(*b*) The determination of the absolute position and the angular diameter of the more intense radio stars in an attempt to identify them with visible objects.

The use of interferometric methods offers important advantages in both types of investigation. When a simple radio telescope is used to detect radio stars, difficulties may arise in distinguishing between the traces produced by the weaker ones and by small irregularities in the background radiation having an angular structure comparable with the resolving power of the telescope. The increased resolving power of an interferometer system makes possible a considerable reduction in the effect of the galactic structure. For observations on a given wavelength and with aerials having a given total area, an interferometric radio telescope may therefore allow a greater number of radio stars to be distinguished, as well as enabling their positions to be found more accurately.

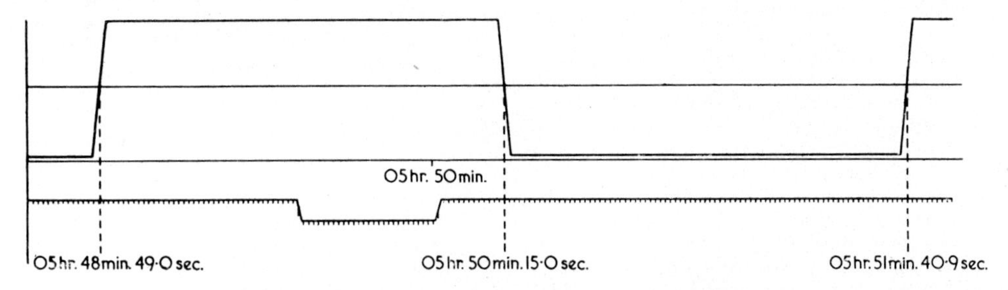

Fig. 5. High-speed record of transit of the intense radio star in Cassiopeia obtained with a phase-switching system on a wavelength of 3·7 m

Both the cliff system (BOLTON and STANLEY, 1948; 1949) and the spaced aerial system (RYLE and SMITH, 1948) have been used for the detection of radio stars. In some of the latter observations (RYLE, SMITH, and ELSMORE, 1950), the phase-switching method was used, and fifty radio stars were located in the northern hemisphere. A more recent survey (MILLS, 1952), using similar methods to observe the southern sky, gave a total of seventy-seven radio stars.

The methods by which observations with spaced aerial interferometers may be used to obtain accurate positions have been discussed by SMITH (1952a), and some of the results obtained (SMITH, 1951) are described in another chapter.

An example of part of a record obtained with a phase-switching interferometer for accurate position finding is shown in Fig. 5.

Observations of the decrease of record amplitude with aerial separation have recently been used to determine the angular diameter of radio stars. An extension of the phase-switching system has been used at Cambridge (SMITH, 1952b, 1952c); observations at Jodrell Bank (HANBURY BROWN, JENNISON, and DAS GUPTA, 1952) and in Australia (MILLS, 1952) have been made with systems employing a radio link between the aerials so as to allow very great separations to be used. The results obtained have shown that some of the more intense radio stars have angular diameters of 1–5 minutes of arc; these observations have been of great importance in confirming the identification of the bodies with visible objects.

The investigation of the scintillation of radio stars has involved simultaneous measurements of the variation of intensity and of the small deviations in the apparent position of the source (RYLE and HEWISH, 1950; HEWISH, 1951, 1952). An account of this work is given in another chapter.

5. The Structure of the General Galactic Radiation

Many of the features of the general galactic radiation can most easily be examined by simple "pencil-beam" radio telescopes; surveys of the distribution of intensity have already been made at wavelengths of 1·5 m and 3 m using systems having a resolution of 10°–15° (Bolton and Westfold, 1950; Allen and Gum, 1950) and measurements with greater resolving power using the 220 ft paraboloid at Jodrell Bank (Hanbury Brown and Hazard, 1953) have made possible the examination of regions near the zenith in greater detail. There are, however, certain problems which require considerably greater resolving power than can be achieved with even the largest of such instruments.

Fig. 6. Photograph of interferometer aerial system used to observe the effect of the H II regions

One such problem concerns the effect of the ionized regions of interstellar matter on the observed intensity of radio emission. Owing to the strong galactic concentration of the early-type stars, the interstellar matter is ionized only in a narrow belt near the galactic plane, and any effect on the radio emission is likely to be restricted to within about 1° of the galactic equator (Westerhout and Oort, 1951). The possibility of detecting the integrated effect of the H II regions, either by their absorption or their emission, therefore depends on observations of the intensity distribution near the galactic equator, with a resolution of better than 1°. Furthermore, it can be shown that a detailed investigation of the temperature and density in these regions requires measurements at wavelengths as great as several metres.

An interferometric method has recently been devised at Cambridge (Scheuer and Ryle, 1953) for studying the H II regions, and observations have been made at wavelengths of 1·4 m, 3·7 m, and 7·9 m. The method is analogous to that used for finding the distribution of "brightness" across the solar disk; two aerials are set up, with a spacing d, on an axis perpendicular to the galactic equator, and they are directed towards a particular galactic longitude. The amplitude of the trace produced with the phase-switching system is then proportional to the term, of angular frequency d/λ, of the Fourier transform of the latitude distribution of intensity. By making observations with a large number of different spacings d, the complete Fourier amplitude curve can be derived, and hence the curve of the variation of intensity with galactic latitude. The resolving power is determined by the maximum aerial spacing used, which in the present experiments was over 200 m. A photograph of the aerial system used is shown in Fig. 6 and an example of the results obtained in Fig. 7.

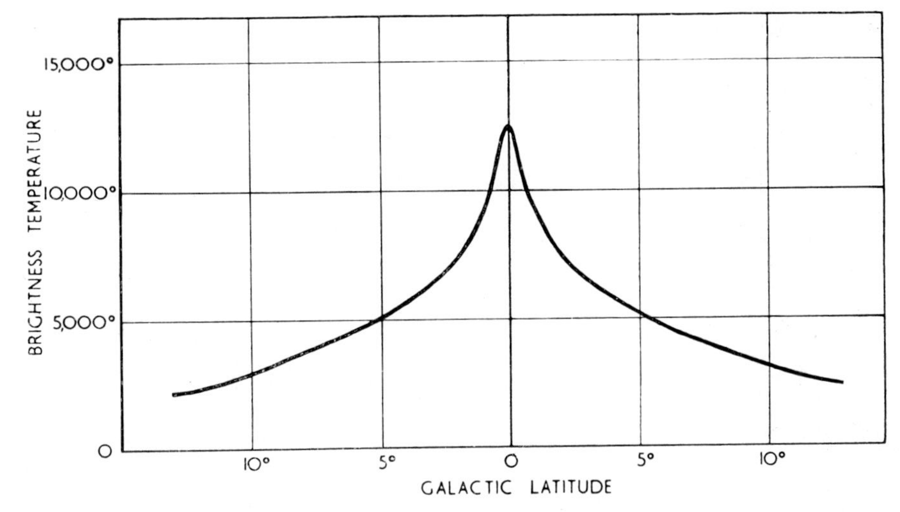

Fig. 7. Variation of "brightness temperature" with galactic latitude at a wavelength of 3·7 m (galactic longitude = 350°)

By comparing the magnitude of the effect on different wavelengths, it is possible to derive both the average temperature and density in the H II regions. The present preliminary results have suggested that the temperature is somewhat greater than the value of 10,000° normally assumed, whilst observations at different galactic longitudes have indicated that the distribution of the regions within the galaxy may show a rather high concentration towards the nucleus.

6. DISCUSSION

Interferometric methods have already been applied to a number of different problems in radio astronomy; in many cases the relatively great resolving power has enabled information to be derived which could not easily have been obtained by other methods.

In some of the problems, such as the derivation of the "brightness distribution" across the solar disk, and the detailed examination of the intensity distribution in regions near the galactic equator, the observations have been made with relatively small aerials, such as those shown in Fig. 6. It seems likely that other problems of this type will also be investigated with similar apparatus.

In other problems, notably the study of radio stars, large interferometric radio telescopes are likely to play an important part. One of the most important problems is related to the location of a large number of radio stars with sufficient accuracy to allow of their identification with visible objects.

The observations (described elsewhere in this volume) which led to the accurate location and eventually to the identification of some of the most intense radio stars were made with an interferometer consisting of two small paraboloids of diameter 27 ft, separated by some 1000 ft.

Fig. 8. One of the four units of the new Cambridge radio telescope

These observations have provided radio astronomy with reference stars which will greatly simplify the location of other weaker sources, whose observation will require the use of very much larger aerials.

In addition to the identification of more radio stars, important information should be derived from a statistical study of the distribution in magnitude and direction of a large number of radio stars.

In order to attack these two problems a new interferometric radio telescope has been built at Cambridge, with an effective area considerably greater than those used previously. As can be seen from Fig. 8, which shows one of the four units of the interferometer, the aerials are built in the form of a cylindrical parabola, which can be rotated about the east-west axis to cover (with the Earth's rotation) the whole sky. The total area of the four units is approximately 4750 m^2.

It seems likely that instruments of this type will play an important part in many of the future observations. The rôles played by large interferometric radio telescopes and the large "pencil-beam" instruments such as the 50 ft paraboloid at Washington and the 250 ft paraboloid now being built at Jodrell Bank are largely complementary; both types of instrument are necessary in the various developments of radio astronomy.

REFERENCES

ALLEN, C. W. and GUM, C. S. . . . . 1950 *Austral. J. sci. Res.*, A **3**, 251.
BLUM, E. J., DENISSE, J. F., STEINBERG, J. L. . 1952 *C.R. Acad. Sci. (Paris)*, **234**, 1597.
BOLTON, J. G. and STANLEY, G. J. . . . . . . 1948 *Nature (London)*, **161**, 312.
1949 *Austral. J. sci. Res.*, A **2**, 139.
BOLTON, J. G. and WESTFOLD, K. C. . . . . 1950 *Austral. J. sci. Res.*, A **3**, 251.
CHRISTIANSEN, W. N., YABSLEY, D. E. and MILLS, B. Y. . . . . 1949 *Austral. J. sci. Res.*, A **2**, 506.
HAIKIN, S. E. and CHICKHACHEV, B. M. . . . 1948 *Bull. Acad. Sci. U.S.S.R.*, **12**, 38.
HANBURY BROWN, R., JENNISON, R. C. and DAS GUPTA, M. K. . . . . 1952 *Nature (London)*. **170**, 1061.
HANBURY BROWN, R. and HAZARD, C. . . . . 1953 *M.N.*, **113**, 109.
HEWISH, A. . . . . 1951 *Proc. Roy. Soc.*, A **209**, 81.
1952 *Proc. Roy. Soc.*, A **214**, 494.
LITTLE, A. G. and PAYNE-SCOTT, R. . . . . 1951 *Austral. J. sci. Res.*, A **4**, 489.
MACHIN, K. E. . . . . 1951 *Nature (London)*, **167**, 889.
1955 In preparation.
MARTYN, D. F. . . . . 1948 *Proc. Roy. Soc.*, A **193**, 44.
McCREADY, L. L., PAWSEY, J. L. and PAYNE-SCOTT, R. . . . . 1947 *Proc. Roy. Soc.* A, **190**, 357.
MILLS, B. Y. . . . . 1952a *Austral. J. sci. Res.*, A **5**, 266.
1952b *Nature (London)*, **170**, 1063.
O'BRIEN, P. A.. . . . . 1953 *M.N.*, **113**, 597.
PAWSEY, J. L. . . . . 1946 *Nature (London)*, **158**, 633.
RYLE, M. . . . . 1952 *Proc. Roy. Soc.*, A **211**, 351.
RYLE, M. and HEWISH, A. . . . . 1950 *M.N.*, **110**, 381.
RYLE, M. and SMITH, F. G. . . . . 1948 *Nature (London)*, **162**, 462.
RYLE, M., SMITH, F. G. and ELSMORE, B. . . 1950 *M.N.*, **110**, 508.
RYLE, M. and VONBERG, D. D. . . . . . . 1946 *Nature (London)*, **158**, 339.
1947 *Nature (London)*, **160**, 157.
SCHEUER, P. A. G. and RYLE, M. . . . . . 1953 *M.N.*, **113**, 3.
SMERD, S. F. . . . . 1950 *Austral. J. sci. Res.*, A **3**, 34.
SMITH, F. G. . . . . 1951 *Nature (London)*, **168**, 555.
1952a *M.N.*, **112**, 497.
1952b *Proc. Phys. Soc. (London)*, B **65**, 971.
1952c *Nature (London)*, **170**, 1065.
STANIER, H. M. . . . . 1950 *Nature (London)*, **165**, 354.
WESTERHOUT, G. and OORT, J. H. . . . . . 1951 *Bull. Astron. Inst. Netherlands*, **11**, 323.

Note added in proof (December, 1954).

A survey of radio stars with the new Cambridge Radio Telescope at a wavelength of 3·7 m has now been completed; a total of 1936 sources has been located between declinations $-38°$ and $+83°$. Some thirty of these are found to be of large angular diameter (20′–2°), and one has been identified with the peculiar galactic nebulosity IC. 443, in the constellation of Gemini. The remainder are of small angular diameter, and good positional accuracies have been obtained for a large number of them; about 500 have been located with errors of about $\pm$ 2 minutes of arc in R.A. and $\pm$ 12 minutes of arc in declination. So far only a few of the sources have been identified with optical objects; a relatively intense source has been found close to the position of Kepler's Nova of 1604. Further search for identifications with optical objects, and a number of statistical investigations of the distribution of the sources are still in progress.

Large Radio Telescopes and Their Use in Radio Astronomy

R. Hanbury Brown and A. C. B. Lovell

University of Manchester, Jodrell Bank Experimental Station

Summary

After referring to the need for directivity and sensitivity in radio astronomical investigations, the paper discusses the fundamental features of radio telescopes. Examples are then given of contemporary methods of obtaining directivity, using radio telescopes of various designs. A description is given of the 218-ft aperture radio telescope at Jodrell Bank. The results obtained with this instrument in the investigation of the radio emission from the Milky Way and also of the extragalactic radio emissions, are described. The paper concludes with a description of the large 250-ft aperture steerable radio telescope which is now being built at Jodrell Bank.

1. Historical Note

The receiving equipment with which Jansky (1932) discovered the radio waves from space in 1932 worked on a wavelength of 15 m, and used a simple aerial system with a wide beam for reception. Even so the small measure of directivity enabled him to conclude that the strongest signals were coming from the direction of the galactic centre. Some years later Reber (1940) used a much shorter wavelength of 1·8 m and a larger aerial of 30 ft diameter which produced a greater measure of directivity. With this equipment Reber was able to find out much more detail about the distribution of the radio intensity over the sky. However, his beam width of 12° was still not sufficiently directive to distinguish individual sources of radiation and, as is well known, he was led to conclude that the radiation must be arising from a diffuse source distributed in the Galaxy, which he suggested might be the interstellar hydrogen gas.

It soon became clear that very much greater directivity and sensitivity would be necessary for the study of the radio emissions from space. In these requirements the needs of radio astronomy are analogous to those of visual astronomy, and the main solutions have likewise been analogous—namely the development of interferometers and large telescopes. The subject of radio interferometers has been treated elsewhere in this volume (Ryle's article, pp. 532–541); this contribution will deal only with the development of very large radio telescopes using pencil beams and with some of the results obtained.

2. The Fundamental Features of Radio Telescopes

The longest wavelengths used in radio astronomy are limited by ionospheric effects and it is not possible to receive extra-terrestrial signals satisfactorily on the Earth on wavelengths much longer than about 15 m. On short waves atmospheric absorption becomes important in the millimetre wave region, but before this extreme is reached the falling off in received power as the wavelength decreases makes investigations difficult. With the notable exception of the study of the powerful radio emissions from the Sun, and of the recently discovered spectral line on 21 cm from the interstellar gas, most investigations in radio astronomy have been carried out in the wavelength region of 1–10 m. In this waveband there are three common methods by which directivity can be readily obtained; by the use of arrays of dipoles, by arrays of

Yagi aerials, or by paraboloidal aerial systems. In each case the basic unit is generally the centre fed half-wave dipole, which has a natural resonant frequency corresponding to a wavelength of twice its length. If G is the power gain over an isotropic receiving element, and if ϕ refers to the equatorial plane, and θ to the plane containing the elements, then for a half-wave dipole it can be shown that

$$G\int_{-\pi/2}^{\pi/2}\int_{0}^{2\pi}\frac{\cos^2(\frac{1}{2}\pi\sin\theta)}{\cos^2\theta}\,.\cos\theta\,.\,d\theta\,.\,d\phi=4\pi,$$

giving $$G=1{\cdot}635.$$

The polar diagram of such a half-wave dipole is shown in Fig. 1. Clearly owing to the lack of directivity and low gain this element alone is not of great value as a

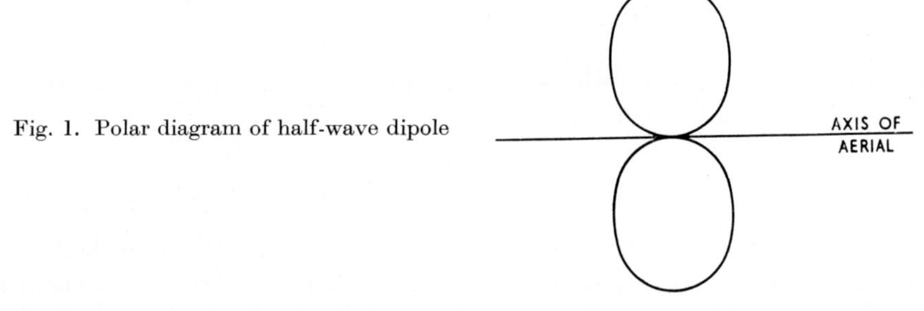

Fig. 1. Polar diagram of half-wave dipole

receiver in radio astronomy. However, if a number of such half-wave dipoles are arranged in a row it is possible to connect them such that the phases of the current in the various elements lead to a reduction of the received power in some directions and an enhancement in others. Consider n equally spaced elements each with a polar diagram $f(\theta)$. A field strength E_r at distance R_r and angle θ will induce a current i_r in the rth element with phase α_r, where

$$E_r=kf(\theta)i_re^{j\left(\frac{2\pi R_r}{\lambda}+\alpha_r\right)}.$$

The corresponding relations for the entire aerial of n elements will be

$$E=kf(\theta)\sum_{r=1}^{r=n}i_re^{j\left(\frac{2\pi R_r}{\lambda}+\alpha_r\right)}.$$

If d is the spacing of the half-wave dipoles, then

$$R_r=R_1+(r-1)d\,.\sin\theta,$$

and for the same current in each element, and for zero α's, summation gives

$$|E|=kf(\theta)\frac{\sin\left(\frac{n\pi d\sin\theta}{\lambda}\right)}{\sin\left(\frac{\pi d\sin\theta}{\lambda}\right)}.$$

The arrangement thus receives maximum power from the direction $\theta=0$ and zero power in directions given by

$$\sin\theta=\frac{N\lambda}{nd}\,(N=1,2,3,\ldots).$$

For a spacing $d < \lambda$ there are principal maxima and a series of side lobes as shown in Fig. 2. In practice such an array is generally mounted in front of a reflecting sheet, so that the radiation is concentrated in the forward direction only. The power

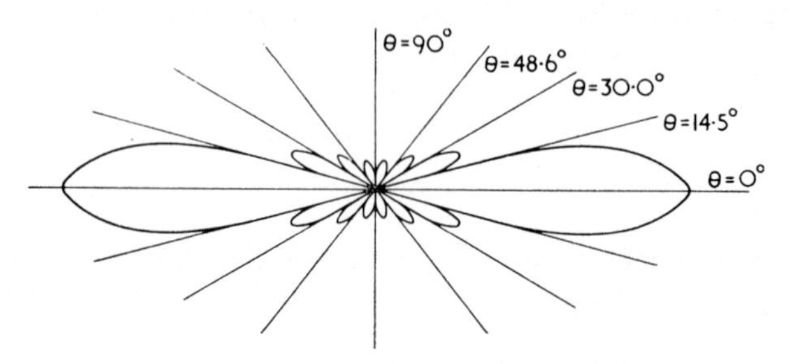

Fig. 2. Polar diagram of linear array with eight equally-spaced elements

received from the main lobe falls to zero at $\sin\theta = \dfrac{\lambda}{nd}$ but if the total aperture of the aerial $(n - 1)d \gg \lambda$ this reduces to

$$\theta = \frac{\lambda}{a}.$$

This arrangement, known as a linear array, produces directivity in the θ direction only. If directivity is also required in the ϕ plane then a series of such linear arrays must be placed side by side to form a broadside array. Fig. 3 shows such an array

Fig. 3. 320-element broadside array, working on a wavelength of 2·4 m. The beam width is $3° \times 10°$ to half-power points

used as a radio telescope in the investigation of the galactic radio emissions. It contains 320 elements on a wavelength of 2·4 m. The overall size is 128 ft $\times$ 41 ft, the beam width $3° \times 10°$ to half power points and the power gain $G = 600$ over a half wave dipole. The mechanical difficulties of building a radio telescope, even on such a comparatively short wavelength as 2·4 m, with appreciable directivity, will be evident from this illustration. This is particularly so if the instrument has to be steerable. In the case illustrated the radio telescope was fixed in azimuth but could be adjusted with difficulty in elevation.

In 1928 a Japanese engineer, Yagi, devised an alternative form of directive aerial in which only the half-wave dipole is fed, but which consists of a parasitic reflector

Fig. 4. An array of Yagi aerials working on a wavelength of 4·1 m. The horizontal polar diagram has a beam width of 10° to half-power points

and a number of parasitic directors. A photograph of a fixed array of Yagi aerials used in radio astronomy is shown in Fig. 4. These work on a wavelength of 4·1 m. The individual arrays are separated by 1·2 λ and are mounted at a height of 1·57 λ above ground. The horizontal polar diagram has a beam width of 10° to half-power points. The power gain over a half-wave dipole was 165. If directivity was also

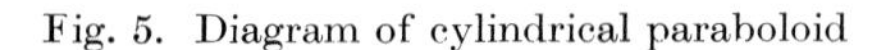

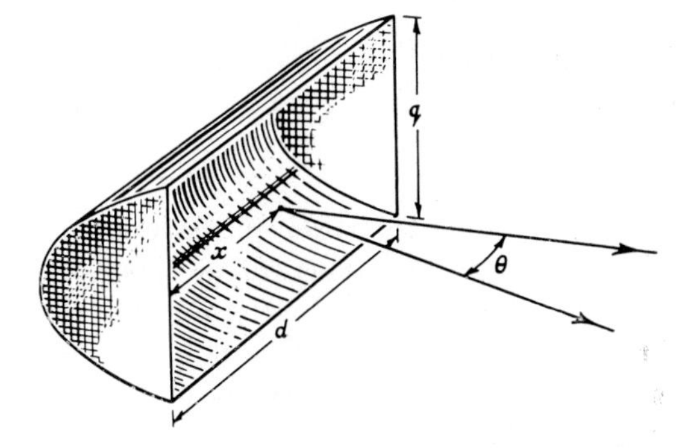

Fig. 5. Diagram of cylindrical paraboloid

required in the vertical plane it would be necessary to place another series of arrays above the existing ones. Yagi aerials of this type have been widely used in radio astronomy, both singly and in arrays. They are somewhat easier to construct and more compact than broadside arrays, but suffer from larger side lobes. The mechanical difficulties of constructing a large and steerable array remain considerable.

In the other main type of radio telescope directivity is achieved by means of a parabolic reflecting sheet with a half-wave dipole or small array mounted at the focus. One great advantage of this type of system is that the wavelength can be readily changed by adjusting the dipole feed system, whereas in the case of an array of dipoles a complete reconstruction would be needed. The simplest form of parabolic aerial is a cylindrical paraboloid, rectangular through the focus as in Fig. 5. If $E(x)$ is the field at a point in the aperture plane, distant x from the edge of the mirror, then by HUYGEN's principle

$$E(\theta) = k \int E(x) e^{\frac{2\pi j x \sin \theta}{\lambda}} . dx,$$

the integral being taken over the aperture. If the distribution of phase and amplitude is uniform across the aperture

$$E(\theta) = \text{const.} \frac{\sin\left(\dfrac{\pi d \sin \theta}{\lambda}\right)}{\dfrac{\pi d \sin \theta}{\lambda}},$$

where d is the size of the aperture.

If $d \gg \lambda$, the first zero occurs at $\theta \simeq \frac{\lambda}{d}$. A similar expression will be appropriate to the ϕ plane. Thus for a rectangular aperture of sides d, q, the first zeroes in the principal planes will occur at $\frac{\lambda}{d}, \frac{\lambda}{q}$, and the received power will be effectively concentrated in a solid angle $\frac{\lambda^2}{dq}$. Thus the power gain will be $\frac{4\pi dq}{\lambda^2}$.

In the case of a full paraboloid with circular aperture of radius a, uniformly illuminated, the beam will have circular symmetry. The polar diagram can be shown (see, for example, SLATER, 1942) to be given in terms of the first-order Bessel function

$$E(\theta) = \frac{2J_1(ka \sin \theta)}{ka \sin \theta}$$

and the power gain

$$G = \frac{4\pi\alpha}{\lambda^2}.$$

where $\alpha = \pi a^2$ is the area of the aperture. The reception beam for a uniformly illuminated circular aperture is therefore 1·22 times wider than that from the uniformly illuminated rectangular aperture with side the same as the diameter of the circular aperture. The field in the first side lobe is, however, reduced to 13 per cent of the main lobe instead of 21 per cent for the rectangular aperture. Whereas the circular paraboloid requires only a single primary feed at the focus, a cylindrical paraboloid will require one for each half wavelength of its length. In actual practice with a circular paraboloid it is impossible to obtain uniform illumination of the aperture, neither is circular symmetry generally achieved because of the differing polar diagrams of the primary feed source in the two planes. The general result is a broadening and asymmetry in the main beam, diminution of gain, and reduction of side lobe intensity relative to the uniformly illuminated aperture. This question is referred to again on page 557.

There are several reasons why aerials of parabolic form have achieved popularity as radio telescopes. They can be used readily over a range of wavelengths by merely changing the dipole system at the focus. The feed arrangements are relatively simple compared with the array aerials, and the paraboloid is somewhat easier to mount in a fully steerable mechanical system. In England there are steerable paraboloids of

Fig. 6. A 30-ft aperture radio telescope in the form of a steerable paraboloid. With the dipole system illustrated the working wavelength was 3·7 m and the beamwidth 26° × 30° to half-power points

aperture 30 ft in use as radio telescopes at Jodrell Bank and Cambridge. One is illustrated in Fig. 6. On a wavelength of 3·7 m the beam width is 26° to half power points in the equatorial plane of the dipole and 30° in the plane of the electric vector. The power gain is 20·6 over a half-wave dipole assuming the primary feed to be 50 per cent efficient. On a wavelength of 21 cm the corresponding figures are 1°6 and 1°8 for the beam widths and 6350 for the power gain.

3. THE LARGE FIXED RADIO TELESCOPE AT JODRELL BANK

During 1946 the construction was commenced at Jodrell Bank of a very large fixed paraboloid with a diameter of 218 ft. The instrument was finished in 1947, and has since been in regular use as a radio telescope for the study of the galactic and extragalactic radio emissions. Some idea of the construction of this instrument may be obtained from the photograph in Fig. 7. The surface of the paraboloid is formed of conducting wires laid on a web of steel cables. This web consists of twenty-four radial runs spaced equally at 15° intervals. These are anchored on a concrete ring at ground level near the centre, and at the perimeter pass over steel posts 23 ft 4 in. high. Each of these radial runs is strained to the ground at equal intervals of

Fig. 7. The 218 ft fixed paraboloid in use as a radio telescope at Jodrell Bank. When this photograph was taken the beam was deflected 14° from the vertical by tilting the mast supporting the primary feed

27 ft 4 in. from the centre in order to pull them down to the parabolic shape. These strain points are joined to the corresponding strain points of adjacent radial runs by means of further steel cables. The maximum deviation of the resultant plano-parabolic surface from the true parabolic shape is nowhere greater than 5 in. and the errors introduced in the construction are less than 2 in. Thin galvanized iron wires are secured to this rigid steel web in order to form the reflecting surface. Although it was originally intended to create a 4 in. section mesh of wire, this has never been done. So far the instrument has operated with a separation of 8 in. between wires laid in the plane of polarization of the dipoles. On the original wavelength of

4 m, the resultant leakage through this surface was only 5 per cent of the incident power. Subsequently the paraboloid has been used on a wavelength of 1·89 m where the loss by leakage through the surface of 37 per cent is much more serious. The departure from the true parabolic shape of 5 in. places a lower limit of approximately 1 m on the working wavelength. At present, therefore, the limit is set by the wire spacing and not by any mechanical inaccuracies.

The focus of this parabolic surface is 126 ft above the ground and the primary feed is carried at this point on a steel tower. It consists of a small array of two folded dipoles with reflectors, arranged to receive maximum power from the region of the mirror surface and as little as possible in the backward direction. A low loss polythene

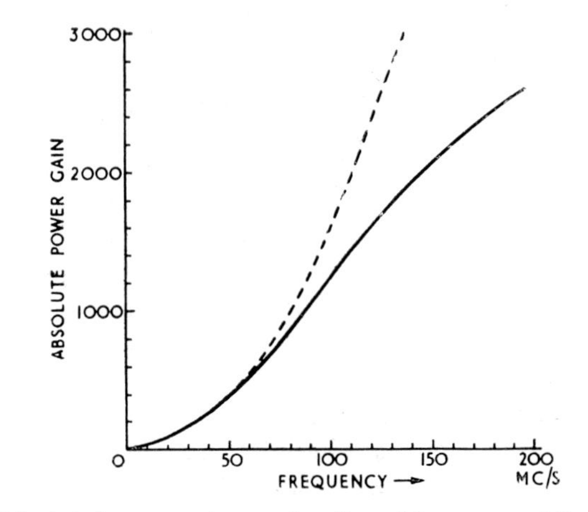

Fig. 8. . . . Calculated power gain as a function of frequency of the radio telescope illustrated in Fig. 7
—— Calculated power gain allowing for the losses due to transmission through the wire reflector

cable connects this primary feed to the receiving equipment at ground level. The steel tower is pivoted at the base so that it can be moved in the north-south direction. This enables the primary feed to be displaced from the focus and is the only method by which the beam may be shifted from the vertical direction. It has been possible to tilt the mast up to 17° from the vertical, giving a beam deflection of about 15°. The shape of the beam remains very satisfactory even under these extreme conditions.

The calculated power gain of this radio telescope as a function of wavelength is given in Fig. 8. This calculation allows for the inefficiency of the primary source, including losses due to back radiation, and the full curve shows the serious effect of the present 8 in. wire spacing on the performance at short wavelengths. A check on the predicted characteristics has been made on two wavelengths. A transmitter carried in an aircraft enabled the characteristics to be plotted on a wavelength of 4 m, and at 1·89 m the radio "star" in Cygnus provided a satisfactory source of emission. The results are as follows:

	4·1 m		1·89 m	
	Theoretical	*Measured*	*Theoretical*	*Measured*
Gain over half-wave dipole .	690	700	2130	—
Beam width to half power .	4° 30′	4° 20′	2° 00′	1° 58′

The technique of using this telescope and some of the results of the measurements of the galactic and extra-galactic radio emissions are described in the following paragraphs.

4. Technique of Observation

The 218 ft paraboloid is at present being used to survey the radio emission from the region of sky which lies within its working field. Since the maximum deflection of the beam from the zenith is about 15° and the latitude of the site is 53° 14′ N, this region of sky lies between declinations 68° N and 38° N. The survey was started in 1950 and at the time of writing (1953) about 70 per cent of the whole field of view of the instrument has been surveyed thoroughly at a wavelength of 1·89 m.

The survey is carried out by fixing the mast which supports the primary feed at a constant azimuth (approximately due south or north) and at an angle of tilt which corresponds to the required elevation of the aerial beam. The position of the primary feed is measured by means of a theodolite and the direction of the beam is then calculated. For each diurnal rotation of the Earth the beam sweeps out a strip of sky 2° wide in declination and 24 hr in right ascension and the power received by the aerial from this strip is recorded continuously on a moving chart. When, for any particular strip, a satisfactory number of records have been made the tilt of the mast is altered and the whole process is then repeated. By making records with the beam at several different elevations it is possible in this way to map the radio emission from a large region of the sky.

The observations are being made at a wavelength of 1·89 m and the receiver has a noise factor of 3·5, a radio-frequency bandwidth of 2 Mc/s and a post-detector time constant of 10 sec. The minimum change of received flux, which it should be possible to detect by inspection of a single record has been estimated theoretically (Hanbury Brown and Hazard, 1951a), and it has been found to correspond to a randomly polarized flux of 10^{-25} W per m^2 per c.p.s. incident on the aerial. Practical experience shows that the minimum value of the flux which can actually be detected depends upon the time of day and the weather. During daylight and in fine weather the performance of the aerial is usually limited by man-made static since it is sited close to a main road and also the primary feed is exposed. The presence of charged rain, thunderstorms, or radiation from the "disturbed" Sun can cause even greater interference and can overload the equipment. During the night the noise level reaches a minimum between midnight and sunrise and the theoretical limits of the equipment are reached on an average for about one-third of the time. On favourable occasions the transit of a source with an intensity of at least 1–2 $\times$ 10^{-25} W per m^2 per c.p.s. can be observed by inspection of a single record. Allowing for losses in the feeder system and in the aerial this corresponds to the detection of an actual power of about 10^{-16} W. This limit can be extended by integrating the results of observations taken on different nights, and a limit of 5 $\times$ 10^{-26} W per m^2 per c.p.s. can be reached by using a number of records which varies from about four, for excellent records, to ten or fifteen for moderate records.

The celestial co-ordinates of a radio source can be found from a set of records taken with the aerial beam at a number of different elevations. The right ascension is determined from the time at which the source is observed to transit the aerial beam and from measurements of the position of the primary feed. There are three main errors in the value of right ascension: (1) the error due to the uncertainty in the position of the electrical centre of the primary feed; (2) the error in measuring the position of the primary feed; (3) the error in observing the time of transit.

The second and third errors can be reduced by taking several independent observations and in practice the value of error (3) will be strongly dependent on the quality of the records and on the strength of the source. In a series of observations made of the intense source in Cygnus during 1950 (HANBURY BROWN and HAZARD, 1951b) the probable error in the value of right ascension was estimated to be $\pm$ 7 sec. due to these two factors alone. Error (1) above is systematic and can only be evaluated approximately. It has been estimated that the probable uncertainty in the position of the electrical centre of the primary feed is about 2 in. and that this may introduce an error of about $\pm$ 20 sec. into the value of right ascension. The final error in the value of the right ascension of the source in Cygnus is believed to be about $\pm$ 25 sec., and this represents about the best that can be done with the aerial without calibrating the direction of the beam on a source of known position.

The declination of a source is obtained by plotting the intensity observed at transit against the angle of the mast, and by finding the angle which corresponds to the greatest value of intensity. The observational errors will again depend upon the quality of the records, the strength of the source and also on the number of elevations at which observations have been made. For the observations of the intense source in Cygnus in 1950 the probable value of these observational errors was estimated to be $\pm 5'$. There are, however, two far greater sources of error which are both systematic. Firstly, in order to find the elevation of the aerial beam it is necessary to know the relation between the angle of tilt of the aerial mast (θ) and the direction of the beam (α). This relation is given by

$$\alpha = k\theta,$$

where α and θ are measured with respect to the axis of the paraboloid and k depends on the ratio of the focal length to the diameter of the aerial. Experiments with models (SILVER and PAO, 1944) suggested that the value of k in this equation should be $0{\cdot}93 \pm 0{\cdot}02$. In practice, this uncertainty in the value of k introduces a systematic error into the values of declination which is zero in the zenith and which increases with the tilt of the mast. To observe the source in Cygnus the mast was tilted to about 14° from the zenith and the value of this error was about $\pm 15'$. A further systematic error of about $\pm 10'$ in the direction of the beam is introduced by the uncertainty in the position of the electrical centre of the primary feed. The total error in the value of declination for the source in Cygnus was therefore about $\pm 25'$ due to systematic errors and $\pm 5'$ due to observational errors.

The effect of these large systematic errors has been reduced by calibrating the aerial on sources whose position is known. In the case of the source in Cygnus the value of the declination was improved by observing the Great Nebula in Andromeda (M 31) (HANBURY BROWN and HAZARD, 1951a), and in this way the error in declination was reduced to about $\pm 15'$. The position of the two most intense sources in the sky (in Cassiopeia and Cygnus) has now been measured by means of interferometers (SMITH, 1951) with an accuracy of about $\pm 1'$, and since both these sources lie within the field of view of the paraboloid it has been possible to reduce the systematic errors by using these two sources as reference points. It is now possible to establish the co-ordinates of any moderately intense source (5×10^{-25} W per m^2

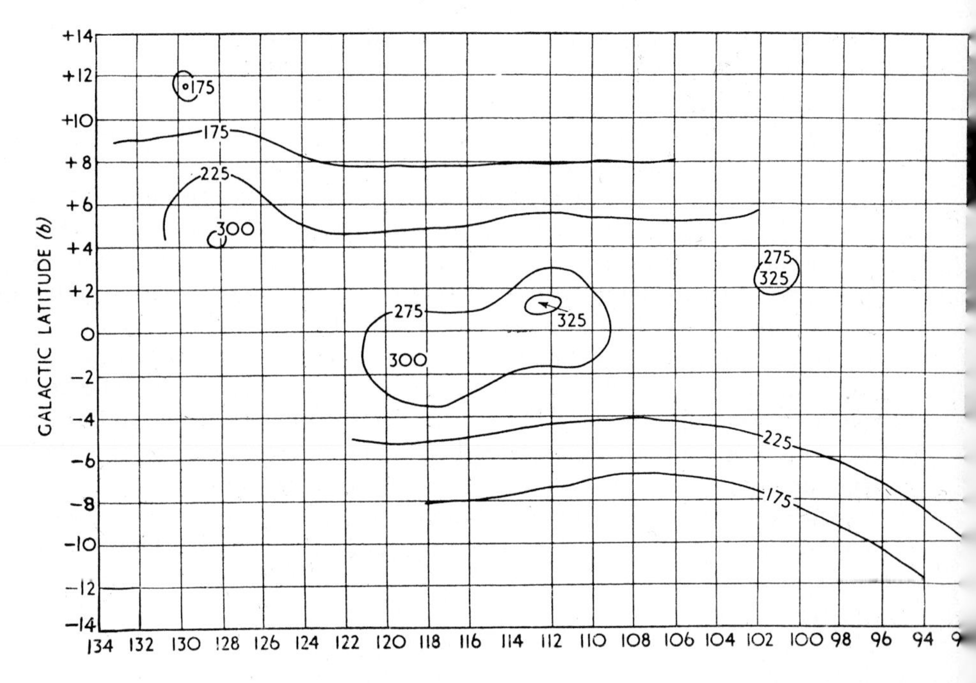

Fig. 9. Isophotes of a region of the Milky Way measur

per c.p.s.) to about $\pm$ 15′. The error in the position of weaker sources is greater and increases to about $\pm$ $1\frac{1}{2}$° for sources near the limit of detection.

5. Results

Isophotes

Fig. 9 shows the isophotes of intensity of the radio emission at 1·89 m observed (Hanbury Brown and Hazard, 1953a) from the region of the Milky Way which lies in the field of view. The isophotes are plotted in galactic co-ordinates (l, b) and it can be seen that the region extends from l = 40° to 130° and b = + 14° to − 14°. The isophotes are calibrated in terms of the equivalent temperature T_A (degrees Kelvin) of the aerial beam. The value of T_A for any point is calculated from the measured value of the power received in that direction and from the constants of the aerial and feeder system. It is assumed that an equivalent value of flux is radiated by a black-body whose temperature is T_A and whose angular diameter is greater than that of the beam and its principal side-lobes.

Inspection of the isophotes shows that they run roughly parallel to the galactic plane (b = 0) and that the intensity of the radiation reaches a maximum in the galactic plane. If the results of other surveys (Reber, 1940; Hey *et al.*, 1948; Reber, 1948; Bolton and Westfold, 1950a; Allen and Gum, 1950) are considered it can be shown that the "radio Milky Way" corresponds closely to the visual Milky Way, and therefore the majority of the observed radiation must be associated with our own Galaxy. The main features of the map are, however, the concentration of intensity in the Cygnus region between l = 65° and 38°, and the highly localized concentrations or so-called "radio stars". The concentration of intensity in the Cygnus region has been noted by several observers and it has been suggested (Reber, 1944;

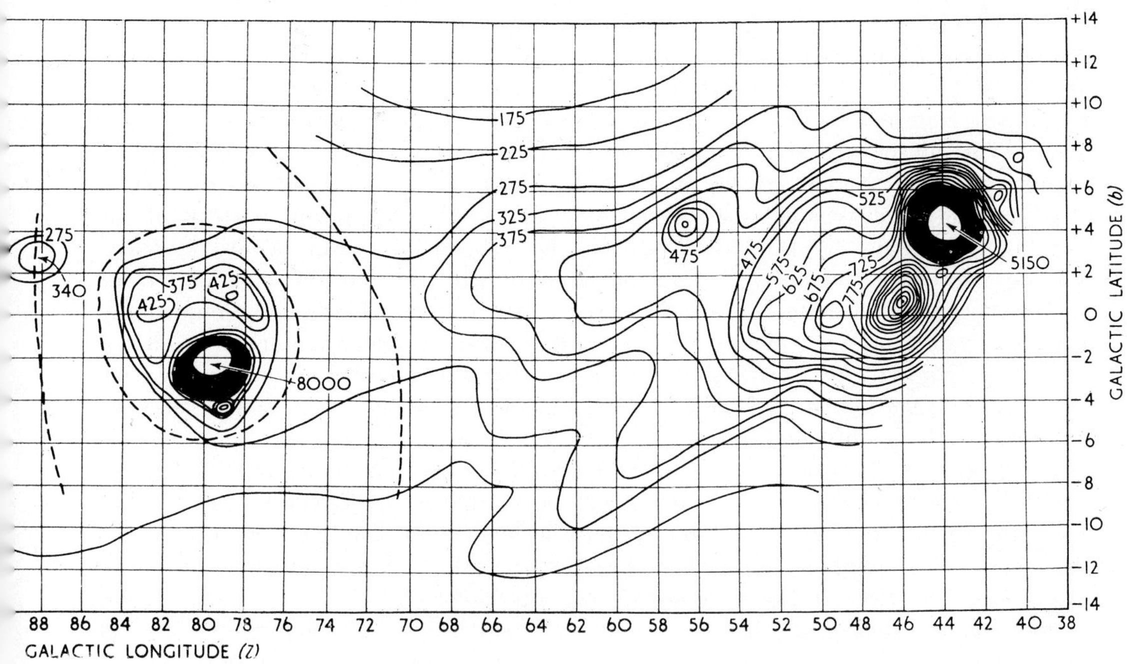

1·89 m using the radio telescope illustrated in Fig. 7

BOLTON and WESTFOLD, 1950b) that it corresponds to the radiation from the aggregation of matter in a spiral arm of the Galaxy. The results shown in Fig. 9 represent a picture of this region obtained with the highest resolving power so far employed, and they suggest that this large and apparently diffuse concentration of intensity in the Cygnus region has several component sources which have not yet been resolved satisfactorily. A further examination of this region must await a survey with a beam of width considerably less than 2°.

Starting from the right hand edge of the map the localized source at $l = 44°$, $b = +4\frac{1}{2}°$ is the well-known (BOLTON and STANLEY, 1948) "radio star" in Cygnus and it will be referred to hereafter as Cygnus (I). It is the second most intense source known and radiates an intensity at 1·89 m of $5{\cdot}7 \times 10^{-23}$ W per m² per c.p.s. So far any attempt to detect variations in this intensity or polarization of the radiation have given negative results. The source has been the subject of a considerable amount of recent work which has suggested some fascinating speculation. Photographs with the 200-in. Hale telescope at Palomar Mountain have shown* that in the position of this source there is the faint image of two extra-galactic nebulae in collision. Furthermore some recent measurements with a radio interferometer have shown (HANBURY BROWN *et al.*, 1952) that the apparent disk of this source has an angular extension of about $150'' \times 30''$, which indicates that it cannot be an object of stellar dimensions.

The source at $l = 46°$, $b = +0{\cdot}8°$ has also been reported in observations at a wavelength of 25 cm where it was found to have a considerable angular extension. Comparison of the intensity observed (PIDDINGTON and MINNETT, 1952), at 25 cm with that at 1·89 m shows that the intensity of this source has not decreased with

* BAADE, W. and MINKOWSKI, R.; Private communication. See also *Ap. J.* **119**, p. 215 (1954).

wavelength in a manner similar to that found for the majority of sources. Further observations are needed to confirm the unusual nature of this source. The source at $l = 80°$, $b = -2°$ is the well-known (RYLE and SMITH, 1948) "radio star" in Cassiopeia (referred to here as Cassiopeia (I)). The curious shape of the isophotes around this source represent the scattering of radiation by the steel guy wires which support the central mast of the paraboloid. The broken lines represent the loci of two systems of weak side-lobes which are part of the diffraction pattern of the paraboloid. Cassiopeia (I) has also been the subject of recent intensive study. Photographs with the 200-in. telescope have shown* that in the position of the source there is a faint nebulosity of extension about $4' \times 4'$ which exhibits unusually high internal motions. Measurements with a radio interferometer have shown (HANBURY BROWN *et al.*, 1952) that the disk of this source corresponds roughly with the photographic image and they suggest strongly that the radio source may be identified with this unusual object.

The source at $l = 88°$, $b = +2°$ corresponds closely in position with TYCHO BRAHE's supernova of A.D. 1572, and it has been suggested (HANBURY BROWN and HAZARD, 1952a) that the radio source may be identified with the remnant of this supernova. This supports an earlier identification (BOLTON and STANLEY, 1949) of a source in Taurus with the remnant of the supernova of 1054 (Crab Nebula), and strengthens the suggestion that the remains of a supernova are a powerful source of radio emission. There are a number of other sources shown on the map but as yet no further identifications with visual objects have been suggested.

The Localized Sources

The data shown in Fig. 9 represent only a selection from the region which has been surveyed. At the present time (January, 1953) about twenty-four localized sources

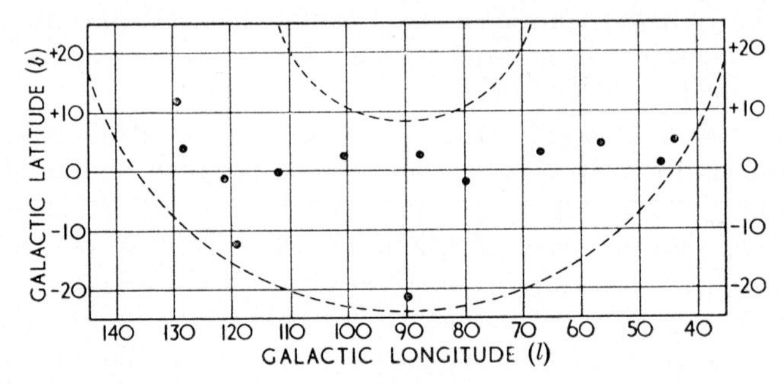

Fig. 10. The distribution of the intense radio sources ($I \geqslant 5 \times 10^{-25}$ W per m² per c.p.s.) The broken lines represent the boundaries of the area surveyed

have been found (HANBURY BROWN and HAZARD, 1953b) in the field of the paraboloid. Although this represents a small sample for the purposes of analysis the results already show that the intense sources are concentrated near the galactic plane. Of the thirteen most intense sources ($\geqslant 5 \times 10^{-25}$ W per m^2 per c.p.s.) ten lie within $\pm 5°$ of the galactic plane and their distribution is shown in Fig. 10. The area of sky which has been surveyed within $\pm 5°$ of the plane is considerably less than the area surveyed away from the plane, and so the results must represent a genuine

* BAADE, W. and MINKOWSKI, R.; Private communication. See also *Ap. J.* **119**, p. 215 (1954).

concentration of the intense sources. This conclusion supports the results of an interferometer survey made in Australia (MILLS, 1952) at a wavelength of 3 m in which a marked concentration of the intense sources near the galactic plane was observed. The distribution of the weaker sources cannot be inferred from the present results since the whole area has not been surveyed with a high sensitivity; furthermore the detection of weak sources near the galactic plane presents great difficulty with a pencil beam, since they tend to be obscured by the large gradient of the background radiation. The results of more extensive surveys (MILLS, 1952; RYLE *et al.*, 1950) with interferometers suggest that the distribution of the weak sources is considerably more isotropic than that of the intense sources.

The high degree of concentration of the intense sources near the galactic plane indicates that they are objects within the Galaxy. Their angular distribution implies that they are rare objects and that they lie extremely close to the galactic plane.

Very little can be said at the present about the weak sources as their distribution is not yet known satisfactorily. There are two obvious possibilities; firstly they may represent the nearest of a group of sources which are distributed densely in the Galaxy, and they may account for the majority of the radiation observed from the Galaxy; secondly they may represent the radiation from extra-galactic objects.

One line of enquiry which promises to throw some light on this question is the investigation of the radiation from extra-galactic nebulae, and a start on this work has been made (HANBURY BROWN and HAZARD, 1952b) with the 218 ft paraboloid. The positions of the brightest extra-galactic nebulae in the field of view has been surveyed and up to the present six radio sources have been detected in the positions of nebulae. The results are shown in Table 1. The identification of these radio

Table 1. *Radio sources associated with extra-galactic nebulae*

Nebula	Type	Apparent photographic magnitude*	Intensity of radio source (W per m² per c.p.s.) at 1·89 m	Apparent radio magnitude‡
NGC 224 (M 31) .	Sb	+ 5·0	$1{\cdot}6 \times 10^{-24}$	+ 6·0
NGC 3031 (M 81) .	Sb	+ 8·9	8×10^{-26}	+ 9·2
NGC 5457 (M 101) .	Sc	+ 9·0	$< 10^{-25}$	> + 9·1
NGC 5194–5 (M 51).	Sc	+ 9·7	$5{\cdot}5 \times 10^{-26}$	+ 9·7
NGC 4258 . .	Sb	+ 10·2	5×10^{-26}	+ 9·8
NGC 2841 . .	Sb	+ 10·5	$3{\cdot}3 \times 10^{-26}$	+ 10·3
NGC 891† . .	S	+ 12·2	10^{-25}	+ 9·0

* These magnitudes are taken from the SHAPLEY-AMES catalogue of nebulae brighter than magnitude + 13·0.
† This nebula is a spiral seen edge-on and showing an intense absorbing lane. Since this lane will absorb a high fraction of the visual radiation from the nebula it may explain why the radio source is more intense than would be expected from the apparent photographic magnitude.
‡ The scale of radio magnitude has been defined so that for most of the nebulae measured the apparent photographic and radio magnitudes are approximately equal. The apparent radio magnitude of a nebula of intensity $5{\cdot}5 \times 10^{-26}$ W per m² per c.p.s. is defined as + 9·7.

sources with nebulae can at present be based only on the agreement of the celestial co-ordinates. The only exception is the Great Nebula in Andromeda (M 31) and in that case it has been possible to show (HANBURY BROWN and HAZARD, 1951a) that not only is the radio source in the correct position, but also it has an angular diameter comparable to that of the nebula. It has also been possible to show that the radiation measured from M 31 at 1·89 m is roughly consistent with that to be expected on the assumption that the ratio of the radio flux to the light flux from M 31 is similar to that observed for our own Galaxy. Examination of the few results in Table 1 suggest that the flux received at 1·89 m from a spiral of type Sb or Sc and of photographic magnitude +10 is about 5×10^{-26} W per m^2 per c.p.s. If this ratio of radio to light flux holds over even a limited range of photographic magnitudes then it is improbable that the radiation from type Sb or Sc nebulae can account for many of the observed sources. There remains the possibility that some other type of extra-galactic nebula or some abnormal state of a nebula may exhibit a far higher ratio of radio flux to light flux, and indeed it may prove that the object discovered in the position of Cygnus (I) represents just such a phenomenon. If this proves to be the case then the identification of the radio sources may prove to be extremely difficult since they may be associated with very faint extra-galactic objects.

It is clear from this brief discussion of the results obtained with the 218 ft radio telescope that many fundamental questions remain to be solved in radio-astronomy. The answers to these questions will only be found after a great deal of further observational data is made available. The present work points the way to what can be expected from even larger and more flexible instruments than the existing paraboloid and it suggests that an increase of resolving power, sensitivity, and field of view would yield a rich reward.

6. A Large Steerable Radio Telescope for Radio Astronomy

It has been mentioned earlier that several fully steerable parabolic reflectors with diameters of 30 ft are in use for radio astronomical investigations. In America a steerable instrument of 50 ft diameter has been built for centimetric observation of solar emissions, and in Holland one of 75 ft diameter is under construction specifically for the investigation of the 21 cm spectral line from interstellar hydrogen. The large fixed radio telescope at Jodrell Bank quickly demonstrated the great potentialities of really large systems and it soon became clear that a completely steerable instrument of this size would be of inestimable value in all aspects of radio astronomy. During 1949 and 1950 the matter was discussed exhaustively and early in 1951 designs were complete for a steerable instrument with an aperture of 250 ft. On account of the steel and financial crisis in Great Britain during 1951 it seemed that the project might have to be abandoned, but early in 1952 the Nuffield Foundation and the Department of Scientific and Industrial Research announced their willingness to share the cost of this enterprise.

The general features of this radio telescope will be evident from the artist's drawing shown in Fig. 11. The paraboloidal surface of 2 in. square mesh is carried on a steel bowl 250 ft in diameter. The depth of the bowl is 62·5 ft so that the focus lies in the aperture plane. This unit, which weighs 300 tons, is carried on steel towers, 180 ft above ground level. These towers rotate on a circular railway track 350 ft in diameter. The engineering difficulties are, of course, considerable. It is necessary to sink 2500 tons of steel and concrete into the ground in order to provide a stable foundation for the railway track, and the superstructure contains a further 1500 tons of steel. The driving system in elevation consists of 100 h.p. metadyne generators working through 28 ft racks and pinions at the top of the steel towers. Similar 100 h.p. generators drive through the bogies on the railway track to give the motion in azimuth. The instrument is controlled electronically and the arrangements are such that almost any form of automatic motion can be obtained. In particular, the radio telescope can be set in automatic sidereal motion with a tracking accuracy of 12 minutes of arc.

In the fundamental design of an instrument of this type the first important parameter to be considered is the focal length of the paraboloid. In the fixed instrument at Jodrell Bank the focal length is long. At the time this was largely determined by the easier mechanical construction involved in a shallow bowl. It has been mentioned that because of the overspill radiation from the primary feed there is a severe loss of

Fig. 11. The 250-ft aperture steerable radio telescope now being built at Jodrell Bank

power gain; on the other hand, the primary feed has a uniform polar diagram over the aperture plane. As the focal length is decreased the overspill is reduced but the uniformity of the polar diagram of the primary feed over the aperture plane decreases, and the beam shape worsens. It can be shown that it is nearly always possible to design a primary feed so that these two effects nearly balance as far as power gain is concerned for changes in focal length. Hence the choice of focal length can be governed by other considerations. In the present case the dominant consideration is that the power received in the overspill radiation must be reduced to the minimum possible. The importance of this in the study of the extra-terrestrial emissions will be obvious, especially in the case where the instrument is used to study regions of low intensity adjacent to intense regions which might be received in the overspill radiation pattern of the primary feed. Other important factors concern the actual design of the primary feed for a given focal length. To achieve the optimum performance the size and complication of the array must increase with increase in focal length. Thus, increases in focal length demand increases in length of the central tower carrying greater loads at the top. This consideration again suggests the desirability of an instrument of short focal length. Criteria such as these explain the ultimate choice of a focal plane design for the radio telescope.

The question of decisive importance in the mechanical design is the lower wavelength limit on which the radio telescope is to operate. It is well known that deviations not greater than $\lambda/8$ from the paraboloidal shape are not serious on the performance of a paraboloidal instrument. For example, departures of 5 in. would not be expected to influence seriously the performance of the radio telescope on a wavelength of 1 m and greater. The specified tolerance has to be maintained in all positions of the bowl, and in designing the stiffness of the structure it is necessary to decide the extent of the departure which can be tolerated with increasing wind speed.

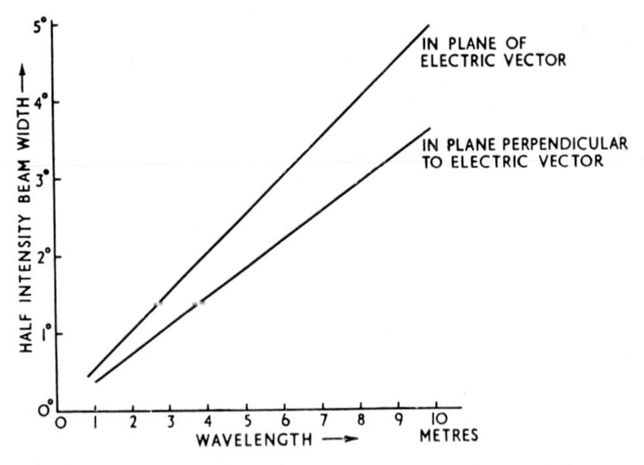

Fig. 12. The calculated relation between the beamwidth and the wavelength for the radio telescope in Fig. 11

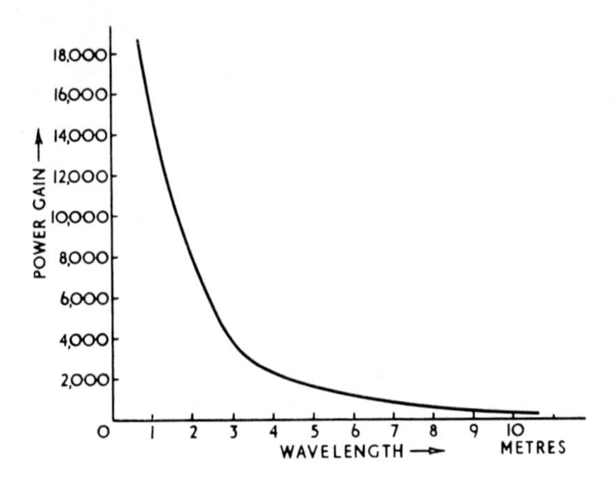

Fig. 13. Calculated relation between the power gain and the wavelength of the radio telescope illustrated in Fig. 11

Wind effects are, in fact, probably the most serious difficulty encountered in the design of this large instrument. These difficulties increase greatly as the wavelength is reduced, firstly because of the increased accuracy required, and secondly because of the need for a closer mesh, which in turn leads to rapidly increasing wind forces. It was necessary to balance these various criteria against tonnage of steel, finance, aperture of paraboloid, and the existing information about the spectrum of the extra-terrestrial radio emissions.

The final decisions were as follows. The shortest wavelength on which the radio telescope would be normally required to operate at full efficiency should be 1 m, involving an accuracy of 5 in. in the figuration of the mesh. At wind speeds of 30 m.p.h. a doubling of this tolerance would be acceptable. For working on this wavelength the surface mesh is to be 2 in. square of 10 gauge wire. With this, the transmission loss is 2·5 per cent at 1 m. It will, of course, be possible to operate the instrument on shorter wavelengths with some sacrifice of efficiency, especially under calm conditions. For the specific study of the 21 cm hydrogen line special arrangements have been made to provide the central portion of the bowl with a $2'' \times 1''$ mesh which will enable the telescope to be used as a 100 ft aperture instrument under these conditions.

The predicted performance of this radio telescope as a function of wavelength is shown in the diagrams of Figs. 12 and 13. Fig. 12 shows the beam width between half-power points assuming a plano-sinusoidal distribution across the aperture. Fig. 13 gives the estimated power gain taking into account the various loss factors such as those due to overspill of the primary feed, and transmission through the mesh and departures of the surface from the true paraboloid as specified under conditions of no wind. At the normal minimum working wavelength of 1 m it will be seen that the beam width is less than $\frac{1}{2}°$ and the power gain about 16,000. The ability to direct such a high gain narrow beam to any part of the sky will clearly be a great asset in radio astronomical investigations. The usefulness is still further increased by the ability to give automatic sidereal motion which will facilitate long period integrations of the power received from particular parts of the sky, thus simulating the long period photographic exposures common in normal astronomical practice.

This contribution has dealt specifically with the use of the radio telescope in the investigation of the extra-terrestrial radio emissions, but the instrument will, of course, be adaptable for use in all aspects of radio astronomy. The radio echo investigation of meteors, the aurorae, the Moon, and perhaps other members of the solar system all fall within its scope. At the time of writing, in January 1953, work is well advanced on the foundations and the 150 deep reinforced concrete piles needed to support the railway track are almost finished. Nevertheless, the major work of construction and the severest engineering difficulties lie ahead. If these are overcome without serious hindrance the radio telescope should be in use by 1955.

References

Allen, C. W. and Gum, C. S.	1950	*Austral. J. Sci. Res.*, A **3**, 224.
Bolton, J. G. and Stanley, G. J.	1948	*Nature (London)*, **161**, 312.
	1949	*Austral. J. Sci. Res.*, A **2**, 139.
Bolton, J. G. and Westfold, K. C.	1950a	*Austral. J. Sci. Res.*, A **3**, 19.
	1950b	*Austral. J. Sci. Res.*, A **3**, 251.
Hanbury Brown, R. and Hazard, C.	1951a	*M.N.*, **111**, 357.
	1951b	*M.N.*, **111**, 576.
	1952a	*Nature (London)*, **170**, 364.
	1952b	*Phil. Mag.*, **43**, 137.
	1953a	*M.N.*, **113**, 109.
	1953b	*M.N.*, **113**, 123.
Hanbury Brown, R., Jennison, R. C. and Das Gupta, M. K.; Smith, F. G.; Mills, B. Y.	1952	*Nature (London)*, **170**, 1061.
Hey, J. S., Philips, J. W. and Parsons, S. J.	1948	*Proc. Roy. Soc.*, A **192**, 425.
Jansky, K.	1932	*Proc. Inst. Rad. Eng.*, **20**, 1920.
Mills, B. Y.	1952	*Austral. J. Sci. Res.*, A **5**, 266.
Piddington, J. H. and Minnett, H. C.	1952	*Austral. J. Sci. Res.*, A **5**, 17.

REBER, G. . . . . 1940 *Proc. Inst. Rad. Eng.*, **28**, 68.
1944 *Ap. J.*, **100**, 279.
1948 *Proc. Inst. Rad. Eng.*, **36**, 1215.
RYLE, M. and SMITH, F. G. . . . . 1948 *Nature* (*London*), **162**, 462.
RYLE, M., SMITH, F. G. and ELSMORE, B. . . . 1950 *M.N.*, **110**, 508.
SILVER, S. and PAO, C. S. . . . . 1944 *Radiation Lab. Report*, 479.
SLATER, J. C. . . . . 1942 *Microwave Transmission.* McGraw-Hill.
SMITH, F. G. . . . . 1951 *Nature* (*London*), **168**, 555.

Positional Radio-astronomy and the Identification of Some of the Radio Stars

F. G. SMITH
Cavendish Laboratory, Cambridge

SUMMARY
A general survey is made of the methods used to determine the accurate positions of the most intense radio stars. The accuracy of the methods is discussed and it is shown that the greatest source of error is due to the "confusion" caused by adjacent radio stars of lower intensity. An account is given of the experimental work which led to the identification of some of the most intense radio stars with certain visible objects. It is found that radio stars are generally associated with visual objects containing diffuse gas exhibiting considerable internal motion. Some of the radio stars are situated within the galaxy and some are found to be extragalactic. The radio source in Cygnus, the most distant radio star yet identified, is believed to represent two galaxies in collision.

1. HISTORY AND METHODS

THE early history of radio-astronomy has followed that of visual astronomy in that it has been largely concerned with the development of methods of finding accurate positions of various heavenly bodies. Since the discovery that there were discrete bodies, which like the Sun, emitted radio waves, there have been remarkable improvements in the various techniques of what may be called "positional radio-astronomy". The recent successes in this field, where the identification of several of these radio stars with visible objects has at last been achieved, make the present time appropriate for a short review of the subject.

When in 1946 HEY, PARSONS, and PHILLIPS (1946) found the radio star in Cygnus, they were unable to determine its direction to better than about $\pm 2°$, since they were using a small aerial array with only limited resolving power. Shortly afterwards, at the Radio-physics Laboratory in Sydney, MCCREADY, PAWSEY, and PAYNE-SCOTT (1947) used an aerial on top of a cliff overlooking the sea to observe radio waves from the disturbed Sun. The resolving power of this aerial is greatly enhanced since it behaves, with its image in the sea, as part of an interferometer. This "Lloyd's Mirror" technique was used by BOLTON and STANLEY (1948), who in 1948 were able to show that the radio star in Cygnus was a discrete source whose position could be accurately determined. Their position was quoted to $2'$ arc in right ascension and $7'$ arc in declination, showing the remarkable improvement obtainable by the use of interferometric techniques.

At Cambridge a different type of interferometer was in use. It had been developed by RYLE and VONBERG (1948) for observations of radio waves from the quiet as well as the disturbed Sun on metre wavelengths, and it was closely analogous to the Michelson interferometer used for the determination of stellar diameters. Two aerials were spaced apart by many wavelengths on an East-West line, and they were connected by radio frequency cables to one receiver. The first observations of radio stars in Cambridge (RYLE and SMITH, 1948) were made by recording the total power from the interferometer as the stars passed through the receptivity pattern of the aerials, but subsequent observations have made use of a more sensitive method of recording (RYLE, 1952) described by RYLE in another chapter of this book (pp. 532–541). It is called the phase-switching interferometer.

The various methods of position finding which have so far been used in radio-astronomy all have parallels in conventional astronomy. The essential difference is that the interferometers used in radio-astronomy have only given a good resolving power in one direction in any one observation. A single observation of the time of transit of a star across the axial plane of an interferometer locates the star on a great circle. A second observation of a transit across the axial plane of another interferometer, or, in some circumstances, an observation of a second transit across the same plane, locates the star on a second great circle and fixes its position.

These great circle observations have been made in various ways. The Lloyd's mirror interferometer records the time when the radio star crosses the horizon. Observations of both rising and setting may be made from the same site if it is fortunately placed, and a complete determination of position may then be made. In Cambridge observations have been made of the upper and lower culminations of a circumpolar radio star, using an interferometer whose axis is considerably inclined to a true East-West line. This again locates the star on the intersection of two great circles. Further similar methods have been suggested (SMITH, 1952a) which involve the use either of two separate interferometers or of one interferometer whose axis can be moved.

Another general principle which has been used is that the angular rate of movement of a star as it crosses the axial plane of an interferometer may be related to its declination. If the axial plane is the meridian the angular rate of movement, which is determined by timing the passage of the radio star through the interference lobes of the receptivity pattern, varies as the secant of the declination. A single observation giving both the time of transit and the rate of movement may then suffice to locate the star, although the accuracy in declination is in practice usually less than the accuracy in right ascension.

The calibration of an interferometer which is to measure angular rate of movement usually depends on measurements of the spacing between the aerials and of the effective frequency of the receiver. The measurement of frequency may be difficult when broad-band receivers are used, but is possible to replace it by a measurement of the electrical length of a section of transmission line. This length of line is used to increase the transmission path from one aerial to the receiver, thereby displacing the collimation plane of the interferometer away from the axial plane. The collimation plane then cuts the celestial sphere in a small circle; when the extra path is placed in the other aerial, this circle occurs on the opposite side of the meridian. The interval between the times of transit across these two small circles is related to the angular rate of movement of the star, and hence to its declination.

2. Experimental Accuracy

The accuracy of these various positional measurements may be limited either by the sensitivity of the receiving equipment or by various external factors; the receiving techniques will not be discussed in detail here, but we shall consider mainly the effects of external factors such as refraction in the atmosphere and ionosphere.

It is possible, using an interferometer, to determine the position of a radio star with an error considerably less than $\lambda/2d$, where λ is the wavelength and d the spacing of the aerials. The method depends upon the measurement of the times of cross-over of the sinusoidal record obtained with a phase-switching receiver, and with an intense radio star a reading accuracy corresponding to an error of less than 1 per cent of $\lambda/2d$ may be obtained. With radio stars of smaller intensity the accuracy may be limited by the presence of irregular fluctuations on the record caused by the noise inherent in the receiver itself, or by the background radiation from the Galaxy.

The extremely good reading accuracy which may be obtained in observation of intense radio stars cannot in practice be fully utilized, since the calibration of the instrument presents special difficulties. The errors of conventional transit telescopes have all to be taken into account: height error, azimuth error, and collimation error. Determination of the axis of the interferometer can only be done to the accuracy with which the electrical centres of the two aerials may be defined in position; for example, the position of the aerials of the 3·7 m interferometer in Cambridge could be defined only to about 2–3 in., whereas the direction of the baseline of another interferometer (on 1·4 m wavelength) which used parabolic reflector aerials could be defined to about ½ in. in 1000 ft spacing, *i.e.* to about 10″ arc. Collimation error results from any differences between the electrical characteristics of the aerials and their associated preamplifiers and transmission lines, since these differences may lead to a change in the phase relation between the two signals. The error may be determined either by electrical measurements or by making observations of both upper and lower culminations of a circumpolar star.

3. External Sources of Error

Further sources of error lie outside the apparatus itself. Atmospheric refraction, of approximately the same magnitude for radio as for visual astronomy, may be considerably less important than refraction in the terrestrial ionosphere and, particularly on the longer wavelengths, ionospheric refraction may constitute the chief limitation to the accuracy of positional measurements.

Ionospheric refraction may cause an apparent displacement of the radio star towards the zenith due to the curvature of the ionospheric layers, or it may cause a displacement in other directions if the total ionization of the ionosphere varies in a horizontal direction. This latter effect, known as "wedge refraction", has been studied recently at Cambridge (Smith, 1952b).

Radio waves from stars may also suffer a random diffraction at irregularities in the ionosphere, and the scintillation which then occurs may seriously limit the reading accuracy of positional observations. Scintillation is the subject of another chapter of this book by Hewish (p. 599).

The most important limitation at the present time is not concerned with any of these difficulties of measurement or with ionospheric refraction. In the foregoing discussion it has been assumed that the signals received by the aerials originate entirely in the radio star under investigation. The addition of a signal from a uniform background would affect none of the arguments, and it would appear therefore that the galactic radiation need not be taken into account. But this radiation is at least in part made up of radiation from other weaker radio stars, and each of these must produce its own trace on the record. The amplitude of the trace resulting from the addition of all these sources may be estimated if their statistical distribution in intensity is known (SMITH, 1952a).

At the longer wavelengths this effect of "confusion" is so great that it may become impossible to distinguish between the various traces made by the weaker radio stars, and only the most intense can be located with an accuracy which is not entirely limited by "confusion error". This is evident in the surveys of radio stars made by RYLE, SMITH, and ELSMORE (1950), and by MILLS (1952a). Although it would be possible to reduce this error by making independent observations with the interferometer aerials at different spacings or in different directions, the extension of accurate position finding to radio stars of lower intensity appears to demand an increase in aerial gain. The confusion error for a star of given intensity may be expressed as an area in which the star probably lies; this area is inversely proportional to the gain of the interferometer aerials.

4. POSITIONAL OBSERVATIONS AT CAMBRIDGE

In this section some account will be given of the positional work on the more intense radio stars which has led to their identification with certain visible objects.

The first tentative suggestions about the identities of some of the radio stars were made by BOLTON, STANLEY, and SLEE (1949). Three stars, in Taurus, Virgo, and Centaurus, were located with the use of the Lloyd's mirror technique to an accuracy of about 50 square minutes of arc. Each position contained unusual visible objects: the Crab nebula (NGC 1952), and the nebulae NGC 4486 and NGC 5128 respectively. The positions of other radio stars showed no clear relation to visible objects. This observation indicated two lines of attack. Either a large number of radio stars should be located to this order of accuracy, when a general correspondence with a particular type of nebula might become apparent; or, alternatively, the positions of a few radio stars of high intensity, including these three, should be redetermined with the greatest possible accuracy so that some individual identifications might be firmly established.

In Cambridge, two radio stars, in Cygnus and Cassiopeia, had been observed with an interferometer receiving at a wavelength of 3·7 m. The positions were again not sufficiently accurate for a detailed search to be made amongst the faintest of the visible objects in the region, but it became clear that no outstanding visible objects coincided with these two radio stars. The first attempt at the problem was therefore to examine as many radio stars as could be detected, and to search in their positions for various classes of objects. This survey of the northern hemisphere (RYLE, SMITH, and ELSMORE, 1950) was carried out at the same wavelength of 3·7 m, and fifty radio stars were located with varying degrees of accuracy. The positions obtained for the stars in Taurus and Virgo tended to confirm the suggested identifications, but the analysis of the remaining stars showed only an expected correlation of four of the weakest with some of the nearest extra-galactic nebulae (M 31, M 33, M 51, M 101).

An interferometer was then constructed at Cambridge which could measure the positions of the four most intense radio stars in the northern hemisphere with the greatest accuracy that has yet been attained. A shorter wavelength (1·4 m) was used in order to reduce the effects of ionospheric refraction; since the flux density from a radio star is roughly proportional to wavelength, and since it becomes increasingly difficult to construct large aerial apertures as the wavelength is reduced, the signal-to-noise ratio for the weaker two of these stars was poor. The aerial gain was comparatively high, while it remained possible to determine their electrical centres accurately; two "Würzburg" aerials (paraboloids having a diameter of 8·5 m) were used, mounted at a fixed azimuth with their bearings aligned along the interferometer axis (Fig. 1).

Fig. 1. The parabolic aerial system used by SMITH to determine the position of radio stars on a wavelength of 1·4 m. The distant section of the interferometer may be seen in the background

The collimation error of this interferometer was determined in two different ways. It was possible to compare the length of the transmission line from the two aerials by direct electrical measurements; the parabolic reflectors were identical, and the dipoles at the foci could be interchanged together with the transmission line transformers associated with them. Alternatively it was possible to observe both the upper and the lower culminations of the radio star in Cassiopeia; the mean of these times gives a value of the right ascension of the star which is unaffected by collimation error. When the right ascension is known, it is possible to find the collimation error in any other transit observations.

Declination was measured by the use of the "displaced collimation plane" technique, in which an extra transmission path of about twenty wavelengths was placed in each side of the interferometer alternately. Observations were here made over periods of about $\frac{1}{2}$ hr, and the angular rate of movement of the four stars was measures to one part in 10^4.

The final values of the positions of the four radio stars (SMITH, 1951a), determined largely by the use of this interferometer, are given in the table. The system of numbering the stars is that which is in use in Cambridge (RYLE, SMITH, and ELSMORE, 1950).

Star	Constellation	Right ascension	Declination
		h m s	
05·01	Taurus	05 31 34·5 ± 3^s	22° 04·0′ ± 5′
12·01	Virgo	12 28 18·0 ± 3	12° 37·0′ ± 10′
19·01	Cygnus	19 57 45·3 ± 1	40° 35·0′ ± 1′
23·01	Cassiopeia	23 21 12·0 ± 1	58° 32·1′ ± 0·7′

By continuing observations over a year, and by searching for changes in the relative positions of the four stars, it was possible to put a very low upper limit on their parallaxes. It was shown that their distances were probably greater than $\frac{1}{2}$ parsec (SMITH, 1951b).

5. THE IDENTIFICATION OF THE RADIO STARS

These measurements gave strong confirmation to the tentative identifications of the radio stars in Virgo and Taurus with the nebulae NGC 4486 (M 87) and NGC 1952 (M 1, the Crab). The two most intense radio stars, in Cygnus and in Cassiopeia, remained at first unidentified.

The first successful search of the region near the Cassiopeia radio star was made by DEWHIRST at the Cambridge Observatories (DEWHIRST, 1951). He found a diffuse nebula near the position of the radio star, but unfortunately outside the limits of accuracy of this position. Soon afterwards BAADE and MINKOWSKI (1954), of Mount Wilson and Palomar Observatories, using the 200″ Hooke reflector, found that a roughly circular region about 5′ arc across and centred on the radio position contained many filamentary patches of nebulosity; the brightest of these was clearly the one found by DEWHIRST, the extent of the region being considerably greater than the limits of accuracy of the radio position (Fig. 2). This nebula is unique. The spectrum shows various lines of high excitation, and these lines show Doppler shifts corresponding to velocities up to 3000 km per sec. The velocity varies at different points in the same filament.

BAADE and MINKOWSKI also examined the spectrum of a nebula in the position of the Cygnus radio star; the nebula had previously not been considered to be of importance because of its extreme faintness (about 15th magnitude). Strong emission lines were found, with large red shifts corresponding to those from a very distant extragalactic nebula. They consider this nebula to be two galaxies in collision. The position has also been determined by MILLS (1952b), whose result agreed well with the Cambridge position.

These two different bodies, and others whose identifications are now fairly certain, appear to exhibit one common characteristic: they contain diffuse gas with large internal velocities. Although no satisfactory explanation has yet been given of the emission of intense radio waves from such bodies, the search for the mechanism of this radiation will be greatly aided by the new information on the physical characteristics of radio stars.

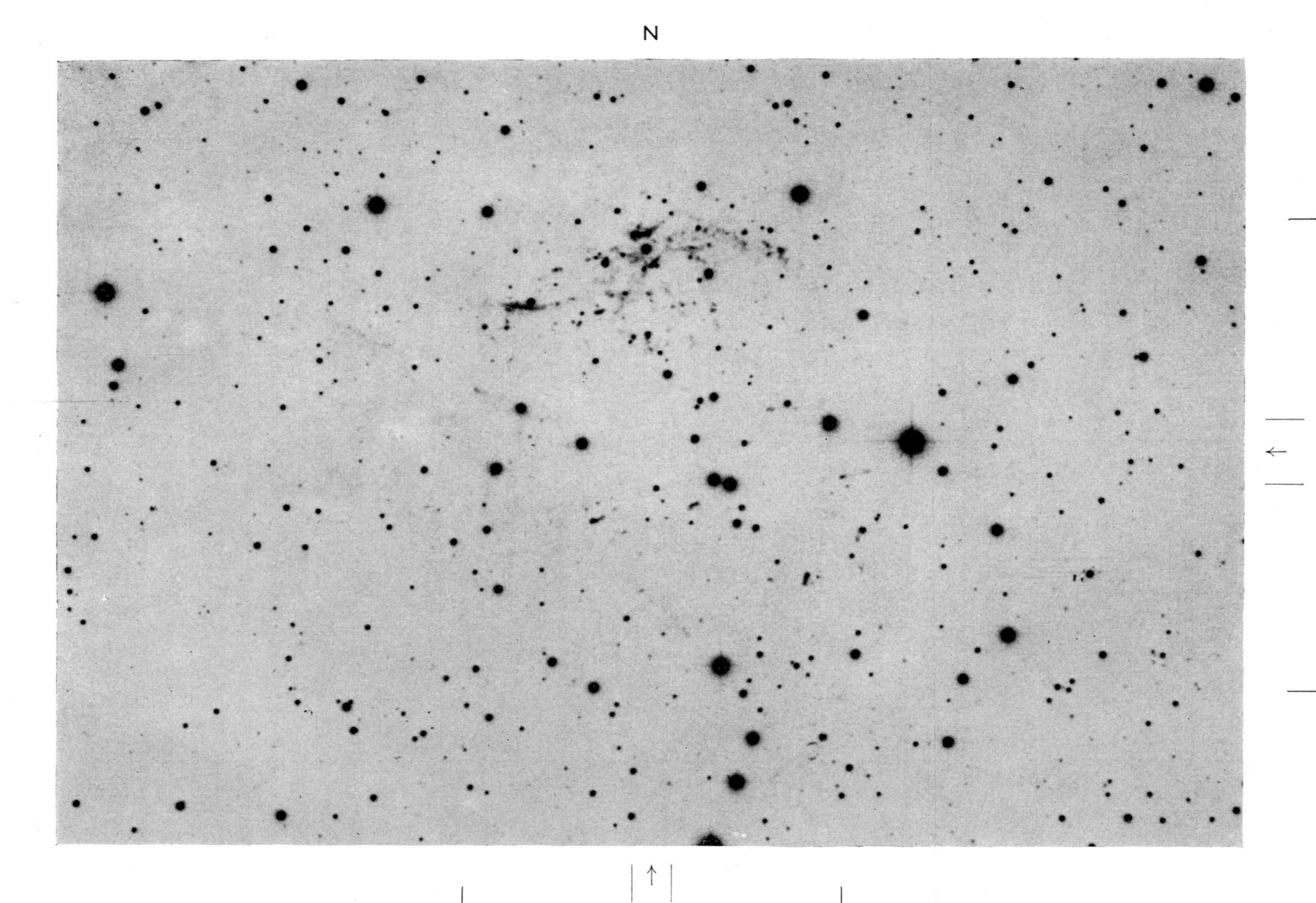

Fig. 2. Radio source in Cassiopeia. Photograph in red light, taken by W. BAADE with the 200-in. Hale Telescope at Mount Palomar in 1951 (scale: 1 mm = 3″.5); north on top. Short fragments of nebulosity cover an approximately circular area of 5′ diameter, and their brighter condensations are most easily seen towards the top of the plate (*i.e.* the northern limit). The position of the radio star is shown by arrows; the parallel strokes indicate the limits of error. The brackets give an indication of the extent of the visual nebulosity; the radio emission

6. THE ANGULAR DIAMETERS OF THE RADIO STARS

The visible objects so far associated with the radio stars are all nebulae with angular diameters of several minutes of arc. There are clear reasons for attempting to measure the angular extent of the sources of radio emission: the results might provide confirmation of the identifications, and secondly, with the measured total intensities, the results would allow apparent brightness temperatures to be assigned to the emitting regions.

It has happened that three separate investigations of the angular extent of radio stars have been made at about the same time (MILLS, 1952c; HANBURY BROWN, JENNISON, and DAS GUPTA, 1952; SMITH, 1952c), in Sydney (Radiophysics Laboratory), in Manchester (Jodrell Bank), and in Cambridge. Interferometric methods were used, and in some experiments aerial spacings were as great as 4000 wavelengths. In Cambridge (SMITH, 1952d) the spacings were all less than 400λ; new methods of observation were devised for this work since observations at small spacings require to be made with great accuracy.

The results of these series of experiments have provided good confirmation for the identifications of the the radio stars in Cygnus and Cassiopeia, since the radio stars appear to have diameters of the order of $3'$ and $5'$ respectively, closely corresponding to those of the nebulae. The Manchester observations show that the Cygnus radio star is considerably asymmetric.

7. CONCLUSIONS

The measurements of the positions of radio stars have reached a turning point. The fundamentals of positional radio-astronomy are now well understood, and their application has already resulted in the identification of several of the more intense radio stars. Since, however, there are apparently several very different types of body represented in these identifications, it still appears as urgent as before to continue the attempts at identification. The experimental problems now become considerably different.

Future positional determinations have the advantage that reference positions are now available for radio stars distributed widely over the sky. Absolute calibrations of interferometers are difficult, but accurate comparisons of the positions of radio stars may often be made simply. The outstanding problem remaining is that of "confusion". Since further identifications have necessarily to be made with weaker radio stars this problem is increasingly important; it can only be countered by the use of aerials with higher gain.

It is with these two points in mind that the large interferometer aerial system recently completed at Cambridge has been designed. The gain is considerably greater than that of previous interferometers, and it is hoped that it will become possible with its use to determine the positions of many radio stars with sufficient accuracy to allow many definite identifications with visible bodies to be made.

REFERENCES

BAADE, W. and MINKOWSKI, R. . . . . 1954 *Ap. J.*, **119**, 215.
BOLTON, J. G. and STANLEY, G. J. . . . . . . 1948 *Nature* (*London*), **161**, 312.
BOLTON, J. G., STANLEY, G. J. and SLEE, O. B. . 1949 *Nature* (*London*), **164**, 101.
DEWHIRST, D. W. . . . . 1951 *Observatory*, **71**, 211.
HANBURY BROWN, R., JENNISON, R. C. and DAS GUPTA, M. K. . . . . 1952 *Nature* (*London*), **170**, 1060.

Hey, J. S., Parsons, S. J. and Phillips, J. W. . 1946 *Nature* (*London*), **158**, 234.
McCready, L. L., Pawsey, J. L. and Payne-Scott, Ruby . . . . 1947 *Proc. Roy. Soc.* A., **190**, 357.
Mills, B. Y. . . . . 1952a *Austral. J. sci. Res.* A., **5**, 266.
1952b *Ibid.*, **4**, 456.
1952c *Nature* (*London*), **170**, 1063.
Ryle, M. and Smith, F. G. . . . . 1948 *Nature* (*London*), **161**, 312.
Ryle, M. and Vonberg, D. D. . . . . 1948 *Proc. Roy. Soc.* A., **193**, 98.
Ryle, M., Smith, F. G. and Elsmore, B. . . . . 1950 *M.N.*, **110**, 508.
Ryle, M. . . . . 1952 *Proc. Roy. Soc.* A., **211**, 351.
Smith, F. G. . . . . 1951a *Nature* (*London*), **168**, 555.
1951b *Ibid.*, **168**, 962.
1952a *M.N.*, **112**, 497.
1952b *J. atmosph. terrest. Phys.*, **2**, 350.
1952c *Nature* (*London*), **170**, 1065.
1952d *Proc. Phys. Soc.* B., **65**, 971.

Australian Work on Radio Stars

J. G. Bolton

Radiophysics Laboratory, Commonwealth Scientific and Industrial Research Organization, Sydney

Summary

The highlights of the Australian investigations on radio stars are briefly reviewed. These include the initial discovery of the first few of these new astronomical objects, determination of their positions, and measurement of their angular diameters. Four of the radio stars have been identified with visible nebulae and an examination of their distribution over the celestial sphere has revealed the existence of two distinct classes. The stars of one class, which are strongly concentrated towards the galactic equator, are presumably galactic objects at relatively great distances. There is insufficient evidence to suggest whether the stars of the other class, which have an isotropic distribution, are galactic or extragalactic. The Australian contributions to the study of the scintillations or "twinkling" of the radio stars are briefly described.

1. Discovery of the Radio Stars or Discrete Sources

The first observations of a radio star in Australia were made in June, 1947. The writer and his associates were following up the pioneer discovery by Hey, Parsons, and Phillips (1946) of short-period fluctuations in the intensity of the galactic noise from a small region in the constellation of Cygnus. Hey, Parsons, and Phillips were able to set an upper limit of two degrees on the size of the region showing fluctuations and pointed out the similarity between these fluctuations and the short-period variations in the intensity of radiation from sunspots.

In order to investigate the region of this discrete source the writer used an interferometer consisting of an aerial situated on a high cliff and its reflection in the sea (Bolton and Stanley, 1948). This technique had previously been used by McCready, Pawsey, and Payne-Scott (1947) in their study of solar radiation on radio frequencies. From the visibility of the interference fringes that were obtained as the source rose above the sea horizon, an upper limit of eight minutes of arc could be placed on the angular width of the source in a direction at right angles to the line of the horizon. The next few months were spent in a study of the fluctuations, or scintillations as they are now called, and in attempts to determine the position of the source. The search for further sources was begun in October, without success in the first two months, but by January, 1948, six had been discovered.

In these observations an aerial of only 4 m^2 effective area was available and detection of the sources was achieved by the use of an extremely sensitive receiver. By maintaining supply voltages constant to within one part in several thousand, receiver output fluctuations were reduced to about one part in a thousand of the total output and thus signals of the same order could be detected.

Early attempts to determine the positions of the radio stars depended on the observation of a small part of the track of the source after it had risen above the horizon. The measurements were severely affected by atmospheric refraction, and failure to apply true corrections for this resulted in the initial positions given by the writer (1948), having errors of several degrees. In order to overcome the difficulty of refraction at low altitudes an expedition was made to New Zealand in June, 1948, with the intention of making observations of the sources both at their rising and setting. With the two sets of observations it is possible to determine both the refraction correction and the position of a source. For a description of the method, see Bolton and Stanley (1949).

The ten weeks of observations from New Zealand were extremely successful in a number of ways. Simultaneous observations of the source in Cygnus from Sydney and New Zealand showed that there was no correlation between the scintillations at two sites separated by 2000 miles and suggested that the scintillations were an atmospheric effect. The results were communicated to workers in Cambridge, who, in conjunction with those at Manchester, confirmed the results in observations made over shorter baselines; Little and Lovell (1950), and Smith (1950).

The New Zealand observations were made from heights of about 300 m or nearly four times the height employed in Sydney. With this improved resolving power the upper limits on angular sizes of the sources were reduced to $1\frac{1}{2}$ minutes of arc in the case of Cygnus and about 7 minutes of arc for three other sources. It is interesting to note that these values are just above the actual sizes determined recently by other observers, although in the case of the Taurus source there was some suspicion of a positive result from the New Zealand observations.

2. Identification of Three Sources with Nebulae

Three out of four of the position measurements resulted in tentative identifications of discrete sources with visible nebulous objects (Bolton, Stanley, and Slee, 1949). The sources were estimated to be within circles of approximately $\frac{1}{4}°$ in diameter and charts covering these regions were searched for unusual objects. In one was the Crab nebula, the remnants of a supernova explosion of 1054 A.D., in another the extragalactic nebula M 87 and in the third the extragalactic nebula N.G.C. 5128. Although the Crab nebula, with its almost unique features of high electron temperature, large velocity of expansion and suggestions of turbulence, immediately appealed to astronomers as a potential source of high radio-frequency emission, the other objects had unusual features which reinforced the identifications. Recent determinations of the positions of these sources (Taurus-A, Virgo-A, and Centaurus-A) by Mills (1952b) and Smith (1951), and the measurement of their angular diameters by Mills (1952c) have put the identifications beyond all doubt. The observations on the Cygnus source were not so successful, the position being some 30 seconds of right ascension (5 minutes of arc at the declination of the source) away from Mills' later result. The reason for this error has never been understood, but it may have been due to a diurnal variation in refraction.

3. Scintillation of the Sources

Observations on the discrete sources were continued during 1949 and by the end of that year about thirty of them had been discovered (Stanley and Slee, 1950). Daily observations on the Cygnus source showed that there was a seasonal variation in the incidence of the scintillations—further confirmation of their atmospheric origin. This seasonal variation showed maxima in midsummer and midwinter and minima at the equinoxes. It was subject to two possible interpretations, either showing a true seasonal effect or a diurnal effect as the local time of observations of the source changed through 24 hr during the year. Observations were made on a second source (Virgo) and these confirmed the former interpretation. From an analysis of five years' data on scintillations the writer has recently concluded that scintillations at very low angles are principally due to sporadic E. However, Mills, from a series of observations of Cygnus made on winter nights of 1949, found a strong correlation between the scintillations of Cygnus at an altitude of 15° and irregularities in the F layer. His result is in agreement with those of Ryle and Hewish (1950), and Little and Maxwell (1951) from observations at near vertical incidence in the northern hemisphere.

4. Identification of the Cygnus Source with Colliding Galaxies

In 1949 Mills and his associates began a long series of measurements on the position of Cygnus. They used an interferometer similar to that used in the Cambridge observations consisting of two aerials spaced along an east-west baseline. Although measurements with such an instrument are not subject to such large refraction corrections, much greater care is necessary in determining the physical constants of the system. Mills and Thomas (1951) list eight different sources of error in the determination of right ascension and seven in the determination of declination. Included in these was the effect of unequal ground reflections at the two aerials. This was particularly important in the Cygnus observations due to the low altitude (15°) of the source at transit and the wide acceptance cone of their aerials. It was the largest single source of error in their determination of the right ascension of the source. Within the probable error rectangle of their position were many faint stars and one extragalactic nebula. (This nebula is on the limit of the probable error rectangle, but even small departures from a probable error rectangle have little significance.) At first the nebula did not greatly attract the attention of astronomers. However, when Smith (1951) obtained a similar position with a smaller estimated probable error, Baade and Minkowski made a close examination of it with the 200-in. telescope. They found that what had appeared to be one galaxy was actually two galaxies in collision. Its spectrum showed emission lines of unusually high excitation, indicating very high gas temperatures.

Mills' initial observations were also made with very small aerial units. In 1950 he constructed three much larger units and started observations using aerial spacings of approximately thirty and ninety wavelengths. The smaller aerial spacing was mainly for use in the determination of the right ascension of a source. Mills (1952b) made accurate measurements of the positions of six sources, including a redetermination of the position of the Cygnus source. The narrower acceptance cones of his new aerials and a more favourable site removed one of the principal sources of error in his previous measurement. His final position is less than 1 minute of arc from the colliding galaxies. He obtained positions of the sources in Taurus, Virgo, and Centaurus which confirmed the writer's suggested identifications for these objects and also of the sources Hydra-A and Fornax-A. These last two have not so far been identified with visible objects.

5. THE DISTRIBUTION OF THE SOURCES

MILLS used his new aerial system to make a survey for sources between declinations $+ 50°$ and $- 90°$, finding evidence for seventy-seven sources of flux density greater than $0{\cdot}5 \times 10^{-24}$ W-m^{-2}-(c/s)$^{-1}$. In making estimates of the approximate positions and flux densities of many of these sources, MILLS was handicapped by the presence of complex interference patterns due to several sources being in the primary acceptance cone of his aerials at one time. As MILLS (1952a) pointed out, some of the information given in his list of sources may be subject to quite large errors because of this effect, but it is sufficiently accurate for the principal object of his survey. This was to make a statistical study of the material. From his results MILLS concluded that there are two classes of source: (I) in which the sources lie close to the galactic plane and (II) in which the sources are more or less uniformly distributed. The sources of Class I have higher flux densities than those of Class II and are assumed to be powerful emitters thinly distributed through the galaxy and at great distances. For the sources of Class II two possibilities exist: either they are extremely powerful emitters at the distances of extragalactic nebulae or relatively feeble emitters fairly close to the Sun—that is, on a galactic scale. A decision on which of these is correct will probably have to wait until a large number of this class has been identified with visible objects, although if the second possibility were correct galactic shape would eventually show in surveys of greater depth.

6. ANGULAR DIAMETERS OF THE SOURCES

Recent work on the sources has mainly been concerned with the measurement of their angular sizes. The first indication of the angular size of any of the sources was given by MILLS' observations with his interferometers of thirty and ninety wavelengths separation. MILLS (1952a) found that three sources had equivalent angular widths of about $\frac{1}{2}°$. However, his measurements, and those with the sea interferometer with an equivalent aerial separation of 180 wavelengths (STANLEY and SLEE, 1950) only set upper limits on the angular sizes of the three sources Cygnus, Virgo, and Taurus. The next step, involving distances of 1–5 km, meant major additions to measuring equipment. At these distances cable connections between the two aerials of the system become expensive and inconvenient. MILLS solved the problem by employing a radio link between a small portable aerial and receiver and the base receiving site. Two transmissions were actually used, as the local oscillator frequency had to be transmitted in order to preserve the phase of the reconstituted signal at the base site. Compensation for the time delay of the radio link was introduced into the base receiver to equalize the propagation times between the two aerials and the mixing point. This time delay was obtained with a mercury-filled acoustic delay line.

MILLS obtained his first positive results on Cygnus with the radio link in July, 1952, shortly before the biennial conference of the International Scientific Radio Union in Sydney. By an amazing coincidence SMITH, of Cambridge, and HANBURY

BROWN, of Manchester, had obtained similar results at the same time. Each observer had made measurements with different aerial spacings, and from a comparison of the results at the conference it became clear that the brightness distribution across the source was quite complex. MILLS, in a preliminary report, gives effective angular sizes for four sources along an east-west line as Cygnus 1′·1, Virgo 5′, Taurus 4′, and Centaurus 6′. These investigations are still in progress but MILLS has recently informed the writer that he suspects that the Virgo and Centaurus sources are considerably elongated. The Taurus source may also be slightly elongated.

At about the time that MILLS began his measurements, the writer found a method of minimizing one of the difficulties in the determination of angular widths by the use of the sea interferometer. This difficulty is associated with the incomplete reflection from the curved surface of the Earth at near grazing incidence. The equipment consisted of two aerials spaced along a cliff edge giving two interference systems mutually at right angles. It was hoped with this system to find whether the Taurus source was greater or less than 3 minutes of arc. The experiment was never carried out, as observations using small aerial spacings, intended to test the equipment, resulted in the discovery of a number of objects with angular widths of more than a degree. These objects had not been previously found, as most interferometers had been operated with their fringe spacings at less than a degree. Some of them have such large angular widths and high total brightness that they must be due to fine structure of the general background of galactic noise and possibly indicate structural features of our galaxy. The others can only be classed as "miscellaneous". One of them appears as a disk of diameter 2° concentric with the smaller source Centaurus-A. MILLS' earlier estimate of 25′ for this object was an effective angular width based on measurements with only two aerial spacings. Another interesting object at $08^h\ 20^m$, $-\ 42\frac{1}{2}°$ has a diameter of about a degree but with some degree of central concentration. In the position of this source MINKOWSKI (private communication) finds an object somewhat similar in appearance to the Cassiopeia source. It consists of a network of filaments, quite different from the emission nebulosities which are abundant in this region. The filaments show no sign of organized motion, but radial velocity measurements show a spread of from + 120 km per sec. to − 30 km per sec.

7. SUMMARY

The main Australian contributions to the study of the discrete sources may be summed up as follows:

(1) The confirmation of HEY's suggestion of a discrete source in Cygnus and the initial evidence for a class of these objects.

(2) The tentative identifications of three of the sources with visible objects.

(3) The statistical work on the distribution of the sources, showing that there are two distinct types.

(4) The first evidence for a finite size (of the order of a degree) for some of the fainter sources.

(5) The measurement of the angular diameters of the sources Cygnus, Virgo, Taurus, and Centaurus.

(6) The accurate determination of the positions of the same sources.

Independent measurements of the angular diameter of the Cygnus source and position determinations of the sources Cygnus, Virgo, and Taurus were made at about the same time in England. In addition, the spaced aerial observations between

Australia and New Zealand pointed the way for the subsequent observations in England which proved that the amplitude fluctuations or scintillations of the sources were of ionospheric origin.

References

Bolton, J. G.	1948	*Nature (London)*, **162**, 141.
	1952	Reported at the General Assembly of the International Scientific Radio Union, U.R.S.I., Sydney.
Bolton, J. G. and Stanley, G. J.	1948	*Nature (London)*, **162**, 462.
	1949	*Austral. J. sci. Res.* A., **2**, 139.
Bolton, J. G., Stanley, G. J. and Slee, O. B.	1949	*Nature (London)*, **164**, 101.
Hey, J. S., Parsons, S. J. and Phillips, J. W.	1946	*Nature (London)*, **158**, 234.
Little, C. G. and Lovell, A. C. B.	1950	*Nature (London)*, **165**, 423.
Little, C. G. and Maxwell, A. B.	1951	*Phil. Mag. xlii*, 267.
McCready, L. L., Pawsey, J. L. and Payne-Scott, Ruby	1947	*Proc. Roy. Soc.* A, **190**, 357.
Mills, B. Y.	1952a	*Austral. J. sci. Res.* A, **5**, 266.
	1952b	*Austral. J. sci. Res.* A, **5**, 456.
	1952c	*Nature (London)*, **170**, 1063.
Mills, B. Y. and Thomas, A. B.	1951	*Austral. J. sci. Res.* A, **4**, 158.
Ryle, M. and Hewish, A.	1950	*M.N.*, **110**, 381.
Smith, F. G.	1950	*Nature (London)*, **165**, 422.
	1951	*Nature (London)*, **168**, 555.
Stanley, G. J. and Slee, O. B.	1950	*Austral. J. sci. Res.* A, **3**, 234.

Outbursts of Radio Noise from the Sun

J. P. Wild

Division of Radiophysics, Commonwealth Scientific and Industrial Research Organization, Sydney, Australia

Summary

The present state of knowledge of the occasional large increases ("outbursts") of radio-frequency radiation from the Sun is briefly reviewed. Outbursts have been detected on frequencies ranging from 9–35,000 Mc/s. They last for several minutes and show a big correlation with solar flares. Recent investigations of their spectrum and position on the solar disk have indicated an origin in corpuscular streams ejected from flares. Current evidence suggests that the intense radio waves are excited at harmonics of the plasma frequency during the passage of the streams through the corona.

---

1. Introduction

The discovery during the war years that the Sun radiates electromagnetic radiation of extraordinary intensity in part of the radio spectrum has opened a new and important field of research in solar physics. The radiation originates entirely in the ionized atmosphere of the Sun, a region almost transparent to light. In principle the outer layers of the Sun can be studied by radio methods with the same directness that the photosphere can be studied by optical methods. It is already becoming evident that radio observations are revealing features and occurrences never observed optically. A striking example is provided by the class of occurrence that forms the subject of this review.

The characteristics of the radio-frequency radiation ("noise") from the Sun are complex. In the absence of disturbed conditions on the Sun, the intensity of the noise remains at a steady level corresponding to the thermal radiation emitted by the solar atmosphere. But when sunspots are visible on the Sun's disk the level may become enhanced and show rapid fluctuations above the basic thermal level. At times sudden increases ("bursts"), lasting some seconds or minutes, may increase the level a thousandfold.

Many of the larger bursts occur at the time of solar flares. Indeed, the converse is now recognized, that large flares are usually accompanied by a particular type of radio burst. Because of their violent effect upon the received level, these flare-accompanying bursts are called *outbursts*. A great deal of the post-war work in radio astronomy has been devoted to their study. The purpose of the present paper is to give a brief review of this work and describe in more detail the results and interpretation of some special experiments on outbursts that have been carried out in Sydney during the last four years.

2. The Association of Solar Radio Noise and Flares

The association of intense solar radio noise with flares was recognized immediately upon the discovery of the former. Appleton (1945) reported that the first hint of excess radio noise from the Sun came from amateur experience in the 10–40 Mc/s band and that the noise "was often the precursor of a catastrophic fade-out associated with a bright eruption on the Sun". The existence of the high-level noise became established when the direction of interference received on numerous British Army radar sets during February, 1942, was traced to the Sun. In reporting this pioneering observation, Hey (1946) concluded that the unusual intensity appeared to be associated with a big solar flare. Early in 1946 Appleton and Hey (1946) recorded six cases of a coincidence between an increase in radio noise and a solar flare or radio fade-out. "We may therefore take as a definite result", they concluded, "that when a solar flare occurs which is accompanied by a radio fade-out there is also in many cases a marked enhancement of solar noise".

Since the early explorations several systematic studies have been made on the relation between solar noise and flares. The first to be published was by Allen (1947). He found no close correlation between short-period 200 Mc/s noise and optical phenomena but that at the time of flares "sudden outbursts of radio noise" could sometimes be detected. Such outbursts would last for a few minutes, fluctuating violently, and then disappear. He could not recognize those that accompanied flares with certainty from the record, but provided the first hint that they may be a special class of disturbance. Allen's descriptive word "outbursts" has now acquired general acceptance.

The statistical correlation between bursts and flares has been investigated by Hey, Parsons, and Phillips (1948) at 70 Mc/s and more recently by Dodson, Hedeman, and Owren (1953) at 200 Mc/s. The former found that the larger flares were the more likely to be accompanied by a burst (50 per cent of class 2 and 3 flares, 32 per cent of class 1); the latter found a generally higher correlation (74 and 79 per cent respectively).

Because bursts show a marked correlation with solar flares, it is natural to expect also an association with the numerous geophysical effects that accompany flares. Apart from short-wave radio fade-outs (Hey, Parsons, and Phillips, 1948), instances have been reported of long-wave enhancements of atmospherics (Laffineur and

Servajean, 1949), D-layer absorption of galactic noise and geomagnetic crochets (Hey, Parsons, and Phillips, 1948), occurring within a few minutes of the start of the outbursts. These effects, due to ultra-violet emission from the flare, usually start a few minutes after the flare onset and a few minutes before the onset of the outburst at 70 Mc/s (Hey, Parsons, and Phillips, 1948). However, Covington (1951) found that at 3000 Mc/s flare and outburst start simultaneously. Cases have also been reported by Payne-Scott, Yabsley, and Bolton (1947), and Wild (1950), of outbursts being followed by great magnetic storms and aurorae (the effects due to corpuscular emission from the Sun at the time of flares) after the elapse of the characteristic delay time of one to two days. Finally, there have been a few instances of an association with world-wide changes in cosmic ray intensity (Edlén, 1946; Hatanaka, Sekido, Miyazaki, and Wada, 1951). Except in the case of magnetic crochets, which tend to accompany only the larger flares and show a very high correlation with bursts (Hey, Parsons, and Phillips, 1948), there is no evidence to suggest that outbursts have a closer connection with any one of the accompaniments of flares than with flares themselves.

3. Characteristics of Outbursts

The majority of observations of solar radio noise are made with simple "radio telescopes" consisting of a directive aerial connected to a radio receiver tuned to a fixed frequency. The receiver output is recorded on a time chart. Although directive, the aerial beam is much wider than the Sun's disk so that the instrument measures the integrated radiation received at the Earth from the whole Sun. The quantity measured is the flux density; its unit in the M.K.S. system is the watt per square meter per cycle-per-second bandwidth.

The early observations of Appleton and Hey were made in the range 20–200 Mc/s; their attempts to detect disturbances at higher frequencies were unsuccessful. The upper frequency of detection of outbursts was subsequently increased to 480 Mc/s by Reber (1946), to 1200 Mc/s by Lehany and Yabsley (1948), to 9500 Mc/s by Schulkin, Haddock, Decker, Mayar, Hagen (1948), and finally to 35,000 Mc/s by Hagen and Hepburn (1952). The lowest frequency of detection known to the author is 9·2 Mc/s from unpublished observations of C. A. Shain. Thus outbursts are observed in a frequency range of four thousand to one

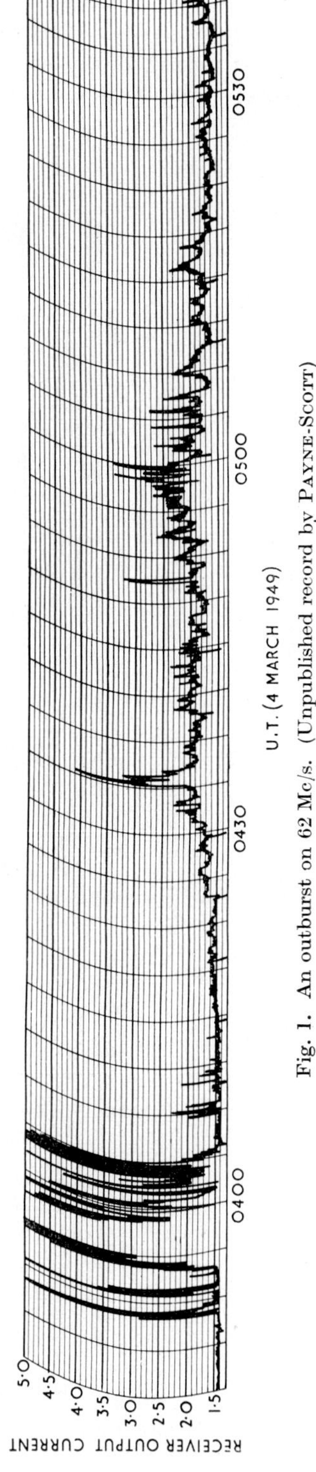

Fig. 1. An outburst on 62 Mc/s. (Unpublished record by Payne-Scott)

extending from the lowest frequencies capable of penetrating the Earth's ionosphere to the highest frequencies used in radio astronomy.

Outbursts vary widely in their characteristics over this range. Their manifestation is most violent in the metre wavelength range, below a few hundred megacycles per second. Fig. 1 shows a typical example at 62 Mc/s. Two distinct phases can be

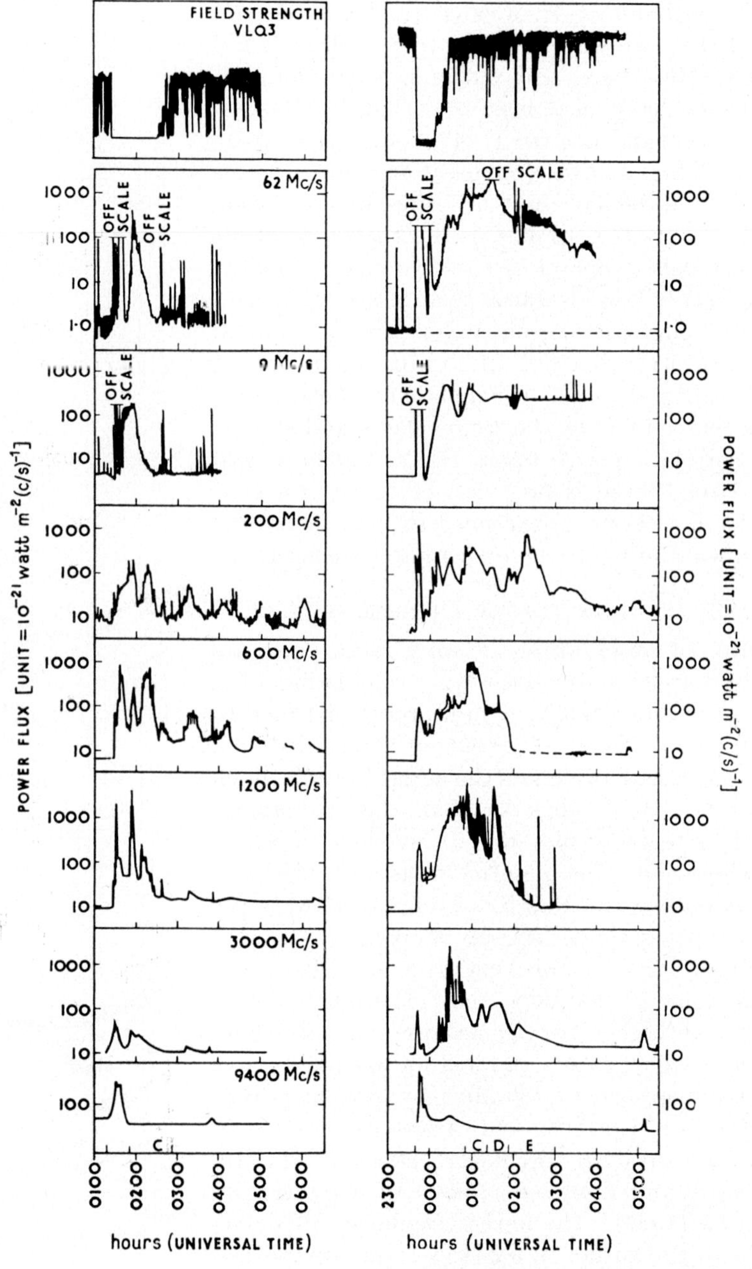

Fig. 2. Two large outbursts recorded on seven frequencies on 17th and 21st–22nd February, 1950. The top record shows the radio fade-out. (Christiansen, Hindman, Little, Payne-Scott, Yabsley, and Allen, 1951)

recognized. In the first or *main* phase the level rises suddenly by a factor of some thousands, fluctuates violently and recedes after about ten minutes. The second or *storm* phase starts soon after with a more gradual rise. The storm may continue for hours or even days showing rapid fluctuations, though both mean and variable components are smaller than those in the main phase. The 200 Mc/s observations of DODSON, HEDEMAN, and OWREN (1953) indicate that in some cases either one of the two phases can occur at the time of flares without the other. Also, observations of PAYNE-SCOTT and LITTLE (1952), described later, have established that the

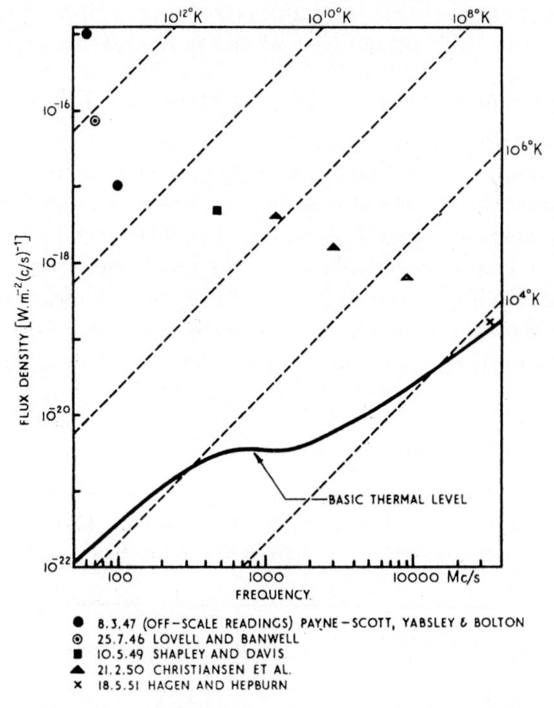

Fig. 3. Peak intensities of the largest outbursts on record. Temperatures on the diagonal grid refer to apparent black-body temperatures of the optical disk of the Sun

polarization of the main phase is random, while that of the storm is circular. These results suggest that the two phases may be due to independent causes.

At higher frequencies (the decimetre and centimetre wavelength bands), the disturbances are of smaller amplitude and show much less fluctuation (COVINGTON, MEDD, 1949). In general the higher the frequency the smaller and smoother is the disturbance. Nevertheless a large disturbance on the decimetre band is sometimes accompanied by only slight activity in the metre band (LEHANY and YABSLEY, 1949).

Fig. 2 shows two large outbursts recorded by CHRISTIANSEN, HINDMAN, LITTLE, PAYNE-SCOTT, YABSLEY, and ALLEN (1951) at frequencies between 62 Mc/s and 9400 Mc/s. The second is perhaps the largest ever recorded on the higher frequencies. In this exceptional case the rapid fluctuations characteristic of the metre wavelengths appear to be in evidence throughout the decimetre range. Peak intensities of this and other large outbursts on record are shown in Fig. 3 as a function of frequency.

Evidence of interest in the interpretation of outbursts has been obtained from the study of the relative times of onset at different frequencies. The first evidence on time delays was given by PAYNE-SCOTT, YABSLEY, and BOLTON (1947). Recording simultaneously on 200 Mc/s, 75 Mc/s, and 60 Mc/s they found that outbursts of exceptionally high intensity occasionally showed systematic time delays of the order of minutes, the high frequencies preceding the low. However, in a subsequent extended series of observations PAYNE-SCOTT (1949) found no further evidence of similar delays and pointed out the difficulty of relating different portions of a complex burst at different frequencies. To resolve the difficulty a technique is required by which the spectrum of the outburst is examined over a continuous range of frequencies. Such a technique is the basis of the first of the investigations described below.

4. OBSERVATIONS USING SPECIAL TECHNIQUES AND THEIR INTERPRETATION

Apart from demonstrating the association of solar noise with flares, evidence obtained from the single-frequency time-charts gave no sure indication of the significance of outbursts. But three other types of observation are fundamentally possible, namely determination of spectrum, direction, and polarization of the incoming radiation. During the past four years special observations have been conducted at the Radiophysics Laboratory, Sydney, covering all three types of measurement. New techniques have been necessary in each case because the rapid variability of outbursts allows us only a brief time interval, of the order of 1 sec., in which to make a measurement.

4.1. *Observations of the Spectrum*

The first observations of the continuous spectrum of solar disturbances covered the frequency range 70–130 Mc/s (WILD and MCCREADY, 1950; WILD, 1950). The receiving instrument consisted of a broad-band aerial connected to a tunable

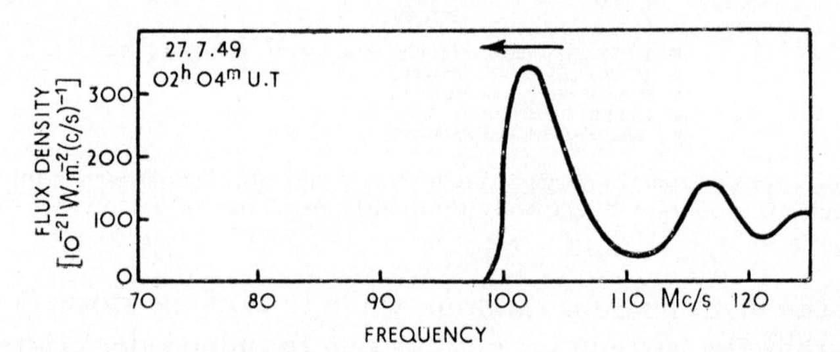

Fig. 4. A typical "instantaneous" spectral profile recorded during an outburst. The arrow shows the direction of frequency drift. (Data from WILD, 1950)

receiver whose frequency was swept rapidly through the range. By this means independent spectra were obtained at the rate of three per second.

The results showed that although bursts differed widely in their spectra it was possible to recognize a distinct class associated with solar flares. In this class the spectrum near the start shows an abrupt "cut-off" in frequency, little or no radiation being received at lower frequencies and one or more intense peaks at higher frequencies (see Fig. 4). With the passage of time the cut-off frequency and other features of the spectrum drift gradually towards the lower frequencies at the rate of about 15 Mc/s per min. Of five examples of this class observed, three accompanied flares or fade-outs. For the first time it was considered possible to recognize with certainty, directly from the record, a type of disturbance associated with flares.

The characteristic spectrum is observed during the first few minutes of the outburst, *i.e.* during the main phase. During the storm phase the enhanced radiation

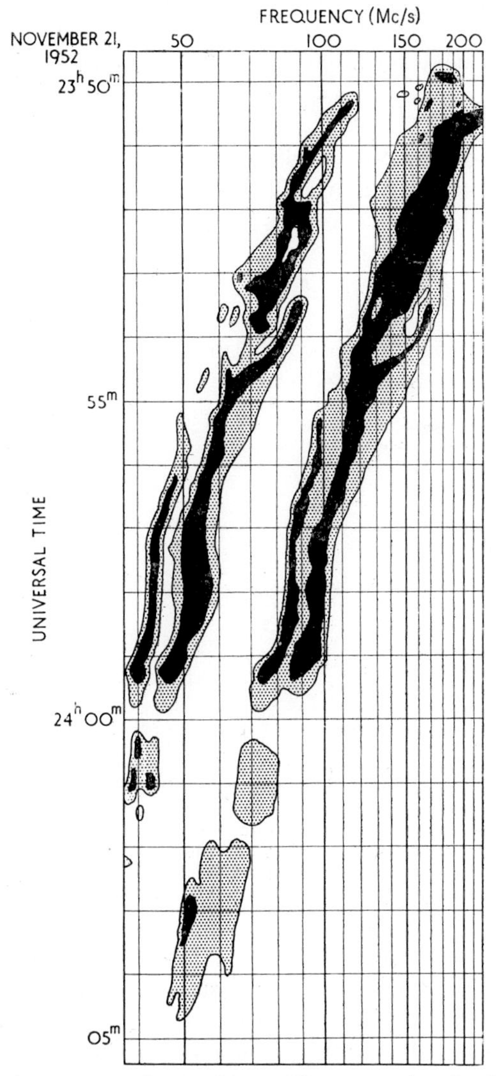

Fig. 5. The dynamic spectrum of an outburst. The two contour levels show intensity as a function of frequency and time and corresponds to levels of approximately 5 and 20 $\times$ 10^{-21} W per m^{-2} per c/s^{-1} (one plane-polarized component). (WILD, MURRAY, and ROWE, 1953)

appears as a relatively smooth continuum above which frequent short-lived narrow-band bursts occur. But if a flare occurs while a storm is in progress the continuum may rise to a high level without showing the spectral features characteristic of the main phase of an outburst (WILD, 1951).

Recently a second series of observations has been started in which the frequency range has been extended to cover 38–220 Mc/s. The dynamic spectrum of the first outburst to be recorded is shown in Fig. 5 (WILD, MURRAY, and ROWE, 1953). In addition to the usual features (cut-off, multiple peaks, and frequency drift), this record shows a new feature which the earlier observations over a narrower frequency range were incapable of revealing. Two widely-spaced "parallel" bands are present, and features in the low-frequency band are simultaneously reproduced in the upper band at almost exactly double the frequency. Although final judgment must await more records, the evidence of this one example seems to indicate that the source of emission radiates the second harmonic in addition to the fundamental frequency.

4.2. *Interpretation of the Spectrum*

It is well known to ionospheric workers that electromagnetic radiation can be propagated through a uniform ionized medium of electron density N only at frequencies above the critical frequency, f_0, given by

$$f_0{}^2 = \frac{e^2N}{\pi m},$$

where e and m are the electronic charge and mass. Thus we should expect that radio noise emitted from within an ionized envelope should show an emergent spectrum

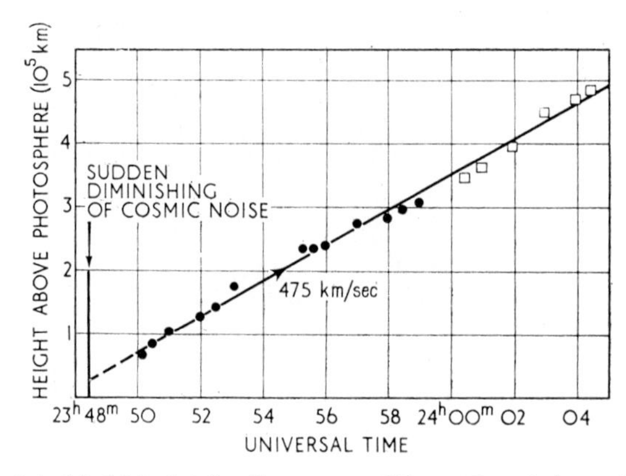

Fig. 6. The calculated height plot for the source of the outburst shown in Fig. 5. Where possible heights are derived from the cut-off frequency of the lower band (small circles). The range is extended to greater heights by using half the frequency of the upper band (small squares). (WILD, MURRAY, and ROWE, 1953)

marked by a cut-off at the critical frequency. The solar atmosphere is a non-uniform ionized medium in which the electron density, and therefore critical frequency, decreases with height. Our interpretation of outburst spectra depends on the hypothesis that the observed cut-off is to be identified with the critical frequency at the locality of the emitting source.* Thus the observed cut-off frequency is taken to be a direct measure of the electron density at the source. Its frequency drift towards the lower frequencies indicates that the source is moving into regions of successively lower density, that is to say *moving outwards through the solar atmosphere.*

Optical measurements of the mean electron-density distribution in the corona allow us to estimate the level of the source at any instant. Fig. 6 shows the passage through the corona of the source of the recent outburst (Fig. 5) assuming the Baumbach-Allen electron-density distribution. The derived velocity is seen to be

* More strictly the cut-off is identified with the lowest frequency of escape. For sources away from the centre of the disk this frequency is slightly above the critical frequency owing to refraction effects. See JAEGER and WESTFOLD (1950).

remarkably constant and extrapolation towards the origin suggests that the outburst may have been initiated near the base of the corona at the onset time of the ultra-violet emission. The value of 475 km per sec. is typical of outbursts previously observed.

The significance of multiple intense peaks above the cut-off frequency is not yet understood. It could be due either to a multiple source or to splitting by magnetic fields. However, the peak of lowest frequency normally occurs immediately above the cut-off frequency and may be due to resonance at the natural plasma frequency, which is identical with the critical frequency f_0. This suggestion is supported by

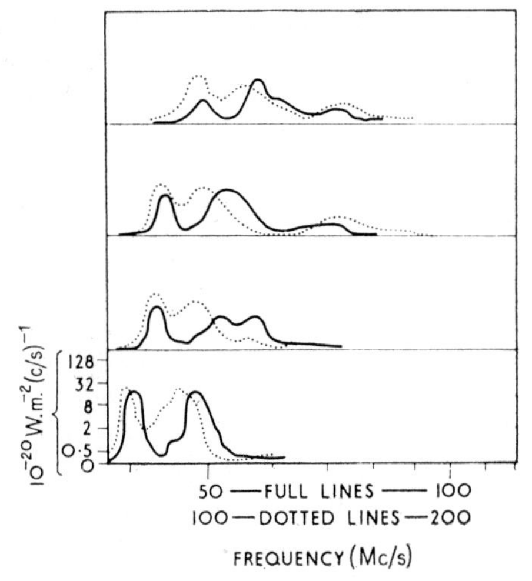

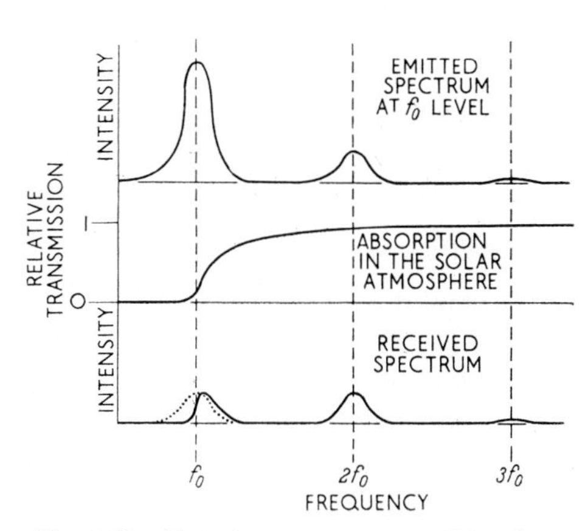

Fig. 7 (*a*). The effect of absorption in the solar atmosphere on harmonics emitted from the f_0 level. In the lowest curve the second harmonic is reproduced with two-to-one frequency shift (dotted line)

Fig. 7 (*b*). Four instantaneous profiles taken at one-minute intervals observed during the outburst shown in Fig. 5. (WILD, MURRAY, and ROWE, 1953)

detailed examination of the harmonic effect present in Fig. 5. The effect of passing a harmonic spectrum, emitted at the f_0-level, through the solar atmosphere is illustrated qualitatively in Fig. 7 (*a*). Owing to the highly selective absorption near the critical frequency, the peak of the fundamental frequency (presumably the strongest emitted) becomes both attenuated and slightly displaced in frequency relative to the second harmonic. These effects are clearly evident in the typical observed profiles shown in Fig. 7 (*b*).

We cannot yet be certain whether the sources of intense radio emission which are presumed to produce outbursts in their outward passage through the solar atmosphere correspond to any previously known phenomenon. Surge prominences sometimes attain outward velocities of 500 km per sec. and reach heights of up to 5×10^5 km above the photosphere before falling back to the Sun (ELLISON, 1951). The unseen solar corpuscles responsible for generating terrestrial magnetic storms and aurorae travel between Sun and Earth at mean velocities of about 800–1600 km per sec. Both are emitted at the time of flares and have velocities comparable with typical derived outburst velocities (500 km per sec.). On present evidence identification with the storm particles seems the more likely of the two because (I) frequency drifts have been observed only in the one direction corresponding to outward motion and

(II), as shown in Fig. 6, the constant outward velocity is still maintained up to a height of 5×10^5 km in the only observation at present available. The discrepancy in velocity by a factor of about two between outburst and storm-generating agencies is not a serious one because calculation of the former is based on an assumed electron-density distribution likely to yield severe under-estimates at times. An independent method of velocity measurement discussed below yields velocities of 500–3000 km per sec., and gives added support to identification with storm particles.

The interpretation of outbursts given above was established as a result of the recognition of the characteristic cut-off frequency and its continuous drift with time (WILD, 1950). This followed tentative suggestions made on the basis of time-delay evidence (PAYNE-SCOTT, YABSLEY, and BOLTON, 1947; HEY, PARSONS, and PHILLIPS, 1948). More recently REBER (1951) has reported an interesting sequence of time-delays which he interpreted as a source moving outwards and thence turning back towards the Sun. Such an explanation, he points out, is fraught with speculation. Indeed the apparent existence in outburst spectra of harmonic bands which start simultaneously emphasizes the hazard of interpreting time-delays at widely-spaced frequencies.

4.3. *Position and Polarization*

The measurement of the position on the Sun's disk of localized sources of radio emission presents a major technical problem. At 100 Mc/s, a radio telescope capable of resolving one-quarter of the Sun's angular diameter would require an aperture of about 1 km. For most purposes in radio astronomy the difficulty may be partially overcome by using the radio equivalent of MICHELSON's stellar interferometer. The directional pattern of two widely spaced aerials consists of a series of narrow "fringes" and position measurements can be made from the undulating record generated as a source drifts, with the Earth's rotation, through the fringes. But outbursts fluctuate so violently that the time taken by the source to pass through a single fringe is normally too great to make a significant measurement.

To meet the problem LITTLE and PAYNE-SCOTT (1951) constructed an interferometer, operated at 97 Mc/s, for which the fringe pattern was artificially swept through the Sun at the rate of 25 times per second. They were thus able to make effectively instantaneous position measurements. Also by taking a second measurement with the polarization axis of the two aerials mutually at right angles, they were able to recognize the polarization of the incoming radiation, whether random, circular, linear, or elliptical.

Using this powerful technique, PAYNE-SCOTT and LITTLE (1952) obtained striking evidence which confirmed the hypothesis discussed above. The results of a typical outburst are shown in Fig. 8. At the start the source is seen to be close to the flare, but subsequently the source moves steadily towards and beyond the limb of the Sun. This limbward movement occurred in five of six outbursts observed and is just that to be expected for a source moving outwards through the solar atmosphere. By assuming that the source moved *radially* outwards from the flare, velocities were found to be in the range 500–3000 km per sec. These velocities, which are in general agreement with, but somewhat higher than, those deduced from spectrum observations, were considered to support the identification of outburst sources with the magnetic-storm generating particles.

The sixth outburst they observed was an exceptional case. The flare occurred close to the limb and the source appeared to move from outside the limb *inwards* to a position just inside the disk. This was interpreted in terms of matter falling back into the Sun, thereby implying a radically different type of outburst. But the phenomenon can be interpreted quite differently. It follows from the theoretical work of JAEGER and WESTFOLD (1950) that when a radio source is viewed from the Earth we see not only the source directly but also its image formed by refraction in the inner strata of the ionized solar envelope which acts rather like a convex mirror. As a source moves radially out from the limb its image appears first just beyond the limb, and then moves *towards the centre of the disk*, eventually crossing the limb. In the case of an isotropic source above the limb, the image appears somewhat smaller than, but nearly as bright as, the source itself. But in the case of an anisotropic source that

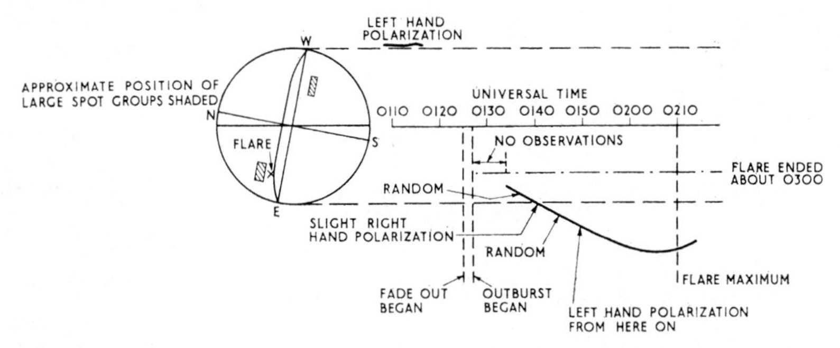

Fig. 8. Observations of source position and polarization during the outburst of 17th February, 1950 (see Fig. 2). (PAYNE-SCOTT and LITTLE, 1952)

radiates most strongly along solar radii the image may be considerably more intense than the source. It is suggested therefore that in this instance of a limb flare the directional observations may have referred to the image rather than the source, and that the outburst conformed to the characteristic pattern.

Measurements of polarization showed that the first surge of an outburst was always randomly polarized but in some cases it was followed by a second rise of intensity showing circular or elliptical polarization. By this observation PAYNE-SCOTT and LITTLE established that some outbursts have the two distinct phases referred to in Section 3 as the main and storm phases. They suggested that the coherent polarization of the latter may be due to the development of coronal magnetic fields during the outburst.

5. CONCLUSION

The significance of outbursts is as yet only partly understood. We can be reasonably sure that the initial violent surge of radiation in the metre wavelength band is due to an outrush of matter from the flare which penetrates to the outer corona, perhaps escaping from the Sun altogether. We may surmise, but cannot be certain, that this is the matter which, in favourable circumstances, causes great magnetic storms and aurorae on the Earth. There is evidence to suggest that in this phase the radio energy at a particular frequency originates in fundamental and harmonic oscillations at the appropriate plasma (f_0) level. However, the detailed mechanism by which the enormous intensities are generated is not yet known. We may speculate that the circularly polarized radiation in the storm phase could be due to a state of turbulence in the corona caused by the corpuscular emission from the flare. The significance of the emission at decimetre and centimetre wavelengths cannot be yet assessed. An extension to higher frequencies of the techniques that have proved so illuminating in the metre wavelength range may provide the key.

ACKNOWLEDGMENT

The author wishes to record that the Australian contribution to the work described here has been due in large measure to the stimulus and guidance of Dr. J. L. PAWSEY, who has led the Radiophysics Laboratory's radio astronomy group since its inception in 1945.

REFERENCES

ALLEN, C. W. . . . . 1947 *M.N.*, **107,** 386.
APPLETON, E. V. . . . . 1945 *Nature* (*London*), **156,** 534.
APPLETON, E. V. and HEY, J. S. . . . . . 1946 *Phil. Mag.*, **37,** 73.
CHRISTIANSEN, W. N., HINDMAN, J. V., LITTLE, A. G., PAYNE-SCOTT, RUBY, YABSLEY, D. E. and ALLEN, C. W. . . . . 1951 *Austral. J. sci. Res.*, A **4,** 51.
COVINGTON, A. E. . . . . 1951 *J. Roy. Astron. Soc. Canada*, **45,** 15.
COVINGTON, A. E. and MEDD, W. J. . . . . 1949 *J. Roy. Astron. Soc. Canada*, **43,** 106.
DODSON, H. W., HEDEMAN, E. R. and OWREN, L. 1953 *Ap. J.*, **118,** 169–196.
EDLÉN, B. . . . . 1946 *Nature* (*London*), **157,** 297.
ELLISON, M. A. . . . . 1951 *Nature* (*London*), **167,** 941.
HAGEN, J. P. and HEPBURN, N. . . . . . 1952 *Nature* (*London*), **170,** 244.
HATANAKA, T., SEKIDO, Y., MIYAZAKI, Y. and WADA, M. . . . . 1951 *Report of Ionospheric Research in Japan*, **5,** No. 1.
HEY, J. S. . . . . 1946 *Nature* (*London*), **157,** 47.
HEY, J. S., PARSONS, S. J. and PHILLIPS, J. W. . 1948 *M.N.*, **108,** 354.
JAEGER, J. C. and WESTFOLD, K. C. . . . . 1950 *Austral. J. sci. Res.*, A **3,** 376.
LAFFINEUR, M. and SERVAJEAN, R. . . . . 1949 *C.R. Acad. Sci.* (*Paris*), **229,** 110.
LEHANY, F. J. and YABSLEY, D. E. . . . . 1948 *Nature* (*London*), **161,** 645.
1949 *Austral. J. sci. Res.*, A **2,** 48.
LITTLE, A. G. and PAYNE-SCOTT, RUBY . . . 1951 *Austral. J. sci. Res.*, A **4,** 489.
LOVELL, A. C. B. and BANWELL, C. J. . . . . 1946 *Nature* (*London*), **158,** 517.
PAYNE-SCOTT, RUBY, YABSLEY, D. E. and BOLTON, J. G. . . . . 1947 *Nature* (*London*), **160,** 256.
PAYNE-SCOTT, RUBY . . . . 1949 *Austral. J. sci. Res.*, A **2,** 214.
PAYNE-SCOTT, RUBY and LITTLE, A. G. . . . 1952 *Austral. J. sci. Res.*, A **5,** 32.
REBER, GROTE . . . . 1946 *Nature* (*London*), **158,** 945.
1951 *Science*, **113,** 312.
SCHULKIN, M., HADDOCK, F. T., DECKER, K. M., MAYAR, C. H. and HAGEN, J. P. . . . . . 1948 *Phys. Rev.*, **74,** 840.
SHAPLEY, A. H. and DAVIS, R. M., Jr. . . . . 1949 *Science*, **110,** 159.
WILD, J. P. . . . . 1950 *Austral. J. sci. Res.*, A **3,** 399.
1951 *Austral. J. sci. Res.*, A **4,** 36.
WILD, J. P. and MCCREADY, L. L. . . . . . 1950 *Austral. J. sci. Res.*, A **3,** 387.
WILD, J. P., MURRAY, J. D. and ROWE, W. C. . 1953 *Nature* (*London*), **172,** 533.

Radio Echo Studies of Meteors

J. G. Davies and A. C. B. Lovell

University of Manchester, Jodrell Bank Experimental Station

Summary

A description is given of the radio-echo techniques for the study of meteors. Particular attention is given to those which are relevant to the work on the astronomy of meteors, such as methods for the determination of meteor radiants and velocities. The discovery and elucidation of the complex series of day-time meteor streams is described as an illustration of the ability of these methods to work under conditions when photographic and visual observations are impossible. A short account is also given of the work on the interstellar meteor problem which has resolved the uncertainty as to whether the sporadic meteors are members of the solar system or come from interstellar space. Brief reference is made to the use of the radio techniques in meteor physics.

1. Historical Note

Although the appearance of a meteor, or shooting star, in the night sky is a familiar sight, the serious study of meteoric phenomena dates back little more than 100 years. During the tremendous display of meteors which occurred in November, 1833, Olmsted, Twining, and others noticed that the meteors appeared to be diverging from a point in the constellation of Leo. It was soon realized that the divergence was the effect of perspective and that the meteors must actually be travelling in nearly parallel paths through space outside the Earth's atmosphere. Interest was further heightened by the return of the great November display thirty-three years later.

At this time Schiaparelli, H. A. Newton, and Adams elucidated the nature of the orbits of some of the prominent meteor streams and the close relation of the Lyrids with Comet 1861 I, of the Perseids with Comet 1862 III, and of the Leonids with Comet 1866 I, was established beyond doubt. Unfortunately the great Leonid shower failed to reappear as predicted in 1899 and this undoubtedly had a most serious effect on the progress of the science. Interest in meteor astronomy waned, and apart from the Harvard expedition to Arizona in 1932 and the work of Hoffmeister in Germany, the main contributions during the first forty years of the twentieth century came from amateur observing organizations. Recently, however, the subject has been revitalized by the application of two new techniques to the observation of meteors. The double camera, rotating shutter, photographic work of Whipple at Harvard is providing fundamental data of high accuracy from which the spatial orbits of individual meteors can be derived, and which is also an important tool for the study of the physical properties of the high atmosphere. In the other technique the observation of radio waves reflected from the ionized trails left when the meteor evaporates provides the only known method of studying meteors in daylight and under cloudy conditions. It is the purpose of this contribution to review some of the achievements of this latter technique.

2. The Radio Echo Technique

The contemporary radio echo methods for the study of meteors have evolved from the radar techniques developed during the Second World War for the location of aircraft and ships. The radio waves are generated in a transmitter T (Fig. 1) connected to an aerial system A, which can either be of simple design to radiate uniformly over the sky, or a large complex structure designed to concentrate the radiation in a specific direction. In the case of large aerial systems it is common practice to use an electronic switching device, S, so that the same aerial can be used to receive the

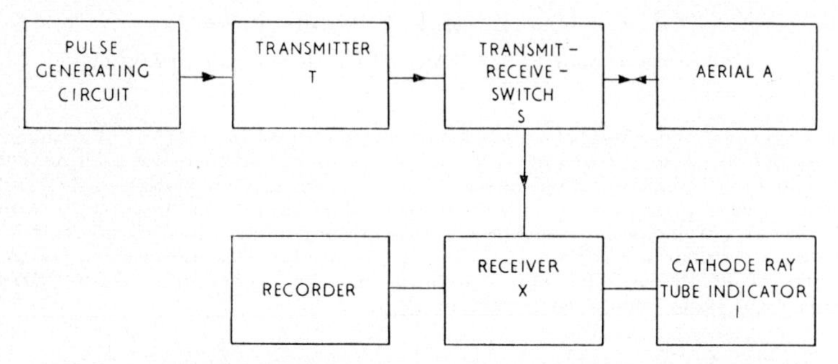

Fig. 1. Block diagram of radio echo apparatus

radiation after scattering from the ionized meteor trail. The output of the receiver X, is connected to an indicator or recording apparatus I.

In many systems T transmits pulses of radio waves with a duration of a few microseconds at a recurrence rate of several hundred per second. The simplest form of display I is a cathode ray tube indicator and the radio waves scattered from the

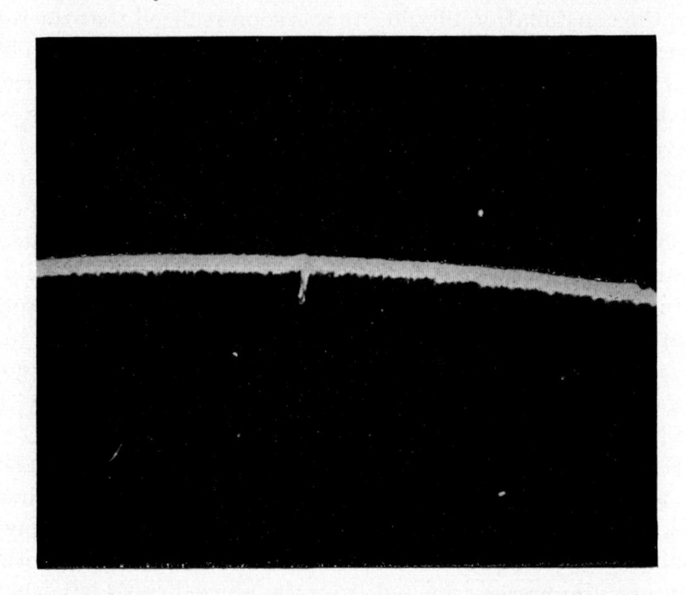

Fig. 2. A radio echo from a meteor trail observed on a range-time cathode ray tube display

meteor trail are then observed as a transient echo as in Fig. 2. This type of display gives immediately the range R of the meteor from the observing station, and the strength of the received signal as measured by the amplitude of the echo. In recent years detailed investigations have been made of the relation between the various parameters of the radio apparatus and the strength of the signal received from a meteor trail of given visual magnitude. The results may be summarized as follows.

If the electron density in the meteor trail is less than the critical density for the wavelength used and if the diameter of the trail is less than the wavelength, then the incident wave penetrates to all the electrons, which scatter freely and in phase giving a received power ε, where

$$\varepsilon = \frac{PG^2N^2\lambda^3}{12\pi^2R^3}\left(\frac{e^2}{mc^2}\right)^2. \qquad \ldots.(1)$$

In this expression, which was first derived by LOVELL and CLEGG (1948), P is the peak transmitter power, G the power gain of the aerial system, referred to as a half-wave dipole (assumed the same for transmission and reception), λ the radio wavelength, N the number of electrons per centimetre in the meteor trail, and (e^2/mc^2) the classical electron radius. When the diameter of the trail is small compared with λ, and the volume density is somewhat greater than the critical density, the nature of the scattering depends on the orientation of the electric vector with respect to the column of ionization, and under certain conditions plasma resonance effects occur. The full analysis has been given by KAISER and CLOSS (1952). When the electron trail is so dense that the radio wave does not penetrate, a different situation arises, and it has been shown by GREENHOW (1952a) and by KAISER and CLOSS (1952) that for $N \gg 10^{12}$ electrons per cm, (1) becomes

$$\varepsilon = \frac{PG^2N^{\frac{1}{2}}\lambda^3}{40\pi^3R^3}\left(\frac{e^2}{mc^2}\right)^{\frac{1}{2}}. \qquad \ldots.(2)$$

In the first case HERLOFSON (1948) has shown that the echo duration to $1/e$ of the initial amplitude is independent of N and is given by

$$T = \frac{\lambda^2}{16\pi^2D}, \qquad \ldots.(3)$$

where D is the diffusion coefficient. In the latter case, however, the duration depends directly on N and is given by

$$T \propto \frac{N\lambda^2}{D}. \qquad \ldots.(4)$$

MILLMAN and MCKINLEY (MILLMAN, 1950) have made a detailed study of the relation between T, for echoes observed on an equipment with $\lambda = 9{\cdot}1$ m, and the visual magnitude of the corresponding meteor. From their results, and by using equation (4), GREENHOW and HAWKINS (1952) have found the electron line density produced by a meteor of given zenithal magnitude. The results indicate that a meteor of zenithal magnitude $+6$ produces approximately 10^{12} electrons per cm path, and a meteor of zero magnitude about 10^{14} electrons per cm path. The value of D (4×10^4 cm^2 per sec.) was determined from echoes of short duration using equation (3).

From these results and the above equations it is possible to determine the parameters of the apparatus necessary to detect meteors down to a given limiting magnitude. Here it is only necessary to mention that with contemporary apparatus working in the wavelength range from about 4 to 10 m, all meteors within the visual range of magnitudes can be readily detected with simple aerial systems, and that echoes of comparatively long duration (determined by (4)) will result. With equipments of higher power, using larger aerial systems, the limit has been extended to about magnitude + 9. In these cases equations (1) and (3) apply and the echo duration is only a fraction of a second.

3. The Measurement of Meteor Radiants and Velocities

Radio echo apparatus employing simple aerials and indicating systems can be used to study the incidence of meteors independently of daylight or cloud. However, significant contributions to meteor astronomy demand the measurement of meteor radiant positions and velocities. Fortunately, appropriate techniques were soon developed and the more important will be described briefly here.

The basis of the method of radiant determination depends on the fact that in the majority of cases the radio reflection from the ionized column is specular (Hey and Stewart, 1947; Lovell, Banwell, and Clegg, 1947). This means that a radio echo can be obtained from a trail only if the perpendicular from the station to the ionized part of the trail lies within the aerial beam. Thus if a narrow beam aerial is used it is possible to establish the plane in which the trails of the meteors lie. This property was utilized by Clegg (1948) in a method for determining the right ascension and declination of an active meteor radiant. The final form of this apparatus (Aspinall *et al.*, 1951) is shown in a preceding article (p. 545, Fig. 4). Two aerial arrays, each producing a narrow beam are erected on either side of the hut, one directed 25° N of W and the other 25° S of W. The output from the transmitter is fed equally into both aerials. An electronic device connects the receiver for alternate pulses to the two aerials in turn and simultaneously the output of the receiver is switched to separate cathode ray tube displays which are photographed continuously on a moving film. By this means continuous records are obtained of the range and time of occurrence of the echoes in each aerial.

Considering for a moment the idealized case of infinitely narrow aerial beams and meteors radiating from a point radiant, it will be evident that in view of the specular reflecting properties of the trails echoes will be observed only when the radiant is at right angles to the beam axis. From the time at which the echoes appeared in each aerial both the right ascension and declination of the meteor radiant could be computed. In practice the beam widths are finite and it is necessary to consider the geometry of the intersection of the aerial beam with the zone in which the meteors ionize. It can be shown that as the Earth rotates, echoes from a given radiant will first appear at short range and that the ranges will increase to a maximum as the radiant moves into a position perpendicular to the beam axis. The echo rate will then fall suddenly. After an interval depending on the declination of the radiant the process will be repeated in the aerial directed to the north west. Examples of the results obtained with this system during the summer daytime streams are shown in Fig. 3. The apparatus has now been in regular use for several years and has given valuable information about the radiant co-ordinates and activity of all the day and night time meteor streams active in the northern hemisphere.

First success in determining velocities by the radio echo technique was achieved by Hey, Stewart, and Parsons (1947) during the great Giacobinid shower of 10th October, 1946. For a small number of very densely ionizing meteors they succeeded in photographing the echo returned from the ionization near the head of the meteor and from the change of range with time the velocity was determined. Such range-time methods are restricted to relatively large meteors which create sufficient ionization to return an echo from the vicinity of the moving head of the

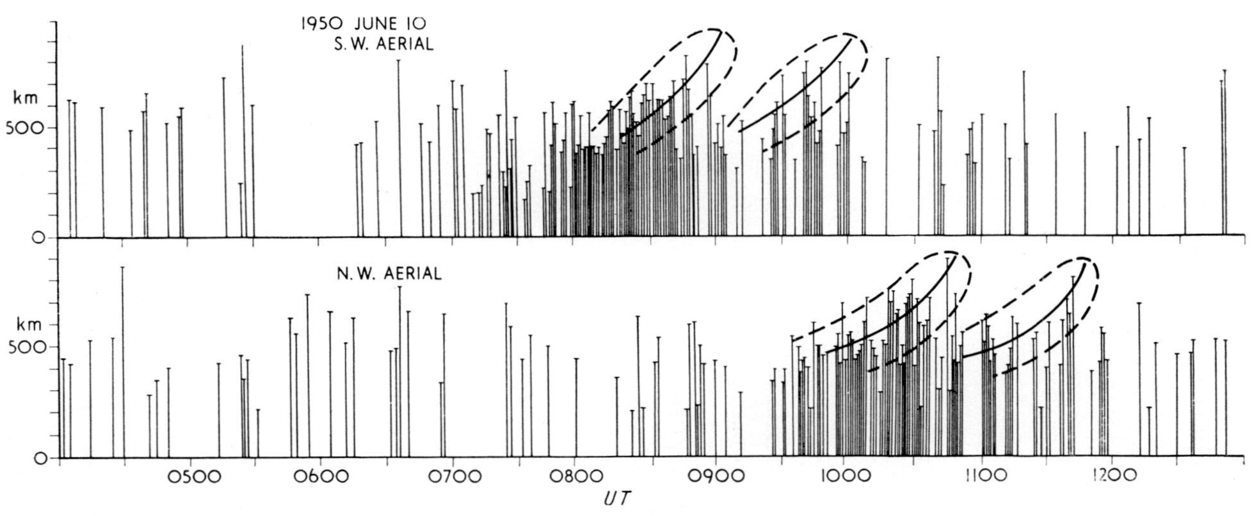

Fig. 3. Examples of the range-time plots obtained on the radiant survey apparatus during the summer day-time meteor streams. The concentration of echoes between 08^h and 09^h in the south-west aerial and between 10^h and 11^h in the north-west aerial was from the Arietid shower. This is followed in each case by a group of echoes from the ζ Perseid radiant

meteor. They have been largely superseded in general use by amplitude-time methods in which the diffraction of radio waves from the trail is observed. The original method was devised by DAVIES and ELLYETT (1949) using a pulsed transmitter, and subsequently McKINLEY (1951) used an alternative method with the transmitter radiating continuous waves. The principle of the method is as follows.

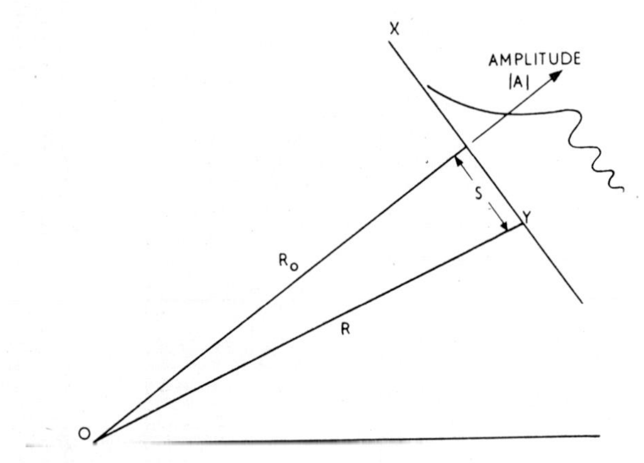

Fig. 4. Diffraction of radio waves from a meteor trail

The amplitude and phase of the wave scattered by an element ds at the point Y on a meteor trail XY, Fig. 4, is given by

$$dA = k \sin\left(\omega t - \frac{4\pi R}{\lambda}\right) ds, \qquad \ldots (5)$$

where k is a constant for a given trail, equation (1), $\omega/2\pi$ is the radio frequency, λ the wavelength, and R the range of Y from the station O on the ground. From this it can be shown that the amplitude received from the whole length of trail XY is

$$|A| = \tfrac{1}{2}k\sqrt{R_0\lambda}\,.\,\sqrt{C^2 + S^2},$$

where C and S are the Fresnel integrals of optical diffraction theory taken between the appropriate limits. The variations in the value of $|A|$ as the point Y moves along the trail are shown in Fig. 4. The values of ∇, the variable of the Fresnel integrals at the maxima and minima can be determined. The length of the trail between these points is then $\frac{1}{2}(\nabla_2 - \nabla_1)\sqrt{R_0\lambda}$ and the velocity is calculated from

$$V = \frac{(\nabla_2 - \nabla_1)}{2T}\sqrt{R_0\lambda},$$

where T is the time taken by the meteor in travelling between points 1 and 2.

In the pulse technique of DAVIES and ELLYETT (1949) the transmitter radiated 600 pulses per second, each of 15 μsec. duration. The first pulse received back from a meteor trail initiated the time bases of two cathode ray tubes, each of which displayed the amplitude of the receiver output. One of these time bases lasted for one-tenth of a second, and thus the first 60 pulses received from the meteor trail were shown side by side. An example of this display is illustrated in Fig. 5. The value of T was then obtained by counting the pulses between successive maxima and minima. The second time base lasted for 7 msec., and was used to determine the range

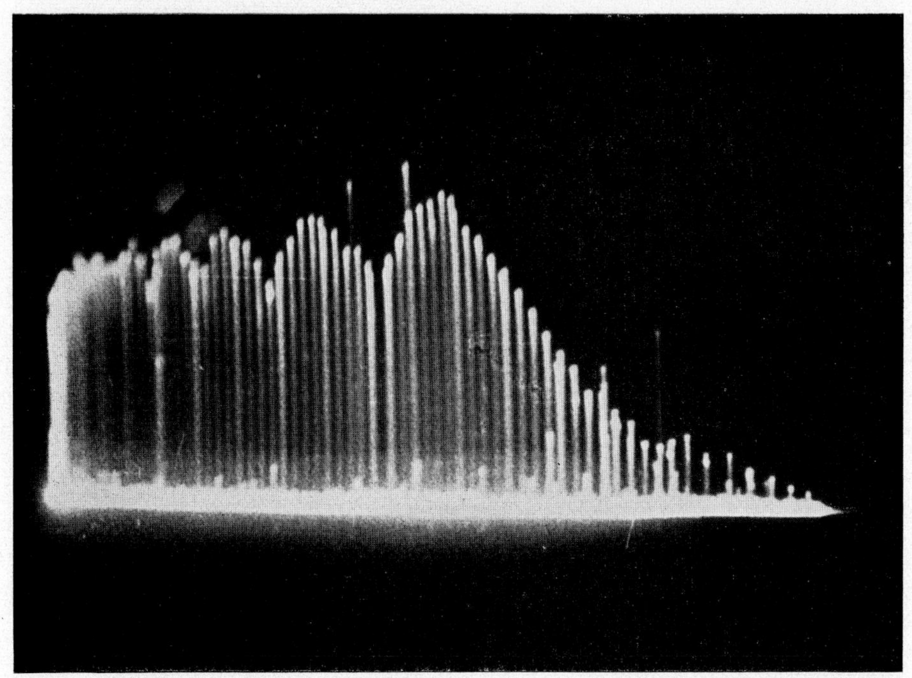

Fig. 5. The diffraction pattern of radio waves scattered from a meteor trail as photographed on a cathode ray tube display. The velocity of the meteor is calculated from the distance between the successive maxima and minima. The range of the meteor is measured on a separate display

Ordinate: Amplitude.
Abscissa: Time, increasing from right to left (each pulse corresponds to an interval of 1/600 sec.).

Fig. 6. The array of Yagi aerials working on a wavelength of 8·2 m used in some of the experiments on the velocity distribution of sporadic meteors

R_0 of the echo. Wavelengths of 4 m and 8 m have been used with this technique, and aerials varying from single dipoles, radiating over a large part of the sky, to arrays of Yagis, as shown in Fig. 6, designed to give a narrow beam, for the observation of faint meteors. In each case the same aerial was used for transmission and reception. With this apparatus velocities have been obtained from meteors of magnitudes down to $+ 9$.

In the continuous wave technique used by McKinley (1951) the transmitter and receiver were separated by several kilometres in order to reduce the direct wave from transmitter to receiver to an amount of the same order as the wave reflected from the meteor trail. In this case the theory is complicated by the addition of the ground wave, but the practical result is that the oscillations in amplitude of the signal can be obtained both before and after the meteor reaches the perpendicular reflecting point.

4. Some Examples of Radio Echo Results

The radiant and velocity measuring techniques described above have been in use at the Jodrell Bank Experimental Station of the University of Manchester since 1948, and have been applied both to shower and sporadic meteors. Some of the more notable results are described briefly below.

(i) *The Daytime Meteor Streams*

A remarkable series of meteor streams active during the daytime in the months of May, June, and July were discovered in 1947, and have since been observed systematically (Clegg *et al.*, 1947; Aspinall *et al.*, 1949; Ellyett, 1949; Aspinall and Hawkins, 1951; Davies and Greenhow, 1951; Almond, 1951; Hawkins and Almond, 1952; Almond *et al.*, 1952). From a large active area in transit during the three hours before noon, four major radiants appear; the first, near *o* Ceti, is active for several days around 15th May, then from 31st May to 16th June, two intense streams appear simultaneously, one in Aries, and the other near ζ Persei. At the end of June and the beginning of July the fourth major radiant is active near β-Tauri. Of these the Arietid radiant, active for three weeks, and reaching an equivalent visual rate of 70 to 80 meteors per hour at maximum constitutes the greatest annual display of meteors at the present time. In the course of its three weeks duration the radiant point moves some 12° across the sky, the right ascension increasing from 39° to 48°, and the declination from $+ 20°$ to $+ 27°$. Several other radiants have also been found in the same general part of the sky, but these have been much less active, and have not reappeared consistently each year.

The velocities of each of the four major daytime streams have been determined over several years, and by combining these with the radiant measurements, orbits have been computed. In each case the inclination of the orbits is small, so the projection of these on to the ecliptic (Fig. 7) gives a fair picture of the true orbit. The orbital periods range from 1·5 years for the *o* Cetids to 3·2 years for the β Taurids, and are amongst the shortest known periods of meteor streams, being closer to those of certain minor planets rather than comets.

Since the inclination of the orbits is small, the streams will pass close to the Earth's orbit again as they approach the Sun. In fact, it is now realized that night time streams with orbits closely similar to the daytime orbits have long been known to visual observers. Thus the Arietid stream, observed in June as it recedes from the Sun is seen again at the end of July as it approaches the Sun, and is then known

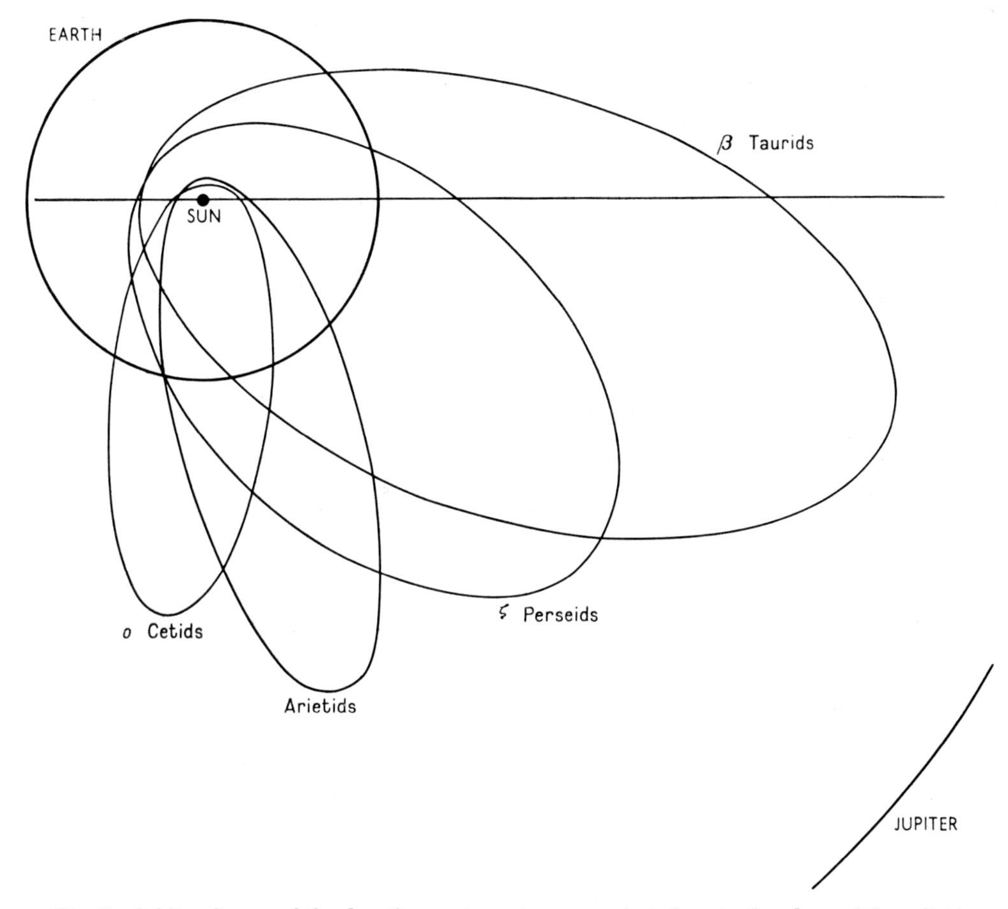

Fig. 7. Orbits of some of the day-time meteor streams projected on to the plane of the ecliptic

as the δ Aquarid stream. Similarly the ζ Perseids correspond to the Southern Arietid stream in October, and the daytime β Taurids to the night time Taurids in November. The β Taurid orbit is compared with the orbits of three individual night time Taurids determined by WHIPPLE's photographic technique (WHIPPLE, 1940; WRIGHT and WHIPPLE, 1950) in Fig. 8.

(ii) *The Interstellar Meteor problem*

One of the major controversies in astronomy during the last fifty years has arisen over the question of the origin of sporadic meteors. One group, led by ÖPIK (1934a, b; 1940; 1941) and HOFFMEISTER (1948), believe that the sporadic meteors have an interstellar origin, and are merely visitors to the solar system. The other group, led by PRENTICE (1948) and PORTER (1943; 1944), believe that they move in elliptical orbits round the Sun. HOFFMEISTER's analysis of the variation of meteor rates throughout the year and throughout the night was based on the assumption, since shown to be false by radio observations, that the meteor orbits were uniformly distributed in space; and visual estimates of velocity, based as they are on durations usually less than a second, are notoriously inaccurate. Even the photographic results, although extremely accurate did not give a clear answer, since they were limited to

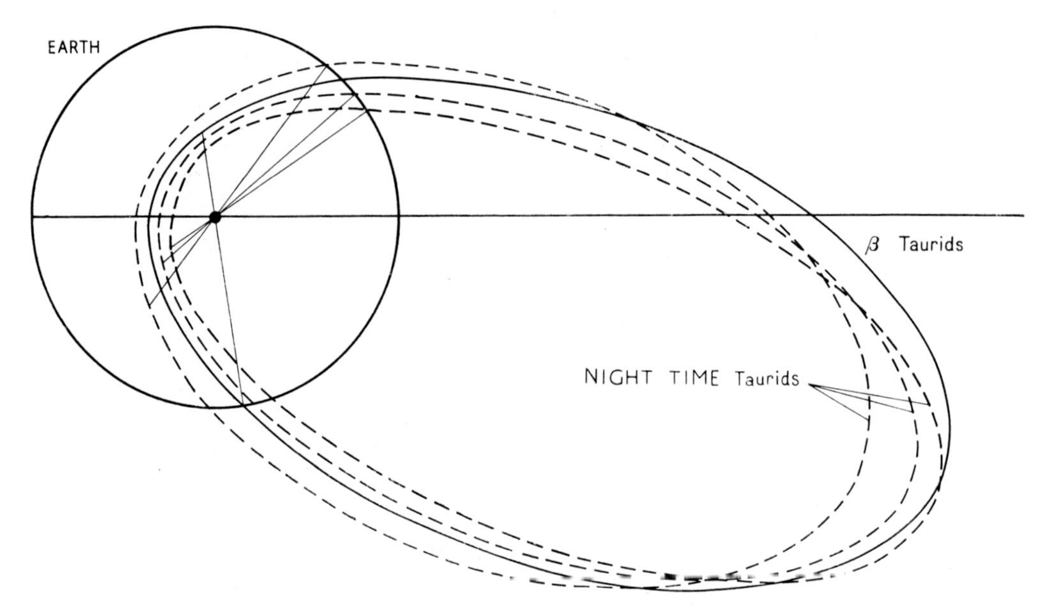

Fig. 8. Comparison of the orbit of the summer day-time β Taurid meteor stream as determined by radio echo measurements with the orbits of three of the autumn night-time Taurids as determined by WHIPPLE's photographic technique

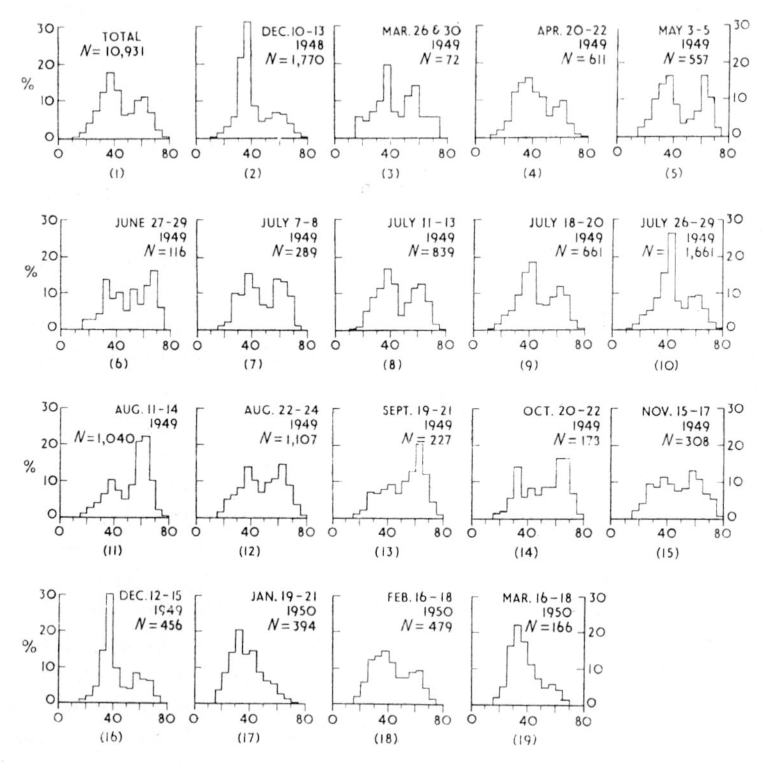

Fig. 9. The velocity distribution of sporadic meteors as measured by McKINLEY using the radio echo technique from December, 1948, to March, 1950

the brightest meteors, and the method is more sensitive to slowly moving particles. Thus in 1948, when the radio echo technique was applied to the problem in England by ALMOND, DAVIES, and LOVELL, and in Canada by MCKINLEY, there were wide divergencies of opinion as to the orbits of the sporadic meteors.

In order to determine whether the orbit of a meteor is elliptical or hyperbolic, it is necessary to know its heliocentric velocity. Unfortunately, no radio technique

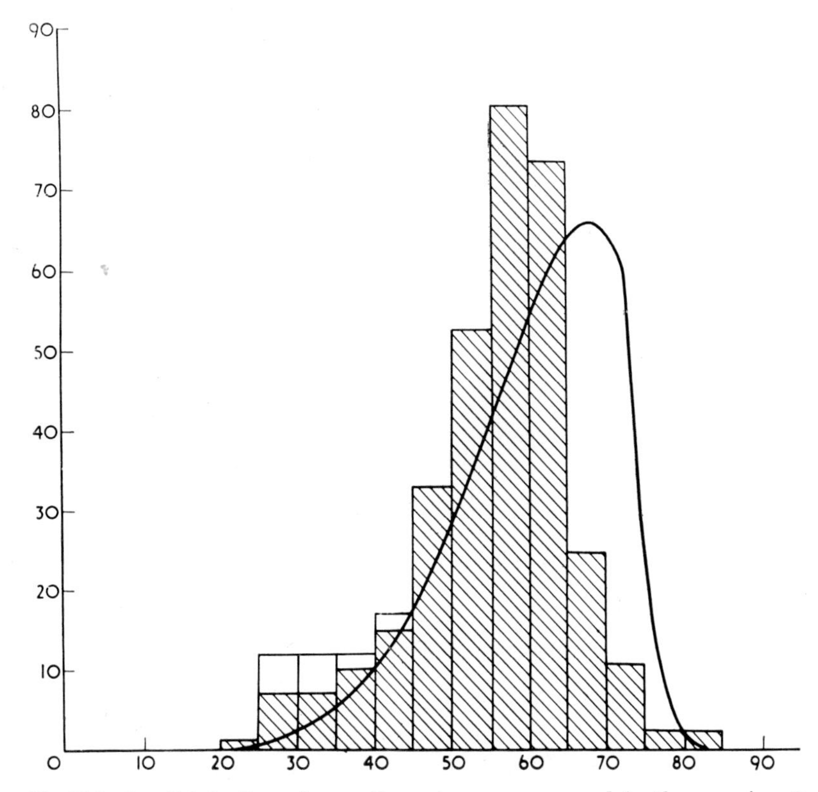

Fig. 10. Velocity distribution of sporadic meteors as measured in the experiments of ALMOND, DAVIES, and LOVELL. This distribution refers to the meteors with radiants close to the apex of the Earth's way, measured during the autumn of 1950. The smooth curve shows the velocity distribution calculated on the assumption that the meteors are travelling in parabolic orbits in random directions

yet used will give this for individual meteors, since the radiant point can only be determined for meteor showers. Therefore statistical experiments have been performed on groups of meteors, and the resulting histograms compared with those expected from different values of the heliocentric velocity. Since heliocentric velocities in excess of 42 km per sec. would indicate hyperbolic orbits, and the Earth's orbital velocity is about 30 km per sec., the greatest geocentric velocity which can be observed for a meteor moving in an elliptical orbit is about 72 km per sec. Velocities in excess of this limit would indicate the existence of hyperbolic meteors.

MCKINLEY (1951), using his continuous wave technique and dipole aerials, measured velocities for periods of 48 hr or 72 hr each month for 15 months. In this time he observed more than 10,000 velocities, the distribution being given in Fig. 9. The effect of the major showers is clear in some of the histograms. Of these only thirty-two meteors had velocities in the range 75–79 km per sec., and there were none with velocities above 80 km per sec. The errors of measurement were such that in no case could any meteor be definitely stated to have a hyperbolic velocity. Simultaneously ALMOND, DAVIES, and LOVELL (1951; 1952; 1953) started a series of experiments, which lasted for four years, using the pulse technique and a beamed aerial system directed to receive echoes from meteors with radiants close to the apex of the Earth's way. In this case the velocity distribution to be expected from meteors travelling in parabolic orbits should show a peak close to 72 km per sec. In all, 860 such meteors were observed, the distribution for one such experiment being given in Fig. 10. A second group of experiments, receiving meteors from the

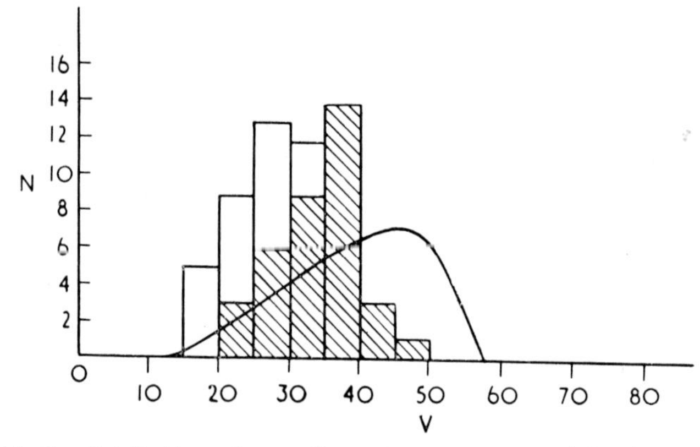

Fig. 11. Velocity distribution of sporadic meteors as measured in the experiments of ALMOND, DAVIES, and LOVELL. This distribution refers to the meteors with radiants close to the antapex of the Earth's way, measured during the spring of 1951. The smooth curve shows the velocity distribution calculated on the assumption that the meteors are travelling in parabolic orbits in random directions

region of the antapex gave the distribution of Fig. 11, and proved that the high velocity cut off observed in the main experiments was not due to instrumental limitations. The peak of the observed distributions is in all cases some 10 km per sec. lower than that derived from the assumption of parabolic velocity, and it is concluded that the majority of the sporadic meteors move in orbits of short periods, similar to those found for the daytime showers. As in MCKINLEY's experiment, no definite case of a hyperbolic velocity was found, and the results are overwhelmingly in favour of the view that sporadic meteors are localized in the solar system.

(iii) *Problems in Meteor Physics*

In addition to astronomical problems, such as those discussed above, the radio echo technique has been applied to the study of meteor ionization and the physics of the upper atmosphere. The measurement of high altitude winds provides an illustration of the value of this technique. Meteors of zenithal magnitude brighter than about + 6, produce echoes whose duration is proportional to the electron density in the trail, and which may last for many seconds (equation (3)). The violent fluctuations in echo amplitude observed under these circumstances have been studied by GREENHOW (1952b). An example, observed on both 4 m and 8 m is shown in Fig. 12. It will be seen that the period of fluctuation is closely proportional to the wavelength. Immediately after formation the meteor trail is subjected to the motion of the winds in the 80 km region where the trail is formed. These winds are both high and turbulent, and the trail rapidly gets distorted out of its initially straight condition. Under these circumstances more than one section of the trail may produce reflection, and these sections may move with different radial velocities relative to the observing station. The signals received from different parts of the trail will then interfere, and produce the amplitude fluctuations observed. From the period of these fluctuations the degree of turbulence can be inferred, and velocity gradients of the order of 5 m per sec. per km are usually observed for points separated by 5–12 km. In addition, by observing the slow drift in range of the long duration echoes, Greenhow was able to measure the radial component of the velocities of individual echoing

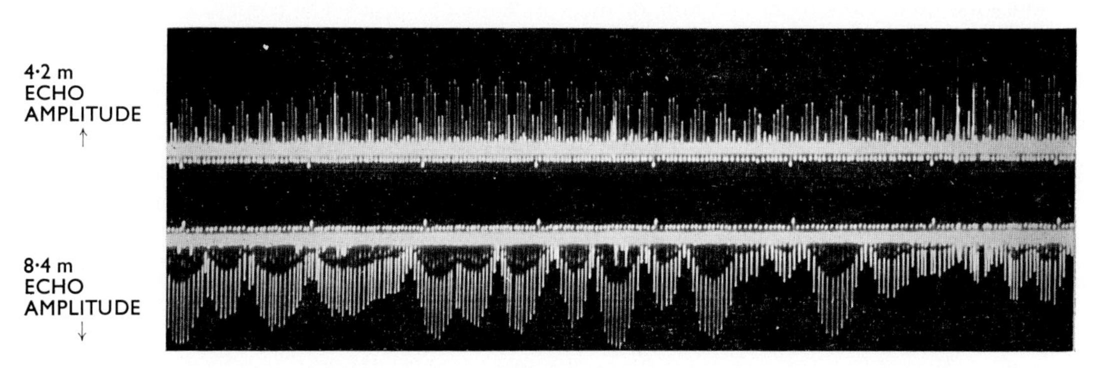

Fig. 12. The fluctuations in amplitude in a radio echo from a meteor trail photographed simultaneously on wavelengths of 4·2 m and 8·4 m. The fluctuations are caused by winds in the high atmosphere

centres. Changes in range of 50 m were observable and drifts of as much as a kilometre in 20 sec. have been recorded. An average value of the winds observed by this method is 50 m per sec.

5. THE FUTURE OF METEOR ASTRONOMY

It must be admitted that in spite of the great advances made by the radio and photographic techniques during the last few years the existence of meteors remains an enigma. The cometary association of the Lyrids, Leonids, and Perseids was established eighty years ago and in the intervening period the list of associations has not increased greatly. In fact, only the association of the night time Taurids and day time β Taurids with ENCKE's Comet, and of the Giacobinids with the Giacobini-Zinner Comet, can be said to be firmly established. The relationship, if any, of the η Aquarids and Orionids with HALLEY's comet remains very problematical. As regards the other major showers the recent measurements of their orbits has created an extremely puzzling situation. Whereas it was believed with some confidence that cometary associations could be found when the orbits were known, this now seems to be impossible. The Geminids, Quadrantids, and the summer daytime streams have orbits with unique characteristics, in some cases of shorter period even than the asteroidal bodies. The parentage of these streams now presents a problem of great interest.

Although the major streams provide the most notable events in meteor astronomy, their contribution to the total number of meteors entering the atmosphere is smaller than that of the sporadic meteors. In view of the recent measurements referred to above, it can scarcely be doubted that these sporadic meteors are moving in closed orbits around the Sun, and, in fact, that a large percentage of the orbits must be of short period. The solution of the problem of the orbits merely serves to raise again the problem of their origin. Are they dispersed from compact streams which were once moving in orbits like the Geminids and the summer daytime streams, or have they an independent origin from the formation of the solar system?

From these remarks it will be appreciated that all the fundamental questions in meteor astronomy remain unsolved. Although some comets and meteor streams are associated, the nature of the association is unknown. Did the comets come before the meteors, do comets form meteors, or meteors comets; or are the comets and meteors moving in the same orbits with a stable relationship? Are the short period streams the remains of a unique family of comets which has now disappeared, or of disintegrated minor planets? Finally, was the large sporadic meteor component originally related to this type of meteor stream?

The radio echo studies of meteors are stimulating a new interest in these fundamental problems of meteor astronomy and the new techniques are opening fresh fields of work in meteor physics and the studies of the high atmosphere.

References

ALMOND, M.	1951	*M.N.*, **111**, 37.
ALMOND, M. BULLOUGH, K. and HAWKINS, G. S.	1952	*Jodrell Bank Annals*, **1**, 13.
ALMOND, M., DAVIES, J. G. and LOVELL, A. C. B.	1951	*M.N.*, **111**, 585.
	1952	*M.N.*, **112**, 21.
	1953	*M.N.*, **113**, 411.
ASPINALL, A., CLEGG, J. A. and LOVELL, A. C. B.	1949	*M.N.*, **109**, 352.
ASPINALL, A., CLEGG, J. A. and HAWKINS, G. S.	1951	*Phil. Mag.*, **42**, 501.
ASPINALL, A. and HAWKINS, G. S.	1951	*M.N.*, **111**, 18.
CLEGG, J. A.	1948	*Phil. Mag.*, **39**, 577.
CLEGG, J. A., HUGHES, V. A. and LOVELL, A. C. B.	1947	*M.N.*, **107**, 369.
DAVIES, J. G. and ELLYETT, C. D.	1949	*Phil. Mag.*, **40**, 614.
DAVIES, J. G. and GREENHOW, J. S.	1951	*M.N.*, **111**, 26.
ELLYETT, C. D.	1949	*M.N.*, **109**, 359.
GREENHOW, J. S.	1952a	*Proc. Phys. Soc.*, **65**, B, 169.
	1952b	*J. atmosph. terr. Phys.*, **2**, 282.
GREENHOW, J. S. and HAWKINS, G. S.	1952	*Nature (London)*, **170**, 355.
HAWKINS, G. S. and ALMOND, M.	1952	*Jodrell Bank Annals*, **1**, 2.
HEY, J. S. and STEWART, G. S.	1947	*Proc. Phys. Soc.*, **59**, 858.
HEY, J. S., STEWART, G. S. and PARSONS, S. J.	1947	*M.N.*, **107**, 176.
HERLOFSON, N.	1948	*Phys. Soc. Rep. Prog. Phys.*, **11**, 444.
HOFFMEISTER, C.	1948	*Die Meteorströme*, Weimar.
KAISER, T. R. and CLOSS, R. L.	1952	*Phil. Mag.*, **43**, 1.
LOVELL, A. C. B., BANWELL, C. J. and CLEGG, J. A.	1947	*M.N.*, **107**, 164.
LOVELL, A. C. B. and CLEGG, J. A.	1948	*Proc. Phys. Soc.*, **60**, 491.
MCKINLEY, D. W. R.	1951	*Astrophys. J.*, **113**, 225.
MILLMAN, P. M.	1950	*J. Roy. Ast. Soc. Can.*, **44**, 209.
ÖPIK, E. J.	1934a	*Circ. Harv. Coll. Obs.*, No. 389.
	1934b	*Circ. Harv. Coll. Obs.*, No. 391.
	1940	*Publ. Tartu. Obs.*, **30**, No. 5.
	1941	*Publ. Tartu. Obs.*, **30**, No. 6.
PORTER, J. G.	1943	*M.N.*, **103**, 134.
	1944	*M.N.*, **104**, 257.
PRENTICE, J. P. M.	1948	*Phys. Soc. Rep. Prog. Phys.*, **11**, 389.
WHIPPLE, F. L.	1940	*Proc. Amer. Phil. Soc.*, **83**, 711.
WRIGHT, F. W. and WHIPPLE, F. L.	1950	*Tech. Rep. Harv. Coll.Obs.*, No. 6.

The Scintillation of Radio Stars

A. Hewish

Cavendish Laboratory, Cambridge

Summary

The intensity and apparent position of radio sources are observed to fluctuate in a manner analogous to the scintillation of visual stars. Experiments are described which show that the scintillation is caused by diffraction of the incoming waves by irregularities of electron density in the upper ionosphere. The origin of the irregularities is not known and the solution may raise problems of astronomical interest.

A phenomenon similar to scintillation occurs when a radio source is viewed through the solar corona, and this fact enables deductions to be made about the irregular structure of the corona at distances in the range 5–15 solar radii.

1. Introduction

The intensity and apparent position of a radio star when observed at metre wavelengths are often found to fluctuate in an irregular manner. Typical records demonstrating the effect are shown in Figs. 1 and 2. These observations were made with

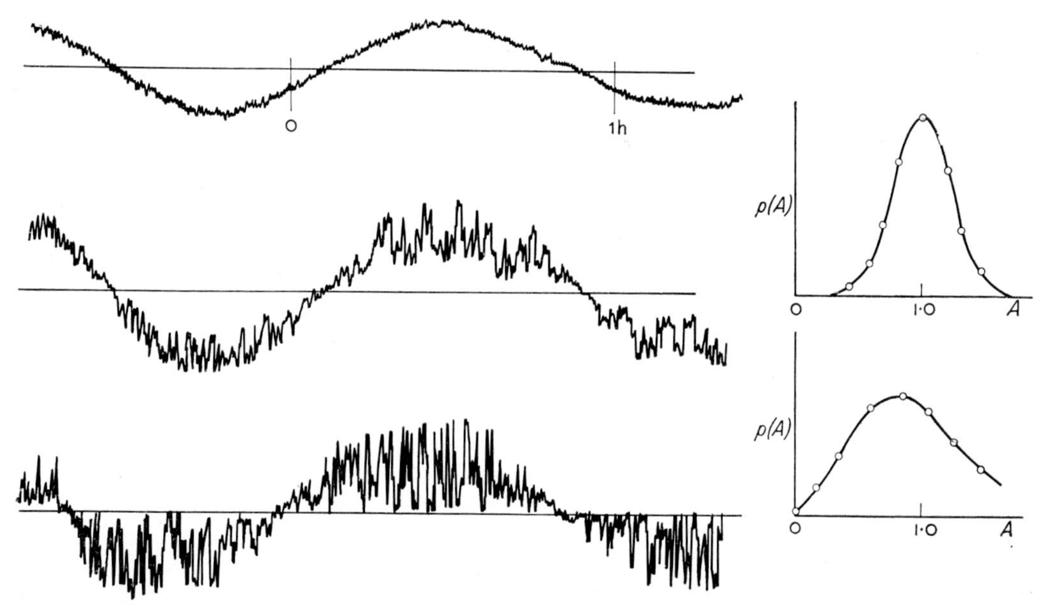

Fig. 1. Observations of the intensity fluctuations of the intense source in Cassiopeia made with a phase-switching receiver on a wavelength of 6·7 m. The probability distribution of amplitude $p(A)$ is shown on the right

interferometer aerial systems in conjunction with a phase-switching receiver where the undisturbed source would be recorded as a steady sinusoidal trace with an amplitude proportional to the intensity of the source.

This phenomenon, which is analogous to the well-known stellar scintillation on visual wavelengths, has been investigated in some detail during recent years. The results of these studies have given new information about the structure and motion of the upper regions of the terrestrial ionosphere, and have also raised fresh problems, the solution of which may be of considerable astronomical interest. In this article a brief survey will be given of the experiments which led to the conclusion that scintillation was produced by the diffraction of galactic radio waves in the ionosphere, and the new problems which have been raised will be outlined.

In addition, recent work has shown that the radiation from a radio star may be considerably modified by its passage through the outer regions of the solar corona, even at distances as great as ten solar radii. By interpreting these results as a scintillation effect it becomes possible to make deductions about the irregular structure

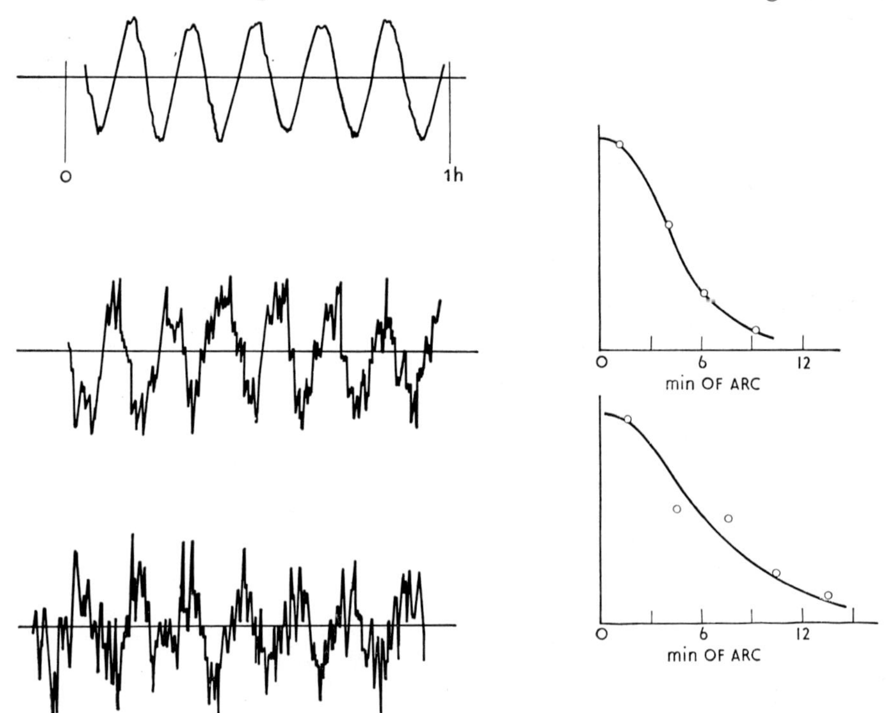

Fig. 2. Typical records of the source in Cassiopeia made with an interferometer of high angular resolution on a wavelength of 8 m. The distribution of the angular deviations is shown on the right

of the outer regions of the corona, and some tentative conclusions derived on this basis will be discussed briefly.

2. Scintillation Introduced by the Ionosphere

(a) *The Identification of the Disturbing Region*

In a series of experiments to determine the origin of the irregular fluctuations of intensity measurements were made with two spaced receivers (Smith, 1950; Little and Lovell, 1950); it was found that the fluctuations, which had an average duration of 30 sec., showed considerable correlation over a distance of about 5 km, but no correlation when the spacing was increased to 20 km. The scale of the intensity variations across the ground was therefore of the order of 5 km and by a simple argument relating this scale to the period of the temporal fluctuations it was possible to show that the disturbing region was associated with the Earth (Ryle and Hewish,

1950). If the fluctuations were caused by interstellar matter the receivers would be moved through the intensity distribution with a velocity of at least 30 km per sec., and fluctuations having a duration of 30 sec. would imply variations with a scale of at least 900 km. The lack of correlation observed at much smaller distances can only be explained if the fluctuations have a local origin.

Later experiments indicated that scintillation was strongly correlated with a particular type of irregularity in the F_2 region which gave rise to scattered echoes in ionospheric pulse sounding investigations (RYLE and HEWISH, 1950; LITTLE and MAXWELL, 1951). It was further concluded, from a detailed analysis of the intensity and angular fluctuations, that the irregularities were situated at a height of the order of 300–500 km (HEWISH, 1952). These results gave conclusive evidence that scintillation was introduced during the passage of the radiation through the ionosphere.

(b) *The Irregular Structure of the Ionosphere*

For wavelengths in the metre region absorption in the ionosphere may be neglected and any irregularities in its structure will give rise to a pure phase variation across the emergent wave front. The irregular wave field across the ground can be conveniently thought of as a diffraction pattern produced by the ionosphere, and the problem then arises of making deductions about the original phase variations from a knowledge of the diffraction pattern at the ground.

A theoretical analysis of this problem has been made and it has been shown how both the lateral extent and the variation of ionization, of the irregularities may be deduced from observations of the phase and amplitude variations across the ground (HEWISH, 1951; LITTLE, 1951). The measurement of these quantities was facilitated by the discovery that the temporal fluctuations were due to the steady drift over the ground of a comparatively unchanging diffraction pattern. It was therefore possible to relate the temporal fluctuations of intensity and apparent position of a radio star to the spatial variations of amplitude and phase in the diffraction pattern.

From the results of a series of observations it was possible to deduce that the ionospheric irregularities had a lateral extent of the order 2–10 km, and a variation of electron content of about 5×10^9 electrons per cm^2 column (HEWISH, 1952). It was also shown that the considerable variations in the rate of fluctuation were largely due to a changing drift velocity, and that the size of the irregularities remained appreciably constant.

(c) *The Origin of the Ionospheric Irregularities*

Before considering the origin of the ionospheric irregularities it is convenient to summarise the evidence obtained from extended observations made during the course of several years (SMITH, 1950; RYLE and HEWISH, 1950; LITTLE and MAXWELL, 1951; LITTLE, 1951; LITTLE and MAXWELL, 1952).

The degree of fluctuation varies irregularly from day to day, and days with large fluctuations tend to occur in groups. It has not been possible to correlate scintillations with the occurrence of meteoric activity, magnetic activity, or aurorae.* During intense magnetic storms the fluctuations are sometimes observed to persist over a greater period of the day, but the most pronounced effect associated with a magnetic disturbance is a marked increase in the rate of fluctuation. It has been shown that the latter is due to an increased drift velocity in the ionosphere. The

* Except for one special occasion (LITTLE, MAXWELL, 1952).

irregularities usually form over a wide region of the ionosphere, but disturbances limited to an extent of about 800 km are sometimes observed.

The most remarkable feature of the irregularities is that they exhibit a pronounced diurnal variation with a maximum around midnight. In Fig. 3 the results of continuous observations over a period of one year are plotted and the effect is immediately apparent.

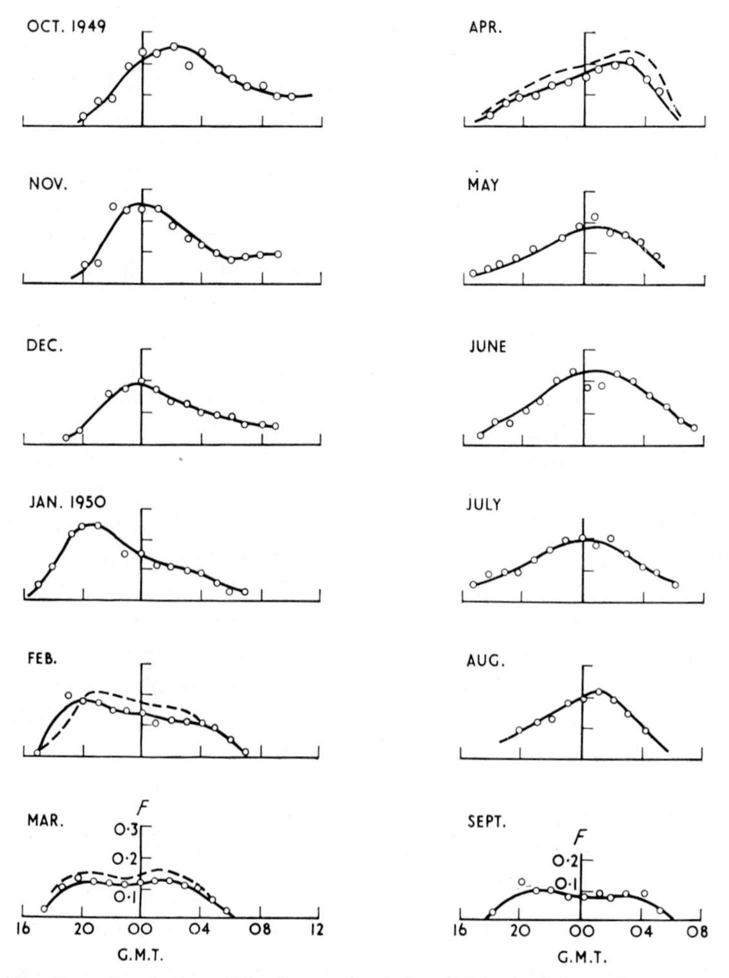

Fig. 3. The diurnal variation of the fluctuation index (F) derived from observations of the source in Cassiopeia. Each curve represents a monthly mean, and the dotted curves were obtained during the following year

Three possible mechanisms for a diurnal variation of the irregular ionization suggest themselves:

(i) A diurnal variation of some additional agency for creating or removing electrons.

(ii) Diurnal changes in the ionosphere which alter the effectiveness of some permanent source of irregular ionization.

(iii) Some disturbance, such as turbulence, of the existing electrons (MILLS and THOMAS, 1951).

Suggestion (ii) does not seem likely in view of the lack of correlation with any of the factors which are known to modify the electron density in the F_2 region and, in particular, the lack of any obvious seasonal variation. (Compare results for June and December in Fig. 3.) It is possible, as in (iii), that an existing uniform electron density becomes irregular due to the onset of turbulence, but it seems difficult to understand why turbulent conditions should only arise during the night. Thus it seems possible that the diurnal variation of the irregularities may be caused by actual variations of some additional ionizing agency.

It has been shown that matter of interstellar origin will fall towards the Sun, under the influence of gravity, from a considerable distance, so that the density of the particles near the Sun may be appreciably greater than that of the undisturbed interstellar matter (HOYLE and LYTTLETON, 1939). Simple calculation shows that the relative motion will be normal to the surface of the Earth at about 0220 hr local time, provided that the general motion of the solar system relative to the interstellar system is small compared to the Earth's orbital velocity. If interstellar particles were responsible for producing irregularities in the ionization of the F_2 region it might therefore be expected that the maximum disturbance would occur at about 0220 hr with a cut off somewhat before 2020 hr and somewhat after 0820 hr. It can be seen that these limits are in approximate agreement with the experimental observations shown in Fig. 3.

If this hypothesis is confirmed by more extended experiments, the results may be of interest in their application to theories of the accretion of matter by the Sun. On this supposition the absence of any appreciable difference in the shape of the diurnal variation at different times of the year does, for example, indicate that the general motion of the solar system relative to the interstellar matter is small compared to the orbital velocity of the Earth.

3. SCINTILLATION INTRODUCED BY THE SOLAR CORONA

(a) *The Occultation of a Radio Star by the Corona*

It was suggested in 1950 that observations of a radio star which passed within a small angular distance from the Sun could be used to provide new information about the electron density in the outer corona (MACHIN and SMITH, 1951). It was shown that rays from the radio star would be refracted away from the Sun during their path through the corona, and when the source was within a certain angular distance from the Sun the rays would no longer be received at the Earth. The effective diameter of the corona for such an occultation was shown to depend upon the wavelength and by making observations at different wavelengths it would be possible to estimate the distribution of electron density in the corona.

(b) *Experimental Observations*

Successful experiments were first made in June, 1952, when the radio star in Taurus, now identified as the Crab Nebula, passed within an angular distance of about 4·6 solar radii from the Sun's centre (MACHIN and SMITH, 1952). The measurements were made using interferometer aerial systems and the phase-switching technique on wavelengths of 3·7 m and 7·9 m, and the results are shown in Fig. 4. It is seen that the apparent intensity of the radio star suffered a gradual decrease, commencing at an angular distance of about 10 solar radii, and being reduced by a factor of about

75 per cent at the distance of closest approach. Additional observations made during 1953 and 1954 have shown effects out to distances of about 20 solar radii.

(c) *Interpretation of the Results*

The ephemerides of the occultation of the radio star in Taurus were computed for several conventional spherical models of the corona based on visual estimates of the electron density (LINK, 1951). It was shown that occultation would occur for some of the models on a wavelength of 7·9 m but that in no case would occultation occur on a wavelength of 3·7 m. In addition it was shown that the onset of the occultation

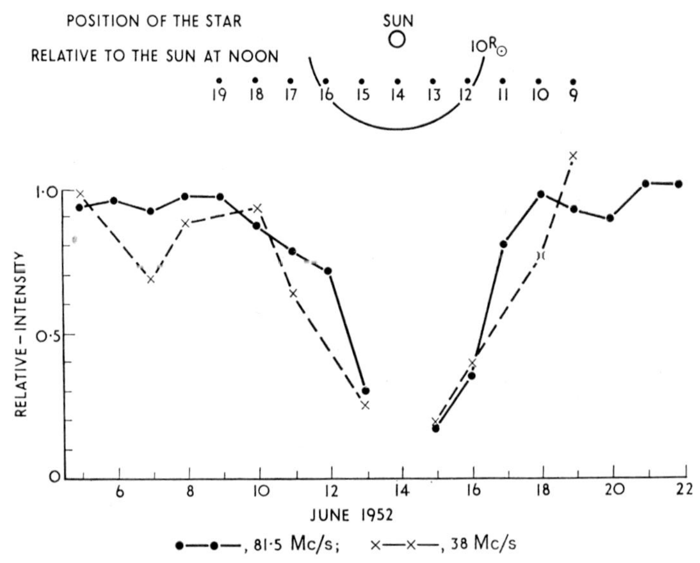

Fig. 4. The variation of apparent intensity of the source in Taurus during the occultation of June, 1952

was essentially a rapid phenomenon, the intensity being reduced to zero in a duration of several hours only.

The observations are seen to be in marked contrast with theory concerning both the time dependence and the wavelength dependence of the occultation, and some alternative explanation must be sought. Some further possibilities will be considered below.

(i) *Absorption*—It is well known that the absorption coefficient of an ionized medium such as the corona depends upon the square of the wavelength, whereas the observations appear to be independent of the wavelength to a first approximation. Again, the value of the absorption, based on visual estimates (VAN DE HULST, 1950) of the electron density, for a ray which passes through the corona at a distance of 5 solar radii is of the order of 1 per cent for a wavelength of 3·7 m. Recent measurements of the radio emission from the corona suggest that the electron density in the outer regions of the corona may be in excess of the visual estimates (O'BRIEN, 1953), which would give rise to an increased absorption, but even if the absorption were appreciable the similarity of the observations on the different wavelengths makes it difficult to explain the results on this basis.

(ii) *Irregularities in the Corona*—If the corona is not uniform, as previously assumed, but contains irregular variations of the electron density, then the rays from a radio star will suffer an irregular deviation while traversing the corona. Such an effect could be regarded as the scintillation of the radio star when viewed through the corona, and the apparent intensity of the radiation at the Earth might be considerably reduced. The precise manner in which the intensity is modified will depend upon the type of coronal irregularity assumed and the angular resolution of the interferometer. It is not possible in an article of this nature to discuss the various mechanisms in detail, but a brief summary will be given of the orders of magnitude required to explain the observations. In all cases the quantities given in the following Table 1 were calculated to give the observed reduction of intensity at a distance of 5 solar radii on a wavelength of 3·7 m. It was assumed for simplicity that the irregularities were distributed isotropically within a distance of 10 solar radii.

Table 1

Type of mechanism	Scale of irregularity (km)	Differential electron density (per cm³)	Distribution	Wavelength dependence
I. Some of radiation scattered into angles wider than the angular resolution of the interferometer ($> 1°$) giving decrease of intensity of unscattered component . .	< 1 10^{-1} 10^{-2}	$3 \cdot 10^2$ $3 \cdot 10$ 3	Contiguous	Decrease of intensity varies as (wavelength)²
II. All radiation scattered into a narrow cone ($\sim \frac{1}{2}°$), giving an apparent increase of source diameter	> 2 10 10^2 10^4 10^5	 $6 \cdot 10^2$ $2 \cdot 10^3$ $2 \cdot 10^4$ $6 \cdot 10^4$	Contiguous	Decrease of intensity varies as wavelength
III. Refraction by large irregularities so that some rays are deviated through angles greater than the angular resolution of the interferometer	$< 10^5$	10^6	About one irregularity in the line of sight through corona	Approximately independent of wavelength

(*d*) *Discussion*

Visual estimates suggest that the electron density in the corona at a distance of 5 solar radii is of the order of 10^4 electrons per cm³ (VAN DE HULST, 1950). With the exception of mechanism III above it may be seen that the variation of density required to produce the observed reduction of intensity is of the same order, or less than, the mean density. Such values would therefore appear to be not unreasonable, but the origin of the irregularities in the corona presents some difficulty since the mean free path of the electrons is considerably greater than the scale of the required irregularities. It is possible that the motion of matter constrained to follow the lines of magnetic force could give rise to irregularities of the same scale as coronal rays (about 7000 km) (VAN DE HULST, 1950), but the generation of smaller irregularities does not seem possible.

The vaporization of small particles falling into the corona might conceivably give rise to small irregularities of the type considered in case I, but it may be shown (RUSSELL, 1929) that vaporization is not sufficiently rapid to maintain the required electron density due to the rapid diffusion of the vaporized atoms.

It is therefore concluded that irregularities having a scale of 10^4–10^5 km, as specified in case II, appear to explain the observations most simply. As a result of additional and more detailed observations, carried out in 1953 and 1954 with interferometers of different spacings, it has been shown conclusively that the measurements are entirely consistent with a scattering mechanism of type II; these results are appended in Table 2, giving size and electron density of the irregularities which would be required to explain the observed scattering at 5, 10, and $15R_{\odot}$ (HEWISH, 1955).

Table 2

$5R_{\odot}$		$10R_{\odot}$		$15R_{\odot}$	
Scale (km)	Electron density (per cm^3)	Scale (km)	Electron density (per cm^3)	Scale (km)	Electron density (per cm^3)
0·27	$7{\cdot}0 \times 10$	1·4	$1{\cdot}5 \times 10$	1·3	$1{\cdot}2 \times 10$
10	$4{\cdot}5 \times 10^2$	10	$3{\cdot}7 \times 10$	10	$3{\cdot}2 \times 10$
10^2	$1{\cdot}4 \times 10^3$	10^2	$1{\cdot}2 \times 10^2$	10^2	$1{\cdot}0 \times 10^2$
10^3	$2{\cdot}5 \times 10^3$	10^3	$2{\cdot}1 \times 10^2$	10^3	$1{\cdot}8 \times 10^2$
10^4	$0{\cdot}8 \times 10^4$	10^4	$0{\cdot}6 \times 10^3$	10^4	$0{\cdot}5 \times 10^3$
10^5	$2{\cdot}5 \times 10^4$	10^5	$2{\cdot}1 \times 10^3$	10^5	$1{\cdot}8 \times 10^3$

Until the advent of more experimental data a detailed discussion of the irregular nature of the corona is not possible, but it is clear that the observation of the occultation of radio stars should provide new information about the nature of the corona in regions which are inaccessible to visual astronomy.

REFERENCES

HEWISH, A.	1951	*Proc. Roy. Soc.* A, **209,** 81.
	1952	*Ibid.*, **214,** 494.
	1955	*Proc. Roy. Soc.* A **228,** 238.
HOYLE, F. and LYTTLETON, R. A.	1939	*Proc. Camb. Phil. Soc.*, **35,** 405.
HULST, H. C. VAN DE	1950	*Bull. Astron. Inst. Netherlands*, **410,** 135.
LINK, F.	1951	*Bull. Astron. Inst. Czechoslovakia*, **3,** 5.
LITTLE, C. G.	1951	*M.N.*, **111,** 289.
LITTLE, C. G. and LOVELL, A. C. B.	1950	*Nature (London)*, **165,** 423.
LITTLE, C. G. and MAXWELL, A.	1951	*Phil. Mag.*, **42,** 267.
	1952	*J. atmosph. terest. Phys.*, **2,** 356.
MACHIN, K. E. and SMITH, F. G.	1951	*Nature (London)*, **168,** 599.
	1952	*Ibid.*, **170,** 319.
MILLS, B. Y. and THOMAS, A. B.	1951	*Austral. J. sci. Res.* A, **4,** 158.
O'BRIEN, P. A.	1953	*M.N.*, **113,** 597.
RUSSELL, H. N.	1929	*Ap. J.*, **69,** 49.
RYLE, M. and HEWISH, A.	1950	*M.N.*, **110,** 381.
SMITH, F. G.	1950	*Nature (London)*, **165,** 422.

Measures of the 21-cm Line Emitted by Interstellar Hydrogen

J. H. OORT

Sterrewacht, Leiden, Netherlands

SUMMARY

The observations of the radiation at 21·1 cm wavelength, emitted by interstellar hydrogen, are discussed. These observations, made mostly by MULLER in the Netherlands and discussed by VAN DE HULST, MULLER, and OORT, show that the interstellar clouds are confined to spiral arms that can be traced to distances that were inaccessible to investigations at the wavelengths of light. From these observations also the velocity of rotation of the galactic system at different distances from the galactic centre could be determined, as well as the distance of the Sun from the centre. The temperature of the interstellar clouds was estimated to be about 100° and their average density in the galactic plane to be about 0·9 hydrogen atoms per cubic centimetre.

1. INTRODUCTION

AN entirely new vista in astronomy was revealed by the results obtained in 1940 and 1943 by GROTE REBER, who was the first pioneer to follow up the revolutionizing, but at that time curiously little appreciated, 1931-observations of JANSKY. During a colloquium on JANSKY'S and REBER'S investigations, organized by the Nederlandse Astronomen Club in 1944, VAN DE HULST suggested that interstellar hydrogen would emit a line at a wavelength of about 21 cm, which might have a measurable intensity. It may well be said that its successful observation, that eventually followed this suggestion, opened by itself a new astronomical vista with previously undreamt possibilities for extending our knowledge of the universe.

The line arises from a reversal of the spin of the electron with respect to that of the nucleus. Its frequency of 1420·4056 Mc per sec (wavelength 21·11 cm) has been accurately determined from atomic beam experiments*. The upper level is triple; the transition probability is known from theory to be $2{\cdot}84 \times 10^{-15}$ sec.$^{-1}$, corresponding to an average life of 11 million years for the atom in its excited stage. The first successful observations were made by EWEN and PURCELL in the spring of 1951, and a few weeks later also by MULLER in Holland and by CHRISTIANSEN and HINDMAN in Australia.†

CHRISTIANSEN and HINDMAN followed up the first measures by making a general survey of the peak intensity of the line in various directions‡, as illustrated in Fig. 1. The irregularity of the contours outside the Milky Way reflects the irregular distribution of the near-by agglomerations of interstellar clouds. We observe, for instance, a bulge in northern latitudes from about 300° to 360° longitude, *i.e.* roughly coinciding with the large masses of dark clouds in Ophiuchus and Scorpius. The large nebulosities south of the Milky Way, in Taurus, are also clearly reflected in these isophotes of 21-cm radiation. On the average, the peak intensity drops to half at about 10° latitude; but this value is evidently quite uncertain on account of the great irregularity. CHRISTIANSEN and HINDMAN also remarked that over a large interval in longitude in the galactic plane the hydrogen line was double.

* Cf. KUSCH and PRODELL; *Phys. Rev.*, **79**, 1009, 1950. For some general physical data concerning the 21-cm line cf. J. P. WILD; *Ap. J.*, **115**, 206, 1952.
† *Nature, London*, **168**, 356–363, 1951.
‡ *Austral. J. sci. Res.*, **5**, 452, 1952.

More refined measures were begun in Holland in the summer of 1952, when MULLER had improved his receiver sufficiently to observe rather accurate line contours in various directions. The observations were made by MULLER or under his direction, using a movable paraboloid of 7·5 m aperture, giving a beamwidth of slightly less than 3°. The reductions were mainly made by VAN DE HULST and myself. For an account of the observations by the Dutch group*, which is supported by the Netherlands Organization for Pure Research, I wish to cite the following statement in the Halley lecture recently delivered by VAN DE HULST in Oxford.†

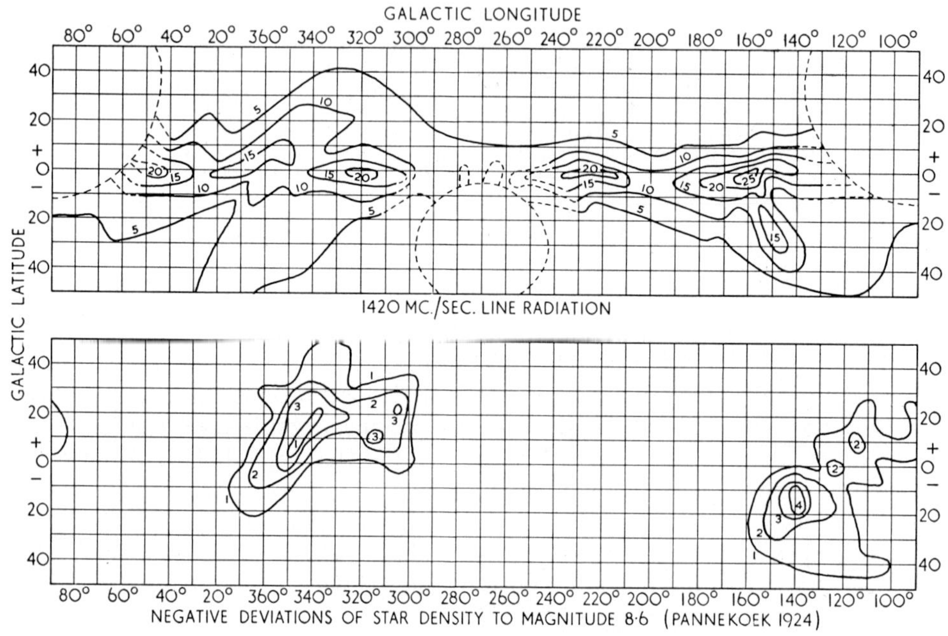

Fig. 1. Maximum intensity in 21-cm line as function of galactic longitude and latitude (from W. N. CHRISTIANSEN and J. V. HINDMAN; *Austral. J. sci. Res.* A., **5,** 452, 1952)

> "Since the improved receiver was ready in July, 1952, about 400 tracings have been obtained, each giving a double line profile for a given point of the sky. Each tracing took two hours or more of observation and roughly twice that amount of reduction, many reductions being made necessary by the still provisional set-up. A full survey would require three-dimensional scanning, for the intensity is a function of l (galactic longitude), b (galactic latitude), and ν (frequency). In this first year we have stuck fairly well to the galactic plane, keeping $b = 0$ and scanning in l and ν only. A point of the galactic circle was selected and followed by manually adjusting the telescope with intervals of $2\frac{1}{2}$ min. In this way three or more double line profiles at many points of the galactic circle have been obtained. They have finally been averaged and brought to a common intensity and frequency scale and the results are represented together in Fig. 2. The abscissa of each figure is the frequency or wavelength, long waves plotted to the right side; the ordinate is the intensity. The

* Cf. *Versl. Amsterdam Akad.*, **61,** 140–143, 1952.

† A more extensive discussion on these observations by H. C. VAN DE HULST, C. A. MULLER, and J. J. OORT appeared in the *Bulletin of the Astronomical Institutes of the Netherlands*, vol. 12, No. 452, p. 117, 1954.

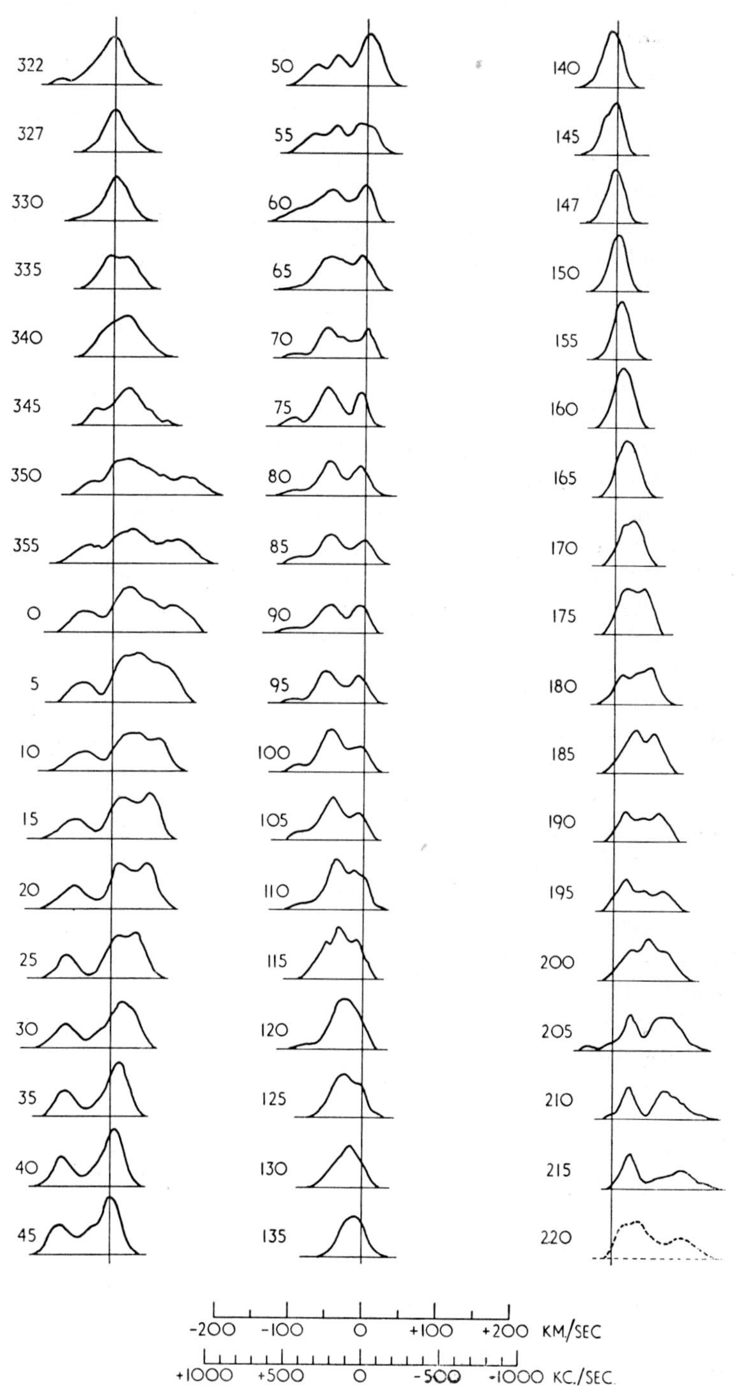

Fig. 2. Line profiles of the 21-cm line at fifty-four points in the galactic equator

vertical line was intended to denote the undisplaced line but corresponds actually to a small velocity of + 1 km per sec.

2. Galactic Rotation and Spiral Arms

Omitting all details of measurement and reduction, I wish to proceed at once with the interpretation of these line profiles. The fact that there is any radiation at all outside the proper frequency of the line is certainly due to Doppler shifts arising from radial velocities. Other broadening effects are insignificant. The known velocities of the Earth in the solar system and of the Sun with respect to the surrounding stars* have been eliminated. The velocities that are left are partly the thermal velocities of the atoms in a cloud; more important are the random velocities of the clouds with respect to each other, which tend to widen the line symmetrically. The most important effect, however, is the differential galactic rotation. This can be made clear by looking at the sign of the displacement. Emission left of the vertical line indicates approach, emission right of the vertical line recession. If we just assume that the angular velocity of rotation is larger for smaller distances from the galactic centre, it is clear that the locus of zero radial velocity consists of (1) the straight line SC, where C represents the galactic centre and S the neighbourhood of the Sun, and (2) the circle with C as centre and going through S. These lines separate four areas, two with positive and two with negative velocities, which for studies close to the Sun appear as four quadrants in a symmetrical arrangement. At $l = 327°$ and $147°$ the line is narrow and undisplaced. This is as predicted. From 220° to 147° the velocities are positive, from 147° to 60° they are negative; in the third quadrant from 55° to 327° both positive and negative velocities occur, all exactly as foreseen. It is most interesting that in the last quadrant negative velocities do occur; this means for sure that we observe parts of the Galaxy farther from the centre of the Galaxy than we are and at distances far above 10,000 parsec.

The line of sight to points in the last-mentioned quadrant passes through regions of positive and negative velocity. The maximum velocity occurs at the point closest to C and near this point the velocity is nearly equal over a long path. This should result in a fairly intense radiation sharply dropping to zero at the plus side. Again the prediction is confirmed, as illustrated very clearly by the profiles for longitudes 10° to 45°. It is possible to measure at what frequency this sharp drop occurs and thus to gain information on the rotational velocity of parts of the Galaxy interior to the Sun. Let R be the distance from C, let $\omega(R)$ be the angular velocity and $\Theta(R)$ the linear velocity, related by $\Theta(R) = R \times \omega(R)$, and let R_0, ω_0 and Θ_0 denote the corresponding values for S. Further let $l' = l + 33°$ be the galactic longitude measured from the direction to the centre.

A point at a distance r from S then has the radial velocity v with respect to S:

$$v = \frac{\mathrm{d}r}{\mathrm{d}t} = R_0\{\omega(R) - \omega_0\} \sin l'. \qquad \ldots.(1)$$

For small r this reduces to the well-known relation

$$v = rA \sin 2l', \qquad \ldots.(2)$$

* A standard value of 20 km per sec towards $\alpha = 18^{\mathrm{h}}.0$, $\delta = +30°$, was used for the solar motion because this agrees well with results derived from interstellar absorption lines.

where $A = -(R/2) \times (d\omega/dR)$ is OORT's constant. A somewhat more general approximation holds for small values of $(R - R_0)$, but arbitrary values of r. This relation reads*

$$v = -2A(R - R_0) \sin l'. \quad \ldots.(3)$$

Relation (3) has been applied to the points closest to the centre on the lines of sight in the range of longitudes $l = 20°$ to $l = 30°$. Here v is measured in the way described, A is known, l' is known and R is replaced by $R_0 \sin l'$. In this way a direct determination of R_0 is made from galactic rotation effects.* So far the estimates of R_0 and Θ_0 were quite uncertain. The values that now seem most probable are: $\omega_0 = 27{\cdot}0$ km per sec.-kps, $R_0 = 8{\cdot}5$ kps, and $\Theta_0 = 230$ km per sec.

The explanation given so far leaves out the most striking feature of the line profiles: the strong maxima and minima. They must indicate an irregular distribution of the densities or of the velocities of the interstellar gas. It seems most likely that most of the maxima and minima are due to a patchy density distribution, although it is already clear that some minor details cannot be explained in that manner. Granting the assumption of a completely regular rotation with a known function $\omega(R)$, the following elegant solution of the density distribution may be made. A certain velocity v observed in a certain direction corresponds by means of equation (1) to a certain value of R. This places the gases at one point (or two or zero) upon the line of sight. Its position in the galactic plane is thereby fixed and can be plotted in a polar diagram with S as the centre.

By means of this method we have obtained the diagram shown in Fig. 3. Each maximum of the line profiles was represented by a dot in the diagram and the dots connected by a cross-hatched band, the width of which corresponds approximately to the width of the maxima. These bands represent the regions in the galactic plane that have a high density of atomic hydrogen gas. These regions have the form of spiral arms and the direction of rotation is with the arms trailing. In viewing Fig. 3 it should be remembered that it is incomplete for three reasons: the sector that is invisible from Holland has been left out; narrow sectors near the centre and anti-centre directions have been left open because the method used is not precise, and the entire portion closer to C than S has been left blank because the solution is ambiguous, one v and R corresponding to two values of r".

So far the account by VAN DE HULST.

In the present article I may elaborate some points a little further.

In Fig. 3 the region where $R < R_0$ has not been filled in, because (as was mentioned) the relation between v and r is not unique. This difficulty is certainly not unsurmountable, but it means that more extensive observations, and preferably a higher resolving power, will be necessary to analyse this most important region. As an example, the relation between v and r is shown for galactic longitude $l = 30°$ in

---

* *Note by the author.* Formula (3) follows directly from (1). For the point at which the line at a given longitude passes closest to the centre, and where the radial velocity becomes a maximum, (3) can be written

$$v_{max} = 2AR_0(1 - \sin l') \sin l'.$$

Measurements of v_{max} thus yield values of the product AR_0. This determination has a considerably greater relative accuracy than the rough estimates of R_0 that have so far been available. The result found in this way (viz. $R_0 = 8{\cdot}26$ kps) agrees very well with BAADE'S estimate of the distance of the galactic nucleus from *RR* Lyrae variables found in the Sagittarius Cloud.

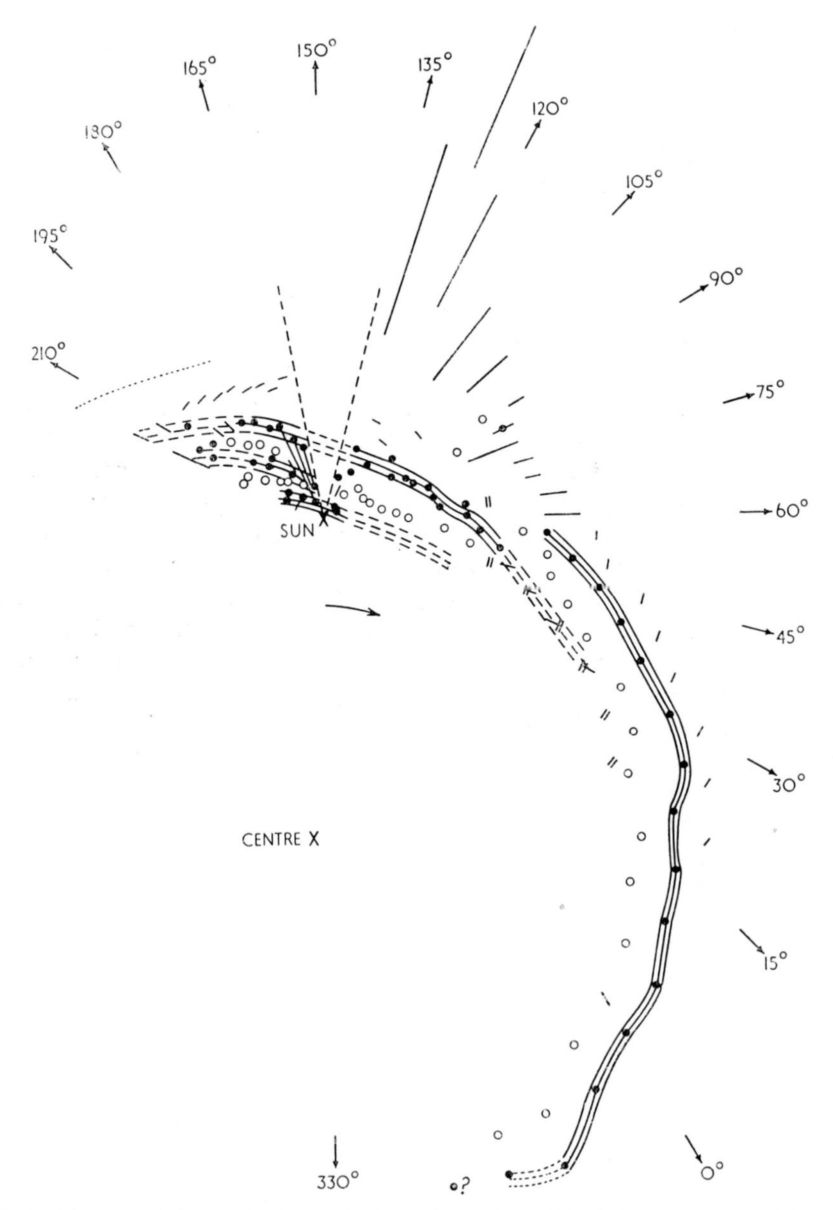

Fig. 3. The more intense spiral arms in the outer region of the Galaxy, as revealed by the 21-cm radiation

Fig. 4. The maximum radial velocity occurs at $r_0 = 3{\cdot}9$ kps, where the line of sight passes closest to the centre. Two points, r_1 and r_2, situated symmetrically with respect to r_0, will give identical radial velocities. In principle, the two regions can, however, be distinguished by the difference in spread in galactic latitude. The unravelling in this case can only be done satisfactorily if the resolving power of the radiotelescope is sufficient to observe the spread in latitude.

In our neighbourhood the total effective thickness of the layer of interstellar clouds may be estimated to about 250 ps. We have no data about its thickness in regions nearer to the centre; but assuming for a moment that it would be everywhere of the same thickness, the apparent thickness would become equal to the beamwidth of the 7·5-m Würzburg at a distance of about 5 kps. The layer may well be thinner in the inner regions. The available observations suffice to show that also in the nuclear part of the Galaxy the gas clouds are strongly concentrated to the galactic plane. So far no indication has been found of clouds at high elevations above the galactic plane, such as have at times been proposed to explain the shadow effects observed in the nuclear parts of spiral nebulae. For a trustworthy distinction between r_1 and r_2

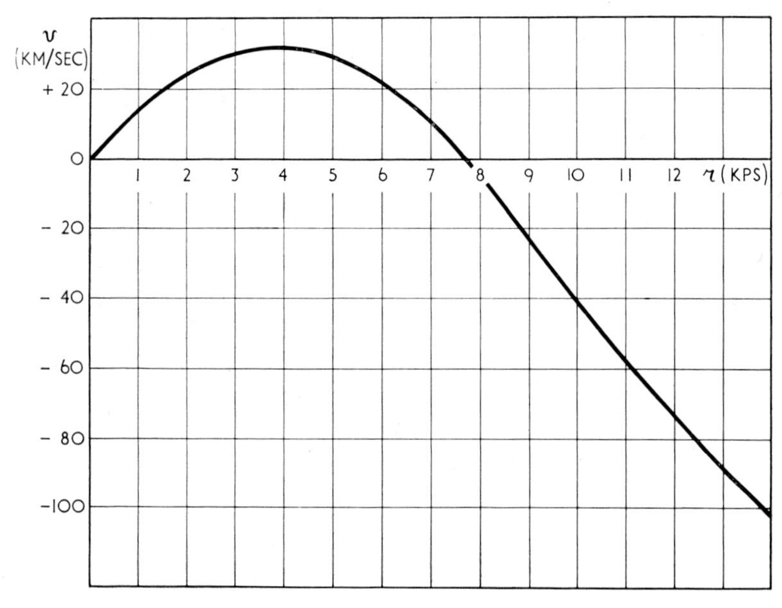

Fig. 4. Relation between radial velocity v and distance from the Sun r for galactic longitude $l = 30°$

we may have to await results with larger instruments. The 25-m paraboloid that is now being constructed in the Netherlands will be suitable for this type of investigation.

Some governing features of the hydrogen distribution in the inner parts can already at a first glance be inferred from the line contours already available; for instance, the presence of a presumably very powerful arm seen in longitudes 350° to 0° between $R = 3{\cdot}3$ and 4·6 kps. The sudden drop in density inside $R = 3{\cdot}3$, shown by the contraction of the line contours (cf. Fig. 2), is quite striking.

From the values of $v_{\max}$ we can get at least an approximate impression of the manner in which the angular rotation varies with distance from the centre. In the above-mentioned inner arm it reaches a value of roughly 55 km per sec.-kps, *i.e.* about twice that near the Sun. The determination of the rotational velocity at different distances from the centre is of importance, because it has to serve as a basis for the relation between v and r, but also because it shows the manner in which the overall density in the Galactic System varies as function of R. Preliminary results indicate that for $R < 5$ kps the density may be roughly six times that in the neighbourhood of the Sun.

The relation between ω and R in the outer regions, used for the construction of Fig. 3, has been taken from *Ap. J.*, 116, 233 (Fig. 5), except that R_0 has been reduced to 8·5 kps (instead to the value of 9·4 kps used in that article). It may give a fair approximation to the true velocities for values of R that do not deviate too much from R_0. For larger values of $R - R_0$ the values of $\omega(R)$ were computed from a rough model of the Galactic System, that could now certainly be improved. For the present the distance-scale for $|R - R_0| > 3$ kps must be considered as quite uncertain.

A more serious uncertainty arises from the fact that the systematic velocities may not always be circular and for the same R may differ at different position angles from the centre. A partial check on the circular character of the motions is furnished by the ordinary observations of differential galactic rotation, which show that the differential motions in our surroundings are practically perpendicular to the direction to the centre. The profiles of the 21-cm line observed in the direction to the centre and opposite to it give another and more important check. The profile in the direction to the centre is symmetrical around zero velocity, indicating that, in this direction at least, there are no large-scale deviations from circular motion. The statement applies more or less to the entire region down to the centre, for in the wings of the line we penetrate to large distances. In the direction of the anti-centre the line is slightly asymmetrical, which causes a displacement of about -1 km per sec. from the true zero. The asymmetry may possibly be due to a part of a distant arm that has a very slight inward motion; this requires further detailed investigation.

3. Random Cloud Motions

As has already been mentioned, the observed Doppler shifts are partly due to the systematic effect of differential rotation, and partly to the random motions of the

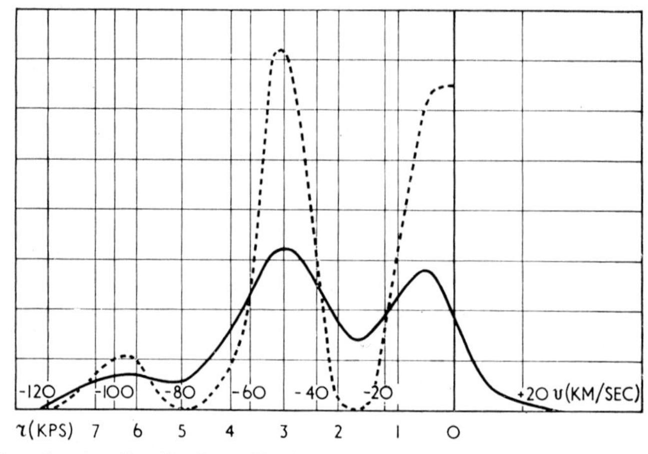

Fig. 5. True density distribution of hydrogen in the direction of 80° longitude (dotted curve). The full curve gives the observed line profile

interstellar clouds. The former is directly related to distance. If we could remove the effect of the random motions the corrected line profile would give the true distance-distribution. This correction for random motions can be effected statistically if we know the distribution of random motions. Fig. 5 shows an example of the result for $l = 80°$. Here the full-drawn line represents the observed line profile, while the broken line gives the true distribution of the hydrogen atoms. The velocity-distribution was assumed to have the form $\frac{1}{2\eta} e^{-|v|/\eta}$, found by Blaauw from interstellar absorption lines. The average radial velocity η was determined from the tail of positive velocities, which must be wholly due to random motions; it was found to be 8·5 km per sec., in good accord with what Blaauw found from the absorption lines.

The line profiles shown in Fig. 5 are typical for the region extending from about 60° to 105° longitude. They show that the principal outer arm in this region has an equivalent width of about 0·8 kps. The regions between the arms are remarkably empty.

4. Temperature and Density; Hydrogen Molecules

In principle, the measurement of the intensity of the 21-cm line permits also to determine both the temperature and the absolute density of interstellar hydrogen. In those directions where the line attains a large optical thickness the intensity in its centre must be equal to that of a black body with the same "temperature" as the interstellar hydrogen. Strictly speaking, the notion of "temperature" in this case applies to the Boltzmann temperature corresponding to the relative population of the two different "spin levels" of the hydrogen atoms. Purcell has shown, however, that this "spin temperature" must be equal to the kinetic temperature in the clouds where hydrogen is non-ionized, because the population of the upper spin levels is sufficiently rapidly replenished by the interaction of passing atoms. Measures of absolute intensities are very difficult. So far we only possess rough estimates, indicating a temperature of about 100°. It is noteworthy that this agrees, within the uncertainties, with the temperature of 60° estimated by Spitzer and Savedoff from theoretical consideration of the different processes of energy exchanges taking place in interstellar space.* It may be remarked that the energy difference between the spin levels is so small that at this temperature they are almost exactly equally populated. The occupation of the upper levels is only 0·07 per cent less than that of the lowest level.

The density of hydrogen can be estimated by measuring the total radiation received from a layer of known linear dimension. The measurements have the same uncertainty as in the case of the temperature. We find for the arm through the Sun the provisional value of 0·9 atoms per cm^3. The density in the arm through h and χ Persei is about the same.

It seems probable that most of the hydrogen is in atomic form—although in very dense clouds H_2 molecules may, apparently, be formed. A first indication of probable combination into molecules has been observed in one of the darkest nebulae in Taurus, where the 21-cm radiation from the cloud was found to be negligible, although it should have been of the same order as in the Milky Way if the hydrogen had been in the atomic state.

The measures of the 21-cm emission enable us to obtain a much better determination of the position of the galactic plane, for which up to the present the principal parts of the Galaxy could not be used. The required measures are now being made.

In the outer parts the arms may deviate considerably from the overall plane of symmetry. Thus, at $l = 50°$, the most distant arm appears to lie at a latitude of

* *Ap. J.*, **111**, 593–608, 1950.

$+ 1^\circ.5$, the gas in this region being situated about 220 ps north of the galactic plane defined by the inner regions. It is interesting to note that the distant δ Cephei variables in the Cygnus cloud are at comparable distances above the galactic plane. These deviations may indicate that the clouds in the thin outer regions have not had time to cancel out entirely the motions perpendicular to the plane of symmetry, which they must have acquired since the System was born from a chaotic mass of gas.

5. Evolution of the Spiral Arms

The observation of spiral arms in the Galactic System, first discovered over a relatively small region by W. W. Morgan, Sharpless and Osterbrock in 1951 (their results agreeing well with the measures of the 21-cm line in this region), opens the door to an important domain of investigations. In the first place we may inquire what the influence of the arms is on the general gravitational field of the Galaxy and on stellar motions. The gas represents about 20 per cent of the total density, so that the arms are relatively massive. Their influence will be especially pronounced in the direction perpendicular to the galactic plane. Their contribution to the components of the force parallel to the galactic plane is limited, because in that case the general force exerted by the more central parts of the Galaxy predominates. This must be the reason why, notwithstanding the presence of strongly inclined arms, the absolute as well as differential rotation in our vicinity is relatively regular and practically perpendicular to the radius vector from the centre.

In the second place, it is interesting to inquire which stars, besides the gas, are concentrated in the arms. Evidently all young stars, like those of types O and B, are situated in the arms. But does the same hold also for stars of type A, and possibly for stars of still later types? And is there a relation between the spiral phenomenon and the deviation of the vertex?

In the third place, it is evidently interesting to inquire into the development and into the past history of the observed spiral arms. In our region of the System the differences in angular velocity are considerable. The derivative $d\omega/dR = -2A/R_0$ is $-0{\cdot}47$ radians per 10^8 years per kps. In the general region around the Sun we observe differences of the order of 2 kps in R over the observed stretches of arms. This means that in a hundred million years the ends would be displaced relative to each other by an amount of the order of a full radian. If that is so, the shape of the arms must have changed entirely in the last half-revolution of the Galactic System. The formation of the present arms should then have taken place within the last 100 or 200 million years. A new vista into the evolutionary process of a galaxy may therefore begin to disclose itself.

The new information obtainable through the radio-emission line of hydrogen does not end in the Galactic System. It is to be expected that it will yield data of fundamental significance concerning the gas content of at least the nearer extra-galactic systems. In the Netherlands we have not yet made any extra-galactic measures. The Australian observers, however, have already announced that they have observed the 21-cm emission from both Magellanic Clouds, and that it has there an intensity comparable to that observed in the Milky Way regions.

SECTION 7

SOLAR PHYSICS

"Wo viel Licht ist, ist starker Schatten!"

JOHANN WOLFGANG VON GOETHE, *Götz von Berlichingen*, Act I, 1773.

Early Solar Research at Mount Wilson

W. S. ADAMS
Mount Wilson Observatory, Pasadena, California

IT is appropriate that a commemorative volume marking an important epoch in the life of our friend and colleague Professor STRATTON should contain a communication, however slight, in the field to which he has devoted a lifetime of fruitful and constructive research, the field of solar and stellar spectroscopy. In addition, it is fitting that it should deal with work of a pioneering character carried on under somewhat difficult physical conditions, because Professor STRATTON is no stranger to such situations, having conducted solar eclipse expeditions to many of the waste places of the Earth. The subject which I have selected has to do with the development of solar research during the early years of the Mount Wilson Observatory in Southern California. It is familiar ground for STRATTON, who in later years has visited it occasionally, and has always maintained a deep interest in its growth and progress.

In 1903 Dr. GEORGE E. HALE, Director of the Yerkes Observatory, spent a winter in Southern California. A distinguished solar investigator and an ardent "sun-worshipper", as he called himself, he enjoyed the many sunny days of the semi-tropical climate and often turned his eyes toward the range of mountains to the northward which bounded the coastal plain. The most accessible of these, although the accessibility was limited to two foot-trails passable by donkeys and mules, was Mount Wilson, which rose to a height of 5700 ft, nine miles in an air-line from the small city of Pasadena. HALE made several trips to Mount Wilson and observed the sun with a 4-in. portable telescope, each time with increasing enthusiasm over the observing conditions.

On his return to the Yerkes Observatory HALE made application to the recently established Carnegie Institution of Washington for a grant to bring a large solar telescope to Mount Wilson upon an expeditionary basis. The grant was made and with three members of the Yerkes staff to accompany him HALE started on his new venture.

The Snow Telescope, as the instrument was named after its donor, was wholly of the reflecting type and consisted of three mirrors. The first and largest of these, a flat mirror 30 in. in diameter, received the direct sunlight. It was mounted in a cell with its axis pointed toward the north pole, and rotated by clockwork to follow the apparent course of the Sun. A second flat mirror, somewhat smaller in size, placed to the south of the large mirror and slightly above it, intercepted the beam of reflected sunlight and sent it northward through a long house to an image-forming concave mirror. This third mirror had a focal length of 60 ft and formed an image of the Sun $6\frac{1}{4}$ in. in diameter. Rotation of this mirror made it possible to throw the image upon any auxiliary instrument within the house.

Fortunately the telescope had no very massive parts, since methods of transportation were primitive, and the limits of size and weight were fixed by what the pack animals could carry over nine miles of twisting mountain trails. The limits of weight were about 200 lb for the powerful mules, and 150 lb or less for the donkeys, or burros, to give them their accepted western name. All these animals were highly temperamental, and their characteristics were doubtless responsible for the development of the now rapidly vanishing group of men of remarkable skill and lurid vocabulary who drove them. It is interesting to realize that in the Snow Telescope building, of steel construction and 180 ft long, no single piece of the framework exceeds 8 ft in length because of these restrictions upon transportation. The problem of the very few pieces of the mounting of the telescope which exceeded the standard limit was solved by HALE through his invention of a narrow truck, with a length of 10 ft and a tread of 20 in., which was drawn by a mule and steered at both ends. This picturesque vehicle was used until the trail was widened into a road to provide for the transportation of the larger telescopes of later years.

Construction of the large stone piers to support the telescope and auxiliary instruments, and of the building to house them, was completed in 1904–5. The building itself was of unusual construction, the inner wall consisting of continuous sheets of canvas, and the outer wall of canvas louvers which permitted circulation of air but shut out direct sunlight. The problem of reducing the injurious effect upon the definition of the solar image of warm currents of air rising from bare ground heated by the Sun led to several experiments. The growth of brush and grass over bare areas near the telescope was encouraged, artificial covering with light white cloth was tried to a limited extent, and some tests were made of the effect of stirring the air inside the telescope building. Probably the most remarkable experiment was that of hoisting an observer and a 4-in. telescope to a height of some 50 ft in the air in a large pine tree, and comparing the quality of the Sun's image observed at this height with that at ground level. It was these experiments which led HALE two or three years later to design the tower telescope, in which the sunlight incident upon the first mirror is received high above the ground and is reflected downward through a shielded vertical tube. A marked improvement was observed in the quality of the image, and this form of telescope has been adopted widely by other observatories.

The auxiliary instruments used with the Snow Telescope were in general three in number. The first was a high-speed photographic shutter which provided direct photographs of the Sun, showing in detail the structure of sunspots and other features of the surface. The second was a spectrograph for analysing the Sun's light and studying the behaviour of the thousands of spectral lines of the elements which compose the Sun's atmosphere. The third was a spectroheliograph, an instrument invented and used extensively by HALE. With this ingenious device the distribution of the white-hot clouds of individual gases over the surface of the Sun may be photographed and their forms analysed.

Two major investigations were undertaken with these instruments, both of them dealing with the nature of sunspots. One was with the large spectrograph, the other with the spectroheliograph. Although quite separate in their beginnings their final results have united to provide an adequate explanation of most of the observed characteristics of sunspots, and at least a partial insight into the phenomena of their origin.

The first of these investigations was a photographic study of the spectrum of sunspots. It had been known to astronomers for a number of years that the spectrum of a sunspot differs from that of the surface of the Sun outside a spot in several respects. The whole spectrum of the spot is fainter, as might be expected from its comparative darkness, but in addition there are two important differences in the spectral lines. First, many lines are strengthened in spots as compared with the disk, and others, fewer in number, are weakened. Secondly, nearly all the lines are considerably widened in spots, some even appearing double. Few, if any, photographic observations, however, had been made, due to the fact that spectrographs of adequate power and size could not be attached to moving telescopes, and forms like the Snow Telescope had not come into general use except for solar eclipses.

The first photographs of the sunspot spectrum showed a wealth of detail which made it possible to catalogue many hundreds of lines according to their behaviour. The most interesting feature was that all lines of the same element did not behave alike. Some of the lines of iron, for example, were greatly strengthened in the spot as compared with the disk, others were little changed, and a few were weakened. These weakened lines were for the most part recognized as belonging to the class of "enhanced" lines, that is, lines which when investigated in laboratory sources are found to be greatly intensified in the spectrum of the electric spark as compared with the electric arc. For the lines which were strengthened, however, no such differentiation was available. This was long before the theory of ionization had provided a logical explanation of the classification of spectral lines according to energy levels in the atom. Hence the Mount Wilson work was of necessity almost wholly on an empirical basis.

As a working hypothesis it was assumed that the temperature of a sunspot is lower than that of the general surface of the Sun. This seemed reasonable on several grounds, and it was soon proved to be correct through the discovery of molecular bands of titanium oxide in the yellow and red portions of the spectrum. The next step was to try to imitate the behaviour of lines in the spot spectrum through observations in laboratory sources. A small spectroscopic laboratory was available on the mountain, and Professor GALE, of the Physics Department of the University of Chicago, who was spending a year at Mount Wilson, took an active part in the observations. The first element examined was iron which has a spectrum rich in lines showing wide differences of behaviour in sunspots. At first the procedure was to photograph the spectrum of a direct-current iron arc, varying the amount of current through as large a range as possible, but later the simple method was adopted of comparing the outer and cooler flame of the arc discharge with the inner and hotter core near the poles. In both cases considerable differences in temperature were attained.

An examination of the photographs showed conclusive results. Many lines were relatively strengthened in the outer flame or cooler portions of the arc, and many others were most intense in the hot central core. The correspondence with sunspots was striking. The lines of iron conspicuously stronger in spots than on the disk were those which were relatively strengthened at low temperatures in the arc, while the lines which were unaffected or sometimes slightly weakened in spots were those which were prominent at high temperatures in the arc. The investigation was later extended to other elements prominent in spot spectra, such as titanium, vanadium, and chromium, with very similar results.

The products of this investigation have had some interesting applications. Not only did they solve many of the complexities of the sunspot spectrum, but they also provided a means for separating spectral lines into classes according to their behaviour with temperature. When the theory of ionization was developed this temperature-classification became of great importance, since it formed a starting point in the problem of analysing the spectral lines of the various elements according to energy levels in their atoms, one of the great accomplishments of modern spectroscopy.

In later years, when the 60-in. reflecting telescope was in operation at Mount Wilson, the spectra of stars were examined in detail with a view to applying the results found in sunspots. Many stars have spectra closely resembling that of the Sun, and many others that of sunspots. In general the agreement was found to be close, but as more and more stars were examined some anomalies were discovered. A few spectral lines seemed to be abnormally strong or weak in spectra which otherwise were nearly identical. These differences proved to be associated with the intrinsic luminosities of the stars, or the total amount of light they emit. The luminosity of a star may readily be calculated from its apparent brightness, if its distance is known, merely by applying the inverse square law of distance. At the time of this investigation relatively few accurate determinations of stellar distances were known. Enough were available, however, to establish a satisfactory correlation between the intensities of the anomalous lines in the spectra and the luminosities of the stars observed. Once such a correlation was established the process could be reversed, and the previously unknown distance of a star could be calculated from its apparent brightness and luminosity.

The final result of this application of the sunspot investigation was the discovery of a new method for determining the distances of stars, and for classifying them into groups according to luminosity and possible order of evolution. It forms an interesting illustration of the ramifications of researches in physical science and of the unforeseen consequences which may follow them.

Another discovery of extraordinary interest came out of the study of sunspots begun with the Snow Telescope. It had its origin in the use of the spectroheliograph. The earlier work with this instrument had consisted in the recording of the calcium flocculi, or clouds of calcium gas, over the surface of the Sun. Such clouds are especially numerous and dense in the vicinity of sunspots, and measurement of their areas furnishes an excellent index of the general state of solar activity. Their forms, however, are irregular, and often they overlie and obliterate the spots, due no doubt to the high level of the calcium gas in the Sun's atmosphere.

Hale had long wished to utilize spectral lines other than those of calcium in the spectroheliograph, more especially those of hydrogen, which is the most abundant element in the Sun and stars. The practical difficulties, however, were considerable at this time. The strongest and most suitable line of hydrogen lies in the red portion of the spectrum, and photographic plates sensitive to red light were not as yet available commercially. Still it was known that certain organic dyes increased sensitivity to red light, and that emulsions treated with these dyes, when dried rapidly, and exposed immediately, might give reasonably satisfactory results. So the experiment was tried, the slit of the spectroheliograph being set on the α-line of hydrogen across a spot group near the centre of the Sun.

The resulting negative had little to commend it from the photographic point of view, but scientifically it was of great interest. Two features characterized the clouds of hydrogen gas around the sunspots. In the first place they were dark against the solar background instead of being bright as were the calcium clouds. In the second place the hydrogen filament showed distinct evidence of vortical structure, with spiral arms centring around the centres of the spots. Later photographs indicated that over areas of intense solar activity portions of the hydrogen clouds might even become bright and be seen as the "solar flares" which affect the transmission of radio waves in the Earth's atmosphere.

It was the vortical structure around spots which interested HALE particularly. He remembered ROWLAND'S observation that a rapidly rotating electrically charged plate produces a magnetic field; and ZEEMAN'S discovery that a magnetic field splits up the spectral lines emitted by a source placed in the field and polarizes them, that is, separates the plane of vibration of the light-waves of the different components according as they are viewed along or across the lines of magnetic force. HALE remembered, too, the widening and occasional doubling of lines observed in the spectrum of sunspots. Could the vortical motion observed indicate rapid revolution within the spot-vortices of streams of electrically-charged corpuscles, corresponding to Rowland's charged plate, and thus producing a magnetic field? If so the field could be detected through the Zeeman effect and the polarization of the sunspot lines. The optical apparatus required to test for the effect was simple to construct, and on a day late in June, 1908, HALE examined a suitable spot with a large spectrograph and the 60-ft tower telescope, which had recently been completed. The results were definite and conclusive in showing that sunspots are the centres of strong magnetic fields, and thus provided the first clear evidence for the existence of magnetic forces outside the Earth.

With the discovery of the Zeeman effect in sunspots the second chief characteristic of their spectrum, the widening and occasional doubling of the lines, found a rational explanation. In later years there followed at Mount Wilson three discoveries of importance: the reversal and return to normal of the direction of the magnetic field in spots in the two hemispheres of the Sun in a period of twenty-three years; the existence of very strong magnetic fields in certain stars; and most recently the presence of changing fields of moderate strength over the surface of the Sun outside of spots. The impetus given to solar research by the early sunspot investigations has persisted throughout nearly half a century.

This brief outline of some of the early solar research at Mount Wilson does not include any attempt to define the contribution of each individual to the investigations undertaken, except in the case of DR. HALE. He was the leader at all times, and his insight and enthusiasm were the driving force and inspiration of those early years. Most of the research was carried on through the collaboration of two or more individuals, and the results were published jointly. These may be found in Volumes 1 and 2 of the *Contributions of the Mount Wilson Observatory* and the *Astrophysical Journal* for the years 1904–1908.

Thirty Years of Solar Work at Arcetri

GIORGIO ABETTI

Osservatorio Astrofisico di Arcetri, Firenze, Italy

SUMMARY

A short review is given of the solar work done at the Astrophysical Observatory in Arcetri from the year 1922 to 1952. The motions and behaviour of bright flocculi and dark filaments have been investigated. Flare intensities have been determined, and compared with those determined by other observatories.

Systematic measures of the height of the chromosphere and of the solar rotation have been made. Researches on the curve of growth, abundance of elements, equivalent width of the spectral lines, motion of gases in sunspots, and other topics, have been included in the programme of work.

THE Arcetri Observatory was founded by G. B. DONATI in the year 1872 and continued under the direction of ANTONIO ABETTI from 1893 until 1921. Astrophysical work began in that year with a regular series of solar observations with the Amici equatorial, with the construction of a solar tower, and of a prismatic reflector for researches in stellar spectroscopy.

In this note, taking great pleasure in this opportunity to honour my eminent colleague and friend Professor F. J. M. STRATTON, I give a short account of the solar work done in Arcetri during about the last thirty years.

Details of this work will be found mainly in *Osservazioni e Memorie dell'Osservatorio di Arcetri* or *Memorie della Società Astronomica Italiana* and in the publications of the *Accademia dei Lincei*.

1. FLOCCULI, FILAMENTS, PROMINENCES

According to the international programme of the International Astronomical Union a series of observations of prominences at the limb has been carried out, since 1921, with a visual spectroscope attached to the Amici equatorial (36 cm aperture, 530 cm focal length). Besides these, when the seeing has been good or fairly good, observations have been made of the height of the chromosphere in the line Hα, using a radial slit at every 30° around the solar limb. The frequency and behaviour of prominences have been studied, their mean heliographic latitude determined, and their migrations towards the poles or equator analysed through the various phases of the cycle.

When the solar tower was completed in the year 1925, daily spectroheliograms were taken in the light of Hα and $K_{2,\ 3}$ and these have been continued regularly on about 150 days every year. From 1932 to 1949 the monthly and yearly means of the projected areas of the bright Hα and $K_{2,\ 3}$ flocculi and of filaments have been measured by RIGHINI and GODOLI. These authors have also made statistical researches on the cycle of the flocculi, finding that the activity of the flocculi has a delay of 0·3 months with reference to the activity of the spots, and also that the mean heliographic latitude of sunspots is smaller than that of the bright flocculi. This difference is variable from about 1°–3°, depending on the monochromatic radiation employed: it seems that this difference is correlated with the observed chromospheric level.

Considering filaments, *i.e.* the prominences projected on the solar disk, and the prominences as seen at the limb, results in better agreement with the present view on the dynamic relations of the various phenomena have been obtained, without separating them into two zones of high and low latitude. In the first stage of their development the filaments are small and irregularly curved, and undergo violent internal motions. Often from their higher parts material is emitted which follows curved trajectories, to be absorbed in regions where there may or may not be spots. Little by little the filaments become more stable, and moving away from the equator

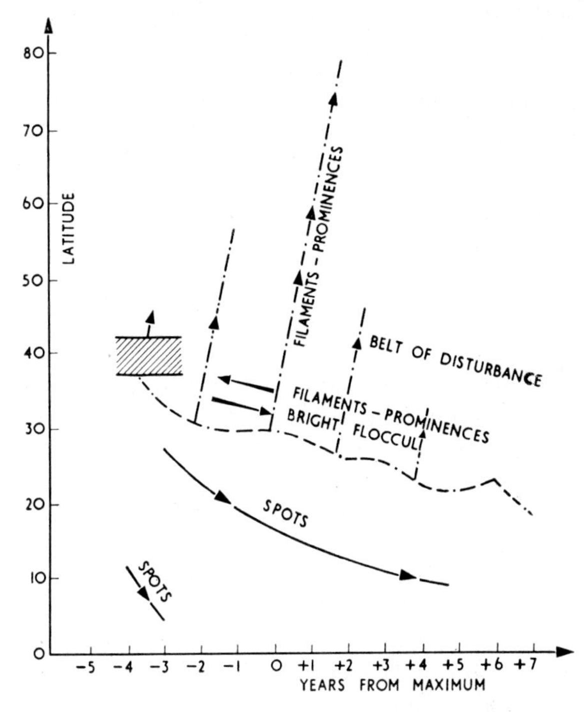

Fig. 1. Hypothesis concerning the migration in latitude of solar phenomena during the eleven-year cycle

they assume a characteristic orientation parallel to solar meridians, changing form in consequence of the polar retardation, and always extending towards higher latitudes. The filaments normally reach their maximum development at the fourth passage across the central meridian, that is in about three months. At that time they present a remarkable curvature because the ends nearer to the poles distribute themselves along, or almost along, the parallels of latitude, so that we see the filaments under various perspective angles. Probably prominences of high and low latitude have the same origin. From the spectroheliograms in hydrogen and calcium, we may conclude that the filament-prominences are probably the same equatorial filament-prominences which little by little emigrate towards higher latitudes, following the general circulation towards the poles and bending little by little along the solar parallels because of the polar retardation of the solar rotation-period. Possibly (Fig. 1) at the beginning of the eleven-year period, that is around the minimum activity of the Sun, perturbations begin to appear in the photosphere in a fairly extended zone of latitudes at around $\pm 40°$; pores and small spots below

$\pm$ 40°, and filaments or prominences above $\pm$ 40°. As the cycle proceeds, the perturbations on one side migrate towards low latitudes with spots and bright flocculi, towards high latitudes with filament-prominences only. The poles are reached, or almost reached, at the epoch of maximum general activity of the Sun. The beginning of the new cycle may arrive from 3–6 years before the maximum, depending on the intensity of the cycle; therefore after maximum the perturbed zone in lower latitudes may last from 8–5 years, continuing to proceed with more or less speed towards the equator, until after a corresponding delay a new perturbed zone at mean latitude $\pm$ 40° shows the beginning of a new cycle.

2. Flares

Precise photometric determinations of the intensity of flares have been made by Ballario on the spectroheliograms taken with the Arcetri solar tower, relative to the scale of importance internationally adopted to-day, with the following results:

Importance	H_α	K_{23}
1	1·6	3·4
2	1·8	4·3
3	2·0	4·8

These intensities are referred to the intensity of the undisturbed chromosphere in $H\alpha$ and K_{23}. The minimum intensity for a bright flocculus to be classified of importance 1 is 0·45 of the continuous spectrum of the centre, both for $H\alpha$ and K_{23}.

The intensities of flares and bright flocculi have also been measured by Ballario on the spectroheliograms taken at the Meudon Observatory and a correlation has been established between the two scales.

3. Height of Chromosphere

The results of a series of observations with the Amici equatorial, aiming at the determination of the height of the chromosphere in $H\alpha$ and extending over 25 years, may be summarized as follows: the mean height of reversal of the line $H\alpha$, observed with the slit of the spectroscope set perpendicular to the Sun's limb, is equal to 10″5 $\pm$ 0″5 for the entire period. Whilst the mean height does not seem to change in relation to the eleven-year cycle, or at most within the limits of the mean errors, it is possible to deduce from the observations a fact which seems well established. In the course of the eleven-year cycle the chromosphere either may be uniformly distributed around the solar disk, or it may be higher at the poles than at the equator. Referring the observed heights in the various heliographic latitudes to those measured at the north pole of the Sun, the percentage values as a function of the latitude for every 30° and for every year are obtained. The resulting curves present a definite sinusoidal shape around the minima, and are flat with small irregularities around the maxima. Giving an empirical index to the sinuosity (namely the sum, in absolute value, of the four differences: north pole – equator, equator – south pole, south pole – equator, equator – north pole), the yearly values of this index have been computed. The result is that they agree perfectly with the Zurich relative numbers and with the numbers of prominences. Curve *C* (Fig. 2) shows the distribution and form of the chromosphere in hydrogen during the eleven-year cycle compared with the other phenomena of solar activity.

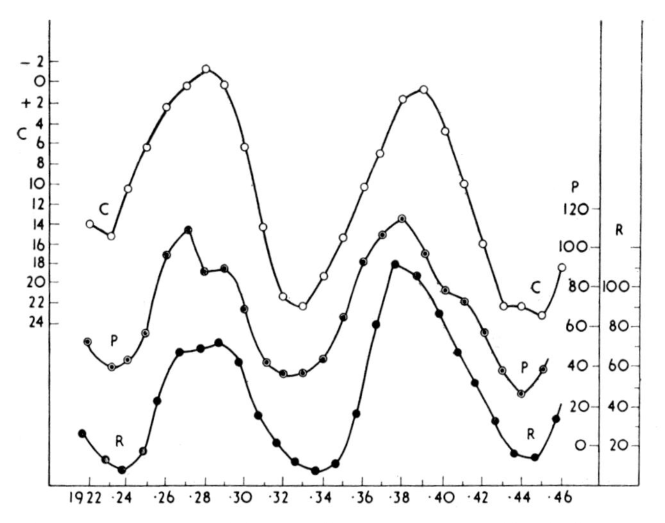

Fig. 2. Correlations between the height of the chromosphere (C), the frequency of prominences (P), and the relative numbers of sunspots (R); derived from the Arcetri observations obtained between 1922 and 1946

4. SOLAR ROTATION

Several series of spectroscopic determinations of solar rotation have been made with the solar tower from 1927 to 1944, in order to make a contribution to the problem of the rotation speeds at different latitudes and of various layers, and of the possible variability of the mean value. The conclusions which may be drawn from these observations are as follows:

(*a*) The various Fraunhofer lines, on account of their different intensities, their different energy levels, the spectral region to which they belong, or other causes due to blends, *etc.*, give widely divergent velocity values.

(*b*) Different instrumental conditions and different observers may give values which are not concordant within the limits of observational errors, indicating the presence of systematic errors due to the size of the solar image used and to the point of the limb observed.

(*c*) With the same instrument and the same observer discordant values may be obtained, probably because of atmospheric conditions (transparency of the sky, scattered light, steadiness of the solar limb).

Under these circumstances it seems more probable that for the time in which observations of this kind have been made, the period of solar rotation at a given latitude and height is constant, at least within the limits of the observational errors.

5. CURVE OF GROWTH

Since the year 1931 G. RIGHINI has made, with the solar tower, a series of researches on the variation of the line intensities from the centre to the limb of the Sun, the curve of growth, the abundance of elements, *etc.* With precise photometric methods the profiles of the intense Mg lines and their variations from the centre to the limb have been investigated. These absorption lines at the limb are produced at an optical depth which is only 4 per cent of the corresponding one at the centre, and the abundance of Mg atoms in the solar atmosphere has been deduced.

Following this research other lines of moderate and faint intensity have been measured, with the result that the faint lines are enhanced at the limb, whilst those of moderate intensity are weakened. It has been possible to conclude that the damping constant of the atoms is greater at the centre than at the limb. The curve of growth method for the analysis of the structure of the solar atmosphere was initiated by RIGHINI in 1933, taking into account all the measurable lines of one element. In this research it has been possible to prove that the damping constant is always greater than the theoretical value given by atomic physics, and this indicates the importance of collisions in the solar atmosphere. The damping constant for the Na lines and their optical depth has been computed, taking into account the simultaneous effect of the collisions with hydrogen atoms and electrons. Comparing these constants with those determined by means of the curve of growth it has been possible to deduce that the mean optical depth at which this absorption originates is 0·08. This result proves that the Na atoms effective in the production of the lines are located in a thin layer and high in the solar atmosphere. It has been proved also that the lines in the ultraviolet and in the infrared give anomalous results which can be explained with an absorption coefficient greater than the normal one.

6. Calibration of King's Scale

The intensities of the spectral lines, estimated by KING years ago on furnace spectra of Fe I, Ti I, and Va I, have been calibrated by RIGHINI, using gf values measured more recently by KING on absorption spectra for some selected multiplets of the same elements. The work, based on about 3000 experimental values, gives the logarithm of the number of atoms involved in the formation of lines derived from their empirical intensities. From KING's scale it is possible to obtain log Nf with an error of $\pm$ 0·37, so that we have the possibility of using rich spectroscopic material for astrophysical purposes.

RIGHINI and BAROCAS have measured the equivalent width of 18 Ti I lines in the spectrum of the centre of the disk, and at four other points along a radius corresponding to cos θ = 0·76, 0·59, 0·32, 0·10. The variation of the equivalent width does not seem to depend on the wavelength, nor on the excitation potential of the atomic levels; the equivalent width attains a maximum at a point corresponding to cos θ = 0·30 approximately. From the line profiles, corrected for the resolving power of the spectrograph, we may conclude that (*a*) the damping constant is much higher than the classical one; (*b*) the line width is greater than the thermal width, so that it is necessary to assume a turbulent motion of the Ti atoms; (*c*) the turbulent velocity has a minimum of 2 km per sec. along the normal to the solar surface, and a maximum of 4·7 km per sec. at an angle of 70°.

7. Motion of Gases in Sunspots

The radial motions of the gases in sunspots, discovered by EVERSHED in 1909, have been investigated with the Arcetri solar tower for twenty-six spot-groups during the years from 1926–1932. The following conclusions have been drawn from these observations:

(*a*) A relation is found between the intensity and excitation potential of the lines and their radial velocities.

(*b*) The motion of the vapours over the spots is irregular and variable, although in the majority of cases it conforms to the Evershed effect. Radial velocities in these observations have been found to vary between almost zero and values which attain a maximum of 6 km per sec.

(*c*) It has been found that there is nearly always a variable tangential motion; this gives the direction of the vortical rotation over the spots at the spectroscopic level of the lines being measured, and the projection of the motion on the plane of the spots is found to be along a logarithmic spiral.

For the metallic vapours it is found that the rotation of the vortex is anticyclonic, that is right-handed, in the northern hemisphere. Comparing this rotation with that deduced from the hydrogen spectroheliograms at a higher level, it is found that the two directions are opposite, the rotation for hydrogen being cyclonic. The direction of motion of the metallic vapours over the spots, right-handed in the northern hemisphere and left-handed in the southern, is that which would be expected as a consequence of the combination of the radial motion with the Sun's rotation. On the other hand, in the chromosphere where the gases move toward and enter the spots, the filaments are curved in a contrary direction to that of the gases at lower levels, as we observe in terrestrial cyclones. The vortices, therefore, are of hydrodynamic character and seem not to depend on magnetic fields.

A theory of the problem of solar rotation and a new set of modern tables to reduce the spectroscopic observations have been developed and computed by F. ZAGAR.

8. INTENSITIES OF FRAUNHOFER LINES

In they year 1933 ABETTI and CASTELLI determined with the solar tower the central intensities of ninety-seven Fraunhofer lines, of neutral and ionized atoms and of cyanogen bands, at the centre and limb. From a discussion of the results it is found that the level at which the lines of ionized atoms are formed is deeper than that of other lines. The central intensities increase with the excitation potential and are a maximum for the lines of ionized elements and the cyanogen bands; the difference in intensity at the centre and limb increases systematically in the same order. It is shown that there is a systematic difference between the intensities at the equatorial and polar limbs, probably in relation to the greater height of the chromosphere at the poles and to a small difference in temperature between the poles and the equator, in agreement with the solar theories.

9. SOLAR CORONA AND ECLIPSES

The shape of the solar corona, expressed in terms of its flattening according to LUDENDORFF, has been investigated by ABETTI and BIOZZI using eclipse observations between 1893 and 1936. The maximum elongation of the corona (equatorial type) arrives 0·7 years before the minimum of the prominences at high latitude, and about 3 years after the minimum of the prominences at low latitude. The minimum elongation (polar type) comes 0·3 years before the maximum of the prominences at high latitude, and 2·3 years after those of low latitude. It may thus be concluded that the shape of the corona follows closely the frequency of the prominences at high latitudes.

With a monochromator, attached to the solar tower, COLACEVICH has determined the profile of Hα and measured its central intensity, which amounts to 15·8, taking the intensity of the adjacent continuous solar spectrum as 100.

From the Arcetri Observatory two expeditions for the purpose of observing total eclipses of the Sun have been sent out and their tasks successfully fulfilled.

The first was to Sara in Baskiria, U.S.S.R., for the eclipse of 19th June, 1936. Results on the photometric characteristic, shape and spectrum of the corona were obtained and published in the Memoirs of the *Accademia d'Italia.*

The second was to Khartoum in the Sudan, for the eclipse of 25th February, 1952. Preliminary results have been published in the *Rendiconti dell'Accademia Nazionale dei Lincei*, while the plates of the corona and of the spectrum of chromosphere and corona are now under measurement and discussion at Arcetri.

During recent years a coronograph of the Lyot type has been built in the workshop of the Observatory. In order to find a suitable locality for this instrument, we have made investigations of the conditions of seeing and atmospheric scattering at various places. A simple instrument planned by FRACASTORO and called a "photo-cyanometer" has been employed to determine the absorption coefficient at Mount Majori (1560 m) in the Tuscan Appennines, at Mount Livrio (3174 m) in the Dolomites, Pic du Midi de Bigorre (2877 m) in the Pyrenees, and at Campo Imperatore (2136 m) in the Abruzzi region. Conclusions have been reached concerning the choice of a suitable place for observations of the solar corona with the coronograph. At present the instrument is undergoing final adjustment in the hands of A. COLACEVICH at the Observatory of Capodimonte at Naples.

10. SOLAR-TERRESTRIAL RELATIONSHIPS

Some work has been done on the problem of solar-terrestrial relationships. We have now eighteen cycles for which we know reliable relative numbers (*Wolf numbers*) of sunspots, and it may even be possible eventually to extend them back to the time of the discovery (1611) of sunspots. There have been many attempts to find, in these numbers, by means of harmonic analysis or other methods, longer or shorter periods superposed on the main eleven-year period. It is interesting to note that in recent work done by VERCELLI and POLLI on terrestrial phenomena, two periods shorter than the eleven-year cycle, respectively of 5 and 8 years, appear very evident. The radio-ionospheric measurements seem to prove that up to a certain degree of solar activity—as indicated by a certain Wolf number—there are no appreciable perturbations in what we can call a "quiet ionosphere": the two periods of quiet Sun and quiet ionosphere may coincide and last from 3–6 years in various eleven-year cycles, whilst, on the other hand, we have remaining 8 and 5 years in which terrestrial phenomena are more affected by solar ones.

Some Problems of the International Solar Survey

K. O. KIEPENHEUER

Fraunhofer Institut, Freiburg i.B., Germany

SUMMARY

The observation and measurement of various solar phenomena are critically reviewed. The relative sunspot numbers are not purely a solar measure but are strongly affected by the Earth's heliocentric position. Reduction to standard conditions is shown to be possible. The survey of faculae, flares, and prominences (filaments), still unsatisfactory, will be greatly improved by the world-wide use of the Lyot filter with automatic cameras and by the application of photo-electric photometry. Also, the daily photometry of the corona, at present mostly made visually, will gain in accuracy and in observing days per year by the use of a photo-electric method as proposed recently by Lyot. Solar magnetic fields in spots and outside the spot-belts will be important components of future solar surveys, suitable instruments already being available. The survey of the Sun's radio emission is already fairly complete. The intensity of solar radiation in the far ultra-violet as revealed daily by the measurable ionospheric electron densities could be obtained on an absolute scale by calibration through a few direct rocket measurements at heights above 100 km. The corpuscular radiation of the Sun is strongly disturbed in the vicinity of the Earth. Terrestrial observations therefore give no reliable solar measure. It seems as if better results can be obtained by observing the effects of this radiation on comet tails and on the brightness of the zodiacal light.

THE Sun is the only star presenting an observable surface. Since the invention of the telescope it has been surveyed with increasing effort. If we saw the Sun without surface and structure as we do other stars, and we could obtain only its total spectrum, we would hardly infer the existence of sunspots, faculae, prominences, and the corona. The contribution of these phenomena to the total spectrum is exceedingly small. Only the chromospheric flares, as observed on the Sun, are bright enough to be detected against the undisturbed spectrum of the other stars (JOY and HUMASON, 1949).

Synopsis of the manifold solar phenomena which are observed on the Sun's visible hemisphere seems to be simple at a first glance, in comparison with the synopsis of terrestrial meteorological conditions. As far as observation is concerned, several well-equipped observatories under good climatic conditions and well-distributed in longitude would serve. Such solar observatories would have at their disposal a number of very different instruments, *e.g.*:

- Refractor for observation and photography of the photosphere (sunspots, *etc.*).
- Monochromatic filter, spectroheliograph or spectrohelioscope for the observation and cinematography of the chromosphere (faculae, flares, prominences).
- Coronagraph for the daily photometry of the inner corona (lines and integrated light).
- Instrument for the measurement of solar magnetic fields.
- Instrument for the simultaneous measurement of solar noise at different frequencies.

There are only a very few observatories which have these instruments available, as well as the experts to build and to run them. The results of a solar survey are therefore based to-day essentially upon the international co-operation of a great number of observatories which mostly concentrate on the observation of a single solar phenomenon.

The reasons which determine that a solar survey be started are by no means always scientific or, in a wider sense, of a cultural nature. In the last 10 years solar surveys have improved enormously in a few countries because of the practical need to have a diagnosis and a prognosis of the quality of radio communications. As is known, these communications are strongly affected by solar activity through the ionospheric layers. Also, other solar-terrestrial relations are pushed into the foreground of the public interest, such as the solar influence on the Earth's magnetic field, on large-scale weather and even on several biological phenomena. Solar surveys do not serve solar physics only but other branches of science as well.

The astronomical goal of the solar survey will always be to record the variable features on the Sun in a complete and quantitative manner, and to eliminate the effect of the heliocentric position of the Earth as much as possible. As will be shown, this meets with great difficulties, because, with a rotating and spherical Sun, the observable phenomena are presented under a constantly varying aspect. Also, the objective synoptic representation of solar phenomena is difficult and still has a rather qualitative character in comparison with the terrestrial weather map.

In general, the data which result from solar surveys are overrated by astronomers and geophysicists. Very often no clear discrimination is made between a genuine property of the Sun and the terrestrial aspect of a solar property. In the following paragraphs some individual solar phenomena will be critically reviewed, their measurement as well as their definition being considered.

1. Sunspots

These may be counted in the form of Wolf's relative sunspot numbers, which are easily obtainable, and are independently available on all continents and without gaps. Apart from the uncertainties which come in by the choice of instrument, by the "seeing" and by the quality of the observer, the variable visibility of spot-groups at different distances from the centre of the solar disk is not taken account of. Also, the spots on the reverse side of the Sun are ignored. In this way, the influence of the heliocentric position of the Earth sometimes surpasses the real fluctuation of the solar spottedness.

The sunspot numbers and, to a small extent, the sunspot areas are not purely a solar measure. Their reduction to the solar centre by an empirical visibility-function can be made but is laborious (Roggenhausen, 1952). Also, the sunspots on the reverse side can be incorporated by extrapolating the development of spot-groups, which have appeared on the east limb or have disappeared on the west limb. The resulting global spot number is then a real solar measure. It has been shown (Becker and Kiepenheuer, 1953) that the 27-day wave has practically disappeared; the influence of the Earth's position has therefore been almost eliminated.

When averaging over several solar rotations, the Wolf numbers and the global spot numbers are of equal value. In this case, however, no information about the day-to-day variations is given. In the study of solar-terrestrial relations by means of the unreduced daily sunspot numbers it is generally assumed that the visibility-function of sunspots and their terrestrial effect vary in the same way with distance from the solar centre. This is certainly not true and may happen only in a very few cases.

2. FACULAE

The recording of faculae, especially of the photospheric faculae, so far is made only in the form of rough sketches. Sometimes a combination of brightness and area is used as an index of facular activity. In none of the cases a visibility-function is applied, covering the variation of brightness with distance from the solar centre. In principle, exact photometry of faculae is possible. It is necessary then to measure photo-electrically the total excess of radiation in Hα, K, or another line within the facular region through a monochromatic filter. Such photometers are under preparation at different observatories.

3. FLARES

In spite of the co-operation of about twenty well-distributed observatories which observe flares with spectrohelioscopes, only half of the flares are obtained. The effective observing-time is decreased by 30 per cent during the winter because most of the observatories are situated in the northern hemisphere. Observation will become much more complete when more observatories in the southern hemisphere contribute and when automatic cameras are used, such as the flare-patrol-cameras in Boulder and on Sacramento Peak. The classification of flares is still made with a rather uncertain scale of estimation, which in practice represents a mixture of brightness and area. The photometry of flares is complicated by the rapid variation of their brightness and their line-profiles, by their filamentary structure and by the complicated shape of the chromospheric background. For the flares as well as for the faculae, it will be necessary to measure the total radiation in Hα by a monochromatic photo-electric photometer. The total number of Hα-quanta radiated by a flare will then be a correct measure of its intensity or importance.

4. PROMINENCES

On the disk prominences appear as dark filaments; on the limb as bright formations. Their daily observation is accomplished without gaps since 1919 (D'AZAMBUJA). The daily values, however, are not usually published. Variations within a day are recorded only occasionally.

On the disk, the true length of the filaments in kilometres is measured; on the limb, the so-called profile-areas are used. Since the limb prominences are visible only for a day or two, while the filaments are visible during half a solar rotation, the latter provide a much better measure of activity than the prominences. As a consequence of their homogeneous structure, their reduction to true length in kilometres can be managed for all distances from the central meridian.

The observation of filaments from minute to minute, especially in the centres of activity, gives a lot of information. The observation is best made by the use of an automatic camera with LYOT's interference filter for Hα. More than ten instruments of this kind are already in use all over the globe. The co-operation of these stations is the aim of Commission 11 of the I.A.U. The composition of a film without gaps is planned, which contains all chromospheric phenomena (faculae, flares, filaments, and prominences) at intervals of a few minutes. With a picture taken every 3 min, there will be 480 per day. This amounts on 35-mm film to a length of 8·5 m per day. A general publication of this material is out of the question. It would be possible, however, to lend copies to other institutions. It is to be hoped that this organization will come into operation soon. The main difficulty seems to be that of obtaining the electronic guiding devices, which are indispensable.

5. Corona

The corona is observable only at the solar limb and only within a limb-distance of about 1′. A regular survey is being made only in the light of the coronal lines $\lambda 5303$ and $\lambda 6374$, and occasionally in $\lambda 5694$. In integrated light (polarized component) only a few observations are available—in spite of the fact that Lyot (1930) has developed a simple photometer for this purpose. The following coronagraphs are in use at present:

Pic du Midi, France (since 1931); by the Observatoire de Paris.

Arosa, Switzerland (since 1939); by the Eidgenössische Sternwarte Zürich.

Wendelstein, Germany (since 1943); attached to the Munich Observatory.

Climax, U.S.A. (since 1946); by the High Altitude Station of Harvard University and the University of Colorado.

Kanzelhöhe, Austria (since 1948); attached to the Vienna Observatory.

Monte Norikura, Japan (since 1951); by the Tokyo Observatory.

Sacramento Peak, U.S.A. (since 1951); by the Air Force Cambridge Research Centre, U.S.A.

The co-operation of the four above-mentioned European coronagraphs gives, on the average, about 3–5 per cent of the days without observation, and about 15 per cent of days with second-rate observations.

In addition, coronagraphic work has been developed by the Stockholm Observatory since 1950 in Abisko and other places in Sweden, and since 1952 in Capri, Italy; furthermore by the Crimean Astrophysical Observatory in Simeis, U.S.S.R., and by the Fraunhofer Institut at Freiburg, Germany. The construction of a number of further instruments is under way, *e.g.* at the Arcetri Observatory in Italy and in India.

The coronal brightness is estimated mostly according to a memory scale, or compared visually with a standard brightness. The values are affected by the variable scattered light of the Earth's atmosphere, by the somewhat uncertain definition of the solar limb, and by the variable criteria of the observer. The differences between simultaneous measurements made with different coronagraphs are still considerable (Behr and Siedentopf, 1952). Since the coronal structure is in general only slowly variable, an isophote map can be made by the extrapolation of the intensities observed at the eastern and western limbs. The accuracy of these maps, which are regularly made for $\lambda 5303$ (Kiepenheuer, 1945 *etc.*) is not high, as inferred from the uncertainty of the intensities. The direct observation of the corona on the solar disk does not seem impossible, but is still far away.

The uncertainties of coronal observations on the limb will be greatly diminished by using the new photo-electric coronagraph proposed by Lyot (1950). This instrument exceeds in accuracy all former procedures. Because of its insensivity to scattered light, it can also be used in the plain. A photometer of this kind is now being developed with the financial help of the I.A.U. and will be provided at different coronal stations.

6. MAGNETIC FIELDS

The regular measurement of the magnetic fields of sunspots is being made at Mount Wilson Observatory by a visual procedure, while in Potsdam a photographic method is in use. The accuracy is of the order of $\pm$ 50 gauss. In this way, the movements of the central fields of larger sunspots from day to day are available. The reduction of these fields to the solar centre can be made by using an empirical foreshortening curve as given by COWLING and others. The co-operation of further stations is urgently needed, as well as improvement in the measuring accuracy and more detailed measurements of the field distribution within the spot-groups. Suitable photo-electric instruments for such measurements are available, and will now be used for solar surveys. Also the fields which are observed outside spots (and even outside the spot-belts), which are of great importance for a diagnosis of the solar activity, will now be surveyed systematically. The method which was applied recently by the BABCOCKS (1952a, 1952b, 1953) for automatic scanning of the solar surface has now become a regular survey programme. (See also this Volume, pages 769, etc.)

7. SOLAR NOISE

The emission of the Sun in the range of centimetre- and metre-waves can be followed independently of weather. Besides the thermal radiations of the corona ($\lambda > 100$ cm) and of the chromosphere and photosphere ($\lambda < 50$ cm), which are almost constant, more or less fluctuating components occur whose intensity in the metre-waves reaches up to a million times the thermal radiation. The slow variations of the short-wave end, *e.g.* at $\lambda = 10$ cm, show in the daily means a remarkable correlation with the sunspot numbers ($r = 0{\cdot}83$). This correlation is increased further by taking account of the spots' magnetic fields ($r = 0{\cdot}9$); BLUM and DENISSE, 1951. Sunspot numbers and intensity of the 10-cm radiation are therefore practically exchangeable measures. Their visibility functions, that is the ratio of the true to the observed spot numbers with respect to the true to the observed intensities of the 10-cm radiation as a function of the distance from the solar centre, must be very similar. A detailed analysis of the observations hitherto obtained, especially of the short intensity variations, will show whether the 10-cm wave can supply more information than the sunspot number or the spot area. The level where the centimetre-radiation originates is probably situated slightly higher than the visible photospheric level.

The intensity fluctuations in the metre waveband are connected with invisible events in the corona. The permanent survey of this radiation in connection with routine observations of the corona and of prominences will bring better understanding of the outer parts of the corona, which can be observed optically only during the few minutes of total eclipses.

The localization of solar noise sources and their mapping can be accomplished only by the use of large interferometers of long base-line and is very difficult. It is doubtful whether such measurements can be made on a regular schedule.

8. SOLAR RADIATION IN THE EXTREME ULTRA-VIOLET

The direct measurement of the Sun's radiation of $\lambda < 2800$ A.U. can be made only from rockets. The components of the solar radiation responsible for the formation of the ionosphere [$\lambda < 1215$ A.U. (L_α), $\lambda < 912$ A.U., and possibly a component in the X-ray region ($\lambda < 30$ A.U.)] become measurable only in heights above 100 km. A permanent survey of this radiation by direct measurement cannot be considered.

However, the extra-terrestrial intensity of this ionizing ultra-violet-radiation can be deduced from the observable electron density of the ionospheric layers (KIEPENHEUER, 1945). It follows without doubt that the intensity of this radiation shows strong fluctuations which are closely correlated with solar activity. The absolute intensity of this radiation is uncertain because of our limited knowledge of the ionization processes involved. It should be kept in mind that this indirect method, which gives the correct relative variation of the extra-terrestrial intensity, can be calibrated on an absolute scale by co-ordinated rocket measurements. Ionospheric physics can in this way contribute to a solar survey in the far ultra-violet and also to the interpretation of the observed ultra-violet excess of solar radiation.

9. Corpuscular Radiation of the Sun

All corpuscles being emitted from the Sun seem to carry an electric charge and then reach the Earth only after deviation in its magnetic field along complicated trajectories. Their greatest penetration-depth in the Earth's atmosphere is around 100 km. Direct detection and measurement of this radiation is therefore almost impossible. Much qualitative information about this radiation can be deduced from geomagnetic perturbations. BARTELS (1941) has even succeeded in isolating perturbations caused by solar corpuscles from those caused by solar wave-radiation. One has to keep in mind, however, that not the emission of solar corpuscles is measured, but only the small part of it which becomes effective around the Earth. The amount of this part is therefore not a solar measure but is dependent in a very complicated way on the magnetic field distribution between Sun and Earth. At best it can be taken as a rough index of the solar corpuscular emission. The same is true for the conclusions derived from the aurorae. More direct information about the corpuscular emission of the Sun, which is not influenced by the Earth's magnetic field, is obtained from the behaviour of comet tails (BIERMANN, 1951, 1952), which are exposed freely to this radiation. Further, the photometry of the polarized component of the zodiacal light will tell us about the electron density in the corpuscular streams (BEHR and SIEDENTOPF, 1953). Such methods, however, are not yet suitable for permanent surveys.

References

BABCOCK, H. W. . . . 1953 *Ap. J.*, **118**, 387.
BABCOCK, H. W. and BABCOCK, H. D. . . . 1952a *Proc. Convegno Volta, Rome* (1953), p. 266.
1952b *Publ. Astron. Soc. Pacific*, **64**, 282.
BABCOCK, H. W. and COWLING, T. G. . . . 1953 *M.N.*, **113**, 357.
BARTELS, J. . . . 1941 *Abhandl. Preuss. Akad. d. Wiss. Berlin, Math. Naturw. Kl.* No. 12.
BECKER, U. and KIEPENHEUER, K. O. . . . 1953 *Z. Astrophys.*, **33**, 132.
BEHR, A. and SIEDENTOPF, H. . . . 1952 *Naturwiss.*, **39**, 28.
1953 *Z. Astrophys.*, **32**, 19.
BLUM, E. J. and DENISSE, J. F. . . . 1951 *C.R. Acad. Sci. (Paris)*, **231**, 1214.
BIERMANN, L. . . . 1951 *Z. Astrophys.*, **29**, 274.
1952 *Z. Naturforsch.*, **7a**, 127.
D'AZAMBUJA, L. . . . 1919 *etc.* *Cartes Synoptiques des Couches Supérieures du Soleil. Annales de l'Observatoire de Meudon.*
JOY, A. H. and HUMASON, M. L. . . . 1949 *Publ. Astron. Soc. Pacific*, **61**, 133.
KIEPENHEUER, K. O. . . . 1945 *Ann. Astrophys.*, **8**, 210. *Sonnenzirkulare des Fraunhofer Instituts*, 1945, *etc.*
LYOT, B. . . . 1930 *C.R. Acad. Sci. (Paris)*, **191**, 834.
1950 *C.R. Acad. Sci. (Paris)*, **231**, 461.

Physical Conditions in the Solar Photosphere

H. H. PLASKETT

University Observatory, Oxford

SUMMARY

From the source-function, recently derived by SYKES, the temperature is found as a function of optical depth in the solar photosphere. The temperature distribution leads in turn to the gas pressure. To find these and other physical parameters it is necessary to assume that the photosphere is in local thermodynamic equilibrium. The uncertainty produced in these quantities by this and other assumptions, as well as by observational error, is briefly discussed. The deep layers of the photosphere, where $(d \ln T/d \ln p) > 0{\cdot}4$, are unstable; the steep temperature gradient may arise from boundary cooling. It is shown that solar granulation probably originates in the zone of observed instability rather than in the deeper-lying hydrogen-convection zone.

INTRODUCTION

THE photosphere is defined to be the visible surface of the Sun. It is the region which emits and absorbs radiation in the wavelength range 3000–10,000 Å, that is the range which carries some 67 per cent of the total energy output of the Sun. The photosphere is thus the seat of interactions between an intense radiation field and a moderately dense gas, and provides an observable example of reactions which, on a more intense scale, must determine the physical conditions in the deep interior. Both on account of the intrinsic interest of these interactions and of their possible application to the theory of stellar interiors, the solar photosphere merits more detailed observational and theoretical investigation than it has yet received.

An essential preliminary to such investigations is a quantitative knowledge of the physical conditions which prevail in the photosphere. This is necessary both to interpret what has already been observed and to decide which problems need further observational study. The present paper represents an attempt to find these physical conditions with the fewest possible hypotheses. In the first section SYKES's recent determination of the source-function from limb-darkening observations is used to find the temperature as a function of optical depth, and then by a method closely analogous to that used by STRÖMGREN the gas pressure, the partial pressure due to free electrons, and other physical variables are computed. The second section contains a brief discussion of the uncertainty in the thus derived values of the physical quantities. The march of temperature and pressure with depth shows that the deep layers of the photosphere are unstable, and in the third section a possible origin of this instability is discussed. The paper concludes with an attempt to locate granulation in the photosphere.

1. DISTRIBUTION OF TEMPERATURE AND PRESSURE

At any wavelength, λ, the limb darkening of the Sun is given by

$$I_\lambda^*(0, \mu) = \frac{1}{\mu}\int_0^\infty B_\lambda^*(\tau)e^{-\tau/\mu}d\tau. \qquad \ldots.(1)$$

Here $I_\lambda^*(0, \mu) = I_\lambda(0, \mu)/I_\lambda(0, 1)$, where $I_\lambda(0, \mu)$ is the surface brightness at a point on the disk where the emerging radiation makes an angle arc cos μ with the outgoing normal and where $I_\lambda(0, 1)$ is the surface brightness at the centre of the disk. Correspondingly $B_\lambda^*(\tau) = B_\lambda(\tau)/I_\lambda(0, 1)$ is the ratio of the source-function at the optical depth τ (defined for the wavelength λ) to the surface brightness at the centre of the disk. This equation is exact for an atmosphere stratified in plane parallel layers, and may be applied (cf. A. UNSÖLD (1948)) to the spherical Sun with an error less than 1 per cent, provided $0{\cdot}2 < \mu \leqslant 1{\cdot}0$. J. B. SYKES (1952) has recently solved this equation to find $B_\lambda^*(\tau)$ from the mean of five determinations of limb darkening at 5485 Å. According to MULDERS (1935) the surface brightness in pure continuous spectrum for this wavelength at the centre of the disk is $I_\lambda(0, 1) = 3{\cdot}73 \times 10^{14}$ erg cm^{-2} sec^{-1} per unit solid angle per centimetre wavelength interval. Hence *in local thermodynamic equilibrium* the temperature may be found from

$$B_\lambda(\tau) = \frac{2hc^2}{\lambda^5} \frac{1}{e^{hc/k\lambda T} - 1} = 3{\cdot}73 \times 10^{14} B_\lambda^*(\tau).$$

The temperatures in the second column of Table I for a number of values of τ are computed from this relation for SYKES'S $B_\lambda^*(\tau)$ (graphically interpolated for these τ's from the values given in the third column of his Table V).

To find the total pressure, p, as a function of optical depth we shall assume, following STRÖMGREN (1944), that the continuous absorption in the photosphere is due to the photoelectric dissociation of the negative hydrogen ion and the photoelectric ionization of the neutral hydrogen atom. If k_λ is the resulting mass coefficient of absorption, then from the definition of optical depth

$$\frac{d\tau}{dz} = -k_\lambda \rho. \qquad \ldots.(2)$$

In hydrostatic equilibrium

$$\frac{dp}{dz} = -g\rho, \qquad \ldots.(3)$$

so that

$$\frac{dp}{d\tau} = \frac{g}{k_\lambda}, \qquad \ldots.(4)$$

an equation which may be integrated to find p provided k_λ is known as a function of optical depth.

Let α_λ^- and α_λ be the atomic absorption coefficients of the negative hydrogen ion and the neutral hydrogen atom respectively. Then if $\phi(T)p_e$ be the ratio of the number of negative hydrogen ions to the number of neutral hydrogen atoms, each per unit volume, the continuous absorption coefficient per neutral hydrogen atom will be

$$\alpha_\lambda^- \phi(T) p_e + \alpha_\lambda.$$

Under photospheric conditions all but a minute fraction of the matter consists of hydrogen in the neutral state, so that the number of hydrogen atoms per gramme is with sufficient accuracy $1/m_H$, where m_H is the mass in grammes of the hydrogen atom. Hence

$$k_\lambda = \frac{1}{m_H} (\alpha_\lambda^- \phi(T) \cdot p_e + \alpha_\lambda). \qquad \ldots.(5)$$

In local thermodynamic equilibrium $\phi(T)$ can be found as a function of T from SAHA's ionization equation, the dissociation potential of the negative hydrogen ion being 0·75 electron volts. CHANDRASEKHAR has made an exceedingly accurate calculation of α_λ^- for the negative hydrogen ion, considering both bound-free and free-free transitions, and in their Table VII CHANDRASEKHAR and BREEN (1946) give values of $\alpha_\lambda^-\phi(T)$, corrected for stimulated emission, as a function of wavelength and temperature. The continuous absorption coefficient per neutral hydrogen atom in the ground state for bound-free and free-free transitions, corrected for stimulated emission, is given by

$$1{\cdot}044 \times 10^{-2}\lambda^3 e^{-u_1}(1 - e^{-u}) \left[\frac{1}{27} e^{u_3} + \frac{1}{64} e^{u_4} + \frac{e^{u_5}}{2u_1}\right] = \alpha_\lambda.$$

In putting this expression equal to α_λ we are neglecting the exceedingly minute fraction of hydrogen atoms which are not in the ground state. In the equation the wavelength is expressed in centimetres, and

$$u_n = R'hc/n^2kT \ : \ u = h\nu/kT = 1{\cdot}4385/\lambda T,$$

where R' is the RYDBERG constant and n the quantum number. Because we require the absorption coefficient at 5485 Å ionizations only from the third and higher quantum states are considered.

We now know everything on the right-hand side of equation (5) except the partial pressure, p_e, of the free electrons. Following STRÖMGREN (1944), we assume that the free electrons come from the ionization of the metals and hydrogen. Then if N_M, N_H, and N_e are the numbers of metal atoms (neutral and ionized), hydrogen atoms (neutral and ionized) and free electrons respectively (each per unit volume), we have

$$N_e = N_M x_M + N_H x_H = N_H \left(\frac{1}{A} x_M + x_H\right).$$

In this equation x_M and x_H are the fractions of positive ions of the metals and hydrogen respectively, and $A = N_H/N_M$ is from STRÖMGREN's investigation a number of the order of 10^4. The total number of particles per unit volume is

$$N = \frac{1}{A} N_H(1 + x_M) + N_H(1 + x_H) \simeq N_H(1 + x_H),$$

the approximation being justified by the overwhelming abundance of hydrogen. Hence

$$\frac{p_e}{p} = \frac{x_M}{A(1 + x_H)} + \frac{x_H}{1 + x_H} \simeq \frac{x_M}{A} + \frac{x_H}{1 + x_H}. \qquad \ldots.(6)$$

The last step follows since the term involving x_M is negligibly small when x_H is effectively different from zero. With this value for p_e, equation (5) takes the form

$$k_\lambda = \frac{\alpha_\lambda^-\phi(T)p}{Am_H}\left[x_M + \frac{Ax_H}{1 + x_H} + \frac{\alpha_\lambda A}{\alpha_\lambda^-\phi(T)p}\right]. \qquad \ldots.(7)$$

Recalling that $p = 0$ when $\tau = 0$, integration of equation (4) gives

$$p = \sqrt{2Am_H g}\left[\int_0^\tau \left[\alpha_\lambda^-\phi(T)\left\{x_M + \frac{Ax_H}{1 + x_H} + \frac{\alpha_\lambda A}{\alpha_\lambda^-\phi(T)p}\right\}\right]^{-1} d\tau\right]^{\frac{1}{2}}. \qquad \ldots.(8)$$

Adopting the value of $A = 10^4$ for the ratio of hydrogen to metal atoms, $\sqrt{2Am_H g} = 3{\cdot}026 \times 10^{-8}$. The values of p may be found by an iterative process. A first approximation, $p^{(1)}$, is found by putting the expression in the curly bracket in the integrand equal to unity and carrying out the integrations to the value of τ in the first column of Table 1 by WEDDLE's rule. A first approximation to the electron pressure, p_e', is from equation (6) given by $p^{(1)}/A$, and from p_e' at the corresponding values of τ, that is T from the second column of Table 1, we may use the convenient Tables 5 and 6 of STRÖMGREN (1944) to find a provisional value of $\{x_M + Ax_H/(1 + x_H)\}'$. We then get an improved value for the partial electron pressure, namely from equation (6),

$$p_e'' = \frac{p^{(1)}}{A}\{x_M + Ax_H/(1 + x_H)\}'.$$

This process, which converges fairly rapidly, may be repeated again and again until a value of the expression in the curly brackets in the integrand of equation (8) is found which corresponds to $p^{(1)}$. Using these values for the curly bracket the set of integrations may be repeated to give us values of $p^{(2)}$, these in turn may be used as before to find improved expressions for the quantity in the curly brackets, and so on. This iterative process for finding the pressure, p, converges rapidly; the maximum difference between corresponding values of $p^{(3)}$ and $p^{(2)}$ is 81 per cent, between $p^{(4)}$ and $p^{(3)}$ is 4 per cent, and between $p^{(5)}$ and $p^{(4)}$ is 1·2 per cent. The values of $p^{(5)}$ have been adopted as those for the total pressure, p, and are given in the third column of Table 1.

Table 1. Physical conditions in the photosphere

τ	T (°K)	p (dyne cm^{-2})	ρ (gm cm^{-3})	p_e (dyne cm^{-2})	x_M	x_H	k_λ (cm^2 gm^{-1})	$z - z_0$ (km)
0·00	4649	0	0	0			0	
0·03	5033	1·7 × 10⁴	4·2 × 10⁻⁸	1·7	0·89	0·1 × 10⁻⁴	0·09	318
0·06	5162	2·5	5·8	2·6	0·90	0·2	0·13	261
0·09	5246	3·1	7·1	3·2	0·91	0·1	0·15	227
0·12	5312	3·6	8·2	4·1	0·90	0·2	0·18	203
0·24	5505	5·1	11·2	6·9	0·93	0·4	0·26	145
0·36	5680	6·2	13·2	10·5	0·93	0·8	0·35	113
0·48	5838	6·9	14·4	17·7	0·93	1·6	0·53	93
0·72	6118	8·0	15·9	27·7	0·95	2·5	0·70	67
0·96	6356	8·8	16·8	46·5	0·96	4·3	1·01	49
1·20	6555	9·4	17·4	70	0·96	6·5	1·35	37
1·80	6959	10·3	17·9	153	0·98	13·9	2·42	19
3·00	7535	11·2 × 10⁴	18·0 × 10⁻⁸	410	1·00	35·6 × 10⁻⁴	5·06	0

The remaining columns of Table 1 are now readily found. The degrees of ionization, x_M of the metals and x_H of hydrogen, are the values $x_M^{(4)}$ and $x_H^{(4)}$ derived from $p^{(4)}$ and used to calculate $p^{(5)}$. They are given in columns 6 and 7. The partial electron pressure in column 5 is computed from

$$p_e = \frac{p^{(5)}}{A}\left[x_M^{(4)} + \frac{Ax_H^{(4)}}{1 + x_H^{(4)}}\right].$$

The density ρ in column 4 is obtained from the equation of state $p = R\rho T/\mu'$ for an ideal gas, where R is the gas constant per mole and μ' is the atomic weight of hydrogen on the chemical scale. The mass coefficient of absorption, k_λ, given in column 8 is calculated from equation (7), where p, x_M, and x_H are taken from columns 3, 6, and 7 respectively. The final column gives

$$z - z_0 = \int_\tau^{\tau_0} \frac{d\tau}{k_\lambda \rho},$$

from numerical integrations by WEDDLE's rule, where the values of k_λ and ρ are graphically interpolated from those given in columns 8 and 4 respectively. Here z_0 is the vertical linear co-ordinate corresponding to an optical depth $\tau_0 = 3$, so that $z - z_0$, expressed in kilometres, is the height in the photosphere above $\tau_0 = 3$.

Apart from minor computational details, the calculation of the total pressure and the associated physical quantities is the same as that used by STRÖMGREN (1944)*. The distinction between his method and results, and those set forth above, lies in the determination of the temperature distribution; upon this temperature distribution, it should be recalled, depends every subsequently calculated quantity. STRÖMGREN assumes that the energy transport is effected entirely by radiation, so that the photosphere on his assumption is in radiative equilibrium with a hypothetical temperature distribution approximately given by $T^4 = T_0^4(1 + 3\bar{\tau}/2)$, where $\bar{\tau}$ is the optical depth corresponding to a ROSSELAND mean absorption coefficient averaged over all wavelengths. To assume this temperature distribution pre-judges, however, the mode of energy transport—one of the very things we wish to discover about the physics of the photosphere. In the present paper, on the other hand, the temperatures are derived from the observed darkening towards the limb, and are, therefore, apart from the assumption of local thermodynamic equilibrium, observed temperatures. A second, though minor advantage, of the present procedure lies in the fact that all computations are carried through with the mass coefficient of absorption for a definite wavelength, rather than with a coefficient averaged to give the constant net flux required for radiative equilibrium. In both respects, therefore, the parameters in Table 1 should represent a closer approach than those of STRÖMGREN'S Table 11 to the physical conditions prevailing in the photosphere.

2. SOURCES OF ERROR

Even if this last statement be conceded, it is only too apparent that the physical conditions now found are subject to at least three major sources of uncertainty. The first of these arises from the error in the observed limb-darkening, which, because of the instability of the integral equation, leads to a magnified error in the observed source-function. Thus from Tables IV and V of SYKES (1952) it will be seen that the r.m.s. error of $I_\lambda^*(0, \mu)$, averaged over all values of μ, is $\pm$ 0·9 per cent,

* *Footnote added in proof.* Of the several model atmospheres which have been computed since STRÖMGREN, the computation of ALLER and PIERCE (1952) most closely resembles the treatment in the present paper. These authors use the Michigan limb-darkening observations and their own solution (1951) of the integral equation to find the temperature, ultimately as a function of the *integrated* optical depth. The effective temperature of the Sun, calculated from this temperature distribution, is, however, much higher than that observed, and to avoid this inconsistency the authors arbitrarily increase the mean absorption coefficient, used in the definition of the integrated optical depth, by some 35 per cent. It is probable that the discrepancy has arisen through the introduction of a mean absorption coefficient, for which the correct definition is not known and for which the evaluation necessarily requires a knowledge of the source function outside the observable range. Once, however, this arbitrary correction has been applied to the integrated optical depth, the finally adopted temperature distribution turns out, according to the authors, to be nearly the same as it would have been if the temperature had been found directly for the monochromatic optical depth at wavelength 5485. If this be so, a comparison may be made between this temperature and that in Table 1 above, and this comparison shows that the temperatures of Table 1 are slightly higher. They are, however, also probably more reliable. Not only is the present determination based on the mean of five independent measures of limb-darkening, rather than on the Michigan measures alone, but also SYKES'S solution of the integral equation, unlike that of PIERCE and ALLER (1951), does not constrain the source-function into a pre-assigned functional dependence on the optical depth.

while the error of $B_\lambda^*(\tau)$, averaged over all values of τ is $\pm$ 6·2 per cent. However, with sufficient accuracy we have from WIEN's law

$$\frac{\Delta B_\lambda}{B_\lambda} = \frac{hc}{k\lambda T}\frac{\Delta T}{T} \text{ or } \frac{\Delta T}{T} = \frac{T}{26{\cdot}2 \times 10^3}\frac{\Delta B_\lambda}{B_\lambda},$$

so that the percentage error of the temperature lies between one-fifth and one-third that of the source-function. In other words, the observational error in limb-darkening produces an error in the temperatures of less than 2 per cent—a satisfactorily small quantity.

The second source of uncertainty, which affects all the physical parameters except the temperature, arises from the assumed abundance of elements in the photosphere. Since the relative abundance can only be determined from the absorption-line spectrum when the physical conditions are known, it is evident that a complete determination of physical conditions in the photosphere can only be reached by successive approximations. The present calculation, in which the ratio of hydrogen to the metals is assumed to be 10^4 and in which the presence of helium and carbon is neglected, is therefore only the first stage in such a process. It may be used, when an adequate theory of line formation is available, to find a further approximation to the abundance of elements from the measured line profiles, and this in turn to give improved values of the physical parameters in Table 1. The present values are valid only in so far as the assumed abundance of elements, on which they are based, itself turns out to be valid.

The third source of uncertainty lies in the assumption of local thermodynamical equilibrium. Since we are dealing with the visible layers of the Sun, this assumption cannot be strictly true, and doubt is therefore thrown upon the calculation of every quantity in Table 1. A less restrictive assumption, however, suffices to justify the calculation. Thus the temperature found in the second column of the table is a "radiation-temperature", that is it is the temperature of equilibrium radiation which has the same specific intensity at the wavelength 5485 Å as the isotropic specific intensity of the source-function. Then if this radiation-temperature, T_r (5485 Å), is equal to the radiation-temperature, $\bar{T}_r$, averaged over all wavelengths, it follows from a theorem of EDDINGTON (1926) that at each depth in the photosphere

$$T_r\ (5485\ \text{Å}) = \bar{T}_r \simeq T_k \simeq T_i,$$

where T_k and T_i are respectively the "kinetic" and "ionization-temperatures". As a consequence it further follows that the pressures, ionizations, and so on in Table 1 are correspondingly correct.

EDDINGTON's theorem applies to evenly-diluted equilibrium radiation. Because of continuous and line-absorption the dilution of photospheric radiation, or in our language its radiation-temperature, varies with wavelength. In the expression for the average energy of the absorbed quantum we may replace the radiation-temperature, variable with wavelength, by the constant average radiation-temperature, T_r, without making a substantial error, and thus following EDDINGTON establish the approximate equality $T_k \simeq \bar{T}_r$ for each depth in the photosphere. Since photo-ionization is determined by T_r and recombination by T_k, the degree of ionization is fixed by these temperatures, or $T_i \simeq T_k \simeq \bar{T}_r$. Absorption-lines in the photospheric radiation field lead to a deficiency of atoms in excited states with the result that the degree of ionization is less than that predicted by the above relation. However, the fraction of excited atoms even under equilibrium conditions at photospheric temperatures is already very small, and a further reduction in this small fraction by absorption lines in the exciting radiation will have a small effect on the degree of ionization. In passing it should be noted that we have assumed a Boltzmann distribution of excited atoms in calculating the continuous absorption due to neutral hydrogen; however, this absorption only becomes significant at optical depths so large that it is justifiable to assume a close approximation to local thermodynamical equilibrium and therefore to the BOLTZMANN distribution. Summarizing, our re-examination of EDDINGTON'S theorem under photospheric conditions appears to establish the approximate equality $\bar{T}_r \simeq T_k \simeq T_i$, but it remains an assumption that T_r (5485 Å) $= \bar{T}_r$. Of this assumption it can only be said that there must be at least one wavelength, λ, for which $T_r(\lambda) = \bar{T}_r$. If 5485 Å $= \lambda$, the calculations of Table 1 are approximately correct.

Nevertheless it remains true that departures from local thermodynamic equilibrium, or from our less restrictive assumption, form the most serious source of uncertainty in our derived physical quantities. Ideally and ultimately we shall have to replace these "equilibrium" calculations by a purely kinetic treatment, using theoretically calculated atomic cross sections (for ionization and excitation) by radiation and collisions, as well as the source-function observed over the whole spectrum. Quite apart from its theoretical difficulties, such a programme can only be carried out when the source-function is found from observations made at a station outside the Earth's atmosphere.

3. INSTABILITY OF THE PHOTOSPHERE

The condition for the mechanical stability of any atmosphere, such as the photosphere, is that

$$\left(\frac{d \ln T}{d \ln p}\right)_{\text{photosphere}} < \left(\frac{d \ln T}{d \ln p}\right)_{\text{adiabatic}} = 0{\cdot}4,$$

where the numerical value for the adiabatic derivative holds for the photospheric gas which is monatomic and relatively unionized. From the values of T and p in Table 1 we may find the required values of $(d \ln T/d \ln p)_{ph}$ from a graphical differentiation of a curve giving $\log T$ as a function of $\log p$. The continuous curve of Fig. 1 shows $(d \ln T/d \ln p)_{ph}$ as a function of p, and of $z - z_0$. It will be noted that when $p > 8{\cdot}4 \times 10^4$ dyne cm^{-2}, or $z - z_0 < 60$ km, the condition for mechanical stability is not satisfied. That there is actual instability may be inferred from the run of ρ (shown as a dotted curve in Fig. 1) which apparently is reaching a maximum at $z - z_0 = 0$. If this be so, dense layers of gas overlie less dense and we shall have EDDINGTON'S catastrophic instability (EDDINGTON, 1942).

STRÖMGREN'S *model* solar atmosphere in radiative equilibrium (STRÖMGREN (1944)) shows an instability of much the same character. His result could, however, be interpreted as meaning that, while an atmosphere in radiative equilibrium is unstable, the actual photosphere would so adjust its temperature distribution, for example, by setting up an adiabatic gradient, as to avoid this instability. The present results, which are a consequence of the observed temperature distribution, cannot be so interpreted, and we must conclude that the deeper layers of the actual photosphere are unstable. Moreover this observed instability has nothing to do with the instability, first predicted by UNSÖLD, which originates from the ionization of the dominant element hydrogen. As can be seen from Table 1 the degree of ionization, x_H, of hydrogen is quite negligible.

How then does the observed instability of the deeper layers of the photosphere originate? Since in hydrostatic equilibrium the pressure p at any level z is given by

$$p = p_0 e^{-\frac{\mu' g}{R}\int_{z_0}^{z} \frac{dz}{T}}$$

it follows that the variation of pressure with depth is determined by the variation of temperature with depth. That $(d \ln T/d \ln p)_{ph} > 0{\cdot}4$ for $z - z_0 \leqslant 60$ km must

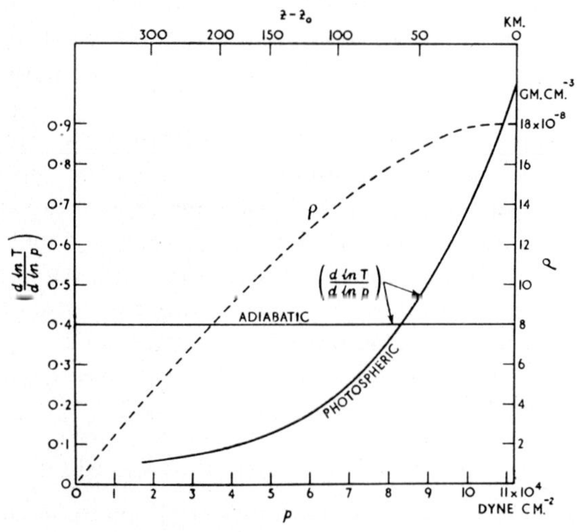

Fig. 1. $(d \ln T/d \ln p)$, photospheric and adiabatic, as functions of p (continuous lines); ρ as a function of p (dotted line)

therefore be a consequence of the very steep temperature gradient, dT/dz, observed in this part of the photosphere. How this gradient steepens for $z - z_0 < 60$ km is well shown in Fig. 2, where T from Table 1 is plotted as a function of $z - z_0$. Now it may be significant that the region where the gradient is steep is just the region where, as we proceed outwards from the interior of the Sun, we first encounter elements of mass which radiate a perceptible fraction of their thermal energy to outside space. For consider an element of mass at optical depth τ in the form of a very thin cyclinder, so that the emitted radiation takes place almost wholly from its top and bottom faces. Then the fraction of the radiation emitted by this element which escapes to outside space is clearly

$$F(\tau) = \frac{\int_0^{2\pi} d\phi \int_0^{\pi/2} e^{-\tau \sec \theta} \sin \theta \cos \theta d\theta}{2\int_0^{2\pi} d\phi \int_0^{\pi/2} \sin \theta \cos \theta d\theta} = \tfrac{1}{2}[e^{-\tau}(1-\tau) + \tau^2 E(\tau)],$$

where $E(\tau)$ is the exponential integral. This fraction, $F(\tau)$, where τ is the optical depth for 5485 Å, is shown as a dotted curve in Fig. 2. In view of the small variation of absorption coefficient with wavelength in the visible region, $F(\tau)$ may be taken as the fraction of the visible energy, that is some 67 per cent of the total, which escapes freely to outside space. Though the reciprocity law cannot be applied to the non-isotropic radiation field of the photosphere, we can at least say that $F(\tau)$ is the fraction of the emitted visible energy of an element which represents an uncompensated loss. From Fig. 2 we note that this uncompensated loss increases from 1 per cent at $z - z_0 = 0$ to some 14 per cent at $z - z_0 = 60$ km, in which region the curve is concave upwards. In other words the zone of steep temperature gradient and instability is just the zone where, as we proceed outwards, we first encounter elements

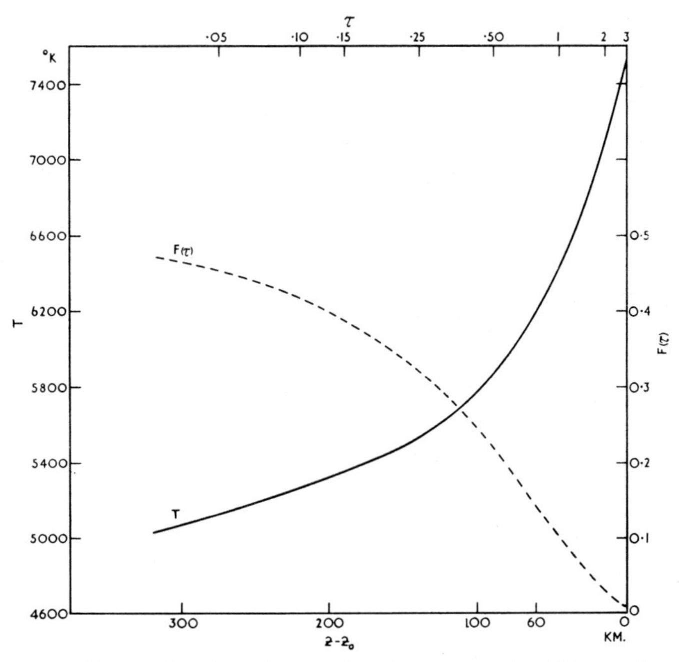

Fig. 2. Temperature, T, as a function of $z - z_0$ (continuous curve): $F(\tau)$ as a function of $z - z_0$ (dotted curve)

of mass which are cooling because of the uncompensated loss of energy. It is at least plausible to suppose that it is this cooling which leads to the steep temperature gradient and the consequent instability.

4. THE ZONE OF GRANULATION

As K. SCHWARZSCHILD pointed out long ago, the presence of photospheric granulation is evidence of motion and instability, a prediction confirmed by his son and R. S. RICHARDSON (1950). The question arises whether granulation occurs in the observed zone of instability, or in the deeper-lying hydrogen-convection zone.

Suppose that the top of the granular elements occurs at an optical depth τ_1 (defined for 5485 Å). Let J_2 and J_1 be the surface brightnesses, constant with direction, at τ_1 of a granule and of an intergranular region respectively. Then the ratio of the observed surface brightnesses at an angle arc cos μ is given by

$$\frac{j_2(\mu)}{j_1(\mu)} = \frac{J_2{}^*e^{-\tau_1/\mu} + \dfrac{1}{\mu}\displaystyle\int_0^{\tau_1} B_\lambda{}^*(\tau)e^{-\tau/\mu}d\tau}{J_1{}^*e^{-\tau_1/\mu} + \dfrac{1}{\mu}\displaystyle\int_0^{\tau_1} B_\lambda{}^*(\tau)e^{-\tau/\mu}d\tau}, \qquad \ldots.(1)$$

where, as in section 1, the asterisks denote that the surface brightnesses and the specific intensities have been divided by $I_\lambda(0, 1)$, the surface brightness at the centre of the disk. According to KEENAN (1939) the ratio of observed surface brightness at $\mu = 1$ is 1·15. Hence for any particular value of τ_1 and for $\mu = 1$ we have from equation (1) one relation between $J_1{}^*$ and $J_2{}^*$. We may get a second relation from the fact that at the centre of the disk the surface brightness, averaged over granule and intergranular region, must be just the surface brightness ordinarily observed, or

$$\tfrac{1}{2}(J_1{}^* + J_2{}^*)e^{-\tau_1} + \int_0^{\tau_1} B_\lambda{}^*(\tau)e^{-\tau}d\tau = 1{\cdot}00. \qquad \ldots .(2)$$

Hence from equations (1) and (2), where the integrals are evaluated by WEDDLE's rule using SYKES's source-functions, we may find $J_1{}^*$ and $J_2{}^*$ for any given optical

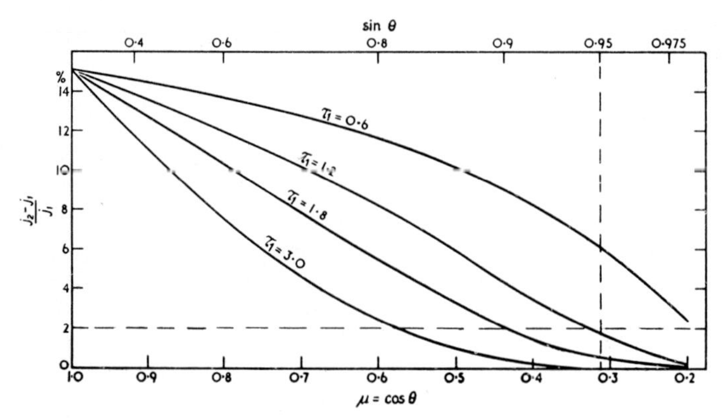

Fig. 3. Contrast, $(j_2 - j_1)/j_1$, of granulation as function of τ_1 and $\mu = \cos\theta$

depth τ_1. The observed contrast, $\dfrac{j_2(\mu) - j_1(\mu)}{j_1(\mu)}$, as a function of μ may then be computed from equation (1) for this optical depth of the top of the granulation. The results of such a computation are shown in Fig. 3 for the cases where the top of the granulation occurs at $\tau_1 = 0{\cdot}6$, 1·2, 1·8, and 3·0.

The figure shows that as the limb is approached the contrast decreases and decreases the more rapidly the deeper the top of the granulation zone. According to TEN BRUGGENCATE (1939) photospheric granulation disappears for sin $\geqslant$ 0·95, that is $\mu \leqslant 0{\cdot}312$. It is not known at what contrast, $\dfrac{j_2(\mu) - j_1(\mu)}{j_1(\mu)}$, granulation becomes invisible, but it will probably be conceded that under ordinary conditions of good seeing detection would become difficult when the contrast is less than 2 per cent. If this criterion be adopted it follows from Fig. 3 that the top of the granulation zone occurs slightly above $\tau_1 = 1{\cdot}2$, corresponding to $z - z_0 = 37$ km and $J_2/J_1 = 1{\cdot}33$. Thus the disappearance of photospheric granulation at the limb of the Sun suggests that the top of the granulation zone occurs in the observed zone of instability, not in the deeper lying hydrogen-convection zone.

It is interesting that granulation, probably occurring in the observed zone of instability, and accompanied by motion, does not set up an adiabatic temperature gradient. That it fails to do so suggests that the onset of uncompensated radiation-loss to outside space is the dominant factor in determining the temperature of the mass elements in this part of the photosphere.

5. CONCLUSION

This preliminary survey of its physical conditions has shown that the photosphere may be considered as composed of two main parts. The upper, where $p < 8{\cdot}4 \times 10^4$ dyne cm^{-2} and $z - z_0 > 60$ km, is the zone of small temperature gradient and the region in which the FRAUNHOFER spectrum originates. As soon as the formation of absorption lines is better understood we should be able to learn much about the physics of this zone and its relation to the chromosphere. The lower part, where $p > 8{\cdot}4 \times 10^4$ dyne cm^{-2} and $z - z_0 < 60$ km, is the zone of high temperature gradient and instability. At present we recognize its existence only by the source-function for the continuous spectrum. However, the instability of the region is so marked that it must be the seat of motions both on a small scale, as in granulation, and on a large scale. If these hypothetical motions should extend into the upper zone, they will be detectable from the resulting DOPPLER displacements and may throw some additional light on what is happening in the deep zone of instability.

These two zones are also recognizable in STRÖMGREN'S model solar atmosphere in radiative equilibrium. This resemblance between the actual photosphere on the one hand and the model in radiative equilibrium on the other, at least suggests that some considerable part of the energy transport in the photosphere is by radiation. Since near the boundary the solar energy has to be carried away to outside space by radiation, it is not surprising that radiative transport plays a part in replacing the thermal energy thus dissipated. However, the observed temperature distribution of the photosphere is not that predicted by radiative equilibrium, and we must recognize that another mode of energy transfer is also at work. This is presumably convective transport. Its amount and effect, in the upper zone at least, will be determinable as soon as we know more about the motions in this region.

Clearly we are still far from an adequate knowledge of the physics of the solar photosphere. However, we can at least begin to see some of the observations and theoretical investigations, which will have to be made to increase that knowledge.

REFERENCES

ALLER, L. H. and PIERCE, A. K. . . . . 1952 *Ap. J.*, **116**, 176.
BRUGGENCATE, P. TEN . . . . . . 1939 *Z. Astrophys.*, **19**, 59.
CHANDRASEKHAR, S. and BREEN, F. H. . 1946 *Ap. J.*, **104**, 430.
EDDINGTON, A. S. . . . . 1926 *Internal Constitution of the Stars*, p. 376; Cambridge University Press.
1942 *M.N.*, **102**, 154.
KEENAN, P. C. . . . . 1939 *Ap. J.*, **89**, 604.
MULDERS, G. F. W. . . . . 1935 *Z. Astrophys.*, **11**, 143.
PIERCE, A. K. and ALLER, L. H. . . . 1951 *Ap. J.*, **114**, 145.
RICHARDSON, R. S. and SCHWARZSCHILD, M. 1950 *Ap. J.*, **111**, 351.
STRÖMGREN, B. . . . . 1944 *Tables of Model Stellar Atmospheres*. Publ. Copenhagen Obs., No. 138.
SYKES, J. B. . . . . 1953 *M.N.*, **113**, 198 = Commun. Univ. Obs., Oxford, No. 38.
UNSÖLD, A. . . . . 1948 *Ann. Phys.* (Leipzig), **3**, 124.

Infra-red Studies of Photospheric Structure

B. E. J. Pagel

The Observatories, Cambridge

Summary

Recent advances in the technique of detecting infra-red radiation by means of photoconductive cells have made it possible to obtain high-resolution tracings of the solar spectrum in the regions 1·0μ to 2·5μ, 2·8μ to 4·2μ, and 4·5μ to 5·5μ, which are not obscured by molecular bands in the Earth's atmosphere. The infra-red region is of particular interest owing to the minimum in the continuous absorption coefficient of the negative hydrogen ion at 1·66μ, and the high excitation potentials of most infra-red Fraunhofer lines. The present paper describes an instrument developed at Cambridge which uses a lead-sulphide cell as detector, and gives a simple analysis of some equivalent-width measurements on the disk and limb in terms of a photospheric model recently proposed by de Jager. To obtain higher resolution, the use of a Fabry-Perot interferometer with variable plate separation is proposed.

1. Introduction

Most of our knowledge concerning the composition and structure of the outer layers of the Sun has been derived from detailed studies of Fraunhofer lines. Such studies in the photographic region have been systematized in the Utrecht Atlas (Minnaert, Mulders, Houtgast, 1940), which gives microphotometer tracings of the entire solar spectrum from λ3332 to λ8771 with effective resolving power 10^5, and in Babcock and Moore's infra-red extension of their Revised Rowland Table, which lists photographically detected Fraunhofer lines in the region λ6600 to λ13,495 (Babcock and Moore, 1947). The wartime development of photoconductive cells which are highly sensitive to infra-red radiation has made it possible to improve considerably our knowledge of Fraunhofer lines up to λ70,000. The commonest infra-red photoconductive cells are of the lead-sulphide type, consisting of a thin layer of PbS either evaporated or chemically deposited on glass in an evacuated tube. The short response time (of the order of microseconds) enables the light to be interrupted at audio-frequency and amplified by a selective amplifier; under these conditions good cells may exceed the Hilger-Schwarz thermocouple by a factor 100 in signal-to-noise ratio in the range λ6000 to λ30,000 (Sutherland, Blackwell, Fellgett, 1946). Deposits of lead-selenide in quartz envelopes are sensitive out to λ45,000 (Blackwell, Simpson, Sutherland, 1947); and lead-telluride cells extend the profitable range still further. Thus it is now possible to record spectra with a resolution limit of about 0·2 cm^{-1} out to 1500 cm^{-1} (*i.e.* 7μ).

The Earth's atmosphere absorbs the infra-red radiation coming from the Sun over a wide range of wavelength (particularly in the H_2O and CO_2 absorption bands for the near infra-red), but the following ranges are "windows" in which the infra-red Fraunhofer spectrum can be observed among the telluric absorption lines (Pettit, 1951):

0·4μ to 1·3μ	2·8μ to 4·2μ
1·5μ to 1·8μ	4·5μ to 5·5μ
2·0μ to 2·5μ	8·0μ to 13·0μ

Fraunhofer lines are distinguishable from telluric lines by the fact that their intensity is independent of the Sun's altitude, and by the rotational Doppler shift between the east and west limbs (BABCOCK and MOORE, 1947).

Since 1947, the entire infra-red region accessible to photoconductive detectors has been explored with high-resolution spectrometers attached to the 75-ft McGregor Solar Tower of the McMath-Hulbert Observatory, Michigan (McMATH and MOHLER, 1949), and to the Snow Telescope on Mount Wilson. Photoconductive detectors have also been used on the Jungfraujoch (MIGEOTTE and NEVEN, 1952) and elsewhere. A photometric atlas of the region $\lambda 8465$ to $\lambda 25,242$, with effective resolving power 3×10^4, has been published at Michigan (MOHLER, PIERCE, McMATH, GOLDBERG, 1950). Further publications give identifications of some of the Fraunhofer lines observed (GOLDBERG, MOHLER, McMATH, 1949; 1950), using as a basis Miss MOORE's tables of atomic energy levels (1949); these lines are due to neutral H, C, Na, Mg, Al, Si, Fe, Ni, and Mn. Most of these lines are relatively faint (residual intensity ~ 90 per cent), since they arise from atomic levels with excitation energy > 5 V (BABCOCK and MOORE, 1947), and therefore originate in relatively deep layers of the solar atmosphere. The continuous absorption coefficient, due predominantly to photo-ionization of H^- below $\lambda 16,600$ and to free-free transitions at longer wavelengths, reaches a minimum at this point (CHALONGE and KOURGANOFF, 1946; CHANDRASEKHAR and BREEN, 1946) and then rises steadily. Correspondingly the frequency of occurrence of atomic Fraunhofer lines is greatest near $\lambda 16,600$ and diminishes rapidly towards longer wavelengths (GOLDBERG, 1950). The high excitation potentials of the lines make them particularly susceptible to pressure broadening owing to the increase in collision cross-section with excitation energy (UNSÖLD, 1938); this fact, combined with the relatively great depth of formation, means that accurate line-profile studies in the infra-red should be particularly interesting for the verification of solar models. Most of the lines are too weak to show damping wings, but the strong Mg line $\lambda 17,108 \cdot 8$ has an equivalent width of about $1 \cdot 0$ Å and should be quite suitable. The resolving power required for such an investigation is of the order of the width of the Doppler core of the line, which is given by

$$k \propto \exp\left[-c^2\left(\frac{\delta\lambda}{\lambda}\right)^2 A/2RT\right],$$

where k is the absorption or scattering coefficient at a distance $\delta\lambda$ from the centre of a line of mean wavelength λ due to an atom of atomic weight A at temperature T. The total half-intensity width is thus

$$\frac{\delta\lambda}{\lambda} = \frac{2}{c}\left(\frac{2RT}{A}\ln 2\right)^{1/2}.$$

If $T = 5000°$K,

$$\frac{\delta\lambda}{\lambda} = 5 \times 10^{-5} A^{-1/2}.$$

For Mg, $A = 25$,

$$\frac{\lambda}{\delta\lambda} = 10^5.$$

Such a resolving power in this region has not been attained with a grating spectrometer, owing to the limited amount of intensity available. With an interferometer, on the other hand, this is much easier, and the development of such an interferometer is one of the chief objects of present work at Cambridge. The conditions of the problem are described in a later section, in which it is shown from intensity considerations that the most promising practical interferometer so far developed, which will give a sufficiently high resolving power in the infra-red, is the Fabry-Perot type with variable optical path between the plates.

The infra-red region has two general advantages over the visible for solar observations.

(1) The Zeeman effect, given by

$$\Delta\nu = \frac{\Delta\lambda}{\lambda^2} = 4{\cdot}7 \times 10^{-5} gH,$$

where $\Delta\nu$ is the splitting in cm^{-1} of a spectral term with Landé factor g caused by a field of H gauss, becomes easier to examine at longer wavelengths owing to the constancy with λ of the relative Doppler width $\delta\lambda/\lambda$. A Fabry-Perot interferometer has a constant resolution limit in frequency units $\delta\nu$ (TOLANSKY, 1947, p. 103); thus a field of 300 gauss should be immediately visible in the high-resolution spot spectrum of a faint line with reasonably large g-factor (say 5), compared with 1000 gauss in the green.

(2) In grating spectrometers, and also in examining details of the Sun's surface, *e.g.* sunspots, difficulties are caused by the scattering of light. The amount of light scattered normally diminishes as the wavelength increases, so that, from this point of view, it is advantageous to work in the infra-red. For photo-electric studies of sunspot spectra, this advantage is off-set by the long time required to scan through a spectral range. Spot spectra have been obtained at Michigan (GOLDBERG, 1950) but not in large numbers, owing to the difficulties of guiding and to poor seeing conditions, which make it necessary to wait for a fairly large spot. At the time of writing there is a sunspot minimum, so that no developments in this particular field are to be expected for some time.

In its present form the Cambridge infra-red spectrometer has an effective resolving power equal to that of the Michigan Atlas, namely 30,000, the limit being set in this case by the amount of energy available, so that its use for spectrum analysis has been confined to measurements of equivalent width. Such measurements were carried out for a few lines around $1{\cdot}7\mu$ at the centre of the solar disk, and also near the limb ($\cos\theta = 0{\cdot}5$ and $0{\cdot}3$), during the summer of 1952, and an account of this work, together with a brief preliminary comparison with theory, is given in Section 3. Unfortunately, the cloudy weather at Cambridge, combined with the necessity for long uninterrupted scanning, made it necessary to take the records early in the morning when there were no clouds (0600 hrs to 0800 hrs U.T.); the resulting high zenith distance of the Sun causes the telluric spectral lines to be very intense, making equivalent width measurements more difficult. Also the number of measurements is rather small owing to weather difficulties and to the large part of the summer spent in adjusting the instrument. Nevertheless, the results are interesting, owing to the special features of the infra-red spectrum that have already been mentioned. A valuable extension of this work would be a comparison of centre-to-limb behaviour in different infra-red regions, which might be correlated with the wavelength variation of the continuous absorption coefficient, bearing in mind the excitation and ionization potentials of the various individual atomic levels, and with the theory of the convective layer (EVANS, 1947).

2. THE GRATING SPECTROMETER

The optical system was designed by Blackwell and installed by him in the dark room formerly used for the concave grating at the Solar Physics Observatory. The electronic recording system was constructed by BLACKWELL, FELLGETT, and the writer, and is similar to a design due to FELLGETT (1951).

(a) *Optics*

The optical system is shown in Fig. 1. It is of the conventional Littrow plane-grating type with two slits using Pfund slotted mirrors (PFUND, 1927), so that the two concave mirrors of the dispersing system are used on axis. The focal length of the

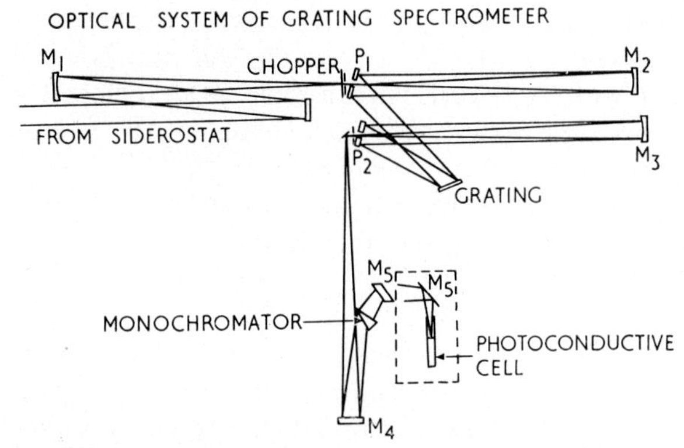

Fig. 1. Optical system of grating spectrometer. Radiation from the Sun is reflected into the laboratory by a siderostat and focused by M_1 through the chopper on to the entrance slit. The light is collimated by M_2 and reflected by the Pfund flat P_1 to the grating, from which it travels to the second Pfund flat P_2 and to the concave mirror M_3, reaching a focus at the exit slit. The radiation is then reflected sideways by a small flat to the concave mirror M_4, from which a convergent beam passes through the small-angle prism and is finally reflected downwards on to the cell, which is mounted vertically

collimator is 304·7 cm (10 ft). The grating finally selected is a Rowland 14,400 lines per inch, 6 in. wide and 4 in. high, so that the aperture of the system is normally about f25. The grating is used in the first order. The detector is a chemically deposited PbS cell jacketed by liquid oxygen in a vertical dewar. It has three parallel graphite electrodes, about 1 cm long and 0·3 cm apart; in the present arrangement, the sensitivity is believed to be uniform over the surface. The aperture of the condensing system M_4 is f12 and the image formed on the cell has about two-thirds the dimensions of the slits, so that the maximum effective slit length is 1·5 cm. The small aperture required to make the cooled air cavity above the cell effective (see below) makes it impossible to increase this figure, but this is a disadvantage in solar work only when one is interested in light integrated over the entire disk, which is not normally the case. It is a disadvantage in recording laboratory spectra.

The wavelength passing through the exit slit is varied continuously by using a geared synchronous motor to rotate the turntable on which the grating is mounted. The maximum permissible rate of rotation is determined by the ratio of the resolving limit to the response time of the detector. During the 1952 observations a rate of 0·18° per hr, or 4 Å per min at 1·7μ was used.

The spectrometer is mounted on steel girders, which rest on brick pillars built directly into the ground, the dewar and the pre-amplifier being given additional protection by mounting them on sponges. The entire instrument is therefore unaffected by floor vibrations. The slits are straight ones, constructed in the observatories workshop. The theory of image curvature in grating spectrometers leads to the result (MINKOWSKI, 1942):

$$\frac{\lambda}{\delta\lambda} = \frac{8f^2}{a^2} = 2 \times 10^5,$$

where $\delta\lambda$ is the total range of wavelength selected by a straight exit slit of total length a as a result of image curvature. As the resolving power is only 3×10^4, image curvature is seen to be quite negligible.

The unwanted orders are removed by a 20° quartz prism placed in the convergent beam from the condensing mirror M_4 (Fig. 1). Owing to the small angle and the low convergence (f12) necessitated by the cell mounting, the prism does not cause any appreciable variation in focal length with wavelength. The cell mounting has to be shifted 1 mm sideways after each wavelength change of 100 Å or so in order to regain the position of maximum output. With a beam of low convergence, this simple monochromator system is more convenient than the usual fore-prism, but it is, of course, not sufficiently selective to remove ghosts.

(b) *Electronics*

The photoconductive cell has already been described. A constant excitation voltage (330 V) is maintained across the two outer electrodes and the incoming radiation interrupted at 750 c/s by a rotating sectored disk, driven by a synchronous motor, in front of the entrance slit, so that the central electrode puts out an a.c. signal with this frequency. The excitation voltage and frequency were found empirically to be optimum for this combination of cell and amplifiers. The resistance of the cell at liquid oxygen temperature is of the order of 50 megohms. The signal is fed through a short cable to an unselective pre-amplifier (maximum gain 300) and then to a selective amplifier (gain 10^4) tuned to the chopping frequency. The tuned signal is then synchronously detected by means of a phase-sensitive rectifier circuit fed from a photocell which is placed above the entrance slit and illuminated through the chopping disk by an auxiliary lamp and lens. The signal and reference wave are thus coherent in phase. The general design of the circuits follows that of FELLGETT (1951). The output is applied to a d.c. amplifier, the current from which (0–5 mA) is continuously recorded. The time-constant of the output circuit is 0·5 sec., so that the bandwidth is only 2 sec.$^{-1}$; this greatly reduces the noise level from the tuned amplifier (bandwidth 50 sec.$^{-1}$).

Cells may be expected on theoretical grounds (FELLGETT, 1949) to work much better in a cooled cavity, though the effect is diminished somewhat by the aperture required to admit the radiation from the spectrometer, which should therefore be as narrow as possible for a given cell area; although it is advantageous in principle to use the smallest available cell, in spite of the increased solid angle over which temperature radiation reaches it. Cooling of the cell itself should also have the advantage of reducing the current and Johnson noise level. The actual effect of cooling varies very much from cell to cell: in general evaporated cells (*e.g.* those produced by the British Admiralty) are not affected appreciably, whereas chemically deposited cells gain a factor of 5–50 over absolute sensitivity at room temperature, while the noise level is diminished by a factor between 2 and 10. Cells of both types have been tried out in the spectrometer and the best performance among those available has been obtained from a chemically deposited cell cooled in liquid oxygen. The cell is mounted in a brass cylinder which extends 7 cm above the cell window and has a thermally insulating bakelite continuation 5 cm long, covered by glass, to prevent condensation of water vapour on the cold cell-window.

(c) *Test Spectra Obtained*

The spectrometer has been used with a tungsten-filament lamp as source to obtain records of the water-vapour spectrum at 1·8μ (5500 cm^{-1}) and 2·5μ (4000 cm^{-1}) with

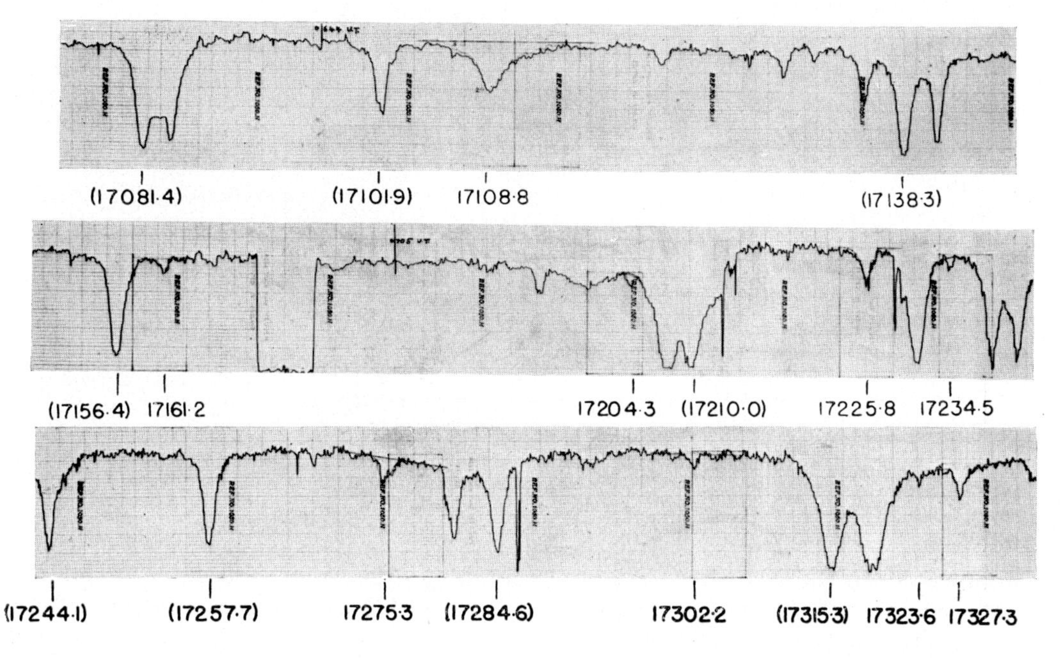

Fig. 2. Photograph of one of the records used in the measurement of equivalent widths. The profiles of the deeper solar lines have been corrected for non-linearity. The figures below the absorption minima are wavelengths in *A*, taken from the Michigan Atlas, those in parenthesis referring to telluric lines. Figures above the spectrum are time marks

slit widths 0·3 cm^{-1} and 0·15 cm^{-1} respectively. The 1·8μ spectrum shows practically all the features discovered by NELSON (1949), who claims a resolving limit 0·15 cm^{-1}, while the well-known doublet at 3977 cm^{-1} in the 2·5μ band (SUTHERLAND, BLACKWELL, FELLGETT, 1946), with separation 0·14 cm^{-1}, is clearly resolved. The solar spectra obtained reproduce exactly the features of the Michigan Atlas between 1·5μ and 1·8μ, where the intense $\nu_2 + \nu_3$ band of H_2O sets in, but are richer in strong telluric lines in the region 2·1μ to 2·5μ. The great intensity of the Sun at 1·5μ, compared with the laboratory source, makes it possible to narrow the slits down to 0·18 cm^{-1}, or 0·6 Å, giving a resolving power 3 $\times$ 10^4. A portion of the solar spectrum is shown in Fig. 2.

The complete blackout obtained in regions of intense water-vapour absorption indicates that ghosts and scattered light are negligible in this apparatus.

3. Equivalent Width Measurements

In order to observe a small portion of the disk at a time, a 5-cm solar image was projected on to the entrance slit, the length of which was restricted to 2 mm. The dispersion on the chart is about 0·6 cm per Å. Wavelengths were taken from the Michigan Atlas and variations (up to 10 per cent) in the chart dispersion were corrected for.

The intensity scale was calibrated by screening portions of the mirror M_4 and comparing the deflection due to a combination of areas with the sum of the deflections

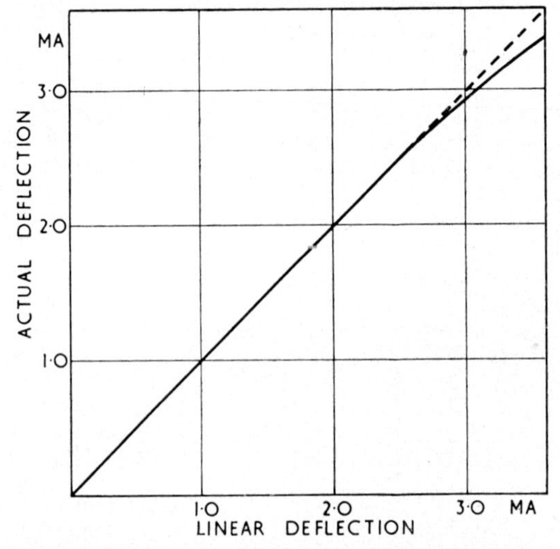

Fig. 3. Intensity calibration curve. The abscissae and the broken line represent the output current that would be obtained if the detecting system were perfectly linear. The solid curve shows the current actually obtained. The deviation from linearity is such that the continuum (about 3·0 mA) has to be elevated by 0·06 mA

obtained when the areas were exposed separately. The calibration curve, which is not far from linear, is shown in Fig. 3. The areas of absorption lines were measured with a planimeter giving the following results.

Line	Equivalent width in milliangstroms at		
	$\cos\theta = 1{\cdot}0$	$\cos\theta = 0{\cdot}5$	$\cos\theta = 0{\cdot}3$
Mg, 17,108·8 Å Transition $4s^1S_0$–$4p^1P^\circ_1$ Excitation 5·39 V (Mohler, Pierce, McMath, Goldberg, 1953)	990 ± 40	1030 ± 30	1110 ± 30
Fe, 17,161·2 Å Transition unidentified (Mohler, Pierce, McMath, Goldberg, 1953)	104 ± 3	104 ± 4	108 ± 10
Fe, 17,338·6 Å Transition $t^5D^\circ_4$–3_4 Excitation 6·33 V (Mohler, Pierce, McMath, Goldberg, 1953)	445 ± 20	463 ± 20	427 ± 20

There is no perceptible change in any of the apparent line profiles, and the changes in equivalent width of the weak lines are below the limit of random error; but the strong Mg line appears to show a significant increase of about 10 per cent at $\cos\theta = 0{\cdot}3$. The weak Fe lines differ in their behaviour from lines of comparable intensity W/λ in the visible region, which are strengthened towards the limb by as much as 30 per cent (ADAM, 1938), but this result is not surprising in view of the high excitation potentials of the lower states of the lines, which are therefore formed at greater mean optical depths than the visible lines. This conclusion is confirmed by detailed calculations according to a recent simple approximate theory due to UNSÖLD (1948), which has been applied to the solar model put forward by NEVEN and DE JAGER (1951), using CHANDRASEKHAR and BREEN'S (1946) values of the

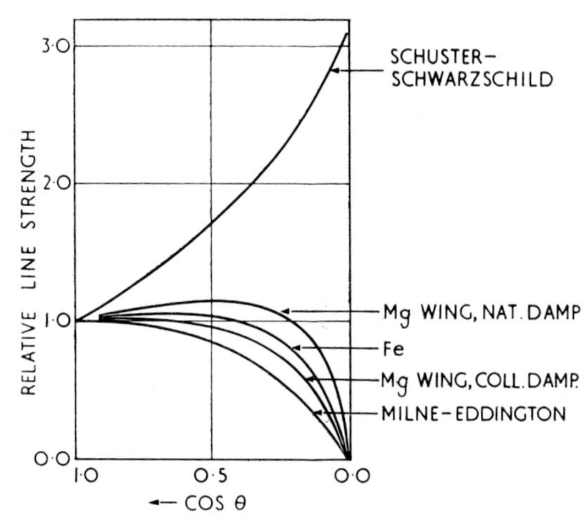

Fig. 4. Predicted centre-to-limb variation in line strengths. The curves indicate the expected centre-to-limb variation in line strength at $\lambda 17{,}100$ for a Schuster-Schwarzschild model; for a faint Fe I line, excitation potential 6·33 V; for the wings of a strong Mg I line, excitation potential 5·39 V, assuming (*a*) natural damping, (*b*) collision damping; and for a Milne-Eddington model; using UNSÖLD's approximate method (1948)

continuous absorption coefficient. The theory is applicable to the residual intensity in weak lines formed by pure absorption and in the absorption wings of strong lines, and its predictions are shown in Fig. 4. The small predicted variation in the weak Fe line between $\cos\theta = 1{\cdot}0$ and $\cos\theta = 0{\cdot}3$ agrees well with the observations. The interpretation of the increase in strength of the Mg line is more doubtful, as the collision damping mechanism is theoretically more likely to predominate than natural damping, so that one might expect a weakening of the wings at $\cos\theta = 0{\cdot}3$. The increase in equivalent width might well be due to an increase in strength of the line core, possibly combined with a correction to $\cos\theta$ due to irregularities in the solar surface (REDMAN, 1943).

4. FUTURE RESEARCH: THE STUDY OF LINE PROFILES

It has already been pointed out that a useful study of line profiles would require an effective resolving power of at least 10^5. An instrument with somewhat smaller resolving power could be used if it were possible to obtain an instrumental profile (REDMAN, 1935), but accuracy requires that the latter should be narrower by a factor of 3 or more than the line itself. The best resolution ever attained with a grating in this region is of the order of 6×10^4 (McMath, 1953), so that some form of multiple-beam interferometer will have to be used. The interferometer will be placed internally in the collimated beam of the spectrograph, which will be used as a monochromator, admitting a range of frequency $\Delta\nu \sim 1\ \text{cm}^{-1}$. An external arrangement would be less convenient, owing to the extra space and equipment needed. The grating and interferometer will scan simultaneously through the spectrum, slight differences of rate being unimportant as long as the small range $\delta\nu \sim 0{\cdot}05\ \text{cm}^{-1}$ admitted by the interferometer remains within the bigger range $\Delta\nu$. The resolving limit of such an arrangement is

$$\delta\nu = \frac{\Delta\nu}{N} = \frac{1\ \text{cm}^{-1}}{N},$$

where N is the effective number of interfering beams.

Interferometers in an internal arrangement generally require entrance and exit slits placed perpendicular to the corresponding slits of the grating monochromator; for the Fabry-Perot and Lummer plates only an exit slit is required to be physically present, as its position automatically defines the direction of the incident beam. It follows that the intensity available for a given resolution $\delta\nu$ is proportional to slit-width, *i.e.* to the dispersion $d\theta/d\nu$. Thus, if I_0 is the intensity falling on the detector when the monochromator is used alone and I the intensity when the interferometer is used, then, provided there is no absorption within the interferometer,

$$\frac{I}{I_0} = \frac{\delta\nu}{\Delta\nu}\frac{F}{l}\frac{d\theta}{d\nu}\,\delta\nu\,\frac{1}{n},$$

where F is the focal length, l the original slit length, and n is 1 or 2, depending on whether single or double orders are obtained. With $\Delta\nu = 1\ \text{cm}^{-1}$, a factor

$$\frac{I}{I_0} \geqslant 10^{-2}$$

would be usable in the present arrangement. Since $l = 1{\cdot}5$ cm, $F = 300$ cm, a dispersion such that

$$\frac{I}{I_0} = \frac{200}{n}\,\delta\nu^2\,\frac{d\theta}{d\nu} \geqslant 10^{-2}$$

is required. At 2μ, the maximum usable $\delta\nu$ is $0{\cdot}05\ \text{cm}^{-1}$.

Therefore, $$\frac{d\theta}{d\nu} \geqslant 4 \times 10^{-2}\ \text{radians per cm}^{-1}.$$

The echelon and Lummer plate (Tolansky, 1947), and also the recently developed echelle grating (Harrison, 1949), have a much smaller dispersion than this ($\sim 10^{-3}$).* The dispersion of a Fabry-Perot etalon used at an angle of incidence θ is

$$\frac{d\theta}{d\nu} = \frac{\cot\theta}{\nu}$$

and $\rightarrow \infty$ at normal incidence. It follows that an etalon used at an angle (as is usual in high-resolution spectroscopy) does not send sufficient intensity on to the cell, but an interferometer used at normal incidence does fulfil the required conditions.

* The echelle grating differs from other multiple-beam interferometers in having a much wider range between successive orders ($\Delta\nu \sim 20\ \text{cm}^{-1}$, compared with $1\ \text{cm}^{-1}$ for the others). Hence, in spite of the small dispersion of the echelle, the monochromator slits could be widened sufficiently to give an observable intensity. However, the echelles produced so far appear to have too much scattered light to be useful in the measurement of line profiles.

With a total path difference 1 cm, a reflecting power of 75 per cent is sufficient to give a resolving limit of 0·05 cm^{-1} (TOLANSKY, 1947, p. 105). The spectrum can be scanned either by increasing the actual separation of the plates or by pumping air into the space between them. A sliding interferometer has been constructed in the observatories workshop at Cambridge, but it is difficult to keep the plates sufficiently parallel during the movement (a resolving power of 10^5 requires that plates with optical path difference 1 cm remain parallel within $\pm 10^{-5}$ cm over their entire surface), so that the air pumping method, which is standard practice in the use of the reflection echelon (TOLANSKY, 1947), appears to be more promising, the pressure in the chamber at any given moment offering a measure of wavelength. The disadvantage of this method is the small range available: 1 atmosphere corresponds to a range of 6 Å at 2μ, but this is sufficient for the study of a single line as the wings can be extrapolated by means of Minnaert's inverse-square law (MINNAERT, 1935). It is therefore suggested that this method is the most suitable approach to the problem of infra-red line profiles.

This work is being done under the supervision of Dr. D. E. BLACKWELL, to whom the writer is most grateful for continual discussion, advice, and encouragement. Thanks are also due to Professor R. O. REDMAN for facilities and for his constant interest and advice, to the D.S.I.R. for a maintenance grant, to Professor R. R. MCMATH for permitting access to unpublished work, and to Professor L. GOLDBERG and Dr. P. B. FELLGETT, who read the manuscript of this paper.

REFERENCES

ADAM, M. G. . . . . 1938 M.N., **98**, 112.

BABCOCK, H. D. and MOORE, C. E. . . . . . . 1947 *The Solar Spectrum λ6600 to λ13,495.* Carnegie Instn. of Washington, Publ. No. 579.

BLACKWELL, D. E., SIMPSON, O. and SUTHERLAND, G. B. B. M. . . . . 1947 *Nature (London)*, **160**, 793.

CHALONGE, D. and KOURGANOFF, V. . . . . 1946 *Ann. Astrophys.*, **9**, 69.

CHANDRASEKHAR, S. and BREEN, F. H.. . . . 1946 *Ap. J.*, **104**, 430.

EVANS, D. S. . . . . 1947 *M.N.*, **107**, 433.

FELLGETT, P. B. . . . . 1949 *J. Opt. Soc. Amer.*, **39**, 970.

1951 *M.N.*, **111**, 537.

GOLDBERG, L. . . . . 1950 *Rep. Progr. Phys.*, **13**, 24.

GOLDBERG, L., MOHLER, O. C. and MCMATH, R. R. 1949 *Ap. J.*, **109**, 28.

1950 *Ap. J.*, **111**, 565.

HARRISON, G. R. . . . . 1949 *J. Opt. Soc. Amer.*, **39**, 522.

MCMATH, R. R. . . . . 1953 Private communication.

MCMATH, R. R. and MOHLER, O. C. . . . . 1949 *J. Opt. Soc. Amer.*, **39**, 903.

MIGEOTTE, M. and NEVEN, L. . . . . . . 1952 *Mem. de la Societé Royale des Sciences de Liège*, **12**, 165.

MINKOWSKI, R. . . . . 1942 *Ap. J.*, **96**, 306.

MINNAERT, M. . . . . 1935 *Z. Astrophys.*, **10**, 40.

MINNAERT, M., MULDERS, G. F. W. and HOUTGAST, J. 1940 *Photometric Atlas of the Solar Spectrum*, Amsterdam.

MOHLER, O. C., PIERCE, A. K., MCMATH, R. R. and GOLDBERG, L. . . . . 1950 *Photometric Atlas of the Near Infra-red Solar Spectrum λ8465 to λ25,242*, Ann Arbor, Mich.

1953 *Ap. J.*, **117**, 41.

MOORE, C. E. . . . . 1949 *Atomic Energy Levels*, Nat. Bureau of Standards, Washington, Circ. No. 467.

NELSON, R. C. . . . . 1949 *Atlas and Wavelength Tables showing the Absorption of Water Vapour in the Regions* 1·33 *to* 1·48 *and* 1·77 *to* 1·98 *Microns.* Northwestern University, Illinois.

NEVEN, L. and DE JAGER, C.	1951	*Bull. Astron. Inst. Netherlands*, **11**, 291.
PETTIT, E.	1951	*Astrophysics, Symposium* (Ed.: HYNEK, J. A.). New York, p. 275.
PFUND, A. H.	1927	*J. Opt. Soc. Amer.*, **14**, 337.
REDMAN, R. O.	1935	*M.N.*, **95**, 742.
	1943	*M.N.*, **103**, 173.
SUTHERLAND, G. B. B. M., BLACKWELL, D. E. and FELLGETT, P. B.	1946	*Nature (London)*, **158**, 873.
TOLANSKY, S.	1947	*High Resolution Spectroscopy*, London.
UNSÖLD, A.	1938	*Physik der Sternatmosphären*, Berlin, p. 280.
	1948	*Ann. Physik.* (Leipzig), **3**, 124.

Temperature at the Poles and at the Equator of the Sun*

A. K. DAS and K. D. ABHYANKAR

Kodaikanal Observatory, India

SUMMARY

The equivalent widths of the g and K lines are measured at the pole and at the equator of the Sun during a period of minimum solar activity. The difference of temperature between pole and equator is evaluated by WOOLLEY's method of the calcium ionization temperature. The pole is found to be hotter than the equator by 96° ± 18° or 86° ± 16°, depending upon the choice of the values of the continuous absorption coefficients. This result lends support to BJERKNES's thermohydrodynamical theory of sunspots. Using BJERKNES's theory, the depth of the sunspot umbral column is calculated from the observed difference of temperature between pole and equator and is found to be 125 km or 140 km.

1. THE PROBLEM

APART from the quantitative theories with a limited objective, such as those due to RUSSELL (1921), ROSSELAND (1926), AMBARTSUMIAN and KOSIREV (1928), MILNE (1930), PETRIE (1930), *etc.*, there are at present two main general theories which aim at explaining the origin of sunspots, one due to BJERKNES (1926, 1937) and the other due to ALFVÉN (1942, 1950). As is to be expected, certain modifications or developments of both these theories have been proposed from time to time. For instance, UNSÖLD (1931) has suggested certain alterations in the original purely hydrodynamical theory of BJERKNES, while WALÉN (1944) has considerably modified and developed ALFVÉN's ideas about the magneto-hydrodynamical origin of sunspots. But the general character of both theories as developed by their original authors remains essentially unchanged. In the present state of our knowledge, in spite of the availability of observational data of sunspots extending over more than a century, it has to be admitted that the mechanism of sunspot formation is still entirely obscure. Theoretical as well as observational objections can be, and have been, raised against both BJERKNES' and ALFVÉN's theories. But taken by and large, BJERKNES' theory offers probably the most extensive explanation of sunspots yet, and appears also to be more amenable to observational test than ALFVÉN's theory.

* Some twenty years ago Professor STRATTON suggested to me (A.K.DAS) this problem for investigation. Although some measurements were made at the time, the results seemed to be very indefinite and I did not pursue the matter further. Recently, however, I have found it necessary to return to this problem as a possible means of clearing, at least partially, some points that have arisen in connection with certain problems of solar physics at present under study in this observatory. In collaboration with Mr. K. D. ABHYANKAR, a Government of India Research Scholar, quite definite results have now been obtained on the difference of temperature between the pole and the equator of the Sun. I am glad to be able to report on our conclusions on this particular question in this volume dedicated to Professor F. J. M. STRATTON.

Without going into the question of the comparative merits and shortcomings of the two theories we consider here only two particular points which emerge rather as by-products of Bjerknes' thermohydrodynamical theory of the origin of sunspots, and which can be tested observationally.

Bjerknes postulates the existence of a general stratified circulation between the poles of the Sun and the equator; the circulatory currents are, in the uppermost layers of the photosphere, directed from the poles to the equator, while in the layers immediately below they flow from the equator to the poles. As in the case of the Earth's atmosphere, the masses moving from the poles toward the equator will be retarded and those moving in the reverse direction will be accelerated with respect to the body of the Sun, thus giving rise to an apparent east wind relative to the body of the Sun. Bjerknes estimates from observational data that this east wind will go round the Sun once in about 20 solar days and therefore will have a velocity of about 130 m per sec. But the pole-to-equator movement will be much slower and will take several years. In the course of this slow drift the superficial layers of the photosphere must cool down by radiation, so that the temperature should be highest at the poles and lowest at the equator. An actual determination of the temperature at the poles and the equator is therefore of very great interest not only as a test of Bjerknes' general hydrodynamical picture, but also as a means of indirectly measuring the depth of sunspot vortices; for in spite of the fact that no rotatory motion has so far been observed with certainty in sunspots, the possibility of the existence of such motion in the deeper levels inaccessible to direct observation cannot be excluded. In fact, Hale's old picture of the generation of the strong magnetic fields in sunspots through vortical motion still remains the simplest and perhaps the most plausible representation of this ill-understood phenomenon.

2. Choice of Method: Calcium Ionization

Bjerknes' theory gives the following simple relationship between the pole-equator difference of temperature and the thickness of the upper circulating stratum above the sub-photospheric zonal vortex in the lower stratum in which sunspots originate:

$$\frac{\Delta\theta}{\theta} = \frac{2v_E \, . \, v}{gh}. \qquad \ldots.(1)$$

In this formula θ = surface temperature of the Sun, v = circumferential velocity at equator = 2000 m per sec., v_E = eastward velocity of the highest circulating stratum, g = solar acceleration of gravity, while $\Delta\theta$ and h are the two quantities mentioned above. To our knowledge, there is no measurement of the quantity $\Delta\theta$ so far available, but there are many conjectural values available for h. From the widely differing estimates of h it would seem that $\Delta\theta$ derived from (1) might be rather small; however, if we note that the internal stability of the Sun increases with a decrease in the depth of sunspot vortices, then a large difference of temperature between the poles and the equator is not precluded. The choice of the method of measurement of the temperature difference, if it exists at all, must therefore depend upon the above considerations. The best values of the effective temperature of the Sun so far available from the measures of Kurlbaum, Abbot, and Fowle, based upon the laws of radiation, vary between 6033° abs. and 6790° abs. For the determination of the supposed pole-equator difference of temperature a procedure capable of much higher precision is therefore desirable. Such a procedure was developed by Woolley (1933) about twenty years ago and was based upon the application of SAHA's formula, as modified by PANNEKOEK (1926, 1930), to the ionization of calcium in stellar atmospheres. It is this method of calcium ionization temperature that has been used in the present work for determining the temperature of the Sun at the pole and at the equator. In the above-mentioned paper WOOLLEY not only worked out completely the theory of the method, but also determined observationally the effective temperature for sunlight integrated over the disk and arrived at the value 6310°A ± 50°. This uncertainty of ± 50° in T_e arises from the probable error in the correcting factor ϕ, which appears in his formula for the ratio of the number of ionized atoms to the neutral atoms per cubic centimetre at a prescribed optical depth, and from the questionable assumption, which WOOLLEY was rather forced to make, namely that the equivalent widths of the g line and the K line measured with widely differing dispersions are affected by the same systematic errors.

3. EXPERIMENTAL DETAILS

The observations were made with a fairly large solar spectrograph designed by one of the present writers (A. K. DAS) and constructed in the workshop of this observatory. The instrument has two independent dispersing trains which are always in position and either system can be brought into operation within a few seconds simply by turning a handle, so that the spectrograph can be used at will either as a grating instrument or as a prismatic one. The grating used is a plane one, ruled by MICHELSON, and has a ruled surface of nearly 7 × 3 in. with 12,500 rulings per inch. For the present work, however, only the prism train was used; this consists of three large 60° prisms and one 30° reflecting prism, each having a clear aperture of 6″ × 4″. The spectrograph is of Littrow construction and has an achromatic collimator camera lens of 21-ft focal length; it is especially designed to reduce stray light to a minimum. With the seven-prism dispersion used the scale of the spectrum is 0·87 Å per mm at λ4227 and 0·74 Å per mm at λ3933. Only violet light was introduced into the spectrograph by the use of a violet filter at the slit. The spectrograph was fed by means of an 18-in. Foucault siderostat through a 12-in. photovisual objectglass of 21-ft focal length, giving a solar image of 59·5 mm diameter in February when the observations were made. The solar image was focussed upon a circular aluminium disk placed immediately in front of the slit plate. This disk had a circle of exactly the same diameter as the solar image marked on it, so that the focussed image could be made to coincide with it. The disk was also mounted in such a manner that it could rotate accurately inside an outer graduated ring capable of both horizontal and vertical movements. A hole of 1 mm diameter which could be placed anywhere on a radius of the metal disk allowed light only from the corresponding point of the solar image to fall on the spectrograph slit. With all these movements it was quite easy to illuminate the slit with the light from the pole or from a corresponding point on the equator. The positions of the accessible pole and the equator for a given time of a given day were determined in advance from the position angles given in the *Nautical Almanac*. On the days on which the spectra were photographed the south pole was 0·5 mm inside the limb on a scale of 59·5 mm to the Sun's diameter; therefore a point 0·5 mm inside the west limb, but exactly on the equator, was selected for photographing the spectra at the equator. The spectroheliograms taken at this observatory on the days on which the polar and equatorial g and K lines were photographed showed that the points concerned were completely undisturbed and that also the general disk of the Sun was practically free from disturbances. Furthermore, the observing conditions on these days were as good as one could have wished.

All the spectra were photographed on $8\frac{1}{2} \times 6\frac{1}{2}$ in. Kodak B20 Process plates. Each plate had eight solar spectra, each of 1 mm width, in addition to the standardization spectra for photometry. If the spectrograph happened to be adjusted for the g-region, then four exposures were made on the pole and another four on the equator immediately afterwards. The exposure times were varied between 1 sec. and 6 sec., but for every polar spectrum exposed for a given time there was a corresponding equatorial spectrum taken with the same exposure time. The spectrograph was then adjusted for the K-region and the exposures were made on another plate for recording the K line at the pole and at the equator in the same way as before. But the exposure times in this case varied between 2 sec. and 100 sec. and also two plates, having eight K spectra each, were taken instead of only one as in the case of the g line. This wide range of exposure times was necessary because of the rapid falling off in the intensity of the spectrum towards the higher frequencies, partly due to absorption by the seven large prisms and partly due to the natural weakening of the solar spectrum in the ultra-violet. Also two separate plates were necessary in this spectral region in order to determine the run of the background to a distance of 80 Å on the shortwave side and to 150 Å on the longwave side of the K line, and in order to determine the interpolated background at the K line. An exposure of 2 sec. sufficed to give a suitable density of the background on the far red side of H, while a 100-sec. exposure was required for the background on the shortwave side of K. An exposure of 50 sec. was sufficient for the centre of the K line. The time required for securing a set of three plates—one for the g region and two for the K region—was about half an hour, but most of this time was spent in changing the spectral region and in bringing the required point of the solar image on the spectrograph slit. However, on the days on which these spectrum plates were taken, the weather was steady and the sky perfectly blue, so that no uncertainty in the quality of the spectra could be feared. Altogether eight sets (twenty-four individual plates) of plates of proper densities were taken for this work, giving 192 solar spectra in addition to an equal number of standardization spectra for photometry.

After exposure of the solar spectra each plate was transferred to a 14-ft plane grating spectrograph which had a permanent arrangement for impressing photometric standards. This auxiliary spectrograph has a very carefully calibrated step-slit (calibrated *in situ*), mounted immediately in front of its normal slit. The step-slit is uniformly illuminated by focussing on it the enlarged image of the ribbon of a tungsten ribbon lamp in quartz envelope, permanently mounted beside the step-slit, with the help of a lens and a plane mirror. For impressing the standardization marks on the plates it is only necessary to open the ordinary slit of the spectrograph to its fullest extent and expose the plates to the ribbon lamp through the step-slit. The ribbon lamp is fed from a high-capacity storage battery at approximately 17 A. The step-slit gives eight standardization spectra of varying densities with one exposure. Exposure times of 15 sec. and 30 sec. respectively were required for the g and K regions for the Kodak B20 Process plates used in this work. There was thus an inequality between the exposure times for the standardization spectra and the solar spectra; this difference was, however, too small to be of any importance, particularly in the case of slow, colour-blind plates. All the plates were developed in

M-Q developer at about 65°F for 6 min in safe red light taking the usual precautions to eliminate EBERHARD effects.

4. Determination of Equivalent Widths of g and K Lines

The spectrum-plates were photometered with the Cambridge recording microphotometer of this observatory and the characteristic curves (D-log I curves) for each plate were prepared in the usual way. For the narrow g line one characteristic curve corresponding to λ4227 for each plate was sufficient; but for the K region characteristic curves for four separate wavelengths at intervals of 50 Å were prepared and each curve was used for about 50 Å so as to ensure that the conversion of densities into intensities was sufficiently accurate. For evaluating the equivalent widths of the g line at the pole and at the equator the following procedure was adopted: out of the eight spectra available on each plate two spectra corresponding to either point (pole or equator) of the solar disk were selected according to their suitability in respect of density, one for the wings and the other for the centre of the line. They were passed through the microphotometer, and the density curves so obtained were converted into intensity curves. The two curves were then joined in the following way. If I_1 and I_2 are the intensities at two points on one curve and I_1' and I_2' are the intensities at corresponding points on the other, then

$$I_1/I_2 = I_1'/I_2' \quad \text{or} \quad I_1/I_1' = I_2/I_2',$$

so that $\log I_1 - \log I_1' = \log I_2 - \log I_2' =$ constant. This constant difference was determined by taking a number of points and thus log I's were obtained on a uniform scale. A graph giving I against λ was then plotted and the background was determined by joining the points of highest intensity by a smooth curve. As this smooth curve was extended to include points up to about 20 Å from the centre of the line on both sides of the centre, the background was reliably determined and proved in practice to be a straight line. The next step was to determine the ratio $r = I/I_0$, where I is the intensity at a point in the line and I_0 is the intensity of the background at the same point. This was done for a sufficiently large number of points and a curve giving r against λ was plotted. The equivalent width of the g line was obtained by determining the area under it by SIMPSON's rule. The procedure for determining the equivalent width of the K line was in principle precisely the same. But in this case all the eight different spectra of appropriate densities for either the pole or the equator available out of each set were photometered in order to reach about 100 Å

Table 1. Equivalent widths in Ångströms

Set No.	Date	Equator		Pole	
		λ4227	λ3933	λ4227	λ3933
10	5–2–53	1·299	11·70	1·254	12·57
11	6–2–53	1·129	10·82	1·227	12·23
12	6–2–53	1·314	11·98	1·256	12·39
13	6–2–53	1·275	11·86	1·252	12·47
14	7–2–53	1·243	11·87	1·276	13·15
15	7–2–53	1·283	11·75	1·299	13·09
18	8–2–53	1·280	11·86	1·286	13·02
20	12–2–53	1·239	12·08	1·227	12·66
Means . .		1·258 ± 0·039	11·74 ± 0·26	1·260 ± 0·016	12·70 ± 0·23

on either side of the line. As in the case of the g line, I for the K line was also obtained on a uniform scale and the background was determined by joining the points of highest intensity by a smooth curve. As is to be expected in this part of the spectrum, this smooth curve representing the background differs greatly from a straight line, but there is no arbitrariness about the determination of the background. The final results of the measurements and computations of the equivalent widths of the g and K lines at the pole and at the equator of the Sun are shown in Table 1. They are based upon eight sets of plates, all of them yielding concordant values.

5. PROCEDURE FOR EVALUATION OF TEMPERATURE

The evaluation of the temperature was done according to the procedure developed by WOOLLEY. WOOLLEY has shown that if k, the absorption coefficient of continuous radiation in the atmosphere, is known and also g, the surface gravity of the Sun, then we can calculate P, the electron pressure at any optical depth τ. The ratio n_1/n_0 of the number of ionized to neutral calcium atoms per cubic centimetre at this depth can be found from the observed equivalent widths of the g line due to the neutral calcium atom and the K line due to the ionized calcium atom and the appropriate absorption coefficients. Knowing n_1/n_0 and P, the temperature at optical depth τ can be calculated by means of the SAHA formula and its modification by PANNEKOEK, and from this we can calculate the temperature of the Sun at any optical depth near the surface. The details of the theory of WOOLLEY's method are given in his paper, but in order to make the purpose of the following tables clear we give here a brief outline of the method of calculation.

The ratio of the number of Ca^+ to Ca atoms per cubic centimetre can be derived in two ways, *viz.* (*a*) by assuming certain temperatures and pressures and using the SAHA formula, and (*b*) from the ratio of the squares of the equivalent widths of the ionized (K) and the neutral (g) lines. By comparing the value obtained from (*b*) with those obtained for various temperatures from (*a*) the temperature corresponding to the ratio n_1/n_0, obtained from the observed equivalent widths of g and K, can be found. We start from a certain effective temperature T_e. Then the boundary temperature T_0 and the effective temperature T at optical depth τ are given by $T_e^4 = 2T_0^4$ and $T^4 = T_0^4(1 + \frac{3}{2}\tau)$. The electron pressure at depth τ is given by

$$P^2 = \frac{2g}{\alpha} \cdot T_0^{11/2} \int_0^\tau (1 + \tfrac{3}{2}\tau)^{11/8} \mathrm{d}\tau, \qquad \ldots .(2)$$

where g, the surface gravity of the Sun $= 2{\cdot}74 \times 10^4$, and α, which appears in CHANDRASEKHAR'S (1932) formula $k = \alpha P\bar{x}/T^{11/2}$ for the coefficient of continuous absorption, has the value $1{\cdot}84 \times 10^{21}$. The ratio $x/(1-x)$ of ionized to neutral atoms is then found from the equations

$$\log\left(\frac{x}{1-x}\right) = -\frac{5040I}{T} + \frac{5}{2}\log T - 6{\cdot}5 - \log P + \log\left(\frac{2B'(\tau)}{B(\tau)}\right) \qquad \ldots .(3)$$

and

$$\frac{Px}{1-x} = \left(\frac{Px}{1-x}\right)_T \times \frac{1}{2}\left\{\frac{T_1}{T}\exp\left\{-\frac{I}{R}\left(\frac{1}{T_1}-\frac{1}{T}\right)\right\} + \frac{3\tau}{2+3\tau}\right\}, \qquad \ldots .(4)$$

due to SAHA and to PANNEKOEK, where I = ionization potential of Ca = 6·09 V. In the first equation $B'(\tau)$ and $B(\tau)$ are partition functions for ionized and neutral atoms. In the case of the calcium atoms $B'(\tau) = B(\tau)$. In the second equation

T_1, the temperature of outgoing radiation, is given by $T_1^4 = T_e^4(1 + \frac{3}{4}\tau)$. The results of the calculation of P and $x/(1-x)$ for $\tau = 0{\cdot}3$ are shown in Table 2 below.

Table 2. Values of P and n_1/n_0 at $\tau = 0{\cdot}3$ for various values of T_e

T_e	6000	6100	6200	6300	6400	6500	6600
T_0	5046	5129	5213	5297	5381	5466	5549
P (in atm $\times 10^{-5}$) .	5·244	5·487	5·738	5·995	6·263	6·534	6·813
n_1/n_0	227	271	322	385	455	533	617

The next step was to deduce the value of n_1/n_0 from the observed equivalent widths given in Table 1. This was easily done by means of the relation

$$n_1/n_0 = \left(\frac{W_1}{W_0}\right)^2 \times \frac{\lambda_0^4 f_0 \omega_0 k_1}{\lambda_1^4 f_1 \omega_1 k_0} \times \phi, \qquad \ldots . (5)$$

given by WOOLLEY. Here W_1 and W_0 are the observed equivalent breadths of the ionized ($\lambda 3933$) and the neutral ($\lambda 4227$) lines respectively, λ_1 and λ_0 are the corresponding wavelengths, f_1 and f_0 are their oscillator strengths, ω_1 and ω_0 their natural widths, k_1 and k_0 are the continuous absorption coefficients at those wavelengths, and ϕ is the correcting factor required to convert the ratio of ionized to neutral atoms found from the observed equivalent breadths to the actual ratio at optical depth $\tau = 0{\cdot}3$. WOOLLEY has calculated the value of this factor and has found it to be $\phi = 0{\cdot}59$ for $\tau = 0{\cdot}3$ in the case of the g and K lines of the calcium atom. He has also deduced the other constants for the calcium lines: $f_0 = 2$, $f_1 = 2/3$, $\omega_0 = 2\omega_{kl(g)}$, $\omega_1 = 1{\cdot}09\omega_{kl(K)}$, so that $\omega_0/\omega_1 = \dfrac{2}{1{\cdot}09} \times \dfrac{\lambda_1^2}{\lambda_0^2}$. WOOLLEY used the value $k_1/k_0 = 1{\cdot}35$ as obtained from PANNEKOEK'S formula for the variation of the coefficient of absorption for continuous radiation with wavelength. But from CHANDRASEKHAR'S (1945) more recent values this ratio comes out to be $k_{3933}/k_{4227} = 2{\cdot}60/2{\cdot}81 = 0{\cdot}925$. Results of the calculations made by using both values are shown in Table 3.

Table 3.

Set-Number	$k_1/k_0 = 1{\cdot}35$					$k_1/k_0 = 0{\cdot}925$				
	Equator		Pole		Difference in Temperature	Equator		Pole		Difference in Temperature
	n_1/n_0	T_e	n_1/n_0	T_e		n_1/n_0	T_e	n_1/n_0	T_e	
10	411	6337	509	6469	+ 132	282	6122	349	6243	+ 121
11	464	6412	503	6462	+ 50	317	6190	345	6237	+ 47
12	421	6351	493	6452	+ 101	289	6135	338	6225	+ 90
13	439	6377	502	6460	+ 83	301	6159	344	6235	+ 76
14	462	6409	540	6508	+ 99	317	6190	369	6275	+ 85
15	425	6357	514	6476	+ 119	291	6140	352	6248	+ 108
18	435	6371	519	6482	+ 111	298	6153	356	6254	+ 101
20	482	6435	540	6508	+ 73	330	6213	370	6276	+ 63
Means .	442 ± 17	6381 ± 23	515 ± 12	6477 ± 14	+ 96 ± 18	303 ± 11	6163 ± 21	353 ± 8	6249 ± 13	+ 86 ± 16

6. CONCLUSION

From Table 3 it is evident that all the values of temperature, whether they are derived from $k_1/k_0 = 1{\cdot}35$ or $k_1/k_0 = 0{\cdot}925$, are systematically higher for the pole than for the equator. Although the individual values of temperatures show some scatter, the mean temperature difference between the pole and the equator is more than five times the probable error and therefore the result can be accepted with a high degree of confidence. WOOLLEY estimated that his method should be capable of yielding the absolute value of temperature within an error of 10 per cent. This included an error of 5 per cent in the value of ϕ and another 5 per cent in the determination of the ratio of the equivalent widths of the ionized and the neutral lines. In the evaluation of the difference of temperature between the pole and the equator, as in the present case, the probable error due to the error in ϕ will be much less than 5 per cent. Furthermore, WOOLLEY used a very much lower dispersion for measuring the equivalent width of the K line than for the g line; but we have used the same spectrograph and the same high-dispersion optical train for both the lines, so that the uncertainty in our values of the ratio of the equivalent widths ought to be considerably less. Thus the results of our determination of the difference of temperature between the pole and the equator of the Sun appear to lend a good deal of probability to BJERKNES' thermohydrodynamical theory that the Sun is a baroclinic cosmic vortex in which the angular velocity decreases with the distance from the equatorial plane and in which, therefore, the temperature increases from the equator to the poles. BJERKNES' theory also leads to a meridional circulation from the poles to the equator at the surface. It is of some interest to note that the same kind of meridional circulation is indicated by simpler and purely dynamical considerations as was shown by one of the present writers (DAS, 1940) in connection with the study of the disk and limb prominences and various other solar phenomena.

From the difference of temperature between the pole and the equator, as determined from the observations described in the present paper, another interesting conclusion also emerges. By using formula (1) and the value of $\Delta\theta$ here derived we obtain $h = 125$ km or 140 km according as we use the mean value in column 6 or column 11 of Table 3. It seems rather significant that this depth is of the same order as that derived from other theoretical considerations by MILNE, PETRIE, and UNSÖLD for the umbral column of sunspots. Therefore the question asked by MILNE many years ago as to "whether the umbral temperature is or is not another solar constant" still remains an open question. The measurement of the umbral temperature is thus an important problem of solar physics; work on this question is in progress in this observatory. It is not without interest to recall that in a paper dealing with the measurement of the radiation flux across sunspots (DAS and RAMANATHAN, 1953), an indication was found of the constancy of the umbral intensity in some sunspots observed under exceptionally good conditions; but this cannot yet be regarded as a definite conclusion.

REFERENCES

ALFVÉN, H. . . . . 1942 *Ark. Mat. Astron. Fys.*, **29,** No. 2.
1950 *Cosmical Aerodynamics*, Oxford.
AMBARTSUMIAN, V. A. and KOSIREV, N. A. . . 1928 *Astron. Nachr.*, **233,** 107.
BJERKNES, V. . . . . 1926 *Ap. J.*, **64,** 93.
1937 *Astrophys. Norveg.*, **2,** 263.

CHANDRASEKHAR, S.	1932	*M.N.*, **91,** 186.
	1945	*Ap. J.*, **102,** 395.
DAS, A. K.	1940	*Indian J. Phys.*, **14,** 369.
DAS, A. K. and RAMANATHAN, A. S.	1953	*Z. Astrophys.*, **32,** 91.
MILNE, E. A.	1930	*M.N.*, **90,** 487.
PANNEKOEK, A.	1926	*Bull. Astron. Inst. Netherlands*, **3,** 207.
	1930	*M.N.*, **91,** 139.
PETRIE, R. M.	1930	*M.N.*, **90,** 480.
ROSSELAND, S.	1926	*Ap. J.*, **63,** 356.
RUSSELL, H. N.	1921	*Ap. J.*, **54,** 293.
UNSÖLD, A.	1931	*Z. Astrophys.*, **2,** 209.
WALÉN, C.	1944	*Ark. Mat. Astron. Fys.*, **30**A, No. 15.
WOOLLEY, R. v. D. R.	1933	*M.N.*, **93,** 691, Supp.

The Lineage of the Great Sunspots

H. W. NEWTON

Royal Greenwich Observatory, Herstmonceux Castle, Sussex

SUMMARY

Early historical records bear witness to the occurrence of great sunspots in past centuries, and telescopic records have contributed more precisely to the story which is taken up by the Greenwich observations started in 1874. Since then, fifty-four great spots, with a mean area for the solar disk-passage of $\geqslant$ 1500 millionths of the Sun's hemisphere have been observed. These spots, of which the first appeared in 1882, are listed in an Appendix. Their characteristics are briefly considered, including a marked tendency for them to be associated with geomagnetic storms. The progress of the observational side of solar physics during the past seventy years is then reviewed against the background of these great sunspots which, by virtue of their size and range of associated phenomena, have presented themselves as unique objects for observational research.

1. INTRODUCTION

THE lineage of the great sunspots emerges from very early historical records from both Asia and Europe with their testimony of a "fleckle" in the Sun, a flying bird, a gourd or an egg, an apple or a coin (TURNER, 1889; SCHOVE, 1950). "Dark spots on the Sun as if nails were driven into it" is the forceful imagery of a fourteenth-century Russian chronicle, when the solar disk appeared blood-red through the smoke of devastating forest fires. In later centuries, naked-eye sunspots were sometimes ascribed to the transits of Mercury or of Venus, such an explanation being offered by KEPLER for a spot that appeared in 1607. However, the first telescopic observations of 1610–11 disposed of this explanation. Sometimes with these early records of dark markings in the Sun, extending back more than 2000 years in the case of the Chinese annals, are records unmistakably of auroral displays at the same epochs as those of the spots. In this country memorable displays are recorded by JOHN OF READING in 1366, by STOW in 1574, and by HALLEY in 1716. Here again is testimony to the presence of active regions on the Sun with a high degree of probability that spots bigger than usual were present. The long lineage of great sunspots appears established, but was it, in fact, unrecognizable in the latter part of the seventeenth century? Concerning a sunspot in 1684, FLAMSTEED comments: "These appearances, however frequent in the days of SCHEINER and GALILEO, have been so rare of late that this is the only one I have seen in his face since December, 1676." (See also MAUNDER, 1889.) And in 1705, CASSINI and MARALDI record that they had never seen a spot in the northern hemisphere. This apparent hiatus in spots receives support from the analysis of tree rings, and also by the absence of reports of auroral displays during this period.

With the opening decades of the nineteenth century, uncertainty arising from imperfect sampling is increasingly reduced. We enter the period, since unbroken, of systematic observation of the Sun's disk—first with the visual work of Schwabe, commencing in 1826: then of Spoerer and Carrington, and passing to the photographic records of Kew and then, in 1874, to Greenwich, which has maintained for many years a virtually complete daily sequence of photographs, with the help of co-operating observatories. No lapse of sunspots, except their marked reduction at solar minima, has been repeated.

2. The Great Sunspots of the Past Seventy Years

To define the lower limit of size for a giant spot is somewhat arbitrary. A spot of area 500 millionths of the Sun's hemisphere when near the centre of the solar disk is just visible to the naked eye. Spots do occur with areas up to six times this size, but

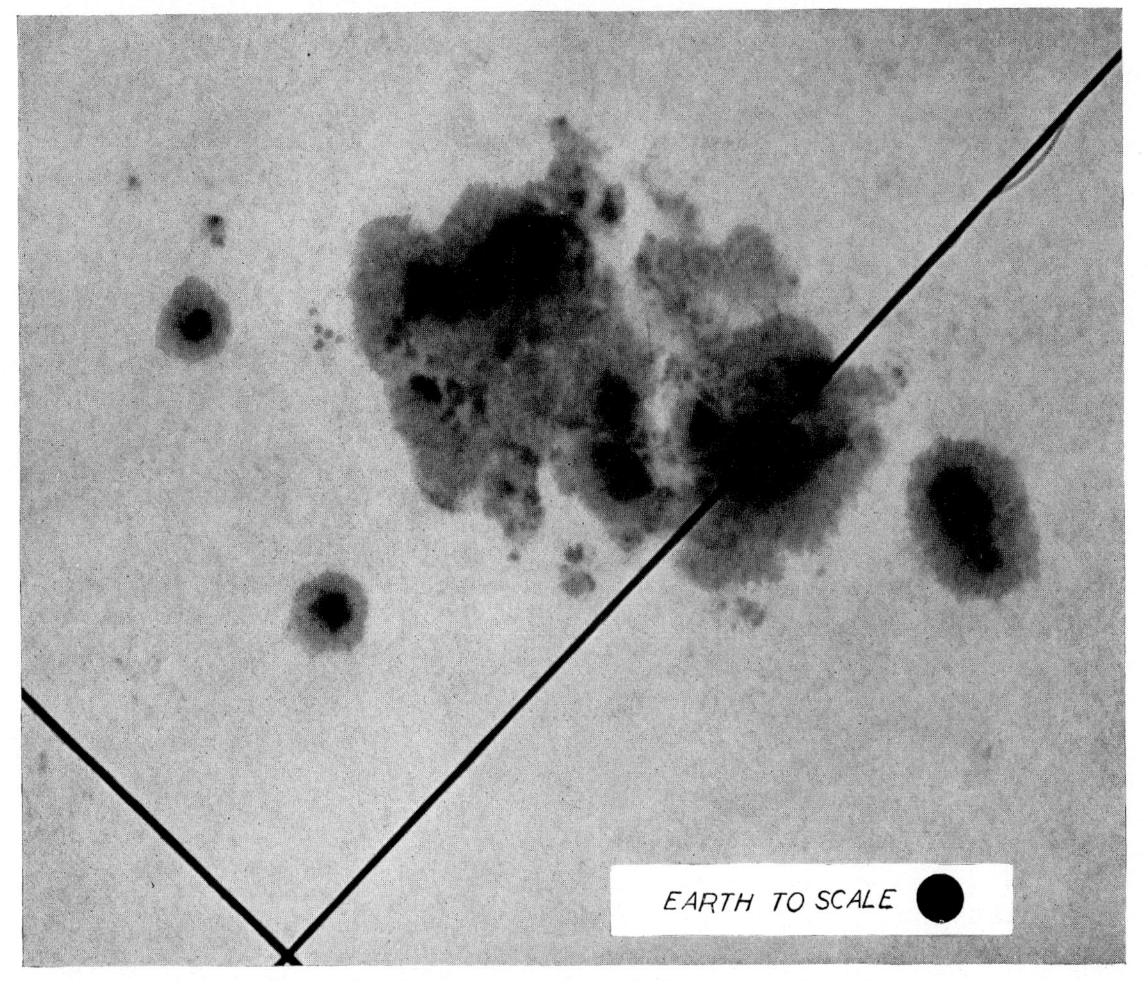

Fig. 1. The giant sunspot of May, 1951 (the fourth in Table 1)
Royal Greenwich Observatory, Herstmonceux

Table 1. Great Sunspots, 1874–1952

Order with respect to		Year	Central Meridian Passage (U.T.)	Area		Latitude	Geomagnetic storms Onset U.T.	
Mean area	Max. area			Mean	Max.		Great	Small
1	1	1947	Apr. 7·2	5520	6132	− 24·4°	—	Apr. 8·9
2	2	1946	Feb. 5·7	4779	5202	+ 26·1	Feb. 7·4	—
3	4	1946	July 26·9	3958	4720	+ 22·2	July 26·8	—
4	3	1951	May 16·0	3743	4865	+ 13·0	—	—
5	5	1947	Mar. 10·2	3637	4554	− 23·1	—	Mar. 7·2
6	6	1926	Jan. 24·5	3285	3716	+ 20·8	Jan. 26·7	Jan. 22·6
7	7	1938	Jan. 18·4	2955	3627	+ 17·1	Jan. 22·2	Jan. 16·9
8	11	1905	Feb. 4·2	2801	3339	− 15·3	—	Feb. 3·1
9	10	1937	Oct. 4·4	2661	3340	+ 9·3	—	Oct. 3·5 Oct. 7·2 Oct. 9·3
10	17	1938	Oct. 11·9	2654	3003	+ 17·0	—	—
11	16	1892	Feb. 11·9	2638	3038	− 28·2	Feb. 13·2	—
12	12	1937	July 28·8	2589	3303	+ 31·5	—	Aug. 1·9
13	21	1925	Dec. 28·6	2587	2934	+ 23·2	—	Dec. 27·6
14	19	1939	Sept. 10·4	2504	2993	− 13·8	—	—
15	13	1917	Aug. 10·1	2389	3178	+ 16·2	Aug. 9·2 Aug. 13·6	—
16	18	1905	Oct. 20·2	2336	2995	+ 13·8	—	—
17	24	1897	Jan. 9·4	2329	2743	− 6·8	—	—
18	34	1910	Jan. 22·7	2242	2471	+ 22·9	Jan. 24·8 Jan. 25·6	—
19	8	1917	Feb. 9·5	2176	3590	− 16·0	—	—
20	20	1947	Feb. 11·3	2131	2944	− 21·1	—	—
21	9	1938	July 15·1	2122	3379	− 12·1	—	July 15·2
22	40	1951	June 18·5	2100	2358	− 12·0	—	June 17·7
23	41	1951	Apr. 18·8	2050	2485	+ 13·0	—	Apr. 18·3 Apr. 20·5
24	44	1938	Nov. 10·8	1990	2245	− 8·5	—	—
25	25	1920	Mar. 21·6	1966	2690	− 4·9	Mar. 22·4	—
26	33	1907	June 19·7	1965	2472	− 14·4	—	—
27	35	1896	Sept. 16·7	1954	2458	+ 13·2	—	Sept. 17·8 Sept. 20·1
28	37	1882	Nov. 18·9	1928	2425	+ 19·2	Nov. 17·4 Nov. 20·0	Nov. 21·7
29	46	1905	July 16·7	1911	2164	+ 13·0	—	—
30	49	1882	Apr. 19·1	1892	2123	− 28·5	Apr. 17·0 Apr. 20·0	—
31	47	1926	Sept. 19·7	1881	2142	+ 24·1	—	Sept. 19·6
32	14	1941	Sept. 16·9	1862	3088	+ 11·6	Sept. 18·2	—
33	29	1907	Feb. 12·4	1860	2555	− 16·6	Feb. 9·6	—
34	23	1950	Feb. 20·2	1850	2856	+ 11·0	Feb. 20·0	Feb. 23·5
35	28	1905	Mar. 7·9	1845	2572	+ 10·5	—	—
36	48	1903	Oct. 11·6	1783	2129	− 21·3	—	Oct. 12·4
37	27	1928	Sept. 27·4	1770	2587	− 15·3	—	Sept. 24·7
38	15	1939	Sept. 1·0	1751	3054	− 14·1	—	Sept. 2·9
39	39	1937	Jan. 31·1	1734	2364	− 10·1	—	Feb. 3·0
40	42	1939	Sept. 28·0	1717	2256	− 12·8	—	—
41	38	1892	July 10·2	1691	2387	+ 11·1	July 12·7	—
42	22	1940	Jan. 5·5	1689	2860	+ 10·5	—	Jan. 3·7
43	50	1946	Dec. 17·0	1675	2099	− 6·4	—	—
44	32	1937	Apr. 24·7	1674	2474	+ 19·5	Apr. 25·7 Apr. 26·7 Apr. 27·8	Apr. 24·5
45	36	1935	Dec. 2·2	1670	2435	− 25·5	—	—
46	26	1893	Aug. 7·4	1640	2621	− 17·8	—	Aug. 6·2
47	45	1898	Sept. 9·2	1636	2235	− 12·1	Sept. 9·5	—
48	30	1948	Dec. 24·0	1592	2513	− 14·4	—	—
49	43	1949	Feb. 5·4	1586	2248	− 9·6	—	—
50	53	1926	Mar. 4·0	1579	1989	− 27·6	—	Mar. 5·4
51	51	1942	Feb. 28·8	1549	2048	+ 7·0	Mar. 1·3	Mar. 5·2
52	31	1894	Oct. 8·1	1542	2511	− 12·2	—	—
53	52	1920	Jan. 28·1	1526	2037	− 6·9	—	—
54	54	1908	Aug. 31·2	1519	1919	+ 6·8	—	Sept. 4·8

NOTES *to Table 1*

1. The sunspot data from the Royal Greenwich Observatory, on which this table is based, extend from April, 1874, to October, 1952. No giant spot was recorded until 1882. Acknowledgment is gratefully made to the Astronomer Royal for permission to use data as yet unpublished, which cover the last few years, and also for permission to reproduce the photograph of the giant sunspot of May, 1951 (Fig. 1). Data after 1948 are subject to slight revision.

2. Areas are corrected for surface foreshortening and are expressed in millionths of the Sun's hemisphere. The mean area given in column 5 is derived from the daily measures covering the disk passage, or part of it when the origin or end of the spot occurs on the disk. Days are excluded when the spot group is not fully in view near the Sun's limb or in any case when it is 80° or more from the central meridian. In accordance with this rule, adopted in 1916, mean areas of spots in the above table, published in the *Greenwich Photoheliographic Results*, have been amended when necessary.

3. The storms listed in the last two columns as having occurred with onset times from $-3^{d}{\cdot}0$ to $+5^{d}{\cdot}0$ (central meridian passage of the spot $= 0^{d}{\cdot}0$) have been abstracted from the lists of storms published in the *Greenwich Photoheliographic Results*, 1927, p. C131–139. These lists have been brought up to date by annual summaries published in *The Observatory*. The lower limit for a small storm at Greenwich-Abinger is 30′ in D or 150γ in H or in V; that for the great storms is 60′ in D or 300γ in H or V.

---

with diminishing frequency. The present cycle has produced the rare example of a group with a mean area of 5500 millionths for its disk passage. Thus a lower limit of 1500 millionths for the mean area during disk passage seems appropriate for the designation "giant spot". According to the Greenwich measures, fifty-four spot groups of this minimum size and upwards have been recorded during the last seven sunspot cycles. The average disk passage area of these giant spots is 2230 millionths and their corresponding peak areas 2879 millionths of the Sun's hemisphere.

This selection of great spots provides an interesting study. A list of the fifty-four largest groups arranged in order of size (mean area during disk passage) is given in Table 1. It is noteworthy that the first five are all recent groups, *i.e.* from the cycle now ending its course.

The distribution of these giant spots, with respect to the epoch of maximum of the respective sunspot cycles to which they belong, is shown in the following Table 2.

Table 2. Distribution of 54 great sunspots during seven sunspot cycles

Years from sunspot maximum	Mean epoch	Total number of spots	Mean area (millionths)	Mean latitude (degrees)	Normal latitude of all spots (degrees)	Latitude range of great spots (degrees)
− 3·9 to − 3·0	− 3·2	1	1783	21°	19°	—
− 2·9 to − 2·0	− 2·2	4	2522	25	21	7°
− 1·9 to − 1·0	− 1·5	10	2273	19	19	18
− 0·9 to 0·0	− 0·4	9	2683	18	17	18
0·0 to + 0·9	+ 0·4	8	2216	17	14	22
+ 1·0 to + 1·9	+ 1·5	7	1958	13	13	16
+ 2·0 to + 2·9	+ 2·5	9	1921	10	11	9
+ 3·0 to + 3·9	+ 3·8	2	2896*	13	10	0
+ 4·0 to + 4·9	+ 4·4	4	1787	11	9	5
+ 5·0 to + 5·9	—	0	—	—	—	—
+ 6·0 to + 6·9	—	0	—	—	—	—

Extreme latitudes + 31°·5 (area 2589) and + 4°·9 (area 1966). 24 spots were in the northern hemisphere and 30 in the southern.

* One of the two spots contributing to this mean value had an area of 3743 millionths. Both spots were related, their latitudes being the same.

Their average size and mean latitude for each of the yearly epochs before and after solar maximum are also given. The latitude progression is essentially that for all spots during the eleven-year cycle, the average interval from minimum to maximum being four years and from maximum to minimum seven years.

The statistical development in area of a giant spot from its origin to maximum is essentially the same as that of medium-sized spot groups, peak area being reached on about the ninth day.* For the smaller groups, *i.e.* less than 500, peak area is usually reached earlier and the total life history is shorter. The decline from maximum to minimum of the giant spots is often very rapid, so that in general the length of life is disproportionately short compared with those spots of area, say, 500 units. In a minority of cases, however, a resurgence of spot formation occurs when the spot is well in decline, thus producing a secondary maximum often reached some three or four weeks after the initial origin of the spots. Usually the second area maximum is below that of the first, but in some cases it is higher. A notable instance is that of the series of giant spots of 1947 culminating in the greatest spot ever recorded. In each case, the renewed surge of spot formation must have started on the invisible hemisphere (Table 3).

Table 3. Three successive maxima of a great sunspot

U.T. of C.M.P. 1947	Longitude	Latitude	Mean area	Maximum area reached on	
Feb. 11·3	86°	21° S	2131	2944	Feb. 10
Mar. 10·2	92	23	3637	4544	Mar. 12
Apr. 7·2	83	24	5520	6132	Apr. 8

Some time after the third maximum the spot must have collapsed, for on its return its average area (made up of two widely separated groups) was only 661 millionths.

Such renewals of spot formation sometimes give rise, as in this recent case, to spots that are apparently of very long duration. But strictly as continued entities it is the smaller spots of circular form (usually the surviving leaders of bipolar groups) that produce the records of longevity up to four months' duration.

The complexity of structure is a noticeable feature of most of the great spot groups. (This complex structure is usually associated with a complex distribution of magnetic polarities classified γ or $\beta\gamma$.) Although a few are giant editions of typical bipolar groups, the great complex groups or single spots rarely appear in miniature.

In spite of this complex structure, the ratio of the mean areas, for a disk passage, of umbra/whole spot (*i.e.* umbra + penumbra) shows a surprisingly small range from spot to spot. For these fifty-four giant spots, the mean ratio is 0·152. (S.D. 0·023.)

3. Giant Spots and Geomagnetic Storms

The association of great spots and geomagnetic storms is one of their most significant attributes (Maunder, 1904; Greaves and Newton, 1928, 1929; Newton, 1949). For example, of the above fifty-four spots, seventeen were associated with at least one "great" storm that began within the period of 3·0 days before and 5·0 days after the central meridian passage of the respective spots. (With four of these spots, there were dual storms and in one case a triple storm.) The chance association for these fifty-four spots would be less than two storms.†

This association of giant spots with geomagnetic disturbance is also illustrated in the adjacent Fig. 2 that is based on international magnetic character figures and the

* See *Observatory*, 1952, **72**, 111. The mean growth curve of the giant spots closely resembles those of the specified area groups of Fig. 2.

† With the greater frequency of small storms in the ratio of nearly 6 : 1 as compared with that of the great storms, the relationship between giant spots and small storms is less convincing. (Indeed an inverse correlation is shown between the small 27-day sequence storms—usually without a "sudden-commencement" onset—and large sunspots.) However, the number of associations of these giant spots with small storms (the larger number *with* S.C. onset) is more than double that to be expected from chance.

disk passage of the spots. The statistical rise of geomagnetic activity is clearly seen as the (mean) spot approaches the central meridian at 0^d. The flattish peak, asymmetrical with respect to the central meridian (because of the involved time of travel of one or two days for the solar particles), is maintained statistically until the spot passes off the disk at day $+ 6\frac{1}{2}$.

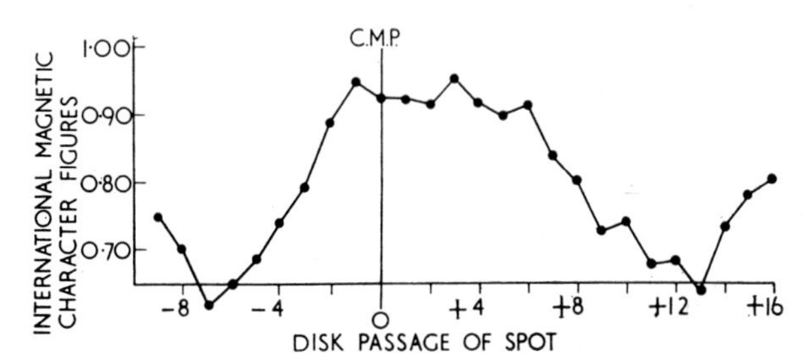

Fig. 2. Large sunspots and geomagnetic storms

The strong but not exclusive tendency for the more intense solar flares to be associated with the larger sunspots is a significant fact in this solar-terrestrial relationship.

4. The Great Sunspots as Time-marks of Progress

Having stated the observational characteristics of great sunspots, let us briefly review their appearances during the last seventy years as time-marks in the progress of solar physics to which, by virtue of their great size and distinctive phenomena, the greatest spots have especially contributed.

The Great Sunspot of November, 1882

The year 1882—the year of the great comet—produced also two giant sunspots, each with an associated geomagnetic storm, one of which was outstanding for its intensity. The interest of this year of astronomical distinction is epitomized by the published observations made at Greenwich on 17 November. From the *Greenwich Spectroscopic Observations*, 1882 (p. 38), we may recapture the strange ethereal quality of that November evening—the Moon in her first quarter near the meridian, the awaited early-morning rising of the great comet, the red and green auroral glow seen above the bank of mist and smoke over London, and finally the apparition of a streak of auroral light that moved majestically across the sky from north-east to west. At noon on 18 November, E. W. MAUNDER (the recorder of the celestial events of the evening before) observed the spectrum of the great spot and saw the C and F lines "exceedingly bright" over the umbra, as YOUNG and LOCKYER had done over other sunspots in the previous sunspot cycle. It cannot be doubted that MAUNDER was observing a phenomenon which more than half a century later (under the designation "solar flare") impinged on the then unimagined field of long distance radio communication. In the light of our present knowledge it is also probable that an intense recrudescence at 0^h U.T. on 20 November of the great magnetic storm (that had begun on 17 November) was related to the brilliant reversals over the spot that MAUNDER observed. He does not seem to have associated these brilliant Hα and Hβ reversals with the unique phenomenon observed in integrated light by CARRINGTON and HODGSON on 1 September, 1859, over a giant spot near the central meridian. The suggestion was to come later from HALE with his important new instrument—the spectroheliograph.

The Great Spot of February, 1892

This epoch was a great one in the history of observational solar physics. HALE had invented his spectroheliograph, capable of recording not only the prominences at the Sun's limb but the disk markings in calcium light. EVERSHED and DESLANDRES quite independently thought out the principle of monochromatic registration and constructed their own instruments, including the unique type by DESLANDRES known as the "spectro-enregistreur des vitesses". The appearance of the great spot of February, 1892, was immediately put to use by HALE who records in his *Study of Stellar Evolution* (1908, p. 86): "The earliest use of the method [*i.e.* the principle of the spectroheliograph] was made in the study of the great sunspot of February, 1892, which, through the great scale of the phenomena it exhibited and the rapid changes that occurred from its exceptional activity, afforded the very conditions required to bring out the peculiar advantages of the spectroheliograph".

One of these advantages was quickly realized five months later. On 15 July, over a spot—a lesser giant that may have been of great size on the Sun's invisible hemisphere—a new phenomenon of brilliant and rapidly changing calcium flocculi was revealed by the spectroheliograph. Within 20 hours after maximum brilliancy, a great geomagnetic storm began. What had been a brilliant reversal of Hα and Hβ in MAUNDER's observation (of 1882) now stood revealed by the spectroheliograph as a brilliant changing cloud of gas over the spot. HALE could, with much certainty, surmise that here was a transient phenomenon most probably akin to the brilliant patches seen in integrated light by CARRINGTON thirty-three years earlier.

Another of these advantages of the new technique that was followed up by HALE (as also by DESLANDRES) was made with the Rumford spectroheliograph of the Yerkes observatory, which gave the requisite greater dispersion for the experiment, *viz.* to record the calcium and hydrogen (Hβ, Hγ; plates sensitive to Hα, not yet being available) flocculi at three merging levels in the chromosphere by isolating a very narrow section of the spectrum line at (*a*) the centre, (*b*) off-centre by a few tenths of an Ångstrom, (*c*) the edge. HALE writes that this experiment was successfully achieved (*loc. cit.* p. 91) on the great sunspot of October, 1903. The results obtained at Yerkes are beautifully depicted in *Publications of the Yerkes Observatory*, Vol. III (1903), and those obtained by DESLANDRES are likewise shown with great perfection in *Ann. de l'Observatoire d'Astronomie Physique de Paris* (*Meudon*), 1910.

The monochromatic records of the Sun's disk have been aided in recent years by the use of the Lyot filter, while the application of cinematography in conjunction with the coronagraph has given us remarkable sequence-records of eruptive prominences at the Sun's limb.

The Great Sunspot of January, 1926

The study of solar flares was greatly advanced by the invention of the HALE spectrohelioscope (1929; 1930; 1931). This delightful solar instrument for the visual observer was still in its experimental form in January, 1926, when a giant sunspot appeared—the largest hitherto in photographic records. Turning his new instrument to the spot, HALE witnessed, on 24 January and again on the following day, brilliant changing areas of hydrogen and dark filaments in motion constituting two intense solar flares, that even from his unique observational experience he described as "the most remarkable solar phenomenon I have ever seen". The papers in which HALE described the spectroheliоscope and the phenomena open to observation with it, stimulated observations of the chromosphere and laid a firm foundation for co-operative work with this instrument under the co-ordination of the International Astronomical Union. His suggestion, supported by several individual cases, that bright chromospheric eruptions (solar flares) might be related to the occurrence of great geomagnetic storms, has since received strong support from statistical papers published from Greenwich (NEWTON, 1943; 1944; 1949; NEWTON and JACKSON, 1951). The recognition of solar flares and their systematic observation were to play in the following decade a fundamental part in correlations between solar and ionospheric phenomena. Perhaps the most spectacular of these is the joint occurrence of intense flares with abrupt fade-outs on long distance short-wave radio transmission on channels within the daylight hemisphere, the solar agency being ultra-violet radiation. About twenty years later, observations of solar flares over the giant spot of February, 1946, were related to intense bursts of solar noise then being observed on metre wavelengths (APPLETON and HEY, 1946). The joint solar and radio phenomena associated with this giant sunspot were thus responsible in no small degree for some of the important developments in radio astronomy that have followed in this country since then.

A second great spot appeared within six months. The close watch kept on it was rewarded on 25 July by the occurrence of an exceptionally intense solar flare. Valuable spectra were obtained (ELLISON, 1946) which showed the Hα line in strong emission broadening to a total width of 20 Å. Many other lines in reversal were listed, and there was evidence of enhancement of the continuum, recalling CARRINGTON'S observation of 1859. An intense burst of solar noise was recorded, closely synchronous in time with the maximum of the visible emission at 16^{h} 27^{m} U.T. The geophysical effects were typical, (1) a synchronous crochet (not recorded, however, in this country); (2) a synchronous fade-out on short-wave radio transmission that was prolonged as was the flare; (3) a great geomagnetic storm with onset 26·3 hours after solar flare maximum.

Sunspots of 1909, 1915, 1921

But, here, we must retrace our steps by forty years. EVERSHED (1909, 1910) had announced the discovery of the radial motions of gases in sunspots, radial to the spot centre and parallel to the photospheric surface. The extended solar minimum, 1911–1913, then intervened, but the new spot cycle found EVERSHED with improved equipment to study this newly-recognized phenomenon (the EVERSHED effect) in greater detail. A minor giant that crossed the Sun's central meridian on 5 April, 1915, though its mean area barely exceeded 1000 millionths of the Sun's hemisphere, was an ideal subject for this detailed study because of its circular form. The radial motions of gases within and around a sunspot itself were furthered by ST. JOHN (1913) working with the powerful equipment on Mount Wilson. Further studies are now in progress at the University Observatory, Oxford.

Although HALE'S fundamental discovery, in 1908, that sunspots were the centres of strong magnetic fields—a discovery not associated with any particular spots—the bigger ones offered some interesting cases for observation. For instance, in May, 1921, a lesser giant spot of mean area 1324 appeared exactly on the Sun's equator. The magnetic polarity observations that were made at Mount Wilson (1938; HALE and NICHOLSON, 1925) showed that the solar equator was a precise line of demarcation for the sign of magnetic polarity, the portions of the spot extending into the northern hemisphere being of opposite polarity to those extending south of the equator.

Many of the giant sunspots, though sometimes not associated with abnormal maximum field strengths, are associated with a mixed distribution of sign for the magnetic field, designated γ or $\beta\gamma$ class. Sunspots most productive of major flares and most clearly associated with geomagnetic storms have a predominant γ classification.

5. THE GREAT SPOT OF FEBRUARY, 1946, AND RADIO ASTRONOMY

The sunspot with central meridian passage 28 February, 1942—a minor giant of mean area 1549 units—gave the first recognizable evidence of solar "noise" received on a wavelength of about 4 m. The solar minimum was then approaching, and it was not until the appearance of the great spot of February, 1946, that the joint phenomena of solar flare occurrence over this spot and solar noise bursts could be studied in detail (APPLETON and HEY, 1946). The statistical study of data provided by the exceptionally high sunspot maximum of 1947–48 left no doubt of the general correlation between sunspots and solar radiation received on metre and centimetre wavelengths, and in particular between the more intense solar flares and notable bursts of solar noise. The new and exciting fields of observation and theory opened up by the discoveries of solar and galactic noise promise much for the future of astronomical research.* The knowledge gained from the detailed study of the Sun—for which the great sunspots provide special opportunities—is thus of basic importance to the new radio-telescopic exploration of the universe itself.

REFERENCES

APPLETON, E. V. and HEY, J. S.	1946	*Phil. Mag.*, **37**, 73.
ELLISON, M. A.	1946	*M.N.*, **106**, 500.
EVERSHED, J.	1909	*Kodaikanal Bull.*, No. 15; *M.N.* **69**, 454.
	1910	*M.N.*, **70**, 217.
F(LAMSTEED), J.	1684	*Phil. Trans.*, **14**, 535.
GREAVES, W. M. H. and NEWTON, H. W.	1928	*M.N.*, **88**, 556; **89**, 84.
	1929	*M.N.*, **89**, 641.
HALE, G. E.	1908	*The Study of Stellar Evolution*, Chicago, p. 86.
	1929	*Ap. J.*, **70**, 265.
	1930	*Ap. J.*, **71**, 73.
	1931	*Ap. J.*, **73**, 379.
HALE, G. E. and NICHOLSON, S. B.	1925	*Ap. J.*, **62**, 270.
LOVELL, A. C. B. and CLEGG, J. A.	1952	*Radio Astronomy*, London, 1952.
MAUNDER, E. W.	1889	*Observatory*, **12**, 193.
	1904	*M.N.*, **64**, 205, 222; **65**, 2.
MOUNT WILSON	1938	*Magnetic observations of Sunspots 1917–1924*, Washington, Pt. 2, 541.
NEWTON, H. W.	1943	*M.N.*, **103**, 244.
	1944	*M.N.*, **104**, 4.
	1949	*M.N.*, *Geophys. Suppl.*, **5**, 321.
NEWTON, H. W. and JACKSON, W.	1951	*7th Report on Solar Terrestrial Relationships*, Paris, 107–130.
SCHOVE, D. J.	1950	*J. Brit. Astron. Assoc.*, **61**, 22.
ST. JOHN, C. E.	1913	*Ap. J.*, **37**, 322; **38**, 341.
TURNER, H. H.	1889	*Observatory*, **12**, 217.

* For a survey of the new astronomy see LOVELL and CLEGG (1952); also the preceding Section 6, pp. 521–616, of this volume.

The Structure of Sunspots

P. A. Sweet

University of London Observatory

Summary

A survey is made of the observations of sunspots. This is used to construct a model of the surface layers. A theoretical interpretation of this leads to a theory of the origin of the Evershed effect. The deeper layers are discussed and an appreciation is given of current theories of the origin of the darkening. The origin of the magnetic field and the relation of sunspots to the solar cycle have not been considered.

1. Introduction

This article deals with sunspots as individuals, and is not concerned with their relation to the solar cycle.

Ever since sunspots were studied with the aid of a telescope by Galileo in 1610, ideas as to their structure have reflected current knowledge of the Sun in general. A. Wilson's (1774) discovery that spot areas decreased towards the limb more than would be expected in normal foreshortening was held to show that spots were depressions in the solar surface. This encouraged analogy with terrestrial meteorology; it was supposed that spots were gaps in a luminous cloudy atmosphere through which the Sun itself could be seen as a dark solid body. What might be called the first theories of the origin of spots were concerned with trying to account for such clearings in the Sun's outer layers. Dawes (1863) suggested a volcanic origin, and Herschel (1864) put foward a meteoritic infall theory.

Visually a sunspot is a dark, and evidently a relatively cool, area of the solar disk. Huggins' (1868) spectroscopic work and the first quantitative bolometric measurements by Langley (1875) gave new evidence for a lower temperature in spots as compared with the photosphere.

By this time it was realized that the Sun's temperature must increase below the surface; this made the problem of the apparent cooling of spots more difficult, quite apart from the question of their origin. As late as 1897 Evershed (1897) was suggesting that spots might in fact be hotter than the photosphere, but appeared cooler on account of the shift in energy to the ultra-violet. Subsequent knowledge of the radiation laws, however, precluded this explanation, and it is now generally agreed that sunspots are regions of local cooling.

The discovery by Hale (1908) of the magnetic fields in spots not only introduced the further problem of their origin, but added a further factor which might be responsible for the cooling. This phenomenon together with the radial outflow of material first detected by Evershed (1909) are essential features in any theory of sunspot structure.

The more recent observations are discussed in the next section, where a preliminary model for the physical conditions in the visible layers is given, based on present knowledge.

This model is then used in a theoretical discussion of the problem of the origin of sunspots.

2. Observations and Model of the Surface Layers

The difficulty in all measurements of sunspots is in correcting for the photospheric light scattered into them. Measurements based on the work of WANDERS (1934, 1935) show that, in bad seeing conditions, half the light from a point source could be distributed outside a circle as large as 15″ in diameter, about the size of the umbra of a fairly large spot. Results thus fall off in accuracy with diminishing spot area; the model set out below therefore represents a large spot of area 350 millionths, as follows.

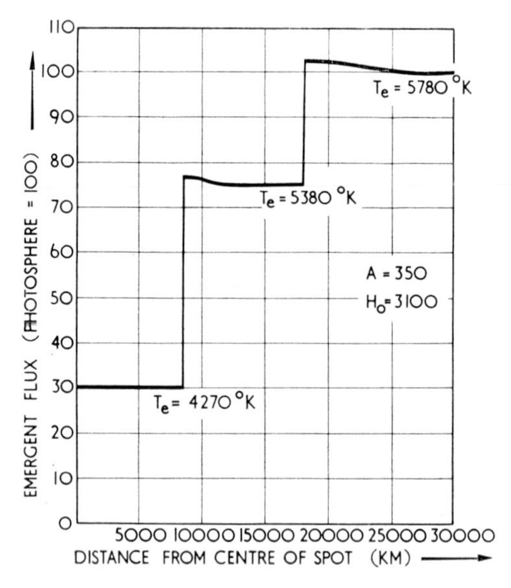

Fig. 1. The total emergent flux distribution across a typical large sunspot

The emergent radiative flux. This is given by the flux profile in Fig. 1, based on the work of WORMELL (1936), WALDMEIER (1938), ANANTHAKRISHNAN (1951), MICHARD (1953), and DAS and RAMANATHAN (1953). Uncorrected profiles do not show sharp boundaries to the umbra and penumbra. WORMELL (1936), however, showed that the boundaries become sharper, and more in keeping with their visual appearance, when corrections are made for seeing and instrumental effects. An important characteristic of the penumbra is the radially elongated granulation. In the umbra little granulation is visible, although THIESSEN (1950) finds evidence of it in larger spots. Recent curve of growth studies by HUBENET (1954) indicate turbulence, which seems to support this evidence.

From spot to spot there is a definite decrease in darkening with decreasing area. Some of this variation is due to scattering, and it is difficult to say how much is real. MICHARD (1953) suggests a real variation of order 1 : 2 in umbral flux with a 10 : 1 range in area.

Temperature and pressure distribution. MICHARD'S (1953) recent model is the most reliable one available at present; the figures in Table 1 are derived directly from this.

The $\tau - T$ relation was derived from centre-limb observations in various wavelengths, and is empirical. The gas pressures P_g were derived from a rediscussion of

Table 1

$\tau_{\lambda 5000}$	Umbra				Photosphere		
	T	log P_g	d* (km)	log W†	T	log P_g	d* (km)
0·02 .	3725	3·79	0	3·79	4513	4·11	0
0·20 .	4340	4·28	1190	5·05	5490	4·77	206
0·67 .	4683	4·45‡	1790‡	5·28‡	6096	4·93	270
2·00 .	5250	4·72‡	2660‡	5·64‡	7100	5·06	332

* d = geometrical depth.
† W = weight per square centimetre of superincumbent gas.
‡ Uncertain values.

TEN BRUGGENCATE and VON KLÜBER'S (1944) line strengths. There is considerable uncertainty in the figures owing to the fact that the electron pressures calculated from the equivalent widths depends strongly on the excitation temperatures adopted. The geometrical depths were calculated from VITENSE'S (1951) absorption coefficients.

A comparison of the geometrical depths at optical depth 0·67 shows that one can see something like 1500 km deeper into the umbra than in the photosphere. The zero levels in the photosphere and umbra were taken arbitrarily at $\tau = 0{\cdot}02$; the geometrical distance between the two levels could only be determined if the mechanical force due to the magnetic field were known. If the field at the zero level in the umbra were vertical and of strength 3000 gauss this level would be about 600 km below the zero level in the photosphere. The actual field fans out in the surface layers, however, and the figure just given is therefore an upper limit. Thus the model would indicate a depression of between 1500 and 2100 km for the umbra in the continuous spectrum. Observational estimates of this depression vary; WILSON'S original figure was about 2000 km, although CHEVALIER (1919) thought that the depth rarely exceeded 750 km. PETTIT (1938) suggested that the Wilson effect might be due to the raised ring of faculae round the spots, and not due to a depression at all.

Since the model makes no assumptions about radiative equilibrium it can be used to examine this. By calculating the flux at various depths MICHARD concluded that the umbra is roughly in radiative equilibrium. This agrees with the previous conclusion of MINNAERT and WANDERS (1932).

Hydrostatic equilibrium also was not assumed in constructing the model. The departure from hydrostatic equilibrium is most clearly shown by comparing the weight W of the superincumbent gas per square centimetre with the gas pressure at any level. The figures in Table 1 were derived by taking the weight above the zero level to be equal to the gas pressure there; any error so introduced would have a negligible effect lower down in the umbra, since the gas pressure is relatively small at the zero level. It is evident that the gas pressure at lower levels is insufficient by a factor of about 6 to support the overlying material. The spot must therefore depend on some other body force, presumably that due to the magnetic field, to prevent it from telescoping on its base.

The magnetic field. From the Mount Wilson Magnetic Observations (1938) and the recent work of MATTIG (1953), the field in the surface layers may be represented roughly in terms of the field strength H_0 at the centre of the umbra, by Table 2: where H is the resultant field strength, θ is the inclination to the vertical and $\tilde{\omega}$ is the distance from the centre of the spot in units of the penumbral radius. These relations cannot be regarded as very accurate owing to the difficulty of disentangling from the observations the effects of scattering and of the variation of field strength with level in the spot.

Table 2.

$H = H_0(1 - \tilde{\omega}^4)e^{-2\tilde{\omega}^2}$

$\tilde{\omega}$	θ
0·00	0
0·15	18°
0·30	49°
0·45	73°
penumbra	~ 90°

The increase of field strength with depth can be assessed roughly in a number of ways. (i) From centre-limb variations of field strength in conjunction with a model atmosphere for determining the variation of mean level of formation of the lines used, HOUTGAST and VAN SLUITERS (1948) gave 5·7 gauss per km for large spots. Used in conjunction with MICHARD's model this would be equivalent to about $1\frac{1}{2}$ gauss per km. (ii) From the surface field model, using the condition div $\mathbf{H} = 0$, the rate of change of the vertical component at the surface in the centre of the spot adopted is approximately $\frac{1}{2}$ gauss per km, if H_0 is taken as 3000 gauss. (iii) From measurements of lines of different strengths, and therefore at different levels, HOUTGAST and VAN SLUITERS (1948) gave a figure of 2·5 gauss per km. Again, this would be reduced to about $\frac{1}{2}$ gauss per km if MICHARD's model were used.

The model given is not complete enough to derive the magnetic force curl $\mathbf{H} \wedge \mathbf{H}/4\pi$. This requires the values of $\partial H_z/\partial\tilde{\omega}$ and $\partial H_{\tilde{\omega}}/\partial z$; the derivative $\partial H_{\tilde{\omega}}/\partial z$, however, is not determined by the model.

From spot to spot the field strength increases with increasing area. HOUTGAST and VAN SLUITERS (1948) find

$$H_0 = 3700\ A/(A + 66) \text{ gauss}$$

for a spot of area A millionths.

The Evershed effect. The variation of the outflow velocity with distance from the centre of the spot has recently been carefully reinvestigated by KINMAN (1952, 1953). He confirmed EVERSHED's conclusions that the velocity is purely radial and horizontal. In a large spot the velocity increased from 1 km per sec. at the inner edge of the penumbra to a maximum of 2 km per sec. about half way across the penumbra, and persisted well outside the spot. In smaller spots the maximum was smaller and occurred relatively further out, in some cases being outside the spot. ABETTI (1932) also found a decrease in velocity with decreasing area from spot to spot; his results were re-analysed by MICHARD (1951).

The variation of velocity with depth can be estimated from centre-limb variations, and from lines of different strength. ST. JOHN's (1913) results indicated an increase in velocity with depth.

The evolution of a spot. COWLING's (1946a) analysis of the Mount Wilson Magnetic Observations showed that for a typical large spot the magnetic field strength increases roughly in step with the area during the growth of the spot; maximum area and field strength are reached in about 12 days. The area then diminishes steadily, while the field strength remains roughly constant until the disappearance of the spot, the decay time from maximum area being about 50 days.

As COWLING remarks, it would seem from this that the field decays by its outer lines of force successively disappearing below the surface, rather than by a general expansion over the whole area. A study of the way the darkening varies during the lifetime of the spot would be extremely valuable.

3. A THEORETICAL INTERPRETATION

The central problem is to explain the darkening, and it has to be decided whether the cooling is caused by the magnetic field or the field by the same cause as is responsible for the cooling. COWLING (1946a) showed that the latter standpoint is difficult owing to the long time required to build up a field *in situ*. ALFVÉN's (1945) explanation of the formation of the field from hydromagnetic waves does not suffer from this objection. COWLING (1946b), however, pointed out a number of further objections. One of these would seem to rule out the process conclusively in that the energy of a hydromagnetic wave would be used up in about one scale height in raising the centre of gravity of the material in it. In recent approaches the magnetic field has been accepted as of unknown origin and regarded as being the cause of the darkening.

The low temperature in a spot could be due to some initial cooling at birth, after which decay took place as the spot was warmed up by the surrounding gas. The difficulty here is that the heat capacity and opacity are too small for a shallow spot to survive long enough. WALÉN (1944) estimated that the depth required for sufficient duration would be of the order of 10,000 km. COWLING (1946b), however, pointed out that the observed sharpness of the umbra precludes a spot as deep as this.

The fact that such reservoir effects can be neglected raises a paradox the resolution of which gives a clue to the origin of the Evershed effect.

The flux paradox. In the absence of reservoir effects the total energy flux is solenoidal. The flux deficit at the surface of a spot therefore means that the energy flowing up underneath the spot is deflected round it. If the spot were deep-seated it is difficult to see at first glance how the umbra and penumbra could be so sharp. On the other hand, if the spot were shallow the flux would be squeezed tightly round the edges, and a bright ring might be expected. The bright rings observed only account for a small fraction of the flux deficit, however.

The resolution of the paradox. Consider the spot quite generally as a disturbed region in and below the Sun's surface. Outside the disturbed region there will be hydrostatic equilibrium. In a gas of uniform chemical composition this means that the pressure and density, and therefore the temperature, are functions of the gravitational potential. The latter is little affected by local disturbances such as sunspots, thus the state of the gas is unaffected by the proximity of the disturbance. But the radiative flux at any point is determined by the state and its neighbouring values within an optical distance of order unity, and is therefore also unaffected over regions large compared with unit optical distance. A spot can therefore be reasonably shallow without producing a bright ring at its boundary. The flux deficit is in fact carried away from the spot by mass circulation. It is reasonable to attribute the Evershed effect to this. The circulation velocity so arising upsets hydrostatic equilibrium slightly through the Bernoulli pressure and viscosity effects. The Evershed velocity is of order 2 km per sec. as compared with about 9 km per sec. for the velocity of sound at the surface. Thus the radiative flux would not be affected by more than a few per cent, the same order of magnitude as the excess brightness of the bright rings.

The foregoing general argument shows that the brightness of the photosphere is only seriously affected in an area where hydrostatic equilibrium is upset by some external mechanical force. A sunspot, with its magnetic field, is just such an area.

The Evershed effect. Fig. 2 (*a*) shows a spot defined by HK at the surface. For simplicity the lowering of brightness in the penumbra will be neglected. AB is the level of the base of the spot, and is defined as the level at which the disturbance in the normal upflow of energy in the Sun is negligible. Considering the energy balance

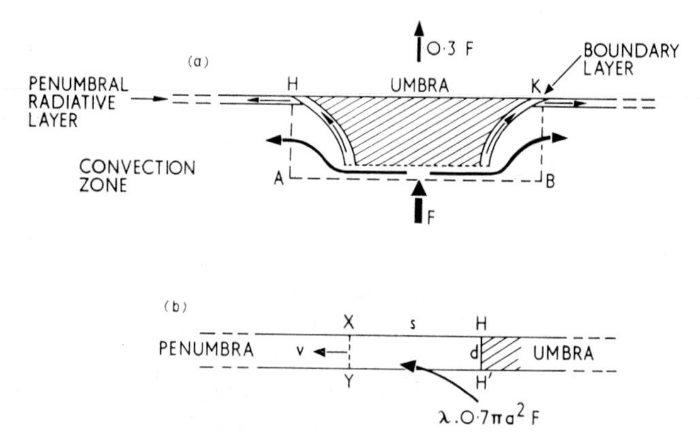

Fig. 2 (*a*). The deflection of energy from below a sunspot. Upflow of material is shown in a boundary layer round the spot
(*b*) Energy deflected from the base of the spot entering the penumbra. λ is the proportion of the defected energy reaching the visible layers within a distance s from the umbra. **v** is the EVERSHED velocity produced thereby

in the region defined by $ABHK$ the inflow of energy across AB is $\pi a^2 F$, where F is the photospheric flux and a is the radius of the umbra. The outflow from the surface HK is $0{\cdot}3\pi a^2 F$, where $0{\cdot}3F$ is the emergent flux from the umbra as given by the model in Fig. 1. There must therefore be an outflow of energy $0{\cdot}7\pi a^2 F$ across the sides defined by AH and BK. The preceding paragraph shows that this energy must be carried by convection. Part of the spot is almost certainly in the hydrogen convection zone, and the energy outflow at these depths will be carried partly by increased turbulent convection and partly by mass circulation. All the energy deflected into the radiative layer at the top must be carried away by mass movement. At the spot boundary in the convection zone the magnetic field might tend to prevent increased turbulent convection, and it is to be expected that an appreciable proportion of the energy is carried by mass movement upwards in a boundary layer as shown in Fig. 2 (*a*). It is reasonable to suppose that this boundary layer motion is responsible for the development of the Evershed velocity at the umbral boundary. The augmentation of the Evershed velocity across the penumbra, as observed by KINMAN (1952), may be due to that part of the energy $0{\cdot}7\pi a^2 F$ which the convective zone is unable to convect away from the spot in the deeper layers. In order to assess what this proportion would have to be, consider the energy balance in a zone of the penumbra as shown in Fig. 2 (*b*). The inflow of energy across the area A defined by YH' is

$0{\cdot}7\pi a^2F\lambda + AF$, where λ is the required proportion of the energy surplus $0{\cdot}7\pi a^2F$; the outflow across XH is AF. The outflow across XY due to the Evershed velocity v is $2\pi(a+s)dc_pT\rho v/\mu$, where ρ, T, and v are mean values for the layer, and c_p is the specific heat at constant pressure. For gas at the surface $c_p = \frac{5}{2}\mathfrak{R}$, where $\mathfrak{R}$ is the gas constant; energy balance therefore shows

$$5\pi(a+s)dPv = 0{\cdot}7\pi a^2F\lambda, \qquad \ldots .(1)$$

where P is the gas pressure. Taking $P \sim 10^5$ dynes per cm^2, $d = 3 \times 10^7$ cm as typical values for the photosphere, together with $a = 10^9$ cm and $F = 6{\cdot}2 \times 10^{10}$ ergs per sec. per cm^2, equation (1) shows that

$$\lambda \sim \frac{1}{30}v(1+s/a), \qquad \ldots .(2)$$

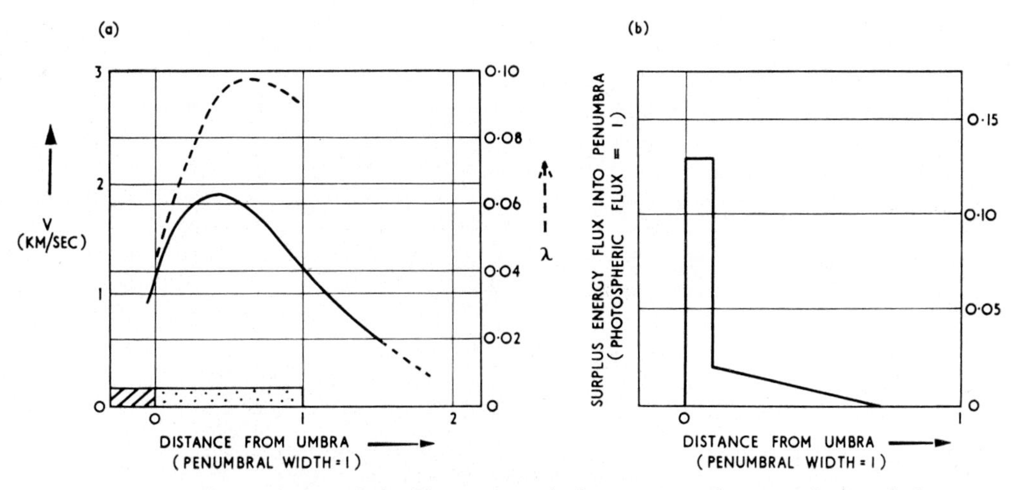

Fig. 3 (*a*). The variation of the EVERSHED velocity **v** across the penumbra, and the proportion λ of the energy deflected from under the spot required to produce the velocity. (*b*) The energy needed to enter the penumbral layers from below in order to produce the observed EVERSHED velocity

where v is expressed in kilometres per second. Fig. 3 (*a*) shows the variation of λ across the penumbra as derived from KINMAN's (1952) values for v. It is seen from the graph that the Evershed effect can be accounted for if about one-tenth of the surplus flux from under the spot is deflected into the visible layers. The decrease in λ near the edge of the penumbra is spurious; it should be remembered that the darkening of the penumbra has been neglected. At the inner edge of the penumbra $\lambda = 1/30$; comparing this with its maximum value of about $1/10$ it is seen that about one-third of the excess energy entering the penumbra from below is concentrated in a transition region between the umbra and penumbra. The width of this region cannot be determined without a detailed study of the boundary layer. From the observed sharpness of the boundary, however, it seems unlikely that it could exceed 1000 km or so. The excess flux distribution for a transition region of this width is sketched in Fig. 3 (*b*). The upward velocity necessary to supply the horizontal outflow of 1 km per sec. would be of order $\frac{1}{3}$ km per sec. This would no doubt be difficult to detect in such a narrow region of the solar disk.

The origin of the darkening. Two likely processes have been considered as possible causes; mass motion of gas in the umbra in which cooling is produced by adiabatic expansion, and inhibition of convection in the convection zone, due to the presence of the magnetic field.

The adiabatic upflow theories of RUSSELL (1921) and ROSSELAND (1926) and others depended on the fact that gas moving upwards in a region in stable radiative equilibrium gets cooler than its surroundings. The discovery of the hydrogen convection zone in the Sun, together with the conclusions of MINNAERT and WANDERS (1932) that at least the surface layers were not in adiabatic equilibrium, threw doubt on this process. The situation is best illustrated with an UNSÖLD entropy diagram as in Fig. 4.

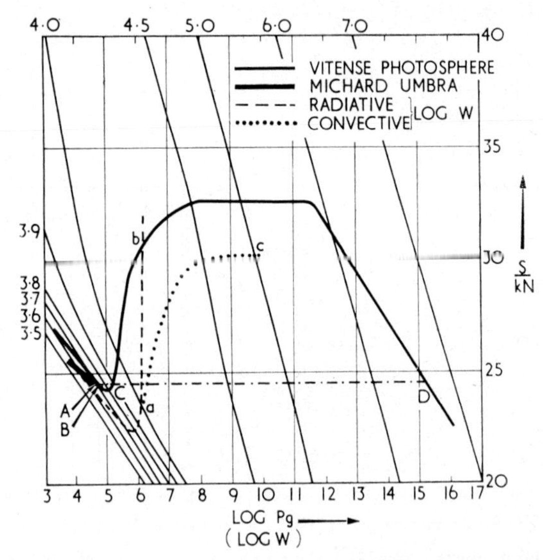

Fig. 4. A comparison of possible entropy-pressure distributions in and below the photosphere and umbra

A gas undergoing an adiabatic change simply moves along a straight line of constant entropy. The figure shows one of the recent models for the sub-photospheric layers, due to VITENSE (1953), together with a plot of MICHARD's model umbra down to its limit of reliability $\tau \sim 1$ at the point A. The most favourable case for an adiabatic upflow process would be where the gas flows adiabatically up to the level corresponding to A. Neglecting for the moment the pressure difference between umbra and photosphere at a given geometrical depth, the base of the spot could then be at levels corresponding to the intersections B, C, or D. COWLING (1935) estimated that at the levels B and C the opacity is too low to allow adiabatic equilibrium unless the upward velocity was impossibly high. The point D is in the stable region well below the hydrogen convection zone, and considerable energy would be required to lift the gas to the surface. The objection of COWLING's just mentioned would again arise for this gas as it approached the surface. This could be overcome by starting from a level appreciably below D. An added objection to the points B and C is that they do not correspond to the same geometrical depth in the umbra and photosphere, owing to the reduction in gas pressure in the umbra effected by the magnetic field. For the umbra, W in Table 1 is a closer estimate of the surrounding gas pressure than the umbral values of P_g. An entropy- $\log W$ plot would come below and to the right of the umbral log P_g plot in Fig. 4, as shown; no adiabatic continuation of it would intersect the photospheric curve except below the hydrogen convection zone.

More promising are the magnetic inhibition theories of BIERMANN (1941) and HOYLE (1949). In the first it is supposed that the magnetic field inhibits convection in the hydrogen convection zone, thereby reducing the flux of energy through the gas. If the umbral model in Fig. 4 were continued to deeper levels it would follow a radiative curve of the general form *ab* if convection were prevented completely. A curve of the form *ac*, similar to the photospheric curve, merging into a convective region would result if convection were unimpeded. The actual curve would presumably lie somewhere between these two. The base of the spot in the former case would correspond to the point *b* in Fig. 4. The point of breakaway, *b*, from the photospheric curve would be determined by the level at which the magnetic field was strong enough to stop convection. It would probably occur further to the right in Fig. 4 for stronger fields, and a darker spot would result, in agreement with observations.

HOYLE'S (1949) suggestion is that convection is affected only to the extent that the convective flux follows the lines of force of the magnetic field. As the field fans out near the surface the flux is therefore reduced.

It is not possible at present to discriminate between the last two theories since no theory is yet available describing the effect of magnetic fields on convection. At first sight it would appear easier for a turbulence element to move along the lines of magnetic force than at right angles to them, as implied in HOYLE'S approach. It should be remembered, however, that a turbulence element has to displace the neighbouring gas at right angles to its path in order to move through it; the magnetic field is therefore highly distorted even in this case. CHANDRASEKHAR (1952) has derived criteria for the effect of a magnetic field on marginal stability; these appear to show that a magnetic field has little effect under the particular conditions obtaining near the solar surface. The motions concerned in this study have low velocities and leak across the field without producing more than first-order distortion in it. These conditions bear little relation to the velocities and field distortions in a region in full-scale convection.

The penumbra. The causes of this phenomenon are not yet understood; the chief difficulty is to explain the sharpness of the outer edge. Again, a detailed study of the effect of a magnetic field on convection may provide an answer. CHANDRASEKHAR (1954) has shown that, for a magnetic field in a rotating medium, the wavelength of a disturbance giving maximum instability in a region of marginal stability can change sharply at a critical field strength. It is possible, therefore, that in a sunspot the differentiation between the umbra and penumbra may be due to a sudden change in the mode of convection. The elongation of the granulation in the penumbra might be connected with this effect. The elongation could, of course, be due simply to the action of the Evershed velocity on the rising granules.

The decay of the spot. COWLING (1946a, 1946b) has suggested that the decay might be due to heated gas at the umbral boundary expanding and pushing the lines of magnetic force below the surface. The picture of the spot just given conforms with this idea; the lines of force, which are nearly horizontal in the penumbra, are pushed slowly downwards by the gas moving horizontally across the penumbra. The spot area diminishes as the lines of force are successively peeled off downwards in this way. Since the field strength in the umbra remains constant during decay, the area is proportional to the total magnetic flux at the surface. Thus the rate of diminution of the area is proportional to the rate of decrease of the magnetic flux. Suppose the penumbra contains a constant proportion p of the total flux K at any stage. If T be the time taken for gas to cross the penumbra then the flux is peeled off at the rate pK/T. The decay time of the spot is therefore of order T/p. With a mean Evershed velocity of 1 km per sec. and a penumbral width 10,000 km, $T \sim 10^4$ sec. In order to give the observed decay time of about 30 days from maximum area to $1/e$ of maximum, p would have to be of order 1/250.

The expression for the resultant field strength given in Table 2 shows that this would involve a mean inclination of the field in the penumbra of about 1° to the horizontal. This is smaller than would be expected from the figure of 17° to the horizontal near the edge of the umbra as shown in Table 2.

4. Conclusion

The arguments developed above can only give an indication of how most of the observational properties of a spot could be accounted for theoretically. Unfortunately, none of the arguments can be followed far enough to be tested quantitatively. What is mainly lacking on the theoretical side is a theory of the effect of magnetic fields on convection, both in the body of a spot and in the boundary layers. On the observational side it is important to confirm Michard's model, particularly at the lower levels, and also to find out more about the magnetic field distribution.

References

Abetti, G. . . . 1932 *Mem. Soc. Astron. Ital.*, **6**, 353.
Alfvén, H. . . . 1945 *M.N.*, **105**, 3; **105**, 382.
Ananthakrishnan, R. . . . 1951 *Nature (London)*, **168**, 291.
Biermann, L. . . . 1941 *Vierteljahrsschrift Astron. Gesellschaft*, **76**, 194.
Bruggenfate, P. ten and von Klüber, H. 1944 *Veröff. Univ. Sternwarte, Göttingen*, No. 78, 139.
Chandrasekhar, S. . . . 1952 *Phil. Mag.*, **43**, 501.
1954 *Observatory*, **74**, 2.
Chevalier, P. . . . 1919 *Ann. Obs. Zô-Sè*, **10**, 11.
Cowling, T. G. . . . 1935 *M.N.*, **96**, 15.
1946a *M.N.*, **106**, 218.
1946b *M.N.*, **106**, 446.
Das, A. K. and Ramanathan, A. S. . . 1953 *Z. Astrophys.*, **32**, 91.
Dawes, W. R. . . . 1863 *M.N.*, **24**, 54.
Evershed, J. . . . 1897 *Ap. J.*, **5**, 249.
1909 *Kodaikanal Obs. Bull.*, **2**, 63.
Hale, G. E. . . . 1908 *Ap. J.*, **28**, 100; **28**, 315.
Herschel, Sir J. F. W. . . . 1864 *Quart. J. Sc.*, **1**, 219.
Houtgast, J. and van Sluiters, A. . . 1948 *Bull. Astron. Inst. Netherlands*, **10**, 325.
Hoyle, F. . . . 1949 *Some Recent Researches in Solar Physics*, p. 11 (Cambridge University Press).
Hubenet, E. . . . 1954 *Observatory*, **74**, No. 880.
Huggins, Sir W. . . . 1868 *Phil. Trans.*, **158**, 529.
Kinman, T. D. . . . 1952 *M.N.*, **112**, 425.
1953 *M.N.*, **113**, 625.
Langley, S. P. . . . 1875 *M.N.*, **37**, 5.
Mattig, W. . . . 1953 *Z. Astrophys.*, **31**, 273.
Michard, R. . . . 1951 *Ann. Astrophys.*, **14**, 101.
1953 *Ann. Astrophys.*, **16**, 217.
Minnaert, M. and Wanders, A. J. M. . 1932 *Z. Astrophys.*, **5**, 297.
Pettit, E. . . . 1938 *Mount Wilson Annual Rep.*, p. 12.

ROSSELAND, S. . . . . 1926 *Ap. J.*, **63,** 356.
RUSSELL, H. N. . . . . 1921 *Ap. J.*, **54,** 293.
ST. JOHN, C. . . . . 1913 *Ap. J.*, **37,** 322.
THIESSEN, G. . . . . 1950 *Observatory*, **70,** 234.
VITENSE, E. . . . . 1951 *Z. Astrophys.*, **28,** 81.
1953 *Z. Astrophys.*, **32,** 135.
WALDMEIER, M. . . . . 1938 *Astron. Mitt. Zürich*, No. 138.
WALÉN, C. . . . . 1944 *Ark. Mat. Astron. Fys.*, **30A,** No. 15.
WANDERS, A. J. M. . . . . 1934 *Z. Astrophys.*, **8,** 108.
1935 *Z. Astrophys.*, **10,** 15.
WILSON, A. . . . . 1774 *Phil. Trans.*, **64,** 1.
WORMELL, T. W. . . . . 1936 *M.N.*, **96,** 736.

The Magnetic Fields of Sunspots

SIR EDWARD C. BULLARD
National Physical Laboratory, Teddington, Middlesex

SUMMARY
The occurrence of sunspots in pairs with opposite magnetic polarities can be explained if it is assumed that lines of magnetic force encircle the axis of rotation in opposite directions a little below the surface of the Sun. Material moving upwards will carry a loop of field through the surface and give a south magnetic pole on the same circle of latitude.

Such a field is to be expected from the interaction of the Sun's non-uniform rotation with a meridional field. The progression of the spots in latitude requires a slow meridional circulation. The dynamics of the non-uniform rotation and the meridional circulation are not understood; it is possible that the equatorial parts of the Sun are coupled magnetically to a more rapidly rotating interior.

1. PROBLEMS

It has been known for forty years that sunspots possess magnetic fields of the order of 2000 gauss. These fields show striking regularities, their main features being:

(*a*) The field of a spot is roughly normal to the Sun's surface.

(*b*) Spots usually occur in pairs close together in the same latitude. When they do so, one of the pair behaves like a north magnetic pole and the other like a south pole so far as the external field is concerned.

(*c*) During a single sunspot cycle, the leading spot in the northern hemisphere is always of a single polarity (south during the present cycle). In the southern hemisphere the polarity is reversed.

(*d*) The polarity of the pairs is reversed in alternate cycles.

(*e*) In the early part of the cycle, the spots are in latitude 20°–30°. Later in the cycle they occur nearer the equator, reaching it at the end of the cycle. As the spots of the old cycle reach the equator, the first ones of the new cycle appear around latitude 25°.

The purpose of this note is to suggest that all these facts can be plausibly explained if some assumptions are made about the motions taking place in the Sun.

2. ORIGIN OF THE FIELD OF A SPOT

The difficulties encountered in previous explanations of the magnetic fields of sunspots have arisen in the attempt to devise a mechanism for producing the field in the spot. As COWLING has pointed out (1946 (*a*)), it may be better to suppose that the spot field is a sample of a field that always exists within the Sun and which is brought to the surface by the rising of material.

The occurrence of spots in pairs on a circle of latitude suggests that the field forms rings round the Sun's axis. The field of a sunspot would then develop as in Fig. 1. The opposite polarity of pairs of spots in the two hemispheres would require rings of field running in opposite directions, one in the northern and one in the southern hemisphere. The field would then be what ELSASSER has called "toroidal" and would be of the type known as a T_2 field.

It is natural, though perhaps not essential, to suppose that the rings of toroidal field are at a depth comparable with the separation of a pair of sunspots, say

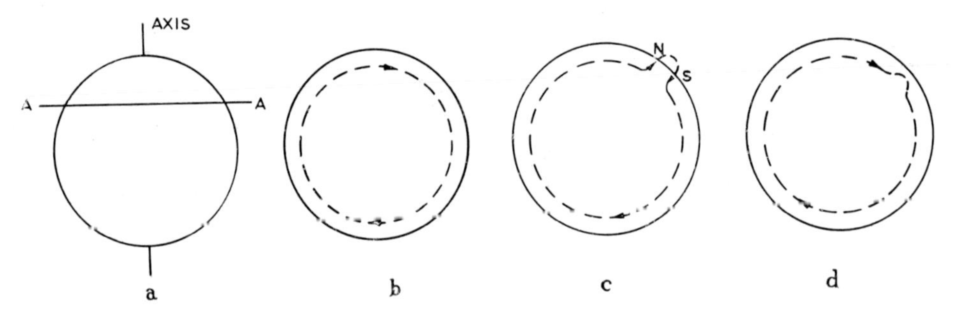

Fig. 1. Formation of a pair of sunspots from the toroidal field: (*b*), (*c*), and (*d*) are sections in the plane AA of (*a*); (*b*) shows the toroidal field when no spot is present; (*c*) shows the formation of a pair of spots; and (*d*) a region liable to spots

20,000 km. A sunspot group develops in a time of the order of a week. The vertical velocity needed for the material to rise 2×10^4 km in this time is only 0·03 km per sec and would not be detected.

The life of a pair of sunspots is normally a few weeks or months. There is no difficulty in maintaining the field for this time since it will decay in a time of the order κd^2, where κ is the electrical conductivity in electromagnetic units and d is the distance between the pair of spots. Taking $\kappa = 1 \times 10^{-6}$ e.m.u. (COWLING, 1945) and $d = 2 \times 10^9$ cm, this gives 4×10^{12} sec or 10^5 years. This is much too long and the difficulty is rather to get rid of the field of the spot than to maintain it. This difficulty has been discussed by COWLING (1946a); he suggests that the rapid fall of the field is due to convection spreading it over a greater area. Another possibility is that the field is broken up by turbulent motions and that the detached bits decay in a time $\kappa d_1{}^2$, where d_1 is the diameter of a typical eddy. A large-scale subsidence of material in the region of a spot would, of course, remove the field also into the interior and provides another method of removing it from view. The existence of regions on the Sun that are liable to a recurrence of spots over a period of several years suggests that after the disappearance of a visible spot and its magnetic field, there is still some peculiarity of the material below the surface that can produce more spots. This might be an upward bulge in the toroidal field, as is shown in Fig. 1 (*d*).

3. THE MAINTENANCE OF THE TOROIDAL FIELD

It is not immediately obvious whether it is necessary to provide a mechanism for the maintenance of the toroidal field. If the Sun were a homogeneous conducting sphere its period of decay would be of order κa^2. With $\kappa = 10^{-6}$ e.m.u. (the value near the surface) and $a = 7 \times 10^{10}$ cm, this gives 5×10^{15} sec or 2×10^{8} years; with $\kappa = 10^{-4}$ (the value half-way to the centre), the period would be 2×10^{10} years, which is probably larger than the age of the Sun. However, if it is assumed that the toroidal field is concentrated in a relatively thin layer, it will decay much more rapidly. It is likely also that there are other processes tending to destroy the field; in particular the formation of sunspots will tend gradually to dissipate the toroidal field by taking parts of it up to the surface into regions of lower electrical conductivity. On an average through the sunspot cycle about 3×10^{-4} of the Sun's surface is covered by spots. If the spots last a month on the average, the whole surface over which spots occur (about half of the Sun's area) will be covered in about 300 years. This suggests that the process of spot formation could produce

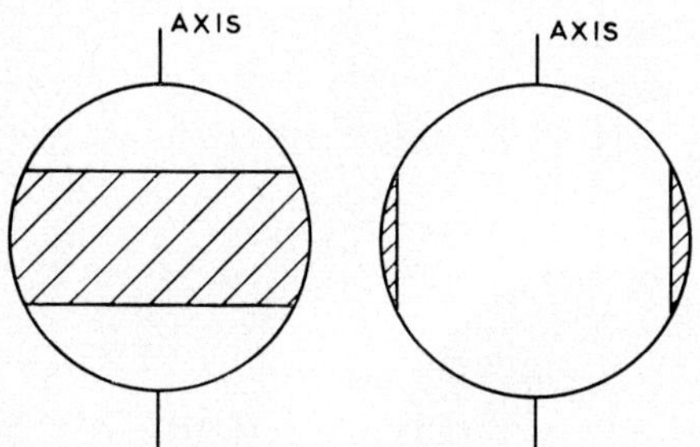

Fig. 2. Axial sections showing regions of rapid rotation. The two would be indistinguishable when seen from the outside

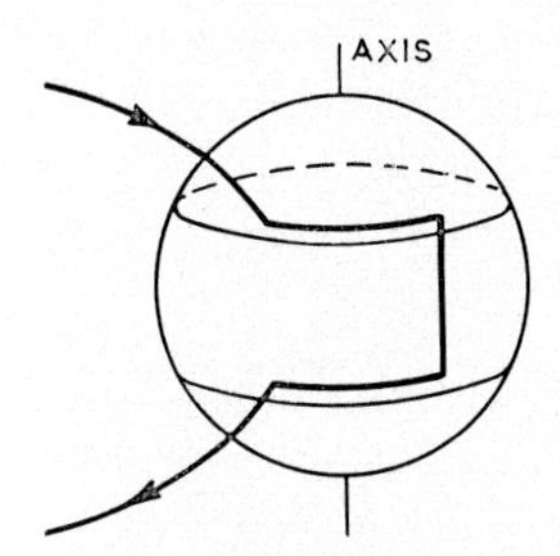

Fig. 3. Formation of the toroidal field

a substantial reduction in the toroidal field in 300 years. It is therefore prudent to provide some mechanism to maintain it.

The variation of the rate of rotation of the Sun with latitude provides a powerful mechanism of the kind required. The equatorial regions rotate in 24·5 days and the region in latitude 30° in 25·8 days. Material on the equator therefore makes a complete rotation relative to that in latitude 30° in about 500 days and moves 0·1 km per sec faster than it would if the Sun rotated as a rigid body with the angular velocity it has at latitude 30°. Only the surface of the Sun can be observed, and it is not known how the angular velocity of rotation varies in the interior. Two extreme views are shown in Fig. 2, and many intermediate arrangements are possible. If the Sun possesses a general magnetic field in any way similar to a dipole field and a line of force crosses the equatorial region of high angular velocity, it will be gripped by the highly conducting material and pulled out along circles of latitude as is shown in Fig. 3. This will give a field of the required kind, going from east to west in one hemisphere and from west to east in the other. The process has been worked out in detail in the case where the two rotating regions are concentric spheres (BULLARD, 1949); the behaviour in other cases will be generally similar. For the case studied it is found that the toroidal field exceeds the inducing field by a factor of about six times the number of rotations made in one decay period. If we take the free decay period of 10^{9}–10^{10} years, the toroidal field would be absurdly large. From this, FERRARO (1937) has deduced that within the Sun the rotation must be arranged in such a way that the angular velocity is very closely constant along each line of force. ALFVÉN (1950) has supposed that this implies that there is no toroidal field; in fact all it proves is that the rotation must be arranged so as not to involve electromagnetic forces large enough to stop the relative rotations, that is that the toroidal field should not be too large.

If the time for the field to decay is determined by sunspot formation rather than by the free decay of the current system, there is no tendency to produce too large a field. Taking the decay period as 300 years and a relative rotation period of 500 days, we get a toroidal field about 1300 times greater than the dipole field. The toroidal field should be of the order of the sunspot field, that is about 2000 gauss; a dipole

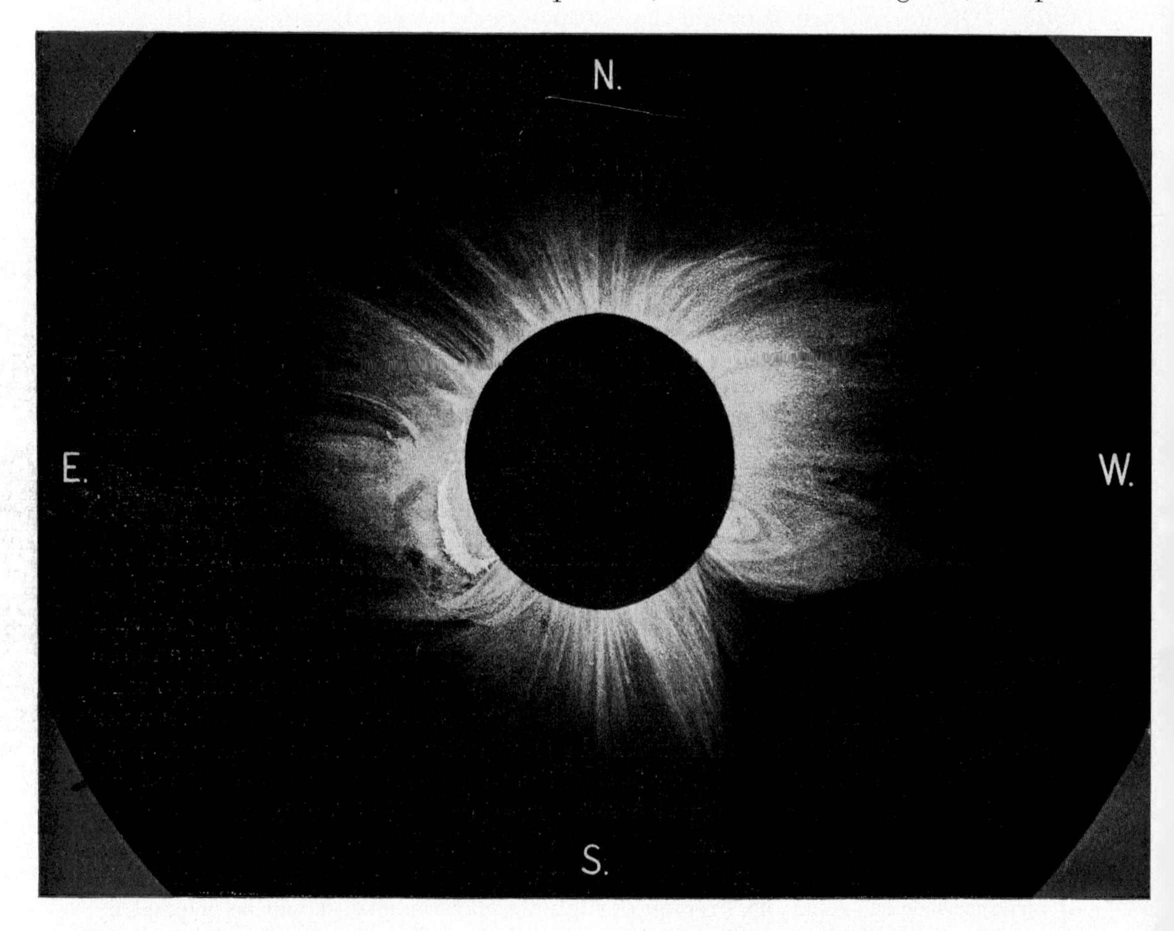

Fig. 4. A drawing of the solar corona on 29th May, 1919 (after F. W. DYSON)

field of about 2 gauss would therefore be required. There is no certain observational evidence for a solar dipole field, but one of 2 gauss would just have escaped detection. Many photographs and drawings of the solar corona (*e.g.* Fig. 4) suggest strongly that such a field exists; it is difficult to think of any mechanism other than a magnetic field that could cause the corona to take such a form as is shown in the figure.

4. Motion of the Sunspot Zone

At any particular stage in the sunspot cycle, spots may occur over a zone of latitude about 20° in width. As the cycle progresses, this zone moves towards the equator, but does not cross it. This suggests that the belt of toroidal field is about 20° wide and moves steadily towards the equator. When it gets near the equator it must somehow disappear to make way for one of the opposite sign appearing near latitude 25° and in its turn moving towards the equator.

Such behaviour strongly suggests that when the material carrying the toroidal field reaches the equator it sinks down and moves poleward beneath the field of the next cycle and emerges again in latitude 25° to start the next cycle but one, as is

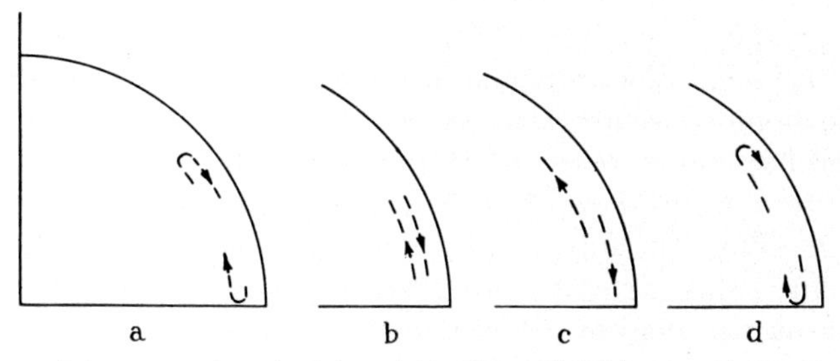

Fig. 5. Progress of the sunspot cycle, (*a*) early in the cycle, (*b*) sunspot maximum, (*c*) nearing the end of the cycle, (*d*) start of a new cycle

shown in Fig. 5. This arrangement explains the asymmetry of odd and even numbered cycles, since the same material and the same rings of field only emerge in alternate cycles; it also explains why the width of the sunspot zone is less at the beginning and end of a cycle than it is in the middle, and why the velocity of the centre of gravity of the spots is less at the start and end of the cycle than in the middle.

The necessity for providing two rings of toroidal field in each hemisphere requires some modification of the mechanism shown in Fig. 3. A loop of field such as that of

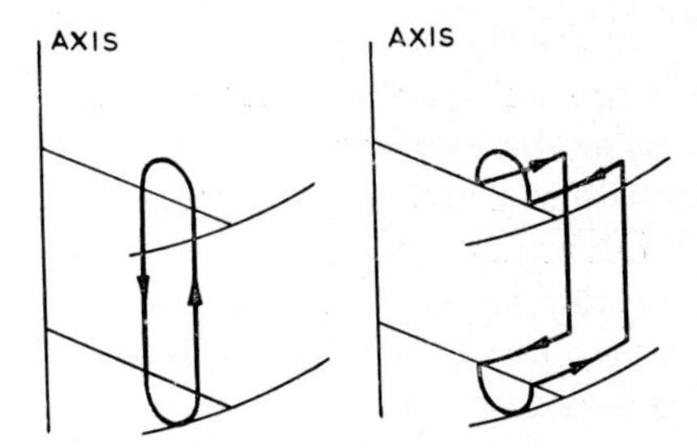

Fig. 6. (*a*) Loop of field in meridian plane, (*b*) loop pulled out by differential rotation

Fig. 6 would produce what is needed, but it is not clear exactly what would happen to such a system when the meridional circulation of Fig. 5 is superposed on it.

The theory resembles that of BJERKNES except that where he has ring vortices we have rings of field. The rings of field are easier to account for than the vortices and are also more closely related to the properties of sunspots, which are not conspicuously associated with rotary motions about the axis of the spot. The theory is quite different from that of ALFVÉN (1950) and WALÉN, who assume that the field within the Sun is predominately in meridian planes and that sunspots are associated with magnetohydrodynamic disturbances propagated along the lines of force of this field. In view of the large differential rotations within the Sun it seems likely that the field will be predominately toroidal.

5. Dynamics of the Process

The proposed mechanism depends on the occurrence of a particular type of motion in the Sun. Part of this, the rotation varying with latitude, is a fact of observation, but the other part, the circulation in meridian planes, is a pure hypothesis. It is difficult to see how the facts can be explained without some assumptions of this kind, but it is not easy to give a dynamical explanation of the motion. The principal difficulty is to account for the fast equatorial rotation. Any radial or meridional flow will tend, by the conservation of angular momentum, to produce a slowing in the equatorial regions. Electromagnetic forces may give large effects. The force per unit volume along a circle of latitude is $H_1 \times I$, where H_1 and I are the field and current in a meridian plane. I is curl $H/4\pi$ and is of the order of magnitude $H_2/4\pi b$, where H_2 is the toroidal field and b is the radial thickness of the region occupied by the toroidal field. Thus the force is of order $H_1H_2/4\pi b$ if $H_1 = 2$ gauss, $H_2 = 2000$ gauss, $b = 10^4$ km; this gives 3×10^{-7} dynes per cm^3. The effect of such forces depends greatly on the density. For material of the mean density of the Sun, it would take 1600 years to produce the observed excess of equatorial velocity over that at 30°, but with the much smaller densities, of the order of 10^{-3} to 10^{-5} gm per cm^3, given by Cowling's and other stellar models (Gardiner, 1951) in the outer part of the Sun, very rapid effects would be produced. Even if the forces are of the right order of magnitude, it is not clear that they can provide a systematic force in the right direction in spite of the reversal of field in the sunspot cycle. The detailed working out of a specific model might be instructive.

This difficulty is a general one that occurs in any theory that requires appreciable mixing of the material in the sunspot zone in the course of a sunspot cycle. During a cycle, the spots, their magnetic fields, and all the associated phenomena change their latitude, and it seems most unlikely that this is not associated with some transport of matter. Jeans (1928) has suggested that the interior of the Sun may rotate more rapidly than the exterior and has suggested a mechanism by which radiation transfers angular momentum from this rapidly rotating core to the equatorial regions. It seems possible that a more powerful version of this theory could be produced in which the momentum was transferred by electromagnetic forces and the rotation of the equatorial regions maintained against the coriolis forces by drawing on the angular momentum from the interior.

The views set out in this note are not so much an explanation of the magnetic fields of sunspots as an indication of a model that might, if it were worked out, provide an explanation. The essential point is the introduction of the toroidal field and the meridional circulation. The detailed working out of these ideas is a problem in magnetohydrodynamics of a kind that has not yet been attempted. It would require the simultaneous solution of Maxwell's equations and the equations of hydrodynamics. So far, we have only solutions of the hydrodynamic equations with arbitrarily assumed fields and solutions of Maxwell's equations with arbitrary velocities. The combination of these two approaches to give self-consistent solutions of both sets of equations is a problem of great difficulty. It may reasonably be hoped that its solution would not only provide a systematic theory of the processes discussed in this paper, but would also provide a dynamo mechanism for the maintenance of the meridional part of the field which we have arbitrarily assumed to exist. It is possible that the motions in the convective core of the Sun provide the dynamo required.

REFERENCES

ALFVÉN, H. . . . . 1950 *Cosmical Electrodynamics*. Oxford: Clarendon Press.
BJERKNES, V. . . . . 1926 *Ap. J.*, **64**, 93–121.
BULLARD, E. C. . . . . 1949 *Proc. Roy. Soc. A*, **199**, 413–443.
COWLING, T. G. . . . . 1945 *M.N.*, **105**, 166–174.
1946a *M.N.*, **106**, 218–224.
1946b *M.N.*, **106**, 446–456.
FERRARO, V. C. A. . . . . 1938 *M.N.*, **97**, 458–472.
GARDINER, J. G. . . . . 1951 *M.N.*, **111**, 94–101.
JEANS, J. H. . . . . 1928 *Astronomy and Cosmogony*. Cambridge: University Press.

On Mean Latitudinal Movements of Sunspots

JAAKKO TUOMINEN
The University, Helsinki, Finland

SUMMARY
An analysis of the motions of sunspots in latitude indicates that (1) between about − 16° and + 16° heliographic latitude the spots are drifting towards the equator, while beyond these parallels they are moving towards the poles, and that (2) this larger system of meridional "vortices" seems to be supermposed by another system which divides the Sun into zones about 10° wide from South to North.

1. INTRODUCTION AND STATEMENT OF THE PROBLEM

IT is well known that the Sun does not rotate as a rigid body but that its angular velocity diminishes gradually from the equator towards the poles. Theoretically, this differential rotation is no longer difficult to understand. If the viscosity of solar matter is supposed to be completely negligible, the forces which must balance each other in a rotating Sun are the force of gas pressure, the gravitational attraction, and the centrifugal force. In a way, the problem is similar to that of the planetary movements, where gravitational attraction and centrifugal force balance each other and a certain law of angular velocity is obtained. MARTIN SCHWARZSCHILD (1947) has derived a theoretical law of solar rotation based on the assumption of vanishingly small viscosity, and qualitatively it agrees well with the observed law. But quantitatively it gives a much too fast variation of the angular velocity with latitude. This fact in itself is easy to understand in a general way, as the viscosity will tend to reduce differences in angular velocity.

If, however, the viscosity of solar matter is not negligible, it will produce an additional viscous force which must be in equilibrium with the three above-mentioned forces. In this case it is easy to understand that movements in the Sun are not purely rotational but that there must also be meridional movements. Under very much simplified assumptions, as has been shown by RANDERS (1941), matter will rise from the centre of the Sun parallel to its axis of rotation, then follow the direction of the surface, and finally return to the centre by descending parallel to the equatorial plane. On the surface, matter will accordingly stream from the poles towards the equator.

SWEET (1950) has shown that this large-scale motion is very slow, and apparently it cannot be observed. In any case, however, we may expect that there will be meridional vortices with observable speeds. These may appear in an outer hydrogen convection zone (BIERMANN, 1942). The only phenomenon by which such currents could possibly be discovered is the systematic drift of sunspots in heliographic latitude. This drift has previously been studied by several authors (CARRINGTON, 1863; TURNER, 1907; STRATTON, 1909 and 1946; DYSON and MAUNDER, 1913; Royal Observatory Greenwich, 1924; TUOMINEN, 1941, 1945, 1952; RICHARDSON and SCHWARZSCHILD, 1953). It can now be regarded as certain that sunspots are drifting towards the equator between the parallels $-16°$ and $+16°$, while beyond these parallels they are drifting towards the poles. TUOMINEN (1945) has suggested that the Sun can be divided into zones about 10° wide from South to North, and that the drift is in the opposite direction in adjacent zones. A new calculation will be presented here, which supports the reality of the existence of such zones. In this calculation the variability of the drift with phase in the solar cycle, which apparently also exists (TUOMINEN, 1952; RICHARDSON and SCHWARZSCHILD, 1953), will be neglected.

2. OBSERVATIONAL MATERIAL

The observational material consists of the data on recurrent groups of sunspots published in the *Greenwich Photoheliographic Results for the Years* 1874–1939.* I have taken the first and the last observation of each spot or group of spots, the total number of which was 1329. The difference in heliographic latitude between these observations is denoted by l, the difference in time by t, and the average latitude is ϕ; l is expressed in degrees, and t in units of 10,000 days. The value of $l(l/t)$ was calculated for each spot. After this, I formed the sums $[t]$, $[l]$, $[l(l/t)]$, and the value of $[l]/[t]$ for zones of 1°, *i.e.* for spots with $\phi = 0° - (\pm 1°)$, $(\pm 1°) - (\pm 2°)$, $(\pm 2°) - (\pm 3°)$, *etc.*, respectively.

3. ELIMINATION OF THE EXPECTED PERIODICITY

In order to derive the latitudinal periodicity of the drift, we first derive the drift-curve from which this periodicity has been eliminated. A first approximation for such a curve can be found by forming averages of $[l]/[t]$ in the zones $(-5°) - (+5°)$, $0° - (\pm 10°)$, $(\pm 5°) - (\pm 15°)$, $(\pm 10°) - (\pm 20°)$, *etc.*, respectively. The resulting values give the drift for $\phi = 0°, \pm 5°, \pm 10°$, *etc.* Next, the curve was made symmetrical by taking averages of the absolute values of the drift on both sides of the equator. The values thus found are denoted by X_a and from them intermediate values were interpolated (extrapolated for outer zones).

The curve was then improved by writing

$$l/t = X_a + \alpha\phi + \beta\phi^3, \qquad \ldots.(1)$$

where α and β were calculated from a least square solution. Evidently the term $\alpha\phi$ corrects the X_a-values near the equator, while the term $\beta\phi^3$ corrects the values farther away from the equator. Each spot was given weight t and, in order to simplify the calculation, each ϕ-value was replaced by the value at the centre of the zone

* When the *Greenwich Photoheliographic Results for the Years* 1940 *and* 1945 appeared, most of the present calculations had been completed; the inclusion of these two years would, however, hardly affect our results.

($\phi = \pm 0\overset{\circ}{.}5, \pm 1\overset{\circ}{.}5, \pm 2\overset{\circ}{.}5, \ldots$). The results of the calculation, with their probable errors, are

$$\alpha = +0{\cdot}6 \pm 1{\cdot}2, \quad \beta = -0{\cdot}0017 \pm 0{\cdot}0025.$$

The large probable errors of α and β are partly due to the periodic variability of the drift with latitude. It is permissible to take the mean drift to be

$$X = X_a + 0{\cdot}6\phi - 0{\cdot}0017\phi^3, \qquad \ldots.(2)$$

in which ϕ is expressed in degrees and X in degrees per 10,000 days. The values of X can be taken from Fig. 1.

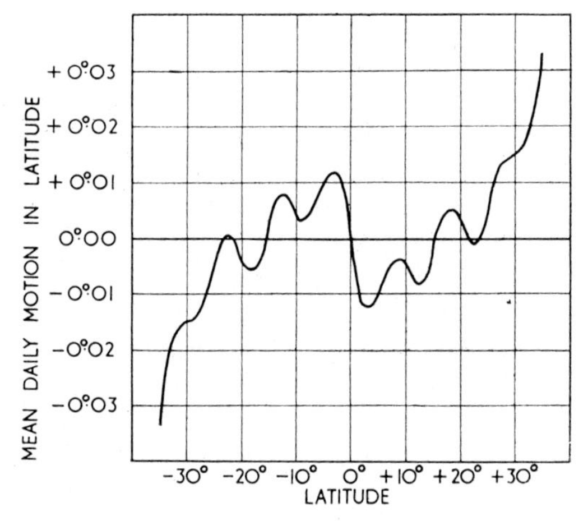

Fig. 1. The drift of sunspots in latitude indicates that there are two systems of meridional "vortices" in the Sun. The limits of the first system are at the equator and at $\pm$ 16°, while the second system divides the Sun into zones about 10° wide

4. DERIVATION OF THE PERIODIC TERM

We write

$$l/t = X + a \sin \nu(\phi + \delta\phi), \qquad \ldots.(3)$$

where $a = a_a + \delta a$ is the amplitude, and $\nu = \nu_a + \delta\nu$ the frequency of the periodic term. a_a and ν_a are approximate values and $\delta\phi$ indicates a possible asymmetry of the drift curve. Neglecting small terms of the second order, the equation can be expressed in the form:

$$\delta a \sin \nu_a \phi + \delta\nu \frac{a_a\phi}{57} \cos \nu_a\phi + \delta\phi \frac{a_a\nu_a}{57} \cos \nu_a\phi = l/t - (X + a_a \sin \nu_a\phi). \qquad \ldots.(4)$$

Each spot was again given weight t and, in order to simplify the calculation, each ϕ-value was replaced by the value at the centre of the zone. The following approximations have been estimated from the diagram published in the author's previous notes (TUOMINEN, 1941, 1945).

$$\nu_a = 36 \quad \text{and} \quad a_a = -40^\circ \text{ per 10,000 days.}$$

$\nu_a = 36$ implies that the latitudinal period is supposed to be about $360°/36 = 10°$.

A least square solution of the whole material of 1329 spots gives the following values and probable errors:

$$\nu = 36{\cdot}1 \pm 1{\cdot}3,$$
$$a = -\,47^\circ \pm 13^\circ \text{ per } 10{,}000 \text{ days},$$
$$\delta\phi = -\,0^\circ\!.2 \pm 0^\circ\!.5.$$

The first relation shows that the period in latitude is equal to $10^\circ\!.0 \pm 0^\circ\!.4$. The amplitude can also be written as

$$a = -\,0^\circ\!.0047 \pm 0^\circ\!.0013 \text{ per day}.$$

The value obtained for $\delta\phi$ indicates that the curve does not show any observable lack of symmetry.

5. Possible Connections with the Study of Solar Structure

The drift of sunspots in latitude might be a consequence of changes in the structure of the groups (see Tuominen, 1941, p. 106–7, and 1952, p. 274). It seems, however, more natural to interpret it as caused by meridional vortices in the Sun. The above results, then, show that there are two systems of meridional vortices: (1) between about $-\,16^\circ$ and $+\,16^\circ$ matter flows towards the equator, while beyond these parallels it flows towards the poles; (2) on this large system of vortices is apparently superimposed a system of vortices of smaller dimensions, as has been shown above.

We may ask: how deep into the Sun's interior do these vortices extend? Is their main cause the instability of the hydrogen convection zone? Also, with what process of viscosity are they connected? As is well known, a very small fraction of the effective viscosity is molecular; a greater part of it is eddy viscosity, and electromagnetic forces probably play an important part in the phenomenon (Alfvén, 1950).

Consideration of electromagnetic forces leads us naturally to the question of the magnetic variability of the Sun, the period of which is about 22 years. In fact the magnetic variability can also be traced in the drift curve of sunspots (Tuominen, 1952; Richardson and Schwarzschild, 1953). The reason why this periodicity has been neglected in the present investigation is that we have been mainly interested in the physical reality of the latitudinal periodicity of about 10° in the drift-curve of sunspots.

The writer wishes to express his gratitude to Professor T. G. Cowling for his helpful comments on Sections 1 and 5 before these reached their final form, and to Dr. St. Temesváry and Dr. L. Mestel for clarifying discussions concerning the same questions during the 1953 Astrophysical Symposium in Liège.

References*

Alfvén, H. . . . . 1950 *Cosmical Electrodynamics* (Oxford University Press).
Biermann, L. . . . . 1942 *Z. Astrophys.*, **21,** 320.
Carrington, R. C. . . . . 1863 *Observations of the Spots on the Sun,* p. 222 and p. 242.
Dyson, F. W. and Maunder, E. W. . . . . . 1913 *M.N.*, **73,** 687.
Randers, G. . . . . 1941 *Ap. J.*, **94,** 109.

* See also the recent paper by U. Becker ("Die Eigenbewegung der Sonnenflecken in Breite") in *Z. Astrophys,* **34,** 129–136, 1954.

ROYAL OBSERVATORY, GREENWICH . . . . . 1924 *M.N.*, **85**, 185.
RICHARDSON, R. S. and SCHWARZSCHILD, M. . . 1953 *Accademia Nazionale dei Lincei, Fondazione Allessandro Volta, Atti dei Convegni*, **11**, Roma, p. 228.
SCHWARZSCHILD, M. . . . . 1947 *Ap. J.*, **106**, 427.
STRATTON, F. J. M. . . . . 1909 *M.N.*, **69**, 659.
1946 *Observatory*, **66**, 300.
SWEET, P. A. . . . . 1950 *M.N.*, **110**, 548.
TUOMINEN, J. . . . . 1941 *Z. Astrophys.*, **21**, 96.
1945 *Observatory*, **66**, 160.
1952 *Z. Astrophys.*, **30**, 261.
TURNER, H. H. . . . . 1907 *M.N.*, **68**, 98.

The Evolution of Solar Prominences

L. D'AZAMBUJA
Observatoire de Paris, Meudon (Seine & Oise), France

SUMMARY
From a study of synoptic charts of the solar chromosphere published by the Observatory at Meudon and of a very large number of spectroheliograms taken there, it has been possible to recognize the essential features of the evolution of quiescent prominences. In their spatial form they are immense bridges with multiple arches, which project onto the disk in the form of long dark filaments. They last from six to eight times as long as sunspots and originate at the same latitudes as these, often in the accompanying faculae. They then move slowly towards the poles, near which they end their existence, orientating themselves parallel to the equator. Their career is nearly always perturbed by the phenomenon of "sudden disappearance", in the course of which they rise sometimes to immense heights and finally vanish. They then frequently re-appear, as if they were the external sign of a more permanent underlying current.

IT is well known that the prominences of the Sun cannot be seen by direct observation except during the brief periods of total eclipses, when the Moon exactly covers the disk, but leaves its external features completely visible. The prominences—as their name implies—appear in the various forms of rose-tinted plumes or flames, rising to various heights above the limb, but usually between the limits of 1/100 and 1/15 of the Sun's diameter (*i.e.* between 15,000 and 100,000 km).

In 1868 a total eclipse gave J. JANSSEN the opportunity to show that the prominences can be observed at any time owing to the properties of their spectrum. Pointing a spectroscope at a well-defined prominence during totality, he noticed that it emitted several bright lines of relatively high intensity. Keeping his instrument in position, he succeeded in following the lines after the eclipse, and thus showed that their intensity is higher than that of the corresponding region in the spectrum of the diffuse light of the sky on ordinary occasions. Sir NORMAN LOCKYER made the same discovery independently at the same time.

From then onwards it was possible to find the prominences situated around the solar disk at any given moment by exploring the limb with a spectroscope attached to a telescope, and by drawing the contours of objects obtained by following the variation in height of one of their emission lines, usually the red Hα-line of hydrogen. This was obviously a long and tedious process. In 1869 HUGGINS had the idea of opening the spectroscope slit sufficiently wide to admit the entire prominence, or at least an appreciable part of it. In this way he obtained a truly monochromatic image of the object. This is the *wide-slit* method which is still used at several observatories. Observation thus became comparatively easy. Through the initiative of Father SECCHI a society of spectroscopists was founded in Italy and began to publish *Spectroscopic Images of the Sun's Limb,* a publication which continued up to the last few years, thus providing statistics of prominences over more than five cycles of solar activity.

The study of these statistics has led to important results. It has indicated in particular the fundamental laws of the distribution of prominences with latitude and as a function of the 11-year cycle.*

These laws may be summarized as follows:

(1) There are four zones around the Sun, two in each of the Northern and Southern hemispheres, where prominences are especially abundant.

(2) The zones of low latitude, between 10° and 40°, are the most important in both the number and the intensity and size of the prominences that appear in them. Their activity increases and diminishes with that of the spots.

(3) The high-latitude prominences begin to appear about three years after sunspot maximum, between 40° and 50°. They continue until spot minimum, at which time they, too, show diminished activity. Activity then increases again very quickly and the zones move upwards towards the poles, which they reach approximately at the time of the next maximum. They disappear shortly afterwards, so that, for a period of three years or so, the second zone practically ceases to exist.

In 1891 H. DESLANDRES and G. E. HALE independently developed an instrument, the *spectroheliograph*, by means of which it became possible to photograph prominences at the Sun's limb. The method used was based on the original system, in which the spectroscope slit "scanned" the regions of the limb occupied by prominences; but a second slit, or *selector slit,* placed in the image plane of the spectrograph, made it possible to select from the spectrum a given emission line. With this arrangement a photographic plate placed behind the second slit and moved in synchronism with the scanning motion of the first slit produced an image of the prominence of higher contrast and therefore more detail than those provided by visual observation with the wide slit. The radiation used for photography was the K-line of ionized calcium which, being in the violet, was much more suitable for photography than $H\alpha$ at a time when the production of plates with a given colour sensitivity was still in an experimental stage. Furthermore, the plates obtained by scanning the solar disk from one limb to the other showed a completely different aspect from that obtained with an ordinary telescope. They revealed in fact the structure of the chromosphere, which consists essentially of a multitude of small brilliant clouds some thousands of kilometres above the surface.

Although the new images of prominences provided the security of photographic documentation and could be studied at leisure after the actual observation, they provided hardly any more information than had been obtained previously from drawings made from visual observation. They still gave only a very incomplete idea of the spatial form of prominences, since only their projection on the apparent limb of the Sun could be observed. No further information was obtained as to the stability and persistence of the effects. These were visible indeed for three or four

* A cycle of solar activity is generally defined as the interval between two consecutive spot minima. Its mean length is 11·2 years.

days at the most, after which the rotation of the Sun about its own axis* caused them to disappear behind the limb, if they were in the West, or onto the disk itself, if they were in the East, after which the spectroheliograph proved unable to detect them.

Not until 1903, when the power of this instrument had been sufficiently increased, did it become possible to follow prominences across the disk by means of their absorption of the light from the chromosphere (see Fig. 1). They then appeared in

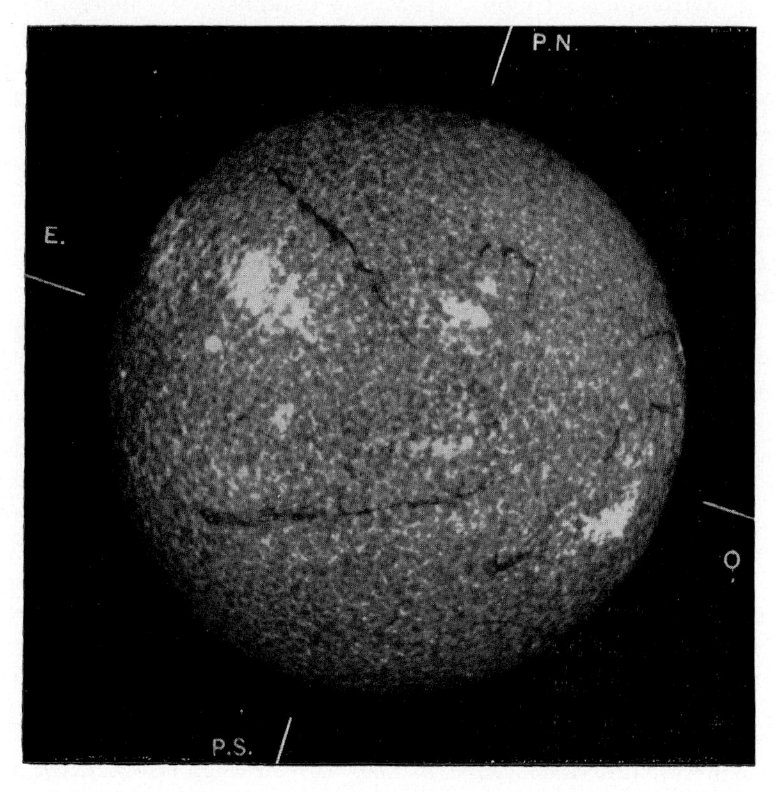

Fig. 1. Prominences projected on the Sun's disk. (Spectroheliogram in the K_3 line, obtained at the Observatoire de Meudon)

contrast against the background as long dark ribbons which H. DESLANDRES called *filaments*, a name which is still in general use.† Since then the possibility of following them like spots from their first appearance on the East limb to their disappearance at the West limb, *i.e.* in the various perspectives they present to a terrestrial observer, has enabled us to establish their usual spatial form. The latter has turned out to be very different from that which might be suggested by their appearance on the limb: the prominences are in fact immense bridges with multiple flattened arches, so narrow in proportion to their other dimensions that some authors have described them as very thin sheets of gas hovering above the chromosphere. One may thus understand how, seen from above, they have the appearance of ribbons or filaments.

* It is known that a point on the Sun's surface on the average takes 27·2 days to return to the same position with respect to a terrestrial observer, after one complete revolution of the Sun about its own axis.

† From the time of the very first observations Father SECCHI had recognized the existence of two distinct classes of prominences, eruptive and quiescent. The eruptive ones, which are connected with the centres of activity around spots, are small and of very short duration. Only in exceptional cases do they go over into filaments on the disk. They will not be discussed in this note.

Another line of approach was available to observers: to study how prominences are formed, develop and finally disintegrate; in other words to determine the normal characteristics of their evolution. However, before embarking on this investigation, maps of filaments had to be prepared to facilitate their identification. This has been the object of the *Synoptic Charts of the Solar Chromosphere* (illustrated in Fig. 2) prepared at Meudon Observatory since 1919 under the auspices of the International Astronomical Union. These now extended over more than 350 solar rotations and the abundant evidence which they provide has enabled the Meudon observers to deduce a certain number of general results which we will now summarize.

The chief zones of highest prominence frequency which are, as we have seen, between latitudes 10° and 40° approximately, on both sides of the equator, coincide very nearly with the regions where spots are formed. It is known that the latter follow a curious law, discovered by CARRINGTON and put in a precise form by SPÖRER: at the beginning of an 11-year cycle, they are formed at the edge of the zone furthest from the equator; as the cycle continues, they appear at increasingly low latitudes which hardly exceed 5° at the end of the cycle. Remarkably enough the birth of prominences follows the same law. At the time when they first appear they usually have the form of a small filament some 50,000 km long, and orientated (and this is a second remarkable fact) approximately along the meridians. On measuring the latitude of the two ends of the object it is found that the end nearer to the equator has the same mean latitude as the spots in the same phase of the 11-year cycle. In consequence, when one cycle gives way to another, the latitude of formation, as in the case of spots, is suddenly increased by 30°. Furthermore, on one occasion in three there is also a connection in longitude between the spots and the formation of filaments. These latter, which form about 25 days after the corresponding spots, appear in the same faculae that accompany the spots, and in almost all cases point directly towards them. However, in their subsequent evolution, these filaments are not distinguished from those whose longitude of formation is independent of spot longitude.

For some days the filament remains practically unchanged, but it already possesses the arched structure described above, as may be seen when the solar rotation brings its profile into view; it then begins to be elongated, especially at the end away from the equator, and acquires an eastward curvature in consequence of the polar retardation in the angular velocity of the Sun's rotation. Indeed, it is known that the Sun does not rotate as a solid body, but that the rotational speed decreases from the equator to the poles, the 70° parallel taking 5 days longer for a complete revolution than the equator.

After its disappearance at the West limb, the filament is almost always seen to re-appear 13 days later at the opposite limb, more developed and more curved towards the East than before. If nothing disturbs its evolution it usually reaches its highest degree of development at its third passage over the disk. At this time it may then have the following dimensions: length 300,000–400,000 km; height 60,000–80,000 km; thickness 6000 km.*

After this the high-latitude end tends to become parallel to the equator, and the filament becomes less conspicuous; it breaks up and finally disappears at the fifth or sixth passage over the disk, losing the equatorial end first. It is seen that these

* It should once again be emphasized that these dimensions are average values only. Filaments have been observed at Meudon which have lasted through more than ten rotations and have reached or exceeded one million kilometres in length,

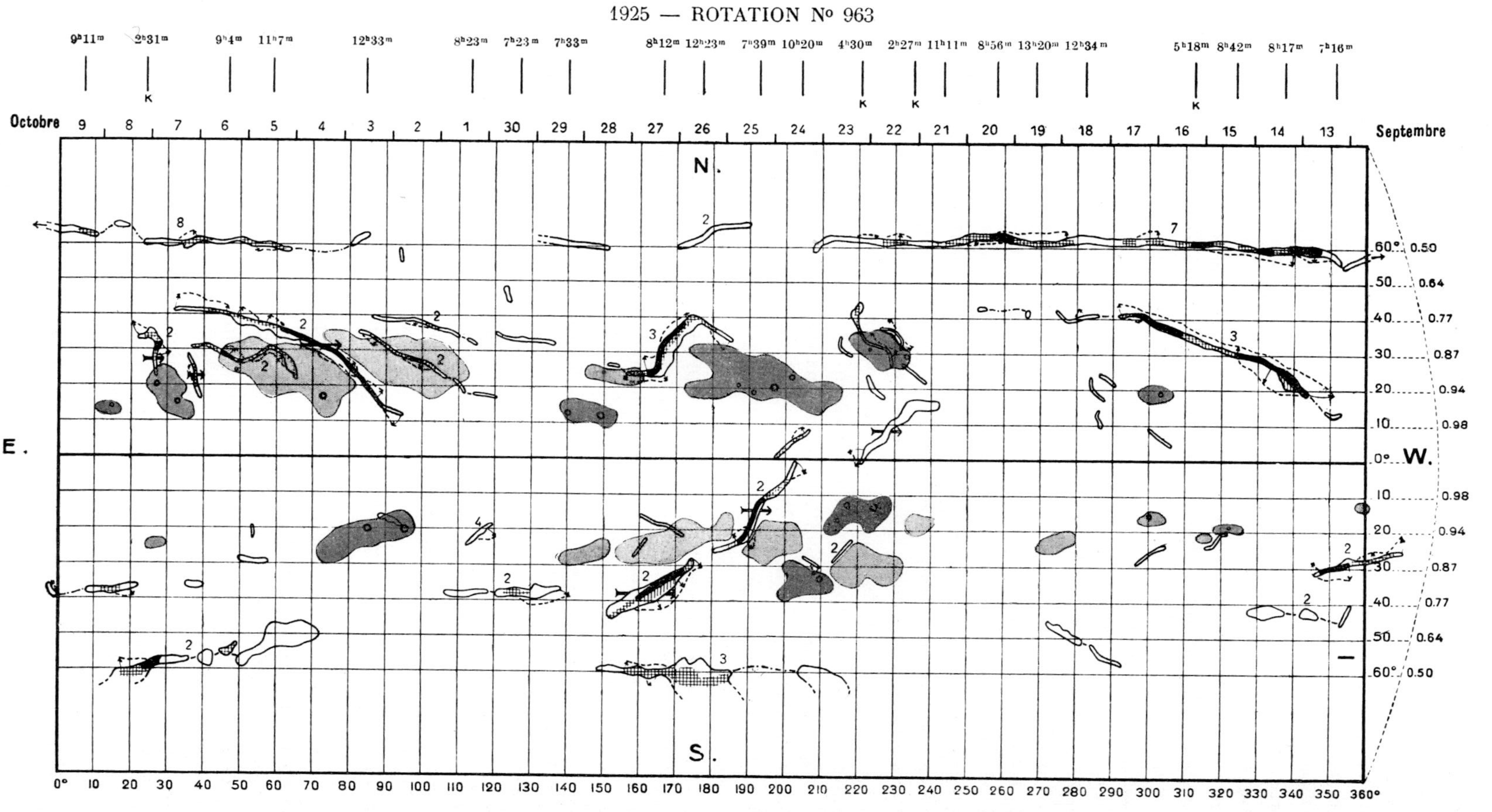

Fig. 2. Synoptic chart of the Sun's chromosphere, showing the centres of activity (spots and faculae)—the more heavily shaded the greater their intensity—and the filaments. (Note that the characteristic orientation of the filaments is oblique in the equatorial region and sensibly parallel to the equator in the polar regions.)

objects are very long-lived in comparison with spots, which usually have a mean life of the order of 10 days. It is also seen that the filaments are formed in the spot belts and are displaced slowly towards the poles in the course of their evolution. Precise measurements obtained from a large number of objects have shown that their motion is of the order of 2° per rotation in low latitudes and is retarded at higher latitudes. It appears also to be more rapid in the rising phase of the 11-year cycle than in the declining phase.

The second zone of prominence activity, in the polar regions, has an appearance on the disk quite different from that of the principal zone. Here the filaments are orientated almost along the parallels of latitude and are lined up end to end, often forming a nearly continuous crown around the Sun. Their evolution is much more complex and it is difficult to determine their beginning and end clearly. It should be mentioned that a filament from the equatorial zone which has already lost its low latitude end often reaches the latitude of this crown towards the end of its life. It is visible there as a distinct element for a certain period of time and is then completely incorporated into the crown. This is taken as an indication that the zone of polar filaments may consist entirely of elements coming from the equatorial filaments. In some way they seem to come to an end here.

The evolution of prominences does not always have the continuous character described above. It is often disturbed by some accident, through which it is shortened or modified. One of the most frequent of these is the coalescence of two filaments which have hitherto evolved independently. Frequently the filaments are separated again after their union. While they are still together, the complex object may differ greatly in its form from normal filaments, which are straight or slightly curved. In this way one may observe filaments of wavy, rectangular, or even U-shape.

But the most remarkable event in the life of a prominence is the sudden destruction of an equilibrium which has persisted for several weeks; this is shown first by violent movements of the mass of vapour, followed by its rapid ascent and then by its complete disappearance. This phenomenon lasts only for a few hours, during which observations of the limb have shown prominences starting at 50,000 km and reaching a height of 1,700,000 km, *i.e.* a little more than the Sun's diameter. In many cases after this sudden disappearance they re-appear gradually after some days and finally have the same aspect as before. This strange phenomenon suggests that prominences may be the external manifestation of a more permanent underlying current travelling slowly from the equator to the poles. This would indicate a circulation inside the Sun which begins in a meridional sense and is then strongly deformed by the polar retardation of the rotational velocity, finally describing a kind of spiral normal to its original direction.

If this hypothesis is correct and it is admitted that the observed interruptions in the visibility of prominences permit them to disappear for three consecutive rotations without ceasing to remain the same object, then the synoptic charts of Meudon show several examples of filaments which pass slowly from the equatorial to the polar zone, and then trace out the spiral described by the underlying current for more than two years. Those that appear at a time of resumption of solar activity after a minimum are formed at a latitude of about 30° and disappear at 70° or 80°. On the other hand, prominences formed during the declining phase when the latitude of formation is down to 15° never reach a higher latitude than 50°.

Thus the high latitude prominences that are visible in the rising phase of the cycle near the maximum of spot activity should first have appeared two or three years previously in the zone of formation of the first spots of the cycle. In the declining phase the secondary zone of prominence activity between 40° and 50°, which lasts until after the minimum, should in the same way be directly linked with the equatorial prominences formed two or three years previously at lower and lower latitudes. This conclusion is clearly provisional, since it is based on a study of two 11-year cycles only; nevertheless it does throw new light on the large-scale movements of the chromosphere which have been discovered from systematic observations of prominences and of their projections on the solar disk.

Investigation of Flares and Prominences at the Crimean Astrophysical Observatory

A. B. SEVERNY

Krimskaya Astrofysicheskaya Observatoria, Simeis, Crimea, U.S.S.R.

SUMMARY

A survey is given of some of the results obtained at the Crimean Observatory during the past five years, referring to the hydrogen spectrum of flares, the development of prominences and flares, and allied problems.

1. THE HYDROGEN SPECTRUM OF FLARES

SPECTRA of more than fifty flares were obtained by means of a spectrohelioscope adjusted for spectrographic work in order to examine the line-contours in the region from H_α to H_ζ. The observed increase of the width of emission with the quantum number of Balmer lines definitely suggests the Stark-effect as the principal mechanism of broadening of the spectral lines in flares. However, the best explanation of the observed line profiles of H_α is obtained for the process of natural damping of radiation, provided the kinetic temperature T_{kin} is 11,000°K and the electron density n_e is about 5.10^{12} (T_{kin} hardly exceeds 10,000–15,000°K, as otherwise the spectra of flares should closely resemble the spectra of O and B stars and we could not observe the emission in the lines belonging to the neutral metals). With such values of T_{kin} and n_e a good agreement of the observed and theoretical Stark contours is also obtained for all the other Balmer lines, such as H_β, H_γ, . . . As the greatest part of the free electrons is produced by hydrogen ionization, the ionization of hydrogen in flares is about ten times greater than the ionization of the undisturbed chromosphere ($\sim 1/100$). Hydrogen in flares is, therefore, mostly in a neutral state and the L_α-emission in flares will be N_2 (flare): N_2 (chrom) times larger than that of the chromosphere. This leads to a L_α-contour with width ~ 30 Å and a flux of L_α-radiation equal to ~ 1 W per cm^{-2} for the flares and $\sim 5.10^{-4}$ W per cm^{-2} for the undisturbed chromosphere. The last value, obtained by E. MUSTEL and A. SEVERNY (1952), appears to be in a good agreement with the recent data concerning L_α-radiation of the undisturbed Sun obtained by the rocket method (W. RENSE *et al.*, 1953a, 1953b).

The rise of the far wings of the Balmer lines and the H- and K-lines is obviously due to the decrease of the number of absorbing atoms and is probably connected with the process of diffusion of L_α-quanta into the deepest parts of the reversing layer. There is strong observational evidence that flares appear at different levels of the solar atmosphere (*e.g.* "hydrogen bombs" observed by ELLERMAN, spectral observations at radio-frequencies at the time of flares, *etc.*). According to MUSTEL and SEVERNY (1950) the mean height of flares is about 2500 km above the photosphere.

To support this view we could also mention the phenomenon of prominences having the brightness of typical flares (such as were, for instance, observed by DODSON and McMATH (1952) and by SEVERNY (1952) in the vicinity of a gigantic group of spots on 8th to 10th May, 1951).

2. Cinematography of Flares

An H_α-birefrigent filter with a 0·5 Å transmission band was used for the examination of the development of more than 100 flares. The minute after minute cinematography of spot groups carried out during 1951–53 showed that even in years of minimum solar activity one flare, on the average, appears during every 7 hr of the life of a group. More than 30 per cent of flares show marked motions which are quite similar to those observed for "electromagnetic" prominences (see below). These motions attain velocities of 300 km per sec. in several cases. In almost all cases the growth of H_α-intensity of flares is accompanied by a similar growth of their area and the larger the velocity of expansion of a flare, the larger is the maximum H_α-intensity. The life of a flare (the "width" of the light-curve) increases as the square of the size of a flare. This last correlation is very similar to that observed by the author in 1949 for fluctuations of brightness in prominences (SEVERNY, 1950). It seems probable that the motions and expansion of a flare in the magnetic field of a spot, H, can be connected with the appearance of the inductive electric field $(1/c)\,\mathbf{v} \times \mathbf{H}$. The larger the velocity $\mathbf{v}$, the larger is the current associated with this field and consequently the larger will be the Joule heat and the intensity connected with the process of recombination (SEVERNY and SHAPOSHNIKOVA, 1954a and 1954b).

3. The Motions and Fluctuations of Brightness in Prominences

Investigations of motions and fluctuations of the intensity of more than 150 prominences with the aid of an H_α filter, carried out in 1948–51, led us to the division of prominences into three classes, according to the type of their dominant motions: (*a*) eruptive (very rare), (*b*) electromagnetic (about 50 per cent of all cases), (*c*) irregular or turbulent prominences (about 50 per cent). The motions of eruptive prominences are well known; however for all the observed prominences of this type we have discovered a temporary splash of brightness, accompanying the sudden accelerations (SEVERNY, 1950). Electromagnetic prominences are characterized by regular, almost uniform, motions of knots and streams along paths similar to channels and resembling magnetic lines of force. There are no regular motions in the irregular prominences and all the changes are rather like metamorphoses similar to those of clouds or smoke on the Earth. Here the concept of turbulence may obviously be applied. A detailed examination of velocities and scales of motions of these prominences has led us to believe that the velocities and the scale of turbulent motion increase with their height in the solar atmosphere. The time of decay for fluctuations of intensity observed in prominences (*b*) and (*c*) is proportional to the square of the linear dimension of a knot. This fact, also observed for flares, is probably the most pronounced evidence in favour of a highly dissipative process acting in both cases. In the case of *c*-prominences the fluctuations of brightness can be explained in terms of the current theory of turbulence (SEVERNY and KHOCHLOVA, 1953). Assuming that the motions in electromagnetic prominences are determined by a horizontal pressure gradient, we concluded from the solution of the fundamental equations of magnetohydrodynamics by means of linear approximations, that the plasma of a prominence can move along the lines of force of a strong external magnetic field of about 10^3 gauss; (SEVERNY, 1953).

4. THE EXCITATION OF HYDROGEN AND HELIUM IN PROMINENCES

Although the cause of the emission in prominences is far from clear, the possibility is not excluded that we are dealing here with the same process of Joule heating as that connected with the motion of plasma in a magnetic field (ALFVÉN, 1950). The ionization and excitation are, however, non-uniform within a single prominence, as follows from a comparison of the shapes of several prominences in the light of H_α and in the light of the line λ10,830 Å of helium (IVANOV-KHOLODNY, 1952). It is also of interest to note that this infra-red helium line is suitable for the evaluation of the helium content in prominences; the corresponding determination based on this line (SEVERNY, 1951) leads to an abundance of the order of 20 per cent.

REFERENCES

ALFVÉN, H. . . . . 1950 *Cosmical Electrodynamics*, Chapter V (University Press, Oxford).
DODSON, H. and MCMATH, R. . . . . 1952 *Ap. J.*, **115**, 78.
IVANOV-KHOLODNY, G. . . . . . . 1952 *Publ. Crimean Astrophys. Obs.*, **8**, 115.
MUSTEL, E. and SEVERNY, A. . . . . 1950 *Publ. Crimean Astrophys. Obs.*, **5**, 3.
1952 *Publ. Crimean Astrophys. Obs.*, **8**, 19.
RENSE, W. *et al.* . . . . 1953a *Phys. Rev.*, **90**, 156.
1953b *Phys. Rev.*, **91**, 299.
SEVERNY, A. . . . . 1950 *Comptes Rendus de l'Acad. Sci. de l'URSS*, **53**, 475.
1953 *Comptes Rendus de l'Acad. Sci. l'URSS*, **91**, 1051.
SEVERNY, A. and KHOCHLOVA, V. . . . 1953 *Publ. Crimean Astrophys. Obs.*, **10**, 9.
SEVERNY, A. and SHAPOSHNIKOVA, E. . . 1954a *Astron. J. of the USSR*, XXXI, N 1.
1954b *Publ. Crimean Astrophys. Obs.*, **12**, 3.

Eclipse Observing

SIR JOHN A. CARROLL
Admiralty, London

SUMMARY
A review of the principal items of interest in modern eclipse observing, surveying the trend of the last fifty years and outlining future developments.

BY the end of the first decade of this century, the observations that could be made during total eclipses of the Sun appeared to be largely exhausted. The chromospheric spectrum had been photographed repeatedly and the wavelengths and intensities of its lines were determined with considerable precision. The corona continued to present the puzzle of its constitution and of the origin of the coronal monochromatic radiation, but apart from the necessary recordings by coronagraphs of its form at each eclipse there seemed little factual information that could be added to the store already accumulated.

The development of physical theory and of new agencies and methods of observations has dramatically changed the picture so that to-day there are a number of important enquiries to which answers may be found during the brief opportunities offered by solar eclipses.

Following approximately a chronological order, the first novel possibility was that of testing whether the proximate stellar universe behaved in the way predicted by EINSTEIN's general relativity theory. This could, in principle, be tested by observing the deflection of light rays passing close to the Sun on their journey to us from a star, and this test formed part, indeed a major part, of the first expeditions to observe eclipses after the war of 1914–18. It is hardly unfair to describe the results then obtained, and at the subsequent attempts, as inconclusive. The deflections to be observed are not very different from the errors of observation, particularly under the difficult conditions prevailing, which are markedly inferior to the conditions for good stellar parallax determinations.

How far it is any longer of interest to obtain such a check on the accuracy to which EINSTEIN's formulation fits the actual universe is perhaps open to question; the present writer doubts the value of any observations that would not be an order of magnitude more precise than has yet been achieved.

The difficulties are very great, not least in respect of the fairly violent temperature change near the Earth's surface during the partial phase of a total eclipse, and instrumentally a very high standard of constancy of rate of rotation of any equatorial or coelostat used is demanded. The writer feels that for success the optical train of the observing system must, by the use of fused quartz mirrors for example, and by thermostatic control, be made immune from the effects of such disturbance. The observation certainly remains, if not perhaps of first importance, an important challenge to the technique of eclipse observing.

The next new type of observation arose partly out of the application of developments of atomic theory to the problems of the constitution of stellar atmospheres; and partly out of the developments of photographic spectrophotometry. The theory of the outer electronic structure of atoms, starting from BOHR's first explanation of the origin of the hydrogen spectrum and extending to all the refinements and additions of multiplet theory and of wave mechanics, enabled theories of stellar atmospheres to be set up by SAHA, MILNE, FOWLER, CHANDRASEKHAR, and others, which might be tested or refined by determining the intensity of monochromatic emission from the chromosphere, a measurement only barely practicable on a very few special lines except during a total solar eclipse.

There is the further interest that certain experiments yielding data concerning atomic electron structure cannot be made in the laboratory, for even were the very low pressures attainable the size of practicable containing vessels would limit the free paths to the largest diameter of the containers, whereas in a stellar atmosphere free paths may be many kilometres, certainly long enough to exceed the lives of some quite long-lived excited states of atoms. During a solar eclipse the atmosphere of one star at least becomes accessible to observation. A celestial laboratory exists where the experiments are made for us if we can devise the means of observing them. Nearly all these observations require spectrophotometry and interest attaches too to the detailed determination of the profiles of lines in the chromospheric spectrum. This last enquiry poses some formidable problems of technique, since the lines are only some tenth of an Ångstrom wide and spectrographs of very great resolving power are called for. Interferometry seems the obvious answer, but it is less obvious how to apply it and most interferometers require a fairly precise thermostatic control for their successful operation.

A further interest emerging in the 1920's arose out of ionospheric studies. Information bearing on the nature and speed of the emissions from the Sun responsible for producing the ionized layers in the Earth's atmosphere may be obtained by observing the changes occurring as the Moon progressively intercepts them on their way to the Earth. These observations share with observations of radio "noise", to be mentioned later, the unusual feature of being possible even though the eclipse be obscured by clouds, a luxury denied the purely astronomical and astrophysical observers.

The advent of radio time signals, by removing the dependence of observers on transported chronometers or telegraphic signals available only in certain places, has introduced the interesting possibility of using eclipses to improve our knowledge of the figure of the Earth. Knowledge of the lunar profile at various phases of libration is now sufficiently precise for timing of the last residue of the partial phase, the rapidly disappearing crescent and the "Baily Beads", to be accurate to a fraction of a second. Cinema records of this stage and of the end of totality can be made with accurate time signals superimposed and the geocentric co-ordinates of the site of observation deduced.

There will always be a need to photograph the corona for purposes of record, although the magnificent pioneering work of LYOT has made some observation of the corona possible at any time without solar eclipses, and probably observation of prominences is little aided by photographs taken during eclipses.

Thanks to GROTRIAN and EDLÉN the origin of the coronal monochromatic radiations is no longer a mystery and the structure of the corona is now largely understood.

The greatest advance has come from the discovery and study of the radio noise emitted from the corona and from sunspots. Solar eclipses offer opportunities for special investigations in this field and such observations have been made at eclipses since 1945.

There are a number of auxiliary observations that are made when an eclipse observing expedition of any size is established, some have geophysical or meteorological interests, and there is nearly always the possibility of adding to existing knowledge by extending or refining older types of observation by applying to them modern techniques or technical developments. An obvious example is the use of infra-red sensitive plates to extend the survey of the chromospheric spectrum.

The use of aircraft for eclipse observing has been considered on many occasions. Clearly there is the possibility of having one's observing platform above the all-too-often obscuring clouds. Unless the aircraft can fly really high, well over 20,000 ft, say, this possibility is not so definite as at first one would hope, but a serious obstacle to most observations is that the equipment to be carried is very heavy. Modern very large aircraft could, however, carry the loads necessary; a far more formidable difficulty is that nearly all observations require the instruments to be pointed at the Sun (either directly or with a coelostat) and to remain so directed during intervals of from a fraction of a second up to the whole of totality and this to an accuracy of a fraction of a second of arc. This is quite beyond the stability, of course, of any existing aircraft, even could vibration be sufficiently reduced. It is, however, well worth remembering that modern aircraft can fly to heights of 30,000 or 40,000 ft; they can carry quite substantial loads; and stability at least good enough to take first-rate photographs of the ground beneath them with exposures of a few seconds duration is a *fait accompli*. For some suitable instrument and special purpose, airborne observation may be very well worth while in the near future. It is even conceivable that observations by equipment carried in high altitude rockets may turn out to be desirable and feasible.

This survey of the trend of eclipse observing in the last fifty years is far from exhaustive. There are many other kinds of observation of interest and value besides those mentioned. The whole field of polarimetry has not been touched upon, and the possibilities of replacing photography by electronic observing and recording can only be hinted at here in passing.

The intention of this survey is rather to expound the somewhat surprising fact that there is so much of real value to be done at future solar eclipses, and to indicate the great variety of techniques involved in eclipse observations as a whole. Fortunately, co-ordination of eclipse observing programmes is good, and one of the paramount needs, to offset the all-too-great likelihood of bad weather by the spreading of different expeditions along the track of totality, can commonly be arranged satisfactorily. Modern transportable electrical generating plants and other "caravan-like" facilities have freed the chooser of a site from restriction to the environs of a fair-sized town that was becoming oppressive.

Eclipse observing is an arduous and exacting task. There is no second chance, the equipment must work at the time of the eclipse and the drill must be perfect. The lost effort, probably eighteen months or more of devising, making, and testing equipment, is most disappointing to those suffering bad weather. In the wider interests of all, willingness of some to go to the less "favourable" sites is very necessary. Nor should anyone be too hopeful or too pessimistic; at other than exceptional places the meteorological hazard can be forecast with no great reliability. A cloudy eclipse is certainly far from a wasted venture if it has enabled new equipment to be proved in the field. It is unfortunately only too true that experience of eclipse observing is a very necessary ingredient of success, and that experience needs to include experience of a clear eclipse before the utmost attainment can be counted on.

A rather pleasing feature of eclipse observing is that compared to many modern physical researches it is not expensive, a few thousand pounds cover the cost of a medium-sized expedition. Most of the equipment is normally used for other purposes, and the most serious increase in cost in recent years is due to the large rises in the charges for transport of men and equipment and the higher living costs on site. So far the rise in cost has not gravely restricted observations and that there has to be a selection of the more far-reaching from among the possibilities is no unhealthy thing.

Finally, a few words of advice, based on experience, to those preparing to observe eclipses. First and foremost make sure that every piece of equipment has been assembled and tried at home before leaving, in the exact way and form it will be used during the eclipse. Secondly, take a very liberal supply of spares and a very complete tool kit and portable workshop, including a liberal assortment of common laboratory equipment. A humble retort-stand can be worth far more than the weight and space it occupies.

In fine as to choice of a site, it is less critical than many suppose. If very good seeing is a prime consideration, a morning eclipse with totality a few hours after sunrise is to be preferred. It is also rather more than wise to be able to live fairly near the site of the instruments. However good the planning and however kind the weather during the period of preparation, there will always be a spell of intense activity only really tolerable if there is no tiring journey to be faced daily.

Thermostatic control of observing instruments is nowadays quite practicable and really essential if the best performance is to be gained from most optical instruments. The diurnal variations to which eclipse cameras, spectrographs, *etc.*, are subject are quite large and it is idle to expect definition, resolving power, and focus to equal that obtained from the same instruments set up and tested at home in a laboratory whose temperature fluctuates slowly and by a mere degree or so.

It is important to realize that proper temperature control is almost sure to demand refrigerating plant as well as heaters. The peak temperatures reached by uncooled arrays, even if shaded from direct Sun, are much higher than it is practicable or desirable to maintain. The ideal is, of course, to maintain equipment at the temperature that will prevail at totality. This can be done well enough in practice by using as a datum 5°C below the temperature normal at the same time on a fine day.

It is an added pleasure in writing this article that it deals with a topic which has so extensively occupied Professor STRATTON'S attention during his long career and to which he has himself contributed so substantially in exploiting older and applying new techniques.

Recent Eclipse Work in Japan

YUSUKE HAGIHARA
Tokyo Astronomical Observatory, Japan

SUMMARY

Eclipse work is summarized for the total eclipses on 1936 June 19, 1941 September 21, 1943 February 4–5, and for the annular eclipse on 1948 May 9—in which photographic and three-colour photoelectric photometry, spectrophotometry, and polarization of the solar corona were studied. The solar limb-darkening was also investigated. In the geodetic programme during the annular eclipse for determining the figure and the dimension of the Earth the large plumb-line deviation of the Japanese Islands was confirmed.

SINCE 1934 Japan has been fortunate in having had several occasions for observing solar eclipses in her own islands. In 1936 Professor F. J. M. STRATTON and his collaborators came to Japan to observe the total solar eclipse at Kamishari in Hokkaido; see Fig. 1 (*a*), (*b*), (*c*), (*d*). This event stimulated the study of eclipses in Japan, and several important papers have since been published.

I. TOTAL SOLAR ECLIPSE OF 1936 JUNE 19 IN HOKKAIDO

MATUKUMA (1940) obtained, at Koshimizu, one EINSTEIN plate with a horizontal camera of aperture 20 cm, focal length 500 cm, and an exposure time of 80 sec. About ten stars were found on the plate, and from them, using two comparison plates of the same field taken at Sendai six months later, he derived an Einstein light deflection of $2''.134 \pm 1''.146$ and $1''.284 \pm 2''.672$, respectively.

HAGIHARA and SAITO (1938) photographed, at Mombetsu, the solar corona with a Dallmeyer camera of 8 cm aperture. A special apparatus was devised for obtaining the intensity calibration of the coronal light, consisting of a cylindrical lens and an optical wedge. The photometric treatment of the plates was carried out with special care. The effect of halation on the photographic plates was studied with auxiliary laboratory experiments, and the effect of sky scattering with auxiliary observations. Both these effects were eliminated by solving FREDHOLM's integral equations (HAGIHARA and SAITO, 1938; HAGIHARA, 1938). The effect of flare was studied by SAITO (1939) by using a similar method, and was shown to be negligible. After the completion of these discussions, the intensity distribution in space of the corona was derived by solving ABEL's integral equation. From the plates, showing the intensity distribution of the light over the Moon's disc, the albedo of the Earth was derived.

KABURAKI (1938) photographed, at Mombetsu, the inner corona through a yellow filter with a Merz visual equatorial of aperture 16·2 cm. He also obtained flash spectra with an objective prism and a Cook lens of aperture 18 cm. KUBOKAWA took photographs of several phases of the eclipse using a horizontal camera of focal length 11 m.

HASIMOTO and OKUDA, at Naka-Tombetsu, obtained spectrograms of the corona with a Zeiss three-prism slit spectrograph, and SOTOME at Memambetsu tried to obtain an Einstein plate with a Cook triplet of aperture 30 cm.

(*a*)

(*b*)

(*c*)

(*d*)

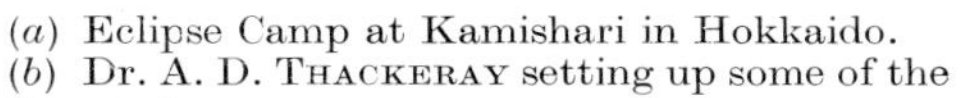

(*a*) Eclipse Camp at Kamishari in Hokkaido.

(*b*) Dr. A. D. THACKERAY setting up some of the instruments.

(*c*) Eclipse observers: the fourth from the left is Prof. F. J. M. STRATTON, the leader of the expedition. The first on the left is Prof. R. O. REDMAN, the third Prof. C. W. ALLEN, the sixth Prof. F. W. ASTON, the ninth Dr. A. D. THACKERAY, the tenth Brigadier R. A. BAGNOLD, and the twelfth Dr. T. ROYDS. The two Japanese members of the group are Kamishari's schoolmaster and the interpreter.

(*d*) Professor STRATTON changing shoes according to local custom on leaving the house of his Japanese hosts (a press comment was that he was rather slow at this operation!).

(*Photographs by courtesy of Prof.* REDMAN *and Dr.* THACKERAY)

Fig. 1. Cambridge Solar Physics Observatory expedition to Japan for the total solar eclipse of 1936 June 19

Sekiguti (1938) and Koiwai, at Memambetsu-Nisshin, secured slitless spectrograms of the corona with an objective four-prism camera of focal length 25 cm, and found (Sekiguti, 1937) new emission lines at 6583·8 Å and 6548·7 Å, *i.e.* nearly coincident with the nebular lines 6583·6 Å and 6548·1 Å of [N II]. Sekiguti also found six new coronal lines with a two-prism slit spectrograph. These lines, however, have not yet been confirmed. Tanaka (Tanaka, Koana, and Kondo, 1945; Tanaka, Oura, and Murakami, 1945), with the assistance of Hattori, took slit spectra of the corona with a Littrow-type three-prism spectrograph at Shari, and obtained the wavelengths of the emission lines and the radial intensity decrement from the limb.

Notuki, at Memambetsu-Nisshin, tried to get flash- and corona-spectra in the longer-wavelength regions with a slitless spectrograph and a plane grating, and Fujita attempted similar observations in the shorter-wavelength regions with a Hilger quartz spectrograph. Oikawa, at Kunneppu, also tried to get slitless flash spectra with a Steinheil 45° prism and a Cook 8°5 prism. The weather was not favourable for these observers.

Yamazaki and Goto (1937) succeeded in taking cinematograph films of the flash spectra, with time-marks, showing the transition from Fraunhofer lines to emission.

Other parties tried to get Einstein plates and corona plates, but were not successful. Several parties went to Manchuria and to Siberia to take direct photographs of the corona, in order to study the changes of its shape during the eclipse in conjunction with the photographs taken with similar equipment at the three above-mentioned stations in Hokkaido. The result has not been published.

II. Total Solar Eclipse on 1941 September 21 in Okinawa Islands

Osawa (1943) made a three-colour photometry of the corona with a simple photo electric photometer. The brightness of the corona in red, green, and violet light, respectively, was 0·27, 0·205, and 0·21, expressed in units of the full moon. He also observed the intensity distribution at the solar limb, using a Steinheil spectrograph and a photo-electric photometer. The output current of the amplifier was recorded cinematographically. He found that the darkening at the extreme limb was much steeper than one would have expected from the well-known law of the intensity distribution over the central zone of the solar disc.

Sekiguti observed the flash- and the corona-spectra by means of a miniature spectrograph of his own design, using glass prisms and a hyperboloid camera mirror. The observation was successful, but no new features were discovered.

Oikawa and Saito took a corona spectrum with a three-prism spectrograph, consisting of Jobin's prisms, a Zeiss collimator lens of focal length 91 cm, and a Steinheil camera-lens of focal length 54 cm. The dispersion was 55·7 Å per mm at λ6374, and 8·7 Å per mm at λ4231. Only one exposure of 180 sec. was made, with the slit tangential to the solar limb. An iron arc at reduced pressure was recorded on the plate. The wavelengths of the coronal lines were measured with great care, giving $\lambda\lambda$6374·057 Å, 5303·010 Å, 5116·014 Å, 4566·412 Å, 4231·135 Å.

Fujita made spectrophotometric observations of the corona with a slitless spectrograph. It had five dispersing prisms and two right-angle reflecting prisms. The camera lens had an aperture of 2·3 cm, and a focal length of 58 cm. The dispersion was 7·0 Å per mm at λ4800. Comparing the coronal light with a standard lamp and with photospheric light, Fujita (1943, 1944) derived the absolute value of the intensity of $\lambda 5303$, the electron pressure in the prominence, the density gradient of hydrogen atoms, and the profile of the Moon's limb.

The above observations were made at Ishigaki Island in the Okinawa Island group. TANAKA went to Paichou, China, to observe corona spectra with a large slit spectrograph, mounted equatorially. MATUKUMA, at the same place, tried to get flash- and corona-plates. UETA tried to obtain an Einstein plate at Hêshêngchiao, Central China. The weather, however, was unfavourable, both at Paichou and at Hêshêngchiao.

III. TOTAL SOLAR ECLIPSE ON 1943 FEBRUARY 4–5 IN HOKKAIDO

SAITO (1948) observed photographically the polarization of the corona, using an equatorial polarigraph with a Rochon prism, and a red filter of effective wavelength $0{\cdot}68\mu$. By measuring the double images of the corona, photographed at various position angles of the Rochon prism, he determined the degree of polarization and the plane of polarization in different parts of the corona. The direction of the polarization was shown to be almost radial to the solar limb, not deviating by more than 10°. The polar polarization exceeded the equatorial polarization near the limb, but the two became nearly equal at 4′ from the limb. The situation was seen to be reversed in the outer corona. SAITO (1950) then proposed an interpretation of this result, based on a theory in which the oblateness of the corona was taken into account.

HURUHATA (1947), on the other hand, measured the polarization of the corona photo-electrically with a 13 cm Cook refractor of focal length 168·4 cm, colour filters, a Nicol prism, a vacuum potassium phototube, a high resistance in vacuo, and a UX 54 tube. The amplified photocurrent was recorded automatically by means of a sliding-plate equipment. The polarization was measured at five points of the corona along a radial direction from the Sun's centre, assuming the plane of polarization to be radial. Care was taken to eliminate the sky light. The results of SAITO and of HURUHATA are in agreement.

OSAWA observed the brightness distribution at the extreme limb of the Sun by means of three-colour photoelectric photometry, but the result was unsatisfactory, owing to the low altitude of the Sun.

FUJITA (1949) observed the brightness distribution at the solar limb by SCHWARZSCHILD's method, using a slitless spectrograph with a rocking-plate camera. Twenty exposures, each of 0·01 sec., were made at intervals of 5 sec. FUJITA solved the well-known integral equations of the problem, and concluded that the brightness distribution was approximately represented by a linear relationship with respect to $\cos\theta$ (for $\lambda 5200$, $\lambda 5300$, or $\lambda 5400$); the brightness was lower than the value given by the linear relation for $0{\cdot}28 > \cos\theta > 0{\cdot}18$, and higher for $0{\cdot}18 > \cos\theta > 0{\cdot}13$.

NOTUKI took flash- and corona-spectrograms, mainly with the intention of measuring the relative strengths of the green and the red coronal lines.

The above observations were made at Akkeshi, Hokkaido.

TANAKA, OURA, MATUKUMA, and MITUNOBU (1946) observed the coronal spectrum at Yubetsu, Hokkaido. They determined accurate wavelengths of coronal lines, including the new line $\lambda 5306{\cdot}40$ which TANAKA observed at the eclipse of 1936. The brightness distribution and the polarization of the corona were also studied by

---

* This eclipse began on 1943 February 4 at $21^{h}\,26^{m}$ G.M.T., and ended on February 5 at $01^{h}\,49^{m}$.

putting a Wollaston quartz-prism behind the collimator lens. The maximum polarization occurred at a distance of about one-half solar radius from the limb.

MATUKUMA obtained at Kushiro a spectrogram of the corona with a three-prism slit spectrograph; he noticed some unknown lines.

IV. Annular Eclipses on 1948 May 9 at Rebun Island in Hokkaido

An Eclipse Committee was organized by the National Research Council, consisting of astronomers, geophysicists, and ionosphere workers. As well as organizing our own observational programme (Nat. Res. Coun. Japan, 1948), the Committee also worked in co-operation with the American National Geographic Society on their geodetic programme.

HIROSE (1948) found, from previous observations of occultations, the systematic deviation of the observations in Japan from those made in Europe and America. He attributed it to the systematic deviation of the vertical in the Japanese islands, due to the presence of the big Asiatic continent to the West, and of the deep Pacific oceanic basin to the East. The amount of this deviation could explain the systematic discrepancy found by the land survey along the boundary of Manchuria and Korea. Thus, the results of these eclipse observations confirmed HIROSE's theory.

KAHO (1949) took direct photographs of the partial phases with a horizontal camera, the objective lens of which had an aperture of 18 cm and a focal length of 500 cm. Sixty-two plates were obtained, and the relative position of Sun and Moon computed (HIROSE, KAHO, and TOMITA, 1950).

OSAWA (1951) observed the brightness distribution at the extreme limb of the Sun using JULIUS' method. A photo-electric photometer with a potassium vacuum-phototube served for recording the variation of the integrated brightness of an effective wavelength of about λ4000. The analysis of the results, solving the well-known integral equation, showed that the intensity dropped rapidly at the extreme limb.

UETA and FUJINAMI photographed the partial and the annular phases. FUJINAMI (1952), with the collaboration of a group of cinematograph technicians, obtained cine films of Bailey's beads. He recorded the heights of the mountains and the depths of the valleys on the lunar surface; the depths of some of the valleys were found to be much deeper than previously determined by HAYN.

References

FUJITA, Y.	1943	*Japan J. Astron. Geophys.*, **20,** 83.
	1944	*Ibid.*, **21,** 1.
	1949	*Publ. Astron. Soc. Japan*, **1,** 23.
FUJINAMI, S.	1952	*Publ. Astron. Soc. Japan*, **4,** 115.
GOTO, S. and YAMASAKI, M.	1937	*Popul. Astron.*, **45,** No. 5.
HAGIHARA, Y.	1938	*Astron. Nachr.*, **266,** 285.
HAGIHARA, Y. and SAITO, K.	1938	*Ann. Tokyo Astron. Obs.*, Ser. II, **1,** 91.
	1938	*Ibid.*, 151.
HIROSE, H.	1948	*Proc. Japan Acad.*, **24,** 35, 51, 56.
	1948	*Tokyo Astron. Obs. Repr.*, Nos. 83, 84, 85.
HIROSE, H., KAHO, S. and TOMITA, K.	1950	*Ann. Tokyo Astron. Obs.*, Ser. II, **3,** 23.
HURUHATA, M.	1947	*Japan J. Astron. Geophys.*, **21,** 173.
KABURAKI, M.	1938	*Ann. Tokyo Astron. Obs.*, Ser. II, **1,** 139.
KAHO, S.	1949	*Tokyo Astron. Obs. Repr.*, No. 39.
MATUKUMA, T.	1940	*Japan J. Astron. Geophys.*, **18,** 51.
	1940	*Nature (London)*, **146,** 264.

Osawa, K.	1943	*Tokyo Astron. Bull.*, Nos. 692–695.
	1943	*Tokyo Astron. Obs. Rep.*, **9**, 143 (in Japanese).
	1951	*Ann. Tokyo Astron. Obs.*, Ser. II, **3**, 53.
Nat. Res. Coun. Japan, Tokyo	1948	Solar Eclipse Committee: *Provisional Report of Observations of the Annular Eclipse on 9th May*, 1948.
Saito, K.	1939	*Ann. Tokyo Astron. Obs.*, Ser. II, **1**, 201.
	1948	*Tokyo Astron. Bull.*, Ser. II, No. 8.
	1950	*Ann. Tokyo Astron. Obs.*, Ser. II, **3**, 1.
Sekiguti, R.	1937	*Nature (London)*, **140**, 724.
	1938	*Ann. Tokyo Astron. Obs.*, Ser. II, **1**, 59.
Tanaka, T., Koana, Z. and Kondo, K.	1945	*J. Phys. Soc. Japan*, **1**, 31.
Tanaka, T., Oura, H. and Murakami, T.	1945	*Ibid.*, 30.
Tanaka, T., Oura, H., Matukuma, T. and Mitunobu, T.	1946	*J. Phys. Soc. Japan*, **1**, 30.

The Widths of Narrow Lines in the Spectrum of the Low Chromosphere, Measured at the Total Eclipse of February 25, 1952.

R. O. Redman

The Observatories, Cambridge

Summary

The spectrum of the chromosphere was photographed at the total solar eclipse of 25th February, 1952, with a slit spectrograph, using the second order of a 6-in. Rowland concave grating. Line widths were measured in both ultra-violet and yellow. Both regions show ξ_0 near 2·5 km per sec. for heavy elements in the lowest chromosphere (average line of sight velocity about 1·8 km per sec.), in close agreement with measures made at the 1940 eclipse. There is some evidence for a small increase of velocity with height.

1. Introduction

The question of the distribution of material velocities within the chromosphere, whether of the nature of turbulence in an essentially dynamic atmosphere, or of a temperature in an approximately static medium, is of fundamental importance to all theories of the structure of this part of the Sun's atmosphere. Perhaps the best way of obtaining evidence on this point is to measure accurately the shapes of lines in the chromospheric spectrum. The method is not free from serious difficulties, but at least it can give an upper limit to the velocity spread at any particular chromospheric level. It can also distinguish to some extent between turbulent motion, where the velocities are the same for all constituents of the gas, and thermal motions where the velocities vary with atomic weight. The chief difficulty is that there are other mechanisms of line broadening, the most important being self-absorption, so that to obtain significant velocity data the lines for measurement have to be selected with great care.

In good seeing and in longer wavelengths, say $\lambda > 5000$ Å, the chromospheric spectrum has been photographed in broad daylight, first by Hale, and later by Adams and others. Dr. H. D. Babcock, who is among those who have made use of this procedure, has recently drawn attention to the possibility of measuring the shapes of the chromospheric lines in this way (Babcock, 1954). However, up till now observers have chosen to measure line widths at eclipses, the inconvenience of an eclipse being off-set by being able to photograph the spectrum of the low chromosphere free of Fraunhofer spectrum, even in the ultra-violet.

At the 1940 eclipse, line widths were measured at both second and third contacts with a three-prism spectrograph (3·5 Å per mm at Hγ), and the narrowest lines were little wider than the instrumental profile (REDMAN, 1942).* Correcting for instrumental broadening by means of photographs of the krypton spectrum, it appeared that $\xi_0 = 3{\cdot}1$ km per sec. (average line of sight velocity about 1·8 km per sec.), inclusive of temperature motions. This result referred only to comparatively heavy elements, and to lines whose counterparts in the Fraunhofer spectrum are weak. Among the narrowest chromosphere lines are those of La II and the singly ionized rare earths, Ce II, Pr II, Nd II, Sm II, Eu II, and Gd II. Lines of these elements, when also weak in the Fraunhofer spectrum, appear to be little distorted by self-absorption, and very suitable for the present purpose, although unfortunately, they all tend to fade away rather rapidly with increasing height in the chromosphere.

Besides showing that the horizontal component of any turbulence must be small in the low chromosphere, the 1940 measures seemed to suggest that the kinetic temperature of the chromosphere may be high ($\approx$ 30,000°K), an interpretation of the observations which has been much criticized (MIYAMOTO, 1951; SUEMOTO, 1951; UNSÖLD, 1952). If the suggested high temperature is not a valid interpretation, either the 1940 line widths are wrong, or we have the problem of explaining them in some other way, *i.e.* we have to explain a chromosphere which has little horizontal turbulence, but yet produces relatively wide lines of helium, and still wider lines of hydrogen. From every point of view it seemed highly desirable to obtain more, and if possible better, observations and the first good opportunity of doing this was at the total solar eclipse of 25th February, 1952 (REDMAN, 1952a). The present paper will discuss results from this eclipse relating only to narrow lines originating near the bottom of the chromosphere. Hydrogen, helium, and other elements giving very strong and generally broad lines are specifically excluded, since they undoubtedly furnish greater difficulties of interpretation, and work on them is not yet completed. Thus it will not be possible to discuss here the kinetic temperature of the chromosphere.

2. OBSERVATIONS

The Cambridge party observed the eclipse at Fort Stanley (15° 36′ 17″ N, 32° 32′ 50″ E), on the outskirts of Khartoum, about 8 km from the centre line of the eclipse track. The surroundings there are sandy desert, broken only by the Nile, and since February falls in the dry season, the chances of a clear sky for the eclipse were believed to be nearly 100 per cent. Although general sky conditions during the preceding weeks were not so good as predicted, sunshine recorders being less discriminating than eclipse observers, totality actually occurred in a cloudless sky, with very good transparency and remarkably little scintillation. The following data for Khartoum are quoted by courtesy of the Sudan Meteorological Service:

Mean diurnal range of temperature, 4th to 24th February, 1952 .	17°·2 C
Absolute maximum, 15th January to 25th February . . .	40·8
Absolute minimum, 15th January to 25th February . . .	10·9
Maximum temperature, 25th February	29·7
Minimum temperature, 25th February	13·2

* The results in this first paper were expressed in terms of the average line of sight velocity; in the present work ξ_0 is used. The distribution of line of sight velocity is $\exp - (\xi/\xi_0)^2$.

The choice of spectrograph was dictated by (*a*) the optical parts available at Cambridge, (*b*) the need for high resolution, (*c*) the limited exposure time for observations made near the centre of the eclipse belt (effectively about 1 sec. for the low chromosphere). Ultimately a 6-in. Rowland concave grating of nominal radius 21 ft was selected; it was used in a Wadsworth mounting, with a 10-in. concave mirror of 12 ft focal length for the collimator. The ruled area of the grating is 5''.7 wide, 2''.2 high, and in the green about 15 per cent of the incident light falls in each second order. For its age the grating is of good quality, and in the second order the strongest Rowland ghost has an intensity about 0·004 of the parent line. Two separate cameras

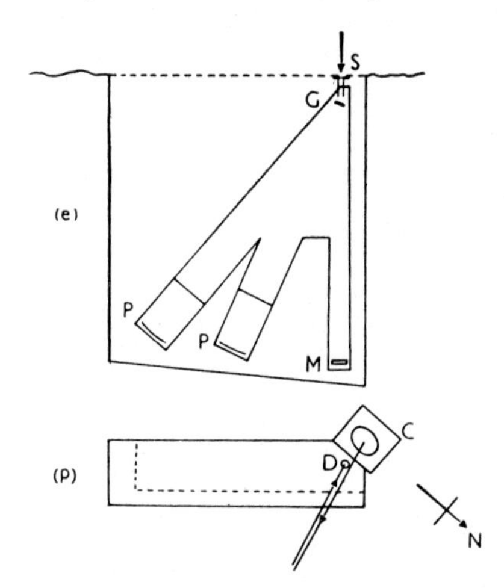

Fig. 1. Diagram of spectrogram pit, (*e*) in elevation, (*p*) in plan. Slit *S*, grating *G*, collimator mirror *M*, photographic plates *P*, *P*, coelostat *C*, diagonal plane mirror *D*

were provided on the spectrograph, allowing a fairly rapid succession of exposures in the second order yellow at second contact, and in the second order ultra-violet at third contact.

In order to obtain thermal protection, the spectrograph was installed in a pit in the ground, the slit at ground level and the collimator axis vertical (Fig. 1), a procedure which had proved very successful in 1940, but which on this occasion gave almost more trouble than it was worth, owing to a less suitable subsoil, and a less favourable optical lay-out. The thermal stability was not specially good, and there was much trouble from an almost incessant rain of fine sand into the pit, some of which inevitably penetrated into the instrument, although this was completely boxed in, and gradually accumulated on the face of the collimator mirror. The grating, held face down, fortunately kept relatively clean.

Preceding the spectrograph was a 16-in. coelostat, rated to follow the Moon, reflecting light horizontally eastwards to a 12-in. concave mirror of 18 ft focal length. The light beam returned almost along itself towards the coelostat, near which it was intercepted by a small plane mirror and passed thence vertically downwards to the slit, which was approximately at ground level. The pit was orientated so that the slit when tangential to, and on the appropriate side of, the solar image, was about 7° in position angle from second or third contact, as the case might be, the cusps being 166° apart. One observer managed the coelostat, the image guiding and various other operations, including opening the mouth of the pit during totality, in order to rotate the grating and alter the spectrograph focus. I. Kendall, of the R.E.M.E. unit stationed at Fort Stanley, operated the camera handles at the bottom of the pit.

The scale of the solar image was 1 mm = 37″.6. The spectrograph slit width was 0·03 mm, and the dispersion 2·2 Å per mm in the yellow, 2·4 Å per mm in the ultra-violet. The plates, Ilford HP3 for the yellow and Ilford Zenith for the ultra-violet, were backed and were on specially thin glass, bent in the plateholders to a radius 4′ 5″. Optical parts were cleaned, and the last focus tests made, about 2 hr before totality. The collimation of the optical train was finally checked at the same time.

Eight exposures were made at second contact, covering the wavelength range 5550–6040 Å, and seven at third contact in the region 3430–4110 Å. Exposures were recorded on a chronograph; the details are in Table 1.

Table 1

5550–6040 Å	mid-exposure U.T. 9^h 9^m $7^s.9$	duration $1^s.42$
	10·1	1·38
	12·2	1·31
	14·4	1·45
	16·6	1·31
	18·8	1·48
	21·1	1·51
	23·4	1·21
3430–4110 Å	mid-exposure U.T. 9^h 12^m $18^s.7$	duration $1^s.65$
	21·0	1·65
	23·3	1·50
	25·8	1·76
	28·2	1·75
	30·9	1·87
	33·5	1·94

Second contact occurred at about 9^h 9^m 13^s. Third contact was marked by two rather conspicuous Baily's beads, the first of which began to appear about 9^h 12^m 23^s.

Each plate was calibrated with 1-sec.-exposures with the help of a Hilger rhodium on quartz step wedge whose transmission has been measured at the National Physical Laboratory. The optical arrangement was similar to that used on a previous occasion (Redman, 1941), with the Sun the source. The wedge exposures were repeated with the wedge reversed, in order to help reduce any trouble from irregular illumination. As a further precaution, with each eclipse plate a second plate from the same box was calibrated with both 1- and 3-sec. exposures, and given as near as possible identical treatment in processing. Each of the spare plates carried also a number of exposures made with a mercury 198 isotope lamp, with the normal 0·03 mm slit. Processing was carried out in total darkness, with Kodak D19b developer at 18°C, the plates being brushed, followed by a hardener stop-bath, and then a hardener fixer.

3. Comments on the Material

The exposures were generally successful and the definition in the spectra appears to be very satisfactory. The yellow plate was found to be cracked when removed from the plateholder. Subsequent tests showed that a small fault in the plateholder itself was responsible, but that the emulsion in all probability nevertheless remained in the focal plane. The crack is near one end of the spectrum and seems not to have affected the definition. Fortunately, the plate remained in one piece during processing.

The calibration exposures could with advantage have been a little heavier, although this point does not affect the present paper, which is restricted to results from weaker lines. The wedge spectra are badly streaked by sand and dust on the slit and on adjacent optical surfaces (wedge, filter, field lens, *etc.*). In the circumstances this trouble was practically unavoidable, and the result is that some of the calibration steps are badly defined. However, since exposures were available on two different plates, with wedge both erect and reversed, and with two slightly different exposure times covering rather different ranges of density, a quite good characteristic curve could be constructed. There is no detectable change of characteristic curve over the whole 700 Å covered by the ultra-violet plate. On the yellow plate a constant characteristic curve has been used, since the measures made so far cover only 100 Å.

The other point on which the material may be criticized concerns the mercury lamp exposures for the instrumental profile. In ideal conditions these would have been made within a few seconds of the eclipse, the spectrograph adjustments remaining untouched, but practical difficulties made this course impossible. Satisfactory ultra-violet exposures were made on the evening of the eclipse. The yellow exposures on that occasion were of poor quality and had to be repeated two evenings later, by which time the spectrograph temperature had changed appreciably and re-focussing had become necessary.

Nevertheless, the material has its strong points too. Compared with 1940 the spectrograph dispersion has been increased on the average by 50 per cent, and the resolution, set primarily by the photographic emulsion, is correspondingly greater. Two fairly widely separated wavelength regions are now available, allowing some test of the assumption that the line shapes are determined chiefly by Doppler broadening. Further, although the principle of the method remains unchanged, most of the practical details have been entirely altered, as follows:

	1940	1952
Instrument	three-prism spectrograph	concave grating
Dispersion	2·7–6·8 Å per mm	2·2–2·4 Å per mm
Region	4030–5000 Å	3430–4110 Å 5550–6040 Å
Plate	Zenith	Zenith, HP3
Calibration	glass wedge: transmission measured by author	rhodium step wedge: transmission measured by N.P.L.
Calibration source	sky	Sun
Instrumental profile source	krypton tube	mercury isotope lamp
Microphotometer	Casella, photographic recording	Cambridge, pen recording

We may add that, although on both occasions the sky transparency was excellent, in 1940 the definition of the solar image was very poor owing to scintillation, while in 1952 it was surprisingly good. No accurate record of the scintillation was possible on either occasion; the impression given by hurried inspection of the image on the slit was of a 10″ blur in 1940, and perhaps only 2″ in 1952. Shadow bands were very strong in 1940, but not seen in 1952.

4. Measures

At the time of writing work is still in progress on the shapes of the hydrogen and helium lines, and of strong lines appreciably distorted by self-absorption, where a fully satisfactory interpretation of the measures is not entirely simple. Here we shall confine attention to the narrowest lines of heavy elements, the principal questions in mind being these: are the 1940 measures substantially confirmed? do the spectrum regions agree with each other, supposing the line shapes to be determined by Doppler broadening? is there evidence of an increase of line of sight velocities with height?

Most of the measures reported now were made on the fourth exposure at second contact (yellow region), and on the third at third contact (ultra-violet). The yellow exposure shows no trace of Fraunhofer spectrum, although there is a fairly strong continuum, with numerous weak absorption lines due to water vapour. (In view of the comparatively high altitude of the Sun, 60°, and the very low humidity of Khartoum the conspicuousness of these lines was rather surprising, but naturally they are more readily seen in the absence of Fraunhofer lines.) Chromospheric emission lines in this region are mostly due to neutral metals and are rather weak and widely separated. The ultra-violet spectrogram (Fig. 2) shows a slight trace of Fraunhofer spectrum in a narrow strip due to the stronger of the Baily's beads mentioned earlier, but for the rest gives a good sample of the spectrum of the low chromosphere. At the lowest levels the spectrum is very crowded, so that many lines are heavily blended. Despite the blends, and despite the smaller scale in kilometres per second per millimetre in the ultra-violet, the yellow region gives no real superiority, for the grain of the Zenith plate is finer than that of the HP3 plate, and it also gives sharper images because the opacity of the Zenith emulsion to ultra-violet light is greater than that of the HP3 emulsion to orange light.

Microphotometer tracings were made at a magnification 100 : 1, with the same slit adjustments for both chromosphere and mercury isotope spectra. In the ultra-violet the instrumental profile was determined separately from the lines near 3650 Å and 4046 Å respectively. The definition is slightly better at 4046 Å. For the yellow lines 5770 Å and 5791 Å were used. The yellow profile was measured over a considerably shorter intensity range than the others, but the figures appear adequate for the purpose in hand. The inferior resolution of the panchromatic emulsion is very clearly shown in Table 2, page 720.

The mercury lamp exposures on which the instrumental profile is based were made through the wedge, with the idea that they should be self-calibrating. It was found difficult, however, to construct a good characteristic curve from them, so the measures were reduced with the calibration used for the chromosphere. This is likely to lead to systematic error, for the duration of the shortest Hg lamp exposure was 1 min and of the longest calibration exposure 3 sec. Although inadequate for a full characteristic curve, the self-calibration was able to provide a general check of the instrumental profile, and this suggested that the ultra-violet intensity scale in Table 2 should be reduced by about 15 per cent. It is difficult to say how reliable this correction is, but we shall see later that our results are little changed by it. The error in

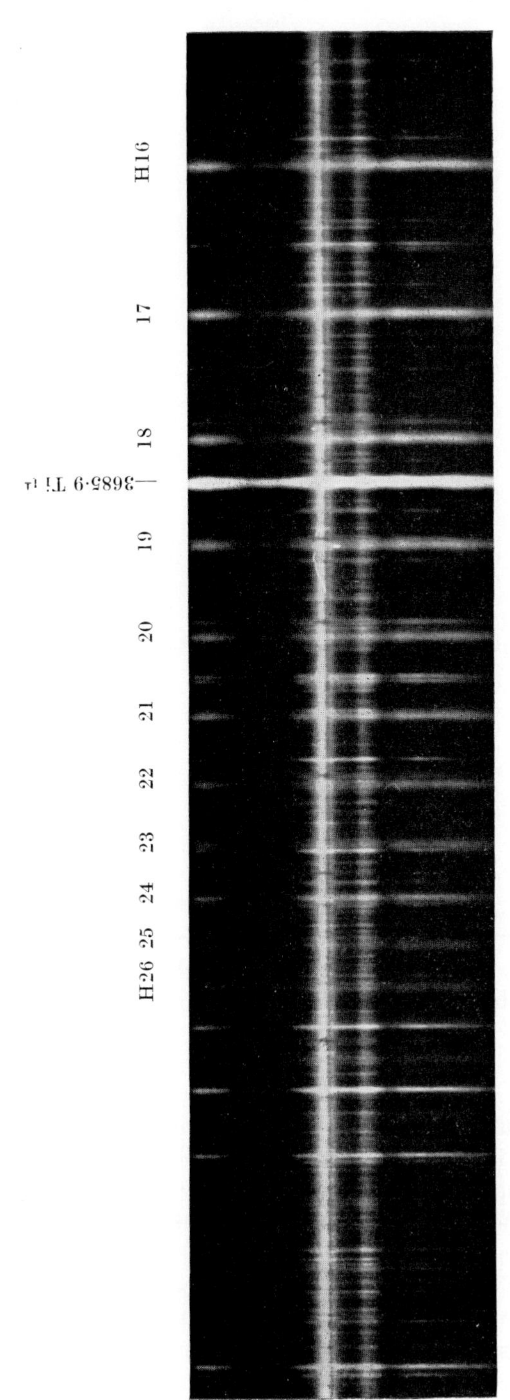

Fig. 2. A section of the chromospheric spectrum near to the head of the Balmer series. Note the comparatively wide hydrogen lines. Irregularity of illumination along the slit is caused in three ways: (1) the place of observation being 8 km from the centre line, the slit was tangential to the solar image 7° from the point of third contact; (2) irregularities of the Moon's limb caused two conspicuous Baily's beads, of which the first traces are clearly seen here; (3) sand and dust on the slit could not be entirely avoided

Table 2. Instrumental profile

Position	Intensities		
	4046 Å	3650 Å	5780 Å
− 0·08 mm	0·0034	0·0081	—
0·075	0·0039	0·0096	—
0·07	0·0046	0·011	—
0·065	0·0055	0·014	—
0·06	0·0068	0·017	—
0·055	0·0087	0·023	—
0·05	0·012	0·032	—
0·045	0·017	0·046	0·17
0·04	0·025	0·078	0·22
0·035	0·039	0·13	0·29
0·03	0·072	0·21	0·36
0·025	0·132	0·33	0·47
0·02	0·26	0·48	0·58
0·015	0·45	0·63	0·72
0·01	0·68	0·79	0·85
0·0075	0·81	0·88	—
0·005	0·92	0·96	0·96
− 0·0025	0·98	0·99	—
0·000	1·00	1·00	1·00
+ 0·0025	0·98	0·94	—
0·005	0·92	0·87	0·97
0·0075	0·78	0·76	—
0·01	0·65	0·63	0·89
0·015	0·42	0·40	0·79
0·02	0·25	0·25	0·69
0·025	0·14	0·14	0·59
0·03	0·091	0·089	0·51
0·035	0·063	0·055	0·43
0·04	0·047	0·036	0·36
0·045	0·037	0·025	0·31
0·05	0·030	0·018	0·26
0·055	0·024	0·014	0·21
0·06	0·020	0·011	—
0·065	0·016	0·0098	—
0·07	0·013	0·0083	—
0·075	0·011	0·0072	—
+ 0·08	0·0098	—	—

any case affects only the correction for instrumental broadening and not the actual measures on the chromospheric spectrum. No check of this kind was possible on the yellow instrumental profile.

5. Heights

In the yellow all the measures quoted in this paper are from one exposure only, that at U.T. $9^{h}\,9^{m}\,14^{s}.4$. There is at present no means of determining closely the effective height at which this was made, but since the preceding exposure contains Fraunhofer spectrum the lowest chromosphere visible at the beginning of the measured exposure could not have been above about 550 km. The strength of the continuous spectrum suggests that the effective height may have been considerably less than this.

In the ultra-violet the exposure chiefly used, that at U.T. $9^{h}\,12^{m}\,23^{s}.3$, was measured along three different strips, whose heights in the first place were simply computed from the corresponding positions on the spectrograph slit, supposing the strip from the brighter Baily's bead to be at zero height. Dr. Z. Suemoto questioned this in a personal discussion, pointing out that the surface brightness gradient in the chromosphere is steep, so that scintillation, scattering by instrument or sky, or poor guiding, all very easily bring serious quantities of spurious low level light into supposedly high levels. Further experience has confirmed that there is indeed trouble of this kind, so that direct computations of heights from distances along the spectrograph slit is practically certain to be misleading.*

If direct geometry is ruled out, one must try to obtain reasonably accurate effective heights directly or indirectly from the progressive motion of the Moon across the chromosphere. Owing to the comparatively slow sequence of exposures, and in part also because with a slit spectrograph one is at the mercy of the seeing and guiding, the amount of information on heights which can be obtained directly from intercomparison of successive images on each plate is rather limited. More information is obtainable from relative line intensities, comparing them with the intensities measured on slitless spectrograms. This has been possible through the kind co-operation of Dr. J. HOUTGAST, who has made available in advance of publication relevant line intensity material obtained by himself and Mr. C. ZWAAN at Khartoum. The Utrecht and Cambridge spectra overlap in the neighbourhood of λ4000 Å, and a comparison of intensities has been made for the lines 4034·5, 4033·1, 4030·8, Mn I; 4028·4, Ti II; 4026, He I.

The intercomparison unfortunately is not without its own difficulties. The slitless measures refer to second contact, whereas we wish to determine heights in third contact spectra. We must assume that the chromospheric spectrum has the same kind of uniformity as the Fraunhofer spectrum, *i.e.* that it has practically constant characteristics, apart from exceptional areas of disturbance near active sunspots, etc. which as far as is known are not involved here. A more difficult point is that the slitless spectra refer to the whole visible chromosphere, integrated from the limb of the Moon outwards; to obtain the spectrum at discrete heights the measures must be differenced in some way. Now the structure, as well as intensity, of most of the stronger chromospheric lines changes greatly with height in the chromosphere, chiefly owing to self-reversal at low levels. To sort out these changes in full detail would require more measurements than are yet available, and even were this done one is still faced with some unknown blurring of heights in the slit spectra. Perhaps for these reasons the slit and slitless line intensities, in so far as they are yet available, do not agree very closely, but effective heights obtained by intercomparing them are the best that can be obtained at present. Data for three heights are given in Table 3

Table 3. Observed line ratios (slit observations) and estimated heights

Line ratio	Height 1	Height 2	Height 3
4034/4028 . .	0·89	0·78	2·37
4033/4028 . .	0·97	0·84	2·56
4030/4028 . .	—	0·90	2·92
4026/4028 . .	0·45	0·36	1·33
Estimated height .	< 50 km	< 100 km	≈ 600 km

(all from the exposure at U.T. 9^{h} 12^{m} $23^{s}.3$). It may be noted that height 2 is certainly greater than height 1 from the general appearance of the spectra.

* Preliminary results (REDMAN, 1952b) mention a height 4000 km derived in this way. This height is identical with height 3 in Table 3, now believed to be about 600 km.

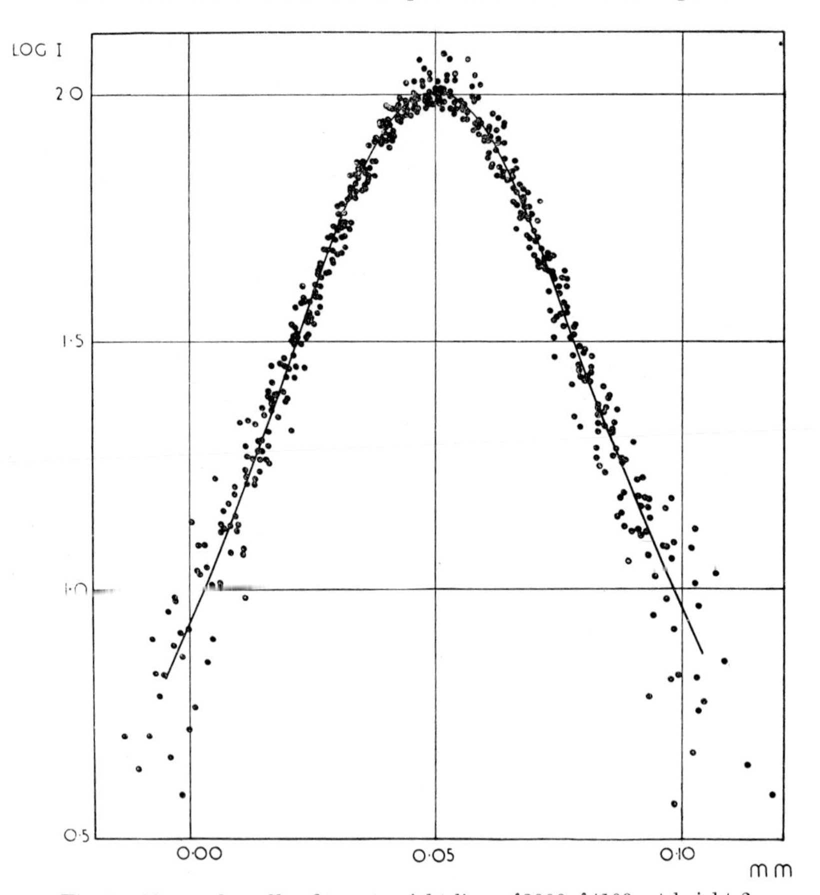

Fig. 3. Observed profile of twenty-eight lines, $\lambda3900$–$\lambda4109$, at height 2

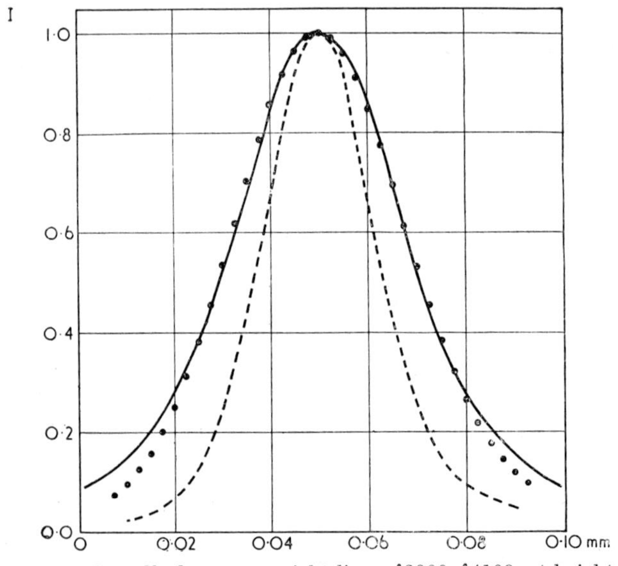

Fig. 4. Mean measured profile for twenty-eight lines, $\lambda3900$–$\lambda4109$, at height 2, compared with points computed for $\xi_0 = 3{\cdot}2$ km per sec. The broken curve is the instrumental profile for $\lambda4046$. Dispersion: 1 mm = 2·4 Å

6. Correction for Instrumental Broadening

Lines in the ultra-violet have been divided into two groups, with wavelengths respectively above and below 3800 Å. Those above have been corrected with the instrumental profile from 4046 Å, those below from 3660 Å. It is sufficiently accurate to assume all profiles symmetrical and to correct the measured lines by the method of the Voigt functions (van de Hulst and Reesinck, 1947). The results in several cases have been confirmed by numerical integration from the actual instrumental profile, with no assumption of symmetry. As an example, Fig. 3 shows the observed profile from twenty-eight lines measured at height 2 in the region 3900–4109 Å, the ordinates being log (intensity). The same profile is given on an intensity scale in Fig. 4, together with the instrumental profile and points computed by direct numerical integration for $\xi_0 = 3{\cdot}2$ km per sec., the velocity given by the van de Hulst method. The true line has been assumed to have a Gaussian profile due effectively to line of sight velocities only. In general lines below and above 3800 Å respectively have given closely agreeing velocities.

7. Results

A summary of results for narrow lines is given in Table 4. Other data will be given in a later publication, when a more complete examination has been made of other causes of line broadening, especially self-reversal and (in hydrogen) the Stark effect.

Table 4. Average line of sight velocities from narrow lines

Region	Height	ξ_0	Number of lines	Mean atomic weight
Ultra-violet .	$<$ 50 km	2·6 km per sec.	38	127
	$<$ 100	3·2	54	125
	$\approx$ 600	3·8	27	63
Yellow . .	$<$ 500	2·2	18	53
1940-violet .	$<$ 500	3·1	30	114

The first point to notice is that the results of 1940 are satisfactorily confirmed. Further, although they came from opposite sides of the Sun, the yellow and ultra-violet spectra agree in giving about the same low line of sight velocity. As already explained, we have some reason to suspect a 15 per cent systematic error in the corrections for instrumental distortion in the ultra-violet. Were such an error accepted and removed from the results, the ultra-violet velocities would be reduced by about 10 per cent, which would bring the value for the lowest chromospheric level into almost exact agreement with the yellow velocity.

Measures of hydrogen lines, not yet ready for publication, show that the kinetic temperature of the chromosphere probably does not exceed 15,000°K. At the other extreme, the lowest temperature other authors have in recent years seriously considered for the chromosphere is about 4500°K. With an atomic weight 120 the mean square line of sight velocity in the absence of turbulence would be 0·6 km per sec. at $T = 4500°$ or 1·0 km per sec. at $T = 15{,}000°$. If we take the resultant mean velocity for temperature plus turbulence to be 1·9 km per sec., turbulence alone would amount to 1·8 km per sec. at 4500°, 1·6 km per sec. at 15,000°. We conclude that in any case the horizontal component of turbulence at the bottom of the chromosphere is small.

There is a small increase of line of sight velocity with height. It is impossible to make much comment on this here, except to say that the work still in progress suggests that this increase may be real. The evidence as given here is rather weak on this point. As height increases the lines available for measurement become rapidly fewer, and for our purposes, less reliable, because those which survive in measurable

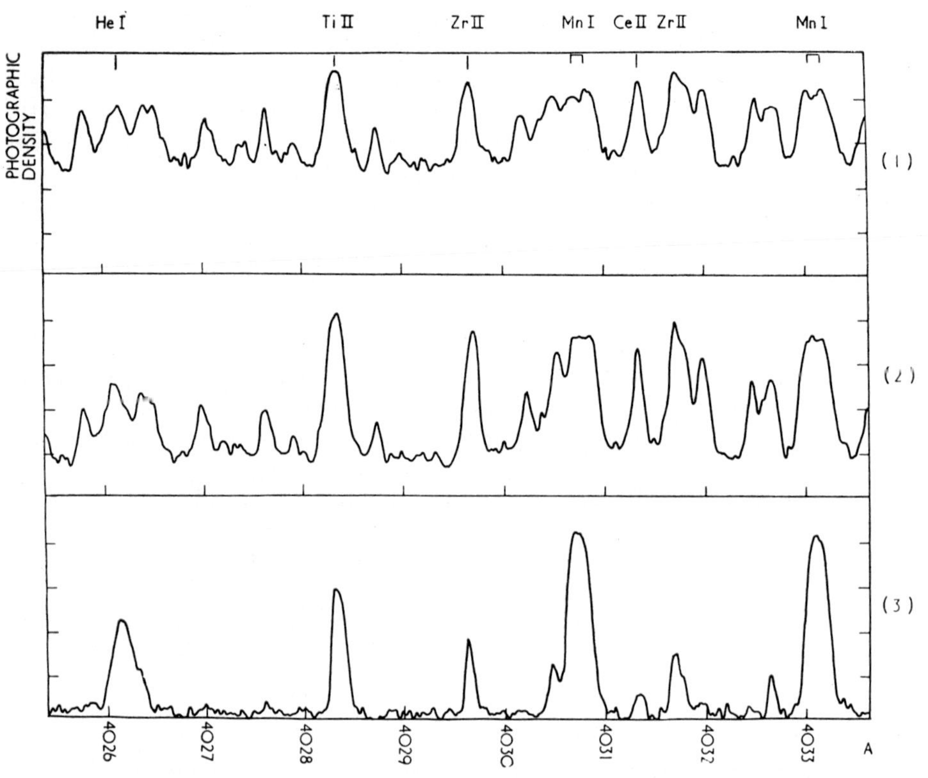

Fig. 5. Reproduction of microphotometer tracings (photographic density only) of the chromospheric spectrum near 4030 Å at heights 1 (< 50 km), 2 (< 100 km), and 3 (≈ 600 km)

strength are precisely those which at low levels show the greatest distortion by self-reversal.

At the same time as these observations were being made, Brück and Jackson (1953) observed the chromospheric line widths with an interferometer. They failed to find interference fringes on any of their lines, although practically simultaneous exposures with a helium discharge tube showed the interferometer to be in good adjustment. They concluded that the line widths could not have been less than 1·2 cm^{-1} for the unblended lines, or 0·5 cm^{-1} for blended lines. At first sight this might appear to contradict the results of this paper. The explanation is that owing to their use of a comparatively small and slow spectrograph, only the strongest chromospheric lines registered on their plate. Nearly half the lines in their Table I are of H, He, or Ca II, all of which are known to be wide lines. Six more lines in their list are on the Cambridge spectrograms. Of these, 3441·4, 3462·6, 3488·0, 3580·0, and 3573·5 are each blends of two to four lines, mostly rather strong and of comparable strength, and at least one of each group is strongly self-reversed in the low chromosphere. The other line, 3685·9, Ti II, is not appreciably affected by blends, but shows heavy self-reversal at heights 1 and 2 on our photographs; at height 3 (600 km), it still has a half-width about 0·5 Å; on an earlier exposure, whose effective height from line intensities may be put tentatively at 1200 km, the half width is 0·35 Å; on yet an earlier exposure, at a still greater height, the half-width is about 0·28 Å. These widths are uncorrected for instrumental broadening, but the correction for such a wide line would be comparatively small. BRÜCK and JACKSON's lower limit for the width of an unblended line at this wavelength is 0·16 Å. This agrees with our results, but tells us only that at least at these levels 3685·9 is an unsuitable line, if our purpose is to study line of sight velocities. The importance of BRÜCK and JACKSON's experiment is that they demonstrated that the interferometer method can be made to work at an eclipse, even under quite unfavourable climatic conditions, and it is to be hoped that at a future eclipse they may be able to record fainter and narrower lines, whose widths are not seriously affected by self-reversal. In that case it should be possible to get results of very great value.

The example of 3685·9, Ti II, which has just been quoted, illustrates rather well the difficulty presented by strong chromospheric lines. Heavily distorted by self-reversal at low levels, the line width *decreases* at first with height, as this self-reversal diminishes, although from the evidence of weaker lines it seems quite certain that there is if anything an *increase* of Doppler width. This rather typical behaviour can also be seen with the Mn I lines in Fig. 5, which reproduces microphotometer tracings at heights 1, 2, and 3 of a short range of spectrum near 4030 Å. What is needed is some test which will show at what height it becomes safe to neglect self-reversal and to take the line profile as a reliable guide to the distribution of velocities in the line of sight.

8. ACKNOWLEDGEMENTS

This work was carried out with the aid of a grant from the Joint Permanent Eclipse Committee, whose Secretary, Professor STRATTON, did a great deal to help the Cambridge party, especially in the earlier arrangements for the expedition. At Khartoum the spectrograph would never have been brought into good working adjustment by the day of the eclipse without much patient aid from Dr. D. E. BLACKWELL. Photographic processing was carried out in an excellent temporary darkroom constructed by Dr. and Mrs. H. VON KLÜBER. Members of the R.E.M.E. unit at Fort Stanley, of whom I. KENDALL should be specially mentioned, gave essential help both before and during the eclipse. The spectrograph was constructed in the Cambridge Observatories by Mr. R. S. OVERHILL. The writer wishes to express his grateful thanks to all these and to others who assisted in making the Cambridge 1952 Eclipse Expedition a success.

REFERENCES

BABCOCK, H. D. . . . . 1954 Private letter; also *Observatory*, January.
BRÜCK, H. A. and JACKSON, D. A. . . . . . . 1953 *Proc. Roy. Soc.*, **216**, 183.
MIYAMOTO, S. . . . . 1951 *Publ. Astron. Soc. Japan*, **3**, 67.
REDMAN, R. O. . . . . 1941 *M.N.*, **101**, 266.
1942 *M.N.*, **102**, 140.
1952a *Nature (London)*, **169**, 686.
1952b *Atti dell'XI Convegno Volta, Firenze*, p. 72.
SUEMOTO, Z. . . . . 1951 *Publ. Astron. Soc. Japan*, **3**, 110.
UNSÖLD, A. . . . . 1952 *Z. Naturforsch.*, **7**, 121.
VAN DE HULST, H. C. and REESINCK, J. J. M. . 1947 *Ap. J.*, **106**, 121.

The Excitation Temperature of the Solar Chromosphere Determined from Molecular Spectra

D. E. BLACKWELL

The Observatories, Cambridge

SUMMARY

The excitation temperature of the solar chromosphere is derived from a study of the rotational structure of the 4216 Å system of CN and the P- and R-branches of the (0, 0) band of the $^2\Delta - {}^2\Pi$ system of CH. The temperatures are 5400°, 7400°, and 5100°.

---

1. INTRODUCTION

DURING recent years the temperature of the chromosphere has been measured by several different methods with apparently discordant results. Observation of line profiles by REDMAN (1942b) gives a kinetic temperature of 30,000° at a height of 1500 km* with a line of sight turbulent velocity of less than 1 km per sec. This is in good agreement with the temperature of 35,000° deduced by WILDT (1947) from the density gradient of hydrogen. Values for the electron temperature are lower than these. An upper limit obtained by KOELBLOED and VELTMAN (1951) using ZANSTRA'S method lies between 10,000° and 16,000° for a height of 400 km. MIYAMOTO and KAWAGUCHI (1949) show that the ionization of the metals is compatible with a temperature of less than 10,000°, and from the curve of growth of the Ca II K line they deduce, with considerable uncertainty, the value of 6000°. WURM (1948), from the non-appearance of forbidden lines, concludes that the temperature is less than 10,000°. Excitation temperatures are even lower. GOLDBERG (1939) finds the excitation temperature of He I to range from 4300° at 760 km to 6700° at 2330 km. WILDT (1947) states that the excitation temperature for Fe I and Ti II lies between 3000° and 4000° at a height of about 1000 km. No direct determinations of the temperature of the lower chromosphere have been made by radio methods, but HAGEN (1951) surmises that the temperature is less than 10,000° at all heights up to 8000 km. Our understanding of the structure of the solar chromosphere is far from complete while these anomalies remain unexplained, but while they exist it is of interest to seek other methods of determining the chromosphere temperature.

Now that high dispersion chromosphere spectra are available it is possible to analyse the intensity distribution in the rotational structure of molecular bands and derive another value for the temperature. Supposing that the distribution of rotational states corresponds to thermal equilibrium at a temperature T, the distribution of intensity among the rotational lines is expressed by the relation

$$I = C\nu^4 i e^{-F(J)/kT},$$

where C is a factor of proportionality, ν the frequency of the rotational transition, and i is the intensity factor which depends upon the type of transition and upon the quantum numbers of the initial and final states. In most applications the factor ν^4

* The measurements on hydrogen and helium lines were made at this height to avoid self-reversal, but measurements on other lines were made at a lower level.

is omitted because often under laboratory conditions a band covers only a small frequency range, but in the solar atmosphere the excitation temperature is high and the band more diffuse, so that it becomes necessary to take account of this factor. The quantity $F(J)$ is the term value of the upper state which can be calculated from the relation

$$F(J) = B\nu J(J+1) - D\nu J^2(J+1)^2 + \ldots,$$

where the coefficients $B\nu$ and $D\nu$ are calculated using lines of low J value.

A plot of log $(I/\nu^4 i)$ against $F(J)$ should give a straight line from the slope of which a value for T can be derived. This is the only reliable method of treating the observations for it shows at once whether there is any deviation from thermodynamic equilibrium. If there is appreciable deviation the value for the temperature becomes meaningless except as a parameter giving the relative populations of different states over a restricted range of term values. If, as is often done, the J value of the line of maximum intensity is noted, or the J values of lines that have equal intensities and lie on opposite sides of a maximum, a value for the temperature can be deduced very easily but at the expense of uncertainty.

The choice of suitable bands for reliable determinations of temperature is rather restricted. The bands must not be too intense as otherwise there is an uncertain correction for self absorption. If a correction is not applied the rotational structure appears to be too diffuse and the deduced temperature therefore too high. On the other hand, the bands must not be too weak for then a large correction for blending with other lines has to be applied, and the true position of the background is uncertain. Under these conditions also it is never possible to be sure that the molecular line is not apparently strengthened by an unseen weak line. Difficulties arising from blending are reduced by avoiding the more complex molecular bands and using a high dispersion spectrum; indeed, the complexity of the chromospheric spectrum is such that a dispersion of at least 4 Å per mm is needed.

Two bands have been chosen for this investigation. One is the (0, 1) band of the $^2\Sigma^+ - {}^2\Sigma^+$ transition in the CN radical with a head at 4216 Å. This band is of simple structure but rather weak. Also investigated is the (0, 0) band of the $^2\Delta - {}^2\Pi$ transition in the CH radical near 4300 Å, a weak and complex band.

2. Observational Data

The spectra used were those photographed by REDMAN (1942a) at the total solar eclipse of 1st October, 1940. The spectrograph had three glass prisms of 3 in. aperture used with a lens of 8 ft focal length, giving a dispersion of 3·6 Å per mm at Hγ. The region covered was $\lambda\lambda$4030–5000 Å and the effective height in the chromosphere 1500 km. It is very difficult to compare the results of two methods of temperature measurement when the observations are made at different eclipses, for there is always the possibility that the observations refer to different heights, or that one set of observations refers to a "hot spot" or even that the chromosphere has changed between the two times of observation. These spectra therefore were of particular interest for the molecular work because they had already been used to derive the kinetic temperature of the chromosphere.

3. Use of the CN Spectrum

Of the CN spectrum, the sequence $\Delta\nu = 0$, including the 3883 Å band, is ideal for the method, but unfortunately it is outside the range of spectra available. The only other useful band is that at 4216 Å and this is used extensively. Both the upper and lower electronic states conform to HUND's coupling case (b), so that the only quantum numbers needed are K, which is integral, and $J = K \pm \frac{1}{2}$. There is a P- and an R-branch but no Q-branch because the transition $\Delta K = 0$ is forbidden. Each rotational level is a spin doublet and hence each line in the band has three components. The lines appear to be doublets rather than triplets, however, because one of the three lines is very faint. The doublet separation increases with increasing K, being unobservable at first and increasing to 0·29 Å at $m = +97$. A valuable series of measurements and identifications has been made for this band by HEURLINGER (1918): in his tables the lines A_1 are those of the R-branch and that part of the P-branch which lies between the band origin and the band head, the lines A_2 are the remaining members of the P-branch. HEURLINGER lists 174 lines, many of which are doublets. A large number of these lines may be identified in the chromospheric spectrum, but only a few are useful in an analysis of this kind. The procedure has been to note all the lines which are relatively unblended and suitable for measurement and then to divide these into long and short wavelength components of doublets in P- and R-branches. When this was done it was found that the only long series of suitable lines was formed by seven short wavelength components of P branch doublets. Identifications of these lines and measurements of their intensity, I, on an arbitrary scale are given in Table 1; for all lines an allowance has been made for blending.

Table 1. *The* 4216 Å *system of* CN

Wavelength (HEURLINGER)	m (HEURLINGER)		Intensity factor	Intensity	Term value of upper state
4213·273 Å	− 9	A_1	8·46 9·46	45/2 = 23	135 cm^{-1}
4214·236	− 33	A_2	33·6	49	1980
4212·215	− 38	A_2	38·5	46	2630
4211·192	− 40	A_2	40·5	57	2920
4210·048	− 42	A_2	42·5	60	3220
4199·675	− 55	A_2	55·5	33	5550
4194·243	− 60	A_2	60·5	34	6610

The first line of this table is an unresolved doublet of which the components should have nearly equal intensities. The intensity of this line has been halved, therefore, to find the intensity of the short wavelength component. Values of the intensity factor have been calculated by MULLIKEN (1927a, 1927b) and in this example,

$$i = \frac{2K(K+1)}{2K+1},$$

where K is the larger of the numbers K' and K''. Fig. 1 shows a plot of $\log_e I/\nu^4 i$ against the term value of the upper state. As the points all lie close to a straight line it is concluded that there is a state of equilibrium. The slope of the line corresponds to a temperature of 5500°.

4. USE OF THE CH SPECTRUM

The only CH lines of appreciable intensity in the chromosphere spectrum are those of the (0, 0) band of the $^2\Delta - {}^2\Pi$ transition near 4300 Å. Laboratory analyses have been made by FAGERHOLM (1941) and GERÖ (1941), and NICOLET (1945) has

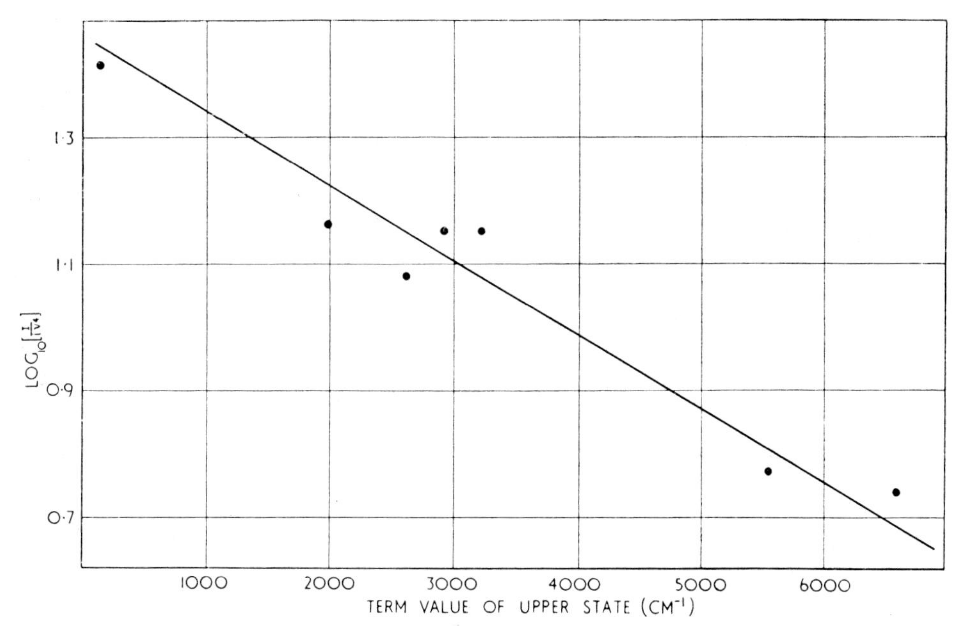

Fig. 1. Intensities of lines of (0, 1) band of the $^2\Sigma^+ - {}^2\Sigma^+$ transition in CN

made a study of the occurrence of the bands in the Fraunhofer spectrum. It is unfortunate that the bands of this system are exceedingly complex, there being twenty branches, but only twelve of these branches are sufficiently intense to be of importance. In this intensity analysis no distinction has been made between the four sub-members of each main P-, Q-, and R-branch, for otherwise there would be insufficient data to form a graph. Many lines of this system can be identified in the chromospheric spectrum, but there are only thirteen reasonably unblended P-branch lines and nine R-branch lines. Data on these lines are summarized in Tables 2 and 3, the values for the intensity factors being those given by MULLIKEN:

$$i_{P_1} = \frac{(K+1)(K-2)(K-1)}{K(2K+1)},$$

$$i_{P_2} = \frac{(K-1)^2(K-2)}{K(2K-1)},$$

$$i_{R_1} = \frac{(K+1)^2(K+2)}{K(2K+1)},$$

$$i_{R_2} = \frac{(K-1)(K+2)(K+1)}{K(2K-1)}.$$

As would be expected in a spectrum of this complexity there are occasional chance coincidences between lines of different branches. Where this occurs it is noted in the tables and the intensity factors for the two lines added. There are also coincidences with lines of the (1, 1) and (2, 2) bands, but the effect of these coincidences has been neglected because the interfering band systems are very weak. In the calculation of the intensity factors no account has been taken of Δ-type doubling.

Table 2. P-branch of (0, 0) *band of* $^2\Delta - {}^2\Pi$ *system of* CH

Wavelength (GERÖ)	K	Branch	Intensity factor	Intensity
4334·662 Å	4	P_1	0·83	31
4334·004	4	P_2	0·64	30
4347·965	7	P_1 P_2	4·26	116
4352·562	8	P_1	2·78	70
4363·322	11	P_1	4·26	82
4374·237	14	P_1 P_2	11·1	110
4377·258	15	P_1 P_2	12·1	86
4389·664	20	P_1	8·76	68
4388·899	21	P_1 P_2	18·1	100
4396·979	24	P_1	10·8	58
4398·633	28	P_2	12·3	60
4412·911	34	P_1	15·8	< 20

Table 3. R-branch of (0, 0) *band of* $^2\Delta - {}^2\Pi$ *system of* CH

Wavelength (GERÖ)	K	Branch	Intensity factor	Intensity
4286·077 Å	4	R_1	4·59	72
4273·791	6	R_2	4·75	65
4267·763	7	R_1	5·95	72
4267·389	7	R_2	5·25	58
4255·000	9	R_2	6·25	50
4172·517	23	R_1	13·8	27
4164·276	24	R_1	14·3	20
4159·179	25	R_1	14·8	20

A plot of log $(I/\nu^4 i)$ against the term value of the upper state is given in Figs. 2 and 3 for the P- and R-branches. The slopes of the lines correspond to the temperatures 7400° for the P-branch and 5100° for the R-branch.

5. Conclusions

Summarizing, the excitation temperatures are:

(*a*) For the P-branch of the (0, 1) band of the $^2\Sigma^+ - {}^2\Sigma^+$ transition in CN . . . . 5500°

(*b*) For the P-branch of the (0, 0) band of the $^2\Delta - {}^2\Pi$ transition in CH . . . . 7400°

(*c*) For the R-branch of the (0, 0) band of the $^2\Delta - {}^2\Pi$ transition in CH . . . . 5100°

(*b*) and (*c*) together: 6300°

It is clear from the scatter of points about the straight lines, and from the difference in the temperature found from the P- and R-branches of the CH spectrum that the method, when applied to these observations, is not accurate. It is, however, striking that the derived temperature differs greatly from the kinetic temperature of 30,000° deduced by REDMAN, and is much lower than all measurements of the electron temperature, but is near to the effective temperature of photospheric radiation. This makes it likely that the molecular bands are excited by radiation and not by collision. If this is true we should expect the intensities of individual lines to show a

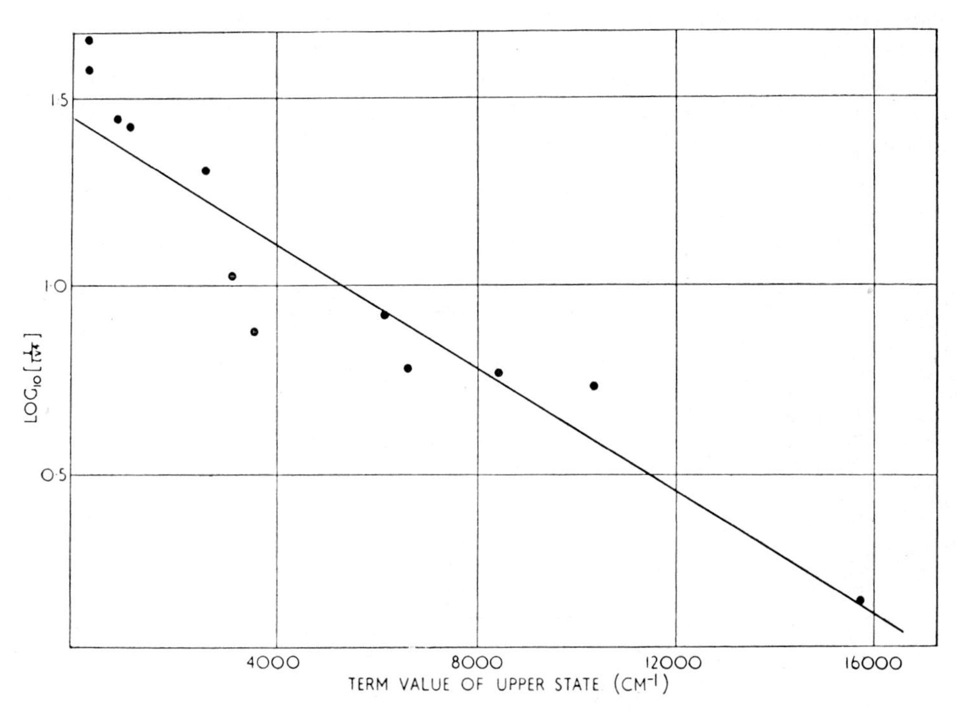

Fig. 2. Intensities of lines of (0, 0) band of the $^2\Delta - ^2\Pi$ system of CH (P-branch)

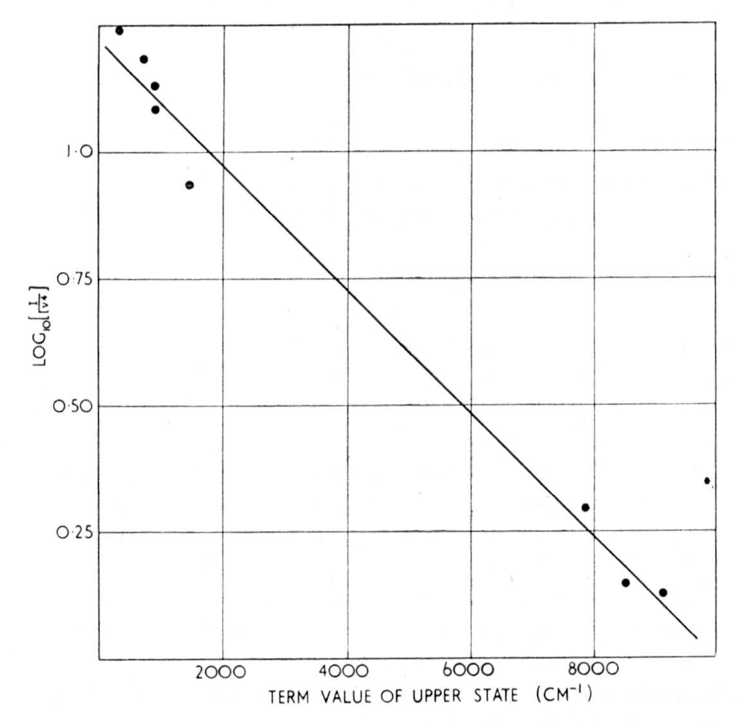

Fig. 3. Intensities of lines of (0, 0) band of the $^2\Delta - ^2\Pi$ system of CH (R-branch)

scatter which would be correlated with the Fraunhofer spectrum. No such correlation is evident, but it might easily be hidden by errors of measurement.

The author is most indebted to Prof. R. O. REDMAN for the use of microphotometer tracings of his eclipse spectra, and to Mrs. CHARLOTTE MOORE-SITTERLY for making available parts of HEURLINGER'S Dissertation.

REFERENCES

FAGERHOLM, E.	1941	*Ark. Mat. Astron. Fys.*, **27**, No. 19.
GERÖ, L.	1941	*Z. Phys.*, **118**, 87.
GOLDBERG, L.	1939	*Ap. J.*, **89**, 673.
HAGEN, J. P.	1951	*Ap. J.*, **113**, 547.
HEURLINGER, J.	1918	Dissertation (Lund).
KOELBLOED, D. and VELTMAN, W.	1951	*Proc. K. Ned. Akad. Wetensch.*, **54**, 468.
MIYAMOTO, S. and KAWAGUCHI, I.	1949	*Publ. Astron. Soc. Japan*, **1**, 114.
MULLIKEN, R.	1927a	*Phys. Rev.*, **30**, 138.
	1927b	*Phys. Rev.*, **30**, 785.
NICOLET, M.	1945	*Misc. Inst. Roy. Meteorol. Belgique*, No. 20.
REDMAN, R. O.	1942a	*M.N.*, **102**, 134.
	1942b	*M.N.*, **102**, 140.
WILDT, R.	1947	*Ap. J.*, **105**, 36.
WURM, K.	1948	*Z. Astrophys.*, **25**, 109.

Solar Work and Plans in Egypt

M. R. MADWAR
Helwan Observatory, Egypt

A PICTURE of present and future astronomical activity in Egypt would hardly be complete without a reference, however slight, to the background of Ancient Egyptian and Arab astronomy in the Nile Valley.

The Ancient Egyptian priests were practical astronomers, as is evident from their astronomical observations, amongst which the heliacal rising of Sirius (Sothis) may be mentioned. In this way they laid the foundation of a solar calendar. Their religious beliefs were closely related to the heavenly bodies; the Sun was the Supreme God "Ra", the stars and planets were divinities. In almost every monument and temple there exists imperishable evidence of high culture in astronomy and its practical application. The Great Pyramids were designed and built by these people, and they were among the earliest geometers. The orientation of the Pyramids with a high degree of accuracy is a testimony to the insight and depth of thought of the Egyptian astronomers as founders and pioneers of practical astronomy. For a long time, it was customary for Ancient Greek philosophers, like THALES, PYTHAGORAS, PLATO, to travel in Egypt, sitting in the temples at the feet of those masters of ancient scientific knowledge, which knowledge they are believed to have jealously and secretly kept to themselves although they permitted its results and practical application to be spread over the then-known world.

Fig. 1. Solar corona, 25th February, 1952. Photographed by M. R. MADWAR, of the Egyptian Eclipse Expedition at Khartoum, with the Worthington camera. Exposure time 75 sec.; North, *etc.* (*i.e.* N, W, S, E), refer to the axis of solar rotation

In the second century A.D. there lived at Alexandria the great astronomer CLAUDIUS PTOLOMAEUS, known as PTOLEMY, whose treatise on astronomy, translated by the Arab astronomers, became known to us as the *Almagest*. It is considered as the foundation stone of mediaeval astronomy. The first astronomical observatory was founded at Alexandria and remained active for 400 years.

During the seventh century Islamic astronomy and science in general was prevalent in the whole Middle East; the first Arab observatory in Egypt was that of IBN YUNIS (940–1008 A.D.) on the Mokattam Hills in Cairo. It was in this epoch that the Hakemites tables of the Sun, Moon, and planets were constructed.

During the early part of the nineteenth century an observatory was founded at Boulak; it was transferred to Abbassia in 1865, and later to Helwan in 1903. The past activity of Helwan Observatory has been fully described elsewhere, but it may be well here to recapitulate briefly some of its achievements during the first half of this century.

Three main sections of astronomy are actively pursued at Helwan; namely the determination of time by stellar observations, the study of bright and large nebulae, and positional astronomy.

A Brunner transit circle was in use until 1930; transit observations necessitated the determination of the various associated errors and it was found that there was a large discrepancy amounting to $0^{s}.13$ in a transit observation when the instrument was reversed on its bearings. In 1930 a Troughton and Simms transit circle, purchased from the Royal Observatory Greenwich, was erected on the same pillar. This instrument has since been used for the determination of the clock error, and was in service during the international longitude operations of 1933.

The presentation of a 30-in. reflector by the late J. H. REYNOLDS in 1903 to the Egyptian Government marks the beginning of astrophysical research activity.* The first important work which this telescope produced was the fine series of photographs of Halley's Comet 1910, taken by Dr. H. KNOX-SHAW, which are still recognized as the first and most complete photographic record of this historic comet. The principal work performed with the telescope, however, was the photography of bright and large nebulae in the zone 0°–40° South of the equator. This zone had not been explored before with such a large instrument, and it provided precious data of the statistical distribution of external galaxies.

The study of the variable nebula in Corona Australis, the position of the eighth satellite of Jupiter, and the planet Pluto were among the outstanding of the other observations carried out by means of the Reynolds reflector.

On 5th May, 1949, an order was given to Grubb Parsons, Newcastle-upon-Tyne, for the construction of a 74-in. telescope similar in design to the one ordered by the Canberra Observatory. It may be recalled that this telescope is one of a series of three (for Australia, France, and Egypt). It has already been pointed out that the Helwan reflector has been used mainly for direct photography of nebulae, comets, *etc.* The physical study of heavenly bodies necessitates spectroscopic and spectrophotometric observations, and in order to reach faint stars a moderately large aperture is required. The large aperture reflectors in the world ranging from 74 in. to 100 in. are able to reach to about the same depth in the universe, whereas the 200 in. telescope at

* *Helwan Observatory Bulletins*, Nos. 1–45.

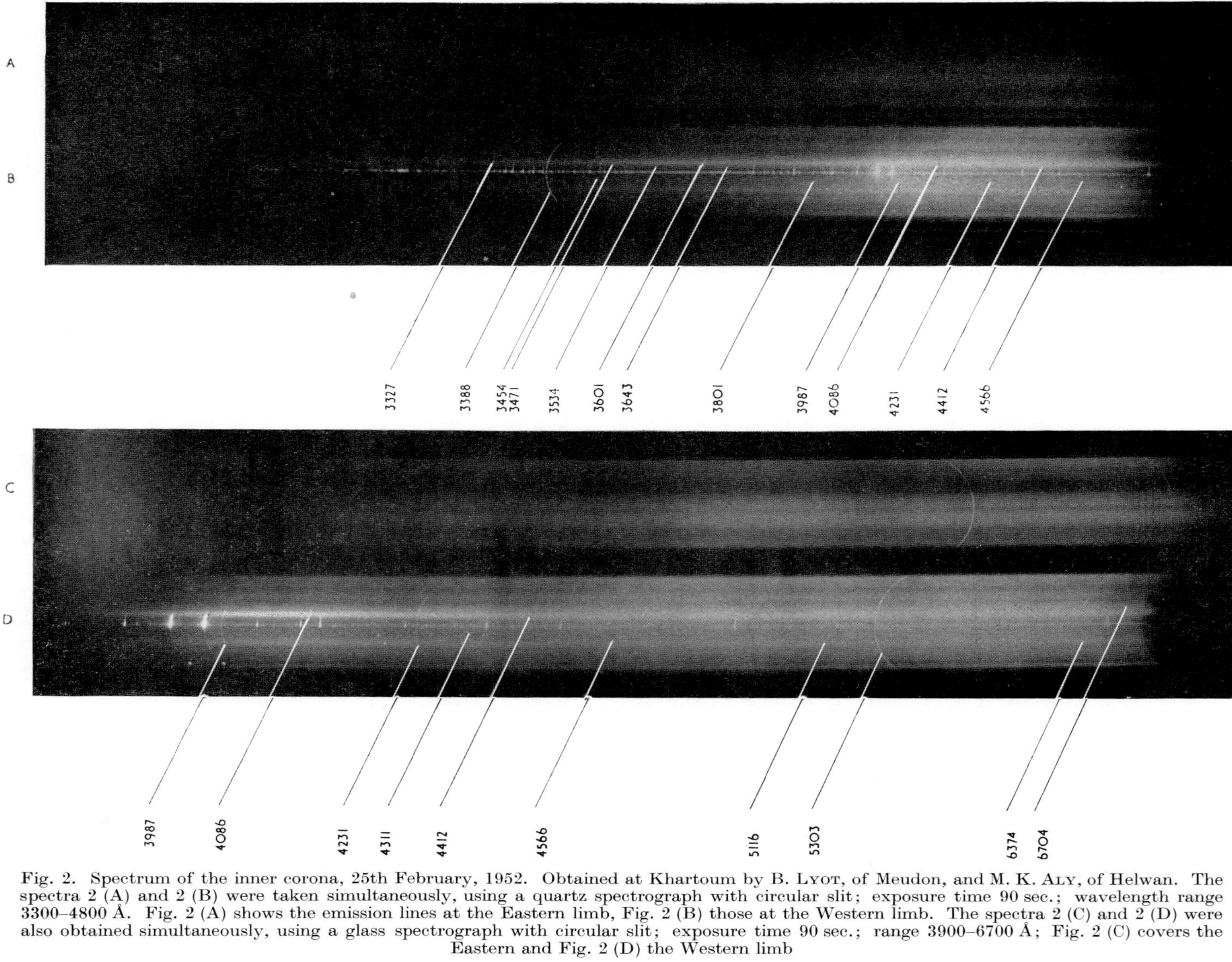

Fig. 2. Spectrum of the inner corona, 25th February, 1952. Obtained at Khartoum by B. Lyot, of Meudon, and M. K. Aly, of Helwan. The spectra 2 (A) and 2 (B) were taken simultaneously, using a quartz spectrograph with circular slit; exposure time 90 sec.; wavelength range 3300–4800 Å. Fig. 2 (A) shows the emission lines at the Eastern limb, Fig. 2 (B) those at the Western limb. The spectra 2 (C) and 2 (D) were also obtained simultaneously, using a glass spectrograph with circular slit; exposure time 90 sec.; range 3900–6700 Å; Fig. 2 (C) covers the Eastern and Fig. 2 (D) the Western limb

Palomar has a range four times as great. However, it is likely that for many years the 200 in. will be used in work for which it alone is suitable, and there will be many useful programmes left to be carried out with the smaller reflectors.

The future programme of work for the new telescope at Helwan cannot yet be given in full, as it is intended to co-operate in an international scheme, as was done with the 30 in. telescope, and the details are still pending. However, the main items of this work may be classified thus:

(1) Radial velocities and spectroscopic features of stars not included in the programmes of other observatories.
(2) Spectra of planetary nebulae.
(3) Spectroscopic binaries of special interest.
(4) Direct photography of nebulae with various filters.
(5) Spectra of bright-line stars.
(6) Proper motion studies in selected fields.

It should again be emphasized that this is a provisional programme.

The site of the new telescope is not yet finally decided upon, but in principle it is proposed to be in the vicinity of Helwan, about 20 km or so distant from existing large cities or small towns likely to expand to a large city in future.

The new Helwan telescope when erected and in full working order will be one of the large African telescopes, which together with the Canberra telescope will make the Southern hemisphere fairly well observable, although as yet not on the same scale as the Northern hemisphere. It is to be hoped that the telescope will be completed by the end of 1954, but it may be well in 1955 before the actual erection is started.

Among the recent activities of the Helwan Observatory was the participation in the observations of the total eclipse of the Sun at Khartoum on 25th February, 1952.

From the onset it may be mentioned that eclipse work is not included in the programme of activity at Helwan Observatory. However, a total eclipse of the Sun at Khartoum was a tempting occasion to participate in solar work.

Preparations were made as early as 1950 when Professor G. van Biesbroeck and myself met at Khartoum in February of that year and began to survey the possible sites on the path of totality; our conclusions were summarized in a short memorandum sent to the Joint Permanent Eclipse Committee of the Royal Society, and it appeared to us that the surroundings of the city of Khartoum would provide the best possible site for observations. The incidence of a dust storm was about 1 per cent, and the fact that such a storm did actually happen 3 days before eclipse day raised our hopes that the eclipse day would be dust-free.

The Egyptian expedition was, indeed, very lucky in having as a guest member the eminent astronomer, the late Dr. B. Lyot, of the Observatoire de Paris at Meudon, whose researches in solar physics, especially on the corona, are outstanding. It was, indeed, the saddest and most regretted event to us all when the astronomical world learnt that Bernard Lyot has passed away a month after his return from Khartoum.

The programme of the eclipse work was outlined in the *Bulletin de l'Institut d'Egypte*, vol. 34, 1951. It comprised direct photography of the corona with the Worthington camera ($f = 13$ m), which was kindly lent to us by the Permanent Eclipse Committee (observer Dr. M. R. Madwar); spectroscopic work of the inner corona in the visible and ultra-violet regions (by Dr. B. Lyot and Dr. M. K. Aly);

photo-electric photometry of the corona reaching outwards up to twenty solar radii (by Dr. I. A. RAHMAN); cinematographic recording of the partial phase of the eclipse at two stations on either side of the totality (jointly with the Greenwich expedition); and, finally, ionospheric observations during the partial phase at Helwan. (See Figs. 1 and 2.)

From preliminary reductions on the two plates, ALY found in the visible and ultra-violet regions several new lines which are listed in Table 1. Greek letters (α, β, γ) opposite each line give the estimated weight of the observation, and for lines designated α also the estimated intensities expressed as a percentage of the well-known line 5303 Å.

Table 1

Wavelength	Weight	Intensity	Wavelength	Weight	Intensity	Wavelength	Weight	Intensity
3471	α	2	4140	α	2	4744	α	3
3502	β	—	4165	α	4	4956	α	2
3533	α	12	4181	β	—	5178	β	—
3575	γ	—	4221	α	3	5188	β	—
3605	γ	—	4256	α	2	5233	β	—
3636	γ	—	4267	γ	—	5251	γ	—
3685	α	4	4363	β	—	5446	α	3
3876	β	—	4376	γ	—	5468	α	2
3911	γ	—	4402	γ	—	5559	α	1
3924	β	—	4418	β	—	5774	β	—
3956	γ	—	4467	β	—	5903	γ	—
3996	α	3	4475	γ	—	5955	γ	—
4020	β	—	4481	γ	—	5974	β	—
4038	γ	—	4484	β	—	6176	γ	—
4074	α	1	4504	α	3	6725	γ	—
4075	γ	—	4512	β	—	6740	β	—
4091	γ	—	4576	γ	—			
			4582	α	1			

It should be remembered before concluding this short account that besides the astronomical department there are two other sections, which deal with geophysics and meteorology: the Helwan Observatory includes a magnetic section for recording the three elements of the Earth's magnetic field (the observations extend as far back as 1900); there is also a seismic section equipped with seismographs for recording distant and near earthquakes (*Bull. de l'Institut d'Egypte*, vol. 36, pp. 1–16).

In 1952 magnetic observations along the thirtieth meridian from the magnetic equator to latitude 15° N. were carried out to investigate the anomalous variation of the diurnal range in the horizontal component.

Observations of the horizontal component of the Earth's field were also made at the intersection of the magnetic equator with the path of totality during the eclipse of February, 1952.

The recent programme of our activities further includes radio-sonde observations and automatic recording of heights and electron density of the ionospheric layers.

Spectroscopic Studies of the Solar Corona

G. RIGHINI

Osservatorio Astrofisico di Arcetri, Firenze, Italy

THE solar corona is an extended atmosphere enveloping the Sun, which normally becomes visible only during the total phase of a solar eclipse. The brightness of the sky near the uneclipsed Sun is many times that of the corona, except at very carefully selected places, generally at high altitudes. If the sky is very clear and the atmosphere free from dust it becomes possible to observe the brightest part of the corona when suitable instruments are employed. The late BERNARD LYOT was a pioneer in this field. In his coronagraph the scattered light coming from the optical parts has been kept to a very low level, so that the inner corona may be observed. LYOT started regular observations of the solar corona at the Observatory of the Pic du Midi in the Pyrenees at an altitude of 2870 m; later WALDMEIER organized another observatory on the Swiss Alps at Arosa, and MENZEL and ROBERTS another at Climax, Colorado. Thanks to this chain of high altitude observatories we now have regular observations of the shape and intensity of the *inner corona* (see Fig. 1 below). But the faint *outer*

Fig. 1. The inner solar corona photographed by the Italian eclipse expedition at Khartoum (Sudan) on 1952 February 25. Exposure time 8 sec.; through red filter. North, *etc.* (N, W, S, E), refer to the axis of rotation; the equator goes through the brightest prominence on the left. The slits of the spectrographs, which furnished the spectra reproduced in Fig. 2 and Fig. 3, were situated to the East of the edge of the Sun and crossed the small prominences visible on this photograph

corona is still invisible outside an eclipse, because of the residual scattering in the atmosphere and in the instrument. Eclipse observations are therefore still very important.

The instruments which are employed in studies of the corona during totality fall into two classes.

(*a*) *Instruments for direct photography of the corona.* These are generally long-focus cameras with a very small *f*-ratio, mounted horizontally and usually fed by a coelostat. Sometimes multiple cameras are employed in order to take photographs through suitable colour filters or polarizing devices.

(*b*) *Slit and slitless spectrographs.* If an image of the eclipsed Sun is projected on to the slit of a conventional spectrograph, one obtains the spectrum of a small part of the corona. The slit has to be placed sufficiently far from the solar limb to avoid including the chromospheric spectrum, which may seriously interfere with the interpretation of the coronal spectrum. This instrument has the serious drawback that it explores only a very small part of the corona; on the other hand, the experimental conditions are well defined and this is a definite advantage if photometry of the spectrum is planned.

At the eclipse of 25th February, 1952, Lyot tried to improve the performances of the slit spectrograph by using, instead of a straight slit, two curved slits which contour the solar disk at a distance of about 1′ from the limb. In order to avoid the superposition of the spectra given by the two slits he shifted them perpendicularly to the dispersion with an ingenious optical device.

In a slitless spectrograph, the thin crescent left uncovered by the Moon at the second and third contact acts as a slit; the spectrum exhibits therefore curved lines which are monochromatic images of the crescent, showing all the fine details of the chromosphere and of the corona. Coronal arches or rings which appear in slitless spectra are fainter than the chromospheric ones and sometimes are perturbed by the continuous spectrum of the chromosphere and of the prominences. In order to avoid such effects, the dispersion of the spectrograph has to be large enough to give a sufficiently faint continuum. For instance, Mitchell (1932) at the eclipse of 1927 obtained very fine monochromatic images of the corona employing a concave grating giving a dispersion of 10 Å per mm.

The coronagraph has proved very useful in the study of the intensity and the shape of the inner corona, and has also yielded many important results when used in conjunction with a spectrograph. In this way Lyot obtained beautiful high dispersion spectra in the visible and infra-red with long exposure.

Another method of observing the corona without an eclipse uses the *electro-coronagraph* invented by Skelett (1934). This instrument takes advantage of television techniques to scan the sky along a spiral around the Sun. As the scattered light is symmetrically distributed around the Sun, its intensity gives no coherent modulation in the photo-electric receiver; faint coronal features and prominences may be detected in this way. The electro-coronagraph is particularly useful when employed in conjunction with a Lyot coronagraph.

The shape of the corona depends on the epoch of the cycle of solar activity. At sunspot maximum, the corona is almost symmetrical around the disk; at minimum, the corona is elongated in the direction of the equator. Lockyer (1922, 1931) classified the corona into four types, and showed that coronal type and solar cycle are strongly correlated. LUDENDORFF (1928, 1934) introduced the more reliable criterion of measuring the ellipticity ε of the isophotes of the corona. The values of ε are correlated better with the number of prominences than with the spot-number. ABETTI (1938) showed in 1936 that the correlation is improved if only high-latitude prominences are taken into account; later BIOZZI (1939) confirmed this result in studies of the correlation between the shape of the corona and the areas of prominences for twenty eclipses which occurred between 1893 and 1937. The fact that the shape of the corona has a minimum ellipticity about three years before the maximum of the sunspot cycle is associated with the behaviour of high-latitude prominences, which have been shown by BAROCAS (1939) to have a maximum which is also just three years before the maximum of the solar cycle.

The brightness of the inner corona shows a sharp decrease within the first 15′ distance from the Sun. BAUMBACH (1937) discussed the more reliable measures of coronal brightness obtained at several eclipses and showed that the brightness may be represented by the simple interpolation formula

$$I(\rho) = 0{\cdot}0532\rho^{-2{\cdot}5} + 1{\cdot}425\rho^{-7} + 2{\cdot}565\rho^{-17}, \quad (1)$$

where I is the brightness in units of 10^{-6} of the brightness at the centre of the disk and ρ the distance from the centre in units of the solar radius.

If the coronal light is solar light scattered by fast-moving electrons, as postulated by K. SCHWARZSCHILD, the electron density $N(r)$ may be readily obtained from formula (1):

$$N(r) = 10^8(0{\cdot}036r^{-1{\cdot}5} + 1{\cdot}55r^{-6} + 2{\cdot}99r^{-16}). \quad \ldots.(2)$$

At the solar surface, $r = 1$, the electron density is $N_0 = 4{\cdot}6 \times 10^8$ electrons per cm^3; for a column of 1 cm^2 cross-section and a height equal to that of the corona one obtains:

$$N = 4 \times 10^{18} \text{ electrons.}$$

VAN DE HULST (1950) has corrected and improved BAUMBACH'S results, taking into account the fact that a certain amount of scattered light is due to the dust particles which are responsible for the zodiacal light.

Photometric, photo-electric, and radiometric measurements of the total brightness of the corona performed between 1886 and 1926 show that the total brightness of the corona is on average $0{\cdot}47 \times 10^{-6}$ times that of the Sun. After 1926 there are only some photo-electric measurements made by STEBBINS and WHITFORD and one photographic observation by RICHTMYER. Both results agree with the average of $0{\cdot}47 \times 10^{-6}$. The general belief was, therefore, that the total light of the corona was constant. But if the above results are corrected, taking into account the masking by the Moon of a significant part of the inner bright corona, the total brightness appears to be variable with a maximum at the time of the sunspot maximum. The ratio of the maximum to the minimum total brightness is about two (NIKONOV, 1943).

The first eclipse observers divided the corona into two parts which they called the *inner* and the *outer corona*; these two parts have not only a very different brightness but also quite different spectra. The inner corona shows a continuous spectrum with some bright lines, whereas the outer corona has a Fraunhofer spectrum. There is, of course, no clear boundary between the two parts, but it is generally assumed that the inner corona ends at 10′–15′ from the solar limb.

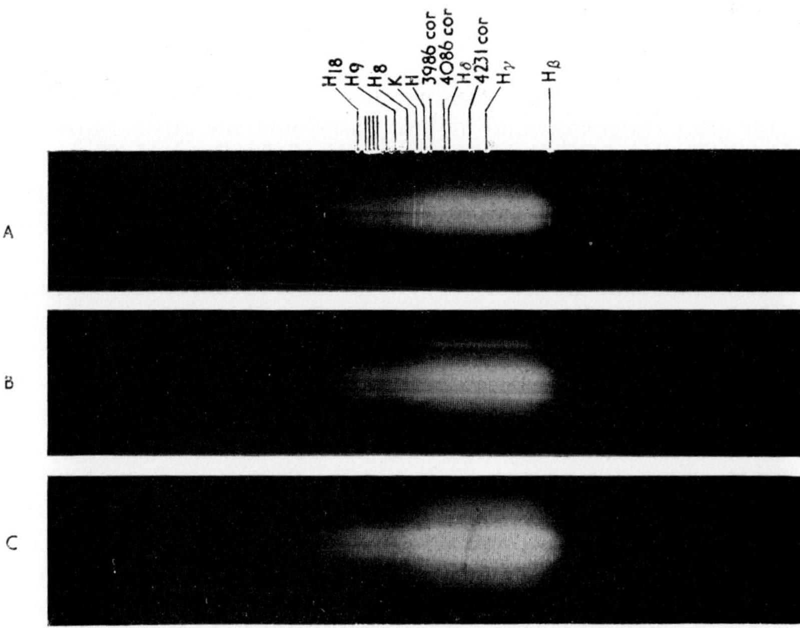

Fig. 2. Spectra of the solar corona of 1952 February 25, obtained by the Italian expedition to Khartoum. Range of wavelengths: 3600–4900 Å

(*A*) Exposed for 10 sec.; many lines of a prominence and some coronal lines are shown.

(*B*) Exposed for 30 sec.; faint trace of the prominence spectrum (the prominence was almost completely eclipsed); several coronal lines are visible.

(*C*) Exposure time 90 sec. The spectrum of the inner corona is over-exposed; some of the coronal emissions are still visible. The Fraunhofer spectrum of the outer corona is clearly shown with the H and K lines in absorption.

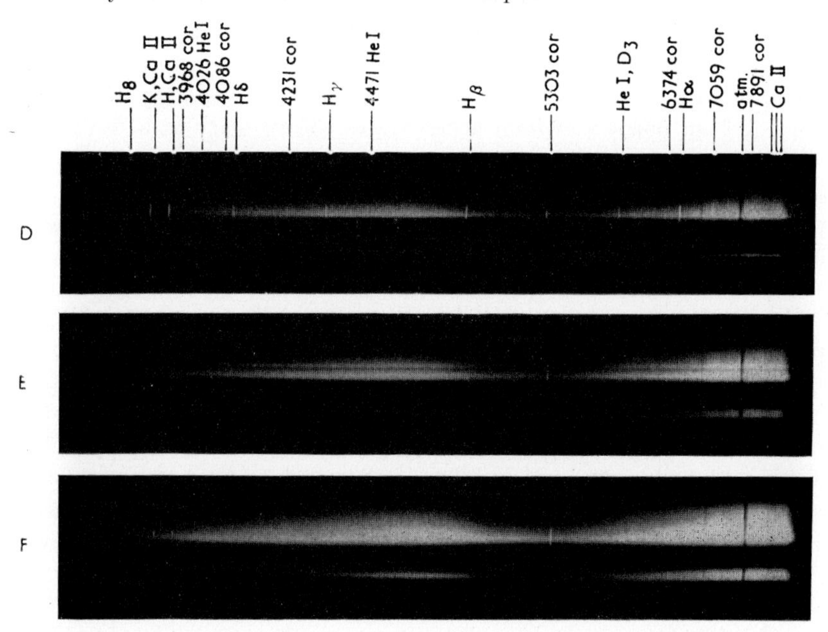

Fig. 3. Spectra of the solar corona of 1952 February 25, obtained by the Italian expedition to Khartoum. Range of wavelengths: 3750–8700 Å

(*D*) Expose time 10 sec.; many prominence lines and several coronal emissions are shown.

(*E*) Exposure time 30 sec.; a few prominence lines are left; the strongest coronal lines are visible, in particular the green line at 5303 Å.

(*F*) Exposure time 90 sec.; faint traces of prominence lines; coronal lines are clearly visible.

Note. The absorptions in the red are terrestrial atmospheric bands. The spectra are divided into two sections, due to the fact that the slit crossed part of the Moon's limb.

The continuous spectrum of the inner corona was first observed by RAYET in 1868; one year later HARKNESS and YOUNG discovered the green emission line at 5303 Å and JANSSEN, in 1871, observed for the first time the Fraunhofer spectrum of the outer corona. K. SCHWARZSCHILD (1906), at the 1905 eclipse, compared the spectra of the inner corona and of the centre of the Sun. He weakened the latter with a neutral diffuser and made careful spectrophotometric measurements. He was able to confirm the fact that the spectral curves of the Sun and of the corona coincide in the wavelength interval 3500–5000 Å. Starting from his results he advanced the hypothesis that the coronal light may be only solar light scattered unselectively by particles. He postulated that these particles were electrons; they must then have a high random velocity in order to blurr out the absorption lines of the solar spectrum.

LUDENDORFF (1925), employing the same method as SCHWARZSCHILD, showed that the identity of the solar and coronal spectrum extends to a distance of 23′ from the limb of the Sun. New and more complete results on this subject were obtained by GROTRIAN (1931) at the 1926 eclipse. He examined the whole coronal spectrum between 3400 Å and 6800 Å taking spectra with a multiple spectrograph. The comparison spectra were photographed with the same instrument weakening the light of the Sun by a metallic wire grating placed in front of the objectives and feeding the instruments with sunlight scattered by a MgO block. GROTRIAN found that the deviations between the spectral distribution of the coronal radiation and that of the Sun are smaller than 10 per cent. These results were later extended to the infra-red by PETTIT and NICHOLSON (1926).

Some minor features of the spectral curve of the corona are still not very clear. GROTRIAN found, for instance, that the wide "absorption" at 3800 Å, due to the incomplete blotting out of very strong absorption lines, is wider and shallower in the corona than in the Sun. From this fact he inferred that the velocity of the electrons must be $7{\cdot}5 \times 10^8$ cm per sec., which is ten times the velocity at a temperature of 6000°K. But this "absorption" was not found by MENZEL at the 1936 eclipse, or by RIGHINI at the 1936 and 1952 eclipses.

It is highly desirable that at the forthcoming eclipses new work should be done to clarify this point and to determine accurate electron temperatures. Very accurate spectrophotometric measurements are required; the solar and coronal spectra should be compared with a standardized terrestrial source, in order to be sure that the absorption of the optical parts does not interfere in the measurements.

The inner corona is sometimes called the K-corona to distinguish it from the outer, or F-corona. K and F are the initials of the German words "Kontinuierliches" and "Fraunhofer" employed by GROTRIAN (1934) in his work on the two types of spectra.

The absorption spectrum of the F-corona begins to be visible at a distance of 5′ from the limb. The lines appear shallow but not broader than those of the solar spectrum. MOORE (1923) at the eclipse of 1922 obtained very good spectra of the F-corona up to a distance of 35′; using an iron arc comparison spectrum, he deduced that a red shift of 26 km per sec. is present on both the east and west sides of the corona. He repeated his work at the 1932 eclipse without obtaining convincing results. ALLEN (1947), however, at the 1940 eclipse measured a red shift of 0·08 Å, corresponding to a velocity of 7 km per sec. A red shift may be interpreted as an outward streaming of the scattering material; the observational results seem to show that if this streaming is present its velocity is very low.

The ratio of the intensities of the K- to F-coronas may be obtained in a simple way by measuring the central depth of the absorption lines, since any superposed K continuum reduces the central intensity of the absorption lines of the F-corona. Calling I_K the intensity of the K continuum, I_F that of the Fraunhofer continuum, and I_c the residual intensity in the centre of an absorption line, we will then observe at the centre of the line an intensity ratio r' given by:

$$r' = \frac{I_c + I_K}{I_F + I_K} = \left(\frac{I_c}{I_F} + \frac{I_K}{I_F}\right) \Big/ \left(1 + \frac{I_K}{I_F}\right). \quad \ldots.(3)$$

As the ratio $r = I_c/I_F$ is readily obtained from the solar spectrum, we can therefore evaluate

$$\frac{I_K}{I_F} = \frac{r - r'}{1 - r'}. \quad \ldots.(4)$$

The ratio r depends on the resolving power attained; therefore the spectra of the corona and of the Sun must be obtained with the same spectrograph. The contribution of the K-corona to the total light according to Grotrian's results is given in Table 1.

Table 1. The intensity of the K-corona at the 1926 eclipse

		Distance from solar limb					
		3'.2	5'	10'	15'	20'	25'
$100 \dfrac{I_K}{I_K + I_L}$	East (per cent) .	85	82	73	60	46	29
	West (per cent) .	—	89	74	56	38	—

Grotrian pointed out that the diffusing medium which produces the F-corona must be different from that producing the K-corona, since in the spectrum of the F-corona the lines are not broadened. This can arise only if the particles of the diffusing medium are at rest or move with very small velocity. Their diameter has to be greater than the wavelength in order to give a neutral scattering. Grotrian postulated the existence of a swarm of dust particles in the outer corona which may be the extension towards the Sun of the cluster of interplanetary dust responsible for the zodiacal light.

The first emission line discovered by Harkness at the eclipse of 1899 was erroneously identified by Young with a line of neutral iron. Thirty years later Lockyer found that the wavelength was 5303·7 Å instead of 5311·7 Å, and that it must therefore belong to another element than Fe. Later, at almost every eclipse, new lines were discovered: Grotrian found one line at 6701 Å at the eclipse of 1929, Dunham another in 1937 at a wavelength of 4412 Å. At the 1936 eclipse Righini (1942) observed a new line having a wavelength of 3533·2 Å; this line was already seen by Lewis at the 1908 eclipse, but was rejected as coronal line because it was believed to be of chromospheric origin and identified as a Ti II line. Lyot, working with a powerful spectrograph attached to his coronograph, has made important discoveries of five new lines in the red and infra-red. Two new lines, at 3997 Å and 4351 Å, have been found by Dollfus (1953) on spectra taken by Lyot during the 1952 eclipse. A recent discovery is also the line at 5445 Å, found by WALDMEIER to be present above large sunspots of exceptional activity.

A list of the emission lines which have been confirmed to date is given in the following Table 2. Columns 2, 3, and 4 contain the intensities referred to the green line 5303 Å (taken as 100).

Table 2. The emission lines of the solar corona

λ	Relative intensity			Element and multiplet	Ionization potential (volts)	Observer
	GROTRIAN 1929	LYOT 1934–36	RIGHINI 1936			
3328·1	1·0	—	—	Ca XII $^2P_{3/2}$–$^2P_{1/2}$	589	LEWIS . . . 1908
3388·10	16·4	—	17·5	Fe XIII $^3P_2$–$^1D_2$	325	NAEGAMWALA . 1898
3453·13	2·3	—	12·9	—	—	NAEGAMWALA . 1898
3533·42	—	—	1·8	—	—	LEWIS . . . 1908
3600·97	2·1	—	1·0	Ni XVI $^2P_{1/2}$–$^2P_{3/2}$	455	LEWIS . . . 1908
3642·87	—	—	0·9	Ni XIII $^3P_1$–$^1D_2$	350	DYSON . . . 1900
3800·77	—	—	1·1	—	—	FOWLER, LOCKYER . 1898
3986·88	0·7	—	2·8	Fe XI $^3P_1$–$^1D_2$	261	FOWLER . . 1893
3997	—	—	—	—	—	LYOT, DOLLFUS . 1952
4086·29	1·0	—	—	Ca XIII $^3P_2$–$^3P_1$	655	FOWLER . . 1893
4231·4	2·6	—	0·0	Ni XII $^2P_{3/2}$–$^2P_{1/2}$	318	FOWLER . . 1893
4311·5	—	—	—	—	—	DYSON . . . 1900
4351	—	—	—	—	—	LYOT, DOLLFUS . 1952
4359	—	—	—	—	—	HILLS and NEWALL . 1896
4412	—	—	—	—	—	DUNHAM . . 1937
4567	1·1	—	—	—	—	HILLS and NEWALL . 1896
4586	—	—	—	—	—	FOWLER, LOCKYER . 1898
5116·03	4·3	2·1	4·8	Ni XIII $^3P_2$–$^3P_1$	350	DYSON . . . 1905
5302·86	100	100	100	Fe XIV $^2P_{1/2}$–$^2P_{3/2}$	355	HARKNESS . . 1869
5445·2	—	—	—	—	—	WALDMEIER . . 1950
5536	—	—	—	—	—	DYSON . . . 1905
5694·42	—	1·3	—	—	—	LYOT . . . 1935
6374·51	8·7	2·3	4·7	Fe X $^2P_{3/2}$–$^2P_{1/2}$	233	CARRASCO . . 1914
6701·83	5·4	2·7	—	Ni XV $^3P_0$–$^3P_1$	422	GROTRIAN . . 1929
7059·62	—	3·3	—	—	—	LYOT . . . 1936
7891·94	—	2·4	—	Fe XI $^3P_2$–$^3P_1$	261	LYOT . . . 1935
8024·21	—	1·1	—	Ni XV $^3P_1$–$^3P_2$	422	LYOT . . . 1936
10746·80	—	200	—	Fe XIII $^3P_0$–$^3P_1$	325	LYOT . . . 1936
10797·95	—	125	—	Fe XIII $^3P_1$–$^3P_2$	325	LYOT . . . 1936

The origin of the coronal lines was for a long time open to question. GROTRIAN (1939) noticed that lines of highly ionized iron are sometimes observed in novae, as well as the most prominent coronal lines. He suggested that probably some coronal lines might be due to forbidden transitions between levels of the multiply ionized Fe-ion, as, for instance, Fe X. He showed that the wave-number difference of the two atomic levels of the ground state of Fe coincided with the wave number of the red line 6374 Å.

Starting from this point, EDLÉN (1941) who had previously carried out extensive researches on spectra of highly ionized Fe, gave a very convincing interpretation of the origin of the coronal spectrum.

The intensity of the lines depends on the density of the emitting ions and that of the continuum, and on the local electron density. According to GROTRIAN (1933) the ratio of the two intensities is constant. RIGHINI (1943), however, found that the equivalent widths of the lines (that is the width in Ångstrom units of a fictitious line, giving complete absorption within a rectangular profile, which would absorb the same amount of energy) do not depend on the continuum intensity but exhibit very wide variations. This fact indicates the existence of local condensations of ions such as have since been observed by WALDMEIER and by LYOT with the coronagraph.

Spectrographic observations performed during a total eclipse are usually unable to give the true shape of the emission line owing to lack of sufficient resolving power; only LYOT at the Pic du Midi and WALDMEIER at Arosa have succeeded in obtaining the resolution needed to reveal the true profile of the coronal lines. The width of the emission lines appears to be very great; this result may be interpreted as showing that high temperatures exist in the corona, or that the coronal material is subjected to turbulent motions with a high velocity. The first assumption leads to a temperature of the order of 10^6 °K, the latter to a velocity of 30 km per sec. This result fits in very well with the hypothesis that the turbulent motions in the outer layers of the Sun have velocities which increase steadily with height; this is illustrated in Table 3.

Table 3. Velocity of turbulence for the outer layers of the Sun

Photosphere	1·8 km per sec.
Low chromosphere	12
High chromosphere	18
Corona	30

It would be possible to decide between thermal and turbulent broadening of the coronal lines if lines arising from atoms or ions having very different atomic weight were observed, since the thermal width of a line increases with decreasing atomic weight and the turbulent width is the same for all lines. Unfortunately, the greater part of the coronal lines whose origin is known are emitted by Fe- and Ni-ions which have almost the same weight. Strong support to the hypothesis of the high temperature is given by the intensity of the radio emission of the Sun, the average intensity of which corresponds to that emitted by a black body having a temperature of 10^6 degrees. Observations of this radiation may yield a great deal of information about the physical state of the corona. If we know, from the optical observations, the electron density of the corona at each level, we may calculate the height at which the corona becomes opaque for any given radio wavelength. All radiation of this wavelength emitted from below this level will be reflected toward the Sun, and thus we can receive only the radiations emitted above this level. The location of the emitting level for each such wavelength has been derived by WALDMEIER (1948).

Radio observations are at present made regularly with many radio telescopes in order to follow the fluctuations in the activity of the Sun. Pioneers in this field have been the English radio-physicists HEY, at Slough, and RYLE, at Cambridge. The latter measured some very high temperatures on perturbed areas of the corona. These temperatures are, of course, "electron temperatures", based on the assumption that emission takes place as a consequence of the deceleration of the electrons which pass near an ion in the coronal plasma.

After EDLÉN's discovery that the visual emission lines are due to highly-ionized metallic atoms, it was believed that the ion densities were sufficient to neutralize the electron charge. That this is not true has been shown by RIGHINI (1943) using the 1936 eclipse. He measured the amount of energy emitted by a square centimetre of the corona, and compared the spectrum of a standard lamp with that of the corona. The energy emitted in each line was obtained immediately in terms of the energy emitted by the neighbouring continuum, and thus the number of ions per square centimetre was finally derived. The number of electrons being known, he deduced that the ratio ions/electrons is of the order of 10^{-7} to 10^{-8}. This means that for the stability of the corona we must postulate the existence of a suitable number of protons in order to neutralize the negative charge of the electrons.

Several hypothesis have been put forward in order to explain the high temperature in the corona. It has been proposed, for instance, that the material forming the corona may come from the interior of the Sun through "volcanoes". Other hypotheses postulate that the heating of the corona may be due to meteoric matter falling on the Sun, or that the high electron temperature may be a consequence of acceleration of the electrons by the variable magnetic fields of the sunspots.

None of these hypotheses seems to be generally accepted; perhaps magnetic acceleration may play a rôle in the formation of local condensations of coronal matter.

An idea which has been developed recently by HOUTGAST, BIERMANN, and M. SCHWARZSCHILD attempts to explain the high temperature of the corona as a consequence of absorption of acoustic shock waves.

Near the surface of the photosphere, there is a layer which is in a state of convective equilibrium, and the movements of solar material in this layer give rise to the granulation which is visible at the top of the photosphere. The granules have a diameter of the order of 1000 km, a vertical velocity of about 1 km per sec., and a mean life of 200 sec. The turbulence that is associated with them gives rise to sound waves which travel with a speed of about 7 km per sec., while the velocity of the material, at the moment when the wave is passing through it, is about 0·5 km per sec. In this case the dissipation of energy is small, and the waves therefore leave the photosphere without hindrance. As soon, however, as the velocity of the material reaches a critical value (which is near to the velocity of sound), the energy dissipation becomes very high owing to the formation of shock waves.

The corona has a density many times smaller than that of the photosphere; the velocity of the sound wave must therefore be greater in order to carry the same amount of energy, and as the speed increases the energy dissipation becomes more and more important and gives rise to a strong heating of the surrounding material.

REFERENCES

ABETTI, G. . . . 1938 *Pubbl. Oss. Astrofis. Arcetri*, **56,** 55.
ALLEN, C. W. . . . 1947 *M.N.*, **106,** 137.
BAROCAS, V. . . . 1939 *Ap. J.*, **89,** 486.
BAUMBACH, S. . . . 1937 *Astron. Nachr.*, **263,** 121.
BIOZZI, M. . . . 1939 *Pubbl. Oss. Astrofis. Arcetri*, **57,** 3.
DOLLFUS, A. . . . 1953 *C.R. Acad. Sci. (Paris)*, **237,** 854–9
EDLÉN, B. . . . 1941 *Ark. Mat. Astron. Fys.*, **28**B, No. 1.
GROTRIAN, W. . . . 1931 *Z. Astrophys.*, **3,** 199.
1933 *Z. Astrophys.*, **7,** 26.
1934 *Z. Astrophys.*, **8,** 124.
1939 *Naturwiss.*, **27,** 214.
LOCKYER, W. J. S. . . . 1922 *M.N.*, **82,** 323.
1931 *M.N.*, **91,** 797.
LUDENDORFF, H. . . . 1925 *S.B. Preuss. Akad. Wiss. Berlin*, V, 83.
1928 *S.B. Preuss. Akad. Wiss. Berlin*, XVI, 185.
1934 *S.B. Preuss. Akad. Wiss. Berlin*, 200.

MITCHELL, S. A. . . . . 1932 *Ap. J.*, **75**, 1.
MOORE, J. H. . . . . 1923 *Publ. Astron. Soc. Pacific*, **35**, 333.
NIKONOV, V. B. . . . . 1943 *Abastumani Astrophys. Obs. Bull.*, **7**, 33.
PETTIT, E. and NICHOLSON, S. B. . . . . . 1926 *Ap. J.*, **64**, 136.
RIGHINI, G. . . . . 1942 *Z. Astrophys.*, **21**, 158.
1943 *Atti Accad. d'Italia*, **XIV**, 113.
SCHWARZSCHILD, K. . . . . 1906 *Astron. Mitt. Sternwarte Göttingen*, Teil 13.
SKELETT, A. M. . . . . 1934 *Proc. Nat. Acad. Sci. U.S.A.*, **20**, 461; and *Phys. Rev.* (2), **45**, 649.
VAN DE HULST, H. C. . . . . 1950 *Bull. Astron. Inst. Netherlands*, **11**, 135.
WALDMEIER, M. . . . . 1948 *Astron. Mitt. Zürich*, No. 155.

For additional references and general discussions of our subject see:

ABETTI, G.; *The Sun.* (Van Nostrand Co., Inc., New York, 1938.)
DYSON, F. and WOOLLEY, R. V. D. R.; *Eclipses of the Sun and Moon.* (Oxford University Press, 1937.)
KUIPER, G. P. (Ed.); *The Sun.* (Volume I of "The Solar System"; 4 volumes): Articles 5 (H. C. VAN DE HULST), 9.3 (G. VAN BIESBROECK), 9.6 (J. W. EVANS). (Chicago University Press, 1953.)
MENZEL, D. H.; *Our Sun.* (Harvard University Press, Cambridge, Mass., 1949.)
MITCHELL, S. A.; *Eclipses of the Sun.* (Columbia University Press, New York, 5th edition, 1951.)
WALDMEIER, M.; *Sonne und Erde.* (Büchergilde Gutenberg, Zürich, 1946.)

Coronagraphic Work at Moderate Altitude Above Sea Level

YNGVE ÖHMAN
Stockholms Observatorium, Saltsjöbaden

SUMMARY
Discussion of the coronagraphic observations of the green corona line, made in Saltsjöbaden and Abisko in Sweden and on Capri, Italy. In spite of the moderate altitude above sea level the intense parts of the corona have often been observed without difficulty in Abisko and on Capri. Estimated intensities obtained on Capri are compared with those on the Pic du Midi. A description is given of a coronagraph-photometer.

1. OBSERVATIONAL BASIS

OF the present coronagraphic stations the Pic du Midi, Climax, Sacramento Peak, and Mt. Norikura are situated at altitudes between 2700 and 3600 m, whereas Arosa. Wendelstein, and Kanzelhöhe are working at altitudes between 1800 m and 2100 m, All these stations have the character of high mountain stations, favoured on clear days by very blue sky right up to the edge of the solar disk. In general it is only in such places that a detailed spectrographic study of both the red and the green corona emission lines is possible.

But several coronagraphs are to be found also at moderate altitude, such as the ones in Freiburg, 1240 m; Partizanskoe, 650 m; Simeis, 360 m; and Capri, 480 m. Even with these instruments spectrographic observations of the λ5303 line can sometimes be made without difficulty, though generally not with the same accuracy as in the high altitude stations.

The last solar instrument constructed by the late Dr. LYOT, called "coronomètre photoélectrique", has made it possible to trace the green corona line even with an ordinary refracting telescope in low altitude places characterized by a less satisfactory sky. Solar physicists round the world look forward very much to the further development and application of this interesting photoelectric method of observation.

In the present article a brief report will be given of coronagraphic observations made in Sweden and on Capri in Italy. As this work has been done near sea level or only a few hundred metres above it, the results obtained show that local meteorological factors may produce sometimes better conditions for coronagraphic work in low altitude stations than has been generally believed.

2. Observations Made in Saltsjöbaden and Abisko

When Dr. Lyot spent a few weeks in North Sweden for observing at Brattås (near the Gulf of Bothnia) the total solar eclipse of 9th July, 1945, he was very much impressed by the blue sky and declared that coronagraphic work no doubt could be done quite often at sea level on this part of the Swedish coast. A contributing factor to the favourable conditions was probably the presence of an extensive and cold water surface and the frequent appearance of polar air. Dr. Lyot's judgment agreed well with our own experience and encouraged me to build later on a coronagraph for work in our country. By the courtesy of Dr. Roberts a sky photometer of Dr. Evans's construction was put at my disposal by the High Altitude Observatory in Boulder and a series of measures were made from the beginning of 1950 both in Saltsjöbaden (altitude 50 m) and during short periods also in Abisko, north of the arctic circle (altitudes around 400 m), and in different places in the mountains of Härjedalen in central Sweden, at altitudes around 650 m. It was found from these measurements that the intensity of the sky at a distance from the solar limb of about 8′ was on a few occasions in Saltsjöbaden as low as about $10 \, . \, 10^{-6}$, expressed in the intensity at the centre of the solar disk as unit. In Abisko and in the mountains of Härjedalen the sky had on rare occasions an intensity of only $3 \, . \, 10^{-6}$ when measured with the same instrument. These results seemed nearly too good to be true and it was of particular value, therefore, to have them checked by Dr. Menzel when he visited Abisko for a few days in 1950. According to his opinion the sky was really of such a quality that coronagraphic work was well possible.

When our coronagraph was ready during the spring of 1951 it was with a great curiosity that it was tried for the purpose of observing the green line. The first very successful result was obtained on 25th April and has been briefly reported previously.* During the summer of 1951 the instrument was stationed in Abisko at the scientific station of the Royal Academy of Sciences. On clear days the green line was fairly regularly recorded, and was in good agreement with the Ursigrammes. The weather was unusually unstable that summer, however, and a disturbing halo, due to ice crystals in the atmosphere, appeared often around the Sun, though the sky looked very blue for the naked eye. In fact this phenomenon often proved to be a presage of the appearance of cirrus-clouds.

Another disturbance was the appearance sometimes of great numbers of mosquitos, gnats, and other flying insects. When a cloud of these insects appeared in the line of sight, the field of view appeared almost as a firework display. These drawbacks made me accept, therefore, a very generous offer of the Italian foundation Centro Caprense through its President Signor E. Cerio, to move the instrument to Capri and install it on Monte Solaro at an altitude of 480 m.

---

* *Observatory*, **71**, p. 110, 1951. It may be mentioned that the sky is not so favourable in Saltsjöbaden during the summer partly because of the appearance of pollen, wind-borne seeds and spores.

3. OBSERVATIONS MADE ON CAPRI IN 1952

During two periods in 1952, one from 1st February to 19th March and the other from 11th July to 24th September, regular observations were made on Capri. The experience collected during these periods shows clearly that observations of the λ5303 line can be made very frequently indeed on Monte Solaro on Capri despite its moderate altitude. That is to say, the recordings cannot be made in the same very complete way as at the high mountain stations, but the intense parts of the corona can generally be recorded without difficulty and with a considerable accuracy.

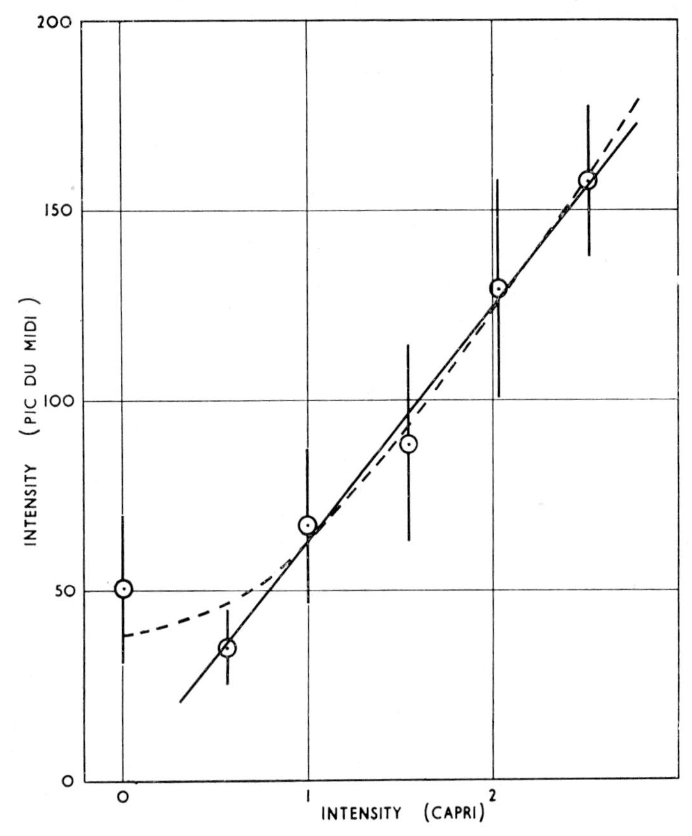

Fig. 1. Comparison of the intensities of the corona line λ5303 between Capri (arbitrary scales) and Pic du Midi (ursigrammes)

This is clear from Fig. 1, showing estimated maximum intensities on different days of the λ5303 line made on Capri and the corresponding Pic du Midi values obtained from the Ursigrammes. The lengths of the vertical lines give twice the dispersion in the individual Pic du Midi values belonging to the estimated group. The total number of individual values is 51. Also the estimated position angles for maximum intensity have been found to agree very well with the values given in the Ursigrammes.

Though the scale of intensity is arbitrary it seems to be remarkably constant and one reason for this may be the stable weather conditions on Capri. During the summer the corona could be seen on almost every day and at its best between one and three in the afternoon, when the wind comes from the North. During the winter the observations in the morning seem to be better. The best sky on Capri compares well with the best sky observed by me in Abisko. But cirrus clouds are very rare on Capri and already at an altitude of a few hundred metres remarkably little dust appears, except when winds from the South are prevailing. The moisture over the sea does not seem to disturb the observations very much, a phenomenon which has been emphasized before by WALDMEIER in a report covering many years' work in Arosa.* On the contrary, the influence of the sea on the stratification of the air seems to be an important factor. As a matter of fact, looking towards the neighbouring continent one can often see on clear days how the dust layer declines from a height of several thousand metres to much smaller heights over the sea.

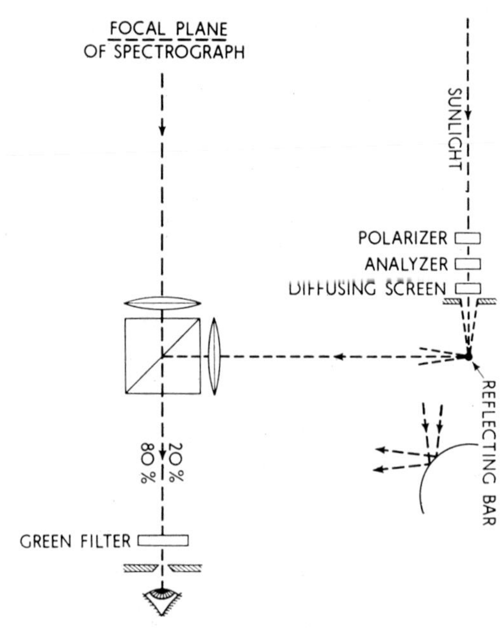

Fig. 2. Coronagraph photometer

During April and May, 1953, the sky has not been found so good on Capri as during the periods of work in 1952. It never reached the top values of 1952 and the number of clear days was smaller than usual; but the general weather situation in Italy has been very unstable during the spring of 1953 and the prevailing winds have been from the South. It may be added that during 1954 the green line of the corona has been extremely faint.

One reason for the successful observations may be the good optical qualities of our coronagraph and of the Siegbahn-grating in the spectrograph. The dispersion used is 21 Å per mm and the slit is circular. Generally it has been found easier to observe the corona visually than photographically because in this way one can go closer to the limb.

Though the estimated intensities have proved remarkably good they have to be checked in order to avoid systematic scale variations. Fig. 2 shows a schematic drawing of a very simple photometer which is used by me for this purpose. It is constructed on the same general principle as many similar devices, that is, it gives an

* *Astron. Mitteilungen*, Zürich, No. 160, 1949.

artificial emission line produced by the sunlight itself which is superimposed on the continuous spectrum of the sky.

In our photometer the artificial line is produced by a thin cylindrical bar of well-polished nickel in which the sunlight is reflected. This gives a very good artificial emission line without sharp edges.

Though a coronagraphic station of the kind we have on Capri may never be expected to give as excellent results as the high mountain stations, it seems to give very reliable observations of the stronger parts of the corona which are of a special geophysical importance. The station has already proved especially valuable in studying prominences; this does not always require the extremely blue sky of the high mountains.

Spectroscopic Measurements of Magnetic Fields on the Sun

H. VON KLÜBER

The Observatories, Cambridge

ABSTRACT

The various methods employed for the measurement of solar magnetic fields, on the basis of the Zeeman effect in the Fraunhofer spectrum, are described. The optical arrangements are discussed and illustrated. A short examination is made of the results obtained for the field strength in sunspots, and of the problem of the general magnetic field of the Sun.

1. SPECTROSCOPIC BASIS

THE investigation of the Zeeman effect of suitable lines in the solar spectrum offers a sensitive method for the determination of magnetic fields in those regions of the solar atmosphere in which the examined lines originate. Fundamental methods and observations of this kind were first introduced by HALE (from 1907–8), and have been developed at Mount Wilson Observatory. This development has continued up to the present, as nearly every *Mount Wilson Annual Report* testifies. It was essentially only during the last decade that research in this field was also begun by other investigators.

The well-founded theory of the Zeeman effect which we possess to-day furnishes, to summarize it in a few words, the following simple basis for the determination of the magnetic field strength (see, for example, BACK and LANDÉ, 1925; CANDLER, 1937). We consider an atom radiating in a magnetic field. An investigator observing parallel to the magnetic lines of force then notices (to take the simplest classical case) that the resulting spectral line is split into two components which are circularly polarized in opposite directions (longitudinal effect). If, however, one observes perpendicular to the line of force, then the line appears split into three components, all of them plane polarized (transverse effect). This is shown in Fig. 1. Viewing obliquely to the lines of force, we observe a mixture of these types of splitting. In very strong magnetic fields, such as have not yet been observed on the Sun, the normal Russell-Saunders coupling breaks down, and the normal Zeeman effect passes over to the more complex Paschen-Back effect. In the simple classical case the relation between the line splitting $\Delta\lambda$, the wavelength λ of the spectral line, and the magnetic field strength H (in gauss) is given by:

$$\Delta\lambda = C \,.\, H \,.\, \lambda^2 \,.\, g,$$

where $C = 4{\cdot}7 \times 10^{-5}$; the Landé factor g has in the simplest case the value 1. The measurement of the amount of splitting $\Delta\lambda$ therefore permits a calculation of the magnetic field strength. In fact, this splitting depends on the magnetic moment of the particular atom. The great majority of all spectral lines therefore split into much more complex patterns. The patterns are always symmetrical about the parent line, both with respect to the amounts of splitting and to the intensities of the components. All the splittings which are of practical importance are very small.

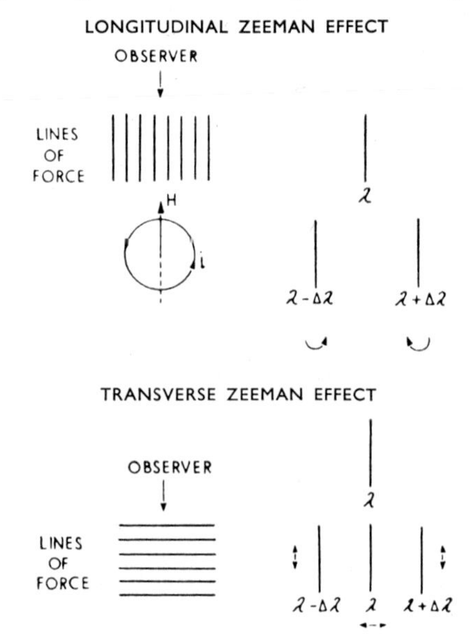

Fig. 1. Sketch of the simple classical Zeeman effect. Arrows indicate sense of polarization. H are the magnetic lines of force, i is the corresponding electric current. The upper part shows a North polarity, as seen by the observer

For the very favourable line $\lambda 6302{\cdot}5$ Fe, $\Delta\lambda$ for a field strength of 10 gauss only reaches 0·0005 Å, which corresponds to a line displacement of 0·0007 mm if observed with a large plane grating spectrograph of 12 m focal length in the second order.

The Landé factor can in general take values between 0 and 3. For $g = 0$ we have no splitting in the magnetic field, and these rather infrequent lines are sometimes very useful for a check of the instrumental arrangement. The theory of the Zeeman effect furnishes simple schemes for the calculation of the amounts of the splitting and the intensities of the components, as soon as the quantum numbers of the particular line are known (CANDLER, 1937; Washington Bureau of Standards, 1928). Examples of such splitting patterns, which are all found frequently in the solar spectrum, are shown in Fig. 2. The Fraunhofer lines of the solar spectrum are absorption lines; nevertheless the theory of the inverse Zeeman effects (JENKINS and WHITE, 1937) justifies the application to them of the considerations which have just been sketched for the emission lines. Of course, it will be practically impossible to resolve the more complicated Zeeman patterns of the wide and diffuse Fraunhofer lines. If, nevertheless, one tries to measure the displacement $\Delta\lambda$ also in these cases one obtains a value corresponding more or less to the centre of gravity of the Zeeman components. For this value, too, the theory gives a simple expression (SHENSTONE and BLAIR, 1929; RUSSELL, 1927). For modern investigations one naturally endeavours to use lines of the large and simple type No. 1 (Fig. 2), which have the Landé factors 3/2, 2 or 3. Such lines can easily be found from term combinations, some of which are indicated in Fig. 3. A survey of the solar spectrum with a view to finding suitable splittings has been carried out by Mrs. COFFEEN (*Mount Wilson Report* 1939–40). A preliminary list of such lines has been compiled by H. v. KLÜBER (1947a).

It becomes obvious that there exist in the solar spectrum only relatively few suitable lines with simple and large splittings (v. KLÜBER, 1947a). Those mainly used

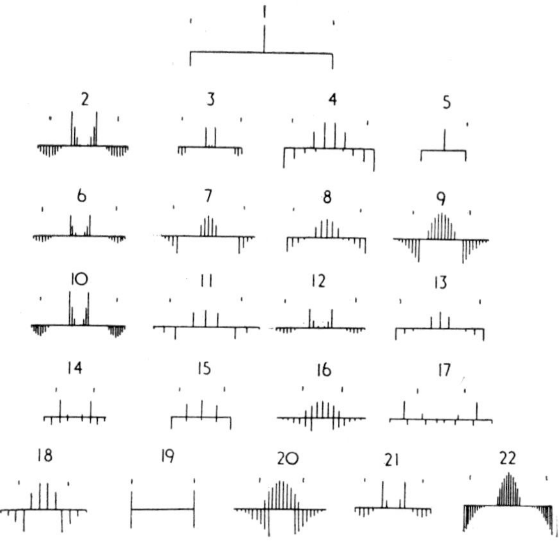

Fig. 2. Typical Zeeman patterns. The π-components, vibrating parallel to the lines of magnetic force are drawn upwards, the σ-components downwards (v. KLÜBER, 1952)

for investigation of the solar field are λ5247·6 Cr, λ5250·2 Fe (EVERSHED, 1939b, 1939c, 1944; v. KLÜBER, 1947a), λ6173·3 Fe, and λ6302·5 Fe (v. KLÜBER, 1947a). Since the splitting increases proportionally to the square of the wavelength, the Fraunhofer lines in the infra-red are most suitable, and thus various investigators have concentrated on the near infra-red region (*Mount Wilson Annual Report*, 1927–8, and 1928–9). Some of the favourable lines in this region have been also tabulated by v. KLÜBER (1947a). The large splitting of the line λ10288·87 Si has been used successfully in the spot spectrum (Cambridge Observatories, 1951).

For the derivation of the magnetic field strength from observed Zeeman splittings we have, in consequence of our considerations above, two possibilities: (1) we can measure the splitting directly as a line displacement and then immediately derive the field strength from this with the help of the formula above: these measurements of Zeeman splittings are particularly suitable for stronger fields and correspondingly larger splittings. (2) On the other hand, however, we can somehow use the accompanying polarization of the Zeeman components to render the splitting more conspicuous. Here, by applying electronic means, particularly sensitive methods have recently been developed. Because of the different sense of polarization on the two sides of the parent line (Fig. 1), the longitudinal effect is much more suitable for this purpose than the transverse effect for very small splittings.

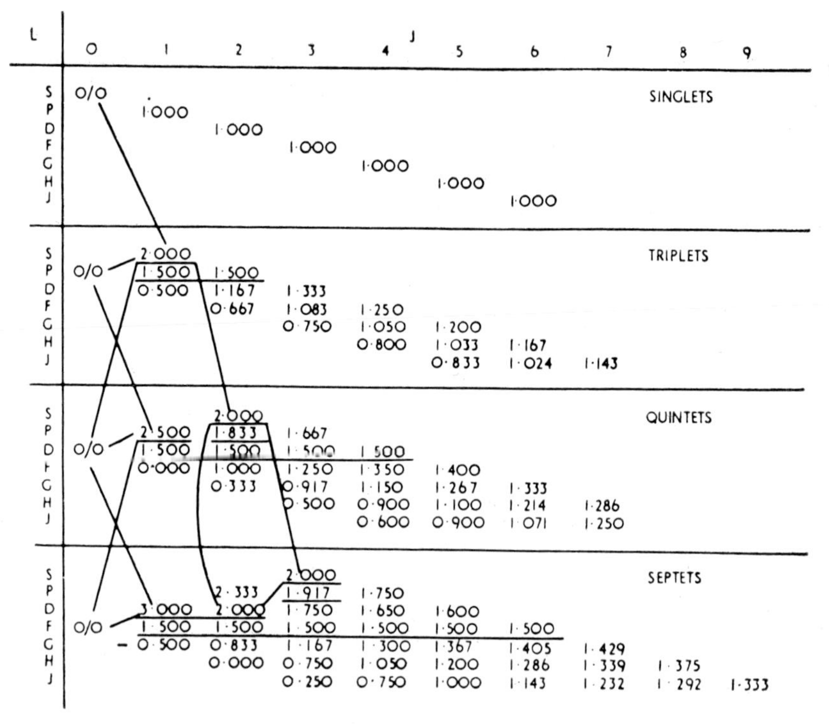

Fig. 3. A table of Landé g-values for odd multiplicities. The connecting lines indicate some transitions with large, simple Zeeman splitting

2. Magnetic Fields of Sunspots

The first investigations by Hale and his collaborators on Mount Wilson, from the beginning in 1907, were concerned with the relatively strong fields which always

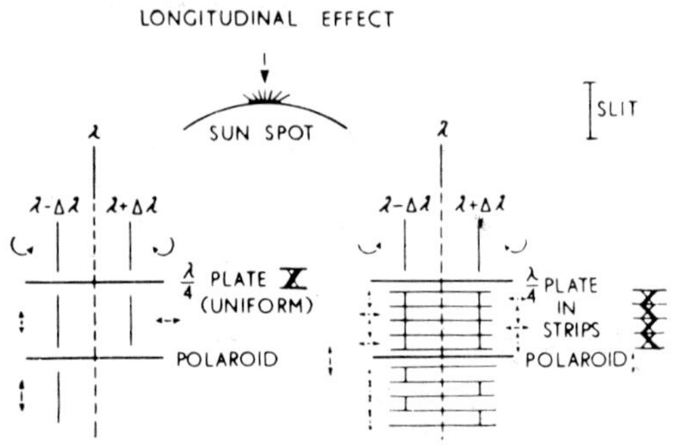

Fig. 4. Explanation of the use of small mica strips for the detection of longitudinal Zeeman effects (v. Klüber, 1947b). Sense of polarization is indicated by arrows

appear in conjunction with sunspots (*Mount Wilson Report*, 1909–10, *etc.*; Hale and Nicholson, 1938). The splitting of the Fraunhofer lines is frequently not complete, and for weak fields we only observe a small widening of the lines which is just noticeable. Hale and his collaborators (*Mount Wilson Report*, 1911–12) therefore used a special trick, illustrated in Fig. 4. The circular polarization of the Zeeman components was converted into a linear polarization by introducing a $\frac{1}{4}$-wave plate. Placing behind the latter a linear polarizer, it is possible to suppress the one or the other of the Zeeman components. In this way the splitting becomes much more conspicuous. In practice the Mount Wilson observers used a number of mica strips, placing them with their optic axes in alternately crossed directions immediately in front of the spectrograph slit. In this way, each of the circularly polarized Zeeman lines is broken up into short sections, which are alternately linearly polarized perpendicular to each other. The subsequent polaroid suppresses the two components

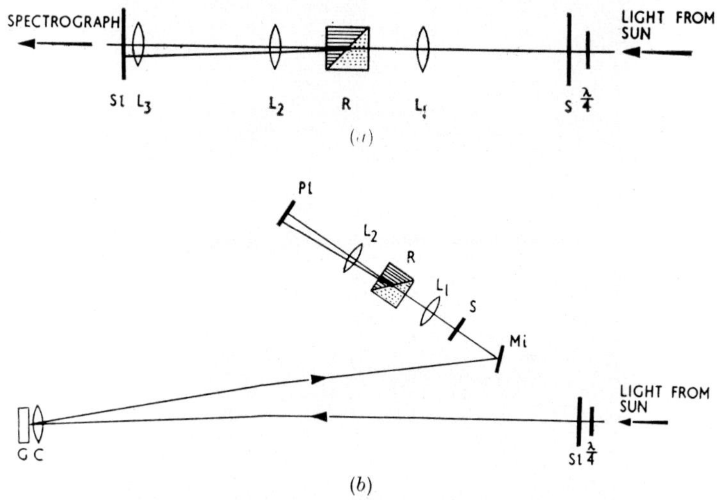

Fig. 5. Optical arrangements using a double image prism R for the investigation of Zeeman effects in spot spectra.
(*a*) Splitting of the image of the spot before the slit.
(*b*) Splitting of the spectrum of the spot behind the image of the spectrum. $\lambda/4$ = quarter-wave plate, S = screen, Sl = slit, L_1 and L_2 = lenses for intermediate imaging, L_3 = field lens, C = collimator, G = grating, Mi = mirror, Pl = photographic plate (v. Klüber, 1948)

alternately. In the spectrum those lines which show a Zeeman effect then display, for a longitudinal magnetic field-component, a "Meander" pattern which is characteristic of these investigations (Fig. 6). It is with this instrumental arrangement that nearly all the fundamental research by Hale and his collaborators on the magnetic fields of sunspots and the general field of the Sun have been carried out. It is possible to use this method for direct visual observations of high sensitivity in the solar spectrum. For example, a rotating $\lambda/2$ plate may be inserted into the optics (Hale and Nicholson, 1938) to reverse the sense of polarization periodically. A longitudinal field then gives rise to a periodic motion of the particular Fraunhofer line.

The observation of magnetic fields on the Sun remained nearly an exclusive privilege of the Mount Wilson Observatory for several decades. However, in 1941 rather extensive photographic observations of solar fields were started at the Potsdam Tower telescope (v. Klüber, 1947a, 1947b, 1947c, 1948). The methods of observation was varied in numerous ways. In one of the optical arrangements, shown in Fig 5 (*a*), a device using a birefringent prism produced two identical images of a sunspot on the slit of the spectrograph, which were polarized perpendicularly to each other. In this simple manner one can study both Zeeman components separately, in any required region of a sunspot, and the influence of local Doppler effects can be eliminated more easily. Furthermore, the useful breaking-up into strips as introduced by HALE can be obtained with this arrangement. An increase in the sensitivity of the measurements, and in particular a complete safeguard against the falsifying influences of local Doppler shifts, was gained by an optical arrangement previously employed for other purposes by WOOD (1914a, 1914b). The principle is illustrated in Fig. 7 (*a*). It requires a quartz plate of a few millimetres thickness, cut parallel to its optic axis

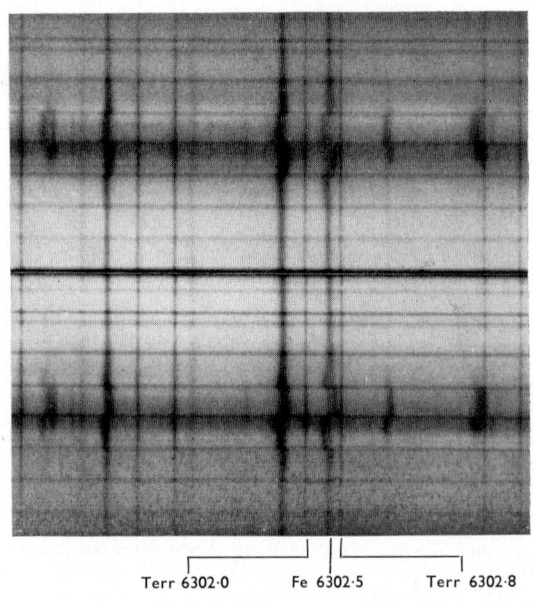

Fig. 6. Sunspot spectra taken with the arrangement of Fig. 5 (*a*), showing strong Zeeman effects and indicating a longitudinal magnetic field component of about 3000 gauss. The horizontal lines are caused by the edges of the mica strips. Upper part vertically, lower part horizontally polarized (v. KLÜBER, 1947c)

in such a way that it acts on one of two suitably chosen close Fraunhofer lines as a $\lambda/4$ plate (of higher order), and on the other at the same time as a $3\lambda/4$ plate. By a linear polarizer (or birefringent prism) the long-wave component of one line and the short-wave component of the other (or vice versa) can be suppressed. A simple measurement of the differences in wavelength of the two Fraunhofer lines then at once gives twice the amount of the Zeeman splitting $\Delta\lambda$, and the result is not affected by any local Doppler effects. In order to detect weak fields, this arrangement can be modified through the use of narrow $\lambda/4$ strips to produce the above-mentioned "Meander" pattern. For one of the two selected spectral lines the resulting Meander pattern is then a mirror image of that produced by the other line; a photograph is shown in Fig. 8. To find such weak fields, both lines can be successfully examined either directly in the solar spectrum or on the photographic plate with a stereo- or blink-comparator (H. v. KLÜBER, 1947c).

Another method for the visual or photographic observation of Zeeman effects has been introduced by ÖHMAN (1950). Here a Savart polariscope (WRIGHT, 1934) is

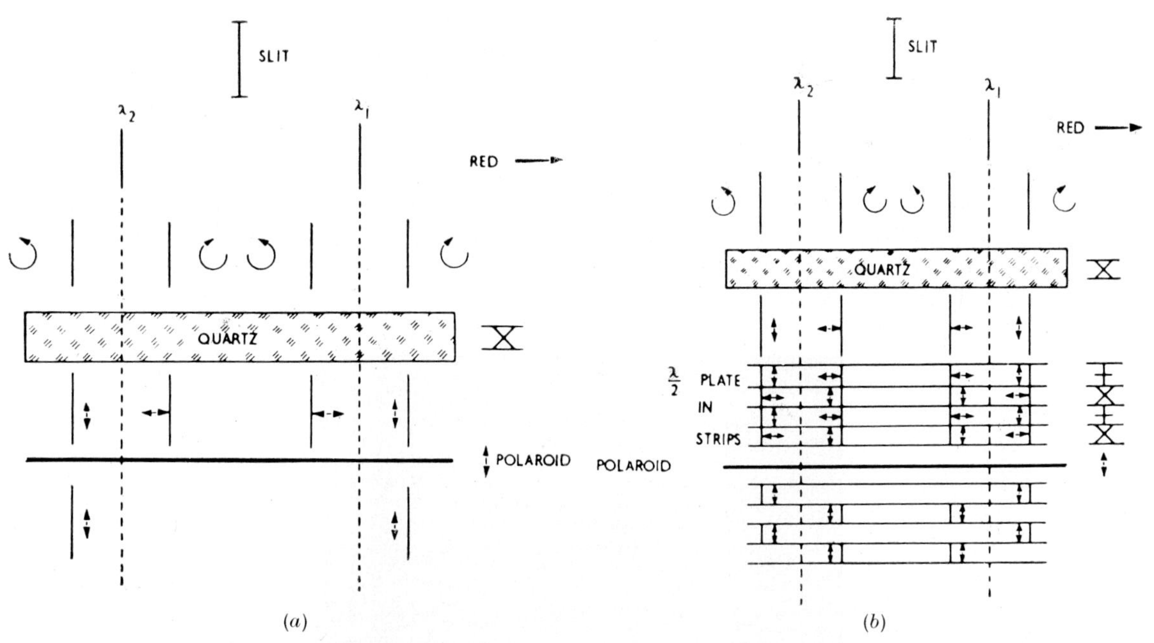

Fig. 7. The use of a thick quartz plate after Wood (1914a, 1914b) for investigating longitudinal Zeeman effects. The sense of polarization is indicated by arrows (v. Klüber, 1947a).
(*a*) Combination of the quartz plate and polaroid only.
(*b*) Combination of the quartz plate with small quarter-wavelength mica strips and polaroid for obtaining the "Meander" pattern of the spectral lines

placed behind the eye-piece of the spectrograph, preferably with a $\lambda/4$ plate in front of the spectrograph slit. Because of their polarization, the Zeeman components then have superimposed on them the fringe-system of the Savart plate; this yields a typical zig-zag line pattern, as illustrated in Fig. 9. As a method of searching visually

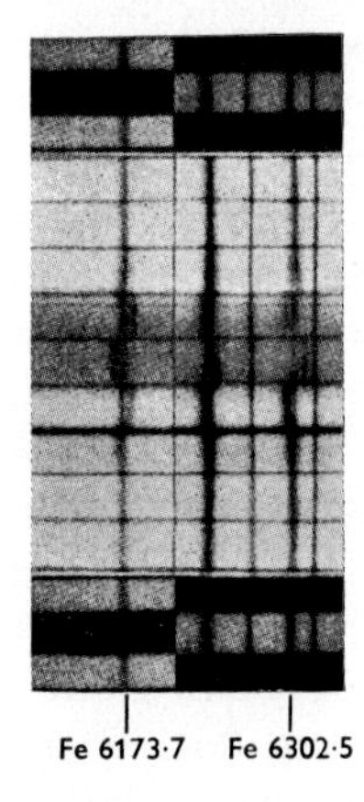

Fig. 8. Photograph of a spot field taken with the optical arrangement of Fig. 7b as seen in a comparator. The horizontal lines are caused by the edges of the mica strips. The two small unaffected lines belong to the oxygen of the Earth atmosphere. Control strips for adjustment of phase (v. Klüber, 1947a) are on the upper and lower edge

for magnetic fields on the Sun, Öhman has proposed the use of two monochromators formed by concave gratings.

The results of extensive observations of sunspot fields carried out on Mount Wilson over a period of many years have been summarized by Hale and Nicholson (1938). Statistical and other analyses of the Mount Wilson material have been published by

Nicholson (1933); Houtgast and van Sluiters (1948); and by Grotrian and Künzel (1949, 1951a, 1951b). This material has led to the recognition of a number of interesting relationships governing the magnetic behaviour of sunspots. In particular the relative polarity of bipolar groups, the change of polarity with the spot cycle, the average distribution of the inclinations of the lines of force, and the dependence of the field strength on the area of the spot. The present magnetic classification of the sunspots developed out of the Mount Wilson observations (Hale, Ellerman, Nicholson, and Joy, 1919; Mount Wilson Summary, in particular 1928; Hale and Nicholson, 1938).

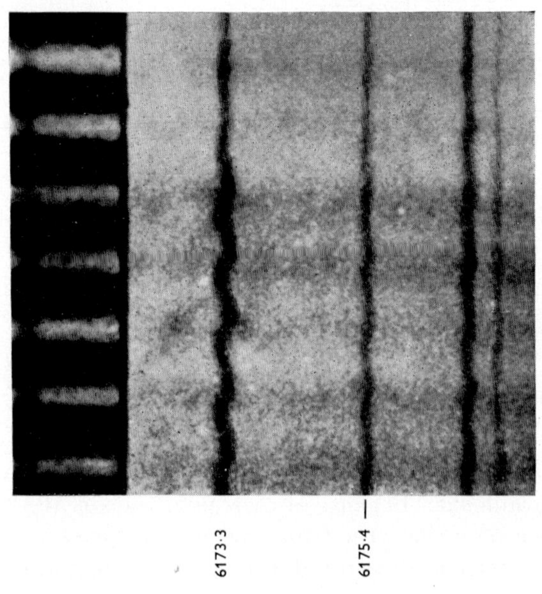

Fig. 9. Spectrum of a sunspot taken with a Savart plate in front of the slit, showing the Zeeman effect. Artificial polarization for control on one side (Öhman, 1950)

Systematic investigations of the Zeeman effect in a large number of Fraunhofer lines of the spot spectrum are due to King (1934), as well as to Tanaka and Takagi (1939). Zeeman effects in the molecular bands of calcium hydride and titanium oxide in the near infra-red have been measured by Nicholson (1938). A preliminary study of the forms of spot fields has been published by v. Klüber (1947a), based on observations obtained since 1942 with the Potsdam tower telescope; Fig. 10 shows an example. Some spectra of flares were also obtained there with large dispersion and polarizing optics, but without leading to any remarkable magnetic properties (Evershed, 1910; v. Klüber, 1947a; Bruce, 1948, 1949a, 1949b; Ellison, 1949). Systematic visual observations of the field strengths of sunspots are carried out at Mount Wilson Observatory, and regularly given in the *Publ. Astron. Soc. Pacific.* Systematic photographic observations of spot fields have been continued regularly at Potsdam since 1943 (v. Klüber, 1947a; Grotrian, 1948; Brunnckow and Grotrian, 1949; Potsdam Jahresberichte since 1950; Mattig, 1953; see also Thiessen, 1953b). The first volume of a series of publications dealing with photographic measurements of sunspot-fields, obtained at the Einsteinturm of the Astrophysikalisches Observatorium at Potsdam, has been published by Grotrian (1953a). Fig. 11

shows as an example a number of approximate types of the fields above rather large unipolar spots, which can be represented by the empirical formula:

$$H = H_0 \left(1 - \frac{r^4}{b^4}\right) . e^{-2\frac{r^2}{b^2}}.$$

Here H stands for the magnetic field strength, H_0 for its maximum value, and b for the distance from the centre of the spot; $b = r$ is the border of the penumbra, where the field strength is too small to be observed (GROTRIAN, 1953b). This formula

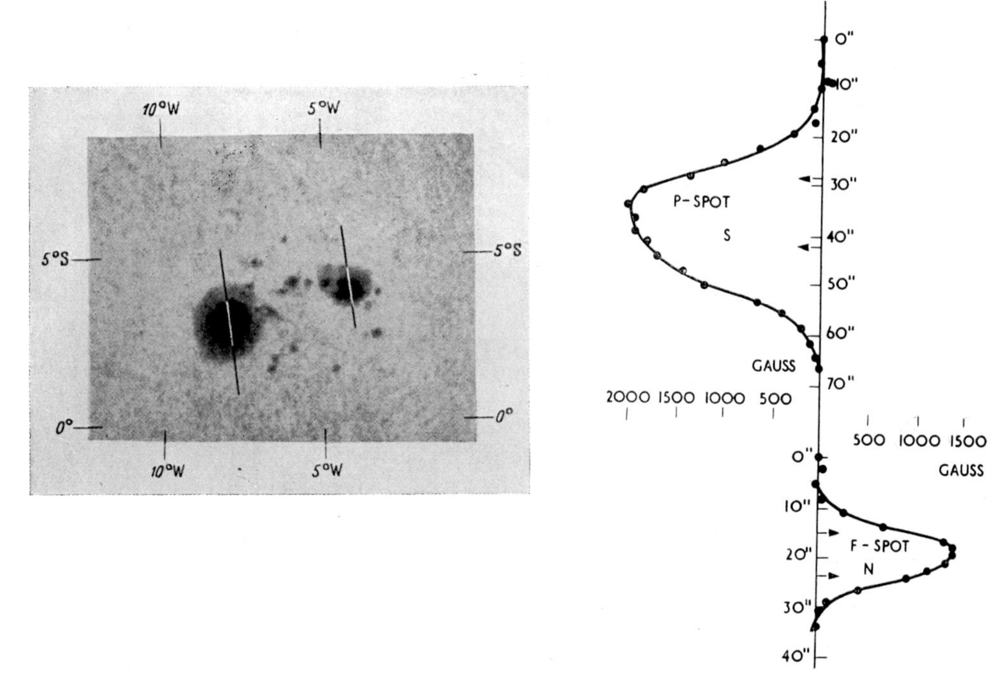

Fig. 10. Example of the determination of magnetic fields (longitudinal component) over sunspots. The position of the slit is indicated by the strokes across the spots. The arrows indicate the limits of the umbra (v. KLÜBER, 1947b)

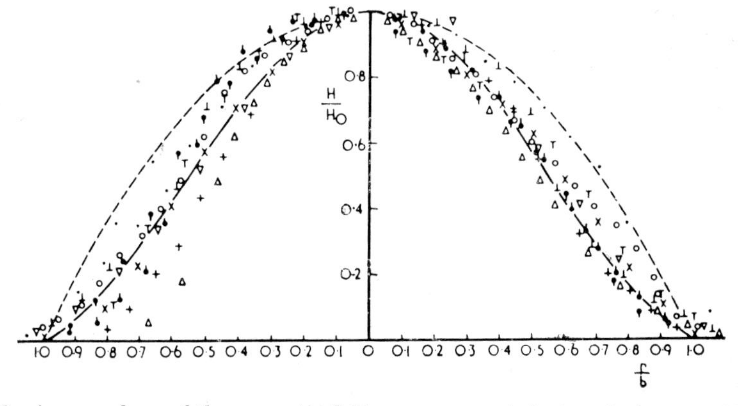

Fig. 11. Average form of the magnetic field over a normal single unipolar spot (MATTIG, 1953). —— mean of ten spots. - - - form of the field according to the empirical formula of BROXON

also furnishes a preliminary basis for the calculation of the magnetic induction through an average sunspot. A similar empirical formula, used by CHAPMAN (1943), has been given by BROXON (1942, 1949a, 1949b).

The structure of the magnetic lines of force above a sunspot has been discussed by SCHRÖTER (1953). HUBENET (1954) developed a theory of profiles of simple Zeeman components of a Fraunhofer line in a spot-field.

RICHARDSON (1939–40) has made a preliminary study of the weak magnetic field which we may expect in the regions farther from the spot, outside its penumbra. It appears that even at five to six diameters from the spot centre a weak field exists, showing about 5 per cent of the maximum spot field strength and an opposite polarity. Some attempts made at Potsdam to detect this field (using WOOD's quartz plates) have not been successful. The limits of measurement were 10–20 gauss, *i.e.* 1 per cent

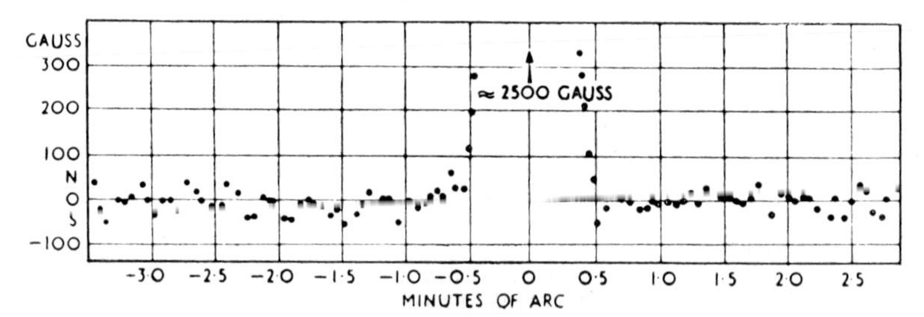

Fig. 12. Investigation of a possible magnetic field in the surroundings of a spot which also indicates the limit of accuracy using the photographic method of Fig. 8 (v. KLÜBER, 1947a)

of the field strength of the main spot (v. KLÜBER, 1947c). The fields to be expected are perhaps still below this limit (Fig. 12).

All these observational results on spot fields have led to a considerable number of theoretical investigations, the description of which is beyond the scope of this article. Most of them can be found in: GUNN, 1929a, 1929b; COWLING, 1932, 1933, 1946a, 1946b; LARMOR, 1934; ALFVÉN, 1942, 1945a, 1945b, 1946, 1947a, 1950; WALÉN, 1944a, 1944b; GUREWITSCH and LEBEDINSKIJ, 1945, 1946a, 1946b, 1948; GIOVANELLI, 1941; BIERMANN, 1950, 1952; BIERMANN and SCHLÜTER, 1951. Papers of a review type are: GROTRIAN, 1948; HALE, 1935; NICHOLSON, 1941; RUBASCHEW, 1949; GROTRIAN, 1949, 1950; BABCOCK and COWLING, 1953.

3. GENERAL MAGNETIC FIELD OF THE SUN

Apart from the study of spot fields HALE and his collaborators have since 1912 also used spectroscopic methods in an extensive series of observations of a general magnetic field of the Sun. Such investigations have again been taken up during the last few years by various investigators. The problem of the existence, form, and strength of a general magnetic field is of great cosmological significance. It has gained particular interest through the recent discovery of very strong stellar magnetic fields by H.W. BABCOCK (1947, 1948, 1949a, 1949b, 1950, 1951a, 1951b) and through BLACKETT's discussion of the relation between rotation and magnetic field (BLACKETT, 1947; see also CHAPMAN, 1948).

If one assumes as a first and obvious assumption that the Sun has a homogeneous magnetic dipole field similar to that of the Earth, having a polar field-strength H_p with the magnetic poles near the rotational poles, then the expression for the field strength at latitude ϕ is $H_\phi = H_0\sqrt{1 + 3 \sin^2 \phi}$. The inclination δ of the lines of force is given by $\tan \delta = 2 \tan \phi$ (see Fig. 13 (*a*)). If one observes the field strength as a function of solar latitude on the central meridian, using the longitudinal Zeeman effect, then one should expect the relationship shown in Fig. 13 (*b*). A maximum effect should occur at the points near $\phi = \pm 45°$ on the central meridian. For this type of work, HALE and his collaborators (HALE, 1913, 1914; SEARES, 1913; HALE,

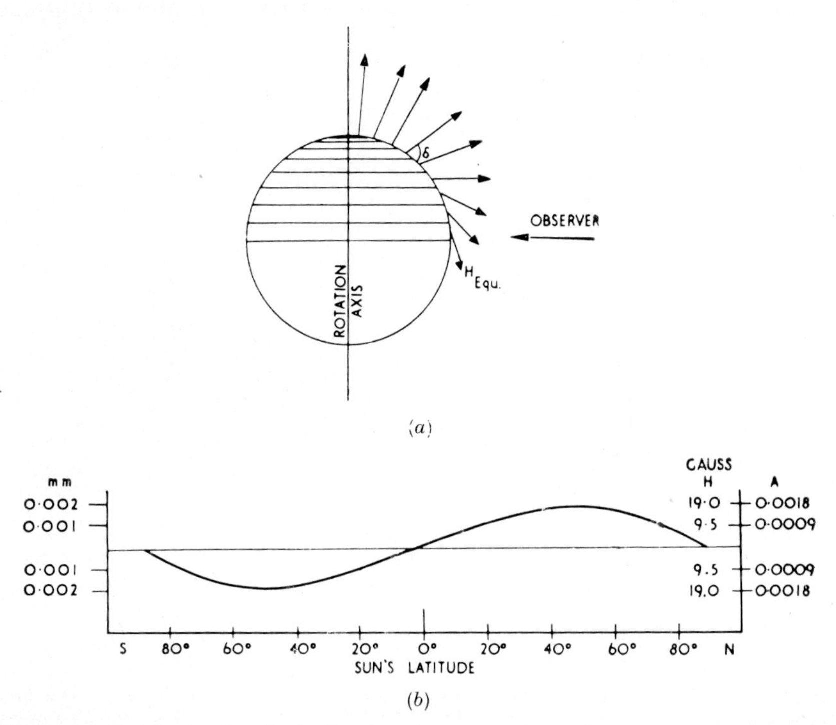

Fig. 13. (*a*) Sketch showing the inclination of magnetic lines of force of a sphere assuming a homogeneous dipole field. (*b*) Sketch representing the amount of the longitudinal magnetic component which exists on the magnetic central meridian of a sphere as a function of the latitude, supposing a dipole field. The right ordinate: on the right is the Zeeman shift $2\Delta\lambda$ (expressed in Å) for the line λ5250·2 Fe, assuming a polar field strength of 25 gauss. The left ordinate on the left: Zeeman shift of same line but expressed in millimetres on the photograph of the spectrum, assuming a plane grating with 12 m focal length in the second order. Maximum effects are to be expected at $\pm$ 45° latitude

SEARES, v. MAANEN, ELLERMANN, 1918) used mainly the 75-ft plane-grating spectrograph of the 150 ft Mount Wilson tower telescope in the second and third order, giving a resolution of about 200,000. Fig. 14 illustrates how, indeed, a number of selected Fraunhofer lines showed the expected behaviour. Surprisingly enough, however, other lines belonging to the same observational material did not show any effect, a phenomenon which has not hitherto been explained. Furthermore, the observers believed that they could deduce from the seasonal variation of their results a certain deviation of the magnetic axis from the rotational axis; they gave the angle between the two as 6°, and the period of mutual revolution as 31·5 days. The value of the polar field strength is given by these authors after a careful discussion of their data as between 25–50 gauss for medium strong Fraunhofer lines. The field shows a southern polarity near the north pole of rotation like that of the Earth.

It also appeared that there exists a clear correlation between the estimated field strength and the Rowland intensity of the line examined, as shown in Table 1, column 7. Originally this result was explained by assuming a rapid decrease of the field with height in the solar atmosphere. Should this intensity effect be real, there is as yet no satisfactory explanation for it. Later observations, which we shall deal with below, do not seem to confirm sufficiently HALE's value of the general field, the existence and form of which still remain an open question. The Mount Wilson work of

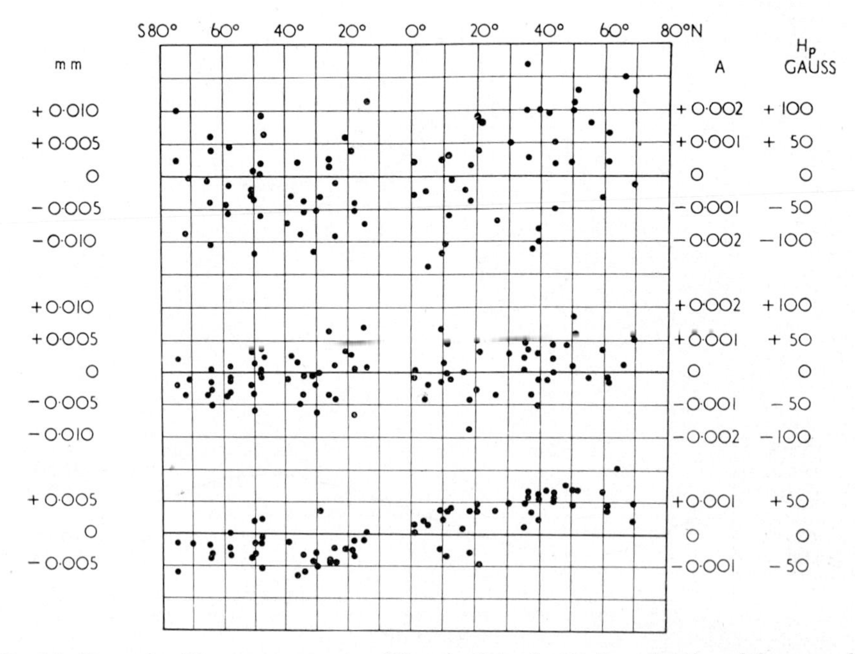

Fig. 14. Example of the measurements of HALE's plates for the investigation of the general magnetic field of the Sun (HALE, *etc.*, 1918). The Zeeman shift of the line λ5856·1 Fe from observations on the central meridian and measured by three different observers. *Abscissa:* heliographic latitudes. *Ordinate, left:* the quantity $2\Delta\lambda$, in millimetres, being the actual measurement on the plates. *Ordinate, right:* the corresponding value of the generally used unit $\Delta\lambda$, expressed in Å. The next column to the right (H_p) gives the polar field-strengths in Gauss, which would apply if the preceding Å-values would have been found for the latitudes $\pm$ 45°.

subsequent years, with various instrumental arrangements, and also the re-measurements of some of the earlier plates, did not give sufficient agreement (*Mount Wilson Report*, 1922–36). This apparently also applies to the attempts made by HALE and his collaborators (*Mount Wilson Report*, 1932–33) to measure the Zeeman effect of the line λ5247·6 Cr directly at the telescope, using polarizing optics and a photocell and amplifier. EVERSHED (1939) has mentioned that HALE's results could perhaps be affected by small local Doppler effects. Recent measures, discussed further below, seem to establish the presence of appreciable variable local magnetic fields, and this fact may have affected most of the observations. As the theory of the Zeeman effect had not yet been developed fully in 1918, which was the time of HALE's measurements, some of the Fraunhofer lines used by the Mount Wilson observers had rather unfavourable Zeeman splittings (*i.e.* all the types pictured as Nos. 2–22 in Fig. 2). In spite of this unfavourable selection, the results are probably not vitiated systematically, as has been shown by v. KLÜBER (1952). The methods described above do not suffice to measure with certainty field strengths smaller than 20 gauss (corresponding to a $\Delta\lambda$ of the order of magnitude of 0·001 Å). A much greater sensitivity of Zeeman measurements with spectrographs of the ordinary type has not seemed possible until recently, even with large gratings, since in practice the resolution remained below 200,000. It was only very recently that progress in the production of

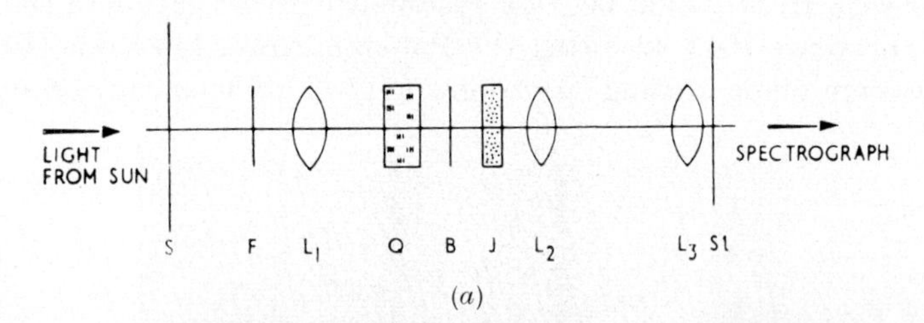

(*a*)

(*b*)

Fig. 15. (*a*) Sketch of an optical arrangement for the investigation of the Sun's general magnetic field by using Wood's (1914a, 1914b) thick quartz plate and a Fabry-Perot single-plate interferometer (v. Klüber, 1947c). S = screen, F = heat filter, L_1 and L_2 = lenses for intermediate imaging, Q = Wood's quartz plate, P = linear polarizer (Bernotar), J = interferometer plate, L_3 = field lens, Sl = slit.
(*b*) Actual arrangement in front of the slit of the spectrograph at the Potsdam Tower Telescope

special diffraction gratings at Mount Wilson and elsewhere (Babcock, 1944; Harrison, 1949a, 1949b) has altered the situation. For the investigation of these weak fields it was obviously necessary to increase the resolving power of the spectroscopic equipment first of all.

Numerous observational series along these lines are due to H. D. Babcock, who, since about 1938, has worked with a Lummer plate (see notes in the *Mount Wilson Reports*). In the polarizing optics he used a thick quartz plate of the kind described above (Babcock, *Mount Wilson Report*, 1939–40, *etc.*; Babcock, 1941, 1944, 1948; *Mount Wilson Report*, 1947–48) for Wood's method (Wood, 1914a, 1914b; v.

KLÜBER, 1947c). He made use chiefly of the line λ6173·3 Fe, but later he also observed some lines in the green spectral region. A preliminary reduction of the plate material from the years 1940–47 gave values of the general field in the range 5–60 gauss for some of the plates, but for the remainder it gave no effect at all. Thus the final result is still uncertain (v. KLÜBER, 1952). Independently, v. KLÜBER (1947b, 1948) together with H. MÜLLER, began a systematic investigation of the general field of the Sun in the years 1943–44, using the Potsdam tower telescope (Einsteinturm). They used the large plane-grating spectrograph (f = 1200 cm) in conjunction with a

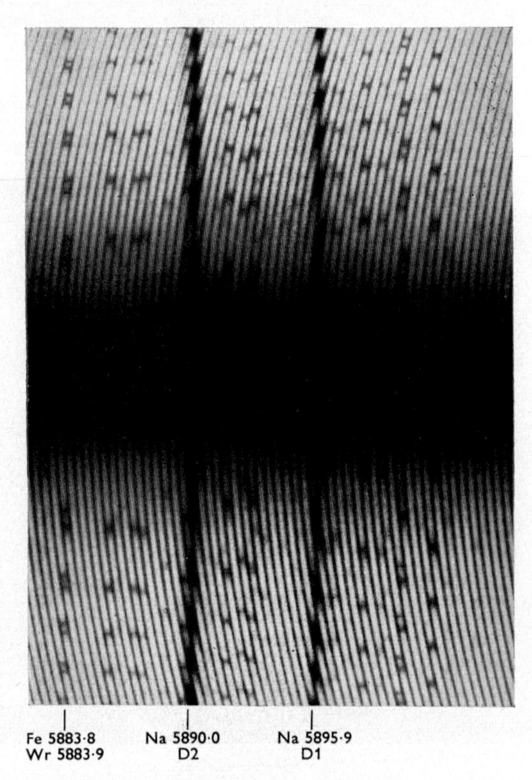

Fig. 16. Example of interference pattern in the solar spectrum, obtained with a Lummer plate and a grating spectrograph (v. KLÜBER, 1951). Resolving power = 600,000; range of interferometer = 0·40 A.U.

one-plate quartz interferometer of the Fabry-Perot type and the Wood arrangement of a thick quartz plate (Fig. 15 (*a*), (*b*)). A region of the Sun's surface was imaged with the tower telescope through a small opening in the screen S, which is at the focus of a lens L_1. The light then passed through L_1, through the Wood quartz plate Q, through a Polaroid B (Zeiss-Bernotar), and through the interferometer plate I, which was silvered on both sides. The lens L_2 imaged the diaphragm in S, together with the interferometer fringes (which are at infinity), on to the slit Sl of the large grating spectrograph; L_3 is a field lens. If the thickness of the interferometer plate and the slit width are chosen suitably (about 6 mm and 0·12 mm respectively), we find in the image plane of the spectrograph the well-known channelled Fraunhofer spectrum with readily measurable interference knots (illustrations are given in v. KLÜBER, 1947c and 1948). The practical resolving power reached about 350,000; the range between orders of the quartz plate interferometer was about 0·14 Å. The diameter of the Sun's image was 25 cm. In Autumn, 1944, the observers obtained about sixty good plates with the combination (which was suited to the Wood quartz plate) of the lines λ5131·5 Fe and λ5250·2 Fe. The line λ5123·7 Fe which does not show any Zeeman effect at all, served as check. The formal accuracy of a single line measurement amounted to about 0·0005 Å. Because of wartime conditions only some of these plates (namely about 2000 interference knots) could be measured completely. All observations referred to the two points at $\pm$ 45° on the central meridian of the Sun. These measurements leave no doubt of the fact that any dipole field with the assumed orientation must have been considerably smaller than 10 gauss at the time of observation. A similar arrangement was used later, in 1949–50, by v. KLÜBER (1951)

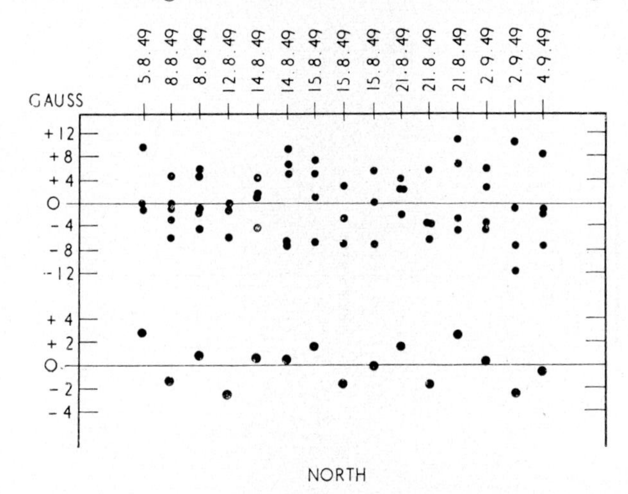

Fig. 17. Results from plates taken at Cambridge in 1949 with a Lummer plate (investigation of a possible general magnetic field of the Sun; v. KLÜBER, 1951). The measurements refer to the point at $\pm$ 45° latitude on the central meridian. Upper part: values for individual pairs of fringes. Lower part: means. Abscissa: date of observation

at Cambridge for two further observational series. This time a quartz Lummer plate, made by A. HILGER of London and of outstandingly good quality, served as interferometer. The range in the green region was 0·33 Å, and the effective resolution reached the high value of about 600,000. The quartz plate of the Wood arrangement had a thickness of 10·222 mm, suitable for the line pair λ5250·2 and λ5263·3 Å, with the Zeeman types (0)3 and (0)5/2. A plate showing the resulting interference system in the solar spectrum is reproduced in Fig. 16. Again, the two points investigated were lying at $\phi = \pm$ 45° latitude on the Sun's meridian. The standard error of one setting on an interference knot was 0·002–0·005 mm. A single value, derived from one pair of plates, had the mean error of $\pm$ 3 gauss. The mean error of all plates was 0·8 gauss for each of these points on the Sun. The scatter of the single observations about their mean is shown in Fig. 17. From the measurement of more than 1000 interference knots on the plates taken in August and September, 1949, *i.e.* the first series, we can conclude that at this time any general field in the form of the assumed dipole field certainly did not exceed 1–2 gauss. The results of the second series, taken in Summer, 1950, indicate a similar result (v. KLÜBER, 1954).

Two other methods, which are visual, and therefore permit much more rapid measurement of line displacements, have been used by THIESSEN (1946, 1947).

Fig. 18 (*a*) illustrates his optical arrangement, which is similar to that of Fig. 15 (*a*); the light from the Sun, imaged on a small opening in the screen S, passes through a heat filter F, is made parallel by the lens L_1, and passes further through a fixed $\lambda/4$ plate, a rotating $\lambda/2$ plate, the polaroid P, and the Fabry-Perot interferometer J, consisting of two plates, and it is finally focused together with the interference pattern on to the slit Sl of the concave grating spectrograph (in the Wadsworth arrangement). In the image plane Sp one thus obtains interference knots again of the kind illustrated in Fig. 15 (*b*). If one rotates the $\lambda/2$ plate, then the sense of polarization of the Zeeman components varies periodically; the knots in the image plane Sp then appear to oscillate by very small amounts. For fields smaller than 100 gauss,

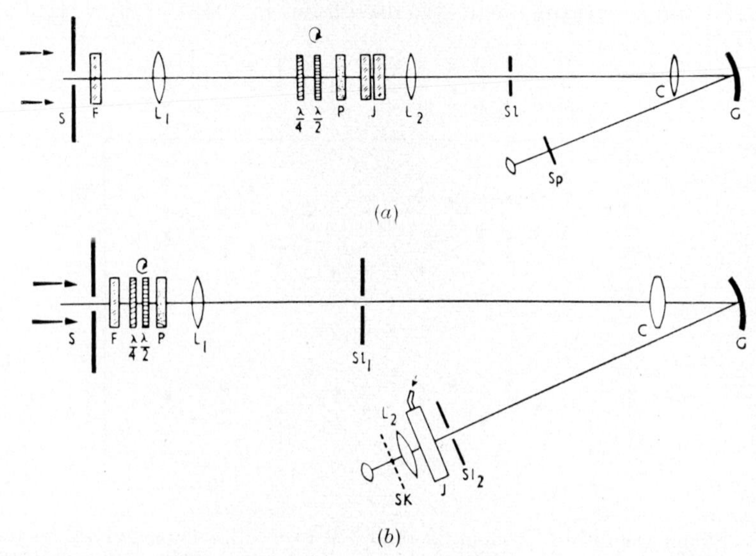

Fig. 18 (*a*) and (*b*). Optical arrangements for the visual observations of small magnetic fields with an interferometer (THIESSEN, 1946). Meaning of notation as given in text

these motions of the knots are so small that they cannot be reliably measured. THIESSEN therefore used an elegant null method to make the measures quantitative; in this method the air pressure in the space enclosed between the two interferometer plates is varied with the frequency of rotation. Thus the periodic movement of the interference knots could be compensated by appropriate pressure variations, and the required pressure is then a measure of the splitting; the calibration was done empirically. This method yielded safe results down to about 70 gauss. Its accuracy was later increased by using an arrangement similar to that frequently used in interference spectroscopy (TOLANSKY, 1947), in which the interferometer is placed behind the image plane of the spectrograph as shown in Fig. 18 (*b*). If the diaphragm Sl_2 is opened by an amount smaller than the range of the interferometer, and provided the dimensions and the light power of the arrangement are properly chosen, then (in the case of a Fraunhofer spectrum) the complete dark interferometric ring system of a Fraunhofer line can be seen in the plane SK of the image on a homogeneously coloured background. If then the $\lambda/2$ plate is rotated, and if there is a Zeeman effect present, the whole ring system vibrates. This vibration can be compensated by pressure variations in the interferometer.

This arrangement was used at first in conjunction with a coelostat, an objective of 900 cm focal length, and a concave grating spectrograph of 457 cm radius of curvature. The accuracy was about 10 gauss for the line $\lambda 6173 \cdot 3$ Fe. The visual observations of 1945 taken at the Hamburg Observatory gave for the two points at $\pm 45°$ a polar field strength of 53 ± 12 gauss. This value, however, could not be confirmed by later observations, and is also in conflict with the slightly earlier Potsdam measures mentioned above. A similar visual series taken with the Hamburg

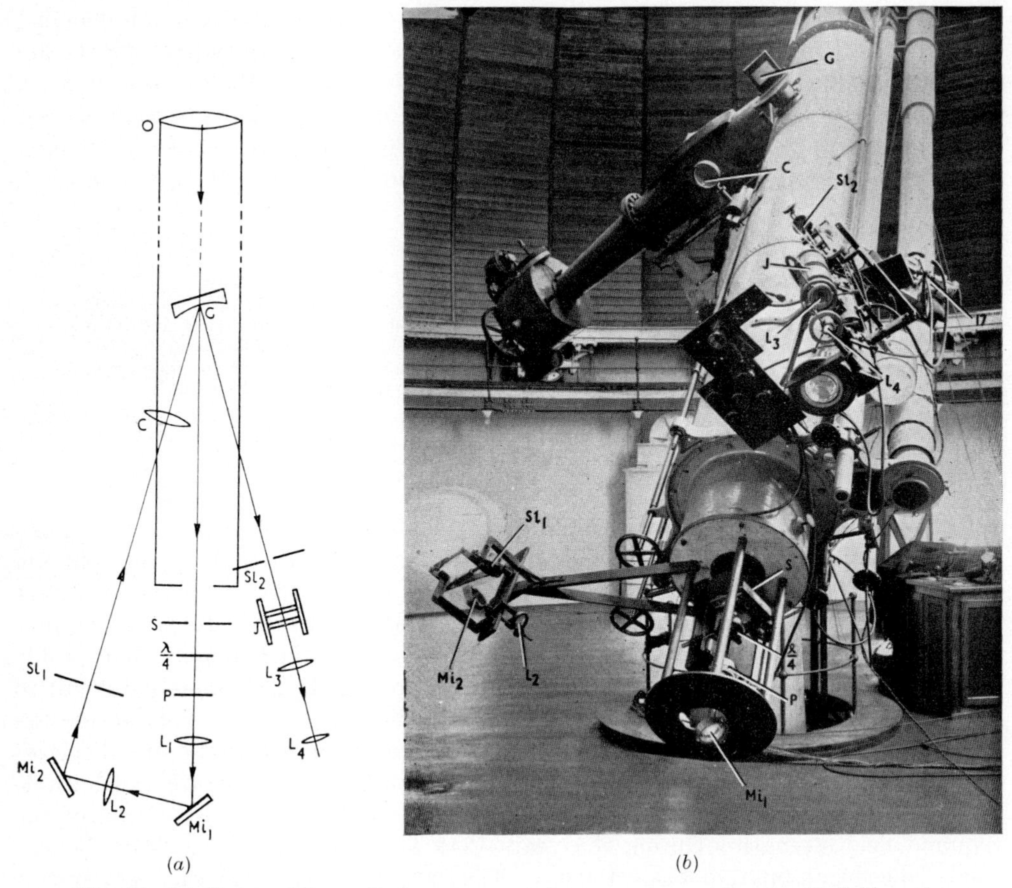

(a) (b)

Fig. 19. (a) Sketch of the optical arrangement for measuring small magnetic fields with a Fabry-Perot interferometer attached to a large telescope (THIESSEN, 1948a); meaning of the letters as before. (b) Actual arrangement attached to the telescope: $f = 900$ cm, $\phi = 60$ cm, $\lambda/4$ = rotating $\lambda/4$-plate, P = polaroid, C = collimator, G = concave grating, I = interferometer, L_4 = eyepiece, 17 = vacuum pump

60-cm refractor ($f = 900$ cm) followed in 1947 and 1948 (THIESSEN, 1949a, 1949b). The optical arrangement was essentially the same as that of Fig. 18 (*b*) and is sketched in Fig. 19 (*a*) and Fig. 19 (*b*). The same two solar points and the following three Fraunhofer lines were used, $\lambda 6173 \cdot 3$ Fe, $\lambda 5247 \cdot 6$ Cr (as was done by HALE), and $\lambda 5250 \cdot 2$ Fe (with the particularly large Landé factor $g = 3$). The accuracy of the method, for suitably chosen lines, was given as about 10 gauss. The results showed that during this observational period there was no magnetic field larger than 5 gauss of the polarity found by HALE. On the contrary, it seemed rather that a very weak field of opposite polarity was indicated, but this is by no means certain. The reduction of the whole material for the years 1947 and 1948 (always assuming a dipole field) led to a polar strength of $-1{\cdot}5 \pm 3{\cdot}5$ and of $-1{\cdot}4 \pm 3{\cdot}3$ gauss, respectively, where the negative sign is chosen to indicate a polarity which is opposite to that of the Earth (with respect to its rotation).

Since 1949 THIESSEN has succeeded in further increasing the accuracy by replacing the measurement of the Zeeman splitting, on which nearly all observations have been based, by measurements of the polarization made with an electron multiplier. For very small fields, and correspondingly small splittings of the Zeeman components, we know that the circularly polarized components of the longitudinal effects are superimposed on each other in the way indicated in Fig. 20. Hence, in each wing of the line, one or the other of the two directions of polarization will more or less predominate according to the size of the splitting and the polarity. By rotating a $\lambda/2$ plate, followed by a fixed polarizer, a longitudinal effect can then be observed,

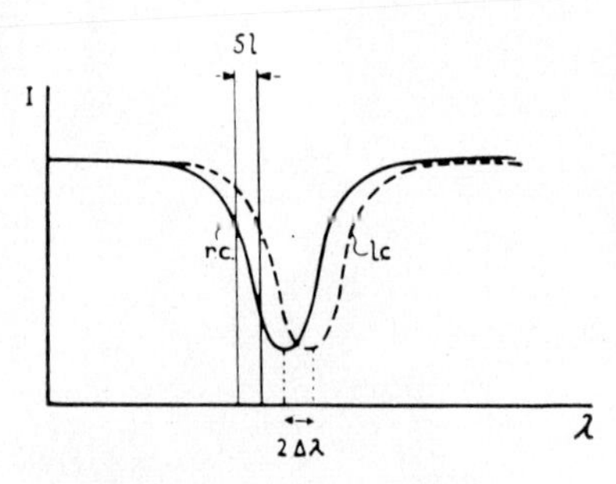

Fig. 20. Sketch of the splitting of a Fraunhofer line by a longitudinal Zeeman effect into two circular polarized components, displaced by $2\Delta\lambda$. The position and width of a slit for measuring the polarization is indicated by *Sl*

as for the previous methods, by a periodic intensity variation. This difference is then measured with a IP21 photomultiplier and furnishes information about the splitting of the two Zeeman components when this is too small to be detected by direct measures of the displacement. To obtain a quantitative calibration from first principles would need a rather accurate knowledge of the instrumental line profile and of the accurate position of the slit *Sl* on this profile. It is extremely difficult to measure these quantities, and therefore the periodic pressure variations inside the interferometer were used to produce artificially a very small periodic line displacement, corresponding to a certain known Zeeman effect. The combination of suitable measurements of our points at latitudes $\pm 45°$ then led to the derivation of the required field strength. During the years 1949 and 1950 the line $\lambda 5247{\cdot}6$ Cr was used. The slit diaphragm was 0·4–0·2 Å. The optical arrangement was very similar to that of Fig. 19 (*a*) and (*b*). The measurements were made with a galvanometer of sharp resonance with a time-constant of decay of 63 sec. The range of the interferometer was 0·5 Å. A small residual polarization (which is produced in all instruments of this kind and is very troublesome) has, as far as possible, been compensated by a rotating sector diaphragm in front of the main objective lens. During the period August–September, 1949, the measurements gave a general solar field of $H_p = -1{\cdot}5 \pm 0{\cdot}75$ gauss. Another series obtained during April–August, 1951, gave $H_p = -2{\cdot}4 \pm 0{\cdot}5$ gauss.

A further attempt to detect weak solar fields by measuring the polarization of Zeeman components was made by KIEPENHEUER in 1950–51 at the Yerkes and Mount Wilson Observatories (KIEPENHEUER, 1953a, 1953b). This attempt, however, is still open to objection, and the author is now engaged in the development of a new set-up at the Fraunhofer Institut at Freiburg i. Br. Instead of a spectrograph it is intended to use an interference filter of very narrow bandwidth, together with an etalon plate, and the aim is to achieve a range of transmission of 0·04 Å, in the expectation that this arrangement will have a greater light-gathering power than large spectrographs.

In 1952 a new and very efficient instrument for regular examination of the whole solar disk, testing the existence of small longitudinal Zeeman effects, was set into operation by H. D. and H. W. BABCOCK (1952, 1955) and H. W. BABCOCK (1953) at the Hale Solar Laboratory in Pasadena. The spectrograph uses an extremely powerful new plane grating having 600 rulings per millimetre over a length of 20 cm. It is blazed in the fifth order and there gives a resolution of 600,000 with a dispersion of 11 mm per A. This high resolving power eliminates the necessity of using the

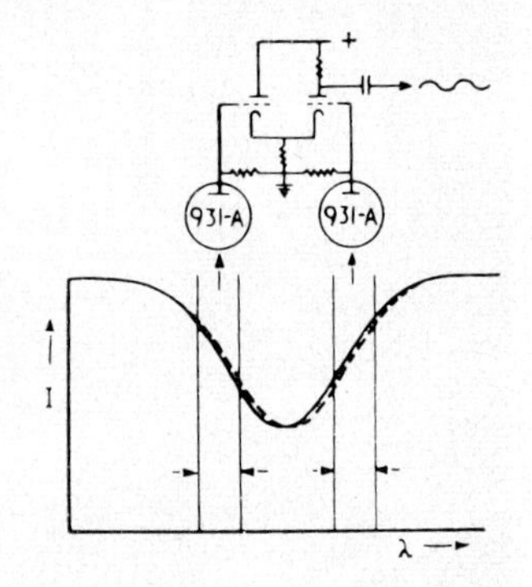

Fig. 21. Sketch of a double-slit arrangement for measuring small Zeeman-shifts (BABCOCK, H. D. and H. W., 1952). Light passing through the two slits Sl_1 and Sl_2 is simultaneously measured by the two 931-A photomultipliers

much more complicated interferometer. The polarizing arrangement in front of the slit consisted of a $\lambda/4$ plate, one $\lambda/2$ plate rotating with 30 r.p.s., and a linear polarizer. An essential improvement of the instrument was recently effected by replacing the two mica plates with an electro-optical crystal (BILLINGS, 1949a, 1949b). An alternating voltage of 9500 V peak is applied to the electrodes of this crystal at a frequency of 120 c/s, and gives rise to a variation in birefringence corresponding to an optical phase delay of $\pm\ \lambda/4$. This set-up has the great advantage of avoiding the use of any rotating parts in the optics, which frequently causes considerable difficulties in such sensitive arrangements. In the image plane of the spectrum a narrow slit is placed across each wing of the Fraunhofer line to be examined (mainly λ5250·2 Fe) (see Fig. 21). A photomultiplier behind each of the slits receives the transmitted light. These photomultipliers are coupled in such a way that only their output difference causes a signal. This has the advantage that the measurements are hardly influenced by accidental variations in visibility, or by a small residual polarization in the optics. This difference signal is applied to a narrow band amplifier, a phase-sensitive rectifier and a smoothing circuit. During the observation the whole solar image of 40 cm diameter was scanned over the 15 mm long first slit. Longitudinal Zeeman splittings were then recorded as vertical deflections on a cathode ray tube. The horizontal deflection on the screen corresponds to the motion of the solar image on the slit. A horizontal line corresponding to "zero-splitting" is also recorded, and the whole pattern is photographed. In 50 minutes, nineteen equally spaced traces over the Sun's disk can be made. The Doppler displacement caused by the Sun's rotation is compensated continuously by tilting a glass plate in front of the first slits. For the calibration a circular polarizer is temporarily introduced (with the solar image at rest) and the compensating glass plate tilted by a known amount, thus introducing an artificial, known displacement. Fig. 22 illustrates the head of this magnetograph, and Fig. 23 gives a record of the Sun's whole disk obtained with this instrument. A deflection

Fig. 22. Magnetograph (H. W. BABCOCK, 1953). Head of the 75-ft spectrograph of the Hale Solar Laboratory, Pasadena, California. A = screen for imaging the Sun; B = electro-optical retardation plate; C = linear polarizer (Nicol) above slit of spectrograph; D = detector arrangement in the image plane of spectrum; E = amplifier; F = cathode-ray tube with recording photographic camera; G = compensator for Doppler-shifts. The large plane grating and the collimator are at the bottom of the 75-ft pit.

of the size of the distance between two successive fiducial traces corresponds to 10 gauss. The accidental noise fluctuations are of the order of 1 gauss. Observations with this instrumental arrangement, which has been in regular use for only a short time, have shown that, even on days with very little solar activity, and even in the absence of visible sunspots, the Sun's surface shows a rather complicated magnetic structure. Even in high latitudes it is still possible to find fields of the order of 5 gauss in the longitudinal component. We may hope that observations with this excellent instrument, if carried out regularly through many years, will lead to a wealth of new insight into the magnetic conditions of the Sun's surface.

In Table 1 are listed some results derived from the more extensive published investigations of the Sun's general magnetic field. There also exist a number of other measurements the details of which have not been published, but which are mentioned in several *Mount Wilson Annual Reports*. It is important to remember that this material has (with very few exceptions) been secured from the beginning in the belief that there are measurable Zeeman effects on the Sun, and that the field is really that

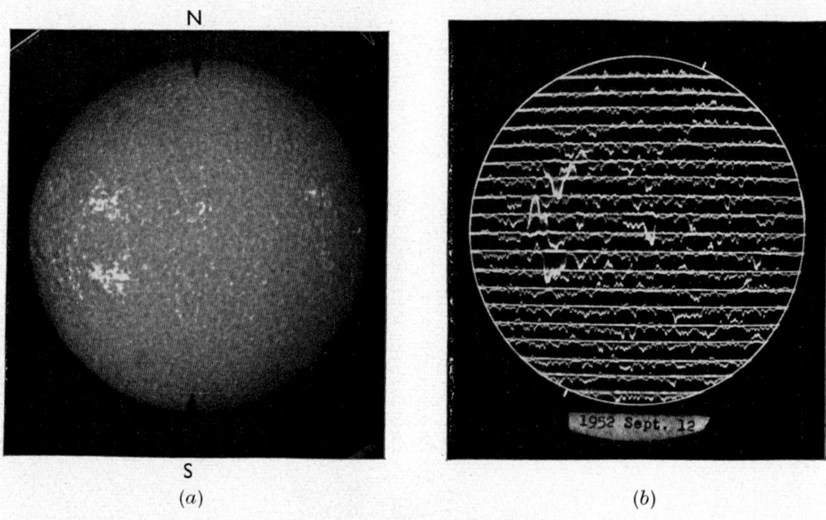

Fig. 23. Example of the records obtained with the arrangement of Fig. 21 (BABCOCK, 1932).
(*a*) Ca II-spectroheliogram of the Sun (1952, 12th September).
(*b*) Magnetic record of the Sun on the same day. A deflection equal to the separation of the fiducial traces corresponds to about 10 gauss.

of a dipole orientated towards the axis of rotation. It was always only the longitudinal component in the line of sight that was measured. From all available material the following conclusions can be drawn:

(*a*) The spectroscopic observations of the kind discussed above are not yet sufficient to confirm the existence of a general magnetic field, although for other reasons the existence of such a field (perhaps smaller than 1 gauss) appears quite probable.

(*b*) Size, polarity, and form of such a field are therefore not yet established.

(*c*) A variable general field is not impossible, but this, too, cannot yet be deduced with sufficient certainty from the present observations (contrary to the views mentioned in the Draft Report of the International Astronomical Union Meeting, 1952).

Many contradictions and difficulties in the available measurements are probably due to small local fields, which may vary relatively rapidly, as has recently been suggested by both observation and theory (H. D. BABCOCK and H. W. BABCOCK, 1952). Of special interest is a note by ALFVÉN (1952). In his opinion, it is possible to find appreciably variable magnetic fields in the granulation elements of the photosphere (ALFVÉN, 1947); turbulence, and the dependence of the absorption of a Fraunhofer line on the temperature, could completely falsify or even altogether prevent measurements of a weak general magnetic field by the Zeeman effect.

Table 1

No.	Year	Author	Reference	Principal instrument	Record	Polar field strength* (in gauss)
1	1913	HALE, *etc.* (Mount Wilson)	HALE, 1913	Grating, Resolving power ~ 200,000	Photographic	$\leqslant +50$
2	1918	HALE, *etc.* (Mount Wilson)	HALE *et al.*, 1918	Grating, Resolving power ~ 200,000	Photographic	+ 40 + 20 + 3 Gauss 0 2 4 Intensity
3	1922, *etc.*	HALE, *etc.* (Mount Wilson)	Mount Wilson, 1922–36; LANGER, 1936	Grating, Resolving power ~ 200,000	Photographic	Not yet conclusive
4	1932–33	HALE, *etc.* (Mount Wilson)	Mount Wilson, 1922–36	Grating, Resolving power ~ 200,000	Photoelectric amplifier	Not yet conclusive
5	1933–34	NICHOLSON, ELLERMAN, HICKOX	Mount Wilson, 1933–34	Grating, Resolving power ~ 200,000	Visual	$-3{\cdot}6 \pm 1{\cdot}7$
6	1939–48	H. D. BABCOCK (Mount Wilson)	Mount Wilson, 1939–40; BABCOCK, 1948	Lummer-plate interferometer, Resolving power = 350,000	Photographic some visual	Not yet conclusive
7	1944	v. KLÜBER, H. MÜLLER (Potsdam)	v. KLÜBER, 1951	Fabry-Perot interferometer, Resolving power = 350,000	Photographic	$\leqslant 10$
8	1945	THIESSEN (Hamburg)	THIESSEN, 1946	Fabry-Perot interferometer	Visual	$+53 \pm 10$
9	1947	THIESSEN (Hamburg)	THIESSEN, 1949a	Fabry-Perot interferometer	Visual	$-1{\cdot}5 \pm 3{\cdot}5$
10	1948	THIESSEN (Hamburg)	THIESSEN, 1949a	Fabry-Perot interferometer	Visual	$-1{\cdot}4 \pm 3{\cdot}3$
11	1948–49	NICHOLSON, HICKOX (Mount Wilson)	BOWEN, 1949	Grating, Resolving power ~ 200,000	Visual	$+2{\cdot}3 \pm 3{\cdot}3$
12	1949	THIESSEN (Hamburg)	THIESSEN, 1952	Fabry-Perot interferometer, Photomultiplier	Galvanometer Visual	$-1{\cdot}5 \pm 0{\cdot}75$
13	1949	v. KLÜBER (Cambridge)	v. KLÜBER, 1951	Lummer-plate interferometer, Resolving power = 600,000	Photographic	< (1 to 2)
14	1950	v. KLÜBER (Cambridge)	v. KLÜBER, 1954	Lummer-plate interofermeter, Resolving power = 600,000	Photographic	< (1 to 2)
15	1951	THIESSEN (Hamburg)	THIESSEN, 1952	Fabry-Perot interferometer, Photomultiplier	Galvanometer Visual	$-2{\cdot}4 \pm 0{\cdot}5$
16	1952	H. D. and H. W. BABCOCK (Mount Wilson)	BABCOCK, 1953	Grating, Resolving power = 600,000, Photomultiplier	Electronic, combined with Photographic	Probably weak, and polarity opposite to that of Earth

* *Note.* The + sign in this last column indicates that the polarity agreed with that of the magnetic field of the Earth (with respect to rotation).

Under certain conditions the magnetic field of the Sun could also have a form which deviates from the dipole character (SCHWARZSCHILD, 1949; GJELLESTAD, 1952; FERRARO and MEMORY, 1952). In this case most of the observations available at present would have to be explained quite differently, and also different observational procedures would be necessary.

The question of the existence of a general magnetic field of the Sun, and the possibility of its detection, has given rise to numerous theoretical investigations, a discussion of which is beyond the scope of this article; we just quote some of the more recent papers in this field: CHAPMAN, 1928a, 1928b, 1948; DESLANDRES, 1929; ALFVÉN, 1942; WALÉN, 1946; FERRARO, 1935, 1937, 1938; FERRARO and UNTHANK, 1949; COWLING, 1945; FALLIN, 1948; SWEET, 1949; KIEPENHEUER, 1953c.

4. FUTURE PROBLEMS

The investigation of solar Zeeman effects, and the derivation of magnetic fields from them, offers a wide field for the future. Investigations of spot fields and their variability, particularly in view of their possible dependence on motions in the spots, are almost completely lacking.

Weak fields farther out of sunspots still need thorough investigation. Extensive systematic observations over the whole solar disk in the search for weak local or general fields are important. In this respect the method developed and applied by H. D. and H. W. BABCOCK (see H. W. BABCOCK, 1953) appears to be particularly hopeful.

In principle it should also be possible to find small magnetic fields by a careful measurement and discussion of the line profiles of certain spectral lines which are influenced by Zeeman effects. Such observations and calculations would, however, be rather involved, and would probably have to be restricted to the study of a limited range of phenomena of special interest.

Attention has been drawn by THIESSEN (1952c, 1952d) to further spectroscopic possibilities for the detection of weak fields, which, however, would also encounter considerable observational difficulties. Assuming pure resonance fluorescence it can be shown that even a very weak magnetic field should cause a weak polarization of, at the most, a few per cent in the hydrogen line H_α and the helium line D_3 of solar prominences. Such an effect appears indeed not improbable, according to the measurements of LYOT (1937). Further, for certain spectral lines of a star with a dipole field we should expect a small polarization of the order of 1 per cent; it is interesting that one could detect in this way a transverse dipole which is not otherwise observable.

Finally, a general magnetic field should also produce a very weak polarization of the so-called F-Corona of the Sun (THIESSEN, 1953), which should be measurable under certain circumstances.

REFERENCES

ALFVÉN, H. . . . . 1942a *Ark. Mat. Astron. Fys.*, **29A,** No. 11 and 12.
1942b *Ark. Mat. Astron. Fys.*, **29B,** No. 2.
1945a *M.N.*, **105,** 3.
1945b *M.N.*, **105,** 382.
1946 *Ark. Mat. Astron. Fys.*, **34A,** No. 23.
1947a *Phys. Rev.*, (2), **72,** 88.
1947b *M.N.*, **107,** 211.
1950 *Cosmical Electrodynamics.* Oxford.
1952 *Ark. Fys.*, **4,** 406.

Babcock, H. D.	1941	*Publ. Astron. Soc. Pacific*, **53**, 237.
	1944	*J. Opt. Soc. Amer.*, **34**, 1.
	1948	*Publ. Astron. Soc. Pacific*, **60**, 244.
Babcock, H. D. and Babcock, H. W.	1952	*Publ. Astron. Soc. Pacific*, **64**, 282.
	1955	*Nature (London)*, **175**, 296.
Babcock, H. W.	1947	*Ap. J.*, **105**, 105.
	1948	*Ap. J.*, **108**, 191.
	1949a	*Publ. Astron. Soc. Pacific*, **61**, 226.
	1949b	*Observatory*, **69**, 191.
	1950	*Publ. Astron. Soc. Pacific*, **62**, 277.
	1951a	*Publ. Astron. Soc. Pacific*, **63**, 81.
	1951b	*Ap. J.*, **114**, 1.
	1953	*Ap. J.*, **118**, 387.
Babcock, H. W. and Cowling, T. G.	1953	*M.N.*, **113**, 357.
Back, E. and Landé A.	1925	*Zeemaneffekt und Multipletstruktur der Spektrallinien.* Berlin.
Biermann, L.	1950	*Z. Naturforsch.*, **5a**, 65.
	1952	*Ann. Phys. (Leipzig)*, **10**, 413.
Biermann, L. and Schlüter, A.	1951	*Phys. Rev.*, **82**, 863.
Billings, B. H.	1949a	*J. Opt. Soc. Amer.*, **39**, 797.
	1949b	*J. Opt. Soc. Amer.*, **39**, 802.
Blackett, P. M. S.	1947	*Nature (London)*, **159**, 658.
Bowen, I. S.	1949	*Publ. Astron. Soc. Pacific*, **61**, 245.
Broxon, J. W..	1942	*Phys. Rev.*, **62**, 521.
	1949a	*Phys. Rev.*, **75**, 349.
	1949b	*Phys. Rev.*, **75**, 606.
Bruce, C. E. R.	1948	*Brit. Elect. and Allied Industries Res. Assoc.*, No. 15.
	1949a	*Nature (London)*, **164**, 280.
	1949b	*Observatory*, **69**, 110.
Brunnckow, K. and Grotrian, W.	1949	*Z. Astrophys.*, **26**, 313.
Cambridge Observatories (Report)	1951	*M.N.*, **111**, 190.
Candler, A. C.	1937	*Atomic Spectra.* Cambridge.
Chapman, S.	1928a	*M.N.*, **89**, 57.
	1928b	*M.N.*, **89**, 80.
	1943	*M.N.*, **103**, 117.
	1948	*M.N.*, **108**, 236.
Cowling, T. G.	1932	*M.N.*, **93**, 90.
	1933	*M.N.*, **94**, 39.
	1945	*M.N.*, **105**, 166.
	1946a	*M.N.*, **106**, 218.
	1946b	*M.N.*, **106**, 446.
Deslandres, H.	1929	*C.R. Acad. Sci. (Paris)*, **189**, 413.
Ellison, M. A.	1949	*Nature (London)*, **164**, 280.
Evershed, J.	1910	*Kodaikanal Obs. Bull.*, No. 22, 265.
	1939a	*M.N.*, **99**, 438.
	1939b	*M.N.*, **99**, 217.
	1939c	*M.N.*, **99**, 438.
	1944	*Observatory*, **65**, 190.
Fallin, J. W.	1948	*Phys. Rev.*, (2), **74**, 1232.
Ferraro, V. C. A..	1935	*M.N.*, **95**, 280.
	1937	*M.N.*, **97**, 458.
	1938	*Observatory*, **61**, 241.
Ferraro, V. C. A. and Unthank, H. W.	1949	*M.N.*, **109**, 462.
Ferraro, V. C. A. and Memory, D. J.	1952	*M.N.*, **112**, 361.
Giovanelli, R. G.	1941	*M.N.*, **107**, 338.
Gjellestad, G.	1952	*Ann. Astrophys.*, **15**, 276.
Grotrian, W.	1948	*Naturwiss.*, **35**, 321.
	1949	*Forschung und Fortschritte*, **25**, 280.
	1950	*Z. angew. Phys.*, **2**, 376.
	1953a	*Publ. Astrophys. Obs. Potsdam*, No. 97.
	1953b	Convegno di Scienze Fisiche Matematiche e Naturali, 14–19 Settembre 1952. (Accademia Nazionale dei Lincei, Fondazione Allesandro Volta, *Atti dei Convegni*, vol. 11, pp. 257–63, Rome.)
Grotrian, W. and Künzel, H.	1949	*Z. Astrophys.*, **26**, 325.
	1951a	*Z. Astrophys.*, **28**, 28.
	1951b	*Z. Astrophys.*, **29**, 173.
Gunn, R.	1929a	*Ap. J.*, **69**, 287.
	1929b	*Terrestr. Magnetism*, **33**, 231.

Gurewitsch, L. E. and Lebedinskij, A. J. . . . 1945 *Acad. USSR*, **49**, 92.
1946a *J. Phys. Moscow*, **10**, 327.
1946b *J. Phys. Moscow*, **10**, 425.
1948 *Jubil. Publ. Lenin University*. Leningrad.

Hale, G. E. . . . . 1907–8 (and following years): *Ann. Report Mt. Wilson Obs.*
1907–8 *Ann. Report Mt. Wilson Obs.*
1913 *Ap. J.*, **38**, 27.
1914 *Publ. Astron. Soc. Pacific*, **26**, 146.
1935 *Nature (London)*, **136**, 703.

Hale, G. E., Ellermann, G. E. F., Nicholson, S. B., Joy, A. H. . . . . 1919 *Ap. J.*, **49**, 153.

Hale, G. E. and Nicholson, S. B. . . . . . 1938 *Magnetic Observations of Sunspots* 1917–1924. Washington.

Hale, G. E., Seares, F. H., v. Maanen, A., Ellerman, F. . . . . 1918 *Ap. J.*, **47**, 206.

Harrison, G. R. . . . . 1949a *J. Opt. Soc. Amer.*, **39**, 413.
1949b *J. Opt. Soc. Amer.*, **39**, 522.

Houtgast, J. and van Sluiters, A. . . . . 1948 *Bull. Astron. Inst. Netherlands*, **10** (No. 388).

Hubenet, H. . . . . 1954 *Z. Astrophys.*, **34**, 110.

International Astron. Union . . . . . 1952 *Draft Report Rome* (I), p. 56.

Jenkins, F. A. and White, H. E. . . . . . 1937 *Fundamentals of Physical Optics*. New York and London.

Kiepenheuer, K. O. . . . . 1953a Contract N9onr 97, 100; and *Ap. J.*, **117**, 447.
1953b *Ap. J.*, **117**, 447.
1953c *Z. Naturforsch.*, **8a**, 225.

King, R. B. . . . . 1934 *Ap. J.*, **80**, 136.

Klüber, H. von . . . . 1947a *Z. Astrophys.*, **24**, 1.
1947b *Z. Astrophys.*, **24**, 121 (with detailed bibliography).
1947c *Fiat Review of German Science* (Naturforschung und Medizin in Deutschland, 1939–1946), Bd. **20**, 208.
1948 *Z. Astrophys.*, **25**, 187.
1951 *M.N.*, **111**, 91.
1952 *M.N.*, **112**, 183.
1954 *M.N.*, **114**, 242.

Langer, R. M. . . . . 1936 *Publ. Astron. Soc. Pacific*, **48**, 208.

Larmor, J. . . . . 1934 *M.N.*, **94**, 469.

Lyot, B. . . . . 1937 *l'Astronomie*, **51**, 217.

Mattig, W.. . . . . 1953 *Z. Astrophys.*, **31**, 273.

Mount Wilson Observatory Annual Reports . . 1909–10, 1911–12, 1922–36, 1927–28, 1928–29, 1932–33, 1933–34, 1939–40, 1947–48

Mount Wilson: Summary of Magnetic Observations *Publ. Astron. Soc. Pacific* (regularly published since May, 1920); see vol. **40**, 51, 1928.

Nicholson, S. B. . . . . 1933 *Publ. Astron. Soc. Pacific*, **45**, 51.
1938 *Publ. Astron. Soc. Pacific*, **50**, 224.
1941 *Publ. Astron. Soc. Pacific*, **53**, 305.

Öhman, Y. . . . . 1950 *Ap. J.*, **111**, 362.

Potsdam, Jahresberichte des Astrophysikalischen Observatoriums Potsdam . . . . 1950 (and following years): *Mitteilungen der Astronomischen Gesellschaft*.

Richardson, R. S. . . . . 1939–40 *Mt. Wilson Obs. Ann. Report.*

Rubaschew, B. M. . . . . 1949 *Priroda, Leningrad*, **38**, 59.

Russell, H. N. . . . . 1927 *Ap. J.*, **66**, 307.

Schwarzschild, M. . . . . 1949 *Ann. Astrophys.*, **12**, 148.

Seares, F. H. . . . . 1913 *Ap. J.*, **38**, 99.

Schröter, E. H. . . . . 1953 *Z. Astrophys.*, **33**, 20.

Shenstone, A. G. and Blair, H. A. . . . . 1929 *Phil. Mag.*, **8**, 765.

Sweet, P. A. . . . . 1949 *M.N.*, **109**, 507.

Tanaka, T. and Takagi, Y. . . . . 1939 *Proc. Phys. Math. Soc. Japan*, **21**, 421 and 496.

THIESSEN, G. . . . . 1946 *Ann. Astrophys.*, **9,** 101.
1947 *Himmelswelt*, **55,** 21.
1949a *Z. Astrophys.*, **26,** 130.
1949b *Observatory*, **69,** 228.
1952a *Z. Astrophys.*, **30,** 17.
1952b *Nature (London)*, **169,** 147.
1952c *Observatory*, **72,** 75.
1952d *Z. Astrophys.*, **30,** 307.
1953a *Z. Astrophys.*, **32,** 173.
1953b *Naturwissenschaften*, **40,** 218.
TOLANSKY, S. . . . . 1947 *High Resolution Spectroscopy*. London.
WALÉN, C. . . . . 1944a *Ark. Mat. Astron. Fys.*, **30A,** No. 15.
1944b *Ark. Mat. Astron. Fys.*, **31B,** No. 3.
1946 *Ark. Mat. Astron. Fys.*, **33A,** No. 18.
WASHINGTON BUREAU OF STANDARDS . . . . 1928 "Tables of Theoretical Zeeman-effects" *Research Paper No.* 23.
WOOD, R. W. . . . . 1914a *Phil. Mag.*, **27,** 524.
1914b *Phys. Z.*, **15,** 313.
WRIGHT, F. E. . . . . 1934 *J. Opt. Soc. Amer.*, **24,** 206

The Name Index and Subject Index for both Volumes are at the end of Volume 2.

[illegible] K. D. Abhyankar Walter S. Adams C. W. Allen Lawrence H. Aller [illegible]

B. [illegible] Edward V. Appleton R. d'E. Atkinson Walter Baade [illegible]

[illegible] V. Barocas G. K. Batchelor [illegible] F. Becker W. Becker

Arthur Beer [illegible] [illegible] P. L. Bhatnagar William P. Bidelman [illegible]

[illegible] D. E. Blackwell [illegible] Bart J Bok John Bolton H Bondi

Max Born [illegible] I. S. Bowen Dirk Brouwer H. A. Brück Máire Brück

E. C. Bullard E M Burbidge Geoffrey Burbidge H. E. Butler G. L. Camm

J. A. Carroll D. Chalonge S Chandrasekhar Sydney Chapman [illegible]

T. G. Cowling J. F. Cox [illegible] [illegible] M. Davidson J. G. Davies

Leverett Davis, Jr. [illegible] J de Kort S.J. [illegible] [illegible] Armin J. Deutsch

G de Vaucouleurs David W. Dewhirst [illegible] A E Douglas J. Dufay

Theodore Dunham Jr. Bengt Edlén D. L. Edwards M. A. Ellison David Evans

J. Evershed [illegible] Peter Fellgett [illegible] Mario G. Fracastoro

E. Finlay-Freundlich [illegible] G. Gamow Cecilia Payne Gaposchkin Sergei Gaposchkin

R H Garstang [illegible] Leo Goldberg R. M. Goody Livio Gratton W. M. H. Greaves

Jesse L. Greenstein Margherita Hack Y Hagihara [illegible] Hanbury Brown

Bob Hardie Daniel L. Harris III G R Harrison [illegible] John [illegible]

O. Heckmann Ejnar Hertzsprung G. Herzberg A. Hewish J. S. Hey W. A. Hiltner

Fred Hoyle C M Huffer Milton L Humason [illegible] [illegible] J Jackson

[illegible] Martin Johnson [illegible] Hans Kienle [illegible]